5TH EDITION

ELECTRICAL PRINCIPLES

5TH EDITION

ELECTRICAL PRINCIPLES

PETER PHILLIPS

Electrical Principles
5th Edition
Peter Phillips

Head of content management: Sophie Kaliniecki
Content manager: Sandy Jayadev
Content developer: Sarah Payne
Project editor: Raymond Williams
Cover design: Leah Ashforth, Watershed Design
Text designer: Rina Gargano (Alba Design)
Project Designer: Linda Davidson
Editor: Julie Wicks
Proofreader: James Anderson
Permissions/Photo researcher: Catherine Kerstjens
Indexer: Max McMaster
Cover: Alamy Stock Photo/Cultura Creative RF

KnowledgeWorks Global Ltd.

Any URLs contained in this publication were checked for currency during the production process. Note, however, that the publisher cannot vouch for the ongoing currency of URLs.

Fourth edition published by Cengage in 2019

For product information and technology assistance,
in Australia call 1300 790 853;
in New Zealand call 0800 449 725

For permission to use material from this text or product, please email
aust.permissions@cengage.com

National Library of Australia Cataloguing-in-Publication Data
ISBN: 9780170458412
A catalogue record for this book is available from the National Library of Australia.

Cengage Learning Australia
Level 7, 80 Dorcas Street
South Melbourne, Victoria Australia 3205

Cengage Learning New Zealand
Unit 4B Rosedale Office Park
331 Rosedale Road, Albany, North Shore 0632, NZ

For learning solutions, visit cengage.com.au

Printed in China by 1010 Printing International Limited.
5 6 7 26 25

BRIEF CONTENTS

CONTENTS

PREFACE

This 5th edition of Electrical Principles has been comprehensively revised. The revisions are based on the feedback provided by many teachers, which I welcome. The revisions have been driven by a number of suggestions, including reduced and, where possible, simpler text, a greater emphasis on a real-world feel and the addition of a basic maths chapter. This new chapter is at the front of the book, and covers the maths students need to master as they work through their trade course. This chapter also includes a section on drawing phasor diagrams. The use of a Casio calculator is also integrated throughout the book, and calculator keystrokes to solve problems are presented in many of the examples as a guide to students. The book retains its previous structure, with some now redundant content removed, and in a few chapters, new content has been added. Many illustrations have been changed to make them clearer and more relevant, with care taken to integrate these illustrations with the text.

The book covers the knowledge component of seven core competency units in the 2020 Electrotechnology Training Package for courses UEE33020 and UEE30820. Units that are fully covered by this book are UEECD0046, UEECD0044, UEEEL0021, UEEEL0019, UEEEL0020, UEEEL0025 and UEEEL0024. The content in the book is sequenced in the order of these units.

Pedagogical features in this book include ensuring all Review Exercises can be answered by reference to the text. Examples are frequently used as a way of explaining the calculations or the development of a phasor diagram. At no stage does the text depart from Certificate III level. There are now more photographs of components and actual circuits, and in Chapter 8, the complexity of series-parallel circuits is explained using an example students can relate to.

Safety messages are included throughout the book, as are margin note references to relevant parts of the Maths chapter to help students deal with transposition, trigonometry, Pythagoras' theorem and so on. During this revision, a few errors have been found, and these have been corrected. Many Review Exercises have been changed to give them a real-world feel, and the number of words has been reduced in all chapters. The language has been changed to give a more direct approach, and is more consistent with how we converse today. In short, this edition has had more revisions than any previous edition, and I believe is a better and more accessible book than ever before.

My sincere thanks to those teachers who have provided the necessary feedback that have driven the revisions in this new edition. Please keep them coming, as I rely on your feedback.

Peter Phillips

ABOUT THE AUTHOR

Peter Phillips trained as an apprentice electrical fitter at a large Federal Government manufacturing plant at Lithgow NSW, after which he spent over eight years working in two power stations. He then joined TAFE NSW, teaching a wide range of electrical subjects in various colleges in that state. He held numerous teaching and supervisory positions and became involved in curriculum development during the 1990s, when the National Curriculum was being introduced. In 1994, after leaving TAFE NSW, he started his own company that offered services including curriculum development and producing training resources. In 2010 he was contracted to write a range of teaching resources for a NSW energy distribution company (Endeavour Energy), that included a number of topics for Certificate III and IV qualifications. This involvement lasted several years and provided considerable insight to the equipment and procedures associated with electrical energy distribution.

Phillips has written extensively for technical magazines over a period of 20 years, and has been considerably involved in developing various Training Packages and their support documentation. This included working with curriculum associated with a number of different areas in the manufacturing sector. As a result, he travelled extensively throughout Australia, visiting a wide variety of manufacturing industries and RTOs. His first textbook, *Electrical Fundamentals*, was released in 1993 in response to a requirement that all trade courses teach this subject. This was followed by a series of books for the Electrotechnology industry. In 2010, Phillips combined the content of these books with new content to produce his first text book that covered the core theory content of various Electrical Trades courses. He remains committed to updating and revising this textbook so it maintains currency with the relevant Training Packages and standards.

Peter Phillips is actively connected with trainers and others involved with the Electrical Training Package. He maintains his connections with TAFE NSW and is widely known through his writings and involvement in tertiary education. He holds a Bachelor of Education (Distinction) and a wide range of technical qualifications including Electrical Trades, Advanced Diploma of Engineering, Industrial Electronics and other Certificate IV qualifications. He was awarded a doctorate by Sydney University in 2017.

ACKNOWLEDGEMENTS

There are many people who have helped me over the years in preparing this book. My particular thanks go to Greg Robinson and Frank Cahill from Miller TAFE and to Alan Birse from Granville TAFE who all provided considerable help and guidance in the original edition of this book, from which this fourth edition has evolved. Others who have been of great help are Col Berry from Dubbo TAFE, and teachers from Ultimo TAFE who provided important feedback when the second edition was first published. Numerous teachers (listed below) participated in a review of the third edition, providing valuable guidance and advice for this new edition.

Thanks in particular to the team at Cengage, publishers of this book. My thanks to Sarah Payne who was the first point of delivery of each revised chapter, also to Sandy Jayadev for her support and belief in this book. As well, thanks to the production team, graphic artists, the sales team and the many others working behind the scenes. It's through the team effort of us all that saw the last edition of this book win two important awards.

The author and Cengage would also like to thank the following reviewers for their incisive and helpful feedback:

Frank Cahill, TAFE NSW
Kris Heel, Tec-NQ
Simone Loone, TasTAFE
Paul Mansfield, TAFE SA
Tristan McNaught, Go TAFE Wangaratta
Greg Moore, TAFE NSW
Geoff Swanson, Tec-NQ

Guide to the text

As you read this text you will find a number of features in every chapter to enhance your study of electrotechnology and help you understand how the theory is applied in the real world.

CHAPTER OPENING FEATURES

Refer to the **Chapter introduction** for a contextualised summary of the chapter.

The **Chapter outline** lists the topics that are covered in the chapter.

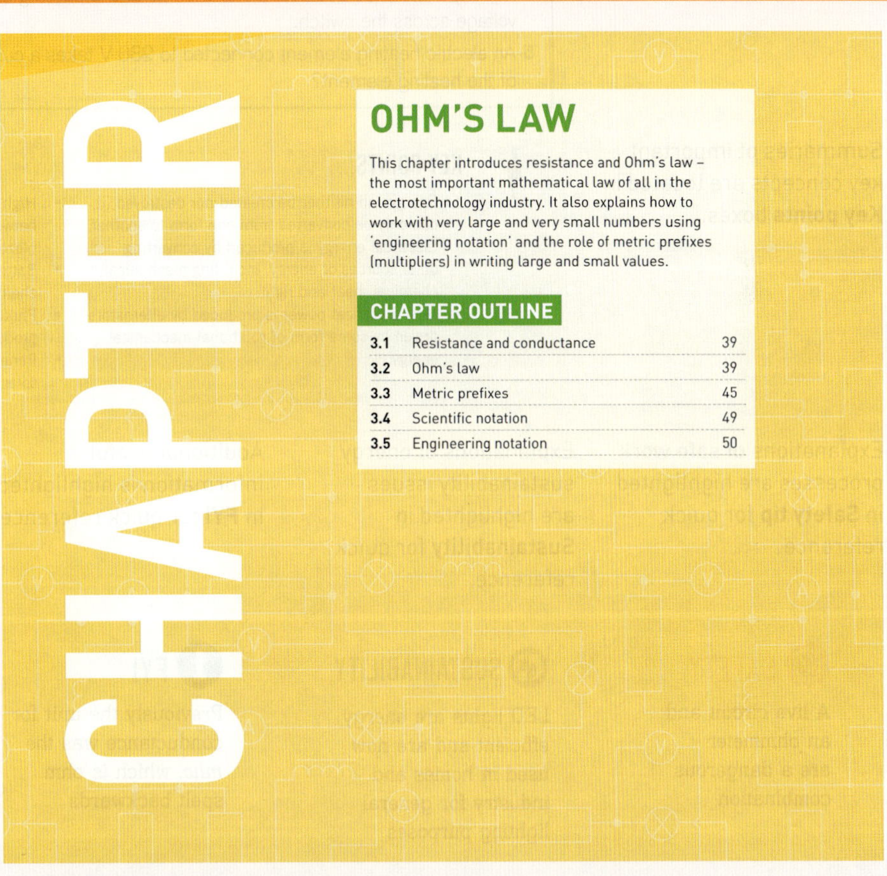

FEATURES WITHIN CHAPTERS

Analyse practical applications of concepts through the **Examples**.

EXAMPLE 3.1

A voltage of 25 volts is measured across a resistor of 15 ohms. Calculate the current flowing in the resistor.

Solution

Values $V = 25$ volts

$R = 15$ ohms

$I = ?$

Equation $I = \frac{V}{R}$

$I = \frac{25}{15}$

Answer $\mathbf{I = 1.67}$ **amperes**

FEATURES WITHIN CHAPTERS

Examine how theoretical concepts have been used in practice through the **Task** boxes.

TASK 3.1

1 You measure 12 V across a resistance and 0.5 A flowing through the resistance. Find the value of the resistance.
2 An ammeter shows a reading of 0.4 A flowing through a 100 ohm resistor. What value of voltage would you measure across the resistor?
3 A long length of electrical cable has a voltage drop across it of 15 V when it is carrying a current of 30 A. Calculate the resistance of the cable.
4 A faulty switch has a resistance of 5 ohms and is passing a current of 6 A. Calculate the voltage across the switch.
5 An electric heating element connected to 230 V takes a current of 5 A. What is the resistance of the heating element?

Summaries of important key concepts are located in **Key points** boxes.

KEY POINTS...

- Energy can neither be created nor destroyed, it can only be converted from one form to another.
- Electrical energy is produced by converting other sources of energy, including mechanical, chemical, heat and light.
- Most electrical power is produced by alternators driven by some form of rotational mechanical energy.
- High voltages can be produced with friction between two materials, or by applying force to a piezo-electric material.
- Batteries produce electricity by converting chemical energy to electrical energy.
- Photo-voltaic devices such as the solar cell produce electricity when exposed to light.
- Renewable energy sources include geothermal, solar, hydro, biogas and wind power.

Explanations of safe work processes are highlighted in **Safety tip** for quick reference.

Explanations of energy sustainability issues are highlighted in **Sustainability** for quick reference.

Additional useful information is highlighted in **FYI** for quick reference.

Explanations of key concepts are highlighted in **Key concepts** for quick reference.

A live circuit and an ohmmeter are a dangerous combination

LED lights are energy efficient and are now used in homes and industry for general lighting purposes

Previously the unit for conductance was the *mho*, which is ohm spelt backwards

KEY CONCEPT

Heat and light are both forms of 'radiant' energy

END-OF-CHAPTER FEATURES

At the end of each chapter you will find several tools to help you to review, practise and extend your knowledge of the key learning outcomes.

Review your understanding of the key chapter topics with the **Summary**.

CHAPTER SUMMARY

- Electrical energy is produced by converting other forms of energy (mechanical, chemical, heat and light).
- Electrical energy is converted to the required form of energy (e.g. heat, light or mechanical energy) by the electric load.
- Large alternators driven by steam turbines (mechanical energy) generate most of Australia's electrical energy.
- An alternator has coils of wire arranged so they intersect with a moving magnetic field, thereby producing electricity.
- Renewable energy sources that can drive an alternator include geothermal, solar thermal, hydro, biogas and wind power.
- Electrolysis is the process of passing current from one electrode to another through an electrolyte. Electroplating is an application of electrolysis.
- Effects caused by an electric current include heating, magnetism, chemical, luminous and physiological.
- Discharge lamps are filled with a gas that produces light directly, or in the case of the fluorescent lamp, ultraviolet light that excites phosphors inside the tube.
- Physiological effects of an electric current include muscle contraction, ventricular fibrillation, burns and death.
- Heat is produced in a conductor because of its resistance.
- Current flowing in a conductor creates a magnetic field around the conductor. Coiling the conductor

END-OF-CHAPTER FEATURES

Test your knowledge and consolidate your learning through **Review exercises** and **Worksheets**.

REVIEW EXERCISES

Check your answers at the back of the book.

1 List four sources of energy that can be used to produce electrical energy.
2 A car alternator produces electricity due to what type of effect?
3 Why does the copper used in electrical cables need to be produced by electrolysis?
4 In electroplating, what is the polarity of the anode?
5 Why does zinc galvanising protect steel from galvanic corrosion?
6 What are the three components of a basic electric cell?
7 Name the two main chemical effects of an electric current.
8 Give three advantages of LED lighting systems.
9 What two things determine the value of current flowing through your body when you get an electric shock?
10 What are the two situations specified by AS/NZS 3000:2018 concerning protection against an electric shock?

ONLINE RESOURCES

COMPLETE WORKSHEET TWO

Check with your instructor for worksheets on this chapter.

Guide to the online resources

INSTRUCTOR RESOURCES PACK

INSTRUCTOR RESOURCE PACK

Premium resources that provide additional instructor support are available for this text, including Mapping Grid, Worksheets, Testbank, Premium PowerPoints, Instructor Manual, and more.

These resources save you time and are a convenient way to add more depth to your classes, covering additional content and with an exclusive selection of engaging features aligned with the text.

The Instructor Resource Pack is included for institutional adoptions of this text when certain conditions are met. The pack is available to purchase for course-level adoptions of the text or as a standalone resource.

Contact your Cengage learning consultant for more information.

FOR THE STUDENT

MINDTAP

Premium online teaching and learning tools are available on the *MindTap* platform - the personalised eLearning solution.

MindTap is a flexible and easy-to-use platform that helps build student confidence and gives you a clear picture of their progress. We partner with you to ease the transition to digital – we're with you every step of the way.

The *Cengage Mobile App* puts your course directly into students' hands with course materials available on their smartphone or tablet. Students can read on the go, complete practice quizzes or participate in interactive real-time activities.

MindTap for Phillips's Electrical Principles is full of innovative resources to support critical thinking, and help your students move from memorisation to mastery! Includes:

- Phillips's Electrical Principles eBook
- Instructor videos
- Worksheets
- Fill in the blank activities
- Revision quiz

MindTap is a premium purchasable eLearning tool. Contact your Cengage learning consultant to find out how MindTap can transform your course.

MATHS AND DEFINITIONS

This chapter explains some of the mathematical processes used in this book. It also lists definitions of the important measurement units, and summarises the Greek letter symbols and metric prefixes used in the book. Refer to this chapter if you need an explanation of a mathematical process as it appears in the book. Calculator examples use the Casio fx-82AU PLUS II 2nd edition calculator. Other models in the fx-82AU range are almost identical.

CHAPTER OUTLINE

1. BASIC MATHEMATICS

1.1 BASIC CALCULATIONS

The four basic maths functions are addition (+), subtraction (–), multiplication (×) and division (÷). Examples:

10 + 2 = 12, 10 – 2 = 8, 10 × 2 = 20, 10 ÷ 2 = 5, also written as $\frac{10}{2}=5$.

An expression involving these four basic functions might have several parts, which should be dealt with as explained in Example 1.

EXAMPLE 1

Solve this calculation: $12+\left(\frac{16}{4}\right)+(4\times3)-5$

Rules:

1 Work out the calculations inside the brackets: $12+(4)+(12)-5$

2 Remove the brackets = 12 + 4 +12 – 5

3 Work out the addition = 28 – 5

4 Work out the subtraction = 23 (Answer)

KEYSTROKES FOR SOLVING WITH A CALCULATOR

The expression in Example 1 can be entered directly as written into an FX-82 AU Casio calculator. Figure 1 shows the keystrokes, which use the bracket keys as in the expression.

Figure 1 Keystrokes to solve the calculation in Example 1

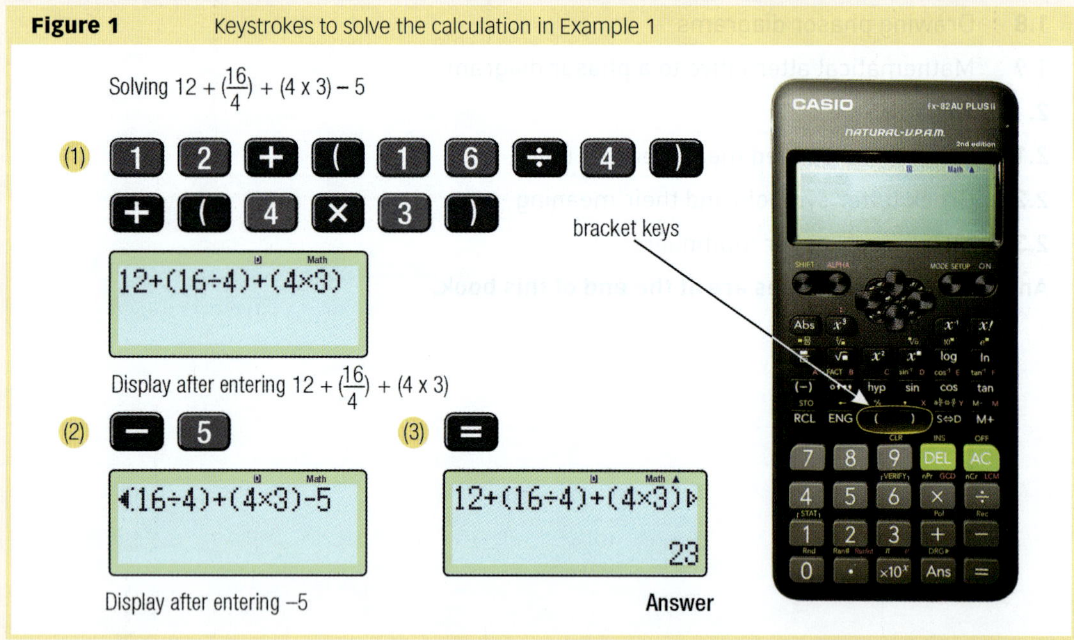

Another way of solving this calculation with a Casio calculator is shown in Figure 2. This method solves the calculation one step at a time, in which the bracketed terms are entered without brackets. Each step is separated by pressing the equal key.

Figure 2 Keystrokes for solving a calculation one step at a time

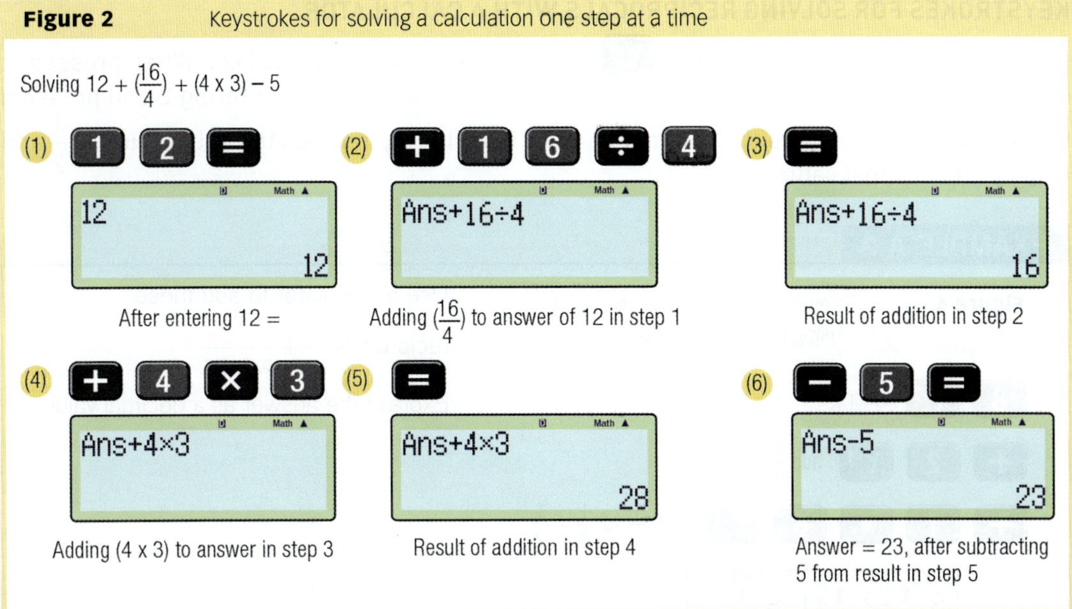

EXERCISE 1

Solve the following calculations. Check your answers at the end of the book.

1 $5 \times (30 - 10)$

2 $\left(\frac{24}{6}\right) \times (8 - 3)$

3 $12 + (6 \times 3) - \left(\frac{25}{5}\right)$

1.2 RECIPROCALS

The reciprocal of a number is that number divided into 1. For example, the reciprocal of 5 is $\frac{1}{5}$, which equals 0.2 as a decimal value. A reciprocal therefore looks like a fraction, except the numerator (top number) is always 1. It is easiest to use a calculator, although simple calculations can be done mentally, such as $\frac{1}{2} + \frac{1}{4}$ which as you probably know is $\frac{3}{4}$. Figure 3 shows how to arrive at that answer using a Casio calculator.

Figure 3 Adding two fractions with a calculator and finding the answer as a fraction

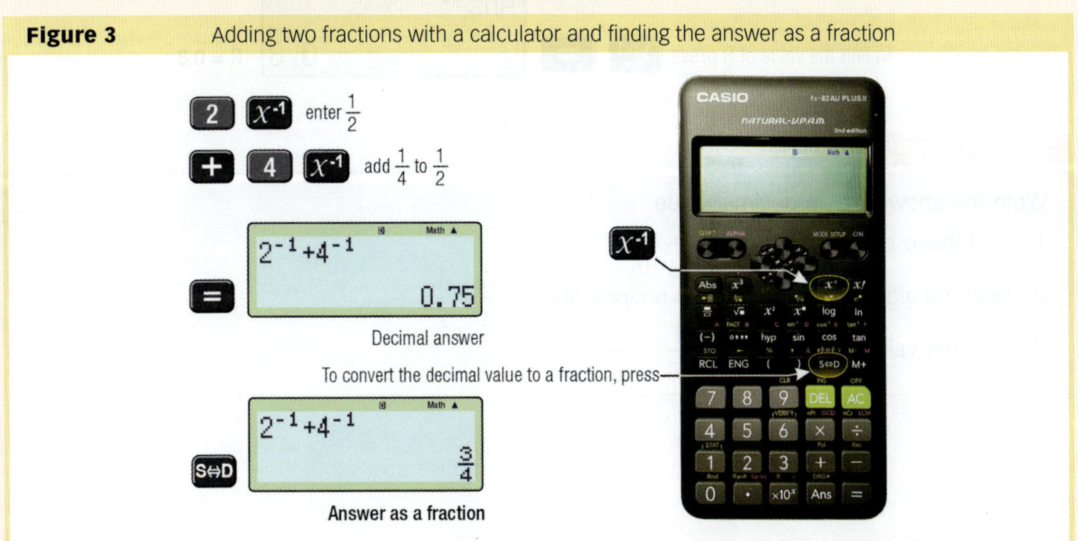

KEYSTROKES FOR SOLVING RECIPROCALS WITH A CALCULATOR

In Figure 3, we used the key marked x^{-1}, which is called the *reciprocal* key. When pressed, it divides an entered number into 1, giving its reciprocal. For example, entering 2 and pressing the reciprocal key will give the decimal number 0.5. That is, it solves the calculation of $\frac{1}{2}$ and displays its decimal value.

EXAMPLE A.2

Use a calculator to add these reciprocals: $\frac{1}{8}+\frac{1}{2}+\frac{1}{1.6}$

Express the answer as a decimal value.

Figure 4 Key presses to solve Example 2 using the reciprocal key

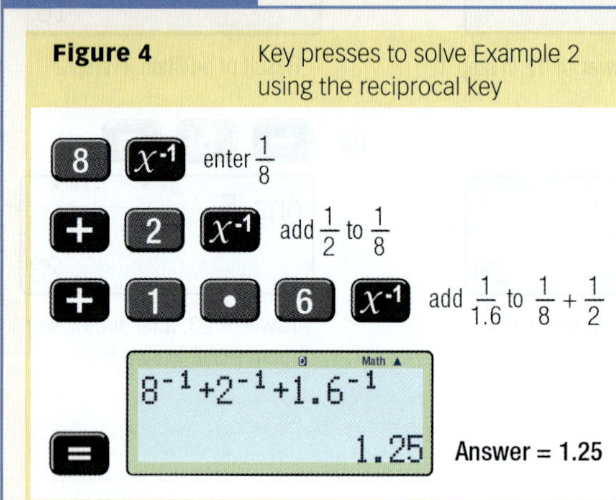

Chapter 7 (*Parallel Circuit*) uses an equation in which the sum of the reciprocals gives the answer as a reciprocal. For example, using the same reciprocals from Example 2, the equation becomes: $\frac{1}{R}=\frac{1}{8}+\frac{1}{2}+\frac{1}{1.6}$

But we really want the value of R, not $\frac{1}{R}$. Figure 5 shows how to solve the equation in terms of R, in which the reciprocal key is used to invert the answer obtained in Example 2.

Figure 5 To solve for R, after adding each reciprocal, press the reciprocal key

Find the value of R when $\frac{1}{R}=\frac{1}{8}+\frac{1}{2}+\frac{1}{1.6}$

8⁻¹+2⁻¹+1.6⁻¹ 1.25

Result of $\frac{1}{R}=\frac{1}{8}+\frac{1}{2}+\frac{1}{1.6}$

$\frac{1}{R}=1.25$

To find the value of R press x^{-1} =

Ans⁻¹ 0.8

R = 0.8

EXERCISE 2

Write the answers as a decimal value.

1 Add these reciprocals: $\frac{1}{5}+\frac{1}{2.5}+\frac{1}{10}$

2 Find the algebraic sum of these reciprocals: $\frac{1}{3}-\frac{1}{4}+\frac{1}{2}$

3 Find the value of R: $\frac{1}{R}=\frac{1}{40}+\frac{1}{30}+\frac{1}{50}$

FYI

Algebraic sum means considering the sign of a number, which can be plus or minus. The algebraic sum of 2, 5 and −9 is −2.

1.3 ROUNDING DECIMAL VALUES

A decimal number might have a decimal point followed by one or more digits. Calculators will often give an answer with many numbers on the right of the decimal point. In most cases, we only require two numbers after a decimal point, which can require *rounding* answers from a calculator. The method is: work from the right, towards the decimal point. If a number is 5 or greater, the number on its left is increased by 1. For example:

a 1.23446 (value displayed on a calculator)

b becomes 1.2345 after rounding to four decimal places

c becomes 1.235 after rounding to three decimal places

d becomes 1.24 when rounded to two decimal places.

EXERCISE 3

Round the following numbers to two decimal places.

1 6.446
2 234.3412
3 12.989
4 10.6145

1.4 TRANSPOSING EQUATIONS

An equation is a mathematical statement that has an equal symbol between two algebraic expressions of the same value. For example, A = B + C. That is, A equals the sum of B and C, or the sum of B and C equals A. Equations can have all types of mathematical operations, such as division and multiplication, as well as addition and subtraction. The four equations in Figure 6 show examples using the four main mathematical functions in which all equations have three terms, and A is the subject. Division is usually shown in an equation with one variable above the other, as in Figure 6.

If we know the values of B and C, the value of A can be found in all these equations. If the values of A and B are known, finding the value of C means rearranging, or transposing, the equation so C is the subject. To do this, both sides of the equation are dealt with so a term can be cancelled out on one side and caused to appear on the other side. The rule is whatever is done to one side of the equation must be done to the other side.

Figure 6 Equation with three terms in which A is the subject

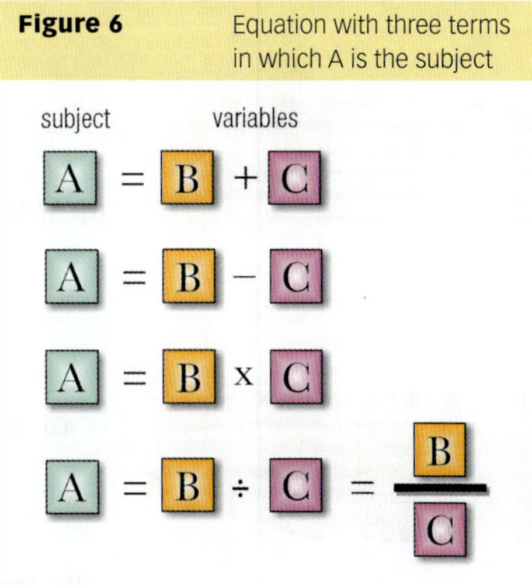

TRANSPOSING EQUATIONS THAT HAVE ONLY PLUS AND MINUS OPERATORS

An equation containing only plus and minus signs will have all terms on the one line. To move a term from one side of the equal sign to the other, apply the opposite maths function to cancel it on one side and place it on the other. Example 3 shows how to transpose an equation that has only plus and minus signs.

EXAMPLE 3

Make D the subject of the equation A = B + C – D

1 Add D to both sides: A + D = B + C – D + D
2 Cancel D on the right-hand side because adding minus D to plus D equals zero, giving: A + D = B + C
3 Subtract A from both sides: A + D – A = B + C – A
4 Cancel A on the left-hand side: D = B + C – A

In step 1, we added D to both sides to move it from the right to the left of the equation, because this cancels the –D on the right side. In step 3, we subtracted A from both sides, as this cancels the +A on the left side.

TRANSPOSING EQUATIONS THAT HAVE ONLY DIVIDE AND MULTIPLY OPERATORS

Most equations in this book have division and multiplication. A common example is the equation known as Ohm's law, introduced in Chapter 3. It always has three terms, I, V and R, and if I is the subject, the equation is $I = \frac{V}{R}$.

EXAMPLE 4

Make V the subject of the equation $I = \frac{V}{R}$.

Figure 7 shows the steps to make V the subject of this equation. As before, to move a term, cancel it on one side with the opposite maths function, and apply the same function to the other side. In this example, a multiplication cancels a division.

In step 2 in Figure 7, we multiply both sides of the equation by R, so in step 3, the R on the right side is cancelled, because $R \times \frac{1}{R} = 1$. This gives the equation shown in step 4, and in step 5, the equation is rewritten with V as the subject on the left.

Figure 7 Transposing an equation by multiplying both sides with the same variable to cancel a division

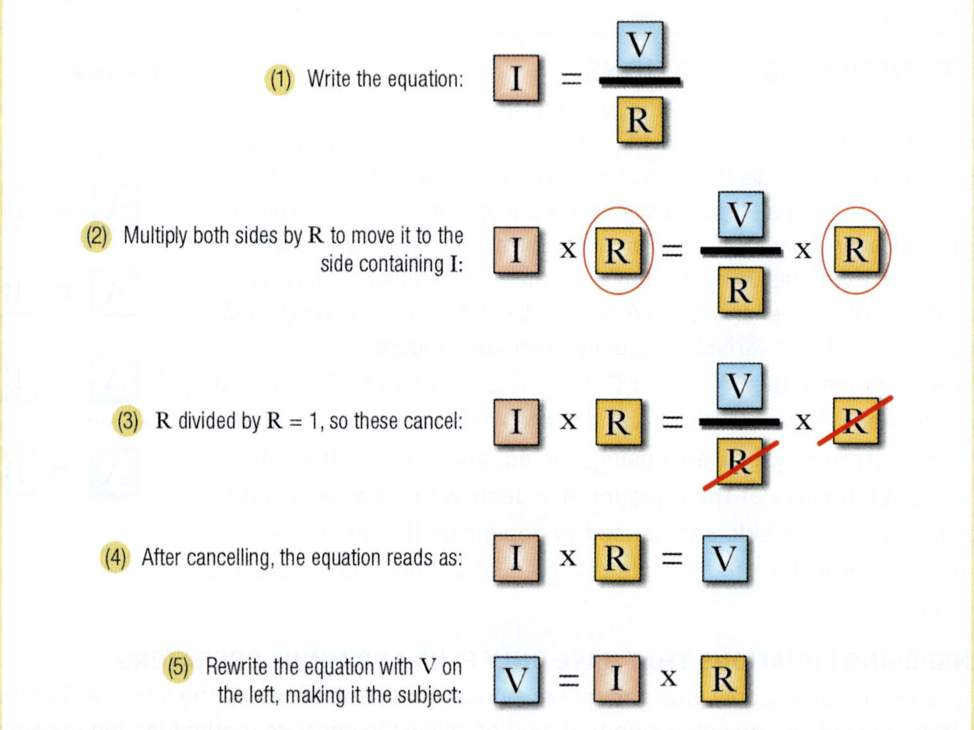

(1) Write the equation: $I = \frac{V}{R}$

(2) Multiply both sides by R to move it to the side containing I: $I \times R = \frac{V}{R} \times R$

(3) R divided by R = 1, so these cancel: $I \times R = \frac{V}{\cancel{R}} \times \cancel{R}$

(4) After cancelling, the equation reads as: $I \times R = V$

(5) Rewrite the equation with V on the left, making it the subject: $V = I \times R$

If an equation being transposed has a multiplication, it is cancelled with a division as explained in Example 5.

EXAMPLE 5

Make R the subject of the equation V = IR.

Figure 8 shows the sequence to transpose an equation that has a multiplier operator. In step 2, we divide both sides by I. In step 3, because I divided by I equals 1, they cancel, giving the equation in step 4. Rewriting the equation as in step 5 gives the equation in which R is the subject.

Figure 8 Transposing an equation by dividing both sides with the same variable to cancel a multiplication

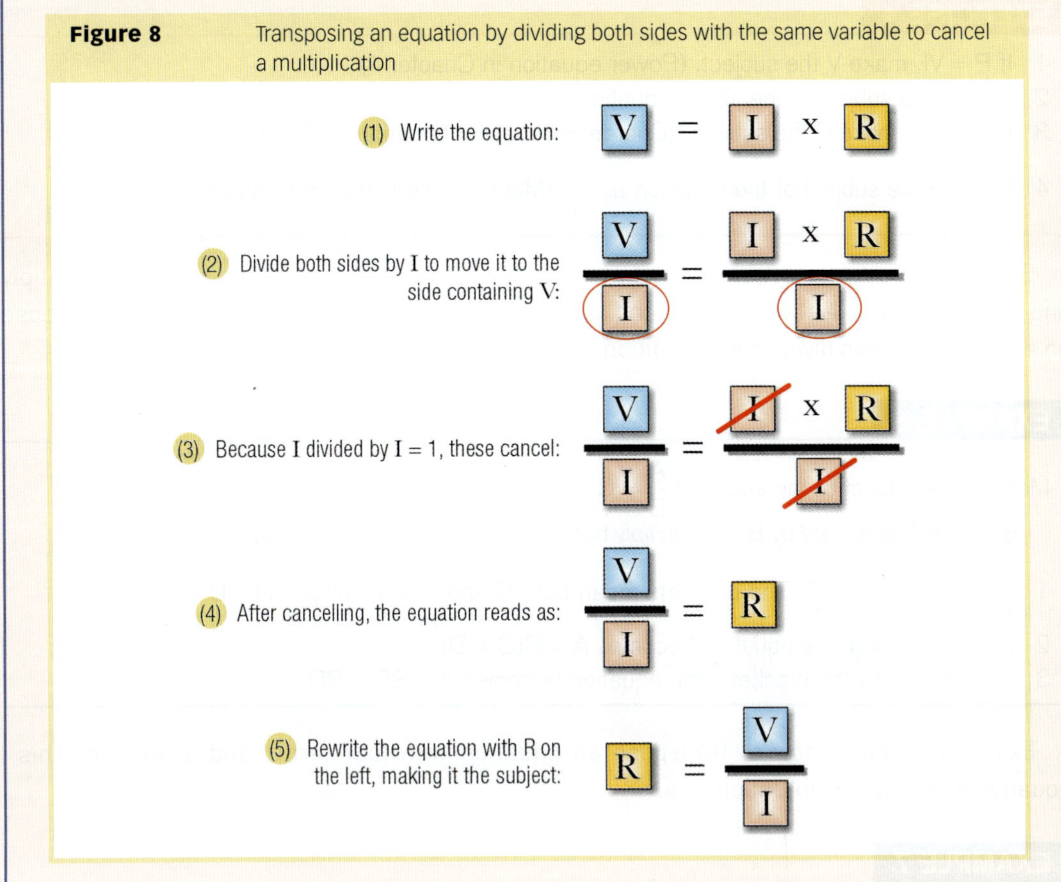

BASIC TRANSPOSING TRIANGLE

An equation with the general form of A = BC can be represented in a triangle, making it easy to transpose the equation. In Figure 9, A (the subject) is in the top half of the triangle, and B and C (values to find A) are in the two sections in the bottom half. To make any term the subject, cover that term and the equation is the other terms. This triangle is used when transposing Ohm's law equations, as shown in Chapter 3.

Figure 9 A = BC represented in a triangle

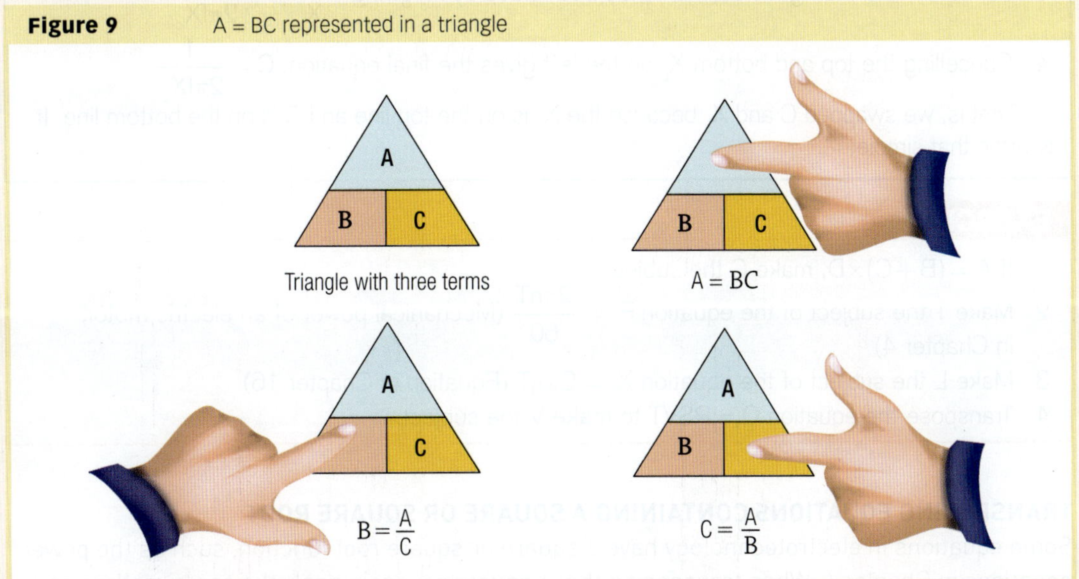

EXERCISE 4

1 If $P = VI$, make V the subject. (Power equation in Chapter 4)
2 Make I the subject in the above equation.
3 If $Q = CV$, make C the subject. (Charge in capacitor equation in Chapter 10)
4 Make H the subject of the equation $\mu = \frac{B}{H}$ (Magnetism equation in Chapter 11)

Some equations in electrotechnology have more than three terms and can include multiply and divide functions, as well as plus and minus operators. Example 6 shows how to transpose an equation that has division and addition, and four variables.

EXAMPLE 6

Make A the subject of the equation $\frac{A}{B} = C + D$

Because A is *divided* by B, we *multiply* both sides by B to remove B from the left side:

1 $\frac{A}{B} \times B = (C + D) \times B$. The brackets mean both C and D are multiplied by B.
2 After cancelling, the equation becomes $A = B(C + D)$.
3 After removing the brackets, the equation becomes $A = BC + BD$.

Example 7 shows how to transpose an equation with four terms and a division. This equation is introduced in Chapter 16.

EXAMPLE 7

Make C the subject of the equation $X_c = \frac{1}{2\pi fC}$

Because C is on the bottom line of the equation, we multiply both sides by C:

1 $X_c \times C = \frac{1 \times C}{2\pi fC}$. We could have written this as $X_c \times C = \left(\frac{1}{2\pi fC}\right) \times C$
2 Cancelling the top and bottom C gives $X_c \times C = \frac{1}{2\pi f}$
3 To move X_c to the right side, we divide both sides by X_c giving $\frac{X_c \times C}{X_c} = \frac{1}{2\pi fX_c}$
4 Cancelling the top and bottom X_c on the left gives the final equation: $C = \frac{1}{2\pi fX_c}$

That is, we swapped C and X_c because the X_c is on the top line and C is on the bottom line. It is often that simple!

EXERCISE 5

1 If $A = (B + C) \times D$, make C the subject.
2 Make T the subject of the equation $P = \frac{2\pi nT}{60}$ (Mechanical power of an electric motor, in Chapter 4)
3 Make L the subject of the equation $X_L = 2\pi\eta T$ (Equation in Chapter 16)
4 Transpose the equation $Q = RSVT$ to make V the subject.

TRANSPOSING EQUATIONS CONTAINING A SQUARE OR SQUARE ROOT

Some equations in electrotechnology have a square or square root function, such as the power equations in Chapter 4. When transposing these equations, again apply the same mathematics to both sides of the equation. That is, *all* terms on the left of the equal sign are treated in the same way as those on the right.

EXAMPLE 8

Figure 10 shows the steps to transpose the equation $A = B^2C$ to make B the subject. Steps 1 to 4 transpose the equation to make B^2 the subject. In step 5, to make B the subject, a square root function is applied to both sides so the square function applied to B can be cancelled.

Figure 10 Steps to transpose the equation $A = B^2C$ to make B the subject

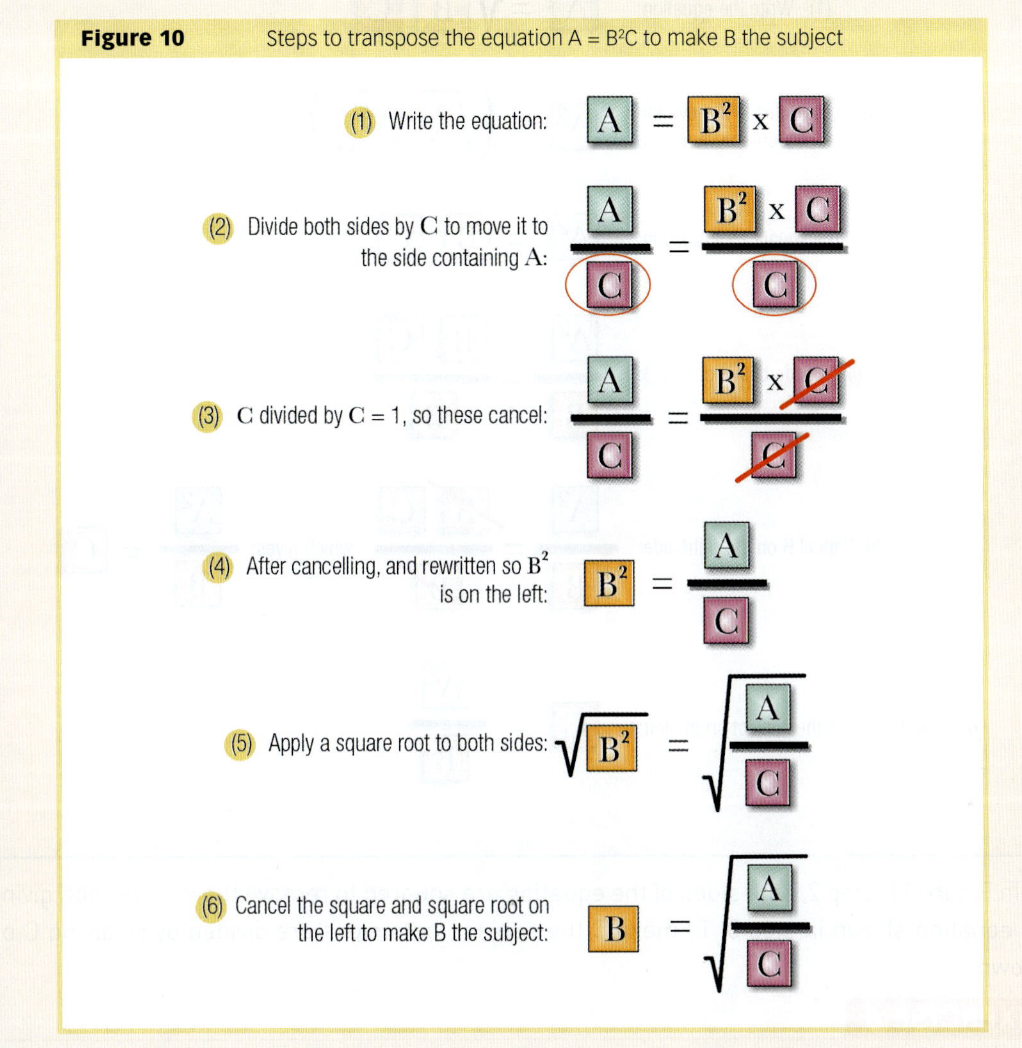

To transpose an equation with a square root function, apply the square function to both sides of the equation to cancel the square root function, as shown in Example 9.

EXAMPLE 9

Transpose $A = \sqrt{BC}$ to make C the subject.

Figure 11 Steps to transpose the equation $A = \sqrt{BC}$ to make C the subject.

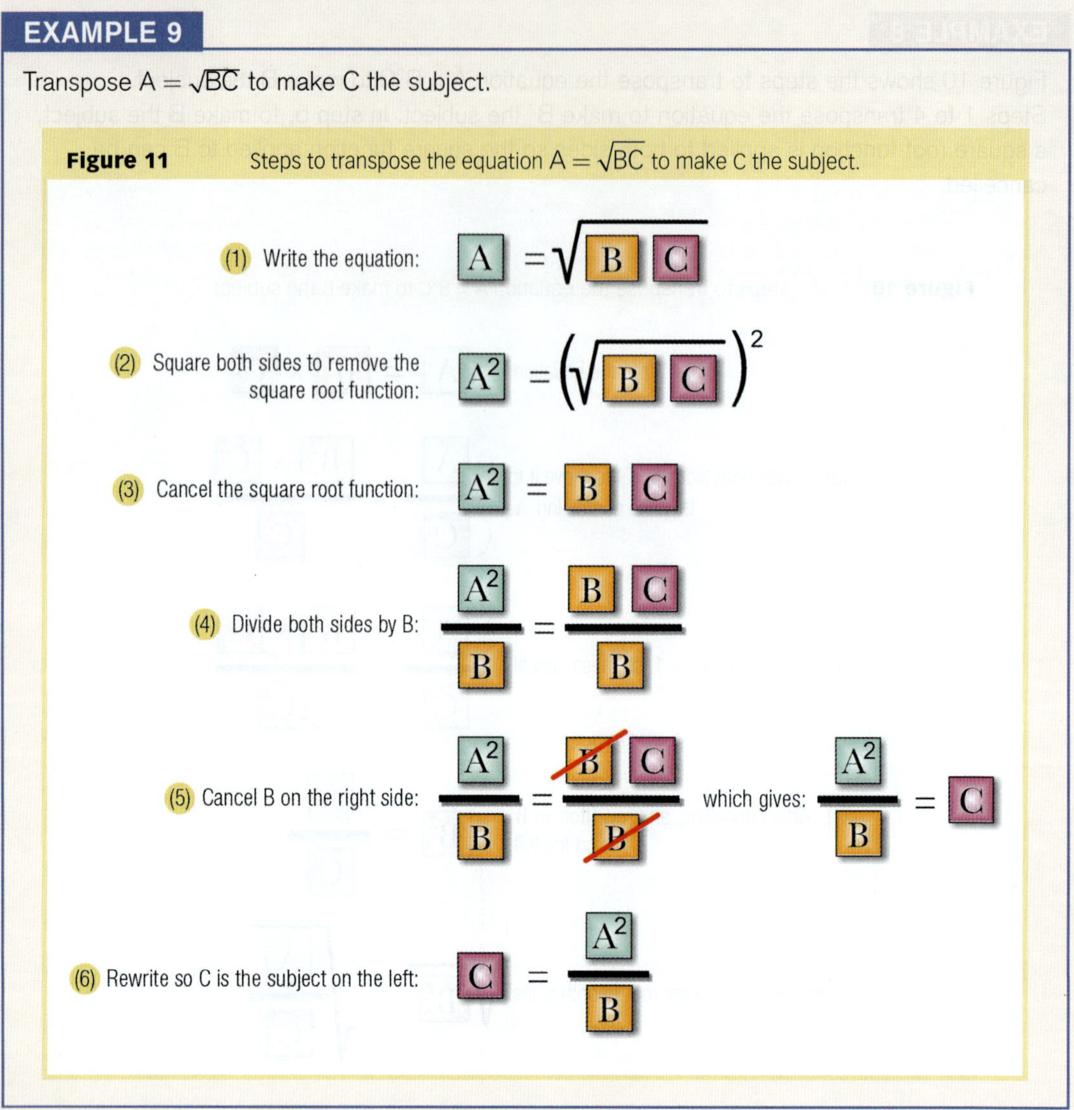

In Figure 11, step 2, both sides of the equation are squared to remove the square root, giving the equation shown in step 3. To make C the subject, both sides are divided by B, giving C on its own.

EXERCISE 6

The equations in this exercise are from Chapter 4. They are used to find power, voltage, resistance and current.

1 Make P the subject of the equation $R = \frac{V^2}{P}$.

2 Given that $P = I^2R$, make I the subject.

3 If $V = \sqrt{PR}$, transpose the equation to make R the subject.

4 Transpose the equation $I = \sqrt{\frac{P}{R}}$ to make P the subject.

The following tables list all the equations in this book that you may need to transpose. Use these tables as a check if you are not sure of your transposition. The tables do not include all the equations in this book, only those you may have to transpose.

Table 1 DC circuits

DC circuits	Transpositions		
Electric charge: Q = coulombs, I = amperes, t = seconds	$Q = I \times t$	$I = \frac{Q}{t}$	$t = \frac{Q}{I}$
Velocity: v = metres/second, d = metres, t = seconds	$v = \frac{d}{t}$	$d = v \times t$	$t = \frac{d}{v}$
Force: F = newtons, m = kilograms, a = metres/second^2	$F = m \times a$	$m = \frac{F}{a}$	$a = \frac{F}{m}$
Work: W = joules, d = metres, m = kilograms, P = watts, t = seconds	$W = F \times d$	$F = \frac{W}{d}$	$d = \frac{W}{F}$
	$W = P \times t$	$P = \frac{W}{t}$	$t = \frac{W}{P}$
Ohm's Law: I = amperes, V = volts, R = ohms	$I = \frac{V}{R}$	$V = I \times R$	$R = \frac{V}{I}$
Efficiency: n% = percentage efficiency, P = watts	$\eta\% = \frac{P_{out}}{P_{in}} \times 100$	$P_{out} = \frac{n\%}{100} \times P_{in}$	$P_{in} = \frac{n\%}{100} \times P_{out}$
Resistance of conductors: ρ = ohm/metres, l = metres, A = meters^2	$R = \frac{\rho \times l}{A}$	$A = \frac{\rho \times l}{R}$	$\rho = \frac{R \times A}{l}$
Electric Power: P = watts, V = volts, I = amperes	$P = V \times I$	$V = \frac{P}{I}$	$I = \frac{P}{V}$
	$P = I^2 \times R$	$V = \sqrt{P \times R}$	$I = \sqrt{\frac{P}{R}}$
	$P = \frac{V^2}{R}$		$R = \frac{V^2}{P}$
Series circuits: R = ohms, V = volts	$R_T = R_1 + R_2 + R_3$	$R_1 = R_T -$ other resistors	
	$V_T = V_1 + V_2 + V_3$	$V_? = V_T -$ other voltages	
Parallel circuits: R = ohms, I = amperes	$R_T = \frac{1}{\frac{1}{R_1} + \frac{1}{R_2} + \frac{1}{R_3}}$	$\frac{1}{R_T} = \frac{1}{R_1} + \frac{1}{R_2} + \frac{1}{R_3}$	
	$I_T = I_1 + I_2 + I_3$	$I_? = I_T -$ other currents	
Capacitance: C = farads Q = coulombs, V = volts	$C = \frac{Q}{V}$	$Q = C \times V$	$V = \frac{Q}{C}$
Capacitors in series: C = farads	$C_T = \frac{1}{\frac{1}{C_1} + \frac{1}{C_2} + \frac{1}{C_3}}$	$\frac{1}{C_T} = \frac{1}{C_1} + \frac{1}{C_2} + \frac{1}{C_3}$	
Capacitors in parallel:	$C_T = C_1 + C_2 + C_3$	$C_? = C_T -$ other capacitors	
RC time constant (τ = secs): R = ohms, C = farads	$\tau = R \times C$	$R = \frac{\tau}{C}$	$C = \frac{\tau}{R}$

Table 2 Electromagnetism

Electromagnetism	Transpositions		
Magnetomotive force (F_m): F_m = ampere turns, I = amperes, N = turns	$F_m = I \times N$	$I = \frac{F_m}{N}$	$N = \frac{F_m}{I}$
Magnetic flux: Φ = webers, F_m = ampere turns (At), R_M = ampere turns/weber	$F_m = \Phi \times R_M$	$\Phi = \frac{F_m}{R_M}$	$R_M = \frac{F_m}{\Phi}$
Permeability: μ_r has no unit μ = henry/metre (H/m), μ_0 = 1.26 x 10^{-6} H/m	$\mu = \mu_o \times \mu_r$	$\mu_o = \frac{\mu}{\mu_r}$	$\mu_r = \frac{\mu}{\mu_o}$
Flux density (B): B = tesla, Φ = webers, A = metres^2, μ = henry/metre (H/m), H = ampere turns/metre	$B = \frac{\Phi}{A}$	$\Phi = B \times A$	$A = \frac{\Phi}{B}$
	$B = \mu \times H$	$\mu = \frac{B}{H}$	$H = \frac{B}{\mu}$
Magnetising force (H): l = length in metres	$H = \frac{F_m}{l}$	$l = \frac{F_m}{H}$	$F_m = l \times H_m$
LR time constant (τ = secs) R = ohms, L = henrys	$\tau = \frac{L}{R}$	$L = \tau \times R$	$R = \frac{L}{\tau}$

Table 3 AC circuits

AC circuits	Transpositions		
Periodic time: t= seconds, f = hertz	$t = \frac{1}{f}$	$f = \frac{1}{t}$	
RMS value of voltage V: V and V_{max} = volts	$V = 0.707 \times V_{max}$	$V_{max} = \frac{V}{0.707}$	
Average value of voltage (half cycle) V_{av} = volts	$V_{av} = 0.637 \times V_{max}$	$V_{max} = \frac{V_{av}}{0.637}$	
Instantaneous voltage: v = volts, θ = degrees	$v = V_{max} \times \sin\theta$	$V_{max} = \frac{v}{\sin\theta}$	$\sin\theta = \frac{v}{V_{max}}$
Ohm's law: I = amperes, V = volts, Z = ohms	$I = \frac{V}{Z}$	$V = I \times Z$	$Z = \frac{V}{I}$
Impedance: $X = X_L - X_C$	$Z = \sqrt{R^2 + X^2}$	$R = \sqrt{Z^2 - X^2}$	$X = \sqrt{Z^2 - R^2}$
Inductive reactance: f = hertz, L = henrys	$X_L = 2\pi \times f \times L$	$L = \frac{X_L}{2\pi \times f}$	
Capacitive reactance: f = hertz, C = farads	$X_C = \frac{1}{2\pi \times f \times C}$	$C = \frac{1}{2\pi \times f \times X_C}$	

Table 3 AC circuits

AC circuits	Transpositions		
Power factor (λ): φ = phase shift in degrees, P = true power in watts, S = apparent power in VA, Z = impedance ohms	$\lambda = \cos\phi$	$\phi = \cos^{-1}$ of λ	
	$\cos\phi = \frac{P}{S}$	$P = \cos\phi \times S$	$S = \frac{P}{\cos\phi}$
	$\cos\phi = \frac{R}{Z}$	$R = \cos\phi \times Z$	$Z = \frac{R}{\cos\phi}$
Single-phase power (P): V = volts, I = amperes	$P = V \times I \times \cos\phi$	$V = \frac{P}{I \times \cos\phi}$	$I = \frac{P}{V \times \cos\phi}$
Single-phase reactive power:	$Q = V \times I \times \sin\phi$	$V = \frac{Q}{I \times \sin\phi}$	$I = \frac{Q}{V \times \sin\phi}$
Single-phase apparent power: P = watts, S = VA, Q = VA_R	$S = V \times I$	$V = \frac{S}{I}$	$I = \frac{S}{V}$
	$S = \sqrt{P^2 + Q^2}$	$P = \sqrt{S^2 - Q^2}$	$Q = \sqrt{S^2 - P^2}$
Voltages – star connected 3-phase system:	$V_L = \sqrt{3} \text{ x } V_P$	$V_P = \frac{V_L}{\sqrt{3}}$	
Currents – delta connected 3-phase system:	$I_L = \sqrt{3} \times I_P$	$I_P = \frac{I_L}{\sqrt{3}}$	
3-phase true power:	$P = \sqrt{3} \times V_L \times I_L \times \cos\phi$	$V_L = \frac{P}{\sqrt{3} \times I_L \times \cos\phi}$	
	$I_L = \frac{P}{\sqrt{3} \times V_L \times \cos\phi}$	$\cos\phi = \frac{P}{\sqrt{3} \times V_L \times I_L}$	
3-phase reactive power:	$Q = \sqrt{3} \times V_L \times I_L \times \sin\phi$	$\cos\phi = \frac{P}{\sqrt{3} \times V_L \times I_L}$	
	$I_L = \frac{Q}{\sqrt{3} \times V_L \times \sin\phi}$	$\sin\phi = \frac{Q}{\sqrt{3} \times V_L \times I_L}$	
3-phase apparent power:	$S = \sqrt{3} \times V_L \times I_L$	$V_L = \frac{S}{\sqrt{3} \times I_L}$	$I_L = \frac{S}{\sqrt{3} \times V_L}$
Two wattmeter power measurement:	$P = W_1 + W_2$	$W_1 = P - W_2$	$W_2 = P - W_1$
$\tan\phi = \sqrt{3} \times \left(\frac{W_1 - W_2}{W_1 + W_2}\right)$	$W_1 = W_2 \times \left(\frac{\tan\phi + \sqrt{3}}{\sqrt{3} - \tan\phi}\right)$	$W_2 = W_1 \times \left(\frac{\sqrt{3} - \tan\phi}{\tan\phi + \sqrt{3}}\right)$	

1.5 PYTHAGORAS' THEOREM

The equation known as Pythagoras' theorem was developed over 2000 years ago and concerns measurements with a right-angled triangle. This equation has many applications in electrical theory. The equation is generally written as $c^2 = a^2 + b^2$ and is shown in Figure 12.

Figure 12 Pythagoras' theorem

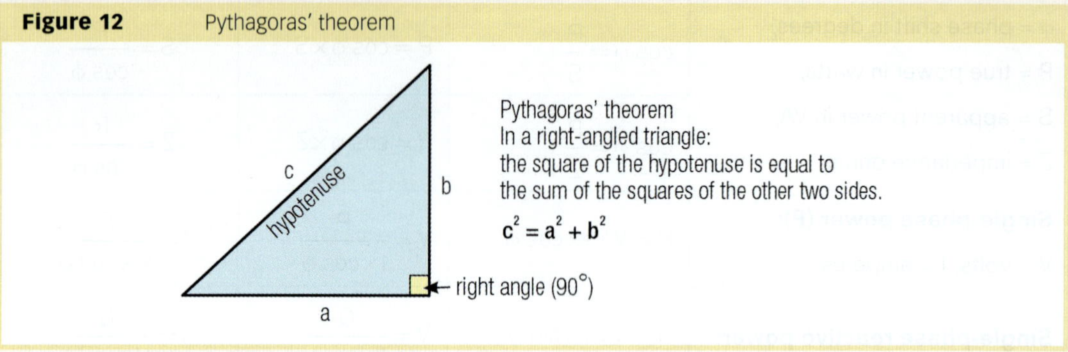

To find c (the hypotenuse), applying a square root function to both sides gives this equation:

$c = \sqrt{a^2 + b^2}$

If a = 3 and b =4, $c = \sqrt{3^2 + 4^2} = \sqrt{9+16} = \sqrt{25} = 5$

Transposing the general equation to make a or b the subject requires rearranging the terms and applying a square root function to both sides, as shown in Example 10.

EXAMPLE 10

Figure 13

c = 12 mm

b

a = 7 mm

For the triangle in Figure 13, find the length of side b.

Method

Step 1. Transpose $c^2 = a^2 + b^2$ to make b the subject

1 Subtract a^2 from both sides: $c^2 - a^2 = a^2 + b^2 - a^2$

2 Cancel a^2 on the right side: $c^2 - a^2 = b^2$

3 Rearrange so b is on the left side: $b^2 = c^2 - a^2$

4 Apply square root function to both sides: $\sqrt{b^2} = \sqrt{c^2 - a^2}$

5 Cancel the square root on the left, as the square of a square root cancels:

$b = \sqrt{c^2 - a^2}$

6 Write the equation by replacing letters c and a with their values:

$b = \sqrt{12^2 - 7^2}$

Step 2. To find b, enter the values into a Casio calculator as in Figure 14.

Figure 14 Calculator keystrokes to solve Example 10

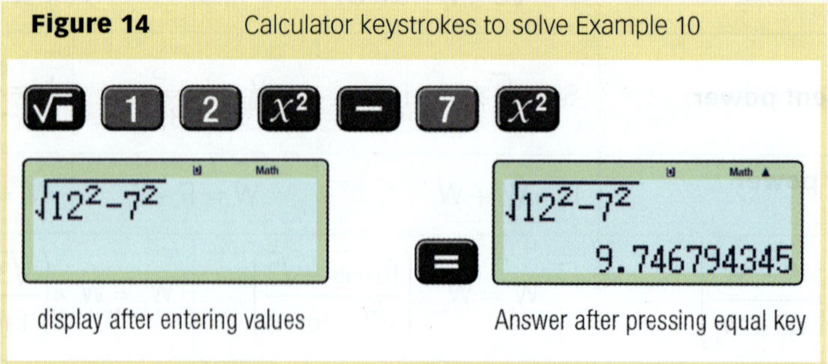

After rounding the calculator result, answer: **b = 9.75 mm**

EXERCISE 7

1 Transpose the equation $c^2 = a^2 + b^2$ to make a the subject.

2 Find the value of side R for the triangle in Figure 15.

3 Make X the subject of the equation $Z = \sqrt{R^2 + X^2}$.

4 Find the length of the hypotenuse of a right-angled triangle if the other sides measure 27 mm and 36 mm.

Figure 15

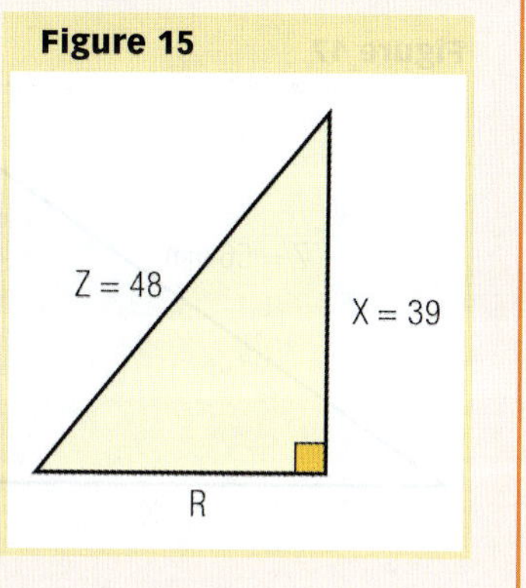

1.6 BASIC TRIGONOMETRY

Trigonometry is used to find the length of any side of a right-angled triangle or any angle within the triangle, providing you know at least two measurements. Trigonometry was first used over 2000 years ago and has its own terminology to describe its functions. Figure 16 shows the three terms used to identify each side of a right-angled triangle, and introduces the terms related to angles. Note that the hypotenuse is *always* opposite the right angle. A way of remembering these three functions is: SOH CAH TOA (try saying it).

Figure 16 Basic trigonometrical functions

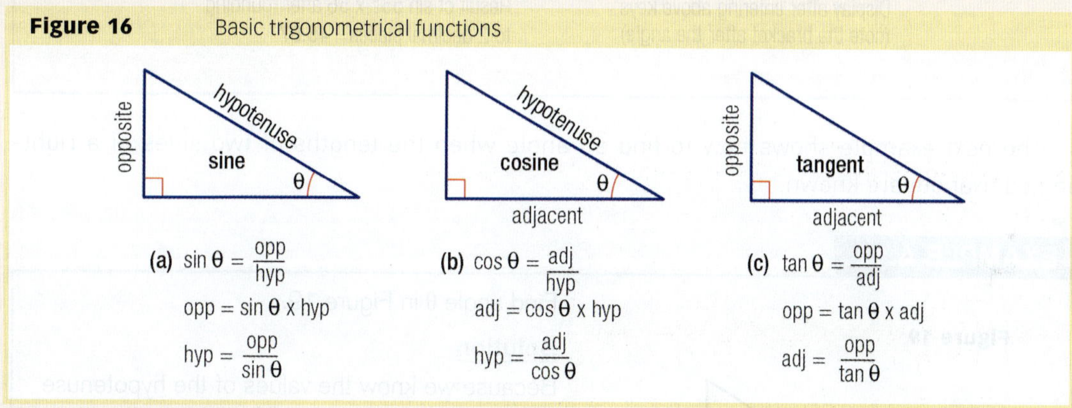

THE SINE FUNCTION

The sine of an angle is found by dividing the length of the side *opposite* the angle by the length of the hypotenuse. Sine is shortened to sin, which is how a calculator key is marked. That is, $\sin\theta = \frac{\text{opp}}{\text{hyp}}$ as shown in Figure 16 (a). The Greek letter theta (θ) is the angle in question. For example, if the opposite side is 20 and the hypotenuse is 40, sin θ equals 20 divided by 40. Therefore, sin θ equals 0.5.

Transposing $\sin\theta = \frac{\text{opp}}{\text{hyp}}$ to make opp the subject gives opp = sin θ x hyp.

Making the hypotenuse the subject gives $\text{hyp} = \frac{\text{opp}}{\sin\theta}$.

EXAMPLE 11

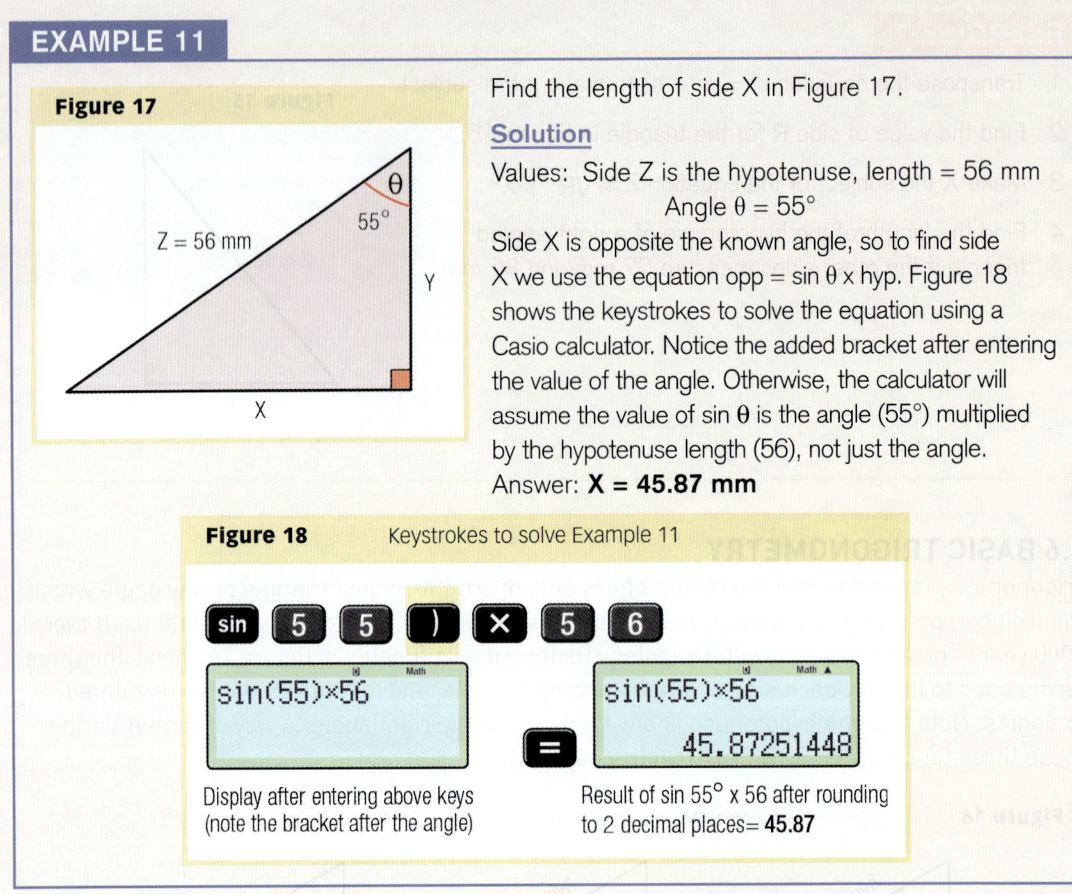

Figure 17

Find the length of side X in Figure 17.

Solution

Values: Side Z is the hypotenuse, length = 56 mm

Angle $\theta = 55°$

Side X is opposite the known angle, so to find side X we use the equation opp = sin θ x hyp. Figure 18 shows the keystrokes to solve the equation using a Casio calculator. Notice the added bracket after entering the value of the angle. Otherwise, the calculator will assume the value of sin θ is the angle (55°) multiplied by the hypotenuse length (56), not just the angle.

Answer: **X = 45.87 mm**

Figure 18 Keystrokes to solve Example 11

The next example shows how to find an angle when the lengths of two sides of a right-angled triangle are known.

EXAMPLE A.12

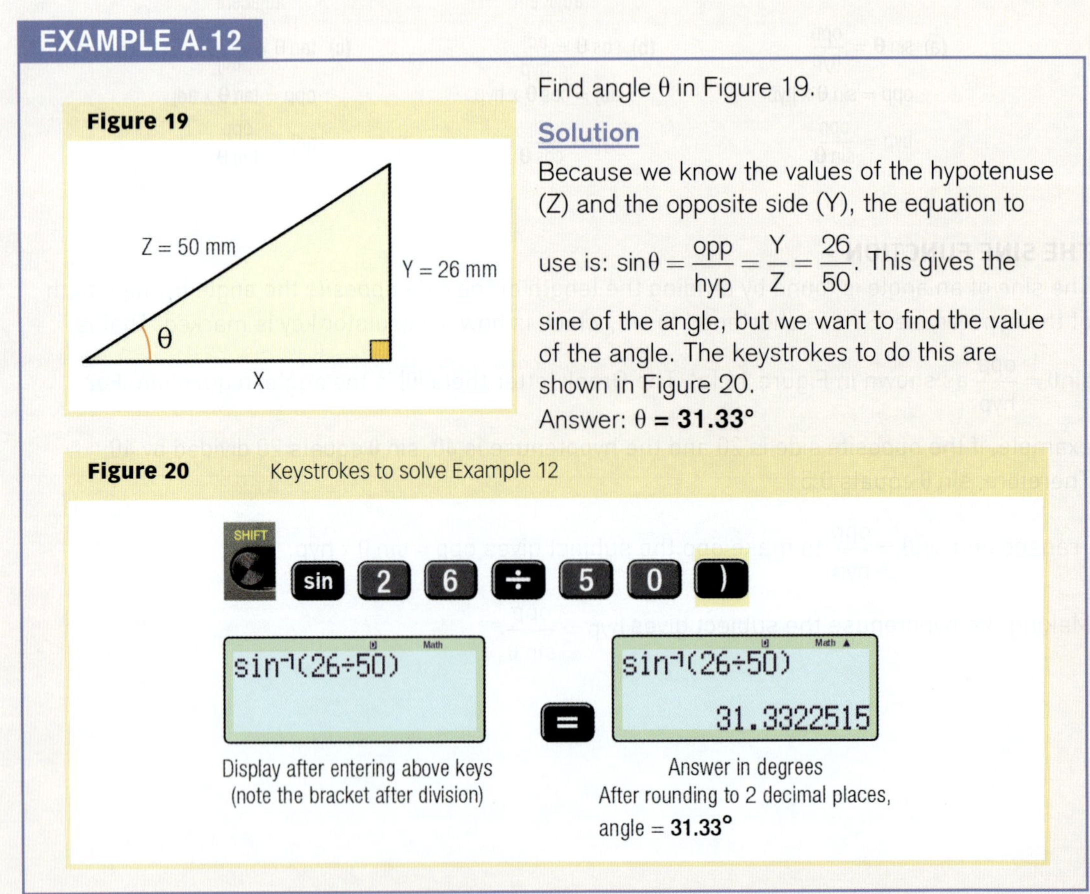

Figure 19

Find angle θ in Figure 19.

Solution

Because we know the values of the hypotenuse (Z) and the opposite side (Y), the equation to use is: $\sin\theta = \frac{\text{opp}}{\text{hyp}} = \frac{Y}{Z} = \frac{26}{50}$. This gives the sine of the angle, but we want to find the value of the angle. The keystrokes to do this are shown in Figure 20.

Answer: **θ = 31.33°**

Figure 20 Keystrokes to solve Example 12

Figure 20 shows how to find the angle whose sine is a particular value. That is, the calculator SHIFT key is pressed so the key marked SIN becomes SIN^{-1}. This means 'the angle whose sine is'. The SIN key is used on its own to find the sine value of an angle.

THE COSINE FUNCTION

The cosine of an angle is found by dividing the length of the side next to the angle (adjacent side) by the length of the hypotenuse. Cosine is shortened to cos, as marked on a calculator key. That is:

$\cos\theta = \frac{adj}{hyp}$.

In any triangle, the values of the enclosed angles add up to 180°. In a right-angled triangle, because one angle is 90°, the other two must add up to 90°. These two angles are *complementary*, which means if one angle is known, subtracting it from 90° gives the value of the other angle. Therefore, the unknown angle in Figure 19 is 90° minus 31.33°, which, as shown in Figure 21 is 58.67°. The 'co' in cosine stands for 'complement', because of this relationship between these two acute angles.

Figure 21

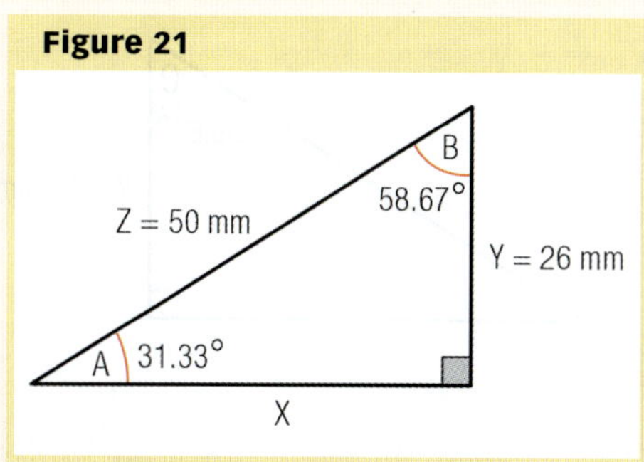

In Figure 21, the cosine of angle $B = \frac{adj}{hyp} = \frac{Y}{Z} = \frac{26}{50} = 0.52$. The sine of angle $A = \frac{opp}{hyp} = \frac{Y}{Z} = \frac{26}{50}$ also equals 0.52, because these angles are complementary.

Transposing $\cos\theta = \frac{adj}{hyp}$ to make either side the subject gives two more equations:

1 $hyp = \frac{adj}{\cos\theta}$
2 adj = $\cos\theta \times hyp$

Figure 22 shows the keystrokes to find angle B in Figure 21. First set the calculator to find the angle whose 'cosine is' by pressing the SHIFT key, then the COS key. The calculator adds a bracket, after which the lengths of the adjacent and hypotenuse sides are entered, separated by the divide sign.

Figure 22 Keystrokes to find angle B in Figure 21 using the cosine function

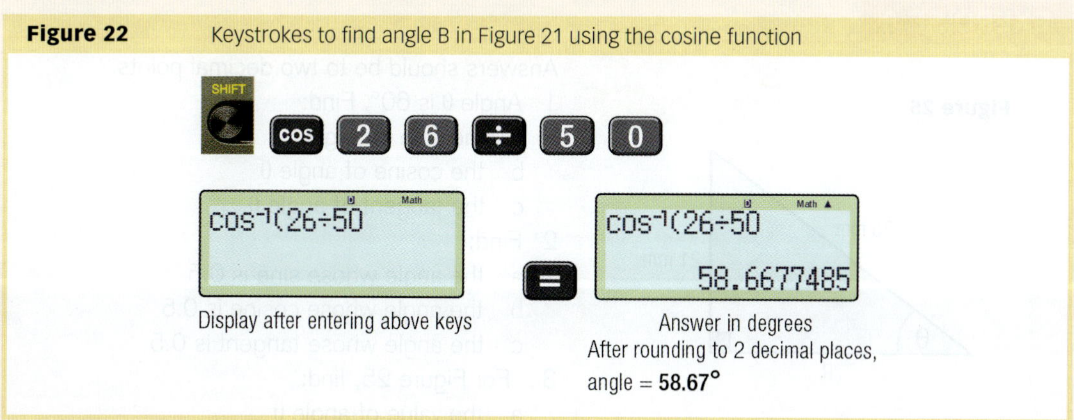

THE TANGENT (TAN) FUNCTION

The tangent of an angle is found by dividing the length of the side opposite the angle by the side next to the angle (adjacent side). That is $\tan\theta = \frac{opp}{adj}$. Transposing this equation to make either side the subject gives:

1 opp = tan θ × adj

2 $\text{opp} = \dfrac{\text{adj}}{\tan\theta}$

Tangent is shortened to tan as marked on a calculator key. Using the tangent function is like using sine or cosine functions, as shown in Example 13.

EXAMPLE 13

Figure 23

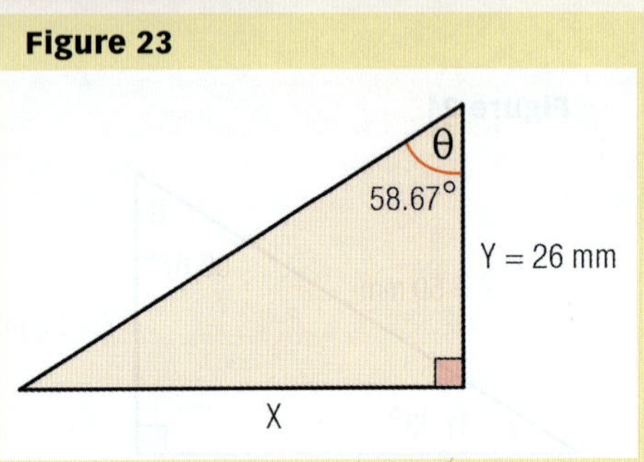

Find the length of side X of the triangle in Figure 23.

Solution

Because we know the angle and the length of the side the angle is next to (adjacent), the tan function can be used to find the length of side X, as it is the opposite side. The equation is opp = tan θ × adj.

The keystrokes to find side X are shown in Figure 24. Notice the bracket after entering the value of the angle. Otherwise, the calculator will assume the value of tan θ is the value of the angle (58.67°) multiplied by the length (26) of the adjacent side.

Answer: **X = 42.71 mm**

Figure 24 Keystrokes to solve Example 13

Display after entering above keys (note the bracket after the angle)

Result of tan 58.67° x 26 rounded to = **42.71**

EXERCISE 8

Figure 25

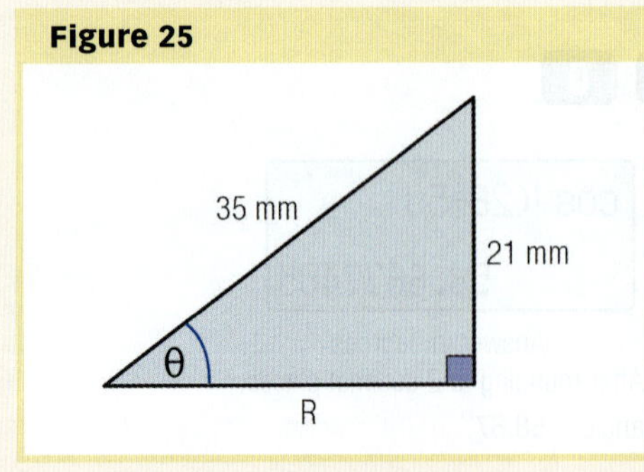

Answers should be to two decimal points.

1 Angle θ is 60°. Find:
 a the sine of angle θ
 b the cosine of angle θ
 c the tangent of angle θ

2 Find:
 a the angle whose sine is 0.5
 b the angle whose cosine is 0.5
 c the angle whose tangent is 0.5

3 For Figure 25, find:
 a the value of angle θ
 b the length of side R
 c the value of the other angle.

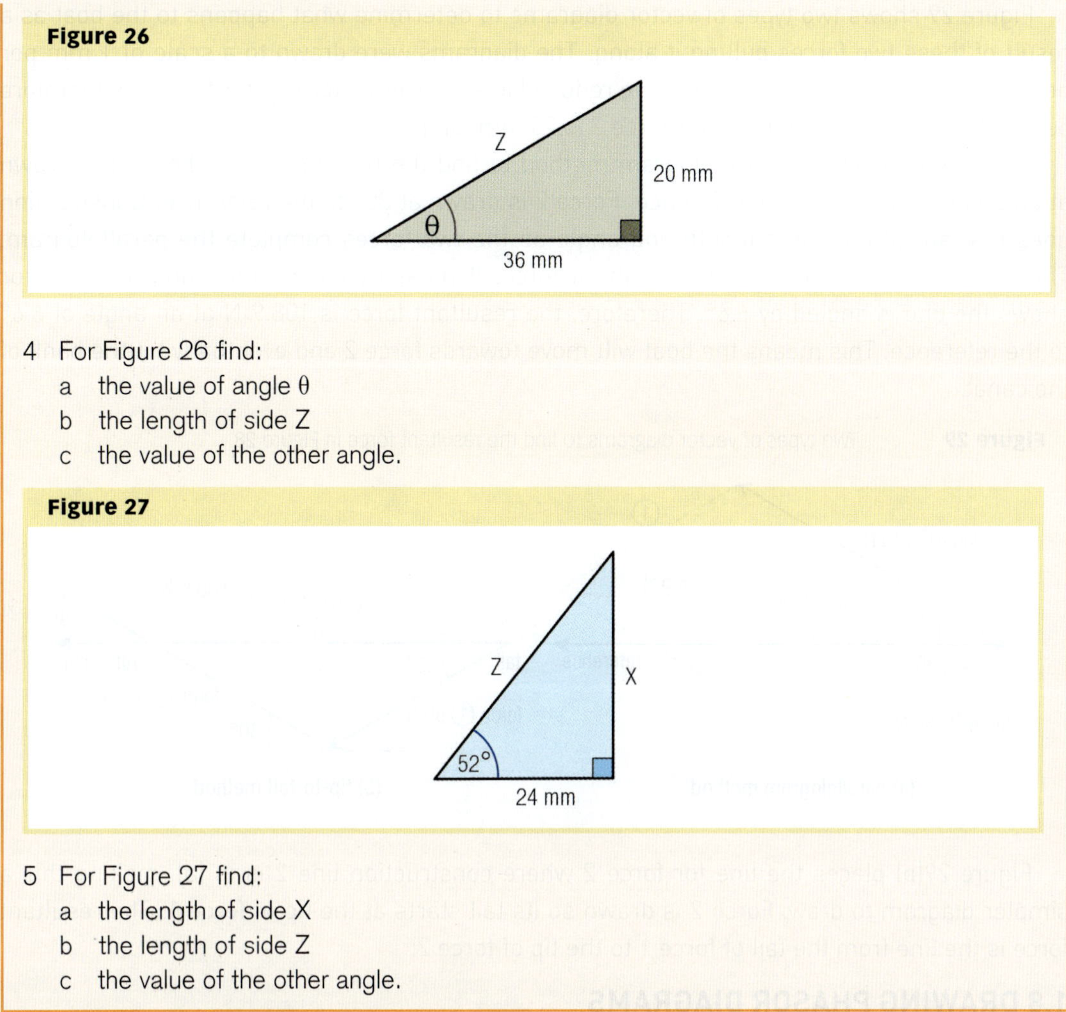

Figure 26

4 For Figure 26 find:
   a the value of angle θ
   b the length of side Z
   c the value of the other angle.

Figure 27

5 For Figure 27 find:
   a the length of side X
   b the length of side Z
   c the value of the other angle.

1.7 VECTOR AND PHASOR DIAGRAMS

A vector is a line with an arrow head that represents a physical force that has size and direction. The length of the arrow represents the size of the force, and the angle of the arrow represents its direction. A vector diagram is used in physics as a graphical way of finding the resulting force when two or more forces are acting on an object. Figure 28 shows a boat being pulled by two people, one pulling with a force of 75 newtons (N), the other with a force of 50 N. Both forces are acting at 30° to the centre line.

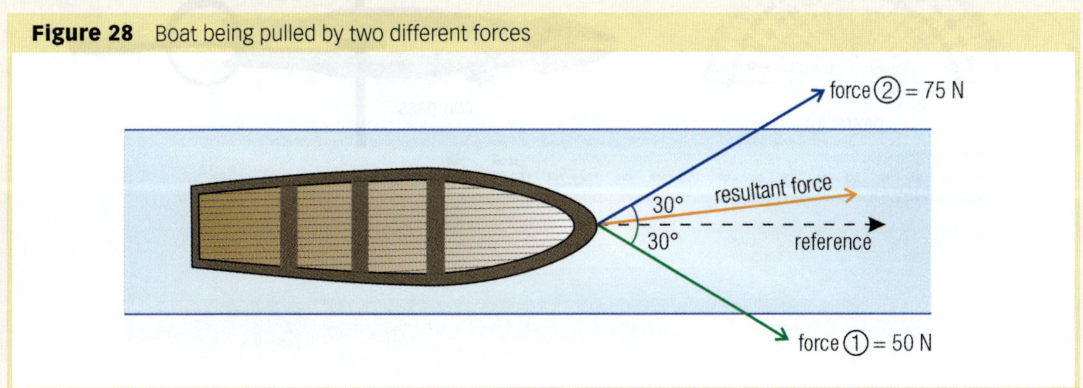

Figure 28 Boat being pulled by two different forces

Figure 29 shows two types of vector diagrams to determine what happens to the boat as a result of these two forces pulling it along. The diagrams were drawn to a scale of 1 mm per newton, although the drawings here are reduced in size. The vector line for force 1 is therefore 50 mm long, and the vector line for force 2 is 75 mm long.

Figure 29(a) uses the parallelogram method to find the resultant force. Force 1 is drawn at an angle of -30° from the reference. Force 2 is drawn at 30° to the reference. Construction lines that are of the same length and angle as the two forces complete the parallelogram. The diagonal line represents the resultant force. It measures 108.9 mm, and a protractor shows the line is angled by 6.6°. Therefore, the resultant force is 108.9 N at an angle of 6.6° to the reference. This means the boat will move towards force 2 and eventually hit the bank of the canal.

Figure 29 Two types of vector diagrams to find the resultant force in Figure 28.

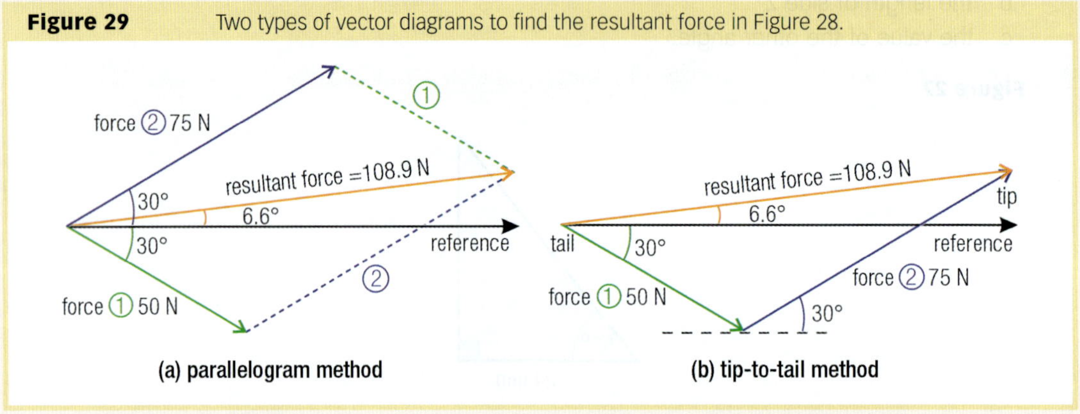

Figure 29(b) places the line for force 2 where construction line 2 is in (a), making this a simpler diagram to draw. Force 2 is drawn so its tail starts at the tip of force 1. The resultant force is the line from the tail of force 1 to the tip of force 2.

1.8 DRAWING PHASOR DIAGRAMS

A phasor diagram is similar to a vector diagram and is used in electrotechnology to solve problems in which voltages or currents in a circuit are out of phase with each other. A phasor is like a vector, except it represents an electrical value. The drawing tools needed to construct a phasor diagram are shown in Figure 30.

Figure 30 Drawing tools to construct phasor diagrams

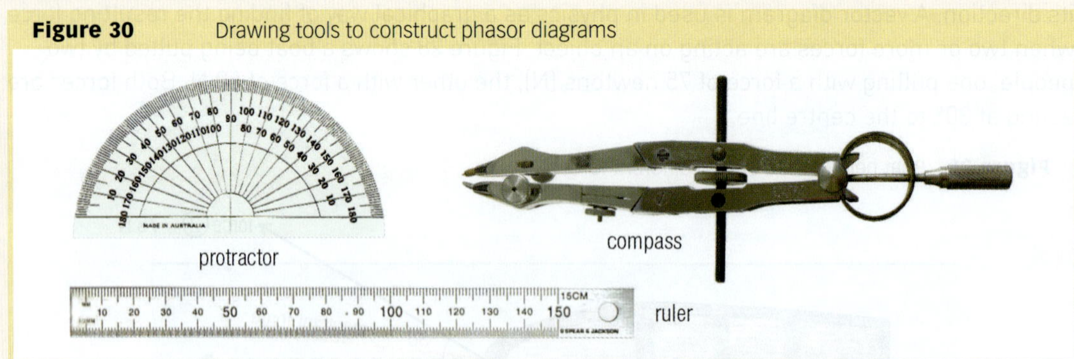

EXAMPLE 14

This example shows how to draw the phasor diagram to solve the problem in Figure 15.24 (Chapter 15). The problem is to find the total current and its phase angle in a parallel circuit, given the values and phase angles of two branch currents, I_1 and I_2. Current I_1 is 7 A at a leading phase angle of 40°, and current I_2 is 8 A at a lagging phase angle of 20°.

Method 1

1 Select a scale. In this example, 1 amp = 10 mm. On graph paper, draw a horizontal line of any length for the reference, which is voltage.
2 Place a protractor as shown in Figure 31(a), and draw an arrow mark to indicate an angle of 40°.
3 Because I_1 equals 7 A, scribe an arc with a compass set to 70 mm. Place the point of the compass at the tail of the reference line. Draw the phasor line for I_1 from the tail of the reference line (0) to the arc, as shown in Figure 31(b). It should measure 70 mm.

Figure 31 Constructing the phasor for current 1_1, 7 A at a leading phase angle of 40°

70 mm radius arc
mark at 40°
0
voltage V

(a) place protractor, make a mark at 40°, draw arc with compass at 70 mm radius

70 mm radius arc
mark at 40°
$I_1 = 7\,A$
0
voltage V

(b) draw 70 mm line for phasor I_1 from 0 point to intersect with arc

4 As explained by comment (a) in Figure 32, for phasor line I_2, draw an arrow mark at 20° and scribe an arc with a radius of 80 mm, because I_2 = 8 A. The compass point is placed at the tail of the reference.
5 Draw an 80 mm construction line at 20°, as explained by (b) in Figure 32.
6 Draw a 70 mm construction line at 40°, shown by (c) in Figure 32.

Figure 32 Constructing the phasor for current I_2, which is 8 A at a lagging phase angle of 20°

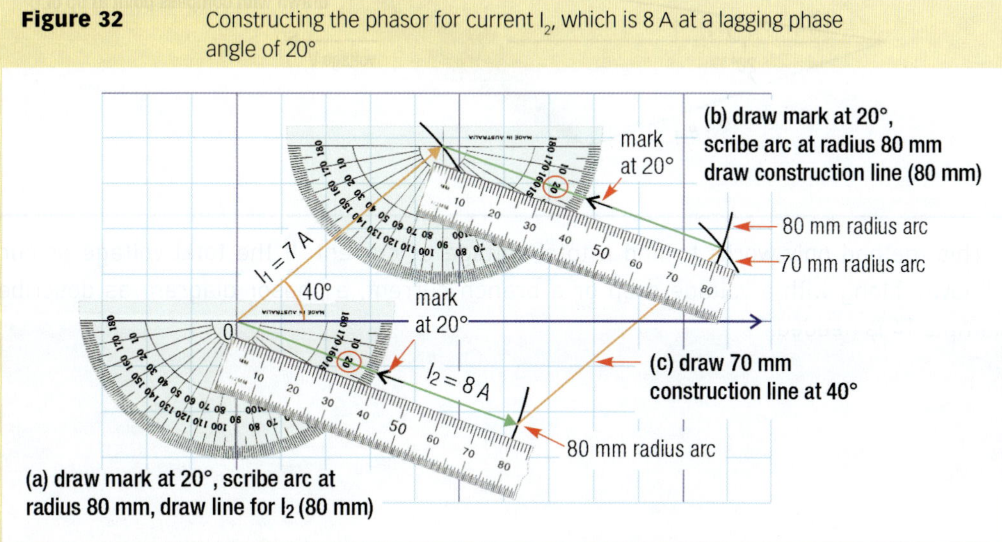

7 Complete the phasor diagram by drawing a diagonal line for I_T from the tail of the reference to the point where both construction lines meet. See Figure 33.
8 As shown in Figure 33, solve the problem by measuring the angle and length of phasor I_T. Answer is 13 A at a leading phase angle of 7.8° (13 A ∠7.8°).

Figure 33 Complete the diagram to find the value and phase angle of current I_T

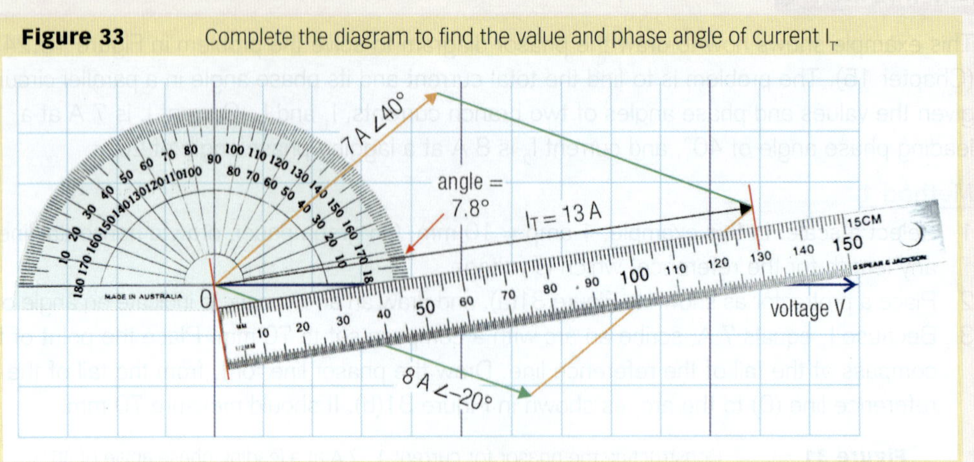

Method 2

A simpler way to construct the phasor diagram to solve this problem is shown in Figure 34. The diagram can be drawn on plain paper, as all angles are related to the reference line.

1. Draw a horizontal line for the reference.
2. Construct the phasors for I_1 and I_2 as shown in Figures 31 and 32.
3. Set the compass to the length of phasor I_1, and place its point at the tip of phasor I_2 and scribe an arc.
4. Set the compass to the length of phasor I_2, and place its point at the tip of phasor I_1 and scribe an arc.
5. Draw the resultant phasor from the tail of the reference to the point where the two arcs intersect. Measure its length and angle as in Figure 33 to solve for I_T.

Figure 34 Phasor diagram on plain paper without construction lines

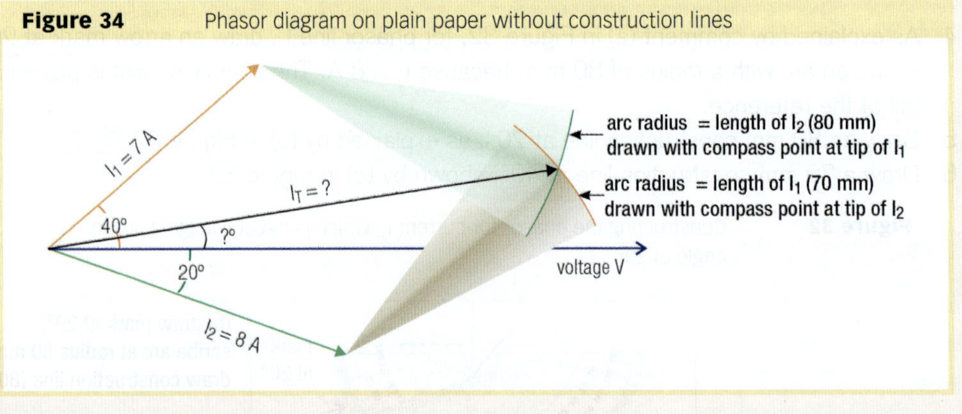

This method only works to find a total voltage or current. If the total voltage or current is known, along with a voltage drop or a branch current, a phasor diagram as described in Example 14 is needed.

EXAMPLE 15

This example shows how to draw the phasor diagram to solve the problem in the series circuit shown in Figure 15.28(d) Chapter 15. The problem is to find an unknown voltage drop (V_1) and its phase angle, given the values and phase angles of one voltage drop (V_2) and the total voltage V_T. Voltage V_2 is 85 V at a lagging phase angle of 30°, the total voltage is 120 V at a lagging angle of 10°.

Figure 35 Tip-to-tail phasor diagram to find a voltage drop in a series circuit

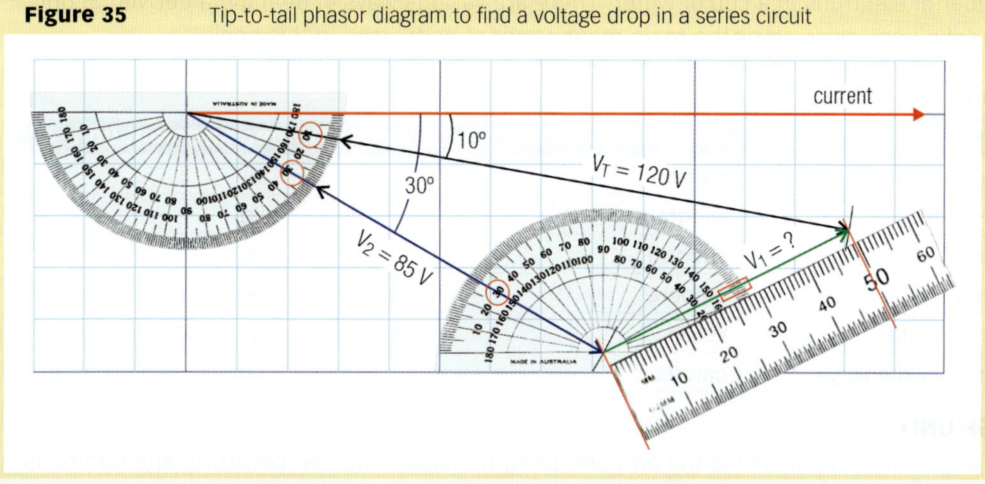

1. Draw the reference line, which is current. We have used a scale of 1 mm per volt.
2. Draw both the known phasors (V_2 and V_T) as explained in Figure 32.
3. Draw a line for unknown phasor V_1 from the tip of phasor V_2 to the tip of phasor V_T.
4. Measure the angle and length of phasor V_2 to solve the problem. By measurement, V_2 = 49.6 V at a leading angle of 25.9°. It is leading because the angle rotates the phasor in an anticlockwise direction.

This method is sometimes called 'tip-to-tip' construction, as the unknown phasor is drawn between the tips of the known phasors. This is an easier construction than the parallelogram method.

1.9 MATHEMATICAL ALTERNATIVE TO A PHASOR DIAGRAM

In some cases, a problem can be solved with trigonometry rather than drawing a phasor diagram. The mathematical solution for Example 14, described in Figures 31 to 34, is shown in Figure 36.

Figure 36 Solving Example 14 with trigonometry

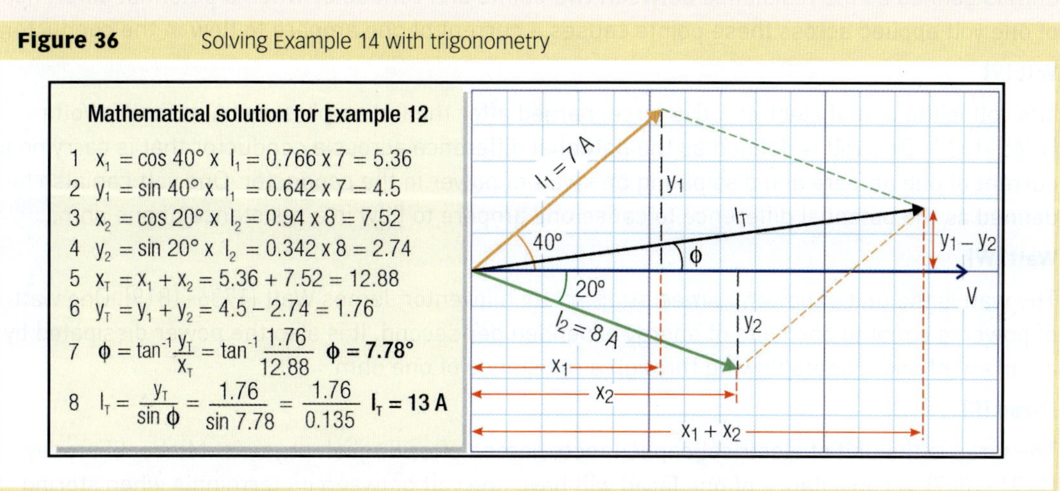

2. Definitions

2.1 SI BASE AND DERIVED MEASUREMENT UNITS

The International System of Units (SI) defines seven units of measurement as a basic set from which all other SI units are derived. Six base units are listed in Table 1; the unlisted unit (which is not used in this book) is the mole, described broadly as a measure of stuff, such as the number of electrons in a cup of water. There are various ways of defining a derived SI unit. The definitions given cover the main measurement units used in this book.

Table 1 SI Base units

Base quantity	Measurement unit	Symbol for measurement unit
length	metre	m
mass (m)	kilogram	kg
time (t)	second	s
current	ampere	A
temperature	kelvin	k
luminous intensity	candela	cd

BASE UNIT

The important SI base unit is the ampere. Length, mass, time, temperature, and luminous intensity are defined using atomic science to establish exact references.

Ampere (A)

Ampere is an SI base unit for current flow, named after André-Marie Ampère (1775–1836). It has the symbol A for amperes, but in an equation is written as I (for Intensity). A current of 10 amperes is written as I = 10 A. One ampere is defined as 6.24×10^{18} (about) electrons flowing past a point in one second, or one coulomb of charge per second. One ampere is also the current produced by one volt acting across a resistance of one ohm.

COMMONLY USED SI DERIVED ELECTRICAL UNITS

Coulomb (C)

The coulomb is the unit of electric charge, named after French physicist Charles-Augustin de Coulomb (1736–1806). One coulomb is defined as 6.24×10^{18} (about) electrons or as that number of electrons passing a point when a current of one ampere is flowing.

Ohm (Ω)

The ohm is the unit of resistance, named after German physicist Georg Ohm (1787–1854). One ohm is defined as the resistance between two points of a conductor when a potential difference of one volt applied across these points causes a current of one ampere to flow in the conductor.

Volt (V)

The volt is the unit of electromotive force, named after the Italian physicist Alessandro Volta (1745–1827). One volt is defined as the potential difference across a conductor that is carrying a current of one ampere and dissipating one watt of power in the conductor. One volt can also be defined as the potential difference to cause one ampere to flow in a resistance of one ohm.

Watt (W)

The watt is the unit of power, named after Scottish inventor James Watt (1736–1819). One watt of power is equal to one joule of energy expended per second. It is also the power dissipated by a current of one ampere flowing through a resistance of one ohm.

Farad (F)

The farad is the unit of electrical capacitance, named after English physicist Michael Faraday (1791–1867). A capacitance of one farad will have one volt between its terminals when storing one coulomb of charge. Or one farad is the capacity of a capacitor to store one coulomb of charge if one volt is applied.

Henry (H)

The henry is the unit of inductance, named after American scientist Joseph Henry (1797–1878). One henry is the amount of inductance that causes an induced voltage of one volt across the inductance when the current in the inductance is changing at one ampere per second. Inductance is the property of a circuit or component that opposes a change in the value of an electric current.

Hertz (Hz)

The hertz is the unit of frequency, named after German scientist Heinrich Hertz (1857–1894). One hertz is one cycle per second of a periodically changing electrical voltage or current.

COMMONLY USED SI DERIVED UNITS OF FORCE AND ENERGY

Joule (J)

The joule is the unit of work and energy. It is named after the English physicist James Prescott Joule (1818–1889). The joule can be defined in various ways. In an electrical sense, a joule is the amount of energy dissipated as heat when an electric current of one ampere passes through a resistance of one ohm for one second. A joule is also the work required to produce one watt of power for one second. In mechanics the joule can be defined as the work done by a force of one newton to move an object by one metre in the direction of the force.

Newton (N)

The newton is the unit of force, named after Isaac Newton (1642–1727). One newton is the force needed to accelerate a mass of one kilogram at the rate of one metre per second per second in the direction of the applied force.

COMMONLY USED SI DERIVED UNITS OF MAGNETISM

Tesla (T)

The tesla is the unit of magnetic flux density (B), named after Nikola Tesla (1856–1943). One tesla is equal to one weber per square metre or flux (Φ) divided by area $\left(B = \frac{\phi}{A}\right)$.

Weber (Φ)

The weber is the unit of magnetic flux, named after German physicist Wilhelm Weber (1804–1891). It is defined as the amount of flux that causes one volt to be produced in a single loop of wire when the current in the loop is reduced at a uniform rate to zero over a period of one second. It can also be defined as the total flux produced by a magnet, in which flux is depicted as lines of force. One weber is said to have 100 000 000 (10^8) lines of magnetic flux. Magnetic lines of flux are an imaginary concept to help us understand magnetism.

Magnetomotive force (F_m)

Magnetomotive force (F_m) is to magnetism as electromotive force (emf) is to an electric circuit. It is the driving force that sets up a magnetic field. The SI unit is ampere-turn (At) and equals the product of the current (I) flowing in a coil multiplied by the number of turns (N) in the coil. That is, $F_m = NI$ ampere-turns.

Magnetising force (H)

Magnetising force is magnetomotive force (ampere-turns) divided by the length of the magnetic circuit. It is officially measured in amperes per metre, as turns does not have a unit, but to avoid confusion most texts use ampere-turns per metre (At/m). $H = \frac{NI}{l}\,\text{At/m}$.

Reluctance (R_m)

Reluctance is resistance to a magnetic field, measured in ampere-turns per weber (At/Wb). It is equal to the magnetomotive force divided by magnetic flux $R_m = \frac{F_m}{\Phi}$. This equation is to magnetism as Ohm's law is to electricity in which $R = \frac{V}{I}$. Reluctance is sometimes given the symbol S. A high permeability material has a low reluctance.

Permeability (μ)

Permeability is a measure of the ability of a material to support, or conduct, a magnetic field. It is measured in henrys per metre (H/m) and has the symbol μ (μ in italics).
The permeability of air (μ_0) is 1.26×10^{-6} (approx) or $4\pi \times 10^{-7}$ (exactly).

Relative permeability (μ_r) is a comparison of the permeability of a material to air (air has a μ_r of 1). Absolute permeability (μ) of a material is its relative permeability multiplied by the permeability of air. That is $\mu = \mu_r \times \mu_0$ henry/metre (H/m).

2.2 GREEK LETTER SYMBOLS AND THEIR MEANING

The Greek letters used in this book are listed and defined in Table 2.

Table 2 Greek letters

Letter	Name	What it represents
Δ	delta	change in (for example Δt means change in time)
ε	epsilon	permitivity of an insulator, value only
η	eta	efficiency
λ	lamda	power factor (= cos ϕ), also wavelength in communications
μ	mu	micro (10^{-6} or one millionth)
μ	mu (in italics)	permeability, unit is henrys per metre (H/m)
Ω	omega	resistance, unit is the ohm
Φ	phi upper case	magnetic flux, unit is the weber (Wb)
ϕ or φ	phi lower case	phase angle between phasors in an AC circuit, measured in degrees
π	pi	constant used in circular measurement = 3.1416 (to 4 decimal places) also = 355/113
ρ	rho	resistivity, unit is the ohmmetre (Ωm)
τ	tau	time constant, unit is the second (s)
θ	theta	angle of rotation or angle in a triangle, measured in degrees

2.3 METRIC PREFIXES (OR MULTIPLIERS)

The four most commonly used metric prefixes are mega, kilo, milli and micro. These and other prefixes are given in Table 3. A prefix is expressed in mathematical form using scientific notation, in which each prefix is treated as a power of 10. Engineering notation only uses powers of 10 that are a multiple of 3. Engineering notation is fully described in Chapter 3.

Table 3 Engineering notation

Term	Symbol	Multiply by	Engineering notation	Example
tera	T	1 000 000 000 000	10^{12}	2 THz = 2×10^{12} Hz
giga	G	1 000 000 000	10^{9}	4.62 GHz = 4.62×10^{9} Hz
mega	M	1 000 000	10^{6}	2.2 MΩ = 2.2×10^{6} Ω
kilo	k	1000	10^{3}	5 kV = 5×10^{3} V
milli	m	0.001	10^{-3}	3.4 mA = 3.4×10^{-3} A
micro	μ	0.000 001	10^{-6}	89 mA = 89×10^{-6} A
nano	n	0.000 000 001	10^{-9}	68 nF = 68×10^{-9} F
pico	p	0.000 000 000 001	10^{-12}	120 pF = 120×10^{-12} F

CHAPTER 1

THE ELECTRIC CIRCUIT

This chapter provides an introduction to the electrotechnology industry and two important laws that cover the entire workforce: workplace safety and environmental legislation. The meaning of terms such as voltage, current and resistance, the difference between a conductor and an insulator, and what is meant by an open-circuit and a short-circuit are also described. The electric circuit and some electrical component symbols are presented, but first we look at how electrical power reaches our homes.

CHAPTER OUTLINE

1.1 The Electrotechnology Industry

Electricity in most countries is generated by a mix of renewable energy sources and coal- and gas-fired power stations. The lower cost of renewable energy is changing the mix and in states like South Australia and Tasmania, over half their electricity is generated by renewables. As shown in Figure 1.1, electricity is transmitted to all parts of the country over high voltage power transmission lines. These lines supply power to substations, which convert the high voltage to a more suitable voltage and distribute the electrical power to homes, industry and commerce.

FIGURE 1.1 Electrical power is transmitted at high voltages, and after stepping down to lower voltages, is distributed to homes and industry

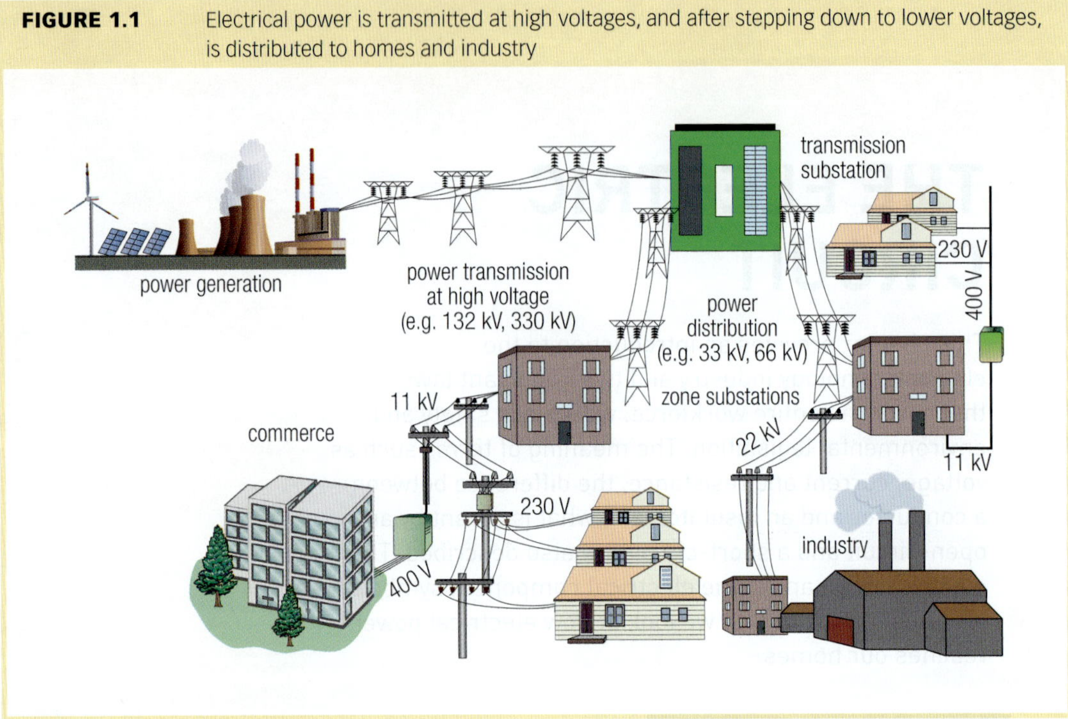

The electrical industry

The electrical industry can be divided into three main areas:

- **Electrical supply** – includes power generation, transmission and distribution. The supply industry includes statutory authorities in Australia such as Transgrid (in New South Wales), Powerlink (Queensland), Western Power (Western Australia), SA Power Networks (South Australia) and AusNet Services (Victoria), which are responsible for operating and managing the high voltage network. The state-owned transmission networks are interconnected and are under the management of Australia's National Electricity Market (NEM) authority. The New Zealand Electricity Authority is responsible for that country's electricity market. Power distribution is done by energy companies, who are responsible for distributing electrical power to commerce, industry and homes.
- **Industrial** – involves installing and maintaining electrical machinery and electrical wiring in factories and industrial complexes. Industries include manufacturing, mining, lifts and refrigeration.
- **Commercial and domestic** – covers installation of solar panels, wiring (lighting, power, data) and the installation and repair of appliances such as stoves, lighting systems and some types of motorised equipment. Electricians who are installing or working on wiring that connects to the electrical supply must have the correct licence.

The electronics industry

The electrical and electronics industries have much in common. An electrician might install data cabling, electronically controlled lighting systems, air conditioners, alarm systems and electronic modules that form part of a solar panel installation. An electronics technician would be trained to repair, but not necessarily install, this type of equipment. Figure 1.2 shows examples of the type of equipment associated with the electronics industry.

FIGURE 1.2 These are some of the main areas in the electronics industry

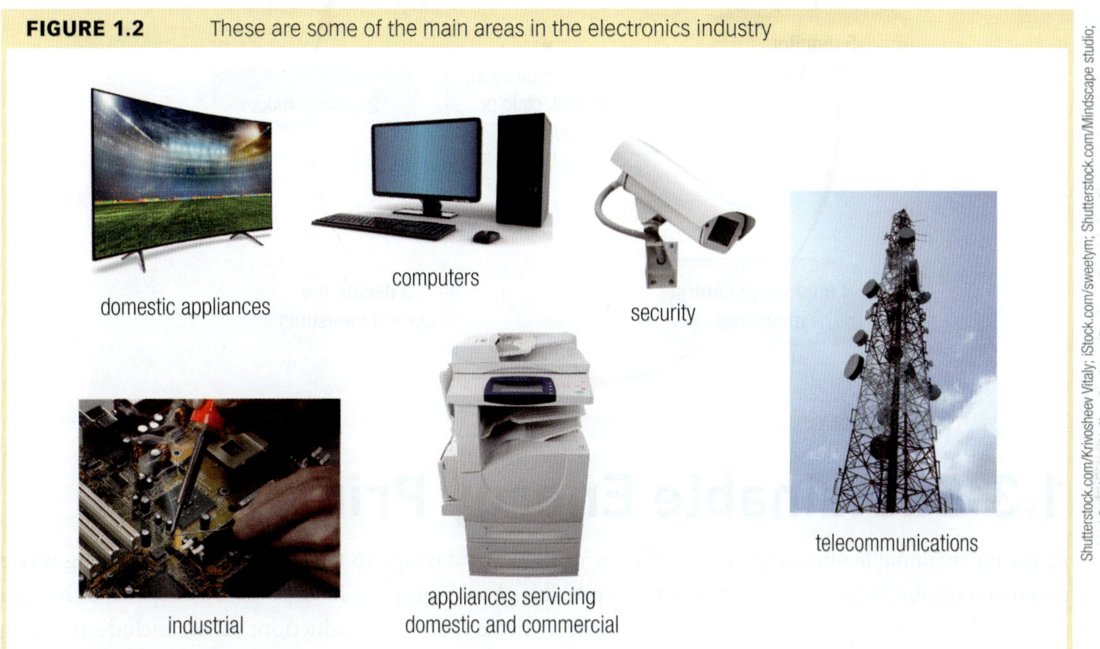

Shutterstock.com/Krivosheev Vitaly; iStock.com/sweetym; Shutterstock.com/Mindscape studio; Getty Images/Jeffrey Coolidge; iStock.com/by_nicholas; Shutterstock.com/drpnncpptak

1.2 Workplace Safety Laws

Workplace safety applies to all areas of the workforce, and all Australian states and territories and New Zealand have legislation that covers the rights and responsibilities of employers and employees. Those undertaking a Certificate III (trades) course will learn about workplace health and safety (WHS) legislation as part of the course. In general, WHS legislation imposes a duty of care on employers and certain obligations on employees. Briefly, employers are required to ensure the health, safety and welfare at work of all employees and others who come into the workplace.

FYI

A hazard is anything that can cause harm to people, plant or the environment

Employees are obliged to take care of the health and safety of co-workers, cooperate with employers in matters of health and safety, work safely and to notify their supervisor of actual or potential hazards. Essentially, this means that the WHS legislation places an obligation on *every* person associated with a workplace in any way to ensure his or her own workplace health and safety.

Working with electricity has its own special hazards, as well as those associated with other industries; for example, working on a roof near overhead power lines, where there's a risk of falling *and* a risk of electric shock. The important thing, before starting any electrical work, is to first identify the hazards and assess the risk associated with each hazard. Some employers require work teams to complete hazard and risk identification paperwork before starting a job. This requires all team members to identify the potential hazards and their associated risk, determine how the risks will be dealt with (controlled), and when everyone agrees, to sign the completed form.

SAFETY

Recommended procedures for safe working practices when working on low-voltage electrical installations are covered by AS/NZS 4836:2011

In the above example, the risk of falling is high and potentially fatal. The control measures would include wearing a safety harness plus other personal protective equipment (PPE), along with being very aware of the hazard and taking appropriate care. The electrical hazard would also need to be controlled in a manner that ensures your safety. As a job proceeds, you might find some of the control measures are not providing sufficient safety, so a review becomes necessary and changes need to be made. The risk management process is shown in Figure 1.3, and applies to every job, whether outdoors or indoors.

FIGURE 1.3 Risk management is a constant process

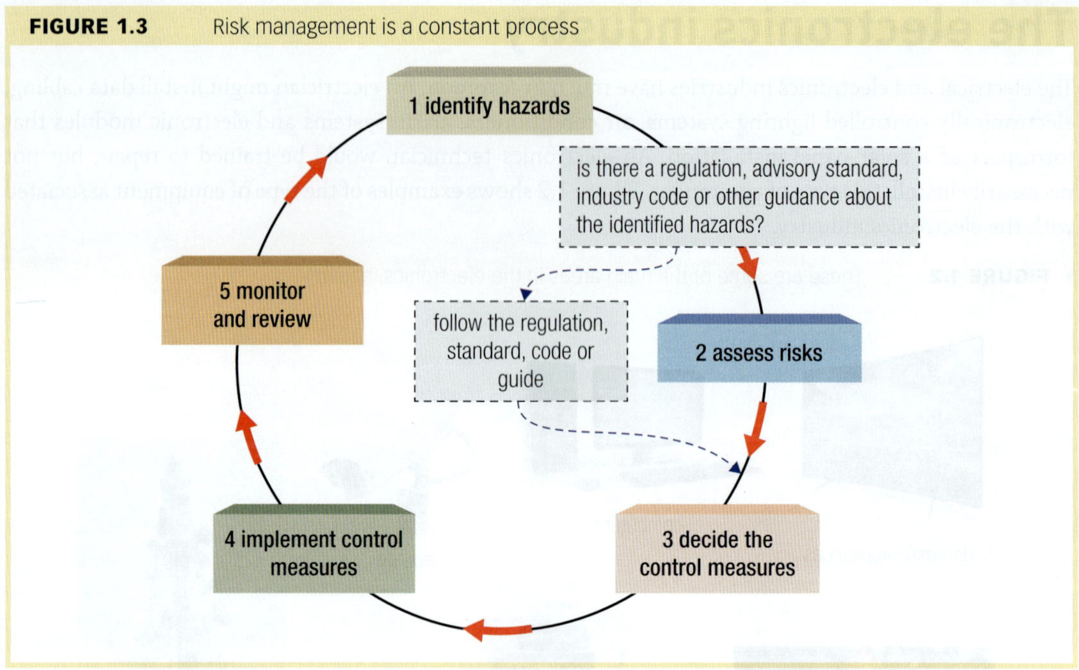

1.3 Sustainable Energy Principles

Applying sustainable energy principles means minimising damage to the environment, reducing waste and using resources in the most efficient way possible. As a concept, it incorporates all aspects of human activity that have an environmental impact. Among these is energy production; others include anything that compromises the ability of future generations to meet their own needs. Sustainable energy is made up of two parts: renewable energy and energy efficiency. It is energy which can be replenished within a human lifetime and causes no long-term damage to the environment.

All Australian states and territories and New Zealand have legislation that covers air, land and water pollution. Environmental legislation usually includes protection for animals and plants that are in danger of extinction and also covers noise pollution, waste disposal and any aspect to do with the environment. Penalties can be very high, for both corporations and individuals.

Industrial organisations will have an *environmental management plan*, which sets out ways of ensuring the organisation's activities have minimum impact on the environment. For example, if a job involves digging a trench to lay cables, the fill taken from the trench must be stored so it cannot run into the stormwater system. This could mean building a sediment fence around the dirt pile to prevent runoff during rain.

A particular environmental and health hazard is asbestos, often used as wall cladding, roofing material and concrete water pipes. Others include oil spillage, chemical waste spillage or anything that enters the stormwater system. It is illegal to discharge anything other than clean water into a stormwater system. Electricity suppliers will also have an *environmental incident management plan* to deal with the many potential situations that could pose a threat to the environment.

Waste disposal is a particular issue for the electrotechnology industry, due to the widespread use of batteries, computers, electronic equipment, oils, chemicals and other environmentally unfriendly items. Because these items are toxic to the environment, they cannot be disposed of as landfill.

In general, sustainable energy principles refers to adopting a lifestyle and work habits that enable the present generation to meets its needs while maintaining a healthy environment for future generations. Sustainable energy technologies are helping make this possible, in which electric power is now being increasingly produced from renewable energy sources. (Energy sources that produce electricity are discussed in Chapter 2.)

FIGURE 1.4 Environmental incidents: fallen power pole with transformer leaking oil, and a pad mount transformer on fire.

Energy *efficiency* refers to using less energy for the same outcome. Products include those with a high efficiency rating such as, energy efficient lamps, in particular, light emitting diode (LED) lamps, which are replacing fluorescent tubes, compact fluorescent lamps and all types of filament lamps. New buildings have to conform to a set of standards designed to reduce energy and water usage. In New South Wales, these are known as BASIX. Other states and territories and New Zealand have similar legislation.

SUSTAINABILITY

Sustainable energy issues are highlighted in chapters where they apply

WHS and environmental legislation have been briefly described before we start looking at electrical theory, as it is very important to be aware of these two items of legislation, and to understand your responsibilities.

KEY POINTS...

- Electrical power is sent from power generators to substations over high voltage transmission lines and distributed to users over lower voltage power lines.
- The electrotechnology industry has over 20 qualifications, each requiring a knowledge of electrical principles.
- The various state and territory WHS and Environmental Protection Acts impose obligations on both employers and their employees.
- Sustainable energy is made up of two parts: renewable energy and energy efficiency.

1.4 Voltage

The three most common electrical quantities are **voltage**, **current** and **resistance**. These three terms are closely related, which means that if any two are present, the third is also present. For instance, if voltage and resistance are both present, so is current.

A voltage is *electrical pressure*. As an example, lightning is caused by a very high voltage (or electrical pressure) between a cloud and another point, such as Earth. When the pressure is high enough, there's a lightning strike. A car battery develops a voltage of 12 volts; there's 230 volts present at a power outlet in a house; a torch battery produces 1.5 volts.

A simple way to describe a voltage is to compare it with the water pressure produced by a water pump, as illustrated in Figure 1.5. A water pump, such as that used in a swimming pool, produces a pressure that forces water through the pool filter. A battery produces electrical pressure (voltage) that causes an electric current to flow when a circuit is connected to the battery.

FIGURE 1.5 Electrical pressure and water pressure both provide a force that causes flow

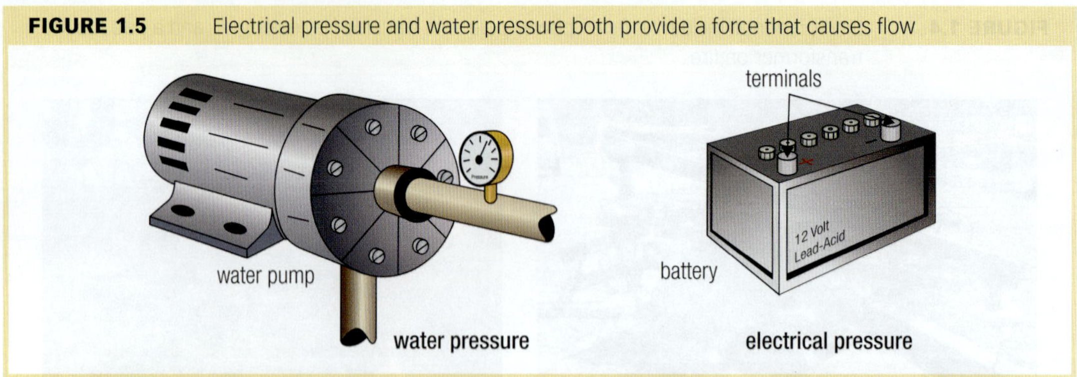

FIGURE 1.6 Electrons orbit around the nucleus of the atom

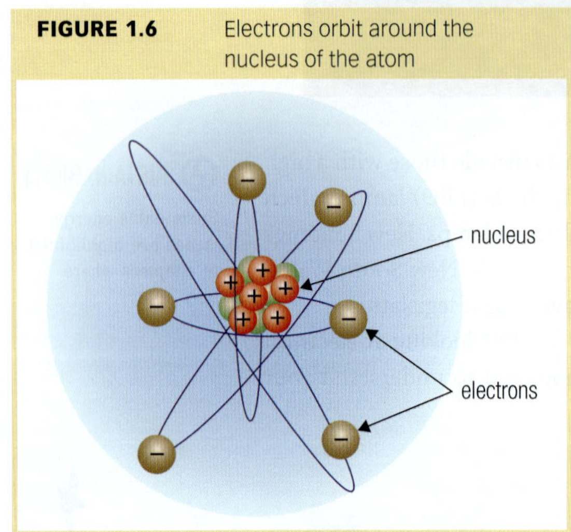

It's important to understand that a voltage always occurs between *two* points. Notice that the battery in Figure 1.5 has two terminals. The value of the voltage produced by a battery is determined by the materials that make up the battery. To understand voltage we need to understand the particles that form atoms. An atom is the smallest part of an element. The simplest atom has an equal number of elementary particles called protons and electrons. More complex atoms have a third particle called a neutron. An element is a substance made entirely from atoms of the same type. There are 92 natural elements (such as copper, oxygen, gold) and (so far) 26 man-made elements. These are produced using particle accelerators or nuclear bombardment. Each element has a number that identifies the number of protons in each atom. The higher the number, the heavier that element. Uranium has 92 protons and is the heaviest natural element. The simplest is element number 1 (hydrogen) with one proton and one electron. Copper has 29 protons, 29 electrons and 35 neutrons, giving it an atomic number of 29 and an atomic weight of 64 (29 protons plus 35 neutrons). The structure of a carbon atom, which has six electrons, six protons and six neutrons, is shown in Figure 1.6.

The nucleus of an atom contains protons (which have a positive charge) tightly bonded with neutrons (no charge). The nucleus, therefore, has a positive charge, balanced exactly by the negative charge of the orbiting electrons. Electrons are about three times larger than protons, but protons are over 1800 times heavier. Electrons can move more easily than the heavier protons, and are attracted by the positive charge of the protons in the nucleus of an atom.

Because an electron has a negative charge, a point with more electrons has a *negative potential* compared to one with fewer electrons. The point with fewer electrons has a *positive* potential. See Figure 1.7. (The term 'potential' is another word for voltage.) There's always a voltage, or potential difference, between two points that have a different number of electrons.

FIGURE 1.7 There is a voltage when one point has more electrons than another

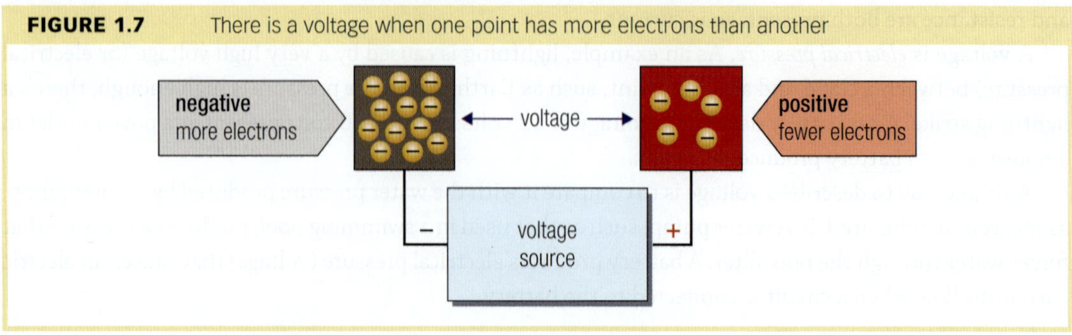

Voltage is measured in **volts** (after Alessandro Volta, who invented the electric cell). The symbol for volts (and for voltage) is the letter V. A car battery has a voltage of 12 volts, which can be written as V = 12 V. The voltage from a voltage source (such as a battery) is also called its **electromotive force** (EMF). The symbol for EMF is the letter E.

KEY POINTS...

- Voltage is electrical pressure between two points.
- Voltage is measured in volts.
- The symbol for voltage (and for volts) is V.

1.5 Current

Current is a *flow of electrons*. You could think of water flowing in a pipe as being like an electric current, just as a water pump is similar to a voltage source, such as a battery. Remember that you need a voltage to make the current flow, just as water pressure is needed to make the water flow. Without a voltage there can be no current flow.

FIGURE 1.8 An electric current is like water flow, except it is a flow of electrons

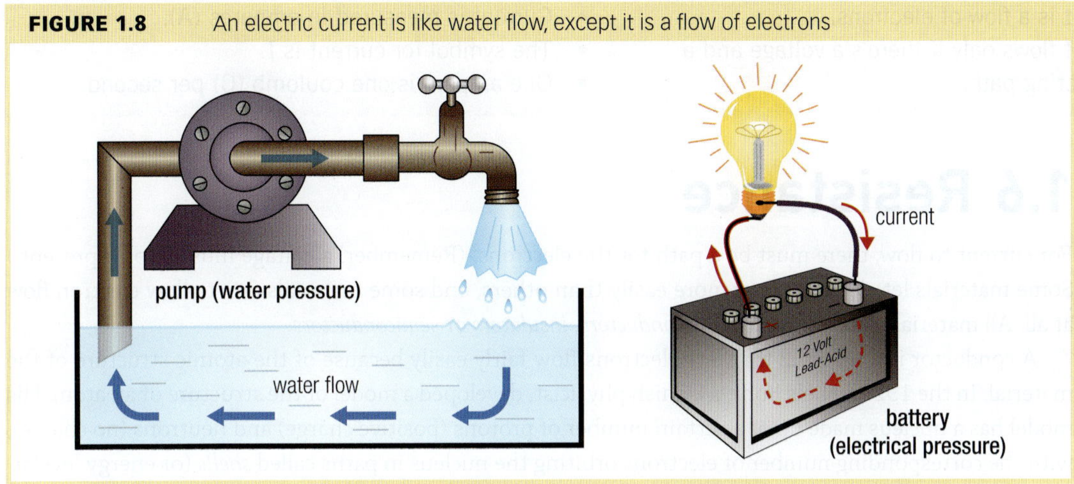

Although electrons are one of the smallest particles known, in quantity they drive motors, make lamps light up and operate electrical (and electronic) appliances. Although an enormous number are needed, it's still impossible to see an electric current, no matter how large the current is. What you can see is the *effect* of an electric current. Lightning is an electric current flowing through ionised air, and the bright flash you see is the ionising effect the electrons have on the air.

Because current is a flow of electrons, the electrons flow from the point where there are more electrons to the point where there are fewer electrons. That is, electrons flow from the negative potential to the positive potential. However, because the nature of electricity was not understood when it was first discovered, it was assumed that current flowed from positive to negative.

So, even though electrons flow from negative to positive, by convention we assume that current flows from *positive to negative*, as in Figure 1.9. (A confusing hangover from the past!) Conventional current flow is used in this book.

FIGURE 1.9 By convention, current flows from positive to negative and electrons flow from negative to positive

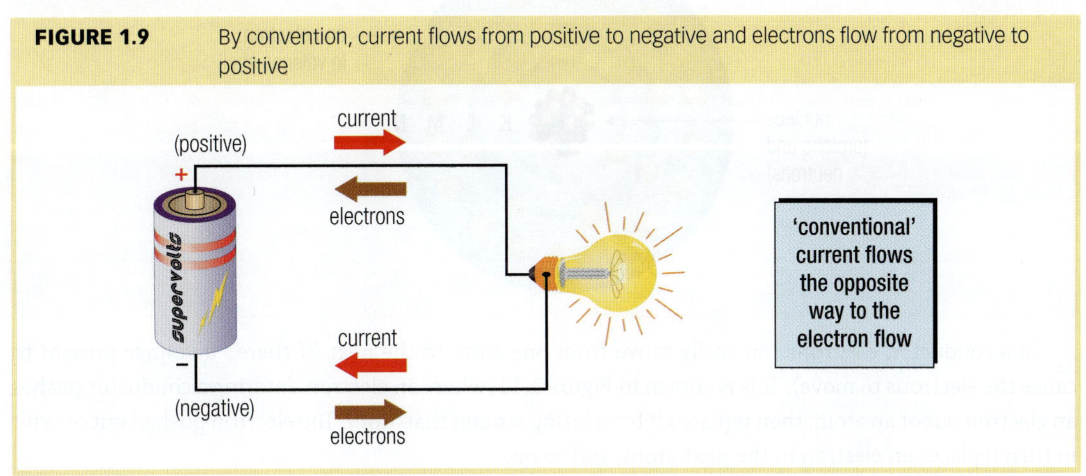

An electric current is measured in **amperes** (after André-Marie Ampère, a French scientist). One ampere is equal to 6.24×10^{18} electrons flowing past a point in one second. As 10^{18} is another way of writing 18 zeros, there are obviously a lot of electrons. This number of electrons is called one *coulomb* of electric charge. The coulomb is the unit for charge, and a flow of one coulomb per second is one ampere. The symbol for charge is the letter Q. The symbol for coulomb is C. An electrical charge is stationary. For example, charge is stored in a cloud before lightning discharges it. So current is *charge in motion*, or electrons (negative charge) moving in a conductor.

The symbol for current is the letter I (which originally stood for Intensity). The symbol for ampere is A. For example, a current of two amperes might be written as 2 A, or as I = 2 A. The starter motor in a car takes about 100 A, while a ceiling fan takes about 0.3 A.

KEY POINTS...

- Current is a flow of electrons.
- Current flows only if there's a voltage and a conducting path.
- Current is measured in amperes (A).
- The symbol for current is I.
- One ampere is one coulomb (C) per second.

1.6 Resistance

For current to flow, there must be a path for the electrons. (Remember, a voltage must also be present.) Some materials let electrons flow more easily than others, and some materials don't allow electron flow at all. All materials can be classified as *conductors*, *insulators* or *semiconductors*.

A conductor is a material that lets electrons flow fairly easily because of the atomic structure of the material. In the 1920s, Niels Bohr, a Danish physicist, developed a model of the structure of an atom. His model has a nucleus made up of a certain number of protons (positive charge) and neutrons (no charge), with the corresponding number of electrons orbiting the nucleus in paths called *shells* (or energy levels).

Each shell can only contain a certain number of electrons, and the outer shell often has less electrons than it is capable of housing. The outer shell is called the *valence* shell, and the number of electrons in this shell gives a *valency number* for that atom. The lower the valency number, the more easily the atom can gain or lose electrons, so a conductor will have a low valency (3 or less), while an insulator will have a high valency (greater than 5). Semiconductors have a valency of 4.

Figure 1.10 shows the atomic model for a copper atom, which has 29 protons in the nucleus, and 29 electrons orbiting the nucleus. The electrons are arranged in four orbits labelled K (inner) to N (valence shell). Because there is one electron in the valence shell, copper has a valency of 1, indicating that copper is a good conductor.

FIGURE 1.10 Copper atom with 29 electrons orbiting the nucleus in four shells: K, L, M and N

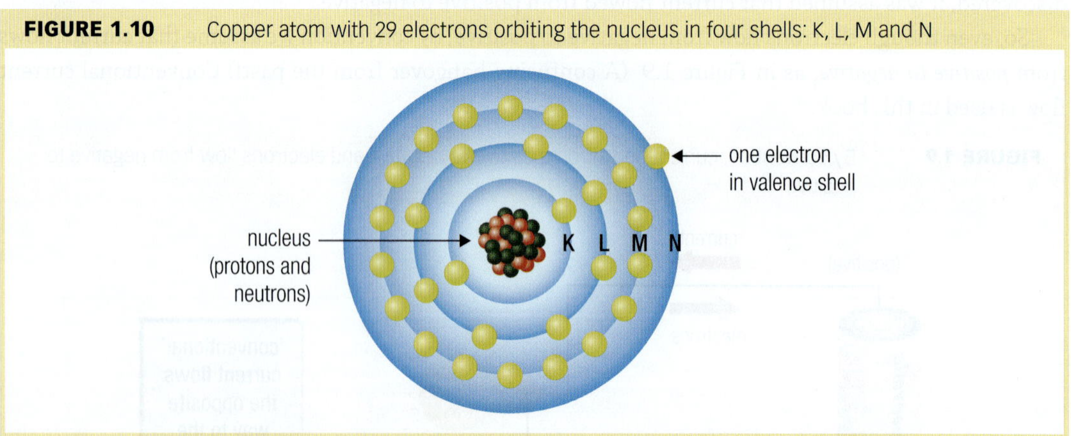

In a conductor, electrons can easily move from one atom to the next (if there's a voltage present to cause the electrons to move). This is shown in Figure 1.11, where an electron entering a conductor pushes an electron out of an atom, then replaces it by orbiting around that atom. The electron pushed out of orbit in turn replaces an electron in the next atom, and so on.

FIGURE 1.11 Electron movement in a conductor

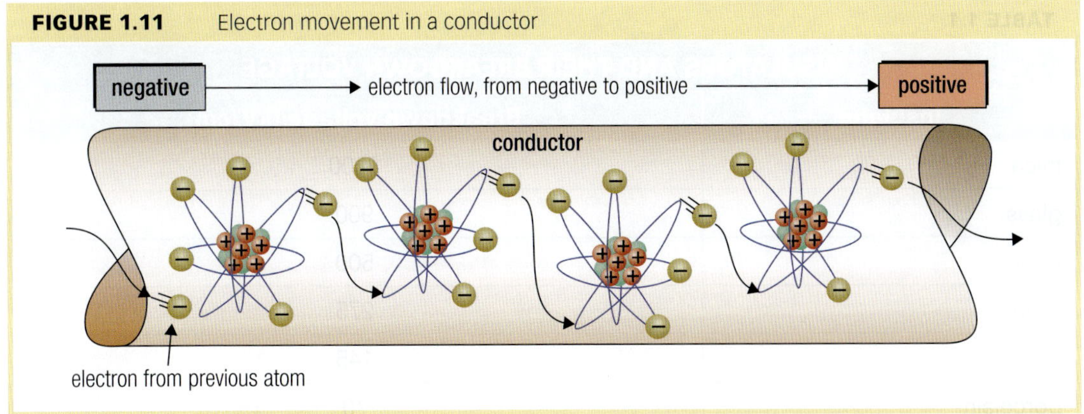

Conductors include most metals, in particular silver, copper, gold and aluminium. Metals are used in various ways in the electrotechnology industry, some of which are described below.

- **Silver** is the best conductor of all the metals, and is also expensive. It is mainly used in switch contacts, sometimes alloyed with copper to give it better wear resistance.
- **Copper** is the next best conductor and is widely used in the electrical field. Copper is relatively weak in its natural state, and is refined to remove impurities and often alloyed with other metals to give it the required physical strength. It is used in most electrical cables, in motor windings and on printed circuit boards.
- **Gold** has less conductivity than copper, but does not easily corrode and is often used as the contact material on plugs and sockets, on some relay contacts and in situations where a corrosion free contact surface is required.
- **Aluminium** has about 60 per cent of the conductivity of copper, but because of its relative lightness, it is used extensively in high-voltage power transmission lines. To get the required strength, these cables have a steel core. Aluminium oxidises easily, which causes an insulating layer of oxide to form on the surface, requiring special methods of connection.
- **Nickel** has around a quarter of the conductivity of copper, but is used as an alloy with other metals such as chromium to make a resistive conductor that produces heat when current is passed through it, as in an electric heating element.
- **Iron** has conductivity 18 per cent that of copper, and it rusts easily. Iron is rarely used in its pure form, and is usually alloyed with other metals to produce steel, which is widely used in the electrical industry to construct motors, transformers, switchboard boxes and tools.
- **Lead** is a poor conductor of electricity, but is often found in solder, a metal alloy that melts at a relatively low temperature. It is also used in lead-acid batteries, such as car batteries.

Some liquids are also good conductors, such as salt water, acids and alkalis. Pure water doesn't conduct electricity, but tap water usually has enough impurities to turn it into a conductor.

An **insulator** is a material in which electron transfer between atoms is generally impossible, due to its high valency (above 4). However, if the voltage is high enough, an insulator will *break down* and conduct electricity. Air is a good insulator, but will conduct electricity if the voltage is high enough. Lightning is the effect of air breaking down and conducting (ionising) due to a high voltage. In many cases, insulating materials are destroyed when they break down, and they go on conducting electricity even when the voltage is reduced to its normal value.

Table 1.1 shows various insulators in the order of their breakdown voltage. For example, air breaks down if a voltage of 30 kV (30 000 V) is applied between two points separated in air by 10 mm (1 cm). Impurities in any of these insulators will reduce their breakdown voltage.

A **semiconductor** is neither a conductor nor an insulator. The most common semiconductor material is silicon. Germanium is another. When these two semiconductors are 'doped' (a special treatment that changes their atomic structure), they can be used to make solid-state components such as transistors and diodes. Carbon is also a semiconductor and is used to make resistors.

SAFETY

Lead is a toxic metal. Wear safety gloves when handling, and regard it as a poison. (The chemical symbol for lead is Pb.)

SAFETY

Be extra careful with wet clothing. Normally cotton or wool won't conduct electricity, but when wet, these materials can become a good conductor. If you are working with electricity, beware of liquids - including tap water

SAFETY

Water and electricity is a dangerous combination

TABLE 1.1

INSULATORS AND THEIR BREAKDOWN VOLTAGE	
Insulator	**Breakdown voltage (kV/cm)**
mica	2000
glass	900
paper	500
rubber	275
oil	145
porcelain	70
air	30

Resistance is a measure of a material's ability to conduct an electric current. A conductor has a low resistance to an electric current, whereas an insulator has a high resistance. As shown in Figure 1.12, silver has the least resistance and mica the most.

FIGURE 1.12 Conductors have much less resistance to electron flow than insulators

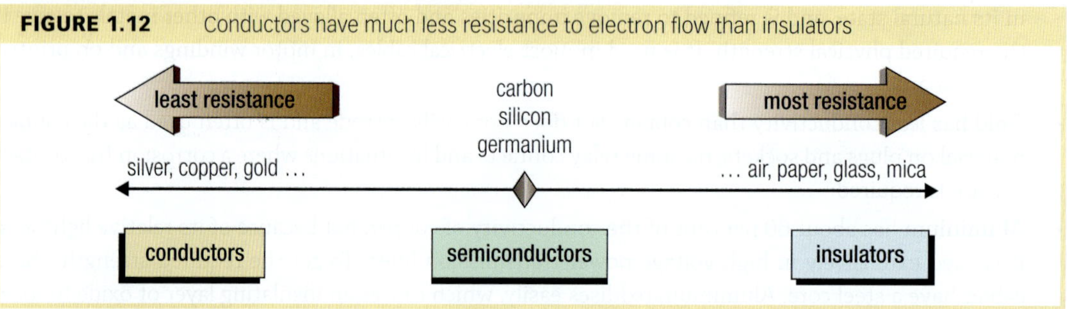

Resistance as an electrical quantity is defined as *opposition to current flow*. It is measured in **ohms** (after Georg Ohm, a German physicist). Its symbol is R. In writing, the term *ohm* is often replaced with the Greek letter Ω (*omega*). A resistance of two ohms is written as R = 2 Ω.

In a plumbing system, a fully open tap offers very little resistance to the flow of water. As the tap is turned off, it offers more resistance to the water flow and the flow of water decreases. Notice that as the resistance increases, the flow decreases.

So, if voltage doesn't change, *the higher the resistance, the less the current*.

KEY POINTS...

- Resistance is opposition to current flow.
- A conductor has a low resistance.
- An insulator has a high resistance.
- Resistance (R) is measured in ohms (Ω).

TABLE 1.2

SUMMARY OF VOLTAGE, CURRENT, CHARGE AND RESISTANCE			
Quantity	**Symbol for quantity**	**Measurement unit**	**Symbol for unit**
EMF	E	volt	V
voltage	V	volt	V
current	I	ampere	A
charge	Q	coulomb	C
resistance	R	ohm	Ω

1.7 Basic Electric Circuit

An electric circuit is a number of electrical components connected with conductors such that current can flow through the circuit when it's connected to a voltage source. The circuit in Figure 1.13 is a simple example in which a lamp (the load) is connected with conductors (wires) to a battery (voltage source). The lamp and the conductors are the path for the current. The battery provides the electrical pressure (voltage) to cause current to flow in the circuit.

In this circuit, the lamp stays on as there's no way to stop the current flow. By adding a switch to the circuit, the current path can be broken by turning the switch off. The switch, therefore, determines whether current is flowing in the load or not, as in Figure 1.14.

It is usual to have some form of electrical protection in a circuit in case a fault occurs. The simplest protection device is a fuse, which in principle consists of a thin wire held in a former of some sort. The fuse wire is much thinner than the circuit conductors, and is selected so the fuse wire heats and eventually 'blows' or operates if the current in the circuit is too high. The diagram in Figure 1.15 shows how a fuse is connected in a basic electrical circuit.

FIGURE 1.13 A basic electric circuit has a voltage source, a path for the current and a load

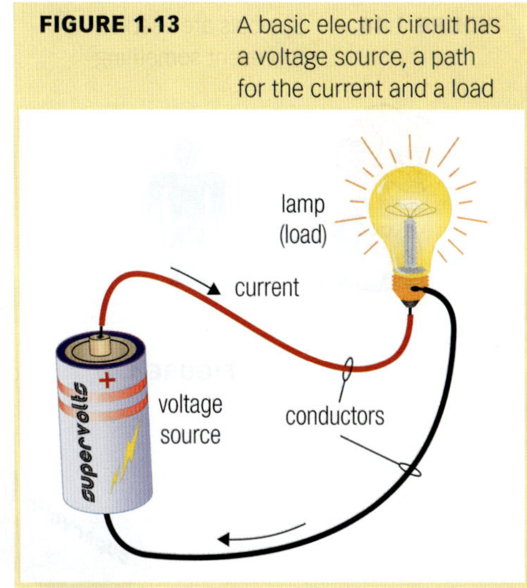

FIGURE 1.14 Adding a switch to the circuit allows the current to the load to be turned on and off

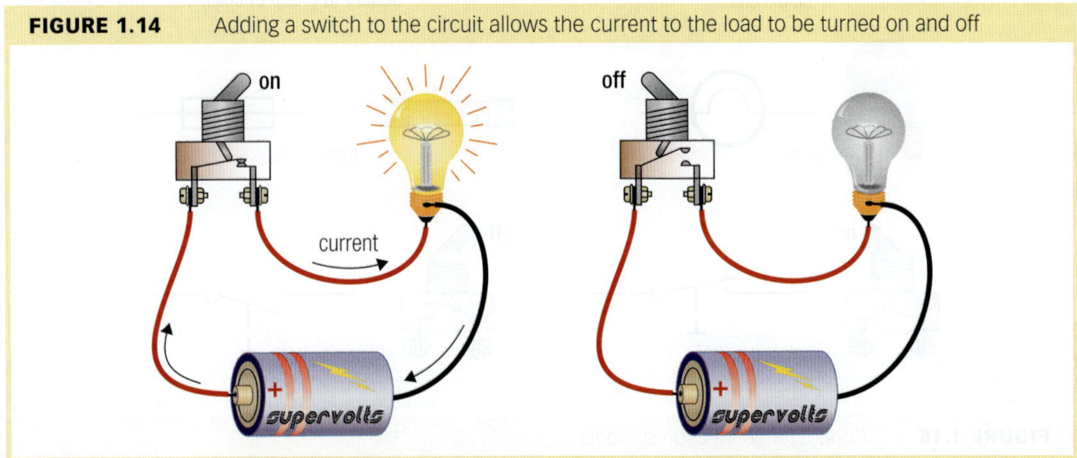

FIGURE 1.15 Adding a fuse to the circuit protects it against overload if a fault occurs

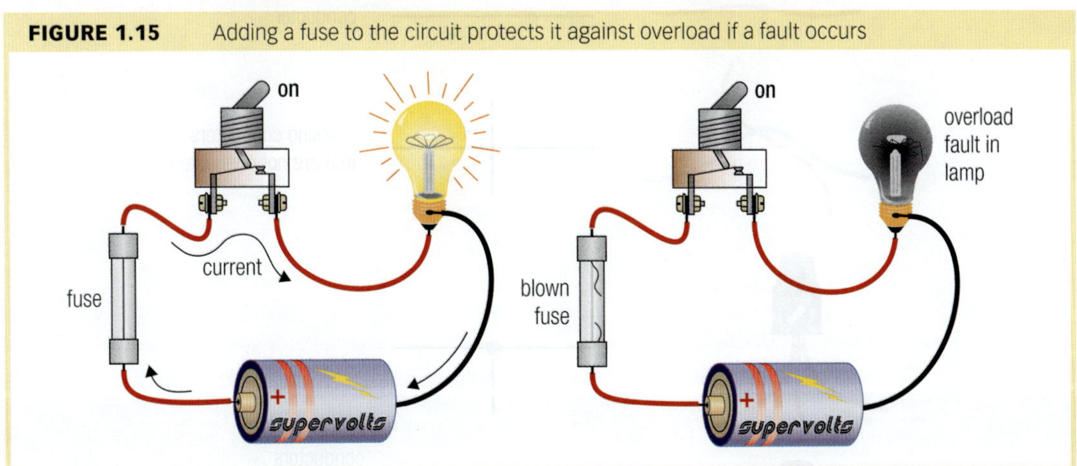

A basic electrical circuit has:

- a voltage source such as a battery or mains power
- a load, such as a lamp, motor or heater
- a switch so the load can be turned on and off by interrupting the current path
- conductors to connect the components and to complete the current path
- a protection device such as a fuse.

1.8 Circuit Diagrams

FIGURE 1.16 Symbols are used to represent something

The circuits shown so far have drawings to represent the components, but we normally use circuit *symbols* to represent each component. There are many electrical and electronic components and they all have their own symbol. Those shown in Figure 1.17 are Australian Standard symbols and are for the devices in the drawings.

Circuit conductors are also drawn in a standard way, as in Figure 1.18. Notice the connection in which no more than three conductors are shown at a connection point. Multiple connections are drawn this way to avoid confusion between crossing and connected conductors.

FIGURE 1.17 Component symbols

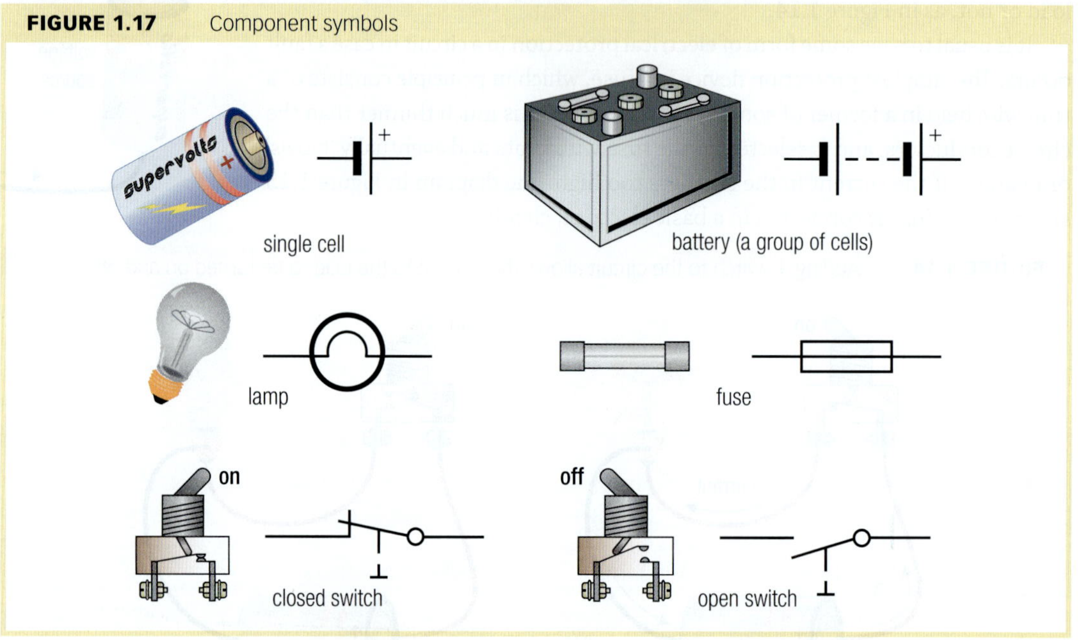

FIGURE 1.18 Conductor connection symbols

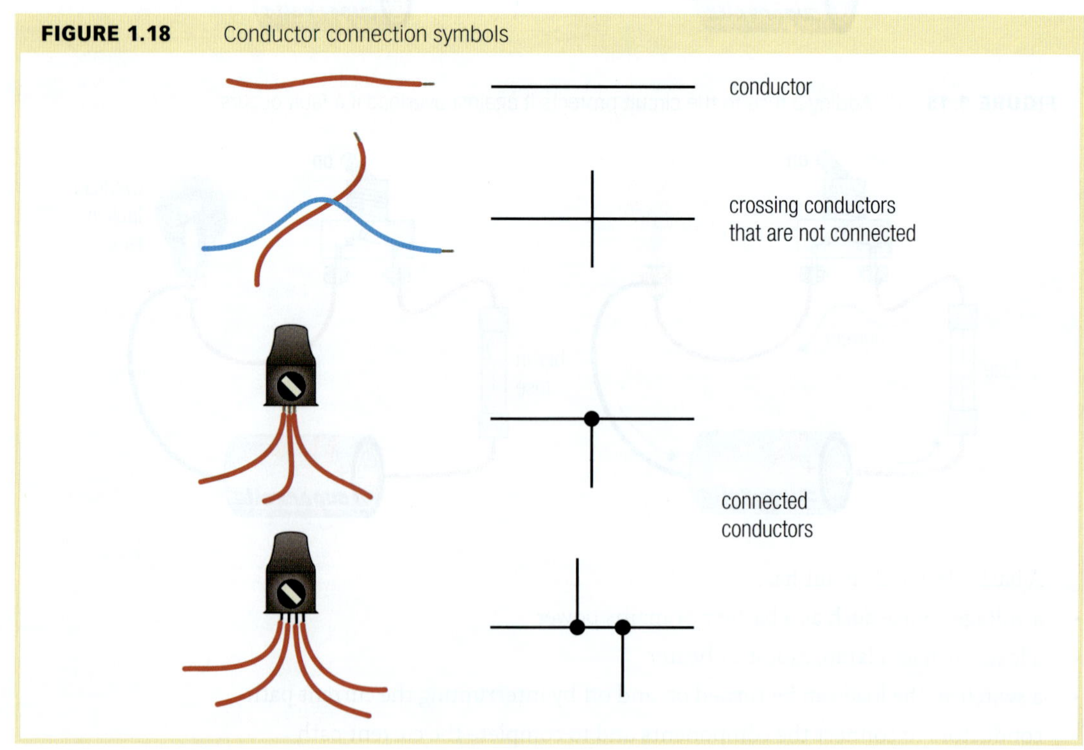

These symbols can be combined in a *circuit diagram* to show how the components are connected together. A circuit diagram only shows connections and not where the components are. The main use of a circuit diagram is to show how the circuit components are connected, and by using standard symbols, anyone who knows the symbols can understand what a circuit represents. Figure 1.19 shows a circuit diagram alongside the actual circuit it is representing.

FIGURE 1.19 Circuit diagrams use symbols and show how the circuit is connected

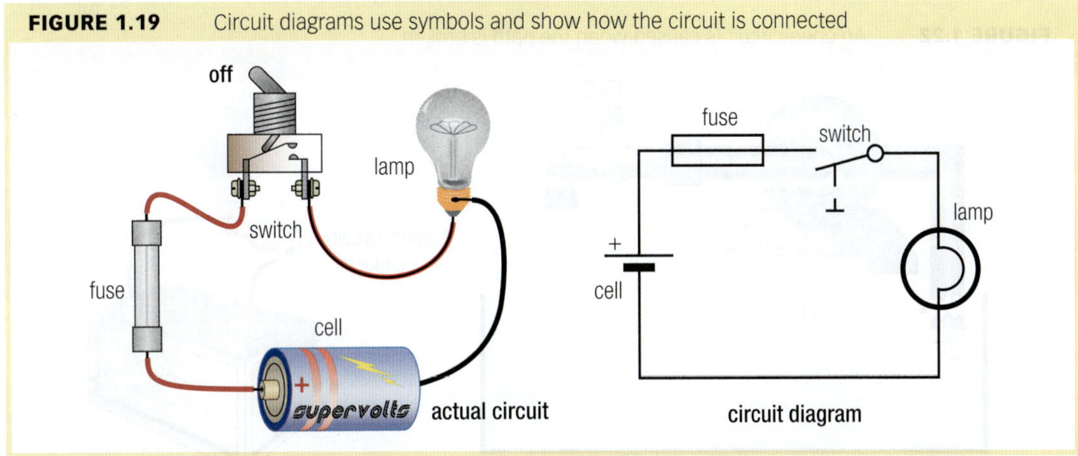

Examples of switches and connectors that are typically used in electrical work are shown in Figures 1.20 and 1.21. There are hundreds of different types of switches, so those shown are a sample only. Cables and connectors also come in a wide variety of sizes, ratings and shapes, so again those shown are typical examples.

FIGURE 1.20 Switches: (a) Miniature toggle (b) Rocker (c) Mechanism from a light switch (d) Toggle (e) Pushbutton (f) Rotary or wafer switch (g) Limit switch which is operated mechanically

FIGURE 1.21 Cables and connectors: (a) Mains power wiring (b) 12 V garden lighting wiring (c) Different types of conductors, multi-strand to single-strand, used in data and telecommunications

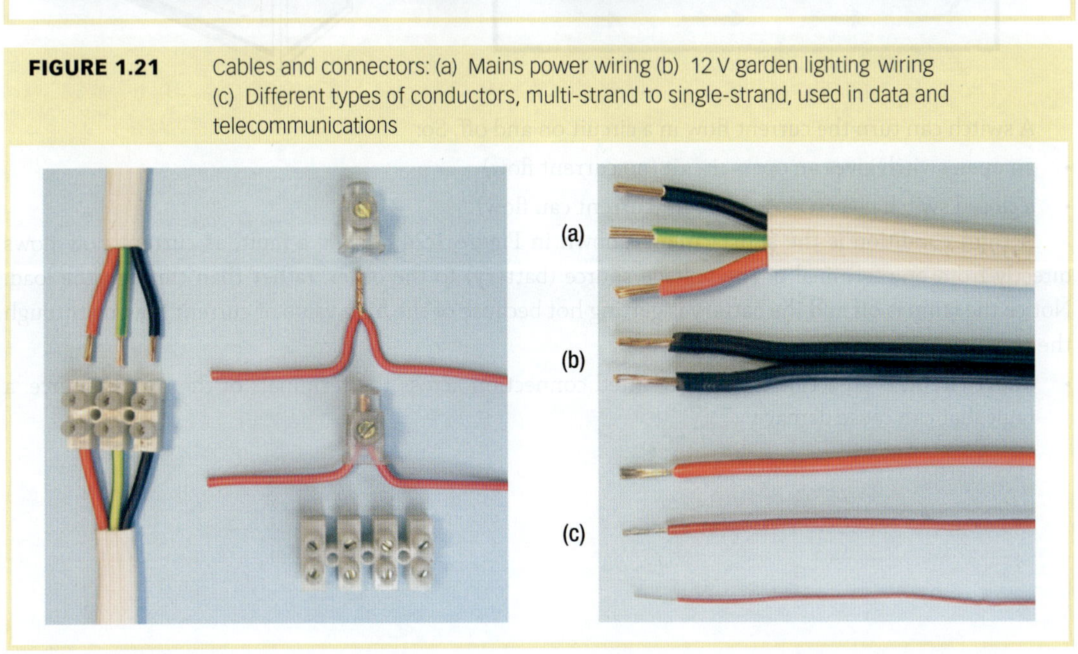

1.9 Open-Circuit and Closed-Circuit

A circuit is said to be *open-circuit* when the conducting path for the current is broken. Turning a switch off or disconnecting a wire gives an open-circuit.

- **Open-circuit** – no current flow because the path is broken

FIGURE 1.22 An open-circuit is caused when the path is broken

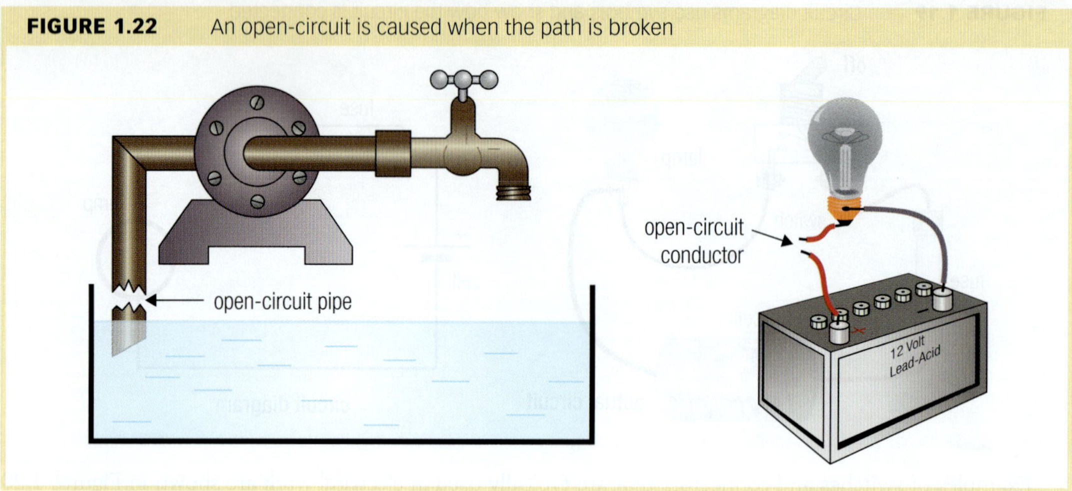

A closed-circuit is a circuit's normal operating condition. Current flows from the voltage source, through the circuit conductors and the load.

- **Closed-circuit** – current path is complete, allowing current to flow (circuit is said to have *continuity*)

FIGURE 1.23 A closed-circuit: the lamp is on and current is its normal value

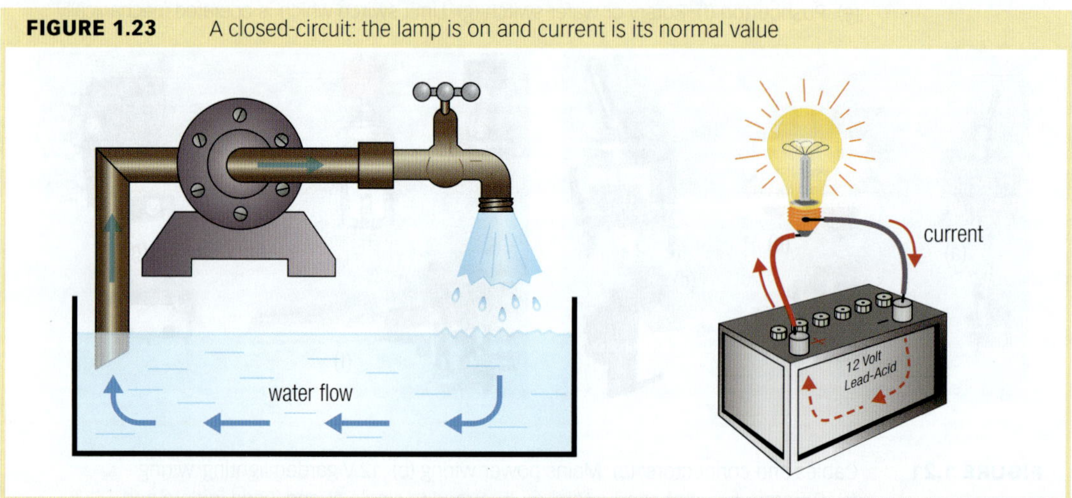

A switch can turn the current flow in a circuit on and off. So:

- an open switch gives an open-circuit (no current flow)
- a closed switch gives a closed-circuit (current can flow).

A third condition is the **short-circuit**, shown in Figure 1.24. This is a fault, as current now flows directly from one terminal of the voltage source (battery) to the other, rather than through the load. Notice the lamp is off and the battery is getting hot because of the high value of current flowing through the short-circuit.

- **Short-circuit** – an accidental or unwanted connection across the terminals of the supply source: a fault that can cause damage

FIGURE 1.24 A short-circuit is a fault that can burn out wiring and cause damage

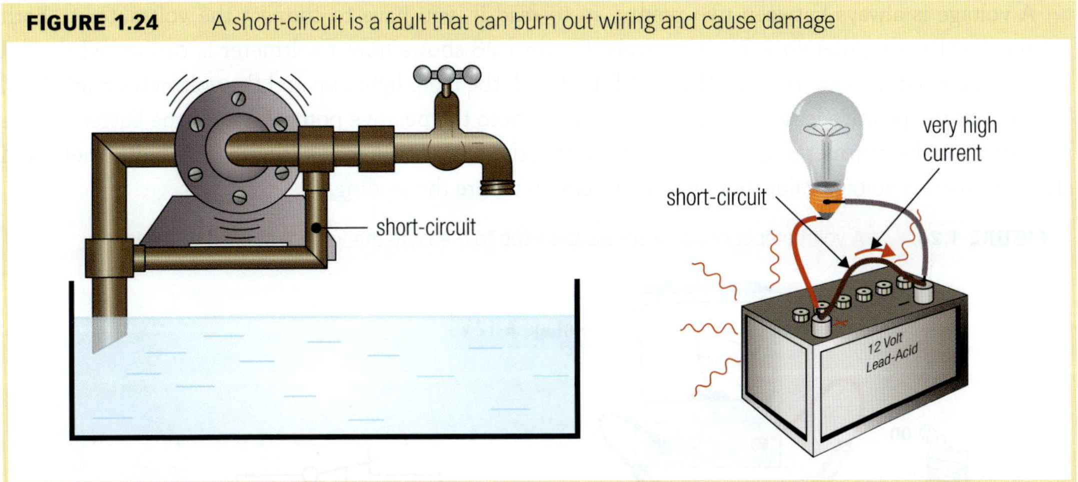

KEY POINTS...

- A basic electrical circuit has a voltage source, switch, conductors, protection device such as a fuse and a load.
- All electrical components have a standard symbol.
- Conductor connections are shown in a standard way.
- A circuit diagram shows how all the components in a circuit are connected.
- An open-circuit prevents current flow; a closed-circuit allows current flow.
- A short-circuit is a fault that allows too much current flow.

1.10 Measuring Voltage

A voltage is measured with a voltmeter like those shown in Figure 1.25. There are two types of electrical measuring meters: analog and digital. An analog meter has a scale and a pointer, and a digital meter has a numerical readout. A voltmeter is usually part of an instrument called a multimeter, but can also be a basic standalone meter. A voltmeter (either analog or digital) has the circuit symbol shown in Figure 1.25.

FIGURE 1.25 A voltmeter, whether standalone or part of a multimeter, is either analog or digital

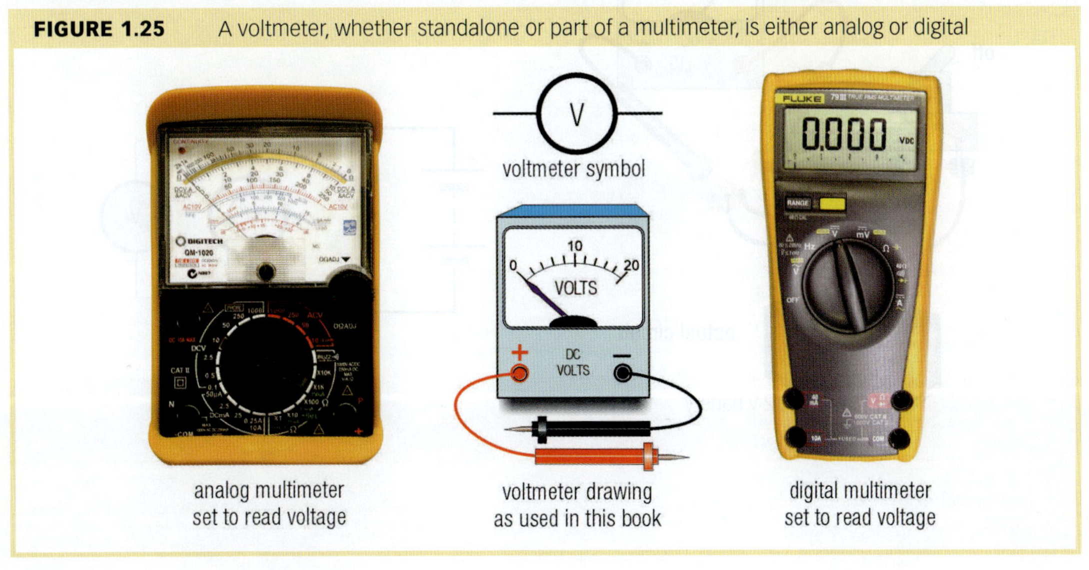

A voltage is always between two points, so to read it, you have to connect the voltmeter to both points. A voltmeter, therefore, has two leads. Figure 1.26 shows how a voltmeter is connected to read the voltage across a lamp. Because the switch is closed, the lamp lights up and the voltmeter reads 12 V. Notice that the positive lead of the voltmeter connects to the positive potential from the battery. If the leads of an analog voltmeter have the wrong polarity, the pointer will read backwards. A digital meter will show the correct voltage value, but with a minus sign before the reading.

FIGURE 1.26 A voltmeter connected across the lamp to measure the voltage applied to the lamp

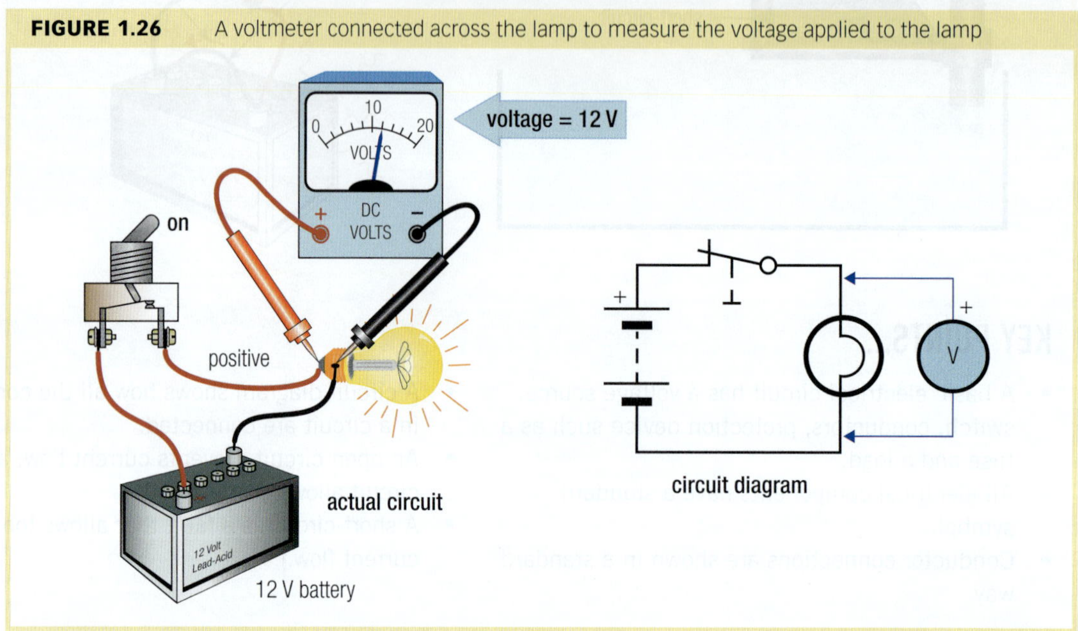

When the switch is turned off (an open-circuit) the lamp is off and, as you would expect, there is no voltage across the lamp. The voltmeter will read zero volts, as in Figure 1.27.

FIGURE 1.27 Voltmeter shows 0 V when it's connected across the lamp and the switch is off

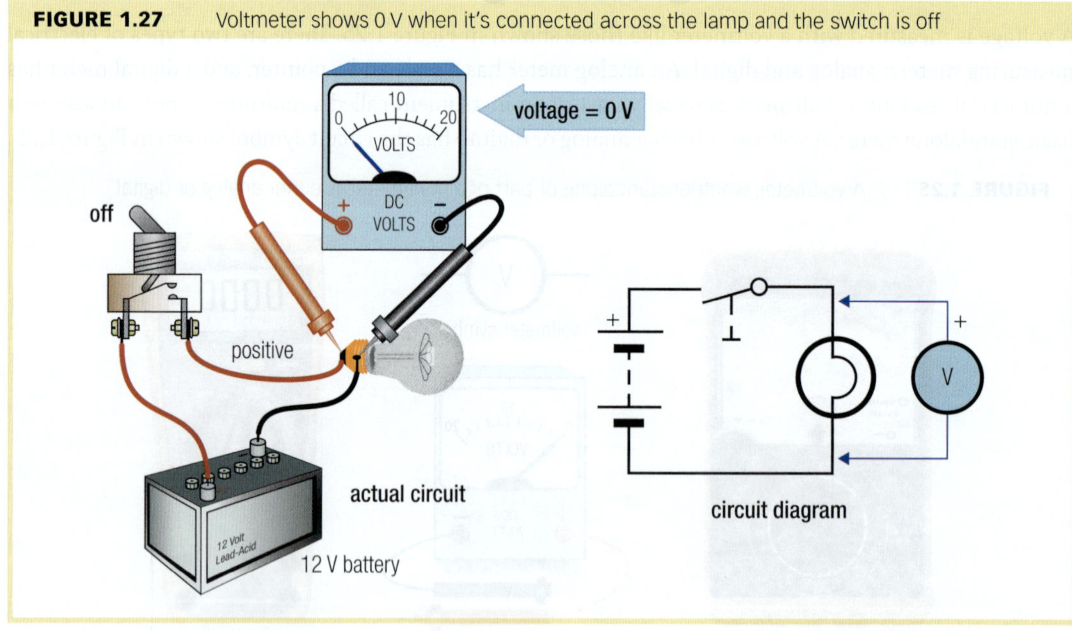

1.11 Measuring Current

Current is measured with an instrument called an *ammeter* (which is short for an 'ampmeter'). An ammeter generally looks like a voltmeter, except the scale has a different calibration. As shown in Figure 1.28, an ammeter (analog or digital) is either a single instrument or part of a multimeter. The symbol for an ammeter is similar to the voltmeter symbol except that the letter A (or sometimes the letter I) is used to show it's an ammeter.

FIGURE 1.28 Ammeters, like voltmeters, are either digital or analog

An ammeter is connected differently from a voltmeter. Remember that current is a flow of electrons, rather like water flowing in a pipe. To measure water flow, a flow meter is fitted in the water pipe so the water passes through the meter. To measure current, an ammeter is connected so the current passes through the meter. When a flow meter is installed, the water pipe has to be cut and the flow meter fitted so that the meter reconnects the pipe. The same applies to the ammeter. The conductor carrying the current has to be cut, and the ammeter connected, so the current passes through the ammeter, as shown in Figure 1.29.

FIGURE 1.29 Measuring current is like measuring water flow – the current has to pass through the meter

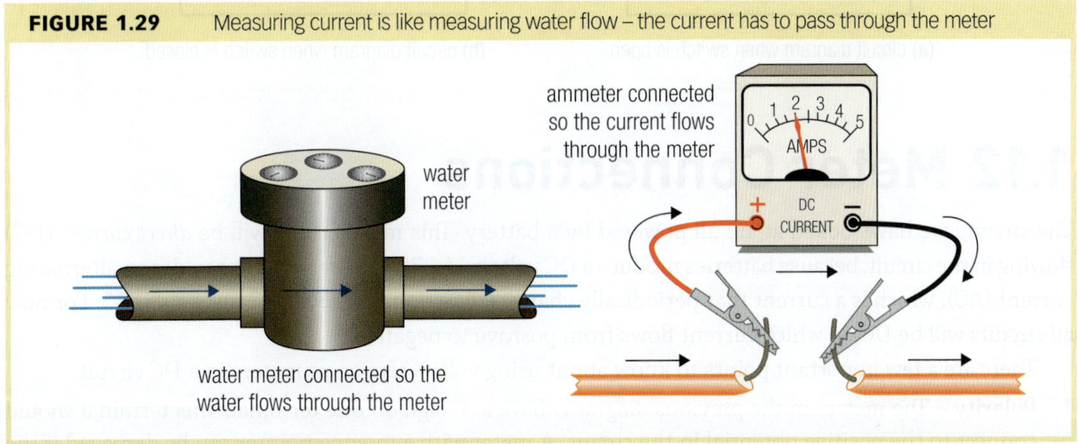

Figures 1.30 and 1.31 show an ammeter connected to measure current flowing through a lamp. As before, when the switch is off, the current is zero. When it's on, a current (in this case 2 amperes, or 2 A) flows.

FIGURE 1.30 An ammeter is connected so the current being measured passes through the meter. When the switch is closed, current flows in the circuit and through the ammeter, which shows the value of the current.

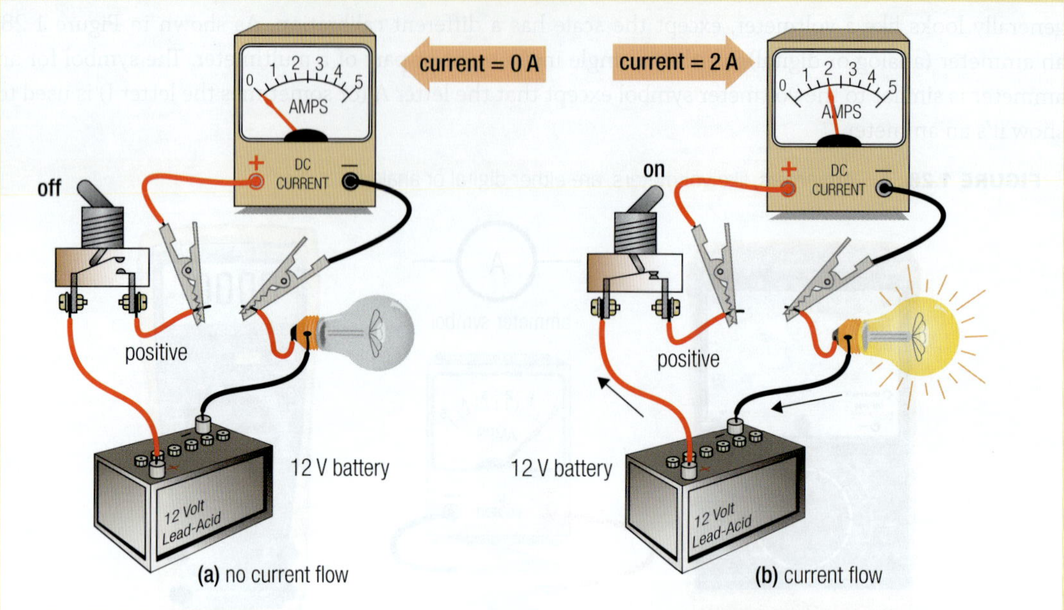

current = 0 A
current = 2 A
(a) no current flow
(b) current flow

The circuit diagrams for the actual circuits of Figure 1.30 are shown in Figure 1.31.

FIGURE 1.31 Circuit diagrams showing (a) switch open, and (b) switch closed

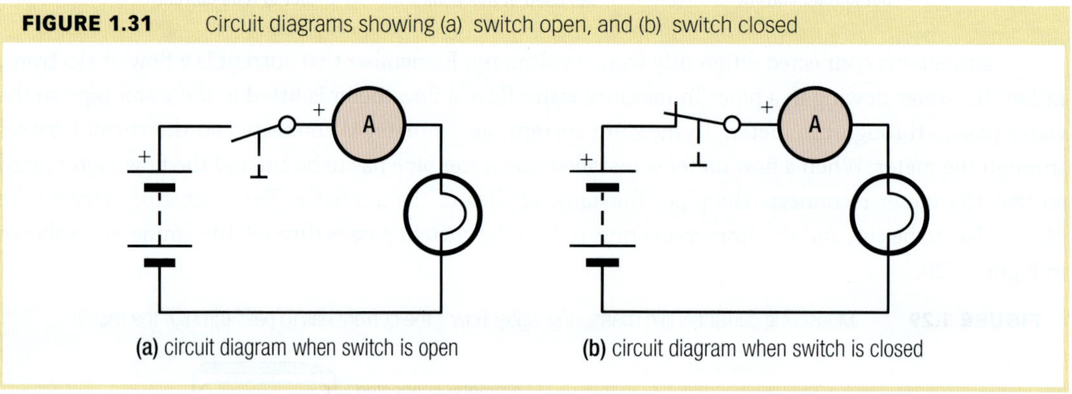

1.12 Meter Connections

The circuits we have looked at are all powered by a battery. This means there will be *direct current* (DC) flowing in the circuit, because batteries produce a DC voltage. A 230 V power outlet provides an alternating current (AC), which is a current that periodically changes direction. We cover AC in Chapter 15. For now all circuits will be DC, in which current flows from positive to negative.

There are a few important points to know about using voltmeters or ammeters in a DC circuit:

1 **Polarity** – The meters in the previous diagrams have a + sign on one terminal. This terminal should connect to the positive potential in the circuit. A meter with a moving pointer can be damaged if the leads are connected the wrong way, while a digital meter will show a minus sign in front of the reading if the leads are reversed.

2 **Range** – Analog and some digital meters have a voltage range switch, which should be set to its highest value when making the first measurement. For a digital meter with auto-ranging, select volts, not millivolts. After connecting the meter (reversing the leads if necessary), select a voltage range that suits the meter. For example, an analog meter should have the pointer between one-quarter to three-quarters of the scale, as shown in Figure 1.32. A digital meter will show 999 or similar if the range setting is too low.

FIGURE 1.32 Choose the correct meter range to avoid damaging the meter

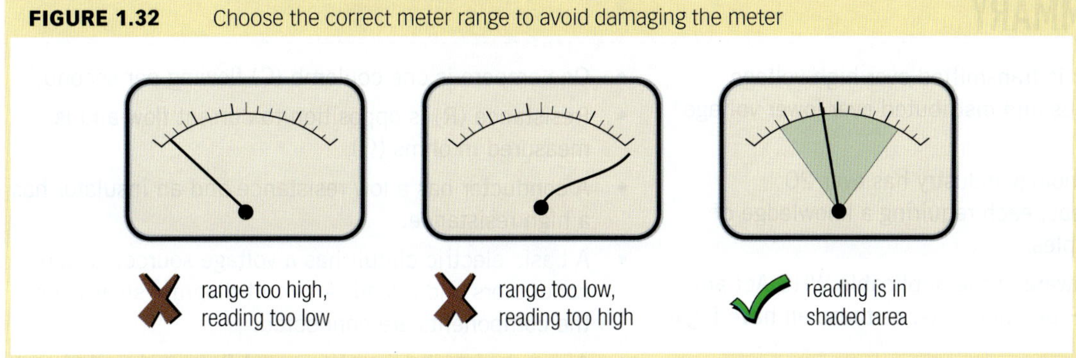

3 **Voltmeter connections** – A voltmeter is always connected across two points in a circuit, such as across the terminals of a battery. We refer to this as connecting a voltmeter in *parallel*, in this case with a battery. (The parallel circuit is explained in Chapter 7.) The circuits in Figure 1.33 show a voltmeter connected in parallel (or across) the lamp, the switch and then the battery. Notice how the positive terminal of the voltmeter is always connected to the positive potential in the circuit.

FIGURE 1.33 The voltmeter connection

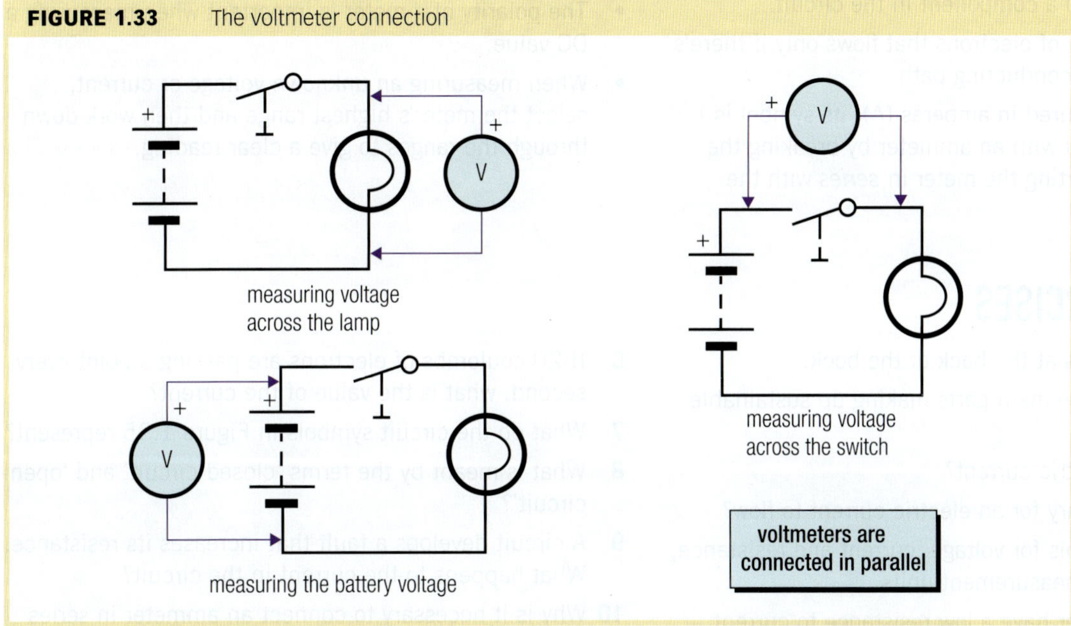

4 **Ammeter connections** – An ammeter is always connected so the current being measured flows through the meter. We call this a *series* connection, in which there is one pathway for the current. (The series circuit is described in Chapter 6.) The circuits in Figure 1.34 show how an ammeter can be connected to measure current. Because there is only one path for the current, it does not matter where in the path the meter is connected.

FIGURE 1.34 These circuits show how an ammeter can be connected to measure the current in the circuit

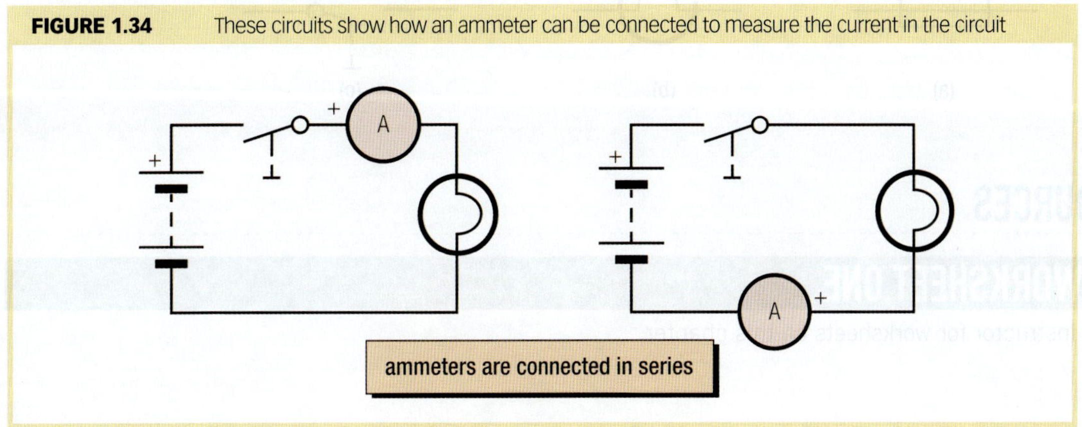

CHAPTER SUMMARY

- Electrical power is transmitted over high-voltage transmission lines and distributed over lower voltage power lines.
- The electrotechnology industry has over 20 qualification areas, each requiring a knowledge of electrical principles.
- You should be aware of the applicable WHS Act and Environmental Protection Acts, as these all have legal obligations.
- Sustainable energy is made up of two parts: renewable energy and energy efficiency.
- Voltage is *electrical pressure* between two points; it is measured in volts and is given the symbol V.
- Voltage is measured by connecting a voltmeter across (in *parallel* with) a component in the circuit.
- Current is a flow of electrons that flows only if there's a voltage and a conducting path.
- Current is measured in amperes (A); its symbol is I and is measured with an ammeter by breaking the circuit and inserting the meter in *series* with the circuit.
- One ampere is one coulomb (C) flowing per second.
- Resistance (R) is opposition to current flow and is measured in ohms (Ω).
- A conductor has a low resistance and an insulator has a high resistance.
- A basic electric circuit has a voltage source, switch, conductors and a load. A circuit diagram shows how the components are connected.
- An open-circuit prevents current flow, a closed-circuit allows current flow, and a short-circuit is a fault that can cause too much current flow.
- Voltmeters and ammeters can be analog or digital, and are either part of a multimeter or are standalone meters.
- The polarity of a meter is important when measuring a DC value.
- When measuring an unknown voltage or current, select the meter's highest range and then work down through the ranges to give a clear reading.

REVIEW EXERCISES

Check your answers at the back of the book.

1. What are the two main parts making up sustainable energy?
2. What is an electric current?
3. What is necessary for an electric current to flow?
4. Write the symbols for voltage, current and resistance, and give their measurement units.
5. Why does copper have a low resistance to current flow?
6. If 20 coulombs of electrons are passing a point every second, what is the value of the current?
7. What do the circuit symbols in Figure 1.35 represent?
8. What is meant by the terms 'closed-circuit' and 'open-circuit'?
9. A circuit develops a fault that increases its resistance. What happens to the current in the circuit?
10. Why is it necessary to connect an ammeter in series with a circuit?

FIGURE 1.35

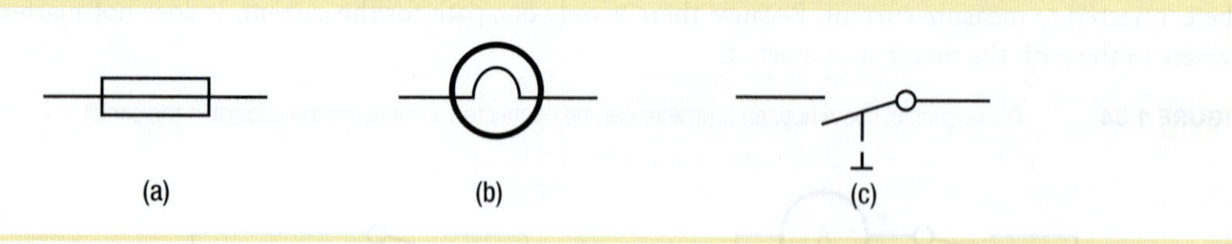

ONLINE RESOURCES

COMPLETE WORKSHEET ONE

Check with your instructor for worksheets on this chapter.

CHAPTER 2

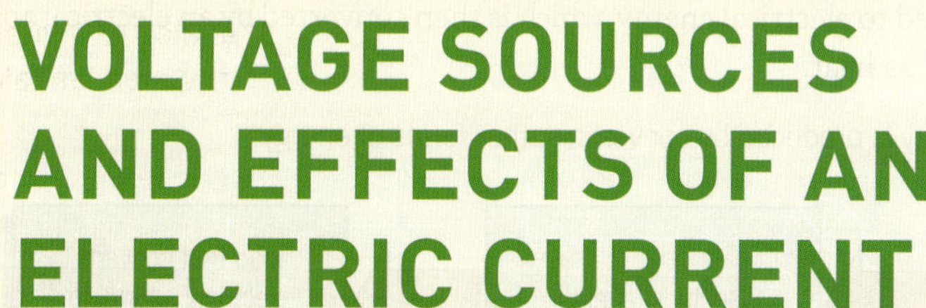

VOLTAGE SOURCES AND EFFECTS OF AN ELECTRIC CURRENT

This chapter outlines how electrical energy is produced from other sources of energy. It also explains the effects of an electric current, and how these effects are used in everyday life. Some effects are not wanted, including the effect of an electric current on the human body, so methods of protecting against the unwanted effects are also given.

CHAPTER OUTLINE

2.1 Producing a Voltage

As explained further in Chapter 4, electricity is a form of energy. A basic law, called the *Conservation of Energy*, says that energy cannot be created or destroyed. However, energy can be converted from one form to another. This means that electrical energy has to come from other sources of energy, by a conversion process. The most commonly used energy sources for this are:

- mechanical
- chemical
- heat
- light.

Heat and light are both forms of 'radiant' energy

Producing electricity often involves several processes, such as first converting heat to mechanical energy, then using the mechanical energy to produce electricity. The basic cycle is shown in Figure 2.1, where an energy source is converted to electrical energy, which is then converted by an electrical appliance into another form of energy, such as heat.

FIGURE 2.1 Electrical energy is produced by converting other forms of energy

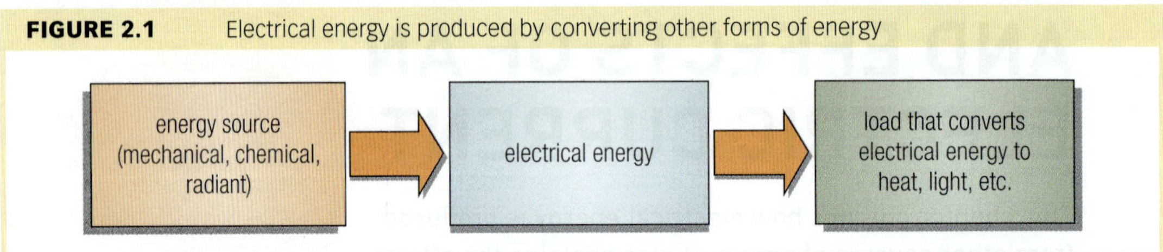

As you saw in Chapter 1, a voltage is 'electrical pressure' and current is a flow of electrons. The current does the work (lights a lamp, makes a motor turn), as a voltage by itself is simply a pressure source. So, electrical energy is a combination of voltage *and* current. That is, electrical energy is used (or more correctly, transformed) only when current flows.

This means the energy source to produce electrical energy is only being used when current flows. For instance, a battery that's not connected to anything is not supplying electrical energy, even though there's a voltage developed across its terminals. Here's a brief look at the various ways of producing electricity. All these methods convert one form of energy into electrical energy.

Mechanical to electrical

Mechanical energy is movement. The three main ways to convert mechanical movement into electrical energy are:

- friction between two materials
- mechanical stress applied to a piezo-electric element
- moving a conductor in a magnetic field.

Friction

You have probably felt the effect of 'static' electricity. When you slide over a fabric-covered car seat you can be 'charged up' to a high voltage. This is because electrons have either been transferred to you or taken from you as you slid over the seat. Touching the car door handle causes the electrons to discharge to the car body, giving the sensation of an electric shock.

Static electricity is produced when clouds move in the atmosphere, caused by friction between the clouds and the surrounding air. Once the charge differential is high enough, the air between two charged clouds ionises, and an electric current in the form of lightning flows between the clouds, causing them to discharge. Rubbing certain types of materials together can also produce static electricity. The available energy from static electricity is rarely useful, although very high voltages can be produced for use in certain types of tests and experiments.

Piezo-electric crystal

The piezo-electric effect was discovered in 1880 by the Curie brothers. (Piezo is Greek for pressure.) They found that when mechanical stress was applied to crystals such as topaz, quartz, Rochelle salt and cane sugar, electrical charges appeared, and that this voltage was proportional to the applied stress. Materials that have this characteristic are called *piezo-electric* materials.

The piezo-electric effect occurs when the charge balance within the crystal lattice of a piezo material is disturbed. When there is no applied stress on the material, the positive and negative charges are evenly distributed so there is no potential difference. When the lattice is changed slightly, the charge imbalance creates a potential difference, often as high as several thousand volts. However, the current is extremely small.

Piezo elements are used in electronic gas lighters, guitar pick-ups, some kinds of microphones and various types of sensors, such as accelerometers. When used in electronic gas lighters, a trigger-operated mechanical mechanism is arranged to apply a short, sharp force to the piezo element, which produces a voltage that is high enough to cause a spark, as in Figure 2.2. In a microphone, the variations in air pressure due to sound waves cause the piezo-electric crystal element in the microphone to produce a very small voltage that changes with the sound waves.

FYI

If a varying voltage is applied to a piezo-electric crystal, it will distort and produce a sound determined by the nature of the varying voltage. Piezo-electric buzzers are commonly used in electrical appliances.

FIGURE 2.2 Piezo-electric effect, where pressure is applied to opposite faces of a quartz crystal to produce a voltage. Pressure can be by compressing, flexing or distorting the piezo element.

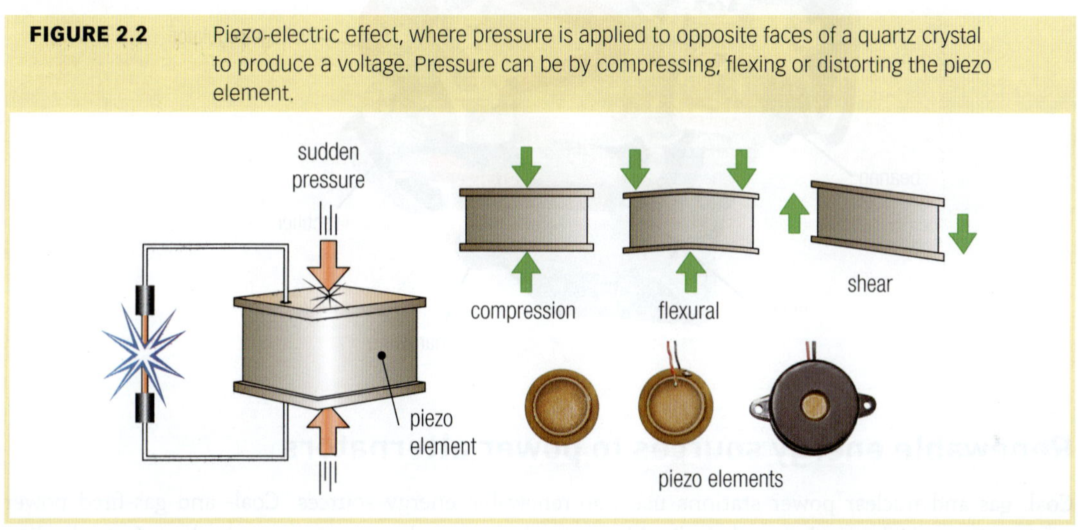

Magnetism

Magnetism plays a big role in the electrotechnology industry, and the topics of magnetism and electromagnetism are covered in Chapters 11 and 12. In principle, if a conductor is moved through a magnetic field, a voltage is *induced* in the conductor. This effect was discovered by Michael Faraday in 1831, a discovery that led to the electrical power generation industry of today. Figure 2.3 shows a single conductor being moved through a magnetic field, although the voltage it produces will be very small (a few millionths of a volt).

Faraday found that the faster the conductor moves through a magnetic field, the higher the voltage. It doesn't matter whether the magnetic field or the conductor moves – a voltage is produced if either moves relative to the other. He also found that a higher voltage would be produced by increasing the strength of the magnetic field and also by forming the single conductor into a coil.

Electricity produced by the magnetic effect is usually achieved with machines called alternators, such as the alternator in a motor vehicle. The cutaway illustration in Figure 2.4 shows the main parts of a typical car alternator. The rotor (3) has a coil that is energised to produce a magnetic

FIGURE 2.3 Producing a voltage by using mechanical energy to move a conductor in a magnetic field

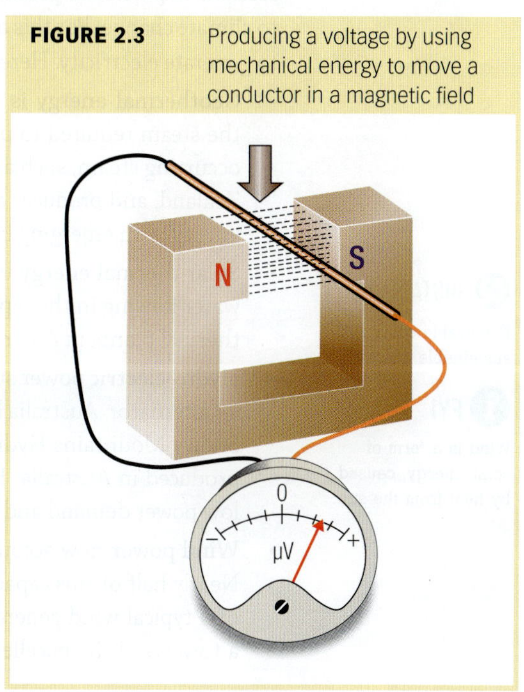

field. Power to the rotor coil is supplied through two slip rings (8). When the rotor is turning, it induces a voltage in stationary coils that are fitted to the stator (2). A regulator (7) adjusts the amount of current supplied to the rotor to regulate the generated voltage induced in the stator coils.

A common way of generating electrical power is with a rotating prime mover driving an alternator, such as a steam-driven, hydro-powered or wind-powered turbine. These range in size, depending on how much electrical power they can produce, but all of them rely on the effect discovered by Faraday.

FIGURE 2.4 An automotive alternator has a rotor driven by the vehicle's engine. Because a vehicle requires 12 V DC, alternators have solid state rectifiers to convert the AC to DC.

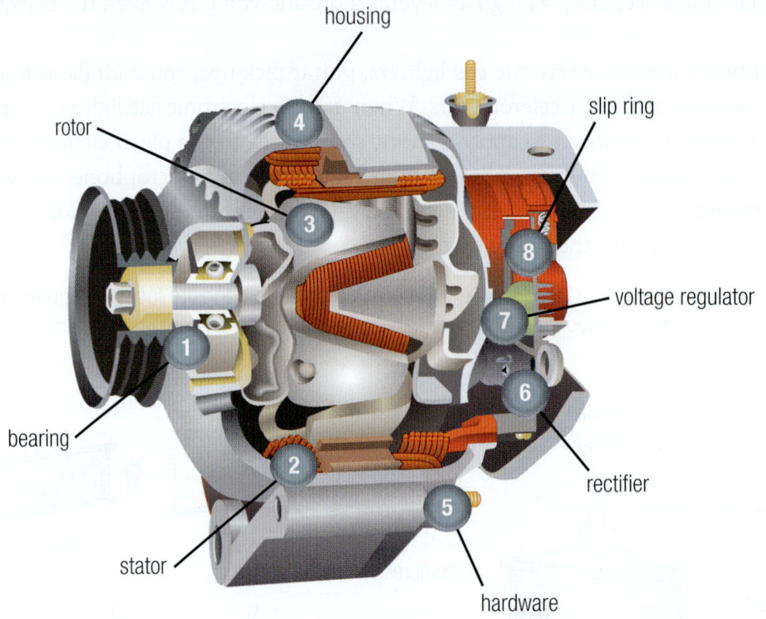

Renewable energy sources to power alternators

Coal, gas and nuclear power stations use non-renewable energy sources. Coal- and gas-fired power stations in particular produce carbon dioxide, a greenhouse gas that is now recognised as a factor leading to climate change. For this and many other reasons, renewable energy sources are being increasingly used to generate electricity. Here's a brief look at some of these energy sources:

- **Geothermal energy** is the heat contained within the Earth, and is used to heat water to produce the steam required to power a turbine driving an alternator. A hydro-thermal system uses naturally occurring steam, such as that produced by a geyser. Geothermal energy is relatively abundant in New Zealand, and produces around 10 per cent of the country's electrical power. In Australia, geothermal energy is an emerging industry, with exploration being conducted in all parts of the country.
- **Solar thermal energy** uses the heat of the sun, which is focused onto specially shaped pipes to heat water flowing in the pipes to give a high pressure to power a turbine. At this stage, there are no solar thermal plants in Australia or New Zealand, due to various limitations with the technology.

Renewable energy is sustainable energy

Wind is a form of solar energy, caused by heat from the sun

- **Hydro-electric power** stations have long been in use, in which flowing water from a dam or river turns an alternator. Australia has more than 100 operating hydro-electric power stations (best known is the Snowy Mountains Hydro-electric Scheme) which generate over 40 per cent of the renewable energy produced in Australia. Hydro systems can store energy by pumping water to a dam during times of low power demand and releasing the water to drive the alternators when there is a high demand.
- **Wind power** now accounts for over seven per cent of Australia's clean energy electricity generation. Nearly half of this capacity is from wind power generators in South Australia. The main components of a typical wind generator are the rotor, usually three blades up to 80 or more metres in diameter, a tower and the nacelle (an aerodynamically shaped housing), which houses the gearbox, alternator

and yaw motors. These motors rotate the nacelle assembly to keep the blades facing the wind. Wind turbine development is driven mainly by the development and exploitation of rare earth magnets. These magnets are contained in the rotor of wind generators and provide a large magnetic field which cuts the conductor coils in the stator.

FIGURE 2.5 (a) Components that make up a wind generator. (b) Gearbox, rotor, shaft and brake assembly being raised into position.

(a) wind generator components

(b) wind generator under construction

Paul Anderson

While most of the electricity we use today is generated by electromagnetic induction, there are other methods, all of which produce a DC voltage (one that always has the same polarity).

FYI

Electromagnetic induction is covered in Chapter 12, and refers to conductors moving in a magnetic field

Chemical to electrical

A battery is the most common application of producing electricity by way of chemical energy. The principle was first discovered by Alessandro Volta in 1800, who noticed that the legs of dead frogs twitched when they were touching two different metals. He went on to find that when two different metals are put into a liquid called an *electrolyte* (an acid or an alkali) a voltage is developed between the two metals. He had discovered the voltaic effect, or the electric cell, the first useful way of producing electricity.

The basic electric cell (shown in Figure 2.6) has three components: the positive electrode, the negative electrode and an electrolyte. The choice of the electrode metals and the electrolyte determine the voltage the cell produces. You can make an electric cell able to produce about 0.8 V by putting a galvanised nail and a piece of copper wire into a lemon, because lemon juice is an acid (citric acid).

FIGURE 2.6 An electric cell converts chemical energy to electrical energy

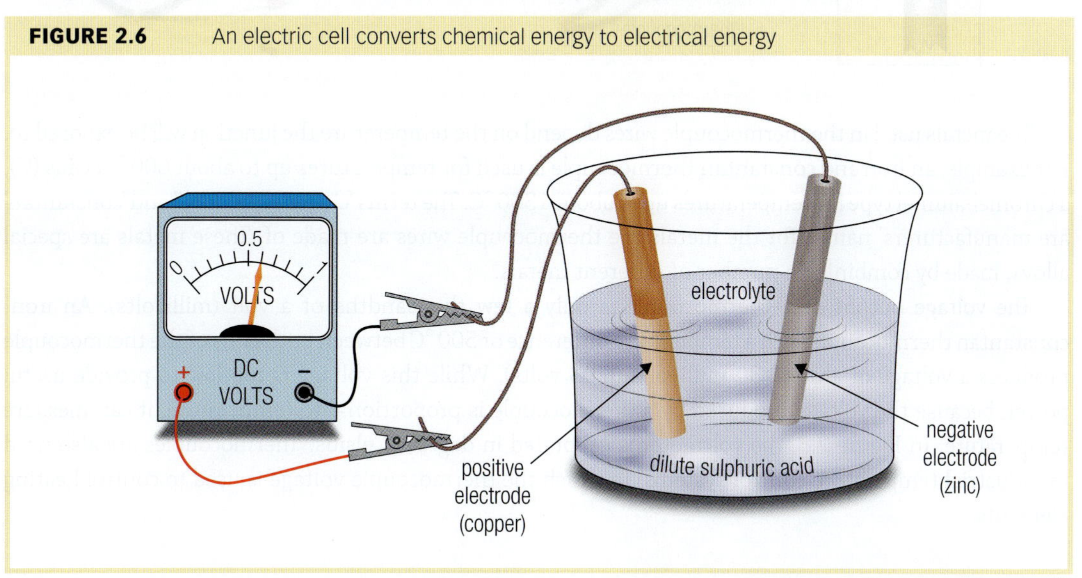

FIGURE 2.7 The most common types of electric cells are the carbon - zinc and the lead-acid

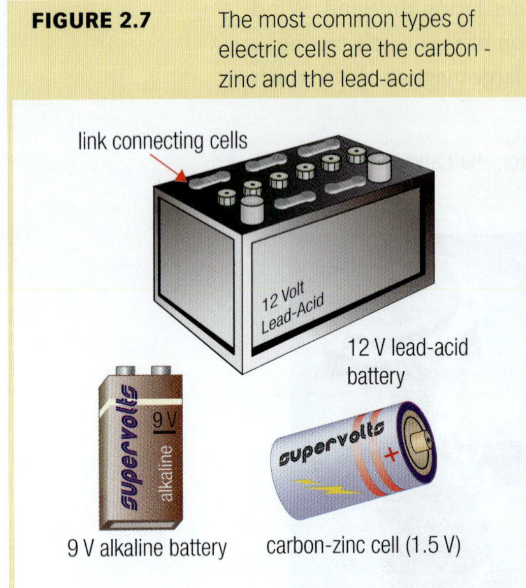

Cells are either rechargeable (called secondary cells) or non-rechargeable (primary cells). Rechargeable types include lead-acid, nickel-cadmium (NiCad), nickel-metal hydride (NiMH) and lithium-ion (Li-ion). Non-rechargeable cells include carbon-zinc (alkaline and conventional) and most types of 'button' cells.

A carbon-zinc cell produces 1.5 V and a lead-acid cell produces 2 V. To get a higher voltage, cells are connected in series to form a battery, so their voltages add together. A 12-volt car battery has six lead-acid cells connected in series to give 12 V.

There are at least 21 700 lithium-ion cells in the storage battery at Hornsdale in South Australia, creating a capacity of 129 MWh when it was built in 2017. Its capacity has since been upgraded to 194 MWh.

Heat to electrical

Although heat is used in coal-fired (and nuclear) power stations, it doesn't directly produce electricity. Instead the heat is used to produce high pressure steam to drive a steam turbine. However, heat can be used to directly produce a voltage using a device called a *thermocouple* which is simply two different metal wires joined at a point called the junction. The principle was discovered in 1821 by German–Estonian physicist Thomas Seebeck, and is now known as the *thermoelectric* effect or Seebeck effect.

If heat is applied to the junction as in Figure 2.8, a voltage is developed by the thermocouple. How much voltage depends on the type of materials used in the thermocouple and the difference in the temperature between the hot and cold ends of the thermocouple.

FIGURE 2.8 A thermocouple is formed by two different metals joined at one point

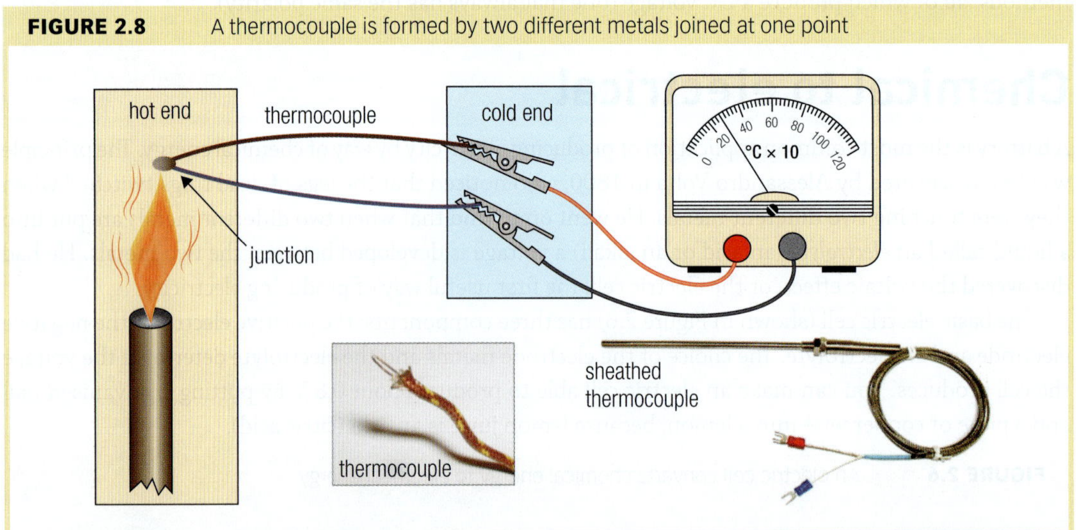

The metals used in the thermocouple wires depend on the temperature the junction will be exposed to. For example, an iron and constantan thermocouple is used for temperatures up to about 600 °Celsius (C), a chromel-alumel type for temperatures up to about 1300 °C. The terms 'chromel', 'alumel' and 'constantan' are manufacturers' names for the metals the thermocouple wires are made of. These metals are special alloys, made by combining a number of different metals.

The voltage output of a thermocouple is only a few thousandths of a volt (millivolts). An iron-constantan thermocouple with a temperature difference of 500 °C between both ends of the thermocouple produces a voltage of around 30 millivolts (0.03 volts). While this voltage is too low to provide useful power, because the voltage produced by a thermocouple is proportional to temperature, it can measure temperature. In Figure 2.8, the voltmeter is calibrated in degrees Celsius. Thermocouples are also used in industrial temperature control systems, in which the thermocouple voltage is used to control heating elements.

Light to electrical

A device that produces a voltage when exposed to light is called *photo-voltaic*. The best known photo-voltaic (PV) device is the solar cell. A simple solar cell has two layers of semiconductor material, such as crystalline silicon (typical) or a type of gallium arsenide. The layers are 'doped' to form an N-type and a P-type layer, as found in transistors and diodes. When the layers are sandwiched together, a PN junction is formed, which creates a barrier that prevents current flow between the layers. When light energy in the form of photons falls on the junction, the energy causes electrons from the P-type layer to jump across the barrier to the N-type layer, creating an electric current. The greater the light intensity, the higher the current. The principle is shown in Figure 2.9.

Doping a semiconductor means adding an impurity such as phosphorus (for N-type) or boron (for P-type)

FIGURE 2.9 A solar cell is a semiconductor in which light energy causes electrons to flow across the PN junction and, if connected, through an external circuit

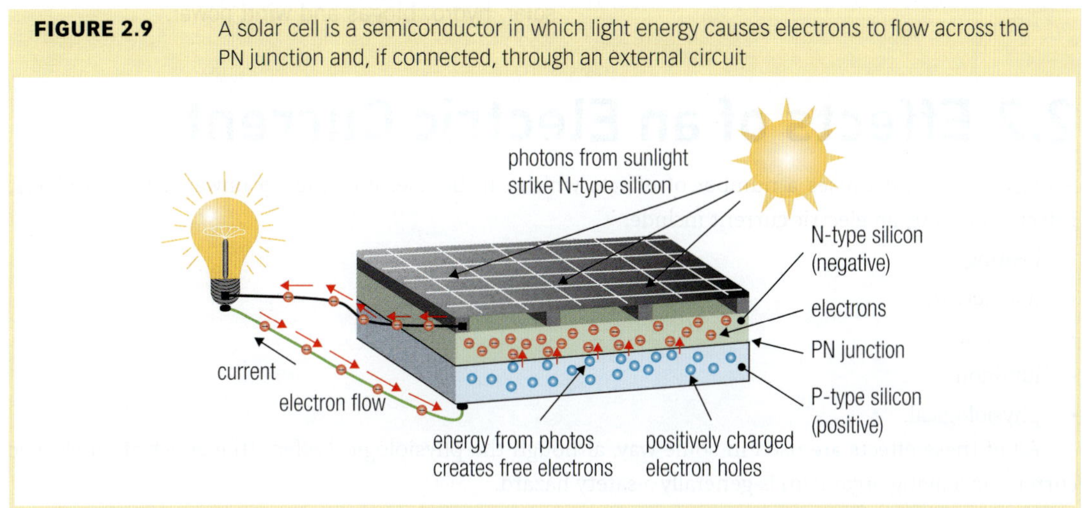

A single solar cell produces about 0.5 V and cells are combined to form a solar panel that typically produces an open circuit voltage of 17 V to 45 V DC and up to 350 watts of power. The power available from a solar panel installation depends on the number, efficiency and size of the panels used. Typical power ratings for a domestic installation range from 1000 watts (1 kilowatt) to 5000 watts (5 kilowatts). Commercial installations produce large amounts of power such as the 352 megawatt solar farm in Western Downs, Queensland. Australia is one of the sunniest countries on Earth, and even larger solar farms are under construction.

Solar farms are rated by their peak power, as their output varies with the sunlight

FIGURE 2.10 (a) A 352 MWp solar farm in Queensland (b) A 200 W solar panel

Solar panels are used to charge batteries that power remotely located roadside signs, telephones and weather stations. Satellites and the international space station are powered by banks of solar panels. Tracking systems are often used in commercial installations to keep sunlight striking the solar panels at all times during the day. A solar cell doesn't need any maintenance (except cleaning) and produces electricity whenever light is present.

KEY POINTS...

- Energy can neither be created nor destroyed, it can only be converted from one form to another.
- Electrical energy is produced by converting other sources of energy, including mechanical, chemical, heat and light.
- Most electrical power is produced by alternators driven by some form of rotational mechanical energy.
- High voltages can be produced with friction between two materials, or by applying force to a piezo-electric material.
- Batteries produce electricity by converting chemical energy to electrical energy.
- Photo-voltaic devices such as the solar cell produce electricity when exposed to light.
- Renewable energy sources include geothermal, solar, hydro, biogas and wind power.

2.2 Effects of an Electric Current

An electric current causes a number of effects which can be useful or merely unwanted by-products. Effects caused by an electric current include:

- heating
- magnetism
- chemical
- luminous
- physiological.

All of these effects are used in some way, although the physiological effect (the effect of an electric current on a living organism) is generally a safety hazard.

Heating

Whenever current flows in a conductor, heat is produced. This effect is used in electric heaters, frypans, cooking ranges and so on. Heat is produced by a conductor because of its resistance. All conductors have a certain amount of resistance, so heat is always produced when current flows. The higher the resistance, the greater the amount of heat produced (for a given value of current). Because of this, the resistance of conductors supplying power to electrical equipment should be as low as possible.

If, however, heat is wanted, as in an electric radiator, the conductor used in the heating element needs to have a certain amount of resistance. The material used is therefore designed to have resistance, and to withstand the heat it produces without burning out. Materials for this purpose include nichrome (a combination of nickel and chrome), which is used in electric radiator elements.

FIGURE 2.11 Heat is developed when current flows through a resistance

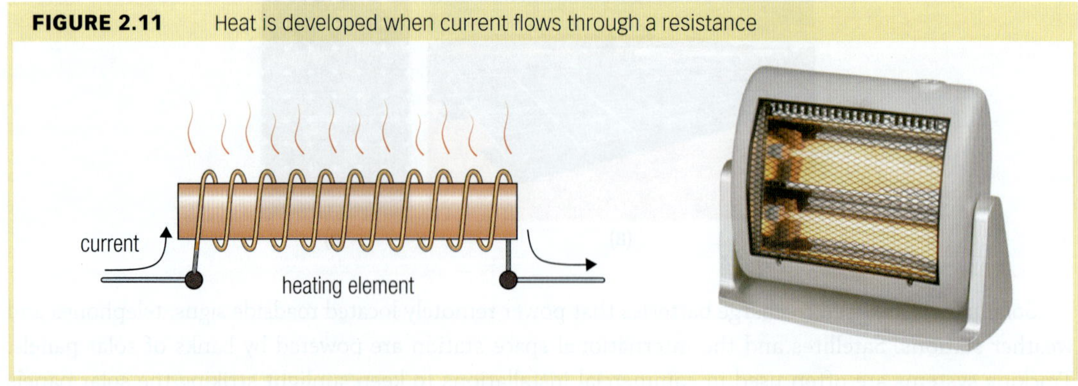

As further explained in Chapter 4, because heat is a form of energy, electrical energy is used to produce the heat. Unless it is wanted, heat in an electric circuit is actually a loss, as energy is needed to produce

it (wasted energy). Efficiency is a measure of power in to power out, so the more unwanted heat that is generated, the lower the efficiency. For example, an electric motor usually gets hot after it has been running for a while. The more efficient the motor is, the less heat it will produce (at its rated load).

FYI

As explained in Chapter 11, a magnetic field is developed around a current-carrying conductor, with a strength proportional to the value of the current

Magnetism

The magnetic effect caused by an electric current is perhaps the most useful of all. The effect was discovered in 1820 when Danish physicist Hans Oersted noticed that a compass needle was deflected when current from a battery was flowing in a nearby wire. On hearing of this discovery, French physicist André-Marie Ampère presented a paper a week later that explained the effect, and thereby opened up the field of electromagnetism. This topic is covered in Chapter 11.

A basic application of the magnetic effect is the *electromagnet*, which has a coil of wire wound around a soft iron core. When current flows in the coil, the iron core becomes magnetised. A common use for the electromagnet is in a device called an electric relay.

A simple relay is shown in Figure 2.12. The moving contact is spring-loaded and is attracted downwards by the magnetised core when the coil is connected to a voltage (energised). When this happens, the moving contact touches the fixed contact. If these contacts are connected between a power source and a load, energising the relay coil will switch power to the load, such as a heater or motor. De-energising the relay coil will turn off power to the load.

FIGURE 2.12 A relay has electrical contacts that close when the relay coil is energised, thereby switching on the load controlled by the relay

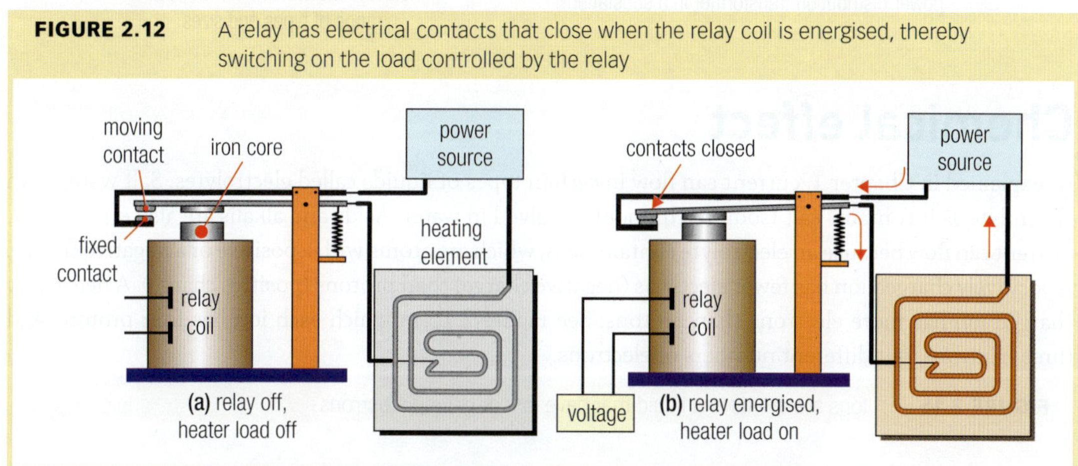

The main application of a relay is to switch loads such as motors, heating elements, lights and so on. A small amount of power is needed to operate the relay coil, but the relay contacts can switch large amounts of power. Relays are available in a range of sizes and ratings. Those shown in Figure 2.13 all have a low voltage coil rated at 12 V. The contacts are rated to handle 230 V AC and currents of 10 A or more.

FIGURE 2.13 Relays used to control a 230 V AC load such as motors, lights and heating elements

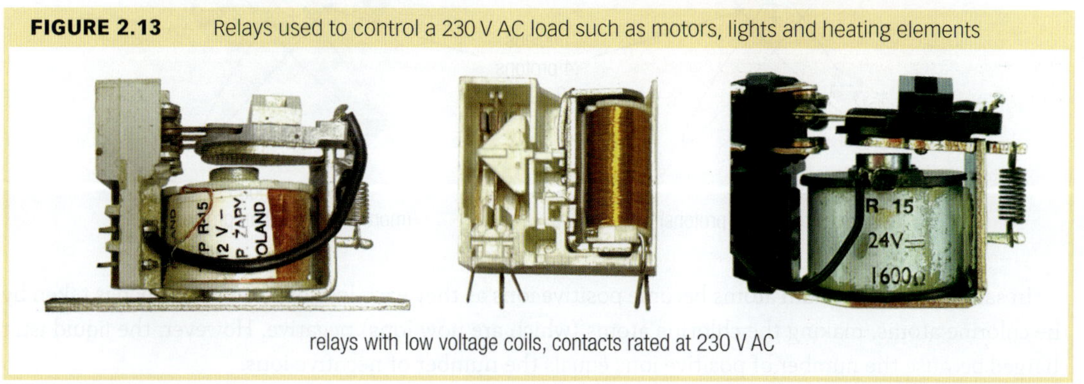

Two other very common applications of the magnetic effect are the electric motor and the transformer. As described in Chapter 21, a transformer converts an alternating voltage from one value to another. Transformers are used extensively in electrical power distribution and transmission, and are widely used

FYI

Larger types of relays are called contactors, which are rated to switch heavy electrical loads

in electronics and telecommunications. Electric motors are also widely used in many forms of transport, machines and appliances, and are explained in Chapters 22 and 23.

Other uses of the magnetic effect are magnetic strips on plastic credit cards or storing computer data on a hard disk. Sensors that detect magnetism, called Hall effect devices, are used extensively in car engines to give timing signals. Large electromagnets are used in metal scrap yards to lift heavy metal objects, or to sort magnetic materials from non-magnetic materials.

FIGURE 2.14 Motors and transformers are common applications of the magnetic effect of an electric current

power distribution transformer in a substation

electric motors come in a wide range of types and sizes

Chemical effect

As explained in Chapter 1, current can flow in certain types of liquids called electrolytes. Salt water is an electrolyte as it contains salt (sodium chloride) dissolved in water. Acids and alkalis are also electrolytes. Current can flow because an electrolyte contains *ions*, which are atoms with a positive or a negative charge. A positively charged ion has fewer electrons (negative charge) than protons (positive charge). A negatively charged ion has more electrons than protons. See Figure 2.15 in which each ion has four protons and three neutrons, but different numbers of electrons.

FIGURE 2.15 Ions are atoms in a liquid that have lost or gained electrons

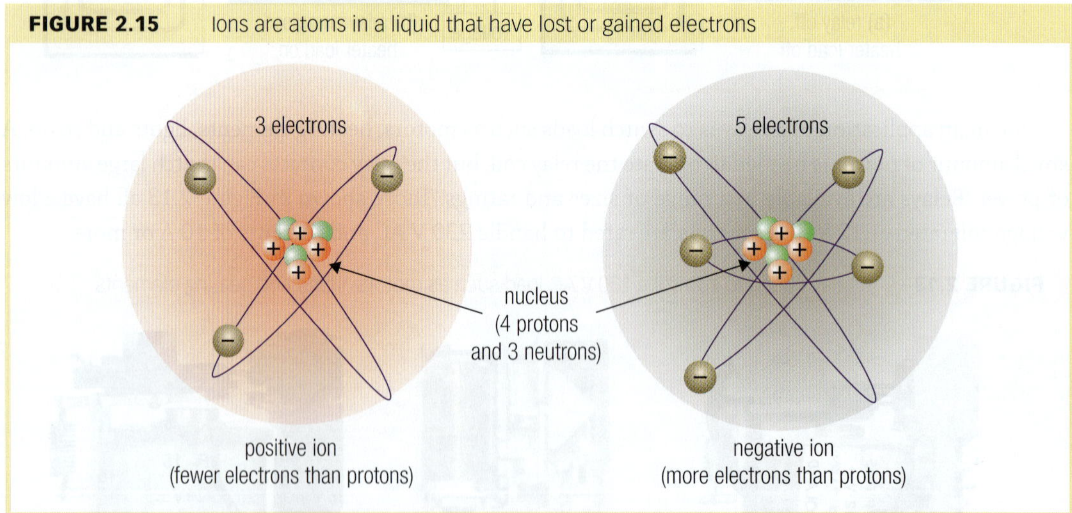

In salt water, the sodium atoms become positive ions as they each lose one electron which is taken by the chlorine atoms, making the chlorine atoms (which are now ions) negative. However, the liquid isn't charged because the number of positive ions equals the number of negative ions.

Because the ions are charged, they can be attracted by a voltage. The rule is: *like charges repel and unlike charges attract*. If two metal electrodes connected to a voltage source are placed in an electrolyte, the positive ions will flow towards (be attracted by) the negative electrode (cathode) and the negative ions will flow towards the positive electrode (anode), as shown in Figure 2.16. As a result, an electric current flows in the liquid.

FIGURE 2.16 Current flow in an electrolyte – the positive ions move to the cathode and negative ions move towards the anode, in this case to produce purified copper

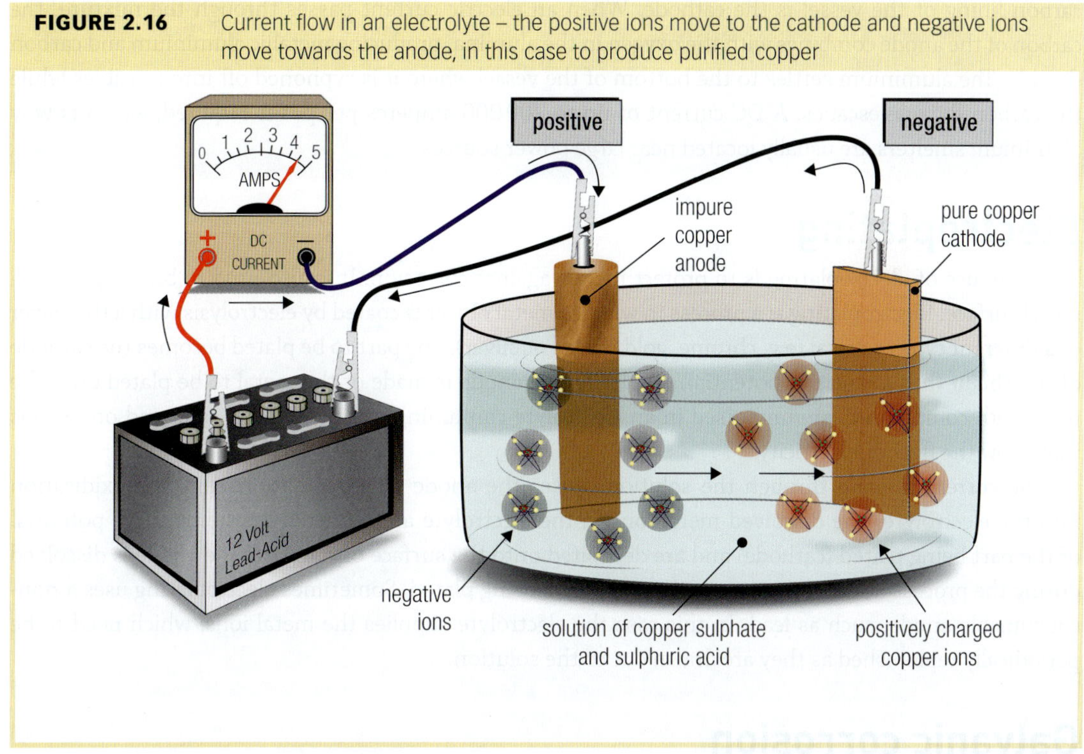

In other words, the electrolyte behaves as a conductor. The more ions there are in the liquid, the better the liquid can conduct an electric current. If more salt is added to the liquid, more ions are created and the resistance of the liquid is less. This is why salt water is a better conductor than fresh water.

The two main useful chemical effects of current and an electrolyte are *electrolysis* and, as already explained, the *voltaic* effect (the electric cell).

Electrolysis

In the electrical industry, an important use of electrolysis is to produce pure copper (greater than 99.95 per cent), called electrolytic copper. High purity is needed because small amounts of impurity metals can seriously reduce copper's ability to conduct electricity. For example, 0.05 per cent of arsenic impurity reduces copper's conductivity by 15 per cent. Electric cables must be made of very pure copper to prevent electrical losses in them, which is particularly important when the cables run for long distances.

To purify copper electrolytically, the impure copper metal is made the anode (the positive electrode) in an electrolytic cell, as in Figure 2.16. A thin sheet of previously purified copper is used as the cathode (the negative electrode). The electrolyte is a solution of copper sulphate and sulphuric acid. When current is passed through the cell, positively charged copper ions are pulled out of the anode into the liquid and are attracted to the negative cathode, where they lose their positive charges and stick tightly as neutral atoms of pure copper metal.

As the electrolysis continues, the impure copper anode dissolves and pure copper builds up as a thicker and thicker coating on the cathode. Impurity metals such as iron, nickel, arsenic and zinc leave the anode and are dissolved by the liquid electrolyte. Impurities such as platinum, silver and gold are also released from the anode, but as they are not soluble, they fall to the bottom, forming a valuable sludge that can often cover the cost of the large amount of electricity used.

Another use of electrolysis is the production of hydrogen, in which water (H_2O) is decomposed into hydrogen and oxygen by passing an electric current through the water. About 4 per cent of hydrogen gas produced worldwide is created by electrolysis, mainly as a by-product of passing current through salt water to produce chlorine gas.

Aluminium is produced by electrolysis, in which powdered alumina, derived from bauxite, is contained within a large vessel (pot) lined with carbon or graphite and fitted with a carbon electrode (anode). The carbon lining of the vessel is the cathode. When an electric current passes through the mixture, the carbon of the anode combines with the oxygen in the alumina, producing metallic aluminium and carbon dioxide. The aluminium settles to the bottom of the vessel where it is syphoned off into crucibles while the carbon dioxide escapes. A DC current of up to 200 000 amperes per pot is required, which is why aluminium smelters are usually located near large power sources.

Electroplating

The main use of electroplating is to protect metals against corrosion. It is also used to build up a worn metal surface. Electroplating is a process in which a metal object is coated by electrolysis with a thin layer of a different type of metal (e.g. chrome, gold, nickel or silver). The part to be plated becomes the cathode of the circuit (has a negative potential). Typically, the anode is made of the metal to be plated onto the part. Both components are immersed in an electrolyte containing dissolved metal salts and other ions that allow the flow of electricity.

The current flowing through the solution causes the anode to release atoms through oxidisation into the electrolyte. The dissolved metal ions in the electrolyte are attracted by the negative potential of the part being plated (cathode) and are deposited onto the surface of the part. The anode is dissolved during the process at the same rate as the cathode is being plated. Sometimes electroplating uses a non-consumable anode, such as lead. In this case the electrolyte supplies the metal ions, which need to be periodically replenished as they are drawn out of the solution.

Galvanic corrosion

FYI

Protection against corrosive or polluting substances is covered in AS/NZS 3000:2018, clause 3.3.2.5

Most metals corrode in some way due to being exposed to oxygen, which causes the surface to oxidise. Rust is the formation of iron oxide and, over time, iron in the presence of moisture will corrode away completely. Other metals such as copper and aluminium form an oxide layer, which helps protect the metal from further oxidisation.

Another form of corrosion is *galvanic corrosion*, caused by electrolysis. This occurs when two different types of metals are in electrical contact and are exposed to an electrolyte, such as rain or sea spray. Corrosion occurs because of the voltaic effect, in which the dissimilar metals and the electrolyte form an electric cell. The current produced by the cell causes the metals to corrode, with one of the two metals corroding before the other. AS/NZS 3000:2018 specifies that different metals likely to cause galvanic corrosion must not be in contact with each other.

An understanding of this process led to the development of *cathodic protection*, a technique to control the corrosion of a metal surface by making it work as the cathode of an electrochemical cell. This is achieved by placing the metal in contact with another more easily corroded metal that acts as the anode of the cell. The first application of this occurred in 1824 when Sir Humphry Davy succeeded in protecting copper sheathing on a ship against corrosion from seawater by the use of iron sacrificial anodes. Today, most storage hot water heaters (gas and electrical) have an internal sacrificial anode to help prevent the boiler from corroding.

Galvanising is a common example of cathodic protection, in which steel sheets are covered with a thin layer of zinc. The zinc not only coats the steel, preventing it from oxidising (rusting), it forms a galvanic cell when the zinc is scratched, protecting the exposed steel by acting as a sacrificial anode. That is, the zinc corrodes before the steel, until eventually the zinc is completely corroded away, allowing the steel to rust. Figure 2.17 lists the commonly used metals in their electrochemical series, which shows that zinc is below iron, hence zinc will corrode in preference to iron.

FIGURE 2.17 Sacrificial anodes in use and the electrochemical series listing metals in the order of their behaviour as a sacrificial anode

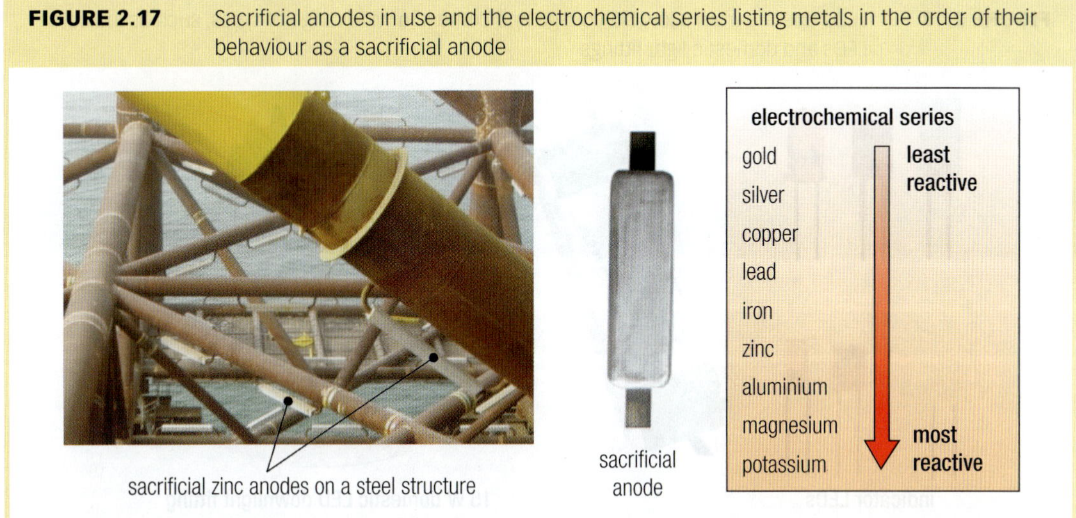

Rechargeable batteries

Charging a battery is another example of the chemical effect, in which the charge current reverses the chemical action caused when the battery is discharged. In effect, electrical energy is stored in the battery during charging, and released by chemical action during discharge. This applies to rechargeable batteries like those used in cars or mobile phones, but not to carbon-zinc batteries, which are not designed to be recharged.

FYI

Large rechargeable batteries are being integrated into the national grid to provide peak power in times of high demand

Luminous effect

One of the first applications of electricity was the development of the arc lamp. This device has two carbon electrodes in open air, arranged so an electric arc is established between the electrodes when they are brought in contact with each other, then separated. While this type of lamp can produce high amounts of light, it also generates high levels of ultraviolet light and a large amount of heat.

The principle of the arc lamp led to the development of the discharge lamp, which has a glass envelope filled with a gas and contains two metal electrodes. Gases used include neon, argon, xenon, sodium, metal halide and mercury vapour. Light is produced by passing an electric current through the gas, which ionises the gas, creating a plasma. A starting circuit provides a high voltage to cause the lamp to light, and a device (called a *ballast*) limits the current flowing through the lamp.

The fluorescent lamp is a common example of a gas-discharge lamp in which the ultraviolet radiation created by the current flowing through the gas (mercury vapour) causes a fluorescent coating inside the tube to glow. These lamps are far more efficient than incandescent lamps, where light is produced by heating a filament.

The light emitting diode (LED) is a solid state device that produces light by way of the effect called *electroluminescence*, which is the result of electrons releasing energy in the form of photons (light). LED lighting is efficient, as not much heat is generated. The first practical visible-spectrum (red) LED was developed in 1962. Today, LEDs can produce red, green, blue, yellow, white, infrared and ultraviolet light. High-power LEDs (over six watts) are used in applications such as traffic lights, vehicle lights and torches. LED light fittings are increasingly being used in homes and industry to provide general lighting. These fittings are designed to manage heat dissipation within a tight tolerance and they have an electronic circuit to control current flowing in the LEDs. The advantage is their high efficiency compared to other forms of lighting. LEDs operate from a low DC voltage, generally around two volts, they have a long life and they maintain a consistent light output even in cold temperatures, unlike traditional lighting methods.

SUSTAINABILITY

LED lights are energy efficient and are now used in homes and industry for general lighting purposes

FIGURE 2.18 LEDs operate from a low DC voltage and come in a wide range of types, such as indicator LEDs and domestic light fittings

electronic control module

heat sink to keep LEDs cool

indicator LEDs

15 W domestic LED downlight fitting

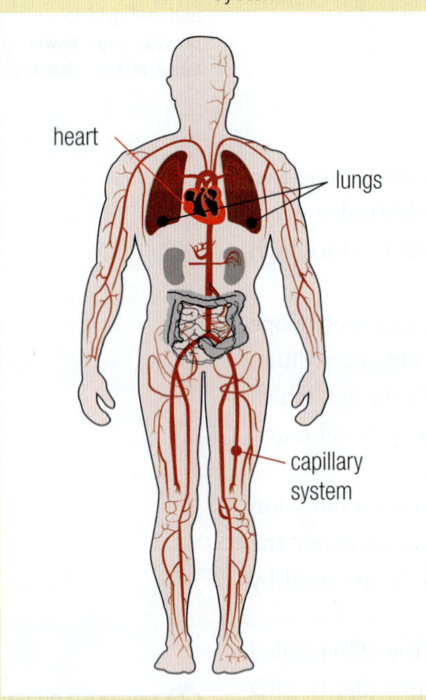

FIGURE 2.19 Blood is a good conductor of electricity and is carried through the body by the capillary system

Physiological effect

This effect is the result of passing an electric current through a living organism, such as a human or an animal. It has a few medical uses, mainly to stimulate a heart by applying a short burst of current that will shock the heart out of ventricular fibrillation (explained shortly).

Otherwise, this effect is limited to stun-guns, cattle prods, electric fences and to those countries that still use the electric chair. The purpose of these items is either to produce a short, sharp electric shock or, in the case of an electric chair, to cause death. So it's important you understand this effect as your life may depend on it!

Life is kept going by two important body functions: breathing (lungs) and blood circulation (heart-beat). If either of these stop for more than a few minutes, the brain is starved of oxygen, and there will be so much damage that death soon follows. If an electric current passes through the body, the current can interfere with the tiny electrical impulses that travel through the body's nervous system to control the heart and lung muscles.

The amount of interference depends on the value of the current, how long it flows and where it flows. Current which flows through the brain, or the chest region where the heart and lungs are, is more dangerous than a current passing through two fingers on the same hand.

If breathing stops after an electric shock, the heart can go into *ventricular fibrillation*. This is when the electrical impulses from the brain controlling the heart are confused and out of step, making the heart quiver and stop pumping blood.

An electric current can also cause terrible burns and make muscles tighten severely. Death and injury can not only result directly from electric shock, but indirectly. Electric shocks have caused people to fall off ladders and power poles, flung them into dangerous chemicals and made them drop something heavy on themselves, such as the live appliance they were carrying at the time.

The value of the current causing the shock can determine whether it's lethal. Table 2.1 shows the effects of various values of current. Remember that a milliamp (mA) is one-thousandth of an ampere, so the current values are quite small. A current as low as three or four-thousandths of an ampere (3 to 4 mA) can cause pain and a current over 20 mA can stop the lungs from functioning. These values of current will hardly even light a lamp, but could kill you. Skin has a relatively high resistance, but blood is a good conductor. If your skin is wet, this reduces its resistance and current can flow through your skin to the low resistance path of the arteries.

FYI

Automatically controlled defibrillators are now found in many workplaces. They can be used by anyone, and can save the life of someone who has had sudden cardiac arrest.

The value of current that flows during an electric shock depends on two things:

- the value of the voltage causing the shock
- the resistance your body offers to the current.

TABLE 2.1

EFFECTS OF AN ELECTRIC CURRENT PASSING THROUGH THE BODY

Current	Effect
up to 2 mA	barely perceptible
2 mA to 8 mA	sensation becomes obvious and more painful
8 mA to 12 mA	muscle spasms and greater pain
12 mA to 20 mA	unable to let go of the conductor, can't control muscles
20 mA to 50 mA	if passing through the chest, breathing might stop
50 mA to 100 mA	if near the heart, causes ventricular fibrillation
100 mA to 200 mA	heart stops beating
above 200 mA	severe burns as well!

There is no such thing as a 'safe' **voltage**. Although rare, there have even been cases of death caused by voltages of less than 10 V, due to electrical contact with the bloodstream by pricking the skin. Always take care.

Resistance is mainly that of the skin. Wet, sweaty hands have less resistance than dry skin. Standing on a wet floor allows current to flow through you to the ground by way of the feet. Standing barefoot is obviously very dangerous – **always wear rubber-soled shoes when working with electricity.**

FYI

Carrying out certain electrical work requires holding the correct electrical licence. It is generally illegal to work on energised electrical equipment.

Wet clothes are another hazard when working with electricity, as wet cloth can conduct an electric current. The resistance of the cloth depends on the type of liquid soaked up by the material. Remember that tap water with all its usual impurities will conduct electricity. As we've said before, electricity and water is a dangerous combination! Here are some basic rules for avoiding an electric shock:

- Think of *any* appliance, conductor, terminal or equipment as live, until you have proved otherwise.
- Don't work on an electrical appliance or circuit with wet hands.
- A wet floor is a good conductor of electricity – be extra careful.
- Don't wear metal jewellery (rings, watch) when working with electricity.
- Isolate and lock out the supply before working on an electrical installation.
- Wear correct PPE, including cotton clothing (not synthetics) and enclosed footwear.
- Remember that 'familiarity breeds contempt'. Always have respect for electricity – *it has no respect for you!*

2.3 Protection against Effects of an Electric Current

FYI

Direct contact is defined in AS/NZS 3000:2018 clause 1.4.38. Indirect contact is defined by clause 1.4.39.

Section 1.5 specifies the requirements for protection against dangers and damage associated with an electrical installation.

The effects of an electric current can sometimes cause dangerous situations, the most obvious being the danger of an electric shock. The Australian and New Zealand wiring rules (AS/NZS 3000:2018) specify ways of minimising these dangers. In regard to the physiological effect, the standards identify two situations that must be protected against:

1. **Direct contact under normal service conditions** with a live terminal, conductor or part of an electrical installation or circuit. Protection against this is achieved by a combination of insulation, barriers or enclosures and placing exposed electrical contacts out of reach. To protect the public, appliances are required to pass safety tests, and electrical wiring installations must be to the standards given in AS/NZS 3000:2018.

FIGURE 2.20 Direct contact with a live conductor can cause current to flow through your body to ground

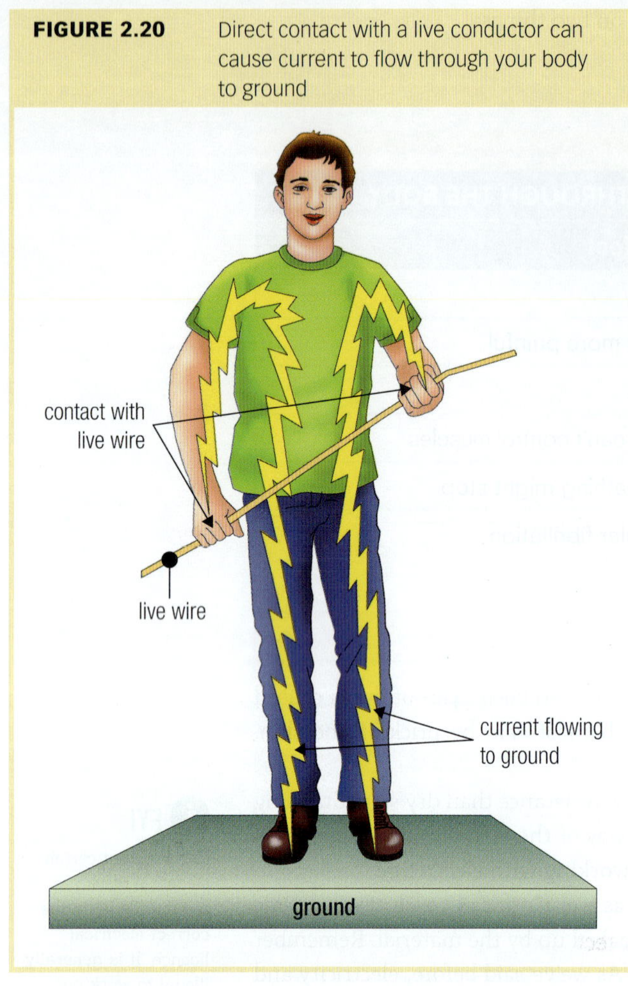

Those working on live electrical circuits are the most likely to receive an electric shock. Normally a circuit is isolated from the electrical supply before doing work, but in some cases this is not possible (e.g. when working on power lines or a live fuse box). Under these conditions, you need to wear protective gloves and other items of personal protective equipment, as well as understand the risk and take steps (control measures) to prevent being electrocuted.

2 **Indirect contact that can occur under a fault condition.** This might happen if you touch a metal-cased appliance that, because of an internal fault, is now live. Other faults are exposed live conductors because the insulation is missing, or water reaching live terminals or electrical parts; any fault that could cause an electric shock to the user.

   Protection against these dangers is usually achieved by automatic disconnection of the electrical supply by way of fuses, circuit breakers and a unit called a residual current device (RCD), generally known as a safety switch. This is in combination with a system of earthing, in which metal parts of an appliance are connected to ground via the earth wire.

The AS/NZS 3000:2018 standards also require electrical installations to provide protection against the thermal effects of an electric current. For example, electrical equipment that produces a lot of heat must be properly ventilated and fireproof enclosures should be used where appropriate. The magnetic effect must also be protected against, particularly when a cable is carrying more than 300 A. This is further explained in Chapter 11.

FIGURE 2.21 Faults that can cause an electric shock include touching a live appliance or exposed live conductors

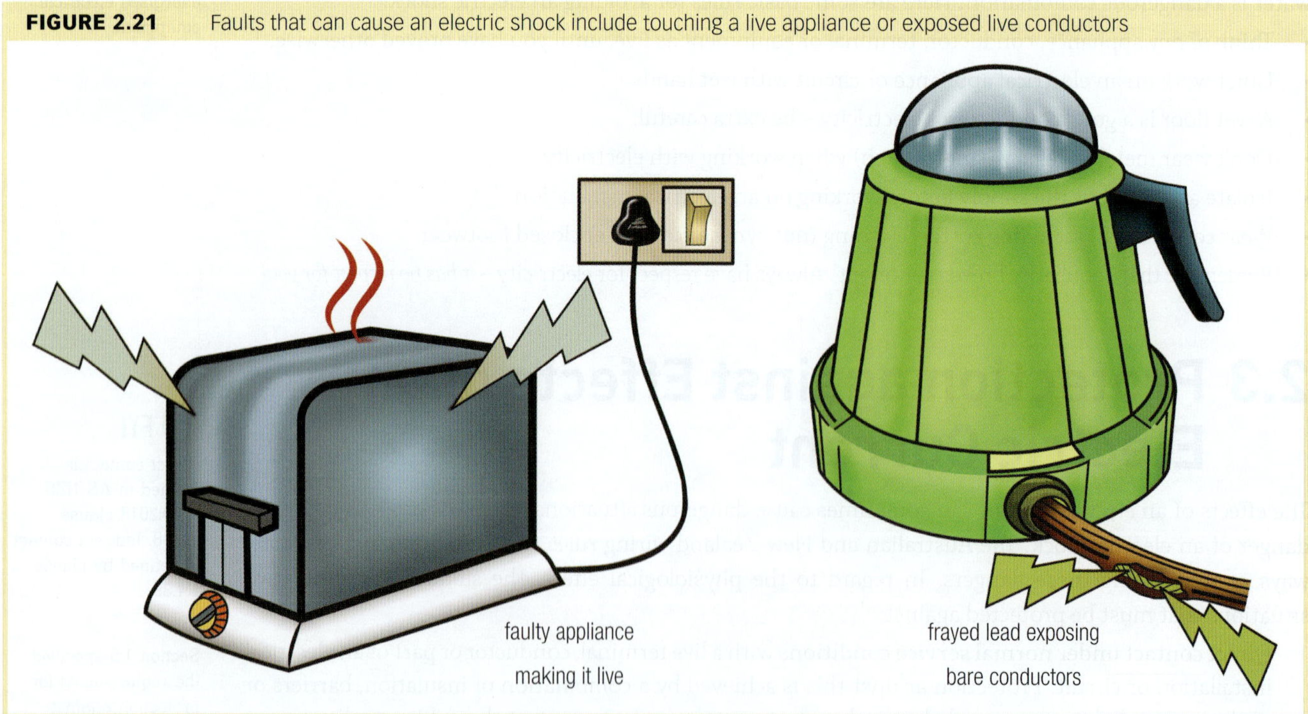

CHAPTER SUMMARY

- Electrical energy is produced by converting other forms of energy (mechanical, chemical, heat and light).
- Electrical energy is converted to the required form of energy (e.g. heat, light or mechanical energy) by the electric load.
- Large alternators driven by steam turbines (mechanical energy) generate most of Australia's electrical energy.
- An alternator has coils of wire arranged so they intersect with a moving magnetic field, thereby producing electricity.
- Renewable energy sources that can drive an alternator include geothermal, solar thermal, hydro, biogas and wind power.
- Solar panels are photo-voltaic devices that convert light energy to electrical energy.
- Friction between two surfaces can produce a high voltage, but not much current.
- The electric cell produces electricity by converting chemical energy to electrical energy.
- Charging a battery (a group of cells) reverses the chemical action that took place when the battery was being discharged.
- Electric motors, transformers and electromagnets work by way of the magnetic effect of an electric current.
- Electrolysis is the process of passing current from one electrode to another through an electrolyte. Electroplating is an application of electrolysis.
- Effects caused by an electric current include heating, magnetism, chemical, luminous and physiological.
- Discharge lamps are filled with a gas that produces light directly, or in the case of the fluorescent lamp, ultraviolet light that excites phosphors inside the tube.
- Physiological effects of an electric current include muscle contraction, ventricular fibrillation, burns and death.
- Heat is produced in a conductor because of its resistance.
- Current flowing in a conductor creates a magnetic field around the conductor. Coiling the conductor makes an electromagnet.
- Chemical effects include electrolysis and voltaic effect (electric cell).
- Galvanic corrosion is caused by electrolysis when two different metals are in electrical contact, creating an electric cell.
- A metal surface can be protected from corrosion with sacrificial anodes that corrode in preference to the metal being protected.
- Fluorescent lamps are an example of the luminous effect of an electric current.
- LED lighting is an example of electroluminescence.

REVIEW EXERCISES

Check your answers at the back of the book.

1. List four sources of energy that can be used to produce electrical energy.
2. A car alternator produces electricity due to what type of effect?
3. Why does the copper used in electrical cables need to be produced by electrolysis?
4. In electroplating, what is the polarity of the anode?
5. Why does zinc galvanising protect steel from galvanic corrosion?
6. What are the three components of a basic electric cell?
7. Name the two main chemical effects of an electric current.
8. Give three advantages of LED lighting systems.
9. What two things determine the value of current flowing through your body when you get an electric shock?
10. What are the two situations specified by AS/NZS 3000:2018 concerning protection against an electric shock?

ONLINE RESOURCES

COMPLETE WORKSHEET TWO

Check with your instructor for worksheets on this chapter.

CHAPTER 3

OHM'S LAW

This chapter introduces resistance and Ohm's law – the most important mathematical law of all in the electrotechnology industry. It also explains how to work with very large and very small numbers using 'engineering notation' and the role of metric prefixes (multipliers) in writing large and small values.

CHAPTER OUTLINE

3.1 Resistance and Conductance

Chapters 1 and 2 explained that resistance in an electrical circuit is *opposition to the flow of current* and that resistance is present (to some extent) in all electrical conductors. Resistance is, therefore, a fundamental property of an electric circuit. It is measured in ohms and is identified by the letter R. The Greek symbol Ω (omega) is often used instead of writing the word ohms, so a resistance of 100 ohms becomes 100 Ω. Resistance and a component known as a *resistor* are covered in Chapter 5. Drawing symbols for a resistance (or resistor) are shown in Figure 3.1.

Conductance, rather than resistance, is a term often used when talking about cables and their current carrying capability. Conductance has the symbol G; it is measured in siemens (symbol S) and is the reciprocal of resistance. For example, if a cable measuring 30 metres has a resistance of 10 ohms, it has a conductance (G) of 1/10, or 0.1 S. If a larger cable of the same length has a resistance (R) of 0.1 ohms, it has a conductance (G) of 1/0.1 or 10 siemens (10 S). Notice that as resistance goes down in value, the conductance increases. That is, the lower the resistance, the higher the conductance, which is what is needed in a current-carrying cable.

Previously the unit for conductance was the *mho*, which is ohm spelt backwards

FIGURE 3.1 Resistor symbols and how a resistor is depicted in a circuit diagram

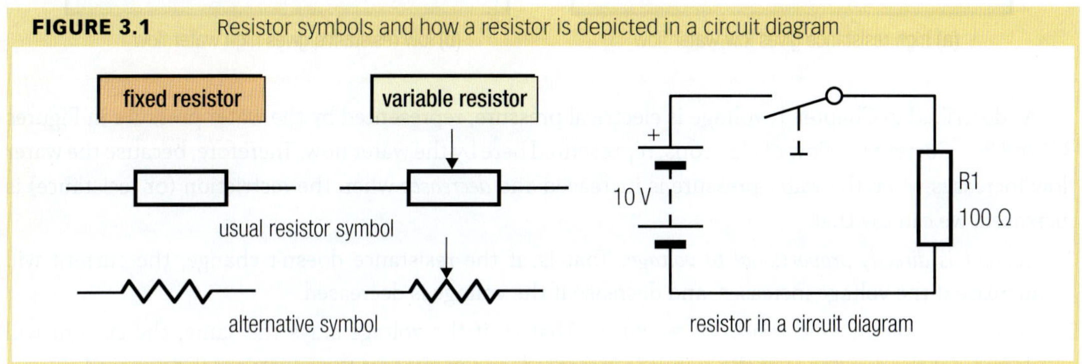

3.2 Ohm's Law

In 1827, Georg Simon Ohm (German physicist, 1789–1854) presented his laws on resistance, voltage and current. At first his laws were dismissed, but in 1881 a meeting of the International Electrical Congress in Paris named the unit of electrical resistance the 'ohm' in his honour.

Ohm's law brings together the three fundamental electrical quantities of voltage, current and resistance. The relationship between these three quantities can be shown with a water pump pushing water through a pipe. Figure 3.2 shows something you already know – if the pressure from a water pump is low, the water flow is also low. Increase the pressure, and the water flow increases.

FIGURE 3.2 Water flow is directly proportional to the water pressure

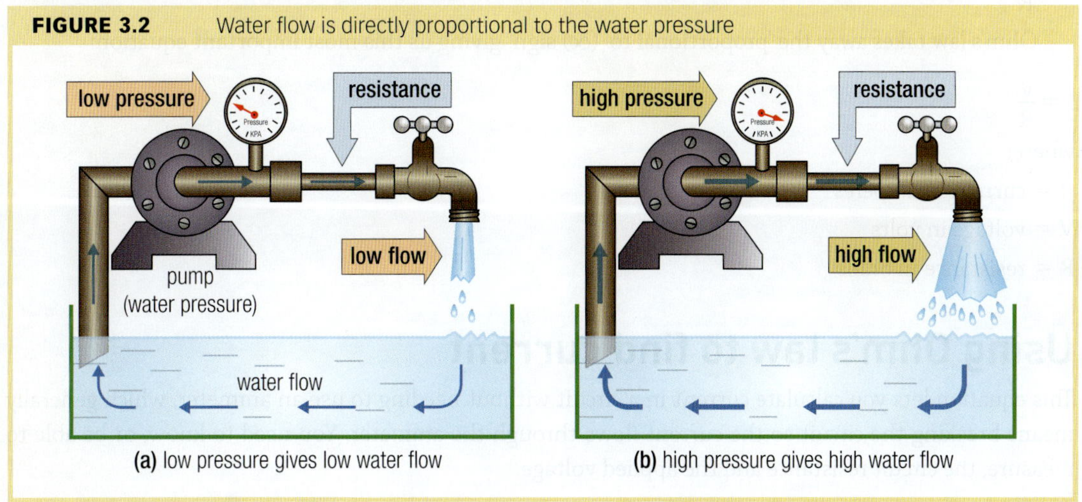

(a) low pressure gives low water flow

(b) high pressure gives high water flow

Figure 3.3 shows something just as obvious. If the pipe carrying the water has a restriction put in it (and the pump pressure stays the same), the water flow reduces. The restriction could be a tap, or as in Figure 3.3 (a), a piece of thin pipe.

FIGURE 3.3 Water flow is inversely proportional to resistance of the pipe

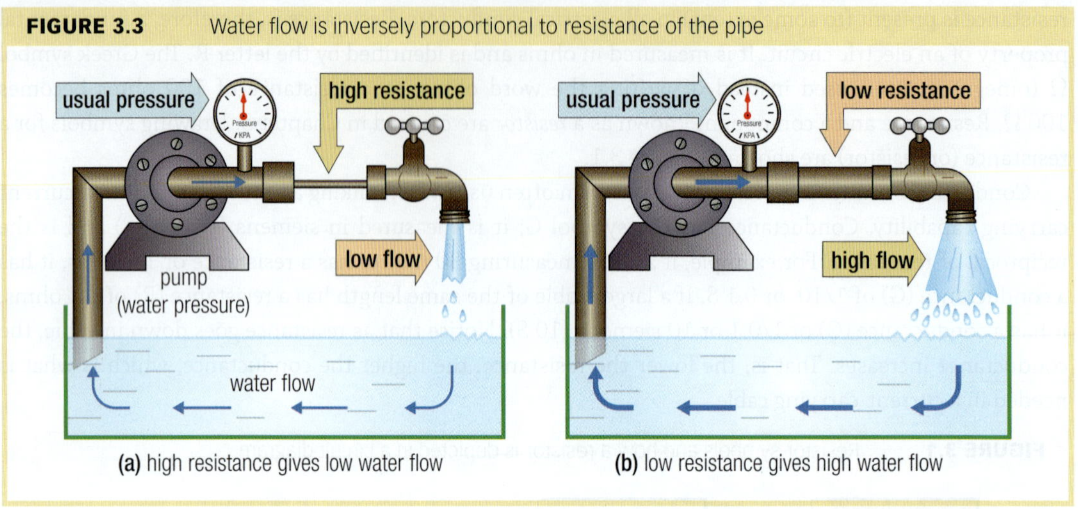

(a) high resistance gives low water flow **(b)** low resistance gives high water flow

As described in Chapter 1, voltage is electrical pressure, represented by the water pressure in Figures 3.2 and 3.3. Current is a flow of electrons, represented here by the water flow. Therefore, because the water flow increases when the water pressure is increased and *decreases* when the restriction (or resistance) is increased, we can say that:

1 *Current is directly proportional to voltage*. That is, if the resistance doesn't change, the current will increase if the voltage increases, and decrease if the voltage is decreased.

2 *Current is inversely proportional to resistance*. That is, if the voltage stays the same, the current will decrease if the resistance of the circuit is increased, and increase if the resistance is decreased.

Ohm's law for current

From the above we can say: The current flowing between any two points in an electric circuit is directly proportional to the voltage between the two points, and is inversely proportional to the resistance of the circuit between the two points.

From this we can write the mathematical relationships between current (I) and voltage (V), and between current (I) and resistance (R):

$I \propto V$ (means I is *proportional to* V)

$I \propto \frac{1}{R}$ (means I is *inversely proportional to* R)

KEY CONCEPT

Ohm's law for current

Ohm's law takes away the 'proportional to' ($\propto$) sign, giving us this most important equation:

$$I = \frac{V}{R}$$

where:

I = current in amperes

V = voltage in volts

R = resistance in ohms

Using Ohm's law to find current

This equation lets you calculate current in a circuit without needing to use an ammeter, which generally means breaking the circuit so the current flows through the ammeter. You need to know, or be able to measure, the circuit resistance and the applied voltage.

Figure 3.4 shows a 10 ohm resistor connected across a 12 V battery. In this diagram, the resistance value (10 Ω) is stamped on the resistor, and the voltage is measured with a voltmeter. Because the values of resistance and voltage are known, the current flowing through the resistor can be calculated.

FIGURE 3.4 The current in the circuit can be found with Ohm's law because resistance and voltage are known

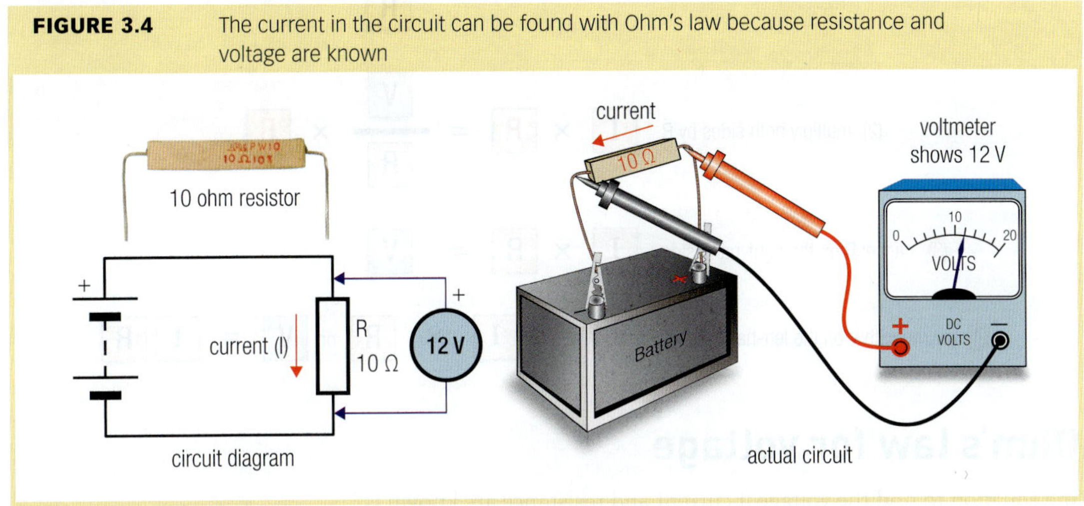

From the measured values in the circuit above, R = 10 Ω and V = 12 V.

Therefore, by Ohm's law: $I = \frac{V}{R} = \frac{12}{10} = 1.2$ amperes

EXAMPLE 3.1

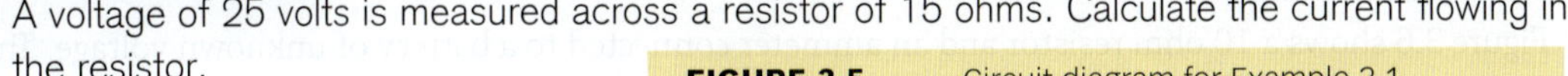

A voltage of 25 volts is measured across a resistor of 15 ohms. Calculate the current flowing in the resistor.

FIGURE 3.5 Circuit diagram for Example 3.1

+
I = ?
R
15 Ω
25 V

Solution

Values $V = 25$ volts

$R = 15$ ohms

$I = ?$

Equation $I = \frac{V}{R}$

$I = \frac{25}{15}$

Answer **I = 1.67 amperes**

Using Ohm's law to find voltage

Sometimes you know the values of the current and resistance in a circuit, but not the voltage. Because Ohm's law is a mathematical equation, it can be rearranged by transposing it to make voltage the *subject*. The subject is the unknown quantity and is on the left-hand side of the equal (=) sign, with the remaining terms on the right-hand side. The process for transposing the equation $I = \frac{V}{R}$ so V is the subject is shown in the equations on the following page.

MATH

See page xxii for help with transposing equations

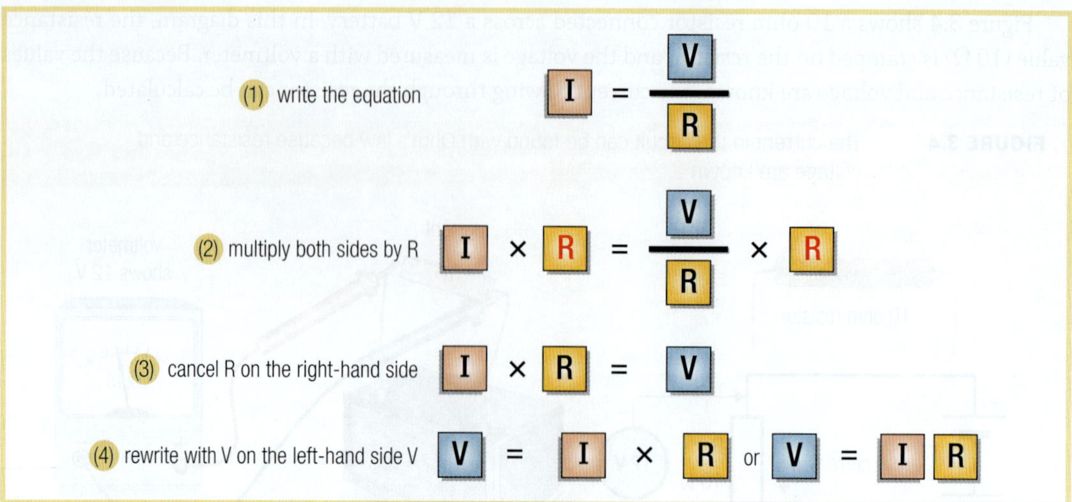

KEY CONCEPT

Ohm's law for voltage

Ohm's law for voltage

The equation to find the voltage if current and resistance are known is:

V = IR

where:

V = voltage in volts

I = current in amperes

R = resistance in ohms

Figure 3.6 shows a 10 ohm resistor and an ammeter connected to a battery of unknown voltage. The ammeter is showing a current of 1.2 A, and the resistance value (10 Ω) is stamped on the resistor. Because the values of resistance and current are known, the voltage across the resistor can be calculated.

FIGURE 3.6 The battery voltage in the circuit can be found with Ohm's law because resistance and current are known

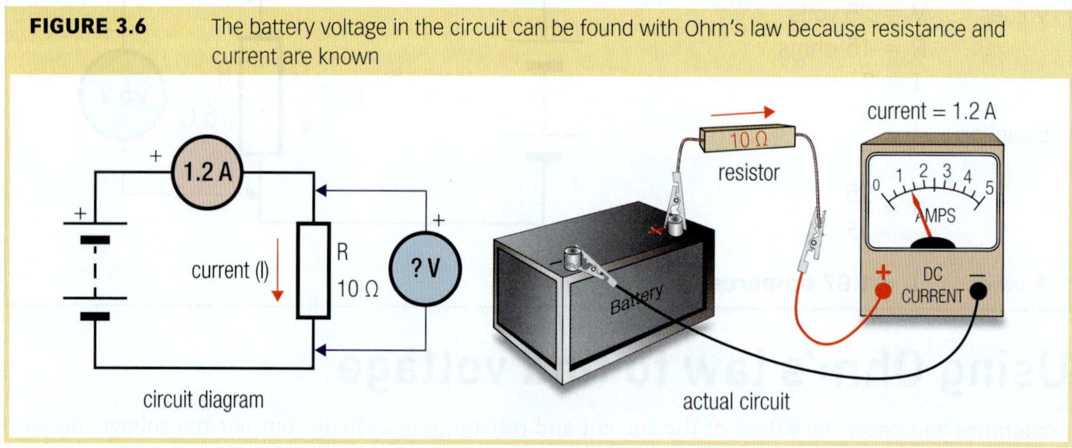

From the measured values in the circuit in Figure 3.6, R = 10 Ω and I = 1.2 A. Therefore, by Ohm's law: V = I × R = 10 × 1.2 = **12 volts**. This is to be expected, as the circuit is the same as in Figure 3.4.

EXAMPLE 3.2

A 50 ohm resistor has a current of 1.5 amperes flowing through it. Calculate the voltage across the resistor.

Solution

Values $I = 1.5$ amperes
$R = 50$ ohms
$V = ?$

Equation $V = IR$
$V = 1.5 \times 50$

Answer **$V = 75$ volts**

FIGURE 3.7 Circuit diagram for Example 3.2

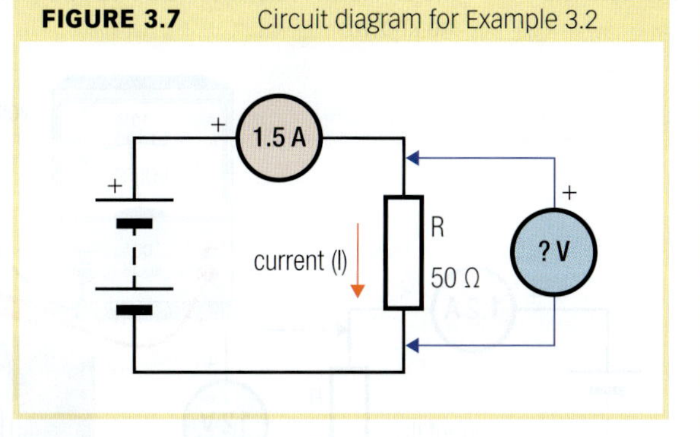

Using Ohm's law to find resistance

MATH

See page xxii for help in transposing this equation

In some cases the voltage and current are known, with resistance the unknown quantity. The sequence to make *R* the subject of the equation starts with V = IR

(1) Write the equation $V = IR$

(2) Divide both sides by I $\frac{V}{I} = \frac{IR}{I}$

(3) Cancel I on the right-hand side $\frac{V}{I} = R$

(4) Rewrite with R on the left-hand side $R = \frac{V}{I}$

Ohm's law for resistance

KEY CONCEPT

Ohm's law for resistance

The equation to find resistance if the voltage and the current are known is:

$$R = \frac{V}{I}$$

where:

V = voltage in volts

I = current in amperes

R = resistance in ohms

Figure 3.8 shows a resistor of unknown value connected to a battery that, according to the voltmeter, is producing 12 V across the resistor. The current in the resistor, as shown by the ammeter, is 1.2 A. Because the values of voltage and current are known, the resistance of the resistor can be calculated.

FIGURE 3.8 The unknown resistance value can be found with Ohm's law because voltage and current are known

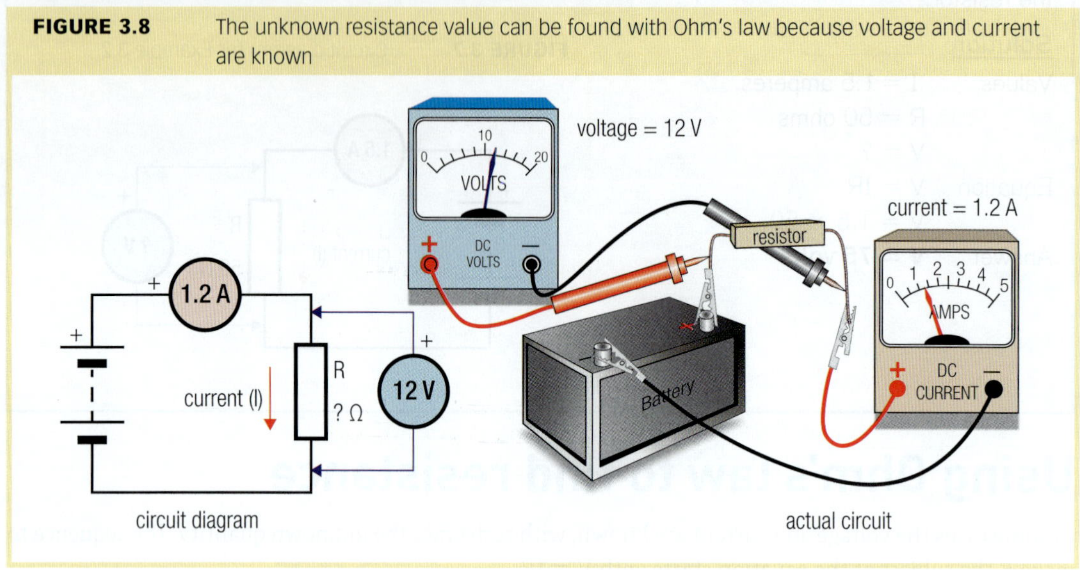

From the measured values in the circuit in Figure 3.8, V = 12 V and I = 1.2 A. Therefore, by Ohm's law: $R = \frac{V}{I} = \frac{12}{1.2} = 10$ ohms. This is to be expected, as the circuit values are the same as in Figures 3.4 and 3.6.

EXAMPLE 3.3

An electric heater connected to 230 V takes a current of 2.3 amperes. Calculate the resistance of the heater.

Solution

Values $V = 230$ volts

$I = 2.3$ amperes

$R = ?$

Equation $R = \frac{V}{I}$

$R = \frac{230}{2.3}$

Answer **R = 100 ohms**

FIGURE 3.9 Circuit diagram for Example 3.3

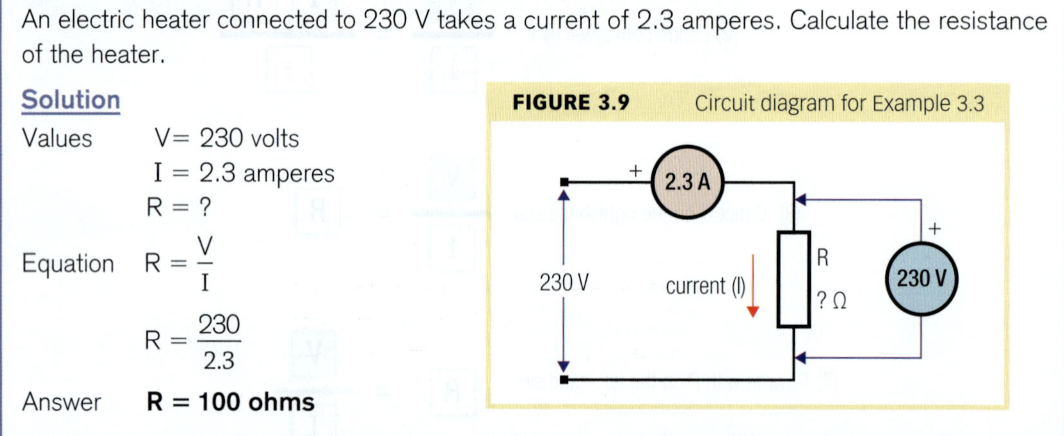

Sometimes E (for EMF) is given instead of V (for voltage). There is no difference between these two as far as the Ohm's law equations are concerned. If E is given, rather than V, substitute E for V in the equations. The equations then become:

$I = \frac{E}{R}$ where E is the applied EMF in volts

$E = IR$

$R = \frac{E}{I}$

The three equations Ohm presented in 1827 are now seen as the most important equations of all in electrical and electronic theory. You need to know these equations and, of course, how to use them.

Ohm's law triangle

The triangle in Figure 3.10 might help you remember the three Ohm's law equations. To find any of the three equations, cover the value you want to find and the equation is the other two terms. Notice that V is near the top vertex of the triangle.

FIGURE 3.10 Ohm's law represented in a triangle. Cover the term you want to find: the equation is the other two terms.

Ohm's law triangle

$V = IR$

$I = \frac{V}{R}$

$R = \frac{V}{I}$

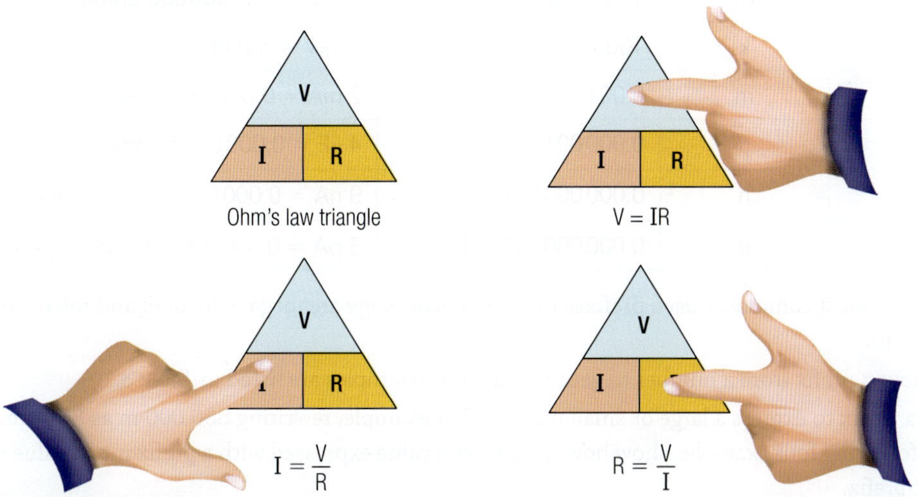

TASK 3.1

1. You measure 12 V across a resistance and 0.5 A flowing through the resistance. Find the value of the resistance.
2. An ammeter shows a reading of 0.4 A flowing through a 100 ohm resistor. What value of voltage would you measure across the resistor?
3. A long length of electrical cable has a voltage drop across it of 15 V when it is carrying a current of 30 A. Calculate the resistance of the cable.
4. A faulty switch has a resistance of 5 ohms and is passing a current of 6 A. Calculate the voltage across the switch.
5. An electric heating element connected to 230 V takes a current of 5 A. What is the resistance of the heating element?

3.3 Metric Prefixes

In the examples given so far, the numbers have been simple to handle as they were neither exceptionally large nor exceedingly small. However, this is often not the case. For example, resistor values can range from 0.0001 ohm to 82 million ohms. The current flowing in a circuit can be from a few millionths of an ampere up to thousands of amperes, and power stations supply millions of watts of electrical power.

In everyday life, we refer to kilometres rather than saying one thousand metres, or millimetres instead of saying one thousandth of a metre. The terms 'kilo' and 'milli' are known as *metric prefixes* or *multipliers* and are given the symbols 'k' and 'm'. That is, rather than write 25 kilometres, you would write 25 km, or 66 mm instead of 66 millimetres. Table 3.1 lists the prefixes used in electrotechnology.

TABLE 3.1

METRIC PREFIXES			
Prefix	**Symbol**	**Multiply by**	**Example**
tera	T	1 000 000 000 000	2 THz = 2 000 000 000 000 hertz
giga	G	1 000 000 000	2 GW = 2 000 000 000 watts
mega	M	1 000 000	39 MΩ = 39 000 000 ohms
kilo	k	1000	33 kV = 33 000 volts
milli	m	0.001	2 mA = 0.002 amperes
micro	μ	0.000 001	4 μF = 0.000 004 farads
nano	n	0.000 000 001	9 nA = 0.000 000 009 amperes
pico	p	0.000 000 000 001	3 pA = 0.000 000 000 003 amperes

The four most commonly used prefixes in electrotechnology are mega, kilo, milli and micro. You need to be able to:

1 Remove a prefix from a number to get its base unit. Examples are listed in Table 3.1.

2 Use a prefix to express a large or small number. For example, rewriting 800 000 m as 800 km.

The following four examples show how to convert a value expressed with a prefix into a value without a metric prefix.

EXAMPLE 3.4

Convert 3.9 MV to volts.

Solution

M (mega) is the prefix for a million, which means multiplying 3.9 by 1 000 000 to replace the prefix with a number. A simple method is to move the decimal point by the number of zeros represented by the prefix, in this case six places. Because the number will become larger, the decimal point is moved to the right, as shown in Figure 3.11.

Answer: **3 900 000 volts**

FIGURE 3.11 Converting 3.9 MV to volts by moving the decimal point to the right

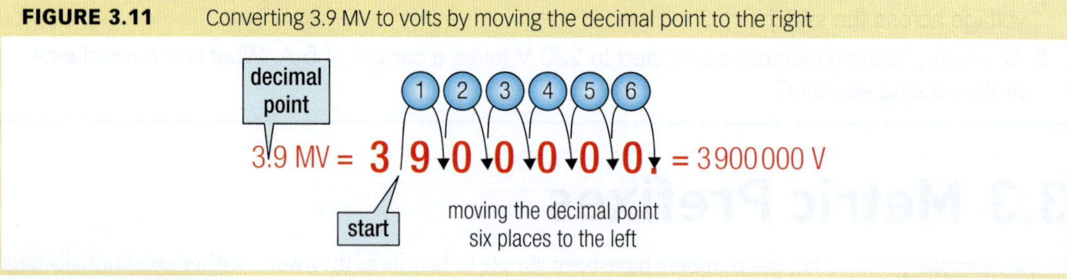

EXAMPLE 3.5

Convert 2.75 kV to volts.

Solution

k (kilo) means *multiply* the number by 1000. This is the same as moving the decimal point *three* places to the right.

Answer: **2750 volts**

To use a calculator for Example 3.4, enter 3.9 and multiply it by 1 000 000. For Example 3.5, enter 2.75, then multiply it by 1000.

The next two examples show how to convert a small value expressed with a metric prefix to a value without the prefix.

EXAMPLE 3.6

Convert 25.8 μA to amperes.

Solution

μ (micro) means *divide* the number by 1 000 000 or move the decimal point *six* places to the left, as shown in Figure 3.12.

Answer: **0.0000258 amperes**

FIGURE 3.12 Converting 25.8 μA to amperes by moving the decimal point to the left

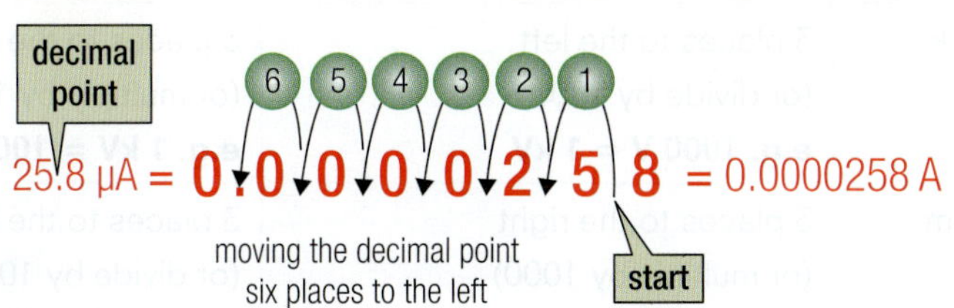

EXAMPLE 3.7

Write 4.76 mA as a value in amperes.

Solution

m (milli) means *divide* the number by 1000 which is the same as moving the decimal place *three* places to the left.

Answer: **0.00476 amperes**

Examples 3.6 and 3.7 can also be done on a calculator by dividing the value in Example 3.6 by 1 000 000 and dividing by 1000 in Example 3.7.

The next two examples show how to express a large or a small value with a metric prefix.

EXAMPLE 3.8

Convert 2 560 000 ohms to MΩ (pronounced megohms).

Solution

Divide the value by 1 000 000 or move the decimal point *six* places to the left. Put the letter 'M' after the new value to show the missing decimal places.

Answer: **2.56 MΩ**

EXAMPLE 3.9

Rewrite 0.0056 A to a value in milliamperes (mA).

Solution

Multiply the value by 1000 or move the decimal point *three* places to the right. Add the letter 'm' after the new value to show the missing decimal places.

Answer: **5.6 mA**

In these examples, the prefix replaces a certain number of zeros. This works only if the zeros are either before or after the value. For instance, a value of 2000.007 can't be simplified with a prefix, as the zeros are between two numbers.

Table 3.2 summarises how to remove or add a prefix. It shows the number of places to move the decimal point, and which way to move it. When using a calculator, the table also shows whether you divide or multiply the value, and by how much. Only the four most commonly used prefixes are listed, but the same rules apply for all prefixes. You just need to know the number of decimal places or the number the prefixes represent.

TABLE 3.2

USING PREFIXES			
Prefix	**Symbol**	**To add the prefix, move the decimal point:**	**To remove the prefix, move the decimal point:**
mega	M	6 places to the left (or divide by 1 000 000) **e.g. 1 000 000 Ω = 1 MΩ**	6 places to the right (or multiply by 1 000 000) **e.g. 1 MΩ = 1 000 000 Ω**
kilo	k	3 places to the left (or divide by 1000) **e.g. 1000 V = 1 kV**	3 places to the right (or multiply by 1000) **e.g. 1 kV = 1000 V**
milli	m	3 places to the right (or multiply by 1000) **e.g. 0.001 A = 1 mA**	3 places to the left (or divide by 1000) **e.g. 1 mA = 0.001 A**
micro	μ	6 places to the right (or multiply by 1 000 000) **e.g. 0.000 001 A = 1 μA**	6 places to the left (or divide by 1 000 000) **e.g. 1 μA = 0.000 001 A**

Decimal point notation

Values expressed with a prefix often have a decimal point. For example, 6 800 000 ohms converted to megohms gives 6.8 MΩ. However, on a circuit diagram or the component itself, the decimal point could be badly printed, even missing. To avoid the problem, the prefix symbol is sometimes placed where the decimal point would otherwise be. So another way of writing 6.8 MΩ is 6M8 Ω. A resistance value of 2700 Ω can be written as 2k7 Ω or 2.7 kΩ.

Figure 3.13 shows a range of resistors where the decimal point is replaced by a prefix. Where a prefix is not required, the decimal point is replaced with the letter *R* (for resistance). There is no standard method. The important thing is to be careful when writing a value with a decimal point, so that anyone reading it has no doubt. For example, write 0.005 instead of .005, as the leading zero helps show there's a decimal point.

FIGURE 3.13 Resistance values where a letter replaces the decimal point

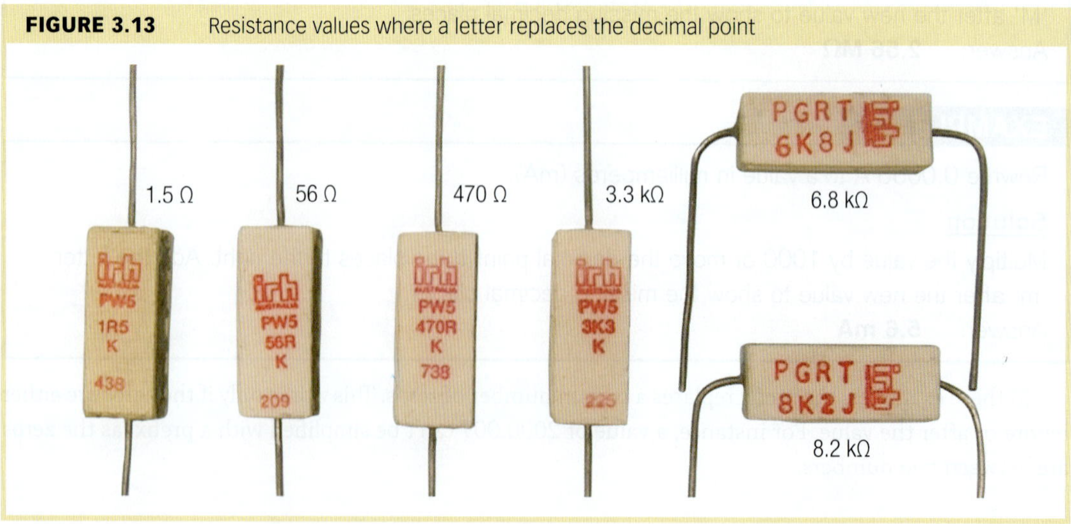

TASK 3.2

1 You measure a resistance value of 2 500 000 ohms. Express the value in megohms.
2 The nameplate on an electrical appliance states the current taken by the appliance is 0.004 A. Write this value in milliamps.
3 A power station has a maximum output power of 2 200 000 000 watts. Express this value using a suitable metric prefix.
4 Convert 6530 millivolts to volts.
5 A resistance value is given as 0.000003 ohm. Express this value in micro ohms.

3.4 Scientific Notation

Metric prefixes, or multipliers, are a convenient way to express large and small numbers but, as already explained, they need to be removed when making calculations. However, another way is to express the prefix in mathematical form using *scientific notation*, in which each prefix is treated as a power of 10.

The term *power of 10* refers to the number of times 10 is multiplied by itself. For example, 1000 equals $10 \times 10 \times 10$. That is, 10 is multiplied by itself three times, or raised to the power of 3. So, rather than write $10 \times 10 \times 10$, we use the notation 10^3. The power (3 in this case) is called the *exponent*.

Scientific notation also gives an easy way of comparing magnitudes, in which a change by 'an order of magnitude' means a change by a factor of 10. For example, if the current in a circuit is 40 A, and it increases by an order of magnitude, the current will be 400 A. A change from 40 A to 4000 A is two orders of magnitude. In scientific notation, 40 A is written as 4×10^1 and 4000 A becomes 4×10^3. The change in magnitude is shown by the difference between the exponents, which in this case is $3 - 1 = 2$.

Each prefix used in the metric system represents a power of 10 and can be written in scientific notation. Therefore, because kilo (k) is 10^3, to enter a value like 2 kΩ into a scientific calculator, follow the keystrokes shown in Figure 3.14, where after entering the value 2, the 'k' is entered as exponent 3.

FIGURE 3.14 Engineering notation is easily handled with a scientific calculator. Some calculators call the exponent key EXP, others mark this key as x10ˣ.

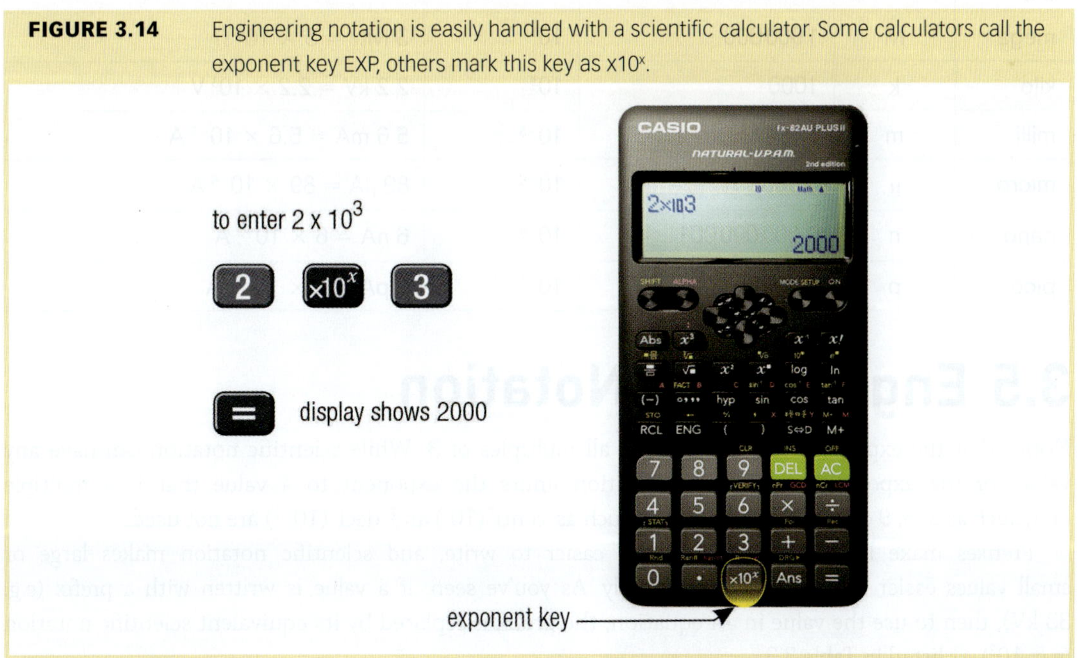

Figure 3.15 shows how to enter a negative exponent into a calculator and what the display will show when 5×10^{-3} is entered.

FIGURE 3.15 How to enter a negative exponent into a calculator and what the display will show when 5×10^{-3} is entered

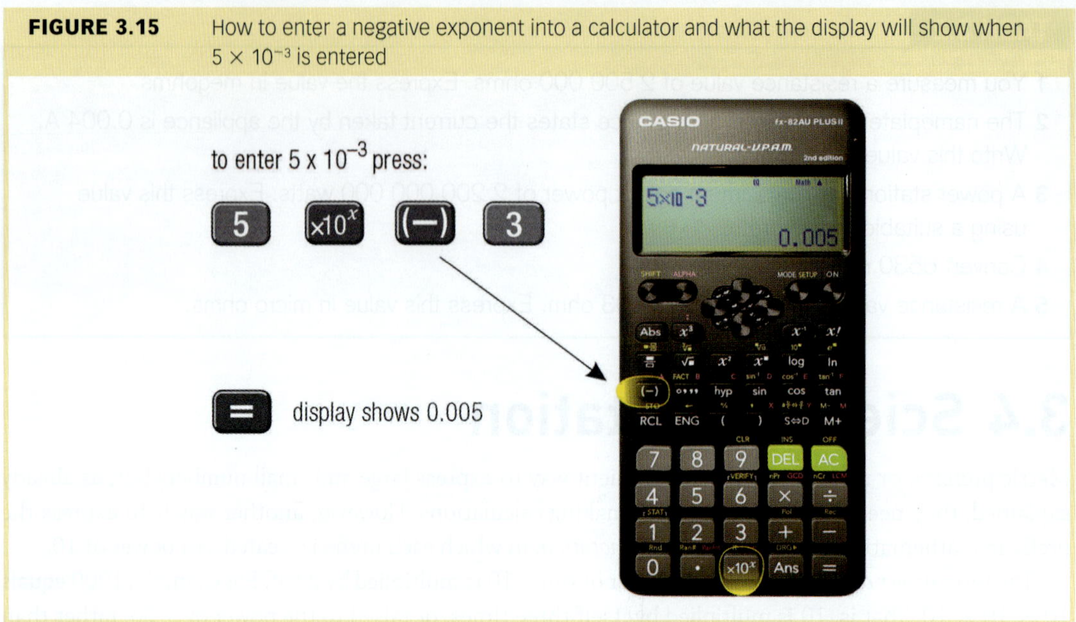

Table 3.3 shows the scientific notation for the prefixes used in electrical work.

TABLE 3.3

SCIENTIFIC NOTATION				
Prefix	**Symbol**	**Multiply by**	**Scientific notation**	**Example**
tera	T	1 000 000 000 000	10^{12}	1.6 THz $= 1.6 \times 10^{12}$ Hz
giga	G	1 000 000 000	10^{9}	4 GW $= 4 \times 10^{9}$ W
mega	M	1 000 000	10^{6}	3 M$\Omega = 3 \times 10^{6}\ \Omega$
kilo	k	1000	10^{3}	2.2 kV $= 2.2 \times 10^{3}$ V
milli	m	0.001	10^{-3}	5.6 mA $= 5.6 \times 10^{-3}$ A
micro	μ	0.000 001	10^{-6}	$89\ \mu$A $= 89 \times 10^{-6}$ A
nano	n	0.000 000 001	10^{-9}	6 nA $= 6 \times 10^{-9}$ A
pico	p	0.000 000 000 001	10^{-12}	5 pA $= 5 \times 10^{-12}$ A

3.5 Engineering Notation

Notice that the exponents in Table 3.3 are all multiples of 3. While scientific notation can have any value for the exponent, engineering notation limits the exponent to a value that is a multiple of 3, such as 3, 6, 9 or 12. Therefore, terms such as 'centi' (10^{2}) and 'deci' (10^{-1}) are not used.

Prefixes make large or small numbers easier to write, and scientific notation makes large or small values easier to handle mathematically. As you've seen, if a value is written with a prefix (e.g. 33 kV), then to use the value in an equation, the prefix is replaced by its equivalent scientific notation (e.g. 10^{3}), as listed in Table 3.3.

In electrical work, you will need to be able to multiply and divide numbers expressed in scientific notation, and also to add and subtract them. When a value is expressed with scientific notation, the value has two components: the *coefficient* and the *exponent*, as shown in Figure 3.16. These two terms are now used to describe how to work with numbers expressed in scientific notation.

FIGURE 3.16 Exponents and coefficients

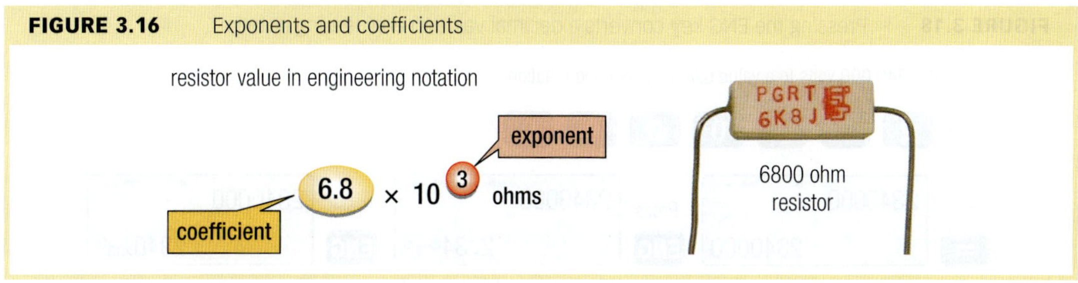

Multiplication

For multiplication – *multiply the coefficients and add the exponents*. Adding the exponents must be done algebraically in which adding a minus number means you subtract that number.

For example:

$(3 \times 10^3) \times (4 \times 10^6) = 12 \times 10^9$ (because 3 plus 6 gives 9)

$(15 \times 10^6) \times (5 \times 10^{-3}) = 75 \times 10^3$ (because +6 added to −3 gives +3)

$(2.2 \times 10^{-6}) \times (1.5 \times 10^{-3}) = 3.3 \times 10^{-9}$ (because −6 added to −3 gives −9)

EXAMPLE 3.10

Find the voltage across a 1.5 MΩ resistor that has a current of 5 mA flowing through it. Express the voltage value with a prefix.

Solution

Values $V = ?$ volts

$I = 5$ mA

$R = 1.5$ MV

Equation $V = IR$

$V = (1.5 \times 10^6) \times (5 \times 10^{-3})$ [replacing prefixes with engineering notation]

$= 1.5 \times 5 \times 10^3$ [adding the exponents in which +6 added to −3 gives +3]

$= 7.5 \times 10^3$ volts

Answer **V = 7.5 kilovolts or 7.5 kV**

Using a calculator to solve Example 3.10 is shown in Figure 3.17. Enter the values with their exponents, perform the calculation and replace 10^3 with the letter k for kilo.

FIGURE 3.17 Calculator keystrokes to solve Example 3.10

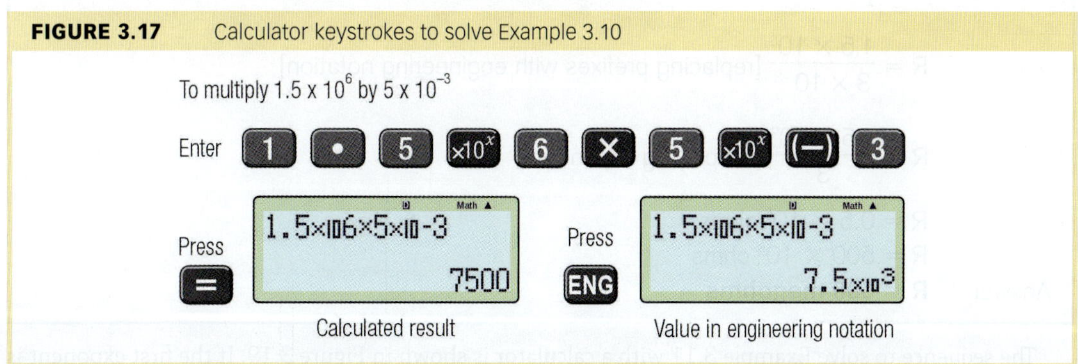

In this example, the ENG key is used to convert the calculated result to a value expressed with engineering notation. Pressing the ENG key again changes to the next lower engineering exponent. Figure 3.18 shows the conversion of 2 340 000 volts to values with engineering notation. After the first press, the voltage is in megavolts (10^6), the next press gives the value in kilovolts (10^3).

FIGURE 3.18 Pressing the ENG key converts a decimal value to engineering notation

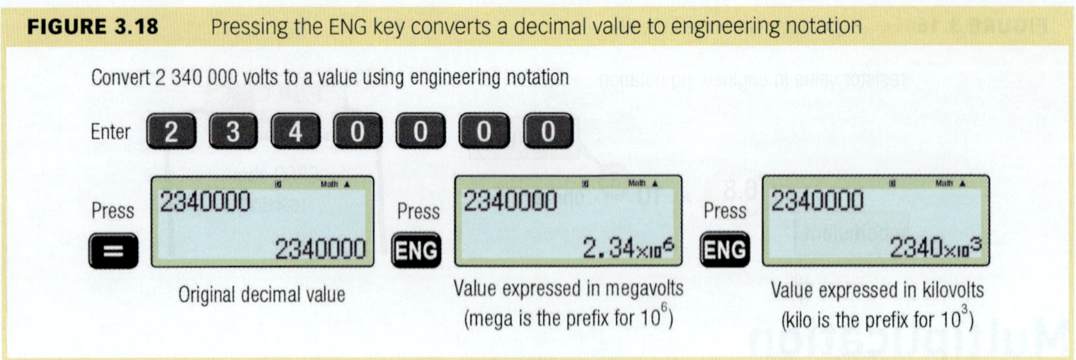

Division

For division – *divide the coefficients, then algebraically subtract the exponent of the number below the line (denominator) from the exponent of the number above the line (numerator).* For example:

$$\frac{6.6\times10^{6}}{2\times10^{3}}=3.3\times10^{3}$$

$$\frac{8\times10^{3}}{4\times10^{6}}=2\times10^{-3}\text{ (6 subtracted from 3 gives }-3\text{)}$$

$$\frac{15\times10^{6}}{3\times10^{-3}}=0.5\times10^{9}\text{ (subtracting a minus gives a plus, giving }6+3=9\text{)}$$

Subtracting the exponents can be done mentally and, if necessary, the coefficients divided on a calculator.

EXAMPLE 3.11

When a voltage of 1.5 MV is applied to a circuit, a current of 3 mA flows. Find the resistance of the circuit and express its value with a prefix.

Solution

Values $V = 1.5$ MV

$I = 3$ mA

$R = ?$

Equation $R = \frac{V}{I}$

$$R = \frac{1.5\times10^{6}}{3\times10^{-3}}\text{ [replacing prefixes with engineering notation]}$$

$$R = \frac{1.5\times10^{9}}{3}\text{ [simplifying exponents]}$$

$R = 0.5 \times 10^{9}$ ohms

$R = 500 \times 10^{6}$ ohms

Answer **R = 500 megohms**

The sequence to solve Example 3.11 with a calculator is shown in Figure 3.19. If the first exponent is not the one you are after, press SHIFT then ENG for the next higher exponent.

FIGURE 3.19 Pressing the ENG key converts the value to engineering notation. Pressing SHIFT then the ENG key changes the exponent to a higher value.

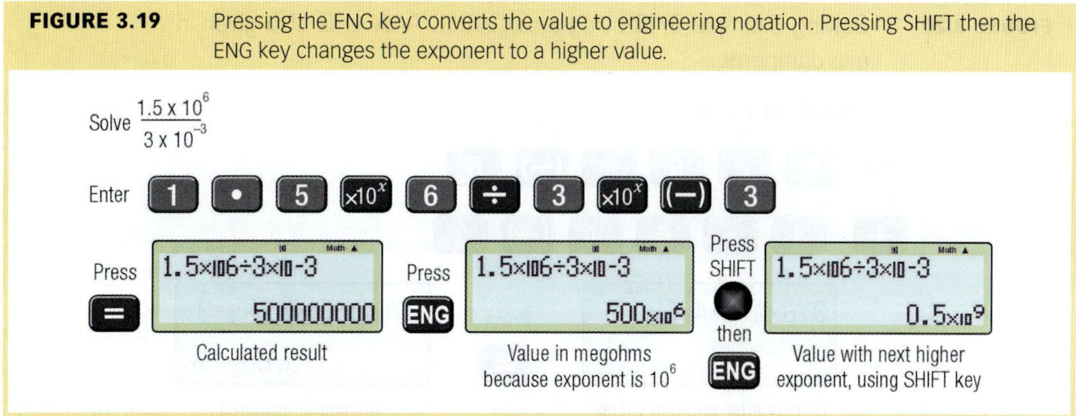

Addition and subtraction

To manually add (or subtract) values written in engineering notation, the coefficients must all have the *same* exponent (or metric prefix). The coefficients are added (or subtracted) and the exponent is left alone. For example, if the exponents are the same, we can add the coefficients, and express the answer with the same exponent: $3.5 \times 10^3 + 4.1 \times 10^3 = 7.6 \times 10^3$.

Manually adding or subtracting values written in engineering notation generally requires that all exponents be the same. The four rules to change from one exponent to another are:

1 *To reduce a positive exponent, move the decimal point to the right.* For example, to convert 12.5 megavolts (MV) to kilovolts (kV), write 12.5 MV as 12.5×10^6 V, move the decimal point by three places right, and reduce the exponent by 3 to give $12\,500 \times 10^3$ V (or 12 500 kV).

2 *To increase a positive exponent, move the decimal point to the left.* For example, to convert 1234 kilohms (1234×10^3 ohms) to megohms move the decimal point three places left to give 1.234×10^6 ohms or 1.234 MΩ.

3 *To increase a negative exponent, move the decimal point to the right.* For example, to convert 0.12 mA (0.12×10^{-3} A) to microamps, move the decimal point three places right to give 120×10^{-6} A (or 120 μA).

4 *To reduce a negative exponent, move the decimal point to the left.* For example, to convert 1234 μA (1234×10^{-6} A) to milliamps, move the decimal point three places left to give 1.234×10^{-3} A or 1.234 mA.

These rules cover all possibilities and also show how to convert from one prefix to another.

FYI

The coefficient is the number; the exponent is the power of 10

EXAMPLE 3.12

Add 876 μA and 4.6 mA and express the answer in milliamperes.

Solution

876 μA = 876×10^{-6} A and 4.6 mA = 4.6×10^{-3} A

Converting 876×10^{-6} A to milliamps requires reducing its negative exponent to −3.

Therefore, move the decimal point three places left and reduce the exponent by 3:

876×10^{-6} A = 0.876×10^{-3} A

Add the two values of current:

0.876×10^{-3} A + 4.6×10^{-3} A = 5.476×10^{-3} A

Answer = **5.476 mA**

Figure 3.20 shows the keystrokes to solve Example 3.12 with a calculator.

FIGURE 3.20 Solving Example 3.12 using a calculator. Answer is also given in microamperes.

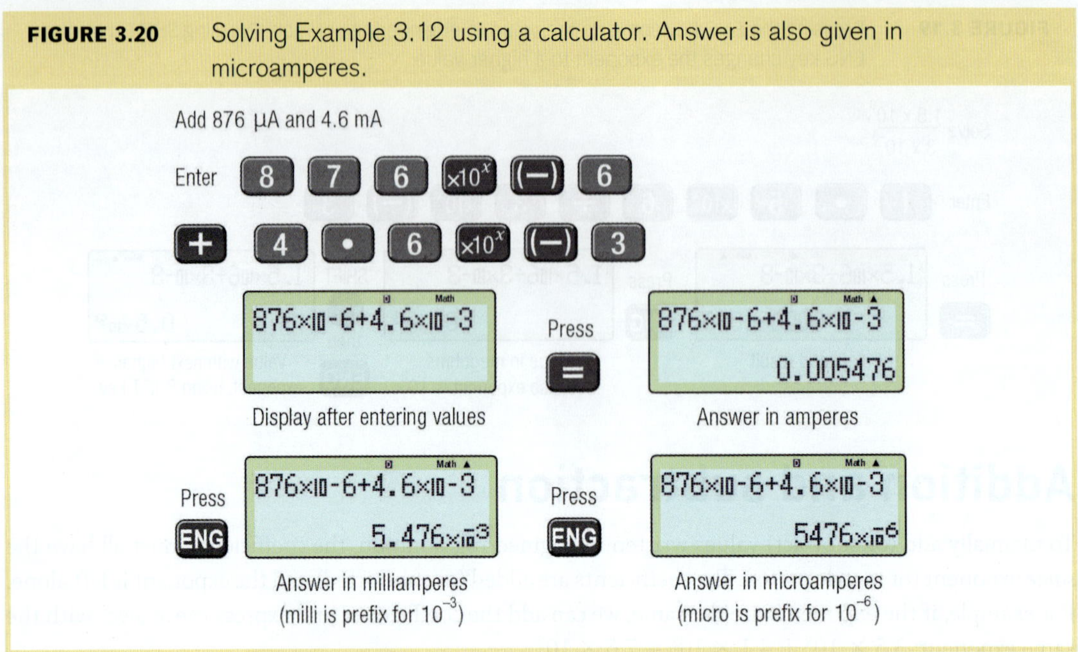

When using a Casio calculator, values are entered as they are written. The calculator takes care of the different exponents and gives the answer without an exponent. Pressing the ENG key displays the answer in engineering notation, pressing it again changes the exponent.

EXAMPLE 3.13

Subtract 120 kilohms from 3.9 megohms.

Solution

$120\ k\Omega = 120 \times 10^3\ \Omega$ and $3.9\ M\Omega = 3.9 \times 10^6\ \Omega$

One of the values must be converted to have the same prefix (exponent) as the other value. In this case we will convert 3.9×10^6 ohms to its equivalent value in kilohms, which means reducing its exponent from 6 to 3 by moving the decimal point right by three places.

$3.9 \times 10^6 = 3900 \times 10^3\ \Omega$

Do the subtraction:

$3900 \times 10^3 - 120 \times 10^3 = 3780 \times 10^3\ \Omega$

Answer = **3780 kilohms**

Figure 3.21 shows the keystrokes to solve Example 3.13 with a calculator.

FIGURE 3.21 Solving Example 3.13 using a calculator

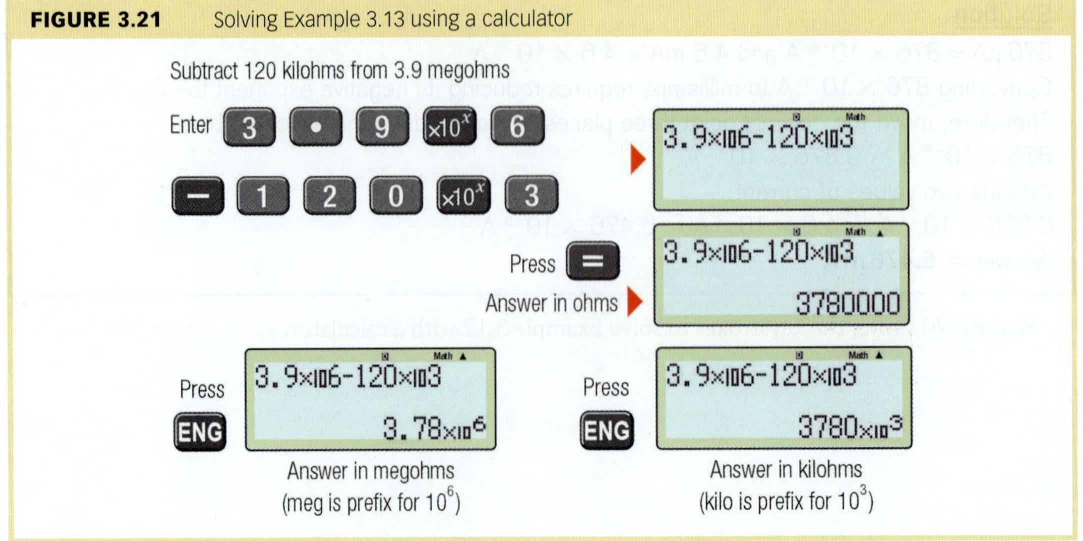

TASK 3.3

1 Express 1.8 GW in megawatts.
2 Convert 1250 mV to volts.
3 Calculate the voltage across a 4.7 kΩ resistor that is passing a current of 100 μA.
4 Add these three resistor values together: 330 Ω, 33 kΩ, 3.3 MΩ.
5 Subtract 250 MW from 4.5 GW.

CHAPTER SUMMARY

- Resistance is opposition to current flow.
- Resistance has the symbol R, and is measured in ohms, which has the Greek symbol Ω.
- Conductance is the reciprocal (inverse) of resistance, and is a measure of how well a conductor can carry an electric current.
- Conductance has the symbol G, and is measured in siemens, symbol S.
- Ohm's law mathematically relates the three fundamental electrical quantities of voltage (V), current (I) and resistance (R).
- Ohm's law says that the current flowing between any two points in an electric circuit is directly proportional to the voltage between the two points, and is inversely proportional to the resistance of the circuit between the two points.
- The three Ohm's law equations are: $I = \frac{V}{R}$, $V = IR$ and $R = \frac{V}{I}$, where V = voltage in volts, I = current in amperes and R = resistance in ohms.
- Metric prefixes are used to abbreviate the written form of a large or small number. The most commonly used prefixes in electrotechnology are kilo (k), mega (M), milli (m) and micro (μ). Others include giga (G), pico (p) and nano (n).
- A metric prefix symbol is sometimes used to replace the decimal point in a value.
- Scientific notation expresses a metric prefix in mathematical form, in which each prefix is treated as a power of 10.
- Engineering notation is a subset of scientific notation, and uses only powers of 10 that are multiples of 3.
- Problems involving values expressed in scientific notation can be solved with a scientific calculator, in which each entered value has two components: the *coefficient* (a number) and the *exponent* (power of 10).

Rules with manual calculations involving engineering notation:

- Multiplication: multiply the coefficients and algebraically add the exponents.
- Division: divide the coefficients, algebraically subtract the exponent of the denominator (number below the line) from the exponent of the numerator.
- Addition and subtraction: convert all values so they have the same exponent.

REVIEW EXERCISES

Check your answers at the back of the book.

1 An oven develops a fault in which the resistance of its element increases to double its previous value. What effect does this have on the current taken by the oven?

2 A conductor has a conductance of 12 siemens. What is its resistance?

3 An appliance takes a current of 12 A when connected to 230 V. Calculate its resistance.

4 A circuit has a resistance of 120 ohms. How much current does it take when the circuit is connected to 12 V?

5 Solve the following:

a $\frac{2.4 \times 10^{3}}{2 \times 10^{-3}} =$

b $(2.2 \times 10^{3}) \times (4 \times 10^{-3}) =$

c $\frac{46 \times 10^{-3}}{2.3 \times 10^{-6}} =$

6 You measure 25 V across a resistance of 10 Ω. How much current is flowing in the resistance?

7 Write the following electrical values in engineering notation:
  a 330 kilovolts
  b 22 milliamps
  c 4.7 megohms.

8 Convert:
  a 0.0012 volts to millivolts
  b 0.000 009 6 amperes to microamps (μA)
  c 35.6 mA to amperes.

9 A conductor has a resistance of 2 milliohms and is passing a current of 4 kiloamps. Calculate the voltage across the conductor.

10 When 2.5 volts is applied between the anode and cathode of an electrolysis tank, a current of 10 kA flows in the electrolyte. Calculate the resistance of the electrolyte. Express the resistance value in engineering notation.

ONLINE RESOURCES

COMPLETE WORKSHEET THREE

Check with your instructor for worksheets on this chapter.

CHAPTER 4

ELECTRICAL POWER

This chapter introduces power, energy and work, starting with a review of mechanical energy and the various forms of energy. This is followed by an explanation of electrical power and how to calculate the power taken by an electrical load. We also look at the effect on power if the voltage or current changes, and how the efficiency of an electrical load affects the amount of power required to achieve a certain task.

CHAPTER OUTLINE

4.1 Energy and Work

The law called 'Conservation of Energy' means that there is no such thing as perpetual motion, as this would require creating energy

Electricity is a convenient form of energy that is obtained from other forms of energy. It is convenient because electrical energy can be produced near a source of energy, such as hydro, wind or solar energy, then transmitted over long distances to where it is needed. In effect, electricity is a means of transferring energy from one point to another. At the receiving end, electrical energy is converted to the required form, such as heat, mechanical, chemical or light, as outlined in Chapter 2. But what is energy?

Energy is the *ability to do work*. It can only exist in a form, such as heat or light, and energy in one form can disappear but the same amount of energy will appear in another form. That is, energy can only be *transformed* from one form to another, as it cannot be created or destroyed.

The unit of energy is the *joule*. When energy is transformed from one form to another, *work* is done. The amount of work done equals the amount of energy transformed, so work is also measured in joules. If no work is done, no energy is transformed. We can see this by looking at the most familiar form of energy: mechanical energy.

Mechanical energy

In 1687, Isaac Newton presented his three laws of motion, which laid the foundation for a study of classical mechanics. Newton's first law says: *An object at rest tends to stay at rest and an object in motion tends to stay in motion with the same speed and direction unless acted upon by an unbalanced (or external) force.* This law is often called 'the law of inertia'.

Pushing a trolley is an example, in which the force you apply overcomes the trolley's inertia and causes it to move against any external forces (e.g. friction or other forms of resistance) that are acting to stop the trolley. If there were no external opposing forces, once it is set in motion, the trolley would keep moving forever.

Another example is lifting a heavy object from ground level to a higher level. In both cases a force is required to move the object a certain distance and work is done (energy is transformed) in the process. That is, work is done when a force applied to an object causes it to move through a certain distance. Work is always against some resistance such as gravity, friction or other opposing forces.

work done = force × distance

where:
work done is measured in joules (J)
force is measured in newtons (N)
distance is measured in metres (m).

KEY CONCEPT

Joules

One joule of work is done when a force of one newton is moved through a distance of one metre, in the direction of the force.

Vector quantities have direction, where scalar quantities have magnitude only. Examples of scalar quantities are time, speed, temperature, resistance.

It's important to understand that force has a direction, which makes it a *vector* quantity. Vectors (and phasors) are covered later in this book.

Force

Newton's second law says: *Acceleration is produced when a force acts on a mass*. The greater the mass of the object being accelerated, the greater the amount of force needed to accelerate the object. We all know that heavier objects require more force to move than lighter objects; however, this law gives us an exact relationship between force, mass and acceleration. It can be expressed as a mathematical equation:

$F = ma$

where:
force (F) is in newtons
mass (m) is in kilograms
acceleration (a) is in metres per second per second (written as m/s^2).

KEY CONCEPT
Force

Force is measured in newtons (N) where one newton is defined as the force required to accelerate a one kilogram mass by a rate of one metre per second per second (or m/s^2).

In many cases work is done against gravity, such as lifting an object. Acceleration in the above equation is, therefore, the acceleration due to gravity (g), which is 9.8 m/s^2. Example 4.1 explains this further.

FIGURE 4.1 Electric forklift raising a 200 kg load through a distance of 2 metres

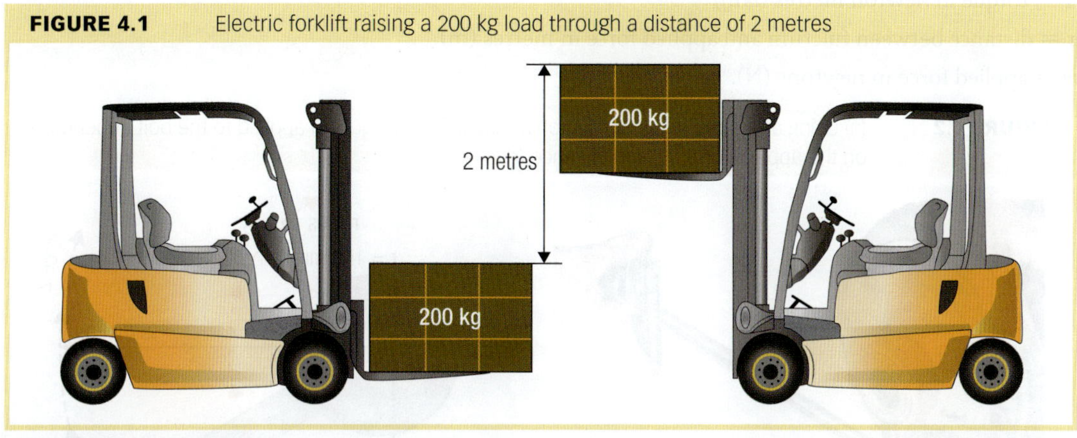

EXAMPLE 4.1

The electric forklift in Figure 4.1 is lifting a 200 kg load vertically through a distance of 2 metres. Find the force required, the work done and the amount of energy that has been transformed.

Solution

Values $m = 200$ kg
$d = 2$ m
a = acceleration due to gravity = 9.8 m/s^2 (g)

Equation $F = ma$ (or mg)
$= 200 \times 9.8$

Force (F) = 1960 N

Work = force × distance
$= 1960 \times 2$

Work = 3920 joules

Energy = work done = 3920 J

Answer **F = 1960 N, work done = energy converted = 3920 J = 3.92 kJ**

The electrical energy used by the forklift in Example 4.1 is at least 3920 J, but as explained later, unless the forklift is 100 per cent efficient, additional energy is needed to overcome losses, which will be expended, usually as heat.

Rotational energy

A familiar form of mechanical force is rotation, known as *torque*. Examples of torque are the force required to turn a steering wheel or to tighten a nut with a spanner. Motors are generally rated by their torque, as well as their power. Figure 4.2(a) shows a torque wrench being used to tighten a bolt. It has a torque indicator which lets you tighten nuts and bolts to a value specified by the equipment manufacturer. (Bolted electrical connections are often given a torque specification.)

Torque is measured in newton metres (Nm). As Figure 4.2(b) shows, the force to rotate a spanner might be applied anywhere along the length of the spanner. As you probably know, the further the distance (r) between the fulcrum and the applied force (F), the greater the torque (T). If the force is applied at right angles we can say that:

$T = Fr$

where:

T = torque in newton metres (Nm)

r = distance between fulcrum and applied force in metres (m)

F = applied force in newtons (N).

FIGURE 4.2 (a) Using a torque wrench to tighten a bolt. (b) The torque (T) applied to the bolt depends on the applied force (F) and distance (r).

(a) torque wrench with scale indicating the torque being applied to the bolt

(b)

EXAMPLE 4.2

A manufacturer of electrical equipment specifies a torque of 40 Nm to correctly tighten the 10 M nuts in Figure 4.3. Calculate the force you need to apply to the torque wrench in Figure 4.2 (b) when (i) the force is applied at point A (r = 100 mm) and (ii) the force is applied at point B (r = 200 mm).

FIGURE 4.3 Busbar attached with 10 mm bolts

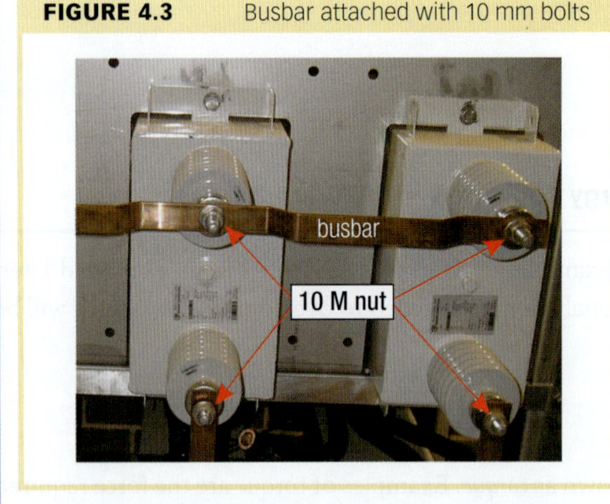

Solution

Values $T = 40$ Nm

$r =$ (i) 100 mm, (ii) 200 mm

$F = ?$

Equation $T = F \times r$

$F = \frac{T}{r}$ (by transposing the above equation)

(i) when r = 100 mm = 0.1 m

$= \frac{40}{0.1}$

Answer (i) = **400 N**

(ii) When r = 200 mm = 0.2 m

$= \frac{40}{0.2}$

Answer (ii) = **200 N**

Potential and kinetic energy

Mechanical energy can be either kinetic energy (energy of motion) or potential energy (stored energy of position). The kinetic energy of wind provides the energy to rotate a wind generator. A raised hammer has potential energy that is converted to kinetic energy when you cause the hammer to strike a nail.

FIGURE 4.4 A raised, stationary mass has potential energy that is converted to kinetic energy when the mass is moving. When the moving mass is suddenly stopped, the kinetic energy is converted to other forms of energy.

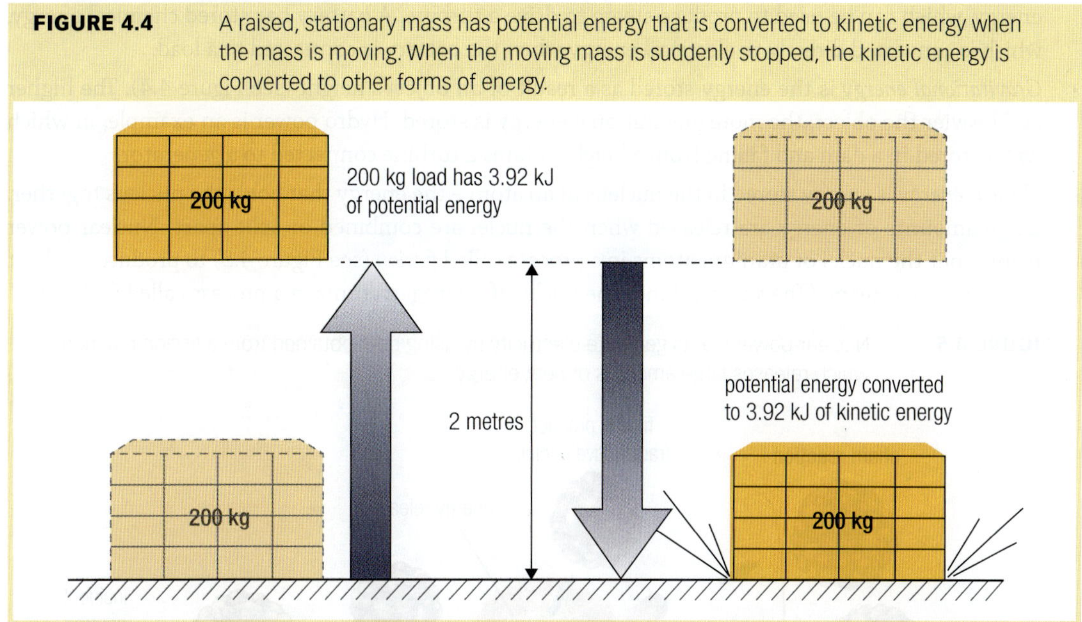

In Figure 4.4, a 200 kg load is raised through a distance of two metres. As calculated in Example 4.1, this takes 3.92 kJ of energy. When the load is held by a support, it has 3.92 kJ of potential energy. Newton's third law of motion says: *For every action there is an equal and opposite reaction*. That is, the downward force exerted by gravity on the 200 kg load is counteracted by an upwards force from whatever is supporting the load.

FYI

All objects, regardless of weight when dropped together, will hit the ground at the same speed. However, a heavy object will dissipate more energy than a light-weight object

If this force (support) is removed, the load will drop from its height of two metres and strike the ground. The 3.92 kJ of potential energy in the load will then be converted to 3.92 kJ of kinetic energy. When the load hits the ground, this energy is dissipated in other forms of energy, including changing the shape of the object, creating sound and heat. The equation to find kinetic energy is:

$$\text{kinetic energy} = \frac{1}{2}mv^2 \text{ joules}$$

where:

m = mass in kilograms

v = velocity in metres per second (m/s^1).

This equation says that an object's kinetic energy equals half its mass multiplied by the square of its velocity. That is, if the velocity is doubled, the kinetic energy increases four times.

KEY POINTS...

- Energy is the ability to do work. It is measured in joules.
- Work is done when energy is transformed, also measured in joules.
- Force is a push or pull on an object that causes it to change shape or to move.
- In mechanics, work = force × distance, and one joule of work is done when a force of one newton is moved through a distance of one metre, in the direction of the force.
- Force = mass × acceleration (Newton's second law).
- Torque is a rotational force (vector quantity) and is measured in newton metres (Nm).

Forms of potential and kinetic energy

Energy has many forms, some of which are used to generate electricity. All forms of energy can be classified as either potential or kinetic. Potential energy is stored energy, waiting to be used. Examples of **potential energy** sources used to generate electricity include:

FYI

Electrical energy is now routinely stored in large batteries and hydro setups, and is released when demand requires it

- *Chemical energy* is the energy stored in the bonds of atoms and molecules. Examples are biomass, petroleum, natural gas and coal. When burned, the stored chemical energy is converted to thermal energy, which can be used to produce steam to drive a turbine. A battery has stored chemical energy, which is converted directly to electrical energy when the battery is connected to a load.

- *Gravitational energy* is the energy stored as a result of an object's height (see Figure 4.4). The higher and heavier the object, the more gravitational energy is stored. Hydro power is an example, in which water stored in a dam and falling from a height rotates a turbine connected to a generator.
- *Nuclear energy* is energy stored in the nucleus of an atom – the energy that holds the nucleus together. Large amounts of energy are released when the nuclei are combined or split apart. Nuclear power plants split the nuclei of uranium atoms in a process called *fission* (see Figure 4.5) to produce the heat that generates steam. (The sun combines the nuclei of hydrogen atoms in a process called fusion.)

FIGURE 4.5 Nuclear power plants generate electricity by using heat obtained from a fission reaction, which releases large amounts of heat energy

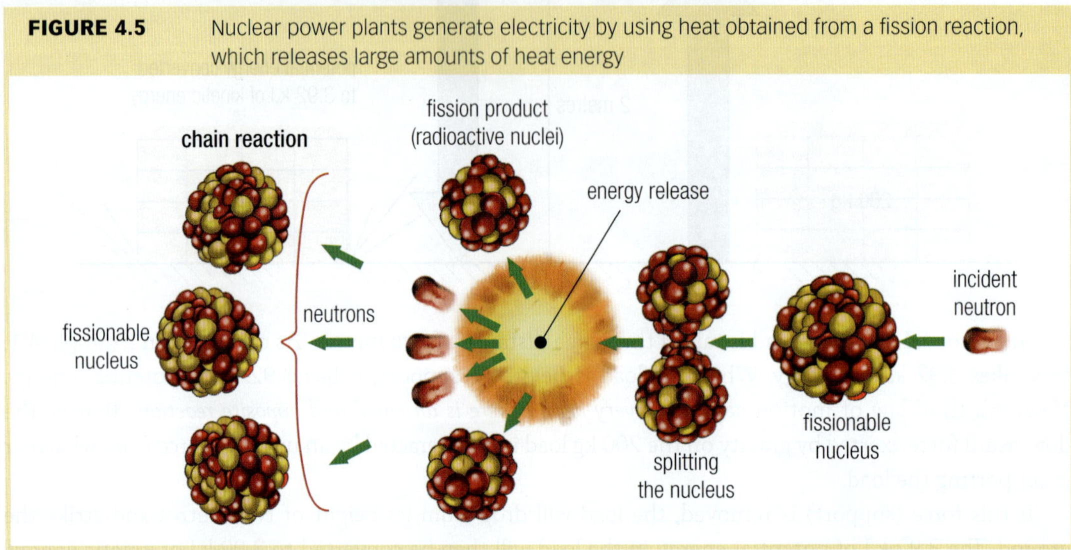

Forms of **kinetic energy** (energy of motion) used to generate electricity include:

- *Radiant energy* is electromagnetic energy that travels in waves. Examples are visible light, x-rays and radio waves. Sunshine is radiant energy which can be converted to electricity using solar panels.
- *Thermal energy* (heat) is the vibration and movement of the atoms and molecules within substances. As an object is heated, its atoms and molecules move and collide more quickly. Geothermal energy is thermal energy in the Earth.
- *Motion energy* is energy in the movement of objects or substances. Examples are wind power from the kinetic energy of moving air, hydro power from moving water and power from rotational forces such as a steam turbine.

Sound is also kinetic energy, as it is the movement of waves through a substance (e.g. air). Sound energy is usually far less than other forms of energy. Electrical energy is the energy due to electrons flowing in a conductor, so it too is a form of kinetic energy.

4.2 Power

As already explained, when energy is transformed, work is done. However, we also need to consider how *quickly* energy is being transformed. In Figure 4.6, two men of equal weight are exercising during lunch. One man walks slowly and takes 20 minutes to walk a distance of one kilometre. The other is running and covers the same distance in five minutes. Who did the most work? Who used the most amount of energy?

KEY CONCEPT
Power

The answer is that the men have used the *same* amount of energy and have done the *same* amount of work. The difference is that the runner has taken less *time* to do so compared to his companion. The explanation is power.

Power is the *rate at which energy is used*, or the *rate at which work is done*. So while the walker is transforming less energy per second, he is doing it over a longer time than the runner. Therefore, they both transform the same amount of energy.

FIGURE 4.6 If both men travel the same distance, who does the most work?

The unit of power is the *watt*, which has the letter W as its symbol. By definition, the power is *one watt if one joule of energy is transformed in one second*. Power is given the symbol P.

The watt is named after James Watt, a Scottish inventor remembered mainly for his steam engines. Watt also introduced the term 'horsepower' to rate the power of his steam engines. A horsepower is equal to 746 watts. Older electric motors have their power rating given in horsepower (hp). The motor in a domestic refrigerator is usually rated at between ¼ hp to ½ hp.

$$\text{power} = \frac{\text{work}}{\text{time}}$$

where:

power is in watts

work is in joules

time is in seconds.

EXAMPLE 4.3

The two electric forklifts in Figure 4.7 are lifting the same 200 kg load vertically through a distance of 2 metres. One does the lift in 2 seconds, the other takes 8 seconds. Find the power of both forklifts.

Solution

Values work = 3920 joules (from Example 4.1)
time = (a) 2 seconds (b) 8 seconds
power = ?

Equation $\text{power} = \frac{\text{work}}{\text{time}}$

(a) $\text{power} = \frac{3920}{2} = 1960 \text{ watts}$

(b) $\text{power} = \frac{3920}{8} = 490 \text{ watts}$

Answer **(a) power = 1960 watts, (b) power = 490 watts**

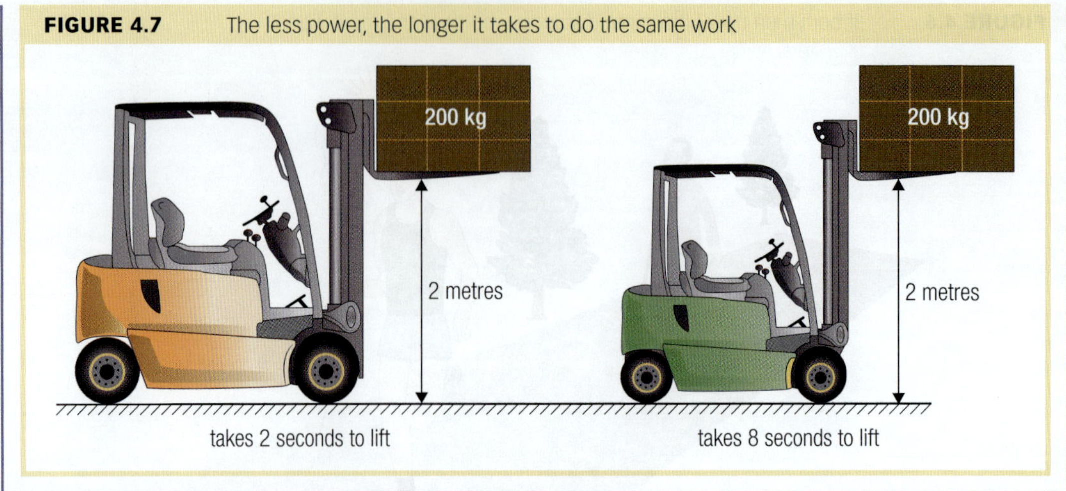

FIGURE 4.7 The less power, the longer it takes to do the same work

Mechanical power

Mechanical power is a rating of how much work a machine such as a motor can do in a certain time. The mechanical power produced by a motor depends on its speed of rotation and the torque it is delivering at that speed. Because we are dealing with a rotational quantity, we need to consider *angular velocity*, which is why the following equation includes the circular reference of pi (π).

$$P = 2\pi NT$$

where:

P = mechanical power in watts

2π is a circular reference equal to 6.28 (approx.)

N = revolutions per second

T = torque in newton metres.

As motor speed is generally given in revolutions per minute (RPM), a more useful equation is:

$$P = \frac{2\pi nT}{60}$$

where:

n = revolutions per minute (RPM).

EXAMPLE 4.4

Find the power (P) of a motor that is delivering 20 Nm of torque when it is running at 1440 RPM.

Solution

Values: torque (T) = 20 Nm
RPM (n) = 1440
power (P) = ?

Equation: $P = \frac{2\pi nT}{60} = \frac{6.28 \times 20 \times 1440}{60}$

Answer: **power = 3014 watts**

Efficiency

The source of energy to power a motor has to come from another energy source. In Example 4.4, the motor is delivering slightly over 3 kW of mechanical power. This is called the output power. The input power is the power taken by the motor from its energy source. If there were no losses in the motor, it would require the same input power as its output power. That is, the motor would be 100 per cent efficient.

It is impossible for any machine to be 100 per cent efficient, as there are always losses in the process of converting energy from one form to another. In the case of an electric motor, the losses appear mainly as heat. Efficiency is, therefore, a measure of power in and power out, and is sometimes given the symbol η (Greek letter *eta*). The general equation to find efficiency is:

$$\eta = \frac{\text{power out}}{\text{power in}} \times 100\%$$

where:

efficiency is expressed as a percentage (%)

power in and power out are expressed in watts.

SUSTAINABILITY

Efficient use of energy helps towards sustainable energy

EXAMPLE 4.5

Find the efficiency of the motor in Example 4.4, if the power taken by the motor is 3500 watts.

Solution

Values power out = 3014 watts (as per Example 4.4)
power in = 3500 watts
efficiency (η) = ?

Equation $\eta = \frac{\text{power out}}{\text{power in}} \times 100\% = \frac{3014}{3500} \times 100\%$

Answer **efficiency = 86.1%**

KEY POINTS...

- All forms of energy can be classified as either potential or kinetic.
- Kinetic energy is energy in motion.
- Potential energy is energy in storage.
- Power is the rate of doing work. It has the symbol P and is measured in watts.
- Efficiency of a machine equals its power out divided by its power in, multiplied by 100.
- Efficiency gives a measure of the losses in a machine and is expressed as a percentage.

TASK 4.1

1 How much force is required to lift a roll of cable weighing 25 kg?
2 Calculate the work done if the cable in question 1 is lifted through a vertical distance of 3 metres.
3 When tightening a bolt, the manufacturer specifies a maximum torque of 60 Nm. How much force can you apply if you hold the spanner at a point 200 mm from the bolt head?
4 Calculate the power needed to climb a 10-metre-long ladder in 20 seconds if you weigh 65 kg and you are carrying tools and equipment weighing 8 kg.
5 A machine requires 500 W of power and delivers 450 W of useful output power. Calculate the losses and the efficiency of the machine.

4.3 Electrical Power

As already explained, the power is *one watt if one joule of energy is transformed in one second.* Electrical energy is used only when a voltage causes a current to flow (kinetic energy). Electrical power is the *rate* at which electrical energy is transformed into another form, such as heat.

Electrical power, like mechanical power, is measured in watts. One watt of electrical power is dissipated in a resistor when a voltage of one volt causes a current of one ampere to flow through the resistor. (By Ohm's law, the resistance is one ohm.)

KEY CONCEPT

Calculating electrical power

The general equation to find electrical power is:

power = voltage × current (P = VI)

where:

P = power in watts

V = voltage in volts

I = current in amperes.

EXAMPLE 4.6

Calculate the power being dissipated by the lamp in Figure 4.8.

Solution

Values V = 24 volts

I = 1.5 amperes

P = ?

Equation P = VI = 24 × 1.5

Answer **P = 36 watts**

FIGURE 4.8

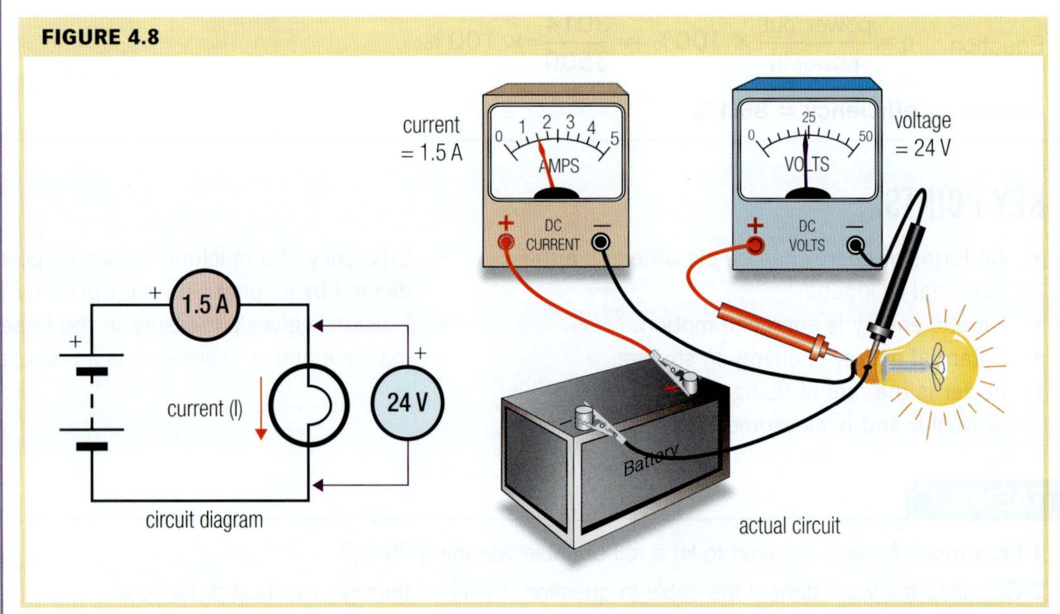

MATH

See page xxii for help in transposing equations

4.4 Transposing the Power Equation

As with Ohm's law, the power equation P = VI can be transposed to make either voltage (V) or current (I) the subject. This gives equations that let you find the voltage (or current) if the power and the current (or voltage) are known.

To make voltage the subject

1 Write the equation: (P = VI)

2 Divide both sides by I: $\left(\frac{P}{I} = \frac{VI}{I}\right)$

3 Cancel out the Is on the right-hand side of the equation: $\left(\frac{P}{I} = V\right)$

4 Rewrite the equation with V on the left-hand side giving: $V = \frac{P}{I}$

To make current the subject

1 Write the equation: (P = VI)

2 Divide both sides by *V*: $\left(\frac{P}{V} = \frac{VI}{V}\right)$

3 Cancel out the Vs on the right-hand side of the equation: $\left(\frac{P}{V} = I\right)$

4 Rewrite the equation with I on the left-hand side giving: $I = \frac{P}{V}$

This gives three equations that relate power, voltage and current:

$P = VI$

$V = \frac{P}{I}$

$I = \frac{P}{V}$

KEY CONCEPT

Equations relating power, voltage and current

These equations, like those for Ohm's law, can be arranged in a triangle. As before, cover the term you want to find, and the equation is the other two terms (see Figure 4.9).

FIGURE 4.9 Power triangle. Cover the term you want to find, and the equation is the other two terms. Notice that P is at the peak of the triangle.

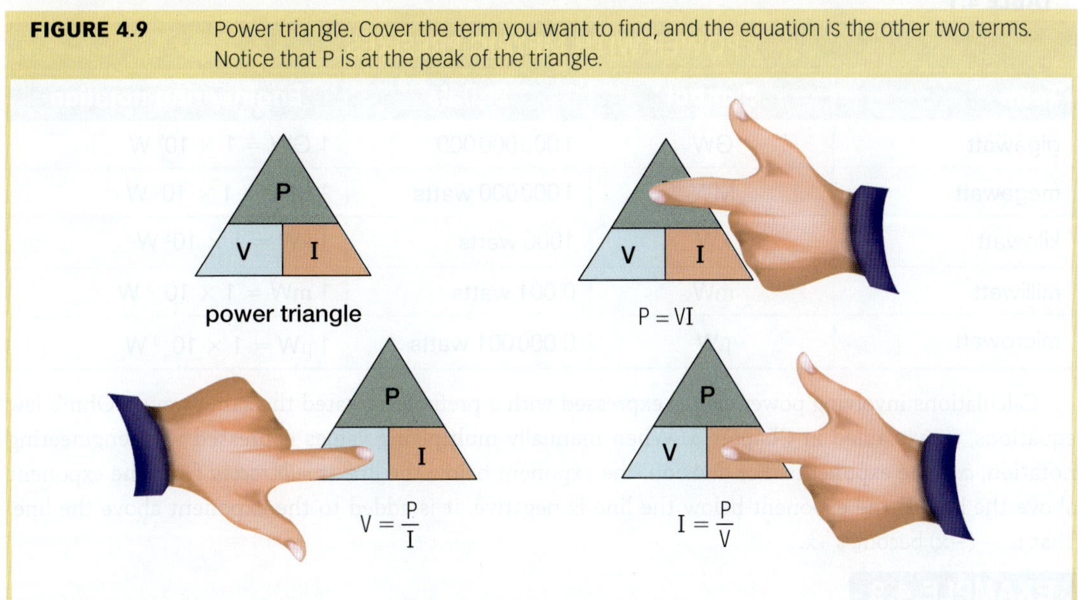

The next two examples show how to use these equations.

EXAMPLE 4.7

Find the voltage across the lamp in the circuit of Figure 4.10.

Solution

Values $P = 40$ watts

$I = 3.2$ amperes

$V = ?$

Equation $V = \frac{P}{I} = \frac{40}{3.2}$

Answer **V = 12.5 volts**

FIGURE 4.10 Circuit for Example 4.7

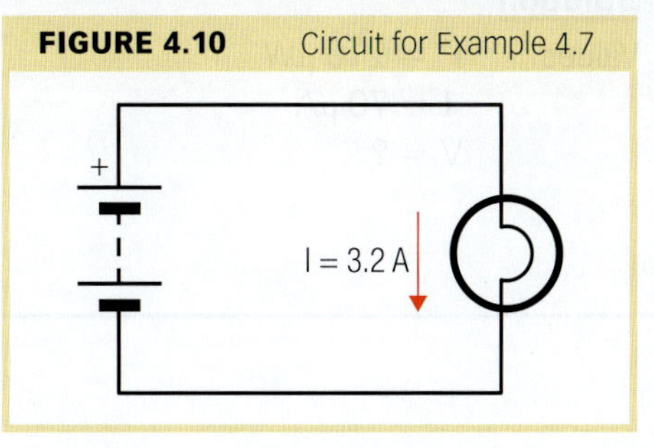

EXAMPLE 4.8

How much current does an electric heater rated at 750 W take when it is connected to a 230 V supply?

Solution

Values $P = 750$ watts
$V = 230$ volts
$I = ?$

Equation $I = \frac{P}{V} = \frac{750}{230}$

Answer **$I = 3.26$ amperes**

Power and metric prefixes

Power values in electrotechnology can range from gigawatts (GW) to less than a microwatt (μW). For example, the Eraring power station in NSW has a combined power capacity of 2880 megawatts (MW), or 2.88 gigawatts (GW). An electric locomotive takes more than 2 MW of power to pull out of a station. Live concert sound systems require kilowatts of power.

On a much smaller scale, mobile phone earphones take only a few milliwatts (mW) of power. The liquid crystal display in a calculator takes only a few microwatts of power from the battery.

Therefore, as with current, voltage and resistance values, prefixes and engineering notation are often used with power values (see Table 4.1).

TABLE 4.1

POWER WITH METRIC PREFIXES			
Term	**Symbol**	**Equals**	**Engineering notation**
gigawatt	GW	1 000 000 000	$1 \text{ GW} = 1 \times 10^{9} \text{ W}$
megawatt	MW	1 000 000 watts	$1 \text{ MW} = 1 \times 10^{6} \text{ W}$
kilowatt	kW	1000 watts	$1 \text{ kW} = 1 \times 10^{3} \text{ W}$
milliwatt	mW	0.001 watts	$1 \text{ mW} = 1 \times 10^{-3} \text{ W}$
microwatt	μW	0.000 001 watts	$1 \text{ μW} = 1 \times 10^{-6} \text{ W}$

Calculations involving power values expressed with a prefix are treated the same way as Ohm's law equations, as explained in Chapter 3. When manually multiplying values expressed with engineering notation, *add* the exponents. For division, the exponent below the line is *subtracted* from the exponent above the line. If the exponent below the line is negative, it is added to the exponent above the line. That is, $-(-x)$ becomes $+x$.

EXAMPLE 4.9

A solar-powered calculator takes 70 μA from its solar panel and uses 210 μW of power. Find the solar panel voltage.

Solution

Values $P = 210$ μW
$I = 70$ μA
$V = ?$

Equation $V = \frac{P}{I}$

$= \frac{210 \times 10^{-6}}{70 \times 10^{-6}}$

Answer **$V = 3$ V**

EXAMPLE 4.10

A manufacturing company is connected to a 3.3 kV supply and takes 1200 amperes when operating at full production. How much power is being taken from the supply?

Solution

Values $V = 3.3$ kilovolts
$I = 1200$ amperes
$P = ?$

Equation $P = VI$
$= (3.3 \times 10^3) \times (1.2 \times 10^3)$

Answer $\mathbf{P = 3.96 \times 10^6}$ **W or 3.96 MW**

FIGURE 4.11 3.3 kV supply to a factory

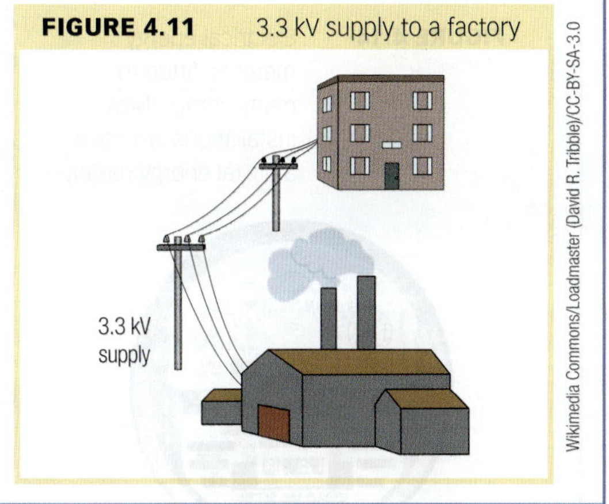

Power and energy rating

Electrical appliances usually have a rating plate showing the power consumption when the appliance is connected to its rated voltage. This is the appliance's *power rating*. For instance, as shown in Figure 4.12, a domestic heater might take 2300 watts when connected to 230 V AC. Electric motor rating plates show the current taken by the motor at full load when operating at its rated voltage. Power consumption must therefore be calculated.

Domestic appliances such as clothes driers, refrigerators, air conditioners and dishwashers are required to display an energy consumption label. As the label in Figure 4.13 shows, the value is given in kilowatt hours used during normal operation over a year.

As explained before, the power is one watt if one joule of energy is transformed in one second. Therefore, if the power remains constant at one watt, 3600 joules of energy will be transformed over one hour (as there are 3600 seconds in an hour). An appliance rated at one kilowatt would use 3 600 000 joules each hour, or put another way, it has used one kilowatt hour (kWh) of energy. That is:

energy in joules = power × time

or

energy in kWh = kW × hours.

FIGURE 4.12 Electric heater rating plate

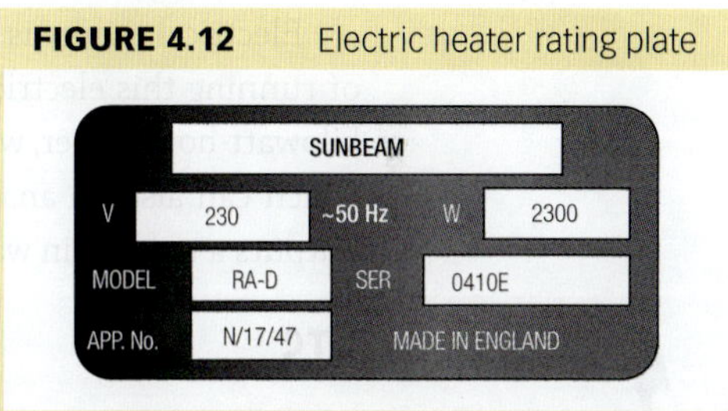

FIGURE 4.13 Appliance energy rating is given in kWh per year

EXAMPLE 4.11

FIGURE 4.14 Electrical energy meter as fitted to many homes. New installations will have a digital energy meter.

Getty Images/UniversalImagesGroup

An electric heater rated at 2500 watts is left switched on for two hours. How much energy does it use (a) in joules and (b) in kilowatt hours?

Solution

(a) Equation energy = power × time (joules)

time = 2 × 60 × 60 = 7200 seconds

power = 2500 watts

energy = 2500 × 7200

Answer (a) = **18 000 000 joules or 18 MJ**

(b) Equation energy = kWh = kW × hours (kWh)

time = 2 hours

energy = 2.5 × 2

Answer (b) = **5 kWh**

Electrical energy is priced by the kilowatt hour. In Example 4.11, if the cost is \$0.30 per kWh, the cost of running this electric heater for two hours is \$1.50 (5 × \$0.30). Electrical energy is measured with a kilowatt-hour meter, which can be either analog or digital. Electrical power is measured with a wattmeter, which can also be analog or digital. A wattmeter measures the circuit current and circuit voltage and outputs a reading in watts. Wattmeters are described in Chapter 25.

KEY POINTS...

- Electrical power (P) is the rate at which electrical energy is transformed.
- Power is measured in watts, where one watt is the power dissipated in a resistance that has one volt across it and one ampere flowing through it.
- Power equals voltage multiplied by current (P = VI) watts.
- By transposition, $V = \frac{P}{I}$ and $I = \frac{P}{V}$.
- Electrical power values are often expressed with a metric prefix.
- Electrical energy is measured in kilowatt hours, where 1 kWh = 3 600 000 J.

TASK 4.2

1. A commercial coffee machine connected to a 230 V supply takes a maximum current of 20 A. How much power is the machine using?
2. An electric water heater rated at 3.6 kW is connected to a 230 V supply. Calculate the current taken by the water heater.
3. A factory connected to an 11 kV supply takes 220 A at full production. How much power is being consumed?
4. An industrial oven takes 25 A when connected to a supply of 400 V. How much power does the oven require and how much energy in kilowatt hours does the oven use in 24 hours?
5. If the cost of electricity is \$0.30 per kilowatt hour, how much does it cost to operate the oven in question 4 per 24-hour period?

4.5 Power and Ohm's Law

Ohm's law gives us a mathematical relationship between current (I), voltage (V) and resistance (R). The three Ohm's law equations introduced in Chapter 3 are:

$$I = \frac{V}{R}$$

$$V = IR$$

$$R = \frac{V}{I}$$

These and the power equations can be combined to give a set of equations that relate the four quantities of power, voltage, current and resistance. For example, if the resistance and current in a circuit are known, you can determine the voltage using Ohm's law. Then, the equation P = VI can be used to calculate the power dissipated in the circuit. By combining the Ohm's law and power equations, power can instead be directly calculated from the resistance and current values. We start by deriving an equation to find power, if resistance and current are known.

Power from resistance and current

To obtain an equation for power in terms of resistance and current we combine the power equation P = VI with the Ohm's law equation V = IR, as shown in Figure 4.15.

FIGURE 4.15 Combining Ohm's law and power equations to get power in terms of current and resistance

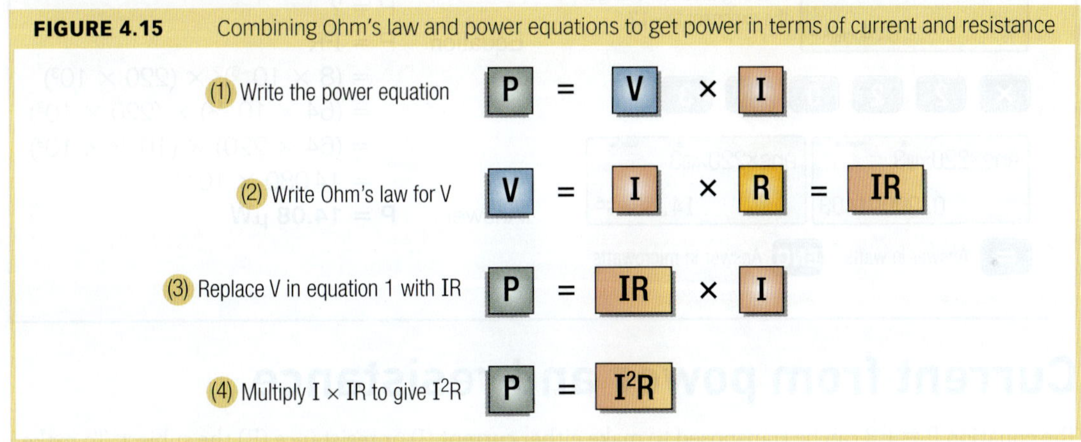

The equation to find power when current and resistance are known, is:

$$P = I^2R$$

where:

P = power in watts

I = current in amperes

R = resistance in ohms.

The next two examples use this equation.

EXAMPLE 4.12

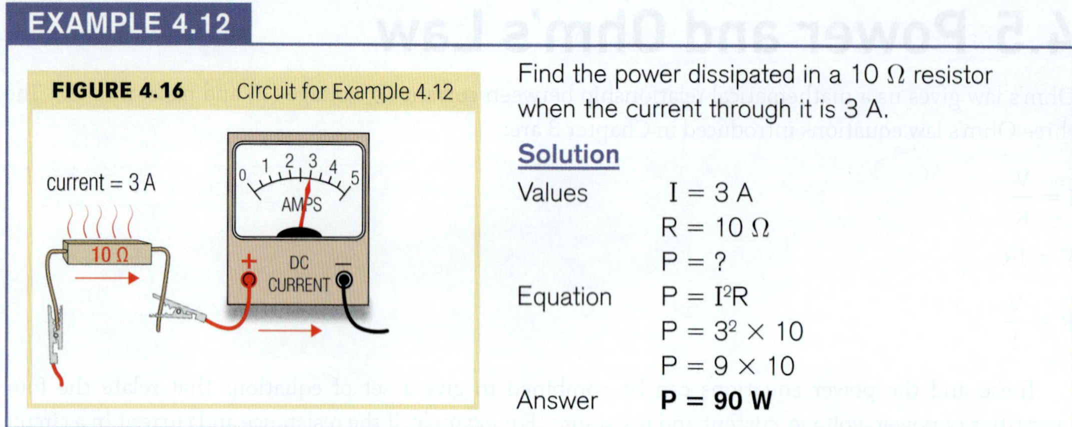

FIGURE 4.16 Circuit for Example 4.12

Find the power dissipated in a 10 Ω resistor when the current through it is 3 A.

Solution

Values	$I = 3$ A
	$R = 10\ \Omega$
	$P = ?$
Equation	$P = I^2R$
	$P = 3^2 \times 10$
	$P = 9 \times 10$
Answer	$\mathbf{P = 90\ W}$

Example 4.13 shows how to use the equation when the values are expressed with metric prefixes.

EXAMPLE 4.13

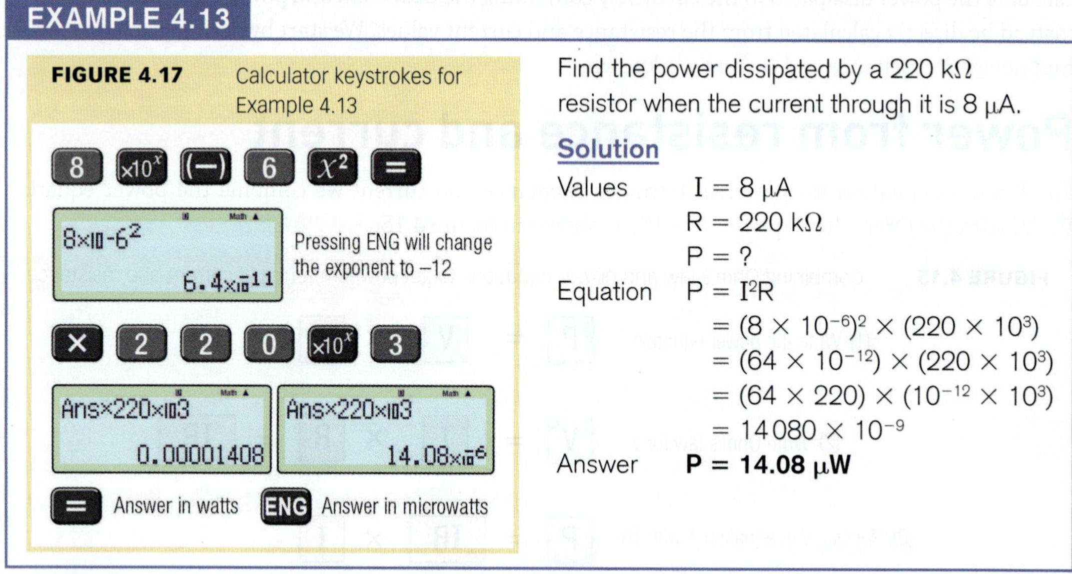

FIGURE 4.17 Calculator keystrokes for Example 4.13

Find the power dissipated by a 220 kΩ resistor when the current through it is 8 μA.

Solution

Values	$I = 8\ \mu A$
	$R = 220\ k\Omega$
	$P = ?$
Equation	$P = I^2R$
	$= (8 \times 10^{-6})^2 \times (220 \times 10^3)$
	$= (64 \times 10^{-12}) \times (220 \times 10^3)$
	$= (64 \times 220) \times (10^{-12} \times 10^3)$
	$= 14\,080 \times 10^{-9}$
Answer	$\mathbf{P = 14.08\ \mu W}$

Current from power and resistance

The equation $P = I^2R$ can be transposed to make either current (I) or resistance (R) the subject. To make I the subject:

1 Write the equation: $P = I^2R$

2 Divide both sides by R: $\frac{P}{R} = \frac{I^2R}{R}$

3 Cancel the Rs on the right: $\frac{P}{R} = I^2$

4 Take the square root of both sides and rearrange so I is on the left: $I = \sqrt{\frac{P}{R}}$

EXAMPLE 4.14

Find the current flowing in a 1 kΩ resistor if it is dissipating 0.5 W of power.

Solution

Values $P = 0.5\ W$

$R = 1000\ \Omega$

$I = ?$

Equation $I = \sqrt{\frac{P}{R}} = \sqrt{\frac{0.5}{1000}} = \sqrt{\frac{0.5}{10^3}}$

$= \sqrt{0.5 \times 10^{-3}}$

$= 0.02236\ A$

Answer **I = 22.36 mA**

FIGURE 4.18 Calculator key strokes for Example 4.14

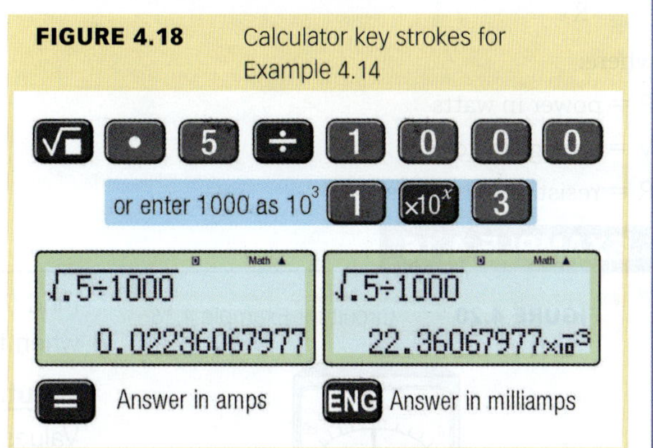

Resistance from power and current

The sequence to make R the subject of $P = I^2R$ is shown below.

1 Write the equation: $P = I^2R$

2 Divide both sides by I^2: $\frac{P}{I^2} = \frac{I^2R}{I^2}$

3 Cancel the I^2s on the right and rearrange the equation so R is on the left: $R = \frac{P}{I^2}$

EXAMPLE 4.15

Find the resistance of a heating element that takes 250 W of power at a current of 3 A.

Solution

Values $P = 250\ W$

$I = 3\ A$

$R = ?$

Equation $R = \frac{P}{I^2} = \frac{250}{3^2}$

$= \frac{250}{9}$

Answer **R = 27.78 Ω**

FIGURE 4.19 Calculator key strokes for Example 4.15

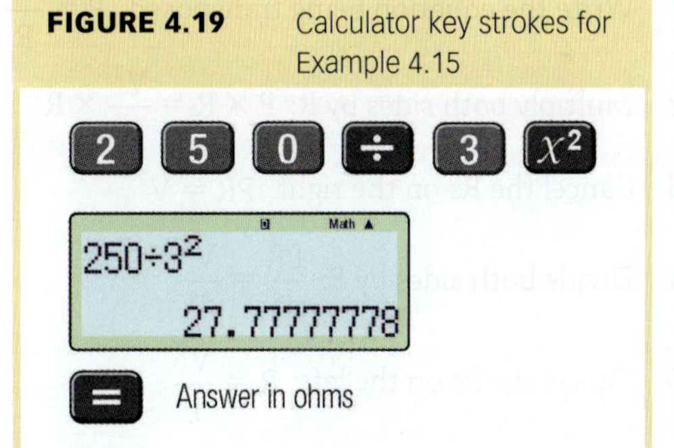

Power from resistance and voltage

To obtain an equation for power in terms of resistance and voltage, combine the power equation P = VI with the Ohm's law equation $I = \frac{V}{R}$ as follows:

1 Write the power equation: P = VI

2 Write the Ohm's law equation for I: $I = \frac{V}{R}$

3 Replace I with $\frac{V}{R}$ in equation P = VI: $P = V \times \frac{V}{R}$

4 Multiply $V \times \frac{V}{R} = \frac{V^2}{R}$

5 Write final equation: $P = \frac{V^2}{R}$.

The equation to find power if the voltage and the resistance are known is:

$$P = \frac{V^2}{R}$$

where:

P = power in watts

V = voltage in volts

R = resistance in ohms.

EXAMPLE 4.16

FIGURE 4.20 Circuit for Example 4.16

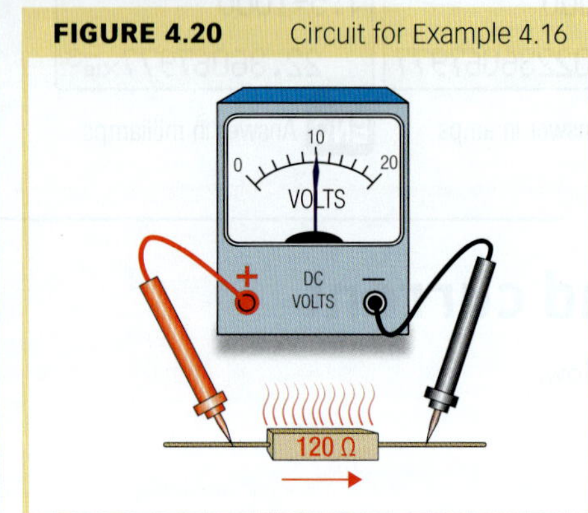

Find the power dissipated by a 120 Ω resistor when the voltage across it is 10 V.

Solution

Values $V = 10\ V$

$R = 120\ \Omega$

$P = ?$

Equation $P = \frac{V^2}{R} = \frac{10^2}{120} = \frac{100}{120}$

Answer **P = 0.833 watts**

Resistance from voltage and power

If resistance is the unknown value when power and voltage are known, transpose the equation $P = \frac{V^2}{R}$ to make R the subject, as shown below:

1 Write the equation being transposed: $P = \frac{V^2}{R}$

2 Multiply both sides by R: $P \times R = \frac{V^2}{R} \times R$

3 Cancel the Rs on the right: $PR = V^2$

4 Divide both sides by P: $\frac{PR}{P} = \frac{V^2}{P}$

5 Cancel the Ps on the left: $R = \frac{V^2}{P}$

The equation to find resistance if the voltage and power are known is:

$$R = \frac{V^2}{P}$$

where:

R = resistance in ohms

V = voltage in volts

P = power in watts.

FYI

This value is the resistance of the lamp filament at its operating temperature. If you measure the filament's resistance when it is cold, the value will be much lower, at around 50 Ω

EXAMPLE 4.17

Find the resistance of a 230 V, 100 W lamp (resistance when the lamp is on).

Solution

Values $V = 230\ V$

$P = 100\ W$

$R = ?$

Equation $R = \frac{V^2}{P}$

$= \frac{230^2}{100} = \frac{230 \times 230}{100} = \frac{52900}{100}$

Answer $\mathbf{R = 529\ \Omega}$

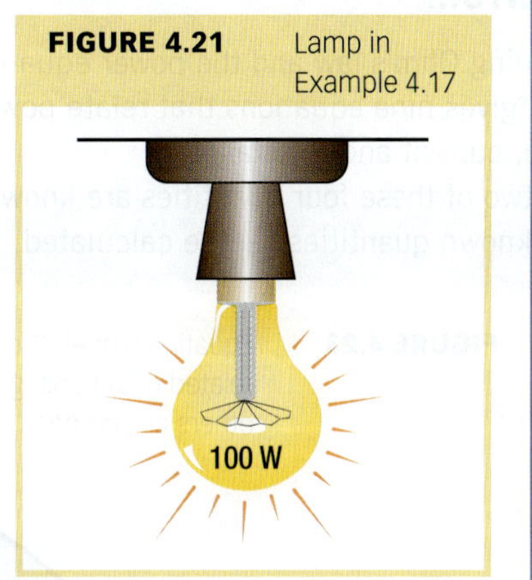

FIGURE 4.21 Lamp in Example 4.17

Voltage from power and resistance

If voltage is the unknown value when power and resistance are known, transpose the equation $P = \frac{V^2}{R}$ to make V the subject, as shown below:

1 Write the equation being transposed: $P = \frac{V^2}{R}$

2 Multiply both sides by R: $P \times R = \frac{V^2}{R} \times R$

3 Cancel the Rs on the right: $PR = V^2$

4 Take the square root of both sides and write equation so V is on the left: $V = \sqrt{PR}$.

The equation to find voltage if the resistance and power are known is:

$V = \sqrt{PR}$

where:

R = resistance in ohms

V = voltage in volts

P = power in watts.

EXAMPLE 4.18

What is the maximum voltage that can be applied across a 330 Ω resistor that has a maximum power rating of 0.5 W?

Solution

Values $R = 330$ ohms

$P = 0.5$ watts

$V = ?$

Equation $V = \sqrt{PR}$

$= \sqrt{0.5 \times 330} = \sqrt{165}$

Answer **V = 12.85 volts**

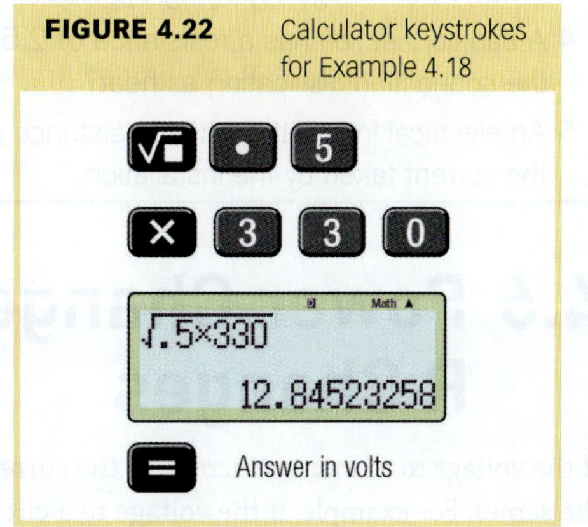

FIGURE 4.22 Calculator keystrokes for Example 4.18

KEY POINTS...

- Combining Ohm's law and the power equation P = VI gives nine equations that relate power, voltage, current and resistance.
- If any two of these four quantities are known, the unknown quantities can be calculated.
- Figure 4.23 summarises the nine power equations and the three Ohm's law equations.

FIGURE 4.23 Equation wheel. The four quantities V, I, R and P are in the centre section near their three related equations, giving 12 equations. Remember some of the equations, and derive the others as you need them.

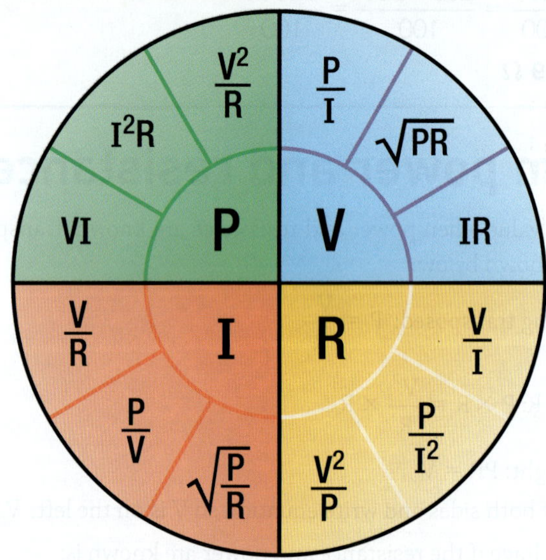

TASK 4.3

1. An appliance has a resistance of 5 Ω and takes a current of 15 A. How much power is being consumed?
2. A commercial oven rated at 10 kW takes 20 A from the supply. Calculate the resistance of the oven.
3. An electrical load has a resistance of 8 Ω and dissipates 20 kW of power when operating. What is the voltage supplying the load?
4. A bad connection has a resistance of 2.5 and is passing a current of 15 A. How much power is the connection dissipating as heat?
5. An electrical installation has a resistance of 20 Ω and is consuming 45 kW of power. Calculate the current taken by the installation.

4.6 Power Change with I, V or R Changes

If the voltage to a circuit is increased, the current increases proportionally (assuming the resistance stays the same). For example, if the voltage to a circuit is doubled, by Ohm's law the current in the circuit will also double. Because power is the product of voltage and current (P = VI), if both voltage and current have doubled in value, the power will be increased by four times. If the resistance of a circuit changes, there will also be a change in the power, but by how much? First a look at what happens when the voltage changes . . .

Effect on power if the voltage changes

As explained earlier, the equation relating power, voltage and resistance is $P = \frac{V^2}{R}$. Notice that the equation shows voltage *squared*. This means that the power taken by the circuit will increase by the *square* of the voltage change (assuming the resistance doesn't change). For instance, as outlined above, if the voltage doubles, the power will increase by four times, as two squared (2^2) equals 4. The following example shows this.

EXAMPLE 4.19

How much power does a 10 Ω resistor take when the voltage across it is (a) 5 V, (b) 10 V?

Solution

Values $V = 5\ V$

$R = 10\ \Omega$

$P = ?$

Equation $P = \frac{V^2}{R} = \frac{5^2}{10} = \frac{25}{10}$

(a) Answer **P = 2.5 W** (when voltage is 5 V)

FIGURE 4.24 Circuit for Example 4.19(a)

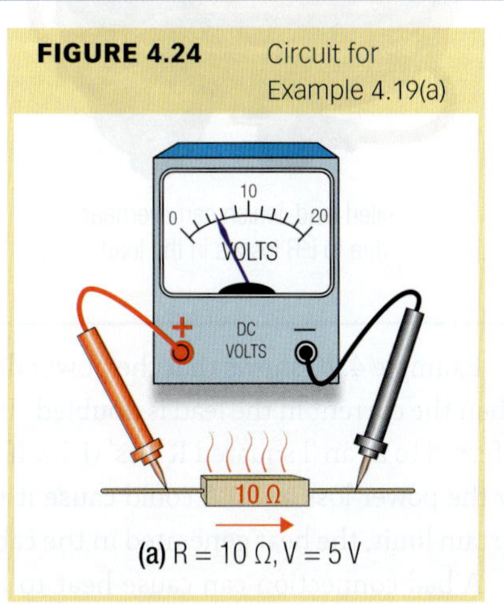

(a) R = 10 Ω, V = 5 V

When the voltage doubles to 10 V, the power is found the same way, as shown below.

Values $V = 10\ V$

$R = 10\ \Omega$

$P = ?$

Equation $P = \frac{V^2}{R} = \frac{10^2}{10} = \frac{100}{10}$

(b) Answer **P = 10 W** (when voltage is 10 V)

FIGURE 4.25 Circuit for Example 4.19(b)

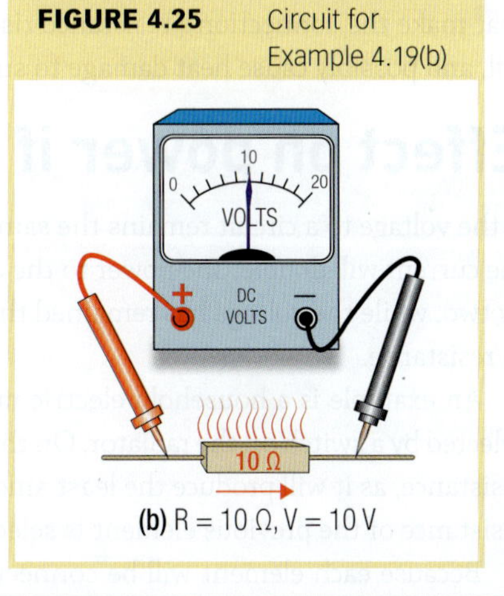

(b) R = 10 Ω, V = 10 V

As you can see, the power has increased by four times (as 2.5 W × 4 equals 10 W). Because power changes by the *square* of the voltage, if the voltage increases by three times, the power taken by the resistor will increase by nine times. If the voltage falls by half, the power will fall to a quarter of its previous value. So, a small change in the voltage causes a much larger change in the power taken by a circuit.

Effect on power if the current changes

The equation $P = I^2R$ shows that the power changes with the *square* of the current (providing the resistance doesn't change). That is, if the current through a circuit doubles (while its resistance remains the same), the power dissipated by the circuit increases by four times. The following example shows this effect.

EXAMPLE 4.20

A 50 m extension lead has a total resistance of 2 Ω. Calculate the power dissipated over the length of the lead when (a) the current in the lead is 5 A and (b) when the current is 10 A.

FIGURE 4.26 Extension lead in Example 4.20

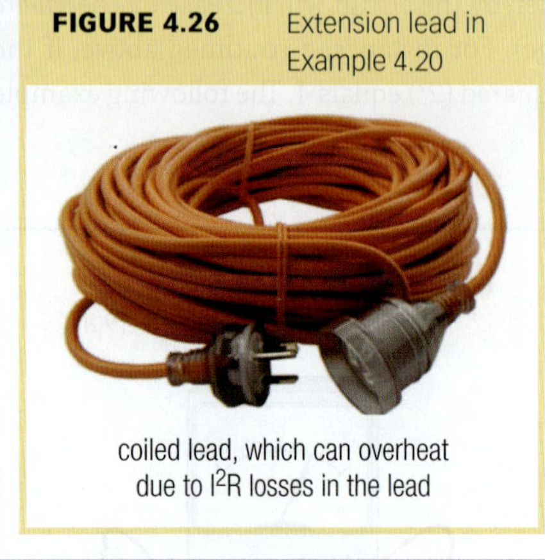

coiled lead, which can overheat due to I^2R losses in the lead

Solution

Values $I = $ (a) 5 A (b) 10 A

$R = 2\ \Omega$

$P = ?$

Equation $P = I^2R$

(a) $= 5^2 \times 2$

$= 25 \times 2$

Answer **(a) P = 50 W** (when current is 5 A)

(b) $P = I^2R$

$= 10^2 \times 2$

$= 100 \times 2$

Answer **(b) P = 200 W** (when current is 10 A)

SAFETY

Bad electrical connections, particularly in roof cavities, are responsible for numerous house fires. Carefully check all connections you make

Example 4.20 shows that the power dissipated by the lead has increased fourfold from 50 W to 200 W when the current in the lead is doubled. The power is dissipated as heat energy, which is a loss, sometimes referred to as an 'I squared R loss' (I^2R). If the extension lead in Example 4.20 is coiled, the heat generated by the power loss at 10 A could cause it to overheat. If the current in an overhead power line exceeds a certain limit, the heat generated in the cable could cause it to sag and come into contact with other cables.

A bad connection can cause heat to be generated at the connection because of the current flowing through a higher than normal resistance. The heat being produced will cause oxidation and other effects that make the connection's resistance rise even further, in turn producing more heat. Eventually it will fail, and possibly cause heat damage to surrounding areas or start a fire.

Effect on power if resistance changes

If the voltage to a circuit remains the same and the circuit resistance falls by half, Ohm's law tells us that the current will double. The power to the circuit will also double, as $P = VI$, and the current has increased by two, while the voltage has remained the same. Therefore, the power changes *proportionally* to a change in resistance.

An example is a household electric radiator that has two or three different sized heating elements, selected by a switch on the radiator. On the low setting, the switch will select the element with the highest resistance, as it will produce the least amount of heat. To obtain twice the heat, an element with half the resistance of the previous element is selected instead.

Because each element will be connected in turn to the same voltage, we can say that if the voltage remains the same, *power dissipation is inversely proportional to the resistance*. That is, the higher the resistance, the less power dissipated for the same voltage.

CHAPTER SUMMARY

- Energy and work are both measured in joules. Energy is the ability to do work, and work is done when energy is transformed.
- Force is measured in newtons (N). One newton is the force needed to accelerate a one kilogram mass by a rate of one metre per second per second.
- Motors are rated by their power and torque, where torque is measured in newton metres (Nm) and power is measured in watts.
- Energy is either kinetic (moving) or potential (stored). Potential energy becomes kinetic energy when used.
- Power is the rate of doing work. That is: $\text{power} = \frac{\text{work}}{\text{time}}$, where power is in watts, work is in joules and time is in seconds. The power is one watt if one joule of energy is transformed in one second.
- Mechanical power of a motor: $P = \frac{2\pi nT}{60}$ watts, where n = RPM, T = torque (Nm).
- Efficiency: $\eta = \frac{\text{power out}}{\text{power in}} \times 100\%$. Efficiency gives a measure of power and energy losses.
- Electrical power (P) to a load equals the product of the current in the load and the voltage across the load (P = VI).
- P = VI is transposed to make either V or I the subject, and is combined with Ohm's law to give 12 equations relating to circuit power, resistance, current and voltage. (See Figure 4.23.)
- $P = \frac{V^2}{R}$ means power changes by the square of the voltage change.
- It also means that power changes inversely with resistance (when the voltage is constant).
- $P = I^2R$ means power changes by the square of the current change.
- 'I squared R losses' refers to the losses generated as heat in a conductor or connection.

REVIEW EXERCISES

Check your answers at the back of the book.

1. What is the relationship between energy and work?
2. A motor delivers a torque of 450 Nm when running at 3000 RPM. Calculate its mechanical power.
3. A motor connected to a 400 V supply takes 50 A when delivering 18.4 kW of mechanical power. Calculate the efficiency of the motor.
4. A lighting circuit takes 1500 W of power at a supply voltage of 230 V. Find the value of the current flowing in the circuit.
5. A floor heating system rated at 1.3 kW is left switched on for 10 hours. Calculate the energy used (a) in joules and (b) in kilowatt hours. What is the running cost if electrical energy is priced at $0.30 per kWh?
6. A commercial oven is rated at 400 V, and has three elements that give a combined resistance (when operating) of 16 Ω. Calculate the power.
7. For the circuit in Figure 4.27, find the resistance (R) and the power (P) dissipated in the resistance for the following meter readings:
   a. Ammeter shows 5 A, voltmeter shows 230 V
   b. Ammeter shows 8 mA, voltmeter shows 24 V
   c. Ammeter shows 200 mA, voltmeter shows 400 V.

FIGURE 4.27

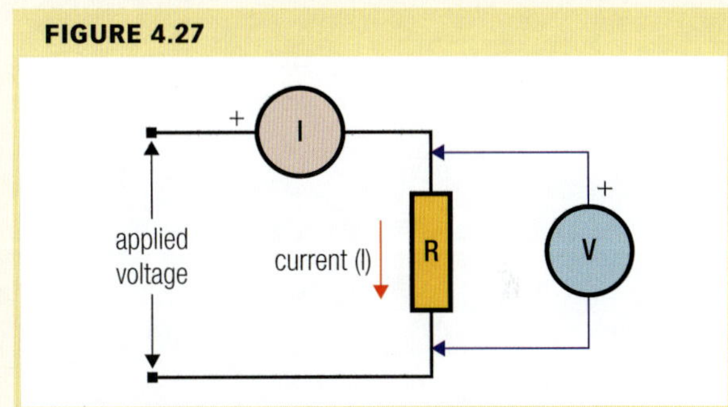

8. For the circuit in Figure 4.28, find:
   a. resistance (R) if the applied voltage is 400 V
   b. current (I) if the resistance is 10 W
   c. voltage (V) if the resistance is 88.2 W.

FIGURE 4.28

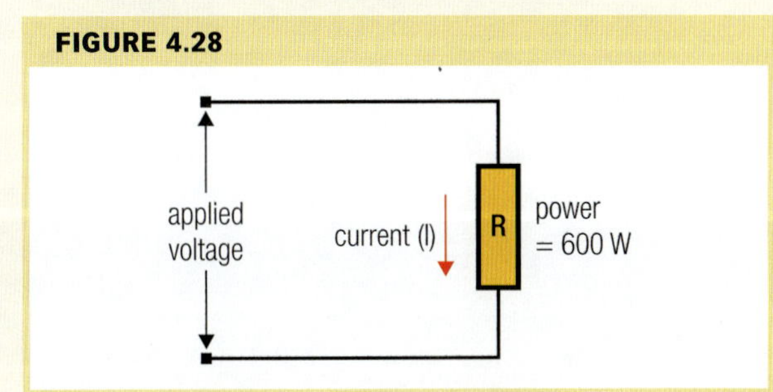

9 An electric stove connected to a 230 V supply has three hot plates, two rated at 900 W and a larger one rated at 1500 W. Calculate the current taken by the stove when:
  a one 900 W hot plate is operating
  b both 900 W hot plates are operating
  c all hot plates are operating.

10 Calculate the resistance of the stove for each of the conditions in Question 9.

ONLINE RESOURCES

COMPLETE WORKSHEET FOUR

Check with your instructor for worksheets on this chapter.

CHAPTER 5

RESISTANCE AND RESISTORS

This chapter explains the factors that determine the resistance of a conductor. It also introduces the resistor, a commonly used component in electrotechnology, and describes a number of other components whose resistance varies with an external influence, such as light or heat. The resistor colour code is explained, along with how to measure resistance with an ohmmeter.

CHAPTER OUTLINE

5.1 Factors that Determine Resistance

Electrical resistance is a property of all conducting materials, and in the case of a cable, resistance should be as low as possible. Resistance is sometimes required, such as in a heating element and also in components called resistors. We look first at conductors and the factors that determine their resistance. The four factors are:

- the material the conductor is made of
- its length
- its cross-sectional area
- the temperature of the conductor.

Material the conductor is made of

The term resistance first appeared in Chapter 1 where it was explained that certain materials are better conductors than others due to their atomic structure. This is a property of the particular material and is referred to as its *resistivity*. All conducting materials have a certain value of resistivity, which is defined as the resistance the material offers between opposite faces of a cube of the material, as shown in Figure 5.1. In the metric system, the cube has one metre dimensions and the resistivity value is expressed in *ohm-metres* (Ω m). The symbol for resistivity is the Greek letter ρ (pronounced rho).

FIGURE 5.1 Resistivity of a material is its resistance measured at a specified temperature between opposite faces of a cube of the material

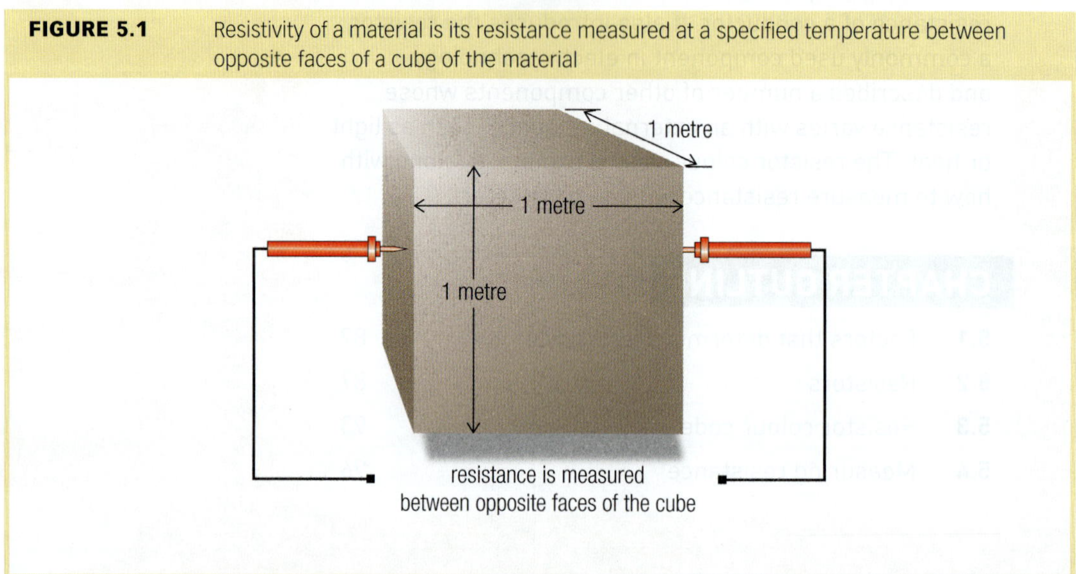

In practice, resistivity is measured with a smaller sample of the material, with the value expressed in terms of a one metre cube. Resistivity, when given for a particular temperature, gives us a value that is independent of all other factors that determine the resistance of a conductor. Table 5.1 lists resistivity values for typical conducting materials used in electrotechnology. These values are for a temperature of 20 °C, because resistivity varies with temperature. The purity of the material also affects the resistivity value, and you may find tables in other books or on the internet give slightly different values.

A cube with one metre dimensions gives very small resistivity values that need to be expressed with scientific notation. The figure for silver is the lowest (as it's the best conductor) and carbon has the highest resistivity. Carbon is found in some types of resistors and is included here for comparison. For example, carbon has a resistivity some 200 000 times higher than silver. Nichrome is only about 69 times greater than silver, but this is high enough to make its resistance suit a heating element.

TABLE 5.1

RESISTIVITY OF CONDUCTORS AT 20 °C		
Material	**Resistivity in ohm-metre (ρ)**	**Notes**
silver	1.63×10^{-8}	= 0.000 000 016 3 ohm-metre
copper	1.72×10^{-8}	value for pure copper
gold	2.44×10^{-8}	
aluminium	2.65×10^{-8}	value for 99.99% pure aluminium
tungsten	5.3×10^{-8}	value varies with impurities
nichrome	112×10^{-8}	used when resistance is required
carbon	3.5×10^{-3}	semiconductor, included for comparison

Table 5.2 gives examples of the resistance of 1 km long cables of different sizes, calculated from the resistivity of the material each is made of. As explained in Chapter 2, only pure copper produced by electrolysis is used in electrical conductors. Aluminium is used in high voltage power transmission lines. Tungsten is often the filament material in an incandescent lamp and nichrome is used in electric radiator elements.

TABLE 5.2

COMPARISON OF CONDUCTOR RESISTANCE AT 20 °C CONDUCTORS ARE 1 KM LONG, RESISTANCE VALUES IN OHMS			
Material	**1 mm dia. (0.79 mm²)**	**2.5 mm²**	**4 mm²**
silver	20.75	6.52	4.08
copper	21.9	6.88	4.30
gold	31.07	9.76	6.1
aluminium	33.74	10.6	6.63
tungsten	67.5		
nichrome	1426		
carbon	4 456 338		

KEY CONCEPT

Resistance is proportional to the resistivity of the material a conductor is made of

Length of the conductor

Electrical resistance is associated with how easily electrons can flow from one atom to another within the material. The easier this is, the lower the energy losses each time an electron moving in a conductor encounters an atom. As the length of a conductor increases, its resistance to the path of electrons becomes greater, as there are now more electron-atom collisions, and therefore increased losses.

KEY CONCEPT

Resistance is proportional to the length of the conductor

The resistance values in Table 5.2 are for cables 1 km long. If they were all 2 km in length, the resistance values would be double the values shown. That is, the resistance of a cable is proportional to its length. The longer the cable, the higher its resistance.

Cross-sectional area

Conductors can be any shape, including circular (wire), rectangular or square. While diameter is a useful measurement for circular cables, it doesn't apply to other shapes. However, all conductors have a cross-sectional area (csa), regardless of their shape. The three conductors in Figure 5.2 all have the same cross-sectional area of 4 mm^2.

FIGURE 5.2 Cross-sectional area of a conductor is the area of the face of the conductor

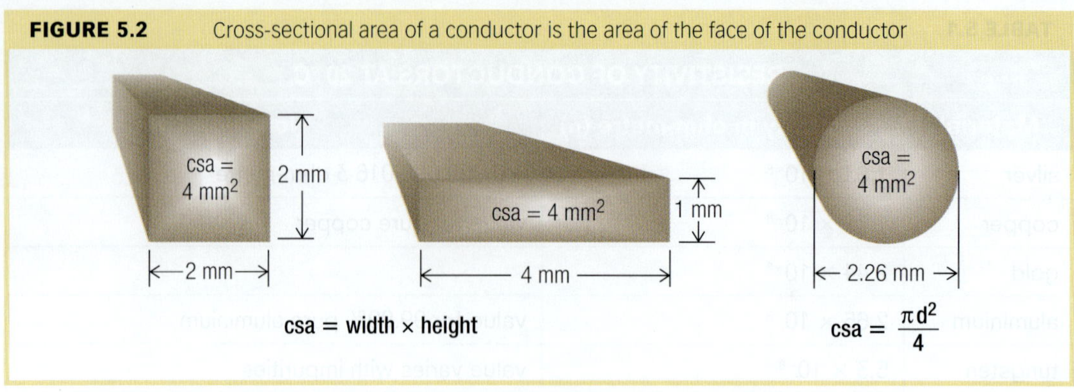

KEY CONCEPT

Resistance is inversely proportional to cross-sectional area

Notice how the cross-sectional area of each of the shapes is determined. For a rectangular or square conductor, the csa is the height multiplied by the width. For a circular conductor, $csa = \frac{\pi d^2}{4}$, which, as you probably recognise, is also an equation to find the area of a circle.

The smaller the cross-sectional area, the higher the resistance, as there are less free electrons to carry the current. If the cross-sectional area is increased, the resistance decreases. Therefore, the resistance of a conductor is *inversely* proportional to its cross-sectional area.

FIGURE 5.3 Electrical cables come in a wide variety of sizes, shapes and types. Cables are usually multi-stranded and made of copper or (less typically) aluminium. Aluminium transmission cables often have an iron core for strength.

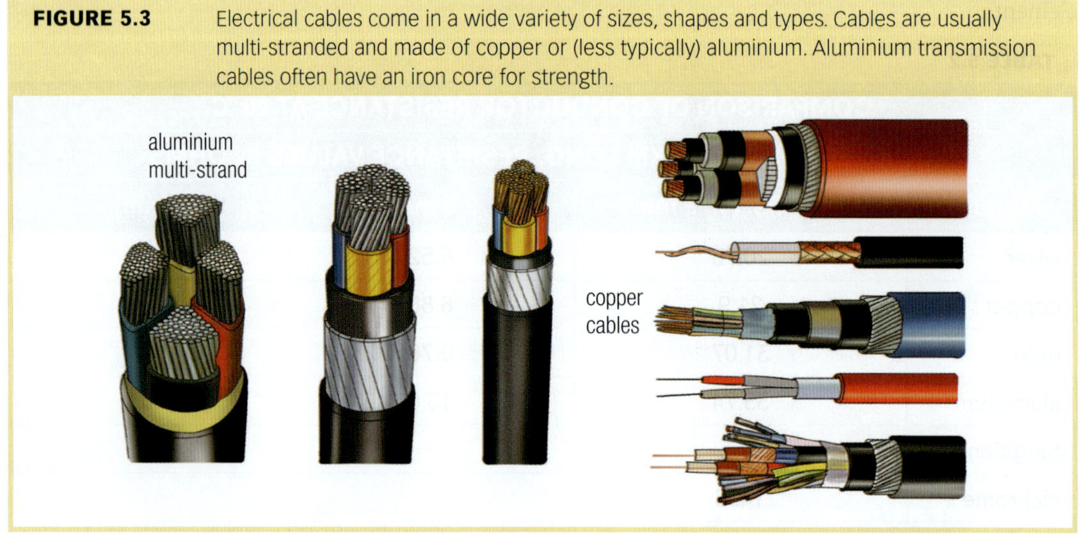

Calculating cable resistance

We have shown that resistance of a conductor (R) is proportional to its cross-sectional area (A), and proportional to the material's resistivity (ρ). It is inversely proportional to the length (l) of the conductor. The equation combining these terms to find R is:

$$R = \frac{\rho l}{A}$$

where:

R = resistance in ohms

ρ = resistivity in ohm-metres

l = length in metres

A = cross-sectional area in square metres.

Typically the cross-sectional area of a conductor is given in square millimetres (mm^2) while the equation requires the value to be in square metres (m^2). One millimetre = 0.001 metres, or 1×10^{-3} metres. A square millimetre converted to square metres is $1 \times 10^{-3} \times 1 \times 10^{-3}$, or 1×10^{-6}. Therefore, to convert from square millimetres to square metres, multiply the value given in square millimetres by 10^{-6} (or divide by 1 000 000).

EXAMPLE 5.1

What is the resistance of a 1 km length of copper cable with a csa of 2 square millimetres (mm^2)?

Solution

Values: From Table 5.1, resistivity (ρ) of copper = 1.72×10^{-8} ohm-metre
csa = 2 mm^2 = 2×10^{-6} square metres (m^2)
length (l) = 1 km = 1000 m

Equation: $$R = \frac{\rho l}{A} = 1.72 \times 10^{-8} \times \frac{1000}{2 \times 10^{-6}}$$

Answer: **R = 8.6 Ω**

EXAMPLE 5.2

Calculate the resistance of a nichrome resistance element made from 1 mm diameter wire. The total wire length is two metres.

Solution

Values: From Table 5.1, resistivity (ρ) of nichrome = 112×10^{-8} ohm-metre
length (l) = 2 m
diameter = 1 mm

Equation 1: $$A = \frac{\pi d^2}{4} = \frac{3.14 \times 0.001^2}{4} = \frac{3.14 \times 10^{-6}}{4} = 7.85 \times 10^{-7} \text{sq metres}$$

Equation 2: $$R = \frac{\rho l}{A} = 112 \times 10^{-8} \times \frac{2}{7.85 \times 10^{-7}}$$

Answer: **R = 2.85 Ω**

Temperature

The resistance of all conductors varies with temperature by an amount that depends on the type of metal. For pure metals, an *increase* in temperature will give an *increase* in resistance.

The change in resistance can be calculated using a value called the **temperature coefficient of resistance** for that metal. Temperature coefficient of resistance is defined as the change in resistance per ohm per degree Celsius of temperature change.

The graph in Figure 5.4 shows how the resistance of a coil of copper wire changes with temperature. If the resistance at 0 °C is 100 Ω, it will be 180 Ω at 200 °C. However, if the temperature drops to −200 °C, the resistance will drop to about 30 Ω.

The graph is a straight line for temperatures above 0 °C. Because the resistance increases with an increase in temperature, the temperature coefficient is *positive*. Most metals have a positive temperature coefficient of resistance. If the temperature is reduced to absolute zero (−273 °C), the resistance of many types of metal conductors falls to virtually zero. This is known as *superconductivity*.

Some materials, such as carbon and silicon (semiconductors), have a negative temperature coefficient of resistance. This means their resistance *drops* as the temperature rises.

FIGURE 5.4 Graph showing the change in resistance of a copper coil with a change in temperature.

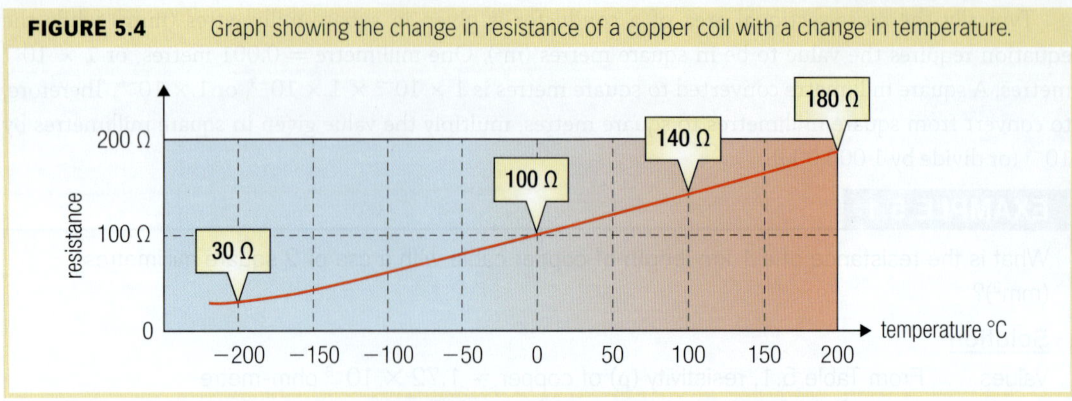

The temperature coefficient for a material is given at a particular temperature, and for most purposes, the value given at 20 °C can be used. For copper, aluminium, gold and silver the temperature coefficient is about 0.004. This means the resistance will increase by about 0.4 per cent for each increase of 1 °C. In Figure 5.4, the resistance of the coil has increased by 40 per cent from 100 Ω to 140 Ω for a temperature rise of 100 °C.

The symbol for temperature coefficient is the Greek letter α (*alpha*). Because α varies with temperature, the reference temperature is usually shown as a subscript; for example α_{20} means the temperature coefficient for the material at a temperature of 20 °C.

Table 5.3 lists the temperature coefficient of resistance for a range of conductors. The purity and the composition of the material will affect these values, and you may find that similar tables in other books or on the internet show different values.

TABLE 5.3

TEMPERATURE COEFFICIENTS AT 20°C (α_{20})		
Material	α_{20}	**Notes**
gold	0.0034	lowest α_{20} of the pure metals
silver	0.0038	approximately 0.004
copper	0.00392	approximately 0.004
aluminium	0.0043	approximately 0.004
tungsten	0.0045	value varies with impurities
nichrome	0.00017	small change in resistance with temperature change
carbon	−0.0005	resistance drops slightly with an increase in temperature

KEY CONCEPT

For pure metals, resistance is proportional to temperature

With spot welding, when a large current is passed through the point being welded, the resistance of the metal generates enough heat to form the weld. Aluminium has a low resistance, so it's harder to spot weld than stainless steel, which has a higher resistance and therefore heats up quickly.

Carbon has various forms and, depending on its composition, temperature coefficients for carbon will vary considerably. Most types of carbon have a *negative* temperature coefficient (resistance falls as temperature rises), although some types have a positive temperature coefficient. The value given is for amorphous carbon. Notice that the value is negative.

The general equation to find the new resistance (R) of a conducting material due to a change in temperature is:

$$R = R_{ref}[1 + \alpha(T - T_{ref})]$$

where:

R = resistance at new temperature (T)

R_{ref} = resistance at reference temperature (T_{ref})

α = temperature coefficient at T_{ref}

EXAMPLE 5.3

A coil of copper wire has a resistance of 30 ohms at 20 °C. Find its resistance at 80 °C.

Solution

Values: From Table 5.3, $\alpha_{20} = 0.00392 = 0.004$

$T_{ref} = 20\ °C$, $R_{ref} = 30\ \Omega$

$T = 80\ °C$

Equation: $R = R_{ref}\,[1 + \alpha_{20}(T - T_{ref})]$

$R = 30\,[1 + 0.004(80 - 20)] = 30\,[1 + 0.24] = 30 \times 1.24$

Answer: $\mathbf{R = 37.2}$ **ohms**

KEY POINTS...

- Resistance is proportional to the length (in metres) of the conductor.
- Resistance is proportional to the resistivity of the material a conductor is made of.
- Resistivity (ρ) is measured in ohm-metres.
- Resistance is inversely proportional to cross-sectional area (csa) of a conductor.
- Cross-sectional area is in square metres (m^2). Values in square millimetres are multiplied by 10^{-6} to convert them to square metres.
- For pure metals, resistance is proportional to temperature.
- Copper and aluminium have a positive temperature coefficient of 0.004.

TASK 5.1

1 Calculate the resistance of a 600 m length of copper cable that has a cross-sectional area of 4 mm^2.
2 A 500 m length of copper cable has a diameter of 3 mm. Calculate its resistance.
3 If the cable in Question 2 is made of aluminium, calculate its resistance.
4 A 5 km long copper cable must have a resistance no greater than 8 Ω. Calculate the minimum required cross-sectional area of the cable.
5 A coil wound with copper wire measures 50 Ω at 20 °C. What is its resistance when the coil temperature has increased to 70 °C?

5.2 Resistors

A resistor is an electrical component with a fixed value of resistance. Resistor values range from less than one ohm up to many millions of ohms. Resistors are also given a *power rating* because they get hot when current is passed through them. The power rating of a resistor determines the maximum current the resistor can pass without overheating.

High power resistors

Resistors with a rating of 5 W or more are generally made by winding resistance wire around a ceramic former, rather like a radiator element. These resistors are known as wire-wound types. Some are fitted with heatsinks to dissipate the heat generated by the resistor, and these must be mounted so heat can be easily dispersed.

Another type is the *grid resistor*, which is designed to handle high currents (400 A or more) and high power dissipation. These have large resistive elements made from a metal alloy, and are used in applications such as dynamic braking and speed control of large electric motors, load banks for testing purposes and in electrical substations. Power resistors are shown in Figure 5.5.

FIGURE 5.5 Power resistors, showing typical construction and types of resistors

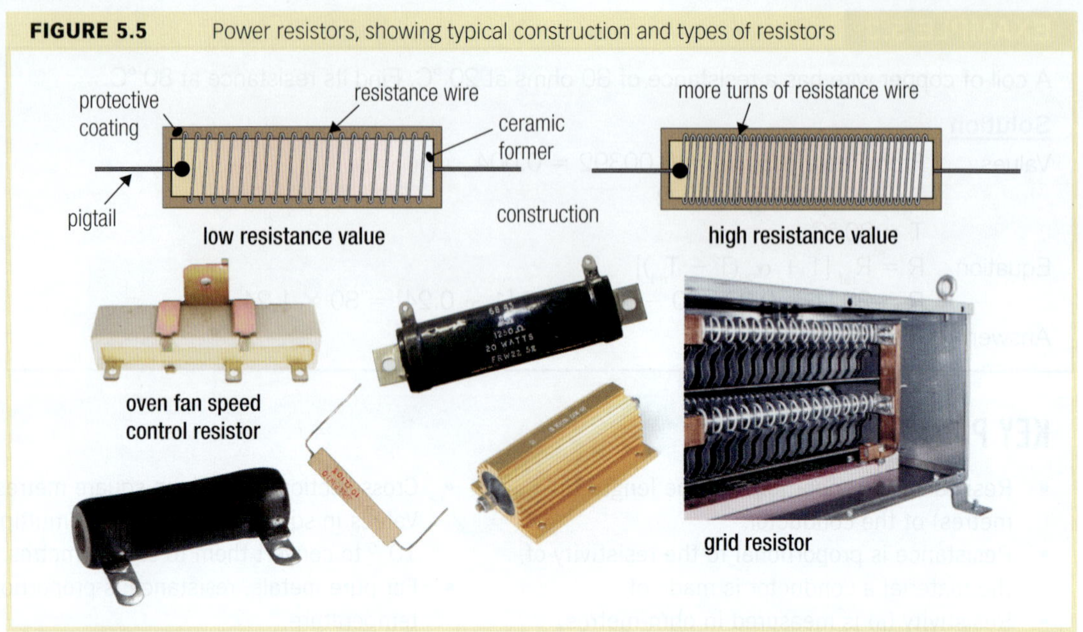

Film resistors

Low-power resistors are made by coating a ceramic cylinder with a film of specially prepared carbon. A spiral is then cut in the film to give a longer path. These resistors are called *carbon-film* resistors and the higher the value, the finer the spiral for a given type of coating. These resistors are used extensively in electronic applications, although they are often found in electrical work. They are the most common type of general purpose resistor.

A variation is the *metal-film* resistor, which has a similar construction to the carbon-film type, except the resistive element is a coating of nickel chromium or some form of metal oxide. These resistors have a higher stability than carbon-film types, and are made using thin-film technology in which the coating is deposited by a vacuum process. Surface mount resistors are made by depositing a metal oxide layer onto a rectangular ceramic substrate, which has contacts at either end. Surface mount resistors (SMD) are widely used in mobile phones, computers and most electronic devices.

In general, the size of a film resistor gives a guide as to its power rating. Typically, these sorts of resistors have power ratings up to 5 W, the most common being 0.25 W, 0.5 W and 1 W. Figure 5.6 shows the construction of a carbon-film resistor, along with typical examples.

FIGURE 5.6 Carbon-film resistors are for low power use. They range in power from 0.25 W to 5 W, and cost a few cents each.

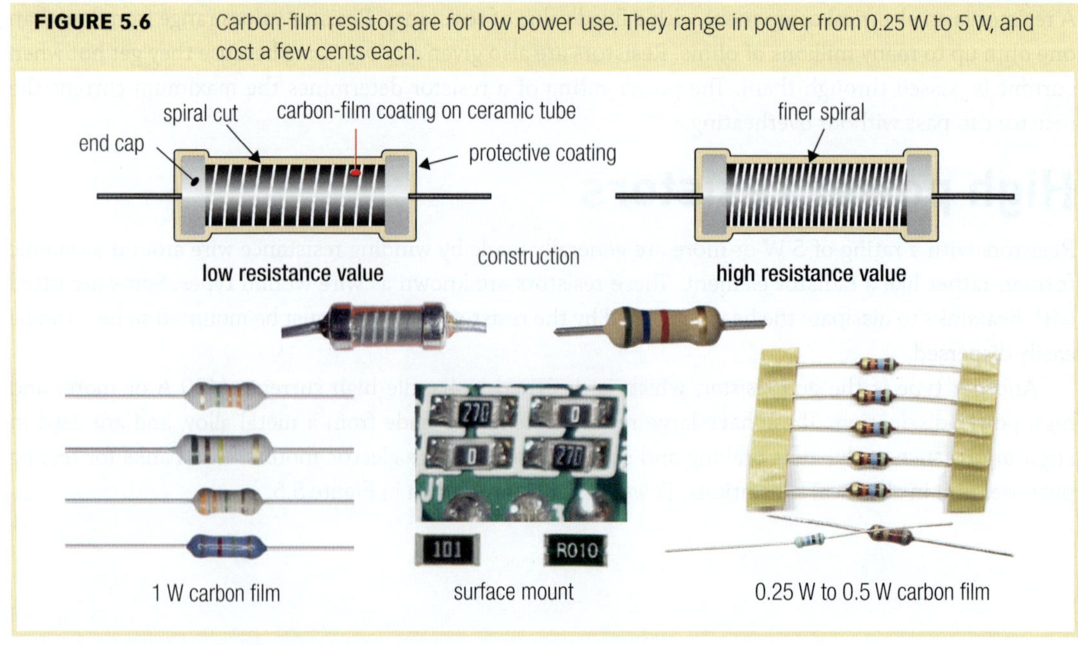

Carbon composition resistors

An older style of resistor is the *carbon composition* type, in which powdered carbon mixed with a filler is compressed inside a phenolic cylinder, with connections (pigtails) at each end. These resistors are used in applications involving high-energy current pulses, as they can withstand this type of stress better than the carbon-film types. However, they tend to change their resistance value over time. Their construction is shown in Figure 5.7.

FIGURE 5.7 Carbon composition resistors were the first types of resistors to be made. They have limited application today, but are still manufactured.

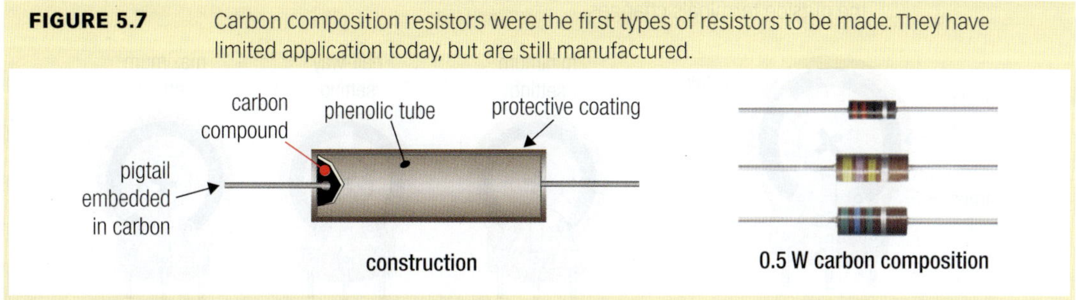

Variable resistors

A *variable resistor* is used in a circuit when you need to be able to adjust the value of a current or a voltage. Variable resistors used to adjust a voltage are often called potentiometers ('pots') as the circuit requires them to have three terminals: one at either end of the resistance element, and another that connects to the moving contact, or wiper. When a variable resistor is connected to adjust a current, it is often called a rheostat, as only two terminals are needed, one to the fixed end of the element, the other to the wiper.

Variable resistors come in a wide range of sizes, resistance values and power ratings. They have a resistance element that is either wire-wound or made of a carbon or metal compound deposited onto a substrate. As in a wire-wound resistor, the resistance element in a wire-wound variable resistor is made from resistance wire wound around a heat resistant ceramic former, allowing it to handle higher values of current and dissipate more power. Most low-power variable resistors have a carbon track resistance element. As shown in Figure 5.8, there are two main types: panel mount and printed circuit board (PCB) mount. The dual-gang panel mount pot has two variable resistors on the one shaft and is used in stereo (two channel) sound systems. Those used on a printed circuit board are often called 'trimpots', as they are adjusted with a screwdriver.

FIGURE 5.8 Typical PCB and panel mount potentiometers

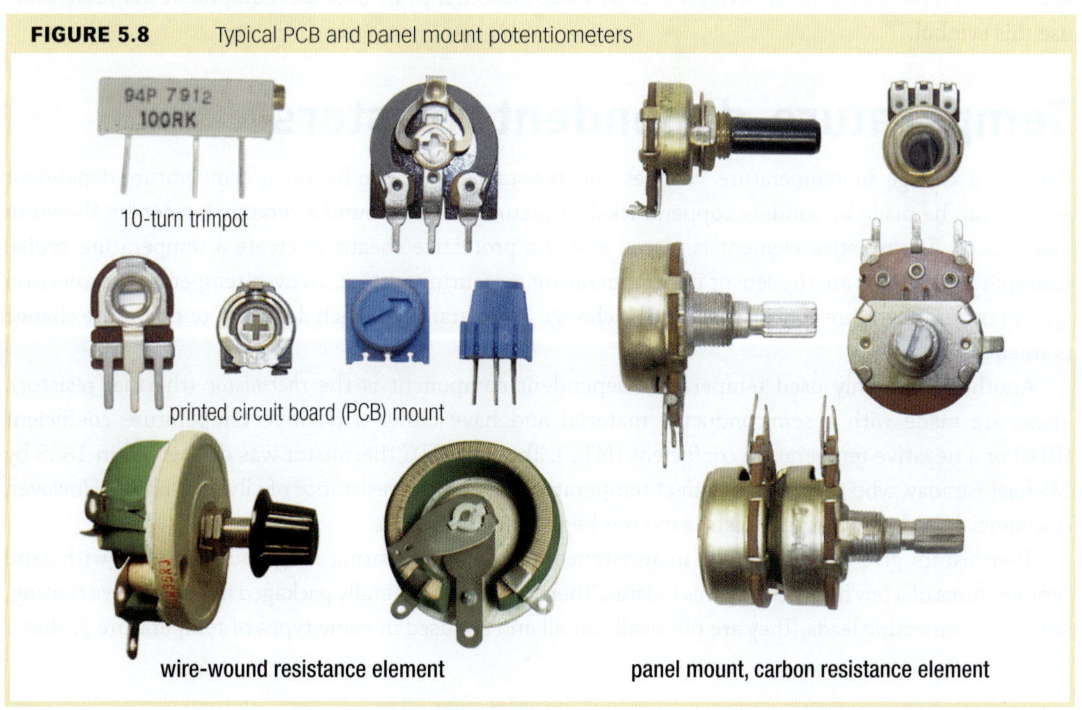

The principle of a variable resistor is shown in Figure 5.9. The resistance element is either a carbon compound deposited onto a Bakelite former or resistance wire wound on a ceramic former. A wiper contacts the resistive track. The wiper is mechanically coupled to a shaft or means of adjustment and is electrically connected to the centre terminal. The ends of the resistive track are connected to the outer terminals.

FIGURE 5.9 A variable resistor (potentiometer) has three terminals, the centre one connected to a moving wiper. As the wiper is moved along the carbon track, the resistance between it and the outside terminals changes.

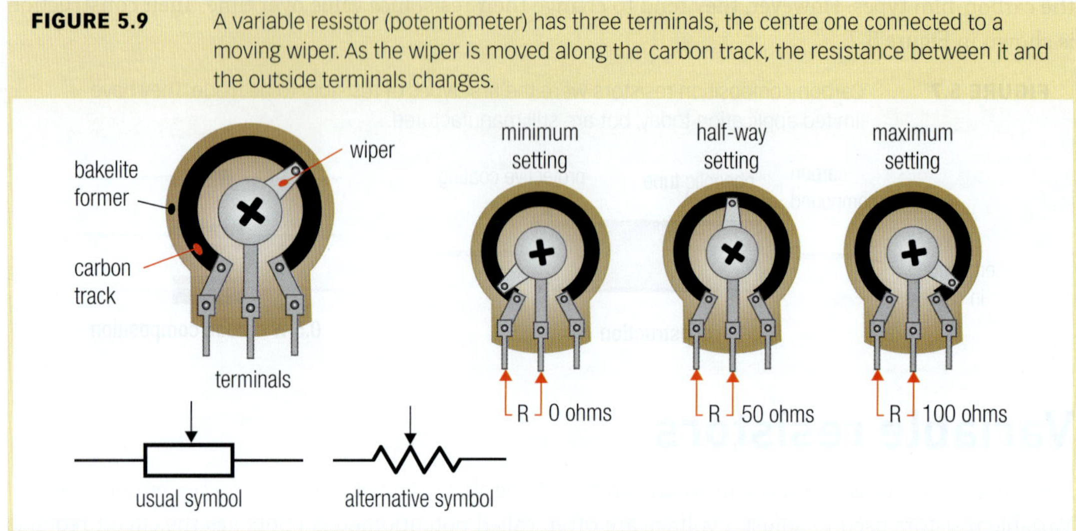

As Figure 5.9 shows, a variable resistance value is obtained by moving the wiper from one side of the resistance element to the other. If the resistive track has a *linear* resistance, when the wiper is halfway, there will be equal resistance between the centre terminal and the outside terminals.

However, in some types of potentiometers, the change in resistance is *non-linear*. That is, when the wiper is halfway, there is a considerable difference in resistance between it and the outside terminals. The most common use for a non-linear variable resistor is the volume control in a radio. Because human hearing is *logarithmic*, the resistance element in a volume control is constructed so its resistance changes in a logarithmic way, to make the change in sound level appear to follow the change in the position of the control.

The Australian Standard symbol for the variable resistor is shown in Figure 5.9. This and the symbol for a fixed resistor were introduced in Chapter 3. As with the resistor, the rectangular symbol for a variable resistor has replaced the older 'wriggly line' symbol, although some overseas equipment manufacturers use this symbol.

Temperature-dependent resistors

Because a change in temperature changes the resistance of a conductor, a temperature-dependent resistor can be made by winding copper, nickel or platinum wire around a ceramic former, as shown in Figure 5.10. The resistive element is placed inside a protective sheath to create a temperature probe. Example applications are the sensor in a temperature measuring system, an over-temperature protection system or a temperature control system. The change in resistance for each degree of temperature change is small but linear.

Another commonly used temperature-dependent component is the *thermistor* (thermal resistor). These are made with a semiconductor material and have either a positive temperature coefficient (PTC) or a negative temperature coefficient (NTC). The first NTC thermistor was discovered in 1833 by Michael Faraday, who reported the effect temperature had on the resistance of silver sulphide. However, commercial production of thermistors did not begin until the 1930s.

Thermistors give a large change in resistance over their operating range but can only withstand temperatures of a few hundred degrees Celsius. Thermistors are generally packaged in a protective coating, with two connecting leads. They are physically small and are used in some types of temperature probes.

FIGURE 5.10 A temperature probe with a resistance winding of thin platinum wire

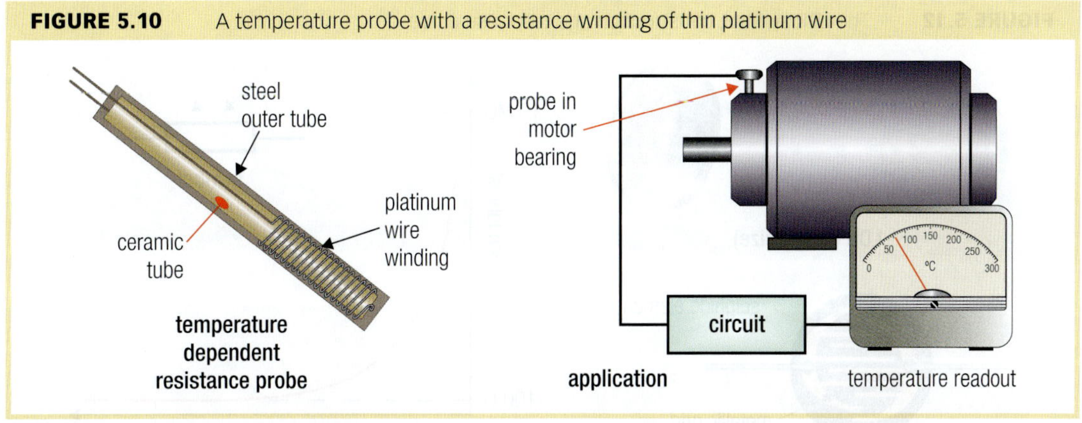

Figure 5.11 shows some typical thermistors, how their resistance changes with temperature and the symbols for NTC and PTC types. The curve for the PTC type shows that the resistance actually drops for a small rise in temperature, then rises rapidly once a certain temperature is reached.

FIGURE 5.11 Thermistors have many applications, and are either NTC or PTC devices packaged in disc or rectangular form

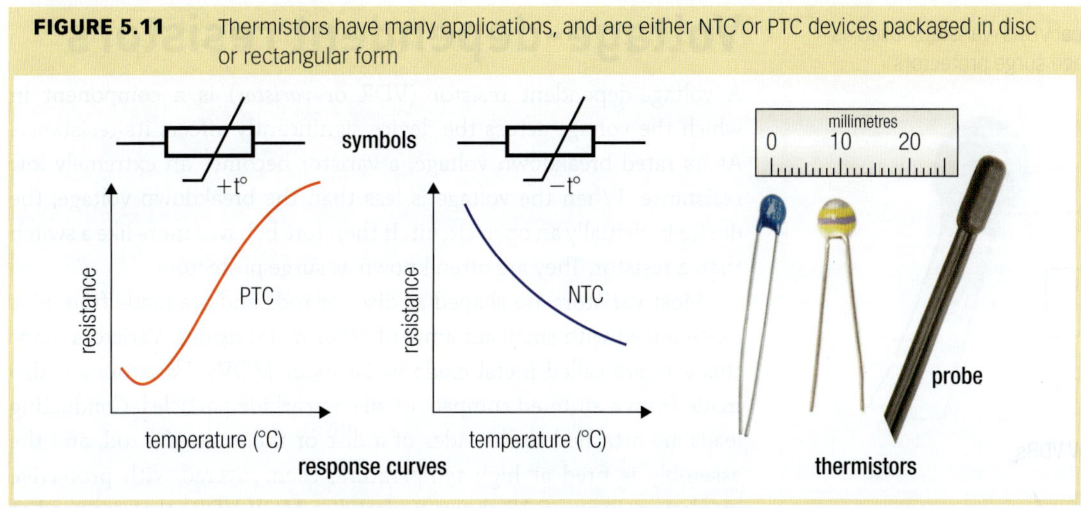

Thermistors are often used to protect an electric motor against overheating. In small electric motors, a PTC thermistor is embedded inside the windings and connected so the motor current passes through the thermistor. If the temperature of the windings rises above a certain value, the resistance of the thermistor will quickly increase (from a low resistance to several thousand ohms), reducing the motor current to almost zero.

In other words, the thermistor behaves like a switch that turns off when the temperature is too high. When the motor cools down, the resistance of the thermistor drops, letting the motor run. In larger motors, the thermistor is connected to a switch that cuts off power to the motor.

Light-dependent resistors

Light-sensitive devices include the light-dependent resistor (LDR), which consists of a thin ceramic disc sintered with cadmium sulphide. (Sintering is the process of using heat to combine two materials.) Cadmium sulphide is a photo-conductive material, in which its resistance is affected by light. A vacuum-deposited metallic grid is applied to the surface of the disc and the whole assembly is covered in clear plastic.

SUSTAINABILITY

LDRs are used to control lighting installations so lights are on only when needed

The resistance of an LDR in complete darkness is over 10 million ohms, which falls to less than 100 Ω in daylight. They are often used as a light sensor in automatically controlled lighting installations in which the LDR is arranged to operate a relay, which in turn switches the lights.

The construction, symbol and response curve of an LDR are shown in Figure 5.12. A typical LDR is about the size of a five-cent piece. As the response curve shows, the LDR is a non-linear resistor.

FIGURE 5.12 The LDR symbol, construction and response curve

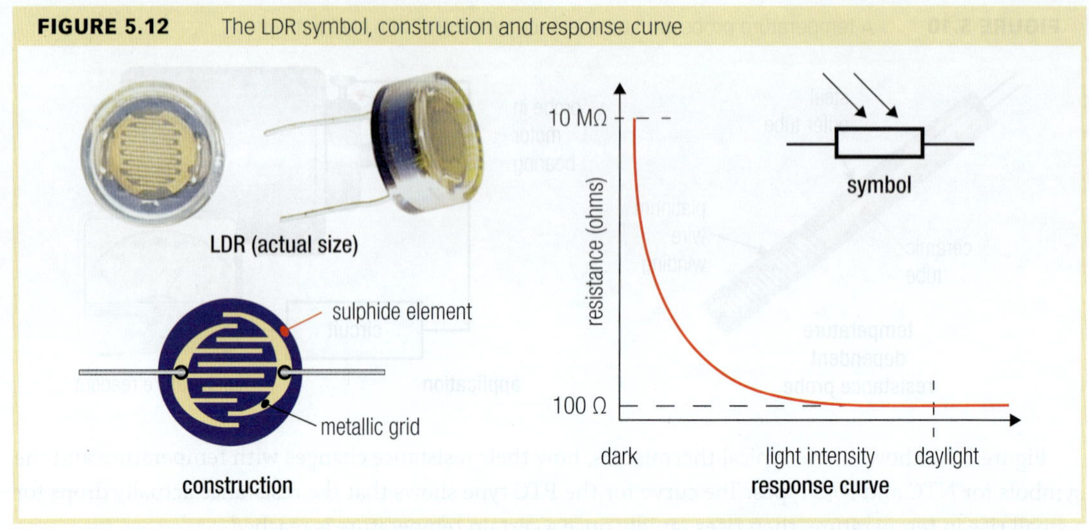

FIGURE 5.13 Typical VDRs (varistors) used as voltage surge protectors

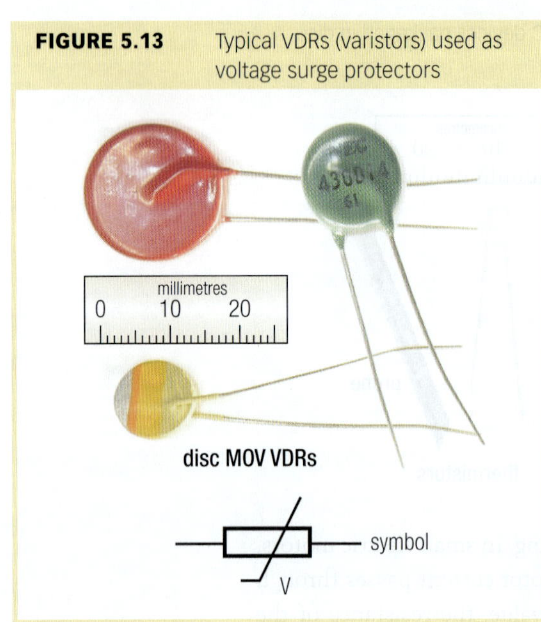

Voltage-dependent resistors

A voltage-dependent resistor (VDR or *varistor*) is a component in which the voltage across the device significantly affects its resistance. At its rated breakdown voltage, a varistor becomes an extremely low resistance. When the voltage is less than the breakdown voltage, the device is virtually an open circuit. It therefore behaves more like a switch than a resistor. They are often known as surge protectors.

Most varistors are shaped as discs or rods and are made from zinc oxide mixed with small amounts of other metal oxides. Varistors made this way are called metal oxide varistors or MOVs. Varistors are also made from a sintered compact of silicon carbide particles. Conducting leads are attached to the sides of a disc or the ends of a rod, and the assembly is fired at high temperature, then covered with protective insulation. Figure 5.13 shows typical disc MOV VDRs that are used as surge protectors in electronic equipment.

The main use of VDRs is to protect a circuit from a voltage surge. When the voltage across a VDR goes over a certain value, its resistance drops within milliseconds to a very low value. If the voltage surge lasts long enough, the large current flowing in the VDR will operate the circuit fuse (or trip a circuit breaker) and isolate the circuit from the supply. If the surge is very brief, the energy in the surge will be dissipated by the VDR before any overload protection operates.

KEY CONCEPT

Surge protector plugs are fitted with a VDR

A VDR is therefore given two ratings: the breakdown voltage and the energy (in joules) it can dissipate. Voltage ratings vary from 5 V to several hundred volts. A typical voltage rating for a VDR connected across the 230 V AC mains supply is 275 V. A VDR is connected across (or in parallel with) the circuit it is protecting, as illustrated in Figure 5.14. Notice that the VDR is connected *after* the fuse.

FIGURE 5.14 A VDR is connected in parallel and the rise in voltage allows a fault current to flow which ruptures the fuse

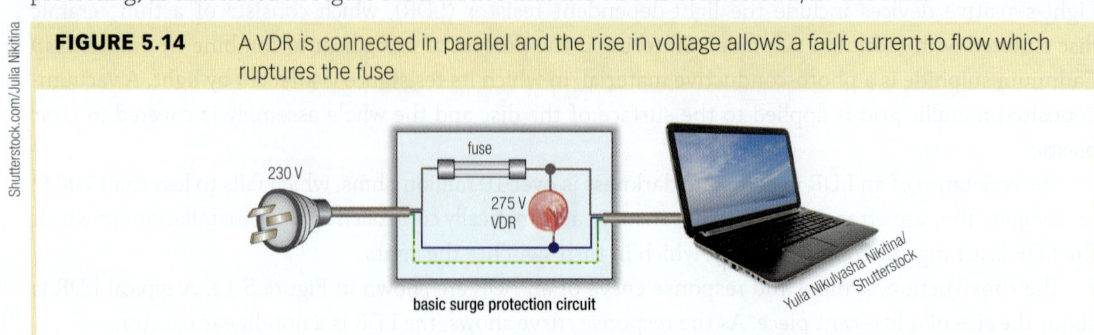

Surge arrestors

In electrical power generation, transmission and distribution, equipment is protected against lightning strikes by devices called lightning or *surge arrestors*. These are similar in construction to the VDRs described above, except they are much larger and can absorb high amounts of energy. They are connected between power lines and a metal stake driven into the ground. If the lines are struck by lightning, the arrestors quickly operate like a switch and divert the voltage surge to the ground, minimising or preventing damage to equipment. The response time of a surge arrestor is very fast – important if it's to provide protection from the voltage surge caused by a lightning strike.

FIGURE 5.15 High voltage power line surge arrestor

KEY POINTS...

- Resistors are rated by their resistance and power handling capability.
- High-power resistors have either a wire-wound or a metal grid resistance element.
- Low-power (up to 5 W) resistors typically have a carbon-film resistance element.
- A variable resistor is adjusted mechanically, and has either a wire-wound or carbon track resistance element, which can be linear or non-linear.
- A temperature-dependent resistor changes resistance with temperature.
- If the resistance *increases* with a temperature increase, the device has a positive temperature coefficient (PTC). If it *decreases*, it has a negative temperature coefficient (NTC).
- Temperature sensors using this effect are either the resistive element wire-wound type (linear PTC) or a thermistor (NTC or PTC).
- A light-dependent resistor has a cadmium sulphide element, which in full darkness has a high resistance. In daylight the resistance falls to around 100 Ω.
- A varistor (VDR) changes from a high resistance to a very low resistance when its breakdown voltage is exceeded. When the voltage is removed, the device returns to being a high resistance.
- Small metal oxide varistors (MOV) are used as surge protectors in electronic equipment. They are rated by breakdown voltage (5 V to over 500 V) and by their ability to dissipate short bursts of energy.
- Large metal oxide varistors are used as surge arrestors in the electrical power industry.

5.3 Resistor Colour Code

The value of a resistor is either printed on its body as a numerical value or, in the case of low-power resistors, as a code of coloured bands. Each colour represents a number and the resistance value is determined by 'reading' the colour code. There are two types of resistors that use colour codes: general purpose resistors and precision resistors. The difference between these two resistors is their *tolerance* value.

Tolerance refers to how close the actual value of the resistor is to its marked value. For example, a resistor might be coded as having a resistance of 1000 ohms with a tolerance of ±10 per cent. This means the maker guarantees the actual value won't be more than 10 per cent higher or lower than 1000 ohms. As 10 per cent of 1000 is 100, the actual value of the resistor could be anywhere between 900 ohms (1000 − 100) to 1100 ohms (1000 + 100).

General purpose resistors are those with a tolerance of ±5 per cent or more. Precision resistors have a tolerance of less than 5 per cent, usually ±2 per cent or ±1 per cent, even as low as ±0.05 per cent.

General purpose resistors show their value with four bands, and precision resistors have five bands. The last band is always the tolerance band, and is spaced further apart from the other bands. Figure 5.16 shows these two types of resistors and what each band represents. There are no standards concerning the body colour of a resistor, although some manufacturers use body colour to identify power rating.

FIGURE 5.16 Colour bands for four-and five-band resistors

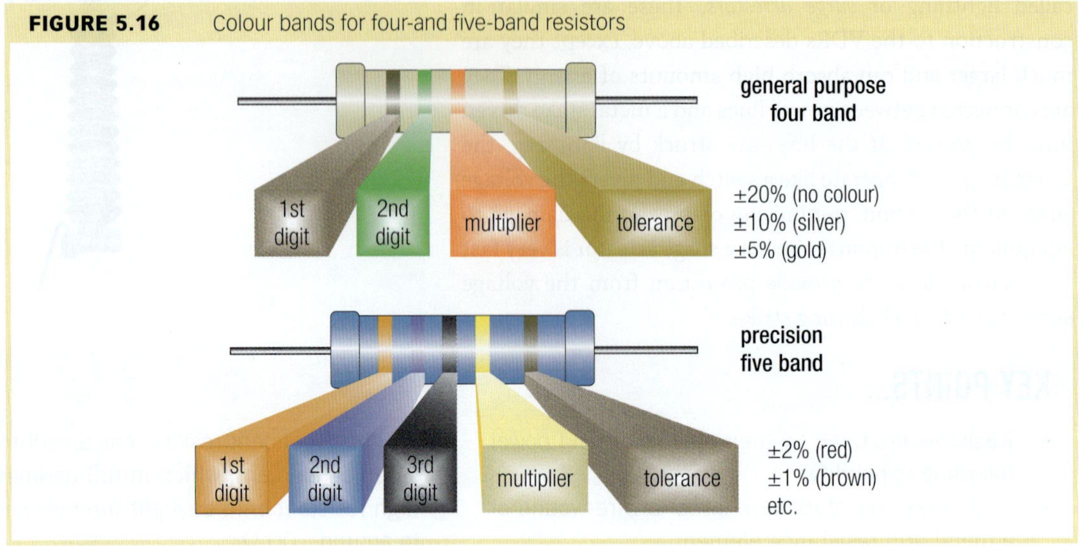

Figure 5.17 shows the 12 colours used in the resistor colour code. Gold and silver are only used to show a tolerance value.

The examples in Figure 5.18 show how to use the colour code. The first two bands (or three for precision resistors) are given a number, the next band is the multiplier and the last (on the right) is the tolerance band.

FIGURE 5.17 Resistor colour code

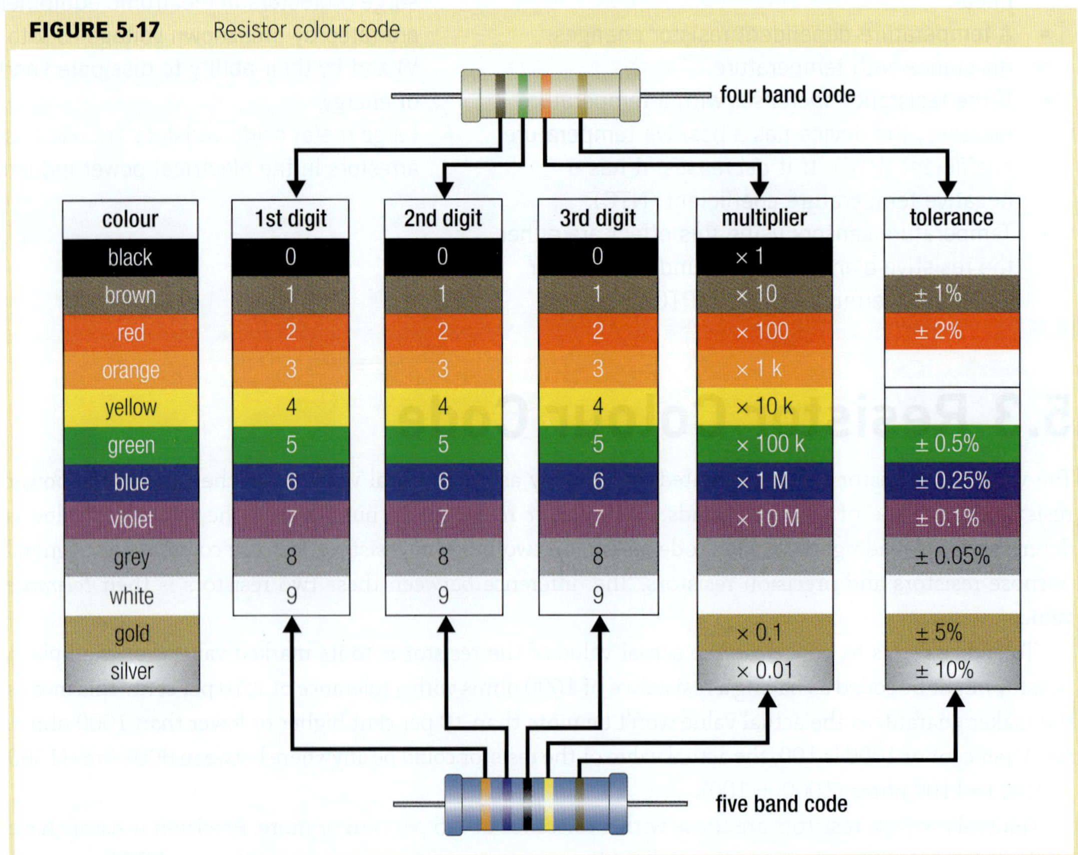

colour	1st digit	2nd digit	3rd digit	multiplier	tolerance
black	0	0	0	× 1	
brown	1	1	1	× 10	± 1%
red	2	2	2	× 100	± 2%
orange	3	3	3	× 1 k	
yellow	4	4	4	× 10 k	
green	5	5	5	× 100 k	± 0.5%
blue	6	6	6	× 1 M	± 0.25%
violet	7	7	7	× 10 M	± 0.1%
grey	8	8	8		± 0.05%
white	9	9	9		
gold				× 0.1	± 5%
silver				× 0.01	± 10%

FIGURE 5.18 Using the resistor colour code

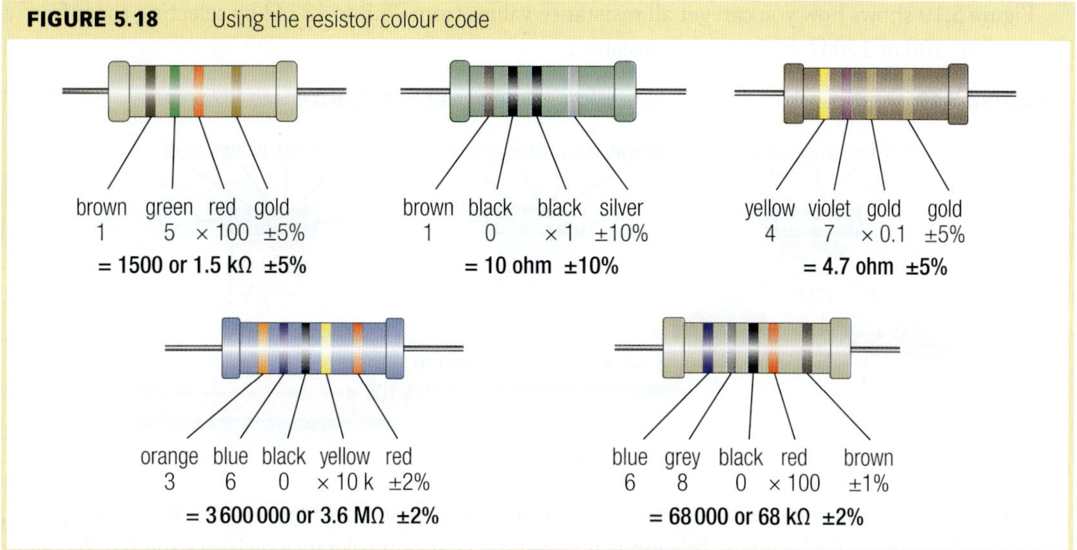

Here are a few points that will help you use the colour code. Practice soon makes the process easy. If in doubt (as is often the case with precision resistors), check the resistance with an ohmmeter!

- The tolerance band is spaced further apart from the other bands.
- The digit bands are generally closer to one end of the resistor. They're read from left to right and the band nearest the end of the resistor is the first digit.
- The first band can never be black, gold or silver.
- Resistors with four bands have either a gold (±5%) or, less often, a silver (±10%) tolerance band.
- Resistors with five bands usually have either a red (±2%) or brown (±1%) tolerance band.
- You might be able to remember the colour sequence with this mnemonic:
  Black **B**erries **R**oam **O**ver **Y**our **G**arden **B**ut **V**iolets **G**row **W**ild

Preferred resistor values

Low-power resistor values range from 0.01 ohms to over 80 million ohms. It's not economical for resistor manufacturers to make every possible value; instead a standard set of values, called the *preferred values*, is made. For a tolerance of ±10 per cent, there are 12 base values per decade. A decade range is, for example, all the numbers between 1 and 9, or between 10 and 99, 100 to 999 and so on. Table 5.4 lists all 120 resistor values in the E12 range. These are commonly available from electronic parts suppliers.

TABLE 5.4

E12 PREFERRED RESISTOR VALUES

Decade	Range of preferred values											
× 0.01	0.01	0.012	0.015	0.018	0.02	0.027	0.033	0.039	0.047	0.056	0.068	0.082
× 0.1	0.1	0.12	0.15	0.18	0.22	0.27	0.33	0.39	0.47	0.56	0.68	0.82
× 1	**1**	**1.2**	**1.5**	**1.8**	**2.2**	**2.7**	**3.3**	**3.9**	**4.7**	**5.6**	**6.8**	**8.2**
× 10	10	12	15	18	22	27	33	39	47	56	68	82
× 100	100	120	150	180	220	270	330	390	470	560	680	820
× 1 k	1 k	1.2 k	1.5 k	1.8 k	2.2 k	2.7 k	3.3 k	3.9 k	4.7 k	5.6 k	6.8 k	8.2 k
× 10 k	10 k	12 k	15 k	18 k	22 k	27 k	33 k	39 k	47 k	56 k	68 k	82 k
× 100 k	100 k	120 k	150 k	180 k	220 k	270 k	330 k	390 k	470 k	560 k	680 k	820 k
× 1 M	1 M	1.2 M	1.5 M	1.8 M	2.2 M	2.7 M	3.3 M	3.9 M	4.7 M	5.6 M	6.8 M	8.2 M
× 10 M	10 M	12 M	15 M	18 M	22 M	27 M	33 M	39 M	47 M	56 M	68 M	82 M

Figure 5.19 shows how you can get all resistance values from 73.8 to 132 Ω by selecting them from a batch of 82, 100 or 120 Ω ±10 per cent resistors.

FIGURE 5.19 These three 10 per cent resistors give all values from 73.8 ohms to 132 ohms

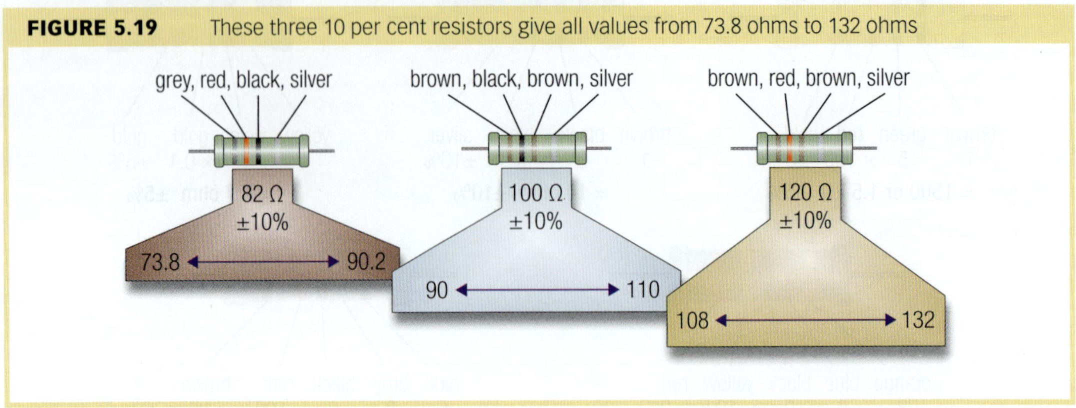

While the E12 preferred resistance values suit most applications, a wider range of resistors is made, called the *E24 preferred value range*. This range is based on 5 per cent tolerance resistors and has 24 values in each decade. The values are shown below. Notice how this range includes the E12 range (in italic) with 12 extra values (underneath them in bold), giving a total of 24 resistance values per decade.

TABLE 5.5

THE E24 RANGE OF RESISTORS												
E12:	*1*	*1.2*	*1.5*	*1.8*	*2.2*	*2.7*	*3.3*	*3.9*	*4.7*	*5.6*	*6.8*	*8.2*
	1.1	**1.3**	**1.6**	**2.0**	**2.4**	**3**	**3.6**	**4.3**	**5.1**	**6.2**	**7.5**	**9.1**

TASK 5.2

1 Use the resistor colour code to determine the values of the resistors in Figure 5.20 below.

FIGURE 5.20 Practice exercise for using the resistor colour code

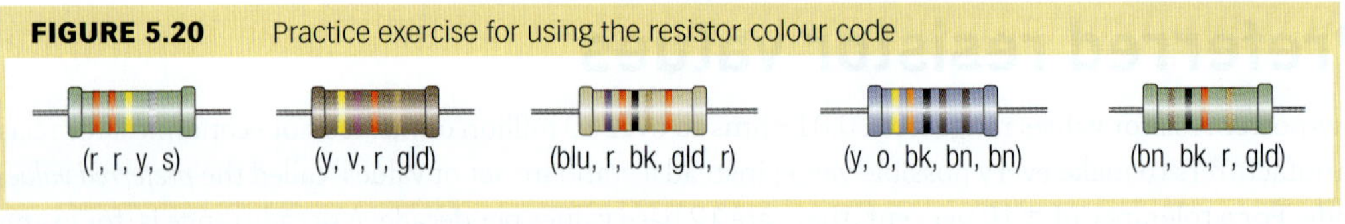

5.4 Measuring Resistance

Resistance is measured with an *ohmmeter*, which can be either an analog or, more typically, a digital meter. An ohmmeter is usually part of a multimeter, in which one or two switches select the range and function (volts, current or resistance). There are two important things to be aware of when measuring resistance with an ohmmeter:

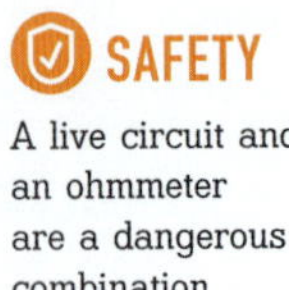
SAFETY

A live circuit and an ohmmeter are a dangerous combination

1 If you are measuring resistance in a circuit, make sure the circuit is not connected to any type of power source, in particular 230 V mains. If there is a voltage in the circuit, it can cause damage to the meter, causing it to possibly explode, catch fire or cause personal injury.
2 While measuring the resistance of a component, do not touch both probes at the same time, as the reading will be affected by your body resistance.

To measure resistance with a multimeter

1 Select the resistance (ohms or Ω) function.
2 For an analog ohmmeter, zero the pointer for both no deflection and full deflection, as shown in Figure 5.21. Make the full deflection adjustment each time you change the range.

FIGURE 5.21 Adjust the mechanical zero if the pointer is not at infinite ohms when the probes are apart. With the probes touching, adjust the meter's 'zero ohm' control for 0 Ω.

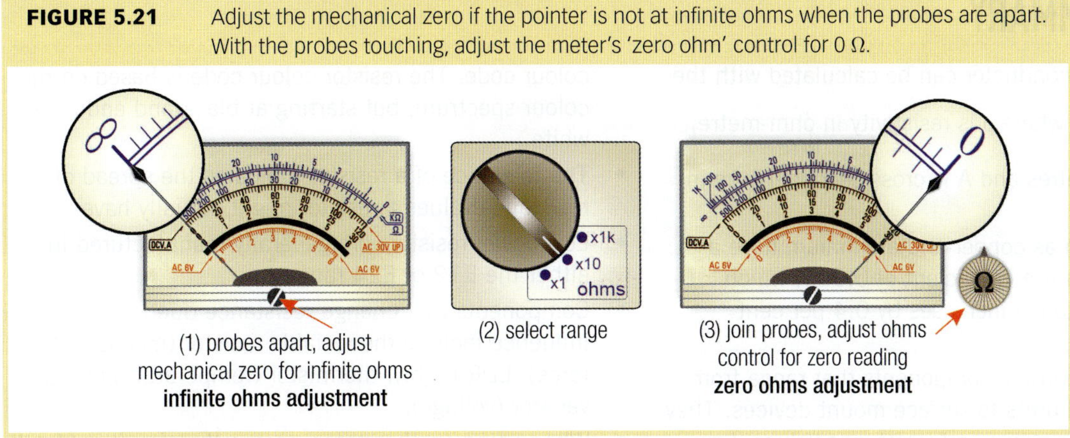

3 Select a range. Digital multimeters are usually auto-ranging, while an analog multimeter always requires a range to be selected. For an analog meter, the best range will give a reading between a quarter and three-quarters of full scale.

4 Connect the ohmmeter probes across the resistor, or to the points in a circuit, and if necessary, change to the range that gives the clearest reading.

5 Read the resistance value as shown in Figure 5.22. For an analog meter, the resistance reading is the scale reading multiplied by the selected range. For example, a reading of 50 on the scale for a range of ohms × 10 means the resistance is 50 × 10, or 500 Ω.

FIGURE 5.22 Analog and digital multimeters measuring the resistance of a 220 Ω resistor

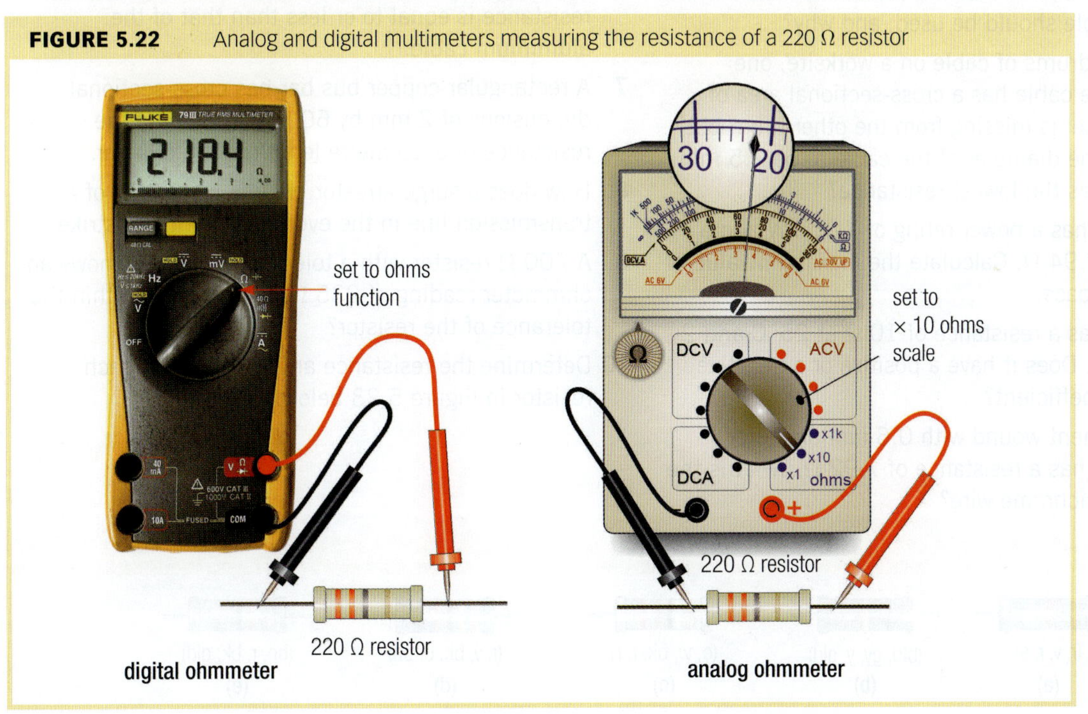

CHAPTER SUMMARY

- Resistance of a conductor can be calculated with the equation $R = \frac{\rho l}{A}$, where ρ is resistivity in ohm-metre, l is length in metres and A is cross-sectional area in square metres.
- Conductors such as copper and aluminium have a positive temperature coefficient of 0.004, which means the resistance increases by 0.4 per cent per 1 °C.
- Resistors are electrical components that range from large, metal grid units to surface mount devices. They are rated by their power and resistance values. Low-power resistors have their resistance identified with a colour code. The resistor colour code is based on the colour spectrum, but starting at black and ending at white.
- The tolerance of a resistor indicates the spread of resistance values a resistor might actually have.
- Low-power resistors are generally manufactured in either the E12 or E24 ranges.
- Components that change resistance due to an external influence include the variable resistor (mechanical force), LDR (light), thermistor (temperature) and the varistor (voltage).
- When using an ohmmeter, ensure the circuit is 'dead' and only touch one metal probe at a time.

REVIEW EXERCISES

Check your answers at the back of the book.

1. A 200 m long copper cable between two buildings must have a resistance of no more than 1.5 Ω. The choice of cable size is 1.5 mm², 2.5 mm² and 4 mm². Which size cable should be used, and why?
2. There are two drums of cable on a worksite, one stating that the cable has a cross-sectional area of 4 mm².The label is missing from the other drum but you measure the diameter of the cable to be 2.5 mm. Which cable has the lowest resistance?
3. A grid resistor has a power rating of 7.35 kW and a resistance of 2.94 Ω. Calculate the maximum value of current it can pass.
4. A thermistor has a resistance of 10 Ω at 25 C and 2 ohms at 80 C. Does it have a positive or a negative temperature coefficient?
5. A heating element wound with 0.6 mm diameter nichrome wire has a resistance of 7.92 Ω. What is the length of the nichrome wire?
6. An aluminium cable with a cross-sectional area of 8 mm² and a length of 2 kilometres is to be replaced with a copper cable. Calculate the minimum csa of the copper cable needed to ensure the cable resistance is equal to or less than that of the aluminium cable.
7. A rectangular copper bus bar has cross-sectional dimensions of 2 mm by 60 mm. Calculate the resistance of a 10 metre length of the bus bar.
8. How does a surge arrestor provide protection of a transmission line in the event of a lightning strike?
9. A 200 Ω resistor with a tolerance of ±10% shows an ohmmeter reading of 235 Ω. Is this value within the tolerance of the resistor?
10. Determine the resistance and tolerance of each resistor in Figure 5.23 below.

FIGURE 5.23

(r, v, r, s)
(a)

(blu, gy, y, gld)
(b)

(o, w, bk, r, r)
(c)

(r, y, bk, o, bn)
(d)

(bn, r, bk, gld)
(e)

ONLINE RESOURCES

COMPLETE WORKSHEET FIVE

Check with your instructor for worksheets on this chapter.

CHAPTER 6

THE SERIES CIRCUIT

Circuits or electrical installations usually have several components that are connected in a particular way. This chapter explains the series circuit, in which each component in the circuit is connected one after the other. It describes how to apply Ohm's law to calculate current, voltage and resistance values in this type of circuit, and to determine the power taken by the whole circuit, and by each component in the circuit. We also look at faults and their effects in a series circuit.

CHAPTER OUTLINE

6.1 The Series Circuit

The term 'connected in series' came up when we explained that an ammeter is connected in series to measure current, and that cells are connected in series to make a battery. Components connected in series are connected one after the other. Figure 6.1 shows four lamps and two cells, all connected in series.

FIGURE 6.1 Lamps and cells connected in series

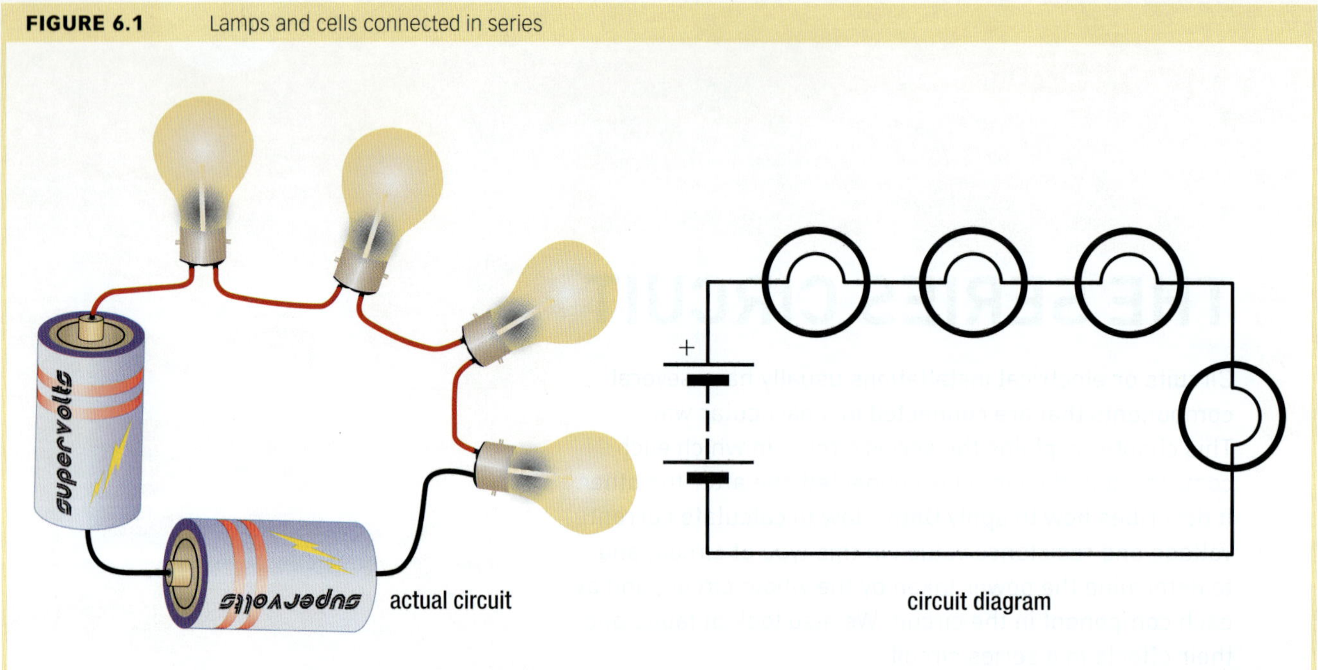

In a series circuit, as Figure 6.1 shows, each component has only *one* conductor connected to each terminal of the component. Because a component must have two terminals – one for the current to enter and the other for the current to flow out of the component – there can only ever be two conductors connected to each component in a series circuit.

FIGURE 6.2 In a series circuit there is only one path for the current

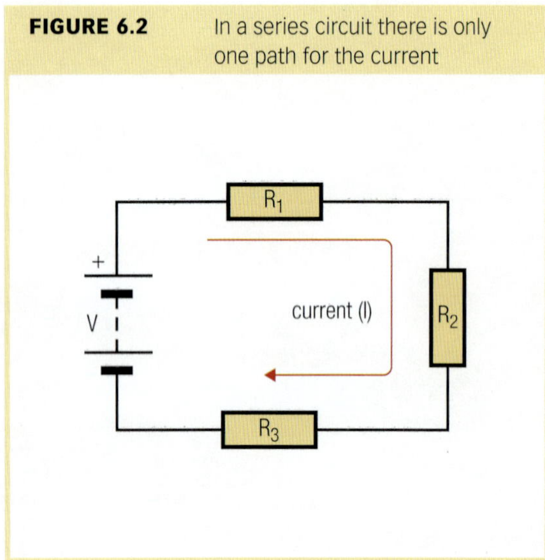

6.2 Current in the Series Circuit

The circuit shown in Figure 6.2 has three resistors connected in series with a battery. Notice that there is only *one* path for the current, which is how a series circuit is defined.

Figure 6.3 shows a series circuit of three resistors, each of different values. An ammeter is connected after each resistor, to measure the current flowing in that resistor. The ammeters are also connected in series, so the whole circuit now has three resistors and three ammeters, all in series, with two cells also connected in series.

As you might expect, the ammeters all read the same value of current. In other words, *current is the same in all parts of a series circuit.* Therefore, it doesn't matter where an ammeter is connected in a series circuit, it will always read the same value of current.

FIGURE 6.3 Current is the same in all parts of a series circuit

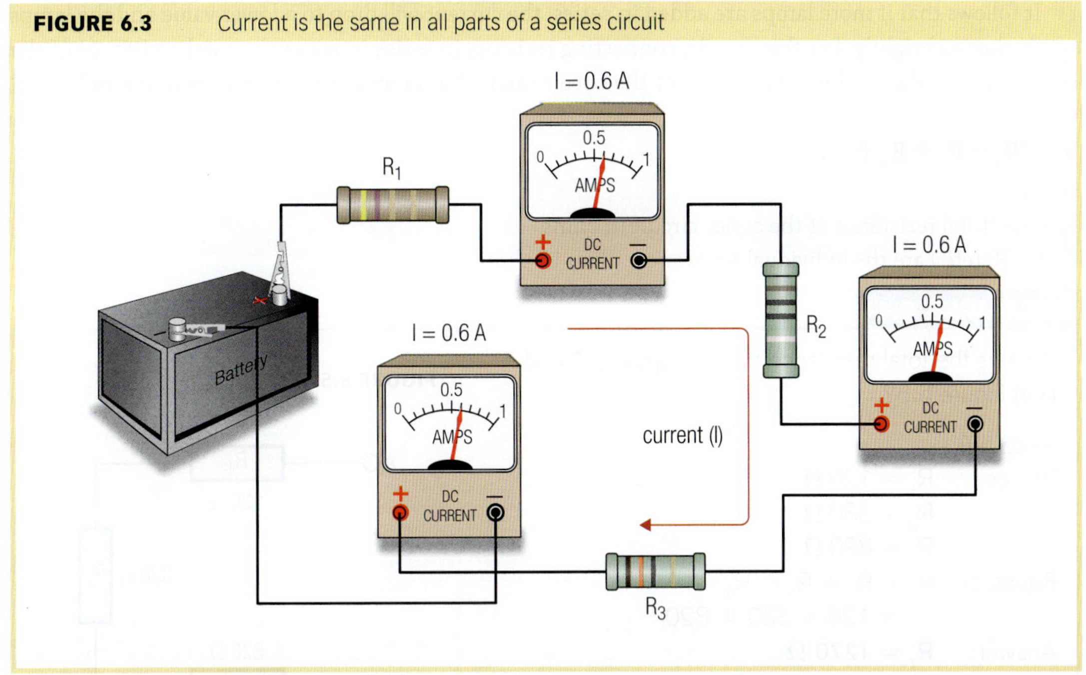

KEY POINTS...

- A series circuit has only one path for the current.
- Current is the same in all parts of a series circuit.

6.3 Resistance in the Series Circuit

In any circuit, the current depends on the resistance of the circuit and the value of the applied voltage. If two identical lamps are connected in series, there will be more resistance in the circuit than in a circuit with one lamp. Therefore the current will be less in the circuit that has two lamps.

This is shown in Figure 6.4, where the current is 1 A when one lamp is in the circuit. The lamp glows at its normal brightness. If a second lamp (with the same resistance as the first) is connected in series with the other lamp, the current falls to 0.5 A. Both lamps glow less brightly as the current is now half its previous value. Because the voltage hasn't changed, the resistance must have doubled.

 **FYI**

For the purposes of explanation, we are ignoring the fact that an incandescent lamp has a higher resistance when it heats up

FIGURE 6.4 Adding resistances in series increases the total circuit resistance

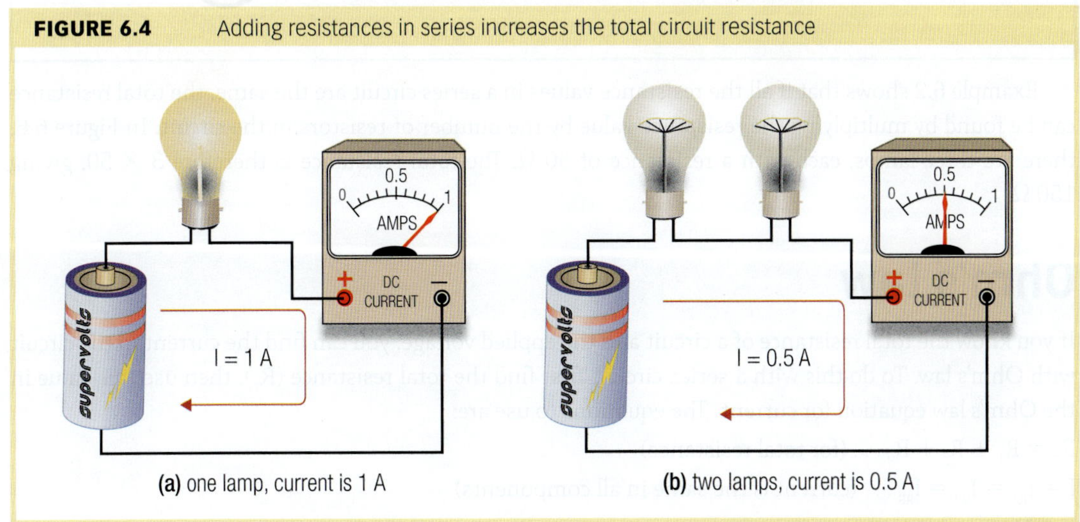

(a) one lamp, current is 1 A

(b) two lamps, current is 0.5 A

KEY CONCEPT

Resistors in series

It follows that if more lamps are added in series, the current will drop to a lower value and the lamps won't glow as brightly. In other words, connecting resistors in series *increases* the total resistance of the circuit. In fact, the total resistance (R_T) of the circuit can be found by adding the values of the individual resistors.

$$\mathbf{R_T = R_1 + R_2 + R_3 + \ldots}$$

where:

R_T is the total resistance of the series circuit in ohms

R_1, R_2, R_3 (etc.) are the individual resistance values in ohms.

EXAMPLE 6.1

What is the total resistance between points A and B in Figure 6.5?

Solution

Values $R_1 = 120\ \Omega$

$R_2 = 330\ \Omega$

$R_3 = 820\ \Omega$

Equation $R_T = R_1 + R_2 + R_3$

$= 120 + 330 + 820$

Answer $\mathbf{R_T = 1270\ \Omega}$

FIGURE 6.5

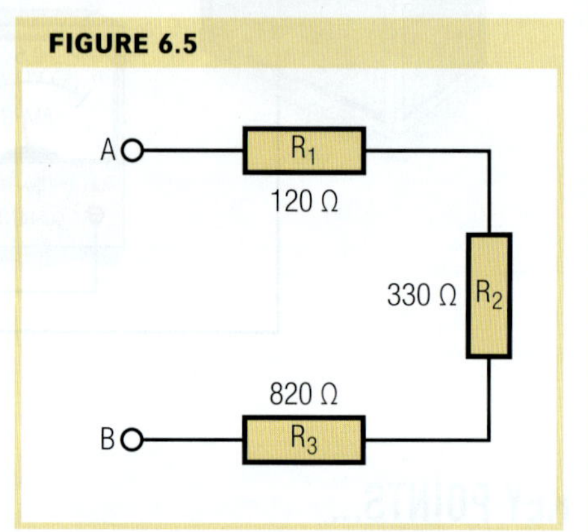

EXAMPLE 6.2

Find the total resistance between points A and B in Figure 6.6.

Solution

Values R_1, R_2 and $R_3 = 50\ \Omega$

Equation $R_T = R_1 + R_2 + R_3$

$= 50 + 50 + 50$

Answer $\mathbf{R_T} = 150\ \mathbf{\Omega}$

FIGURE 6.6

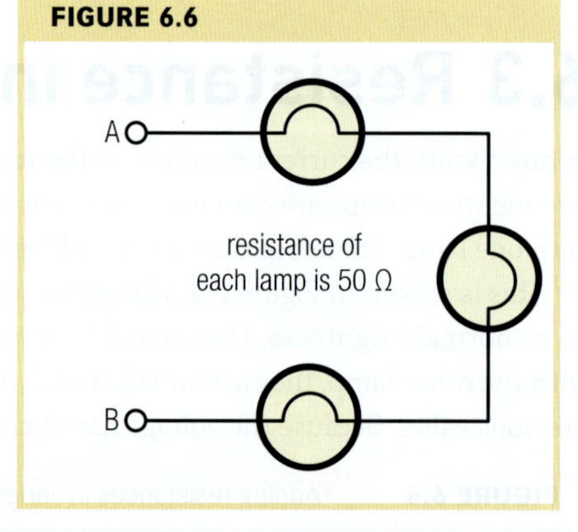

Example 6.2 shows that if all the resistance values in a series circuit are the same, the total resistance can be found by multiplying the resistance value by the number of resistors in the circuit. In Figure 6.6, there are three lamps, each with a resistance of 50 Ω. The total resistance is therefore 3 × 50, giving 150 Ω.

Ohm's law

If you know the total resistance of a circuit and the applied voltage, you can find the current in the circuit with Ohm's law. To do this with a series circuit, first find the total resistance (R_T), then use this value in the Ohm's law equation for current. The equations to use are:

$R_T = R_1 + R_2 + R_3 \ldots$ (for total resistance)

$I = I_{R1} = I_{R2} = I_{R3} \ldots$ (current is the same in all components)

$I = \dfrac{V_T}{R_T}$ (to find the current).

EXAMPLE 6.3

Calculate the current in Figure 6.5, if the applied voltage is 230 V.

Solution

Values $R_T = 1270\ \Omega$ (as found in Example 6.1)
$V_T = 230$ V

Equation $I = \frac{V_T}{R_T} = \frac{230}{1270}$

Answer $I =$ **0.18 A**

6.4 Voltage in the Series Circuit

The circuit in Figure 6.7 shows three resistors connected in series with a 20 V DC power supply. In this circuit:

- V_T is the applied voltage (the voltage of the voltage source) which causes a current to flow in the circuit.
- V_1, V_2 and V_3 are voltage *drops*. A voltage drop is caused by current flowing through a resistance. Its value is directly proportional to the value of the resistance (the higher the resistance, the higher the voltage drop for the same value of current).

To calculate the voltage drop across each resistor, first find the current (I) in the circuit, then find the voltage drop across each resistor with Ohm's law. Example 6.4 explains.

EXAMPLE 6.4

Calculate the voltage drops across the three resistors in Figure 6.7.

Solution

Values $V_T = 20$ V
$R_1 = 20\ \Omega$
$R_2 = 30\ \Omega$
$R_3 = 50\ \Omega$
$V_1, V_2, V_3 = ?$

Equations $R_T = R_1 + R_2 + R_3$

$I = \frac{V_T}{R_T}$

$V_1 = IR_1, V_2 = IR_2, V_3 = IR_3$

Working $R_T = 20 + 30 + 50 = 100\ \Omega$

$I = \frac{20}{100} = 0.2$ A

Answers $V_1 = 0.2 \times 20 =$ **4 V**
$V_2 = 0.2 \times 30 =$ **6 V**
$V_3 = 0.2 \times 50 =$ **10 V**

FIGURE 6.7

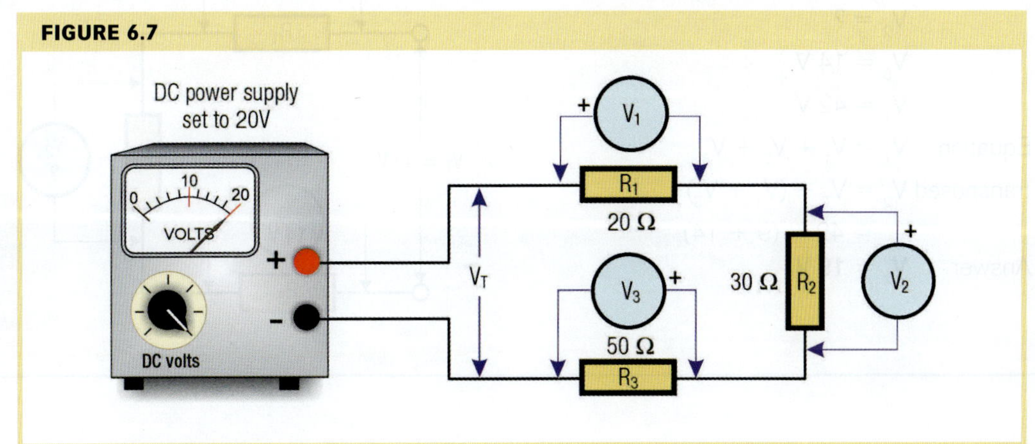

Kirchhoff's voltage law

KEY CONCEPT

Kirchhoff's voltage law

As you would expect, when you add up the voltage drops in Example 6.4, you get 20 V, which is the supply voltage. This is called Kirchhoff's voltage law (after Gustav Kirchhoff, a German physicist), which says:

The algebraic sum of the voltage drops in a series circuit equals the applied voltage.

This law is as important as Ohm's law. Like Ohm's law, it can never be wrong. The voltage drops around a series circuit can *never* total any more, or any less, than the applied voltage. The next two examples show how to use Kirchhoff's voltage law.

EXAMPLE 6.5

What is the value of the supply voltage (V_T) in the circuit shown in Figure 6.8?

Solution

Values $V_1 = 11$ V

$V_2 = 23$ V

$V_3 = 6$ V

$V_T = ?$

Equation $V_T = V_1 + V_2 + V_3$

$= 11 + 23 + 6$

Answer $\mathbf{V_T = 40\ V}$

FIGURE 6.8

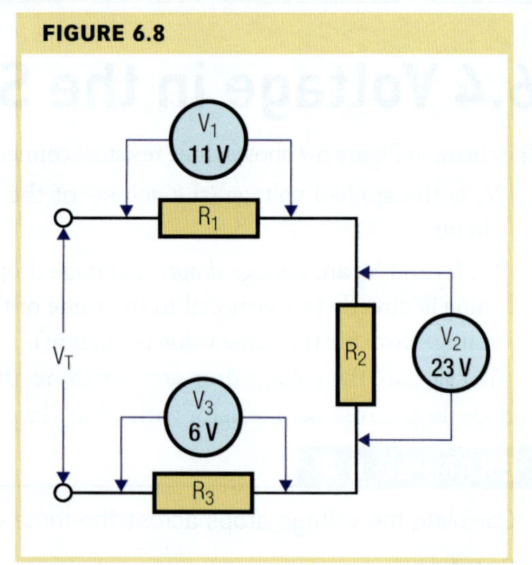

If the supply voltage and all but one of the voltage drops in a series circuit are known, you can calculate the unknown voltage drop by adding the known voltage drops and subtracting their sum from the value of the supply voltage.

EXAMPLE 6.6

Calculate the voltage drop across resistor R_2 in the circuit of Figure 6.9.

Solution

Values $V_1 = 9$ V

$V_2 = ?$

$V_3 = 14$ V

$V_T = 42$ V

Equation $V_T = V_1 + V_2 + V_3$

transposed $V_2 = V_T - (V_1 + V_3)$

$= 42 - (9 + 14)$

Answer $\mathbf{V_2 = 19\ V}$

FIGURE 6.9

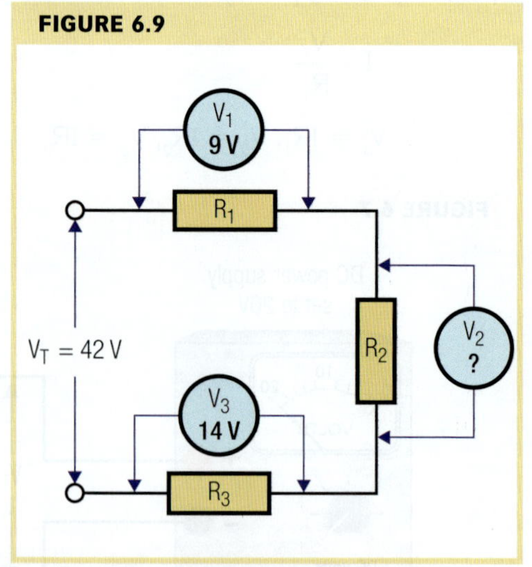

Ohm's law and Kirchhoff's voltage law

If you know (or measure) the voltage drop across a known resistance value in a series circuit, Ohm's law can be used to find the current flowing in this resistance. This also gives the circuit current, because the current in a series circuit is the same in all parts. Knowing the current and the supply voltage means you can calculate the total resistance of the circuit. The following examples demonstrate this.

EXAMPLE 6.7

Find the total resistance of the circuit in Figure 6.10.

FIGURE 6.10

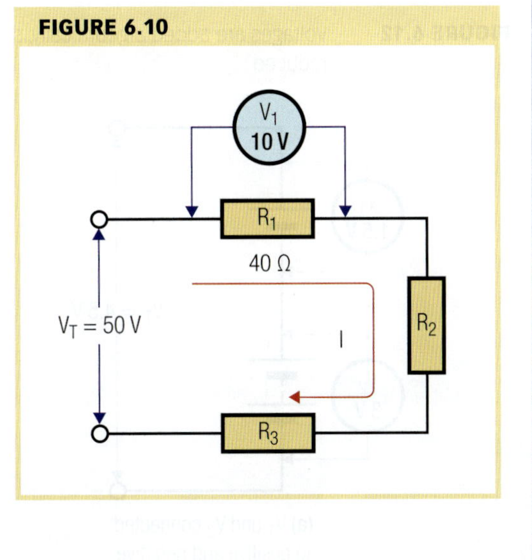

Solution

Values $V_T = 50\ V$

$R_1 = 40\ \Omega$

$V_1 = 10\ V$

$R_T = ?$

Equations $I = \frac{V_1}{R_1}$

$R_T = \frac{V_T}{I}$

Working $I = \frac{10}{40} = 0.25\ A$

$R_T = \frac{50}{0.25}$

Answer $\mathbf{R_T = 200\ \Omega}$

The values of R_2 and R_3 in Figure 6.10 can also be found, providing you know the voltage drop across either one of these resistors. Example 6.8 shows how to apply Kirchhoff's voltage law and Ohm's law to find these values. It assumes the same values of current and total resistance from Example 6.7.

EXAMPLE 6.8

If the total resistance of the circuit in Figure 6.11 is 200 Ω, what are the values of R_2 and R_3? Also, what's the voltage (V_2) across R_2?

FIGURE 6.11

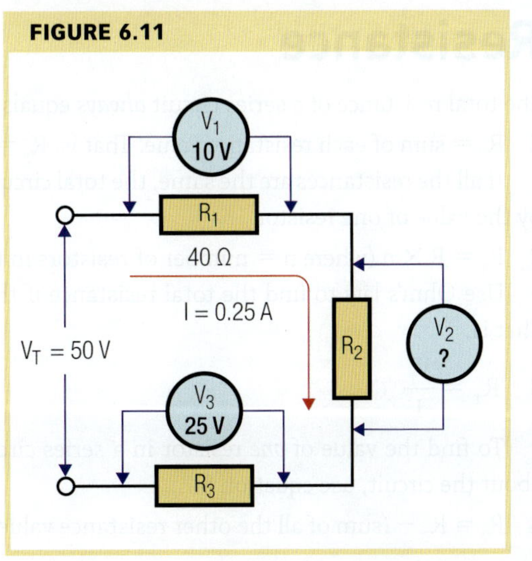

Solution

Values $R_T = 200\ \Omega$, $R_1 = 40\ \Omega$

$I = 0.25\ A$ (as found in Example 6.7)

$V_T = 50\ V$

$V_1 = 10\ V$, $V_3 = 25\ V$

Equations $R_3 = \frac{V_3}{I}$

$R_2 = R_T - (R_1 + R_3)$

$V_2 = V_T - (V_1 + V_3)$

Working $R_3 = \frac{25}{0.25}$

Answer $\mathbf{R_3 = 100\ \Omega}$

$R_2 = 200 - (40 + 100)$

Answer $\mathbf{R_2 = 60\ \Omega}$

$V_2 = 50 - (10 + 25)$

Answer $\mathbf{V_2 = 15\ V}$ (V_2 also equals $IR_2 = 0.25 \times 60 = 15\ V$)

Voltage sources

Kirchhoff's voltage law says that voltages are added *algebraically*, which means you have to consider the *polarity* of the voltage. For example, as shown in Figure 6.12(a), when cells are connected in series, they are arranged so the positive terminal of a cell connects to the negative terminal of the next cell. The total voltage is the sum of the individual cell voltages.

Figure 6.12(b) shows the effect if a cell is connected in reverse to the others. As expected, the total voltage is reduced, as the voltage of the reversed cell is subtracted from the other cell voltages.

FIGURE 6.12 Voltages are added algebraically, so a reversed cell will cause the applied voltage to be reduced

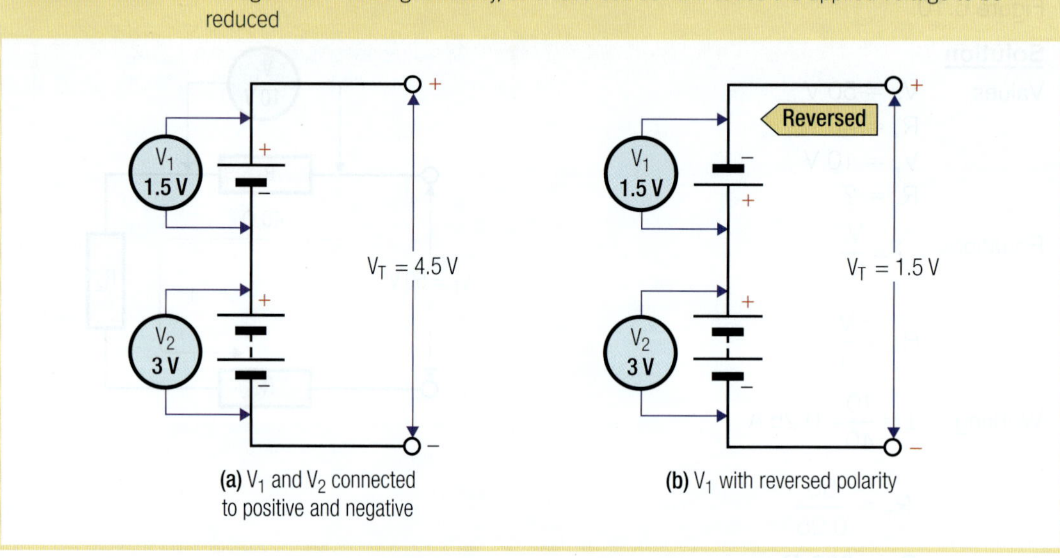

6.5 Summary of the Series Circuit

As we've shown, current, voltage and resistance values in a series circuit can be calculated in various ways, depending on what you know about the circuit. Here's a summary.

Resistance

KEY CONCEPT

Total resistance of a series circuit is the sum of the individual resistances

The total resistance of a series circuit *always* equals the sum of the individual resistance values.

1 R_T = sum of each resistance value. That is, $R_T = R_1 + R_2 + R_3 + \ldots$

If all the resistances are the same, the total circuit resistance equals the number of resistors multiplied by the value of one resistor.

2 $R_T = R \times n$ (where n = number of resistors in the circuit and R = value of one resistor).

Use Ohm's law to find the total resistance if the applied voltage and the circuit current are known. That is,

3 $R_T = \dfrac{V_T}{I}$

To find the value of *one* resistor in a series circuit use equation 4, or depending on what you know about the circuit, use equation 5.

4 $R_? = R_T -$ (sum of all the other resistance values)

5 $R_1 = \dfrac{V_1}{I}$, $R_2 = \dfrac{V_2}{I}$ and so on

Current

The current in a series circuit is always *the same* in all parts of the circuit and can be found with Ohm's law. Use equation 6 if you know the applied voltage and the total resistance of the circuit.

6 $I = \frac{V_T}{R_T}$

If you know the voltage across any *one* resistance in the circuit and the value of that resistance, use equation 7 to find the current.

7 $I = \frac{V_1}{R_1}, = \frac{V_2}{R_2}$ and so on

To measure circuit current, connect an ammeter in *any part* of a series circuit. If you know the current in one component, you then know the current in all others in the circuit.

KEY CONCEPT

Current is the same in all parts of a series circuit

Voltage

A series circuit has an applied voltage (V or V_T for total voltage) and a number of voltage *drops* (V_1, V_2 and so on). A voltage drop is caused by current flowing in a resistance – the applied voltage causes the current in the first place. By Kirchhoff's voltage law, the sum of the voltage drops in a series circuit *always* equals the applied voltage. Use Kirchhoff's voltage law or Ohm's law to find either the applied voltage or a voltage drop, depending on what you know about the circuit. Use equations 8 or 9 to find the applied voltage (V_T).

8 $V_T = V_1 + V_2 + V_3 + \ldots$ (Kirchhoff's voltage law)

9 $V_T = I \times R_T$ (Ohm's law)

To find a voltage drop ($V_?$) with Kirchhoff's voltage law, equation 8 is transposed in terms of the unknown voltage. That is:

10 $V_? = V_T -$ (sum of all the other voltage drops in the circuit)

Ohm's law can also be used, where:

11 $V_1 = I \times R_1, V_2 = I \times R_2$ and so on

KEY CONCEPT

The sum of the voltage drops in a series circuit equals the applied voltage

KEY CONCEPT

The highest voltage drop in a series circuit is across the highest resistance

TASK 6.1

1 In Figure 6.13(a), find current I.

2 In Figure 6.13(b), find resistance R_2.

3 In Figure 6.13(c), find the voltage drop across resistor R_2.

FIGURE 6.13

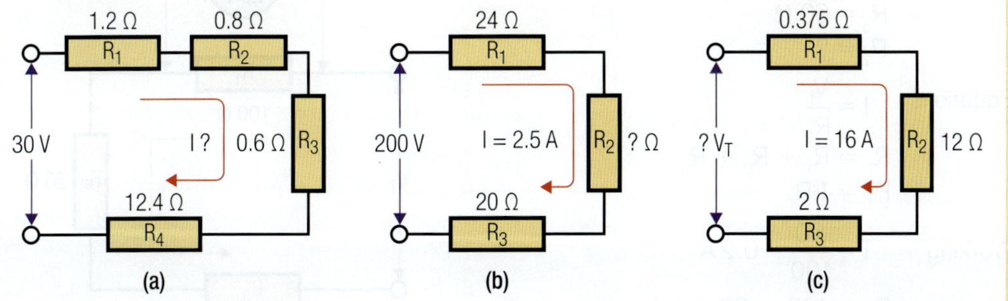

4 In Figure 6.14(a), find the supply voltage V_T.

5 In Figure 6.14(b), find the values of resistances R_2, R_3 and R_4 and the total resistance (R_T) of the circuit.

FIGURE 6.14

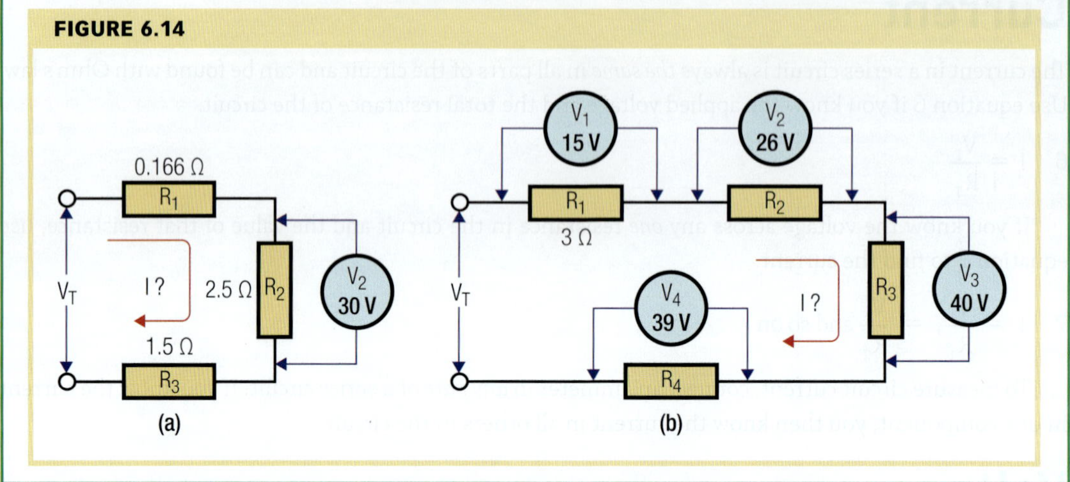

6.6 Power in the Series Circuit

The total power, or the power taken by an individual component in a series circuit, can be found with the power equations given in Chapter 4, combined with Ohm's law and the equations introduced in this chapter.

Calculating total power

If the circuit current (I) and the supply voltage (V) are known, the total power (P_T) taken by a circuit is the product of the current I and the supply voltage. That is, $P_T = VI$ watts. This applies to all types of circuits. Another way, if you know the current and resistance of the circuit, is with the equation $P_T = I^2R_T$. As Example 6.9 shows, sometimes you also have to find the circuit current and total resistance from values associated with one component in the circuit.

EXAMPLE 6.9

Find the total power taken by the circuit in Figure 6.15.

FIGURE 6.15

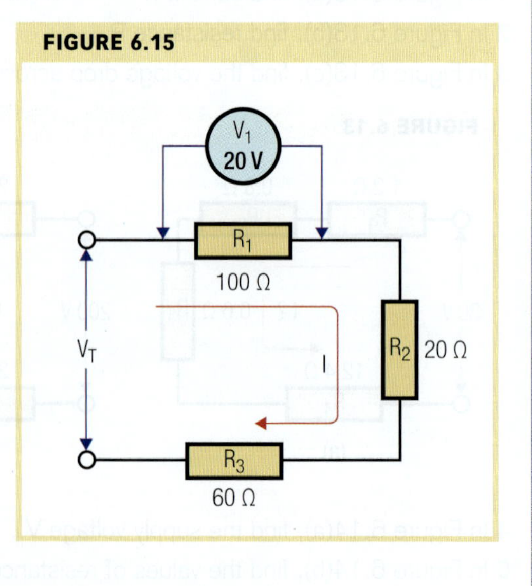

Solution

Values $V_1 = 20\ \text{V}$

$R_1 = 100\ \Omega$

$R_2 = 20\ \Omega$

$R_3 = 60\ \Omega$

$P_T = ?$

Equations $I = \dfrac{V_1}{R_1}$

$R_T = R_1 + R_2 + R_3$

$P_T = I^2R_T$

Working $I = \dfrac{20}{100} = 0.2\ \text{A}$

$R_T = 100 + 20 + 60$

$R_T = 180\ \Omega$

$P_T = 0.2^2 \times 180$

Answer $\mathbf{P_T = 7.2\ W}$

Calculating power taken by individual components

Example 6.10 shows how to find the power taken by each resistor in Figure 6.15. Figure 6.16 is the same as Figure 6.15 except it shows the value of the current, which was found to be 0.2 A. The power taken by R_1 is found using the square of the voltage drop V_1 divided by R_1. For the other resistors, power is found with the equations I^2R_2 and I^2R_3.

EXAMPLE 6.10

Calculate the power taken by each resistor in the circuit of Figure 6.16.

FIGURE 6.16

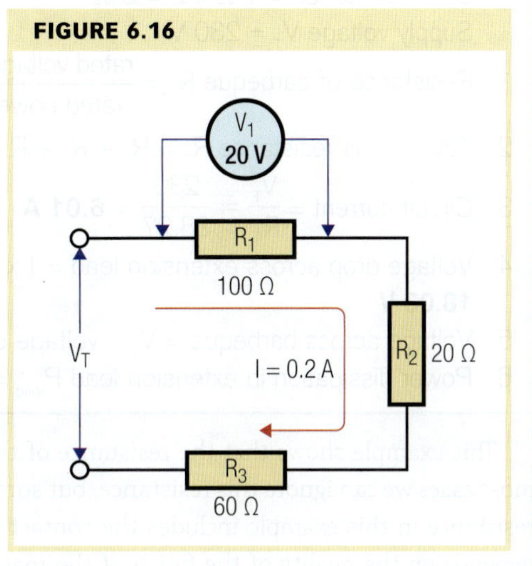

Solution

Values $V_1 = 20$ V
$R_1 = 100\ \Omega$
$R_2 = 20\ \Omega$
$R_3 = 60\ \Omega$
$I = 0.2$ A (from Example 6.9)
$P_1, P_2, P_3 = ?$

Working $P_1 = \frac{V_1^2}{R_1} = \frac{20^2}{100}$

Answer $\mathbf{P_1 = 4\ W}$

Working $P_2 = I^2R_2 = 0.2^2 \times 20$

Answer $\mathbf{P_2 = 0.8\ W}$

Working $P_3 = I^2R_3 = 0.2^2 \times 60$

Answer $\mathbf{P_3 = 2.4\ W}$

Putting it all together

Adding the individual power values found in Example 6.10 gives, as you would expect, the total power calculated in Example 6.9. Therefore, in a series circuit:

$P_T = P_1 + P_2 + P_3 + \ldots$

If you know the total power and the power taken by every component except one, the unknown power consumption can be found by transposing this equation in terms of the unknown power (P). That is,

$P = P_T -$ (sum of the remaining individual values of power)

Example 6.11 uses a number of equations to find all the unknowns in the series circuit shown in Figure 6.17. In this circuit the resistance caused by an extension lead powering an appliance is shown as two resistance values. These represent the individual resistance of the two wires providing power to the appliance.

EXAMPLE 6.11

A temporary setup has two 30 metre light duty extension leads connected in series to power a 1.5 kW electric barbeque. The equivalent circuit is shown in Figure 6.17, in which the resistance of the extension leads and the contact resistance of each plug and socket are considered.

Calculate:

1 resistance of the barbeque R_2

FIGURE 6.17

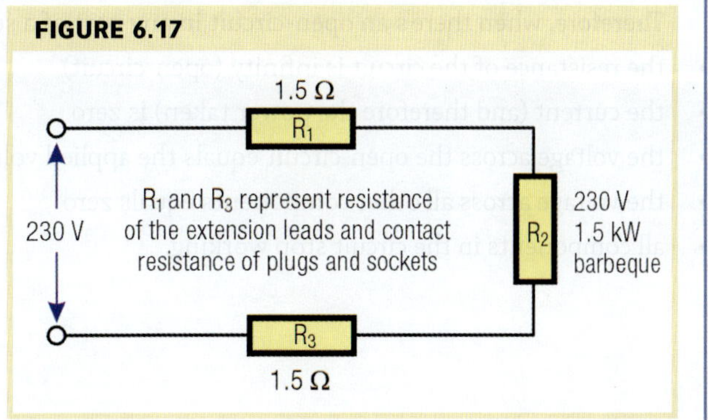

SAFETY

Excessively long extension leads can cause a significant voltage drop, and their resistance might be high enough to prevent a protection device from operating in the event of a fault

2 total resistance of the circuit R_T
3 circuit current I
4 voltage drop across extension lead
5 voltage across the barbeque
6 power dissipated by extension lead.

Solution

Values

Barbeque power rating = 1.5 kW
Barbeque voltage rating = 230 V
Lead resistance = 1 Ω x 2 = 3 Ω
Supply voltage V_T = 230 V

1 Resistance of barbeque $R_2 = \frac{\text{rated voltage}^2}{\text{rated power}} = \frac{230^2}{1500} = \mathbf{35.27\ \Omega}$

2 Total circuit resistance $R_T = R_1 + R_2 + R_3 = 1.5 + 35.27 + 1.5 = \mathbf{38.27\ \Omega}$

3 Circuit current $= \frac{V_T}{R_T} = \frac{230}{38.27} = \mathbf{6.01\ A}$

4 Voltage drop across extension lead $= I \times (R_1 + R_3) = 6.01 \times (1.5 + 1.5) = 6.01 \times 3 =$ **18.03 V**

5 Voltage across barbeque $= V_T -$ voltage drop across lead $= 230 - 18.03 =$ **211.97 V**

6 Power dissipation in extension lead $P_{lead} = I^2R = 6.01^2 \times 3 =$ **108.36 W**

This example shows that the resistance of the supply cables forms a series circuit with the load. In most cases we can ignore this resistance, but sometimes it needs to be considered, as in the example. The resistance in this example includes the contact resistance of the plugs and sockets. Contact resistance depends on the quality of the finish of the mating surfaces, the tension between the plug and mating socket, and how much contact is made between the two surfaces. When drawn as an equivalent circuit, as in Figure 6.17, an apparently simple, if non-ideal, setup becomes a series circuit that can be analysed using Ohm's law and other equations.

6.7 Faults in the Series Circuit

The two most common faults in any circuit are an open-circuit or a short-circuit. A third type of fault is a change in the value of a resistance within the circuit. We look now at the effects these faults have in a series circuit.

Open-circuit

The effects of an open-circuit are shown in Figure 6.18. In this diagram, one lamp has a blown filament, and is therefore an open-circuit. As a result, the current is zero, the voltage across the open-circuit equals the supply voltage and the voltage across all other lamps is zero.

Therefore, when there's an open-circuit in any part of a series circuit:

- the resistance of the circuit is infinity (open-circuit)
- the current (and therefore the power taken) is zero
- the voltage across the open-circuit equals the applied voltage (a useful fault-finding hint)
- the voltage across all other components equals zero
- all components in the circuit stop working.

FIGURE 6.18 Effects of an open-circuit in a series circuit

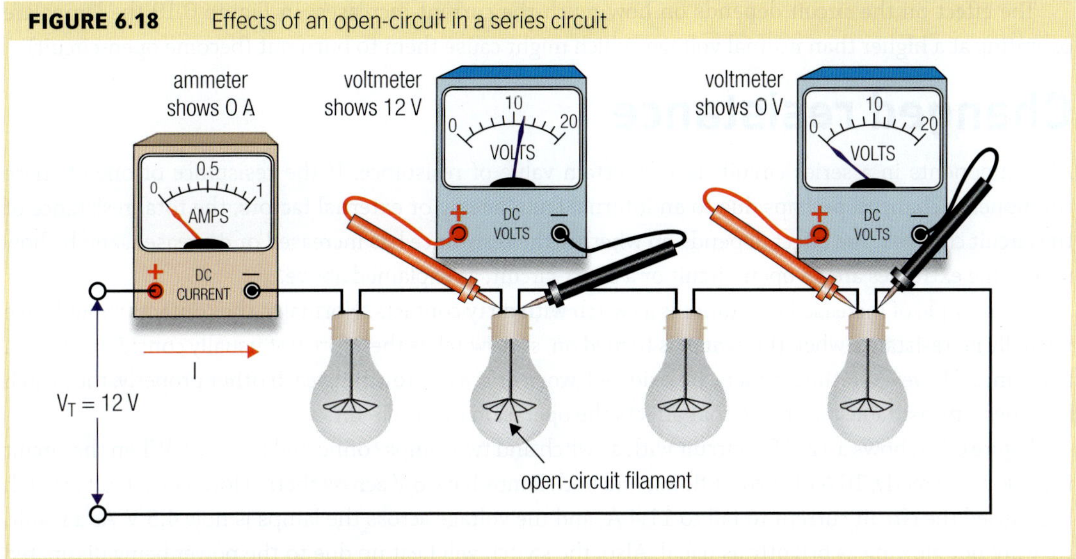

Short-circuit

A short-circuit across a component in a series circuit will *always* cause the current to increase, because the total resistance of the circuit drops.

In Figure 6.19, one lamp has an internal short-circuit. If each lamp has a resistance of 3 Ω, the total resistance of the circuit (without the short-circuit) is 12 Ω. By Ohm's law, the current is 1 A (I = 12 V/12 Ω). The voltage across each lamp is 3 V (V = 1 A × 3 Ω).

The short-circuit will cause the total resistance of the circuit to drop by 3 Ω (from 12 Ω to 9 Ω) and the current is now 1.33 A (12 V/9 Ω). The voltage across the lamp with the short-circuit is zero, as the resistance of the short-circuit is virtually zero. The voltage drop across the other three lamps is now 4 V (1.33 A × 3 Ω), instead of 3 V. These lamps will glow more brightly because of the increased voltage across them.

Therefore, when there's a short-circuit across a component in a series circuit the:

- circuit resistance drops
- circuit current (and therefore the power taken) increases
- voltage drop across the remaining components increases
- voltage drop across the short-circuit equals zero
- components in the rest of the circuit work harder because of the increased voltage drop across them.

FIGURE 6.19 A short-circuit in a series circuit always causes the current to increase

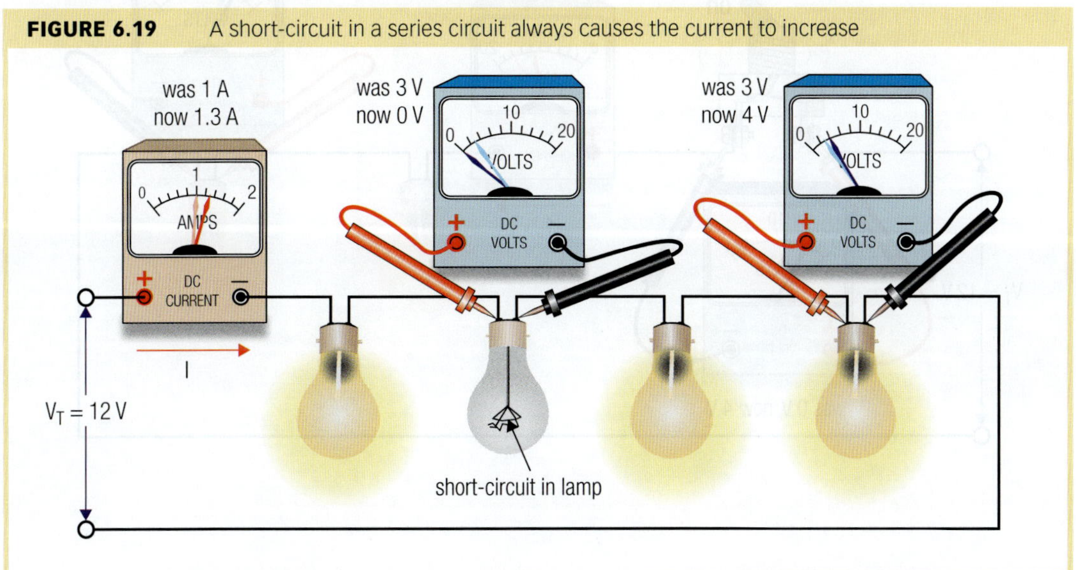

The effect on the circuit depends on how much the current increases. In Figure 6.19 the lamps are operating at a higher than normal voltage, which might cause them to burn out (become open-circuit).

Changed resistance

All components in a series circuit have a certain value of resistance. If the resistance of one of these components changes, perhaps due to an internal fault, ageing or external factors, the total resistance of the circuit changes. The effect depends on whether the resistance has increased or decreased, and by how much. The extremes are an open-circuit or a short-circuit, as explained above.

An example of increased resistance is a switch with dirty contacts. Normally, the contacts should have virtually no resistance when the switch is turned on, so a switch is therefore not usually considered to be a resistance. However, if the contacts are oxidised, worn or are not touching each other properly, the switch becomes a measurable resistance that affects the operation of the circuit.

Figure 6.20 shows a 12 V DC circuit with a switch and two lamps connected in series. When the circuit is working correctly, 16 A of current flows, and both lamps have 6 V across them. However, a faulty switch has caused the circuit current to fall to 11.4 A, and the voltage across the lamps is now 4.5 V. As a result, they are not glowing as brightly as usual. Also, the switch will heat up due to the power being dissipated by the resistive contacts.

When the fault is an *increased* resistance in a series circuit, the:

- circuit resistance increases
- circuit current (and therefore the power taken) decreases
- voltage drop across the remaining components decreases
- voltage drop across the increased resistance increases
- components in the rest of the circuit receive less power.

When the fault is a *decreased* resistance in a series circuit, the:

- circuit resistance decreases
- circuit current (and therefore the power taken) increases
- voltage drop across the remaining components increases
- voltage drop across the reduced resistance decreases
- components in the rest of the circuit receive more power.

FIGURE 6.20 A faulty switch with an increased resistance between its contacts affects the rest of the circuit

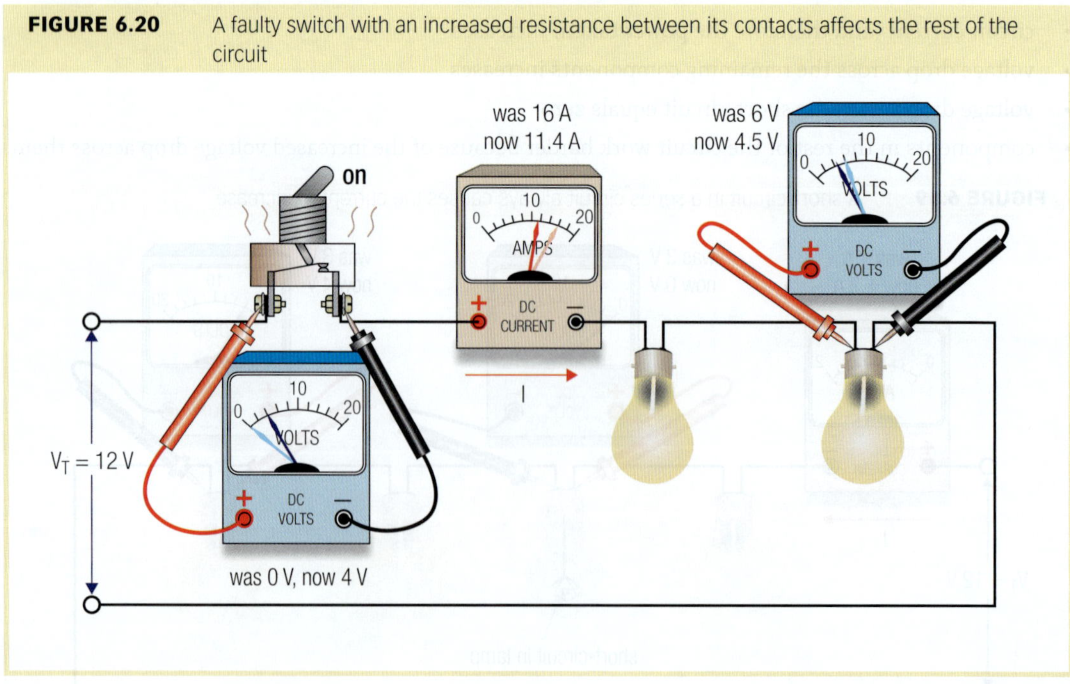

CHAPTER SUMMARY

In a series circuit:

- current is the same in all parts
- total resistance (R_T) is the sum of the individual resistances
- the algebraic sum of the voltage drops in the circuit equals the supply voltage V_T (Kirchhoff's voltage law)
- Ohm's law is used to find unknown voltage, resistance and current values in the circuit
- total power (P_T) taken equals the supply voltage times the current ($P_T = V_T I$)
- total power can also be found with the equations $P_T = I^2 R_T$, or $P_T = \frac{V_T^2}{R_T}$
- the sum of the power taken by each component equals the total power
- an open-circuit anywhere in the circuit prevents any component from receiving power, as the circuit current is zero
- the voltage drop across an open-circuit equals the supply voltage
- a short-circuit will cause an increase in the circuit current
- the voltage drop across a short-circuit equals zero
- effects caused by faults that cause a change in resistance will be somewhere between the extremes of an open- or a short-circuit.

REVIEW EXERCISES

Check your answers at the back of the book.

1 Four heating elements are connected in series. Two elements have a resistance of 15 Ω, the other two have a resistance of 10 Ω. The supply voltage is 230 V. Calculate:
  a total circuit resistance
  b circuit current
  c voltage across the 15 Ω elements
  d voltage across the 20 Ω elements
  e total power.

2 A toaster with four identical elements connected in series is rated at 1.6 kW when operating from a 230 V supply. Calculate:
  a current taken by each element
  b resistance of each element.

3 If one element in the toaster in Question 2:
  a becomes open-circuit, what is the voltage across that element and across the other elements?
  b becomes short-circuit, what is the voltage across that element and across the other elements?

4 A light display has 20 LEDs connected in series. Each LED has 3 V across it when a current of 2 A is passing. Calculate:
  a the value of a series resistance to allow the display to operate from a 100 V DC supply
  b the power dissipated by the resistance.

5 A 230 V, 4.5 kW commercial oven has four identical tubular heater elements connected in series. Calculate the:
  a resistance of the oven
  b current taken by the oven
  c resistance of each element
  d power taken by each element
  e voltage across each element.

6 If one element in Question 5 becomes a short-circuit, determine:
  a the voltage across each element
  b current taken by the oven
  c total power taken by the oven.

7 For the circuit in Figure 6.21, find the:
  a circuit current
  b resistance of resistor R_6
  c total resistance of the circuit
  d resistance of resistor R_2.

FIGURE 6.21

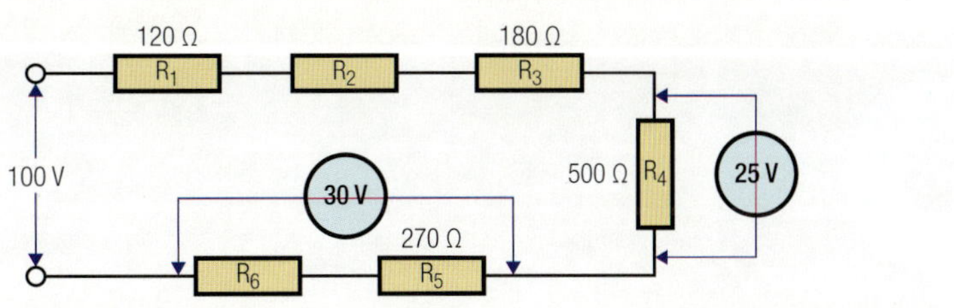

ONLINE RESOURCES

COMPLETE WORKSHEET SIX

Check with your instructor for worksheets on this chapter.

CHAPTER 7

THE PARALLEL CIRCUIT

In a parallel circuit, all components connect directly to the applied voltage. This chapter looks at ways of determining the total resistance of any type of parallel circuit, and how to calculate current in each branch of the circuit. The chapter also describes power in the parallel circuit, and the effects of an open- or short-circuit.

CHAPTER OUTLINE

7.1 The Parallel Circuit

When you plug two appliances into the same power outlet, you are connecting the appliances in *parallel*. In a parallel circuit, all the components connect directly to the supply voltage. In a car, the lights and electrical accessories are all connected in parallel with the car battery – although they are individually controlled with their own on-off switches. In a house, every appliance is connected in parallel with the mains supply.

Figure 7.1 shows three lamps connected in parallel with a battery. There are several ways the wires can be run to do this, such as wiring each lamp directly to the battery as in (a). A more typical way is with a common set of wires that run the length of the circuit, as shown in (b). Regardless of how the wiring is arranged, the circuit diagram is the same.

FIGURE 7.1 Three lamps connected in parallel with different ways of running the wiring

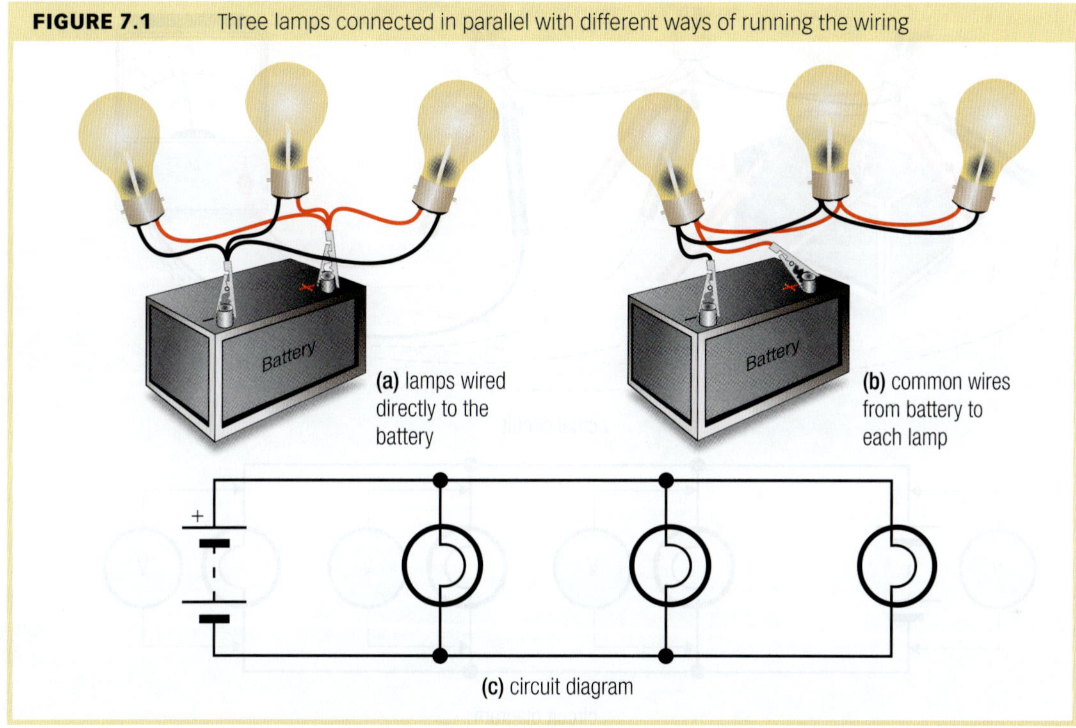

A parallel circuit can be drawn in a number of ways. In Figure 7.2(a), three resistors are connected in parallel with a battery. Regardless of how the parallel circuit is drawn, each component connects directly to the battery. Circuit (b) shows a single resistor connected to a battery, with a voltmeter connected across, or in parallel with, the resistor. As we've explained, a voltmeter is always connected in parallel with a component to measure the voltage across that component.

FIGURE 7.2 More examples of the parallel connection

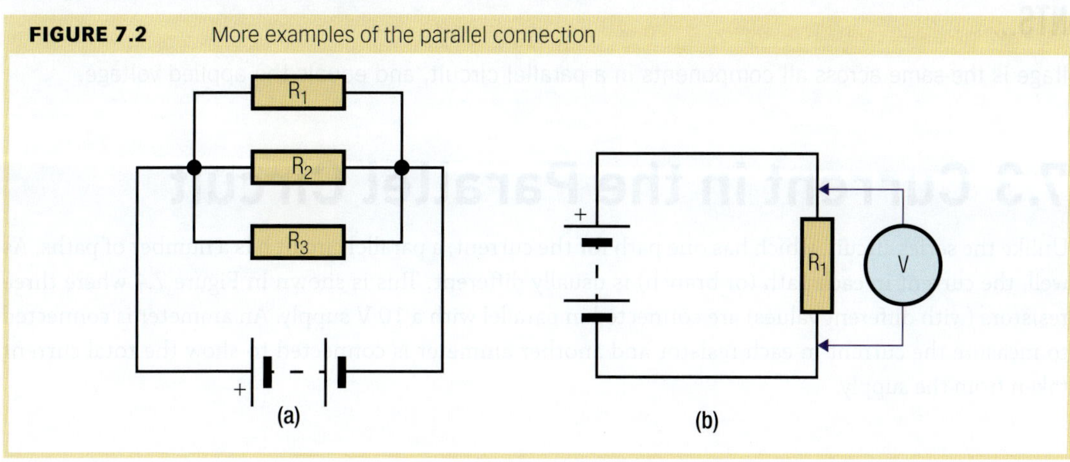

7.2 Voltage in the Parallel Circuit

You can already see that because each component in a parallel circuit connects directly to the voltage source, the voltage across each component must be the same as the voltage of the voltage source.

This is shown in Figure 7.3 in which three lamps are connected in parallel with a 12 V battery. As the diagram shows, the voltmeter will read 12 V if it's connected across any of the three lamps, or across the battery.

FIGURE 7.3 The voltage across each component in a parallel circuit is equal to the source voltage

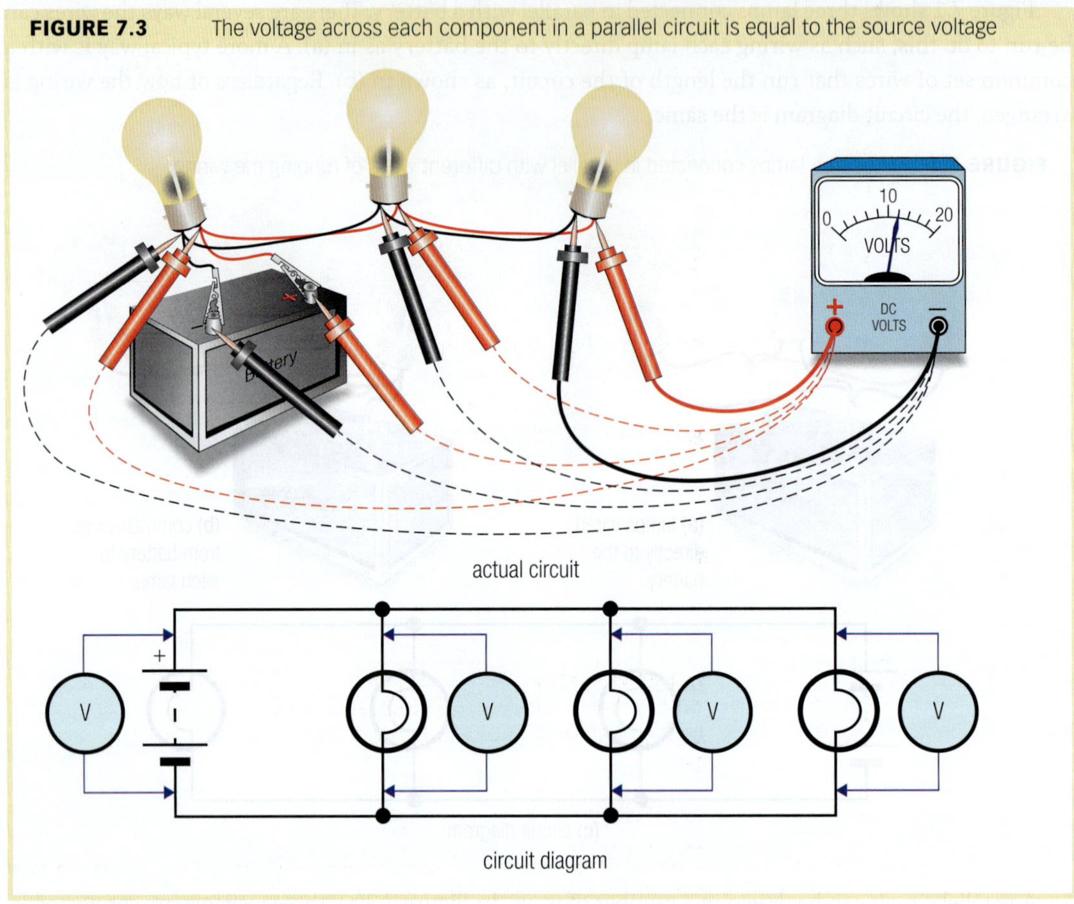

The circuit diagram in Figure 7.3 shows the same thing. Here, all voltages are the same, and equal to the applied voltage (V). This is different to the series circuit, where the voltage drops across each component depend on the value of its resistance and the current flowing in the circuit.

KEY POINTS...

- The voltage is the same across all components in a parallel circuit, and equals the applied voltage.

7.3 Current in the Parallel Circuit

Unlike the series circuit, which has one path for the current, a parallel circuit has a number of paths. As well, the current in each path (or branch) is usually different. This is shown in Figure 7.4 where three resistors (with different values) are connected in parallel with a 10 V supply. An ammeter is connected to measure the current in each resistor and another ammeter is connected to show the total current taken from the supply.

FIGURE 7.4 Measuring current in a parallel circuit

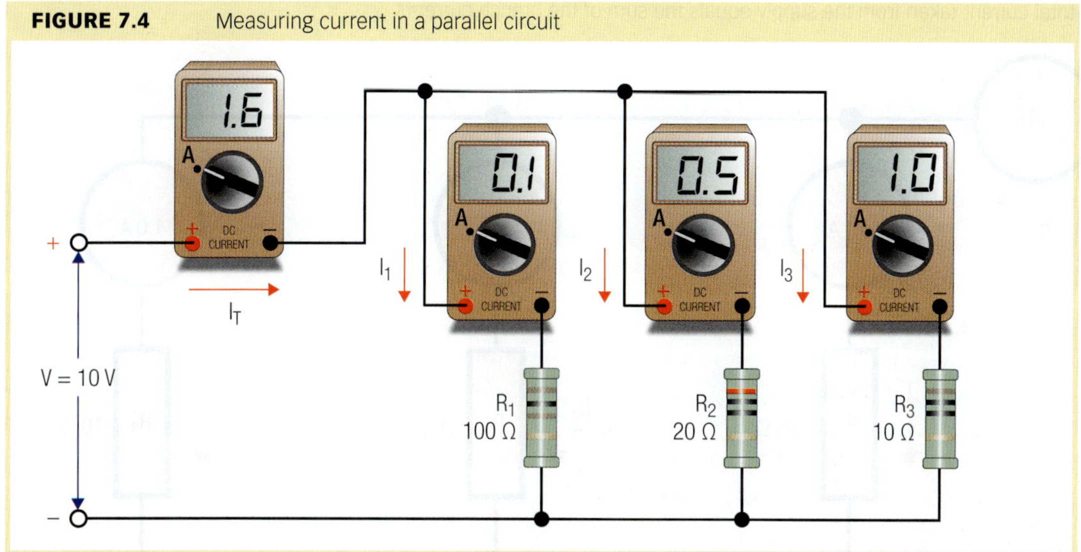

The values of each branch current can be found with Ohm's law by looking at each branch separately. The circuit in Figure 7.5 shows the conditions for resistor R_1 only.

FIGURE 7.5 Looking at R_1 only

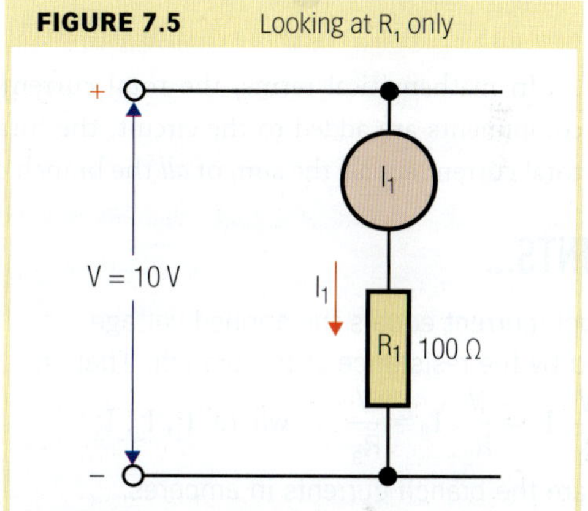

$$I_1 = \frac{V}{R_1} = \frac{10}{100} = 0.1 \text{ A}$$

Notice that this is the value indicated by the ammeter for I_1 in Figure 7.4. The same method is used to find the current in R_2. Figure 7.6 shows the circuit for R_2 only, where I_2 is also found with Ohm's law.

FIGURE 7.6 Looking at R_2 only

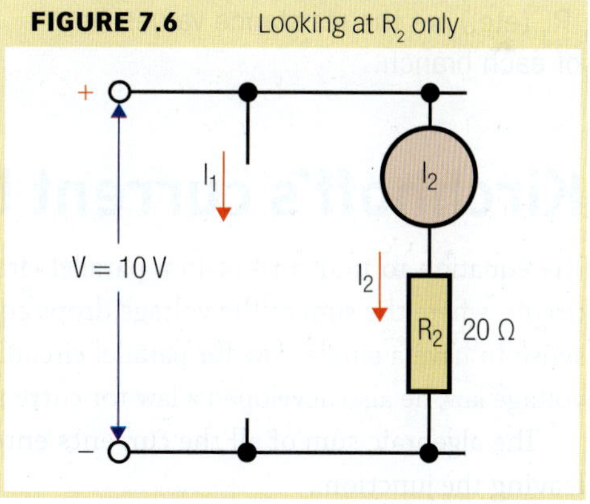

$$I_2 = \frac{V}{R_2} = \frac{10}{20} = 0.5 \text{ A}$$

The current in R_3 is also found with Ohm's law.

$$I_3 = \frac{V}{R_3} = \frac{10}{10} = 1 \text{ A}$$

When calculating a branch current, each branch is treated on its own, as if the other branches don't exist. This applies no matter how many branches there are in a parallel circuit.

If you compare the resistance values in each branch and the current in that branch, you can see that the higher the branch resistance, the lower the branch current. In other words, the current in a branch of a parallel circuit is *inversely* proportional to the resistance of the branch. This is simply an application of Ohm's law, which says that current is inversely proportional to resistance.

The next thing to consider is the total current taken from the supply. Figure 7.7 is the circuit diagram of Figure 7.4, and shows that the branch currents flow from the positive potential of the supply, through each resistor then back to the supply. Because each branch current comes from the supply, the total current (I_T) taken from the supply is therefore the sum of the individual branch currents.

FIGURE 7.7 The total current taken from the supply equals the sum of the branch currents

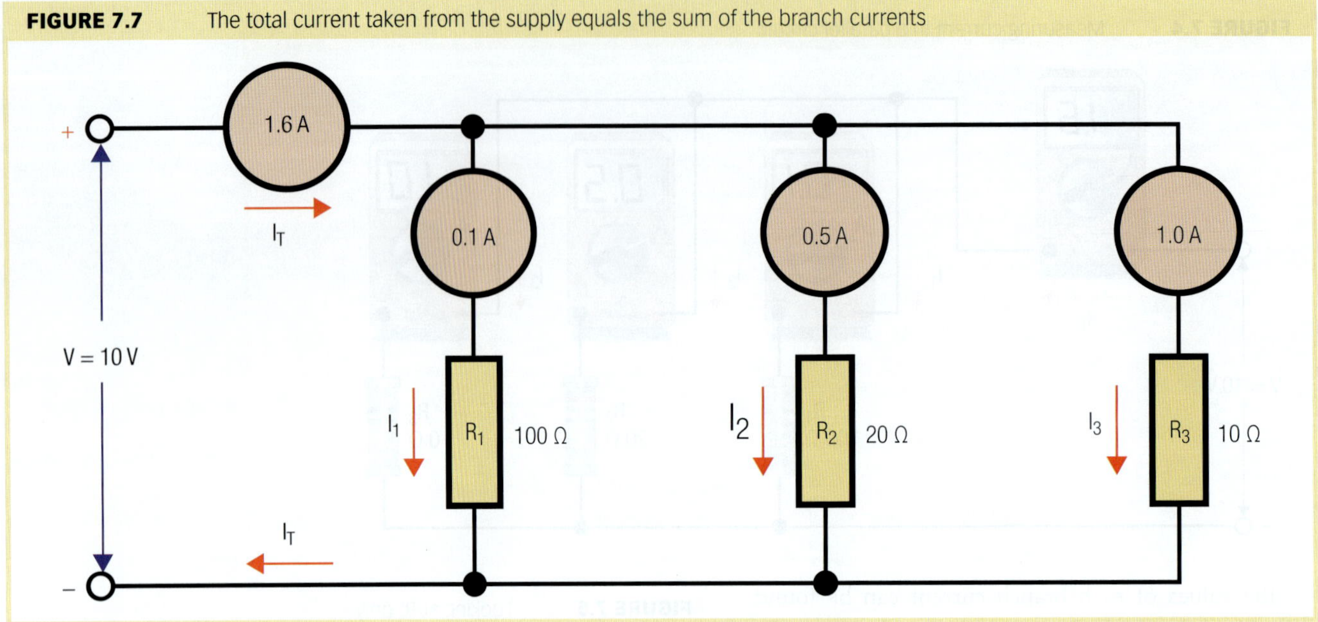

In mathematical terms, the total current (I_T) in Figure 7.7 equals $I_1 + I_2 + I_3$. If more parallel components are added to the circuit, the total current will increase. Therefore, in a parallel circuit the total current equals the sum of *all* the branch currents.

KEY POINTS...

- A branch current equals the applied voltage divided by the resistance of the branch. That is: $I_1 = \frac{V}{R_1}, I_2 = \frac{V}{R_2}, I_3 = \frac{V}{R_3} \ldots$ where: I_1, I_2, I_3 (etc.) are the branch currents in amperes.
- V is the applied voltage in volts.
- R_1, R_2, R_3 (etc.) are the resistance values (in ohms) of each branch.
- Branch current is inversely proportional to the resistance of the branch.
- The total current equals the sum of all the branch currents. That is:
- $I_T = I_1 + I_2 + I_3 + \cdots$ where: I_T is the total current in amperes taken from the supply.

Kirchhoff's current law

KEY CONCEPT

Kirchhoff's current law

The equation to find current in a parallel circuit is similar to that for finding voltage drops in a series circuit, where the sum of the voltage drops equals the applied voltage (Kirchhoff's voltage law). It makes sense to have a similar law for parallel circuits, this time for current. When Kirchhoff came up with his voltage law, he also developed a law for current which says:

The algebraic sum of all the currents entering a junction equals the algebraic sum of the currents leaving the junction.

Figure 7.8 shows how the currents divide for the circuit of Figure 7.7. The total circuit current (I_T) enters junction 1, and two currents leave the junction: I_1 which flows through R_1, and $I_2 + I_3$ which flows into junction 2.

Notice that 1.6 A flows into and out of junction 1. There is 1.5 A entering junction 2, and 1.5 A leaves it, divided into two paths: I_2 and I_3.

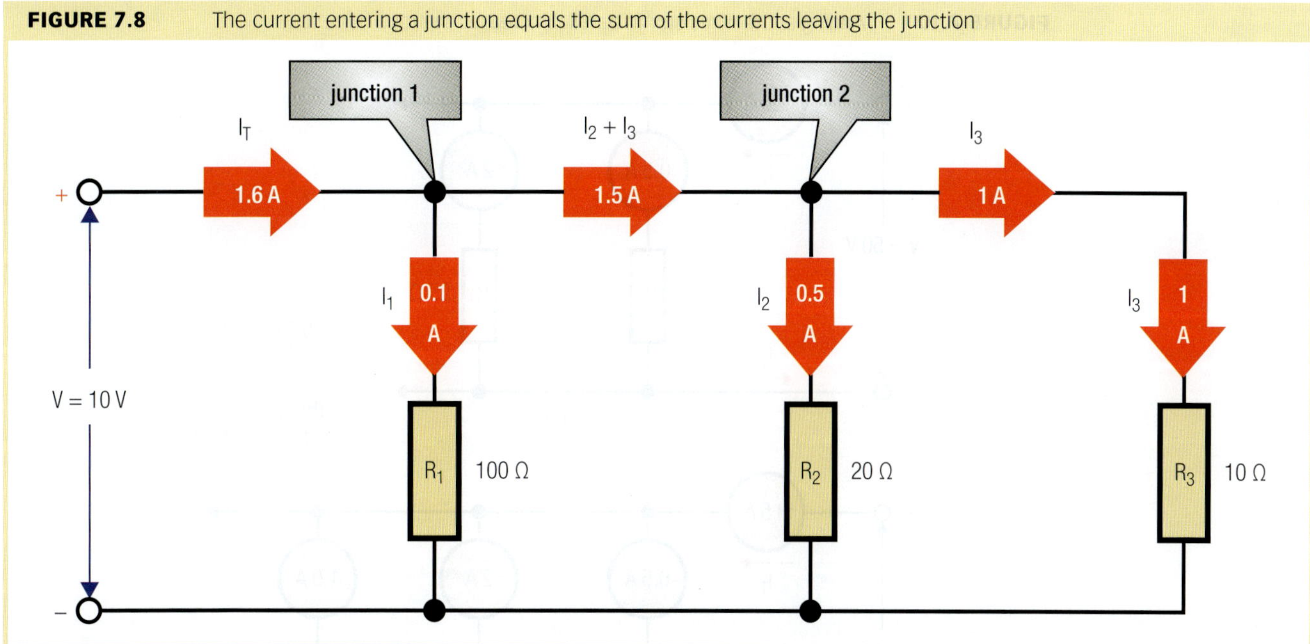

FIGURE 7.8 The current entering a junction equals the sum of the currents leaving the junction

The following example uses Ohm's law and Kirchhoff's current law to find the branch currents in a parallel circuit. Ohm's law is used to find I_1. Kirchhoff's current law is used to find I_2. (The equation has to be transposed in terms of I_2.)

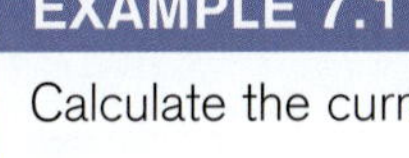

Calculate the currents flowing in resistors R_1 and R_2 in Figure 7.9.

Solution

Values $R_1 = 50\ \Omega$

$I_T = 1.2\ \text{A}$

$V = 25\ \text{V}$

To find I_1

Equation $I_1 = \frac{V}{R_1} = \frac{25}{50}$

Answer $I_1 = 0.5\ \text{A}$

To find I_2

Equation $I_2 = I_T - I_1 = 1.2 - 0.5$

Answer $\mathbf{I_2 = 0.7\ A}$

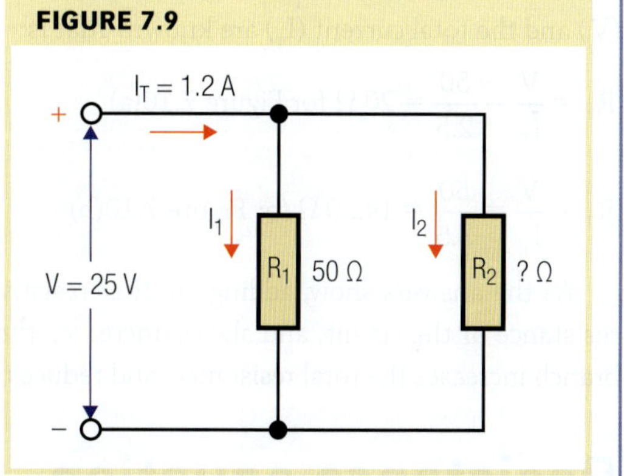

FIGURE 7.9

7.4 Resistance in the Parallel Circuit

We've shown that a parallel circuit has a number of branches, with each branch taking current from the voltage supply. The more branches, the more current. For example, Figure 7.10(a) has two branches, each taking a particular value of current. If a branch is added, as in (b), there will be more current taken. That is, adding another resistor in parallel will cause the total current to increase, which means the total resistance of the circuit must have decreased.

FIGURE 7.10 Adding another resistor in parallel increases the total current

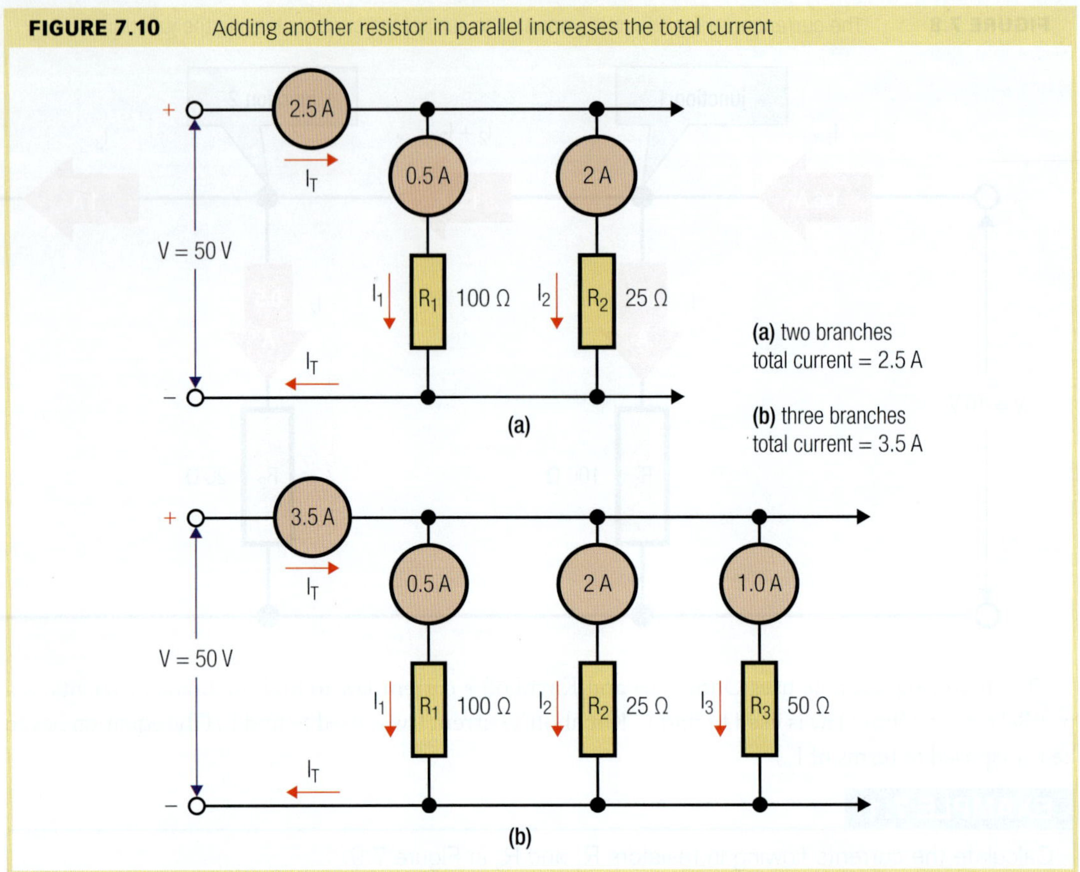

The total resistance (R_T) of Figure 7.10(a) and (b) can be found with Ohm's law, as the applied voltage (V) and the total current (I_T) are known. That is:

$$R_T = \frac{V}{I_T} = \frac{50}{2.5} = 20\ \Omega \text{ for Figure 7.10(a)}$$

$$R_T = \frac{V}{I_T} = \frac{50}{3.5} = 14.29\ \Omega \text{ for Figure 7.10(b)}$$

As the answers show, adding another resistive branch to a parallel circuit always reduces the total resistance of the circuit, and always increases the total circuit current. On the other hand, removing a branch increases the total resistance, and reduces the total current.

Resistance equation

When you don't know the total current or the applied voltage, the total resistance of a parallel circuit must be found in other ways. The following shows the derivation of an equation to find the total resistance (R_T) from the individual resistance values. It starts with Ohm's law, which can be used with any circuit if you know the applied voltage and the total current.

1 $R_T = \frac{V}{I_T}$ where V is the applied voltage and I_T is the total circuit current

This equation can be transposed in terms of I_T:

2 $I_T = \frac{V}{R_T}$

The total current in a parallel circuit equals the sum of the branch currents. Each branch current can be determined by Ohm's law, as in Example 7.1. Therefore, the total current can be found by adding the Ohm's law equations for each branch current. That is,

3 $$I_T = \frac{V}{R_1} + \frac{V}{R_2} + \frac{V}{R_3} + \cdots$$

Because equations 2 and 3 both equal I_T, the right-hand side of these equations must equal each other, giving equation 4.

4 $$\frac{V}{R_T} = \frac{V}{R_1} + \frac{V}{R_2} + \frac{V}{R_3} + \cdots$$

Because V is common to all parts of equation 4, its value doesn't matter as it affects both sides equally. Therefore, by dividing both sides by V, you get a *general* equation to find the total resistance of a parallel circuit. This equation is:

5 $$\frac{1}{R_T} = \frac{1}{R_1} + \frac{1}{R_2} + \frac{1}{R_3} + \cdots$$

Equation 5 gives the *reciprocal* of the total resistance. In other words, we can say that the reciprocal of the total resistance of a parallel circuit equals the sum of the reciprocals of each branch resistance. Equation 6 gives the actual total resistance (R_T) of a parallel circuit, and is derived from equation 5.

6 $$R_T = \frac{1}{\frac{1}{R_1} + \frac{1}{R_2} + \frac{1}{R_3} + \cdots}$$

KEY CONCEPT 

Resistance of a parallel circuit is always less than the lowest value branch resistance

KEY POINTS...

- The resistance of any parallel circuit: $\frac{1}{R_T} = \frac{1}{R_1} + \frac{1}{R_2} + \frac{1}{R_3} + \cdots$
- The total resistance of a parallel circuit is always less than the smallest value of branch resistance.

Example 7.2 shows how to use this equation, with suggestions on using a calculator given in Figure 7.12.

EXAMPLE 7.2

Calculate the total resistance of the circuit in Figure 7.11.

FIGURE 7.11

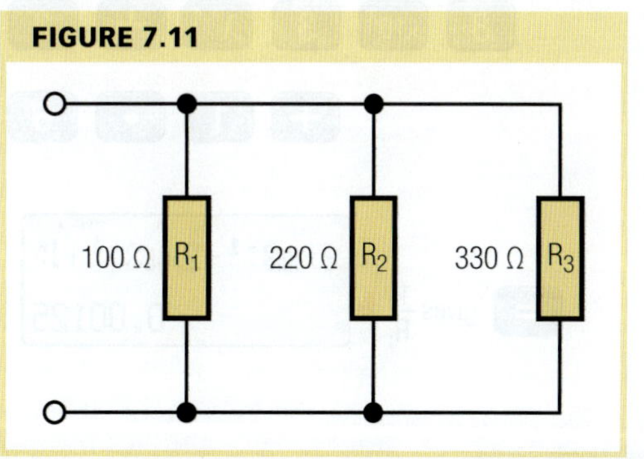

Solution

Values $R_1 = 100\ \Omega$, $R_2 = 220\ \Omega$, $R_3 = 330\ \Omega$, $R_T = ?$

Equation $$\frac{1}{R_T} = \frac{1}{R_1} + \frac{1}{R_2} + \frac{1}{R_3}$$

Working $$\frac{1}{R_T} = \frac{1}{100} + \frac{1}{220} + \frac{1}{330}$$

(add the reciprocals of each resistance)

$$\frac{1}{R_T} = 0.017575757$$ (the sum of the reciprocals)

Answer **R_T = 56.89655172 or 56.9 Ω** (rounded off to two decimal places)

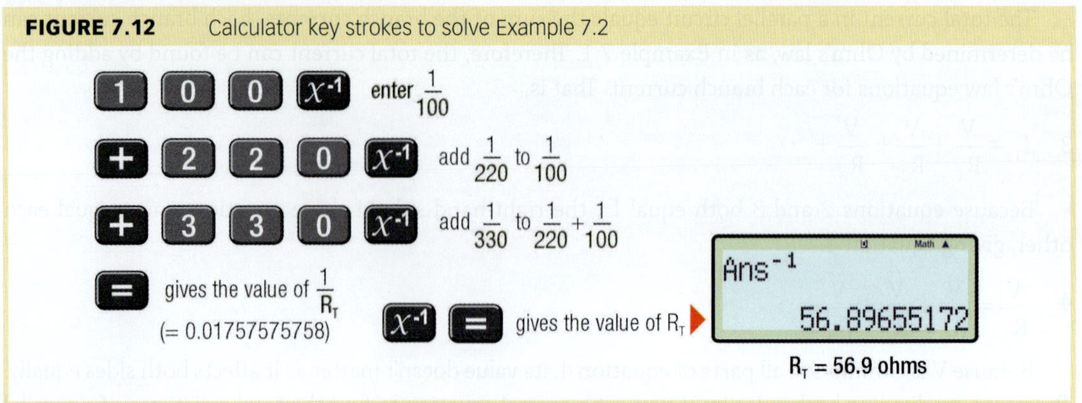

FIGURE 7.12 Calculator key strokes to solve Example 7.2

The next example has resistor values expressed with engineering notation. These values are entered into a calculator as described in Chapter 3 and shown in Figure 7.14.

EXAMPLE 7.3

What is the total resistance of the circuit in Figure 7.13?

Solution

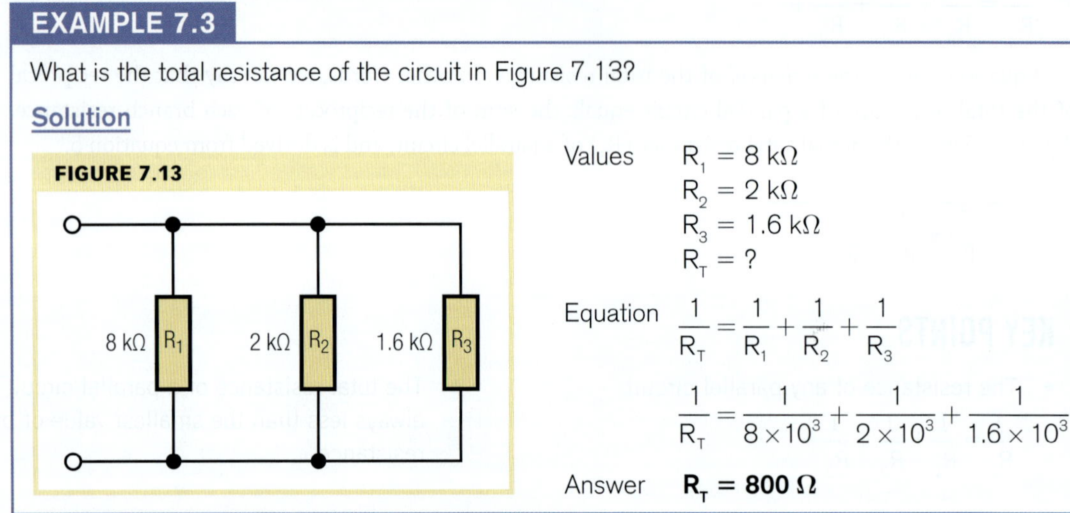

FIGURE 7.13

Values $R_1 = 8\ \text{k}\Omega$

$R_2 = 2\ \text{k}\Omega$

$R_3 = 1.6\ \text{k}\Omega$

$R_T = ?$

Equation $$\frac{1}{R_T} = \frac{1}{R_1} + \frac{1}{R_2} + \frac{1}{R_3}$$

$$\frac{1}{R_T} = \frac{1}{8 \times 10^3} + \frac{1}{2 \times 10^3} + \frac{1}{1.6 \times 10^3}$$

Answer $\mathbf{R_T = 800\ \Omega}$

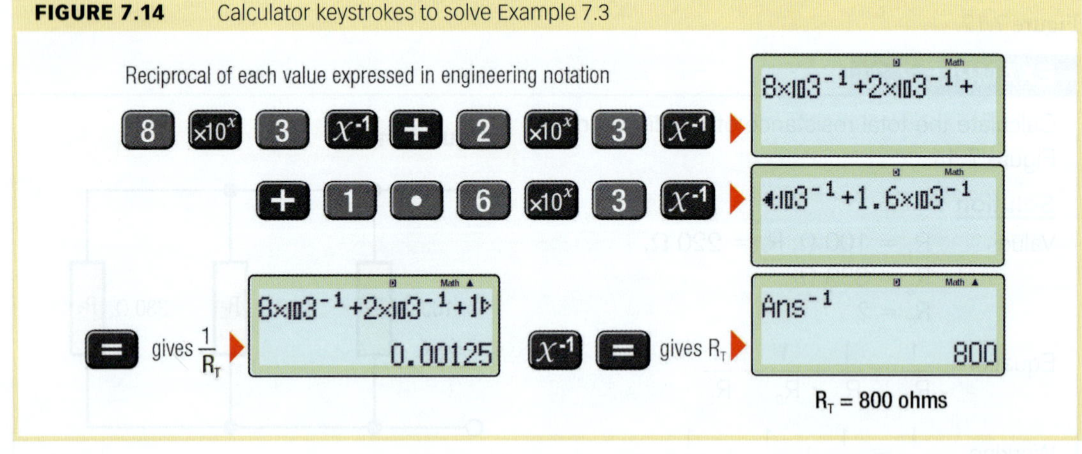

FIGURE 7.14 Calculator keystrokes to solve Example 7.3

Two resistors in parallel

In many cases, there are only two resistors in parallel. The general equation can be used to find their combined resistance, but a simpler way is to use an equation derived from the general equation. This equation (shown below) works out to be the *product of the two resistance values divided by their sum.*

$R_T = \frac{R_1 \times R_2}{R_1 + R_2}$ where R_1 and R_2 are two resistors connected in parallel.

EXAMPLE 7.4

Find the total resistance R_T and the total current I_T for the circuit in Figure 7.15.

Solution

Values $V_T = 12$ V

$R_1 = 60\,\Omega$, $R_2 = 40\,\Omega$

I_T and R_T = ?

To find R_T

Equation $R_T = \frac{R_1 R_2}{R_1 + R_2}$

$\frac{60 \times 40}{60+40} = \frac{2400}{100}$

Answer $\mathbf{R_T = 24\,\Omega}$

To find I_T

Equation $I_T = \frac{V}{R_T} = \frac{12}{24}$

Answer $\mathbf{I_T = 0.5\ A}$

FIGURE 7.15

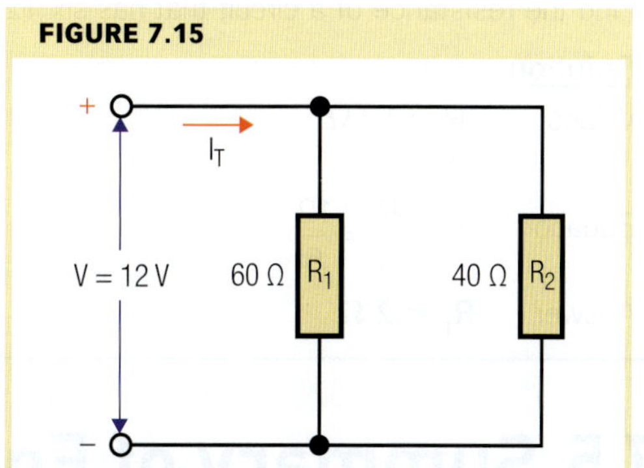

FIGURE 7.16 Calculator keystrokes to solve Example 7.4

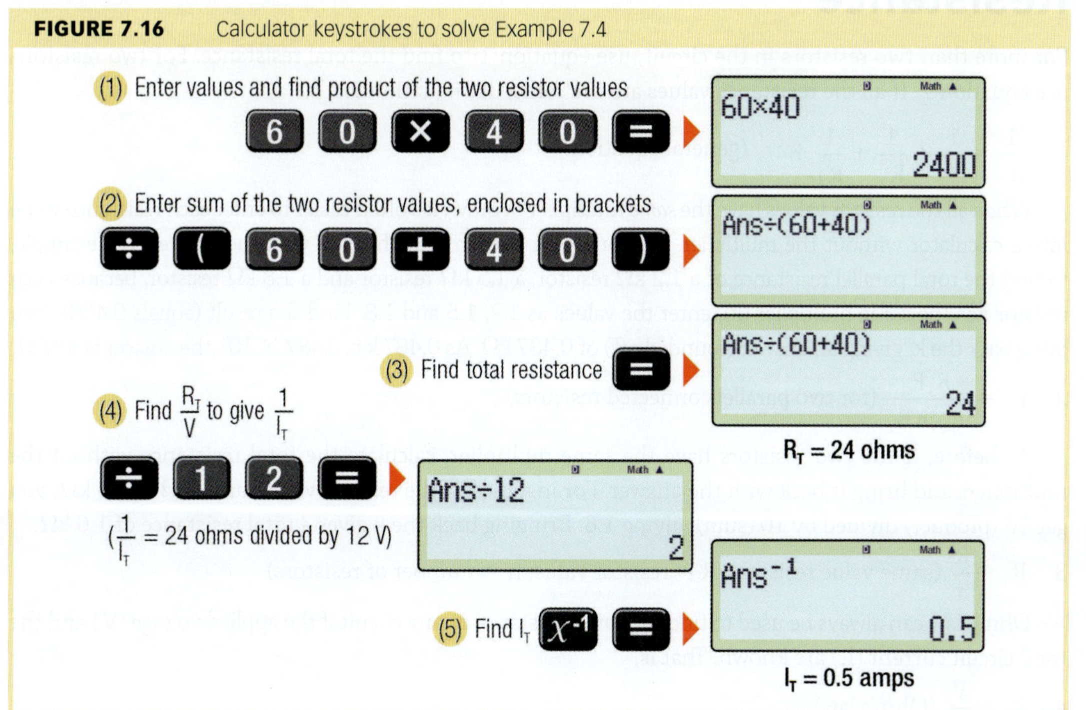

Resistors of the same value

Another special case is when all the resistors in the circuit have the same value. The equation is:

$$R_T = \frac{R}{n}$$

where:
R is the value of each resistance
n is the number of resistances in the circuit.

EXAMPLE 7.5

Find the resistance of a circuit that has six 12 Ω resistors connected in parallel.

Solution

Values $R = 12\ \Omega$
$n = 6$

Equation $R_T = \frac{R}{n} = \frac{12}{6}$

Answer $\mathbf{R_T = 2\ \Omega}$

7.5 Summary of Equations

Like the series circuit, current, voltage and resistance values in a parallel circuit can be found in various ways, depending on what you know about the circuit. Apart from the equations to find parallel resistance, Ohm's law and Kirchhoff's current law will also be useful. The following summarises the important equations, with hints on using them.

KEY CONCEPT

The total resistance of a parallel circuit is always less than the smallest branch resistance and gets smaller when other branches are added to the circuit

Resistance

For more than two resistors in the circuit, use equation 1 to find the total resistance. For two resistors, use equation 2. If all the resistance values are the same, use equation 3.

1 $\frac{1}{R_T} = \frac{1}{R_1} + \frac{1}{R_2} + \frac{1}{R_3} + \cdots$ (general equation)

When all the resistor values have the *same* multiplier (k or M) it's often easier to enter each resistance value into a calculator without the multiplier. The multiplier is then brought back with the answer. For example, to find the total parallel resistance of a 1.2 kΩ resistor, a 1.5 kΩ resistor and a 1.8 kΩ resistor, because each resistor has the same multiplier (k), enter the values as 1.2, 1.5 and 1.8. Find the result (equals 0.486) then bring back the k, giving an answer (rounded off) of 0.487 kΩ. As 0.487 k is 0.487×10^3, the answer is 487 Ω.

2 $R_T = \frac{R_1 R_2}{R_1 + R_2}$ (for two parallel connected resistors)

As before, if the two resistors have the same multiplier, calculate the total resistance without the multipliers and bring it back with the answer. For instance, if the resistor values are 2 kΩ and 8 kΩ, you get 16 (product) divided by 10 (sum), giving 1.6. Bringing back the k gives a total resistance of 1.6 kΩ.

3 $R_T = \frac{R}{n}$ (same value resistors, R = resistor value, n = number of resistors)

Ohm's law can always be used to find the total resistance of any circuit if the applied voltage (V) and the total circuit current (I_T) are known. That is,

4 $R_T = \frac{V}{I_T}$ (Ohm's law)

The resistance of a branch in a parallel circuit is found with Ohm's law, where V is the applied voltage and I_1, I_2 and so on are the currents in the particular branch.

5 $R_1 = \frac{V}{I_1}, R_2 = \frac{V}{I_2}, R_3 = \frac{V}{I_3}$ and so on

Voltage

The voltage (V) in a parallel circuit is *the same* across all components in the circuit. Voltage is found with either of the next two Ohm's law equations. Use equation 6 if you know the total current and the total resistance of the circuit.

6 $V = I_T \times R_T$

If you know the current flowing in any *one* resistance in the circuit and you also know the value of that resistance, use equation 7 to find the applied voltage.

7 $V = I_1 \times R_1$, $V = I_2 \times R_2$ and so on

KEY CONCEPT
The voltage across any component in a parallel circuit equals the applied voltage

Current

Depending on what you know about the circuit, you can use both Kirchhoff's current law and Ohm's law to find either the total current or a branch current. Equations 8 and 9 are to find the total current (I_T).

8 $I_T = \frac{V}{R_T}$ (Ohm's law)

9 $I_T = I_1 + I_2 + I_3 + \cdots$ (Kirchhoff's current law)

To find a branch current with Kirchhoff's current law, transpose equation 9 in terms of the unknown current (I). That is:

10 $I = I_T -$ (sum of all the other branch currents in the circuit)

Remember that Kirchhoff's current law says that the current flowing into a junction equals the current flowing out of the junction. Use this law to find an unknown current flowing into or out of a junction if you know the other currents flowing into or out of that junction.

A branch current can also be found with Ohm's law, where:

11 $I_1 = \frac{V}{R_1}$, $I_2 = \frac{V}{R_2}$ and so on

Because the voltage is the same across each branch it follows that the lower the branch resistance, the higher the branch current.

KEY CONCEPT
Current into a junction equals the current out

KEY CONCEPT
The lower the branch resistance, the higher its current

TASK 7.1

1 In Figure 7.17, find:
   a supply voltage V
   b current in resistor R_2
   c total current I_T
   d total resistance R_T
   e resistance value of R_1

FIGURE 7.17

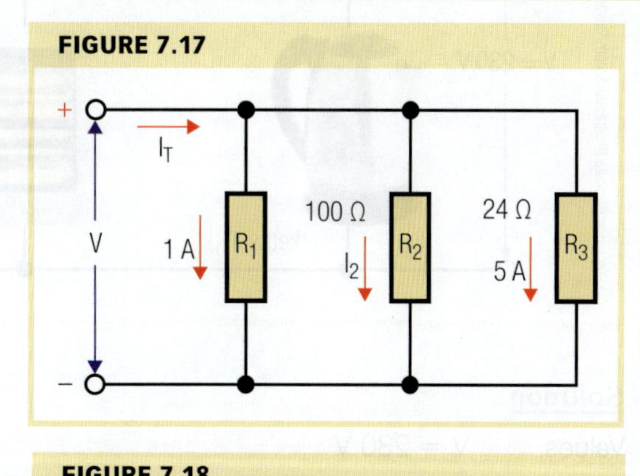

2 In Figure 7.18, find:
   a resistance value of R_1
   b current in resistor R_1
   c current in resistor R_2
   d total resistance R_T
   e resistance value of R_2
   f resistance value of R_3

FIGURE 7.18

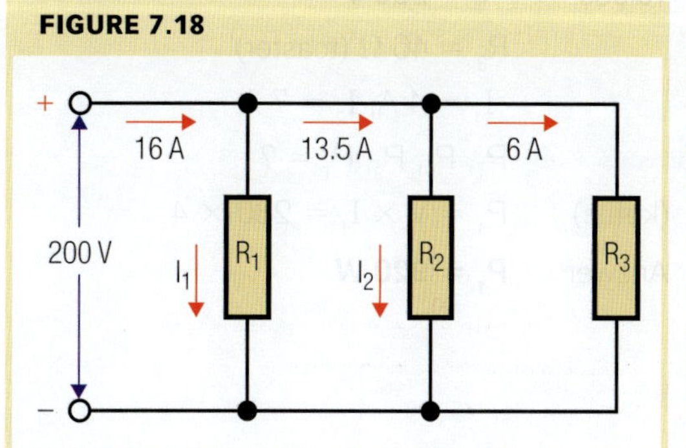

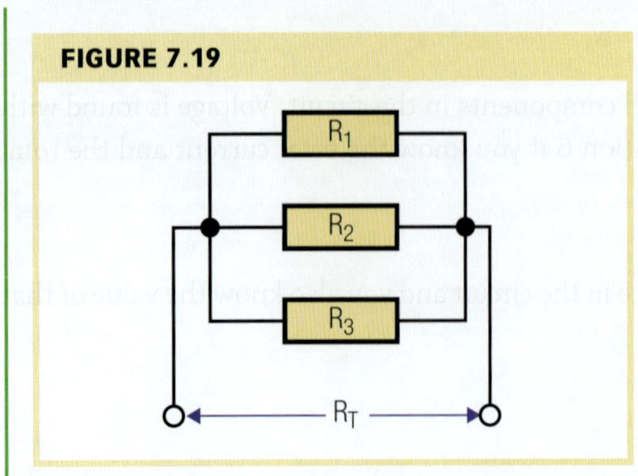

FIGURE 7.19

3 For the circuit in Figure 7.19 find the total resistance when:
   a $R_1 = 48\ \Omega$, $R_2 = 120\ \Omega$, $R_3 = 72\ \Omega$
   b $R_1 = 400\ \Omega$, $R_2 = 630\ \Omega$, $R_3 = 550\ \Omega$
   c $R_1 = 4.7\ k\Omega$, $R_2 = 6.8\ k\Omega$, $R_3 = 5.6\ k\Omega$
   d all resistors = $24\ \Omega$

7.6 Power in the Parallel Circuit

Like the series circuit, each component in a parallel circuit takes power from the supply. But unlike a series circuit, each component is independent of the others. So when looking at the power taken by a branch, ignore the other branches and treat the branch as a single component connected to the supply. The total power taken by the circuit is the sum of the power taken by each branch.

As for the series circuit, total power or power taken by a branch is found using the power equations given in Chapter 4. The equations to use depend on what you know about the circuit, as shown in Example 7.6.

EXAMPLE 7.6

Calculate the power taken by each appliance in Figure 7.20.

FIGURE 7.20

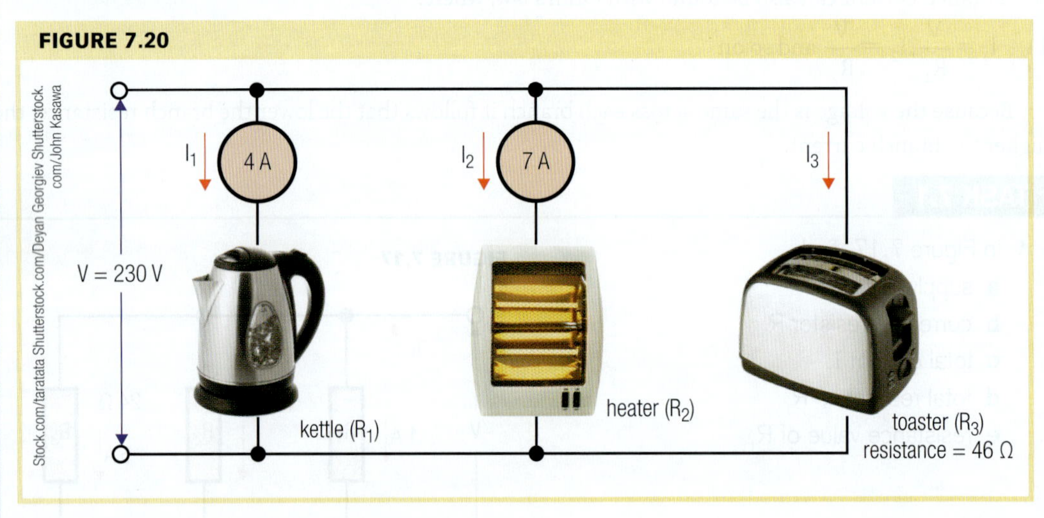

Solution

Values $V = 230\ V$

$R_3 = 46\ \Omega$ (toaster)

$I_1 = 4\ A$, $I_2 = 7\ A$

$P_1, P_2, P_3, P_T = ?$

(kettle) $P_1 = V \times I_1 = 230 \times 4$

Answer $\mathbf{P_1 = 920\ W}$

(heater) $P_2 = V \times I_2 = 230 \times 7$

Answer **P_2 = 1610 W or 1.61 kW**

(toaster) $P_3 = \frac{V^2}{R_3} = \frac{230^2}{46}$

Answer **P_3 = 1150 W or 1.15 kW**

Like the series circuit, the total power equals the sum of the powers taken by each component. That is, $P_T = P_1 + P_2 + P_3$

In Example 7.6, the total power taken from the supply is 920 + 1610 + 1150, which gives a value of 3680 W or 3.68 kW.

There are two ways to find the total resistance of a parallel circuit if you know the value of the supply voltage and the resistance of each branch. The first is to calculate the power taken by each branch, then add these to get the total power. Another is to calculate the total resistance of the circuit, and use this value to calculate the total power, as shown in Example 7.7.

EXAMPLE 7.7

Calculate the total power taken by the circuit in Figure 7.21.

Solution

Values $V = 30\ V$

$R_1 = 33\ \Omega$

$R_2 = 47\ \Omega$

$R_3 = 56\ \Omega$

$P_T = ?$

To find R_T

Equation $\frac{1}{R_{eq}} = \frac{1}{R_1} + \frac{1}{R_2} + \frac{1}{R_3}$

Working $\frac{1}{R_{eq}} = \frac{1}{33} + \frac{1}{47} + \frac{1}{56}$

R_T = 14.4 Ω

To find total power

Equation $P_T = \frac{V^2}{R_{eq}} = \frac{30^2}{14.4}$

Answer **P_T = 62.5 W**

FIGURE 7.21

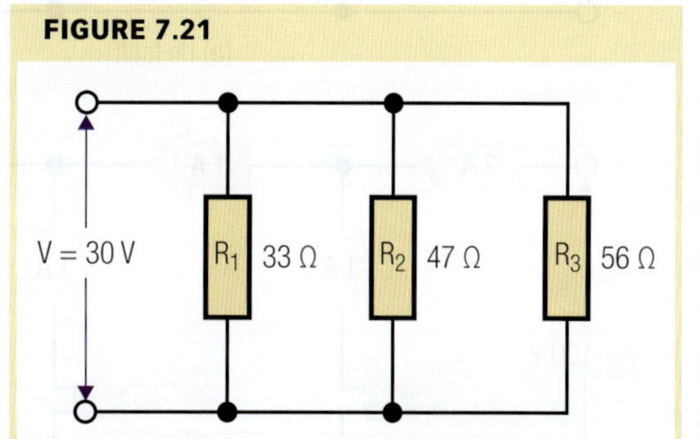

FIGURE 7.22 Equivalent circuit of Figure 7.21

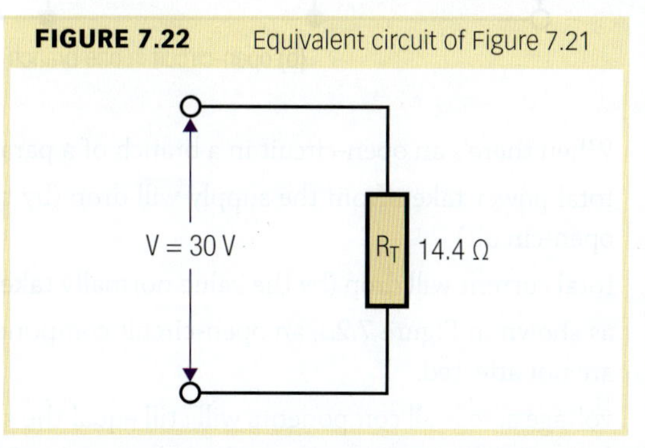

Because each branch in a parallel circuit is considered separately, if you know the power taken by a branch you can find the:

- applied voltage if you know the branch current
- branch current if you know the applied voltage.

7.7 Faults in the Parallel Circuit

As with any circuit, the two basic faults in a parallel circuit are an open-circuit and a short-circuit. Because each branch of a parallel circuit is independent of the others (although they all share the same supply voltage), an open-circuit in one branch won't affect the others.

This is shown in Figure 7.23(a) and (b), which shows a parallel circuit with three lamps, (a) with all lamps operating and (b) with one lamp open-circuited. Notice that there is no current in this branch and that the total current has dropped from 3 A to 2 A. However, as the diagrams show, the voltmeter across the lamp reads the supply voltage in both cases.

FIGURE 7.23 An open-circuit in one component of a parallel circuit causes the total current to drop

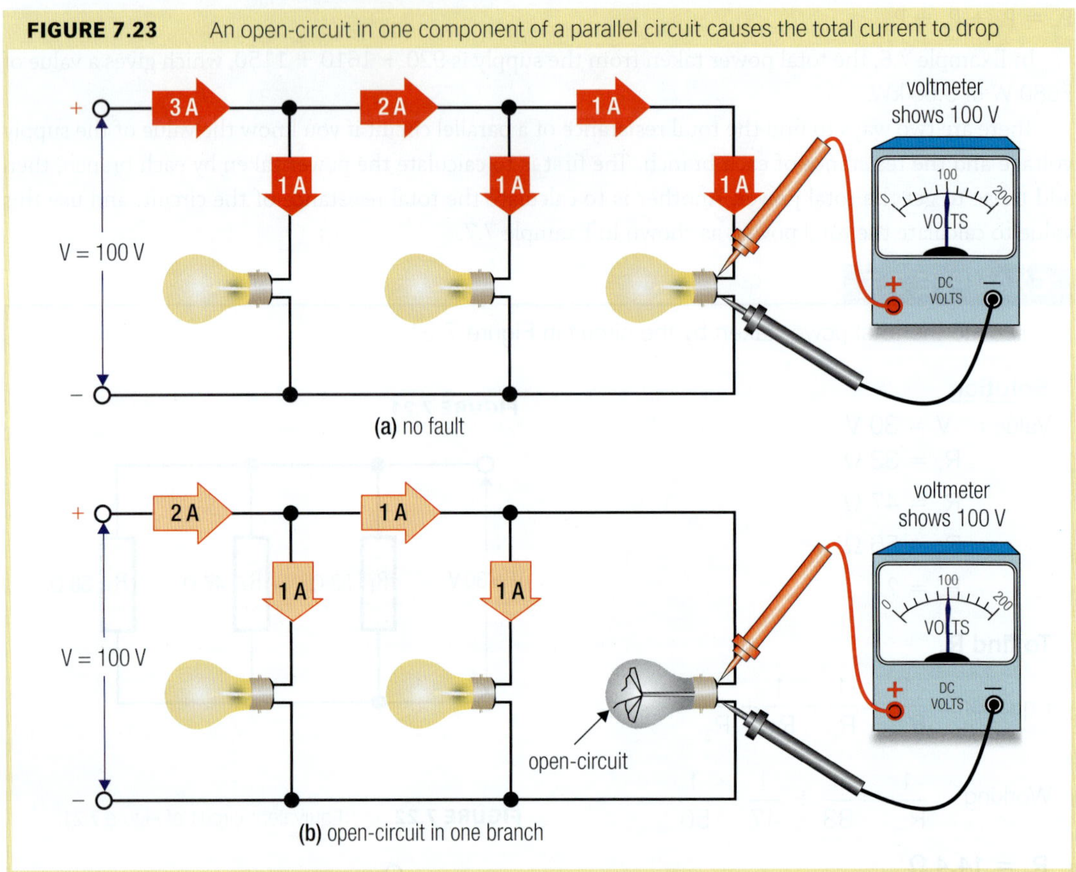

When there's an open-circuit in a branch of a parallel circuit, the:

- total power taken from the supply will drop (by the value normally taken by the branch that is now open-circuit)
- total current will drop (by the value normally taken by that branch)
- as shown in Figure 7.23, an open-circuit component will no longer work, but all others in the circuit are not affected.
- voltage across all components will still equal the supply voltage.

Obviously, an open-circuit in a *conductor* between junctions will have a different effect. For instance, all branches after the open-circuit will be isolated from the supply. The voltage across these branches will therefore be zero. However, the voltage across the open-circuit will equal the supply voltage.

A short-circuit in a parallel circuit means there's a short-circuit across the supply voltage. This will cause a large current to flow, limited only by the capacity of the supply source and the total resistance of the circuit conductors and the short-circuit itself. Usually the circuit protection device (fuse or circuit breaker) will operate and isolate the circuit. Therefore, a short-circuit across a component in a parallel circuit causes:

- the total circuit resistance to drop to a very low value
- the total current to increase by a large amount

- the voltage drop across all branches to drop to almost zero
- all components to stop working, as there's no longer any supply voltage.

7.8 Comparison to the Series Circuit

A parallel circuit has components that are all directly connected to the voltage source. Because each component is independent of the others, there can be any number of components in the circuit, limited by the ability of the supply source to power the components and of the circuit conductors to carry the current.

In a series circuit, the limit on the number of components is determined by the value of the supply voltage. Adding an extra component in a series circuit creates an additional voltage drop across the component, reducing the available voltage to the rest of the components. However, unlike the series circuit where a single switch controls all the components, each component in a parallel circuit can have its own on-off switch.

The main differences between the series and the parallel circuit are given in Table 7.1.

TABLE 7.1

COMPARISON OF SERIES AND PARALLEL CIRCUITS

Function	Series circuit	Parallel circuit
Voltage	unequal	equal
Current	equal	unequal
Adding resistance	total resistance increases	total resistance decreases
Open-circuit in one component	current falls to zero in all components	current falls to zero in that component only
Short-circuit in one component	current increases, to a limit set by the other components in the circuit	current increases, limited by the resistance of the conductors and the current the power source can supply

CHAPTER SUMMARY

In a parallel circuit:

- voltage is the same across all components
- total resistance (R_T) is found with the equations:
  - $\frac{1}{R_T} = \frac{1}{R_1} + \frac{1}{R_2} + \frac{1}{R_3} + \ldots$ (general equation)
  - $R_T = \frac{R_1 R_2}{R_1 + R_2}$ (for two parallel-connected resistors)
  - $R_T = \frac{R}{n}$ (same value resistors, R = resistor value, n = number of resistors)
  - $R_T = \frac{R}{n}$ (Ohm's law)
- total current taken from the supply is found with the equations:
  - $I_T = \frac{V}{R_T}$ (Ohm's law)
  - $I_T = I_1 + I_2 + I_3 + \cdots$ (Kirchhoff's current law)
- Ohm's law is used to find unknown voltage, resistance and current values in any branch of the circuit
- total power (P_T) taken equals the supply voltage times the total current ($P_T = VI_T$)
- total power can also be found with the equations $P_T = I_T{}^2R_T$, or $P_T = \frac{V^2}{R_T}$

- the sum of the power taken by each component equals the total power
- an open-circuit in a branch prevents only that component from receiving power
- a short-circuit will cause a considerable increase in the circuit current and cause protection devices to operate (if present)

REVIEW EXERCISES

Check your answers at the back of the book.

1 A lighting installation has four branches connected to a 230 V supply. The branch currents are I_1 = 1.5 A, I_2 = 3 A, I_3 = 2 A and I_4 = 1 A. Determine:
  a total current taken by the installation
  b power taken by each branch
  c total circuit power.

2 A 230 V, 10 A power board with four outlets has an appliance plugged into each outlet. Two appliances each take 700 W of power and the other two each take 400 W of power. Determine:
  a total power when all appliances are operating
  b current taken by each appliance
  c total current
  d total resistance when all appliances are operating.

3 A workshop has a 1 kW lathe, a 500 W drill press, a 1.5 kW compressor and a 300 W grinder all connected to the same 230 V, 15 A supply. Calculate:
  a the current taken by each machine
  b total current when all machines are operating
  c resistance of each machine
  d total resistance when all machines are operating.

4 For the circuit in Figure 7.24, calculate:
  a supply voltage V
  b current in R_2
  c current in R_3
  d resistance of R_3
  e total circuit power.

FIGURE 7.24

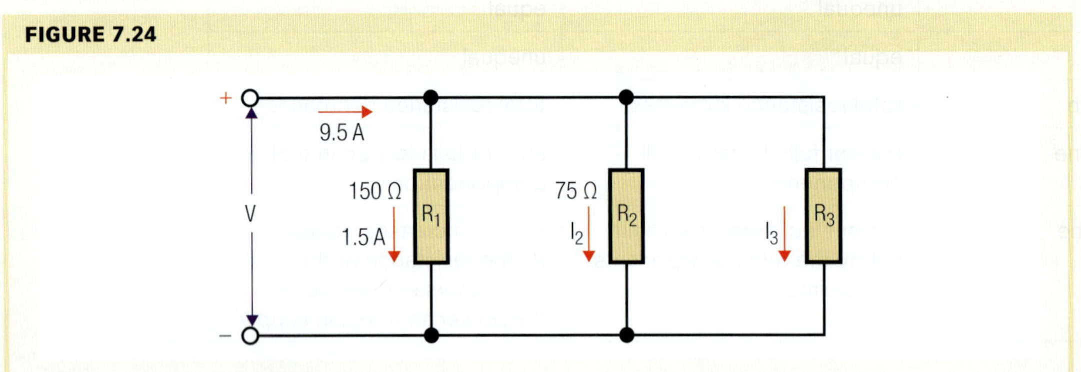

5 Three appliances with resistances of (R_1) 100 Ω, (R_2) 46 Ω and (R_3) 115 Ω are connected in parallel to a 230 V supply. Determine:
  a total circuit resistance
  b current in each appliance
  c power taken by each appliance
  d total circuit power.

6 A 230 V electric stove has five independent heating elements. Four of these have the same resistance of 90 Ω, the fifth element has a resistance of 50 Ω. Determine:
  a total resistance when all elements are operating
  b current taken by a 90 Ω element
  c current taken by the 50 Ω element
  d total current when all elements are operating.

ONLINE RESOURCES

COMPLETE WORKSHEET SEVEN

Check with your instructor for worksheets on this chapter.

CHAPTER 8

THE SERIES–PARALLEL CIRCUIT

Many electrical or electronic circuits have components connected in both series and parallel, giving a more complex arrangement than shown so far. This chapter explains how to calculate voltage, current and resistance values in a series–parallel circuit containing resistors, and how to reduce the circuit to a single equivalent resistance. It also covers ways of calculating the total power taken by the circuit and the power taken by each component.

CHAPTER OUTLINE

8.1 Introduction

A parallel circuit, as explained in Chapter 7, has a number of branches connected to the supply source. It may be that a branch has two or more components connected in series, such as the circuit shown in Figure 8.1. Here, three lamps connected in series are powered from the same supply as two other individual lamps, giving a circuit that has both series-connected components and parallel-connected components: a series–parallel circuit.

FIGURE 8.1 A series–parallel circuit with three lamps in series connected in parallel with two other lamps

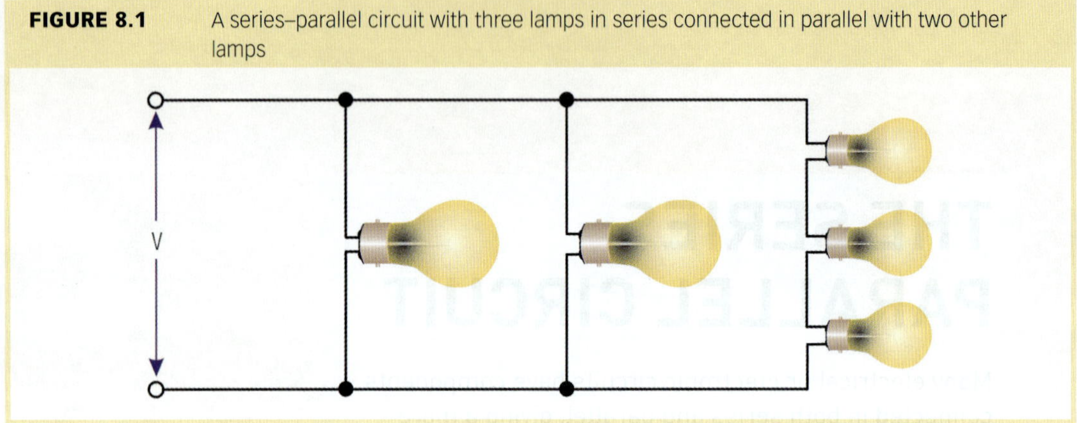

8.2 Resistance in the Series–Parallel Circuit

Figure 8.1 is one possible arrangement of a series–parallel circuit, which in practice can have any arrangement. Figure 8.2 returns to the temporary setup discussed in Chapter 6, in which a long extension lead connected to an electric barbeque gives a series circuit that includes the resistance of the extension lead. Adding to this arrangement, as shown in Figure 8.2(a), the extension lead is now connected to a power board, and the barbeque and a hot water urn are plugged into the power board, giving a series–parallel circuit. The equivalent circuit is shown in Figure 8.2(b). R_1 and R_4 represent the resistance of each cable in the extension lead, and the contact resistance of plugs and sockets. This circuit is used in Example 8.1, which shows how to reduce the entire circuit to a single resistance value.

FIGURE 8.2 A series–parallel circuit of a long extension lead powering two appliances.

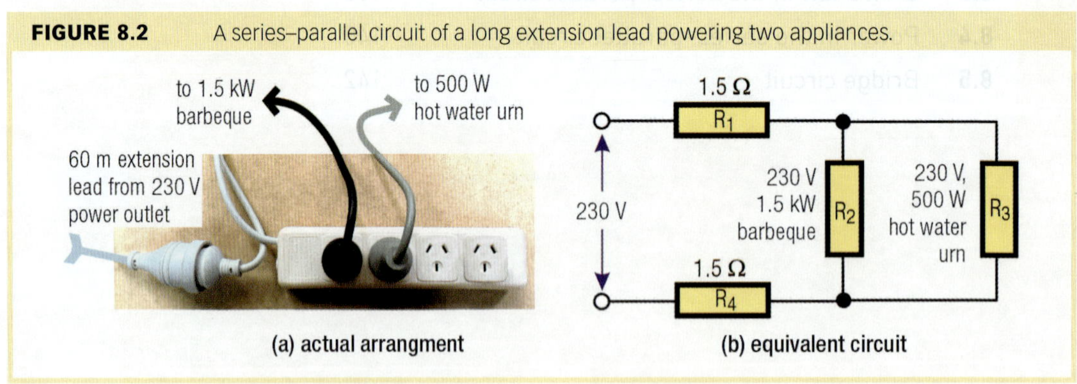

EXAMPLE 8.1

Find the total resistance of Figure 8.2.

Solution

1 Determine the resistance values of R_2 and R_3 using the equatiaon $R = \frac{V^2}{P}$.

$$R_2 = \frac{230^2}{1500} = 35.2\ \Omega, R_3 = \frac{230^2}{500} = 105.8\ \Omega$$

2 Determine the total resistance of R_2 in parallel with R_3.

$$R_{2-3} = \frac{R_2 \times R_3}{R_2 \times R_3} = \frac{35.27 \times 105.8}{35.27 \times 105.8} = 26.45\ \Omega$$

The circuit so far is shown in Figure 8.3.

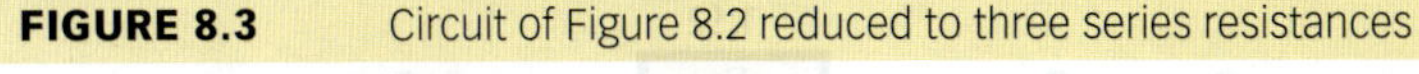
FIGURE 8.3 Circuit of Figure 8.2 reduced to three series resistances

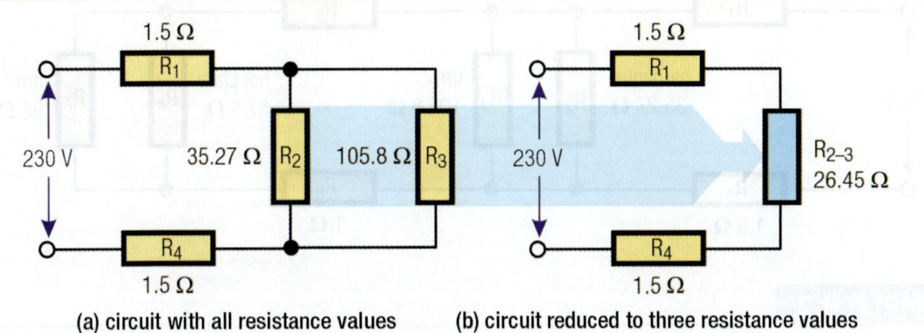

3 Combine all series-connected resistances into a single value by adding their individual resistance values. This gives the total resistance of the circuit, and the equivalent circuit shown in Figure 8.4(b).

FIGURE 8.4 Equivalent circuit of Figure 8.2 reduced to one resistance value, which is the total resistance of the circuit

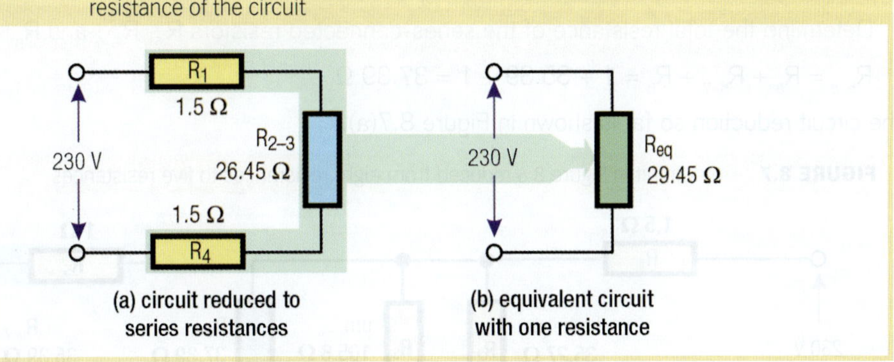

From Figure 8.4(a) $R_T = 1.5 + 1.5 + 26.45$

Answer: $\mathbf{R_T = 29.45}\ \Omega$

The setup shown in Figure 8.2 can become more complicated if a long extension lead is plugged into the power board to provide power to more appliances. The arrangement is shown in Figure 8.5, and the circuit diagram is in Figure 8.6. Notice that the actual wiring and the circuit diagram look quite different.

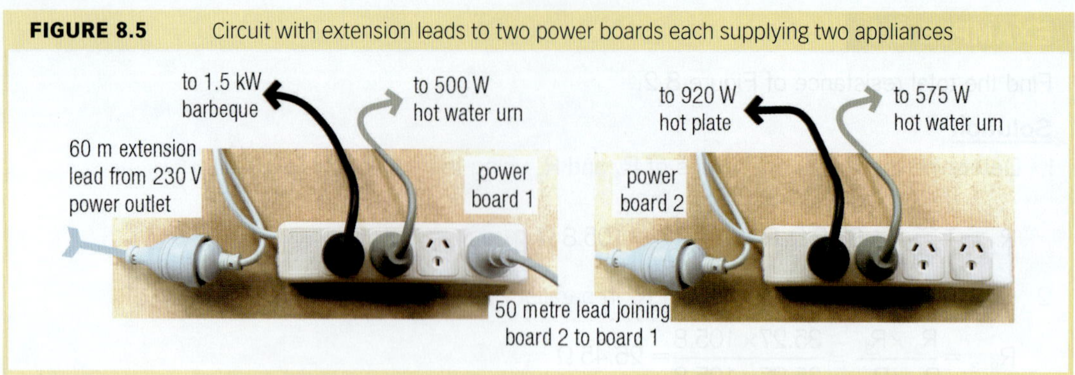

FIGURE 8.5 Circuit with extension leads to two power boards each supplying two appliances

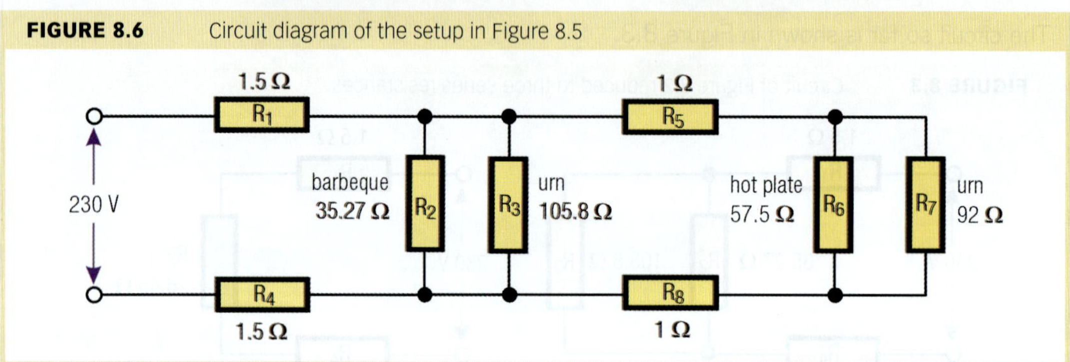

FIGURE 8.6 Circuit diagram of the setup in Figure 8.5

EXAMPLE 8.2

Find the total resistance of Figure 8.6.

Solution

1 Determine the total resistance of R_6 in parallel with R_7

$$R_{6-7} = \frac{R_6 \times R_7}{R_6 \times R_7} = \frac{57.5 \times 92}{57.5 \times 92} = 35.39\ \Omega$$

2 Determine the total resistance of the series-connected resistors R_5, R_{6-7} and R_8

$$R_{5-8} = R_5 + R_{6-7} + R_8 = 1 + 35.39 + 1 = 37.39\ \Omega.$$

The circuit reduction so far is shown in Figure 8.7(a).

FIGURE 8.7 Circuit in Figure 8.6 reduced from eight resistances to five resistances

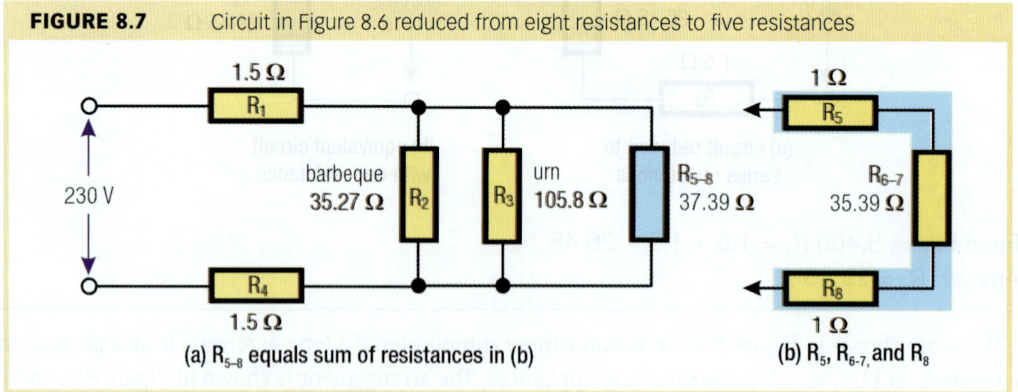

3 Determine the total resistance of R_2 in parallel with R_3.

$$R_{2-3} = \frac{R_2 \times R_3}{R_2 \times R_3} = \frac{35.27 \times 105.8}{35.27 \times 105.8} = 26.45\ \Omega \text{ (as in Example 8.1).}$$

The circuit reduction so far is in Figure 8.8(b).

FIGURE 8.8 Circuit in Figure 8.6 reduced in (b) to four resistances

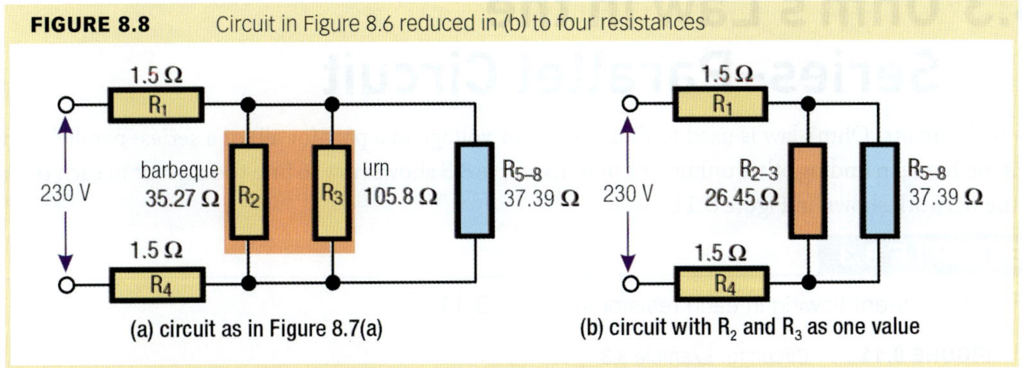

4 Determine the total resistance of R_{2-3} in parallel with R_{5-8}. We will call this value R_{eq1}.

$$R_{eq1} = \frac{R_{2-3} \times R_{5-8}}{R_{2-3} \times R_{5-8}} = \frac{26.45 \times 37.39}{26.45 \times 37.39} = 15.49\ \Omega.$$

This gives the ciarcuit in Figure 8.9(b).

FIGURE 8.9 Circuit in Figure 8.6 reduced to three series resistances

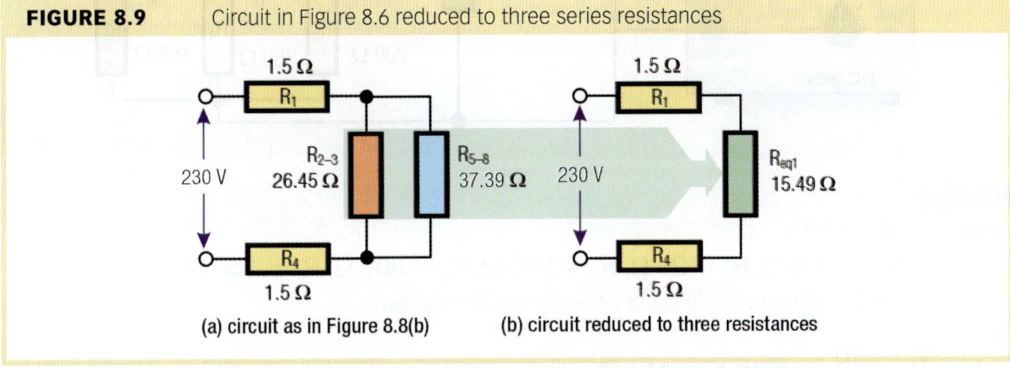

5 Calculate the total resistance of the original circuit (Figure 8.6) by adding the resistance values in Figure 8.9(b).

$R_T = R_1 + R_{eq1} + R_4 = 1.5 + 15.49 + 1.5 = 18.49\ \Omega$

Answer: $\mathbf{R_T = 18.49\ \Omega}$

These two examples show that to find the total resistance of a network of resistances requires simplifying the circuit one section at a time. Resistances in series are combined by adding their values to obtain one resistance value. Resistances in parallel are resolved to one value using the various equations for parallel-connected resistances. Drawing a circuit showing each simplification often helps the process.

TASK 8.1

1 Find the total resistance of both circuits in Figure 8.10.

FIGURE 8.10 Circuits for Task 8.1

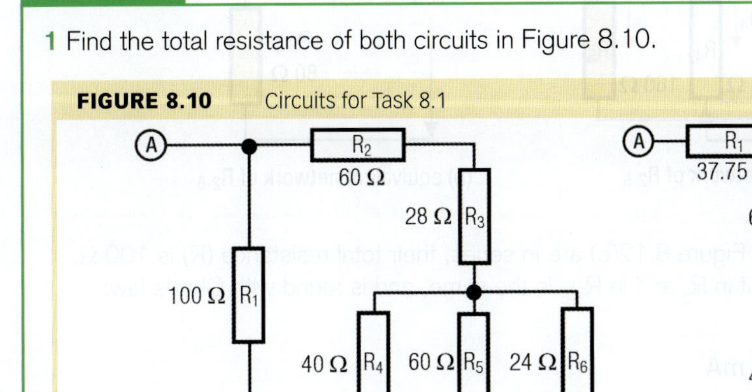

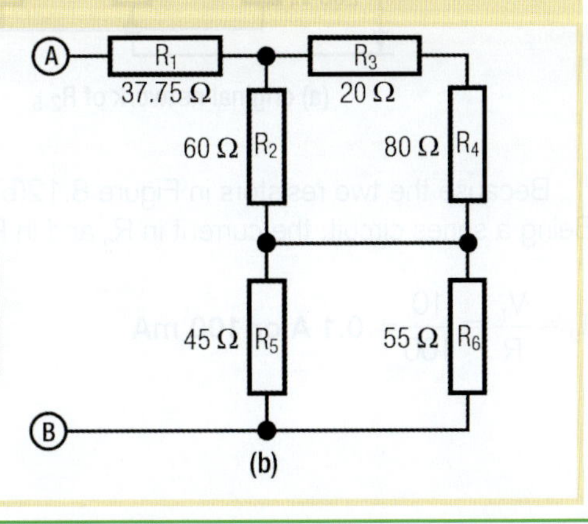

8.3 Ohm's Law in the Series–Parallel Circuit

As in all circuits, Ohm's law is used to find current or voltage in a part (or all) of a series–parallel circuit. This might mean finding other unknowns first. Example 8.3 shows how to find the current in each resistor of the network shown in Figure 8.11.

EXAMPLE 8.3

Find the current flowing in each resistor in Figure 8.11.

FIGURE 8.11 Circuit for Example 8.3

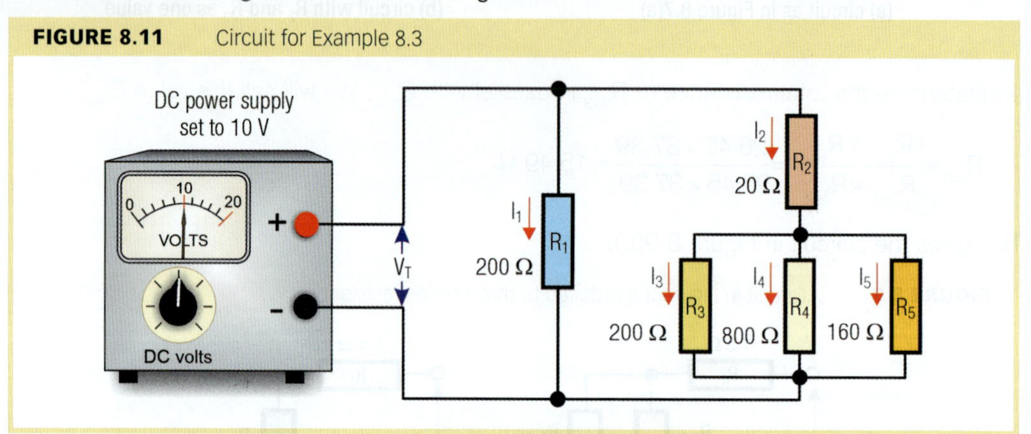

Solution

Values $V_T = 10$ V

$R_1 = 200\ \Omega$, $R_2 = 20\ \Omega$, $R_3 = 200\ \Omega$, $R_4 = 800\ \Omega$, $R_5 = 160\ \Omega$

1 The current in R_1 is a simple case of applying Ohm's law:

$$I_1 = \frac{V_T}{R_1} = \frac{10}{200} = \mathbf{0.05\ A\ or\ 50\ mA}$$

2 To find the current in R_2, simplify the network of R_2 to R_5. The parallel combination of R_3 to R_5 is 80 Ω so the whole section can be drawn as two resistors connected in series with the 10 V supply as shown in Figure 8.12(b).

FIGURE 8.12 Simplifying network of R_2 to R_5

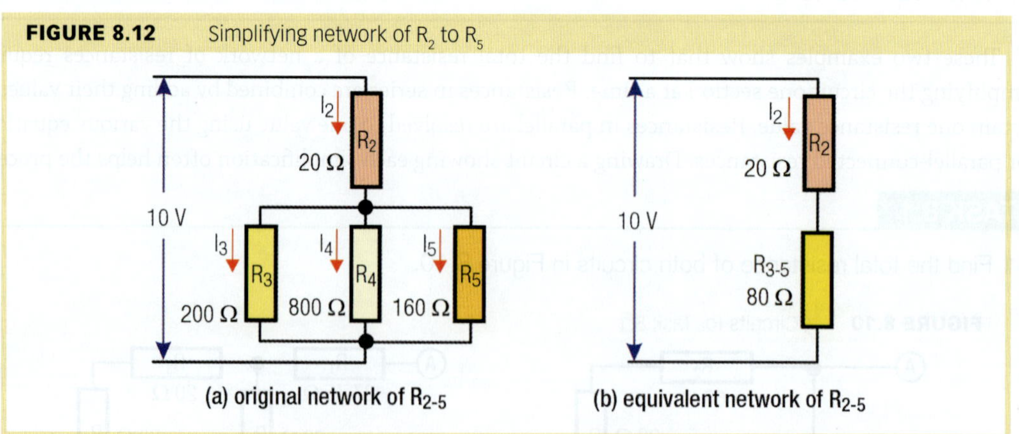

Because the two resistors in Figure 8.12(b) are in series, their total resistance (R) is 100 Ω. Being a series circuit, the current in R_2 and in R_{3-5} is the same, and is found with Ohm's law:

$$I_2 = \frac{V_T}{R} = \frac{10}{100} = \mathbf{0.1\ A\ or\ 100\ mA}$$

3 Calculating the current in resistors R_3, R_4 and R_5 requires finding the voltage drop across them, starting with Ohm's law to find the voltage drop (V_{R2}) across R_2:

$V_{R2} = I_2 \times R_2 = 0.1 \times 20 = \mathbf{2\ V}$

by Kirchhoff's law, $V_{R3-5} = V_T - V_{R2} = 10 - 2 = \mathbf{8\ V}$

Figure 8.13 shows the current and voltage values found so far.

FIGURE 8.13 Currents and voltages found so far in Example 8.3

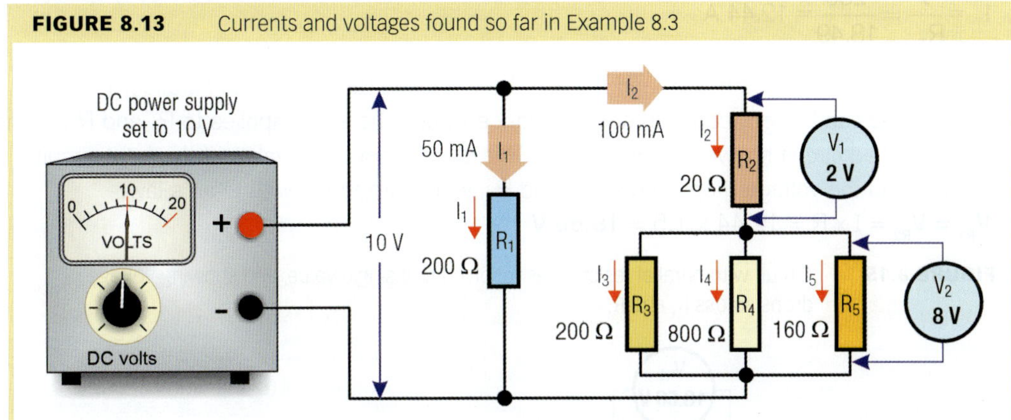

4 Ohm's law is now used to find the remaining currents:

Current in R_3: $I_3 = \frac{V_2}{R_3} = \frac{8}{200} = \mathbf{0.04\ A\ or\ 40\ mA}$

Current in R_4: $I_4 = \frac{V_2}{R_4} = \frac{8}{800} = \mathbf{0.01\ A\ or\ 10\ mA}$

Current in R_5: $I_5 = \frac{V_2}{R_5} = \frac{8}{160} = \mathbf{0.05\ A\ or\ 50\ mA}$

This checks, because the sum of these currents, by Kirchhoff's law, equals I_2.

Example 8.4 shows how to find the voltage across each resistance in the series–parallel circuit from Example 8.2. It will illustrate the importance of using the correct cable size in situations involving long extension leads. The extension lead resistance values are based on a cable size of 1.5 mm^2, as is typical in light-duty leads.

EXAMPLE 8.4

Find the voltage across each resistance in Figure 8.14.

FIGURE 8.14 Circuit as used in Example 8.2

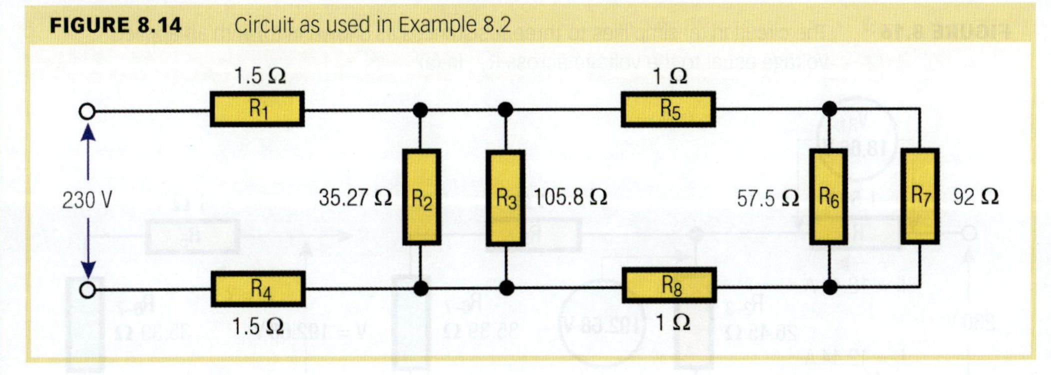

Solution

To find the voltages across each resistance in Figure 8.14 requires finding the current flowing in each part of the circuit. The total resistance of the circuit was found in Example 8.2 to be 18.49 Ω, so:

1 The total current flowing in the circuit is found with Ohm's law.

$$I_T = \frac{V}{R_T} = \frac{230}{18.49} = 12.44 \text{ A}$$

2 Because the voltage across R_2 and R_3 must be the same (as they are in parallel), these resistances can be replaced with one resistance value. The same applies to R_6 and R_7, giving the simplified circuit shown in Figure 8.15. This circuit shows the value of the total circuit current and the voltage drops across R_1 and R_4, which are found with Ohm's law:

$\mathbf{V_{R1} = V_{R2}} = I \times R = 12.44 \times 1.5 = \mathbf{18.66\ V}$

FIGURE 8.15 Circuit with parallel resistances resolved to a single value, and showing voltage drops across R_1 and R_4

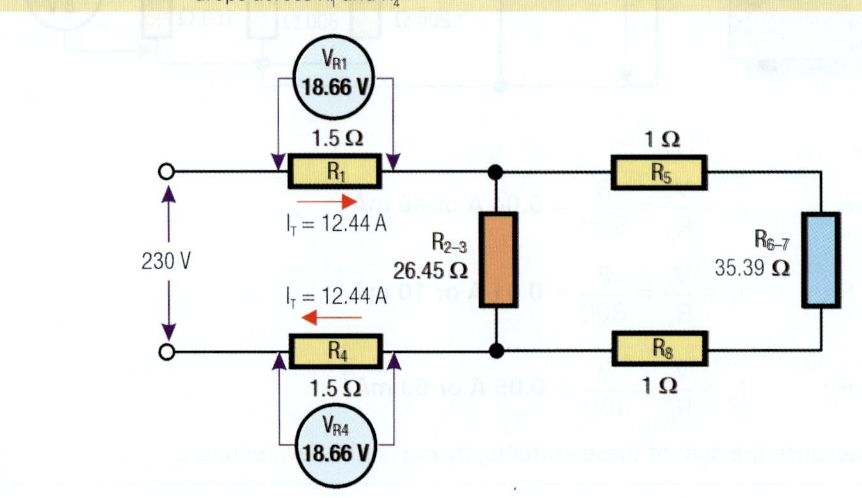

3 The voltage across R_{2-3} can now be found using Kirchhoff's voltage law. This voltage is the applied voltage minus the sum of the two voltage drops:

$\mathbf{V_{R2-3}} = 230 - 18.66 - 18.66 = \mathbf{192.68\ V}$

Figure 8.16(b) shows that R_5 to R_8 can be considered as three series-connected resistances supplied by 192.68 V. The total resistance of R_5 to R_8 is the sum of the three values. We can therefore use Ohm's law to find the current flowing in the loop, and the voltages across each resistance.

FIGURE 8.16 The circuit in (a) simplifies to three resistances as shown in (b) with an applied voltage equal to the voltage across R_{2-3} in (a)

(a) voltage drops found so far

(b) section yet to analyse

4 The total resistance of R_5 to R_8 = 1 + 35.39 + 1 = 37.39 Ω. The current (I) flowing in this section is found with Ohm's law:

$$I = \frac{V}{R} = \frac{192.68}{37.39} = 5.15\text{ A}$$

The voltage across each resistance in (b) is found with Ohm's law:

$V_{R5} = I \times R_5 = 5.15 \times 1 = \mathbf{5.15\ V}$
$V_{R6\text{-}7} = I \times R_{6-7} = 5.15 \times 35.39 = \mathbf{182.37\ V}$
$V_{R8} = I \times R_8 = 5.15 \times 1 = \mathbf{5.15\ V}$

There are other ways of arriving at some of these voltages. If the current in R_{2-3} is calculated with Ohm's law, the current in R_5 can be found with Kirchhoff's current law, rather than Ohm's law. The voltages across R_5 to R_8 are then found with Ohm's law. There is no best method to use, just work through one step at a time. Figure 8.17 shows all the voltage drops in the circuit.

FIGURE 8.17 Resistance R_{2-3} represents the appliances receiving 192.68 V. Resistance R_{6-7} represents the appliances receiving 182.37 V.

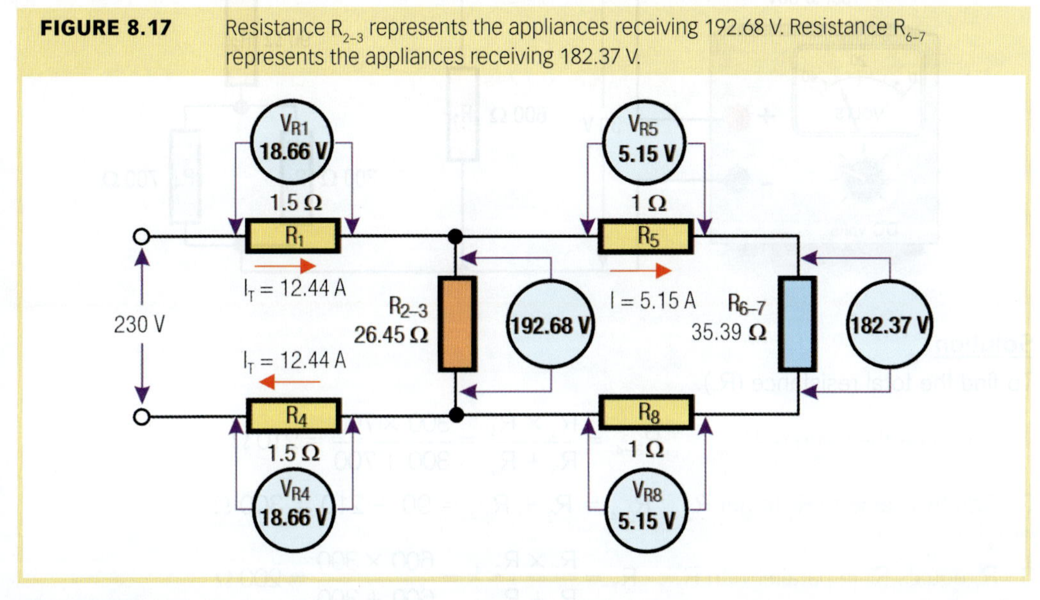

TASK 8.2

1 For the circuit in Figure 8.18 (a), find:

- **a** total resistance of the circuit
- **b** current flowing in R_1
- **c** voltage across R_4

2 For the circuit in Figure 8.18 (b) find:

- **a** total resistance of the circuit
- **b** voltage across R_2
- **c** current flowing in R_6

FIGURE 8.18 Circuits for Task 8.2

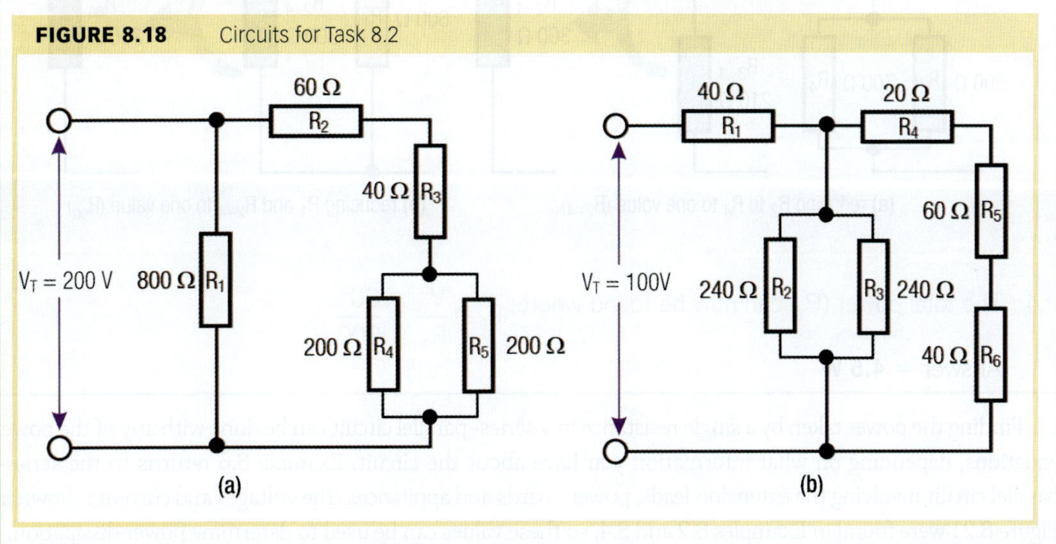

8.4 Power in the Series–Parallel Circuit

Calculating power in a series–parallel circuit uses the same equations as for all circuits. Therefore, it is often necessary to determine current, resistance or voltage values first. Example 8.5 shows a network of resistors in which the power taken by the circuit is found by first calculating the total resistance of the circuit.

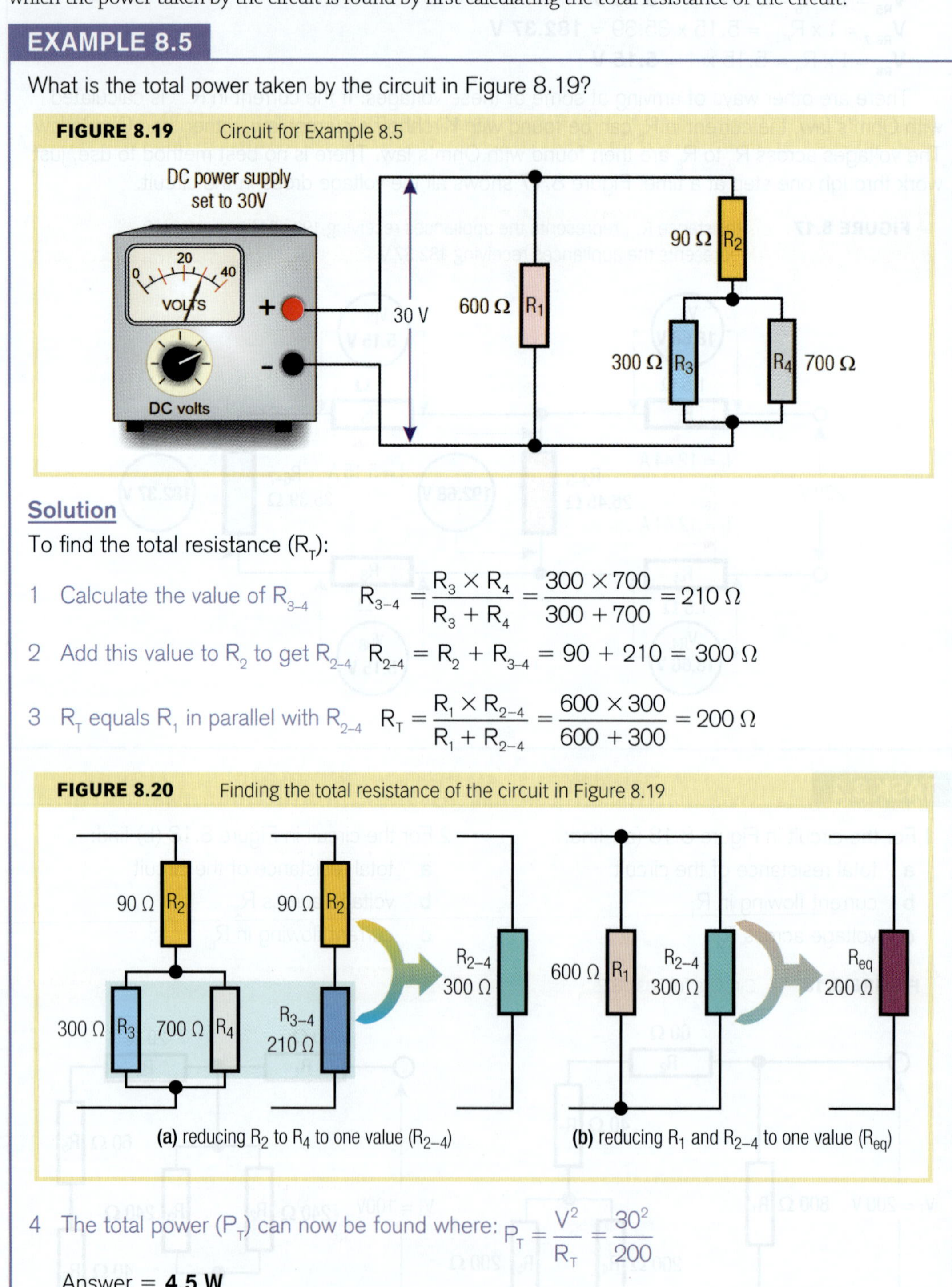

EXAMPLE 8.5

What is the total power taken by the circuit in Figure 8.19?

FIGURE 8.19 Circuit for Example 8.5

Solution

To find the total resistance (R_T):

1 Calculate the value of R_{3-4} $\quad R_{3-4} = \dfrac{R_3 \times R_4}{R_3 + R_4} = \dfrac{300 \times 700}{300 + 700} = 210\ \Omega$

2 Add this value to R_2 to get R_{2-4} $\quad R_{2-4} = R_2 + R_{3-4} = 90 + 210 = 300\ \Omega$

3 R_T equals R_1 in parallel with R_{2-4} $\quad R_T = \dfrac{R_1 \times R_{2-4}}{R_1 + R_{2-4}} = \dfrac{600 \times 300}{600 + 300} = 200\ \Omega$

FIGURE 8.20 Finding the total resistance of the circuit in Figure 8.19

4 The total power (P_T) can now be found where: $P_T = \dfrac{V^2}{R_T} = \dfrac{30^2}{200}$

Answer = **4.5 W**

Finding the power taken by a single resistance in a series–parallel circuit can be done with any of the power equations, depending on what information you have about the circuit. Example 8.6 returns to the series–parallel circuit involving the extension leads, power boards and appliances. The voltages and currents shown in Figure 8.21 were found in Examples 8.2 and 8.4, so these values can be used to determine power dissipation.

EXAMPLE 8.6

Find the:

1 total power taken by the circuit in Figure 8.21
2 power taken by each resistance in Figure 8.21.

FIGURE 8.21 Complete circuit with all voltage values found previously

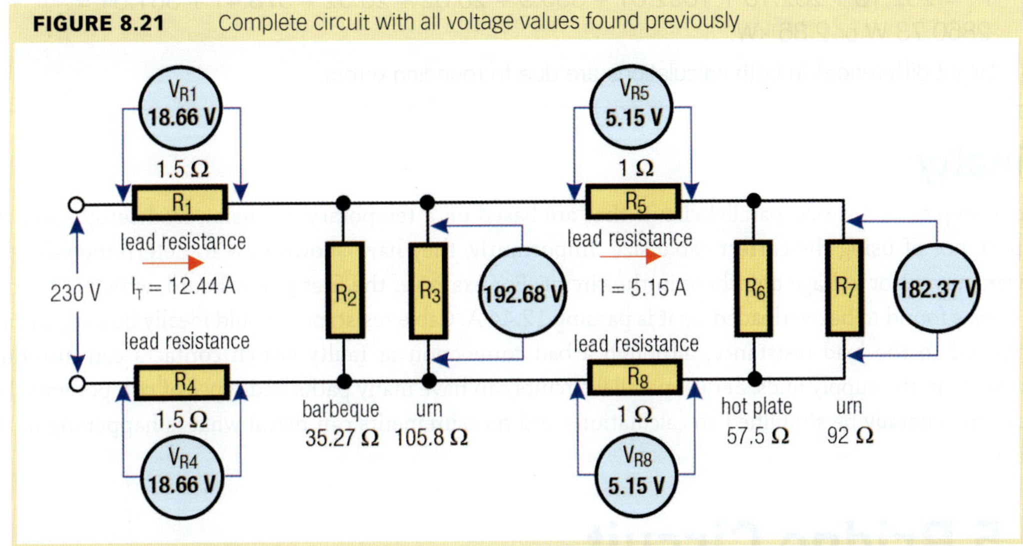

Solution

1 To find total power, because the total circuit current and applied voltage are known, we can use the following equation:

$\mathbf{P_T} = V \times I_T = 230 \times 12.44 =$ **2861.2 W** or **2.86 kW**

2 To find the power dissipated by each resistance:

$$P_{R1} = P_{R4} = \frac{V_{R1}^2}{R_1} = \frac{18.66^2}{1.5} = \mathbf{232.13\ W}$$

Therefore, as P_{R1} and P_{R4} are the power dissipation values in each cable, 464.26 W of power is dissipated in the 60 m extension lead, meaning it will get hot.

$$P_{R2} = \frac{V_{R2}^2}{R_2} = \frac{192.68^2}{35.27} = \mathbf{1052.61\ W\ or\ 1.052\ kW}$$

R_2 is the equivalent resistance of the 1.5 kW barbeque, which is now operating at about 70 per cent capacity.

$$P_{R3} = \frac{V_{R3}^3}{R_3} = \frac{192.68^2}{105.8} = \mathbf{350.9\ W}$$

R_3 is the equivalent resistance of the 500 W hot water urn, which is now operating at about 70 per cent capacity.

$$P_{R5} = P_{R8} = \frac{V_{R5}^2}{R_5} = \frac{5.15^2}{1} = \mathbf{26.52\ W}$$

Therefore, 53.1 W of power is dissipated in the 50 m extension lead.

$$P_{R6} = \frac{V_{R6}^2}{R_6} = \frac{182.37^2}{57.5} = \mathbf{578.41\ W}$$

R_6 is the equivalent resistance of the 920 W hot plate, which is now operating at around 63 per cent capacity.

$$P_{R7} = \frac{V_{R7}^2}{R_7} = \frac{182.37^2}{97} = \mathbf{361.51\ W}$$

R_7 is the equivalent resistance of the 575 W hot water urn which is now also operating at around 63 per cent capacity.

As a check, the sum of the individual powers should equal the total power:

$P_T = P_{R1} + P_{R4} + P_{R2} + P_{R3} + P_{R5} + P_{R8} + P_{R6} + P_{R7}$

$P_T = 232.13 + 232.13 + 1052.61 + 350.9 + 26.52 + 26.52 + 578.41 + 361.51 = 2860.73$ W or 2.86 kW.

Slight differences in both calculations are due to rounding errors.

Finally

The examples of a series–parallel circuit that are based on a temporary extension lead setup show the importance of using the correct cable size. Importantly, they have shown how to determine values of power, current or voltage anywhere in the circuit. For example, the first power board, which is rated at 10 A, was found to be overloaded, as it is passing 12.44 A. Cable resistance should ideally be insignificant compared to the load resistance, although a bad connection or faulty switch contacts can introduce resistance in the supply lines. Series–parallel circuits can have many paths and types of components, but they can generally be simplified so calculations and measurements can reveal what is happening in the circuit.

8.5 Bridge Circuit

The bridge circuit is a special case of a series–parallel circuit. It is used in some types of measuring equipment and is generally drawn in the diamond-like arrangement shown in Figure 8.22(b).

FIGURE 8.22 A bridge circuit has two branches or legs, bridged by, in this case, a voltmeter

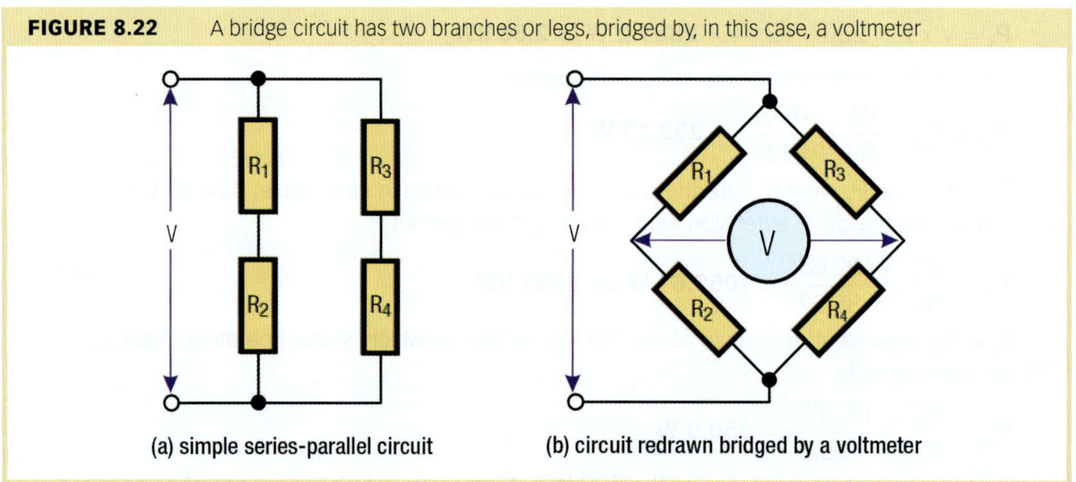

(a) simple series-parallel circuit

(b) circuit redrawn bridged by a voltmeter

The main application of a bridge circuit containing only resistors is the Wheatstone bridge, an instrument developed by Charles Wheatstone to measure low values of resistance.

Figure 8.23(b) shows the basic circuit of a Wheatstone bridge. Using this circuit to find the value of an unknown resistance relies on knowing the ratio of the resistor values in the two legs. The circuits in Figure 8.23 show that R_1 and R_2 have a ratio of 100 to 200, or 1 to 2. Resistors R_3 and R_4 have a ratio of 10 to 20, or 1 to 2. That is, both sides have the same ratio but with different resistor values.

Currents I_1 and I_2 can be found with Ohm's law once the resistance of each leg is determined. Then, voltages V_1 and V_2 can be calculated.

1 To find V_1.

$R_1 + R_2 = 100 + 200 = 300\ \Omega$

$I_1 = \frac{12}{300} = 0.04$ A or 40 mA

$V_1 = I_1 \times R_2 = 0.04 \times 200 = 8$ V

FIGURE 8.23 Bridge circuit in which both legs have resistors with same ratio but with different values

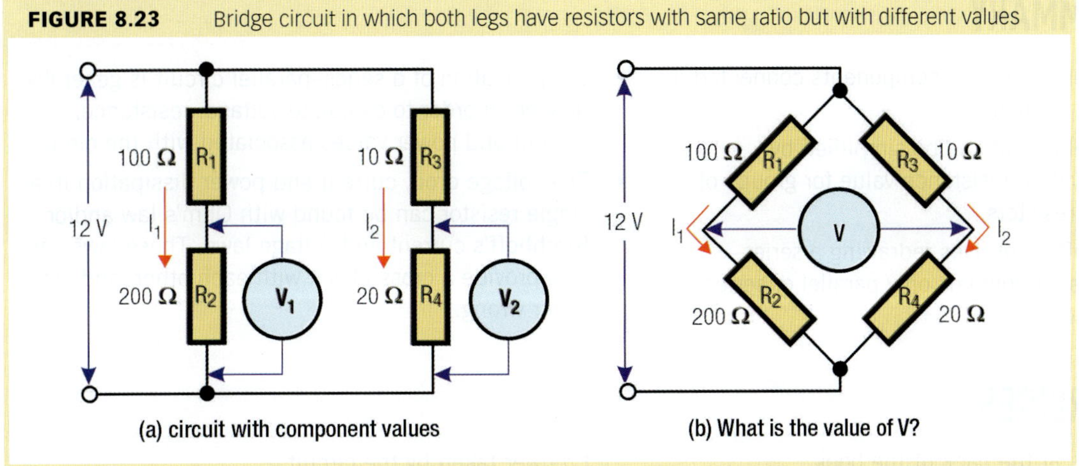

2 To find V_2

$R_3 + R_4 = 10 + 20 = 30\ \Omega$

$I_2 = \frac{12}{30} = 0.4$ A or 400 mA

$V_2 = I_2 \times R_4 = 0.4 \times 20 = 8$ V

Because both R_2 and R_4 have the same voltage across them, the voltmeter in Figure 8.23(b) will read zero, as there is no potential difference between the two points. If the voltmeter shows a reading other than zero, it means there is a difference in the ratio of the resistances in each leg.

Figure 8.24 shows a bridge circuit with a sensitive centre-zero voltmeter connected to indicate the difference in voltages across a known resistance (R_2) and a calibrated variable resistor (R_4). The only resistance with an unknown value is R_3, which is the resistance being measured.

To balance the bridge, the variable resistance R_4 is adjusted until the voltmeter reads zero. The value of R_4 is indicated by a scale, and we know the ratio of resistances in each leg are 1 to 2. That is, the value of R_4 at balance will be twice the value of the unknown resistance R_3. So, a reading of 12 ohms for R_4 means the unknown resistance is half that value, or 6 ohms.

FIGURE 8.24 Concept of using a Wheatstone bridge circuit to measure an unknown resistance

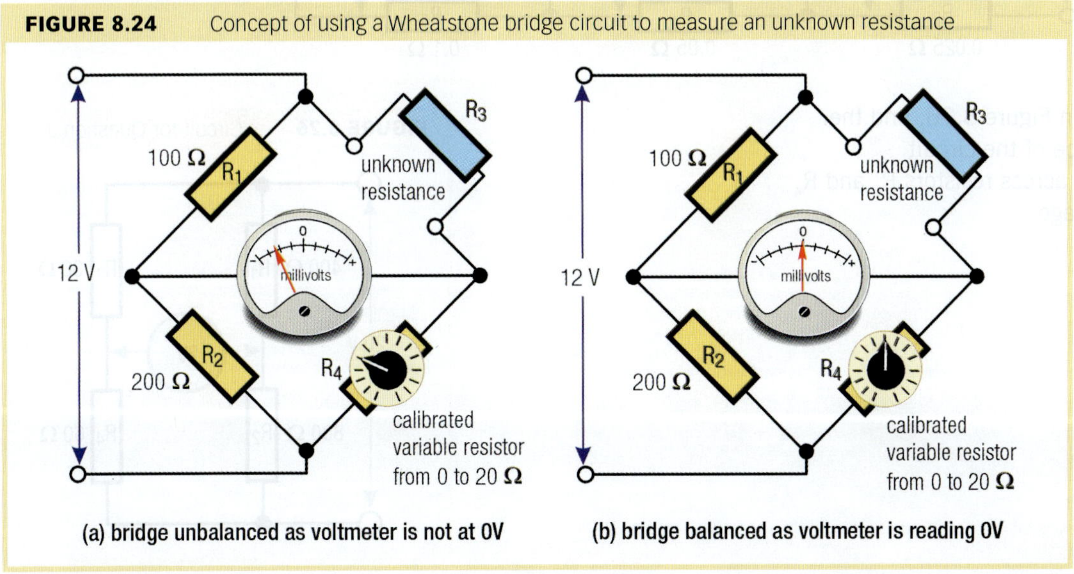

Bridge circuits occur in various forms, and typically have an input and an output. In the example shown, the input is a 12 V supply, and the output is connected to a voltmeter. Another example of a bridge circuit is a rectifier that converts alternating current (AC) to direct current (DC). This circuit has four diodes arranged as a bridge circuit, so an AC voltage source is applied to the input, and the rectified voltage (DC) is taken from the output. This type of circuit is known as a bridge rectifier.

CHAPTER SUMMARY

- A series–parallel circuit has components connected in both series and parallel.
- A series–parallel circuit can be simplified by finding an equivalent resistance value for groups of interconnected resistors.
- Simplification often involves redrawing a series–parallel circuit so it contains only parallel or series-connected resistors.
- Simplification of a series–parallel circuit is generally needed in order to calculate voltage, resistance, current and power values associated with the circuit.
- The voltage drop, current and power dissipation in a single resistor can be found with Ohm's law and/or Kirchhoff's current and voltage laws. These laws can also provide a cross-check with each other, and are never wrong.

REVIEW EXERCISES

Check your answers at the back of the book.

1 The garden lighting circuit shown in Figure 8.25 has three 50 W quartz-halogen lamps connected to a 12 V supply with long cables. From the information given in the circuit, determine:
   a current I_T
   b current I_1
   c current I_2
   d total resistance of the circuit
   e current in each lamp
   f power taken by the circuit
   g power dissipated in the cable.

2 If each lamp in Figure 8.25 is replaced with a 6 W LED lamp, determine:
   a total resistance of the circuit (hint: determine the equivalent resistance of each LED lamp)
   b current I_T
   c voltage across each LED lamp
   d power taken by the circuit
   e power dissipated in the cable.

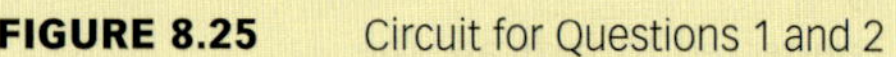

FIGURE 8.25 Circuit for Questions 1 and 2

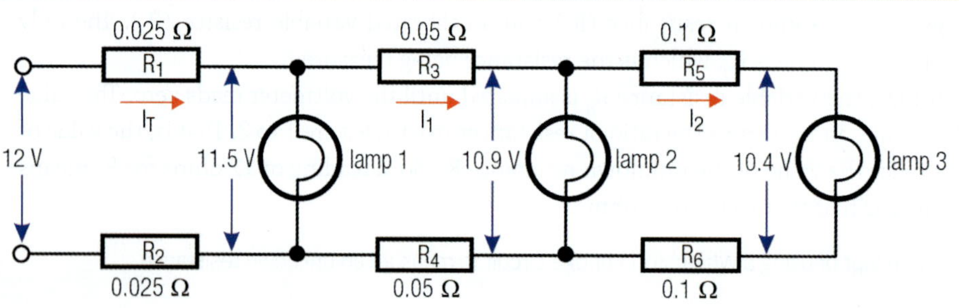

3 For the circuit in Figure 8.26, find the:
   a total resistance of the circuit
   b voltage drops across resistors R_2 and R_4
   c unknown voltage.

FIGURE 8.26 Circuit for Question 3

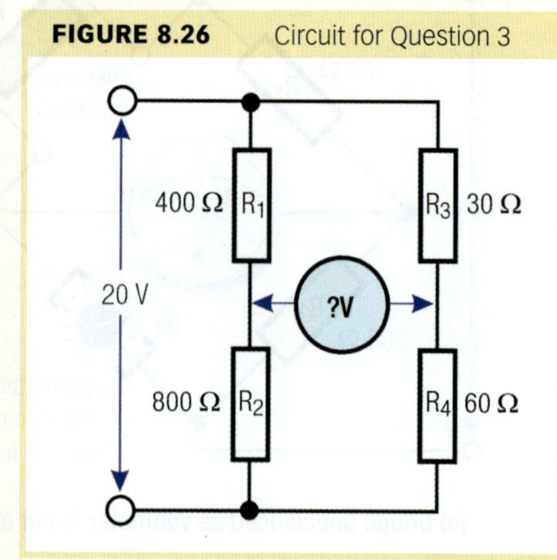

ONLINE RESOURCES

COMPLETE WORKSHEET EIGHT

Check with your instructor for worksheets on this chapter.

CHAPTER 9

BASIC METERS

Electrical quantities such as voltage, current and resistance often need to be measured. This chapter describes the operating principles of an analog and digital ammeter, voltmeter and ohmmeter. It also explains 'meter loading', which is the effect a voltmeter has on the voltage value it is measuring. The chapter discusses the use and safety aspects of a typical analog and digital multimeter, an instrument commonly used in the field and the workshop.

CHAPTER OUTLINE

9.1 Introduction

Voltage, current and resistance can all be measured with a multimeter, which combines an ammeter, voltmeter and ohmmeter in the one case. Most multimeters are digital, although analog multimeters have their advantages and are still used. An analog meter has a meter movement driving a pointer, while a digital meter has an electronic module with a display, powered by a battery. Meters of all types can be classified by their:

- **accuracy** – determined by the quality of the meter, calibration and maintenance
- **resolution** – the number of decimal places shown by a reading
- **sensitivity** – how much power a meter takes from a circuit during a measurement.

In Figure 9.1, the digital meter is showing 1.597 V, giving a resolution to three decimal places. The analog meter is showing 1.62 V, a resolution to two decimal places. A digital multimeter is typically more accurate than an analog multimeter.

FIGURE 9.1 Digital and analog multimeters reading the voltage of a 1.5 V cell

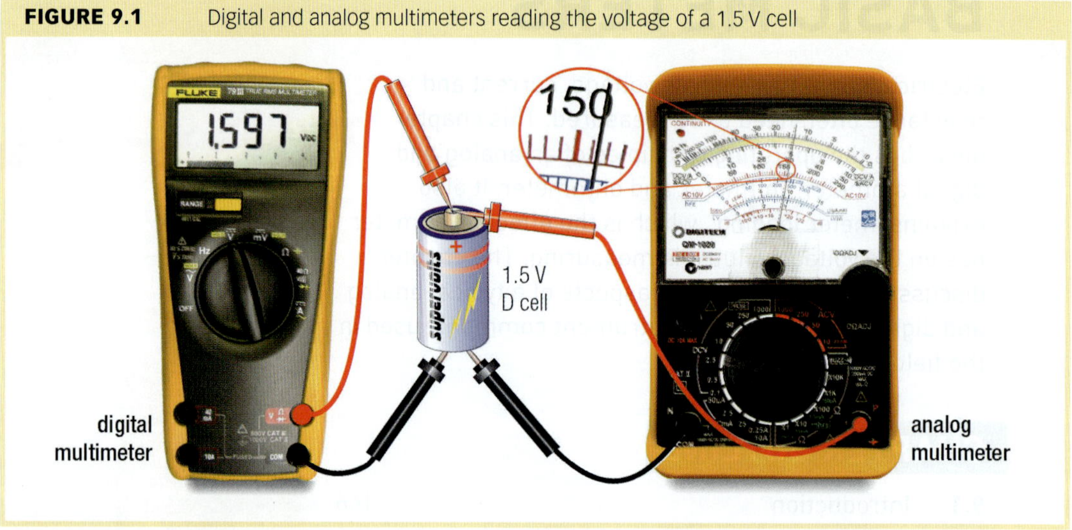

9.2 Analog Meter Movement

An analog meter movement has an assembly that rotates in a magnetic field and moves a pointer over a scale. The d'Arsonval movement, known today as a moving-coil meter, is the most common type, invented by Jacques-Arsène d'Arsonval (1851–1940). See Figure 9.2.

The meter movement develops a turning torque between the field of a permanent magnet and the magnetic field created by current flowing in the coil. The torque is counterbalanced by hair springs, so the assembly moves in proportion to the coil current.

FIGURE 9.2 The moving-coil meter movement, in which the coil is deflected by an amount proportional to the current flowing in the coil

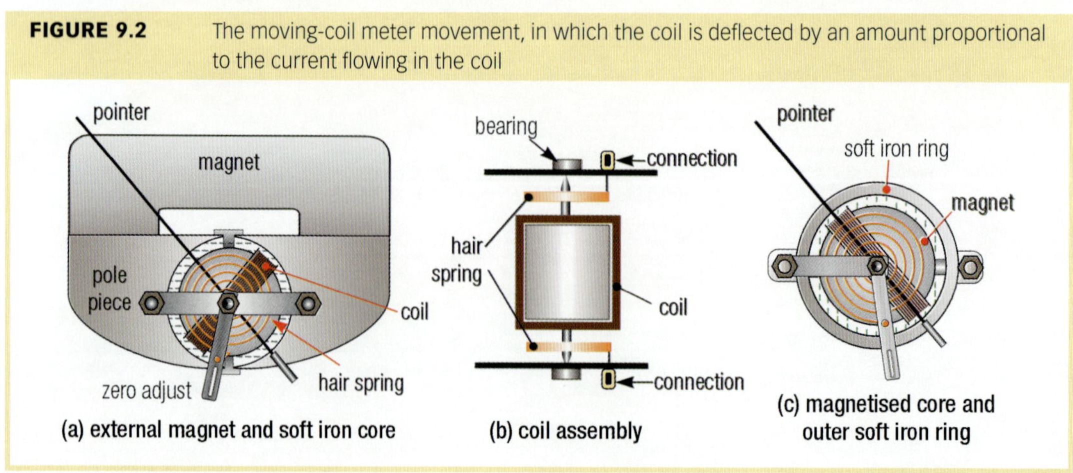

9.4 The Ammeter

An ammeter is connected so the current being measured flows through the ammeter. It therefore must have a low resistance, so it does not affect the value of the current being measured. This applies to an analog and a digital ammeter.

Analog ammeter

A moving-coil meter movement needs as little as 10 μA to be fully deflected. So as an ammeter can measure currents greater than its full-scale deflection current, a resistor called a shunt is connected in parallel with the movement. The shunt bypasses (or shunts) all current away from the meter movement, except the small amount needed to drive the movement.

Figure 9.5 (a) shows a 10 A panel-mount ammeter that has a 5 mA meter movement and a shunt resistor connected internally across its two terminals, shown in (c). The shunt resistor is a piece of thick wire made from manganin (a copper-nickel-manganese alloy). This alloy has a low temperature coefficient, so its resistance is not greatly affected by temperature changes. It also has a higher resistivity than copper, allowing a small but sturdy assembly.

FIGURE 9.5 A panel mount 0 to 10 A ammeter. The meter movement requires 5 mA for full scale deflection, and the shunt inside the case carries the rest of the current.

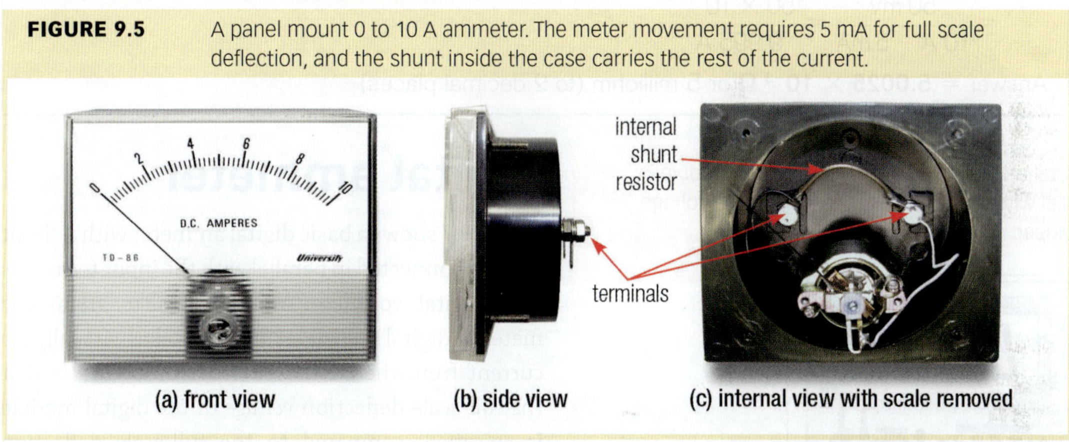

The resistance of an ammeter shunt is always much lower than the resistance of the coil in the meter movement. The lower the shunt resistance, the more current taken by the shunt. The resistance of the ammeter is the parallel combination of its shunt resistor and the resistance of the meter movement.

The value of a shunt resistor for a particular value of ammeter current can be calculated if you know the resistance of the meter movement, and its full scale deflection current. Example 9.1 is based on the meter movement in Figure 9.5.

EXAMPLE 9.1

Calculate the value of the shunt resistor required to make a 10 A ammeter based on a meter movement with a resistance of 10 ohms and a full-scale deflection current of 5 mA.

Solution

Values $I_{FSD} = 5\ \text{mA}$
$R_{meter} = 10\ \Omega$
$I_{ammeter} = 10\ \text{A}$

FYI

In most cases, the current taken by the meter can be ignored when calculating the value of a shunt resistor

To find the value of the shunt resistor, we first determine the voltage that causes full-scale deflection of the meter movement. As shown in Figure 9.6, because they are in parallel, this voltage is the same across the shunt resistor and the meter movement. Therefore:

$$V_{FSD} = I_{FSD} \times R_{meter}$$
$$= 5\ \text{mA} \times 10\ \Omega$$
$$= 50\ \text{mV}$$

The current passing through the shunt resistor is 10 A minus the meter movement current, therefore:

$$R_{shunt} = \frac{V_{FSD}}{I_{shunt}} = \frac{V_{FSD}}{I_{ammeter} - I_{FSD}}$$
$$= \frac{50\ \text{mV}}{10\ \text{A} - 5\ \text{mA}} = \frac{50 \times 10^{-3}}{9.995\ \text{A}}$$

Answer = $\mathbf{5.0025 \times 10^{-3}\ \Omega}$ or 5 milliohm (to 2 decimal places)

FIGURE 9.6 Circuit for Example 9.1

FIGURE 9.7 A digital ammeter has a digital voltmeter module in parallel with a shunt resistor. It displays the voltage developed across the shunt resistor.

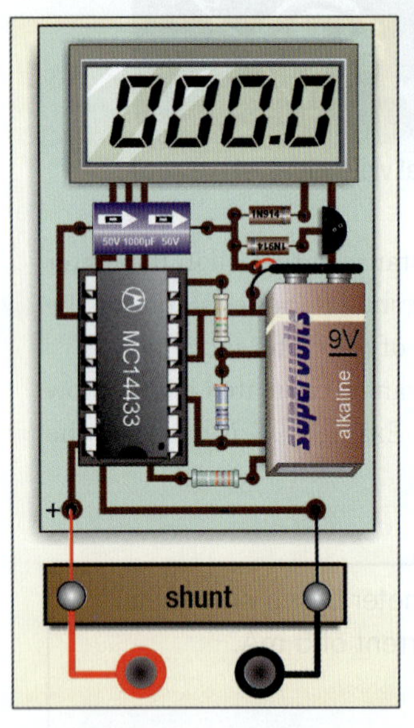

Digital ammeter

Figure 9.7 shows a basic digital ammeter with a shunt resistor connected in parallel with the input terminals of a digital voltmeter module. Unlike an analog meter, a digital voltmeter module takes virtually no current from the circuit. Another difference is that the full-scale deflection *voltage* of the digital module is specified, compared to the full-scale deflection current for a meter movement.

A typical digital voltmeter module has a maximum display value of 199.9, which occurs when the voltage at its inputs is 199.9 mV. If the input voltage to the module is 10 mV, the display will show 10.0. Therefore, so the meter will display 10.0 when 10 A is flowing through a shunt resistor, the value of the resistor must cause a 10 mV drop across it when 10 A is flowing through it. This is shown in Example 9.2 and in Figure 9.8.

EXAMPLE 9.2

Calculate the value of the shunt resistor required to make a 10 A ammeter based on a digital voltmeter module that has a full-scale input voltage of 199.9 mV.

Solution

Values V_{FSD} = 199.9 mV
R_{module} = open-circuit (infinite Ω)
$I_{ammeter}$ = 10 A

Given that the maximum input voltage to the module is 199.9 mV, the input voltage required to show a value of 10.0 on the display = 10 mV.

$$R_{shunt} = \frac{10\ \text{mV}}{10\ \text{A}} = \frac{10 \times 10^{-3}}{10}$$

Answer = **1 × 10⁻³ Ω** or 1 milliohm

FIGURE 9.8 Circuit for Example 9.2

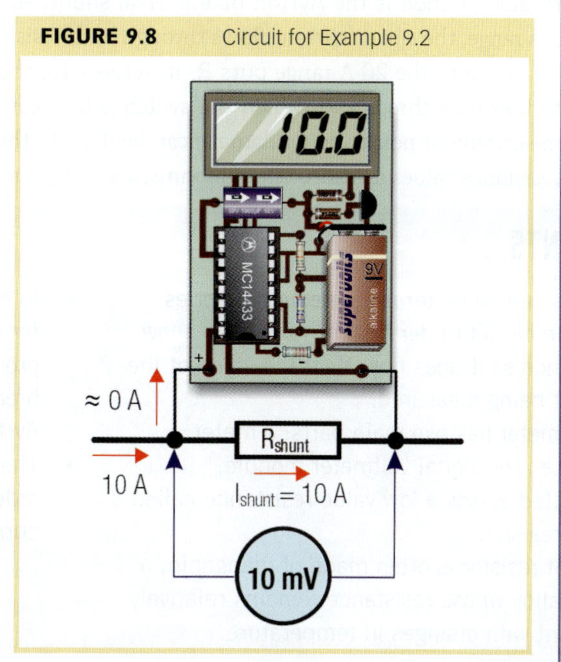

To increase the resolution of the ammeter reading, the resistance of the shunt resistor could be increased by a factor of 10 (to 10 milliohms), so the voltage across it at 10 A becomes 100 mV. The position of the decimal point on the display would be switched to move one place left, so when the current in the shunt resistor is 10 A, the display shows 10.00.

A multi-range ammeter

The most expensive part of an ammeter is the display (analog or digital). Therefore, ammeters often have a number of shunt resistors so the ammeter can read a range of currents. Each range could be selected with a switch, or the ammeter might have a separate terminal for each range.

The diagrams in Figure 9.9 show two ways of connecting the shunt resistors for an ammeter with two ranges. In (a), each shunt is a separate component selected by a switch. However, when the switch is operated, it creates a momentary open-circuit while moving between contacts. If this happens while the ammeter is connected to a live circuit, the full circuit voltage is applied to the meter movement or digital module, which neither can withstand.

FIGURE 9.9 A multi-range ammeter can have as many ranges as required. Those shown have a 2 A and a 20 A range, selected with a switch.

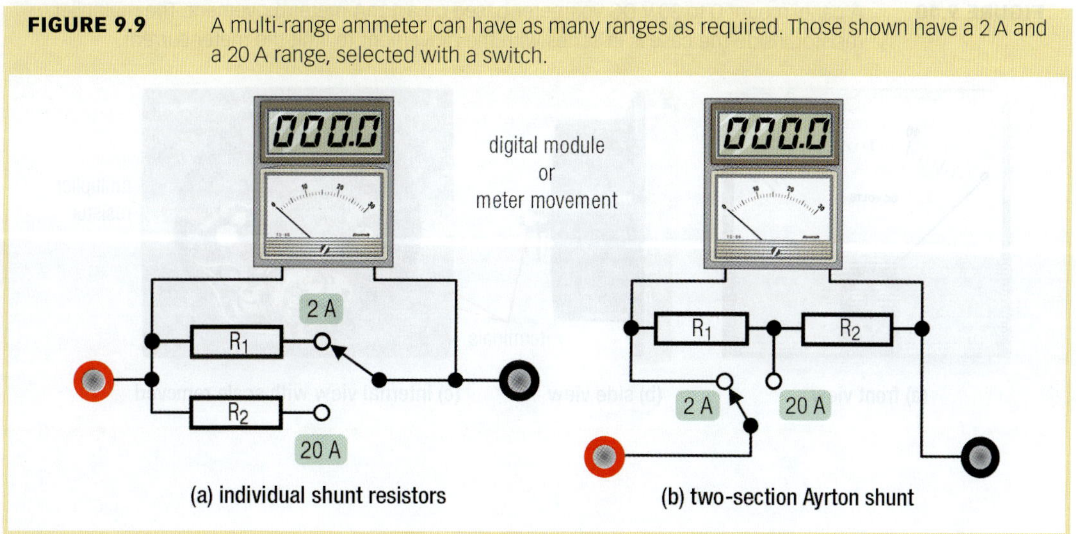

To prevent damage to the meter module, some ammeters have a make-before-break switch. A more reliable method is the Ayrton or universal shunt, shown in Figure 9.9(b). When the switch is on the 2 A range, the circuit current flows through both resistors R_1 and R_2, so the shunt resistance is their sum. Changing to the 20 A range puts R_1 in series with the meter movement and causes the circuit current to flow only through R_2. When the switch is between contacts, the ammeter is disconnected from the measurement points, so no damage can be done to the meter module. Also, R_1 and R_2, which both have resistance values measured in milliohms, are always in parallel with the meter.

KEY POINTS...

- Because an ammeter is connected in series with the circuit under test, it must have a low resistance so it does not affect the value of the current being measured.
- An ammeter has two main parts: a meter movement or digital voltmeter module, connected across a low value resistance called a shunt resistor.
- A shunt resistor is often made of manganin, a metal alloy whose resistance remains relatively constant with changes in temperature.
- In a multi-range ammeter, the shunt resistor for each range can be a separate component, providing the selector switch has a make-before-break action, or a tapped resistor called an Ayrton or universal shunt.
- The resistance of a shunt is generally in the order of a few milliohms, depending on the current range it is designed for.

9.5 The Voltmeter

A voltage in a circuit is either a voltage source (e.g. a battery) or a voltage drop across a component. To prevent it affecting the voltage it is measuring, a voltmeter should have a high resistance. That is, it should not load the circuit, as explained later.

Analog voltmeter

The deflection current of a moving-coil meter movement is supplied by the circuit under test. To limit the voltage across the meter movement to its full-scale voltage, a resistor called a *multiplier* is connected in series with the meter. Its value is chosen so the meter movement is fully deflected at the maximum voltage the voltmeter is designed to measure.

Figure 9.10 shows a 30 V panel-mount voltmeter based on a 1 mA meter movement. The multiplier resistor is a 30 kΩ metal film ½ W resistor connected internally in series with the meter movement.

FIGURE 9.10 A panel mount 0 to 30 V DC voltmeter based on a 1 mA meter movement. The multiplier resistor inside the case is in series with the movement to limit the meter current.

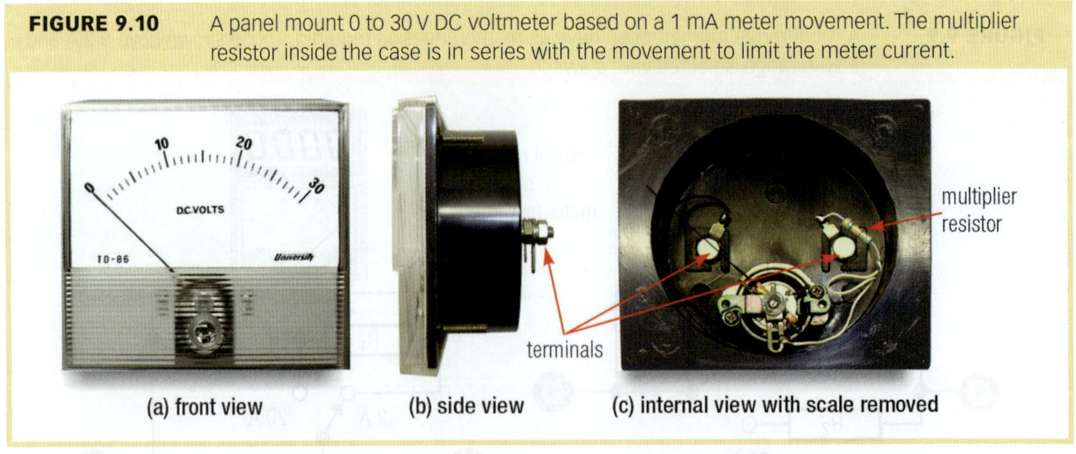

The value of a multiplier resistor for a particular voltage range can be calculated if you know the full-scale deflection current of the meter movement. The resistance of the meter movement is generally low enough to be ignored, which makes the calculations quite easy, as shown in Example 9.3. This example is based on the meter movement in Figure 9.10.

EXAMPLE 9.3

Calculate the value of the multiplier resistor required to make a 30 V voltmeter based on a meter movement with a resistance of 100 ohms and a full-scale deflection current of 1 mA.

Solution

Values $I_{FSD} = 1\text{ mA}$
$R_{meter} = 100\ \Omega$
$V_{voltmeter} = 30\text{ V}$

By Ohm's law, we can find the total resistance required to limit the current to 1 mA when 30 V is applied:

$$R_t = \frac{V_{voltmeter}}{I_{FSD}} = \frac{30\text{ V}}{1\text{ mA}}$$

$$= \mathbf{30\ k\Omega}$$

The meter movement resistance is 100 Ω, which means the multiplier resistor should ideally be 30 kΩ less 100 Ω, giving a value of 29 900 Ω. For practical purposes, the error will be insignificant if we ignore the meter movement's resistance. Therefore:

Answer $= \mathbf{30\ k\Omega}$

FIGURE 9.11 Circuit for Example 9.3

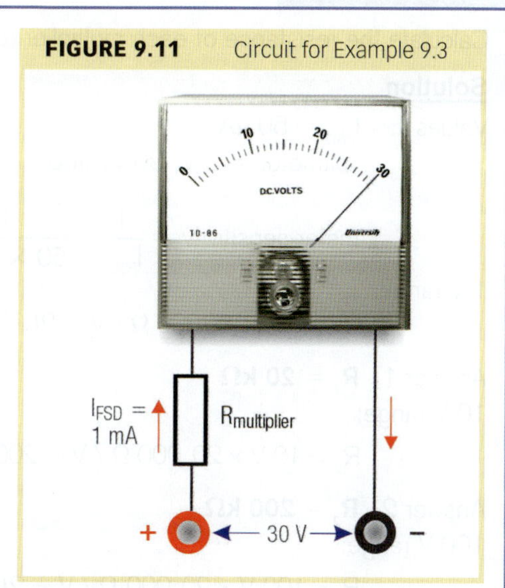

The sensitivity of a moving-coil meter movement can be expressed in ohms per volt, which is the reciprocal of its full-scale deflection current. A 10 μA meter movement has a sensitivity of $1/(10 \times 10^{-6})$, which is 100 000 ohms per volt. A 1 mA meter movement has a much lower sensitivity of $1/(1 \times 10^{-3})$, or 1000 ohms per volt. If you know the sensitivity of the meter movement, the multiplier resistor value can be found by multiplying the sensitivity in ohms/volt by the required full-scale voltage.

In Example 9.3, the multiplier resistor can be determined by multiplying the meter sensitivity of 1000 Ω/V by 30 V, giving 30 kΩ, as found in the example. If a 10 μA meter movement is used, the value of the multiplier resistor will be 100 000 Ω/V × 30 V, which gives 3 MΩ.

A multi-range analog voltmeter

A multi-range analog voltmeter has a number of multiplier resistors selected by a switch. The voltmeter in Figure 9.12 has three ranges, a single scale and a 50 μA meter movement. On the 1 V range, the scale reading is divided by 10; on the 100 V range the scale reading is multiplied by 10.

FIGURE 9.12 A multi-range voltmeter with three multipliers selected with a switch

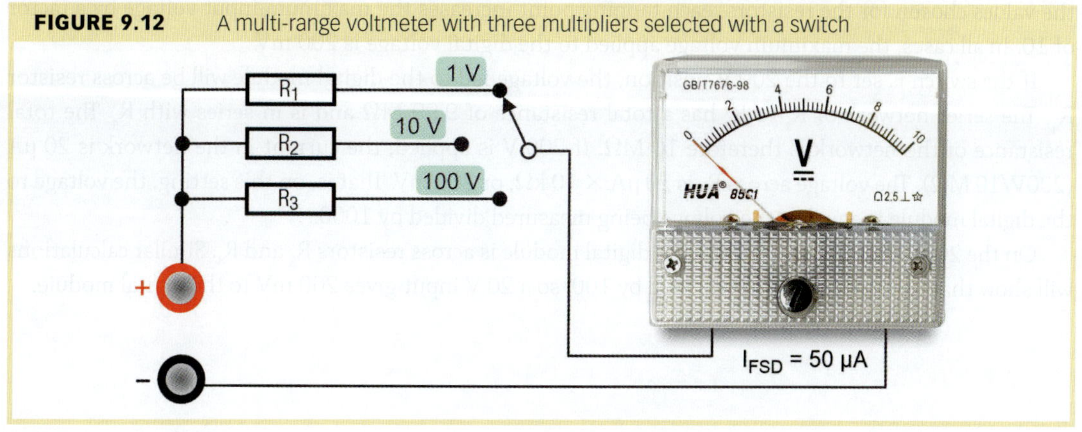

As in all analog voltmeters, the total resistance between the voltmeter terminals is the sum of the selected multiplier resistor and the resistance of the meter movement. Example 9.4 uses the meter sensitivity in ohms per volt to calculate the resistance of each multiplier resistor in Figure 9.12. The resistance of the meter movement is 1 kΩ, which is low enough to be ignored.

EXAMPLE 9.4

Calculate the resistance of each multiplier resistor in Figure 9.12.

Solution

Values $I_{FSD} = 50\ \mu A$

voltmeter = 3 V, 30 V, 300 V

$$\text{meter sensitivity} = \frac{1}{I_{FSD}} = \frac{1}{50 \times 10^{-6}} = 20\,000 \text{ ohms/volt}$$

1 V range:

$$R_1 = 1\text{V} \times 20000\ \Omega/\text{V} = 20\ 000\ \Omega$$

Answer 1 $\mathbf{R_1 = 20\ k\Omega}$

10 V range:

$$R_2 = 10\ \text{V} \times 20\ 000\ \Omega/\text{V} = 200\ 000\ \Omega$$

Answer 2 $\mathbf{R_2 = 200\ k\Omega}$

100 V range:

$$R_3 = 100\ \text{V} \times 20\ 000\ \Omega/\text{V} = 2\,000\,000\ \Omega$$

Answer 3 $\mathbf{R_3 = 2\ M\Omega}$

The important thing this shows is that the resistance of an analog voltmeter changes with each voltage range. As explained later, the resistance of a voltmeter can affect the voltages it's measuring. This is called *voltmeter loading*.

Digital voltmeter

A basic circuit for a digital voltmeter is shown in Figure 9.13. It is different to an analog voltmeter circuit and has two main parts:

- a digital voltmeter module
- a potential divider section.

As already described, the digital voltmeter module converts the analog input voltage at its inputs and shows its value on a four-digit display. The maximum voltage that can be applied to the inputs of the module is usually a few hundred millivolts. In Figure 9.13, the maximum input voltage to the module is 200 mV.

In Figure 9.13, resistors R_1 to R_4 are connected in series across the voltmeter's input terminals, with connections to a four-position selector switch. This arrangement is called a *potential divider*. Because of the values chosen for the resistors, each tapping point increases the maximum input voltage by a factor of 10. In all cases, the maximum voltage applied to the digital voltage is 200 mV.

If the switch is set to the 200 V position, the voltage fed to the digital module will be across resistor R_4. The series network of R_1 to R_3 has a total resistance of 9.99 MΩ and is in series with R_4. The total resistance of the network is therefore 10 MΩ. If 200 V is applied, the current in the network is 20 μA (200V/10 MΩ). The voltage across R_4 is 20 μA × 10 kΩ, or 200 mV. That is, on this setting, the voltage to the digital module is equal to the voltage being measured divided by 1000.

On the 20 V setting the voltage to the digital module is across resistors R_3 and R_4. Similar calculations will show that the input voltage is divided by 100, so a 20 V input gives 200 mV to the digital module.

FIGURE 9.13 A digital voltmeter has a potential divider connected between the measured voltage and the digital module

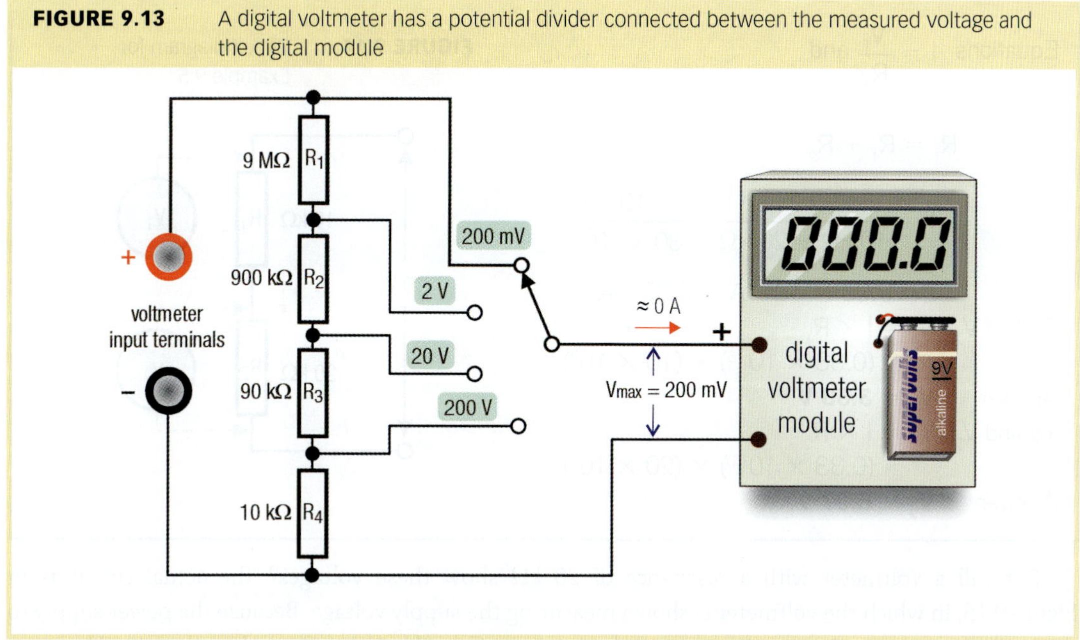

When the switch is at the 2 V position, the voltage to the digital module is the voltage drop across resistors R_2, R_3 and R_4. That is, the potential divider has reduced the applied voltage by a factor of 10. The resistor values in the potential divider network are determined by Ohm's law calculations. The values shown in Figure 9.13 suit a typical digital voltmeter module and assume the resistance between the module's input terminals is almost an open-circuit.

Regardless of the position of the switch, the input voltage is always across a resistance of 10 MΩ. This is a significant difference to an analog meter, in which the resistance of the meter depends on the setting of the range switch. Most digital voltmeters have an input resistance of 10 MΩ.

This high resistance minimises loading on a circuit, but it can also cause 'ghost' readings. Ghost voltages occur due to nearby energised wiring or circuits and appear on a digital voltmeter as a varying value that can range from a few hundred millivolts to several volts. Some digital meters have a facility to lower the input resistance to minimise ghost voltages. An analog meter, due to its lower input resistance, will not usually show ghost voltages.

A DMM will often indicate an AC voltage when a circuit is de-energised. This is called a *ghost* voltage and is due to pick-up from nearby energised wiring

9.6 Voltmeter Loading

The ideal voltmeter should have an infinitely high resistance and therefore take no current from the circuit it is measuring. All voltmeters have some resistance, as already explained. So, when you connect a voltmeter to a circuit, you are connecting a resistance to that part of the circuit. If this resistance value is similar to other resistances in the circuit, the addition of the voltmeter will affect the voltages in the circuit. That is, the voltmeter has *loaded* the circuit and affected the readings.

To explain the effect of voltmeter loading, Example 9.5 starts by calculating the voltage drops in a simple series circuit of two resistors. It then continues by calculating the same voltage drops when a 10 V DC voltmeter with a resistance of 20 k Ω is connected across one of the resistors.

EXAMPLE 9.5

Find the values of V_1 and V_2 in the circuit in Figure 9.14.

Solution

Values $V_T = 10$ V
$R_1 = 10$ kΩ
$R_2 = 20$ kΩ

Equations $I = \frac{V_T}{R_T}$ and

$$R_T = R_1 + R_2$$

$$I = \frac{10}{10\ \text{k}\Omega + 20\ \text{k}\Omega} = \frac{10}{30 \times 10^3}$$

$$I = 0.33 \times 10^{-3}\ \text{A} = 0.33\ \text{mA}$$

To find V_1 $V_1 = I \times R_1$

$= (0.33 \times 10^{-3}) \times (10 \times 10^3)$

Answer $\mathbf{V_1 = 3.33\ V}$

To find V_2 $V_2 = I \times R_2$

$= (0.33 \times 10^{-3}) \times (20 \times 10^3)$

Answer $\mathbf{V_2 = 6.67\ V}$

FIGURE 9.14 Circuit diagram for Example 9.5

But will a voltmeter with a resistance of 20 kΩ show these voltages? The actual circuit is in Figure 9.15, in which the voltmeter is shown measuring the supply voltage. Because the power supply to the circuit has a low resistance, its output voltage is not affected by connecting the voltmeter. However, as Example 9.5 shows, connecting the voltmeter across R_1 will alter the voltage across this resistor.

FIGURE 9.15 Actual circuit for Example 9.5, with a 20 kΩ voltmeter measuring the supply voltage

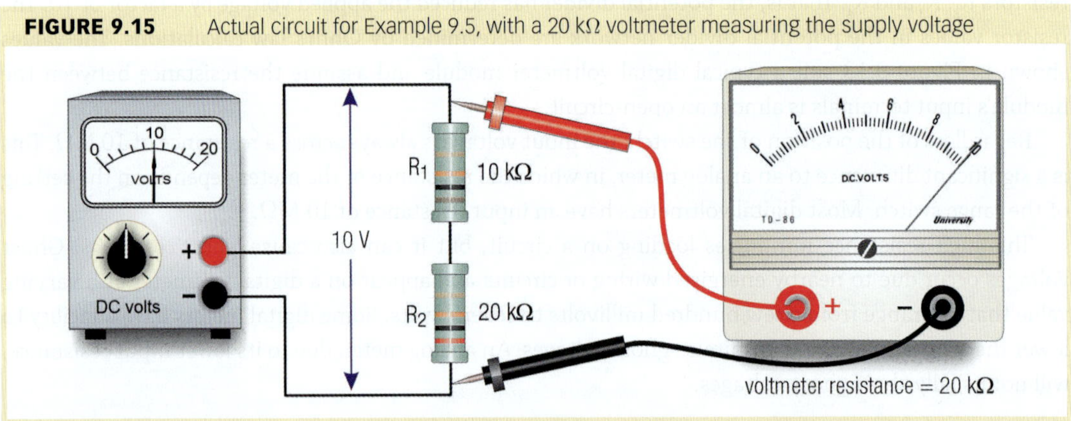

To see the effect the meter has on the circuit voltages when it's connected across R_1, we next calculate the voltage across both resistors when the voltmeter's equivalent resistance of 20 kΩ is in the circuit.

EXAMPLE 9.6

Find the voltage across resistor R_1 in the circuit of Figure 9.16.

Solution

Values $V_T = 10$ V

$R_1 = 10$ kΩ

$R_2 = 20$ kΩ

$R_{VM} = 20$ kΩ

The equivalent resistance (R_{eq1}) of R_1 and R_{VM} is found using the product/sum equation for parallel resistors:

$$R_{eq1} = \frac{R_1 \times R_{VM}}{R_1 + R_{VM}} = \frac{10\ \text{k}\Omega \times 20\ \text{k}\Omega}{10\ \text{k}\Omega + 20\ \text{k}\Omega}$$

$$R_{eq1} = 6.67\ \text{k}\Omega$$

FIGURE 9.16 Equivalent circuit when voltmeter is measuring the voltage across R_1

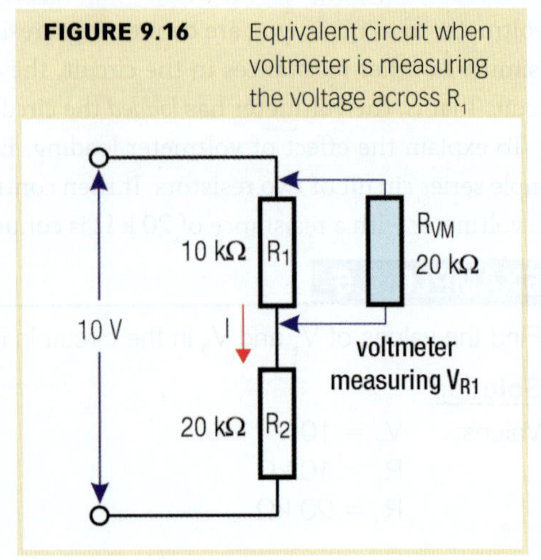

To calculate the voltage across R_1, we first use Ohm's law to find the circuit current:

$$I = \frac{V_T}{R_{eq1} + R_2} = \frac{10\ V}{6.67\ k\Omega + 20\ k\Omega}$$

$I = 0.375\ mA$

The voltage across R_1 and R_{VM} is the same (as they are in parallel), and is found with Ohm's law:

$$V_{R1} = V_{Req1} = I \times R_{eq1} = 0.375\ mA \times 6.67\ k\Omega$$

Answer $\mathbf{V_{R1} = 2.5\ V}$

As shown in Figure 9.17 and calculated in Example 9.5, before the voltmeter was connected, the voltage across R_1 was 3.33 V. With the voltmeter connected to the circuit, this voltage has now fallen to 2.5 V.

FIGURE 9.17 Connecting a 20 kΩ voltmeter has loaded the circuit. Before the voltmeter was connected, the voltage across R_1 was 3.33 V, not 2.5 V as indicated by the voltmeter.

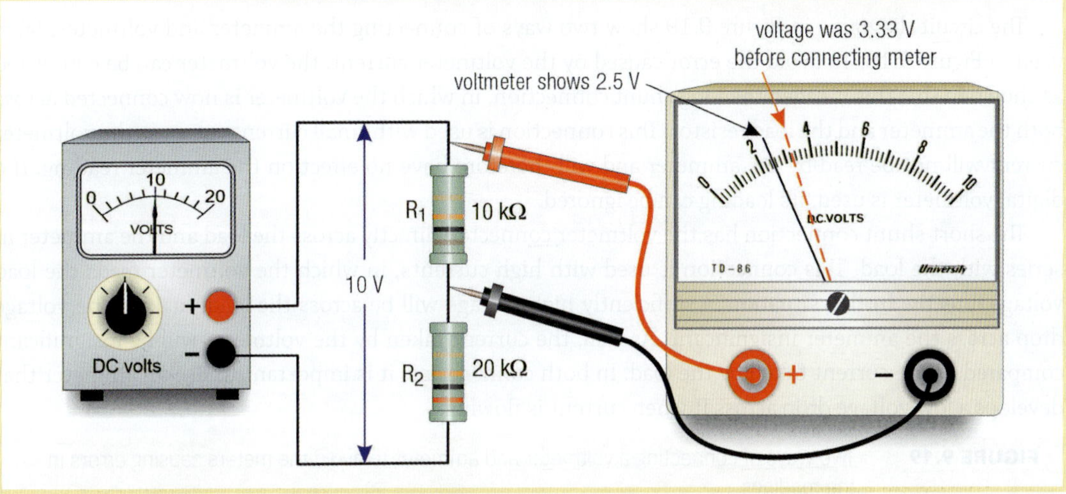

Because a digital voltmeter has a resistance of around 10 MΩ, it would not load the circuit of Figure 9.17 and would therefore display the correct voltages. As a general rule, the voltmeter resistance should be two orders of magnitude (100 times) higher than the resistance values associated with the circuit under test.

Long shunt and short shunt meter connections

Measuring the current in a load and the voltage across the load requires connecting an ammeter and a voltmeter to the circuit. To obtain the highest accuracy, the meters should be connected so one does not affect the readings of the other.

The circuit in Figure 9.18 shows a 1 kΩ resistor connected to a 10 V DC power supply with a voltmeter connected across the resistor, and an ammeter reading the current taken by the resistor. By Ohm's law, the current in this resistor is 10 V/1000 Ω, which equals 10 mA. However, the voltmeter is drawing 1 mA from the circuit, so the ammeter is reading 11 mA, not the expected 10 mA. In this case, the voltmeter is loading the ammeter.

FIGURE 9.18 The voltmeter current is large enough compared to the load current to cause the ammeter to show a 10 per cent error

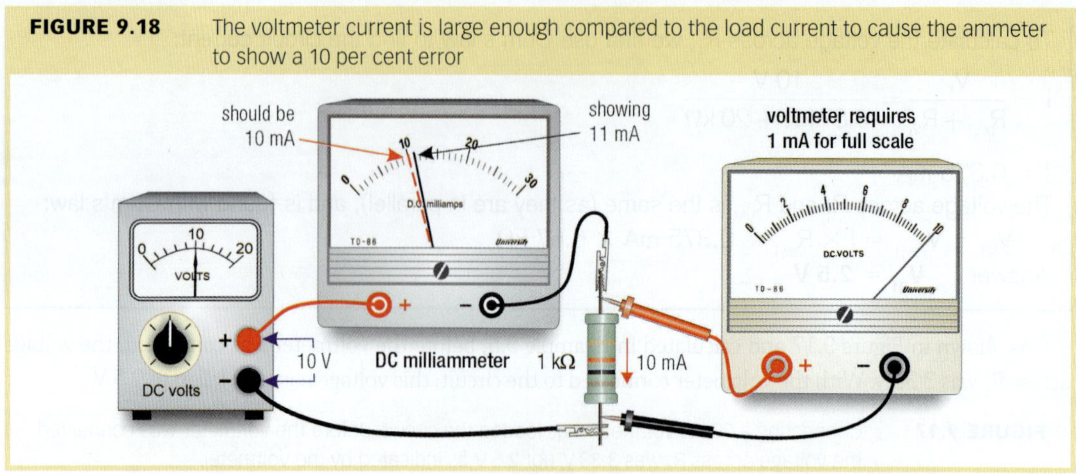

The circuit diagrams in Figure 9.19 show two ways of connecting the ammeter and voltmeter; (a) is used in Figure 9.18. To avoid the error caused by the voltmeter current, the voltmeter can be connected as shown in (b). This is called the long shunt connection, in which the voltmeter is now connected across both the ammeter and the load resistor. This connection is used with small current values as the voltmeter current will not be read by the ammeter and will therefore have no effect on the ammeter reading. If a digital voltmeter is used, its loading can be ignored.

The short shunt connection has the voltmeter connected directly across the load and the ammeter in series with the load. This connection is used with high currents, in which the voltmeter reads the load voltage directly. In this situation, a sufficiently high voltage will be across the load, making the voltage drop across the ammeter insignificant. As well, the current taken by the voltmeter will be insignificant compared to the current taken by the load. In both connections, it is important to use an ammeter that develops a low voltage drop across it when current is flowing.

FIGURE 9.19 Two ways of connecting a voltmeter and ammeter to avoid the meters causing errors in the readings

I, V, 20 A, R_L, V

I, V, V, 100 mA, R_L

(a) short shunt, for high currents, when voltmeter current is insignificant

(b) long shunt, for small currents, when voltmeter current can affect ammeter reading

KEY POINTS...

- An analog voltmeter has a moving-coil meter movement and a resistor called a *multiplier* in series with movement.
- The resistance value of the multiplier can be found with Ohm's law, or with the product of the movement's sensitivity in ohms/volt and the voltage range.
- The resistance of an analog voltmeter changes with each voltage range. The higher the range, the higher the resistance.
- A digital voltmeter has a potential divider network rather than individual multiplier resistors.
- The resistance of a digital voltmeter is the same on all ranges. It is usually 10 MΩ.
- A voltmeter can 'load' a circuit, and therefore give misleading readings.
- The higher the resistance of the voltmeter, the less the loading effect.
- When measuring both load current and voltage, use the short shunt connection for most purposes. For small values of load current, use the long shunt connection to avoid the ammeter giving a misleading reading due to the voltmeter current.

TASK 9.1

1. A digital voltmeter module from a parts supplier has a full-scale input voltage of 199.9 mV. Calculate the value of a shunt resistor so the module displays 25.0 when 25 A is passed through the shunt resistor.
2. Determine the value of a multiplier resistor to make a 100 V voltmeter using a meter movement that has an I_{FSD} of 100 μA.
3. A multi-range analog voltmeter is specified as having a loading of 100 kilohms per volt (Ω/V). What is its resistance when set to the 200 V scale?

9.7 The Ohmmeter

An analog and a digital ohmmeter operate by passing a current through the resistance being measured. The current is supplied by a battery inside the meter case. While both meters operate in a similar way, their internal circuit is different.

The analog ohmmeter

A basic single-range analog ohmmeter is shown in Figure 9.20. The current flowing through the resistance being measured is limited by resistors R_1 and VR_1, which are both connected in series with the battery and the meter movement.

In Figure 9.20, when a resistance is being measured, current flows in the circuit, causing the meter movement to show a reading that depends on the value of the test current. The two extremes of resistance measurement are a short-circuit (zero resistance) and an open-circuit (when the probes are held apart). An open-circuit is indicated on the ohmmeter scale by the infinity (∞) symbol. If not, adjust the position of the pointer with the meter's mechanical adjustment.

FIGURE 9.20 A basic analog ohmmeter, which has an internal battery, and an adjustable and a fixed resistor in series with a meter movement

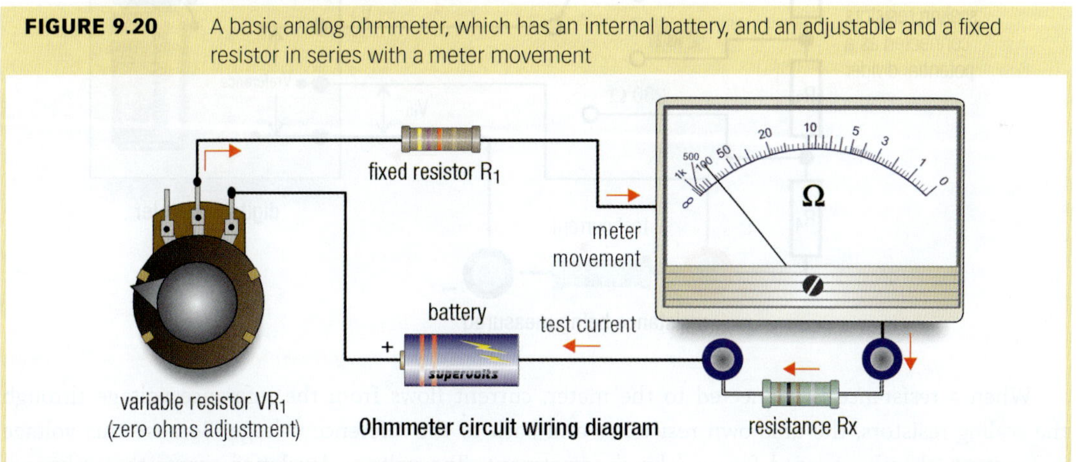

When the meter probes are joined, the resistance is zero and the pointer should be at full-scale (zero ohms). The 'zero ohms' potentiometer VR_1 adjusts the meter current so the pointer is aligned with the 0 ohms indication on the scale. This adjustment compensates for changes in the internal battery voltage and should be done before taking a resistance measurement and when changing ranges on a multimeter.

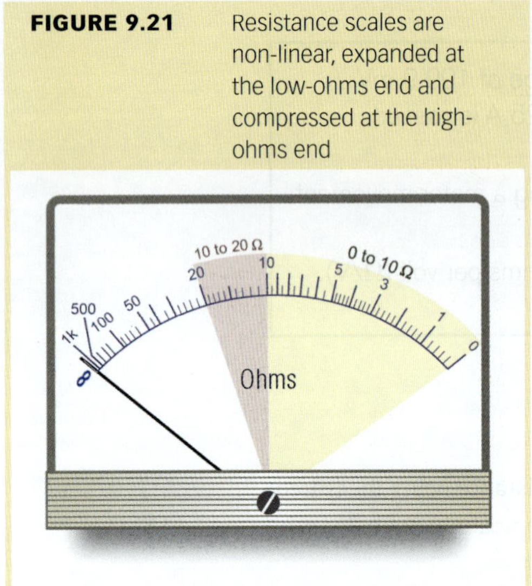

FIGURE 9.21 Resistance scales are non-linear, expanded at the low-ohms end and compressed at the high-ohms end

Because current is inversely proportional to resistance, the ohmmeter scale follows an inverse law, giving a scale that is expanded on the right and compressed on the left. This is shown in Figure 9.21, where the pointer moves nearly halfway for the first 10 Ω (shaded in yellow), but far less for the next 10 Ω (shaded in pink). For this reason, the best measurement accuracy is obtained when the pointer is in the yellow shaded section.

Digital ohmmeter

The internal circuit of a digital ohmmeter is different to an analog meter because a digital module is a voltage-driven device, which means it responds to a DC voltage at its input terminals. A moving-coil meter movement is current driven, as a current is needed to drive the pointer upscale. However, in both types of ohmmeters, a current is passed through the resistance being measured. As shown in Figure 9.22, to supply this current, a typical digital ohmmeter uses a reference voltage, which is produced by the battery supplying the meter. This voltage is held at a constant value by a regulator IC in the digital module, and is applied across the series network of R_1 to R_4 and the unknown resistance.

FIGURE 9.22 An internal voltage source causes a current to flow in the unknown resistance, producing a voltage that's measured by the digital module

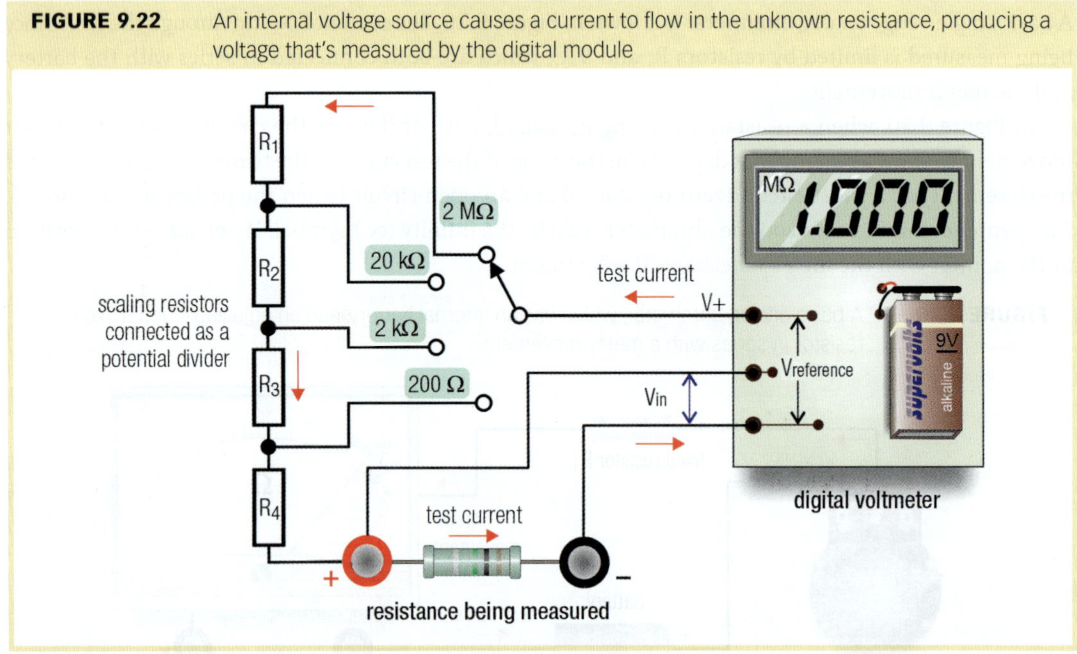

When a resistance is connected to the meter, current flows from the reference voltage through the scaling resistors, the unknown resistance and back to the reference voltage. Because this voltage is constant, there's no need for an 'ohms' adjustment. The voltage developed across the unknown resistance is applied to the input of the digital voltmeter module. By selecting suitable values for the scaling resistors, the reading on the module will show the resistance in ohms. The resistance range switch selects the amount of resistance in series with the unknown resistance. The greater the scaling resistance, the lower the test current. The high-resistance range has the highest scaling resistance in series with the unknown resistance, and therefore the lowest value of test current.

9.8 Multimeters

Digital multimeters, and to a lesser extent analog multimeters, are widely used in electrical measurement. Both types are readily available, and each type has an advantage that makes it preferable for certain types of measurements. It is important that any multimeter used to measure values in a mains power circuit has the correct category rating, as explained later. Both types of multimeter have one or more switches to select:

- **measurement function**, such as volts, current or resistance
- **AC or DC** for current and voltage measurements
- **measurement range**, such as up to 10 V, 100 V etc. Many digital multimeters are auto-ranging, so you simply select the measurement function such as DC volts and let the meter do the rest.

The main difference between both types of multimeters is how readings are displayed. We look first at the analog multimeter and how to read a typical scale.

Reading an analog multimeter scale

Analog multimeters, such as those shown in Figure 9.23, range in cost from \$10 to several hundred, even thousands of dollars. The AVO-8 multimeter shown ceased production in 2008, but it and its predecessors are often found in electrical or electronic workshops. The AVO-8 shown has an accuracy of ±1 per cent of FSD on DC ranges and ±2 per cent on AC ranges. The multimeter in Figure 9.23(b) is typical of a low - cost unit costing \$50 or so.

FIGURE 9.23 Analog multimeters range in price and the functions they offer

(a) AVO multimeter, model 8, mark 7

(b) typical low cost multimeter

An analog multimeter scale combines all the measurement functions of the meter. Therefore, the scale will have markings for various ranges of voltage, current and resistance. Reading an analog scale requires making sure you are reading the right scale and interpreting it correctly. Scales are usually made up of major markings, identified by a thick line or a number, with minor markings in between. The value of a minor mark is determined by noting how many divisions there are between two major marks.

Figure 9.24 shows a typical 0 to 10 V DC voltage scale, with each major division divided into five sections, giving four minor marks between any two major marks. The value of a minor mark in this scale is therefore one fifth of the difference between any two major marks. The difference between any two major marks is 1, so each minor mark equals 1/5 or 0.2. In general:

$$\text{value of a minor scale marking} = \frac{\text{difference between adjacent major marks}}{\text{number of sections between adjacent major marks}}$$

FIGURE 9.24 Meter scale with 10 major divisions that are further divided by five minor divisions

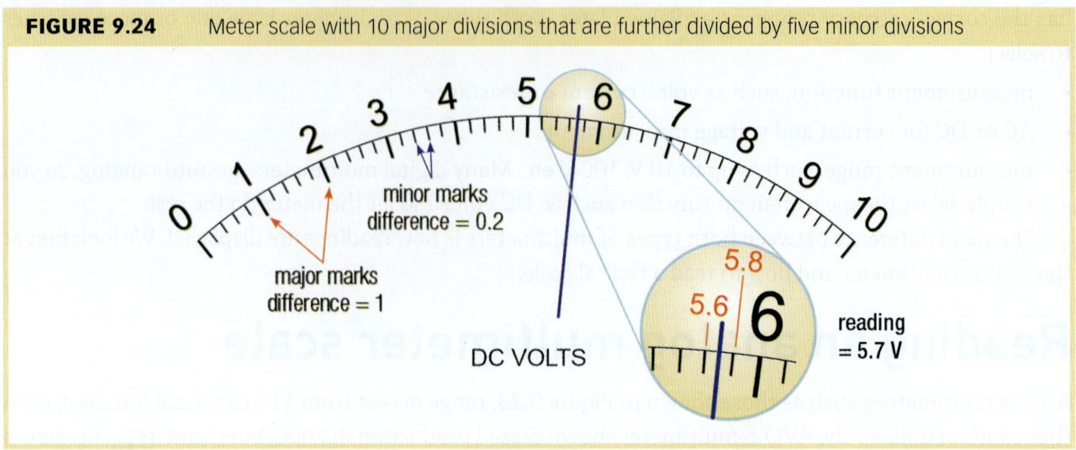

In Figure 9.24, if the meter is set to its 10 V scale, the pointer is reading 5.7 V. Note that the pointer is halfway between two minor marks, and as each minor mark represents 0.2 V, halfway between two minor marks represents 0.1 V. Most analog multimeters use the same scale for each range, so if the range setting is 100 V, the value read in Figure 9.24 becomes 57 V.

When you look at the pointer, you should align yourself so the reflection of the pointer is directly beneath the pointer and is therefore hidden by the pointer. This prevents incorrect readings of the scale caused by *parallax error*, in which the distance between the pointer and the scale can cause an error unless you are viewing the pointer from directly above. Figure 9.25 shows the effect of parallax error in which a reading of anywhere between 42 V to 48 V (top scale) is read, depending on the position of the eye.

FIGURE 9.25 There's parallax error if you can see the reflection of the pointer in the mirror

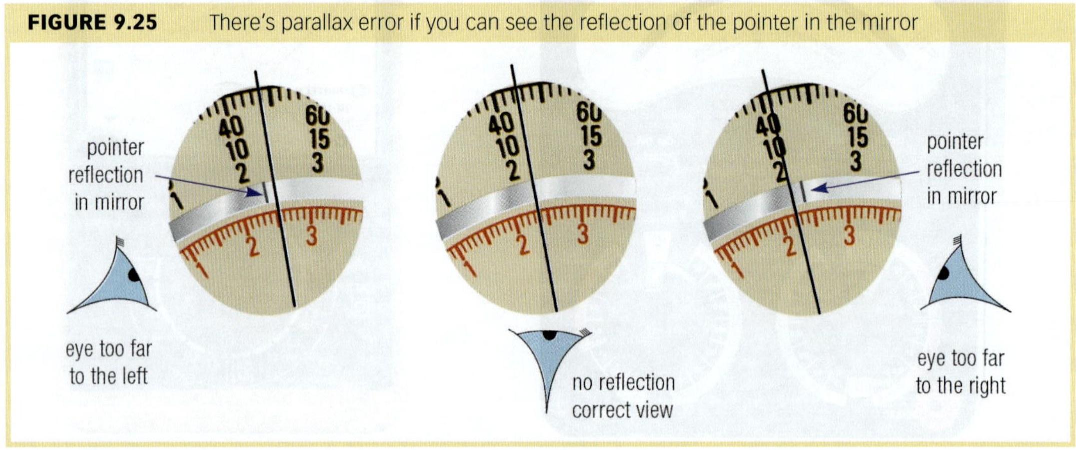

Digital multimeter display

The display in a typical digital multimeter (DMM) shows the selected measurement function, a multiplier (e.g. mV or mA), as well as the value being measured. Functions and multipliers are indicated by way of *annunciators* that are designed into the display. Many displays also have a bar graph to give a clearer indication of a changing or varying value. Although there are considerable variations between DMM makes and models, their displays all achieve much the same thing. The digital meter displays in this book are from a Fluke DMM, which is typical of an auto-ranging instrument likely to be used by electricians in the field. Here's a general explanation of what the displays are showing for each basic function of the meter.

Reading voltage

Figure 9.26 shows the displays when this meter is reading AC or DC volts. The DC millivolts function is not auto-ranging, indicated by the MANUAL annunciator. The meter can read up to 399.9 mV before the display indicates an overload. The AC volts and DC volts functions are both auto-ranging, in which the AC volts function ranges from millivolts to volts. If required, the auto-ranging can be switched off (with the RANGE button) and individual ranges selected manually.

FIGURE 9.26 DMM displays for AC and DC voltage measurement

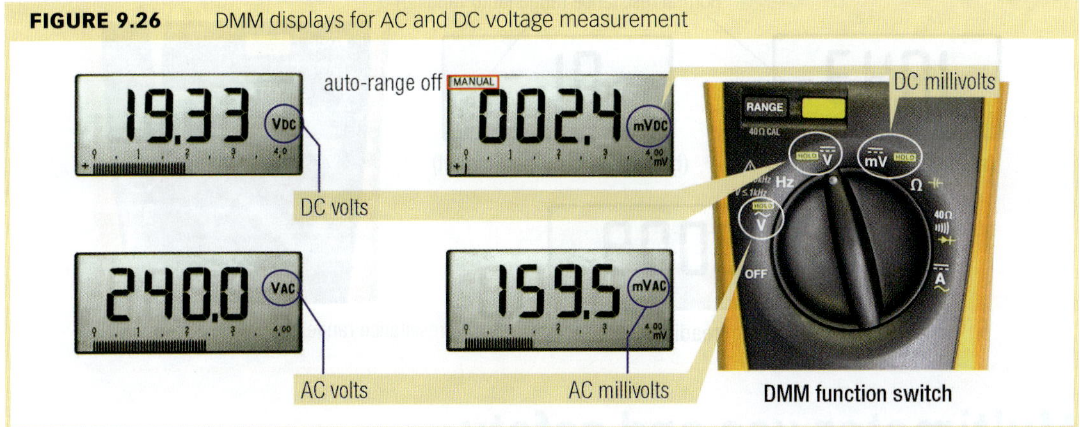

Reading current

Unlike some DMMs, this meter has only one setting for current measurement, as shown in Figure 9.27. It can measure up to 10 amps, AC or DC, in which AC current measurement is selected by pressing the yellow button. The annunciators used with current measurement only show whether AC or DC is selected. The displays shown are for zero current, so the meter has auto-selected the most sensitive range, as shown by the position of the decimal point. When the meter shows a current value, the bar graph also gives an indication.

FIGURE 9.27 DMM displays when set to read AC or DC current

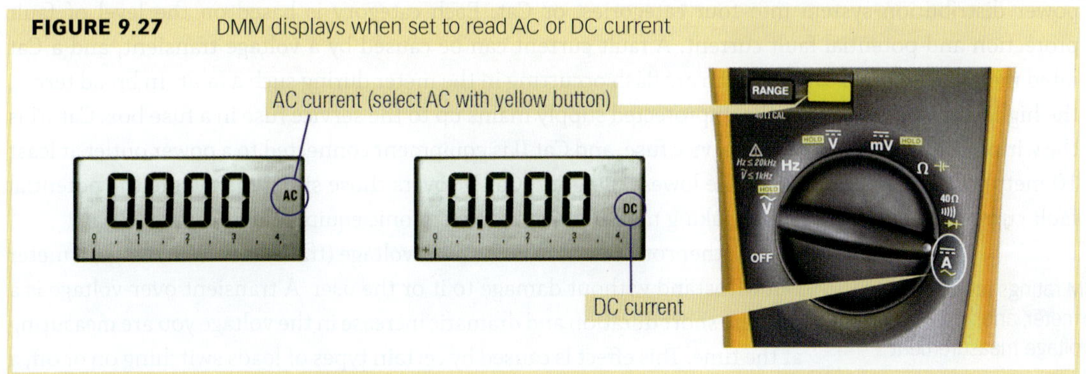

Reading resistance

Most digital multimeters have settings for continuity (or low ohms) as well as standard resistance measurement. The continuity setting will usually cause a beeper to sound if the meter probes are connected across a low resistance, which for the meter in the diagrams is around 40 ohms. Auto-ranging is off when the continuity function is selected.

The resistance function is auto-ranging, and the multiplier that is selected by the meter is indicated on the display. When the meter probes are not connected to a resistance, the meter auto-ranges to its highest range (MΩ) and gives the indication shown in Figure 9.28(b). The bar graph gives an indication on the resistance function, but not on the continuity function.

Multimeters of any type are usually not able to accurately measure resistance values below a few ohms. Lead resistance and probe contact resistance become significant, adding to the resistance value. Probe contact resistance occurs between the probe tip and the point of contact with the resistance being measured due to surface film caused by oxidisation. A minimum value of current, called the wetting current, is required to overcome or reduce this resistance. When set to the low ohms range, a DMM uses a higher test current for this reason.

FIGURE 9.28 DMM displays when set to read resistance

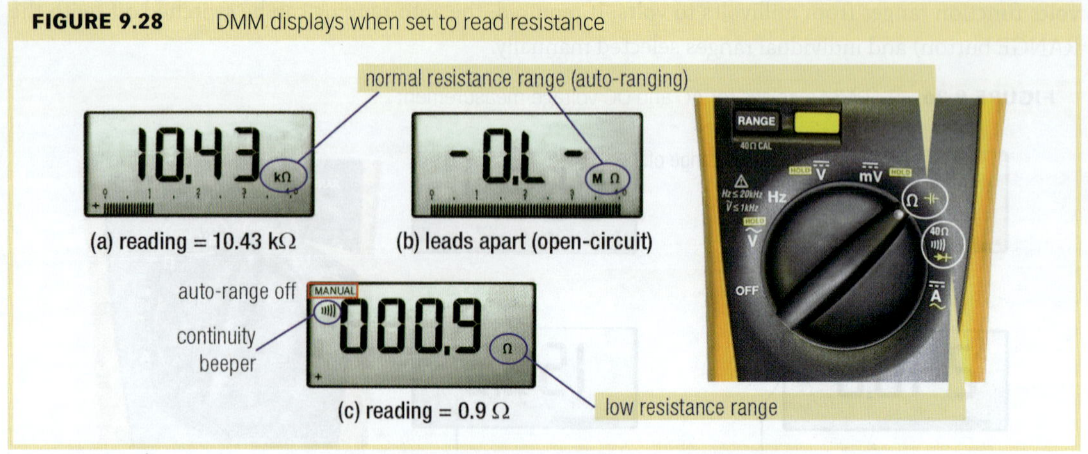

Multimeter use and safety

SAFETY

For mains voltage measurement, use a multimeter with a 300 V Cat III rating or higher

Safety is essential when using a multimeter in a live circuit, including selecting the right type of multimeter and probes. Electrical instruments, digital multimeters in particular, are usually given a category rating by their manufacturer, as explained below. Products that state 'tested to relevant standard' should be selected rather than products that state 'designed to relevant standard'.

Instrument category ratings

Instrument ratings come from the international safety standard IEC 61010, which divides the electrical power distribution system into four categories, or Cat. Each category is based on the level of fault protection and potential fault current. A fault current can be caused by a voltage transient, and a Cat rated meter is designed to prevent an arc flash occurring in the meter during such a fault. In broad terms, the highest is Cat IV, which is the unprotected supply mains up to the service fuse in a fuse box. Cat III is the wiring on the load side of the service fuse, and Cat II is equipment connected to a power outlet at least 10 metres from a Cat III source. The lowest category (Cat I) covers those situations where the potential fault current level is low, such as making measurements on electronic equipment.

Another consideration is the peak voltage (transient over-voltage) a meter can withstand without damage to it or the user. A transient over-voltage is a sudden, short duration and dramatic increase in the voltage you are measuring at the time. This effect is caused by certain types of loads switching on or off, a lightning strike or as a result of a fault condition in another part of the power circuit. The higher the voltage rating of the meter, the higher the transient over-voltage it should be able to withstand.

FIGURE 9.29 DMM ratings are printed on the meter, and apply only to voltage measurement

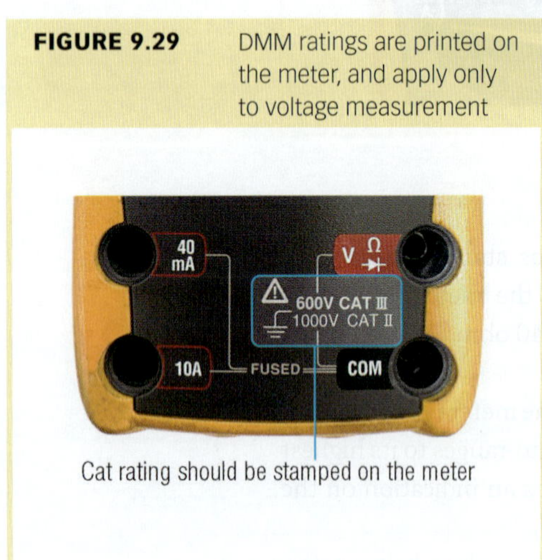

Figure 9.29 shows the ratings of a digital multimeter suitable for use in a Cat III situation (wiring after service fuse). To meet this standard, the meter must be able to withstand an over-voltage of 6000 V (6 kV). This over-voltage value applies also to the Cat II/1000 V rating, hence the Cat II rating on the instrument. However, despite the same voltage rating, a Cat III rated meter can handle a fault current six times that of a Cat II rated meter.

An important consideration is the Cat and voltage rating of the probes and leads being used, which should have the same or higher Cat rating as the multimeter. As shown in Figure 9.30, the Cat rating of a probe should be stamped on the body of the probe. The UL logo means the probe design has been tested to UL standards (Underwriters Laboratory, a standards body in the USA). Some electrical supply authorities specify the use of probes fitted with internal fuses. This gives greater protection in a variety of fault situations.

FIGURE 9.30 Probes should show a Cat rating that is equal to or higher than the meter Cat rating

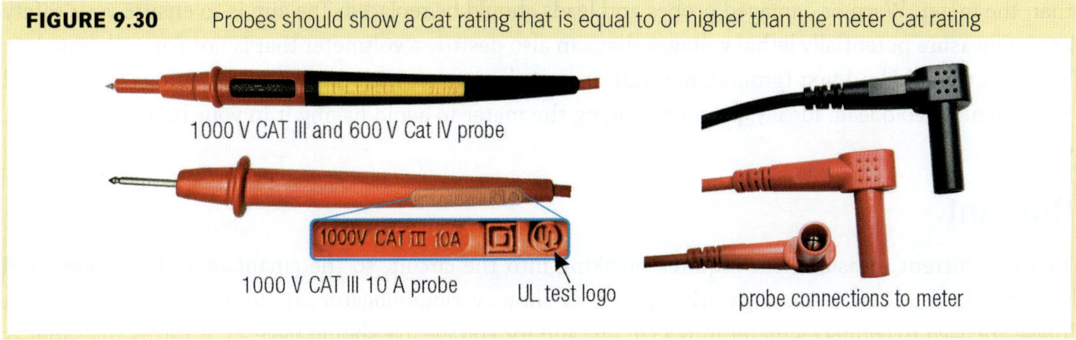

Instrument in-built protection

Category-rated multimeters that are able to measure current have internal fuses to protect the meter against current overload. The type of fuse is important. Cat II meters will often have glass fuses, while Cat III meters should have an HRC (high rupturing capacity) fuse. This type of fuse is capable of rapidly clearing a fault current up to 100 kA (100 000 amps). A glass fuse acts more slowly and does not provide the same level of protection.

FIGURE 9.31 Ammeters with glass fuses are a safety hazard. The correct type of fuse is a high rupturing capacity (HRC) type that is specifically matched to the multimeter.

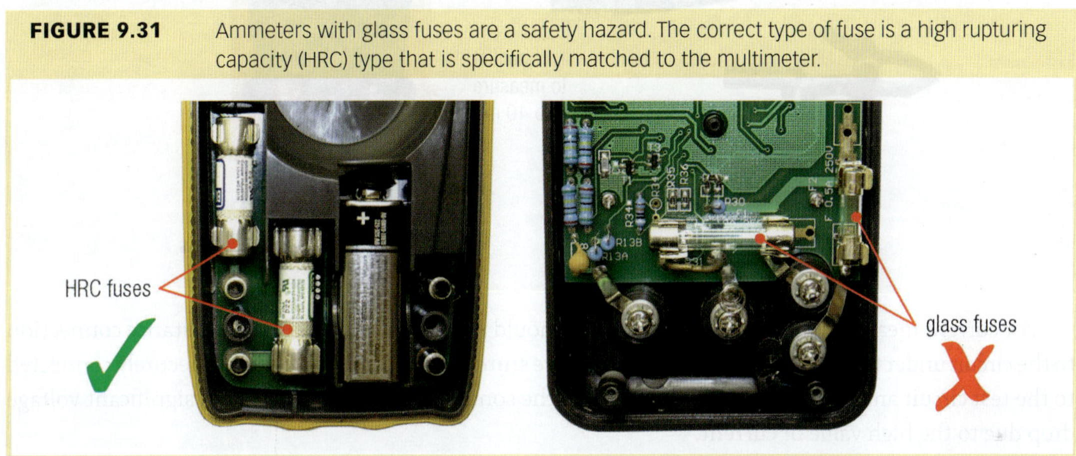

The right type of protection is essential when using a meter to measure current in a 230 V mains circuit. An accidental contact to the wrong point in the circuit can cause an extremely high current to flow through the leads and the meter. A glass fuse can become a plasma fireball under this type of overload condition, while an HRC fuse designed for the job will operate almost instantly, without damage to the meter or its user. Fluke fits fuses with a minimum interrupt rating of 10 kA or 17 kA in their multimeters.

PPE

When using a multimeter to take measurements on mains equipment or wiring, you should always use the correct personal protective equipment (PPE). This includes eye protection, correctly rated gloves and insulated rubber mats. If possible, avoid taking measurements in humid or damp environments and make sure there are no nearby flammable dust or vapour hazards. Remove jewellery, watch and any metal object you might be wearing.

SAFETY

Safety AS/NZS 4836 (Section 9) provides a selection guide for PPE for various types of electrical work

Taking measurements

Voltage

Confirm that your multimeter or voltmeter has the correct Cat rating. For most electrical work, choose a 300 V Cat III or higher rated instrument. Some types of analog voltmeters have Cat ratings, but many do not and are therefore a safety hazard. Confirm also that the probes have the same or higher Cat rating than the meter. Worn or damaged probes and leads should be replaced. The aim is to ensure your safety as you measure potentially lethal voltages that can also destroy a voltmeter that is not correctly rated.

Hook the cold lead first (ground, neutral), then make contact with the hot lead. Remove the hot lead first, then the cold lead. Ideally, prop up or hang the meter to avoid having it in your hand in the event of a fault.

Current

Taking a current measurement requires breaking into the circuit, so the circuit must be isolated and proven to be dead before starting work. Figure 9.32 shows various alligator clips for use with a multimeter. These are used to connect a meter into a circuit, and for voltage use should have a Cat rating the same as or higher than the meter rating.

FIGURE 9.32 Current measurement requires correctly rated clips and using the correct meter sockets

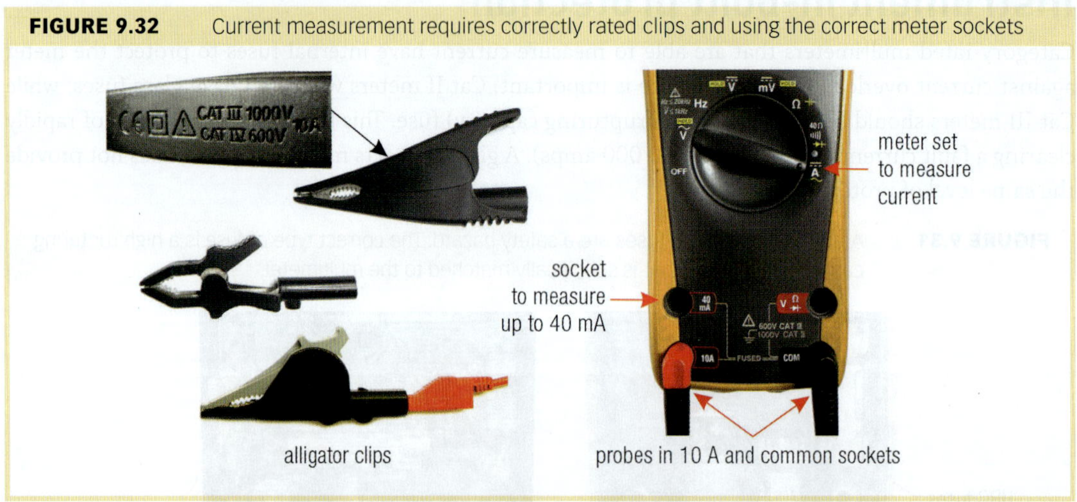

SAFETY

After measuring current with a multimeter, put the meter leads back into the sockets for voltage measurement

For current measurement, the clips being used should provide an extremely low resistance connection to the circuit under test. High values of current require sturdy, undamaged clips that are securely connected to the test circuit and the meter leads. This ensures the connections do not introduce a significant voltage drop due to the high value of current.

Resistance measurement

Resistance values in electrical work range from micro ohms up to the high megohms. Most general-purpose digital or analog multimeters are suited to only reading values above a few ohms. Some digital meters have a 'calibrate' function to zero the meter reading when the probes are joined to cancel out probe lead resistance. However, as previously explained, contact resistance between the probes and the device under test becomes an increasing problem for resistance values less than one ohm.

To read resistance values in the order of milli or micro ohms requires a specialised ohmmeter. The difference between measuring high and low resistance values concerns the test current the meter passes through the resistance being measured. The Fluke digital multimeter in Figure 9.32 passes a test current of 0.45 mA when measuring a resistance of one ohm. This compares to test currents of 10 A, and up to 600 A when using an instrument called a micro-ohmmeter, also called a *ductor* tester. These instruments are used to measure the contact resistance in switching equipment, which is typically specified as 25 μohms or less. They are also used to measure contact bonding resistance in switchyard earthing systems, in which readings should be less than 20 milliohms for each bond between the earth cable and the item of equipment.

When measuring resistance, make sure the circuit is 'dead'

Following are some points about using a multimeter to measure resistance.

- **Safety** – Resistance measurement in a circuit must be done only when the power is removed from the circuit. A common error is to forget, or not be aware, that a circuit is live when making a resistance measurement. This can destroy the meter and is a safety hazard to the user.

- **Probes** – While taking a resistance measurement, be careful not to touch the probe tips, as this will reduce the reading because you are in parallel with the resistance. Also, when measuring the resistance of a component that is in a circuit, isolate at least one leg of the component, to prevent false readings due to other components in the circuit.
- **Continuity** – Most digital meters have a continuity function in which a beeper sounds when the meter is measuring a resistance less than a certain value. A continuity test only indicates a circuit is not open-circuit and is not always indicating a low resistance.

Other digital multimeter functions

Most digital multimeters have additional functions, such as data storage, and the ability to measure capacitance, frequency and temperature (with a matched probe). A useful function is testing a solid-state diode.

Among many applications, these are used to convert AC to DC in a process called *rectification*. A diode is a semiconductor that allows current to flow in one direction only. It has two terminals called anode and cathode. When the anode is positive with respect to the cathode, current flows through the diode, causing a voltage drop across the diode of around 0.5 to 0.6 V. When the anode is negative with respect to the cathode, the diode is an open-circuit.

FIGURE 9.33 Diodes have a marking to identify the anode and cathode

The power diodes shown in Figure 9.33 can pass around 15 A, and withstand a reverse voltage of 600 V. There are many types of diodes, and all have a way to identify the cathode, such as a band around the body at one end, as on the 5 A diode in Figure 9.33. The diode symbol shows that the direction of current is from the anode to the cathode.

A diode can be tested with an ohmmeter to see if it conducts with the probes connected one way, but not when the probes are reversed. An analog ohmmeter will show a low resistance reading one way and infinite resistance when the leads are reversed. However, the polarity of the leads is reversed on the ohms setting, so the black lead is now the positive potential.

Most digital multimeters have a setting called 'diode test', usually marked with the diode symbol shown in Figure 9.33. For the meter in Figure 9.34, this setting is selected by pressing the yellow button, which selects a suitable test voltage and causes the display to show the measurement as a *voltage*, not resistance. As Figure 9.34 shows, a good diode will show around 0.5 V when the probes are connected one way, and open-circuit when connected in reverse. If the diode is a short-circuit, the meter will show around 0 V for both probe polarities; for an open-circuit, the meter display will be OL for both probe polarities.

FIGURE 9.34 A diode or other semiconductor device can be checked with a digital multimeter set to its 'diode test' setting. The reading is given in volts DC, usually around 0.5 V.

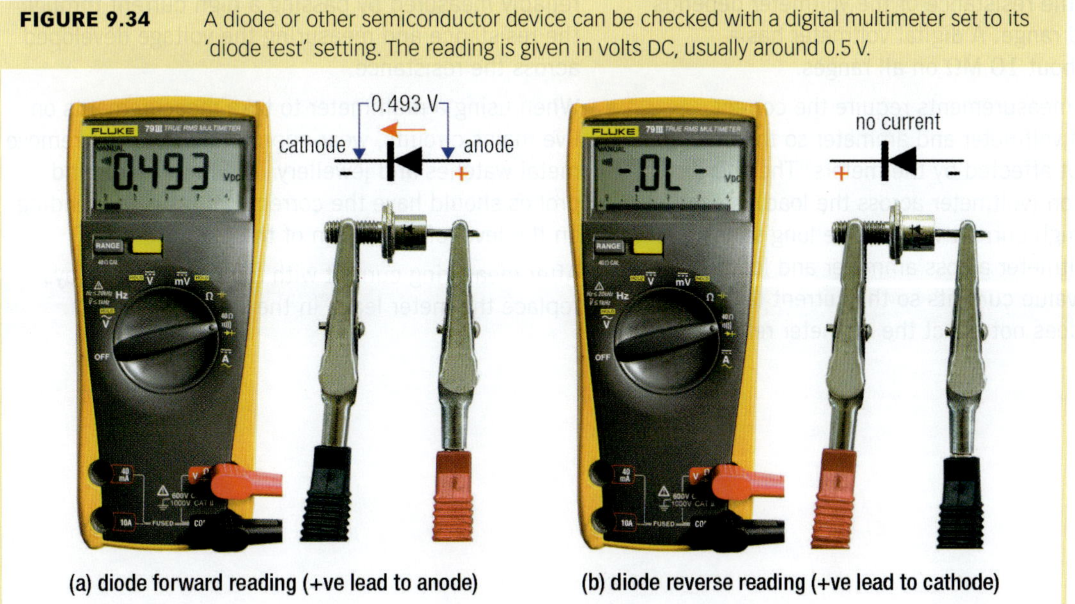

(a) diode forward reading (+ve lead to anode)

(b) diode reverse reading (+ve lead to cathode)

The multimeter is one of the most important items of test equipment in electrical work. It should have the appropriate Cat rating for the work being undertaken and ideally be a quality brand that can prove the meter was tested to ISO specifications. Other items of test equipment you will encounter in the trade are described in Chapter 25.

KEY POINTS...

- An ohmmeter passes a test current through the resistance being measured. The value of the current depends on the range setting.
- Analog multimeters have a pointer and various scales that cover all the meter functions, in which a scale has major and minor divisions to give a resolution of two decimal places or less.
- Digital multimeters usually have a four-digit display to show a reading and annunciators to indicate the selected meter function and multiplier (k or M for ohms, mA for amps). The maximum display value of a digital meter is typically 199.9 or 399.9.
- Meters used in electrical work should have the correct Cat rating, in which a Cat III meter suits any work carried out on an installation protected by a service fuse. Probe Cat ratings should match or exceed the meter Cat rating.
- A multimeter should not be used to measure resistance values less than a few ohms, instead use a micro-ohmmeter (ductor tester).
- A quality DMM will have properly rated HRC fuses to protect against overload when measuring current.

CHAPTER SUMMARY

- An ammeter is connected in series with the circuit under test. Ammeters contain a meter movement or a digital module that reads the voltage developed across a low resistance called a *shunt* resistor. Some types have a universal shunt resistor that allows range changing without damage to the meter.
- A voltmeter is connected in parallel with the circuit under test and should have a high resistance so it does not load the circuit. Analog voltmeters have a resistor called a *multiplier* in series with the meter movement, so the resistance of the voltmeter depends on the selected range. A digital voltmeter has a resistance of about 10 MΩ on all ranges.
- Some types of measurements require the correct arrangement of voltmeter and ammeter so the readings are not affected by the meters. The short shunt connection (voltmeter across the load) is used when reading high current values. The long shunt connection (voltmeter across ammeter and load) is used with low value currents so the current taken by the voltmeter does not affect the ammeter reading.
- Multimeters are typically digital, but analog multimeters have several advantages such as being able to follow a slowly changing current or voltage value. As well, they are not so prone to reading ghost voltages, in which a DMM might indicate several volts on a circuit disconnected from the supply. This is due to the high resistance of the meter responding to electrical fields from an adjacent energised circuit.
- A low resistance, such as circuit wiring or contact resistance of a switch or contactor, can only be reliably measured by passing a high current through the resistance and measuring the voltage developed across the resistance.
- When using a multimeter to take measurements on live mains circuits, wear appropriate PPE and remove metal watches and jewellery. The multimeter and probes should have the correct Cat rating depending on the level of protection of the circuit.
- After measuring current with a multimeter, always replace the meter leads in the voltage sockets.

REVIEW EXERCISES

Check your answers at the back of the book.

1 The 100 V analog panel meter in Figure 9.35 has a defective multiplier resistor. The meter has a full-scale deflection current of 1 mA. Calculate the value of a replacement resistor.

FIGURE 9.35

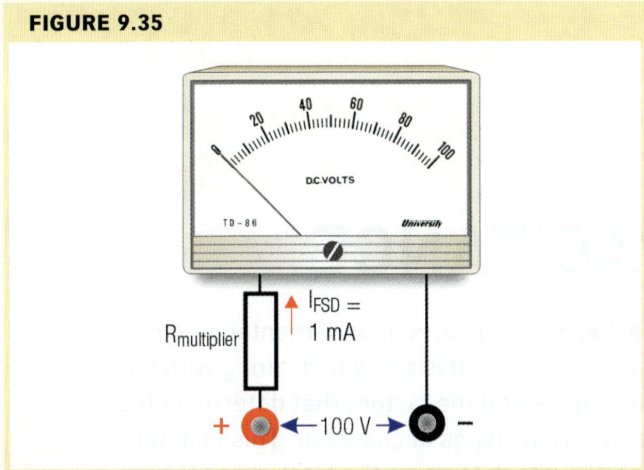

2 Calculate the value of the shunt resistor required to make a 30 A ammeter based on the meter movement in Figure 9.36. The meter movement has a resistance of 50 Ω and a full-scale deflection current of 5 mA.

FIGURE 9.36

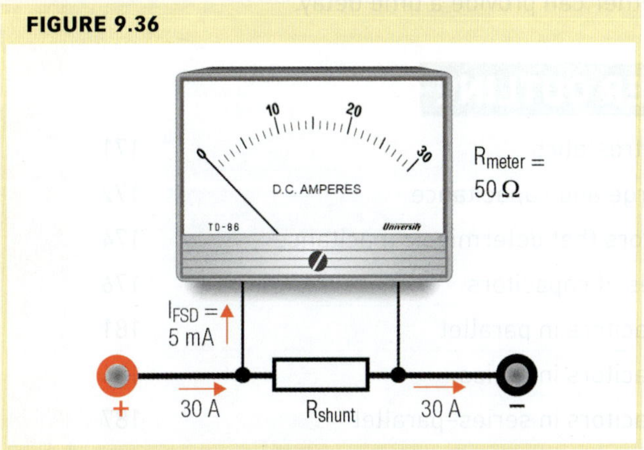

3 Calculate the maximum voltage that can be applied to the input terminals of the digital voltmeter circuit in Figure 9.37 before the meter indicates an overload condition.

4 Which type of voltmeter causes the least loading on a circuit, a typical digital voltmeter, or an analog voltmeter with a 10 μA FSD meter movement, assuming both are set to their 200 V range?

FIGURE 9.37

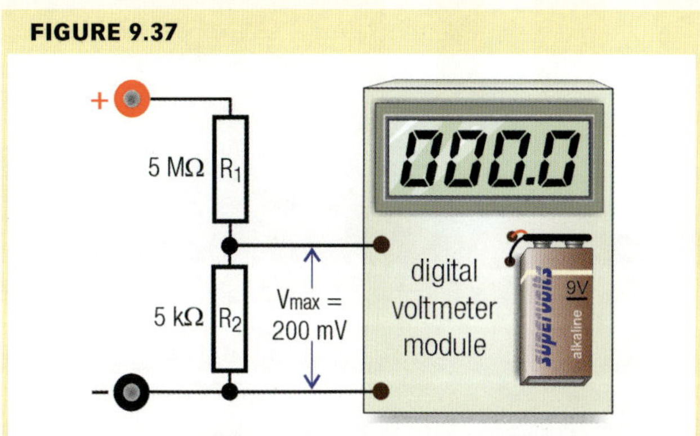

5 When measuring a high current taken by a load while also measuring the voltage across the load, should the long or short shunt meter connection be used, and why?

6 For the circuit in Figure 9.38, calculate the voltage across R_2 before and after the voltmeter is connected.

7 If the voltmeter in Figure 9.36 is replaced with a typical digital voltmeter, calculate the voltage this meter would read across R_2.

FIGURE 9.38

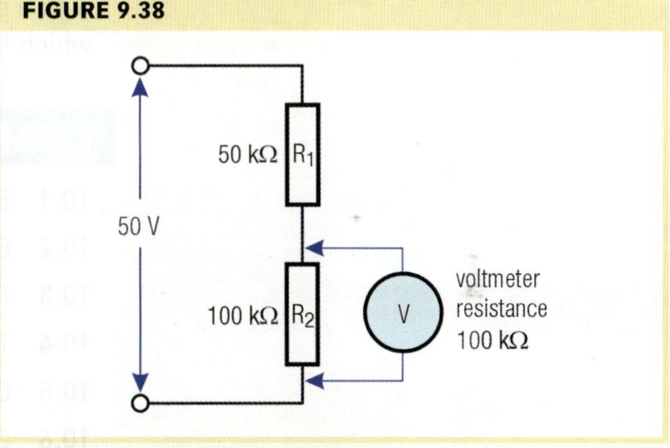

8 Why is a typical DMM or analog ohmmeter unsuitable for measuring resistances below one ohm?

9 A Cat II and a Cat III meter have the same 1000 V rating. What is the difference between these meters?

10 You get two different readings when measuring the same voltage in a circuit with two different voltmeters. Assuming both meters are accurate, explain what is likely to be happening, and which of the two readings is the most likely to be correct.

ONLINE RESOURCES

COMPLETE WORKSHEET NINE

Check with your instructor for worksheets on this chapter.

CHAPTER 10

CAPACITANCE

Capacitance, like resistance, is a fundamental electrical property. Capacitance itself is explained, along with how capacitors are made and the factors that determine the capacitance of various types of capacitor. This chapter also describes how to determine the total capacitance of a group of capacitors connected in series or parallel, or in series–parallel, and the reasons these connections are used. Capacitor testing and safety is described, along with a simple circuit involving a resistor and a capacitor, which together can provide a time delay.

CHAPTER OUTLINE

10.1 Electrostatics

Electrostatics is a branch of physics that studies electric charges at rest. Capacitance is that property of a circuit or a component that stores an electric charge. A capacitor is formed whenever two parallel conductors are separated by an insulator. A long length of twin-core cable has capacitance, as the two cables are close to each other but separated by their insulation. Capacitors are used in all areas of electrotechnology and, like the resistor, are a basic electrical component.

FYI

Charge separation occurs in clouds because rain in a cloud carries electrons from one part to another

Electric charge, as explained in Chapter 2, is measured in coulombs (symbol Q). Static electricity is an electric charge caused when friction robs from or adds electrons to an object. A voltage exists between two objects or points within an object that have a different number of electrons. That is, electrical energy is stored in the form of an electric charge.

As shown in Figure 10.1, lightning is a spark discharge between centres of positive and negative charge, and can occur within clouds, between clouds and between clouds and the ground. It occurs when the voltage between any two points is enough to break down the air so it can conduct a flow of electrons.

FIGURE 10.1 Charge separation in clouds builds up a high voltage that causes current flow in the form of lightning

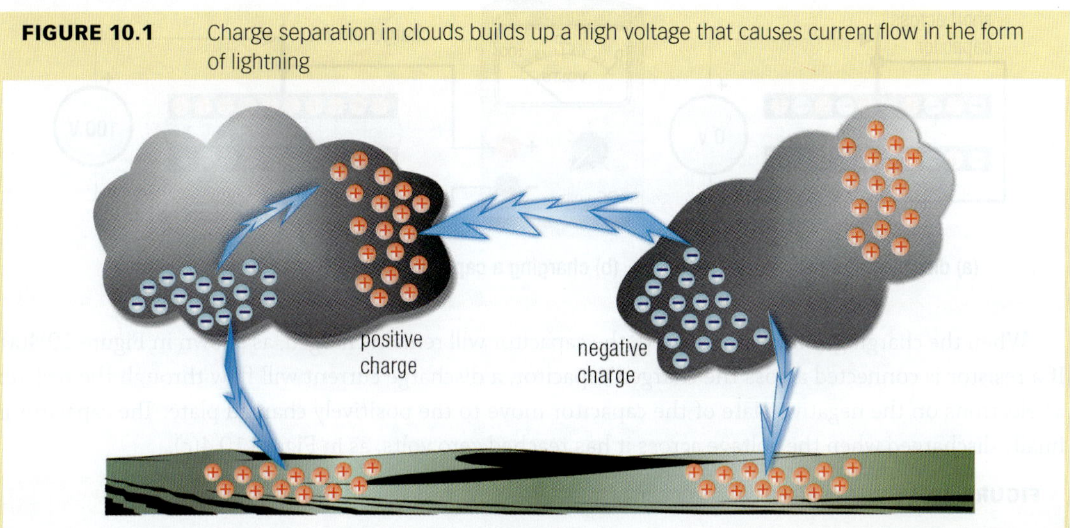

A basic law of electrostatics says that *like charges repel and opposite charges attract*. When there are two oppositely charged objects close to each other, an electrostatic (or electric) field exists between them. This is similar to magnetism, in which a magnetic field exists between two magnetic poles, and like magnetic poles repel while unlike poles attract each other.

An electric field can be thought of as lines of force, shown as dotted lines in Figure 10.2. The intensity of an electric field (number of electrostatic lines of force in a given area) depends on the voltage between the charged objects, the distance between them and the type of insulating material separating them.

Figure 10.2 shows a charged capacitor with a potential difference of 100 V between its plates. In a charged capacitor, a certain number of electrons are stored on the negative plate of the capacitor and the same number of electrons are repelled by the electric field from the positive plate.

The electric field is stored in the insulating material between the two charged plates. The insulating material is called the *dielectric*. In Figure 10.2, the dielectric is air, but most capacitors have a solid material or a chemical dielectric. If the 100 V charging source is removed, the capacitor will remain charged at 100 V, unless there is a path for the electrons between the plates.

FIGURE 10.2 An electric field is set up between two charged objects

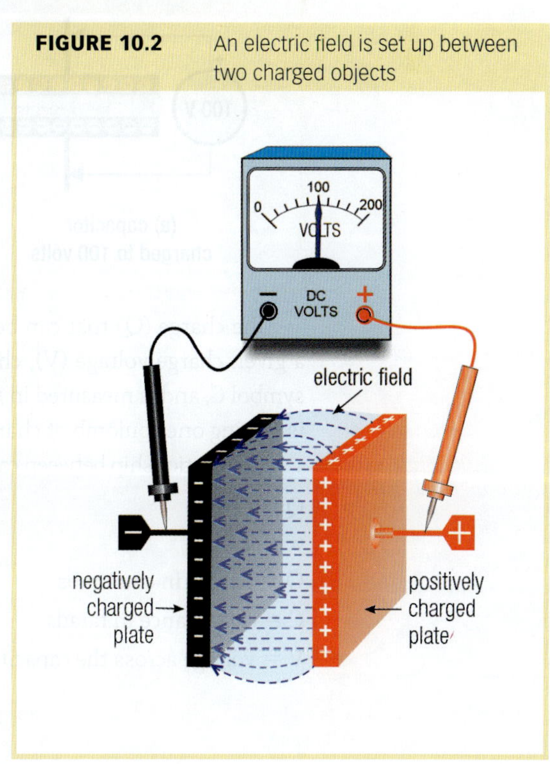

10.2 Charge and Capacitance

A capacitor has two plates made of a metallic material separated by a dielectric. When a capacitor is connected to a voltage source, electrons flow into one plate and electrons are drawn from the other until the voltage between the plates equals the applied voltage. Although a charge current has flowed, electrons have not flowed through the dielectric between the plates. Instead, as electrons enter one plate, the electric field in the dielectric repels electrons from the other.

Figure 10.3 shows a capacitor being charged from a 100 V DC voltage supply. Before charging, there are an equal number of electrons and protons on each plate of the capacitor, so no charge is stored, as shown in Figure 10.3(a). When the 100 V DC supply is connected to the capacitor, a charge current flows until the voltage across the capacitor is 100 V.

FIGURE 10.3 When a capacitor is being charged, a charge current flows until the voltage across the capacitor equals the charging voltage

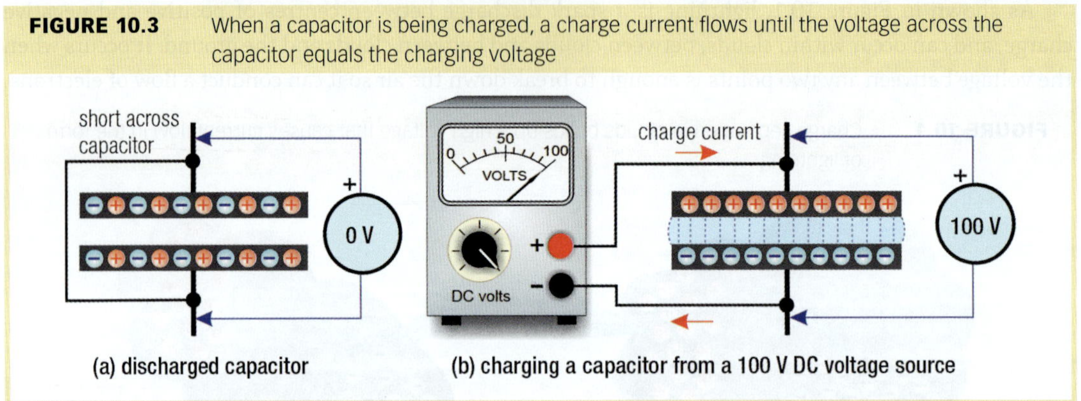

When the charging voltage is removed, the capacitor will remain charged, as shown in Figure 10.4(a). If a resistor is connected across the charged capacitor, a discharge current will flow through the resistor as electrons on the negative plate of the capacitor move to the positively charged plate. The capacitor is finally discharged when the voltage across it has reached zero volts, as in Figure 10.4(c).

FIGURE 10.4 A charged capacitor can be discharged by connecting it across a resistor

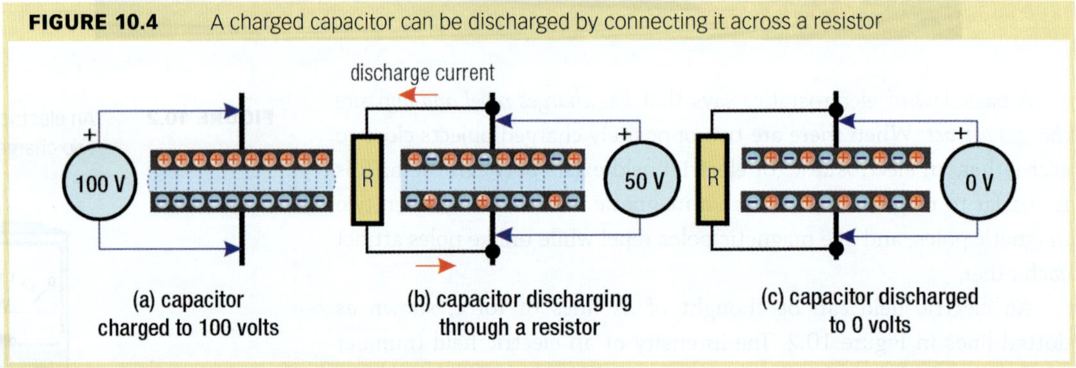

The charge (Q) that can be stored in a given size capacitor depends on the charging voltage. Or, for a given charge voltage (V), charge Q depends on the capacitance of the capacitor. Capacitance has the symbol C, and is measured in *farads*. A capacitance of one farad will have one volt across its terminals if it is storing one coulomb of charge (6.24×10^{18} electrons).

The relationship between capacitance, voltage and charge can be expressed with this simple equation:

$$Q = C \times V$$

where:

Q = charge in coulombs

C = capacitance in farads

V = voltage across the capacitor in volts.

KEY CONCEPT
Charge, capacitance and voltage

This equation can be transposed to make either C or V the subject, giving:

$V = \frac{Q}{C}$ and $C = \frac{Q}{V}$

Most practical capacitors have a capacitance measured in millionths of a farad, referred to as *microfarads*. The symbol for the farad is the letter F, and the term microfarad is written as μF. The amount of charge stored is therefore millionths of a coulomb (μC). Examples 10.1 and 10.2 show how to use these equations.

EXAMPLE 10.1

Find the charge in a 2 μF capacitor that has 400 V across its terminals.

Solution

Values $C = 2\ \mu F$

$V = 400\ V$

$Q = ?$

Equation $Q = CV$

$Q = 2 \times 10^{-6} \times 400$

Answer $\mathbf{Q = 800\ \mu C}$

FIGURE 10.5

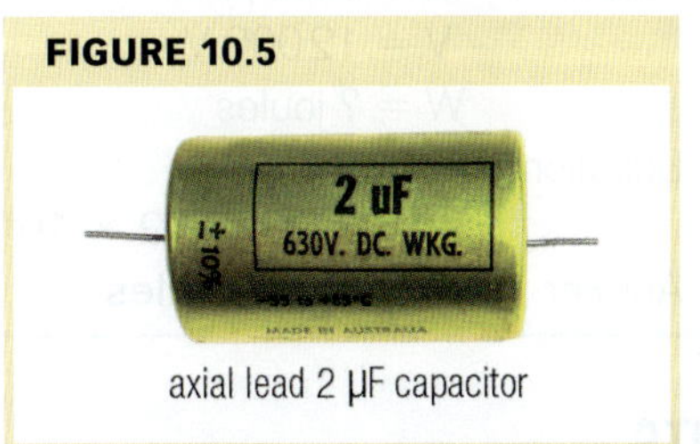

axial lead 2 μF capacitor

EXAMPLE 10.2

How much capacitance is required to store 40 μC of charge at a voltage of 200 V?

Solution

Values given $C = ?$

$Q = 40\ \mu C$

$V = 200\ V$

Equation $C = \frac{Q}{V} = \frac{40 \times 10^{-6}}{200}$

Answer $\mathbf{C = 0.2\ \mu F}$

Energy stored in a capacitor

The energy stored by a charged capacitor is found by multiplying *half* the value of the capacitance by the *square* of the voltage across the capacitor. That is:

$W = ½CV^2$

where:

W = energy in joules

C = capacitance in farads

V = voltage across the capacitor in volts.

EXAMPLE 10.3

Calculate the energy stored in a 100 μF capacitor charged to 200 V.

Solution

Values $C = 100\ \mu F$

$V = 200\ V$

$W = ?$ joules

Equation $W = ½CV^2$

$W = 0.5 \times 100 \times 10^{-6} \times 200 \times 200$

Answer **W = 2 joules**

FIGURE 10.6

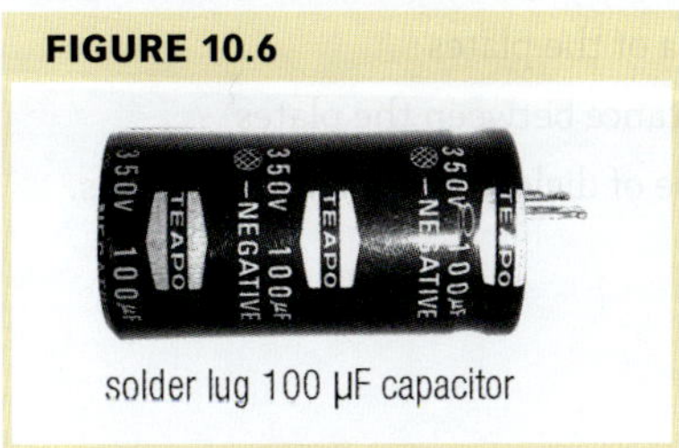

solder lug 100 μF capacitor

As explained later, capacitors able to store high amounts of energy are used in power transmission and distribution. While they have a small capacitance, these special capacitors can be charged to an extremely high voltage, thereby storing a substantial amount of energy. Example 10.4 uses typical values that would be found in a substation.

EXAMPLE 10.4

Find the energy stored in a 0.22 μF capacitor charged to 12 kV.

Solution

Values $C = 0.22\ \mu F$

$V = 12000\ V$

$W = ?$ joules

Equation $W = \frac{1}{2}CV^2$

$W = 0.5 \times 0.22 \times 10^{-6} \times 12000 \times 12000$

Answer **W = 15.84 joules**

KEY POINTS...

- Capacitance is formed between two conductors (plates) separated by an insulator.
- When a capacitor is charged, it will have a voltage across its terminals and an electric field between its plates.
- The unit of capacitance is the farad (F). Most capacitors have values in the order of microfarads (μF) or less.
- The charge (Q) stored in a capacitor equals the capacitance (C) in farads times the voltage (V) across the capacitor. That is $Q = CV$.
- The energy stored in a capacitor equals half its capacitance times the square of the voltage across the capacitor. That is $W = \frac{1}{2}CV^2$.

TASK 10.1

1 A 10 μF capacitor has 200 V across its terminals. Calculate the charge held by the capacitor.

2 A 20 μF capacitor is holding a charge of 12 millicoulombs (mC). What is the voltage between the capacitor terminals?

3 How much energy is stored in a 0.3 μF capacitor charged to 11 kV?

4 Determine the amount of capacitance needed to store 12 millicoulombs (mC) at a voltage of 22 kV.

5 A 10 μF capacitor holds a charge of 6 mC. Find the energy stored in the capacitor.

10.3 Factors that Determine Capacitance

The three things that determine the capacitance of a capacitor are:

1 area of the plates

2 distance between the plates

3 type of dielectric between the plates.

Area of the plates

The larger the area of the plates the higher the capacitance. A large plate area provides more storage for electrons, and the larger dielectric area can store a stronger electric field. Figure 10.7 shows the concept, in which two capacitors are compared, each with plates of different sizes. The plates are separated by the same distance and have the same dielectric. The only difference between the two capacitors is the size of their plates. That is, capacitance is *directly* proportional to the area of the plates.

FIGURE 10.7 The larger the plate area, the larger the capacitance for a given distance between the plates

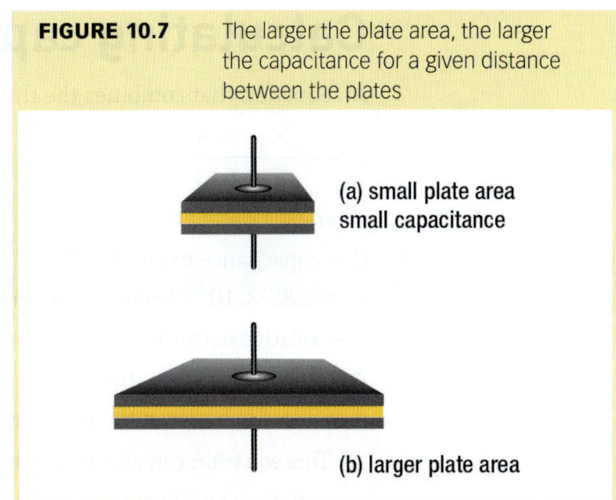

Distance between the plates

The closer the plates, the higher the capacitance. In other words, *the* thinner the dielectric, the higher the capacitance.

A limitation is obtaining a high working voltage with an extremely thin dielectric. Figure 10.8 shows two capacitors with the same plate area but with a different dielectric thickness. A thin dielectric will have a higher concentration of electrostatic lines of force, and therefore a stronger electrostatic field strength. The limit is the maximum voltage the capacitor can withstand before it breaks down and becomes a conductor. That is, capacitance is *inversely* proportional to the distance between the plates.

FIGURE 10.8 A thin dielectric gives a higher capacitance for the same size plate area

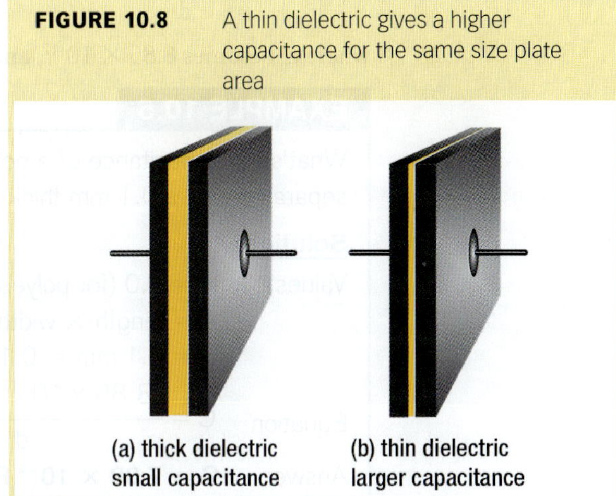

Type of dielectric

The dielectric between the plates of a capacitor supports the electrostatic field (electric lines of force). Its ability to do this is referred to as its *permittivity*. Measured in farads per metre (F/m), the symbol for permittivity is ε, the Greek letter epsilon. The permittivity of a vacuum has the symbol ε_0, and is approximately 8.85×10^{-12} F/m. A more useful term is relative permittivity (ε_r) which is a measure of how much better a material can support an electric field compared to air or a vacuum. Relative permittivity is usually referred to as a material's *dielectric constant* (K). Table 10.1 lists the dielectric constant of various materials used in capacitor manufacture.

Table 10.1 shows, for example, that polyester is four times better at supporting an electric field than air. Therefore, for the same plate area and distance between them, a capacitor with a polyester dielectric has four times the capacitance of a capacitor with an air dielectric.

TABLE 10.1

DIELECTRIC CONSTANTS AND TYPICAL APPLICATIONS

Material	Dielectric constant (K)	Type of capacitor and applications
air	1	variable, used in communications equipment
paper	3 to 6	older style tubular capacitors
oil impregnated paper	2 to 4	high voltage, used in substations
polyester	4	film capacitors, general and high voltage use
mica	6 to 8	low loss, used in high frequency circuits
aluminium oxide	9.6	large capacitance, low and mains voltage use
tantalum oxide	26	large capacitance in small size, low voltage use
ceramic (low-K)	12 to 40	high stability, low loss, high frequency circuits
ceramic (high-K)	200 to 14 000	small size, high capacitance, low voltage use

Calculating capacitance

An equation that combines the three factors determining capacitance is:

$$C = \frac{\varepsilon_0 \times \varepsilon_r \times A}{d}$$

where:

C = capacitance in farads (F)

$\varepsilon_0 = 8.85 \times 10^{-12}$ (value to convert relative permittivity to absolute permittivity)

ε_r = relative permittivity of the dielectric

A = area in square metres (m^2)

d = distance between the plates in metres (m).

This equation can also be written as:

$$C = \frac{(8.85 \times 10^{-12}) \times K \times A}{d}$$

where ε_0 becomes 8.85×10^{-12}, and ε_r becomes the letter K for dielectric constant.

EXAMPLE 10.5

What's the capacitance of a polyester capacitor with foil plates measuring 10 mm by 2 m, separated with a 0.1 mm thick dielectric?

Solution

Values $K = 4.0$ (for polyester dielectric, from Table 10.1)

$A = \text{length} \times \text{width} = 10\text{ mm} \times 2\text{ m} = 0.01\text{ m} \times 2\text{ m} = 0.02\text{ m}^2$

$d = 0.1\text{ mm} = 0.1 \times 10^{-3}\text{ m}$

Equation $C = \frac{(8.85 \times 10^{-12}) \times K \times A}{d} = \frac{(8.85 \times 10^{-12}) \times 4.0 \times 0.02}{0.1 \times 10^{-3}}$

Answer $\mathbf{C = 7.08 \times 10^{-9}\ F = 7.08\ nanofarad\ (nF)}$

As previously mentioned, the farad is too large a unit to be practical, and most capacitance values are in microfarads ($\mu F = F \times 10^{-6}$), nanofarads ($nF = F \times 10^{-9}$) or picofarads ($pF = F \times 10^{-12}$).

KEY POINTS...

- Capacitance is directly proportional to the area of the capacitor's plates, and inversely proportional to the distance between them.
- Capacitance also depends on the type of dielectric between the plates. Dielectric materials are rated by their permittivity and breakdown voltage.
- All dielectrics are compared to air (or vacuum), as both have an absolute permittivity (ε_0) of 8.85×10^{-12} farads per metre.
- The relative permittivity of a dielectric (ε_r) indicates the ability of that dielectric to support an electric field, compared to air (or vacuum).
- Relative permittivity is also called dielectric constant (K). The higher the value of K, the higher the capacitance, if other factors remain the same.

10.4 Types of Capacitors

Capacitors are widely used in all areas of electrotechnology, with physically small capacitors found mainly in electronics and communications equipment. Physically large capacitors with small capacitance are used extensively in electrical power supply and distribution as well as in industry, where voltages and power handling requirements are much higher.

Capacitors are given various ratings, the main two being the capacitance value and its working voltage. Another rating is the reactive power handling capability of a capacitor, as explained in Chapter 19. Other specifications include the temperature range the capacitor can work within, the type of construction of the capacitor, stability over time and the losses the capacitor introduces into a circuit.

Plastic-film capacitors

This type of capacitor is used extensively in a range of applications. Most types have a film dielectric made of polyester, polycarbonate or polypropylene. The most common is the polyester type. Older types used paper as the dielectric.

Two construction methods are shown in Figure 10.9, in which both types have a thin plastic (or paper) film as the dielectric. In Figure 10.9(a), the dielectric film has a metallised coating, and layers are stacked so alternate layers are connected together and linked to a terminal. The assembly is coated in a protective epoxy or plastic covering. Film capacitors generally use a metallised film, as this type of film has self-healing properties, in which a breakdown point between two layers is burned away, usually in a few microseconds.

Another construction, shown in Figure 10.9(b), is to form a roll of two metal foils separated with two layers of an insulating film. Metallised film is also used in the rolled construction. Connection to the two layers of foil is made by the end caps, in which the metallised layers are offset so one endcap will contact one layer, while the other endcap contacts the other layer. After winding, the assembly is enclosed in a protective package that supports the connecting leads

FIGURE 10.9 Two ways of obtaining a large plate area using metallised plastic film, or foil and plastic film

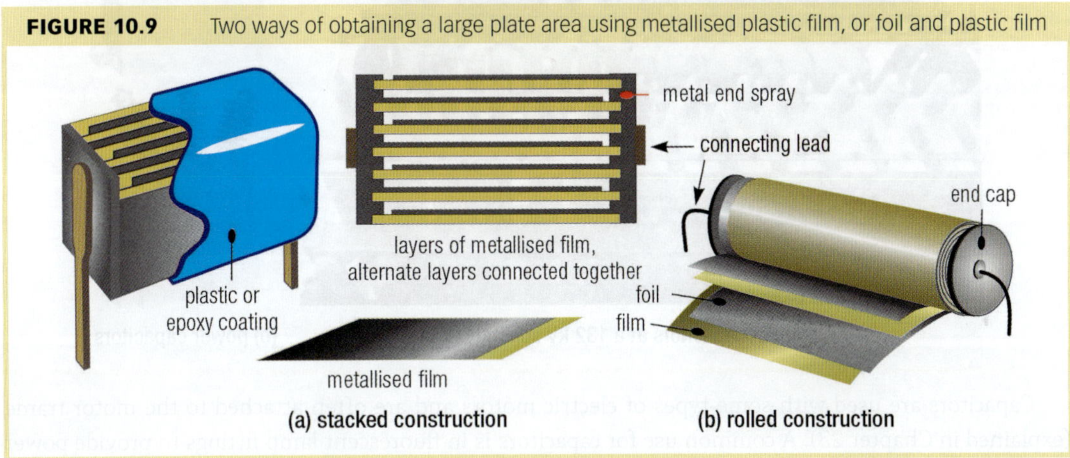

FIGURE 10.10 Examples of plastic film capacitors, shown full size

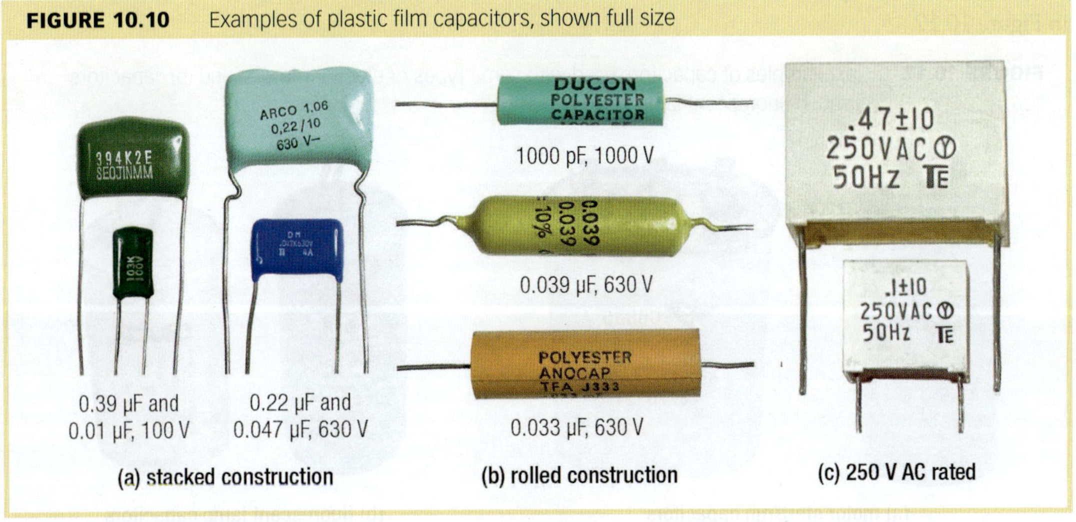

Plastic-film capacitors range from one nanofarad to several microfarads. They have good temperature stability and have a stable capacitance value over time. They can operate from (typically) -40 °C to +85 °C and their DC working voltage is usually from 50 V to 630 V DC. Those used in mains-rated electrical equipment are given an AC voltage rating, typically 250 V AC or higher. Figure 10.10 shows photos of typical film capacitors. Unless stated on the capacitor body, voltage ratings are DC.

Power capacitors

Capacitors are used extensively in electrical power transmission and distribution. Their main use is to provide voltage regulation and to increase the efficiency of the power supply network, as explained in Chapter 19. The types of capacitors shown in Figure 10.11 are physically large and capable of withstanding high voltages. High power capacitors, such as those shown, typically have an oil-impregnated paper or a plastic film dielectric.

FIGURE 10.11 (a) One section of a large bank of capacitors connected to a 132 kV power transmission line, and (b) typical high power capacitors used in electrical power transmission applications (courtesy ABB)

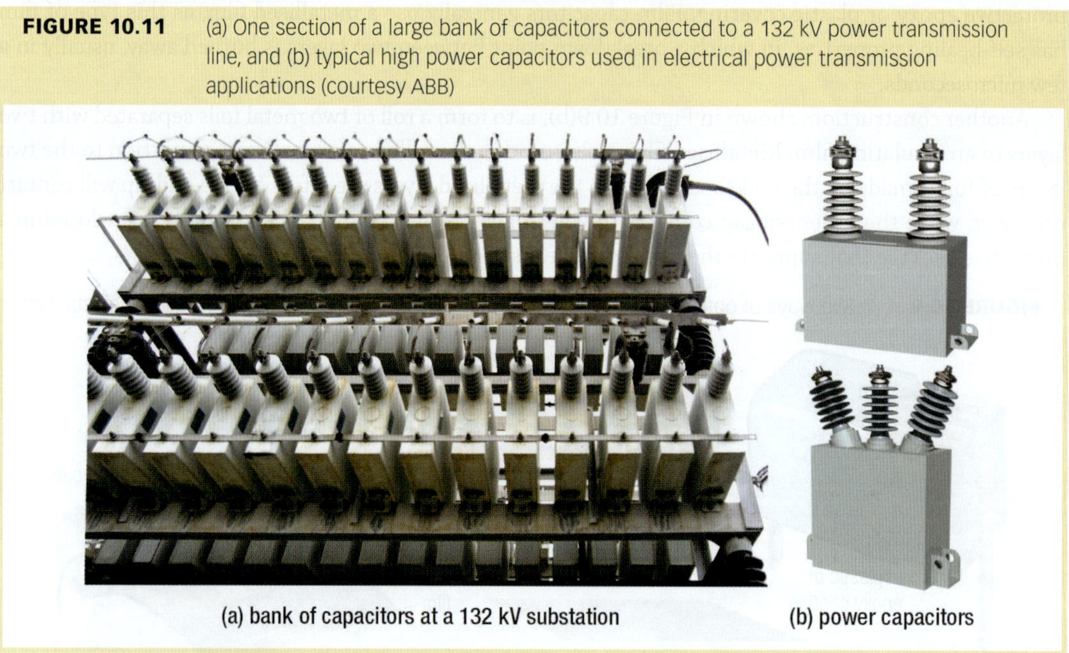

(a) bank of capacitors at a 132 kV substation

(b) power capacitors

Capacitors are used with some types of electric motors and are often attached to the motor frame (explained in Chapter 23). A common use for capacitors is in fluorescent lamp fittings to provide power factor correction, a topic covered in Chapter 19. Typical examples of these types of capacitors are shown in Figure 10.12.

FIGURE 10.12 (a) Examples of capacitors used with some types of electric motors, and (b) capacitors used in fluorescent lamp fittings

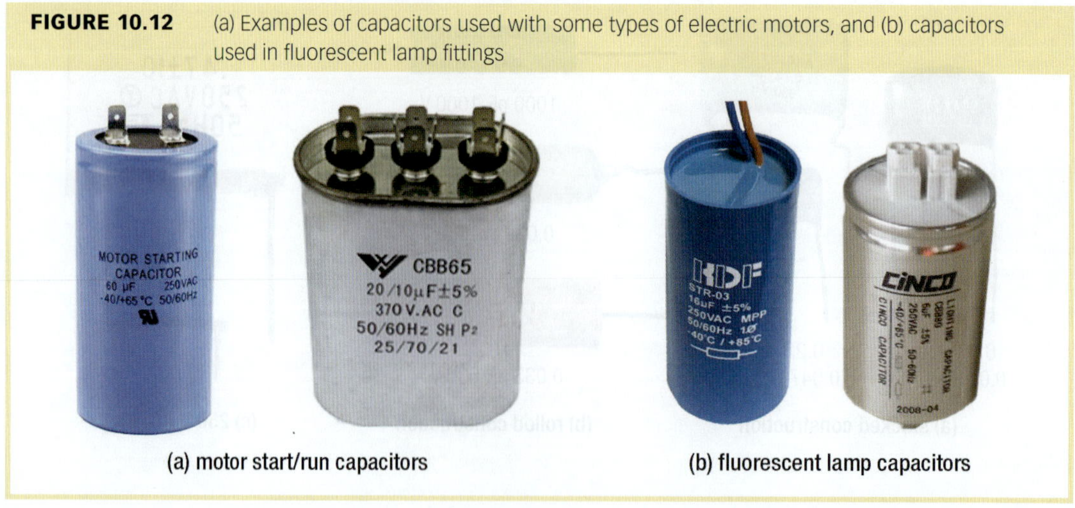

(a) motor start/run capacitors

(b) fluorescent lamp capacitors

Ceramic capacitors

There are two different types of ceramic capacitors, classified by the IEC as class 1, or low-K, and class 2, or high-K, in which K refers to the dielectric constant of the ceramic material. Class 1 capacitors have a dielectric constant of around 12 to 40, which means they are larger for a given capacitance than a class 2 capacitor, which can have dielectric constants of 10 000 or more. Class 1 capacitors have high stability and low losses and can be manufactured to tight tolerance values. They generally have a disc construction, as in Figure 10.13(a).

FIGURE 10.13 Ceramic capacitor construction is either disc or stacked

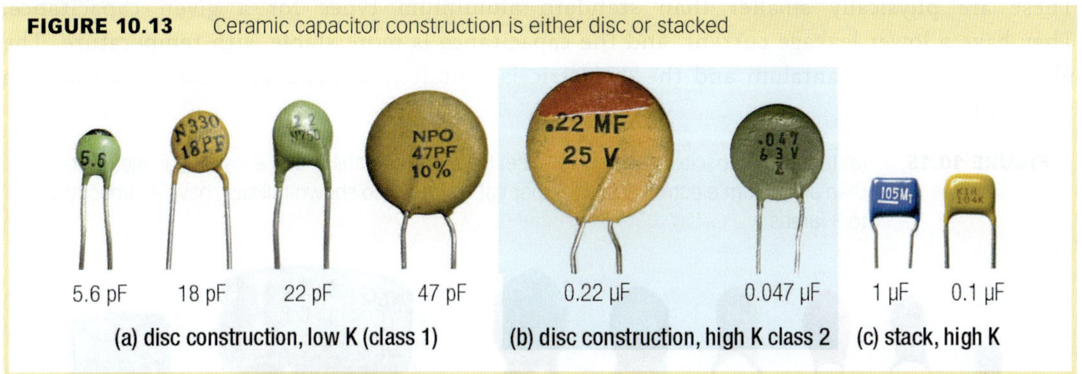

Class 2 capacitors have much smaller package sizes for a given capacitance but lack the stability of the class 1 types. Construction methods are disc and stacked. Disc capacitors are made by coating a ceramic disc metallised with silver on both sides. The stacked construction is similar to that described for plastic-film capacitors and provides a large capacitance in a small package, as shown in Figure 10.13(c).

Ceramic capacitors are used in high voltage applications, with voltage ratings greater than 15 kV and capacitance values up to 2 nF, or 0.002 µF. These are used in a range of industries. The capacitors in Figure 10.13 are generally used in electronic equipment.

Electrolytic capacitors

Electrolytic capacitors have a dielectric formed by chemical action. Because the dielectric is extremely thin, these capacitors have a large capacitance in a small size. The most common type of electrolytic capacitor is the polarised wet aluminium type. The dielectric is formed during manufacture by passing a DC current through the capacitor, so these types must be connected with the right polarity. The polarity of the terminals is marked on the case and the negative lead always connects to the aluminium casing of the capacitor. Examples are shown in Figure 10.14(a) and (c). Bi-polar electrolytic capacitors contain two capacitors in series, with a positive and negative terminal connected together. These capacitors therefore do not have a polarity. Examples are shown in Figure 10.14(b).

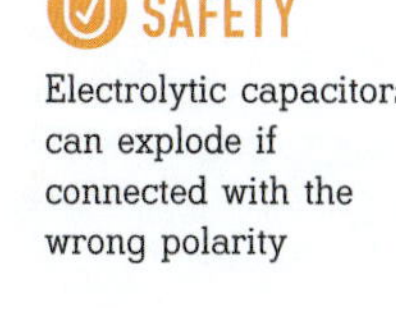

Electrolytic capacitors can explode if connected with the wrong polarity

FIGURE 10.14 Electrolytic capacitors have a large capacitance due to the thin chemically produced dielectric

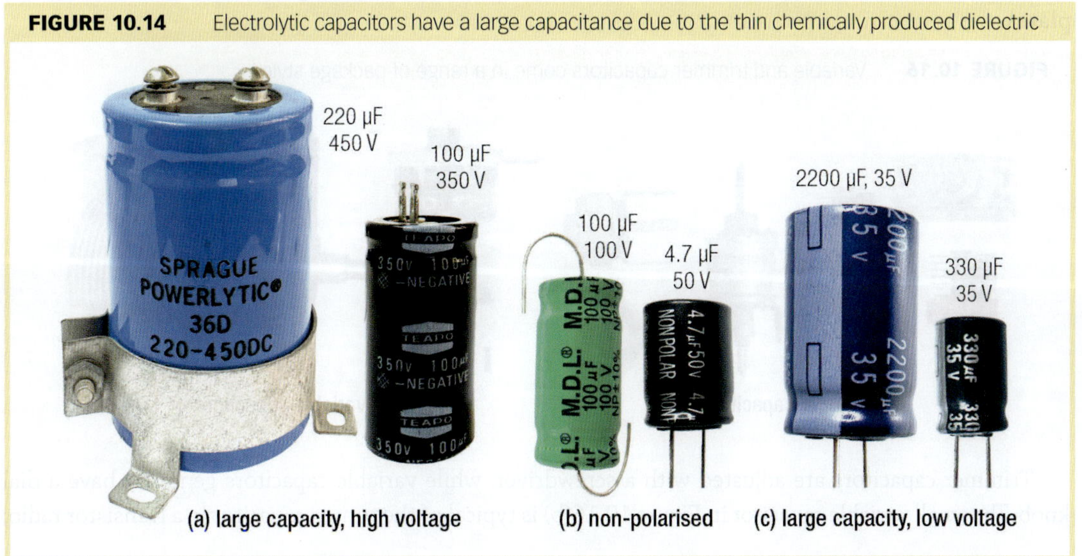

Electrolytic capacitors are made in a range of case styles as shown in Figure 10.14. Some types have a mounting bracket, other styles are mounted directly onto a printed circuit board or an appliance. Some types have several capacitors in the one can.

Capacitance values range from 0.1 μF to over 10 000 μF and have a manufacturing tolerance rarely better than ±20 per cent. Working voltages range from 3 V to over 500 V, and the higher the working voltage for a given capacitance the larger the capacitor.

Another type of electrolytic capacitor is the *tantalum* capacitor, shown in Figure 10.15(a). These are physically smaller than standard aluminium types for a given capacitance. They have a lower leakage current, and the capacitance is more stable with temperature. The plates are made of tantalum and the dielectric is tantalum dioxide. They range in value up to 200 μF.

FIGURE 10.15 (a) Tantalum capacitors (shown full size) have better stability and a lower leakage current than aluminium electrolytics; (b) Super capacitors, also shown full size, have a capacity up to 3 farads

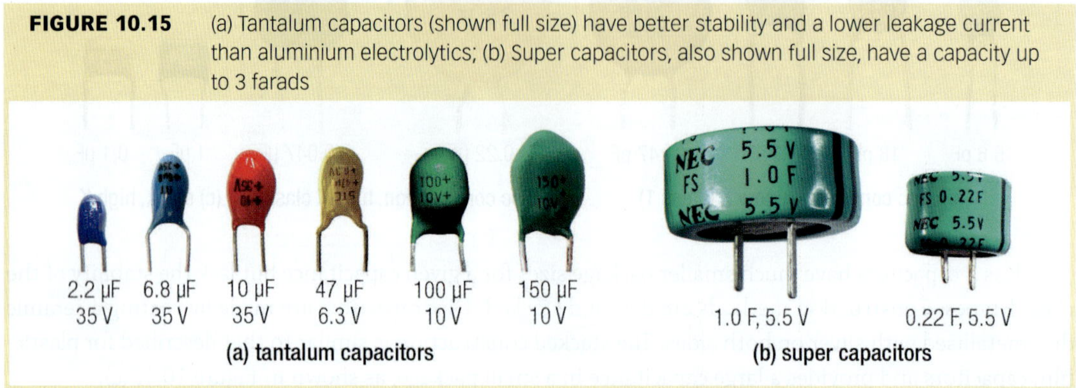

A special type of polarised capacitor is the so-called super capacitor. These have a capacitance up to several farads and are made with activated charcoal plates immersed in an electrolyte. The effect is that a *double layer* of charge builds up between the plate surface (which is very large because the charcoal is porous) and the electrolyte. A one farad 5.5 V rated capacitor, as shown in Figure 10.15(b), can store five coulombs of charge, enough to supply an electronic device in sleep mode for many days. Their main role is to act as a backup supply in computer systems to maintain settings when the equipment is turned off.

Variable capacitors

Variable capacitors are used mainly in telecommunications equipment. Typical types are shown in Figure 10.16. These have a capacitance value measured in *picofarads* (10^{-12}), which is varied either by changing the effective *area* of the plates, or by changing the *distance* between the plates.

FIGURE 10.16 Variable and trimmer capacitors come in a range of package styles

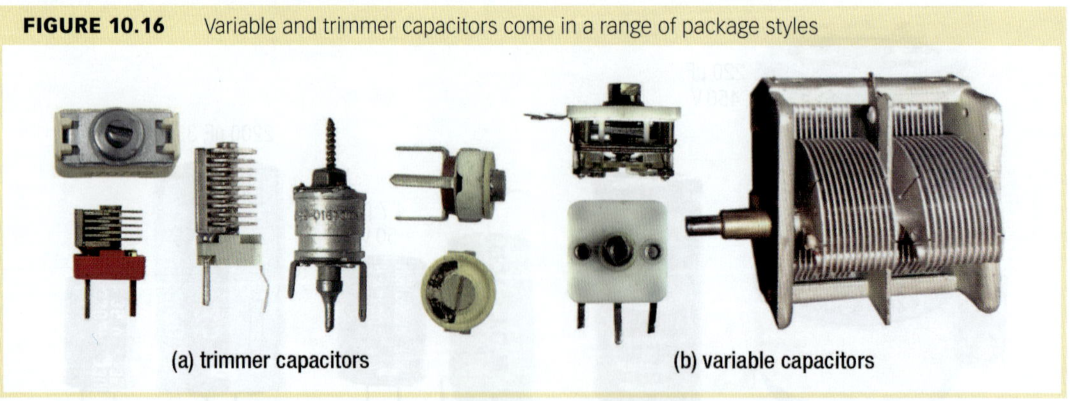

Trimmer capacitors are adjusted with a screwdriver, while variable capacitors generally have a dial knob. The small variable capacitor in Figure 10.16(b) is typical of the tuning capacitor in a transistor radio;

the larger example is more likely to be found in a valve radio. The capacitor at the top left of Figure 10.16 is adjusted by moving the distance between the top plate and the bottom plate. All others are adjusted by interleaving movable plates with fixed plates.

Capacitor symbols

There are four types of capacitors, as listed below; their symbols are in Figure 10.17:

- fixed, non-polarised (plastic-film, ceramic, paper, mica)
- fixed, polarised (electrolytic, tantalum)
- variable (adjustable by hand)
- trimmer (adjustable with a screwdriver).

FIGURE 10.17 Capacitor symbols

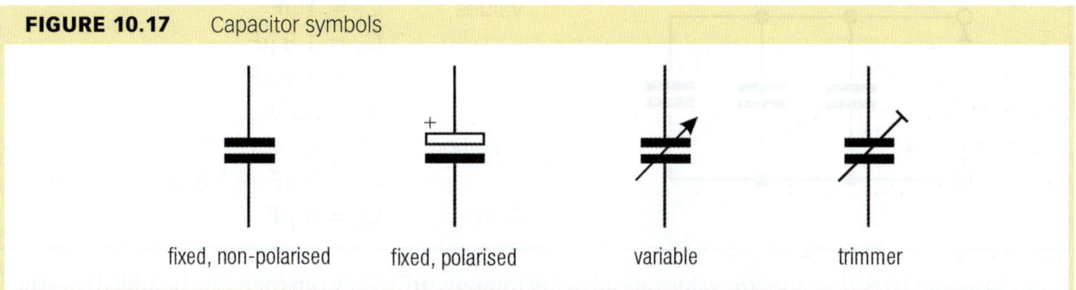

KEY POINTS...

- Capacitors can be broadly grouped into three categories: high voltage, high reactive power (e.g. power transmission applications); medium voltage, medium power (e.g. electric motors, fluorescent lamps and other electrical applications); and low power (e.g. electronics and communications).
- Capacitor dielectrics can be non-polarised (e.g. plastic film) or polarised (e.g. electrolytic). Polarised capacitors must be connected with the right polarity as the dielectric is produced by chemical action. Reversing the polarity can cause some types of capacitors to explode.
- All capacitors have a maximum working voltage that is often printed on the body of the capacitor. If this voltage is exceeded, the dielectric can break down and cause the capacitor to become a short-circuit.
- Capacitors have various types of dielectrics to suit their intended application. High-power capacitors often have oil-impregnated paper as the dielectric.

10.5 Capacitors in Parallel

Figure 10.18 shows the effect of two capacitors connected in parallel. The result is an increase in the total effective plate area. Because capacitance is directly proportional to the area of the plates, the total capacitance is therefore increased.

FIGURE 10.18 Connecting capacitors in parallel gives more capacitance

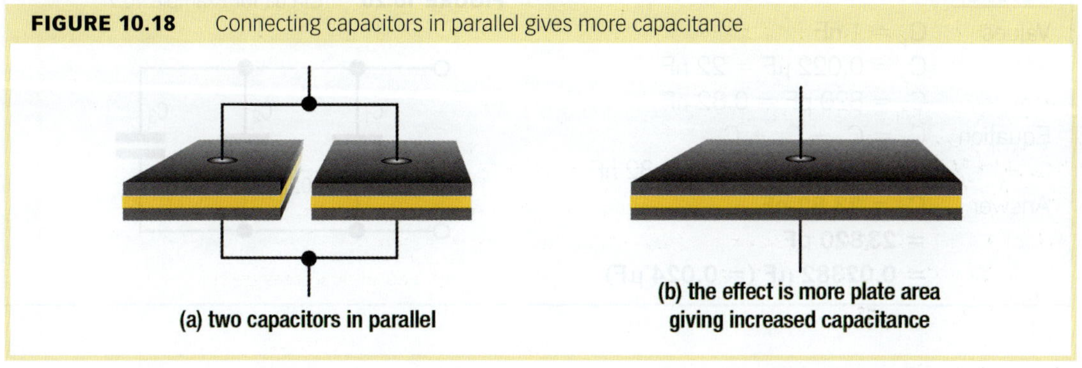

KEY CONCEPT

Capacitors in parallel

When capacitors are connected in parallel, their total capacitance equals the sum of the individual capacitance values. The equation is written as:

$C_T = C_1 + C_2 + C_3 + \cdots$

where:

C_T = the total capacitance

$C_1, C_2, C_3 \ldots$ are the individual capacitor values.

EXAMPLE 10.6

Calculate the total capacitance of the circuit in Figure 10.19.

FIGURE 10.19 Circuit for Example 10.6

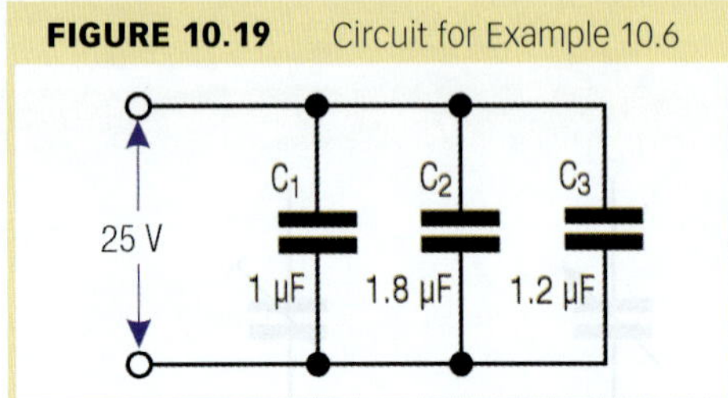

Solution

Values $C_1 = 1\ \mu F$

$C_2 = 1.8\ \mu F$

$C_3 = 1.2\ \mu F$

$V_T = 25\ V$

Equation $C_T = C_1 + C_2 + C_3$

$C_T = 1\ \mu F + 1.8\ \mu F + 1.2\ \mu F$

Answer $\mathbf{C_T = 4\ \mu F}$

In Example 10.6, the capacitor values are all in microfarads (µF). They can therefore be added together, and the answer expressed in µF. Adding capacitor values with different metric multipliers requires them to be converted to the same metric multiplier. Table 10.2 shows how to convert between picofarads (pF), nanofarads (nF) and microfarads (µF)

TABLE 10.2

CONVERTING CAPACITOR VALUES				
pF (10^{-12}F)	**nF (10^{-9}F)**	**µF (10^{-6}F)**	**to convert from**	**move decimal point**
100	0.1	0.0001	pF to nF	3 places left
1000	1	0.001	nF to µF	3 places left
10000	10	0.01	pF to µF	6 places left
100000	100	0.1	µF to nF	3 places right
1000000	1000	1.0	nF to pF	3 places right
	10000	10	µF to pF	6 places right

EXAMPLE 10.7

Calculate the total capacitance of the circuit in Figure 10.20. Express the answer in nF, pF and µF.

Solution

Values $C_1 = 1\ nF$

$C_2 = 0.022\ \mu F = 22\ nF$

$C_3 = 820\ pF = 0.82\ nF$

Equation $C_T = C_1 + C_2 + C_3$

$C_T = 1\ nF + 22\ nF + 0.82\ nF$

Answer $\mathbf{C_T = 23.82\ nF}$

$\mathbf{= 23820\ pF}$

$\mathbf{= 0.02382\ \mu F\ (= 0.024\ \mu F)}$

FIGURE 10.20 Circuit for Example 10.7

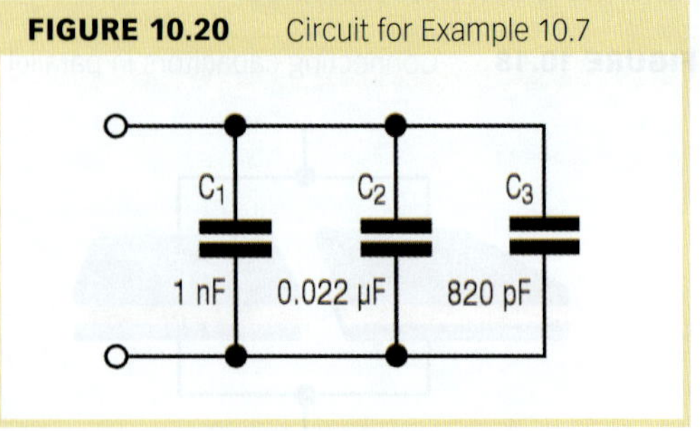

In Example 10.7, the pF and μF values were converted to nF. However, they could have all been converted to pF or μF instead.

To solve Example 10.7 with a calculator, enter each capacitance value using engineering notation. The metric prefix of the result can be changed to suit, as shown in Figure 10.21.

FIGURE 10.21 Keystrokes to solve Example 10.7

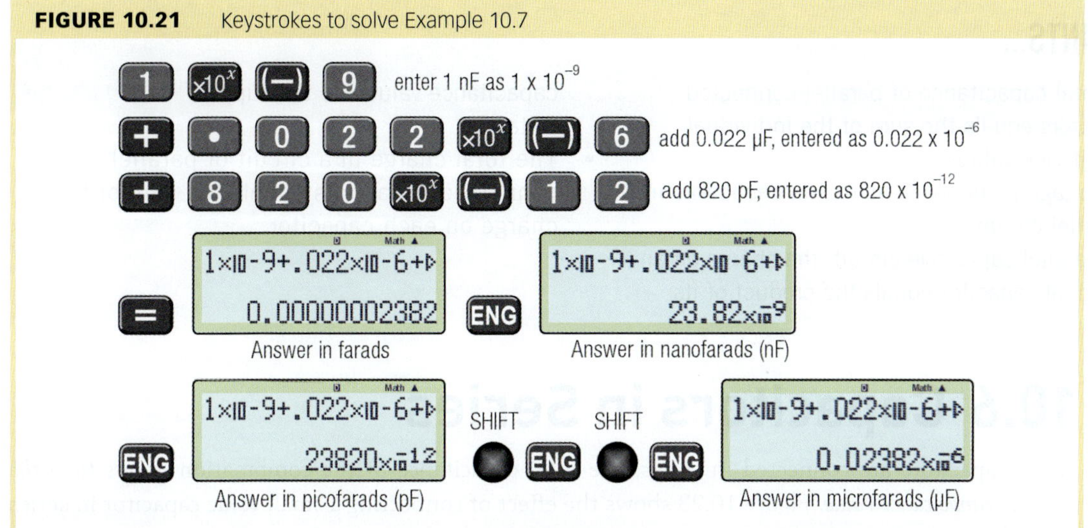

Charge in a parallel capacitive circuit

KEY CONCEPT
Charge in a parallel capacitive circuit

Capacitors in parallel all connect to the same supply voltage, so each capacitor charges independently. The charge taken by each capacitor depends on its value and the supply voltage. The total charge taken by the circuit is the sum of the charge in each capacitor.

$Q_T = Q_1 + Q_2 + Q_3$

$Q_T = C_T \times V$

$Q_1 = C_1 \times V$, $Q_2 = C_2 \times V$, $Q_3 = C_3 \times V$ and so on.

Example 10.8 uses these equations.

EXAMPLE 10.8

For the circuit of Figure 10.22, calculate (a) the charge taken by each capacitor, (b) the total capacitance and (c) the total charge taken by the circuit.

Solution

FIGURE 10.22 Circuit for Example 10.8

40 V; C_1 12 μF; C_2 18 μF; C_3 20 μF

Values $C_1 = 12\ \mu F$

$C_2 = 18\ \mu F$

$C_3 = 20\ \mu F$

$V = 40\ V$

Equation (a) $Q_1 = C_1 \times V$

$Q_1 = (12 \times 10^{-6}) \times 40$

Answer (a) $\mathbf{Q_1 = 480\ \mu C}$ (charge on C_1)

Charge on C_2 and C_3 is calculated the same way, so:

Answer (a) $Q_2 = C_2 \times V = (18 \times 10^{-6}) \times 40\ V =$ **720 μC** (charge on C_2)

Answer (a) $Q_3 = C_3 \times V = (20 \times 10^{-6}) \times 40\ V =$ **800 μC** (charge on C_3)

Total capacitance is the sum of the individual capacitance values:

Equation (b) $C_T = C_1 + C_2 + C_3 = 12\ \mu F + 18\ \mu F + 20\ \mu F$

Answer (b) $\mathbf{C_T = 50\ \mu F}$

Equation (c) $Q_T = C_T \times V = (50 \times 10^{-6}) \times 40\ V$

Answer (c) $\mathbf{Q_T = 2000\ \mu C}$ **or 2 millicoulomb (mC)**

Example 10.8 shows that the sum of the individual charges gives the total charge. Capacitors are connected in parallel to obtain a larger capacitance than possible from a single capacitor of the required working voltage. Another reason is to achieve a higher charge or discharge current, as in high-power applications.

KEY POINTS...

- The total capacitance of parallel-connected capacitors equals the sum of the individual capacitance values.
- The voltage is the same across all capacitors in a parallel circuit.
- In a parallel capacitive circuit, the charge on an individual capacitor equals the product of its capacitance value and the supply voltage.
- The total charge in a circuit of parallel-connected capacitors equals the sum of the charge on each capacitor.

10.6 Capacitors in Series

When capacitors are connected in series, the total capacitance of the combination is *less than the smallest capacitance value*. Figure 10.23 shows the effect of connecting a small value capacitor in series with a large one.

FIGURE 10.23 Connecting capacitors in series gives less capacitance as the plate area is reduced and the equivalent thickness of the dielectric is increased

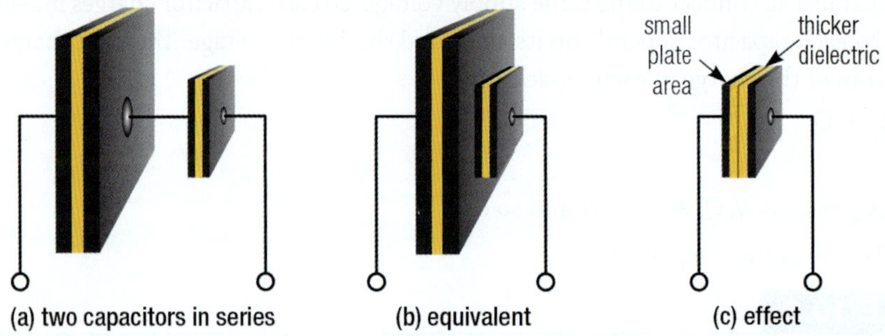

small plate area

thicker dielectric

(a) two capacitors in series

(b) equivalent

(c) effect

Capacitance in series

As Figure 10.23(c) shows, the effect of connecting a small capacitor in series with a larger capacitor is a capacitor with a thicker dielectric and small plate area. This means the total capacitance is less than the value of the small capacitor. This is like resistors in parallel, in which the total resistance is less than the smallest value resistor. As well, the equation to find the total capacitance of series-connected capacitors is similar to the equation for parallel resistors.

$$\frac{1}{C_T} = \frac{1}{C_1} + \frac{1}{C_2} + \frac{1}{C_3} + \ldots$$

where:

C_T = the total capacitance

$C_1, C_2, C_3 \ldots$ = the individual capacitor values.

EXAMPLE 10.9

Find the total capacitance of the circuit in Figure 10.24.

Solution

Values $C_1 = 1\ \mu F$

$C_2 = 2\ \mu F$

$C_3 = 4\ \mu F$

Equation $$\frac{1}{C_T} = \frac{1}{C_1} + \frac{1}{C_2} + \frac{1}{C_3}$$

$$\frac{1}{C_T} = \frac{1}{1} + \frac{1}{2} + \frac{1}{4}(\mu F)$$

$$\frac{1}{C_T} = 1.75\ \mu F \text{ therefore } C_T = \frac{1}{1.75}\ \mu F$$

Answer $\mathbf{C_T = 0.57\ \mu F}$

FIGURE 10.24 Circuit for Example 10.9

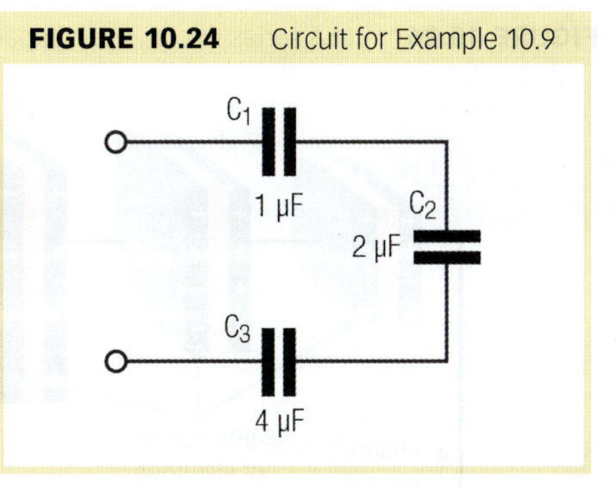

In Example 10.9, because all the capacitance values are given in μF, their numerical values only were entered into the equation, and the answer expressed in μF. A calculator can accept any value, so if the capacitor values have different metric prefixes, enter them as written. After each entered value, press the reciprocal key. When each value has been entered, press the equal key to find the sum of each reciprocal, then press the reciprocal key to get the answer. See Chapter 7, Figure 7.14.

FYI

In a series capacitive circuit, the total capacitance is always less than the smallest value of capacitance in the circuit

Like the equations for resistors in parallel, two simpler equations can be used with series-connected capacitors. If there are only two capacitors in series, use the 'product divided by the sum' equation.

$$C_T = \frac{C_1 \times C_2}{C_1 + C_2}$$

If all the series-connected capacitors have the same value, divide that value by the number of capacitors connected in series:

$$C_T = \frac{C}{n}$$

where:

n is the number of capacitors in the circuit

C is the capacitance value of one capacitor.

Charge in a series capacitive circuit

In a series circuit, each capacitor will hold the same charge, regardless of its value, for reasons explained after Example 10.10. This example uses the circuit from Example 10.9, in which we calculated a total capacitance of 0.57 μF. By connecting the circuit to a 50 V supply, we can now work out the total charge taken by the circuit using the equation for charge: $Q_T = C_T \times V$.

EXAMPLE 10.10

How much charge does the circuit of Figure 10.25 take when it is connected to a 50 V DC supply?

Solution

Values $C_T = 0.57\ \mu F$ (as calculated in Example 10.9)

$V = 50\ V$

$Q_T = ?$

Equation $Q_T = C_T \times V = (0.57 \times 10^{-6}) \times 50$

Answer $\mathbf{Q_T = 28.5\ \mu C}$

FIGURE 10.25 Circuit for Example 10.10

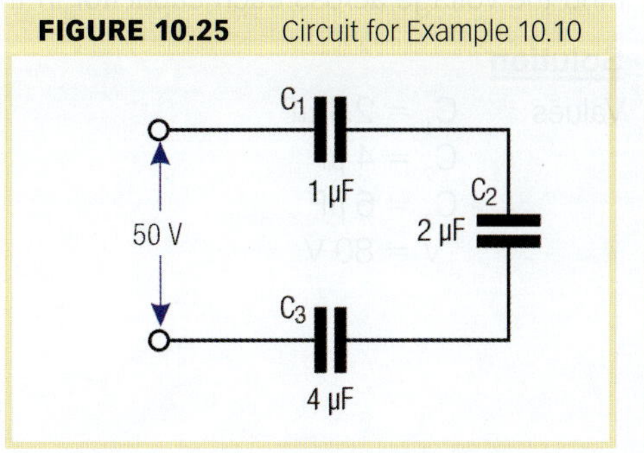

In Example 10.10 the total charge has been found using the total capacitance value and the applied voltage. But how much charge does *each* capacitor hold? The answer can be seen when looking at what happens when the circuit is first connected to the 50 V supply, as shown in Figure 10.26.

FIGURE 10.26 In a series circuit, each capacitor stores the same number of electrons

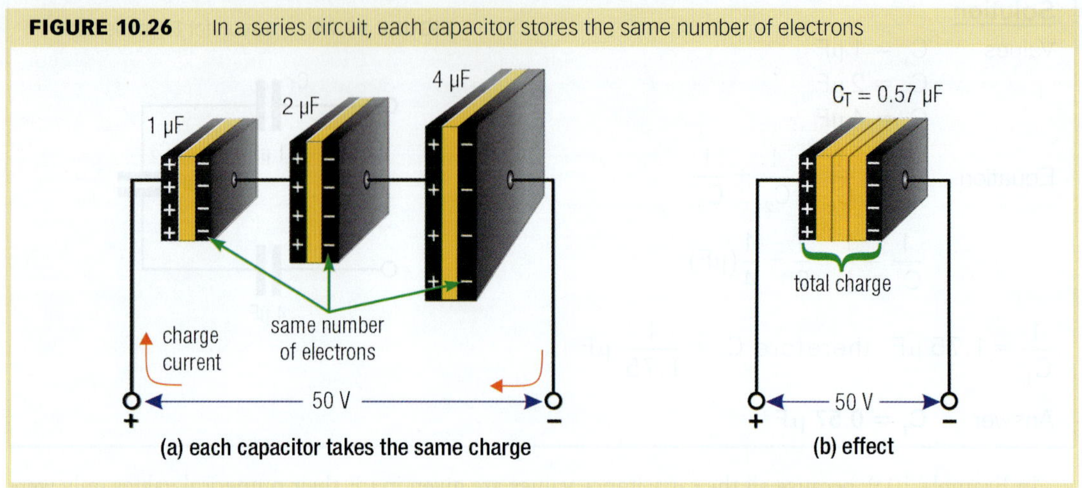

Figure 10.26 shows the three capacitors from Example 10.10 connected in series with a 50 V DC supply. When the supply is first switched on, a current flows as the capacitors charge. However, the current *does not flow through* the dielectric, as it is an insulator. Instead, the electrons stored on one plate of a capacitor repel an equal number from its other plate because of the electric field. These electrons are then stored on a plate of the next series-connected capacitor, in turn repelling an equal number from its other plate. No matter how many capacitors there are in the circuit, this is what happens for them all.

Therefore, when the circuit charges, the number of electrons taken *from* the power supply equals the number of electrons returned *to* the power supply, and each capacitor holds this same number of electrons. That is, *each capacitor has the same charge*, and that value of charge equals the total charge stored by the circuit.

In a series capacitive circuit, total charge Q_T and therefore the charge in each capacitor can be found with these equations:

KEY CONCEPT

Charge (Q) in a series capacitive circuit

$Q_T = Q_1 = Q_2 = Q_3 = \ldots$

$Q_T = C_T \times V_T$

$Q_T = Q_1 = C_1 \times V_1 = Q_2 = C_2 \times V_2 = Q_3 = C_3 \times V_3$, and so on.

DC voltage across capacitors in a series circuit

Because each capacitor in a series circuit has the same charge, the voltage across each capacitor will depend on its capacitance. Example 10.11 shows how to find the DC voltage across each capacitor in a series circuit using the charge equations Q = CV and V = Q/C.

EXAMPLE 10.11

Find the voltage across each capacitor in the circuit in Figure 10.27.

Solution

Values $C_1 = 2.4\ \mu F$

$C_2 = 4\ \mu F$

$C_3 = 6\ \mu F$

$V = 80\ V$

Equation 1 $\frac{1}{C_T} = \frac{1}{C_1} + \frac{1}{C_2} + \frac{1}{C_3}$

$$\frac{1}{C_T} = \frac{1}{2.4} + \frac{1}{4} + \frac{1}{6} (\mu F)$$

$$C_T = 1.2\ \mu F$$

Equation 2 $Q_T = C_T \times V$

$$Q_T = (1.2 \times 10^{-6}) \times 80$$

$$Q_T = 96\ \mu C = Q_1 = Q_2 = Q_3$$

Equation 3 $V_1 = \frac{Q}{C_1},\ V_2 = \frac{Q}{C_2},\ V_3 = \frac{Q}{C_3}$

Answer 1 $\mathbf{V_1} = \frac{96 \times 10^{-6}}{2.4 \times 10^{-6}} = \mathbf{40\ V}$

Answer 2 $\mathbf{V_2} = \frac{96 \times 10^{-6}}{4 \times 10^{-6}} = \mathbf{24\ V}$

Answer 3 $\mathbf{V_3} = \frac{96 \times 10^{-6}}{6 \times 10^{-6}} = \mathbf{16\ V}$

FIGURE 10.27 Circuit for Example 10.11

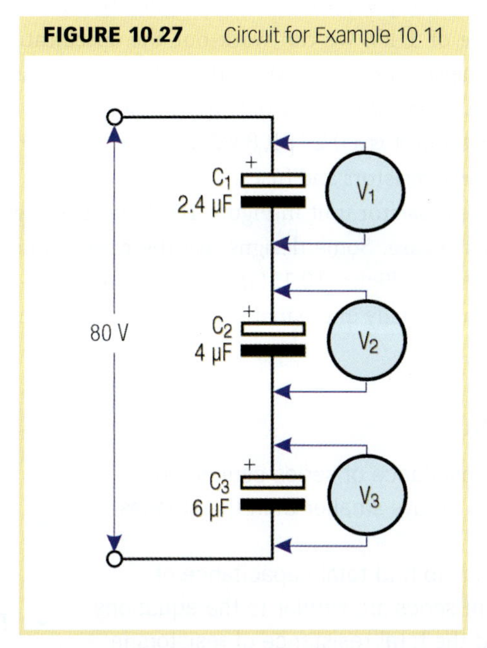

As you would expect from Kirchhoff's voltage law, the three voltages in Example 10.11 add up to the supply voltage. Notice also that the highest voltage drop is across the smallest capacitance. That is, the voltage drop across each capacitor in a series circuit is *inversely proportional* to its capacitance.

FYI

The highest voltage drop is across the smallest capacitance

10.7 Capacitors in Series–Parallel

Capacitors are connected in series to achieve a high working voltage, and in parallel to obtain a higher capacitance and/or a higher charge–discharge current capability. Capacitors are therefore connected in series–parallel to obtain a capacitance of a specified value that can withstand a specified voltage and handle a specified charge–discharge current.

High voltage capacitors are typically made up of small capacitors connected in series–parallel and are typically referred to as capacitor *units*. The example in Figure 10.28 shows six groups of eight parallel-connected capacitors connected in series.

FIGURE 10.28 High voltage capacitors are made up of many smaller capacitors

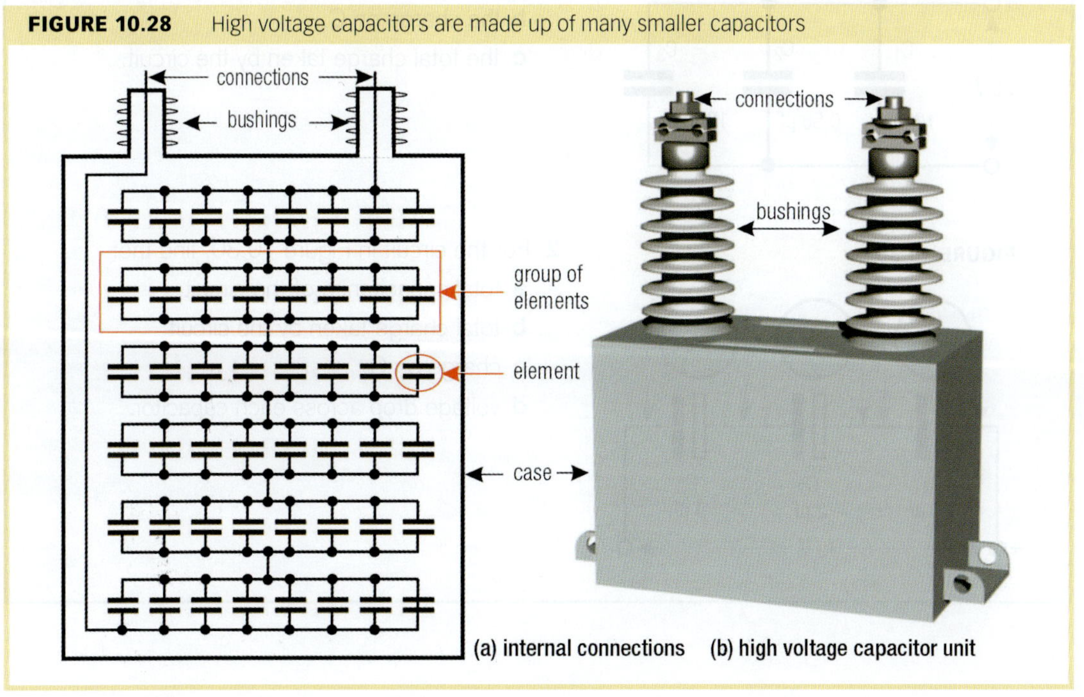

If each capacitor element in Figure 10.28 is identical with a capacitance of 0.22 μF and a working voltage of 1 kV, we can work out the specifications of the capacitor unit. Each parallel group has a total capacitance of 8 × 0.22 μF, or 1.76 μF. The six series-connected groups have a total capacitance of 1.76 μF divided by 6, or 0.29 μF. The working voltage of the capacitor unit is 8 × 1 kV, or 8 kV, so the capacitor unit is a 0.29 μF, 8 kV capacitor. There are many arrangements used, and most include internal discharge resistors and fuses.

The capacitor unit in Figure 10.28 has two high voltage bushings to insulate the connection points from the case. Some designs use the case as one terminal, and therefore have one bushing, like the capacitors in Figure 10.11(a). Others are for three-phase systems and will have three bushings. Capacitor units are usually impregnated with a protective fluid, such as mineral oil, and typically have a plastic-film dielectric.

KEY POINTS...

- The total capacitance of series-connected capacitors is always smaller than the smallest capacitor.
- The equations to find total capacitance of capacitors in *series* are similar to the equations used to find the total resistance of resistors in *parallel*.
- The total capacitance of capacitors in series can be found with these equations:
  - Any number of capacitors: $\frac{1}{C_T} = \frac{1}{C_1} + \frac{1}{C_2} + \frac{1}{C_3} + \ldots$
  - Two capacitors only: $C_T = \frac{C_1 \times C_2}{C_1 + C_2}$
  - All equal value capacitors: $C_T = \frac{C}{n}$
- The voltage drop across individual capacitors in series is indirectly proportional to the capacitance value.
- The sum of the voltage drops across series-connected capacitors equals the supply voltage.
- In a series circuit, each capacitor stores the same charge. This charge is also the total charge stored by the circuit.
- High power capacitor units are made up of many smaller capacitors connected in series–parallel to achieve high voltage and high current specifications.

TASK 10.2

FIGURE 10.29

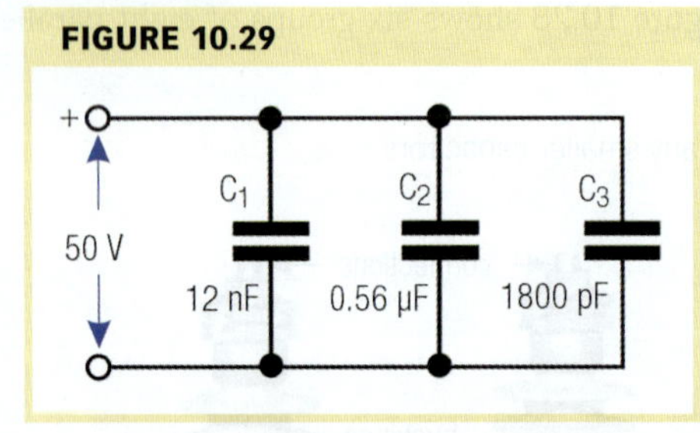

1 For the circuit in Figure 10.29, find:
   a the total capacitance of the circuit
   b the charge on C_1
   c the total charge taken by the circuit.

FIGURE 10.30

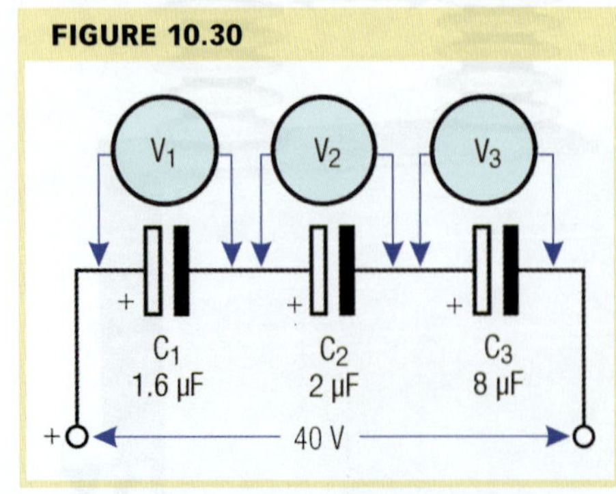

2 For the circuit in Figure 10.30, find the:
   a total capacitance of the circuit
   b total charge taken by the circuit
   c charge on C_1
   d voltage drop across each capacitor.

10.8 Capacitor Safety and Testing

Some older types of capacitors contain polychlorinated biphenyls (PCB), a toxic man-made organic chemical that does not readily break down. First manufactured in 1929, PCBs were banned in 1979. Capacitors in old fluorescent light fittings may contain PCBs, as might a capacitor attached to an old electric motor. Typically, these types of capacitors have a metal case and may have a date marking. They are especially dangerous if they are leaking.

When handling a leaky PCB capacitor, avoid contact with the chemical by wearing personal protective equipment, including gloves that are made of materials resistant to PCBs. Do not use gloves made of polyvinyl chloride (PVC) or natural rubber (latex).

Beware the charged capacitor

Significant energy can be contained in a charged capacitor, enough to give a substantial electric shock and arc burns due to arc flash. Arc flash occurs when an electric current flows through an air gap. (Arc flash incidents cause more deaths than electrocution.) Capacitors greater than 1 μF charged to more than 100 V can deliver a strong momentary arc when discharged by shorting the terminals together.

SAFETY

Discharge a capacitor before touching its terminals to avoid possible electric shock

It is therefore essential to be sure a capacitor is discharged before handling. Even though a circuit is de-energised, it is possible for capacitors to develop a charge, due to chemical action. In a high-voltage work environment, workplace procedures will specify the method of discharging capacitors. In low-voltage situations involving 230 V mains equipment, there is no standard way of discharging a capacitor. Ideally, a capacitor should be discharged through a resistance to limit the peak current and prevent an arc. Then, as a final test, measure the terminal voltage of the capacitor, which should be zero.

Testing capacitors

In low-voltage electrical work, the most common use of capacitors is with electric motors. Other uses are in fluorescent lamp fittings and in power line filters supplying electronic equipment. In all cases, these capacitors are rated for use with 230 V AC. Capacitor faults are an open-circuit, a short-circuit or a low resistance dielectric.

AS/NZS 4836:2011 Section 3.5 covers the requirements when working on live conductive parts such as charged capacitors

There are two tests that most workshops are equipped to do: measuring capacitance and measuring dielectric leakage resistance. Most digital multimeters can measure capacitance, which should be done without touching the capacitor terminals. When measuring an electrolytic capacitor, the positive meter lead connects to the positive terminal of the capacitor. If a capacitor gives a capacitance reading less than the specified variation as printed on the capacitor can (typically ±10%), it should be replaced.

A good capacitor should appear as an open-circuit when its resistance is measured. When a digital ohmmeter is first connected, the capacitor will begin charging as a result of the voltage applied by the ohmmeter. This will cause a resistance reading that rises until the capacitor is charged to the meter voltage. A 250 V capacitor is faulty if it shows less than an open-circuit when measured with a digital ohmmeter. If the meter does not show a rising resistance reading, but instead remains at infinity, the capacitor is probably open-circuit, as will be found when measuring its capacitance. Small values of capacitance (0.01 μF or less) will often show an open-circuit almost immediately.

Capacitors rated at 250 V or higher should be tested for leakage resistance using an insulation resistance tester. This instrument is described in Chapter 25. A test voltage of 250 V or 500 V can be used. As with the ohmmeter test, the capacitor will charge, causing the meter to show a resistance that should rise to infinity (open-circuit). If a reading of slightly less than infinity is shown, the capacitor may still be useful, but is beginning to fail. After testing, the capacitor will be charged to a high voltage, requiring it to be discharged. This can often be done by leaving the probes of an insulation resistance tester connected while turning off the measure switch. In any case, be sure the capacitor is discharged after this test, as it will have a lot of stored energy otherwise.

10.9 The RC Circuit

A circuit that has a resistor and a capacitor connected in series is called an RC circuit. There are two reasons to connect a resistor in series with a capacitor. One is to limit the amount of charge current a capacitor takes when it's first connected to a voltage source. A second reason is to introduce a *time delay*.

Figure 10.31 shows a 10 Ω resistor connected to a 10 V battery in series with a pushbutton and an ammeter. By Ohm's law, the maximum current in the circuit is 10 V/10 Ω = 1 A. So, when the pushbutton is pressed, the current instantly rises to 1 A, and stays there while the pushbutton is held on, as shown in Figure 10.31(b).

FIGURE 10.31 Current in a resistive circuit reaches its maximum value instantly

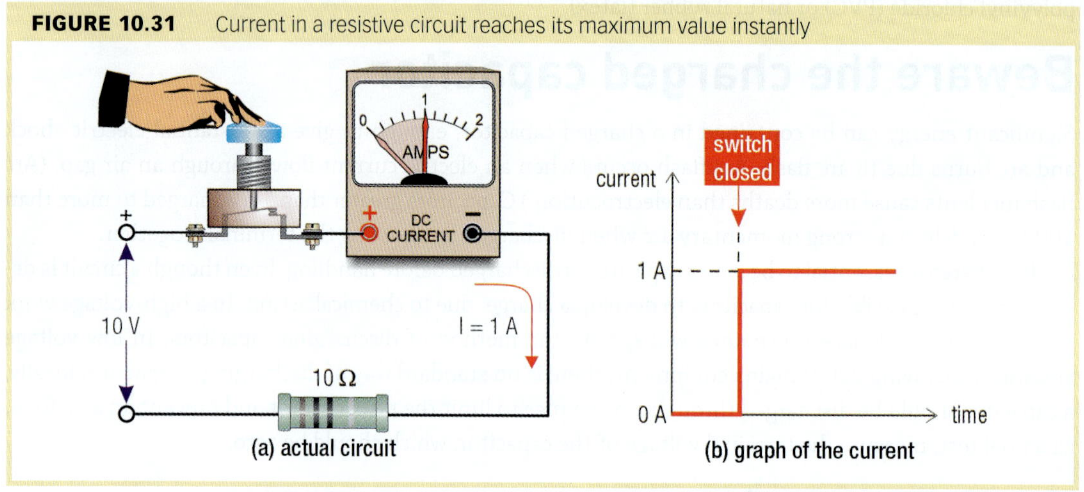

When a capacitor is added to the circuit, as in Figure 10.32, current can only flow when the capacitor is charging or discharging. When the pushbutton is first pressed, the capacitor is discharged, so the voltage across it is zero and the capacitor appears as a short-circuit. Current flows in the circuit, at first limited only by the 10 Ω resistor.

As it charges, the voltage across the capacitor increases, opposing the current. When the capacitor voltage equals the supply voltage, the current falls to zero.

FIGURE 10.32 The charge current in an RC increases instantly to a maximum, then dies away to zero

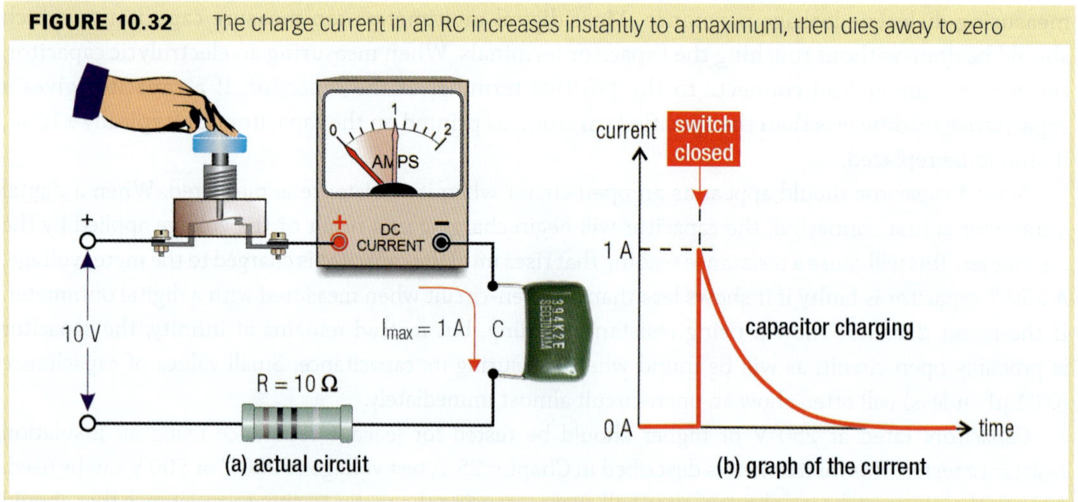

By adding resistance in series with the capacitor, a time delay is introduced, shown by the graph in Figure 10.32(b), in which time has elapsed before the current finally falls to zero. This is shown in more detail in Figure 10.33, which is the same circuit as in Figure 10.32, except C is a 0.1 F capacitor. Note that the capacitor takes 5 seconds to fully charge.

FIGURE 10.33 Current flows in an RC circuit until the capacitor voltage equals the supply voltage

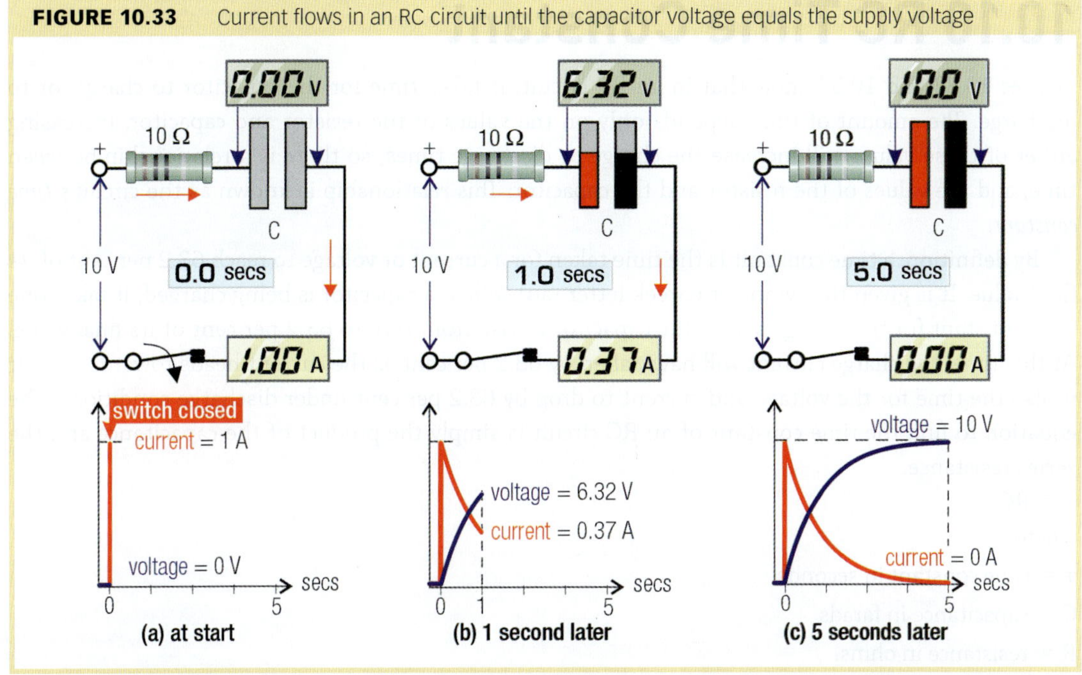

The graphs in Figure 10.33 (and in Figure 10.32(b)) are called exponential curves, in which the values change at first quickly, then more slowly as the capacitor becomes charged. As Figure 10.33 shows, it took one second for the voltage across the capacitor to reach 6.32 V, and a further 4 seconds to reach 10 V.

When a charged capacitor is discharged through a resistor, the graphs of voltage and current are also exponential, as shown in Figure 10.34. The graphs for voltage and current are now the same and are also the same as the graph for the charge current in Figure 10.33.

FIGURE 10.34 When a capacitor is being discharged, the graphs of current and voltage have the same shape

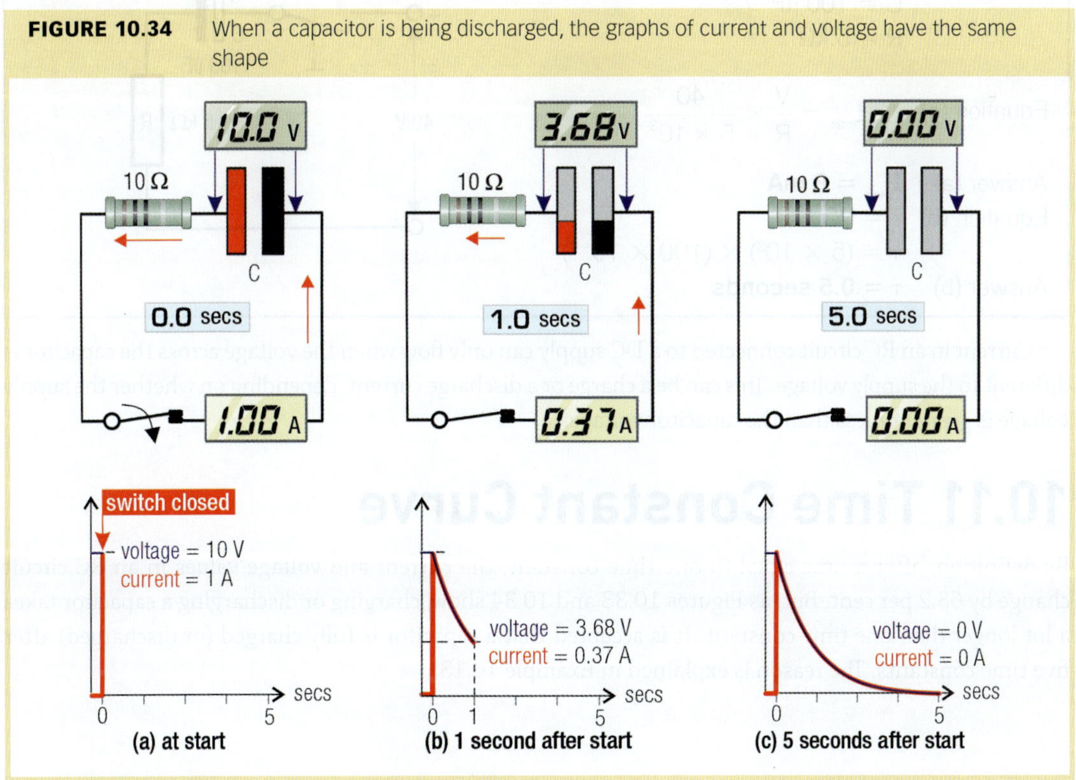

10.10 RC Time Constant

Figures 10.33 and 10.34 show that in an RC circuit, it takes time for the capacitor to charge, or to discharge. The amount of time depends only on the values of the resistor and capacitor. Increasing either of these values will increase the charge or discharge times, so there is a relationship between time, and the values of the resistor and the capacitor. This relationship is known as the circuit's *time constant*.

By definition, a time constant is the time taken for a current or voltage to reach 63.2 per cent of its final value. It is given the symbol τ (Greek letter tau). When a capacitor is being charged, it takes one time constant for the voltage across the capacitor to rise from zero to 63.2 per cent of its final value. At this time, the charge current will have fallen by 63.2 per cent as the current heads towards zero. It is also the time for the voltage and current to drop by 63.2 per cent under discharge conditions. The equation to find the time constant of an RC circuit is simply the product of the capacitance and the series resistance:

KEY CONCEPT

RC time constant

$\tau = RC$

where:

τ = time constant in seconds

C = capacitance in farads

R = resistance in ohms.

EXAMPLE 10.12

For the circuit in Figure 10.35: (a) Calculate the value of the current (I_{max}) that flows the instant the switch is closed. (b) Determine how long it takes for this current to fall by 63.2 per cent. (That is, find the time constant of the circuit.)

Solution

Values $V = 40$ V
$C = 100\ \mu F$
$R = 5\ k\Omega$

Equation (a) $I_{max} = \frac{V}{R} = \frac{40}{5 \times 10^3}$

Answer (a) $\mathbf{I_{max} = 8\ mA}$

Equation (b) $\tau = RC$

$\tau = (5 \times 10^3) \times (100 \times 10^{-6})$

Answer (b) $\boldsymbol{\tau} = \mathbf{0.5\ seconds}$

FIGURE 10.35 Circuit for Example 10.12

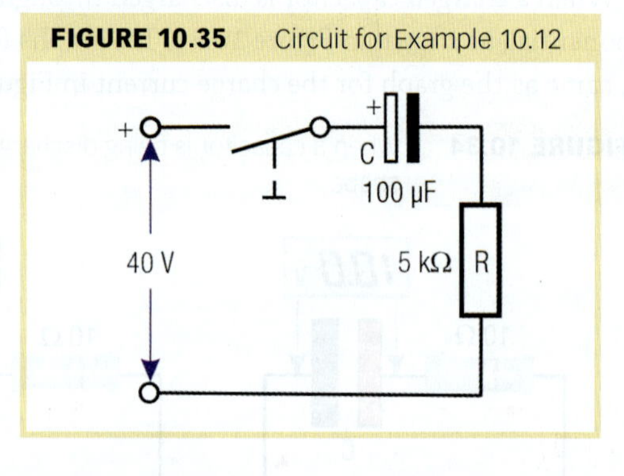

Current in an RC circuit connected to a DC supply can only flow when the voltage across the capacitor is different to the supply voltage. This can be a charge or a discharge current, depending on whether the supply voltage is greater or less than the capacitor voltage.

10.11 Time Constant Curve

By definition, after a time equal to one time constant, the current and voltage values in an RC circuit change by 63.2 per cent. But, as Figures 10.33 and 10.34 show, charging or discharging a capacitor takes a lot longer than one time constant. It is accepted that a capacitor is fully charged (or discharged) after five time constants. The reason is explained in Example 10.13.

EXAMPLE 10.13

For the RC circuit in Figure 10.36, calculate:
(a) the circuit time constant
(b) the voltage across the capacitor after two time constants.

(a) Solution

Values $V = 100\ V$
$C = 1\ \mu F$
$R = 1\ M\Omega$

Equation $\tau = RC$
$\tau = (1 \times 10^6) \times (1 \times 10^{-6})$

(a) Answer τ = **1 second**

(b) Solution

First time constant. After one time constant the capacitor voltage will have changed by 63.2 per cent as it charges from zero, heading towards the applied voltage of 100 V. The total possible change is 100 – 0, or 100 V. To find 63.2 per cent of this (or any) value, simply multiply it by 0.632. So, after one time constant, V_C reaches a voltage of 100 V × 0.632, or 63.2 V.

Second time constant. The capacitor voltage will now change by a further 63.2 per cent, this time from a start voltage of 63.2 V. Because 100 V – 63.2 V = 36.8 V, after the second time constant, V_C will have changed by 0.632 × 36.8 V = 23.26 V.

So, after the second time constant, the capacitor voltage is 63.2 V + 23.26 V:

(b) Answer V_C **= 86.5 V (to one decimal place)**

FIGURE 10.36 Circuit for Example 10.13

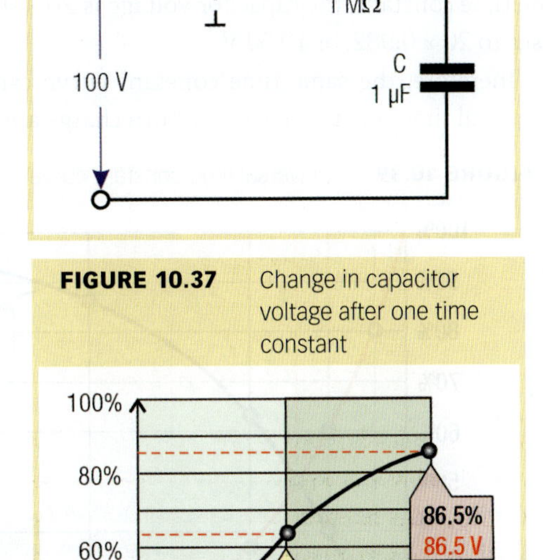

FIGURE 10.37 Change in capacitor voltage after one time constant

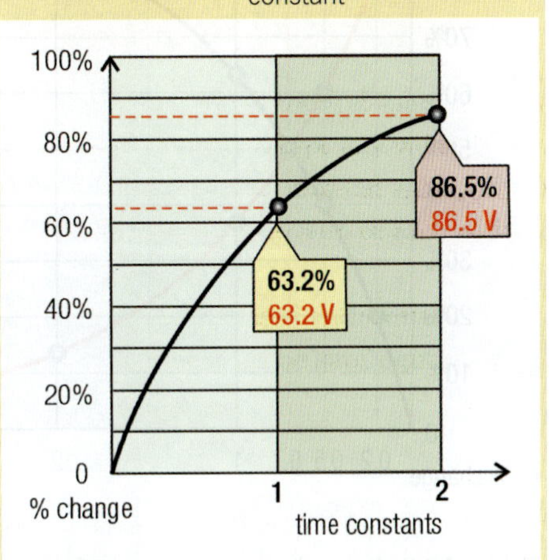

Using the same method shown in Example 10.13, the capacitor voltage can be found after each of the remaining three time constants. This gives the graph shown in Figure 10.38, which shows the percentage change and the voltages for each of the five time constants.

FIGURE 10.38 Change in capacitor voltage over five time constants

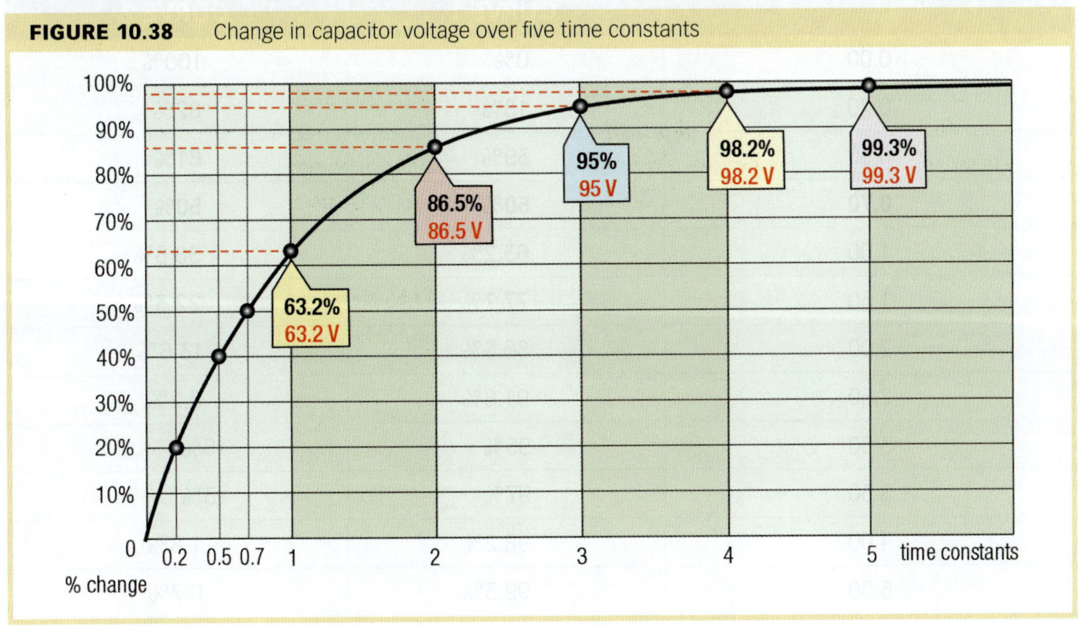

10.12 Universal Time Constant Curve

The curve in Figure 10.38 shows the percentage change and capacitor voltage at each time constant. By knowing the percentage change at each time constant interval, the capacitor voltage can be found by multiplying the supply voltage by the percentage change. For example, if the supply voltage is 20 V, after one time constant, the capacitor voltage is 20 × 0.632, or 12.64 V. After four time constants, the voltage rises to 20 × 0.982, or 19.64 V.

Therefore, the same time constant curve can be used with any RC circuit. Figure 10.39 shows a universal time constant curve for both charge and discharge of a capacitor in an RC circuit.

FIGURE 10.39 Universal time constant curve

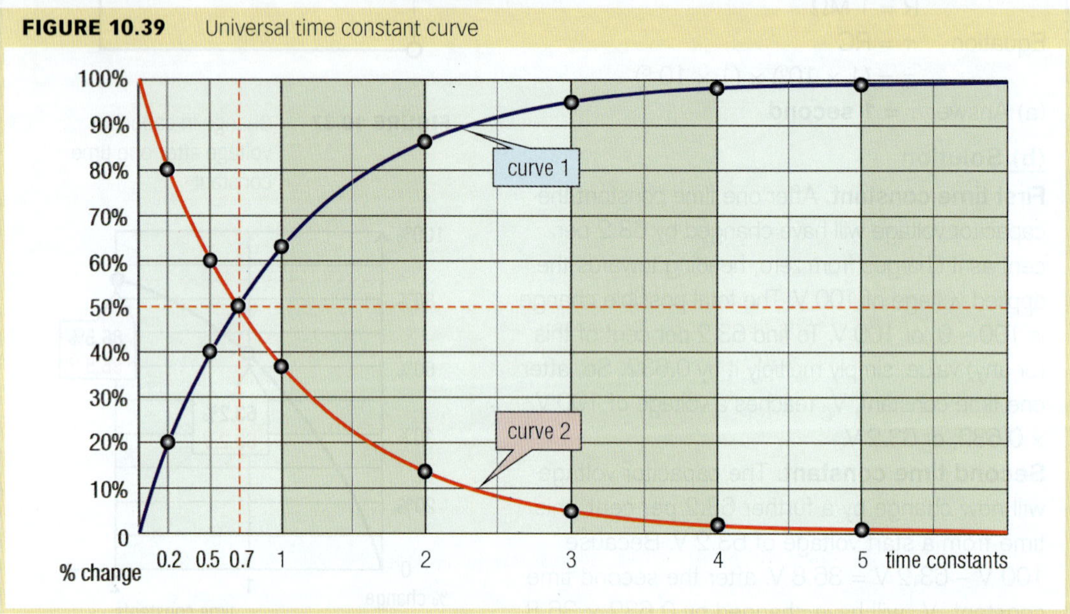

Table 10.3 lists the percentages for various time constants for both curves. Notice that at 0.7 time constants, both curves are at 50 per cent.

TABLE 10.3

TIME CONSTANT (TC) PERCENTAGES		
Time constants	Percentage	
	CURVE 1	CURVE 2
0.00	0%	100%
0.20	18%	82%
0.50	39%	61%
0.70	**50%**	**50%**
1.00	63.2%	36.8%
1.50	77.7%	22.3%
2.00	86.5%	13.5%
2.50	91.8%	8.2%
3.00	95%	5%
3.50	97%	3%
4.00	98.2%	1.8%
5.00	99.3%	0.7%

To find a voltage or current value from this table, multiply its maximum value by the percentage value divided by 100. For example, if the maximum voltage is 30 V, after 1.5 time constants, from the table the voltage will be 30 divided by 0.777 (or 77.7%/100), giving a voltage of 38.61 V. At this time, the capacitor current will be 22.3 per cent of its maximum value, or I_{max} divided by 0.223. The universal time constant curve in Figure 10.39 can be used to find the current and voltage in any RC circuit at any time interval, as shown in Example 10.14.

EXAMPLE 10.14

For Figure 10.40:

(a) Find the voltage across the capacitor two seconds after the switch is closed.

(b) Find the current flowing in the circuit at this time.

FIGURE 10.40 Circuit for Example 10.14

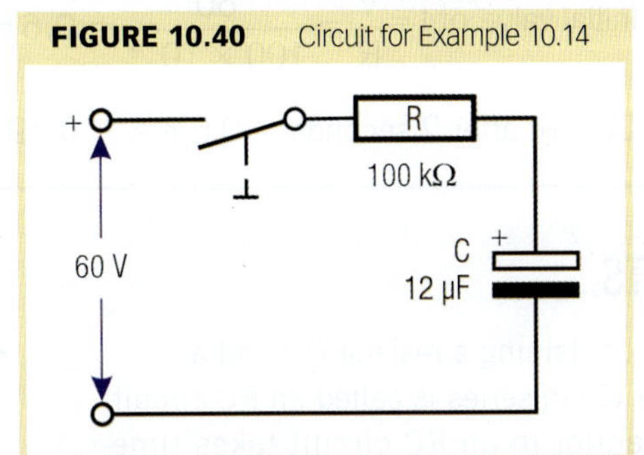

Solution

Values V = 60 V

C = 12 μF

R = 100 kΩ

Equation $\tau = RC$

$\tau = (12 \times 10^{-6}) \times (100 \times 10^{3})$

$\tau =$ **1.2 second**

To find the capacitor voltage, use the universal time constant curve. That is:

1 Replace each time constant (TC) with its time value. That is, 1TC = 1.2 s, 2TC = 2.4 s and so on to calibrate the time axis in seconds.

2 Determine where two seconds is on the time axis by drawing a time scale. From this point, draw a vertical line to cut the curves, then draw a horizontal line across to the % change axis.

3 Read the percentage change and use this to calculate the voltage.

FIGURE 10.41 Universal curves showing percentage values of voltage and current after 2 seconds have elapsed, for the circuit in Figure 10.40

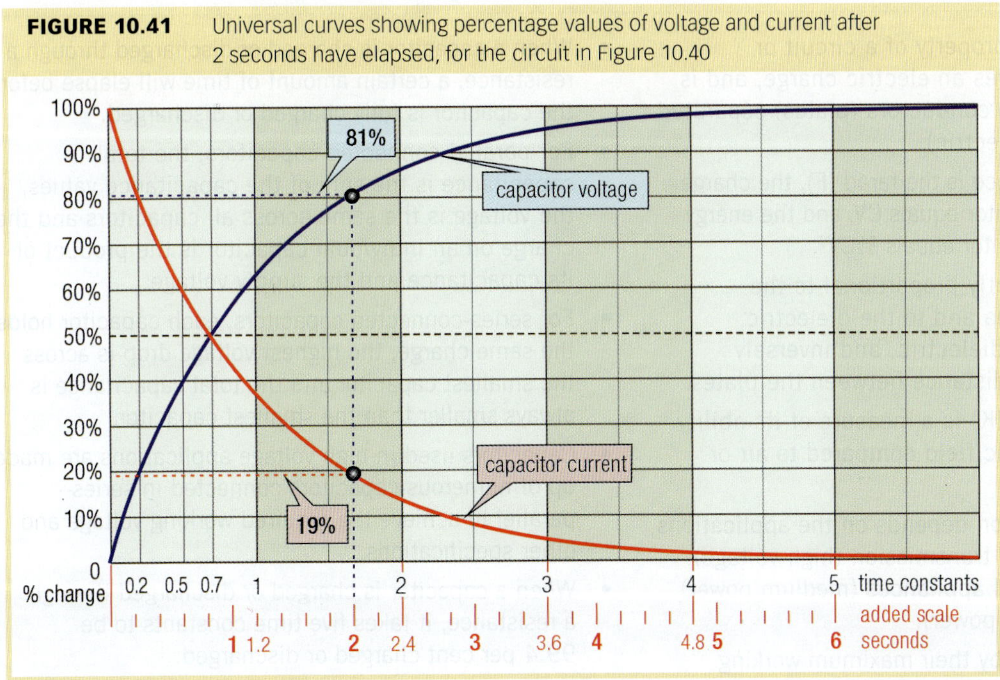

Answer (a) From Figure 10.41, the percentage change of the capacitor voltage after 2 seconds is about 81 per cent. So the voltage across the capacitor = 60 V × 0.81 = 48.6 V.

$\mathbf{V_C}$ = **48.6 V** (after 2 seconds).

Answer (b) From Figure 10.41, the current after 2 seconds equals 19 per cent of its initial value. The initial current is found by Ohm's law:

$$\text{Initial value of } I = \frac{V}{R} = \frac{60}{100 \times 10^{-3}} = 0.6\,\text{mA}$$

Current after 2 seconds = 0.6 mA × 0.19 = **0.114 mA**

KEY POINTS...

- A circuit containing a resistor (R) and a capacitor (C) in series is called an RC circuit.
- The capacitor in an RC circuit takes time to charge and discharge because the resistor limits the maximum current that can flow.
- The time taken for current or voltage in an RC circuit to change by 63.2 per cent is the circuit's time constant.
- The time constant (τ) of an RC circuit equals the product of the component values. That is τ = RC.

CHAPTER SUMMARY

- Capacitance is the property of a circuit or component that stores an electric charge, and is formed between two conductors (plates) separated by an insulator (dielectric).
- The unit of capacitance is the farad (F), the charge (Q) stored in a capacitor equals CV, and the energy (W) stored in a capacitor equals $\frac{1}{2}CV^2$.
- Capacitance is directly proportional to the capacitor's plate area and to the dielectric constant (K) of the dielectric, and inversely proportional to the distance between the plates.
- Dielectric constant (K) is a measure of its ability to support an electric field compared to air or vacuum.
- Capacitor construction depends on the applications, which include power transmission (high voltage), motors and electrical appliances (medium power), and electronics (low power).
- Capacitors are rated by their maximum working voltage, and in power applications, by the reactive power they can handle.
- When a capacitor is charged or discharged through a resistance, a certain amount of time will elapse before the capacitor is fully charged or discharged.
- For parallel-connected capacitors, the total capacitance is the sum of the capacitance values, the voltage is the same across all capacitors and the charge on an individual capacitor is the product of its capacitance and the supply voltage.
- For series-connected capacitors, each capacitor holds the same charge, the highest voltage drop is across the smallest capacitor and the total capacitance is always smaller than the smallest capacitor.
- Capacitors used in high voltage applications are made up of numerous capacitors connected in series–parallel to achieve the required working voltage and other specifications.
- When a capacitor is charged or discharged through a resistance, it takes five time constants to be 99.4 per cent charged or discharged.
- Charged capacitors can cause serious injury or death. Make sure a capacitor is discharged before touching its terminals.

REVIEW EXERCISES

Check your answers at the back of the book.

1 For the circuit in Figure 10.42, calculate the:
   a total capacitance of the circuit in nanofarads and microfarads
   b total charge taken by the circuit
   c charge on each capacitor.

FIGURE 10.42

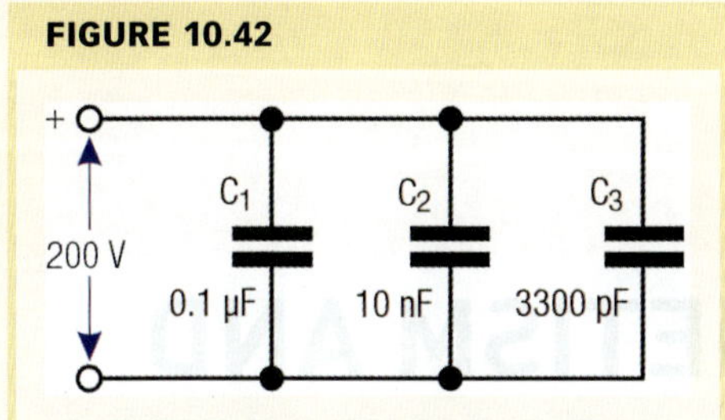

2 For the circuit in Figure 10.43, calculate the:
   a total capacitance of the circuit in nanofarads and microfarads
   b total charge taken by the circuit
   c voltage across each capacitor.

FIGURE 10.43

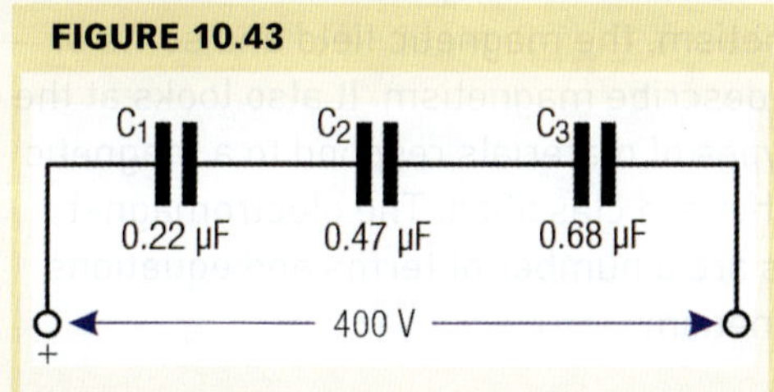

3 For Figure 10.44, calculate:
   a the total capacitance of the circuit
   b the voltage across each capacitor.

FIGURE 10.44

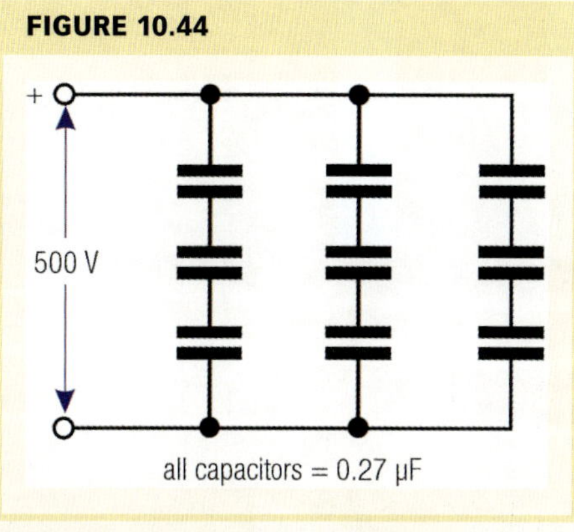

4 Calculate the energy stored in a 3.3 µF capacitor charged to 11 kV.

5 Calculate the capacitance of a capacitor with foil plates measuring 8 mm by 5 m, separated with a 0.006 mm thick dielectric that has a dielectric constant of 5.

6 A 1.5 µF capacitor is holding a charge of 20 millicoulombs (mC). Find the voltage between the capacitor terminals.

7 A 100 V DC power supply has a 2000 µF capacitor connected across its output terminals. A 100 kΩ resistor is connected across the capacitor to discharge it when the supply is turned off. Calculate the time taken for the capacitor to discharge to 99.4 per cent.

8 A capacitor is marked 33 µF, ±10%, but it measures 25 µF. Should the capacitor be replaced, and why?

9 A high voltage capacitor is rated at 0.1 µF, 5 kV. How many 0.1 µF, 500 V capacitors are needed to make this capacitor?

10 Calculate the value of resistance to obtain a time constant of 20 seconds with a capacitor of 4700 µF.

ONLINE RESOURCES

COMPLETE WORKSHEET TEN

Check with your instructor for worksheets on this chapter.

CHAPTER 11

MAGNETISM AND ELECTROMAGNETS

Magnetism and electricity are closely related. In fact, most of the electricity we generate relies on this relationship, as do electric motors and many other electrical components. This chapter starts with a review of magnetism, the magnetic field and some of the terms that describe magnetism. It also looks at the way different types of materials respond to a magnetic field and how they are classified. The electromagnet is explained, as are a number of terms and equations related to magnetism.

CHAPTER OUTLINE

11.1 Magnetism

Magnetism is a force created by a magnetic field. Magnetism was first written about over 2500 years ago, after it was found that a 'lodestone' attracted iron. A lodestone is created when lightning strikes a material called magnetite, an oxide of iron (Fe_3O_4). Chinese sailors were using lodestone compasses by 1100 AD and other countries soon followed. In 1600 AD William Gilbert concluded that the world itself had a magnetic field, which explained why compasses pointed north.

FIGURE 11.1 A lodestone is a naturally occurring magnet, caused when lightning strikes magnetite

Shutterstock.com/Vitaly Raduntsev
Getty Images/DEA/A. RIZZI

A magnet such as the pointer in the compass shown in Figure 11.2 has two poles, called the north and the south poles. If a magnet can move freely, it will align itself so its north pole points towards the Earth's geographic North Pole. However, like magnetic poles repel and unlike poles attract, so the Earth's magnetic north pole is actually near the geographic South Pole. For convenience, we refer to the Earth's 'North' magnetic pole as the pole pointed to by the north pole of a compass. The Earth's magnetic field is caused by large circulating currents in the core of the planet, which shows there is a relationship between an electric current and the production of a magnetic field.

FIGURE 11.2 The Earth's magnetic field causes the north pole of a magnet to point towards geographic North

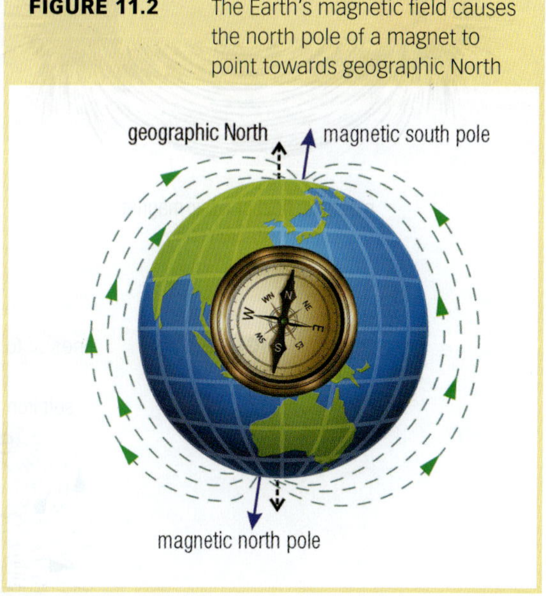

Compass image: Pixabay/ariesjay

Magnetic field

A magnetic field comes from two atomic sources: the spin and orbital motions of electrons. Electrons spin as they orbit their associated nucleus. The combined effect of lots of spinning electrons produces *magnetic domains*. A magnetic domain is the smallest possible magnet, and these are scattered throughout all materials. When a material is magnetised, its magnetic domains are aligned so their magnetic fields combine, rather than cancel. That is, the material now has a north and south pole, and an associated magnetic field.

A magnetic field is assumed to consist of *lines of force* that flow from one pole of a magnet to the other. Figure 11.3 shows the magnetic field (or *flux*) of a bar magnet in which each line of force is a dotted line forming an endless loop, no matter how large the loop.

Magnetic lines of force are assumed to flow from the north to the south pole of a magnet, they never cross each other and lines flowing in the same direction repel each other. For this reason, the magnetic field becomes weaker as the distance from the magnet increases. Lines of force become more spread out in air or other non-magnetic materials, and become denser when flowing in a magnetic material, such as iron.

As shown in Figure 11.3, the north pole of a compass needle placed anywhere along a line of force points to the south pole of the magnet because *opposite magnetic poles attract*. The direction of lines of force outside a magnet is assumed to be from its north pole to its south pole because a north pole is used as the reference.

FIGURE 11.3 A magnetic field is made up of lines of force that travel from the north to the south pole of a magnet

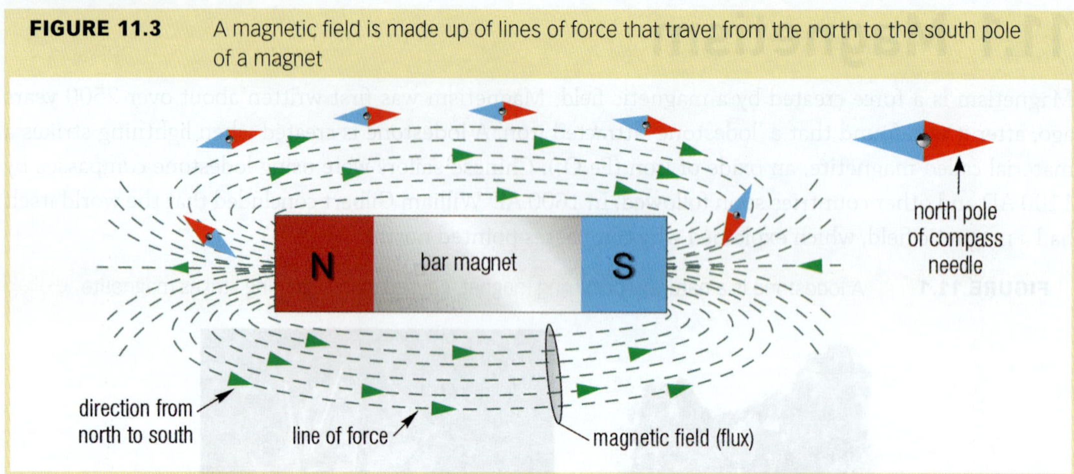

FIGURE 11.4 Iron filings give an impression of lines of force around a magnet

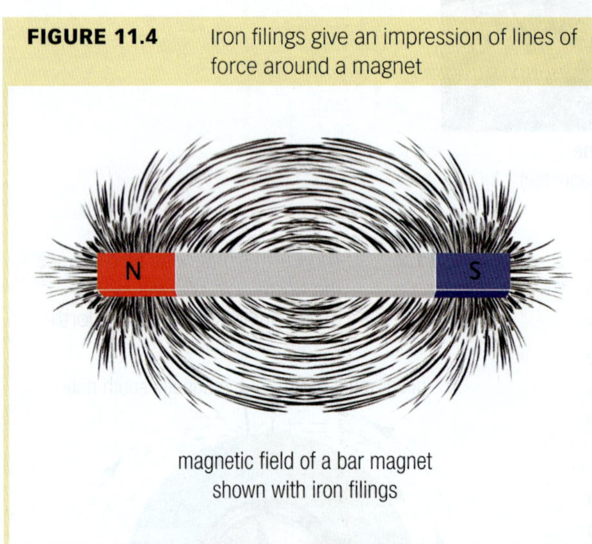

Figure 11.4 shows the field pattern of a bar magnet created with iron filings. The magnetic field is concentrated at the *poles* of the magnet, showing that the lines of force are at their most dense at the north and south poles of the magnet.

Lines of magnetic force always take the path of least magnetic resistance. Therefore, if a piece of iron is placed in the path of a magnetic field as shown in Figure 11.5, the lines of force will distort to pass through the iron because it has a lower magnetic resistance.

When a ferromagnetic material is magnetised, it tends to change its dimensions. This is called *magnetostriction*, an effect that causes the hum associated with a transformer. In a transformer, as explained further on, the alternating current flowing in its windings changes direction every 10 milliseconds (for a 50 Hz supply). This causes the magnetic field to reverse every 10 milliseconds, giving a 100 Hz hum.

FIGURE 11.5 Lines of force take the path of least magnetic resistance

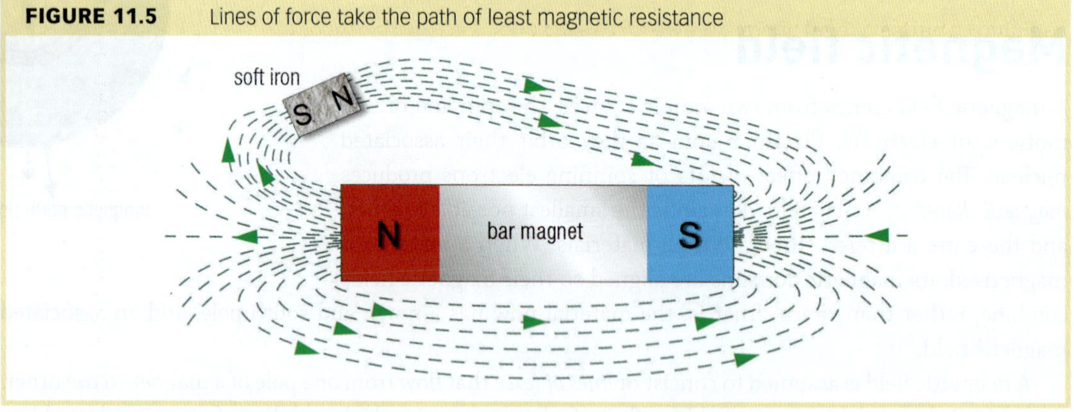

Magnetic induction

When a magnetic material is in a magnetic field, as in Figure 11.5, it becomes temporarily magnetised. This is called *magnetic induction* because magnetism is *induced* in the material by the magnetic field of the magnet. The polarity of the induced magnet is such that lines of force enter the south pole and exit from the north pole, as in the bar magnet.

If the magnetic field in Figure 11.5 is removed, the piece of iron will remain slightly magnetised. Therefore, exposing a magnetic material to a magnetic field provides a way of making a magnet. The field can come from a permanent magnet or, as explained later, an electromagnet, which is a coil of wire carrying a DC current.

A piece of magnetic material can be partially magnetised by using a permanent magnet. Figure 11.6(a) shows the single touch method, in which the material is stroked with a magnet from end to end several times by the same pole of a magnet, and in the same direction. The polarity of the induced magnet depends on the polarity of the magnetising force and the direction of stroking. The induced magnet will have the opposite polarity to the magnetising force at the end where the magnetising force leaves the induced magnet.

FIGURE 11.6 Soft iron can be magnetised by induction by stroking the iron with one or two strong magnets

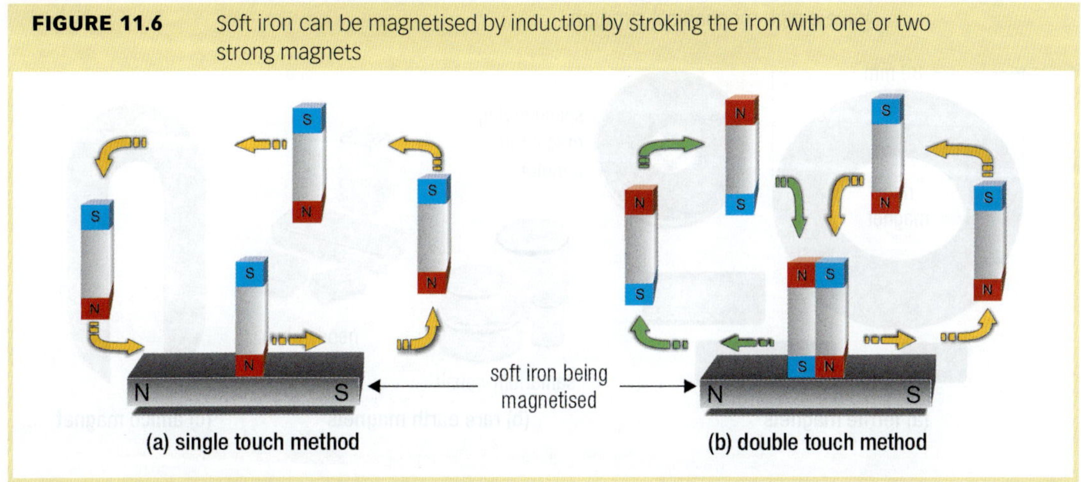

With the double touch method, shown in Figure 11.6(b), the material being magnetised is stroked with two magnets arranged with opposite poles that are moved outwards from the centre of the material. They are lifted clear of the material, then returned to its centre where the process is repeated at least six times. Small magnets can be made by placing a piece of magnetic material across the poles of a powerful horseshoe magnet. Gently tapping the material while it's in position will increase the strength of the induced magnet.

Soft iron is relatively easy to magnetise, but when the magnetising force is removed it will retain only a small amount of magnetism, called *residual magnetism*. This magnetism can often be removed by knocking the magnetised object against a hard surface. The ability of a material to retain its magnetism after magnetisation is called *retentivity*. Soft iron has a low retentivity, while hard steel, although more difficult to magnetise, has a higher retentivity. In general, if a material is easily magnetised, it is also easily demagnetised.

Magnetic materials

Materials are either non-magnetic or magnetic. Non-magnetic materials have no reaction to or effect on a magnetic field. All magnetic materials react in some way to a magnetic field and are classified according to the type of reaction. There are three classifications: ferromagnetic, paramagnetic and diamagnetic.

Ferromagnetic materials are attracted by a magnetic field, can be magnetised and can affect the shape of a magnetic field. A ferromagnetic material has a high *permeability* (or low resistance to lines of force) and is therefore a good conductor of a magnetic field. Materials with a high permeability include iron, steel, nickel, cobalt and their alloys.

The magnetic alloy alnico was developed in the 1940s and combines iron, aluminium (Al), nickel (Ni) and cobalt (Co). Although this alloy is still used, stronger magnets are obtained with *rare earth* alloys. The first of these appeared in the late 1970s, when a material called Sm-Co was developed. Made of samarium, cobalt and iron, plus additives, Sm-Co magnets are up to 10 times stronger than an alnico magnet.

An even better magnetic alloy made of neodymium, iron and boron (Nd-Fe-B) was developed in the 1980s. This alloy is cheaper to produce than Sm-Co, and produces magnets that are nearly twice as strong as Sm-Co magnets. Neodymium magnets are usually zinc or nickel plated to prevent corrosion, and are so strong they can cause an injury due to the strong attraction between the magnet and a magnetic material. Before alloys were developed, magnets were generally made of carbon steel and later tungsten steel. Early alloys to be used include Honda steel (named after its inventor), and in the 1930s, HK steel.

A *ferrimagnetic* material has similar properties to a ferromagnetic material, but is a poor conductor of electricity. An example is ferrite, a ceramic material derived from iron oxides such as magnetite, which is also a ferrimagnetic material. Ferrite was first used in magnets in the 1960s, and is still widely used. The advantage of ferrite is that it's easy to produce magnets in various shapes, such as the ring magnets shown in Figure 11.7.

FIGURE 11.7 Ferrite ring magnets are used in small electric motors. Alnico magnets are the weakest but have a high mechanical strength.

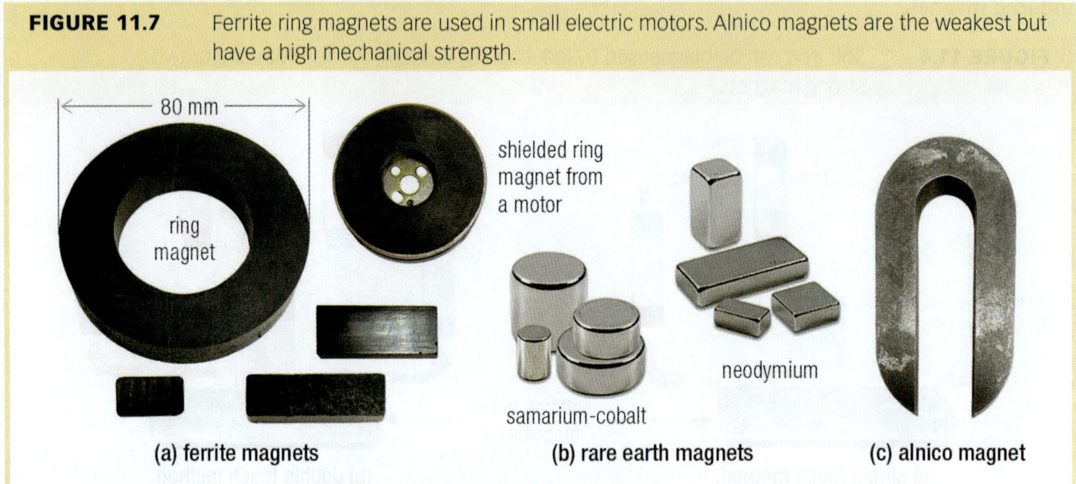

Paramagnetic, diamagnetic and non-magnetic materials

Paramagnetic materials are weakly attracted to a magnet and include aluminium, platinum, manganese, chromium and oxygen. These substances have a permeability slightly higher than one and become weakly magnetised in the same direction as the magnetising force. They have far less strength than a ferromagnetic magnet, and lose their magnetism when the magnetising force is removed. The effect increases when the temperature of the material is reduced. Because oxygen is paramagnetic, oxygen content of a gas can be measured with a paramagnetic oxygen analyser.

Diamagnetic materials also become weakly magnetised, but in the *opposite* direction to the magnetising force. They have a permeability *less* than one. Bismuth and carbon graphite are the strongest diamagnetic materials. Others that are less strong include mercury, copper, gold, silver, water, diamonds, wood and living tissue. The last three materials are carbon-based.

 FYI

A magnet strongly attracts ferromagnetic materials, weakly attracts paramagnetic materials and weakly repels diamagnetic materials

An application is magnetic levitation, where an object made of a diamagnetic material can be made to float above a strong magnet. This requires an extremely strong magnetic force. Another use is with magnetic resonance imaging (MRI) in which carbon-based atoms in the body are exposed to a strong magnetic field, causing them to be slightly repelled by the field. This movement can be detected and used for analysis.

Non-magnetic materials have a permeability of one. They include air, wood, plastic, water, paper and cloth fabrics. They are not affected by a magnetic field (don't become magnetised) and have no effect on the field. For most purposes, paramagnetic and diamagnetic materials are considered non-magnetic.

Preserving a magnet

A magnet of any type will lose some or all of its magnetism if it's exposed to a high temperature. The loss is complete at the *Curie temperature*, which for iron is about 770 °C. Neodymium magnets have a much lower Curie temperature of around 300 °C. Hammering or jarring a magnet will also affect its magnetic strength.

A magnet can lose its magnetic strength if it's exposed to another magnetic field. Figure 11.8 shows the law of magnetic attraction and repulsion. When *unlike* poles are next to each other as in Figure 11.8(a), they attract, and the magnetic field is maintained and the magnetic strength of each magnet is not affected.

FIGURE 11.8 Law of attraction and repulsion – unlike poles attract, like poles repel

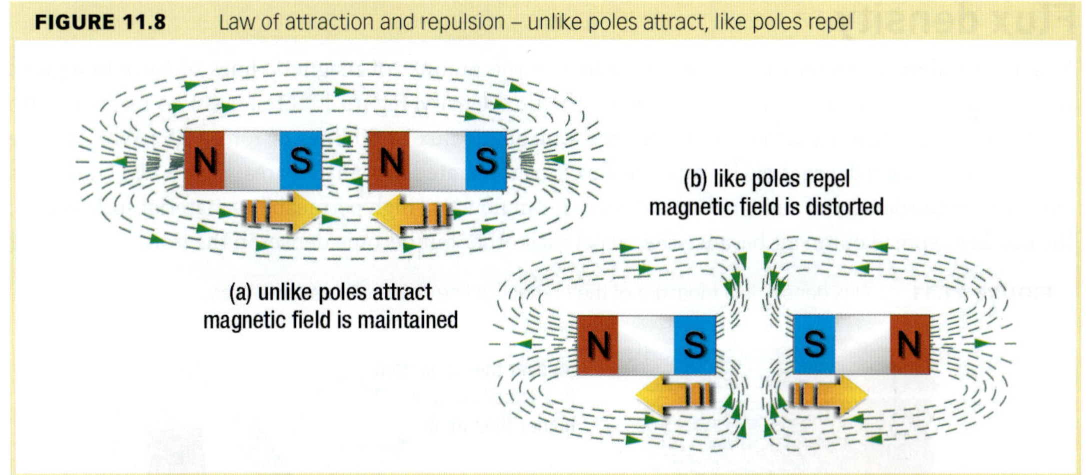

Figure 11.8(b) shows that when *like* poles are brought near each other, they repel each other and the magnetic field is distorted. The stronger of the two magnets will attempt to magnetise the other magnet, but in the opposite polarity. This therefore weakens the field of the weaker magnet even further.

To prevent loss of magnetic strength, an unused magnet or a magnet removed from an item of equipment should be stored with a *keeper* across its poles.

As shown in Figure 11.9, a keeper is a piece of soft iron placed across the poles of a horseshoe magnet or pair of bar magnets. It provides a high permeability path for the magnetic flux, helping the magnet retain its strength.

FIGURE 11.9 Magnets are stored with an iron keeper to preserve their magnetic strength

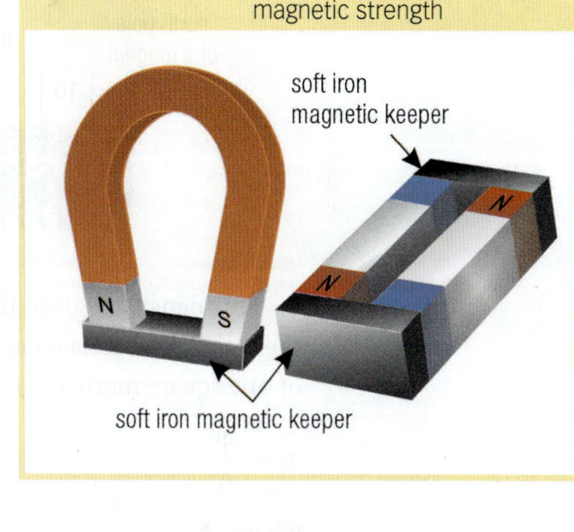

Magnetic shielding

There is no material that can stop a magnetic field from passing through it. Therefore, the only way to stop magnetic flux reaching an area is to divert the lines away with a material that has a high permeability, as shown in Figure 11.10. Some measuring instruments are magnetically shielded to prevent a magnetic field affecting their accuracy.

FIGURE 11.10 A magnetic shield keeps magnetic flux away from an area by providing a high permeability path around the area

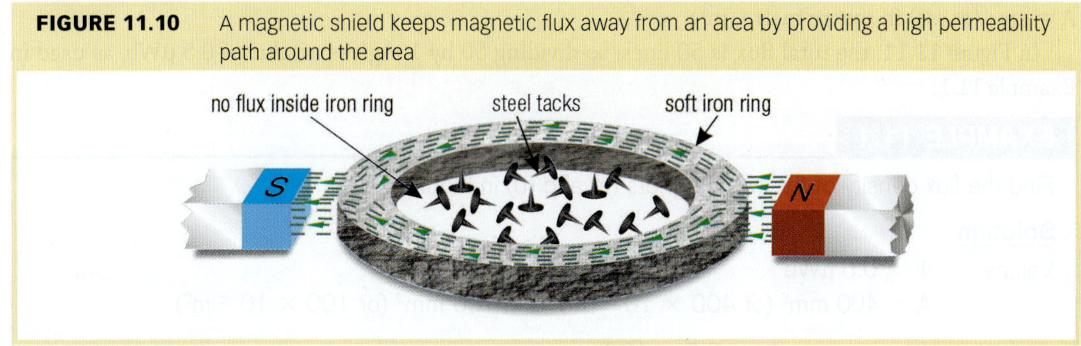

Magnetic flux

Magnetic flux refers to the total magnetic field around a magnet and has the symbol Φ (phi). The SI unit of measurement is the *weber* (abbreviated to Wb), named after Wilhelm Weber (1804–1891), a German physicist. It is defined as the amount of flux that causes one volt to be produced in a single loop of wire when the current in the loop is reduced at a uniform rate to zero over a period of one second. One weber has 100 000 000 (10^8) lines of force. Magnets are often specified by their 'pull force', or the weight the magnet can lift. The ferrite ring magnet in Figure 11.7 can lift about 1 kg.

Flux and flux density can be measured with a magnetometer, also called a gauss meter

Flux density

A more useful measurement is *flux density*, which is the number of magnetic lines of force in a given area. The more lines in a given area, the higher the flux density. Figure 11.11 shows two magnets with a different cross-sectional area (csa). Both have a magnetic flux of 50 lines of force (0.5 μWb), but the smaller magnet has a denser field because the flux is concentrated in a smaller area. Figure 11.11(c) shows the effect on the density of a magnetic field when it can flow in a high permeability material such as iron. The flux lines spread out in air because they repel each other but become denser in the iron.

FIGURE 11.11 Flux density is a measure of the number of lines of force in a given area

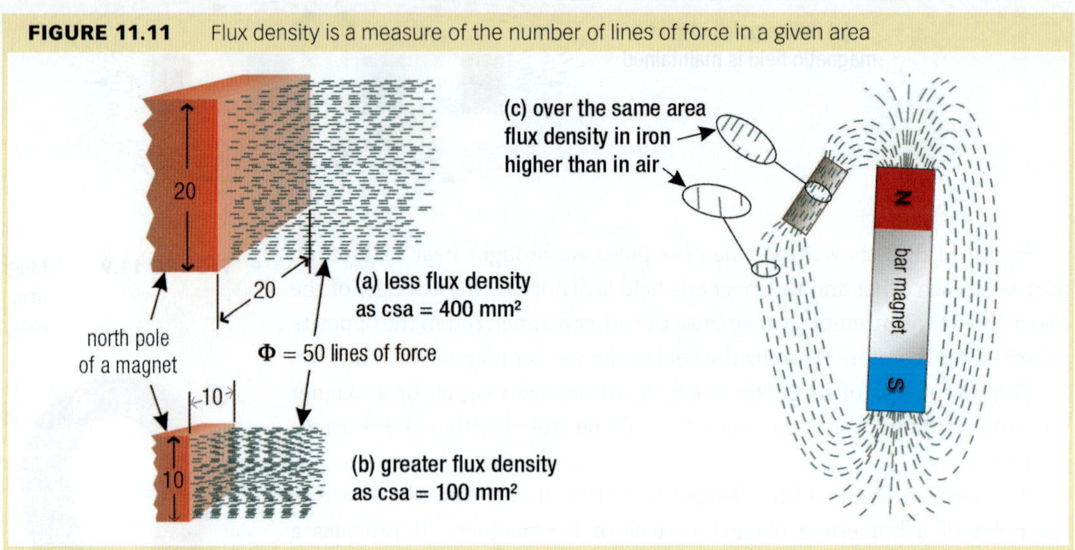

Flux density is given the symbol B, and equals flux (Φ) divided by area (A). The unit of flux density is the tesla (T). A tesla equals one weber per square metre. That is, one tesla is 10^8 lines of force in an area of one square metre (m^2). So:

$$B = \frac{\Phi}{A}$$

where:

B = flux density in tesla (or weber/m^2)

Φ = flux in weber

A = area (or csa) in square metre (m^2).

In Figure 11.11, the total flux is 50 lines, so dividing 50 by 10^8 gives a value of 0.5 μWb, as used in Example 11.1.

FYI

$1\ mm = 1 \times 10^{-3}\ m$ and $1\ mm^2 = 1 \times 10^{-6}\ m^2$

EXAMPLE 11.1

Find the flux density of the two magnets (a) and (b) in Figure 11.11.

Solution

Values Φ = 0.5 μWb

A = 400 mm² (or $400 \times 10^{-6}\ m^2$), and 100 mm² (or $100 \times 10^{-6}\ m^2$)

Answer magnet (a) $B = \frac{\Phi}{A} = \frac{0.5 \times 10^{-6}}{400 \times 10^{-6}}$ = **0.00125 T or 1.25 mT**

Answer magnet (b) $B = \frac{\Phi}{A} = \frac{0.5 \times 10^{-6}}{100 \times 10^{-6}}$ = **0.005 T or 5 mT**

Applications

Permanent magnets have many industrial applications, such as those shown in Figure 11.12. Magnetic chucks and faceplates are used to hold a ferromagnetic object so it can be machined in a lathe, grinder or milling machine.

FIGURE 11.12 Permanent magnets are used in a range of devices and tools

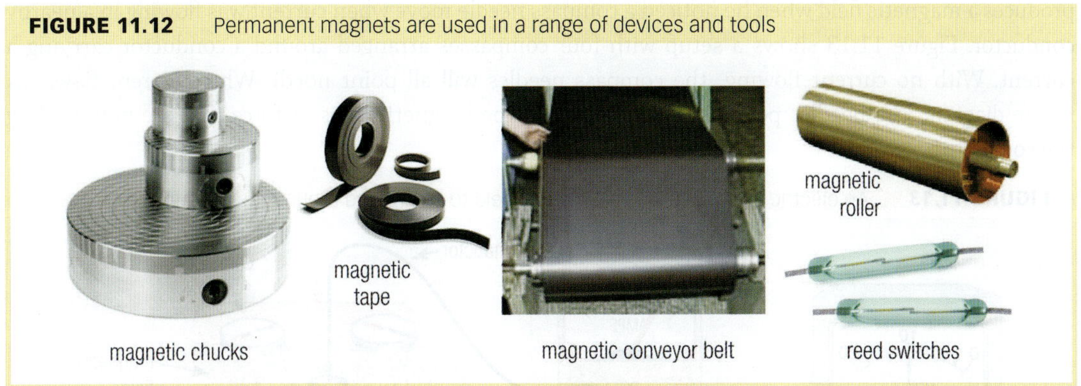

An application involving a permanent magnet is the reed switch, shown in Figure 11.12. A reed switch has a pair of flexible, metal reeds made from a ferromagnetic material that are hermetically sealed in a glass tube. Contacts are attached to the ends of the reeds. Under normal conditions (no magnetic field), the reeds and their contacts remain separated. If a magnetic field from a permanent magnet is brought close to the switch, the reeds are attracted to each other, thereby closing the contacts and making a circuit. Removing the magnetic field causes the metal reeds to open the circuit.

In a house alarm, reed switches are mounted near windows or doors which have magnets attached to them. Closing a door will close its reed switch, so when all protected doors and windows are closed, the alarm system can be armed. Because all the reed switches are connected in series, if a door or window is opened, its associated reed switch will open, triggering the alarm. Other uses of a permanent magnet are in some types of electric motors and generators. Loudspeakers have a strong permanent magnet.

KEY POINTS...

- A magnet can be created by exposing a ferromagnetic material to a strong magnetic field. The field can be from another magnet or an electromagnet. The process is called magnetic induction, as the magnet is inducing magnetism in the ferromagnetic material.
- A magnet has two poles called the north and south poles. The north pole (or north-seeking pole) of a magnet points towards Earth's geographic North Pole.
- Unlike magnetic poles attract; like poles repel.
- A magnetic field (flux) consists of lines of force that flow from the north pole of a magnet to its south pole.
- Lines do not cross each other, they all have the same magnetic strength and although they have a direction, they don't actually move.
- Materials are generally classified as ferromagnetic (strongly attracted to a magnetic field), paramagnetic (weakly attracted by a strong magnetic field, e.g. oxygen) and diamagnetic (weakly repelled by a strong magnetic field, e.g. living tissue).
- A magnet can lose its magnetism as a result of excess heat, physical vibration and stress, and through exposure to a magnetic field of the opposite polarity. A soft iron keeper should therefore be applied across the poles of a magnet to help it retain its magnetism.
- Magnetic flux (symbol Φ) is measured in webers, where one weber (Wb) is 100 million (10^8) lines of force. Most magnetic fields have a flux measured in milliwebers (mWb) or microwebers (μWb).
- Flux density (symbol B) refers to the number of lines of force (Φ) in a given area (A). It is measured in teslas, where one tesla equals one weber of flux per square metre. That is:

$$B = \frac{\Phi}{A}$$

11.2 Magnetic Effect of Electric Current

In 1820, Danish physicist Hans Christian Oersted (1777–1851) discovered that an electric current produces a magnetic field when he noticed a compass needle move when current was flowing in a nearby conductor. Figure 11.13 shows a setup with four compasses arranged around a conductor carrying a current. With no current flowing, the compass needles will all point north. When current flows, as in the diagram, the needles point in the direction of the magnetic lines of force that are now circling the conductor.

FIGURE 11.13 An electric current causes a magnetic field to be created around the conductor

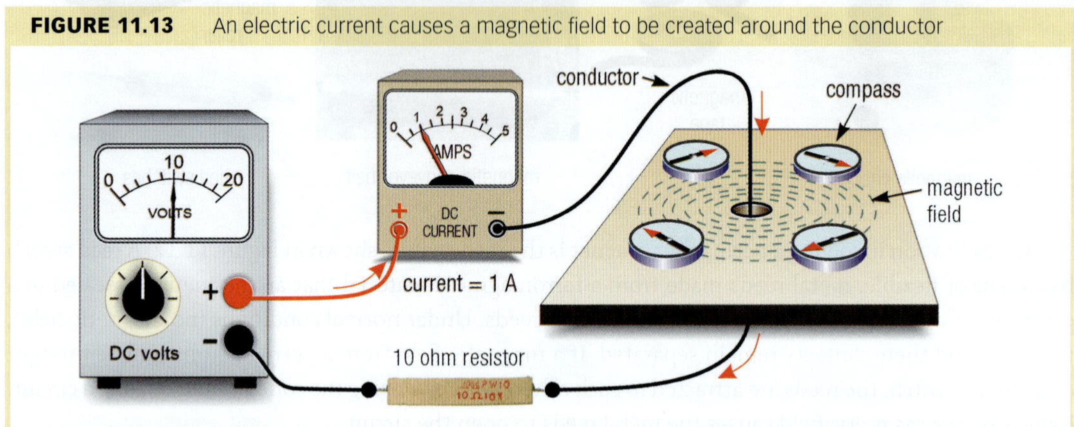

Figure 11.14 shows the field patterns around two conductors. In (a), the current is flowing into the page, and the flux direction is clockwise, the same as shown by the compasses in Figure 11.13. The symbol to show current flowing into the page is a circled cross, like the back of an arrow heading away from you. In (b), the current is flowing out of the page, so the symbol for this direction is a dot, like the point of an arrow.

FIGURE 11.14 The direction of the magnetic field depends on the direction of the current

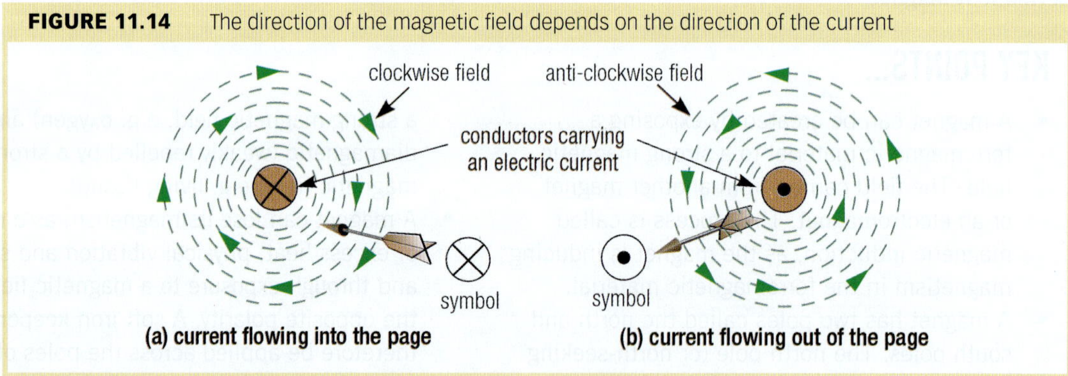

Right-hand rule for a conductor

The *right-hand rule for conductors*, as shown in Figure 11.15, is used to find the direction of the magnetic field around a conductor. That is, point your thumb in the direction of the current and wrap your fingers around the conductor, which then point in the direction of the magnetic field. This rule only works with your right hand.

Magnetic field around two conductors

The magnetic fields around adjacent current-carrying conductors interact and create a force between the conductors. If the currents are in the same direction, the fields combine, and the flux lines try to pull the conductors together, because flux lines seek the shortest path. Therefore, the two conductors are attracted to each other, as shown in Figure 11.16.

FIGURE 11.15 Right-hand thumb rule to find the direction of the magnetic field around a conductor

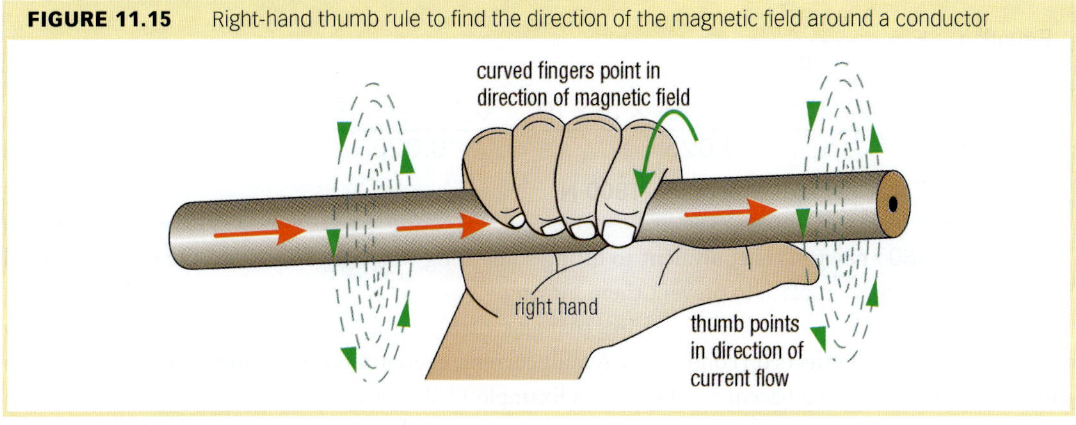

FIGURE 11.16 When the current is in the same direction, adjacent conductors are attracted to each other

(a) current flowing away in both conductors

(b) current flowing towards in both conductors

When the currents in adjacent conductors are flowing in opposite directions, the flux lines between the conductors are travelling in the same direction, as shown in Figure 11.17. Because flux lines travelling in the same direction repel each other, the magnetic force acting on the conductors therefore tends to push them apart.

FIGURE 11.17 Currents in the opposite direction make adjacent conductors repel each other

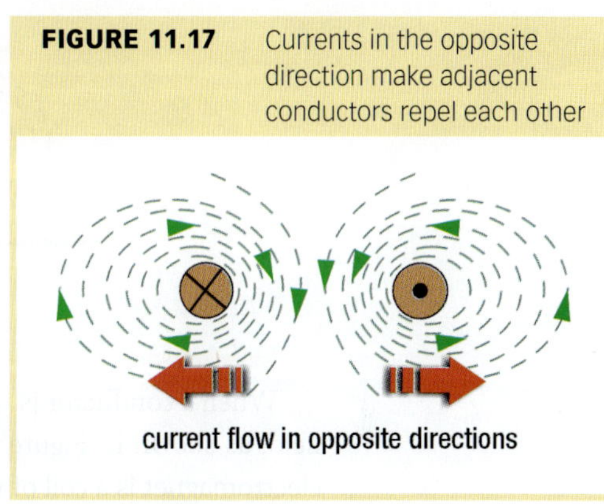

Force between two conductors

Conductors carrying a high current produce strong magnetic fields. If two such conductors are located close to each other, there can be a considerable force acting between them. This is especially important in a substation, where high fault currents can occur, but is also important in many types of electrical installations.

By definition, if one ampere of current passes through two infinitely long parallel conductors placed one metre apart in a vacuum, a force exactly equal to 2×10^{-7} newton per metre of length will be produced between them. If the currents are in the same direction, the conductors will attract each other; otherwise they repel each other with the same force. An equation to find this force (F) is:

$$F = \frac{2 \times 10^{-7} I_1 I_2}{d}$$

where:

F = force in newtons (N) per metre length

I_1 = current in one conductor, I_2 = current in the other conductor

d = distance in metres between the conductors.

EXAMPLE 11.2

Two copper bus bars spaced 25 millimetres apart experience a fault current of 15 000 A that flows in each bus bar. What is the force acting between them over a one metre length?

Solution

Values $I_1 = I_2 = 15\,000 \text{ A} = 15 \times 10^3 \text{ A}$

$d = 25 \text{ mm} = 0.025 \text{ m}$

Equation $F = \frac{2 \times 10^{-7} I_1 I_2}{d}$

$$F = \frac{2 \times 10^{-7} \times 15 \times 10^3 \times 15 \times 10^3}{0.025} = \frac{0.2 \times 15 \times 15}{0.025}$$

Answer Force per metre length = **1800 N**

A force of 1800 newtons is equivalent to lifting something weighing 183 kg, so the bus bars would need to be firmly fixed in place to prevent them moving.

AS/NZS 3000:2018, clause 2:10.6 requires that all switchboard wiring be designed to withstand any thermal and magnetic effects on conductors. A fault current, even in a domestic installation, can create a significant force between adjacent conductors, as Example 11.2 shows.

11.3 The Electromagnet

When a current-carrying conductor is formed into a loop (as in Figure 11.18), the magnetic fields around both sides of the loop tend to push the conductors apart, as shown in Figure 11.17. However, inside the loop are flux lines travelling in the same direction, creating a magnetic field double the intensity of a magnetic field around a straight piece of wire.

FIGURE 11.18 The magnetic field around a loop of wire

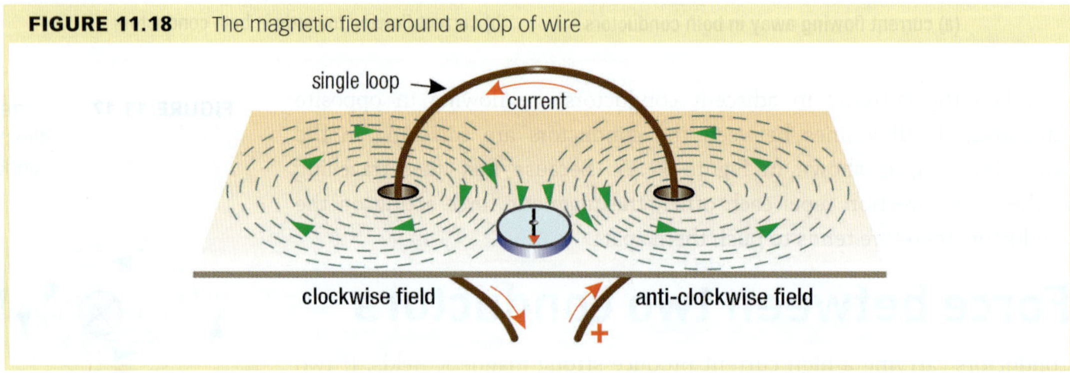

When a conductor is wound into a coil, each loop creates its own magnetic field, creating a combined field, as shown in Figure 11.19. This magnetic field is the same as that of a permanent magnet, so an electromagnet is a coil of wire carrying a current.

FIGURE 11.19 The magnetic field around a coil is the same as the field around a permanent magnet

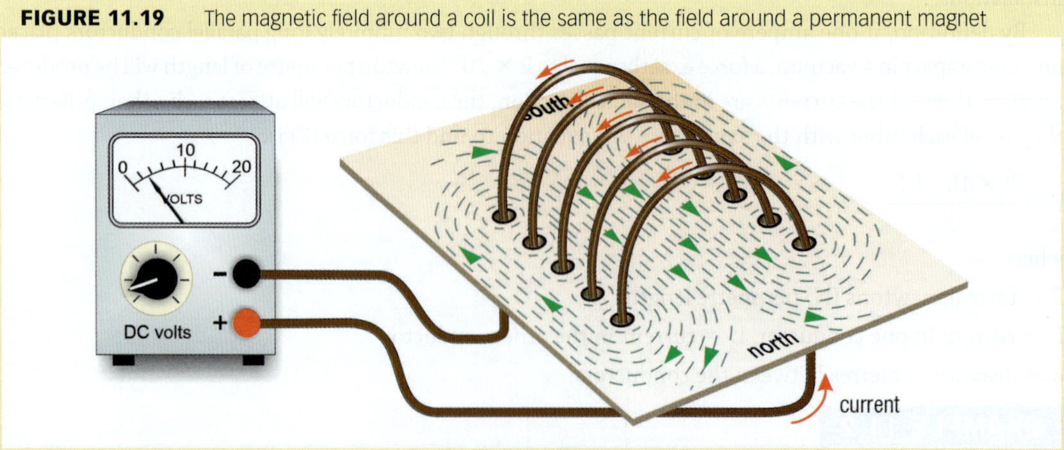

Electromagnets have either an air core or a core made of a magnetic material. An air-cored electromagnet that pulls an armature made of a magnetic material into the centre of the coil is called a solenoid. An example is shown in Figure 11.20 (a). An electromagnet with a ferromagnetic core acts like a magnet that can be turned on or off.

FIGURE 11.20 Examples of electromagnets

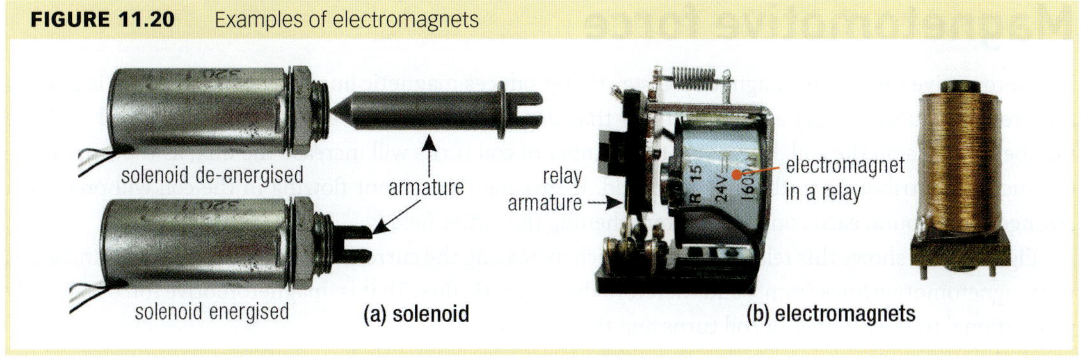

Applications of electromagnets

Electromagnets can produce extremely strong magnetic fields, in which the strength of the magnetic field can be controlled. The solenoid is an application of an air-cored electromagnet, although externally, the coil is encased in a magnetic material. A common application is the starter motor solenoid in a motor vehicle; other applications are as actuators in machinery and in a wide range of domestic and industrial equipment.

Electrically-operated water valves have an encapsulated coil that fits over a watertight brass tube. A soft iron plunger inside the tube is arranged to lift and unplug a hole, letting water through when the coil is energised.

Strong electromagnets are used in recycling yards to separate iron and steel materials from non-magnetic materials as they flow past the electromagnet on a conveyor belt. Electromagnets powerful enough to hold several tonnes of steel during lifting find application in a range of heavy industries.

Permanent magnets are made by exposing them to the field produced by an electromagnet. An extremely high current is passed through the electromagnet for a few milliseconds, giving an intense but short-duration magnetic field. This current is typically provided by a large capacitor, in which the capacitor is charged slowly, then discharged quickly through the electromagnet.

The principle of electromagnetism forms the basis of electrical equipment such as transformers, electric motors and generators. This is why an understanding of magnetism and its relationship with electricity is so important.

Right-hand rule for a coil

The *right-hand thumb rule* can be used to find the north pole of an electromagnet. As shown in Figure 11.21, hold the electromagnet with your right hand so your fingers point in the direction of the current. Your outstretched thumb then points to the north pole of the magnet.

The magnetic polarity of an electromagnet depends on the direction of current and also the direction of the winding. In Figure 11.21, the coil is wound in a clockwise direction when looking at the positive end of the coil (south pole end). Winding it in an anticlockwise direction without changing the direction of the current will reverse the north and south poles of the electromagnet.

FIGURE 11.21 Finding the north pole of an electromagnet

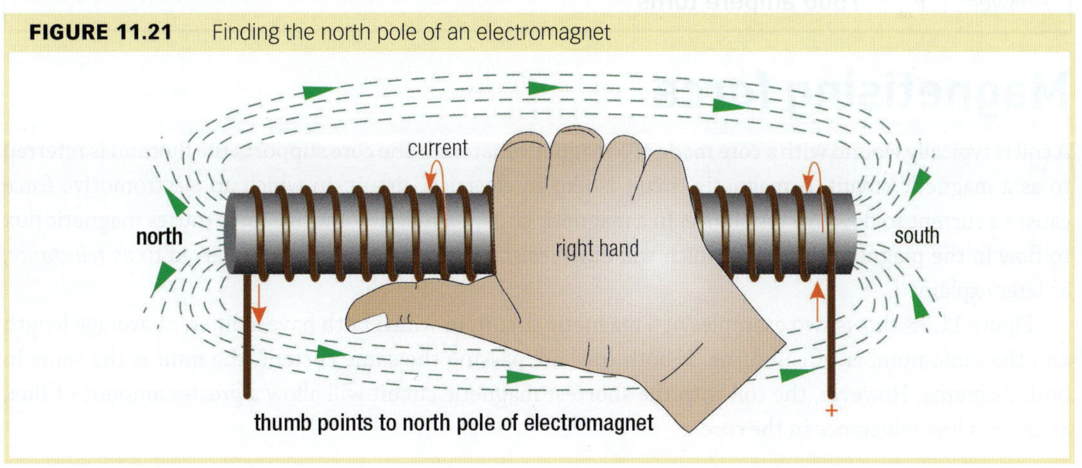

Magnetomotive force

Magnetomotive force is the magnetic *pressure* that produces magnetic lines of force. A coil that is passing a current produces a magnetomotive force that depends on the current flowing in the coil and the number of turns on the coil. Increasing the number of coil turns will increase the flux, as there are more conductors contributing to the magnetic field. Increasing the current flowing in the coil will produce a stronger flux around each conductor, strengthening the entire field.

Figure 11.22 shows this relationship, in which increasing the current or the number of turns increases the magnetomotive force (mmf) and therefore the magnetic flux. That is, magnetomotive force is directly proportional to the number of coil turns and the current.

FIGURE 11.22 The magnetic flux produced by an electromagnet is proportional to the current and the number of turns

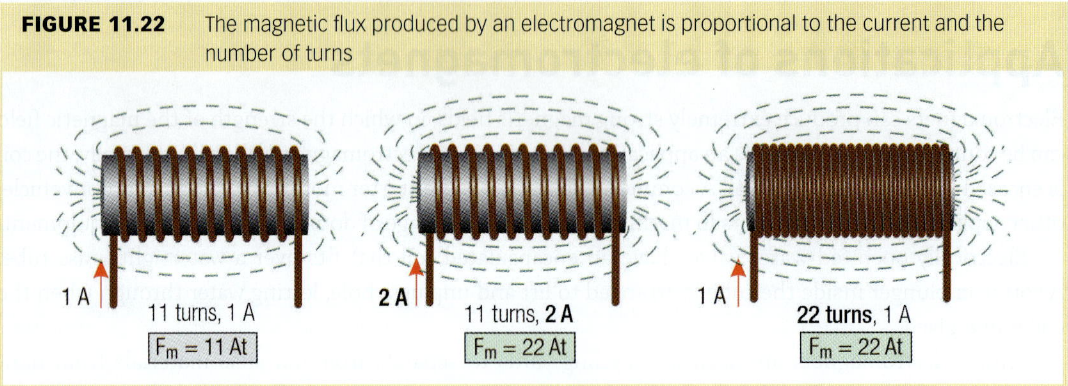

Magnetomotive force (or mmf) is referred to as F_m and is the product of the number of turns and the current flowing in the turns. That is:

$$F_m = I \times N$$

where:

F_m = magnetomotive force in ampere-turns (At)

I = current in amperes

N = number of turns.

EXAMPLE 11.3

What is the magnetomotive force developed by an air-cored coil of 250 turns passing a current of 30 amps?

Solution

Values $N = 250$ turns

$I = 30$ amps

Equation $F_m = I \times N$

$F_m = 30 \times 250$

Answer $\mathbf{F_m = 7500}$ **ampere turns**

Magnetising force

A coil is typically wound with a core made of a magnetic material. The core supports the flux and is referred to as a magnetic circuit. A magnetic circuit is like an electrical circuit, in which an electromotive force causes a current to flow in a resistance. In a magnetic circuit, a magnetomotive force causes magnetic flux to flow in the magnetic material, which will have resistance to the flux. This is referred to as *reluctance*, as later explained.

Figure 11.23 shows two examples of a magnetic circuit, in which both have a different average length and the same number of coil turns. If both coils are passing the same current, the mmf is the same in both diagrams. However, the coil with the shortest magnetic circuit will allow a greater amount of flux, as there is less reluctance in the core.

FIGURE 11.23 Magnetising force depends on the ampere turns of the coil and the length of the magnetic circuit

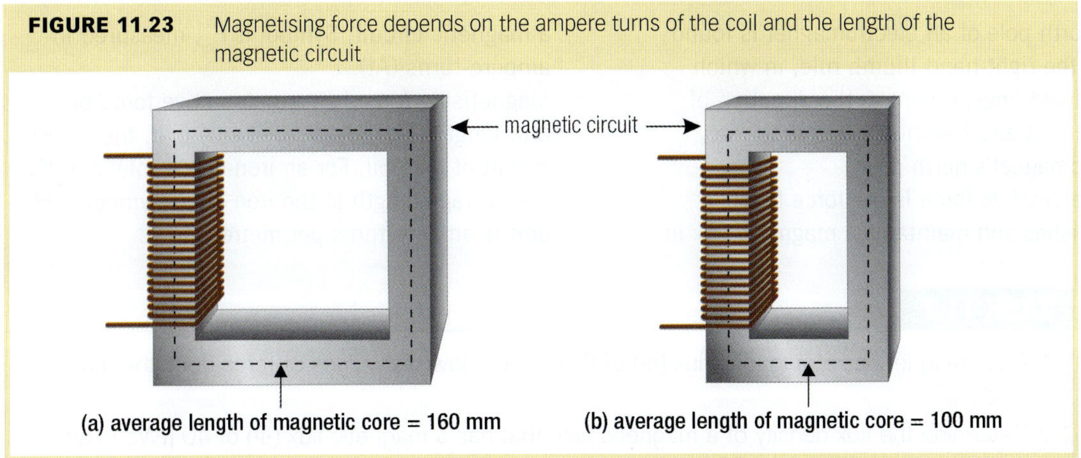

Therefore, the magnetic field strength in Figure 11.23(b) will be greater than in (a), due to its shorter core length. That is, the strength of the field depends on the ampere-turns and the length of the magnetic circuit. Magnetic field strength is called magnetising force and is defined as *magnetomotive force divided by length*. Because the standard unit of length is the metre, the measurement unit for magnetising force is the *ampere-turn per metre*, or At/m. The symbol for magnetising force is H, and the equation to find it is:

$$H = \frac{F_m}{l} \text{ or } \frac{NI}{l}$$

where:

H = magnetising force in ampere-turns per metre (At/m)

F_m (NI) = magnetomotive force in ampere-turns (At)

l = length of magnetic path in metres (m).

This equation gives the magnetising force (H) at any point in the core. The magnetic field around a conductor comprises circular flux lines in air. At a distance of say, 0.05 metres, the magnetic circuit length is the circumference of a circle with a radius of 0.05 m, which is 0.31 m. In a straight conductor, the mmf is the value of the current, so the magnetising force at 0.05 m from a cable carrying a current of 50 A is 50 divided by 0.31 m, giving 159 At/m.

EXAMPLE 11.4

A coil of 1000 turns wound on a 75 mm long iron core has a current of 0.2 amperes passing through the coil. What is the magnetising force (H) in the core?

Solution

Values $N = 1000$ turns
$I = 0.2$ A
$l = 0.075$ m

Equation $H = \frac{NI}{l} = \frac{1000 \times 0.2}{0.075}$

Answer **H = 2666.67 At/m**

KEY POINTS...

- The direction of the magnetic field around a conductor is found by the right-hand thumb rule, in which the thumb points in the direction of the current, and the curved fingers point in the direction of the flux lines.
- Adjacent current-carrying conductors will have a force acting to repel them (currents in different directions) or attract them (current in the same direction).
- Adjacent conductors carrying a high current will have a strong magnetic force between them.
- An electromagnet is a current-carrying conductor wound into a coil.

- The north pole of an electromagnet is found using the right-hand thumb rule, in which the curved fingers point in the direction of the current and the thumb points to the electromagnet's north pole.
- Magnetomotive force is the force that establishes and maintains a magnetic flux in a magnetic circuit. Symbol is F_m, measured in ampere-turns (At).
- Magnetising force is magnetomotive force per unit length. For an air-cored solenoid, the length is that of the coil. For an iron-cored solenoid it's the average length of the iron core. Symbol is H, unit is ampere-turns per metre (At/m).

TASK 11.1

1 A bar magnet has a magnetic flux (Φ) of 0.5 mWb. How many lines of force does the magnet produce?
2 Determine the flux density of a magnetic field that has a magnetic flux (Φ) of 40 μWb in an area measuring 20 mm by 20 mm?
3 Find the force acting over a metre length of two parallel conductors separated by 20 mm that are both carrying a current of 5000 A.
4 Calculate the magnetomotive force developed by a coil of 500 turns when it's carrying a current of 25 amps.
5 If the coil in Question 4 is wound on a 250 mm long iron core, calculate the magnetising force in the core.

11.4 Magnetic Circuit

As previously explained, a magnetic circuit is like an electric circuit, in which voltage (electrical pressure) causes current to flow, providing there is a circuit for the current. The current is opposed by the resistance of the circuit. In a magnetic circuit, magnetomotive force (F_m) is the pressure, and magnetic flux (Φ) is equivalent to current. The magnetic circuit is the path taken by the flux, and its ability to conduct magnetic flux is determined by the material it is made of.

Permeability, as already explained, is a measure of the ability of a material to conduct and concentrate a magnetic flux. This is similar to the conductance (G) of a conductor, in which the higher the conductance, the better the ability of the material to conduct an electric current. A material with high permeability, such as iron, is a better conductor of magnetic flux compared to one with low permeability, such as air. That is, iron has a higher permeability relative to air, just as copper has a higher conductivity compared to carbon.

Permeability of air (μ_0)

In air (or a vacuum), the ratio of the flux density (B) to the magnetising force (H) producing the flux gives a value called the *permeability of air (free space)*. The measurement unit is henry/metre (H/m). This value is constant regardless of the magnetising force, and is called μ_0, where:

$\mu_0 = 4\pi \times 10^{-7}$ (exactly) or for practical purposes, 1.26×10^{-6} henry/metre (H/m)

Relative permeability is a measure of the ability of a material to support a magnetic field compared to air

Relative permeability (μ_r)

A coil with an iron core will have a higher flux density (B) in the core compared to the same coil with an air core. The ratio of the two values of flux density gives the *relative* permeability of the iron being used in the core. The relative permeability of air is one, while ferromagnetic materials can have a relative permeability of over 100 000. Relative permeability does not have a measurement unit, but has the symbol μ_r and can be found with the equation:

$$\mu_r = \frac{\mu}{\mu_0} = \frac{\text{B (in the substitute material)}}{\text{B (in air)}}$$

Absolute permeability (μ)

The absolute permeability of a material is calculated by multiplying its relative permeability by the permeability of air. The symbol is μ, the measurement unit is henry/metre (H/m) and the equation to find it is:

$\mu = \mu_r \times \mu_0$ henry/metre (H/m)

Table 11.1 compares the absolute permeability of a number of typical ferromagnetic materials. Relative permeability μ_r is not a constant value and changes with flux density (*B*). The relative permeability values in the table give an approximate range. The values of absolute permeability μ are minimum and maximum values calculated from the range of relative permeability values.

TABLE 11.1

PERMEABILITY OF TYPICAL FERROMAGNETIC MATERIALS

Material	Permeability of air μ_0	Relative permeability μ_r	Absolute permeability $\mu = \mu_r \times \mu_0$	Comments
Nickel	1.26 × 10^{-6} H/m	100–600	0.126 × 10^{-3} H/m – 0.756 × 10^{-3} H/m	99% pure
Iron		200–5000	0.25 × 10^{-3} H/m – 6.3 × 10^{-3} H/m	99.8% pure
Silicon-iron grain-oriented		1500–40 000	1.9 × 10^{-3} H/m – 50.4 × 10^{-3} H/m	used in transformer cores
Ferrite		8–10 000	0.01 × 10^{-3} H/m – 12.6 × 10^{-3} H/m	μ = 3.78 × 10^{-3} for N41 ferrite (μ_r = 3000)
Permalloy		8000–80 000	10.1 × 10^{-3} H/m – 0.101 H/m	comes in various grades
Mu metal		20 000–100 000	25.2 × 10^{-3} H/m – 0.126 H/m	magnetic shield material

EXAMPLE 11.5

What is the absolute permeability of an iron core, if its relative permeability is 5000 at a certain flux density?

Solution

Values $\mu_r = 5000$

Equation $\mu = \mu_r \times \mu_0$ where $\mu_0 = 1.26 \times 10^{-6}$ H/m

$\mu = 5000 \times 1.26 \times 10^{-6}$

Answer $\boldsymbol{\mu = 6.3 \times 10^{-3}}$ henry/metre (H/m)

In Example 11.5 the iron has a relative permeability 5000 times higher than air. This means the flux density in the iron is 5000 times higher than it would be in an air core, for the same value of magnetising force (H). The absolute permeability (μ) of a material can also be calculated with the equation:

$$\mu = \frac{B}{H}$$

where:

μ is the material's absolute permeability (H/m)

B is the flux density in the material (tesla)

H is the magnetising force (At/m).

This equation can be transposed in terms of either B or H, giving:

$$B = \mu H$$

$$H = \frac{B}{\mu}$$

EXAMPLE 11.6

Calculate the magnetising force (H) and the flux density (B) in the core of the electromagnet shown in Figure 11.24.

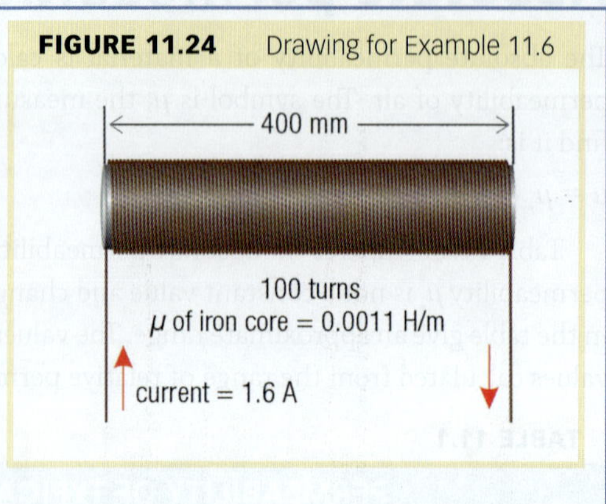

FIGURE 11.24 Drawing for Example 11.6

Solution

Values $\mu = 0.0011$ henry/metre (H/m)
$N = 100$ turns
$I = 1.6$ A
$l = 40$ cm $= 0.4$ m
$H = ?$
$B = ?$

Equation $H = \frac{NI}{l} = \frac{100 \times 1.6}{0.4}$

Answer $\mathbf{H = 400\ At/m}$

Equation $B = \mu H$
$B = 0.0011 \times 400$

Answer $\mathbf{B = 0.44\ T}$

Because the core is made of a ferromagnetic material, the flux density calculated in Example 11.6 is present in all parts of the core.

Reluctance

Reluctance in a magnetic circuit is like resistance in an electric circuit. Any magnetic circuit will have some reluctance, which limits its ability to conduct a magnetic field. The symbol for reluctance is R_m and the measurement unit is the ampere-turn per weber (At/Wb).

In an electric circuit, Ohm's law relates voltage, current and resistance. In a magnetic circuit, there's a similar relationship between F_m, Φ and R_m:

$$\Phi = \frac{F_m}{R_m} \left(\text{equivalent to the Ohm's law equation } I = \frac{V}{R}\right)$$

$$R_m = \frac{F_m}{\Phi} \left(\text{equivalent to } R = \frac{V}{I}\right)$$

$$F_m = \Phi \times R_m \text{ (equivalent to } V = IR)$$

where:

Φ is flux measured in webers (Wb)

F_m is magnetomotive force measured in ampere-turns (At)

R_m is reluctance measured in ampere-turns per weber (At/Wb).

The reluctance of a magnetic circuit depends on three things:

1 *length* – the longer the magnetic circuit, the higher the reluctance
2 *cross-sectional area* – the smaller the csa, the higher the reluctance
3 *permeability* – the lower the permeability, the higher the reluctance.

The reluctance of a magnetic circuit can be found with an equation that relates these three factors:

$$R_m = \frac{l}{\mu_0 \mu_r A}$$

where:

R_m is reluctance in ampere-turns per weber (At/Wb)

l is length of magnetic circuit in metres (m)

μ_0 is permeability of air = 1.26×10^{-6} henry/metre
μ_r is relative permeability
A is the cross-sectional area in square metres (m^2).

If you know the absolute permeability (μ) of the material, this equation becomes:

$$R_m = \frac{l}{\mu A}$$ where μ is the permeability of the material in henry/metre.

The effect of reluctance in a magnetic circuit can be shown by examining the operation of an electric relay. A relay has a coil wound around a former made of a magnetic material. When the coil is energised, the magnetic field in the former attracts a metal armature, causing it to close and bringing the two contacts together. When the coil is de-energised, a spring pulls the armature back to the open position, opening the contacts. This allows a relay to switch an electrical load by energising or de-energising the relay coil.

Figure 11.25 depicts a relay with its armature open, then with the armature closed. When the armature is open, the magnetic path is the iron core of the coil, the armature and the air gaps between the core and the armature. When the armature is closed, there are no air gaps, and the entire path is through high permeability iron. Because air has a low permeability, the air gap will increase the reluctance of the magnetic circuit.

FIGURE 11.25 A relay has an air gap in the magnetic path when the armature is open, which increases the reluctance of the magnetic circuit

(a) armature open, high reluctance
(b) armature closed, low reluctance

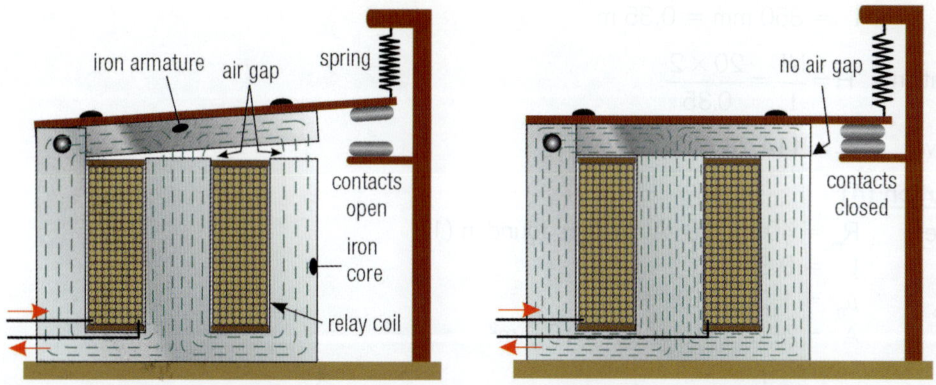

Because the reluctance of the magnetic path drops when the armature is closed, more flux is produced by the coil. For this reason, the magnetomotive force (and, therefore, current in the coil) to make the relay operate is greater than that needed to hold the relay closed. That is, it takes more electrical power to operate the relay than to keep it closed.

In effect, an air gap in a magnetic circuit is like having a high resistance in series with a low resistance in an electric circuit. The reluctance of an air gap can be calculated in the same way as the reluctance of the magnetic material, except the value of μ is now μ_0. The total reluctance of the magnetic circuit is the sum of the two reluctance values.

Example 11.7 shows the relationships between reluctance, magnetomotive force, flux, flux density, magnetising force and permeability.

EXAMPLE 11.7

The core of the electromagnet in Figure 11.26 has an average length (l) of 350 mm and a cross-sectional area (csa) of 400 mm^2. The coil has 20 turns. When a current of 2 A is passed through the coil, a total flux of 100 μWb is established in the core.

1 What is the reluctance (R_m) of the core?
2 What is the flux density (B) in the core?
3 Determine the magnetising force (H).
4 Calculate the relative permeability (μ_r) of the core.

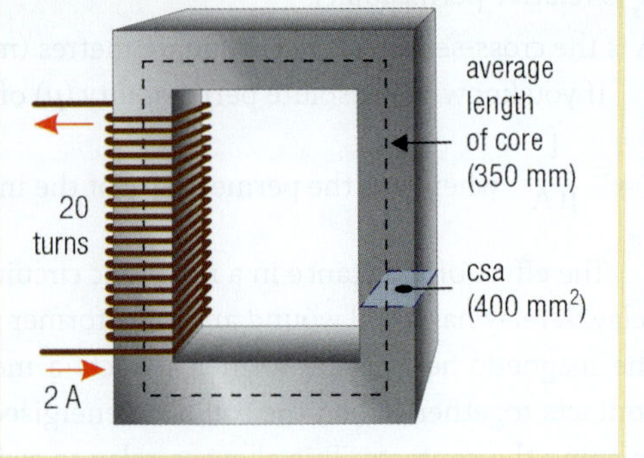

FIGURE 11.26 Drawing for Example 11.7

Solution (1)

Values $N = 20$ turns

$I = 2\text{ A}$

$\Phi = 100\ \mu\text{Wb} = 100 \times 10^{-6}\text{ Wb}$

Equation $R_m = \dfrac{F_m}{\Phi} = \dfrac{NI}{\Phi}$

$$R_m = \frac{40}{100 \times 10^{-6}}$$

Answer $\mathbf{R_m = 400\,000 \text{ or } 400 \times 10^3 \text{ At/Wb}}$

Solution (2)

Values $\Phi = 100\ \mu\text{Wb} = 100 \times 10^{-6}\text{ Wb}$

$A = 400\text{ mm}^2 = 400 \times 10^{-6}\text{ m}^2$

Equation $B = \dfrac{\Phi}{A} = \dfrac{100 \times 10^{-6}}{400 \times 10^{-6}}$

Answer $\mathbf{B = 0.25\text{ T}}$

Solution (3)

Values $N = 20$ turns

$I = 2\text{ A}$

$l = 350\text{ mm} = 0.35\text{ m}$

Equation $H = \dfrac{NI}{l} = \dfrac{20 \times 2}{0.35}$

Answer $\mathbf{H = 114.28\text{ At/m}}$

Solution (4)

Values $R_m = 400 \times 10^3$ At/Wb (as found in (1))

$l = 0.35\text{ m}$

$\mu_0 = 1.26 \times 10^{-6}\text{ H/m}$

$A = 400\text{ mm}^2 = 400 \times 10^{-6}\text{ m}^2$

Equation $R_m = \dfrac{l}{\mu_0 \mu_r A}$, when transposed becomes $\mu_r = \dfrac{l}{R_m \mu_0 A}$

$$\mu_r = \frac{0.35}{400 \times 10^3 \times 1.26 \times 10^{-6} \times 400 \times 10^{-6}}$$

Answer $\mathbf{\mu_r = 1736}$

If there was an air gap in the core of Figure 11.26, the reluctance of the magnetic circuit would increase, so the flux (Φ) and the flux density (B) in the core would reduce. To compensate, a higher value of coil current would be needed to produce a higher magnetising force (H), which would increase the flux density.

KEY POINTS...

- Permeability (symbol μ), also known as absolute permeability, is the ability of a material to conduct and concentrate a magnetic flux. It is measured in henrys per metre.
- Absolute permeability (μ) of a material can be calculated by multiplying its relative permeability μ_r by the permeability of a vacuum μ_0. It also equals flux density (B) divided by the magnetising force (H). That is $\mu = B/H$.
- Permeability of a vacuum (μ_0) is a constant and equals exactly $4\pi \times 10^{-7}$ or approximately 1.26×10^{-6} henrys per metre (H/m).

- Relative permeability (symbol μ_r) is a measure of the increase in flux density a magnetic material causes compared to air, for the same magnetising force. For ferromagnetic materials, μ_r is always greater than one.
- Reluctance (symbol R_m) is the resistance a magnetic path offers to magnetic lines of force (Φ) and is measured in ampere turns per weber (At/Wb).
- Reluctance can be found with either of these equations:
  - $R_m = \frac{F_m}{\Phi}$ where F_m = NI= amps × turns
  - $R_m = \frac{l}{\mu_0 \mu_r A}$ where A = csa in m^2, l = length of magnetic circuit in metres

TASK 11.2

1. The iron core of a relay has a relative permeability (μ_r) of 2600 at the flux density created when the relay coil is energised. Calculate the absolute permeability (μ) of the core.
2. A coil of 100 turns is wound around an iron core 250 mm long with an absolute permeability (μ) of 4.5×10^{-3} H/m. Calculate the flux density (B) in the core when a current of 4 A flows in the coil. (Hint, first find the magnetising force H.)
3. The coil in Question 2 produces a total flux (Φ) of 800 µWb. What is the reluctance (R_M) of the core?
4. The iron core of an electromagnet has a cross-sectional area (A) of 450 mm^2 and a total flux (Φ) of 150 µWb. Calculate the flux density (B) in the core.
5. A 200 mm long iron core has a cross-sectional area of 300 mm^2 and a reluctance (R_m) of 300×10^3 At/Wb. Calculate the relative permeability (μ_r) of the iron core.

11.5 Magnetisation Curves

The magnetic characteristics of a ferromagnetic material are measured with specialised equipment so curves can be constructed. The starting point is a material's B–H curve.

B–H curve

The B–H curve is a measure of the flux density (B) produced in a material for various values of magnetising force (H). Magnetising force is ampere-turns per metre, so changing the current flowing in a coil changes the magnetising force. If the core of the coil is a non-magnetic material such as air, the change in flux density is proportional to the change in the magnetising force, giving the straight line shown in Figure 11.27. If the core is made of a ferromagnetic material, the graph is a curve, called the B–H curve (or the magnetisation curve) of the material, shown in Figure 11.27.

FIGURE 11.27 The B–H curve shows how the flux density changes with magnetising force

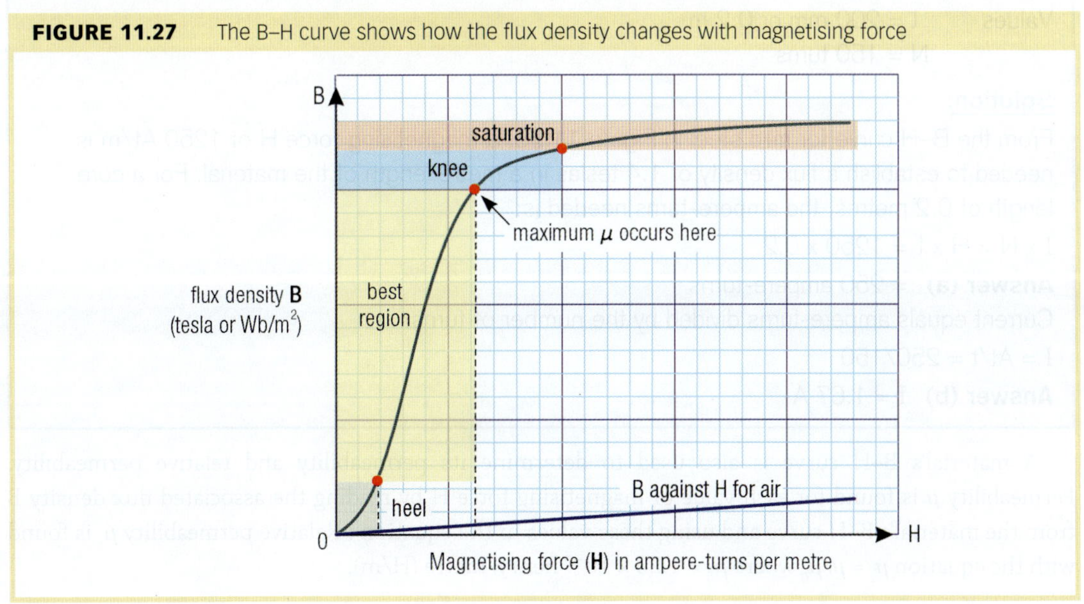

FIGURE 11.28 B–H curves for various types of ferromagnetic materials

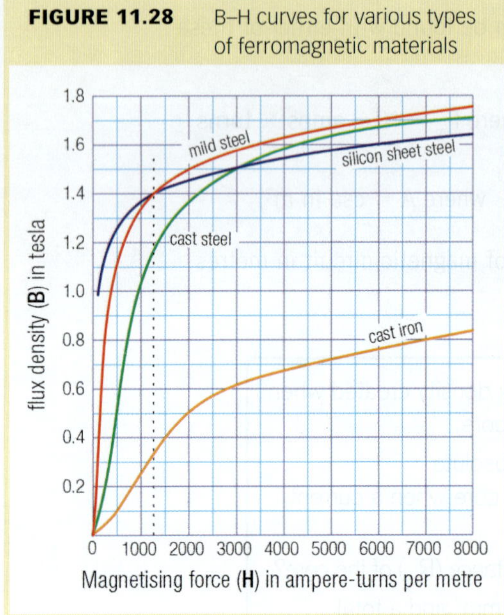

The B–H curve in Figure 11.27 is not for any particular metal and is drawn to highlight certain aspects of a B–H curve, which include:

1. The B–H curve for any magnetic material shows that the ratio of B to H is non-linear.
2. Permeability (μ) of a magnetic material reaches a maximum before saturation occurs.
3. The *heel* region at the bottom of the curve is where the flux density increases, at first slowly, then quite rapidly, as the magnetic flux becomes established in the core.
4. The linear part of the curve is between the heel and the knee. In this region, a small change in magnetising force causes a large and nearly linear change in flux density.
5. After the knee, flux density no longer increases in proportion with the magnetising force, and the material starts to become saturated.

Figure 11.28 shows B–H curves for four ferromagnetic materials. The curve for cast iron shows it has a flux density B of about half that of the other metals, for the same magnetising force. Silicon sheet steel is used in transformer cores and has a similar B–H curve to mild steel. Saturation of these four materials starts when the magnetising force reaches around 2000 At/m. For mild steel, increasing the magnetising force by a further 6000 At/m achieves a flux density increase from 1.5 tesla to only 1.75 tesla.

A material's B–H curve is essential to solve most magnetic problems, such as that shown in the next example.

EXAMPLE 11.8

FIGURE 11.29 Toroid coil for Example 11.8

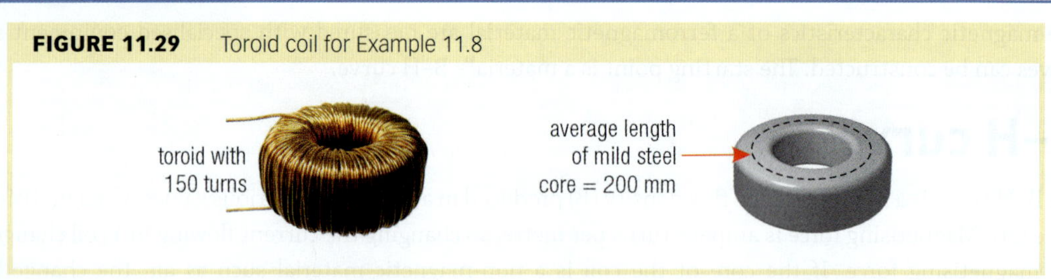

For the toroid coil in Figure 11.29. Find:

(a) the ampere-turns needed to produce a flux density of 1.4 tesla in the core

(b) the required coil current.

Values l = 200 mm or 0.2 m

N = 150 turns

Solution:

From the B–H curve for mild steel in Figure 11.28, a magnetising force H of 1250 At/m is needed to establish a flux density of 1.4 teslas in a metre length of the material. For a core length of 0.2 metres, the ampere-turns needed is:

$I \times N = H \times l = 1250 \times 0.2$

Answer (a) = 250 ampere-turns

Current equals ampere-turns divided by the number of turns:

I = At/t = 250/150

Answer (b) I = 1.67 A

A material's B–H curve is also used to determine its permeability and relative permeability. Permeability μ is found for each value of magnetising force H by reading the associated flux density B from the material's B–H curve and using these values in the equation . Relative permeability μ_r is found with the equation $\mu_r = \mu/\mu_0$ where $\mu_0 = 1.26 \times 10^{-6}$ henry/metre (H/m).

Table 11.2 lists the values of B at various levels of H as read from the B–H curve for mild steel in Figure 11.30. The material's permeability μ and relative permeability μ_r are calculated from each set of these readings and the results for relative permeability are shown as a graph in Figure 11.30.

TABLE 11.2

RELATIVE PERMEABILITY FOR MILD STEEL FROM B AND H			
H (At/m)	B (tesla)	$\mu = B/H$	$\mu_r = \mu/\mu_0$
250	0.9	**0.0036**	2857
500	1.14	0.00228	1810
750	1.26	0.00168	1333
1000	1.35	0.00135	1071
1250	1.4	0.00112	889
1500	1.44	0.00096	762
1750	1.47	0.00084	667
2000	1.51	0.000755	599
2250	1.52	0.000676	536
2500	1.54	0.000616	489
3000	1.58	0.000527	418
3500	1.6	0.000457	363
4000	1.63	0.000408	323
4500	1.65	0.000367	291
5000	1.67	0.000334	265
5500	1.68	0.000305	242
6000	1.7	0.000283	225
6500	1.72	0.000265	210
7000	1.73	0.000247	196
7500	1.74	0.000232	184
8000	1.75	0.000219	174

Table 11.2 shows that the permeability (μ) and therefore the relative permeability (μ_r) of mild steel (and any ferromagnetic material) changes considerably with the flux density. This explains why a B–H curve for a ferromagnetic material is not a straight line.

After the saturation point is reached at around 2000 At/m, permeability changes by smaller amounts for each value of magnetising force. Table 11.2 (and Figure 11.30) show that the maximum permeability of mild steel occurs at 250 At/m, which is also the knee of the B-H curve.

FIGURE 11.30 Relative permeability and B–H curves for mild steel

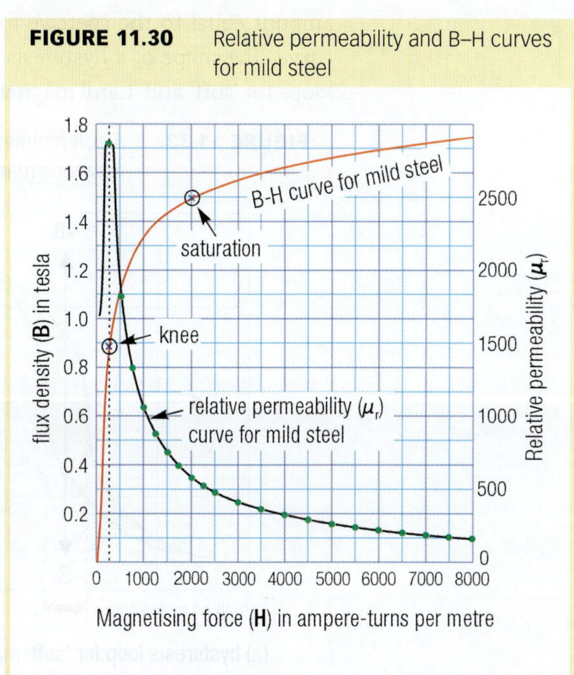

Magnetic hysteresis

A B–H curve shows the relationship between flux density and magnetising force, in which it is assumed the magnetic field does not change direction. Magnetic materials are also used in electrical equipment that involves an alternating current, such as AC motors and transformers. In this case, the magnetic field changes polarity every time the current changes polarity. The type of curve that shows what happens in the material under these conditions is called a *hysteresis loop* (or curve), as shown in Figure 11.31.

FIGURE 11.31 A hysteresis curve shows how flux density changes in strength and direction with a magnetising force that periodically changes direction

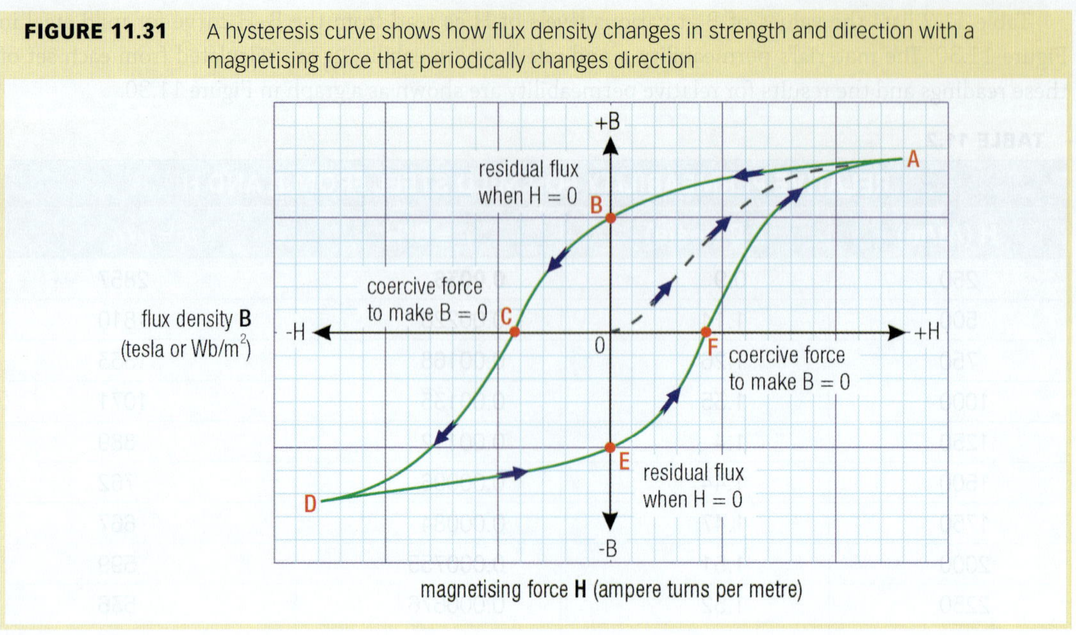

When the magnetising force H is first applied, it's assumed the coil has no residual magnetism. Therefore, the flux density increases from zero to the maximum value *A*, which is determined by the value of the current flowing in the coil. From point A to point B, the current is reduced and is zero at point B. However, the core is now partially magnetised at point B. This is its *residual* magnetism and depends on the *retentivity* of the material.

After the current (and therefore magnetising force H) reaches zero, it begins increasing in value in the opposite direction. At point C, the flux density has fallen to zero (material is demagnetised), and the magnetising force to achieve this is called the *coercive force*. The current reaches its maximum negative value at point D.

At point E, the current and therefore magnetising force H are zero, leaving the material magnetised as before, but with a reverse polarity. The current now increases in the positive direction, and the material is fully demagnetised at point F. The current reaches its maximum positive value at point A, and the cycle repeats.

The flux density in a material, and therefore its magnetisation, lags the magnetising force by an amount equal to the coercive force. The term *hysteresis* means lagging behind, hence the name of the curve. The shape of a hysteresis loop depends on the magnetic material. Figure 11.32 shows hysteresis loops for 'soft' and 'hard' magnetic materials.

FIGURE 11.32 A soft magnetic material is more easily magnetised and demagnetised than a hard magnetic material

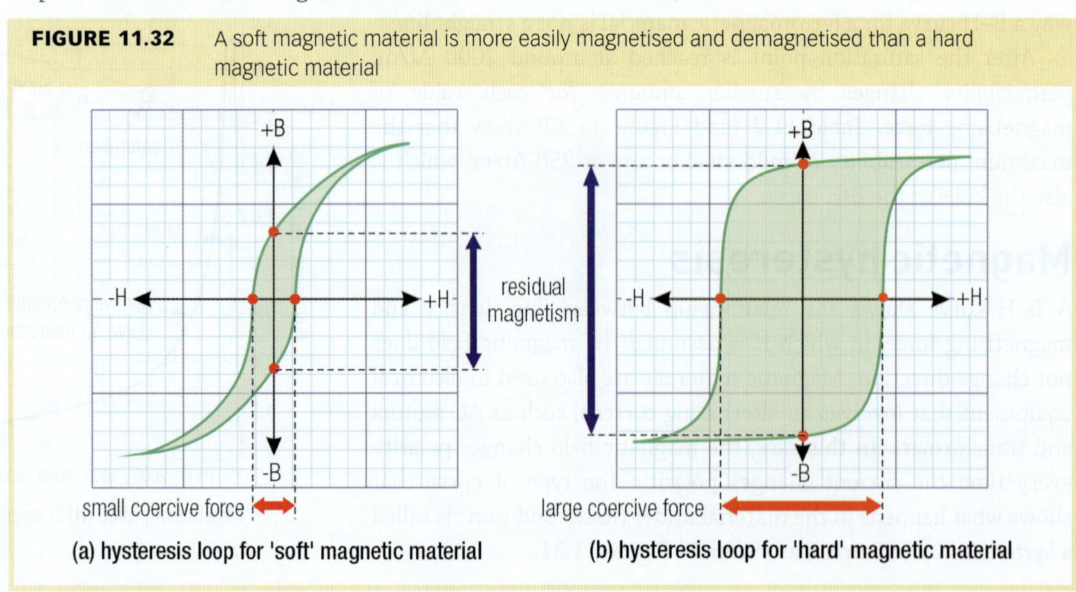

The hysteresis curve in Figure 11.32(a) is for a material that is easily magnetised and demagnetised, such as silicon steel used in transformer cores. Compared to the curve in (b), a soft material requires less coercive force to reduce the flux to zero, and there is less residual flux in the material when the magnetising force is zero. This makes soft magnetic materials such as soft iron and silicon steel suitable for electromagnetic applications involving an alternating current.

The hysteresis curve for a magnetically hard material shows a high level of flux density remaining when the magnetising force is zero (residual flux). As well, a high coercive force is needed to demagnetise the material. A magnetically hard material is therefore suitable as a permanent magnet because most of the magnetism remains after the magnetising force is removed. As well, it's difficult to demagnetise it. Hard magnetic materials include alnico, rare earth alloys, ferrite and carbon steel.

Magnetic losses

When an electromagnet is first energised, energy is expended in aligning the magnetic domains in the core material so they form a magnet and thereby create a magnetic field. In a DC circuit, this energy loss is insignificant, as it only occurs when the electromagnet is first energised. When an electromagnet is supplied with an alternating current, energy is lost in rearranging the magnetic domains each time the current reverses polarity.

The hysteresis loops in Figure 11.32 give a comparison of the energy expended during a cycle, in which (a) takes less energy to change magnetic polarity over a cycle than (b). This means less heat is generated in the core. The loss caused by a material's hysteresis is called its *hysteresis loss*. The smaller the area enclosed by the hysteresis loop, the lower the loss.

Another loss that occurs with an alternating magnetic field is that due to *eddy currents*. As explained further in the next chapter, an alternating magnetic field will induce an alternating current in a nearby conductor. Because a magnetic core is usually made of a conductive material (unless it is ferrite), its own alternating magnetic field will generate a current that circulates in the core. Because a solid metal core has a low resistance, there is little to restrict this current, which will heat up the core.

The resistance of a ferromagnetic core can be increased by building it with laminations that have a layer of insulating oxide between each lamination, as shown in Figure 11.33. The core material of AC motors and transformers is made this way. Although a small circulating current will flow in each lamination, their combined value and heating effect will be far less than for a solid steel core.

FIGURE 11.33 Electromagnets operating with an alternating current have a laminated core to minimise eddy currents flowing in the core

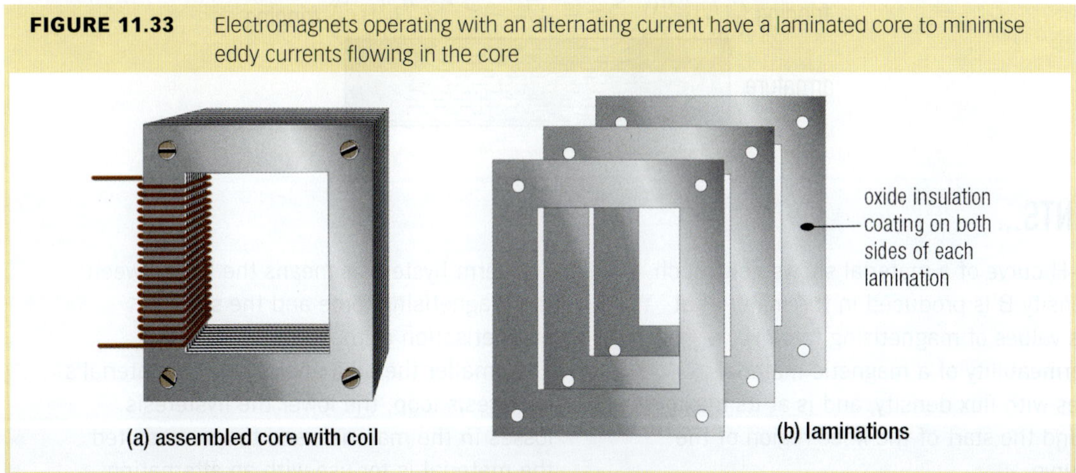

The combined effects of hysteresis and eddy current losses are called *iron losses*, as they occur in the iron core. Iron losses are minimised by the right choice of magnetic core material, and the quality of the insulation between laminations.

FIGURE 11.34 Transformers with laminated magnetic core

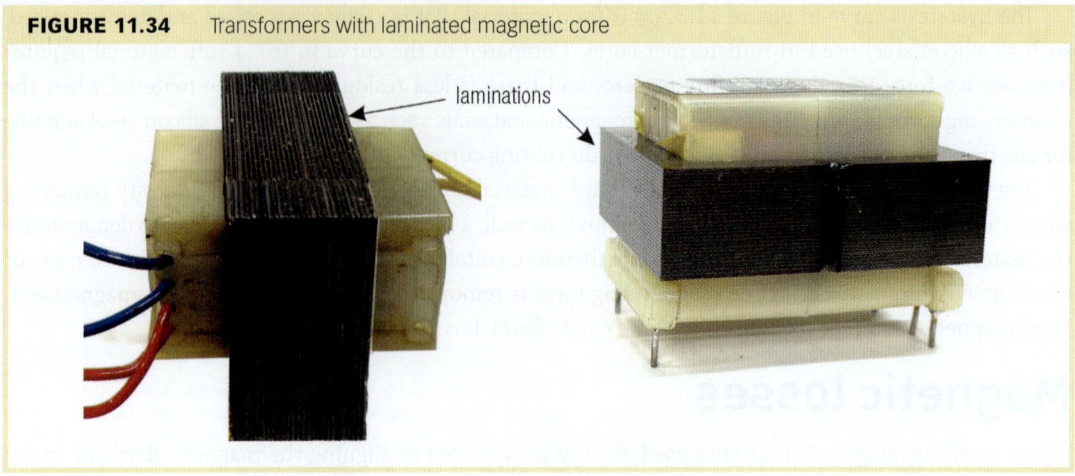

Another source of loss in a magnetic circuit is caused by magnetic flux *leakage* and *fringing*, shown in Figure 11.35. Leakage flux bypasses part of the magnetic circuit, and so does not contribute to the strength of the field attracting the armature. Fringing reduces the field strength in the air gap between the core and the armature, thereby reducing the magnetic attraction. All magnetic circuits have some leakage and fringing, which must be accounted for in the design.

FIGURE 11.35 Magnetic fringing and leakage contribute to the losses in a magnetic circuit

KEY POINTS...

- The B–H curve of a material shows how much flux density B is produced in the material at various values of magnetising force H.
- The permeability of a magnetic material changes with flux density, and is at its maximum at around the start of the knee region of the B–H curve.
- A hysteresis loop (or curve) of a magnetic material shows how the flux density changes when the magnetising force is periodically reversed by passing an alternating current through the magnetising coil.
- The term hysteresis means the lag between the magnetising force and the state of magnetisation of the material.
- The smaller the area enclosed by a material's hysteresis loop, the lower the hysteresis losses in the material, and the more suited the material is for use with an alternating magnetic field.
- All magnetic circuits have losses, which include hysteresis loss, eddy current loss, magnetic fringing and leakage.

CHAPTER SUMMARY

All the equations introduced in this chapter are listed below.

- $B = \frac{\Phi}{A}$ where B = flux density in teslas or webers/m^2, Φ = flux in webers, A = area (or csa) in square metres (m^2).
- $F = \frac{2 \times 10^{-7} I_1 I_2}{d}$ where F = force in newtons (N) per metre length, I_1 = current in one conductor, I_2 = current in the other conductor, d = distance in metres between the conductors.
- $F_m = N \times I$ where F_m = magnetomotive force in ampere-turns (At), I = current in amperes, N = number of coil turns.
- $H = \frac{F_m}{l}$ or $\frac{NI}{l}$ where H = magnetising force in ampere-turns per metre (At/m), l = length of magnetic path in metres (m).
- $\mu_r = \frac{\text{B (in the substitute material)}}{\text{B(in air)}} = \frac{\mu}{\mu_0}$ where μ_r = relative permeability in H/m.
- $\mu_0 = 4\pi \times 10^{-7}$ (exactly) or for practical purposes, 1.26×10^{-6} henry/metre (H/m) where μ_0 = permeability of air (free space, vacuum).
- $\mu = \mu_r \times \mu_0$ where μ = absolute permeability of a material measured in henry/metre (H/m).
- $\mu = \frac{B}{H}$ where values for B and H can be derived from a material's B–H curve to calculate the μ of material at a particular value of H.
- $R_m = \frac{F_m}{\Phi} \left(\text{equivalent to } R = \frac{V}{I} \right)$ where R_m is reluctance measured in ampere-turns per weber (At/Wb).
- $R_m = \frac{l}{\mu_0 \mu_r A}$ where l is length of magnetic circuit in metres, A is cross-sectional area in square metres.

REVIEW EXERCISES

Check your answers at the back of the book.

1 For Figure 11.36, identify the magnetic polarity of:
  a the pole marked A
  b the end marked X on the soft iron.

FIGURE 11.36

2 What is the difference between a ferromagnetic material, paramagnetic material and diamagnetic material?

3 What is the purpose of a magnetic shield and what type of material is the shield made of?

4 A magnet has a total flux of 2 μWb and a cross sectional area of 80 mm^2. Calculate the flux density in the core of the magnet.

5 List three causes that can reduce the magnetism of a permanent magnet.

6 An extension lead has two cables spaced 3 mm apart, and is carrying a current of 10 A. The currents in the cables are in opposite directions. Determine the force and its direction between the cables.

7 Calculate the force between the conductors in Figure 11.37 if the current in one conductor is 550 A and the other has a fault current flowing of 12 000 A.

FIGURE 11.37

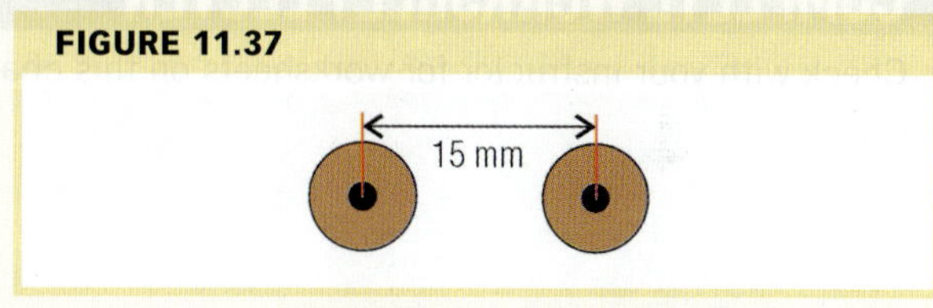

8 If the currents in both conductors in Figure 11.37 are 12 000 A, and one current is flowing in the opposite direction, what is the force between them and in which direction?

9 What is the polarity of the magnetic pole marked A in Figure 11.38?

FIGURE 11.38

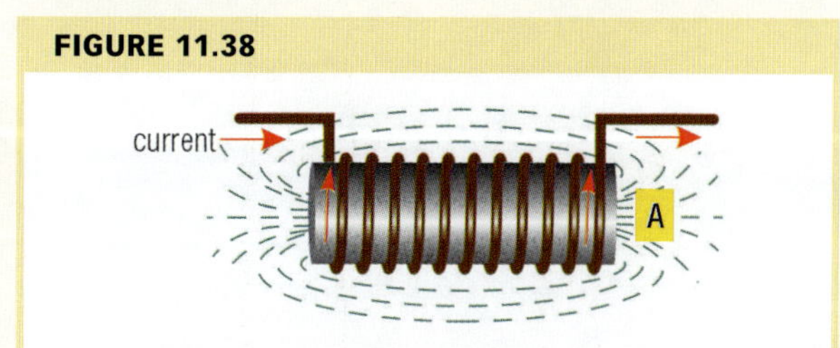

10 A coil has 35 turns and is passing a current of 24 A. Calculate its magnetomotive force (F_m).

11 The electromagnet in Figure 11.39 has 2000 turns wound on a mild steel core. Calculate the magnetising force (H) produced in the core when the coil current is 0.1 A.

FIGURE 11.39

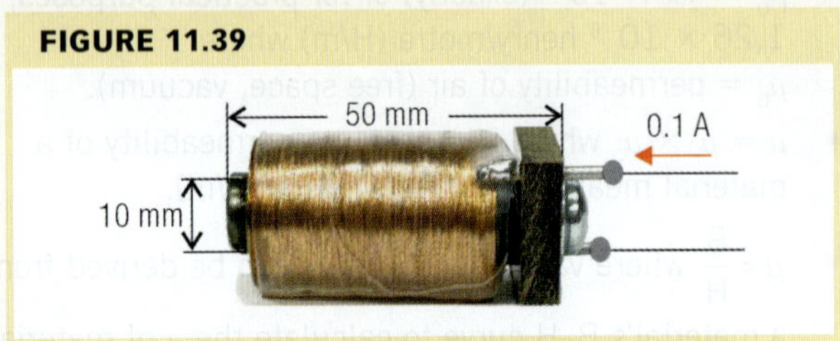

12 From Table 11.2, for the value of H found in Question 11, determine the flux density (B) produced by the electromagnet in Figure 11.39 and the relative permeability (μ_r) of its mild steel core.

13 For the coil in Figure 11.39, calculate the:

a cross sectional area of the core

b total flux (Φ) in the core.

14 Calculate the absolute permeability (μ) of a material that has a relative permeability (μ_r) of 35 000.

15 Calculate the flux density (B) in the core of the coil in Figure 11.40.

FIGURE 11.40

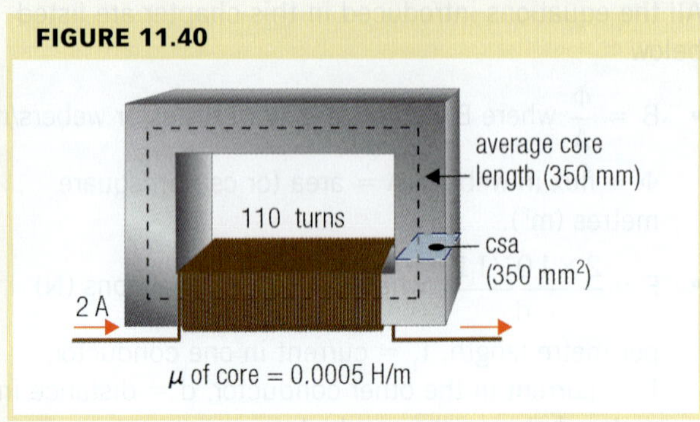

16 Calculate the total flux (Φ) in the core of the coil in Figure 11.40.

17 Determine the reluctance (R_m) of the core in Figure 11.40.

18 What is meant by 'residual flux' and 'coercive force' when talking about the hysteresis loop of a magnetic material?

19 List three losses in a magnetic circuit.

20 A magnetic circuit has an average length of 200 mm, a cross sectional area of 150 mm^2 and a relative permeability at a flux density of 830 T. Calculate its reluctance.

ONLINE RESOURCES

COMPLETE WORKSHEET ELEVEN

Check with your instructor for worksheets on this chapter.

CHAPTER 12

ELECTROMAGNETIC INDUCTION

An electric current produces a magnetic field, and as this chapter explains, a moving magnetic field can produce an electric current. This is called electromagnetic induction and is how most electrical power is generated. This chapter also introduces the inductor, a component which is usually a coil of wire wound around a ferromagnetic core. Inductors are widely used in electrotechnology, and we examine their electrical properties when used in a DC circuit, especially the property called inductance, which, like resistance and capacitance, is a fundamental electrical quantity.

CHAPTER OUTLINE

12.1 Introduction

In 1820, Danish physicist Hans Christian Oersted (1777–1851) discovered the link between electricity and magnetism. In 1831 Michael Faraday (British physicist) made an even more important discovery when he found that a moving magnetic field causes a current to flow in a conductor. Here was a way of producing electricity! This effect, called *electromagnetic induction*, was studied by many notable people of the time including American scientist Joseph Henry (1797–1878). The first DC generator was produced in 1832, a hand-cranked device that paved the way for much larger machines. Today, most of our electricity is produced by electromagnetic induction – thanks to Faraday's discovery.

12.2 Electromagnetic Induction

Figure 12.1 shows a conductor moving between the poles of a magnet. The conductor is connected to a centre-zero microvoltmeter (μV), also known as a galvanometer. This type of meter is able to register very small voltages of either polarity because the pointer can move left or right. As the diagram shows, a voltage is *induced* in the conductor when it moves in the magnetic field.

FIGURE 12.1 A conductor moving in a magnetic field causes a voltage to be induced in the conductor

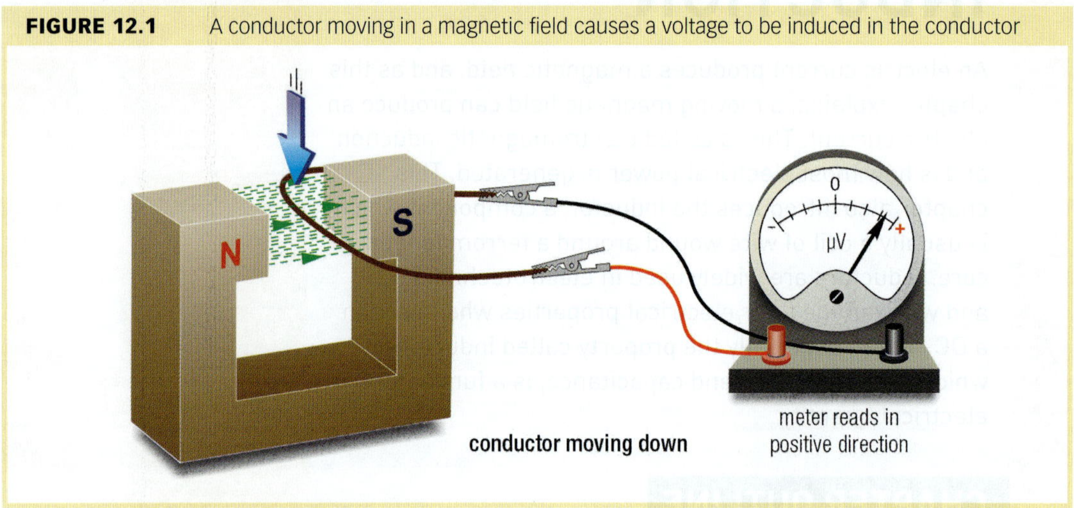

If the conductor is moved in the other direction, the polarity of the voltage will change, as shown in Figure 12.2. Notice that the meter is now registering a negative voltage.

FIGURE 12.2 The polarity of the induced voltage changes with the direction of movement

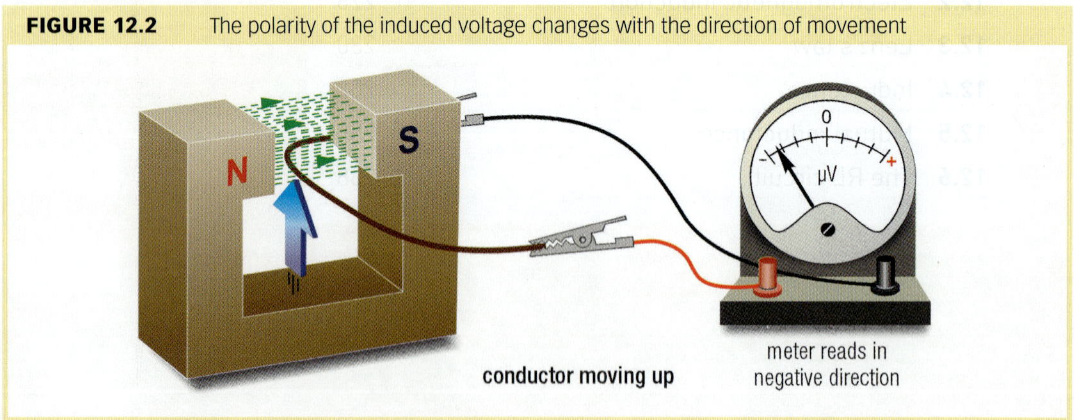

If the polarity of the magnet is changed by swapping the north and south poles, the polarity of the induced voltage will also change.

Fleming's right-hand rule

The direction of the current when a conductor is moving in a magnetic field is found with *Fleming's right-hand rule*.

- Arrange the thumb, first finger and second finger of your right hand so they're at right angles to each other.
- Point the thumb in the direction the conductor is moving relative to the magnetic field (thumb = travel).
- Point the first finger in the direction of the flux (first finger = flux).

The second (or centre) finger then points in the direction the induced current will flow (centre = current). It also points to the positive end of the conductor.

KEY POINTS...

- The polarity of an induced voltage depends on the:
  - direction of movement between field and conductor
  - polarity of the magnetic field.

FIGURE 12.3 Fleming's right-hand rule

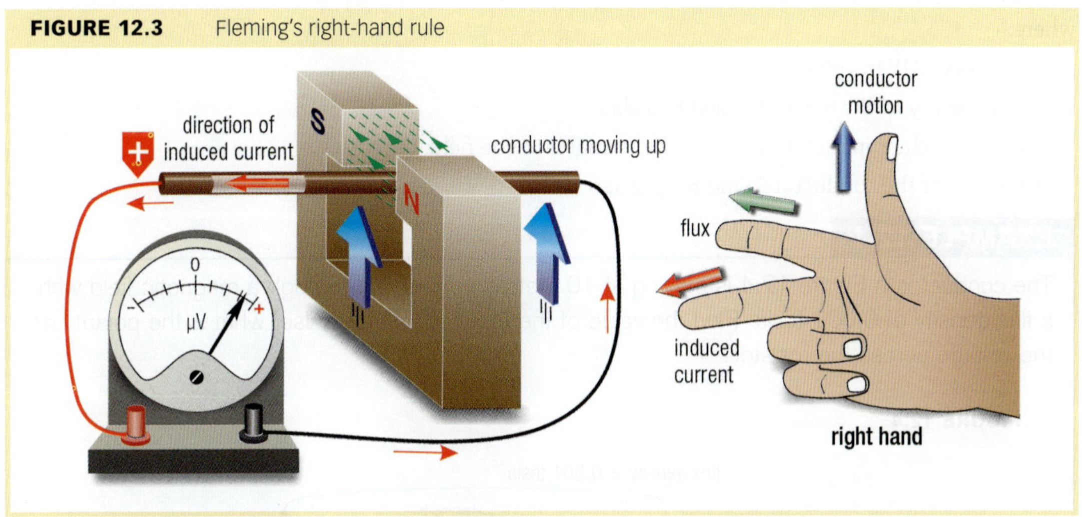

It doesn't matter whether the conductor or the magnetic field moves, providing one of them moves – giving *relative* motion between them. Therefore, if the magnetic field is moving instead of the conductor, you point your thumb in the relative direction of the conductor. For example, if the field is moving down while the conductor is stationary, you point your thumb up.

Fleming's right-hand rule is for *generators*, devices that produce electricity by relative motion between a coil of wire and a magnetic field. The right hand is used where conventional current flow is assumed (as in this book).

Because a generator produces an electric current, the current *inside* the generator flows from negative to positive. Outside the generator it flows from positive to negative as shown in Figure 12.3. This is like a battery, where current flows from the positive terminal, through the circuit to the negative terminal. Inside the battery, the current flow is from negative to positive.

Therefore, your second finger points in the direction of the current and also to the positive end of the conductor because the current in the conductor flows towards the positive end.

Voltage induced in a conductor

The value of the voltage induced in a conductor depends on four things:

1 **Length** of the conductor moving in the magnetic field. The longer the conductor, the higher the induced voltage.
2 **Speed** of movement (or velocity) between the conductor and the magnetic field. The faster the relative velocity, the higher the induced voltage.
3 **Strength** of the magnetic field. The stronger the field, the higher the induced voltage.
4 **Angle** at which the conductor passes through the magnetic field. For the purposes of this chapter, this is assumed to be a right angle (90°) at all times.

The first three factors can be combined into a simple equation that lets you calculate the value of the induced voltage. This equation assumes that the angle of intersection between the conductor and the magnetic field is always 90°.

 FYI

If the angle of intersection between the field and conductor is not 90°, this equation becomes: $e = Blv \sin \theta$ where θ is the angle of intersection

$e = Blv$

where:

e = induced EMF in volts

B = flux density of the magnetic field in teslas

l = length of the conductor moving at right angles to the field, in metres

v = velocity of the conductor in metres per second.

EXAMPLE 12.1

The conductor in Figure 12.4 is moving at 10 metres per second through a magnetic field with a flux density of 0.001 tesla. Find the value of the induced voltage. Also, what is the polarity of the voltage across the resistor?

FIGURE 12.4

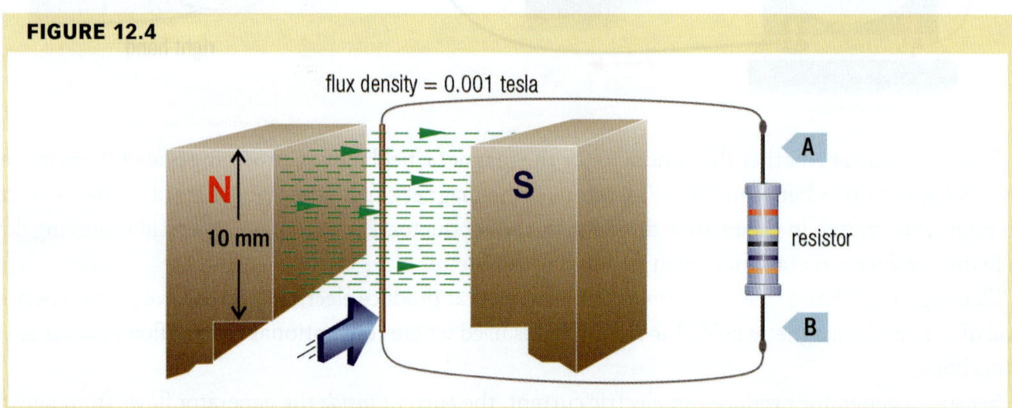

Solution

Values $B = 0.001$ tesla

$l = 10$ mm or 0.01 m (length of conductor exposed to the field)

$v = 10$ m/s

$e = ?$

Equation $e = Blv$

$e = 0.001 \times 0.01 \times 10 = 0.0001$ V

Answer $\mathbf{e = 100\ \mu V}$

Using Fleming's right-hand rule, the current induced in the conductor will flow from top to bottom, so point B is positive in relation to point A.

Notice that the length of the conductor in Example 12.1 is taken as the length of the magnetic field (10 mm). Any part of the conductor outside the field has no effect on the induced voltage. To increase the induced voltage, the conductor length can be increased by winding it into a coil.

Figure 12.5 shows the effects of moving a bar magnet into and out of a coil. In (a) and (c) the magnetic flux cuts the coil windings as the magnet is moved, giving a deflection on the meter that reverses with the direction of movement. When the magnet is not moving, as shown in (b), there is no induced voltage. (This arrangement is how Faraday investigated electromagnetic induction.)

FIGURE 12.5 A voltage is induced in a coil only when the magnet is moving

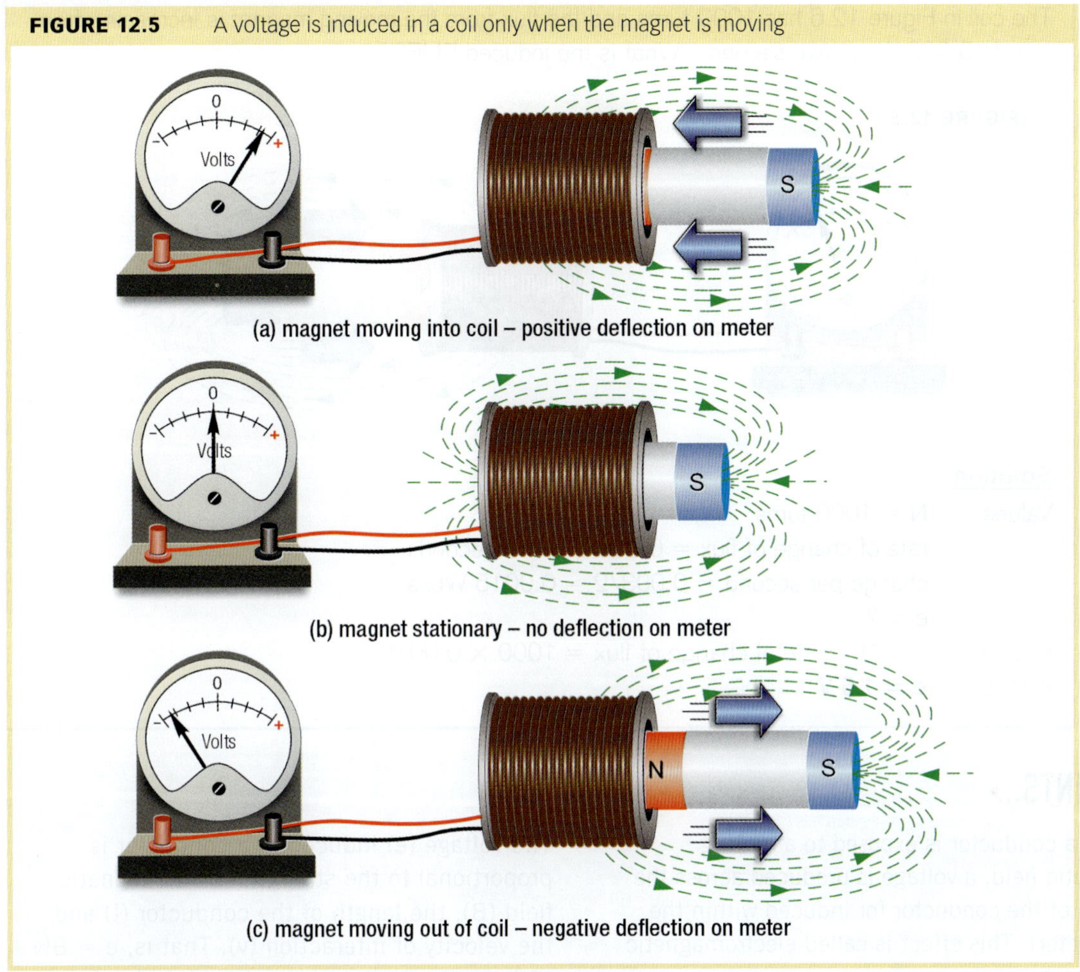

Voltage induced in a coil

A voltage is induced in a conductor only if there is relative movement between the conductor and the magnetic field. When a conductor is wound into a coil, its length is increased, giving, as you'd expect, a higher induced voltage. The voltage induced in a coil depends on the:

- strength of the magnetic field
- number of turns on the coil
- speed of the cutting action between the magnetic field and the conductor.

KEY CONCEPT
Faraday's law

The equation $e = Blv$ includes all these factors, but calculating the length of a conductor that's wound into a coil is rather difficult to do. Another way is to use the equation first described by Faraday, now known as Faraday's law. That is:

$e = N \times$ rate of change of flux

where:

e = induced EMF in volts

N = number of turns on the coil

rate of change of flux = flux in webers cut per second.

Sometimes this equation is written as $e = N\frac{\Delta\Phi}{\Delta t}$ where $\frac{\Delta\Phi}{\Delta t}$ equals the rate of change of flux. The Δ symbol (Greek letter *delta*) stands for 'change in'.

From this equation, we can say that the value of the voltage induced in a coil is determined by the number of conductors in the magnetic field and the rate of change of the flux cutting the conductors.

EXAMPLE 12.2

The coil in Figure 12.6 has 1000 turns and the flux from the moving magnet is increasing from zero to 0.003 Wb in two seconds. What is the induced EMF?

FIGURE 12.6

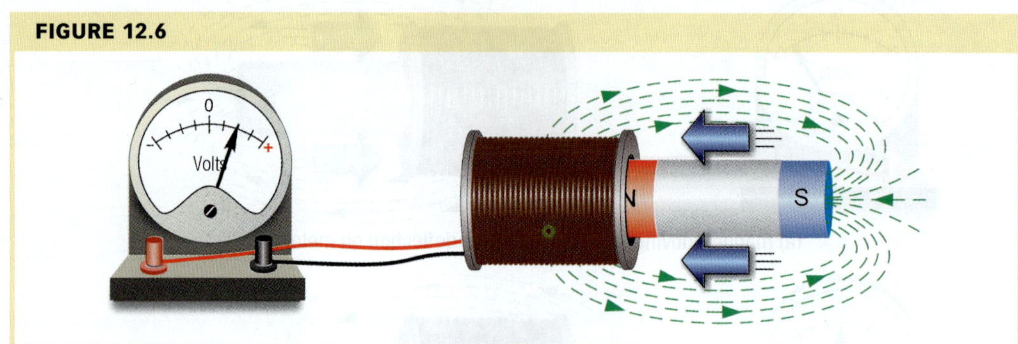

Solution

Values N = 1000 turns
rate of change of flux = 0.003 Wb in 2 seconds
change per second = 0.003/2 = 0.0015 Wb/s
e = ?

Equation e = N × rate of change of flux = 1000 × 0.0015

Answer **e = 1.5 V**

KEY POINTS...

- When a conductor is exposed to a moving magnetic field, a voltage is produced across the length of the conductor (or induced within the conductor). This effect is called electromagnetic induction.
- The polarity of the induced voltage depends on the direction of relative motion and the direction of the magnetic field.
- The direction of the induced current caused by the induced voltage can be found with Fleming's right-hand rule for a conductor.
- The voltage (e) induced in a conductor is proportional to the strength of the magnetic field (B), the length of the conductor (l) and the velocity of interaction (v). That is, e = Blv (B in teslas, l in metres, v in metres/sec).
- The voltage (e) induced in a coil is proportional to the number of turns (N) and the rate of change of the magnetic flux. That is, $e = N\frac{\Delta\Phi}{\Delta t}$

12.3 Lenz's Law

While Faraday was investigating the interaction between magnetism and an electric current, so too was German physicist Heinrich Lenz (1804–1865). As a result he came up with a most important law, now called Lenz's law, which says:

KEY CONCEPT
Lenz's law

- the current induced in a conductor will set up a magnetic field that opposes the magnetic field causing the current.

This basic electrical law is like Newton's third law of motion which says: *Every action has an equal and opposite reaction*. To explain, consider a conductor moving in a magnetic field, as shown in Figure 12.7. In this diagram, the conductor is connected to a milliammeter and is moving downwards through the magnetic field of a permanent magnet. From Fleming's right-hand rule, we can determine that the current induced in the conductor will flow in the direction shown.

The right-hand rule for conductors gives the direction of the magnetic field around the conductor, which is anticlockwise because the current is coming towards you. Because the lines of force around the conductor are in the same direction as those of the permanent magnet, the fields repel each other. The fields therefore try to push the conductor up, against the force that's pushing it down.

FIGURE 12.7 (a) Fleming's right-hand rule gives the direction of the induced current, which in (b) is flowing towards you, causing an anticlockwise field around the conductor

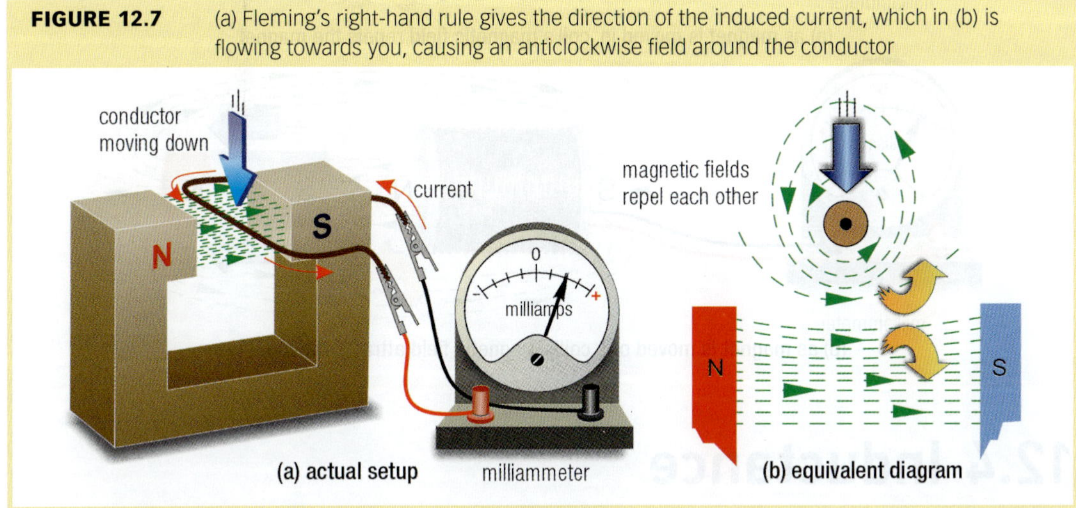

The magnetic field developed around the conductor is due to the current induced in the conductor as it moves. A current can only flow if there is a complete circuit, which is the milliammeter in Figure 12.7. Opposition to the movement of the conductor only occurs when current is flowing in the conductor, so without a complete circuit there is no opposition. Instead, a voltage will be induced across the conductor.

When a magnet is moved towards a stationary coil, the effective length of the conductor is far greater than a single loop, giving higher values of induced voltage and current. In Figure 12.8(a), the moving magnetic field induces a voltage in the coil that causes a current to flow in the circuit. This current sets up a magnetic field around the coil that opposes the movement of the magnet creating the induced current in the first place.

When the magnet is withdrawn from the coil, the current induced in the coil will change direction because the direction of relative motion has changed. The magnetic field set up by the coil now changes direction and attracts the magnet, opposing its withdrawal.

In Figure 12.8, the motion of the magnet is opposed by the magnetic field set up in the coil by the induced current. In (a), as the south pole of the magnet is moved towards the coil, the induced current in the coil develops a south pole that opposes the magnet's south pole. In (b), the reverse happens, in which the coil develops a north pole as the magnet is withdrawn, attracting the magnet's south pole.

FIGURE 12.8 Lenz's law – the coil's magnetic field opposes the movement of the magnet

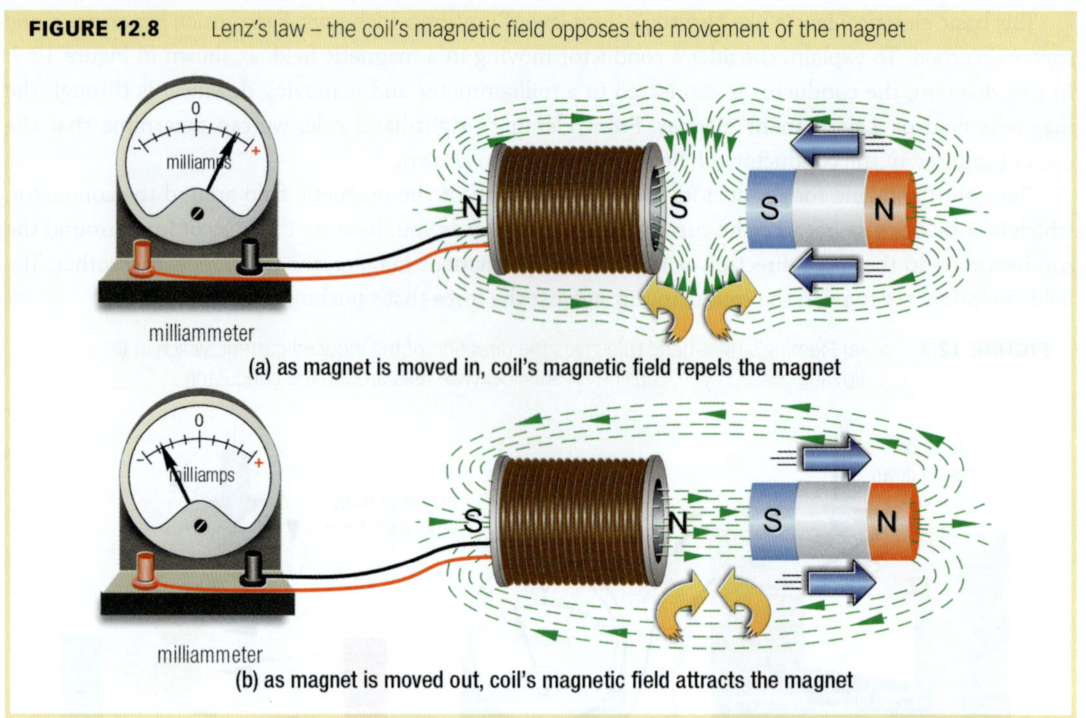

12.4 Inductance

A moving magnetic flux induces a voltage in a conductor (or a coil), and an induced current will flow if there's a complete circuit. When current starts flowing, it causes a magnetic field around the conductor to build up. The size of the field depends on the value of the current. Figure 12.9(a) shows how the magnetic field looks not long after current has started to flow, and in (b), when the field is fully established.

FIGURE 12.9 The magnetic field set up by current flowing in a conductor builds up from the centre and opposes the current flowing in the conductor

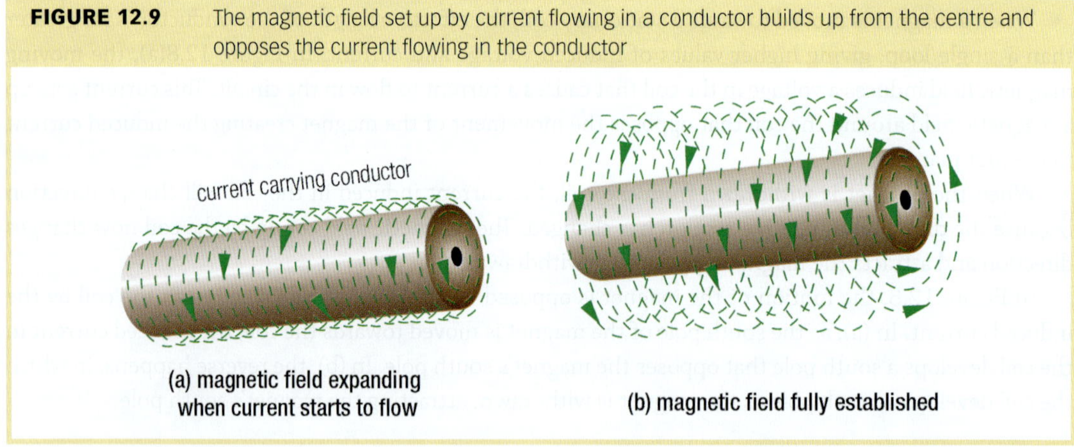

As the magnetic field expands from the centre it cuts, or intersects, the conductor. This continues until the magnetic field is fully established. Because a moving magnetic field induces a voltage in a conductor, there must be a voltage induced in the very conductor producing the magnetic field in the first place.

Lenz's law says that this voltage will oppose the current producing the magnetic field. In other words, the act of the current starting to flow is itself opposed. This opposition is called *inductance*.

The same thing happens when the current is switched off. Here the magnetic field around the conductor collapses, inducing an EMF that will try to maintain the original current. Because the induced EMF opposes the current, it's often called a back-EMF.

Therefore, inductance is a property that opposes a *change* in current. Inductance is present to some extent in all components – even conductors.

KEY POINTS...

- Inductance is the property of a circuit, component or conductor that opposes a change in the value of an electric current.

Winding a conductor into a coil increases its inductance because there will be a higher value of induced EMF to oppose a change in current. If the coil has an iron core, the magnetic field is further concentrated and the induced EMF is even higher. The inductance of such a coil is therefore very much higher than a single conductor. In fact, you can usually ignore the inductance of a single conductor. An exception is power transmission cables, as their great length can result in an inductance that has to be considered.

The effect of the induced EMF in a coil is shown in Figure 12.10. When the switch is first closed, current starts to flow in the circuit, producing a magnetic field that expands as the current increases. As the field expands, it induces an EMF in the coil that opposes a further increase in the value of the current. Because the coil is acting as a generator, the induced EMF is positive at the end of the coil connected to the positive side of the DC supply. The potential difference to cause current flow is the supply voltage minus the induced voltage.

FIGURE 12.10 A coil produces a back-EMF while the magnetic field builds up, which opposes the current in the coil

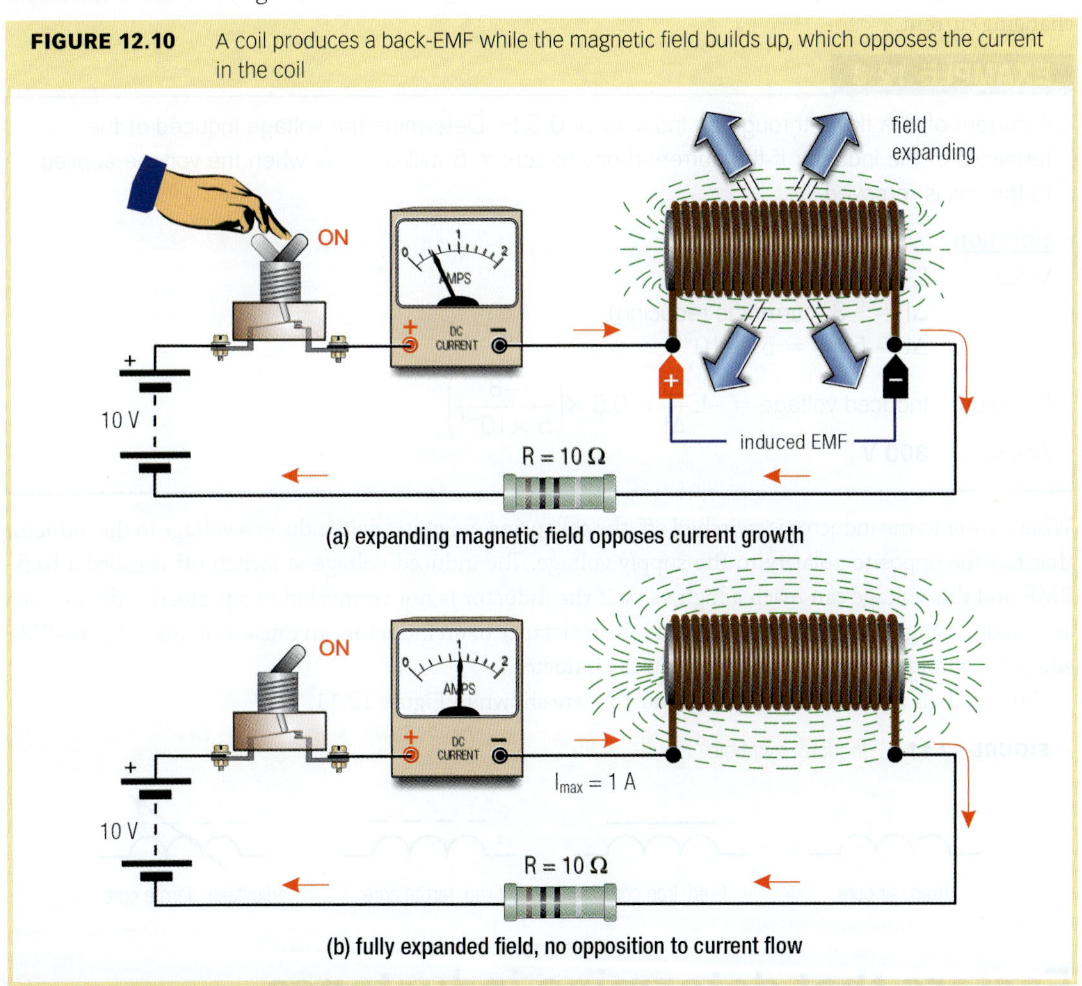

The magnetic field continues to expand, but more slowly as it reaches its maximum. The back-EMF reduces and finally falls to zero when the field is stationary. The current in the circuit is now limited only by the resistance of the circuit. It can take several seconds or more for the current to reach its maximum, depending on the resistance and the inductance in the circuit.

KEY CONCEPT

The unit of inductance is the henry

The inductor

A coil has quite a lot of inductance, so coils are often called *inductors*, depending on the use. For example, although an electromagnet has inductance, it's not usually called an inductor because its purpose is only to produce a magnetic field. Inductors used because of their inductance come in a range of shapes and sizes, and are often referred to as a choke or ballast. The unit of inductance is the *henry*, where:

- an inductance of one henry causes an induced voltage of one volt when the current is changing at a rate of one ampere per second.

The value of the voltage induced in an inductor caused by a change in current flowing in the inductor can be determined with the equation:

Induced voltage $= -L\frac{\Delta I}{\Delta t}$ volts, where:

L = inductance in henrys
ΔI = change in current in amperes
Δt = change in time in seconds.

The minus sign in the equation indicates the induced voltage will have a polarity that acts against the changing current.

EXAMPLE 12.3

A current of 3 A flows through an inductor of 0.5 H. Determine the voltage induced at the terminals of the inductor if the current drops to zero in 5 milliseconds when the voltage applied to the coil is removed.

Solution

Values $L = 0.5$ H
$\Delta I = -3$ (current is reducing)
$\Delta t = 5$ ms $= 5 \times 10^{-3}$ seconds

Equation Induced voltage $= -L\frac{\Delta I}{\Delta t} = 0.5 \times \left(\frac{-3}{5 \times 10^{-3}}\right)$

Answer **300 V**

When power to the inductor is switched off, the collapsing magnetic field induces a voltage in the inductor that has the opposite polarity to the supply voltage. The induced voltage at switch-off is called a back-EMF, and this voltage can reach a high value if the inductor is not connected to a protective device such as a diode. Using an ohmmeter to measure the resistance of an inductor can cause a significant back-EMF when the ohmmeter is disconnected from the inductor.

The symbols for different types of inductors are shown in Figure 12.11.

FIGURE 12.11 Inductor symbols

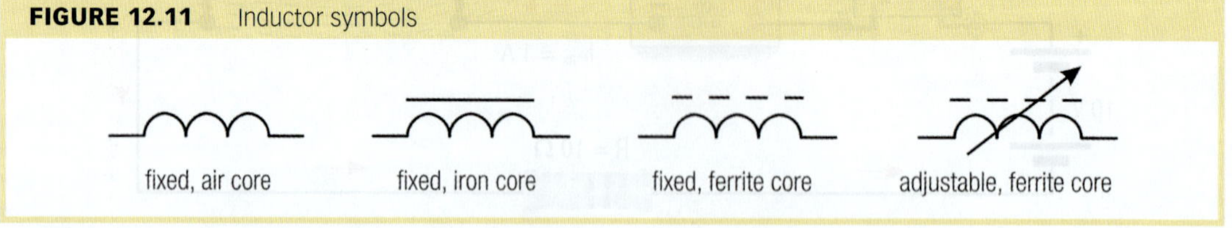

Factors that determine inductance

The inductance of a coil is determined by four factors:

1 number of turns
2 length of the coil
3 cross-sectional area of the coil
4 type of core.

The higher the number of turns, the higher the inductance. This is because the magnetic field produced by the coil is increased if more turns are added. The stronger the magnetic field, the higher the back-EMF, as there are more magnetic lines of force to produce a back-EMF. This is illustrated in Figure 12.12. In fact, the inductance of a coil is proportional to the *square* of the number of turns.

FIGURE 12.12 Inductance is proportional to the square of the turns, so doubling the number of turns gives four times the inductance

2 A
11 turns
less turns, less inductance
2 A
22 turns
double the turns, four times the inductance

For the same number of turns, a long coil has less inductance than a short coil. Figure 12.13 shows two coils of different lengths, but with the same number of turns, type of core and cross-sectional area (csa). The longer coil has less inductance because not as much flux combines to form the magnetic field around the coil. When the turns are packed more closely, as in the shorter coil, the magnetic field is concentrated into a smaller area, giving a higher flux density. The inductance of a coil is therefore *inversely* proportional to the length of the coil, for the same number of turns.

FIGURE 12.13 For the same number of turns, inductance is inversely proportional to the length of a coil

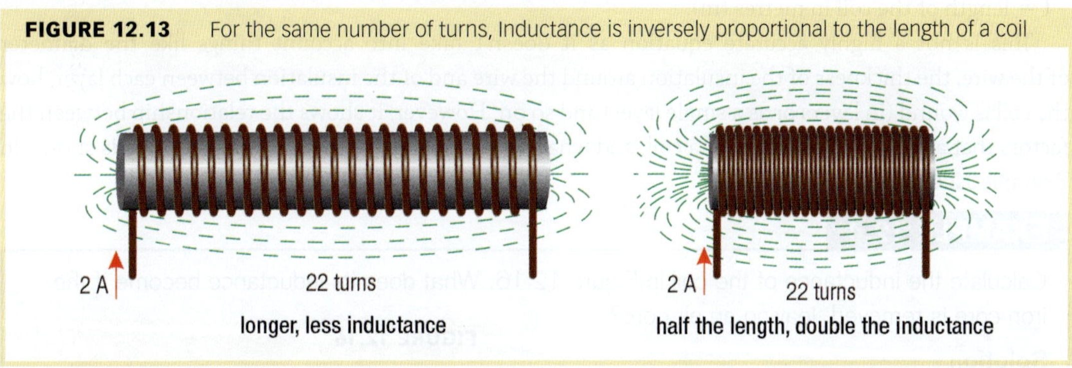

The larger the cross-sectional area (csa) of a coil, the higher its inductance. While coils can be any shape they are usually cylindrical, so the csa of a coil is determined by its diameter. (The larger the diameter, the larger the csa.)

Figure 12.14 shows two coils identical except for their csa. The coil with the larger csa has more wire wound around it because its diameter is greater than that of the smaller coil. Therefore, a stronger magnetic field is produced by the coil with the larger area, giving it a higher inductance. That is, inductance is proportional to the cross-sectional area of the coil, for the same number of turns.

FIGURE 12.14 Inductance is directly proportional to the area of a coil

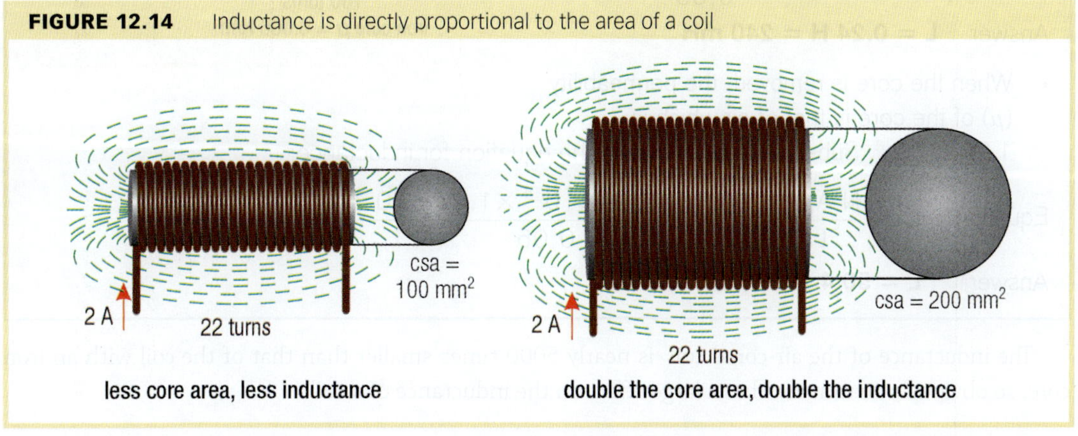

To get a stronger magnetic field, most inductors have a ferromagnetic core. So because there's a stronger magnetic field: the higher the permeability of the core material, the higher the inductance of the coil.

FIGURE 12.15 Inductance is proportional to the permeability of the core

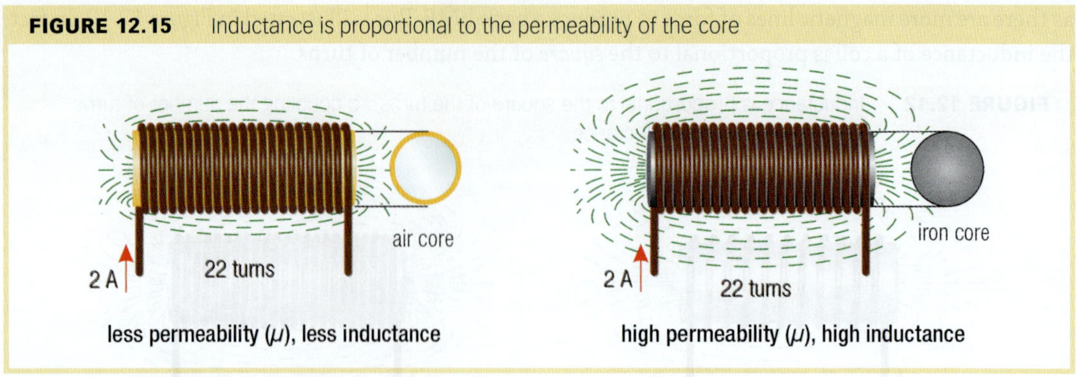

These four factors are combined into an equation to find the inductance of a coil:

$$L = \frac{N^2 \times A \times \mu}{l}$$

where:

L = inductance in henrys

N = number of turns

A = cross-sectional area in square metres (m^2)

μ = permeability of the core ($= 1.26 \times 10^{-6} \times \mu_r$)

l = length of the coil in metres (m).

This is not a highly accurate equation as it doesn't take into account things like the diameter of the wire, the thickness of the insulation around the wire and of the insulation between each layer, how the coil is wound (in layers or as a single layer) and so on. However, it shows the relationship between the factors that affect the inductance of a coil, and what happens if one of them is changed. This is shown in Example 12.4.

EXAMPLE 12.4

Calculate the inductance of the coil in Figure 12.16. What does its inductance become if the iron core is removed, leaving an air core?

FIGURE 12.16

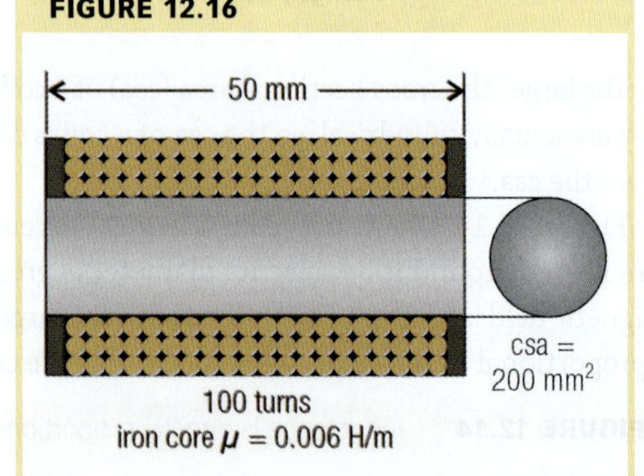

Solution

Values $N = 100$ turns

$A = 200\ mm^2 = 200 \times 10^{-6}\ m^2$

$l = 5\ cm = 0.05\ m$

$\mu = 0.006\ H/m$

Equation

$$L = \frac{N^2 A \mu}{l} = \frac{100^2 \times 200 \times 10^{-6} \times 0.006}{0.05}$$

Answer **$L = 0.24\ H = 240\ mH$**

- When the core is removed, the permeability (μ) of the core is that of air which is 1.26×10^{-6}. Substituting this value in the equation for inductance:

Equation $$L = \frac{N^2 A \mu}{l} = \frac{100^2 \times 200 \times 10^{-6} \times 1.26 \times 10^{-6}}{0.05}$$

Answer **$L = 50.4\ \mu H$**

The inductance of the air-cored coil is nearly 5000 times smaller than that of the coil with an iron core, so obviously an iron core has a large effect on the inductance of a coil.

Inductor cores

There are three main types of core materials used with inductors:

1 air (or other non-magnetic materials)

2 soft iron

3 ferrite.

An inductor without a ferromagnetic core is assumed to have an air core, and will have a low inductance, usually measured in microhenrys. Examples are shown in Figure 12.17(b). Air-cored coils are used mainly in communications equipment and where high frequency signals are involved. Large air-cored coils are used in transmitters where they might handle many kilowatts of power. Small air-cored coils are usually wound on a former made of ceramic, phenolic, plastic or other non-magnetic materials.

Iron-cored inductors are typically used in low-frequency applications, such as mains power (50 hertz) and audio frequencies. As explained in Chapter 11, soft silicon steel laminations are used as the core in these types of inductors. Two examples are shown in Figure 12.17(a), in which their inductance is measured in henrys.

Ferrite cores are used in inductors that operate at relatively high frequencies. These range in shape, and include toroids, pie wound, encapsulated and adjustable types. They range in inductance from millihenrys to several henrys. Typical examples are shown in Figure 12.17(c).

FIGURE 12.17 Low and medium power inductors, typically used in electronic equipment

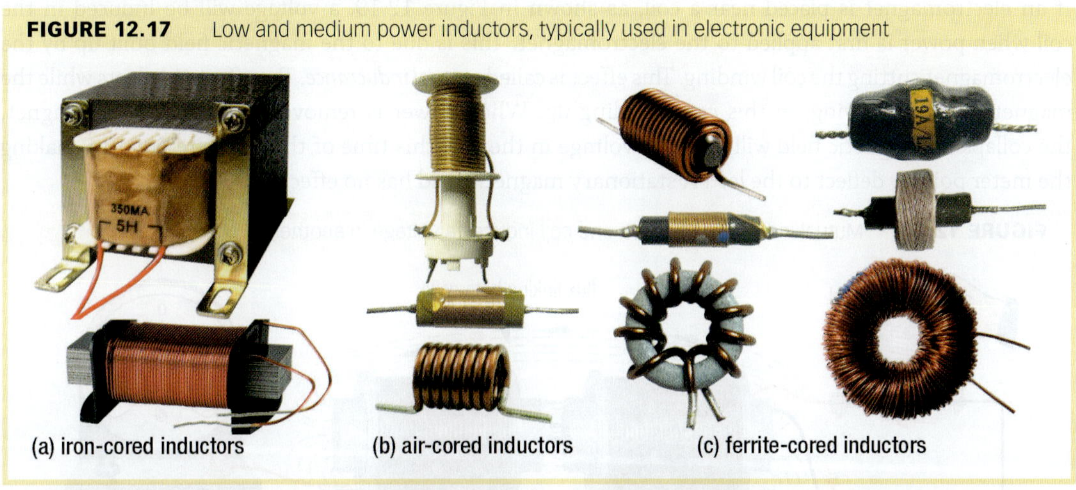

Inductors with high current ratings have numerous applications in the electrical power industry. The blue three-phase filter inductor shown in Figure 12.18 filters out the signal tones that are sent over power lines, allowing the 50 Hz current to pass but preventing the higher frequency signal tones. This device is rated at 11 kV/150 A per phase and has an inductance (per phase) of 11.58 mH. The in-rush inductor in Figure 12.18 is in series with a capacitor bank and limits the surge current when power is first applied. Lamp ballasts are used to limit the current flowing in high-intensity discharge (HID) lamps, such as mercury vapour, fluorescent and sodium lamps.

FIGURE 12.18 Power inductors are used in the power industry, or in power applications such as HID lamps

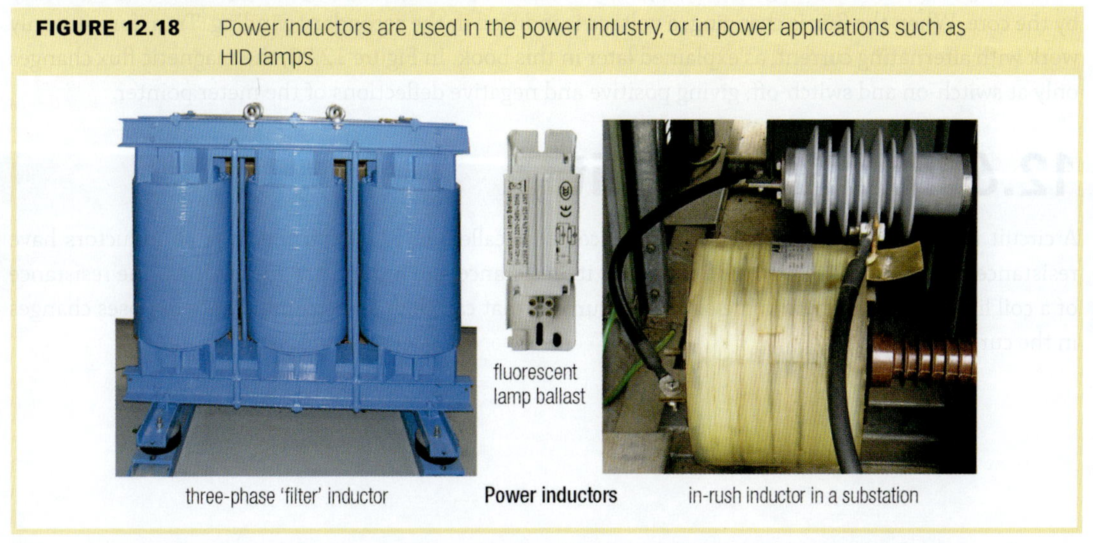

KEY POINTS...

- Lenz's law states that the current induced in a conductor will set up a magnetic field that opposes the magnetic field causing the current. This opposition is called inductance.
- Inductance is measured in henrys (H). A coil has an inductance of 1 H if a voltage of 1 V is induced in the coil when the current in the coil is changing by 1 A per second.
- The inductance (L) of a coil is proportional to the square of the turns (N^2), the cross-sectional area (A), the permeability of the core (μ), and is inversely proportional to the length (l) of the core. That is: $L = \frac{N^2 \times A \times \mu}{l}$
- Inductors can have a ferromagnetic or an air core. Inductors used with an alternating current have a laminated core.
- High power handling inductors are used in electrical power substations, whereas low power inductors are used in electronics and telecommunications.

12.5 Mutual Inductance

If an electromagnet is placed near a coil, as shown in Figure 12.19, a voltage will be induced in the coil when power is first applied to the electromagnet. This is due to the magnetic field built up by the electromagnet cutting the coil winding. This effect is called *mutual inductance*. The effect only lasts while the magnetic field is moving, in this case building up. When power is removed from the electromagnet, the collapsing magnetic field will induce a voltage in the coil, this time of the opposite polarity, making the meter pointer deflect to the left. A stationary magnetic field has no effect.

FIGURE 12.19 Mutual inductance – when one coil induces a voltage in another

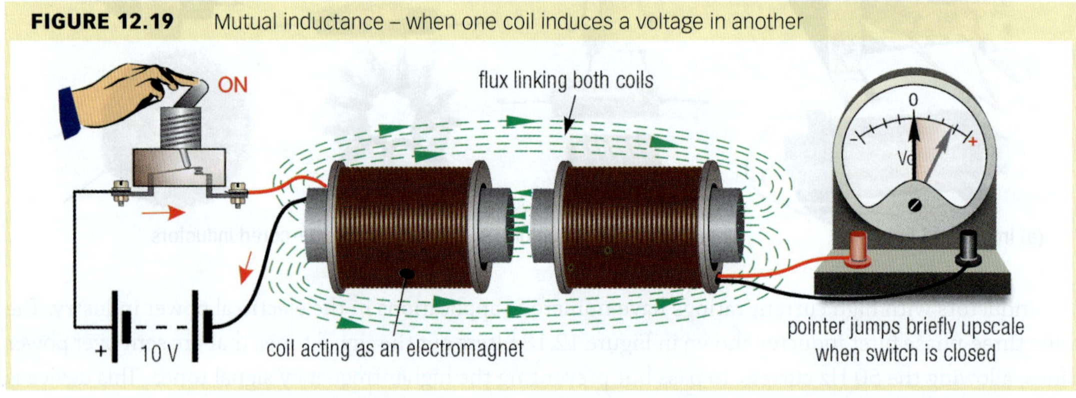

Transformers are explained in Chapter 21. They are used only with an alternating voltage

The effect of mutual inductance can be a problem in some situations, such as power lines inducing voltages in nearby telecommunication cables. However, an essential application of mutual inductance is the transformer, shown in Figure 12.20. Two windings, called the primary and secondary, are wound on a ferromagnetic core. The magnetic field produced by the primary winding is coupled to the secondary by the core. When the flux is changing, a voltage is induced in the secondary winding. Transformers only work with alternating current, as explained later in this book. In Figure 12.20, the magnetic flux changes only at switch-on and switch-off, giving positive and negative deflections of the meter pointer.

12.6 The RL Circuit

A circuit that has resistance (R) and inductance (L) is called an RL circuit. Because all inductors have resistance, an inductor itself is an RL circuit, as its resistance and inductance are in series. The resistance of a coil limits the maximum value of the DC current that can flow, and its inductance opposes changes in the current.

FIGURE 12.20 A transformer works by mutual inductance

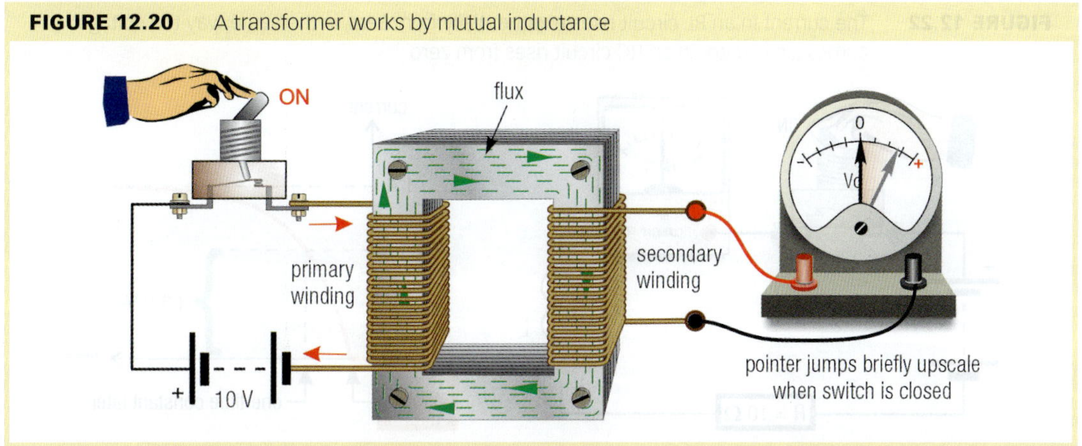

This is illustrated in Figure 12.21 where (a) shows a coil in series with a DC supply, a switch and an ammeter. The equivalent circuit in (b) shows the resistance and the inductance as two separate components connected in series. If the circuit includes another series resistor, its value is added to the resistance of the coil.

Inductance, as previously explained, is *the property of a circuit or component that opposes a change in an electric current* and is caused by the magnetic field produced when a current starts to flow in an inductor. When power is first applied, the field expands, cutting the coil windings and inducing a back-EMF that opposes the current. Eventually the magnetic field stops expanding and the current is then limited only by the resistance of the circuit.

FIGURE 12.21 An RL circuit has inductance and resistance in series, where the resistance is often that of the inductor alone

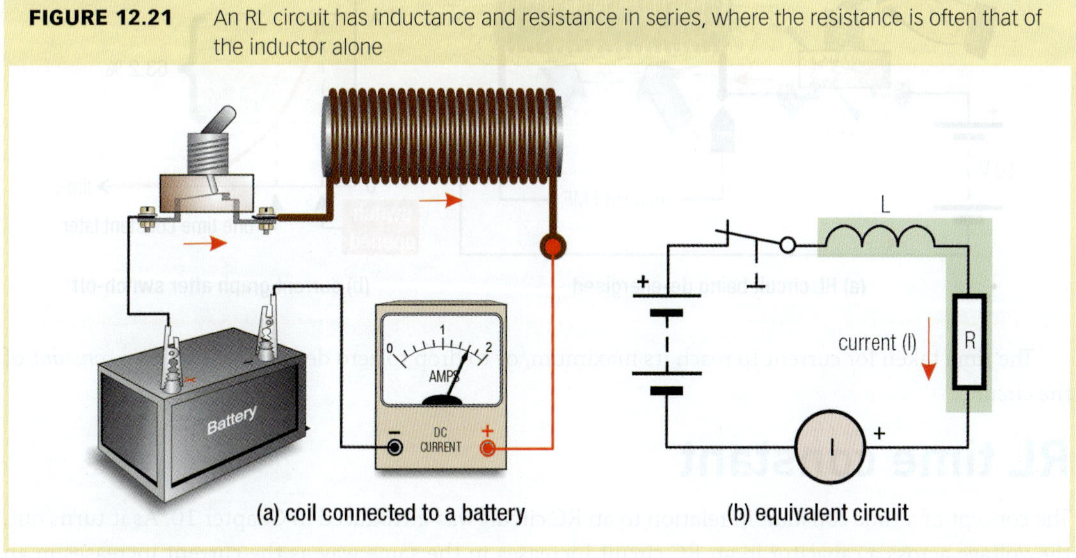

In Figure 12.21, when the switch is closed, current will start flowing, creating a magnetic field around the coil that opposes the increasing current. The final value of the current is determined by the resistance of the inductor. Therefore, the current in an RL circuit takes a certain amount of time to reach its final value, which is shown by the graph in Figure 12.22. The time taken depends on the values of the inductance and resistance in the circuit.

The graph in Figure 12.22 shows that the current rises quickly at first, then ever more slowly until it reaches its maximum value. The maximum value of current (I) is found with Ohm's law, where I equals the applied voltage (V) divided by the resistance (R) of the circuit. For the circuit in Figure 12.22, current equals 10 V divided by 10 Ω, giving 1 A as the maximum current. Once the current reaches its maximum (or steady-state) value, the magnetic field stops expanding, no longer opposing the changing current.

As shown in Figure 12.23, when the switch is opened, the magnetic field collapses, inducing a voltage in the coil that acts to prevent the current from reducing in value. Because the coil is now isolated from the circuit (as the switch is open), the induced voltage has nothing to limit it, allowing it to reach a very high value, typically producing a spark across the switch contacts. The spark is ionised air conducting an electric current, and the sparking will corrode the switch contacts. It is usual to have some sort of protective component in an inductive circuit to prevent sparking and corrosion of the switch contacts.

FIGURE 12.22 The current in an RL circuit rises exponentially from zero, in the same way the voltage across a capacitor in an RC circuit rises from zero

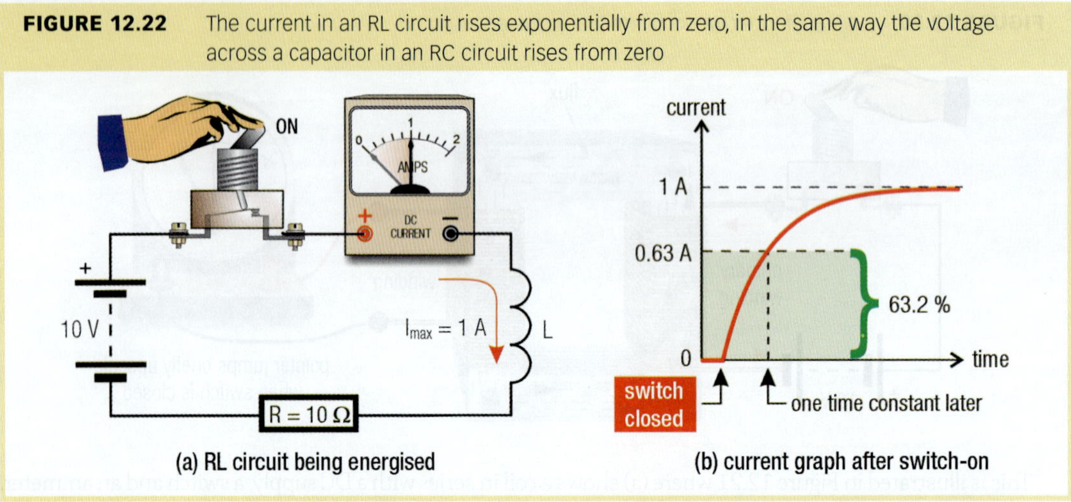

(a) RL circuit being energised

(b) current graph after switch-on

FIGURE 12.23 The back-EMF of a coil causes the current to continue after the supply is disconnected, causing sparking across the opened switch contacts

(a) RL circuit being de-energised

(b) current graph after switch-off

FYI

The ignition system in motor vehicles depends on a coil generating a high value of induced EMF to create a spark

The time taken for current to reach its maximum, or to drop to zero depends on the *time constant* of the circuit.

RL time constant

The concept of a time constant in relation to an RC circuit was introduced in Chapter 10. As it turns out, the voltage across a capacitor in an RC circuit increases in the same way as the current increases in an RL circuit. That is, both circuits have a time constant that determines how long it takes for a current or voltage to reach 63.2 per cent of its final value. As explained in Chapter 10, the symbol for time constant is τ (Greek letter tau – pronounced tor) and the measurement unit is the second. In an RL circuit, the time constant τ is found with:

$$\tau = \frac{L}{R}$$

where:

τ = time constant in seconds

L = inductance in henrys

R = resistance in ohms.

EXAMPLE 12.5

Calculate how long it takes for the current in Figure 12.24 to reach 63.2 per cent of its final value after the switch is closed. That is, find the time constant of the circuit. Also calculate the final value of the current.

FIGURE 12.24 Circuit for Example 12.5

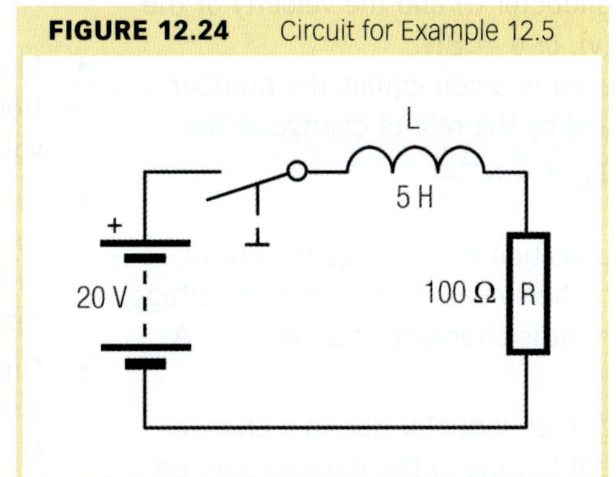

Solution

Values $L = 5\text{ H}$
$R = 100\ \Omega$

Equation 1 $\tau = \dfrac{L}{R} = \dfrac{5}{100}$

Answer 1 $\boldsymbol{\tau = 0.05}$ **seconds**

Equation 2 $I = \dfrac{V}{R} = \dfrac{20}{100}$

Answer 2 $I =$ **0.2 A**

EXAMPLE 12.6

Calculate the time taken for the current in Figure 12.24 to drop by 63.2 per cent when the switch is opened. Assume the resistance of the opening switch is 10 MΩ. That is, find the time constant of the circuit under these conditions.

Solution

Values given $L = 5\text{ H}$
$R = 10\text{ M}\Omega$ (ignore the 100 Ω resistor)

Equation $\tau = \dfrac{L}{R} = \dfrac{5}{100 \times 10^6}$

Answer $\boldsymbol{\tau = 0.5}$ **microseconds (μs)**

These examples show that the *higher* the resistance, the *shorter* the time constant. Because the resistance in series with the inductor is very high when the switch is opened, the back-EMF produced by the coil lasts a very short time. That is, the flux collapses very quickly, inducing a high value of voltage, which can deliver a nasty, even lethal shock. Coils sometimes have a solid-state high-speed diode connected across them to provide a short circuit across the coil to limit the back-EMF. A diode only conducts in one direction, so the diode is connected such that it conducts only during the interval when a back-EMF occurs across the coil. Under normal energised conditions, the diode is an open-circuit.

A high voltage can occur across the terminals of a coil when power is removed, even that supplied by an ohmmeter

It takes around five time constants for the current in an RL circuit to reach its final value. In Example 12.4, it would take about 0.25 seconds before the current reached its final value and the magnetic field around the inductor was fully established.

The universal time constant charts presented in Chapter 10 can be used to find the current in an RL circuit after any time interval. The process is explained for an RC circuit, and the only difference between it and an RL circuit is that the curves refer to current, not voltage.

CHAPTER SUMMARY

- The voltage (e) induced in a conductor by a moving magnetic field equals the product of the flux density (B), length of the conductor (l) and the velocity of the relative movement (v), or e = Blv.
- The voltage (e) induced in a coil equals the number of turns (N) multiplied by the rate of change of the magnetic flux That is, $e = N\frac{\Delta\Phi}{\Delta t}$
- Inductance (L) is opposition to a change in current. An inductance of one henry causes an induced voltage of 1 V when the current is changing at a rate of 1 A per second.
- The voltage induced in an inductor due to a change in the value of current flowing in the inductor can be found with the equation: Induced voltage $= -L\frac{\Delta I}{\Delta t}$
- The inductance (L) of a coil can be found with the equation $L = \frac{N^2 \times A \times \mu}{l}$
- Mutual inductance occurs when the magnetic field from one coil interacts with another coil, inducing a voltage in that coil.
- In an RL circuit, the time taken for the current to reach its maximum value is proportional to the inductance L, and inversely proportional to the resistance R in the circuit.
- One time constant, by definition, is the time taken for a value to change by 63.2 per cent. The time constant (τ) in an LR circuit equals inductance L divided by resistance R. That is, τ = L/R seconds.
- A high voltage (back-EMF) is induced in a coil when it is de-energised. It is prevented by connecting a diode across the coil terminals.

REVIEW EXERCISES

Check your answers at the back of the book.

1 Determine the direction (clockwise or anticlockwise) of the induced current in Figure 12.25.

2 What is the polarity of point A with respect to point B in Figure 12.25?

3 The conductor in Figure 12.25 is moving at 45 metres per second, the flux density is 1 mT, and the effective conductor length is 40 mm. Calculate the value of the induced voltage (e) in the conductor.

4 Calculate the value of the induced EMF in a coil of 2200 turns that is exposed to a magnetic flux that is increasing at a rate of 0.1 Wb every five seconds.

FIGURE 12.25

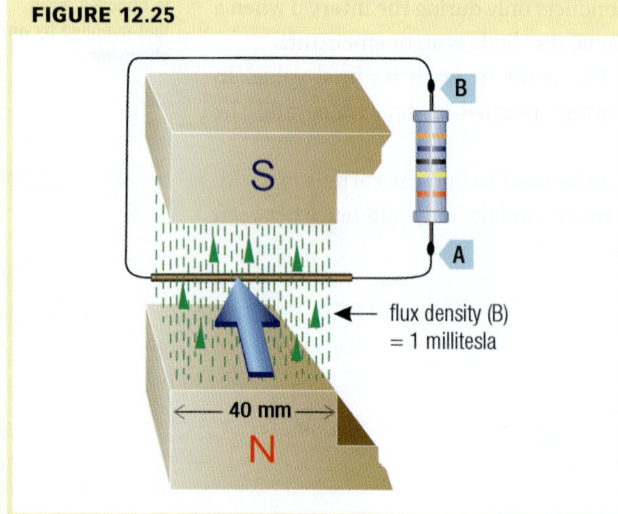

5 Determine the voltage induced at the terminals of a 100 mH inductor if the current rises from zero to 6 A in 10 milliseconds.

6 Calculate the inductance of a 40 mm long coil that has 250 turns wound on an air-filled core with a diameter of 10 mm.

7 Determine the inductance of the coil in Question 6 if the air core is replaced with a core of soft iron with a permeability (μ) of 0.006 H/m.

8 For the circuit in Figure 12.26, determine the:

a steady-state current

b time constant.

9 When the switch in Figure 12.26 is set to position B, in what direction does the current flow before it falls to zero?

10 The field coil of a DC generator has a resistance of 8 Ω and a time constant of 0.3 seconds. Calculate the inductance of the field coil.

FIGURE 12.26

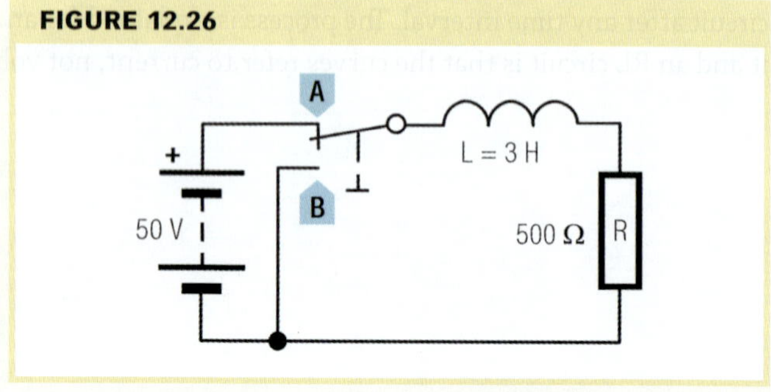

ONLINE RESOURCES

COMPLETE WORKSHEET TWELVE

Check with your instructor for worksheets on this chapter.

CHAPTER 13

DC GENERATORS

Although Faraday discovered the principles of electromagnetic induction in 1831, it took a further 40 years before a commercially viable DC generator was developed by Gramme, later improved by Siemens and others. The invention of the electric light meant larger and more powerful generators were needed, and by 1900 the design of the DC generator was much as we see it today. This chapter describes the operating principles and construction of various types of DC generators. We also look at how each type performs when providing power to a load, along with their typical applications and limitations.

CHAPTER OUTLINE

13.1 Introduction

AC power is the most common form of electrical power, and is produced by machines called alternators. DC power is used in electric traction networks, in which AC power is usually converted to DC by solid-state devices. Car alternators convert AC to DC with solid-state diodes. While the development of solid-state converters, devices and diodes has reduced the need for traditional DC generators, there are numerous situations where these are still used. The main difference between a traditional DC generator and an alternator with solid-state rectifiers is the use of a *commutator* to convert AC to DC.

A DC generator has an almost identical construction to a DC motor, and both are called DC *machines*. In principle, a DC generator converts mechanical energy into electrical energy. The mechanical energy is provided by a prime mover that rotates the generator shaft. Prime movers include diesel or petrol engines, hydro-power, a steam turbine or any form of mechanical energy that can rotate the generator shaft. The amount of electrical power produced by a DC generator depends on the power of the prime mover, and the design and size of the generator.

FIGURE 13.1 A DC generator set, powered by a diesel prime mover

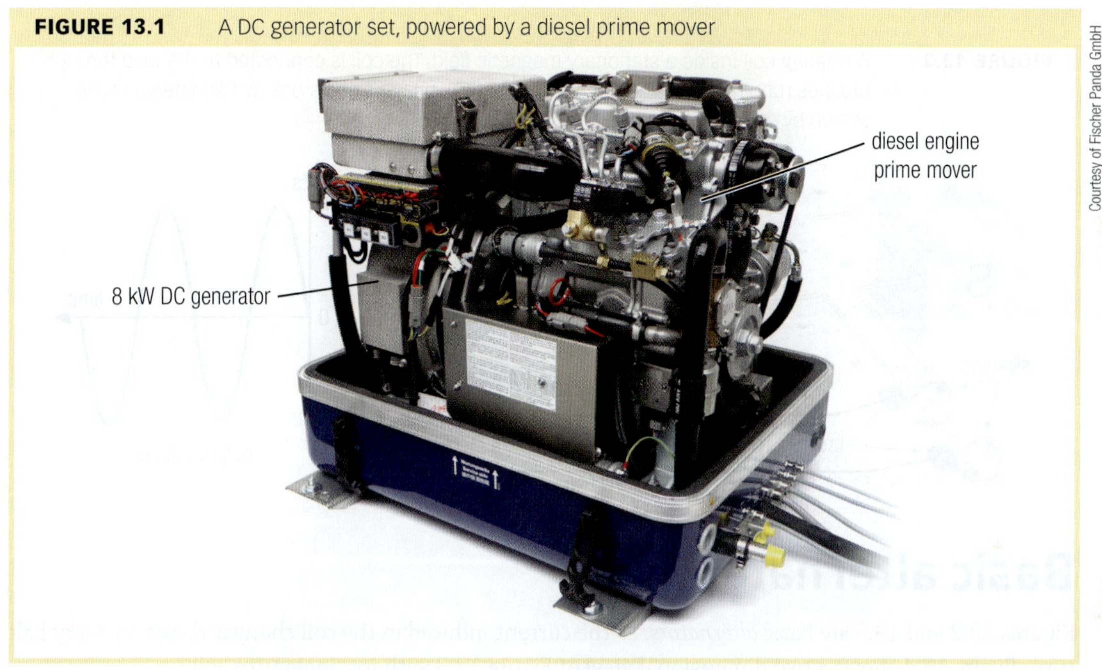

13.2 DC Generator Operating Principles

As explained in the previous chapter, voltage is induced in a coil when the coil is exposed to a moving magnetic field; for example, by moving a bar magnet in and out of a coil. If the coil forms part of a closed circuit, current will flow in a direction that depends on the polarity of the magnetic field and the direction of relative motion. A basic AC alternator is therefore an arrangement of a coil (or coils) and a magnetic field, so that rotary action of one or the other induces a voltage in the coil. A simple setup is shown in Figure 13.2, in which a magnet is arranged so it can rotate inside a coil of wire.

A more usual arrangement in larger machines is to have a stationary magnetic field, with the coil being rotated. Because it's rotating, electrical contact to the coil is made with slip rings and brushes that rub against the slip rings. Figure 13.3 shows a simplified diagram, in which the coil rotates while the magnetic field is stationary. The output voltage is a sinewave, as shown in this diagram, in which the voltage reverses polarity once per revolution. The alternator in Figure 13.2 produces the same output voltage.

FIGURE 13.2 A rotating magnet arranged to induce a current in a stationary coil

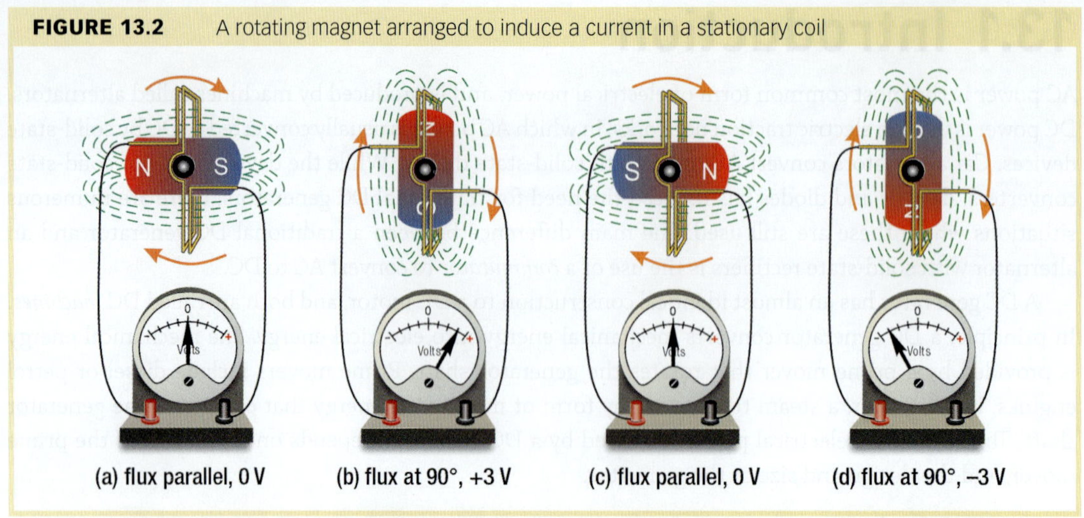

FIGURE 13.3 A rotating coil inside a stationary magnetic field. The coil is connected to the load through brushes rubbing against slip rings. The output voltage swings positive and negative, as shown by the waveform.

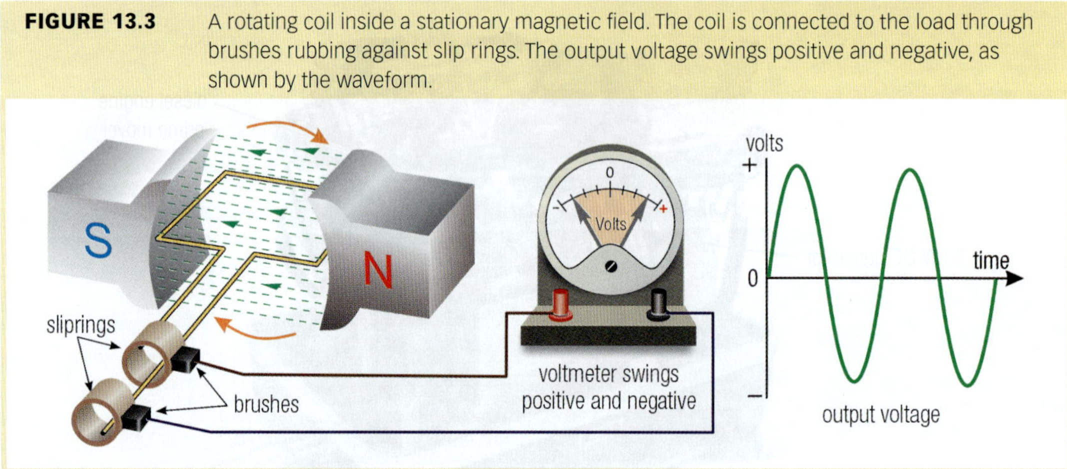

Basic alternator

Figures 13.2 and 13.3 are basic *alternators*, as the current induced in the coil changes direction every half turn. Figure 13.4 shows a two dimensional view of Figure 13.3 with its single turn coil.

FIGURE 13.4 When a coil rotates in a magnetic field, the voltage induced in the coil changes polarity

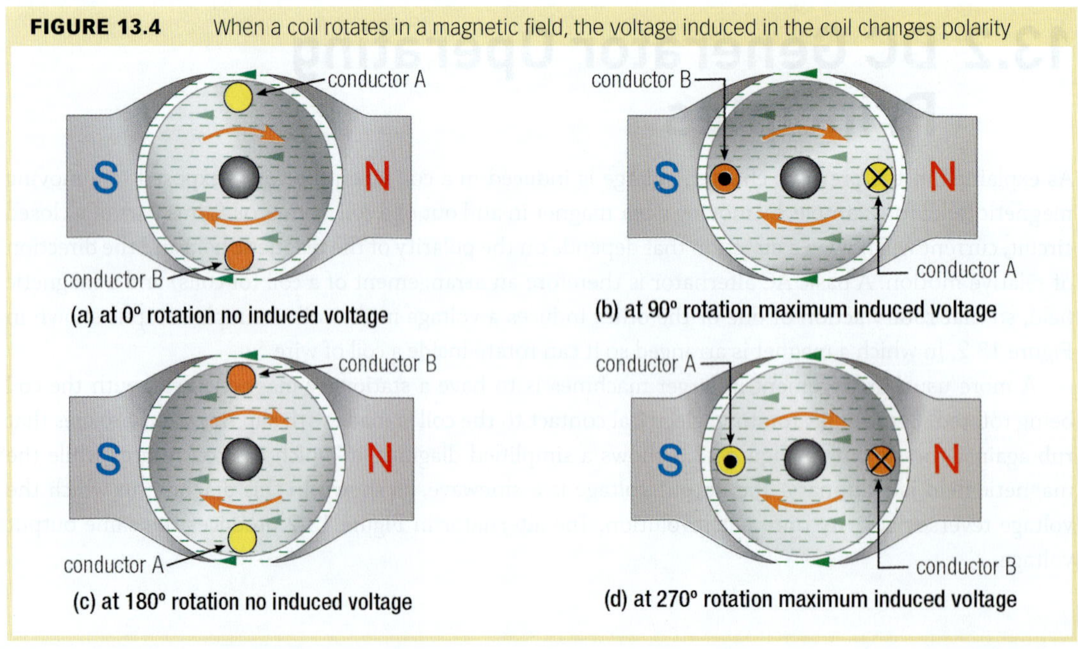

In Figure 13.4(a) the conductors are moving parallel to the magnetic flux, so there is no cutting action between them, and therefore no induced voltage in either conductor. After 90° of rotation, as in (b), the conductors are cutting the flux at right angles, giving maximum induced voltage. The direction of the current (by Fleming's right-hand rule) for conductor A is into the page, and in conductor B the current direction is towards you.

After 180°, as shown in (c), the conductors are again moving parallel to the flux (intersecting at 0°), so there is no induced voltage. Then, as shown in (d), after the coil has rotated by 270°, the conductors are again cutting the flux at right angles, giving maximum induced voltage. This time, the direction of the current in conductors A and B is the opposite of what it was at 90°, so the induced voltage at 270° has the opposite polarity to the voltage produced at 90°. This sequence is also shown in Figures 13.2 and 13.3, in which an alternating voltage (swings positive and negative) is produced each full turn of the magnet or the coil.

Basic generator

To convert the simple alternator in Figure 13.3 into a DC generator requires adding a way of switching the ends of the coil so the load current always flows in the same direction. This can be done as shown in Figure 13.5, in which the slip rings are replaced by a commutator.

FIGURE 13.5 The commutator in this simplified drawing switches the coil connections every half turn, so the load current always flows in the same direction

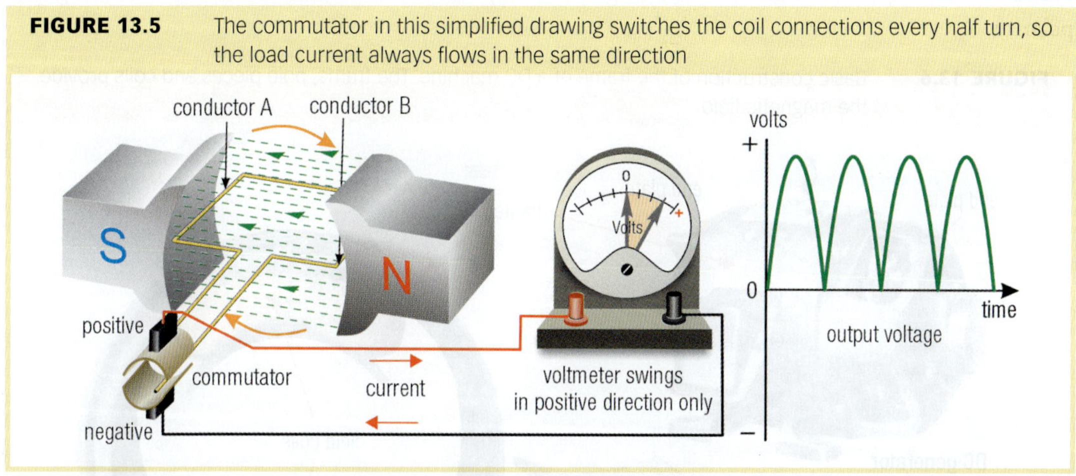

In Figure 13.5, conductor A is passing the south pole of the magnetic field. By Fleming's right-hand rule, the current flows in the direction shown, making the top segment of the commutator the positive terminal. When conductor B passes the south pole, its commutator segment is now at the top, and current flows in the same direction as when conductor A passed the south pole. Therefore, the generated current always flows in the same direction, but varying from zero to maximum positive twice per revolution of the coil, as shown by the voltage waveform.

As explained in Chapter 12, the voltage induced in a conductor depends on its length (l), speed of interaction between the conductor and the magnetic field (v), the strength of the magnetic field (B) and the angle at which the conductor passes through the magnetic field. At this stage, we are assuming the angle is either 0° (no voltage is induced) or 90° (maximum voltage is induced). Combining the first three factors gives the equation to find the induced voltage in a conductor:

$$e = Blv$$

where:

e = induced EMF in volts

B = flux density of the magnetic field in teslas

l = length of the conductor moving at right angles to the field, in metres

v = velocity of the conductor in metres per second.

As this equation shows, the longer the conductor, the greater the induced voltage. A practical generator has *coils* of wire to increase the length of the conductor. The strength of the magnetic field and the speed of rotation of a generator also contribute to the value of the output voltage. All of these factors are considered in the design and construction of a DC generator.

KEY POINTS...

- A DC motor and a DC generator have the same construction, and are referred to as DC machines.
- A simple alternator has a coil of wire rotating in a magnetic field, with the coil connected to the load by way of slip rings and brushes.
- An alternator produces a current in the load that periodically changes polarity, called an alternating current (AC).
- A DC generator has a commutator, a device that switches the coils as they rotate so current in the load is always in the same direction (DC).

13.3 DC Machine Construction

A DC machine (motor or generator) has stationary (stator) parts built into a frame and a rotary part, called the armature. Figure 13.6 shows the frame of a typical two-pole DC machine. The pole pieces are constructed separately from steel laminations, and after assembly are bolted to the frame. To create the magnetic field, coils are wound around the pole pieces. When supplied with an electric current, the pole pieces become magnetised, one with a south magnetic polarity, the other with a north polarity. Some machines have four, six or eight poles, arranged so each pole has an alternate north and south magnetic polarity.

FIGURE 13.6 Basic construction of the frame of a DC machine. The frame, pole pieces and coils provide the magnetic field.

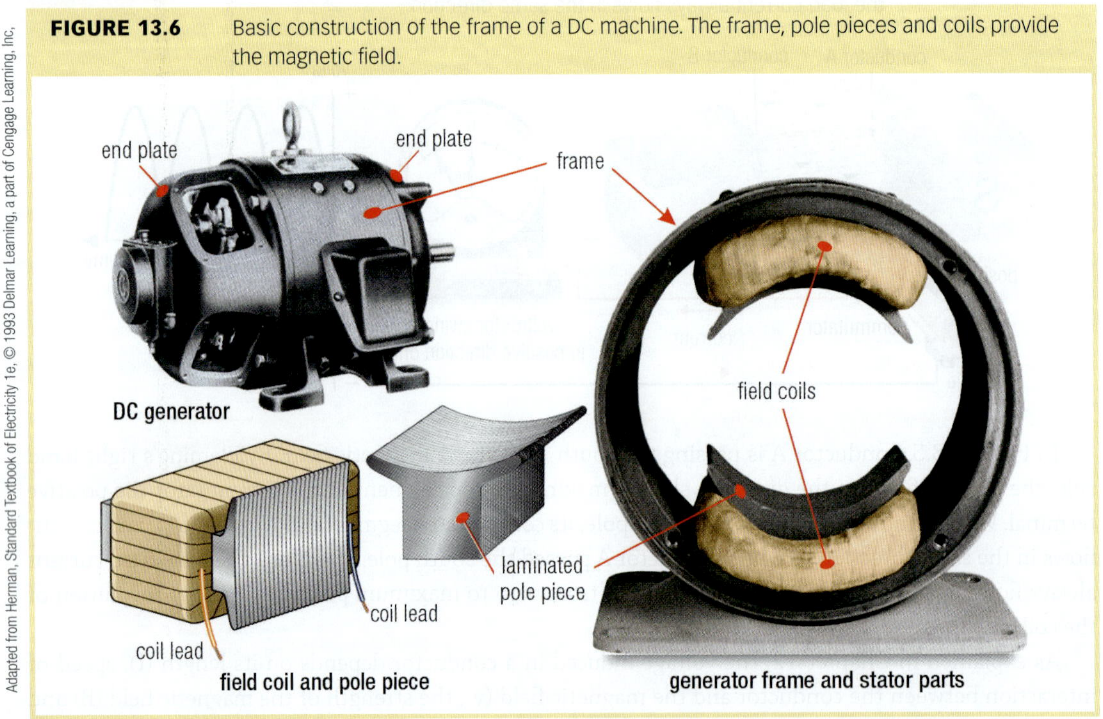

Stator parts

The machine frame is made of a high permeability metal, such as cast steel, which gives a high mechanical strength and also a good magnetic path between the poles. Large machines sometimes have a laminated frame, which improves the frame's magnetic qualities. Motor and generator frames are usually built to standard sizes. They are designed to carry the required magnetic flux and to also support the end shields, pole pieces, feet and other attachments.

Pole pieces support the field coils, and are made of steel laminations (between 1 to 1.5 mm thick) that are riveted together. A pole piece has the same length as the core of the armature and has rounded edges to give best fit with the field coil. Field coils are discussed further in this chapter, but they always consist of a number of turns of insulated copper wire wrapped in an outer layer of protective covering. Coils are coated in a protective lacquer and, if required, baked so the lacquer hardens to gain maximum durability.

During assembly of the machine, coils are fitted to each pole piece, and the pole pieces are then attached to the frame with machine screws.

The end shields provide support for the armature by way of centrally placed bearings. In a large machine, these are usually sleeve bearings. In smaller machines ball bearings are used. The end shield at the commutator end might also support the commutator brush gear.

Armature

Figure 13.7 shows an example of an armature and its component parts. These are from a relatively small machine, but the principle is the same in all machines.

FIGURE 13.7 An armature core is made of steel laminations, which are shaped to provide slots for the windings

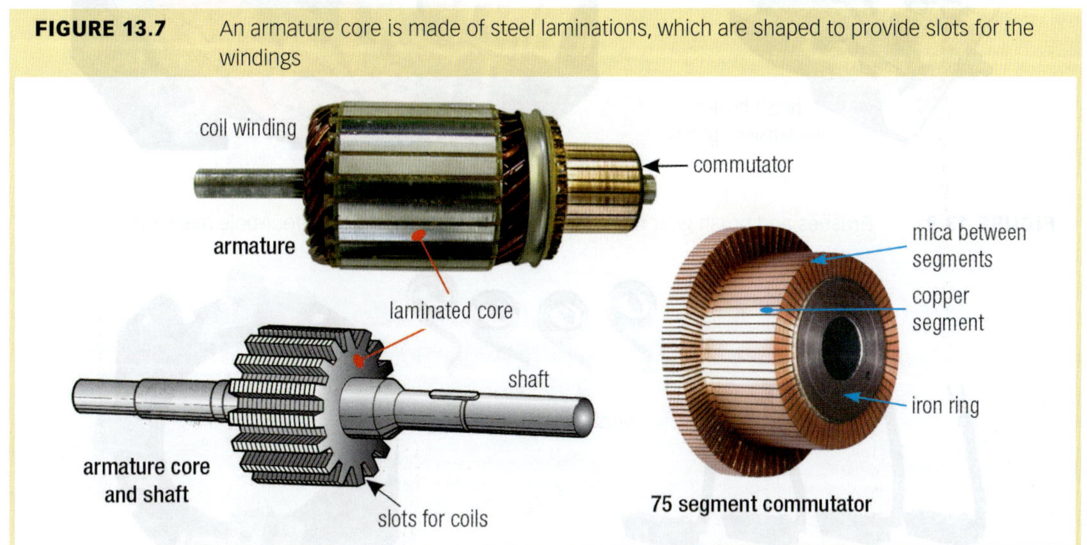

The armature of a DC machine has a central shaft made of steel. The point where the shaft supports the armature core usually has a greater diameter than other sections of the shaft. This gives increased torsional strength and a greater ability to withstand an out-of-balance armature. The commutator is keyed to the shaft, and the whole armature assembly rotates in bearings fitted to the end shields. The drive end of the shaft is coupled to a prime mover.

A commutator has a number of hard-drawn copper segments (or bars) that are separated and insulated from each other by strips of mica. The construction of a commutator is complex, and involves specially shaped copper segments arranged and clamped to form the complete unit, such as the 75-segment commutator in Figure 13.7. Each segment has a notch cut into it at one end to provide a connection to the coil windings. The mica insulation strips between each segment are undercut so they don't protrude and cause unnecessary wear on the brushes.

Brushes

Brushes and brush gear are shown in Figures 13.8 and 13.9. Brushes are made of graphite, sometimes with added copper. They wear with use, requiring routine replacement. The brushes shown in Figure 13.8 are for a large machine. The brush holder applies spring tension to the brushes, which are inserted into the holder. Notice that the machine in Figure 13.8 has six sets of brushes, as it's a six-pole machine.

The brushes in Figure 13.9 are for smaller machines, and again have spring-loaded brushes that fit into a brush holder. Brush gear in smaller machines is usually fixed in place. However, the brush gear in large machines can generally be rotated to some extent around the commutator. This allows best adjustment for the load conditions to reduce sparking between the brushes and commutator.

FIGURE 13.8 Brushes and brush gear for large machines

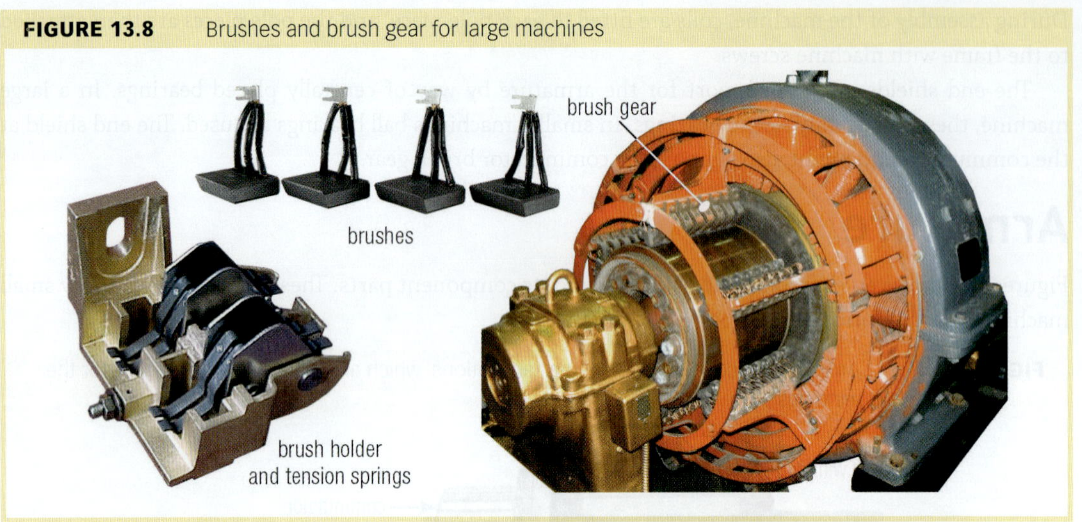

Adapted from Herman, *Standard Textbook of Electricity* 1e, © 1993 Delmar Learning, a part of Cengage Learning, Inc, reproduced by permission www.cengage.com/permissions

FIGURE 13.9 Brushes and brush gear for small machines, in this case for a four-pole machine

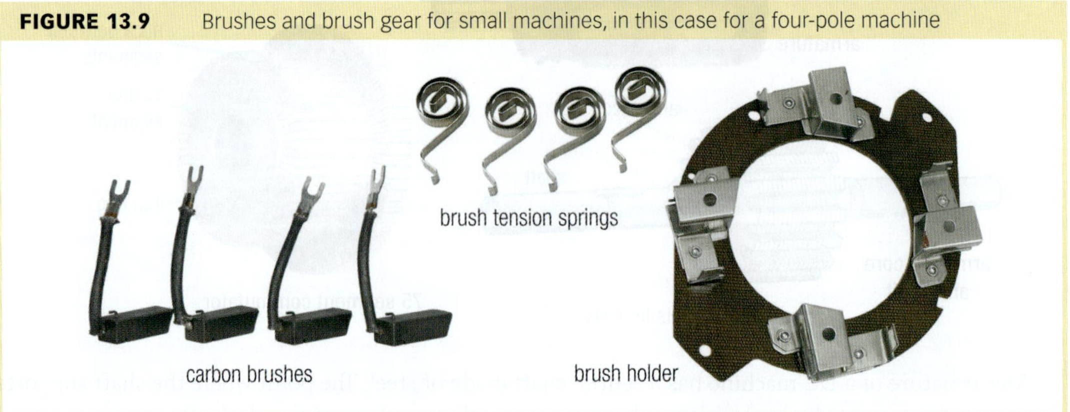

Adapted from Herman, *Standard Textbook of Electricity* 1e, © 1993 Delmar Learning, a part of Cengage Learning, Inc, reproduced by permission www.cengage.com/permissions

Armature windings

The armatures shown in Figure 13.7 are clearly different to the simplified version in Figure 13.5. In the first place there are far more coils, and also each coil has many turns, rather than the single turn shown in our simplified diagrams. A simple DC generator will produce a unidirectional output voltage, but this voltage will vary continually between maximum and zero. To get a smoother output voltage (and therefore load current), armatures are wound in various ways. In all cases, practical machines have a number of armature coils, sometimes with three or more separate coils per slot. It's beyond the scope of this book to describe armature windings in detail, as there are numerous variations.

Figure 13.10 shows a basic two-coil armature in a magnetic field. Generators usually have many coils and a commutator with numerous segments to obtain a smooth output voltage.

FIGURE 13.10 Increasing the number of coils on an armature results in a smoother DC output voltage from the generator. In this two-coil arrangement, the voltage changes polarity twice per revolution.

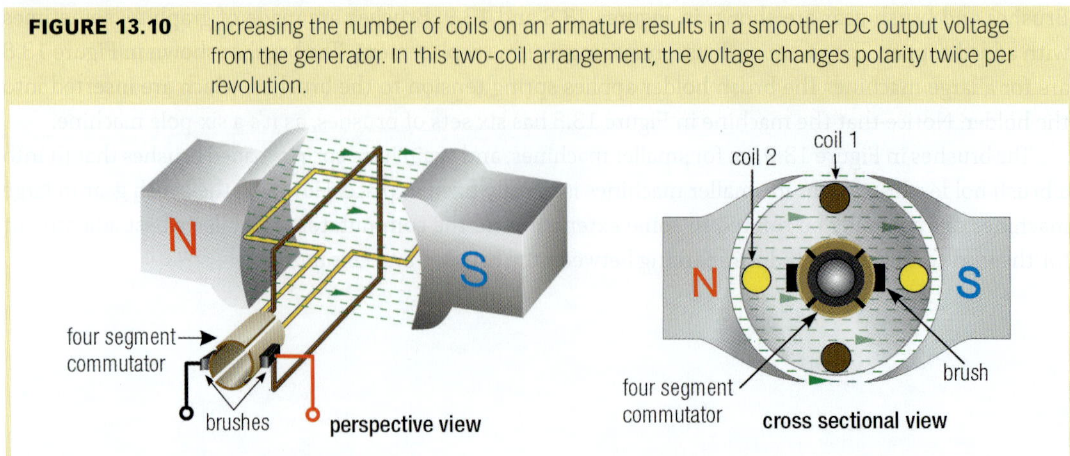

Armature windings with many coils are either lap or wave, with variations. Figure 13.11 shows a lap winding for an armature with an 11-segment commutator. Because the machine has four poles, there are four brushes, connected as pairs.

FIGURE 13.11 Example of a lap winding in a four-pole machine, in which each coil connects to adjacent commutator segments, and the coils span three armature slots

The coils in Figure 13.11 are placed in the armature core so each slot has two coil sides, with each coil spanning three slots, referred to as a coil span of three. One coil side is in slot 1, the other side in slot 4, placed beneath the other coil sharing slot 4. Coil 1 connects to commutator segments 1 and 2, coil 2 connects to segments 2 and 3 and so on.

The schematic diagrams in Figure 13.12 show a radial view and an equivalent circuit for the brush position shown. There are four parallel paths, in which coil 10 is short-circuited at this position. The number of parallel paths equals the number of poles and brushes.

The lap winding in Figure 13.11 is *progressive*, in which the coil connections progress around the commutator in the same direction as the coils themselves progress. A *retrogressive* lap winding has the coil connections crossed and progressing around the commutator in the opposite direction to how the coils are laid in the armature.

FIGURE 13.12 Four-pole lap winding as a circuit diagram in which there are four parallel paths feeding the output, thereby providing a high current at a relatively low voltage

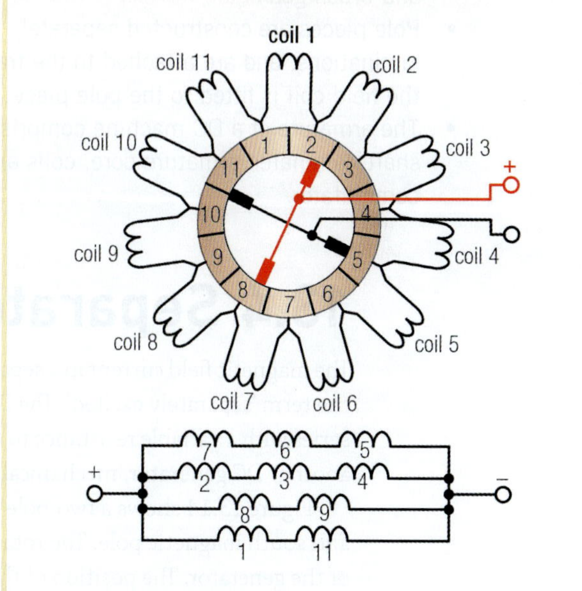

A progressive wave winding is shown in Figure 13.13, again with an 11-segment commutator, a coil span of three and a four-pole stator. The start of coil 1 connects to segment 1, and its end connects to segment 6 (span or pitch of 5). Although only two brushes are required, in practical machines it's usual to have as many brushes as there are poles.

A wave winding has only two parallel paths between brush connections, with windings in series as shown in Figure 13.13. This gives a larger number of conductors contributing to the output voltage, but because there are only two paths, the output current is less than provided by a lap wound armature with its greater number of parallel paths. Both types of armature windings produce the same amount of output power, in which a wave winding provides a higher voltage at a lower current. A lap winding provides a higher current at a lower voltage.

Like the lap winding, a wave winding can be progressive or retrogressive, and both the lap and wave windings shown here are called *simplex* windings. All two-pole DC machines have a lap wound armature. In nearly all cases, a lap wound armature has an even number of commutator segments. Wave wound armatures usually have an odd number of commutator segments. DC machines have one, two (typical) or three coils per slot.

FIGURE 13.13 An example of a wave winding in a four-pole machine, in which each coil spans three armature slots and connects to commutator segments with a spacing of five. The circuit diagram shows the two paths between the brushes.

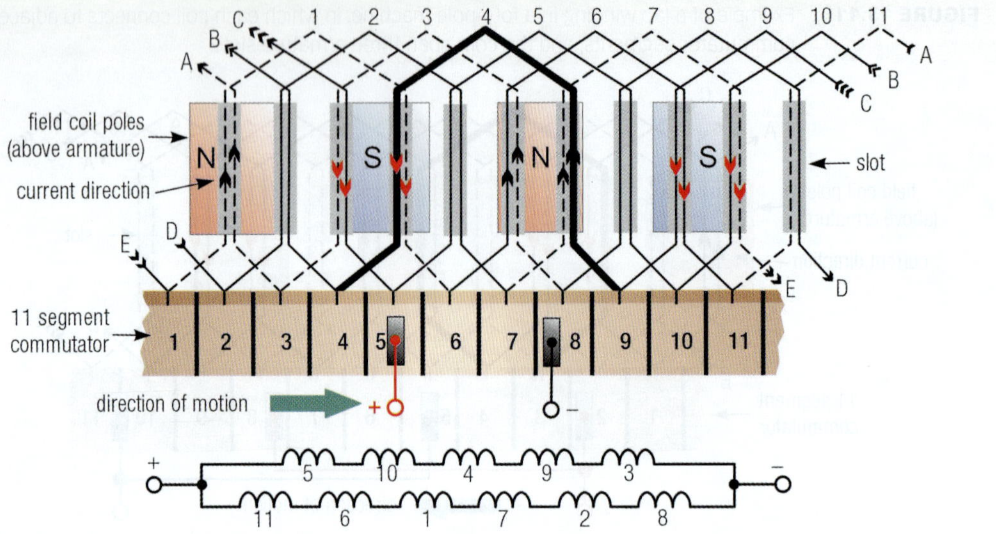

KEY POINTS...

- DC machines have a cast steel or laminated frame that supports the end shields, pole pieces and brush gear.
- Pole pieces are constructed separately from laminations, and are attached to the frame after the field coil is fitted to the pole piece.
- The armature of a DC machine comprises a shaft, laminated armature core, coils and a commutator.
- Carbon brushes are pressed with springs against the commutator and provide a path for the armature current to flow in the load connected to the generator.
- DC machines usually have as many brushes as poles.
- Armatures are either wave or lap wound, and both types produce the same amount of output power.

13.4 Separately Excited DC Generator

The magnetic field current in a separately excited DC generator is provided by an external DC supply, hence the term 'separately excited'. The DC supply can be adjustable, or for a fixed supply voltage, connected in series with a variable resistance to control the value of the field current. The DC supply itself is sometimes another DC generator, mechanically connected to the main DC generator shaft.

Figure 13.14 shows a two-pole DC generator with its two field coils connected in series to give a north and south magnetic pole. The rotating armature has two brushes which connect to the output terminals of the generator. The position of the brushes is shown as being in the centre of the pole pieces to conform with previous diagrams. Brush position depends on how the armature is wound, along with other factors.

The circuit diagram of the generator is shown in Figure 13.15. The DC supply is connected to the field coils through a device called a field *rheostat*. This is simply a variable resistor with a power rating sufficiently high so it can handle the required value of field current. A rheostat needs only two terminals and its resistance value is chosen to give the required range of field current values.

FIGURE 13.14 A basic two-pole separately excited DC generator

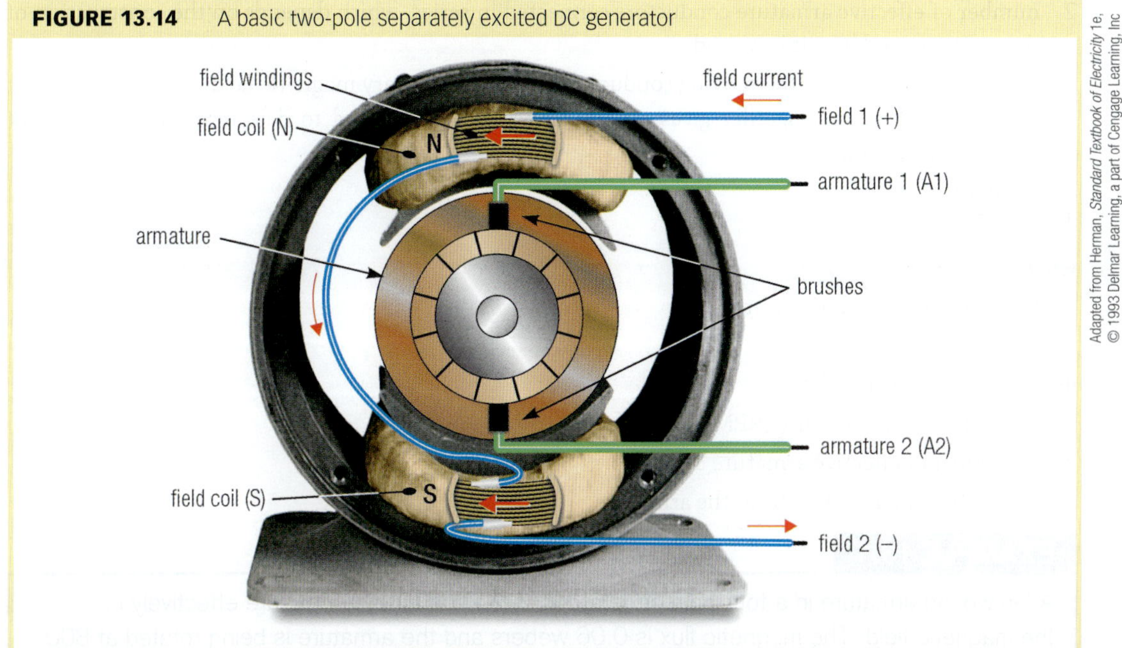

An equivalent circuit diagram of a separately excited generator is in Figure 13.16. The field coils are now shown as a resistance (R_F) rather than as an inductance, as it's their resistance that determines the value of the DC supply voltage needed to cause the required value of field current.

The armature is shown as a resistance (R_A) in series with a DC voltage shown as a battery. This voltage is the generated EMF (E_G) produced within the armature windings. When a load is connected to the armature, current flows through the armature resistance to the load. The voltage drop across the armature resistance (R_A) causes the terminal voltage (V_T) to be less than the armature's generated voltage.

FIGURE 13.15 Circuit diagram of a separately excited DC generator

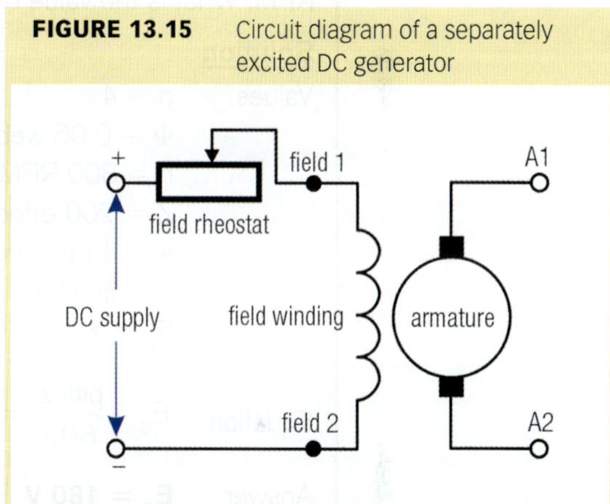

FIGURE 13.16 Equivalent circuit of a separately excited DC generator

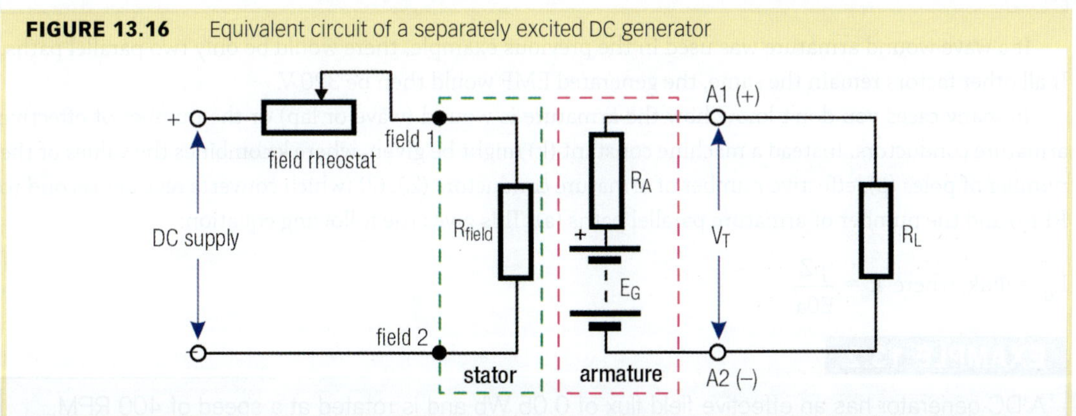

Open-circuit characteristics

The equivalent circuit of a DC generator (Figure 13.16) shows that if there is no load connected to the generator, the terminal voltage of the generator (V_T) will equal the EMF generated in the armature (E_G). The value of this voltage depends on three factors:

1 **strength** of the magnetic field created by the field coils. Each pole typically produces the same amount of flux, where flux has the symbol Φ and is measured in webers

2 **number** of effective armature conductors connected in series, which depends on the construction of the armature and how it is wound

3 **relative speed** between the moving conductors and the stationary magnetic field.

Putting these factors together gives an equation that can be used to determine the open-circuit voltage of a DC generator:

$$E_G = \frac{p\Phi nZ}{60a}$$

where:

E_G = voltage generated by the armature

p = number of poles

Φ = flux per pole in webers

n = revolutions per minute (RPM)

Z = number of effective armature conductors

a = number of parallel paths in the armature.

EXAMPLE 13.1

A lap wound armature in a four-pole generator has 200 conductors that are effectively in the magnetic field. The magnetic flux is 0.06 webers and the armature is being rotated at 800 RPM. What is the value of the voltage generated in the armature?

Solution

Values $p = 4$
$\Phi = 0.06$ webers
$n = 800$ RPM
$Z = 200$ effective conductors
$a = 4$ (as number of parallel paths in a lap wound armature equals the number of brushes, which equals the number of poles)
$E_G = ?$

Equation $E_G = \frac{p\Phi nZ}{60a} = \frac{4 \times 0.06 \times 800 \times 200}{60 \times 4}$

Answer $\mathbf{E_G = 160\ V}$

If a wave wound armature was used in the previous example, there would be only two parallel paths. If all other factors remain the same, the generated EMF would then be 320 V.

In many cases you don't know how the armature is wound (wave or lap) or the number of effective armature conductors. Instead a machine constant (k) might be given, where k combines the values of the number of poles (p), effective number of armature conductors (Z), 60 (which converts revs per second to RPM) and the number of armature parallel paths (a). This gives the following equation:

$$E_G = \Phi nk, \text{ where } k = \frac{pZ}{60a}$$

EXAMPLE 13.2

A DC generator has an effective field flux of 0.05 Wb and is rotated at a speed of 400 RPM. The machine constant is 5. Find the voltage generated by the armature.

Solution

Values $\Phi = 0.05$ webers
$n = 400$ RPM
$k = 5$
$E_G = ?$

Equation $E_G = \Phi nk = 0.05 \times 400 \times 5$

Answer $\mathbf{E_G = 100\ V}$

These equations show that the open-circuit voltage of a given DC generator depends on the amount of magnetic flux produced by the field coils and the rotational speed of the armature. Of particular interest is the strength of the magnetic flux, which is produced by current flowing in the field coils. As explained in Chapter 11, all magnetic materials respond in a particular way to a magnetising force (H). Too much magnetising force will saturate the material, so there is a point at which increasing the current in the field coils gives little or no increase in the magnetic field.

The open-circuit characteristic of a DC generator is found by measuring the field current with an ammeter, and the generator's open-circuit output voltage with a voltmeter. Figure 13.17 shows the circuit of the test setup.

FIGURE 13.17 Circuit when determining a DC generator's open-circuit characteristic

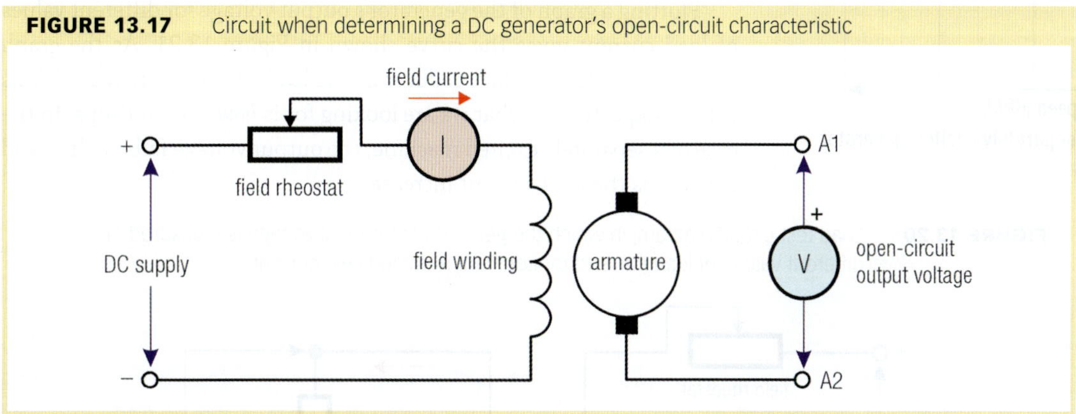

To determine the generator's open-circuit characteristic, the rotational speed of the generator is held constant while the field current is increased from zero to a maximum value. The voltage produced by the generator is measured for each value of field current. These values are then plotted as a graph, shown in Figure 13.18(a). This curve is identical to the magnetisation curve of the ferromagnetic material used in the generator's magnetic circuit, as the graph in (b) shows. When the field current is zero there is still an output voltage due to the residual magnetism in the core.

FIGURE 13.18 The open-circuit voltage curve of a DC generator is the same as the magnetisation (B–H) curve of the machine's magnetic circuit

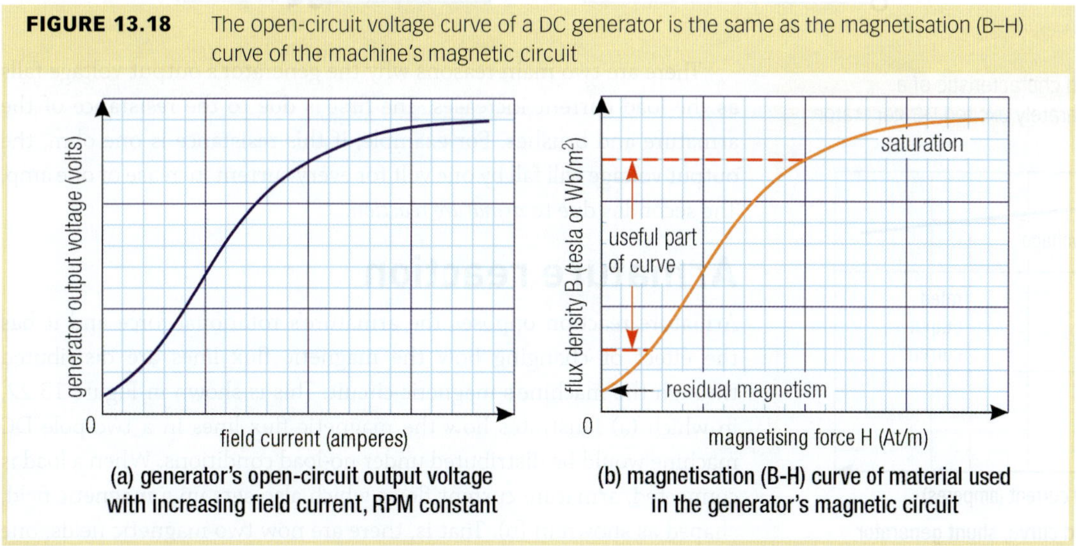

(a) generator's open-circuit output voltage with increasing field current, RPM constant

(b) magnetisation (B-H) curve of material used in the generator's magnetic circuit

The above equation for generated EMF (E_G) tells us that, for a set value of field current, the open-circuit (or no-load) voltage will vary proportionally with the rotational speed. That is, as shown in the graph in Figure 13.19, the open-circuit voltage will increase as the generator speed increases.

FIGURE 13.19 Voltage versus speed curve for a DC generator on no load

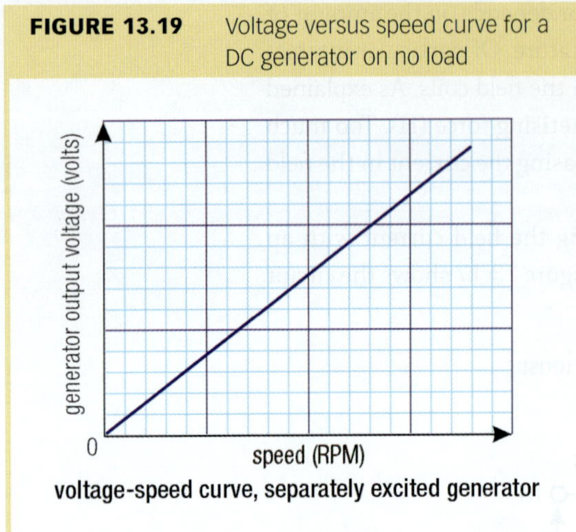

Load characteristics

A DC generator is intended to provide electrical power to a load. So it makes sense to know how the generator performs under load conditions. This is done with a *load test* in which the generator is run at its rated speed and field current while the load current is varied. The circuit is shown in Figure 13.20, in which the load is a rheostat in series with an ammeter. Varying the rheostat resistance varies the load current from zero to around 125 per cent of the generator's rated value.

Plotting a graph of the generator's output voltage for different values of load current gives the curve shown in Figure 13.21. As the graph shows, the output voltage reduces as the load current is increased. This is to be expected, so what we are looking for is *how much* it drops. In the case of a separately excited machine, the output voltage falls slightly and steadily as the load current increases.

FIGURE 13.20 Load test circuit diagram, in which the generator's terminal voltage is measured at different values of load current, for constant RPM and field current

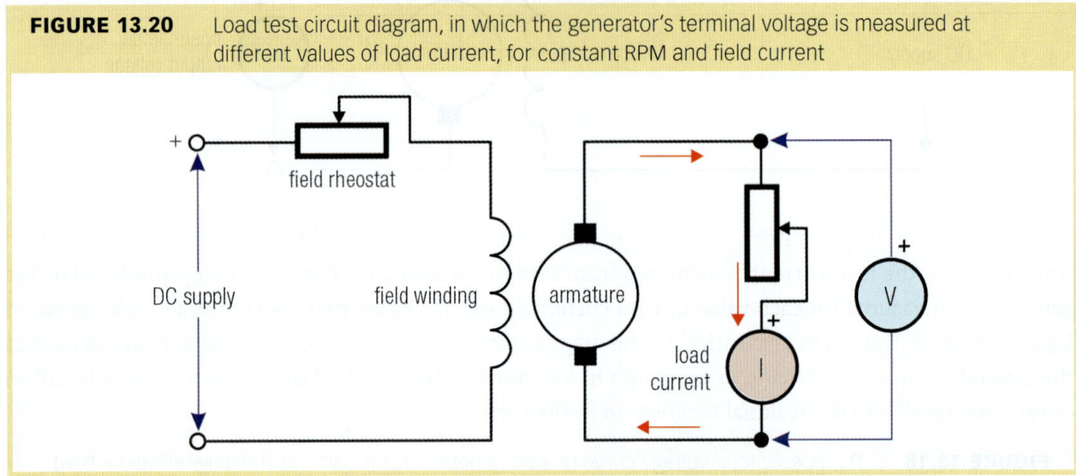

FIGURE 13.21 Load characteristic of a separately excited DC generator

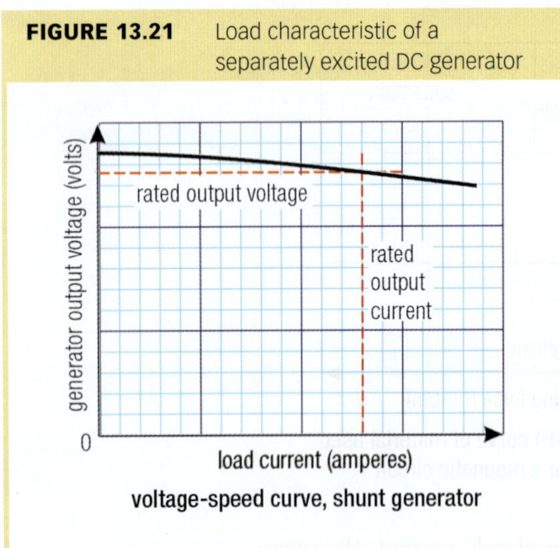

There are two main reasons why the generator's output voltage falls as the load current increases. The first is due to the resistance of the armature and brushes. For example, if this resistance is one ohm, the output voltage will fall by one volt for every current increase of one amp. The second is due to *armature reaction*.

Armature reaction

Armature reaction opposes the armature's rotational force and it has the effect of changing how the magnetic flux lines are distributed through the machine's magnetic circuit. This is shown in Figure 13.22, in which (a) illustrates how the magnetic flux lines in a two-pole DC machine would be distributed under no-load conditions. When a load is connected, armature current flows which also sets up a magnetic field, shaped as shown in (b). That is, there are now two magnetic fields, one caused by the field coils, the other by the armature current. The result of these two interacting magnetic fields is shown in (c).

One of the effects is to cause the flux to concentrate at the trailing pole tips and to reduce the flux at the leading pole tips. (A leading pole tip is the section of the pole piece that the armature moves towards, a trailing tip is the section the armature is moving away from.) This flux redistribution can cause saturation at the trailing pole tips, and in any case, the magnetic flux is now distributed unevenly. As a result, the magnetic circuit in the machine is not being used effectively, and the generated voltage will be reduced.

FIGURE 13.22 Armature reaction causes the machine's magnetic field to twist in the direction of rotation

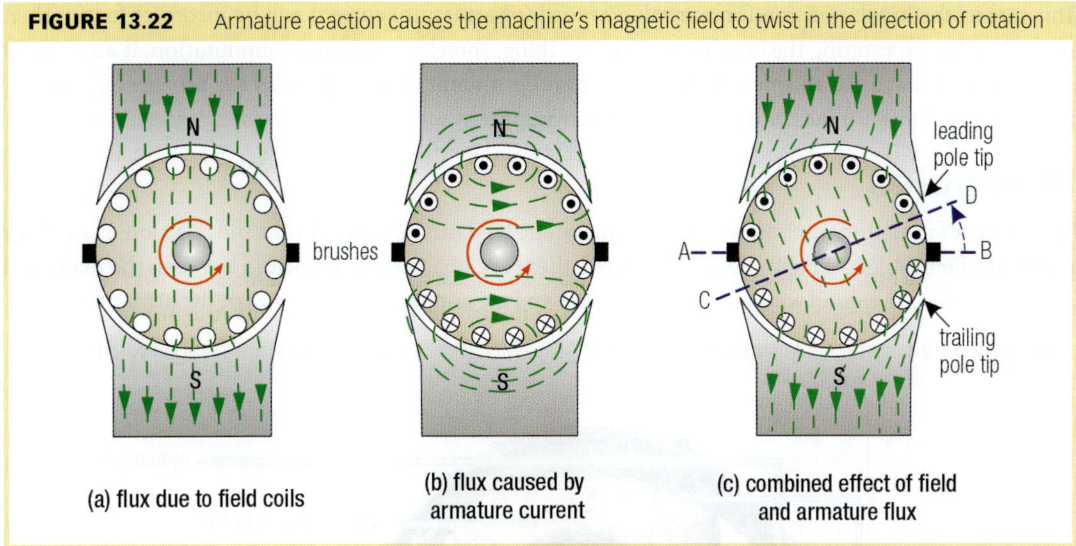

Another effect is sparking at the brushes. So far, we have considered the commutator simply as a switch, but we also need to consider *when* the switching occurs. Figure 13.23 shows a simple DC generator. In Figure 13.23(a) the armature coil is travelling parallel to the magnetic field and therefore no voltage is induced in the coil. It's at this point where the commutator changes segments, and for a few degrees of rotation, both segments are connected together by the brushes. Because there is no current flowing, shorting the segments together has no effect. In Figure 13.23(b), the coil is travelling at 90° to the flux, so maximum voltage is induced in the coil. The brushes are now in the centre of the commutator segments.

FIGURE 13.23 Brushes are positioned so switching between commutator segments occurs when there is no induced current in the coils being switched

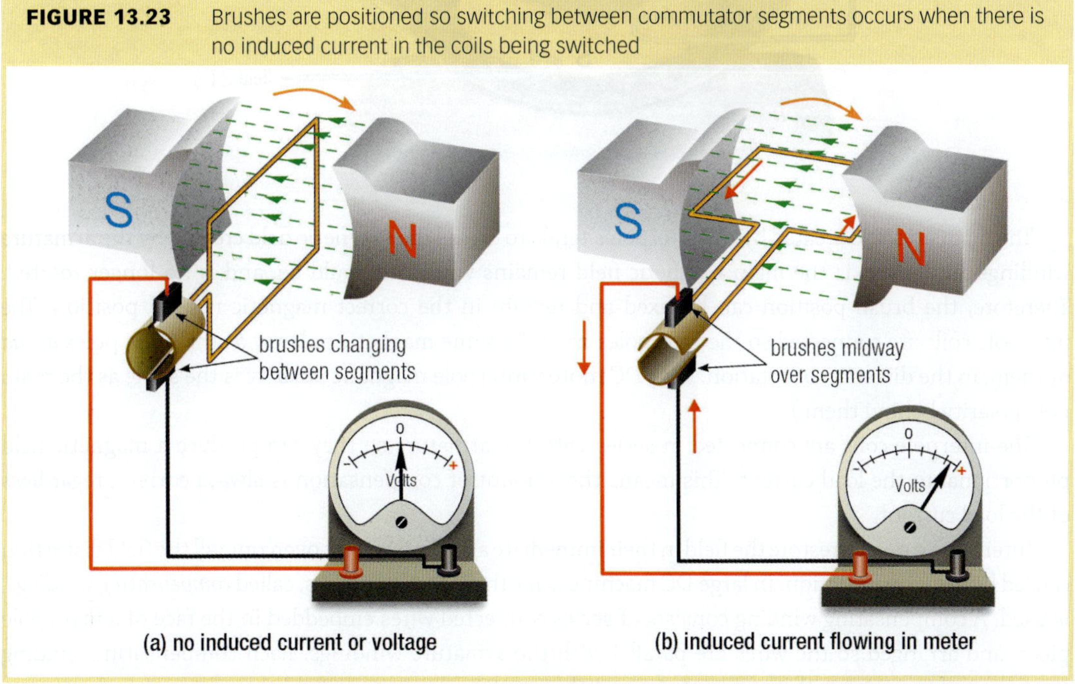

The brush position in Figure 13.23 is aligned to the machine's *magnetic neutral zone*, so segment switching occurs when the induced current is zero in the coils being switched. However, if the magnetic field is distorted due to armature reaction, as in Figure 13.22(c), the magnetic neutral zone shifts, which in the case of a generator is in the direction of armature rotation. The result is that coils are now being switched by the commutator while current is still being induced in them. This causes sparking at the commutator. (In a DC motor, the magnetic neutral zone moves against the direction of rotation.)

A solution is to rotate the brushes so they align with the new magnetic neutral zone. This is shown in Figure 13.22(c), in which the neutral zone was originally central to the pole pieces, indicated by line A–B. Because of armature reaction, the magnetic neutral zone has moved anticlockwise to line C–D. However, the extent to which the magnetic field is rotated depends on the current flowing through the armature.

This means the brush position for one value of load (or armature) current will be incorrect for another. As well, although rotating the brushes reduces sparking, thereby improving commutation, it also causes a reduction in the useful magnetic flux. The greater the amount of 'lead' given to the brushes, the greater the reduction in useful flux, and therefore output voltage.

Interpoles

The effects of armature reaction on the main magnetic field can be reduced with the addition of auxiliary poles called *interpoles*. As Figure 13.24 illustrates, these are small field poles placed between the main field poles. The coils are wound with a heavy gauge wire and are connected in series with the armature.

FIGURE 13.24 Two-pole separately excited DC generator with interpoles, which are in series with the armature

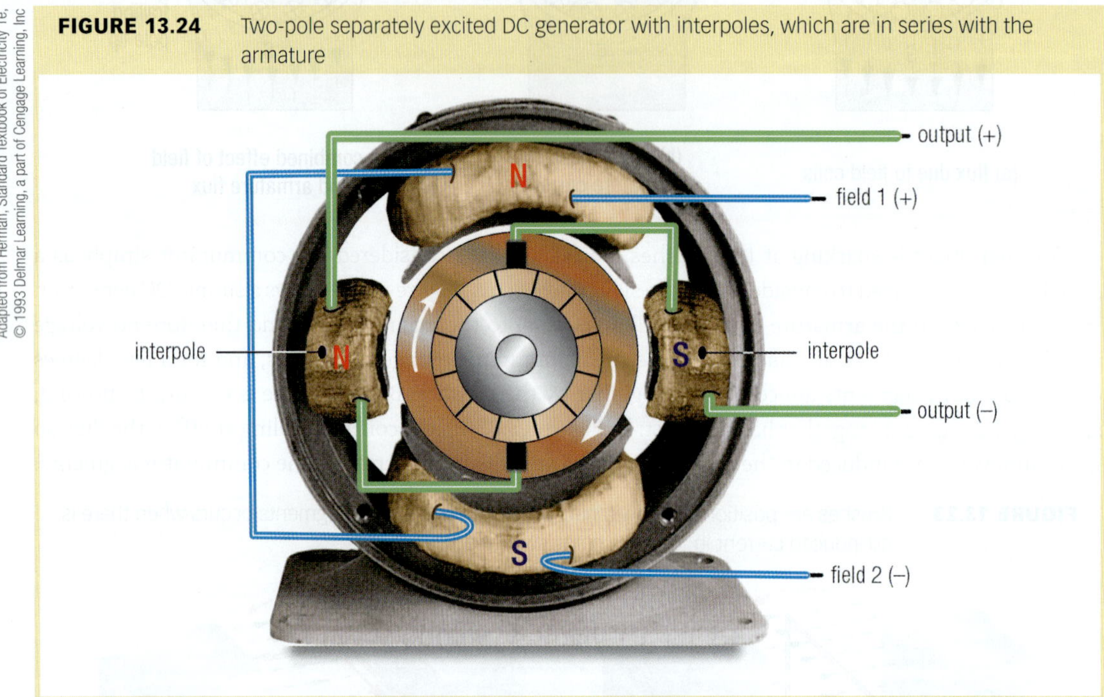

The magnetic field created by the interpoles tends to cancel the magnetic field created by the armature windings. The result is the main magnetic field remains where it should be, and is no longer rotated. Therefore, the brush position can be fixed and remain in the correct magnetic neutral position. The interpole coils are connected so the interpoles have the same magnetic polarity as the main poles *ahead* of them, in the direction of rotation. (In a DC motor, interpole magnetic polarity is the same as the main pole polarity behind them.)

The interpole coils are connected in series with the armature so they can produce a magnetic field proportional to the load current. This means the amount of compensation is always correct, regardless of the load current.

Interpoles can only restore the field in their immediate area, and cannot overcome all the field distortion caused by armature reaction. In large DC machines, another set of windings, called *compensating windings*, is used. A compensating winding consists of series connected wires embedded in the face of a main pole piece, and arranged so the wires are parallel with the armature windings. Each compensating winding is connected in series with the armature, and the result is a magnetic field that is equal and opposite to the armature field. As a result, the effects of armature reaction can be almost completely overcome. Compensating windings are expensive to manufacture, so they are only found in large machines.

KEY POINTS...

- A separately excited DC generator requires an external DC power source to energise the field coils to establish the generator's magnetic field. Smaller generators often have permanent magnets instead of field coils.
- The open-circuit terminal voltage of a DC generator is the same as the armature-generated EMF, as there is no voltage drop across the armature resistance.
- The open-circuit output voltage of a DC generator equals the product of the effective field flux (Φ), its RPM and its machine constant. The machine constant is determined by the machine's construction, such as the number of poles, the effective number of armature conductors and how the armature is wound.
- The open-circuit characteristic of a separately excited DC generator is found by holding the rotational speed constant while measuring its unloaded terminal voltage at different values of field current.
- The load characteristic of a DC generator is a graph that shows how its terminal (output) voltage changes with load current.
- The full-load voltage of a generator is less than the generated EMF due to the resistance of the armature and brushes. Unless compensated for, the full-load voltage is also reduced by the effects of armature reaction.
- Armature reaction occurs when the armature's magnetic field distorts the main field. This reduces the generator's terminal voltage and can cause sparking at the brushes.
- The effects of armature reaction can be reduced with interpoles, which are small poles between the main poles, with the interpole coils in series with the armature.

13.5 Self-Excited Generators

Separately excited generators have the disadvantage of requiring an external source of direct current to supply the field coils. They are therefore only used in special cases, such as when a wide range of terminal voltages is required. Most DC generators are self-excited, which means they provide their own field current. There are various ways this can be done, and each method is named after the manner in which the field coils are connected. They can be in series with the armature, in parallel (shunt) or a combination of both.

All self-excited generators need the pole pieces to retain some degree of magnetism after shutting down, so when starting there is enough magnetism to allow the generator to build up a voltage. At start-up, the armature windings interact with the magnetic field due to the residual magnetism, which must be enough to induce a voltage to cause field current to flow. The field current must flow in a direction that adds to the existing magnetism. If there is enough field current, the magnetic flux will increase, causing the terminal voltage of the machine to increase. This in turn increases the field current until normal operating conditions are established.

Shunt generators

A shunt generator has its field winding connected in parallel with the armature. That is, the voltage developed by the armature is applied across the field windings, which causes a current to flow in the windings. The field current is therefore independent of the load current, but dependent on the load voltage. In general, the field current of a shunt generator is no more than a few per cent of the load current. Figure 13.25 shows how a shunt generator is connected.

The field coils in a shunt generator are wound with numerous turns of relatively fine wire and are connected in series. A two-pole machine is shown in Figure 13.25; others have four or more poles which are connected to give alternate north and south poles.

Figure 13.26(a) shows the circuit diagram of a shunt generator. The field rheostat controls how much current flows in the field coils, giving limited adjustment of the output voltage.

FIGURE 13.25 Two-pole shunt-connected DC generator, in which the field winding is in parallel with the armature and therefore the output terminals

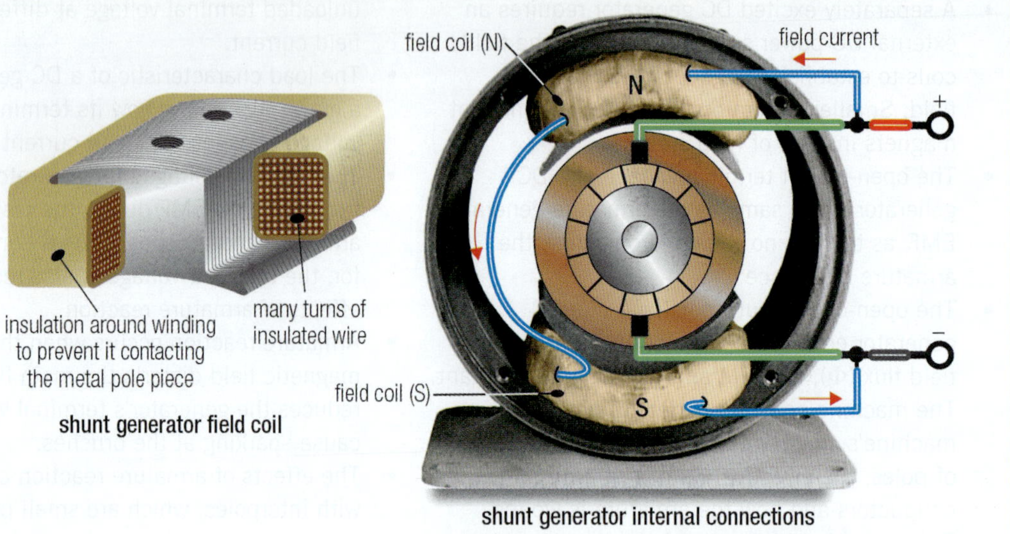

Adapted from Herman, Standard Textbook of Electricity 1e, © 1993 Delmar Learning, a part of Cengage Learning, Inc

FIGURE 13.26 The terminal voltage of a shunt DC generator falls dramatically under overload conditions, shown by the red section of the curve

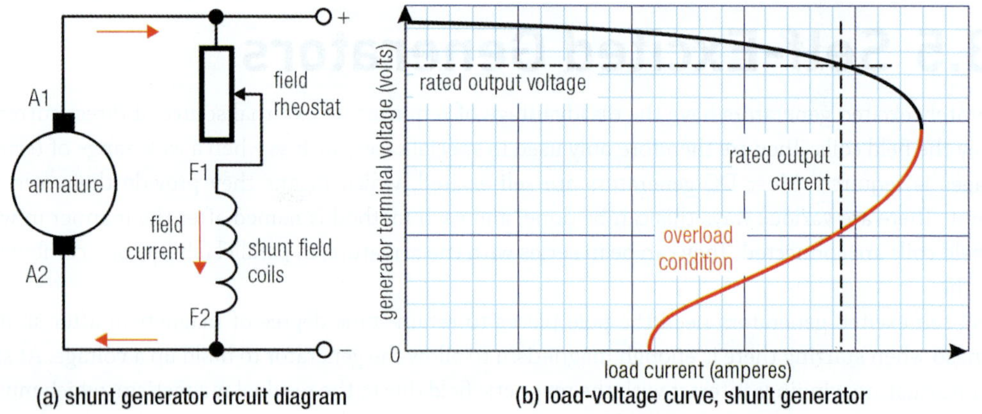

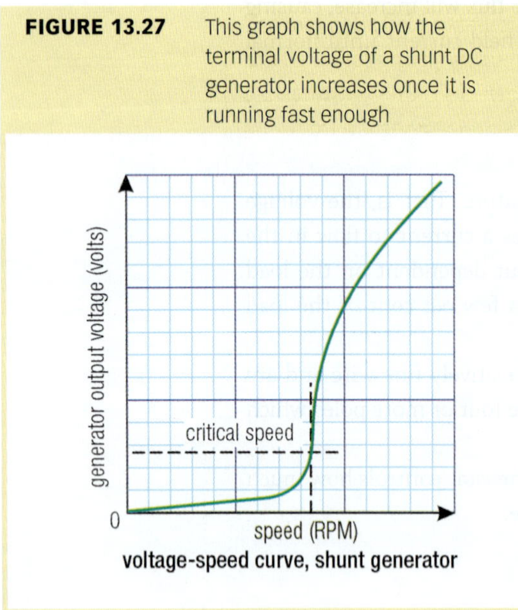

FIGURE 13.27 This graph shows how the terminal voltage of a shunt DC generator increases once it is running fast enough

The load characteristic curve (also called the external characteristic) is shown in Figure 13.26(b). This curve, like that for the separately excited generator, shows how the terminal voltage of a shunt generator behaves for various values of load current. The load characteristic curves for both types of generator are similar, but only up to where the generator is supplying its rated output current.

In a shunt generator, because the generator is supplying its own field current, as the load current increases, the generator's terminal voltage drops, reducing the field current. This reduces the strength of the generator's magnetic field, so when the load current exceeds a certain value, the terminal voltage drops significantly. Further overload will cause the terminal voltage to fall even more.

This means the shunt generator will protect itself from damage if there's a short-circuit across its terminals. Under this condition, there will be no field current (as the field coils are shorted out), and the armature voltage will be due only to the machine's residual magnetism. Because of the much reduced magnetic field, the current flowing in the short-circuit will not be enough to damage the machine.

Figure 13.27 shows that the voltage produced by a shunt generator varies almost proportionally to its speed of rotation, once the *critical speed* is reached. Below this speed, the voltage produced by the generator is due mainly to its residual magnetism. The generated voltage will cause a small field current to flow, but not enough to increase the magnetic flux much above the residual value. Once the critical speed is reached, the terminal voltage increases because there is now a stronger field current flowing.

Like the separately excited generator, for a given speed the terminal voltage of a shunt generator drops under load due to its armature resistance and the effects of armature reaction, with the added effect of reduced field current. Despite this, shunt generators are used where a nearly constant voltage is required over a somewhat varying, but predictable load.

Applications are limited, and include electroplating and charging batteries. Another use is to provide the field current for a larger, separately excited generator or alternator. In some cases the shunt generator might be fitted with an automatic voltage regulation system, in which sections of resistance in its field rheostat are short-circuited to increase the field current in response to a falling terminal voltage.

Series generators

A series generator has its field winding connected in series with the armature. This means the load current flows through the field windings, which need to have a very low resistance. The internal connections of a series generator and the construction of a series field coil are shown in Figure 13.28. Although there are far fewer turns in a series field coil compared to the coils in a shunt generator, the field current is considerably higher, and typically equals the load current. In a shunt generator, the field current is around 3 per cent of the load current.

FIGURE 13.28 The field coils in a series DC generator are wound with a few turns of heavy gauge wire, and are connected in series with the armature

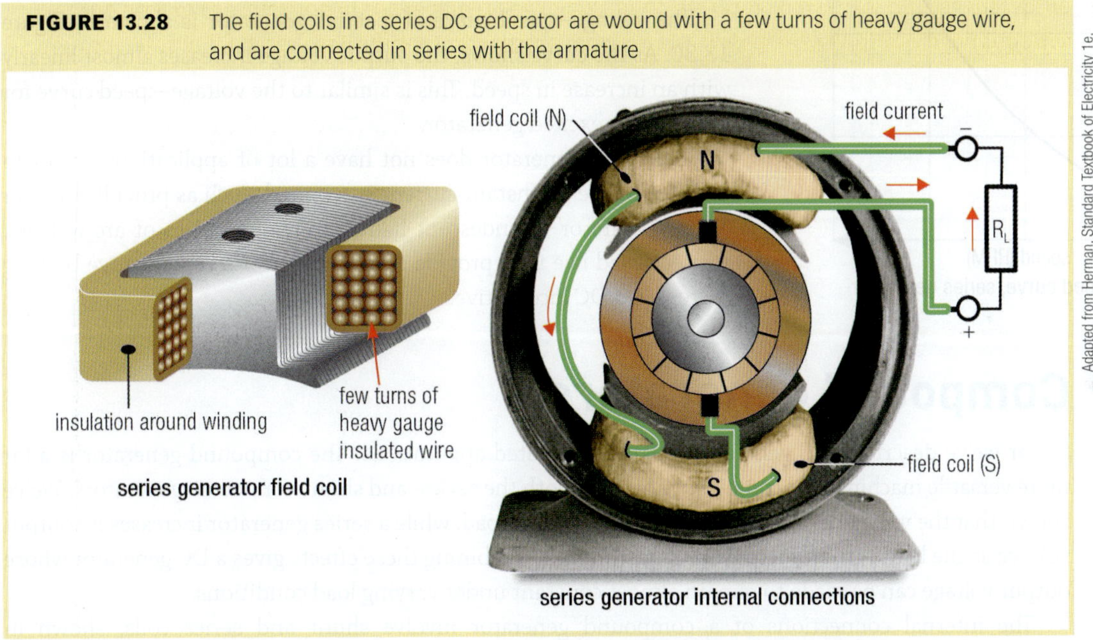

When a series generator is started, it must have the load connected so field current can flow, whereas a shunt generator is usually run up to full speed before connecting the load. Like the shunt generator, there must be some residual magnetism in the poles to allow EMF build-up in the series generator. Because the field current depends on (or equals) the load current, the magnetic field strength varies with the load current. Therefore, the terminal voltage of the series generator increases as the load current increases. Figure 13.29 shows the circuit diagram and load voltage characteristic curve of a series generator.

The field diverter rheostat in Figure 13.29(a) is fitted to some machines to give a degree of voltage control. By changing the resistance of the rheostat, more or less current is diverted through the rheostat, away from the field coils, thereby altering the magnetic field strength and hence the generated voltage.

A series generator does not have an open-circuit characteristic curve, as the unloaded voltage is that due to the machine's residual magnetism. The load characteristic curve in Figure 13.29(b) shows that as the load current increases, the terminal voltage increases. However, if the load current is too low (load resistance too high), the magnetic field created by the field coils will hardly exceed that of the machine's residual magnetism, and the voltage will not increase.

FIGURE 13.29 The terminal voltage of a series DC generator increases as the load current increases

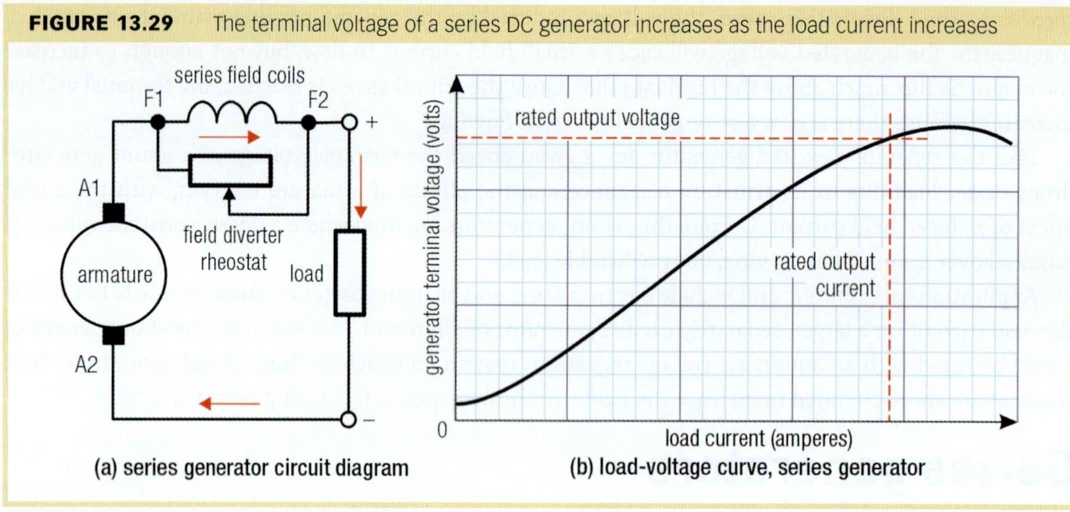

(a) series generator circuit diagram

(b) load-voltage curve, series generator

FIGURE 13.30 The terminal voltage of a series DC generator increases with its rotational speed

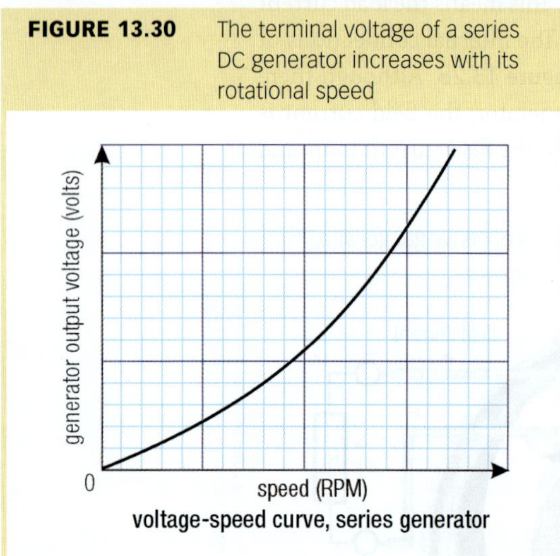

voltage-speed curve, series generator

If the load current exceeds the generator's rated value, the pole pieces tend to saturate and armature reaction increases. Together these effects limit the terminal voltage, and therefore the load current. As the load characteristic curve clearly shows, the series generator is not suitable for applications requiring a constant voltage with a varying load.

The voltage–speed curve of a series generator is shown in Figure 13.30. As the curve shows, the output voltage increases almost linearly with an increase in speed. This is similar to the voltage–speed curve for a separately excited generator.

The series generator does not have a lot of applications. Its main uses are where a constant current is required, such as providing power for series arc or incandescent lights, and some types of arc welders. A specialised use is to provide field current for a regenerative braking system in a DC locomotive.

Compound generators

So far we've described DC generators that have limited applications. The compound generator is a far more versatile machine, and combines aspects of both the series- and shunt-connected generators. We've shown that the voltage of a shunt generator falls under load, while a series generator increases its output voltage as the load current increases (within limits). Combining these effects gives a DC generator whose output voltage can be made to remain almost constant under varying load conditions.

The internal connections of a compound generator involve shunt and series coils, shown in Figure 13.31. The field coils have two windings, a shunt winding made up of many turns of relatively fine wire and a series winding comprising fewer turns of heavy gauge wire. The series winding is connected in series with the armature (and therefore the load), and the shunt windings are connected to the output terminals of the generator.

Typically, the winding connections are arranged so both shunt and series coils produce the same magnetic polarity in their respective pole piece. That is, their magnetic fields add. When connected this way, the compounding is said to be *cumulative*. If the coils are arranged so their magnetic fields oppose, the compounding is *differential*. In this type of generator, the output voltage drops in a similar way to the series generator, but unlike the series generator, it develops full voltage at no load. It has limited applications, unlike the cumulatively compound generator, which we'll now refer to as simply a compound generator.

When a compound generator is running without a load, it behaves like a shunt generator and produces a terminal voltage equal to the armature's generated voltage. The magnetic field is provided only by the shunt field coils, which as shown in the circuit diagrams in Figure 13.32, are connected in series with a field rheostat. The only difference between the two circuits is the point of connection for the shunt coil circuit.

FYI

Long and short shunt connections also refer to how an ammeter and a voltmeter are connected. See Chapter 9

FIGURE 13.31 The compound generator has two separate windings in each field coil, a series-connected coil and a shunt-connected coil

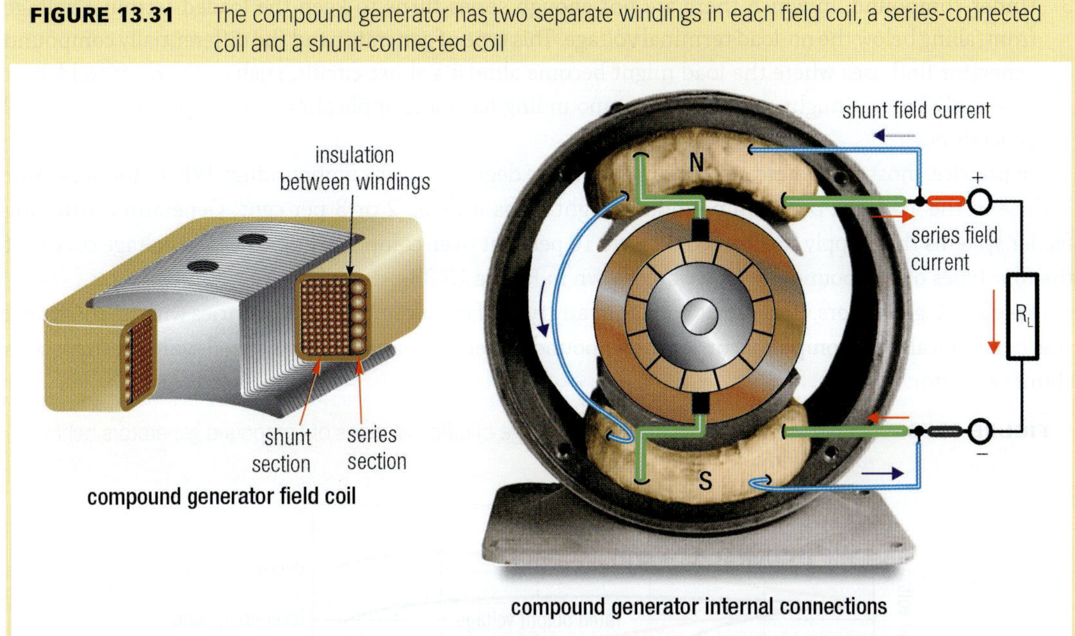

FIGURE 13.32 Two ways of connecting the shunt field coils in a compound generator

(a) compound generator (long shunt)

(b) compound generator (short shunt)

In Figure 13.32(a), the shunt coils and rheostat are connected across the output terminals. This is the so-called long shunt connection, which refers to the apparent extra length of the connecting leads. When connected this way, the voltage drop across the series field coils affects the voltage applied to the shunt coils. Where this is a problem, the shunt coils are connected directly across the armature terminals, giving the short shunt connection. In practice, there is little difference between the performance of either connection.

The shunt field rheostat provides a way of controlling the shunt field coil current, and therefore the generator's terminal voltage. Once the machine is running and a load is connected, current flows in the series field coils as well as the shunt field coils. This increases the strength of the generator's magnetic field, and causes the terminal voltage to increase, offsetting the fall in the voltage that normally occurs in a shunt generator. The amount of increase depends on the number of turns in the series field, giving three distinct types of compound generators:

1 **Level-compound,** where the number of turns in the series coils is proportioned so the generator's terminal voltage remains practically constant from no load to full load. This type of generator is used where the load is relatively close to the generator.

2 **Over-compound,** where there are more than the necessary turns in the series coils, causing the loaded terminal voltage to rise above its no-load value. This is useful where the load is connected some distance from the generator, as it compensates for the voltage drop in the connecting cables. That is, the voltage at the load end of the circuit remains almost constant, regardless of the load conditions.

3 **Under-compound,** in which there are not enough series turns to keep the loaded terminal voltage from falling below the no-load terminal voltage. This type of generator and the differentially compound generator find uses where the load might become almost a short-circuit, such as in an arc welder, or where a DC motor might stall. Under-compounding has more applications in DC motors than in DC generators.

In practice, most compound generators have some degree of over-compounding. Where the generator is close to the load, the over-compounding might be as little as 2 or 3 per cent. Generators with long feeder systems that supply the load could have 10 per cent over-compounding. The load voltage curves of the four types of compound generator are shown in Figure 13.33.

Like all DC generators, the output voltage of any type of compound generator varies with its rotational speed. A typical over-compound or level-compound generator has the same speed–voltage curve as a shunt generator.

FIGURE 13.33 Curves showing how the output voltage of different types of compound generators behave under load

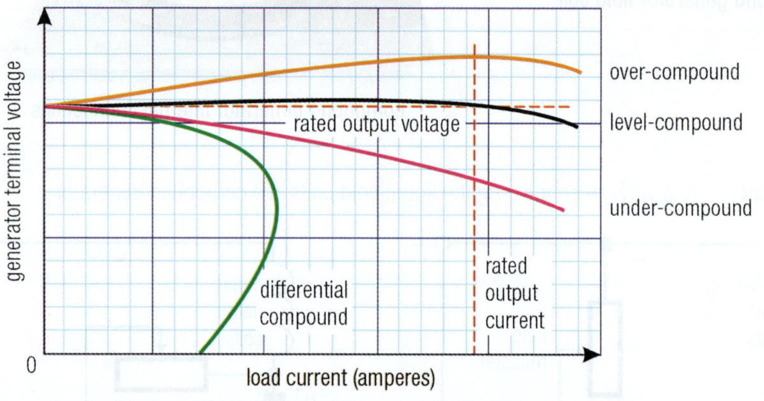

generator terminal voltage
over-compound
level-compound
rated output voltage
under-compound
differential compound
rated output current
0
load current (amperes)

load voltage characteristic curves – compound generators

KEY POINTS...

- The output voltage of all DC generators depends on the rotational speed. For self-excited and permanent magnet generators, the output voltage is proportional to the speed (linear response). In all other generators the relationship is relatively non-linear (their graphs are not straight lines).
- There are five types of DC generators:
  - Separately excited shunt generator in which field current is supplied from an external source. This type of generator is used in Ward Leonard drive systems, in which a DC generator drives a DC motor. The motor speed can be smoothly controlled by adjusting the generator's field current and therefore its output voltage of the generator.
  - Self-excited, which are connected in four possible ways: shunt, series, cumulatively compound (over, level and under) and differential compound. So they can start generating a voltage, all self-excited DC generators require a residual magnetic field.
  - Shunt generators provide a relatively constant output voltage. Their applications include providing excitation current for large alternators in power stations, electroplating and battery charging. A shunt generator is also used where a wide range of DC voltages are needed.
  - A series generator has a few turns of heavy wire connected in series with the armature. Its output voltage increases (up to a point) when the load current increases. Under no load, the output voltage is virtually zero. Series wound generators are used to boost the voltage of a long supply line that supplies large current loads, and in applications that require a constant current, such as a lighting circuit.
- A compound generator has both shunt and series field coils, usually connected so their magnetic fields are cumulative. The number of turns on the series coils determines the degree of compounding which can be over, level or under. Compound generators are used in shipboard DC power systems, in aeroplanes, and as electric traction generators in diesel-electric locomotives.
  - A level-compound generator is used where the load is close to the generator and a constant voltage is needed.

- Over-compound generators are suited to applications when DC power is transmitted over a long distance, as the rise in its terminal voltage compensates for the voltage drop over the lines.
- Under- and differential-compound generators are used where intrinsic overload protection is essential, such as powering electric winches or dredges, in which an overload causes the generator voltage to drop and to limit the current to a safe value. Arc welding is another application, in which the terminal voltage must fall once the arc is struck.

13.6 Calculations

A DC generator, of whatever type, provides electrical power to a load. The generator is driven by a prime mover, and the load is connected to the generator with cables. In many cases, the total load is made up of a number of individual loads. In this type of installation, a feeder cable connects the DC generator to a distribution board, with further cables going from the board to the various loads.

For our purposes, we'll look at an installation in which a single 40 kW load is connected directly to a shunt DC generator located 200 m away. See Figure 13.34. We will work from the load back to the generator, starting with the load requirements in terms of its voltage and current. This allows us to then calculate what the generator has to produce at its output terminals to supply the load.

FIGURE 13.34 Circuit of a shunt generator connected to a load located 200 m from the generator

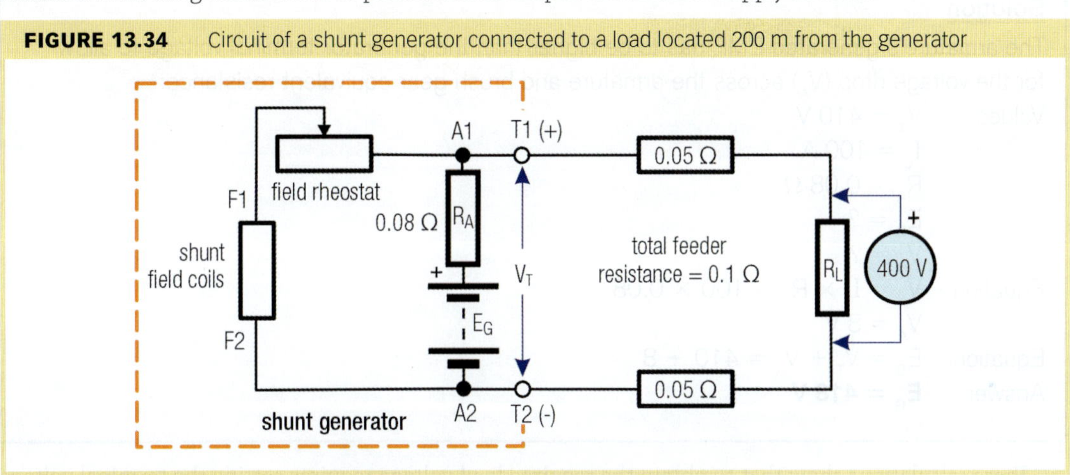

Voltage drops due to resistance

The first thing is to determine the voltage drops due to the feeder resistance.

EXAMPLE 13.3

The shunt generator in Figure 13.34 is connected to a 400 V, 40 kW load with a feeder cable that has a total resistance of 0.1 Ω (0.05 Ω per cable). The armature, brush and brush contact equivalent resistance is 0.08 Ω.

1 Find the load current, assuming the voltage across the load is 400 V.

Solution

Current is found using the given power and voltage values:

Values $V_L = 400$ V

$P_L = 40$ kW

$I_L = ?$

Equation $I_L = \frac{P_L}{V_L} = \frac{40 \times 10^3}{400}$

Answer $\mathbf{I_L = 100\ A}$

2 Find the voltage at the generator terminals (V_T) when the load current is 100 A and the load voltage is 400 V.

Solution

The generator's terminal voltage has to be higher than the load voltage to allow for the voltage drop across the feeder. We use Ohm's law to find the feeder voltage drop, then add this to the load voltage:

Values $V_L = 400$ V

$I_L = 100$ A

$R_{feeder} = 0.05 + 0.05 = 0.1\ \Omega$

$V_{feeder} = ?$

$V_T = ?$

Equation $V_{feeder} = I_L \times R_{feeder} = 100 \times 0.1$

$V_{feeder} = 10$ V

Equation $V_T = V_L + V_{feeder} = 400 + 10$

Answer $\mathbf{V_T = 410\ V}$

3 Find the value of the armature's generated EMF (E_G), assuming the only losses are due to the armature and brush resistance.

Solution

The armature's generated EMF has to be higher than the generator terminal voltage to allow for the voltage drop (V_A) across the armature and brush gear equivalent resistance:

Values $V_T = 410$ V

$I_L = 100$ A

$R_A = 0.08\ \Omega$

$E_G = ?$

$V_A = ?$

Equation $V_A = I_L \times R_A = 100 \times 0.08$

$V_A = 8$ V

Equation $E_G = V_T + V_A = 410 + 8$

Answer $\mathbf{E_G = 418\ V}$

These calculations show that to obtain the required load voltage requires setting the terminal voltage of the loaded DC generator high enough to allow for the voltage drops across the connecting cables. They also show that the generated EMF is higher than the terminal voltage to allow for the voltage drop across the resistance of the armature circuit.

Voltage regulation

There are many factors that can affect the voltage appearing across the load in Figure 13.34. If the load current and the generator speed remain constant, the load voltage will remain constant. But because it's a shunt generator, if the load current is reduced, the generator's terminal voltage and therefore the load voltage will rise (see Figure 13.26). The maximum voltage that will appear at the generator terminals occurs when it is unloaded.

If we know the full-load and no-load voltages of the generator, we can determine its *voltage regulation*. Generators are rated at their full-load output voltage, so regulation is the difference between the no-load and full-load voltages, divided by the full-load value. The answer is multiplied by 100 to express it as a percentage. So:

$$\text{Voltage regulation percentage} = \frac{V_{NL} - V_{FL}}{V_{FL}} \times 100$$

where:

V_{NL} = no-load voltage

V_{FL} = full-load voltage.

EXAMPLE 13.4

The shunt generator in Figure 13.34 has an open-circuit voltage of 450 V and a full-load voltage of 410 V. Calculate its percentage voltage regulation.

Solution

Values $V_{NL} = 450$ V

$V_{FL} = 410$ V

% Vreg = ?

Equation $$\% \text{ Vreg} = \frac{V_{NL} - V_{FL}}{V_{FL}} \times 100 = \frac{450 - 410}{410} \times 100$$

Answer **% Vreg = 9.76 per cent**

Because the output voltage falls under load, the generator's voltage regulation is a negative value. That is, the voltage regulation of the generator is −9.76 per cent. If the output voltage increases under load (as in an over-compound generator), the voltage regulation is a positive figure.

To overcome this poor voltage regulation, an over-compound generator could be used. As the graphs in Figure 13.33 show, with the right amount of over-compounding the increase in the generator's terminal voltage under load would cancel out the voltage drop across the feeder. In our example, this would need to be a rise of 10 V above the no-load voltage.

A low voltage regulation figure means the generator installation has a more stable output voltage than one with a high voltage regulation figure. For example, a generator with -5 per cent voltage regulation would have a full-load voltage only 5 per cent less than its no-load value. If the no-load voltage is 420 V, the full-load voltage is 400 V.

Power loss and efficiency

To deliver 40 kW of electrical power to the load in Figure 13.34 requires a prime mover powerful enough to drive the generator. The question is, how powerful? Because energy cannot be created or destroyed, only transformed, we can say that the prime mover must provide at least 40 kW of mechanical power when the generator is delivering 40 kW of electrical power.

But this assumes the generator is 100 per cent efficient, which we know it is not, as no machine is capable of such efficiency. To see why, we look now at some of the electrical power losses in the circuit and in particular in the generator, which is shown again with greater detail in Figure 13.35.

FIGURE 13.35 Circuit for Example 13.5

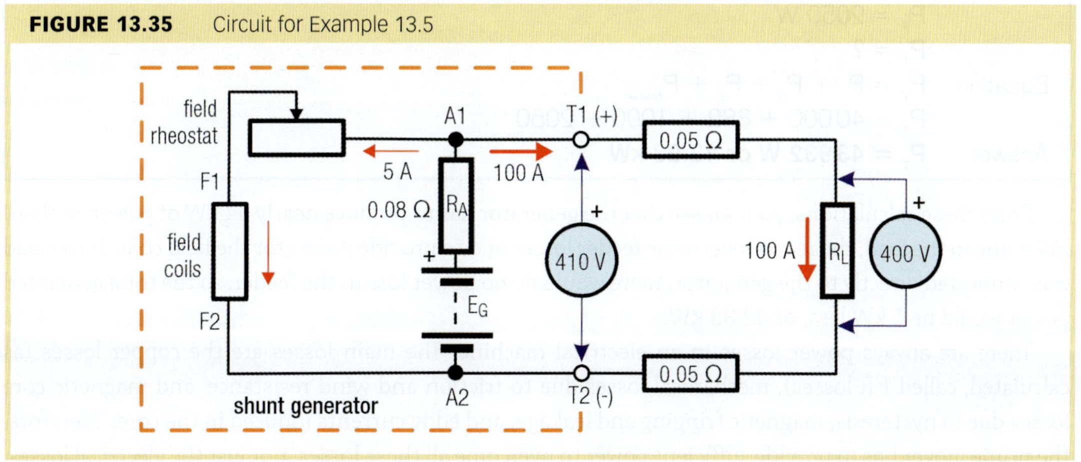

EXAMPLE 13.5

Find the total electrical power (P_T) produced by the generator in Figure 13.35.

Solution

The total generated power is the sum of the load power (P_L), the power losses due to series resistances including the armature, brushes and feeder resistances (P_R) and the power consumed by the field coils (P_F).

To find the power dissipated in the armature and brush resistance we first determine the total armature current. Looking at the circuit shows there are two currents, the field coil current and the load current. By Kirchhoff's current law, these currents add, giving 105 A flowing in the armature.

Values $I_A = I_L + I_F = 105\ \text{A}$
$R_A = 0.08\ \Omega$
$P_A = ?$

Equation $P_A = I_A^2 \times R_A = 105^2 \times 0.08$

Answer $\mathbf{P_A = 882\ W}$

The power dissipated in the feeder is found in a similar way, in which the feeder current equals the load current of 100 A:

Values $I_{feeder} = I_L = 100\ \text{A}$
$R_{feeder} = 0.1$

Equation $P_{feeder} = I_L^2 \times R_{feeder} = 100^2 \times 0.1$

Answer $\mathbf{P_{feeder} = 1000\ W}$

The power required by the field coils is the product of the applied voltage (V_T) and the field current. The applied voltage under full-load conditions, as previously found, is 410 V.

Values $V_{field} = V_T$ (at full load) $= 410\ \text{V}$
$I_F = 5\ \text{A}$
$P_F = ?$

Equation $P_F = I_F \times V_T = 5 \times 410$

Answer $\mathbf{P_F = 2050\ W}$

The total power (P_T) produced by the generator is the sum of the individual powers dissipated in the load, armature, field coils and feeder: That is:

Values $P_L = 40\ \text{kW}$
$P_A = 882\ \text{W}$
$P_{feeder} = 1000\ \text{W}$
$P_F = 2050\ \text{W}$
$P_T = ?$

Equation $P_T = P_L + P_A + P_F + P_{feeder}$
$P_T = 40\,000 + 882 + 1000 + 2050$

Answer $\mathbf{P_T = 43\,932\ W\ or\ 43.93\ kW}$

From these calculations, you can see that the generator has to produce nearly 44 kW of power, with 40 kW going to the load, the rest to overcome feeder losses and to provide power for the field coils. If the load was connected directly to the generator, there would be no power loss in the feeder, so the total generated power would be 1 kW less, or 42.93 kW.

There are always power losses in an electrical machine. The main losses are the copper losses (as calculated, called I^2R losses), mechanical losses due to friction and wind resistance, and magnetic core losses due to hysteresis, magnetic fringing and leakage, and eddy currents induced in the core. Therefore, the prime mover has to provide sufficient power to overcome all these losses, not just the electrical losses.

The ratio of the power output and the power input gives a measure of a machine's efficiency. The higher this value, the better. Efficiency (symbol η, Greek letter eta) is usually expressed as a percentage, and is found with this equation:

$$\eta\% = \frac{P_{out}}{P_{in}} \times 100$$

where:

$\eta\%$ = efficiency expressed as a percentage

P_{out} = output power in watts

P_{in} = input power to the machine when the machine is producing its rated output power.

EXAMPLE 13.6

Find the efficiency of a generator that delivers 40 kW of output power when the prime mover input power is 48 kW.

Solution

Values $P_{in} = 48$ kW
$P_{out} = 40$ kW
$\eta\% = ?$

Equation $$\eta\% = \frac{P_{out}}{P_{in}} \times 100 = \frac{40\,\text{kW}}{48\,\text{kW}} \times 100$$

Answer **efficiency = 83.3 per cent**

Getting back to the question of the required power output of the prime mover, we can say that if the 40 kW generator is 83.3 per cent efficient, the prime mover must deliver 48 kW of mechanical power to the generator. However, this amount of mechanical power is only required when the generator is delivering full electrical power to its load. Under full load, the armature current creates a strong magnetic field (armature reaction) that opposes the main field. The prime mover has to provide the torque to keep the armature rotating against this magnetic opposition. Under no-load conditions, the armature reaction is considerably less, so the prime mover does not have so much work to do.

SUSTAINABILITY

Increasing the efficiency of electrical machines and appliances is one of the principles of sustainable energy

CHAPTER SUMMARY

- DC machines (motors and generators) have the same construction.
- The frame and pole pieces in a generator provide a path for the magnetic field created by its field coils. This field interacts with the rotating armature to induce a voltage in the armature coils. The alternating armature current is converted by the commutator to direct current in the load.
- A lap wound armature has many parallel paths and suits a high current, low voltage machine. A wave wound armature has two parallel paths and suits a high voltage, low current machine.
- The magnetic field in a generator can be produced by permanent magnets (smaller machines) or field coils supplied by either an external DC source or by the generator itself.
- Self-excited generators can be: shunt, where the field coils are in parallel with the generator; series, in which the armature (load) current flows in the field coils; or compound, which combines shunt and series field coils.
- Generators are rated by their output current at a rated output voltage, and have a specific direction of rotation. If the direction of rotation is reversed, a self-excited machine will tend to cancel its residual magnetism and not produce an output voltage.
- The mechanical energy required to drive a generator is proportional to, but greater than, the electrical energy being delivered by the generator.
- A DC generator has resistive (I^2R) losses, core losses (magnetic effects) and mechanical losses (friction, etc.).
- Resistive losses are those due to any resistance in series with the armature's generated EMF (E_G) and the output terminals of the generator. These include the resistance of the armature and associated brushes, and the resistance of series field coils (where used).
- DC machines (motors and generators) operate at a certain efficiency, which is the ratio of the power delivered by the machine (P_{out}) and the power required to drive the machine (P_{in}).

- The percentage voltage regulation of a DC generator is a measure of how much its output voltage drops under load.
- A prime mover driving a DC generator has to overcome the opposing magnetic fields created by the armature and the generator's main field. The armature field strength increases with load current, requiring more mechanical power from the prime mover.

REVIEW EXERCISES

Check your answers at the back of the book.

1 What is the purpose of the field coils in a DC generator?

2 A separately excited DC generator has a faulty rheostat that causes the field current to drop from 10 A to 5 A. How does this affect the output voltage of the generator?

3 A DC generator has a machine constant of 7.5. Its magnetic flux is 250 mWb, and the armature is rotating at 1000 RPM. Calculate the no-load output voltage.

4 The rotational speed of the prime mover driving a DC generator suddenly increases. What is the effect on the output voltage of the generator?

5 What is the effect on starting if there is little or no residual magnetic field in a self-excited DC generator?

6 An armature is generating an EMF of 400 V and is providing a current of 60 A to a load. The resistance of the armature and brushes is 0.36 Ω. Calculate the terminal voltage of the generator.

7 For the conditions shown in the compound generator in Figure 13.36, calculate the:
   a terminal voltage V_T
   b power dissipated in the series field coils
   c voltage between the armature terminals A1 and A2
   d power dissipated in the shunt field coils
   e power dissipated in the armature resistance
   f power dissipated in the load (assume zero feeder resistance)
   g total power being produced by the generator.

8 A 25 kW generator has the following losses:
   - Excitation 650 W
   - Windage and friction 920 W
   - Coupling losses 700 W

   Calculate the full load efficiency of the generator.

9 The armature of a DC generator produces a no load voltage (E_G) of 330 V. Calculate the resistance of the armature if the terminal voltage is 300 V when the armature current is 120 A.

10 The output voltage of a DC generator is 212 V at no load, and 197 V at full load. What is its percentage voltage regulation?

FIGURE 13.36

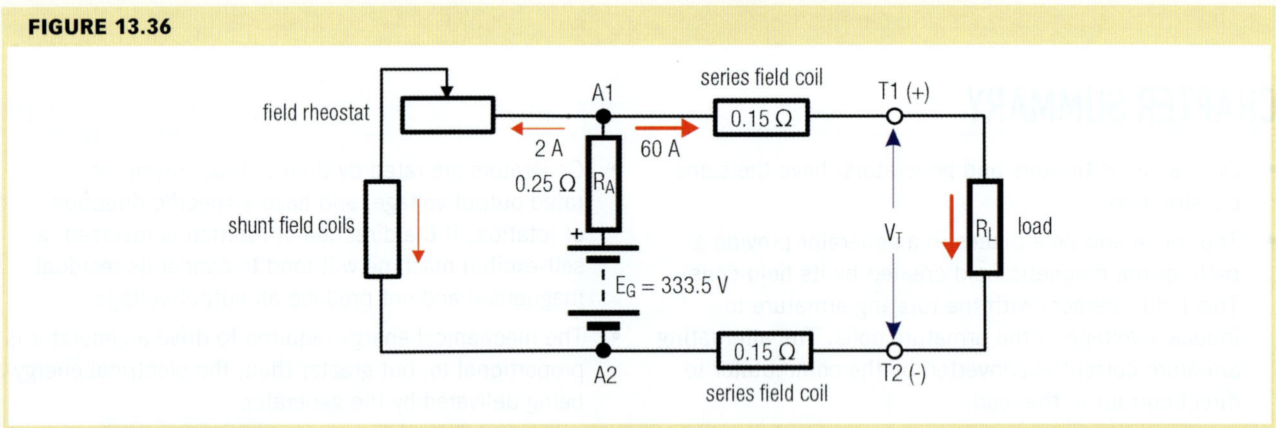

ONLINE RESOURCES

COMPLETE WORKSHEET THIRTEEN

Check with your instructor for worksheets on this chapter.

CHAPTER 14

DC MOTORS

The modern DC motor was invented by accident in 1873, when Gramme connected the DC generator he had invented to a second similar unit. He found the second unit behaved as a powerful motor when the first was driven by a prime mover. Further development was done by Sprague and by the late 1880s DC electric motors were being used in lifts, trolley cars and trains. DC motors and DC generators share much in common, as Gramme discovered. This chapter describes the operating principles and construction of various types of DC motors. We also look at how each type performs when providing power to a load, along with their typical applications and limitations.

CHAPTER OUTLINE

14.1 Introduction

A DC motor converts electrical energy into mechanical energy – a DC generator in reverse. Like the DC generator, DC motors are classified by how the main magnetic field is produced. This gives the same five types as for generators. That is, permanent magnet and separately excited, and the three types of field coil connections used in self-excited generators: shunt, series and compound.

DC motors come in a wide variety of sizes and power ratings, from tiny motors in toys to large industrial motors rated in hundreds of kilowatts. Like the generator, a typical DC motor has an armature that rotates in a stationary magnetic field. Power to the armature is through the same type of commutator and brush gear, the difference being the direction of power transfer. In a generator electric power is delivered; in a motor it is consumed.

While conventional DC motors are the main topic in this chapter, we also look briefly at other types. These include motors that have an electronic circuit instead of the usual commutator and brush set. We start by explaining the operating principles of a conventional DC motor.

FIGURE 14.1 Examples of DC motors, both industrial and automotive

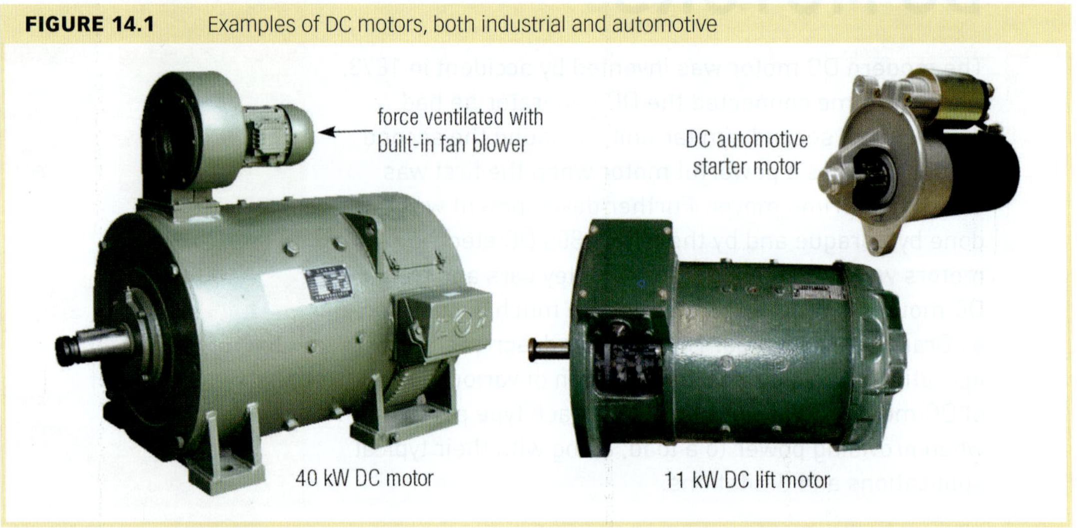

14.2 DC Motor Operating Principles

Chapter 11 explained that the current in a conductor creates a magnetic field around the conductor. If that conductor is in a magnetic field, the two fields interact and the conductor will move. A simple DC motor is shown in Figure 14.2, in which a single coil of wire, free to rotate in a magnetic field, is connected to a battery so current flows in the coil.

FIGURE 14.2 A basic DC motor in which current flowing in the coil produces a magnetic field that interacts with the main field, causing the loop to rotate

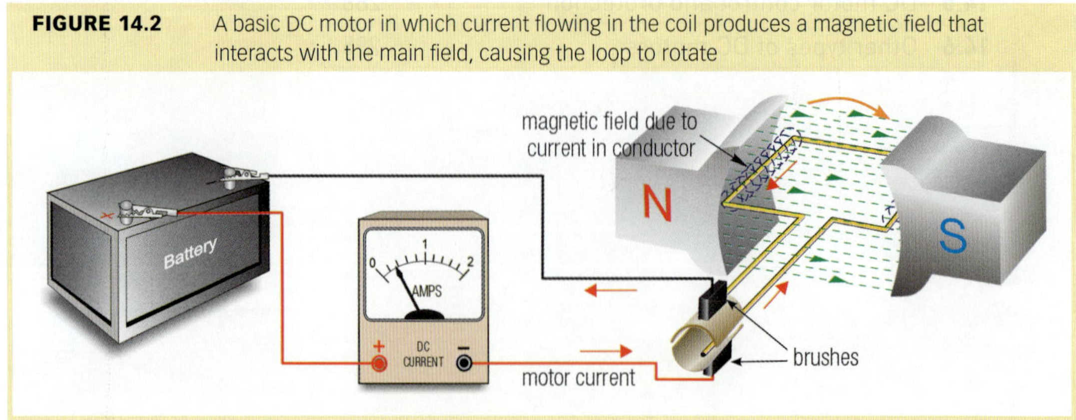

The motor in Figure 14.2 is identical to the basic DC generator, except the current flowing in the coil is from an external supply. The commutator, as in the generator, switches the ends of the coil so the current flow in the coil is always in the correct direction. That is, conductors near the north pole of the field *always* have current flowing in one direction, and those near the south pole have current in the opposite direction. The current in the coil creates a circular magnetic field around the conductor, with a direction that depends on the direction of the current.

The effect of the interacting magnetic fields is shown in Figure 14.3. Fleming's right-hand grip rule, shown in (a), gives the direction of the magnetic field around each conductor. Because lines of force travelling in the same direction repel each other, the main magnetic field is distorted by the fields around the conductors, as shown in (b). The resulting magnetic forces acting on the conductors tend to push the left-hand conductor upwards and the right-hand conductor downwards, causing the coil to rotate.

FIGURE 14.3 The motor effect, which is caused by an electric current flowing in a coil that is free to rotate in a magnetic field

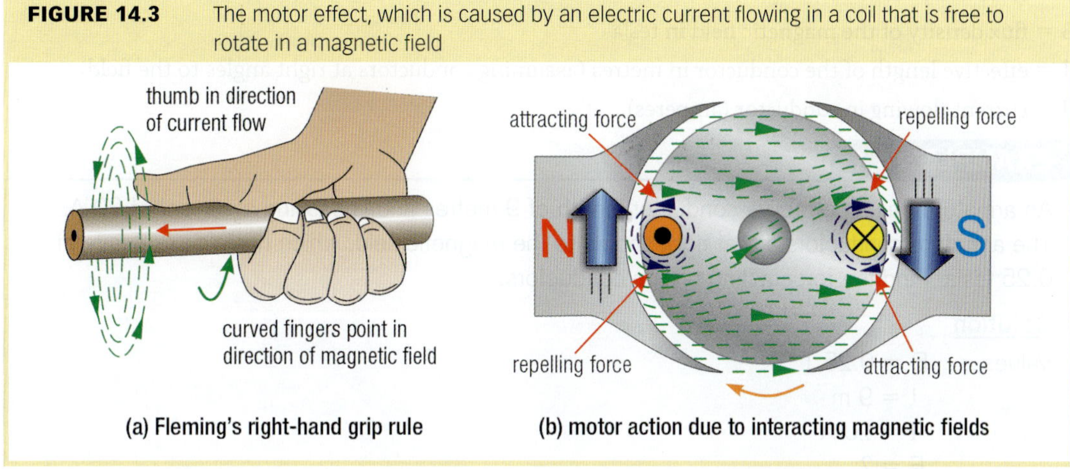

The rotational direction of a motor depends on the polarity of the magnetic field and the direction of the coil current. Reversing either one of these reverses the direction of rotation. Fleming's *left-hand rule* is a way to determine the rotational direction of a DC motor and is similar to his right-hand rule explained in Chapter 12. The right-hand rule is for generators, shown in Figure 14.4(b). This rule is used to determine the direction of an *induced* current when you know the direction of motion and the direction of the magnetic field.

In a motor, the direction of the current and the magnetic field are known. To find the direction of rotation, arrange the thumb and fingers of your *left* hand so they are at mutual right angles, as shown in Figure 14.4(a). The thumb points in the direction of the magnetic force acting on the conductor, and therefore its direction of rotation.

FIGURE 14.4 Fleming's left-hand and right-hand rules apply only when current is assumed to flow from positive to negative. Change hands if current is assumed to flow from negative to positive.

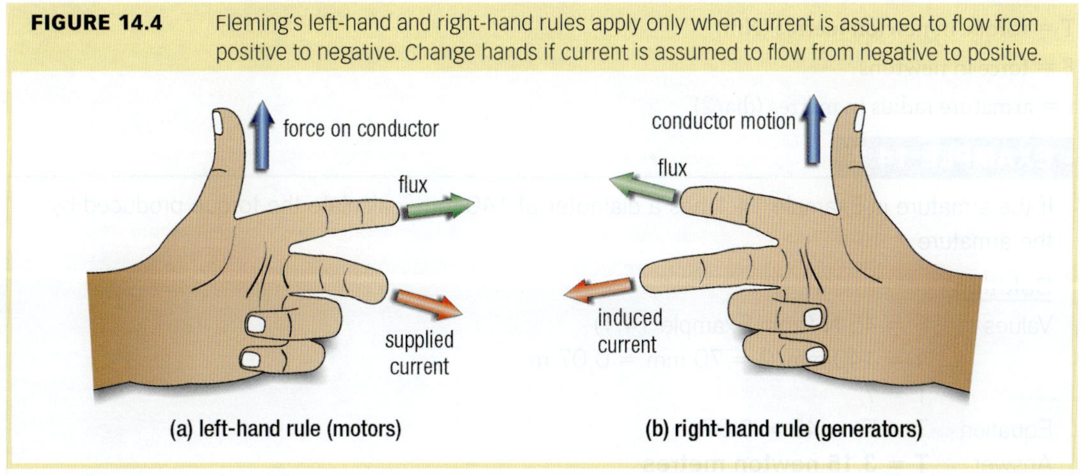

Force and torque

The force acting on a current-carrying conductor depends on the strength of the magnetic field, the current in the conductor and the effective length of the conductor. The effective length is that part of the conductor in the main magnetic field. In a practical armature, as explained in Chapter 13, the effective conductor length depends on the number of coils in the armature and the turns per coil. For our purposes, when using the term 'effective length', we are including all of these factors. The force acting on the armature conductors can be found with this equation:

$F = BlI$

where:

F = force in newtons

B = flux density of the magnetic field in tesla

l = effective length of the conductor in metres (assuming conductors at right angles to the field)

I = current flowing in conductor (amperes).

EXAMPLE 14.1

An armature has an effective conductor length of 9 metres and is passing a current of 20 A. The armature conductors are at right angles to the magnetic field, which has a flux density of 0.25 tesla. Find the force acting on the conductors.

Solution

Values $B = 0.25$ T

$l = 9$ m

$I = 20$ A

$F = ?$

Equation $F = BlI = 0.25 \times 9 \times 20$

Answer **F = 45 newtons**

Of more interest is the *torque* a motor produces. Torque, or turning effort, as explained in Chapter 4, is a force that produces rotational motion and is measured in newton metres (Nm). If the force is known (as in the above example), torque can be determined if we also know the diameter of the armature, or more specifically, the radius of the armature (half the diameter). Torque is found with this simple equation:

$T = Fr$

where:

T = torque in newton metres (Nm)

F = force in newtons

r = armature radius in metres (dia/2).

EXAMPLE 14.2

If the armature in Example 14.1 has a diameter of 140 mm, calculate the torque produced by the armature.

Solution

Values $F = 45$ N (from Example 14.1)

$r = 140\text{ mm}/2 = 70\text{ mm} = 0.07$ m

$T = ?$

Equation $T = Fr = 45 \times 0.07$

Answer **T = 3.15 newton metres**

We discuss torque again later in this chapter. First we need to look further at a motor's electrical characteristics, in particular the factors determining the value of the armature current.

Motor back-EMF

When a coil of wire rotates in a magnetic field, a voltage is induced in the coil. This occurs whether the coil is driven by a prime mover or whether the coil is moving as a motor. Either way produces a voltage in the coil, and if there is a current path, current flows.

Figure 14.5(a) shows a simple DC motor rotating in a magnetic field, in which the coil current is travelling towards you in the conductors near the north magnetic pole, and into the page at the south pole. This current causes the coil to rotate as a motor. However, because the coil is rotating, a current is *induced* in the coil that, by Fleming's right-hand rule, flows in the direction shown in (b). The induced (generated) current flows in the *opposite* direction to the supplied current. As a result, the total current in the armature is reduced.

FIGURE 14.5 The induced voltage in a motor has the opposite polarity to the applied voltage

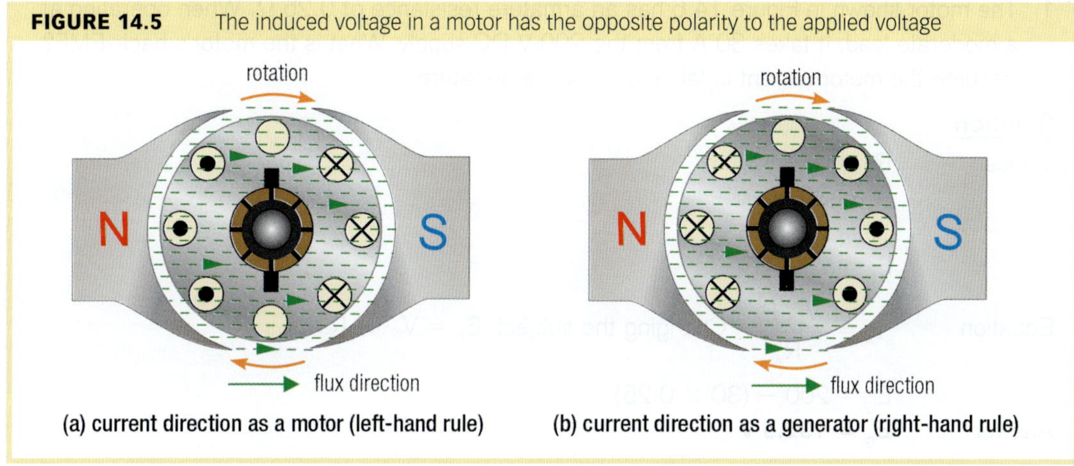

(a) current direction as a motor (left-hand rule) **(b) current direction as a generator (right-hand rule)**

The voltage generated by an armature (E_G), as explained in the previous chapter, can be calculated with the equation $E_G = \Phi nk$. That is, the generated armature voltage equals the product of the flux (Φ), speed of rotation (n) and the machine constant (k). This voltage has the opposite polarity to the motor's applied voltage and is therefore referred to as the motor's *back-EMF*.

The voltage available to cause current to flow in the armature is the difference between the applied voltage (V) and the back-EMF (E_G). If the armature resistance is known, applying Ohm's law gives a useful equation to find the armature current (I_A):

$$I_A = \frac{V - E_G}{R_A}$$

where:

I_A = armature current in amps
V = applied voltage (volts)
E_G = armature generated EMF
R_A = armature resistance (ohms).

Because the back-EMF is proportional to the speed of armature rotation, if the motor tends to slow down due to increased load, the back-EMF will reduce. This allows more current to flow in the armature, thereby providing the extra power as torque needed to drive the increased load.

Example 14.3 shows the effect of back-EMF on motor current under different loads. Although the motor shown would require additional power to provide field current, this is not included in the example.

FIGURE 14.6 Circuit for Example 14.3

EXAMPLE 14.3

1 The motor shown in Figure 14.6 has an armature resistance of 0.25 Ω. When operating at a moderate load, it takes 30 A from the 200 V DC supply. What is the motor's back-EMF? Assume the motor current is taken only by the armature.

Solution

Values $R_A = 0.25\ \Omega$

$I_A = 30\ A$

$V = 200\ V$

$E_G = ?$

Equation $I_A = \frac{V - E_G}{R_A}$. Changing the subject, $E_G = V - I_A R_A$

$E_G = 200 - (30 \times 0.25)$

Answer $\mathbf{E_G = 192.5\ V}$

2 Find the motor's back-EMF if the load increases, causing the motor to take an armature current of 50 A.

Solution

New values $I_A = 50\ A$

Equation $E_G = V - I_A R_A$

$E_G = 200 - (50 \times 0.25)$

Answer $\mathbf{E_G = 187.5\ V}$

Example 14.3 shows that the motor takes more armature current when the load increases because the back-EMF reduces under load. That is, the current taken by a DC motor increases as the load on the motor increases, which is to be expected. But what about the torque being developed?

Torque

The force producing torque is due to the interaction of the armature's magnetic field and the main magnetic field. If either field is increased in magnetic strength, the force between them will also increase. Because the strength of the armature's magnetic field is proportional to the armature current, torque is therefore proportional to magnetic field flux (Φ) and armature current (I_A).

A general equation to calculate torque needs to include other factors, such as the number of poles, how the armature is wound and the number of effective conductors, plus a conversion constant of 0.318. For our purposes, we can put all these together into the machine's constant (k) to give this equation:

$T = k\Phi I_A$

where:

T = torque in newton metres (Nm)

k = machine constant

Φ = main field flux in webers (Wb)

I_A = armature current.

EXAMPLE 14.4

The motor in Figure 14.6 has a main field flux of 0.035 Wb. Calculate the torque delivered by the motor when the armature current is (a) 30 A and (b) 50 A. Assume a machine constant (k) of 100.

Solution

Values $\Phi = 0.035$ Wb

I_A = (a) 30 A (b) 50 A

$k = 100$

$T = ?$

Equation $T = k\Phi I_A = 100 \times 0.035 \times 30$

Answer (a) **T = 105 Nm**

Equation $T = k\Phi I_A = 100 \times 0.035 \times 50$

Answer (b) **T = 175 Nm**

This example shows that as the armature current increases in response to an increased load, the motor torque also increases. This is to be expected, as armature current is a variable in the equation $F = BIl$, which we used in Examples 14.1 and 14.2.

Power output

The mechanical output power of a DC motor can be calculated in two ways. The first uses an equation based on the machine's mechanical performance, namely torque and speed, which we introduced in Chapter 4. The equation is:

$$P_{out} = \frac{2\pi nT}{60} = \frac{nT}{9.55}$$

where:

P_{out} = output power in watts

n = speed in RPM

T = torque in newton metres.

EXAMPLE 14.5

If the motor in Example 14.4 is running at 500 RPM, calculate its power output when the armature current is 50 A.

Solution

Values $n = 500$ RPM

$T = 175$ Nm at 50 A (from Example 14.4)

$P_{out} = ?$

Equation $P_{out} = \frac{2\pi nT}{60} = \frac{nT}{9.55} = \frac{500 \times 175}{9.55}$

Answer **P_{out} = 9162 W or 9.162 kW**

The power output of a DC motor also approximately equals the product of the back-EMF and the armature current. This doesn't take into account all the losses in the machine, but is often accurate enough. That is:

$$P_{out} = E_G \times I_A$$

where:

E_G = armature generated voltage (volts)

I_A = armature current (amps).

EXAMPLE 14.6

The motor in the above examples has a back-EMF of 187.5 V when the armature current is 50 A. Find its output power (ignore losses).

Solution

Values	$E_G = 187.5$ V
	$I_A = 50$ A
	$P_{out} = ?$
Equation	$P_{out} = E_G \times I_A = 187.5 \times 50$
Answer	$\mathbf{P_{out} = 9375}$ **W or 9.375 kW**

Speed

The speed of a DC motor is directly proportional to its terminal voltage (V) and inversely proportional to the field flux (Φ). This means there are two ways to increase the speed of a motor: increase the supply voltage, or reduce the strength of the motor's main magnetic field. We showed that the power output of a motor is the product of its speed and torque, divided by 9.55, so rearranging that equation in terms of speed (n) gives:

$$n = \frac{9.55P_{out}}{T}$$

where:

n = speed in RPM

P_{out} = output power in watts

T = torque in newton metres.

EXAMPLE 14.7

If the motor in Example 14.6 is producing 175 Nm of torque, calculate its speed.

Solution

Values	$P_{out} = 9375$ W (from Example 14.6)
	T = 175 Nm
	n = ?
Equation	$n = \frac{9.55P_{out}}{T} = \frac{9.55 \times 9375}{175}$
Answer	**n = 511.61 RPM**

KEY POINTS...

- A DC motor has the same construction as a DC generator.
- Use Fleming's left-hand rule to find the direction of rotation of a motor when the direction of the current and magnetic flux are known.
- When a DC motor is running, the armature produces a back-EMF that opposes the supply voltage. This limits the value of the armature current.
- When the load on a DC motor is increased, its speed and therefore its back-EMF reduce, causing the armature current to increase. This provides the additional torque to drive the increased load.
- The back-EMF of a motor equals its terminal voltage minus the armature voltage drop (I_AR_A).
- The speed of a DC motor is directly proportional to its terminal voltage and inversely proportional to the strength of the main magnetic field.
- The output power of a DC motor is approximately equal to the product of its back-EMF and armature current. Power also equals the product of torque and speed divided by 9.55.

14.3 Types of DC Motors

As explained at the start of this chapter, there are five basic types of DC motors. These have an identical construction to their equivalent as a generator, and they all have their own characteristics. We look at each type in terms of its basic construction, connections and performance, along with the applications each type of motor is best suited for.

Permanent magnet DC motor

The simplest DC motor is the permanent magnet type. Originally only made for low-power use, their power output capability has increased significantly since the development of rare earth magnetic materials and improved construction techniques. Permanent magnet motors (and generators) are now available with power outputs of over 300 kW, although the majority are low power, around 500 W or less. They have a high efficiency, as there is no power required to maintain the magnetic field. Figure 14.7 shows a 375 W DC permanent magnet motor and its basic construction.

FIGURE 14.7 DC permanent magnet motors come in a wide range of sizes and power ratings. The 375 W motor shown here could be used to power an electric utility vehicle, such as a golf buggy.

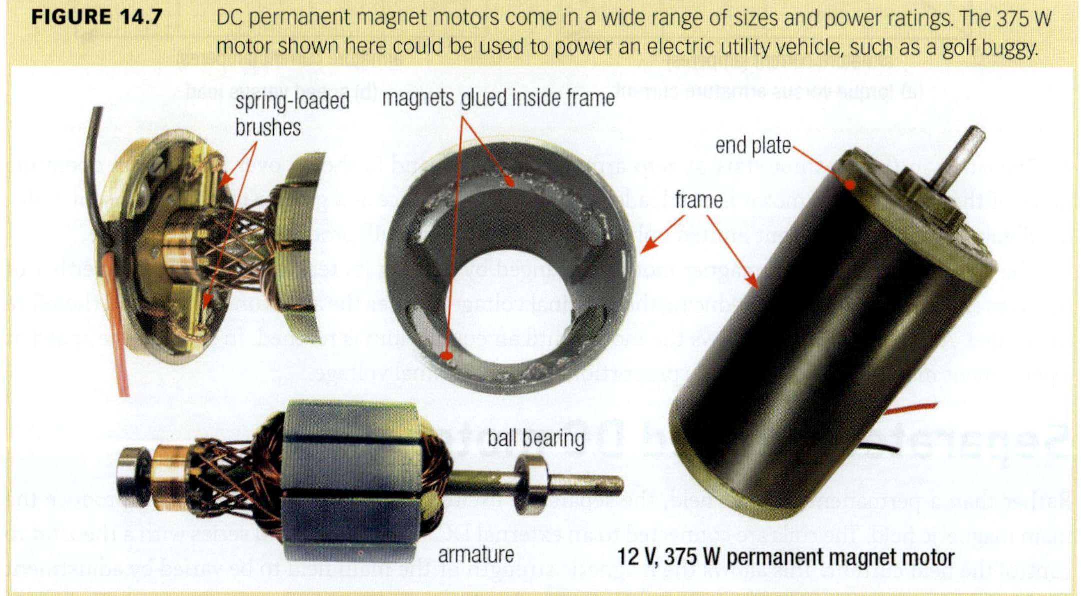

The construction of a permanent magnet motor involves an armature and brushes, with the armature free to rotate inside the magnetic field provided by the magnets. The construction shown in Figure 14.7 is typical of small motors. Because the magnetic field strength is fixed, speed control is only available by varying the armature current. As shown by the circuits in Figure 14.8, this can be done by varying the DC voltage applied to the motor, or if the DC supply is fixed, by inserting a rheostat in series with the motor.

FIGURE 14.8 Because the magnetic field strength is fixed, the speed of a permanent magnet motor is adjusted by varying the armature current

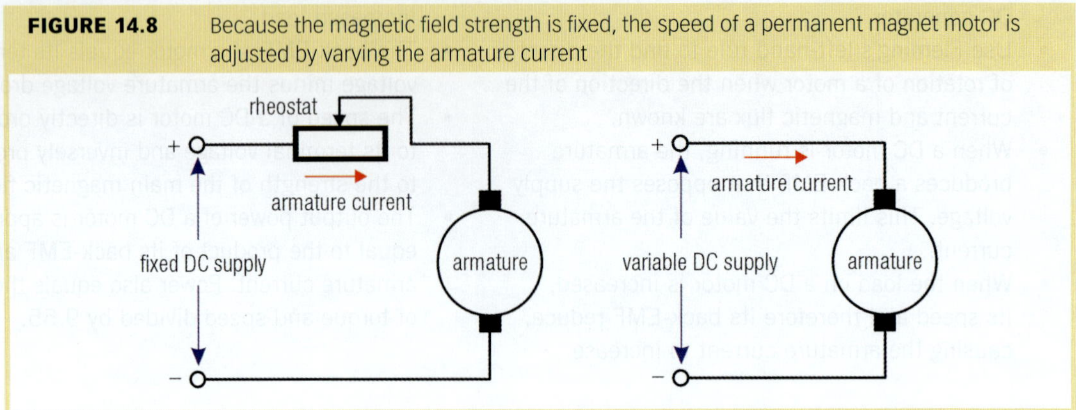

The graphs in Figure 14.9 show some of the characteristics of a permanent magnet motor. In (a), the motor torque is plotted against armature current and shows that the torque increases proportionally with the armature current. Graph (b) shows that the speed of the motor varies by a relatively small amount as the load is increased. Although the speed falls by a small amount as the load increases, it's enough to cause the armature current to increase and provide the additional torque required by the load.

FIGURE 14.9 Torque and speed characteristics for a permanent magnet motor

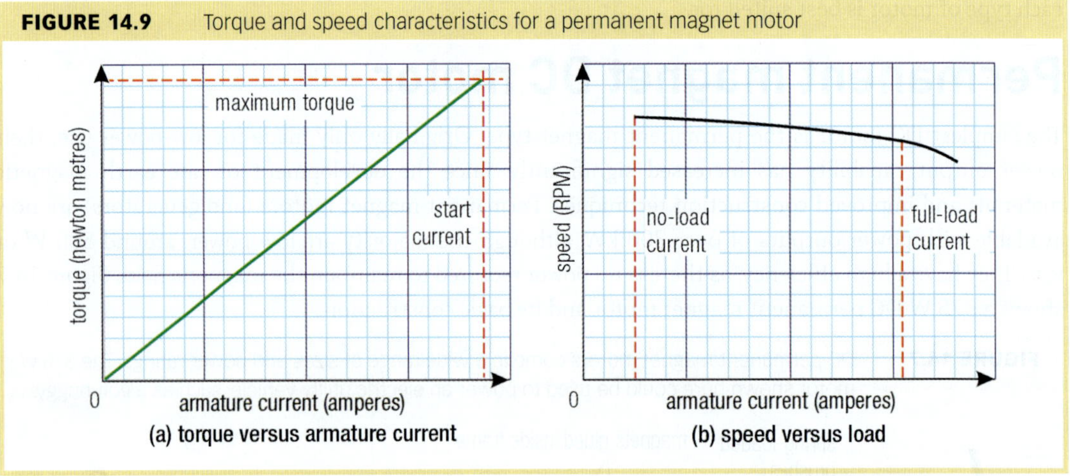

FYI

DC motors have their maximum torque at start up because the armature current is a maximum

The curve in (b) does not start at zero armature current, and is shown over the normal operating range of the motor. If the motor is overloaded, its speed will reduce at a greater rate. If the motor stalls, it will take an armature current limited only by its resistance, and will produce maximum torque.

The speed of a permanent magnet motor is changed by varying its terminal voltage, using either of the ways shown in Figure 14.8. Reducing the terminal voltage reduces the armature current and therefore the torque. As a result, the load slows the motor until an equilibrium is reached. In general, the speed of a permanent magnet motor is directly proportional to its terminal voltage.

Separately excited DC motor

Rather than a permanent magnet field, the separately excited DC motor has field coils to produce the main magnetic field. The coils are connected to an external DC supply, usually in series with a rheostat to control the field current. This allows the magnetic strength of the main field to be varied by adjustment of the field current. Changing the armature terminal voltage also varies the motor speed. These two methods can also be used together to give more control over the speed.

The circuit diagram of a field-controlled separately excited DC motor is shown in Figure 14.10. The DC supply to the armature is usually from a different source to that for the field. Varying the field rheostat changes the field current. This type of motor is used if a wide range of speed control is important, such as in traction applications. Its speed and torque characteristics are the same as for a permanent magnet motor (see Figure 14.9).

FIGURE 14.10 The speed of a separately excited motor can be varied by changing the field current or the armature current, or a combination of both methods

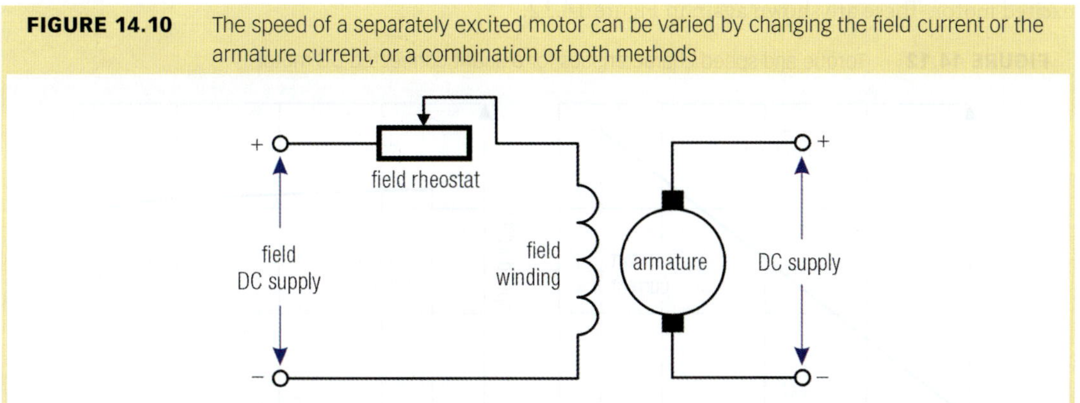

The speed of a DC motor is *inversely* proportional to the field flux. Increasing the resistance of the rheostat in Figure 14.10 reduces the field current and therefore the strength of the magnetic field, so the motor speed increases. When the main field flux is reduced, the armature has to spin more quickly to generate enough back-EMF to oppose the terminal voltage and limit the armature current. If the field current is reduced below a certain level and the motor is unloaded, it can reach a very high, even dangerous speed. This applies to all DC motors.

To reverse the direction of a separately excited motor, reverse either the armature connections or the field connections. Reversing both will not change the direction. This also applies to all DC motors.

SAFETY

If the field (excitation) current in a DC motor is too low, the motor can reach dangerously high speeds

Shunt DC motor

The shunt DC motor, like the shunt connected generator, has its field coils connected in parallel (shunt) with the armature. Unlike the separately excited motor, the field current and the armature current come from the same DC supply. Speed control can be achieved with a rheostat connected in series with the field coils. Figure 14.11 shows the internal connections of a shunt motor.

FIGURE 14.11 Two-pole shunt motor connections

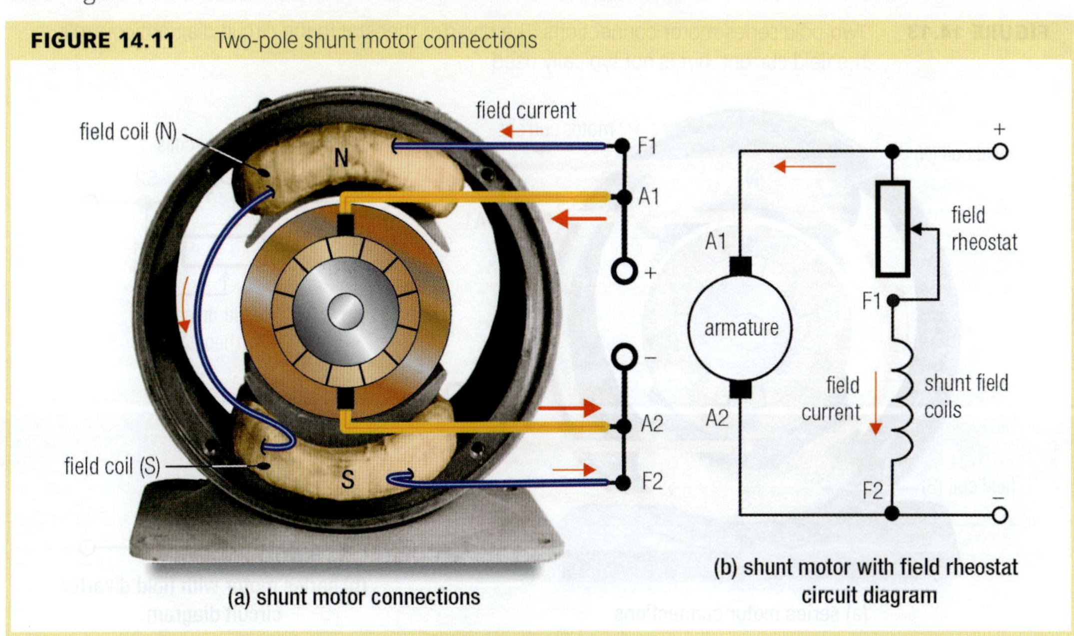

(a) shunt motor connections

(b) shunt motor with field rheostat circuit diagram

The usual way of controlling the speed of a shunt motor is with a field rheostat. As already pointed out, reducing the main field flux below a certain level can cause the motor speed to increase dramatically. This effect can actually help maintain the loaded speed, in which armature reaction caused by the increased load weakens the main field, which tends to increase and therefore stabilise the motor speed. The speed and torque characteristic curves for a shunt motor are the same as for a permanent magnet and separately excited motor. These are shown again in Figure 14.12.

FIGURE 14.12 Torque and speed characteristics for a shunt connected DC motor

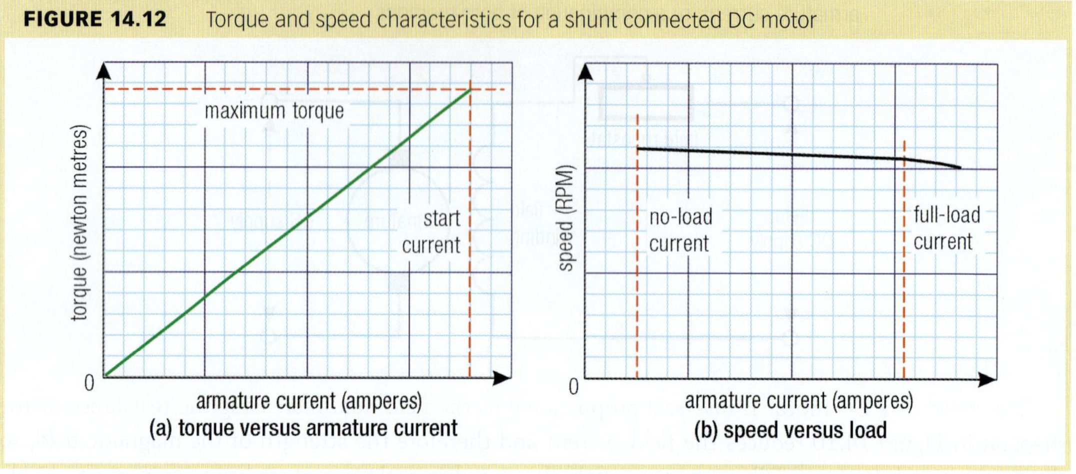

A shunt DC motor operates at a nearly constant speed, regardless of load. It suits applications where a constant speed is required. It also suits a situation where the load speed must be variable, with the new speed remaining constant regardless of load variations. Applications include powering lathes, large blowers, pumps, lifts and elevators. Motor reversal is achieved by reversing either the armature or field connections.

Series DC motor

The series motor has its field coils connected in series with the armature, as shown in Figure 14.13. The field coils, as in the series generator, are wound with a few turns of heavy gauge wire. If required, speed control can be obtained with a diverter rheostat connected as shown in Figure 14.13(b). Adjustment of the rheostat diverts more or less current from the field coils, giving some control over the speed.

FIGURE 14.13 Two-pole series motor connections. The diverter rheostat in the circuit diagram controls the field current, but is not typically used.

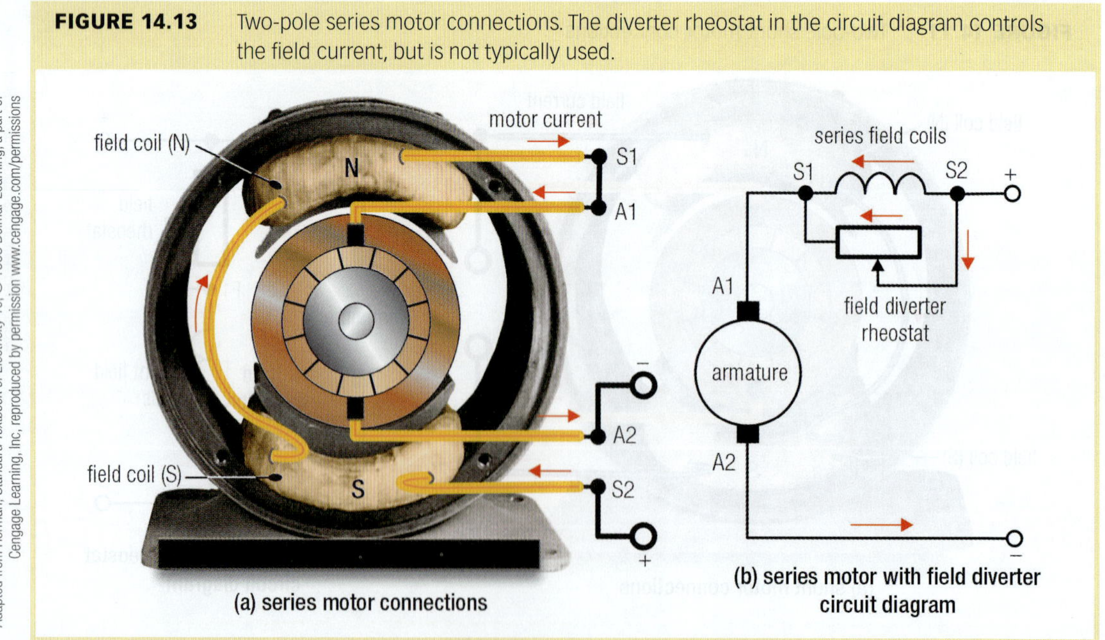

The field strength in a series motor depends on the armature current. If a series motor is unloaded and connected to a DC supply, the only load is friction caused by bearings and air. Therefore, the armature and the field current is small, giving a small field strength. To produce sufficient back-EMF to limit the motor current means the armature has to rotate at a high speed, possibly too high.

SAFETY

An unloaded series motor can reach dangerously high speeds

This is a fundamental issue with a series motor, as it can reach a no-load speed that might cause it to self-destruct. This is especially critical in a large motor, as the centrifugal forces being produced by the high speed can tear the armature coils from their slots. For this reason, a series motor should always be directly coupled to its load. This can be through gearing, but not with a belt drive as belts can break.

At start-up, the series motor takes maximum current, which decreases rapidly as the motor gains speed. The flux developed is approximately proportional to the motor current (up to full load), so the torque of a series motor is nearly proportional to the *square* of the motor current. Magnetic saturation starts to occur above full load, so the torque does not increase so rapidly. The characteristic curves of a series motor are shown in Figure 14.14.

FIGURE 14.14 Torque and speed characteristics for a series motor

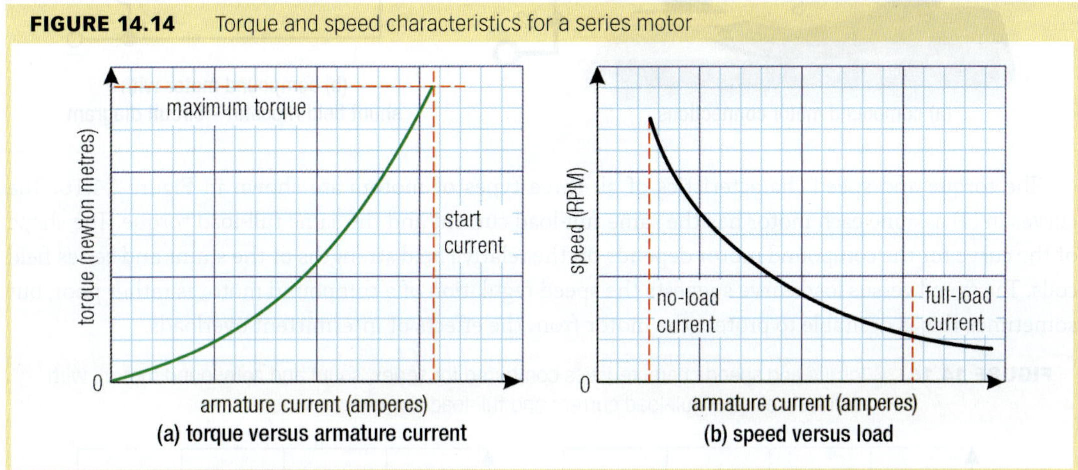

The torque of a series motor at start-up can be up to eight times the rated torque, due to the high current flowing when the motor is just starting to rotate. This makes the series motor ideal for situations where a high starting torque is needed. An example is the starter motor in a car.

A shown by the curve in Figure 14.14(b), the speed of a series motor varies considerably with load. Speed control is therefore not really applicable, except where the motor is driving a constant load. In this case, speed control can be achieved by diverting current from the field coils by the diverter rheostat in the circuit of Figure 14.13.

Because of its ability to start against a heavy load, the series motor is used in applications such as traction motors in trains and electric vehicles (cars, buses, etc.). The motors are also used in cranes, lifts and whenever a high starting torque is needed. To reverse the direction of a series motor, swap the connections of either the series coils or the armature.

Compound motor

A shunt motor has good speed regulation with load variations, but a comparatively low starting torque. However, its torque remains relatively constant over its working range. A series motor has a high starting torque, but poor speed regulation with load changes. A compound motor has characteristics of both these motors and therefore has shunt *and* series field coils, as shown in Figure 14.15. The circuit in Figure 14.15 is 'long-shunt', with the rheostat connected to the positive terminal. The short-shunt connection would have the rheostat connected to the armature terminal A1.

The characteristics of the compound motor depend on the strength of the shunt or series fields. If the field due to the series coil is relatively weak compared to the shunt coil field, the motor will have better torque characteristics than a shunt motor, and with only a minor loss of speed regulation. Making the series field stronger than the shunt field makes the compound motor behave more like a series motor. However, the presence of the shunt field ensures the motor speed does not reach a dangerous level under no load.

FIGURE 14.15 Two-pole compound motor connections, which are made at the motor's terminal box

(a) compound motor connections

(b) compound motor with shunt field rheostat – circuit diagram

The torque and speed characteristics of all three types of motors are shown in Figure 14.16. The curves in (a) assume each motor has the same full-load current and the same full-load torque. The shape of the curve for the compound motor depends on the relative field strengths of the shunt and series field coils. The speed versus load curve suggests the speed regulation of a compound motor is rather poor, but sometimes this is desirable to protect the motor from the effects of intermittent overloads.

FIGURE 14.16 Torque and speed characteristics compared for series, shunt and compound motors with identical values of full-load current and full-load torque

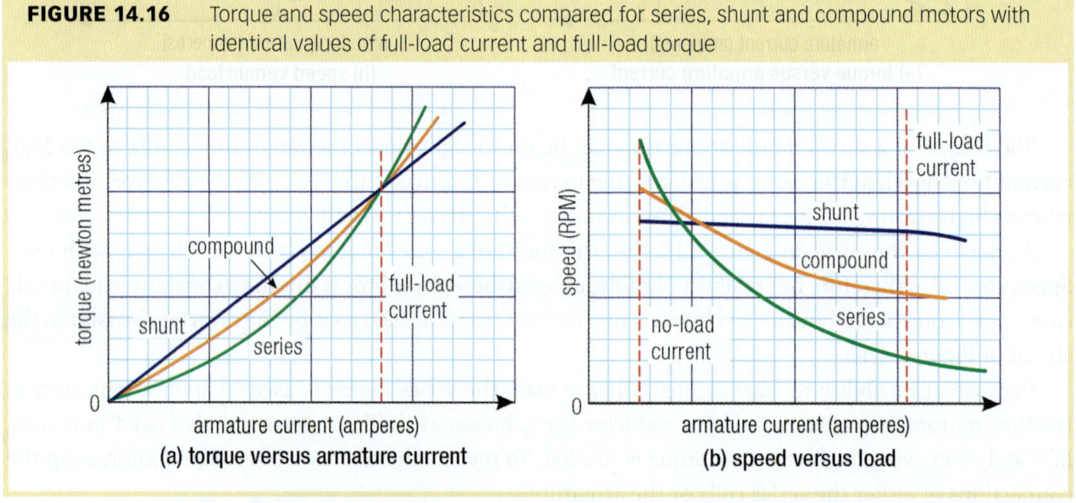

The compound DC motor has many applications. It suits a load that needs a high starting torque, but unlike the series motor, its unloaded speed remains within safe limits. Uses include driving machinery such as powered metal punches, metal cutting guillotines, traction systems, rolling mills and machines that are subject to sudden increases in loads.

Reversing the rotation of a compound motor requires some care. As with all motors, either the field coils or the armature connections are reversed. For a compound motor, this means reversing the connections of both the series and the shunt coils. It's usually simpler to reverse the armature connections. To do this, in Figure 14.15 connections A1 and A2 would be reversed so A1 goes to F1 and A2 goes to S1.

If the connections to the series coils are reversed without also reversing those to the shunt coils, the motor becomes a *differential* compound motor. This type of motor has little application because the speed of the motor increases under load due to the series field coils tending to cancel the field created by the shunt coils, thereby weakening the field as the load current increases. However, with careful design for a particular load situation, the speed of a differentially compounded motor can be held constant over its load range. It is therefore only used in special cases where a very constant speed is important.

KEY POINTS...

- Permanent magnet motors produce a linear torque and have good speed regulation. Their applications are increasing due to larger and more powerful motors becoming available.
- Separately excited DC motors require an external DC supply to provide the field current. They have similar characteristics to the permanent magnet motor, with the advantage that their speed can be controlled by varying the field current.
- A shunt DC motor has good load-speed regulation, and its speed is readily controlled by varying the current in the shunt field coils. It has similar characteristics to the separately excited and permanent magnet motors, but if the field current is too low, the motor speed can become excessive.
- The series motor has a high starting torque, but can reach a dangerously high speed if unloaded.
- A cumulatively compounded DC motor has series and shunt field coils connected so their magnetic fields add to each other. The characteristics of the motor depend on the relative strength of the series and shunt fields. Typically, a compound motor has a high starting torque due to the series windings, but has an acceptable no-load speed due to the shunt coils.
- All DC motors exhibit maximum torque at start-up (zero speed).
- A DC motor will run in reverse if the polarity of the field coils or the armature terminals are reversed, but not both.

14.4 Losses and Efficiency

All practical motors have losses that reduce their efficiency. Like the DC generator, the losses in a DC motor can be grouped as mechanical, electrical or magnetic. Mechanical losses include friction, due to bearings and brushes rubbing against the commutator, and windage, which is the friction air imposes on the rotating armature and cooling fan on the armature shaft.

Magnetic losses are caused by the hysteresis of the armature's iron core, plus eddy currents that are induced in the core, along with magnetic losses in the air gap between the field poles and the armature. These are called *iron losses* and are nearly constant over the full operating range of the motor. Eddy currents are reduced by making the armature core from insulated iron laminations. This stops induced eddy currents flowing lengthwise along the core. The laminations are made of a low hysteresis metal, usually silicon iron.

Electrical losses are due to the resistance of the copper windings, and are called *copper losses*. These plus the iron losses represent the main losses in the machine. Figure 14.17 illustrates the losses in a DC motor.

FIGURE 14.17 Losses in a DC motor cause the motor to heat up

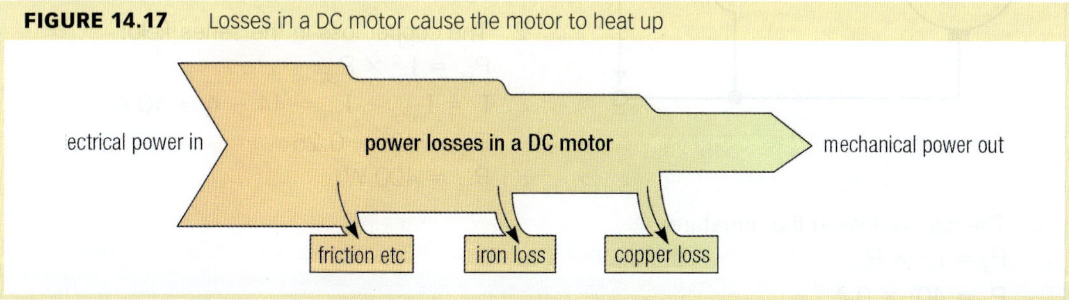

When a motor is loaded, its armature current increases. This increases the losses due to the armature resistance, so the copper losses in a motor increase with load. The difference between the input power and the output power gives the total power losses in the motor. For example, if a 10 kW rated motor takes 12 kW of power at full load, the power loss is 2 kW.

EXAMPLE 14.8

A motor is delivering a torque of 70 Nm at a speed of 1400 RPM while taking a current of 30 A from a 400 V DC supply. Calculate (1) motor input power, (2) output power and (3) total losses in the motor.

Solution

Values $V = 400$ V

$I = 30$ A

$T = 70$ Nm

$n = 1400$ RPM

$P_{in} = ?, P_{out} = ?$ losses = ?

1 Equation $P_{in} = VI = 400 \times 30$

Answer $\mathbf{P_{in} = 12\ kW}$

2 Equation $P_{out} = \frac{2\pi nT}{60} = \frac{nT}{9.55} = \frac{70 \times 1400}{9.55}$

Answer $\mathbf{P_{out} = 10.26\ kW}$

3 Equation Power loss $= P_{in} - P_{out} = 12$ kW $- 10.26$ kW

Answer **Power loss = 1.74 kW**

Copper losses can be calculated using resistance and current values, as in the following example:

EXAMPLE 14.9

Calculate the copper losses in the motor shown in Figure 14.18.

Solution

The total copper loss is the sum of the individual losses.

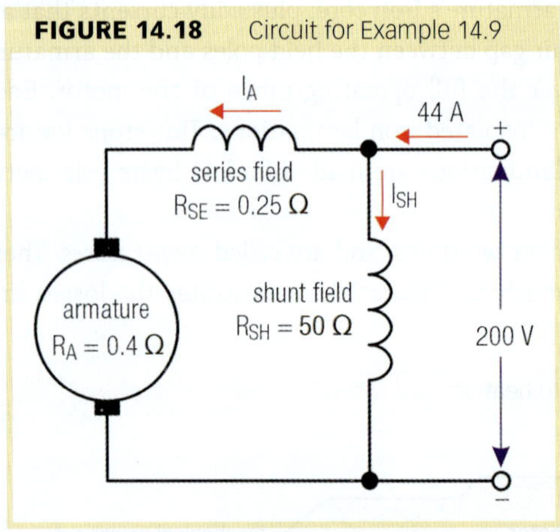

FIGURE 14.18 Circuit for Example 14.9

Values $V = 200$ V

$I_{motor} = 44$ A

$R_{SH} = 50\ \Omega$, $R_{SE} = 0.25\ \Omega$,

$R_A = 0.4\ \Omega$

1 The copper loss in the shunt field:

$$I_{SH} = \frac{V}{R_{SH}} = \frac{200}{50} = 4\text{ A}$$

$P_{SH} = V \times I_{SH} = 200 \times 4$

$P_{SH} = 800$ W

2 The copper loss in the series field:

$P_{SE} = I_A^2 \times R_{SE}$

$I_A = I_{motor} - I_{SH} = 44 - 4 = 40$ A

$P_{SE} = 40^2 \times 0.25$

$P_{SE} = 400$ W

3 The copper loss in the armature:

$P_A = I_A^2 \times R_A$

$P_A = 40^2 \times 0.4$

$P_A = 640$ W

Total copper loss $= 800 + 400 + 640$

Answer $= \mathbf{1.84\ kW}$

Iron and frictional losses in a motor can be determined by measuring the power it takes when unloaded (assuming it is safe to run the motor without a load). The relatively small armature copper losses at full load can be calculated from measurements taken of the armature resistance and full load current.

Efficiency

Chapter 13 explained that the ratio of the power output and the power input gives a measure of a machine's efficiency. Efficiency is given the symbol η (Greek letter *eta*) and is expressed as a percentage:

$$\% \eta = \frac{P_{out}}{P_{in}} \times 100$$

where:

η = efficiency

P_{out} = output power in watts

P_{in} = input power in watts.

EXAMPLE 14.10

Find the efficiency of a 250 V 15 kW DC motor that is taking a full-load current of 70 A.

Solution

Values $V = 250\ V$

$I_{FL} = 70\ A$

$P_{out} = 15\ kW$

% efficiency = ?

Equation $P_{in} = VI_{FL} = 250 \times 70$

$P_{in} = 17.5\ kW$

Equation $\%\eta = \frac{P_{out}}{P_{in}} \times 100 = \frac{15\ kW}{17.5\ kW} \times 100$

Answer **η = 87.7 per cent**

The efficiency of a motor varies according to its load. When a motor is loaded, its armature current increases. This increases the losses due to the armature resistance, so the copper losses in a motor increase with load. Motors are designed to give good efficiency at their rated load, although, as the curves in Figure 14.19 show, maximum efficiency does not always occur at full load.

FIGURE 14.19 Efficiency of a DC motor compared to speed and torque

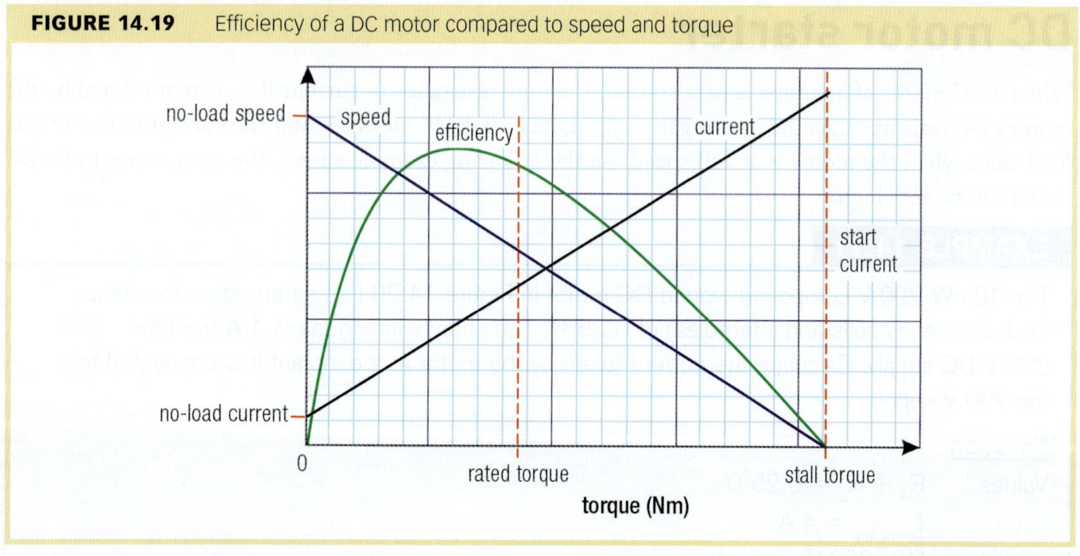

The speed curve (blue line) shows the speed characteristic of a typical DC motor with a constant terminal voltage and varying load. Its end points are the no-load speed and zero, which occurs at the stall torque, or starting torque. The current curve (black line) shows the relationship between current and torque. Its end points are the no-load current and the starting current.

The curve for efficiency (green) shows the relationship between the mechanical power output and the electrical power input. Maximum efficiency in a motor occurs when the fixed losses (iron, friction, windage, shunt coil copper losses, etc.) equal the variable losses. Variable losses are the copper losses in the armature and series windings. These losses are proportional to the square of the armature current (I^2R losses) and vary because the current is changing as the load changes.

The efficiency is zero at start-up (zero speed) as no mechanical power is produced. It quickly rises as the motor output exceeds its fixed losses, peaking when the variable and fixed losses are equal. At full load, depending on the design of the motor, the efficiency is slightly less than its maximum value. Above full load, the efficiency drops due to the rapidly increasing variable I^2R losses.

Minimum energy performance standards (star rating) currently require some types of motors to have a minimum efficiency of 96 per cent

In general, large DC motors have a higher efficiency than small motors. The search for high motor efficiency is driven by governments through programs such as MEPS (minimum energy performance standards). Also called an appliance's 'star rating', it is now mandatory for specified appliances to comply with the requirements.

KEY POINTS...

- A DC motor has resistive (I^2R) losses, core losses (magnetic effects) and mechanical losses (friction, etc.).
- Resistive losses due to the resistance of the armature, brushes and series coils increase with load; other losses remain relatively constant.
- Efficiency (symbol η) of a DC motor is highest when the constant losses equal the resistive losses.
- Maximum efficiency in a motor occurs when the fixed losses (iron, friction, windage, shunt coil copper losses, etc.) equal the copper losses in the armature and series windings.

14.5 DC Motor Control and Protection

Operating a large DC motor requires a means of starting it, controlling its speed and protecting it against various types of fault conditions. Some situations also require the motor to be reversed, or to be stopped quickly. DC motor control and protection can be a complex topic, with many of today's controllers based on computer and solid-state technology. In this section, we look at the principles behind protecting and controlling a DC motor.

DC motor starter

When a DC motor of any type is first started, the only resistance to current flow is that offered by the motor's DC resistance. As the motor gains speed, its back-EMF increases and the current reduces to its final value when the motor is at full speed. As the following example shows, the start current of a DC motor can be very high.

EXAMPLE 14.11

The 10 kW 200 V compound wound DC motor in Figure 14.20 has an armature resistance (includes series coils and interpoles) of 0.25 Ω. The shunt winding takes 4 A from the 200 V DC supply. Calculate the current taken by the motor at the instant it is connected to the 200 V supply.

Solution

Values $R_A + R_S = 0.25\ \Omega$

$I_{shunt\ field} = 4\ A$

$V = 200\ V$

$I_{start} = ?$

Equations $I_{start} = I_{shunt\ field} + I_{armature}$

$$I_{armature} = \frac{V}{R_A + R_S} = \frac{200}{0.25} = 800\ A$$

$I_{start} = 4 + 800$

Answer **Motor start current = 804 A**

If the motor in Example 14.11 is 90 per cent efficient, when delivering 10 kW to its load, it will be taking about 11 kW of power from the 200 V DC supply. Therefore the full-load current of the motor is

FIGURE 14.20 Circuit for Example 14.11

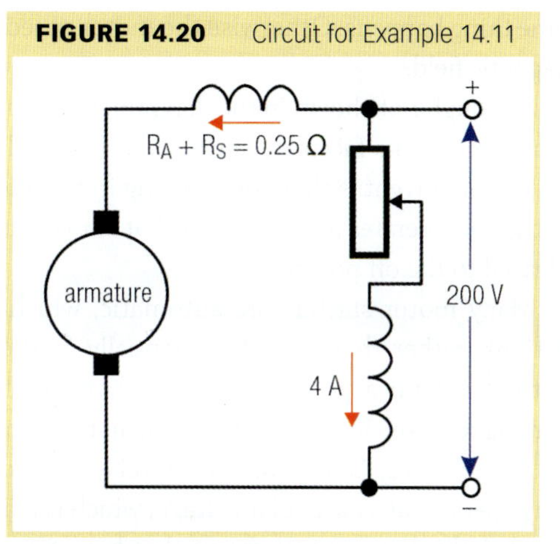

around 55 A. A DC supply capable of delivering 55 A is unlikely to be able to handle 804 A, so fuses will operate, or circuit breakers will trip. To prevent this kind of overload, most DC motors have a device called a *motor starter* which introduces extra resistance in series with the motor to limit its starting current.

The earliest form of motor starter was a manually operated device controlled by a trained operator. At start-up, the operator would move the starter's handle to the start position, thereby introducing the starter's maximum resistance into the motor circuit. As the motor gains speed, the operator moves the handle to reduce the amount of series resistance, the final position having no series resistance. Figure 14.21 shows a typical arrangement of a manual (faceplate) starter.

FIGURE 14.21 Four position, three terminal motor starter

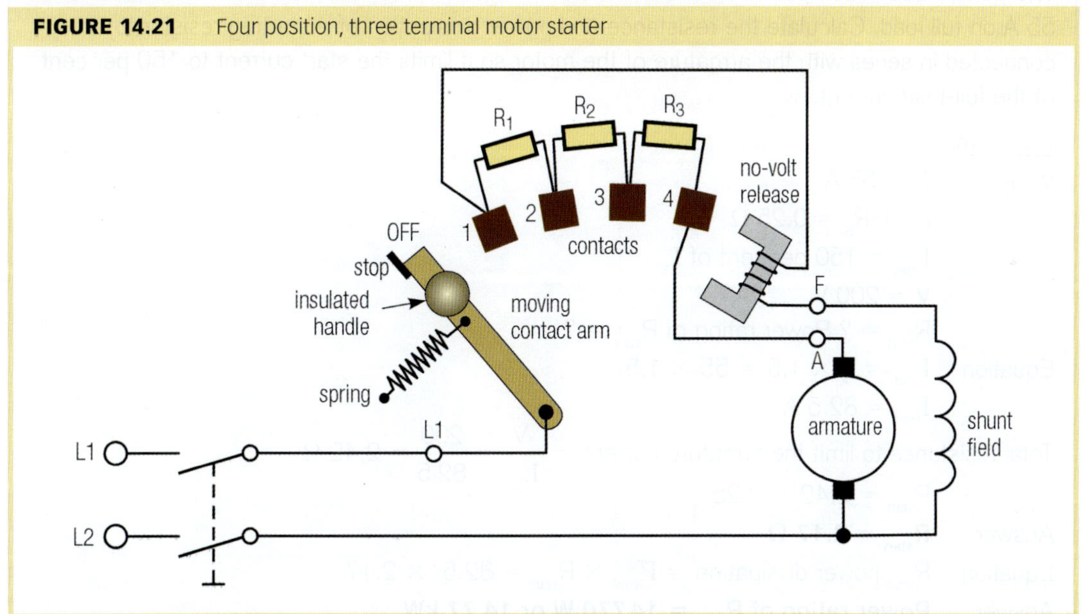

The motor starter in Figure 14.21 has three terminal connections, one to the DC power supply (L1) and two to the motor, one of these to the shunt field, the other to the armature. When the handle is moved to position 1, full voltage is applied to the shunt coils, while all three resistors are in series with the armature. The armature current is limited by the resistors, but the shunt field is receiving full power. As the motor speed rises, the operator moves the arm progressively through each position until, at position 4, full power is applied to the motor. At this point, the shunt field is in series with the starter resistors, but these have a small value compared to the resistance of the shunt coil.

The contact arm is held in position by the magnetic field created by the shunt field current flowing in the no-volt release coil. This device serves two purposes. If the DC supply is interrupted, the contact arm is released and, by spring tension, pulled back to the start position. This ensures the motor is disconnected from the line when power is restored. If the shunt field current fails for any reason, the same thing happens. Otherwise the motor speed could increase to a dangerous level due to the weak main magnetic field.

Although not shown in the diagram, most motor starters also have overload protection. A traditional arrangement in a faceplate starter has a second coil in series with one of the lines to the motor, so the full motor current is therefore flowing in this coil. If it exceeds a preset value, the magnetic field created by the coil operates a mechanism that shorts out the no-volt release coil. As a result, the contact arm is released to the off position.

Many motor starters are automatic, which allows the motor to be controlled remotely by on-off stations, rather than operated manually at the motor. Automation is achieved in several ways. One method is with a preset timing mechanism that operates relays that progressively short-circuit the starting resistors in series with the armature. There could be different delay times between each step.

A second way is to have a set of relays that operate progressively as the armature voltage rises. Each relay operates at a particular voltage, which is reached only when the motor has accelerated to a particular speed. A third way is with relays that have two operating coils, one connected to the power supply, the other in series with the armature current. As the motor accelerates, the armature current reduces. At a particular point, the coil connected to the power supply takes over against the now weaker second coil, causing the relay to operate contacts that reduce the resistance in series with the armature.

Motor start current is usually limited to around 150 per cent of its rated current. This allows the motor to start under full load and accelerate reasonably quickly. Because of the high current involved, motor starting resistors need to be able to dissipate considerable heat.

EXAMPLE 14.12

The 10 kW 200 V compound wound DC motor in Figure 14.20 takes an armature current of 55 A on full load. Calculate the resistance and power dissipation of a starting resistor to be connected in series with the armature of the motor so it limits the start current to 150 per cent of the full-load current.

Solution

Values $I_A = 55$ A

$R_A + R_S = 0.25\ \Omega$

$I_{start} = 150$ per cent of I_A

$V = 200$ V

$R_{start} = ?$ Power rating of $R_{start} = ?$

Equation $I_{start} = I_A \times 1.5 = 55 \times 1.5$

$I_{start} = 82.5$ A

Total resistance to limit the armature current $= \dfrac{V}{I_{start}} = \dfrac{200}{82.5} = 2.42\ \Omega$

$R_{start} = 2.42 - 0.25$

Answer $\mathbf{R_{start} = 2.17\ \Omega}$

Equation R_{start} power dissipation $= I^2_{start} \times R_{start} = 82.5^2 \times 2.17$

Answer **Power rating of R_{start} = 14 770 W or 14.77 kW**

The resistor in Example 14.12 needs to dissipate this amount of heat only for a short time. Even so, starting resistors for large DC motors, such as those used for electric trains and other high-power situations, are generally made in the form of a grid for best cooling, like those shown in Figure 14.22. Metals used include cast iron and stainless steel. In large industrial installations, motor starting resistors consist of electrodes dipped into a vat containing a conductive liquid. These are called liquid starting resistors and the resistance is determined by the depth of electrode immersion.

FIGURE 14.22 Motor starter resistors have to dissipate considerable heat

Speed control

The two methods of controlling the speed of a DC motor are by varying its terminal voltage or varying the strength of the motor's magnetic field. The second method is the most popular because the field coil current is much smaller, and therefore easier to control than the motor current. The simplest method of controlling the field current, as already explained, is with a rheostat in series with the shunt field coils. See Figure 14.23(a), which shows the circuit of a compound motor with a shunt field rheostat.

The resistance and power rating of a shunt motor rheostat depends on the shunt coil resistance and voltage. Example rheostats are shown in Figure 14.23(b).

FIGURE 14.23 A rheostat is a wire wound variable resistor designed to dissipate heat. Those shown are rated at 100 W, although some types have a much higher power rating.

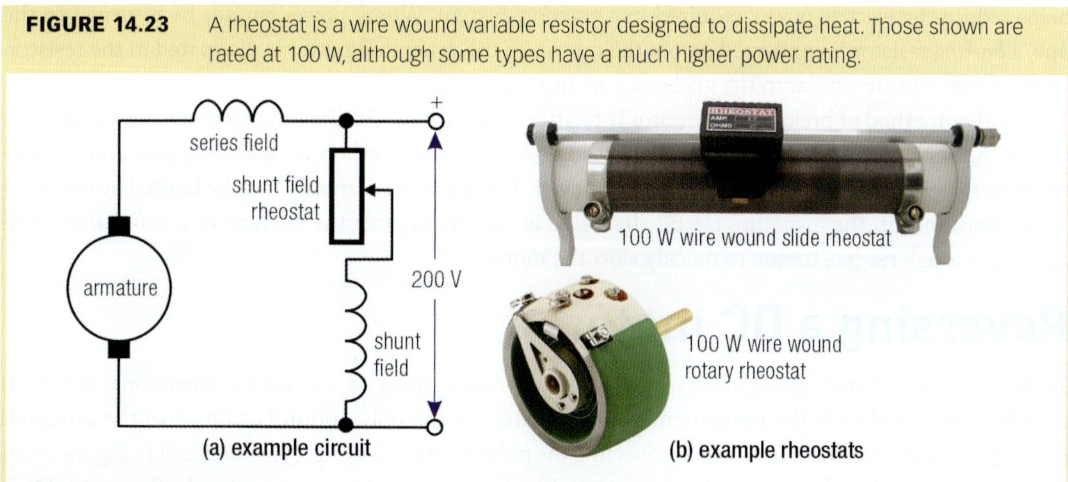

EXAMPLE 14.13

For the motor circuit in Figure 14.23, find the resistance and power rating of a rheostat that on maximum resistance limits the shunt current to 1 A.

Solution

Values $I_{SHmax} = 4$ A

$I_{SHmin} = 1$ A

$V = 200$ V

$R_{rheostat} = ?$ Power rating of $R_{rheostat} = ?$

$$\text{Resistance of the shunt coils} = \frac{V}{I_{SHmax}} = \frac{200}{4} = 50\ \Omega$$

Total resistance to obtain $I_{SHmin} = \frac{V}{I_{SHmin}} = \frac{200}{1} = 200\ \Omega$

Resistance of rheostat = 200 − 50

Answer **$R_{rheostat}$ = 150 Ω**

The power dissipated by the rheostat depends on its resistance setting. At a current of 1 A, power dissipated (I^2R) = 1 × 1 × 150 = 150 W. Maximum power dissipation occurs when the rheostat resistance equals the shunt coil resistance of 50 Ω. When the rheostat is set to 50 Ω, the total resistance is then 100 Ω, and the current is 200/100 = 2 A. The power dissipated in the rheostat is 2 × 2 × 50, so:

Answer **Power rating of rheostat = 200 W**

Braking a DC motor

FYI

Electric vehicles use dynamic braking, thereby extending the life of brake linings

When the power is disconnected from a large DC motor that is coupled to a heavy inertia load, it can take a long time to come to a stop. One way to brake the motor is by mechanical friction, like the brakes in a car. A better method is to circulate a reverse current in the armature, so the motor brakes electrically. This can be done in two ways.

The first and most common method is called *dynamic braking*. When the motor is being driven by its load, the motor becomes a DC generator. If the generated power is fed to an electrical load, like a resistor, it will act as a load on the motor (now acting as a generator) and tend to slow it down. In some situations, the generated power is fed back to the DC supply, such as in electric trains. This is known as *regenerative braking*, as rather than waste the generated power, it is returned to the power supply. This helps reduce the electric power required in a traction system. Where power cannot be returned to the line, a *braking* resistor is connected across the motor, so the generated power is dissipated in the resistor. Braking resistors are similar to the grid resistors in Figure 14.22.

Another method of braking a DC motor is to attempt to reverse its direction by reversing the polarity of the DC supply to either the field coils or the armature. This method is called *plugging* and is more effective than dynamic braking, but uses more power. The armature current has to be limited by inserting series resistance in the armature circuit. Its value is chosen to limit the current to a safe value while allowing enough reverse torque to quickly slow the motor.

Reversing a DC motor

To reverse a DC motor, either reverse the armature connections or the field connections, not both together. The simplest is the permanent magnet motor, as the only option is to reverse the armature connections (in effect this is the same as reversing the polarity of the DC supply). The circuit in Figure 14.24 shows a reversing switch connected to the armature of a permanent magnet motor. In Figure 14.24(a), the switch is set to cause the motor to rotate in a clockwise direction. Armature connections A1 and A2 connect to L1 and L2 respectively.

FIGURE 14.24 A permanent magnet motor is reversed by changing the polarity of the DC supply to the armature connections

(a) armature current from A1 to A2

(b) armature current from A2 to A1

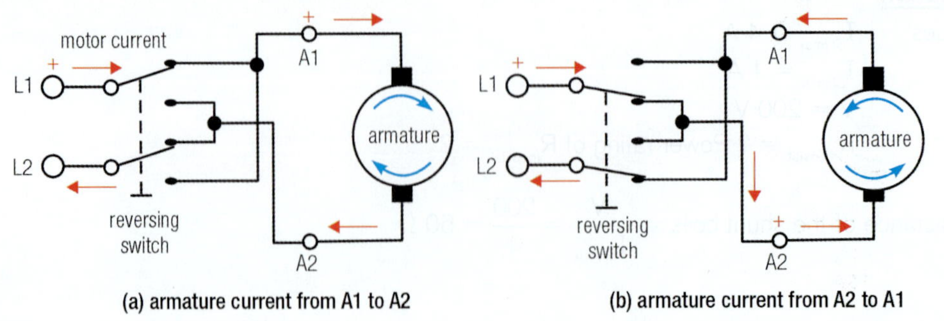

Throwing the switch to its other position reverses the polarity of the armature connections, so A1 is now connected to L2 and A2 to L1. The armature current is now flowing in the opposite direction to before and therefore it rotates anticlockwise. The switch in Figure 14.24 could be replaced with relays or contactors, with their contacts connected in a similar way.

A separately excited motor is similar to the permanent magnet motor, except there is also the option of reversing the motor by reversing the polarity of the field supply. This is often a preferable way of reversing a motor, as less power is handled during switching.

A reversing switch (or contactor) is used when a motor is regularly reversed as part of its normal function. In other cases, a motor is reversed by changing connections at its terminal box. Figure 14.25 shows a shunt motor and its terminal box. In Figure 14.25(a), the connections produce a clockwise rotation. In Figure 14.25(b) and (c), the rotation is reversed by changing links between the supply terminals and either the armature or field connections.

FIGURE 14.25 A shunt motor is reversed by changing the polarity of the supply voltage to either the armature or the field. This can usually be done at the motor's terminal box.

A1
F1
armature
shunt field coils
F2
A2
shunt motor circuit

(a) initial rotation

(b) opposite rotation by reversing shunt field

(c) opposite rotation by reversing armature

Reversing a series motor is done in the same way as for a shunt motor, in which either the armature connections or the series field connections are reversed at the terminal box, as shown in Figure 14.26. Because the field coils are in series with the armature, the connections to change motor direction are not as obvious as in the shunt motor.

FIGURE 14.26 A series motor can sometimes be reversed by changing links in the terminal box to reverse the direction of current in either the armature or the series field coils

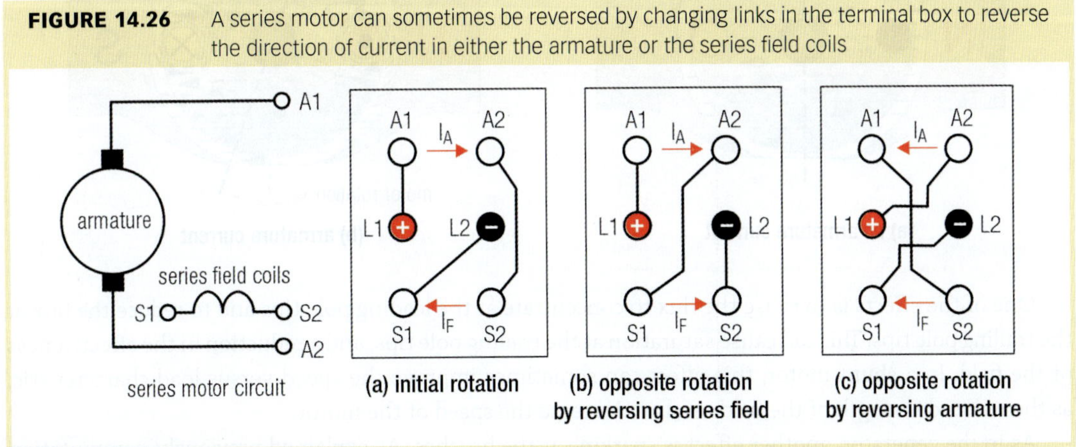

A compound motor has a series and a shunt field. To reverse the motor, either the armature connections are reversed, or the series and shunt coils are reversed. This is difficult to show at the terminal box, so instead we've shown the circuit diagram in Figure 14.27. In Figure 14.27(a), the motor connections cause clockwise rotation. In Figure 14.27(b) the armature connections are reversed, and in (c) the series and shunt coils are reversed. If only the series *or* the shunt coils are reversed, the motor becomes a differentially compounded motor.

FIGURE 14.27 When reversing the direction of a compound motor, either reverse the armature connections, or both the series and shunt coil connections

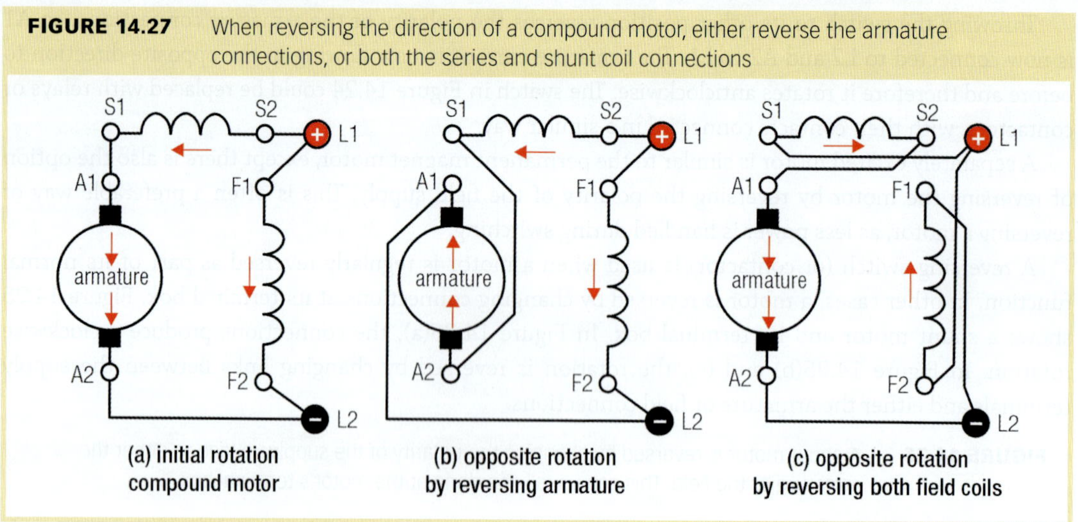

Armature reaction

Armature reaction is discussed in Chapter 13, where we explained that it has the effect of changing how the magnetic flux lines are distributed through the machine's magnetic circuit. This is shown in Figure 14.28, in which (a) illustrates how the magnetic flux lines in a two-pole DC motor would be distributed when there is no current in the armature (motor stationary). When power is applied to the armature, armature current flows which also sets up a magnetic field that interacts with the main field, as shown in Figure 14.28(b).

FIGURE 14.28 In a motor, armature reaction causes the magnetic field to twist in the opposite direction to the direction of rotation

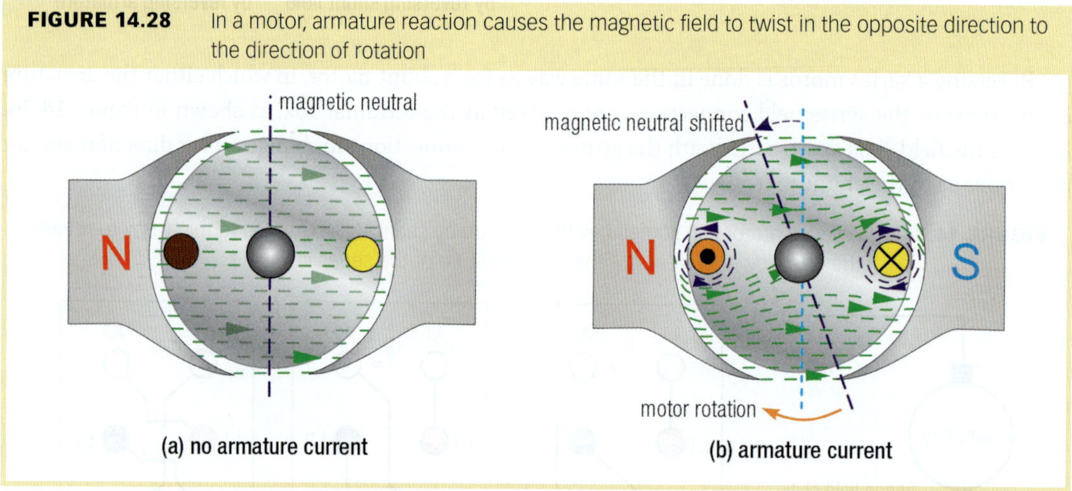

One of the effects is to cause the flux to concentrate at the leading pole tips and to reduce the flux at the trailing pole tips. This can cause saturation at the trailing pole tips, and a reduction in the effectiveness of the field. In a shunt motor, this effect can sometimes improve the speed versus load characteristic, as the reduced strength of the field tends to increase the speed of the motor.

As in the generator, another effect is sparking at the brushes. As explained previously, commutation should occur when the coils being switched are passing through the motor's magnetic neutral zone. That is, the coils should be switched at a time when little or no current is flowing in them. A simple solution is to move the brushes to the new neutral position. This works when the load is relatively constant, but does not suit a varying load.

A common way of compensating for armature reaction effects in a motor is the addition of *interpoles*. These are small field poles placed between the main field poles. The coils are wound with a heavy gauge wire and are connected in series with the armature. A four-pole shunt wound motor with interpoles is shown in Figure 14.29.

FIGURE 14.29 Four-pole shunt wound motor with interpoles. Once the polarity of the interpoles is set correctly the motor can work equally well as a generator.

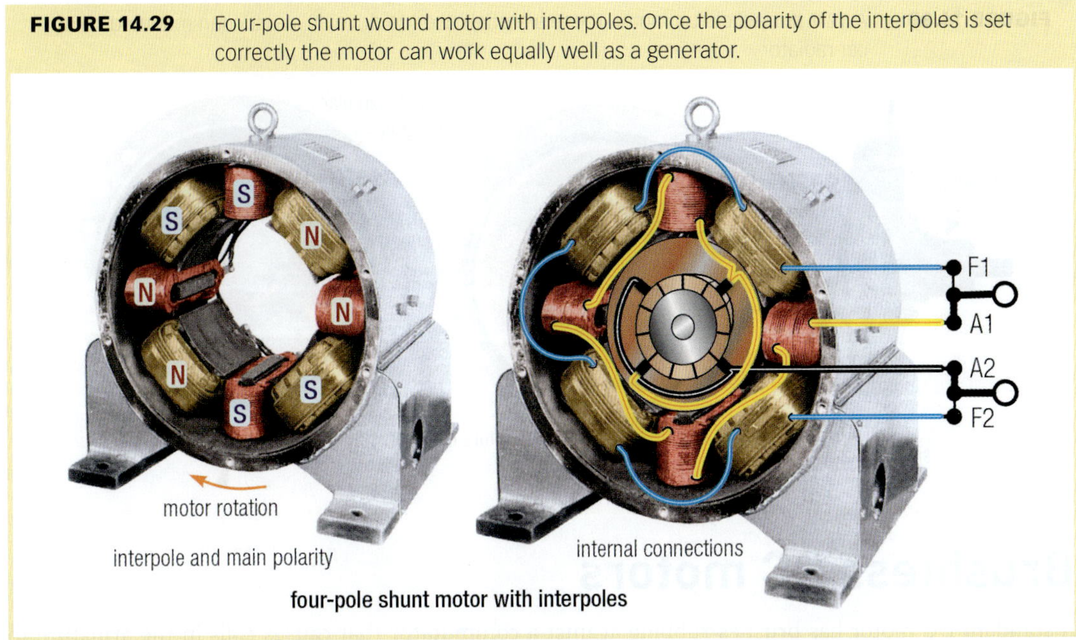

The magnetic field created by the interpoles tends to cancel the magnetic field created by the armature windings (see also Chapter 13). The result is the main magnetic field remains where it should be, and the magnetic neutral is no longer rotated. Therefore, the brush position can be fixed. The interpole coils are connected so the interpoles have the same magnetic polarity as the main poles *behind* them, in the direction of rotation.

The interpole coils are connected in series with the armature so they can produce a magnetic field proportional to the load current. This means the amount of compensation is always correct, regardless of the load current. In large DC motors, compensating windings are also often embedded in the main pole pieces. As a result, the effects of armature reaction can be almost completely overcome. Compensating windings are expensive to manufacture, so they are only found in large machines.

14.6 Other Types of DC Motors

The DC motors described so far have been around for many years, and could be considered as the workhorses of industry. Smaller motors also have an important role, and some of these are described in this section. The only common thing about these motors is that they all work from a DC supply. Some of them operate on different principles to traditional motors. It's beyond the scope of this book to explain their detailed operation, so we give an overview instead.

Printed circuit motors

The printed circuit motor has an armature constructed on a thin piece of circular insulating material laminated on both sides with copper. By a process of chemical etching, a complex winding is produced. Figure 14.30 shows a side view of the motor, and how it looks when pulled apart. The armature winding is around 2 mm thick and has a diameter of 90 mm (in this example). When assembled, the strong eight-pole ceramic magnet sits very close to the armature.

Because of its large diameter and low mass armature, the printed circuit motor has excellent torque characteristics. It also has very little sparking at the brushes because the armature coils have a low inductance due to being air-cored. The lack of iron in the armature has other advantages, including smooth running at low speeds as there is no 'cogging' effect. The light weight of the armature allows it to be rapidly accelerated. Power rating is usually less than 200 W. These motors are used extensively in the automotive industry, but find application in many areas, such as in servo systems.

FIGURE 14.30 A 12 V printed circuit motor. These are often used to drive an electric fan mounted on a car radiator.

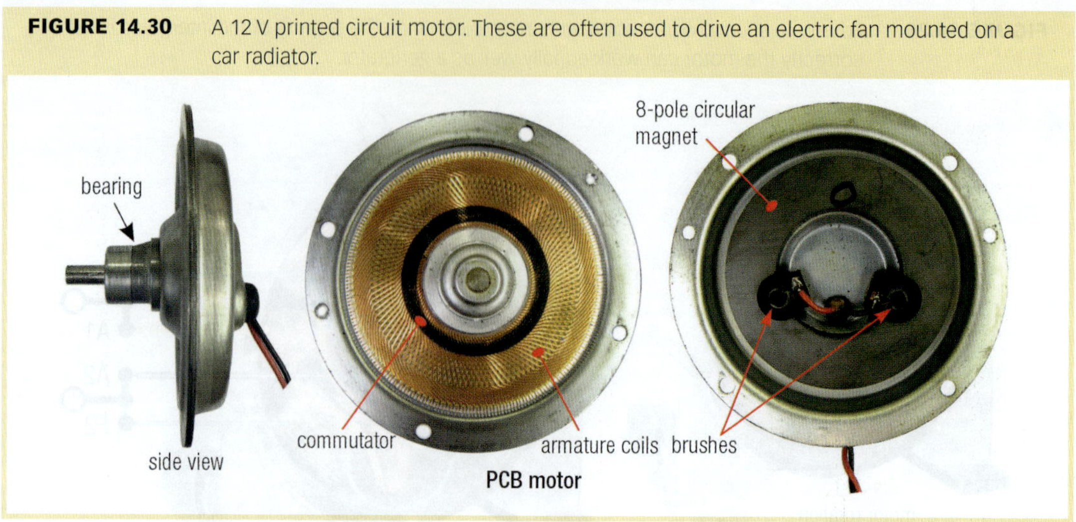

Brushless DC motors

A standard DC motor has brushes rubbing against a commutator that causes coils in the armature to be connected at the right moment in time as the armature rotates. In brushless DC (BLDC) motors the commutation (or switching) is controlled by an electronic circuit that switches current as required to coils arranged in a stator. A common example of a BLDC motor is the 12 V brushless DC fan motor, typically found in computers. This type of motor is shown in Figure 14.31, in which a magnetised rotor rotates in the rotating magnetic field created by the coils. The rotating magnetic field is produced by the electronic switching circuit.

FIGURE 14.31 Examples of brushless motors. The magnetised rotor sits over the coil assembly in these examples, while other types have the rotor rotating inside the coil assembly as in a conventional motor.

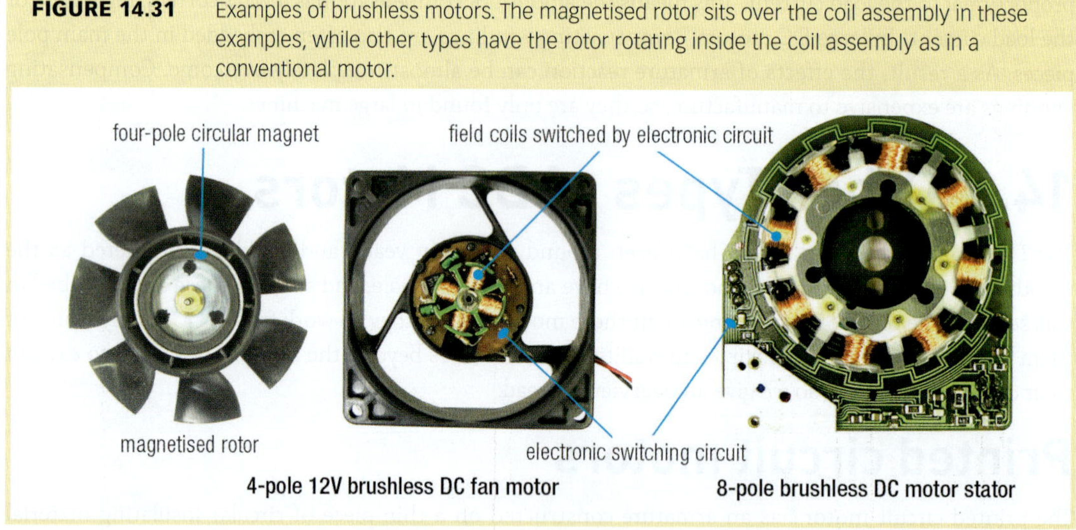

A BLDC motor sometimes has position sensors (Hall effect devices) installed in the motor that can be used by the electronics to determine the position of the rotor. In a simpler type of motor, the electronics uses the back-EMF generated in the motor coils to determine when coil switching should occur. Position sensors are used where a high initial torque is needed, or where the starting torque varies greatly. Sensorless BLDC control is often used when the initial torque does not vary a lot, such as in a fan motor.

BLDC motors have many advantages, despite their increased cost and complexity. Advantages include smooth operation, low maintenance (no brushes or commutator to wear out), silent running, high level of control and improved efficiency. For these reasons, BLDC motors are being increasingly used. For example, British technology company Dyson uses high speed BLDC motors in some of their products (vacuum cleaners, hand driers). In refrigeration, to reduce energy consumption, BLDC motors to drive cooling fans are being increasingly used.

A BLDC motor often has the stator coils connected to form a three-phase circuit (explained in Chapter 20). As shown in Figure 14.32, the basic three-phase BLDC motor has six coils to form two sets of three coils, referred to as U, V and W. This differs from the smaller motors in Figure 14.31, in which all coils are independently connected, not in pairs. As Figure 14.32 shows, some BLDC motors have more than two sets of coils, to give smaller rotation steps and smaller torque ripple.

FIGURE 14.32 Different types of BLDC motors. A basic motor with two sets of coils and a four pole magnetised rotor is shown in (a); the motor in (b) has three sets of coils and eight poles, and (c) has four sets of coils and eight poles.

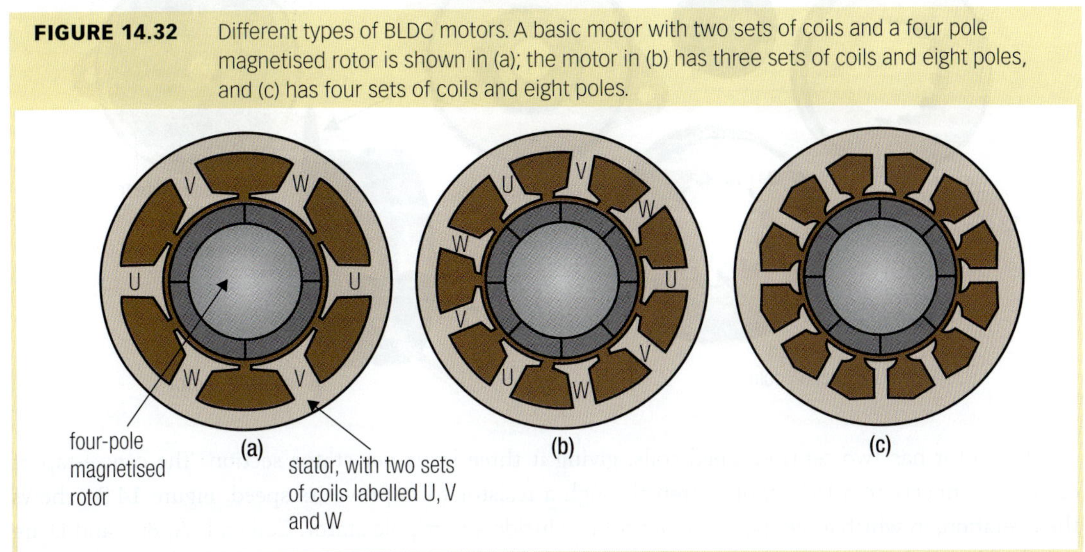

The rotor in a BLDC motor contains an even number of permanent magnets. The number of magnetic poles in the rotor also affects the step size and torque ripple of the motor. More poles give smaller steps and less torque ripple.

To make the motor rotate, the stator coils are energised in a particular sequence to make the motor turn in the required direction. Reversing the sequence makes the motor run in the opposite direction. As already explained, rotation is achieved by changing the current flow in the coils (and therefore the polarity of the magnetic fields) at the right time and in the right sequence, a process referred to as commutation.

A three-phase BLDC motor has six states of commutation. When all six states have been performed the sequence is repeated to continue the rotation. Note that this sequence is one full *electrical* rotation, not necessarily mechanical rotation. For example, a four-pole BLDC motor requires two electrical rotation cycles for one mechanical turn of the motor shaft.

The simplest method used to drive BLDC motors is to switch coils on and off as required. That is, a coil is either conducting (in one or the other direction) or not conducting. This is called square wave commutation or block commutation. The electronics associated with a BLDC motor has to therefore switch the coil currents in the right sequence. This requires suitable solid-state switching devices that are controlled by a microprocessor or other type of controller. Therefore, a BLDC motor always has an electronic circuit to drive it, which may be inside the motor, or an external unit. A full explanation of this type of circuit is beyond the scope of this book.

Stepper motors

This type of motor also requires an electronic switching circuit but works in a different way to the brushless motor. A stepper motor is rated by its *angular resolution*, which is the number of discrete steps the motor can achieve per revolution. The stepper motor in Figure 14.33 has 24 magnetic poles, with 12 each on the upper and lower sections of the stator. The two sections are offset so the 12 poles in each section interleave with the centre section, giving 24 poles. The rotor has 12 poles and sits so it is free to rotate in the centres of the three stator sections.

FIGURE 14.33 A 7.5° stepper motor, shown assembled and when dismantled. The rotor is a ceramic magnet with 12 poles and sits inside the stator assembly.

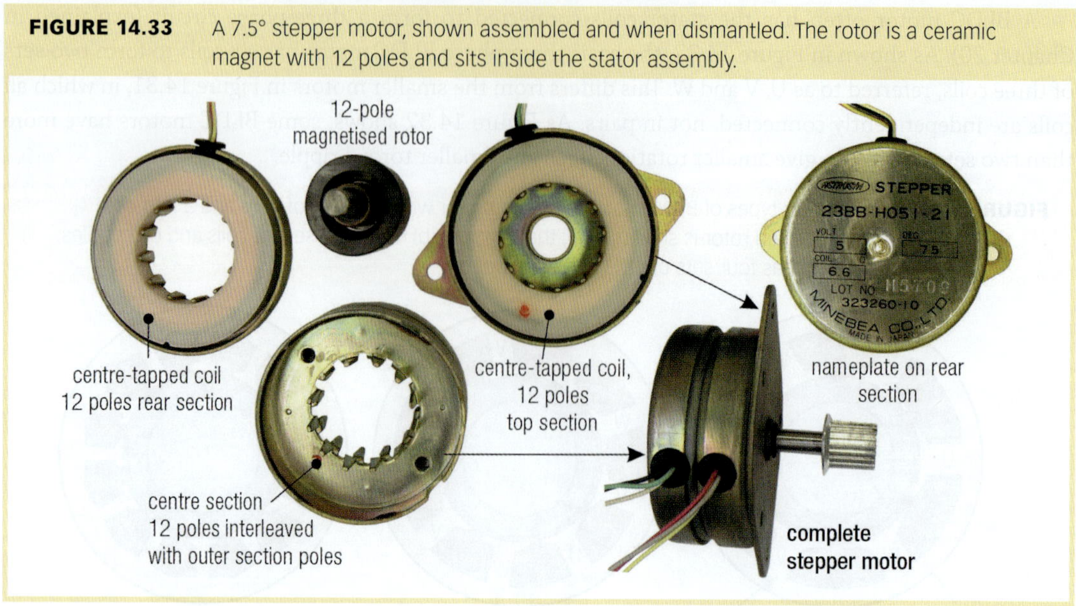

The motor has two centre-tapped coils, giving it three leads per stator section. The centre-tap of each coil connects to a DC supply, often through a resistor to improve its speed. Figure 14.34 shows the operation, in which a two-pole magnet rotates inside a four-pole stator. Coil ends A, B, C and D are switched by an electronic circuit that energises the coils in a particular sequence. The switch sequence is controlled by a program or an integrated circuit that causes the motor to rotate by one position on arrival of a signal pulse. A constant series of pulses will make it rotate continuously at a speed determined by the rate at which pulses are received.

FIGURE 14.34 Sequence showing how four cycles of operation for a two-pole stepper motor rotates it by 90° per cycle. Coils A to D are energised as required by connecting them with an electronic switch to the negative side of the DC supply.

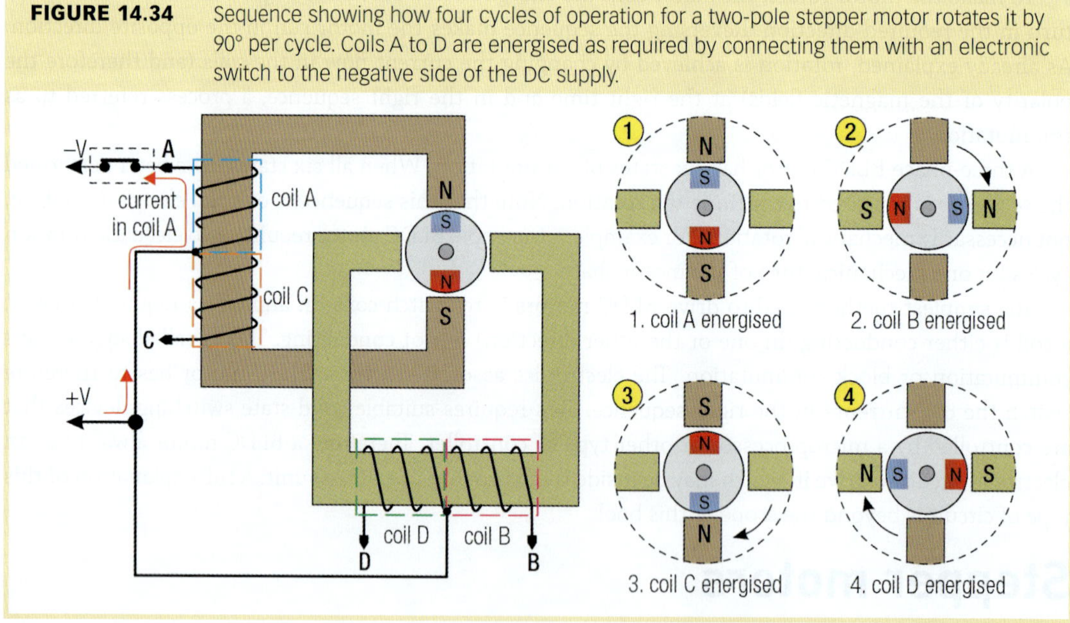

The sequence shown in Figure 14.34 starts with coil A energised, which makes the magnetic rotor align itself with the two vertical poles. When coil B is energised, the rotor is aligned with the horizontal poles. Energising coil C and then coil D causes the magnet to rotate clockwise as shown. This type of drive is called *wave* drive. Another way of driving the motor is to energise two coils at the same time, called *full-step* or *two-phase* drive. Either method gives a 15° step resolution for the 24-pole motor in Figure 14.33.

Combining the wave and full-step drive methods doubles the number of positions the rotor can take up, aligned with the poles or between them. Therefore, a motor with 24 poles can have 48 steps, which is a resolution of 360°/48, or 7.5°.

Another type of stepper motor is shown in Figure 14.35. This motor has a resolution of 1.8°, achieved by the type of construction shown. The rotor has two circular magnets of opposite polarity, in which each magnet has 50 'teeth', or serrations. The teeth on one magnet are offset from those on the other, so one set of teeth sits halfway between the other set.

FIGURE 14.35 Stepper motor with a serrated magnet rotor that aligns with the serrations on each pole piece. Each step moves the rotor by one serration.

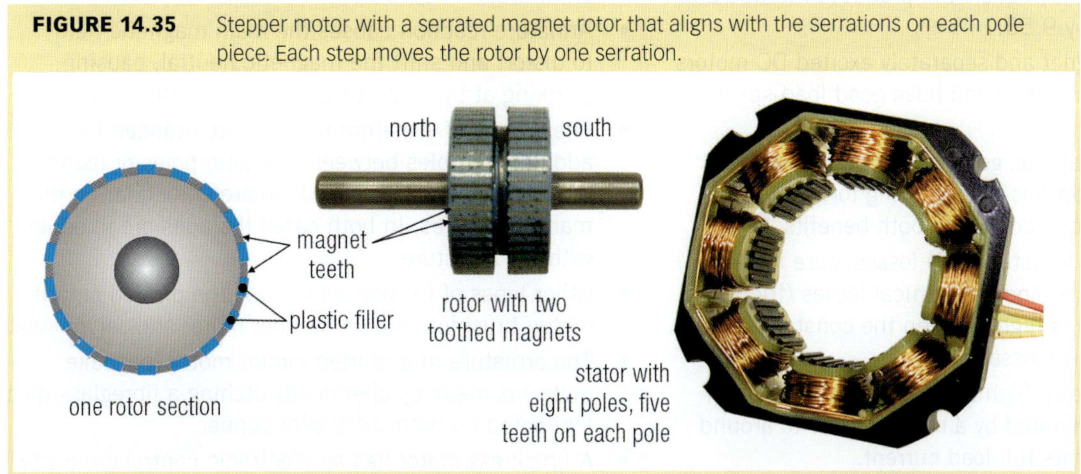

This type of stepper motor has four wires, unlike the six in the previous stepper motor. The coils are arranged in two series groups, each alternate coil forming part of its group. Coils are energised sequentially as explained, in which the teeth on the energised pole pieces attract teeth on the rotor. The design of the motor causes it to rotate by half a tooth at a time, giving it 200 steps per revolution, or 1.8° angular resolution.

Servomotors

Servomotors are used in applications that require very fast speed response and accuracy. For example, a motor might need to accelerate from zero to 6000 RPM in a few hundredths of a second, and to slow down or reverse just as quickly.

Typically, servomotors are long and narrow, with low inertia armature assemblies so they can be accelerated quickly. Servomotors that are used in positioning applications have some sort of encoding device fitted to the shaft to send information to the system controlling the motor.

The operating principles of a permanent magnet servomotor are the same as for a standard permanent magnet motor. The difference comes in its physical size and shape, as well as its performance and speed characteristics. A stepper motor is sometimes called a servomotor, as it too is used in positioning tasks. However, as explained, it operates differently to the servomotors just described.

The tacho generator is another application that uses a permanent magnet DC motor, this time as a DC generator. This type of tachometer gives an output DC voltage proportional to the rotating speed of its armature. Tacho generators are used in motor speed control applications, in which the tacho generator's output voltage indicates the speed of the motor shaft it is attached to.

CHAPTER SUMMARY

- The speed of a DC motor is directly proportional to its terminal voltage and inversely proportional to the strength of the main magnetic field.
- The output power of a DC motor is approximately equal to the product of its back-EMF and armature current. Power also equals the product of torque and speed divided by 9.55.
- Permanent magnet and separately excited DC motors produce a linear torque and have good load-speed regulation.
- A shunt DC motor has good load-speed regulation; a series DC motor has good starting torque. A compound motor combines both benefits.
- A DC motor has resistive (I^2R) losses, core losses (magnetic effects) and mechanical losses (friction, etc.). Efficiency is highest when the constant losses equal the resistive losses.
- A DC motor takes a high starting current, which in large motors is limited by a motor starter to around 150 per cent of its full-load current.
- A motor starter can be manual or automatic, and consists of a number of resistors that are connected in series with the motor armature. As the motor gains speed, the starter progressively reduces the amount of series resistance.
- The speed of a DC motor is usually controlled by varying the current in the shunt field coils. Another method is to limit the armature current with a series resistor between the armature and the power supply.
- Methods to bring a large DC motor to a halt include applying a mechanical brake to its shaft, passing the motor's generated current through a load (dynamic braking), injecting the generated current back to the supply (regenerative) and reversing the DC supply to reverse the motor (plugging).
- A DC motor can only be reversed by changing the direction of current in either the field coils or the armature.
- Armature reaction causes the main magnetic field to distort and shift the magnetic neutral, causing sparking at the brushes and reduced efficiency.
- The effects of armature reaction are reduced by adding interpoles between the main poles or using compensating windings, which are embedded in the main field poles. In both cases the coils are in series with the armature.
- Other types of DC motors include the printed circuit motor, brushless motor, stepper motor and servomotor.
- The armature in a printed circuit motor (pancake motor) is made by chemically etching a fibreglass disc laminated on both sides with copper.
- A brushless motor has an electronic control circuit to replace the conventional commutator and brushes. The control circuit switches power to coils in the stator to create a rotating magnetic field that the magnetised rotor is compelled to follow.
- A stepper motor rotates by a certain angle each time the control electronics receives an electrical pulse. This type of motor is used in positioning applications.
- A servomotor is a type of permanent magnet motor designed to rapidly accelerate and decelerate.
- A DC tachometer is a permanent magnet motor acting as a generator, where the output voltage is proportional to the shaft speed.

REVIEW EXERCISES

Check your answers at the back of the book.

1 A 300 mm diameter armature with an effective conductor length of 12.5 metres is passing a current of 42 A. The conductors are at right angles to a magnetic field with a flux density (B) of 0.75 tesla. Find:
   a the force acting on the conductors
   b the torque produced by the armature.

2 An armature with a resistance of 0.2 ohms is passing a current of 32 A when connected to a 400 V DC supply. Calculate the back-EMF (E_G) produced by the armature.

3 A DC traction motor produces a torque of 260 Nm when rotating at 500 RPM. Calculate its power output.

4 Which type of motor has the greatest starting torque, series or shunt wound? Why?

5 A motor is producing a torque of 220 Nm at a speed of 620 RPM while taking a current of 40 A from a 400 V DC supply. Calculate:
   a motor input power
   b output power
   c total losses in the motor.
   Express your answers in kW.

6 Calculate the copper losses in the motor shown in Figure 14.36.

FIGURE 14.36

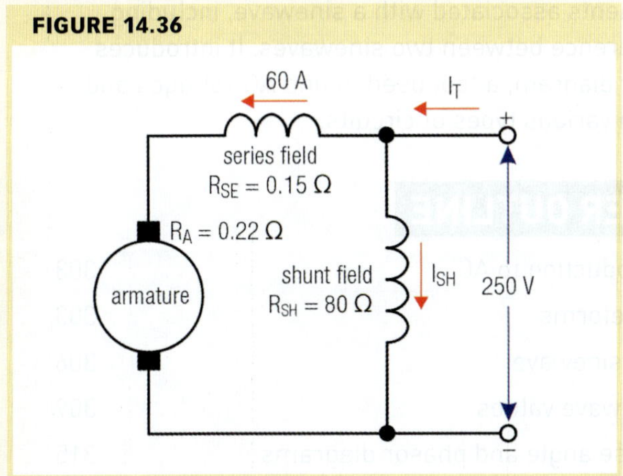

7 When the motor in Figure 14.36 is running unloaded, it takes 6 A from the 250 V DC supply. Determine the approximate value of the iron and mechanical losses in the motor.

8 A traction motor running at 1200 RPM is producing a torque of 200 Nm. If the input power to the motor is 27 kW, find the efficiency of the motor.

9 A shunt motor has an armature resistance of 0.25 Ω and a shunt winding resistance of 100 Ω. Calculate the start current if the motor is connected directly to a 250 V DC supply.

10 Find the resistance and power rating of a resistor that could be connected in series with the armature in Question 9 to limit the start current to 100 A. Ignore the shunt field current.

ONLINE RESOURCES

COMPLETE WORKSHEET FOURTEEN

Check with your instructor for worksheets on this chapter.

CHAPTER 15

AC FUNDAMENTALS

While Thomas Edison was promoting the use of DC as a source of electrical power, others such as Nikola Tesla and Sebastian Ferranti were promoting an alternating current (AC) power system. Today, AC power is standard practice, with the first AC power station going online in the early 1890s. This chapter introduces alternating current and voltage. It describes the sinewave (the shape of an alternating current waveform) and the measurements associated with a sinewave, including phase difference between two sinewaves. It introduces the phasor diagram, a tool used to find AC voltages and currents in various types of circuits.

CHAPTER OUTLINE

15.1 Introduction to AC

An alternating voltage or current *periodically changes polarity*. That is, it regularly changes from a positive to a negative value, back to positive and so on. A DC voltage or current never changes polarity, although its value might vary, giving a varying DC. Alternating current is the form in which electrical power is distributed in most parts of the world.

FIGURE 15.1 An alternating voltage can be transmitted at a high voltage, then stepped down with a transformer to a low voltage

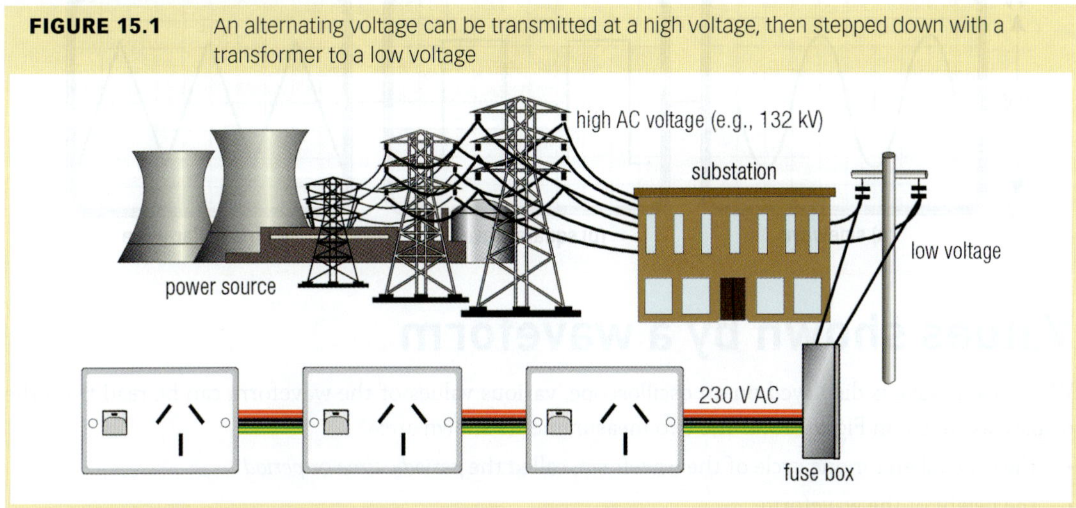

AC power distribution is used because it is easy to change a voltage from one value to another with a transformer. In Figure 15.1, the power source produces electrical power at a particular voltage, which is stepped up with a transformer to typically 132 kV, so the electrical power can be transmitted many hundreds of kilometres with best efficiency. Substations are connected to the 132 kV transmission lines, where the voltage is stepped down to a lower value for local distribution.

15.2 Waveforms

The usual way of seeing the shape of an alternating voltage is with an instrument called an oscilloscope. Briefly, an oscilloscope, like that in Figure 15.2, displays an alternating voltage as a waveform. Controls are used to adjust the instrument for the best display.

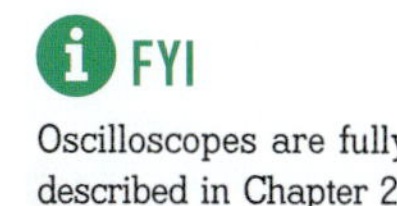

Oscilloscopes are fully described in Chapter 25

FIGURE 15.2 The shape of an alternating voltage can be seen on an oscilloscope

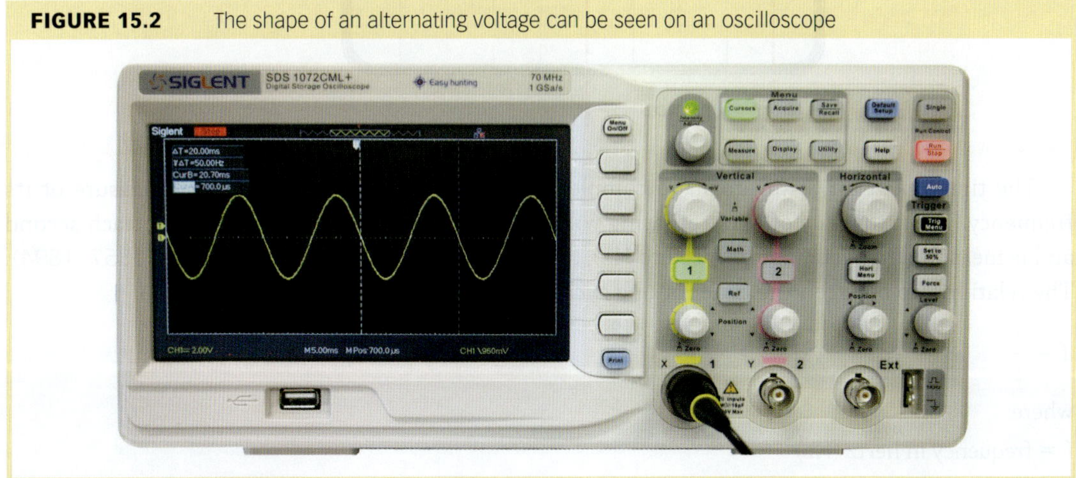

The three waveforms in Figure 15.3 show that an alternating voltage can have any shape. The waveform for AC power is the sinewave, shown in Figure 15.3(a). The graticule line across the centre of the screen is zero volts, with the trace swinging above and below it.

FIGURE 15.3 A waveform shows how a voltage changes with time. A DC voltage doesn't change, so its waveform is a straight line.

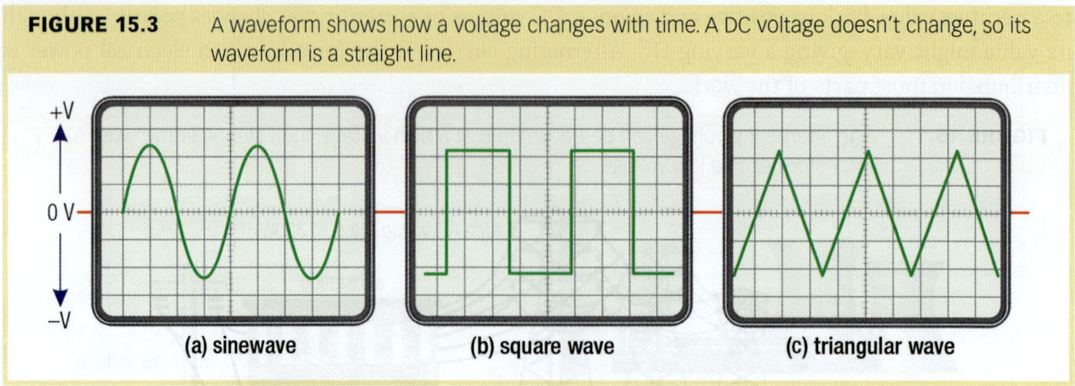

Values shown by a waveform

When a sinewave is displayed on an oscilloscope, various values of the waveform can be read from the display, as shown in Figure 15.4. The two measurements shown are:

- the time taken for one cycle of the waveform, called the *periodic time* or *period*
- the height of the waveform.

FIGURE 15.4 The time taken for a complete cycle and the peak values reached by the waveform can be read from an oscilloscope

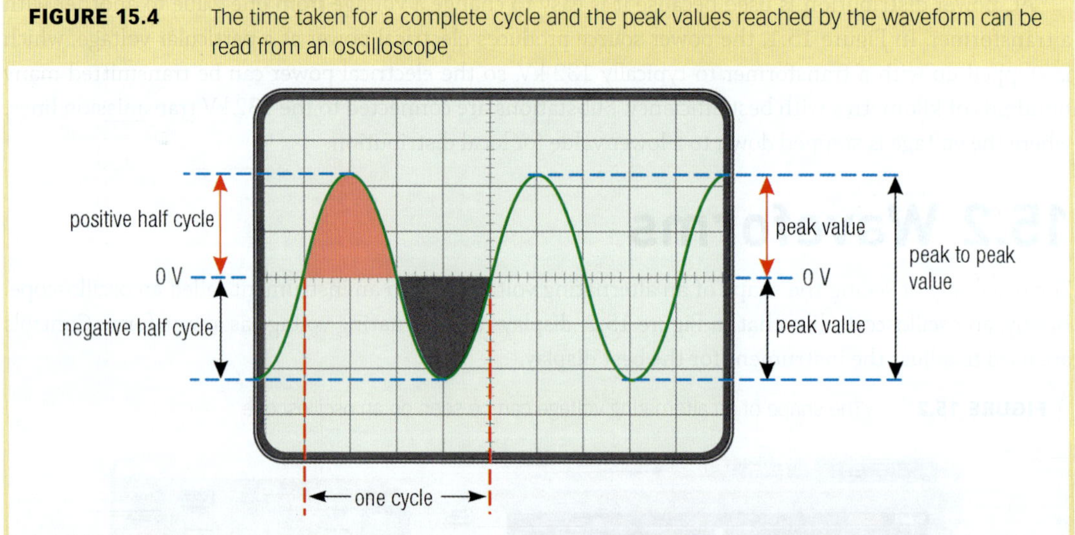

The time taken for a waveform to complete one cycle (periodic time) gives a measure of its frequency. The frequency of an alternating voltage is the number of cycles that occur each second and is measured in hertz (abbreviated to Hz), after German physicist Heinrich Hertz (1857–1894). The relationship between the time for one cycle and frequency is::

$$f = \frac{1}{t}$$

where:

f = frequency in hertz (Hz)

t = period of the waveform in seconds.

EXAMPLE 15.1

The period of a waveform is 20 milliseconds. What is its frequency?

Solution

Values $t = 20 \text{ ms} = 20 \times 10^{-3}$ seconds

$f = ?$

Equation $f = \frac{1}{t} = \frac{1}{20 \times 10^{-3}} = \frac{10^3}{20}$

Answer $\mathbf{f = 50\ Hz}$

As shown in Figure 15.4, the peak voltage of a symmetrical waveform occurs twice per cycle. The positive peak value and the negative peak value are equal if the waveform is symmetrical around the zero line. The peak-to-peak voltage is, therefore, twice the peak value.

EXAMPLE 15.2

Given the oscilloscope settings provided in Figure 15.5, for the displayed sinewave, determine its: (a) period and frequency, and (b) peak voltage and peak-to-peak voltage.

FIGURE 15.5 Sinewave display with known oscilloscope settings

oscilloscope set to 1 ms per horizontal division

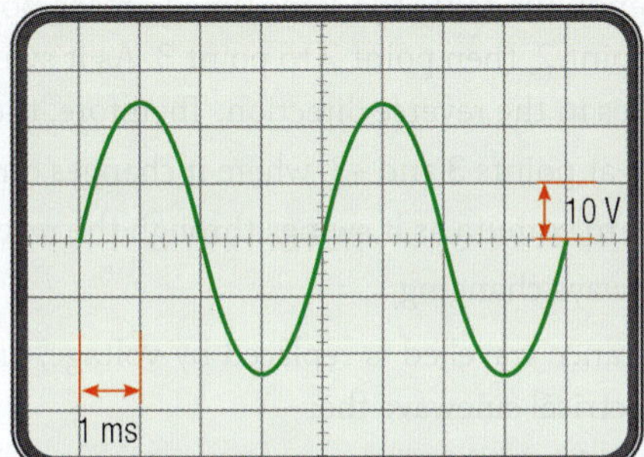

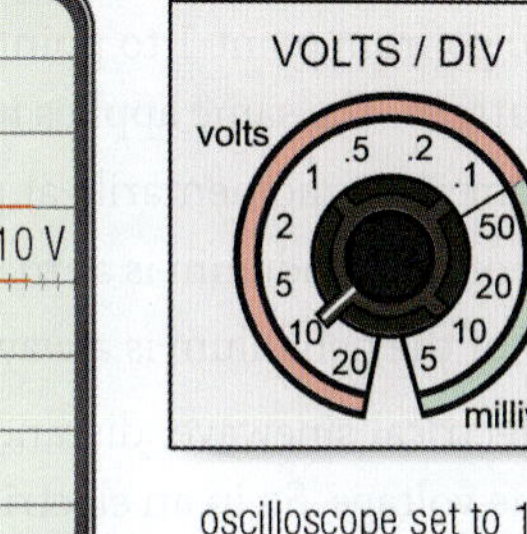

oscilloscope set to 10 V per vertical division

Solution

(a) The sweep time is 1 millisecond per major division. One cycle of the sinewave takes 4 major divisions. Therefore:

Period (t) = number of divisions × time per division

$= 4 \times 1 \text{ ms} = 4 \times 10^{-3}$ seconds

Answer $\mathbf{t = 4\ ms}$

Frequency (f) $= \frac{1}{t}$

$= \frac{1}{4 \times 10^{-3}} = \frac{10^3}{4}$

Answer $\mathbf{f = 250\ Hz}$

(b) The height setting is 10 V per major division. Height of the positive half cycle = 2 divisions, height of the negative half cycle = 2 divisions.

Peak voltage (V_{pk}) = number of divisions × volts per division = 2 × 10 V

Answer $\mathbf{V_{pk} = 20\ V}$ for both half cycles

Peak-to-peak voltage ($V_{p\text{-}p}$) = number of divisions × volts per division = 4 × 10 V

Answer $\mathbf{V_{p\text{-}p} = 40\ V}$

FYI

Sweep time is the time taken for the oscilloscope trace to move by one major division

15.3 The Sinewave

The sinewave is a fundamental waveform, and all other waveforms are many sinewaves of different frequencies and voltages added together. It's therefore useful to examine sinusoidal motion to learn about the nature of a sinewave. A simple example is a pendulum, discovered by Galileo around 1600 while he watched a light swinging during a church service. Figure 15.6 shows a pendulum swinging and its relationship in time to a sinewave.

FIGURE 15.6 The motion of a pendulum traces a sinewave

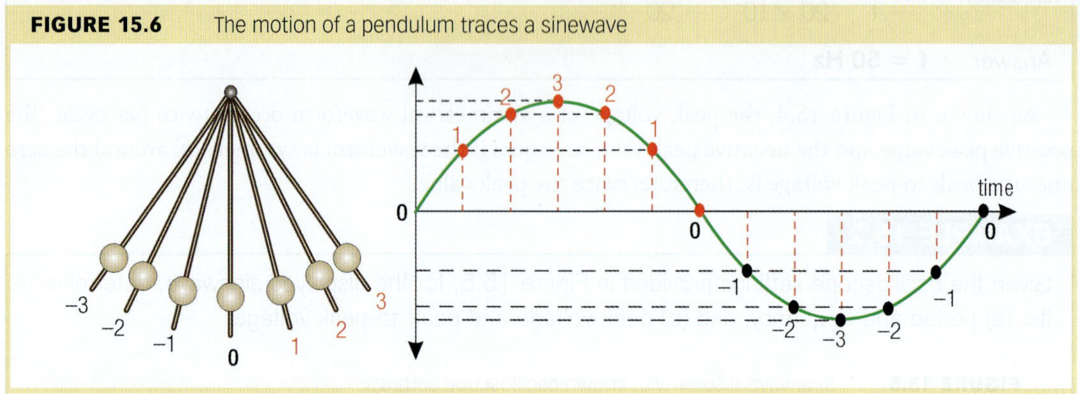

Figure 15.6 shows that the pendulum takes the same time to reach point 1 from the rest position (0), as it does to go from point 1 to point 2, then point 2 to point 3. As it swings back towards zero, it traces the same pattern. The same applies in the reverse direction. Therefore, the waveform shows that the:

- pendulum stops momentarily at points 3 and −3, where it changes direction
- velocity of the pendulum is at maximum as it swings through the rest (or zero) position
- velocity of the pendulum is always changing

In an electrical sinewave, distance travelled is replaced by voltage, and velocity becomes the *rate of change* of the voltage. So in an electrical sinewave the:

- rate of change of the voltage is always changing
- voltage changes more quickly as it approaches zero
- voltage stops changing momentarily when it reaches its maximum value.

Generating a sinewave

As explained in Chapter 13, a coil of wire in a rotating magnetic field produces an alternating voltage. The polarity depends on the direction of rotation, and the direction of the magnetic flux. The generated voltage value (e) depends on the effective length of the coiled conductor (l), the speed of interaction between the conductor and the magnetic field (v), the strength of the magnetic field (B) and the angle at which the conductor passes through the magnetic field. So far we have ignored the angle and have assumed it to be either 0° (no voltage is induced) or 90° (maximum voltage is induced). That is, ignoring the angle, the equation is $e = Blv$.

Figure 15.7 shows a simplified alternator in which a magnet rotates inside a stationary coil. Chapter 13 showed that this arrangement produces an alternating voltage that varies between zero and a maximum positive or negative value. Of interest is what happens *between* these values. As Figure 15.7 shows, the waveform of the voltage produced by the coil is a sinewave.

FIGURE 15.7 A magnet rotating inside a coil produces a sinusoidal alternating voltage

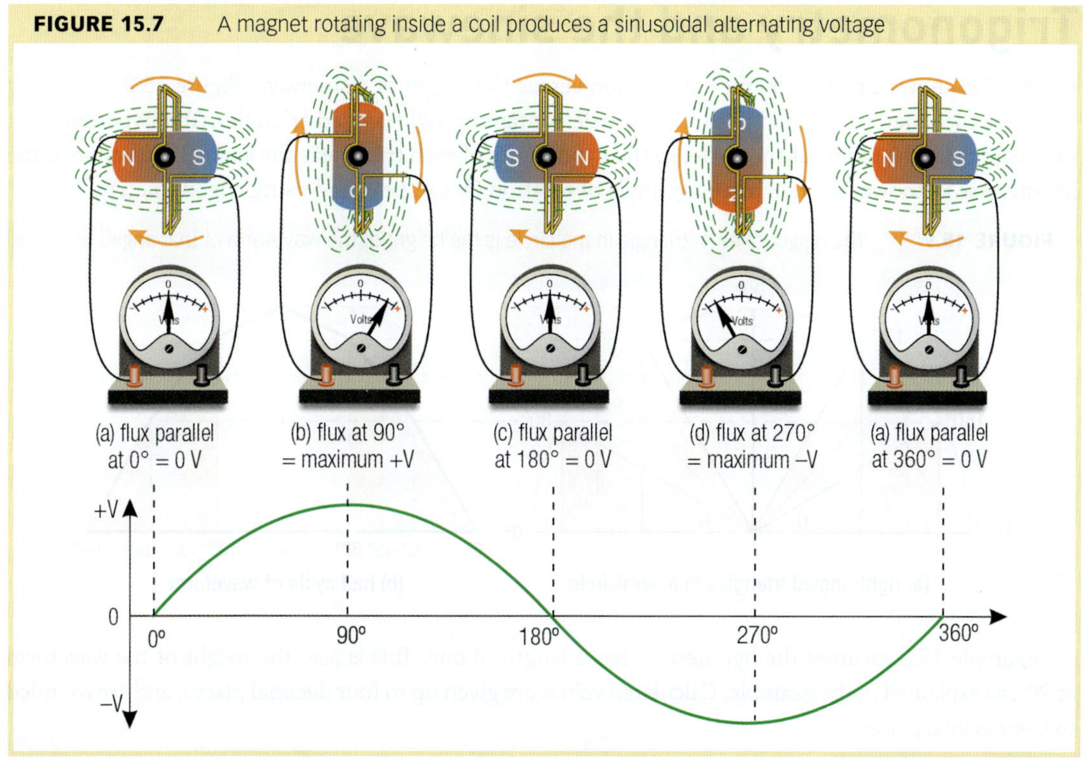

The circle and the sinewave

A pendulum traces an arc, and a rotating magnetic field intersecting a fixed coil produces a sinewave, as in Figure 15.7, showing there's a relationship between a circle and a sinewave.

The waveform in Figure 15.8 (which is almost a sinewave) is constructed from points around the circumference of a circle. Each point is where the radius of the circle intersects its circumference after rotating by a known number of degrees. The horizontal axis of the waveform shows degrees of rotation, and the vertical axis is the height of the waveform.

Figure 15.8 proves the relationship between a sinewave and a circle. Although the waveform only has nine plot points in each half cycle, it still looks like a sinewave. If enough points of rotation are used to plot the waveform, it becomes a perfect sinewave. Note that the horizontal axes for the waveforms in Figures 15.7 and 15.8 are calibrated in degrees of rotation.

All repetitive waveforms take 360° for a complete cycle. Therefore, the horizontal axis can always be calibrated in degrees, giving a convenient way of referring to any point on the waveform. For example, the waveform in Figure 15.8 has a height of zero at 0°, 180° and 360°. It has a maximum height at 90° and 270°.

FIGURE 15.8 A sinewave can be constructed from a circle

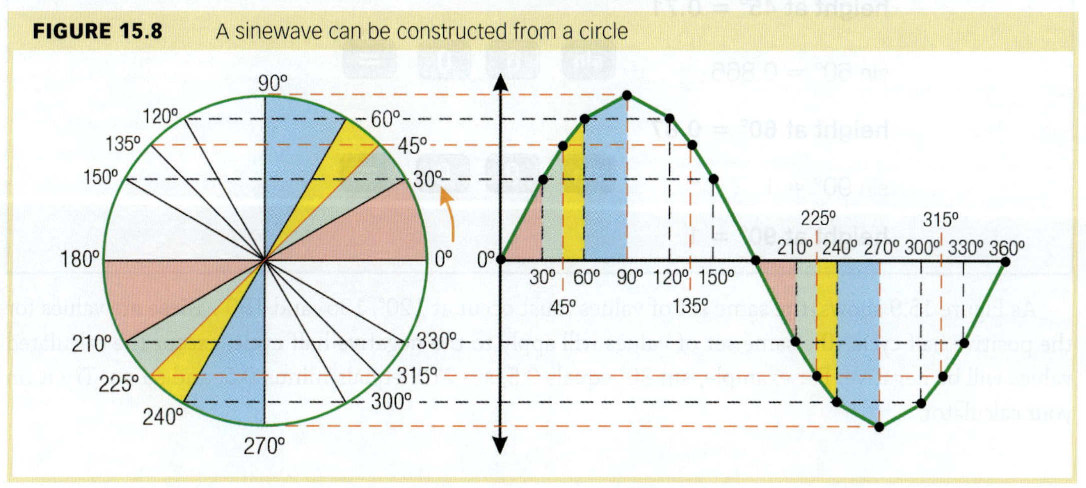

Trigonometry and the sinewave

MATH

For help with trigonometry, see page xxxi

Figure 15.8 shows a plot of degrees of rotation versus the height of a sinewave. Figure 15.9 shows *half a cycle* of the sinewave aligned with a semicircle containing right-angled triangles. The hypotenuse of each triangle is the radius of the circle, so the hypotenuse of each triangle is the same. If the value of the hypotenuse is known, the height at any angle of rotation (θ) can be found using trigonometry.

FIGURE 15.9 The height of each triangle in the circle is the height of the waveform at that angle

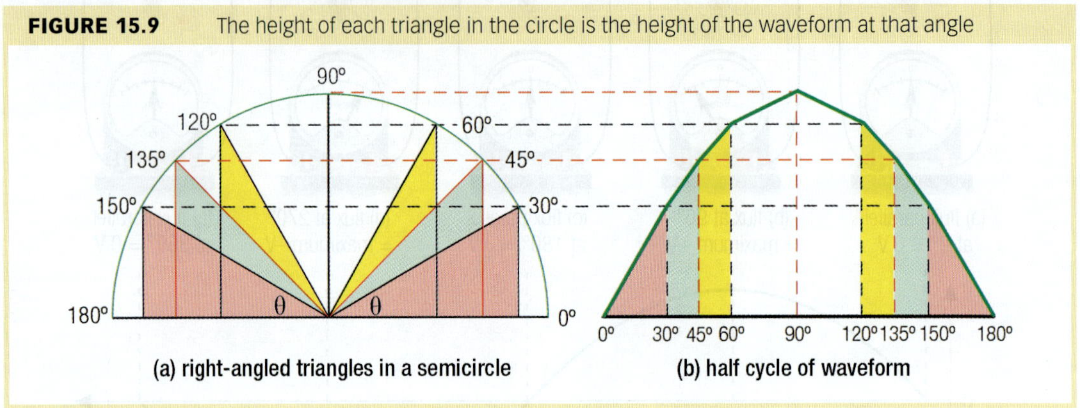

Example 15.3 assumes the hypotenuse has a length of one. This is also the height of the waveform at 90°, as explained in the example. Calculated values are given up to four decimal places, and are rounded to two decimal places.

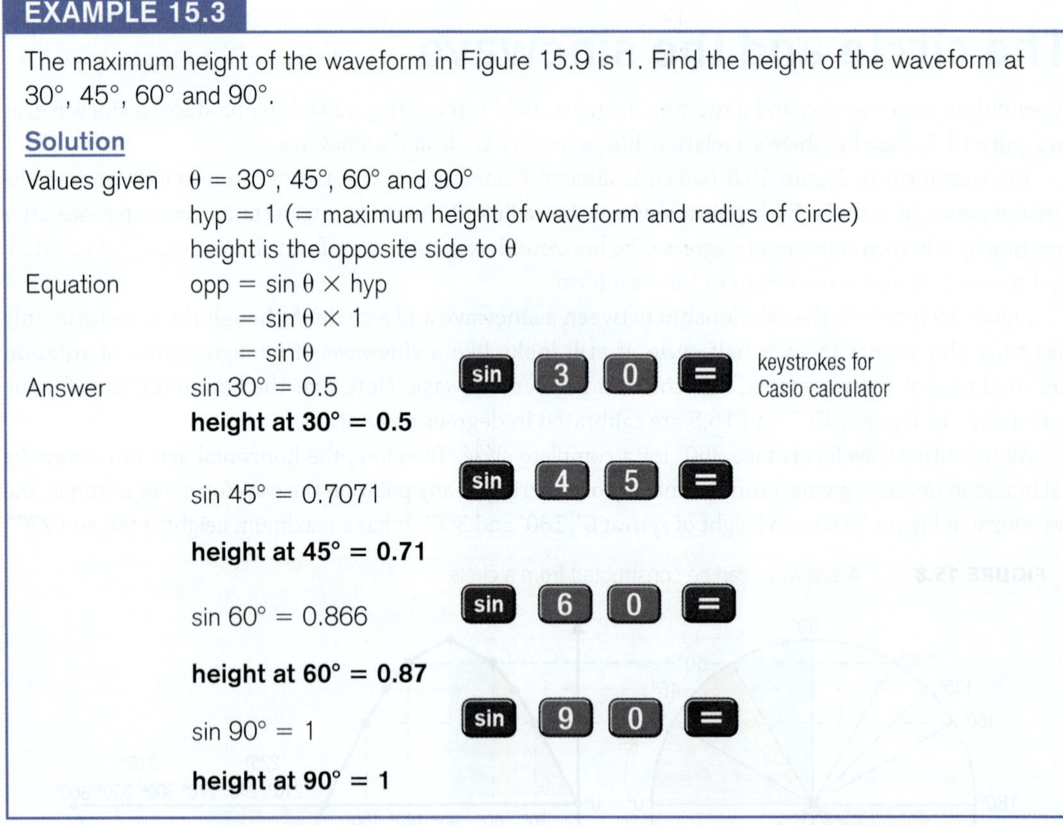

EXAMPLE 15.3

The maximum height of the waveform in Figure 15.9 is 1. Find the height of the waveform at 30°, 45°, 60° and 90°.

Solution

Values given $\theta = 30°, 45°, 60°$ and $90°$

hyp = 1 (= maximum height of waveform and radius of circle)

height is the opposite side to θ

Equation opp = sin θ × hyp

= sin θ × 1

= sin θ

Answer sin 30° = 0.5

height at 30° = 0.5

sin 45° = 0.7071

height at 45° = 0.71

sin 60° = 0.866

height at 60° = 0.87

sin 90° = 1

height at 90° = 1

As Figure 15.9 shows, the same set of values must occur at 120°, 135° and 150°. These are values for the positive half cycle. The same set of values will apply to the negative half cycle, except the calculated values will be negative. For example, sin 30° equals 0.5, sin 210° equals minus 0.5, and so on. Try it on your calculator.

KEY POINTS...

- An alternating current periodically changes direction because an alternating voltage periodically changes polarity.
- An alternating voltage has a waveform, which can be displayed for analysis on an oscilloscope.
- A waveform is made up of cycles, in which each cycle is usually identical in shape and one cycle is one complete trace of its shape.
- The time taken for one complete cycle is the period (t) of the waveform.
- Frequency (f) is the number of cycles that occur per second, measured in hertz (Hz). Frequency equals the periodic time divided into one ($f = 1/t$).
- The sinewave is the fundamental wave shape and is the shape of the alternating voltage produced by an alternator.
- A sinewave and a circle are closely related and calculations to find certain values of a sinewave require trigonometry.

15.4 Sinewave Values

There are two aspects to any waveform: its height and time to complete one cycle. As already shown, to determine the height at any point in a sinewave requires knowing the angle of rotation. A full cycle always takes 360°, regardless of how long a waveform takes for one cycle. Therefore, a waveform cycle can be divided into degrees, even if the horizontal axis is calibrated in time. The vertical axis is usually calibrated in volts or amperes. To show a waveform of an alternating *current* on an oscilloscope requires passing the current through a low-value resistor. The oscilloscope is connected to display the voltage developed across the resistor.

The vertical axis

The height or amplitude of a sinewave can be expressed in at least four ways, as shown in Figure 15.10. Values can be read from a calibrated vertical axis, but are generally calculated using the waveform's maximum value, as explained below.

FIGURE 15.10 The voltage or current value of a sinewave can be given as peak, peak-to-peak, RMS and an instantaneous value

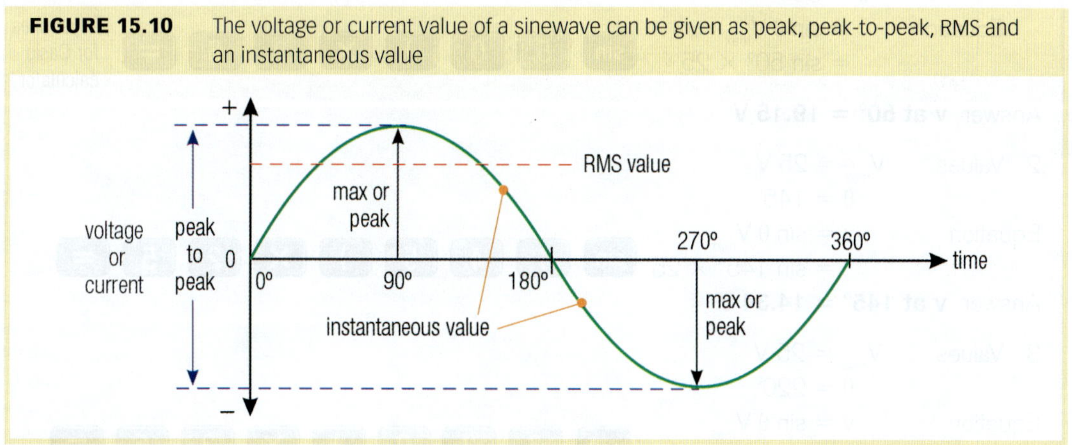

Maximum or peak value

For a sinewave, the maximum value occurs at 90° (positive peak) and 270° (negative peak), assuming the waveform starts a positive cycle at 0°. The voltage rating of components is often given as a maximum or peak value. It is written as either V_{max} or V_{pk}.

Peak-to-peak

The peak-to-peak value of a sinewave is twice its maximum (or peak) value, written as $V_{p\text{-}p}$. A waveform with a peak-to-peak value of 10 V is written as 10 $V_{p\text{-}p}$.

Instantaneous value

The instantaneous value of a sinewave is its value after a number of degrees of rotation. It is written as v (not a capital V). It is found with this equation (which only applies to a sinewave):

$v = \sin\theta\, V_{max}$

Example 15.4 uses this equation. See also Example 15.3.

EXAMPLE 15.4

For the waveform in Figure 15.11 find the instantaneous voltage at:

1 50°
2 145°
3 220°
4 335°

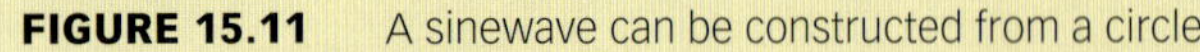

FIGURE 15.11 A sinewave can be constructed from a circle

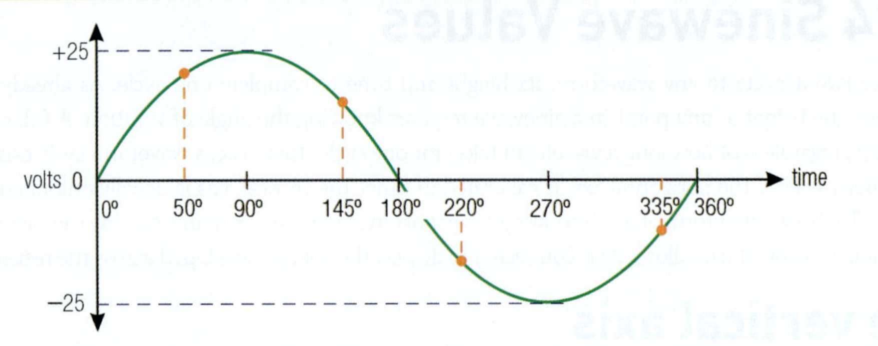

Solution

1 Values $V_{max} = 25$ V
$\theta = 50°$

Equation $v = \sin\theta\, V_{max}$
$= \sin 50° \times 25$

sin 5 0) × 2 5 = keystrokes for Casio calculator

Answer **v at 50° = 19.15 V**

2 Values $V_{max} = 25$ V
$\theta = 145°$

Equation $v = \sin\theta\, V_{max}$
$= \sin 145° \times 25$

sin 1 4 5) × 2 5 =

Answer **v at 145° = 14.34 V**

3 Values $V_{max} = 25$ V
$\theta = 220°$

Equation $v = \sin\theta\, V_{max}$
$= \sin 220° \times 25$

sin 2 2 0) × 2 5 =

Answer **v at 220° = –16.07 V** (note the negative value)

4 Values $V_{max} = 25$ V
$\theta = 335°$

Equation $v = \sin\theta\, V_{max}$
$= \sin 335° \times 25$

sin 3 3 5) × 2 5 =

Answer **v at 335° = –10.57 V** (also a negative value)

RMS value

KEY CONCEPT

RMS value and heating effect

RMS stands for *root mean square* and is described as the *effective* value of a waveform. This is how most sinusoidal AC voltages (like the mains voltage) are given. A digital multimeter usually displays the RMS value of an alternating voltage. An RMS value relates an AC value to an equivalent DC value.

If an AC voltage is applied across a resistor, an alternating current flows. That is, it flows one way for the positive half cycle, then the other during the negative half cycle. However, both half cycles cause the same value of current, regardless of its direction. When an AC or DC current flows through a resistor, heat is produced. Therefore, the RMS value of an alternating current causes the same heating effect as an equal value of direct current, as depicted by Figure 15.12. That is, because a kettle is a resistance, a 230 V kettle can be powered by either 230 V RMS or 230 V DC.

FIGURE 15.12 In a kettle or any heater with a resistive element, the same heat is produced by 230 V AC or 230 V DC

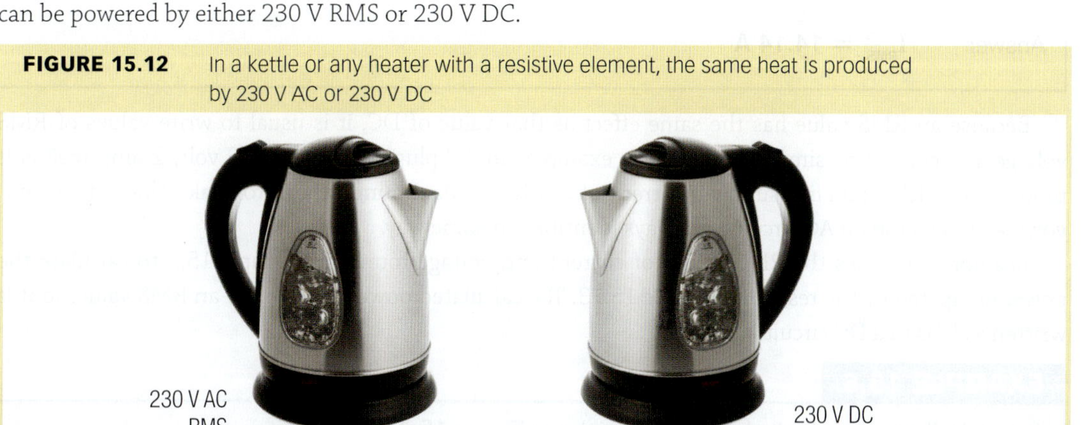

The RMS value for *any* waveform can be found by:

- calculating a number of instantaneous values for the waveform
- calculating the square of each of these values
- adding the squared values
- dividing their sum by the number of values to get an average (or mean)
- finding the square root of the mean.

Fortunately, there's an easier way to find the RMS value of a sinewave:

- $V_{RMS} = \frac{1}{\sqrt{2}} \times V_{max}$ which simplifies to $0.707\ V_{max}$

For a sinewave, the RMS value equals the maximum value multiplied by 0.707.

RMS value of a sinewave

Example 15.5 uses an RMS voltage and Ohm's law to find the value of the current. Current flowing in a resistor is *always* voltage divided by resistance. If the voltage is an RMS value, the current will be an RMS value. If the voltage is a maximum value, the current will be a maximum value.

EXAMPLE 15.5

Find the RMS voltage and current in the circuit in Figure 15.13.

FIGURE 15.13

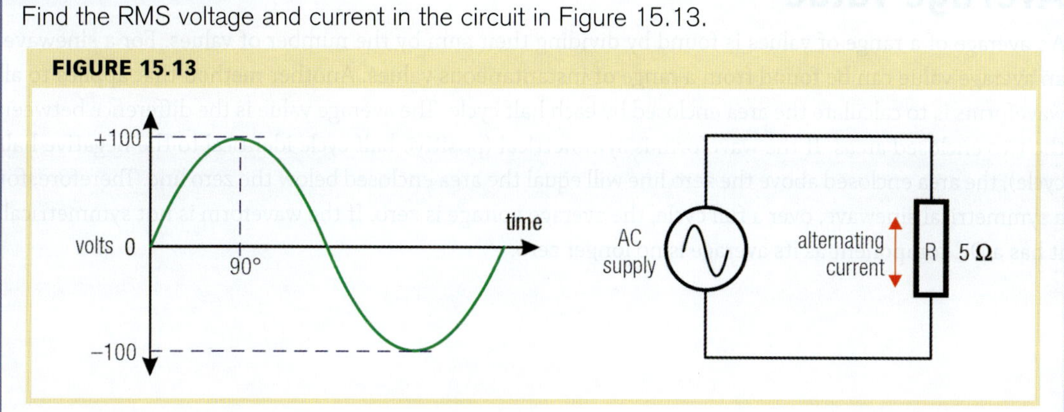

Solution

Values $V_{max} = 100\ V$

$R = 5\ \Omega$

Equation 1 $V_{RMS} = 0.707\ V_{max}$

$= 0.707 \times 100$

Answer $\mathbf{V_{RMS} = 70.7\ V}$

Equation 2 $I_{RMS} = \frac{V_{RMS}}{R}$

$= \frac{70.7}{5}$

Answer $\mathbf{I_{RMS} = 14.14\ A}$

Because an RMS value has the same effect as that value of DC, it is usual to write values of RMS voltage and current as simply V and I. For example, an AC plug-pack rated at 6 volt, 2 amp means it produces 6 V RMS and can supply a current of 2 A RMS. Maximum or peak-to-peak values of current, voltage or power in an AC circuit are always identified in some way.

Example 15.6 uses the RMS values of current and voltage found in Example 15.5 to calculate the power dissipated by the resistor in Figure 15.13. The calculated power is therefore an RMS value, so it is written as P, as in a DC circuit.

EXAMPLE 15.6

Calculate the power dissipated by the resistor in Figure 15.13.

Solution

Values $V_{RMS} = 70.7\ V$

$I_{RMS} = 14.14\ A$

Equation $P = V \times I$

$= 70.7 \times 14.14$

Answer $\mathbf{P = 999.7\ W}$

MATH

For help with transposing, see page xxi

The value of an alternating current or voltage is usually given as an RMS value. It is also the value displayed by a digital multimeter, so maximum or peak-to-peak values are calculated from the RMS value. The equation to find V_{RMS} from a maximum value was previously shown to be:

$$V_{RMS} = \frac{1}{\sqrt{2}} V_{max} = 0.707\ V_{max}$$

Transposing this equation to make V_{max} the subject gives:

$$V_{p\text{-}p} = 2\ V_{max} \text{ or } 2\sqrt{2}\ V_{RMS}$$

Average value

An average of a range of values is found by dividing their sum by the number of values. For a sinewave, an average value can be found from a range of instantaneous values. Another method that applies to all waveforms is to calculate the area enclosed by each half cycle. The average value is the difference between the two enclosed areas. If the waveform is symmetrical (positive half cycle identical to the negative half cycle), the area enclosed above the zero line will equal the area enclosed below the zero line. Therefore, for a symmetrical sinewave, over a full cycle, the average voltage is zero. If the waveform is not symmetrical, it has a DC component as its average is no longer zero.

However, as shown in Figure 15.14, looking at one half cycle only, the area enclosed above the zero line is not counteracted by an enclosed area below the line because the negative half cycle is missing. This applies to all waveforms. Because a sinewave is related to a circle, the average voltage of a half cycle is found with this equation:

$$V_{av} = \frac{2}{\pi} \times V_{max} = 0.637 \times V_{max}$$

This equation only applies to a sinewave, and then only for one half cycle.

FIGURE 15.14 The average of one cycle of a symmetrical sinewave is zero, the average of a half cycle is the maximum voltage multiplied by 0.637

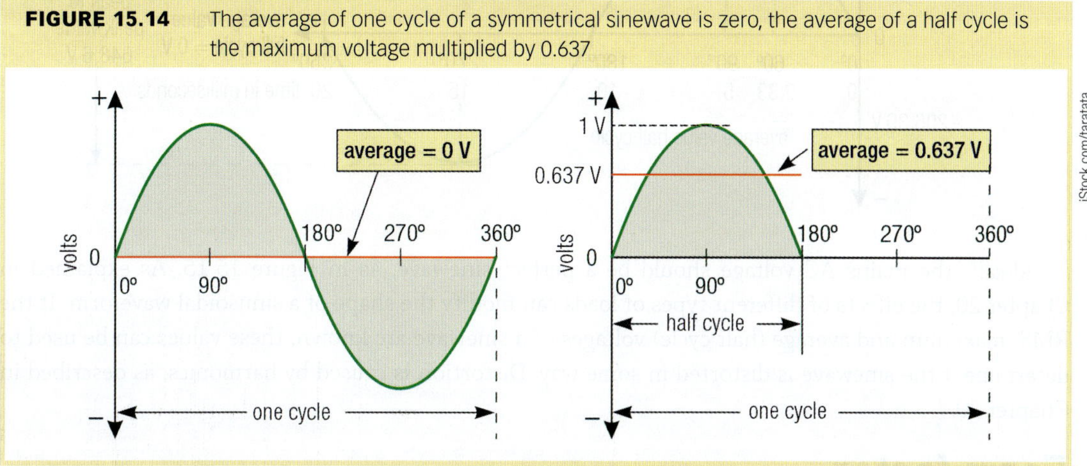

When an AC voltage is converted to a DC voltage by a circuit called a rectifier, both halves of the cycle have the same polarity. That is, one cycle of the AC voltage gives two positive (or negative) half cycles, giving a varying DC voltage that always has the same polarity. The average value of the rectifier's output voltage is therefore $0.637 \times V_{max}$. A typical application is a battery charger. Example 15.7 applies the voltage equations so far discussed for a 230 V AC sinewave.

EXAMPLE 15.7

A circuit is supplied with 230 V AC mains power. Calculate the:

(a) maximum voltage of the supply
(b) peak-to-peak voltage of the supply
(c) average voltage over a full cycle
(d) average voltage over half a cycle
(e) the instantaneous voltage of the supply at 60° from the start of the cycle.

Solution

Value $V = 230\ V_{RMS}$

(a) Equation $V_{max} = 1.41\ V_{RMS}$
$= 1.41 \times 230$

Answer $\mathbf{V_{max} = 324.3\ V}$

(b) Equation $V_{p\text{-}p} = 2\ V_{max}$
$= 2 \times 324.3$

Answer $\mathbf{V_{p\text{-}p} = 648.6\ V}$

(c) Equation $V_{av\ full\ cycle} = 0\ V$

Answer $\mathbf{V_{av\ full\ cycle} = 0\ V}$

(d) Equation $V_{av\ half\ cycle} = 0.637\ V_{max}$
$= 0.637 \times 324.3$

Answer $\mathbf{V_{av\ half\ cycle} = 206.39\ V}$

(e) Equation $V = \sin\theta \times V_{max}$
$= \sin 60° \times 324.3$

Answer **V at 60° = 280.85 V**

Figure 15.15 shows the voltages for the 230 V RMS sinewave as calculated in Example 15.7.

FIGURE 15.15 Voltages of a 230 V RMS, 50 Hz sinewave as found in Example 15.7. Horizontal axis is shown in degrees and time

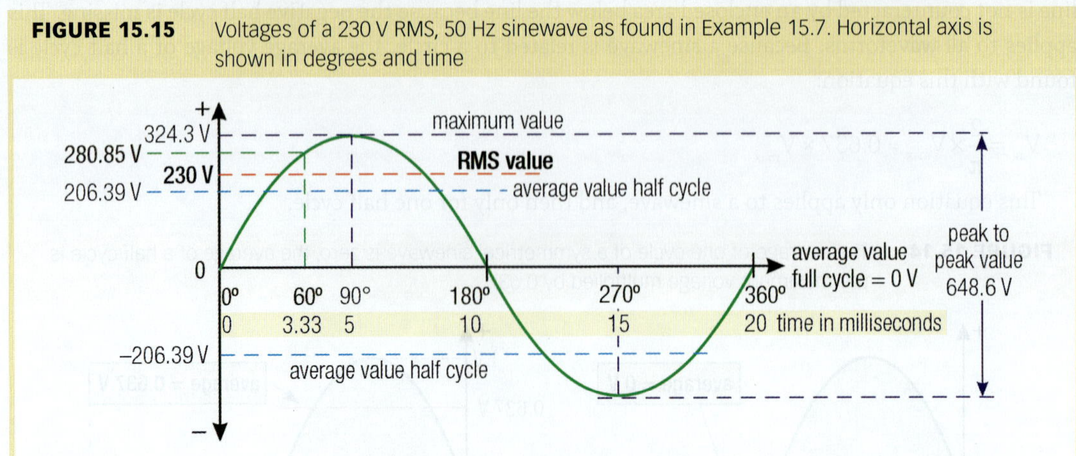

Ideally, the mains AC voltage should be a perfect sinewave, as in Figure 15.15. As explained in Chapter 20, the effects of different types of loads can modify the shape of a sinusoidal waveform. If the RMS, maximum and average (half cycle) voltages of a sinewave are known, these values can be used to determine if the sinewave is distorted in some way. Distortion is caused by harmonics, as described in Chapter 20.

Form factor

A sinewave's form factor is defined as the ratio of its RMS value to its average value (half cycle):

$$\text{Form factor} = \frac{\text{RMS value}}{\text{average value}} \text{ which for a sinewave} = \frac{0.707}{0.637} = 1.11$$

If a sinewave has a different form factor to 1.11, it indicates it is no longer a pure sinewave. A form factor higher than 1.11 means the sinewave is peaky, like a triangular wave. A form factor below 1.11 indicates it has a flattened top, like a square wave. A square wave has a form factor of 1 because its RMS and average values are the same.

Crest factor

The crest factor of a sinewave is defined as the ratio of its maximum value to its RMS value:

$$\text{Crest factor} = \frac{\text{Maximum value}}{\text{RMS value}} \text{ which for a sinewave} = \frac{1}{0.707} = \sqrt{2} = 1.414$$

A crest factor value higher than 1.414 means the waveform has a higher maximum value than for a sinewave. For example, if a waveform has a crest factor of 1.5, it means its maximum (peak) voltage is 1.5 times the RMS value. Crest factor is taken into account when determining insulation requirements.

KEY POINTS...

- Figure 15.15 shows the various voltage values for a 230 V AC sinewave
- Table 15.1 lists the equations to convert from one value to another. These only apply to a sinewave.

TABLE 15.1

LIST OF EQUATIONS TO CONVERT BETWEEN SINEWAVE VALUES			
Unknown value	**Equations**	**Polarity of value**	**Comments**
V_{max}	$= \frac{V_{p\text{-}p}}{2} = 1.41\ V_{RMS}$	positive or negative	also called peak value
$V_{p\text{-}p}$	$= 2\ V_{max} = 2.82\ V_{RMS}$	no polarity	twice the peak value
V_{RMS}	$= 0.707\ V_{max}$	no polarity	an RMS value has the same heating effect as that value of DC
V_{av} (half cycle)	$= 0.637\ V_{max}$ (= 0 for full cycle)	positive or negative	average value usually only applies to half a cycle
v	$= \sin\theta\ V_{max}$ (θ is the angle at the point in the cycle where v is being calculated)	positive or negative	v = an instantaneous value anywhere over the full sinewave cycle
Form factor	$= \frac{V_{RMS}}{V_{av}} = 1.11$	no polarity	value for a pure sinewave
Crest factor	$= \frac{V_{max}}{V_{RMS}} = 1.414$	no polarity	value for a pure sinewave

15.5 Phase Angle and Phasor Diagrams

A DC circuit can have two or more voltages, such as one battery connected in series with another. If a battery is reversed, one voltage is subtracted from the other; that is, their voltages are added algebraically. In an AC circuit, as in a DC circuit, there can be two or more voltages in the circuit. However, the voltages may not be changing polarity at the same time because there is a *phase difference* between them, as shown in Figure 15.16(b).

FIGURE 15.16 Two AC voltages as displayed on an oscilloscope

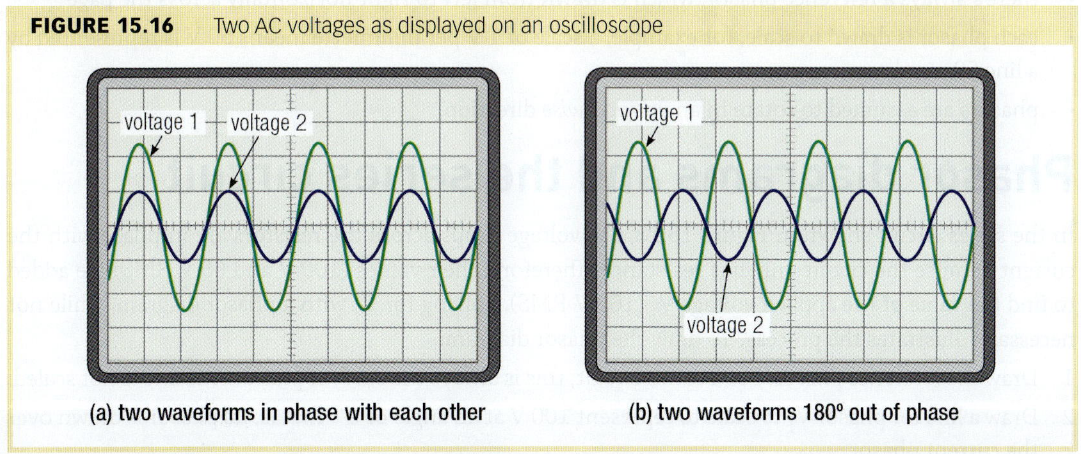

In Figure 15.16(a) the two voltage waveforms are in phase, so they both reach a positive peak and a negative peak together. If voltage 1 is 100 V AC (RMS) and voltage 2 is 65 V AC (RMS), the total voltage is 165 V RMS. The waveforms in Figure 15.16(b) are in opposition to each other, or 180° out of phase. As voltage 1 reaches a positive peak, voltage 2 reaches a negative peak, so for the same voltages as before, the total voltage is now 100 – 65, or 35 V AC (RMS).

In a resistive AC circuit, voltages and currents are always in phase, so calculations are similar to those for a DC circuit. However, in many AC circuits, voltages and currents may not be in phase with each other.

FYI

Phase difference is denoted with phi (ϕ), angle of rotation is denoted with theta (θ)

In Figure 15.17, two alternating voltages are out of phase by an amount denoted by the angle ϕ (Greek lower case letter phi).

In Figure 15.17(a), voltage 1 passes through zero in a positive-going direction before voltage 2 does the same thing, but ϕ° later. Therefore, voltage 1 *leads* voltage 2 by ϕ°. In Figure 15.17(b), voltage 2 is now leading voltage 1, or put another way, voltage 1 *lags* voltage 2 by ϕ°

FIGURE 15.17 Two AC voltages out of phase with each other by ϕ°

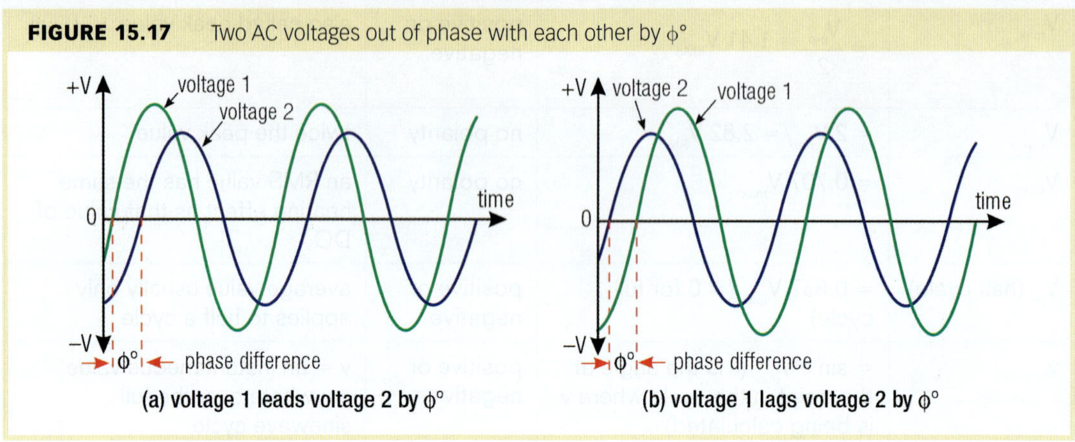

MATH

See page xxxv for information about vector and phasor diagrams

In an AC circuit, if voltages or currents are in phase or 180° out of phase, the total value is their algebraic sum. Otherwise, their total value has to be found graphically, using a *phasor diagram*, which is similar to a vector diagram as used in physics.

Phasor diagrams

A phasor diagram, like a vector diagram, uses lines with arrow heads to represent a value and a direction, in this case of an alternating voltage or current. The length of the line is scaled to represent the value. Its angle compared to a reference line is its phase difference. Each line is called a *phasor*. There are several conventions to follow when drawing phasor diagrams:

- voltage phasors are shown with an open arrow head⟶
- current phasors are shown with a closed arrow head ➞
- the arrow head of a phasor is called the tip, the start end of the phasor is called the tail
- there's always a reference phasor, which is drawn from left to right horizontally across the page
- each phasor is drawn to scale, for example, a scale of 1 V per millimetre means 50 V is represented by a line 50 mm long
- phasors are assumed to rotate in an *anticlockwise* direction.

Phasor diagrams and the series circuit

In the series circuit shown in Figure 15.18, the voltage drops across the resistors are in phase with the current because the circuit only has resistance. Therefore, their values (100 V and 65 V RMS) are added to find the value of the applied voltage V_T (165 V RMS). Solving for V_T with a phasor diagram, while not necessary, illustrates the process. To draw the phasor diagram:

1. Draw the reference phasor. For a series circuit, this is always the current phasor. The line is not scaled.
2. Draw a line for phasor V_1 to scale to represent 100 V at an angle of 0°. That is, its phasor is drawn over the current phasor.
3. Draw a line to scale for phasor V_2, starting at the tip of V_1, also drawn over the current phasor.
4. Determine the total voltage by measuring the total length of the two voltage phasors.

FIGURE 15.18 Phasor diagram to find the applied voltage V_T in a series AC circuit with two resistors

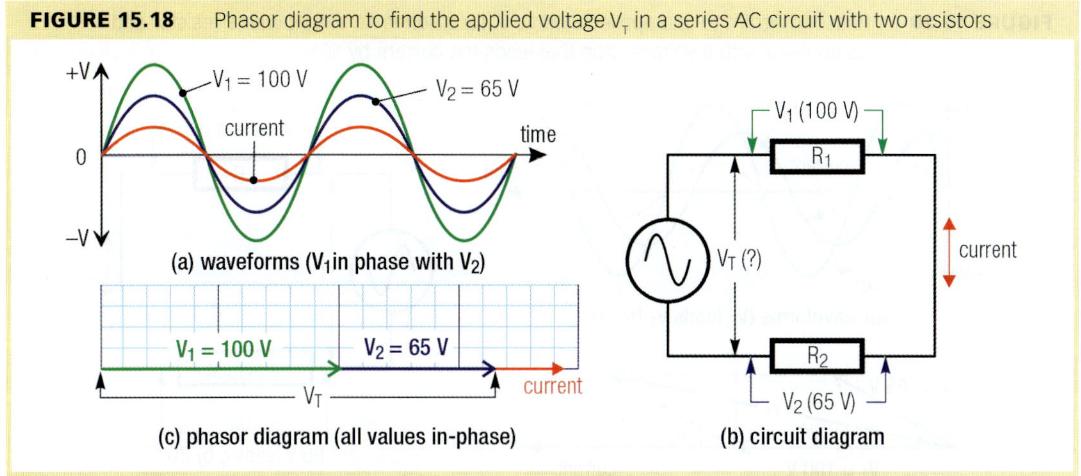

The series circuit in Figure 15.19 has two AC voltage sources that are 180° out of phase with each other. Again, as previously explained, this means their total voltage is their algebraic sum, or 100 V – 65 V, giving 35 V. Solving for V_T with a phasor diagram, while also not necessary, illustrates the process in which there is now a phase difference to deal with.

FIGURE 15.19 Phasor diagram to find the applied voltage V_T in a series AC circuit in which one voltage is 180° out of phase with another voltage

Solving for V_T with the phasor diagram in Figure 15.19:

1 Draw the reference phasor (current, as it's a series circuit). The line can be any length.
2 Draw a line to scale over the current phasor for phasor V_1 to represent 100 V at 0°.
3 Draw a line to scale over the current phasor for phasor V_2, starting at the tip of V_1. Because it is 180° out of phase with the current, V_2 is drawn in the opposite direction to phasor V_1.
4 Measure from the tail of phasor V_1 to the tip of phasor V_2 to find V_T (35 V RMS)

Figure 15.20 shows a series AC circuit with a resistor and a component with a voltage drop (V_2) that *leads* the current by 30°. V_1 is a voltage drop across a resistor, so V_1 is in phase with the current. Solving for V_T in which a leading phasor is rotated *anticlockwise*:

1 Draw the reference phasor (current, because it's a series circuit).
2 Draw a line to scale for phasor V_1 over the current phasor because V_1 is in phase with the current.
3 Draw a line to scale for phasor V_2, starting at the tail of V_1 and drawn at an angle of 30°, rotating *anticlockwise* from the reference because it's a *leading* phase shift.
4 Complete the parallelogram by drawing construction lines (shown dotted).
5 Draw a diagonal line to represent the phasor addition of V_1 and V_2 (gives V_T).
6 Measure the length of the diagonal line to find the applied voltage V_T (159.6 V).
7 Measure the angle (ϕ°) between the reference (current) and the V_T phasor to find the phase difference between the applied voltage V_T and the circuit current (11.8° lead).

FYI

Components that cause current and voltage to be out of phase are said to be 'reactive', as explained in Chapter 16

FIGURE 15.20 Phasor diagram to find applied voltage V_T for a series AC circuit with a resistor and a component with a voltage drop that leads the current by 30°

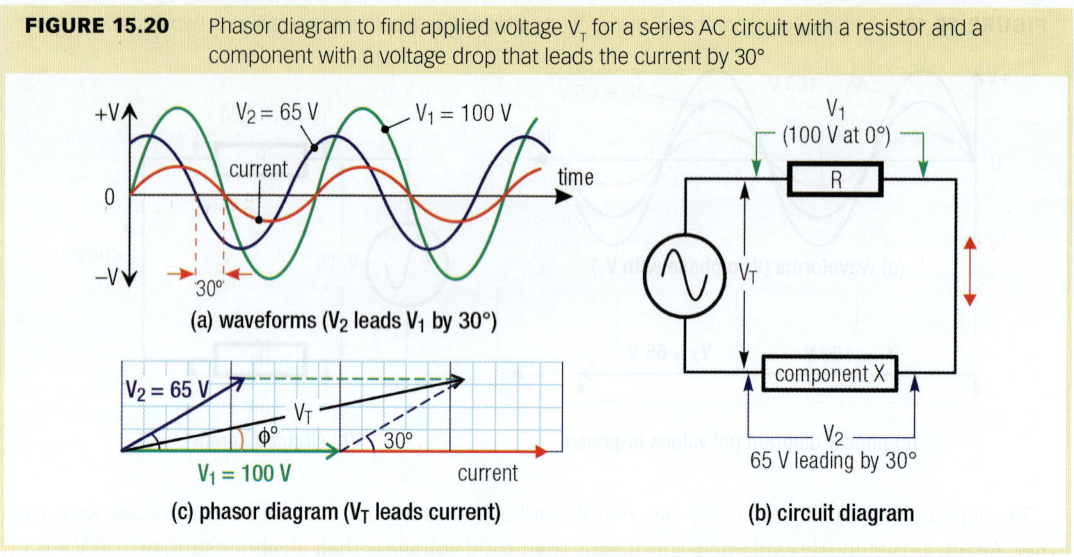

Figure 15.21 shows the waveforms for the applied voltage (V_T) and the current for the circuit in Figure 15.20. It also shows an enlarged view of the phasor diagram with divisions on the angled phasors for a scale of one division per 10 V.

FIGURE 15.21 Waveforms of the applied voltage and current for the circuit in Figure 15.20 and an enlarged view of the phasor diagram in Figure 15.20

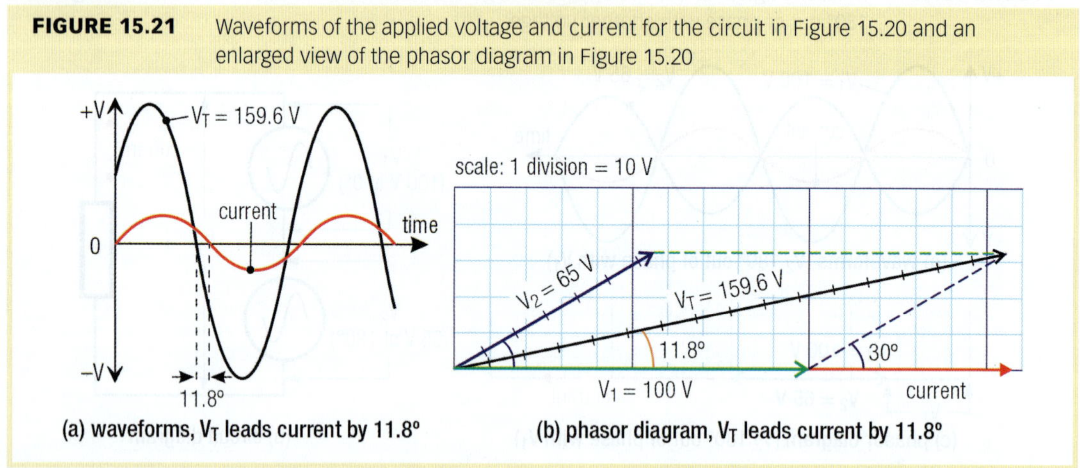

Figure 15.22 shows a series AC circuit with a resistor and a component with a voltage drop (V_2) across it that *lags* the current by 30°. (A lagging phasor is rotated *clockwise*.) V_1 is a voltage drop across a resistor, so V_1 is in phase with the current.

FIGURE 15.22 Phasor diagram for a series AC circuit with a resistor and a component with a voltage drop that *lags* the current by 30°

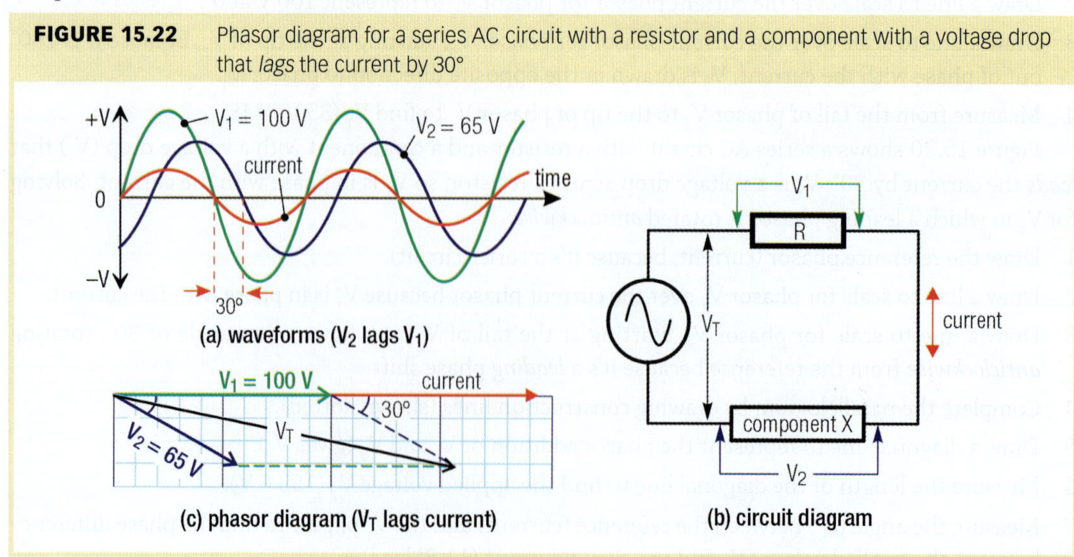

The phasor diagram in Figure 15.22 is drawn in the same way as for Figure 15.20, except phasor V_2 is rotated clockwise:

1 Draw the reference phasor (current).
2 Draw a line to scale for phasor V_1 over the current phasor.
3 Draw a line to scale for phasor V_2, starting at the tail of V_1 and drawn at an angle of 30°, rotating *clockwise* from the reference (because it's a lagging phase shift).
4 Complete the parallelogram.
5 Draw a diagonal line to represent the phasor addition of V_1 and V_2 (gives V_T).
6 Measure the length of the V_T phasor to find the applied voltage (159.6 V).
7 Measure the angle between the reference and the V_T phasor to find the phase difference between V_T and the circuit current (11.8° lag).

Figure 15.23 shows the waveforms for the applied voltage (V_T) and the current for the circuit in Figure 15.22, and an enlarged view of the phasor diagram in Figure 15.22.

FIGURE 15.23 Waveforms of the applied voltage and current for the circuit in Figure 15.22, and an enlarged view of the phasor diagram in Figure 15.22

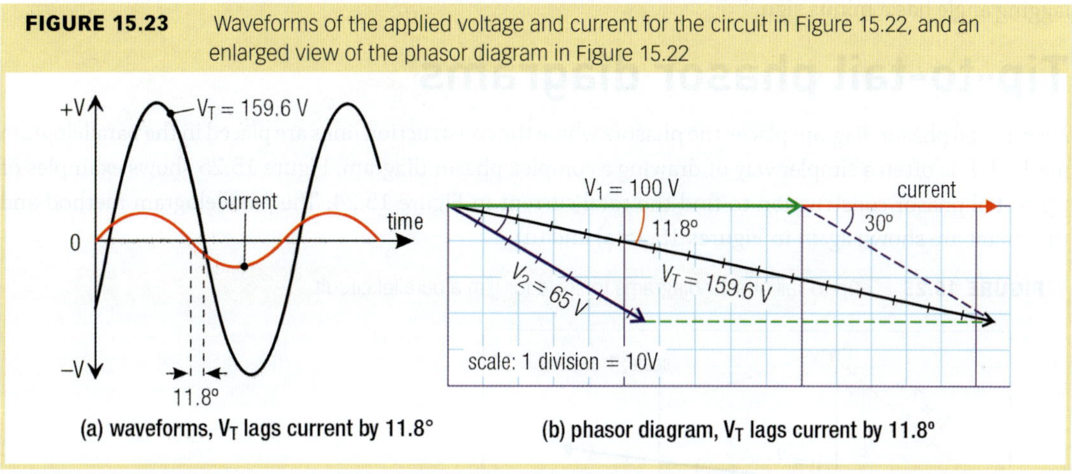

(a) waveforms, V_T lags current by 11.8°

(b) phasor diagram, V_T lags current by 11.8°

Phasor diagrams and the parallel circuit

In a parallel circuit, voltage is common to all branches in the circuit, so the phasor diagram now deals with branch currents referenced to the applied voltage. Figure 15.24 shows a parallel circuit with two branches supplied by applied voltage V. Each branch has a reactive component that causes a phase shift between the applied voltage and the branch current. The phasor diagram to find the total circuit current I_T is constructed in the same way as previously described, except the applied voltage is the reference, and the phasors are of branch currents.

 **MATH**

Pages xxxvii–xxxviii show two ways to construct the phasor diagram in Figure 15.24(a)

FIGURE 15.24 Parallel circuit with two branch currents out of phase with the applied voltage, and the phasor diagram to find the total current and its phase angle

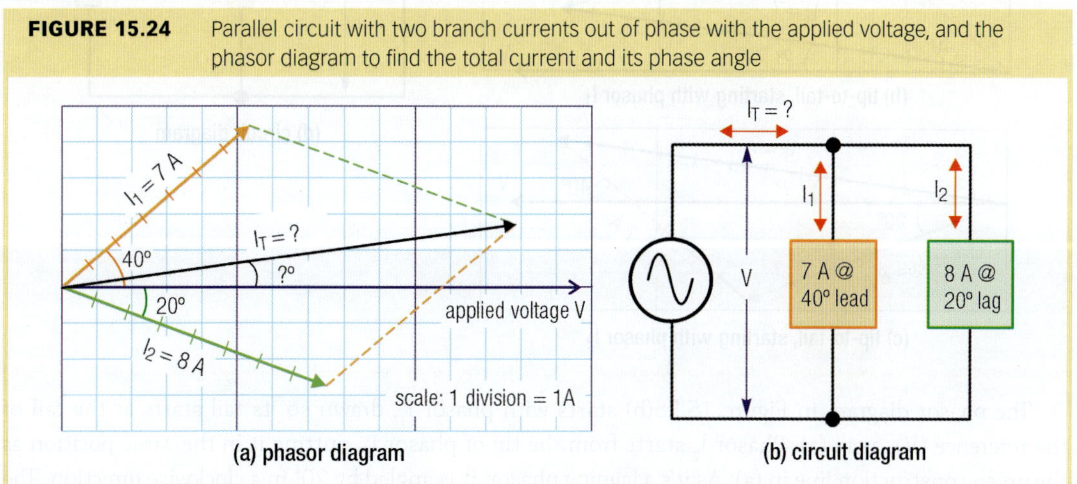

(a) phasor diagram

(b) circuit diagram

Solving for I_T with the phasor diagram in Figure 15.24:

1 Draw the reference phasor (applied voltage, because it's a parallel circuit).
2 Draw a line for I_1 to scale to represent a current of 7 A. Because I_1 is leading the voltage by 40°, its phasor is drawn at an angle of 40°, rotated anticlockwise compared to the reference.
3 Draw a line to scale for phasor I_2, starting at the tail of I_1 and drawn at an angle of 20°, rotated clockwise (because I_2 is lagging the voltage).
4 Complete the parallelogram.
5 Draw a diagonal line to represent the phasor addition of I_1 and I_2 (gives I_T).
6 Measure the length of the I_T phasor to find the total current (13 A).
7 Measure the angle between the reference (V) and the I_T phasor to find their phase difference (I_T leads V by 7.8°).

The two currents in Figure 15.24 are given as 7 A @ 40° lead, and 8 A @ 20° lag. A mathematical shorthand way is to write 7 A ∠40° and 8 A ∠–20°. The ∠symbol indicates an angular measurement; a lagging angle has a minus sign.

Tip-to-tail phasor diagrams

A tip-to-tail phasor diagram places the phasors where the construction lines are placed in the parallelogram method. It is often a simpler way of drawing a complex phasor diagram. Figure 15.25 shows examples of tip-to-tail phasor construction to find the total current in Figure 15.24. The parallelogram method and the circuit are shown again in Figures 15.25(a) and (d).

FIGURE 15.25 Tip-to-tail phasor diagrams to solve for I_T in a parallel circuit

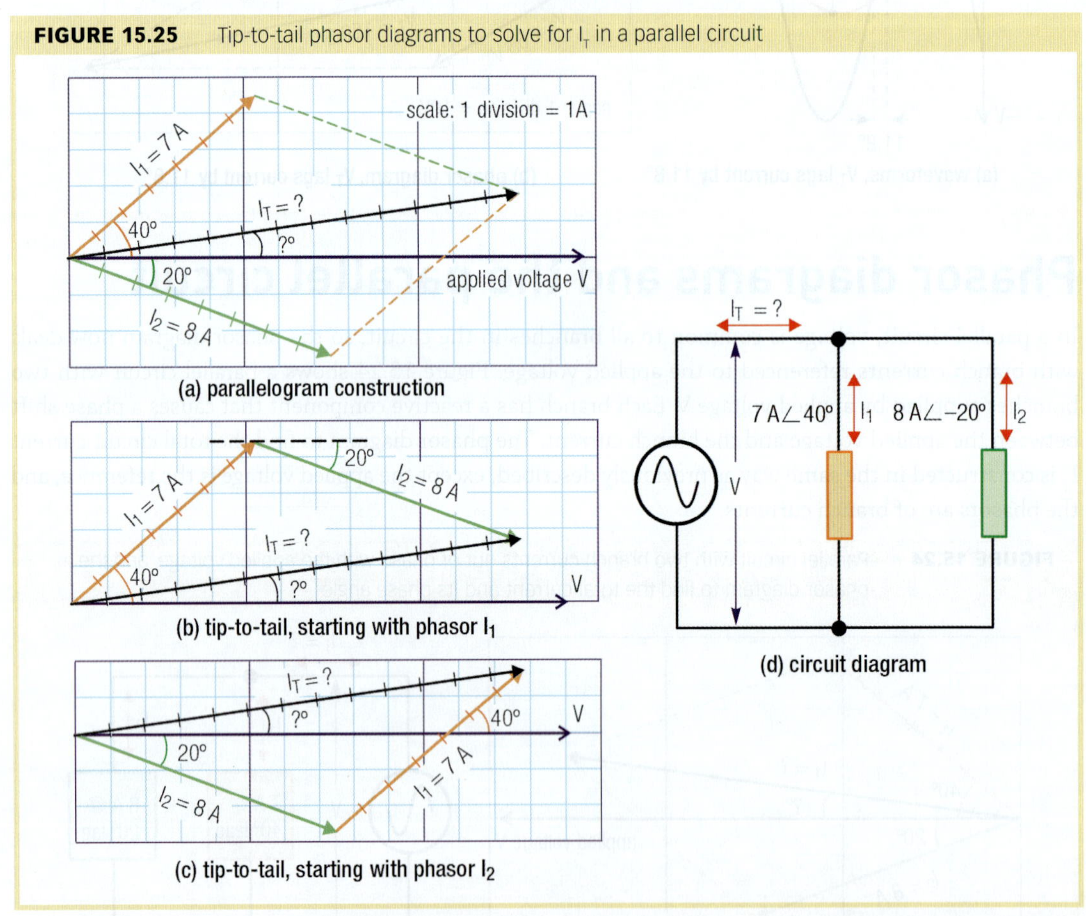

The phasor diagram in Figure 15.25(b) starts with phasor I_1, drawn so its tail starts at the tail of the reference (V), as in (a). Phasor I_2 starts from the tip of phasor I_1, putting it in the same position as the green construction line in (a). As it's a lagging phasor, it is angled by 20° in a clockwise direction. The phasor for I_T is drawn from the tail of phasor I_1 to the tip of phasor I_2. Its length and angle to the reference gives its value and phase angle, as before (13 A ∠7.8°).

The phasor diagram in (c) starts with I_2, the other known value. Because I_2 is lagging the voltage by 20°, it is angled in a clockwise direction. Phasor I_1 leads the voltage by 40°, so it is rotated in an anticlockwise direction, putting it in the same position as the orange construction line in (a). The phasor for I_T is drawn from the tail of phasor I_2 to the tip of phasor I_1. Gives 13 A ∠7.8°, as before.

Phasor diagrams (b) and (c) show it doesn't matter which of the known phasors is drawn first, as the same result is obtained for phasor I_T.

FIGURE 15.26 Tip-to-tail phasor diagrams to solve for unknown voltage drop V_1 in a series circuit

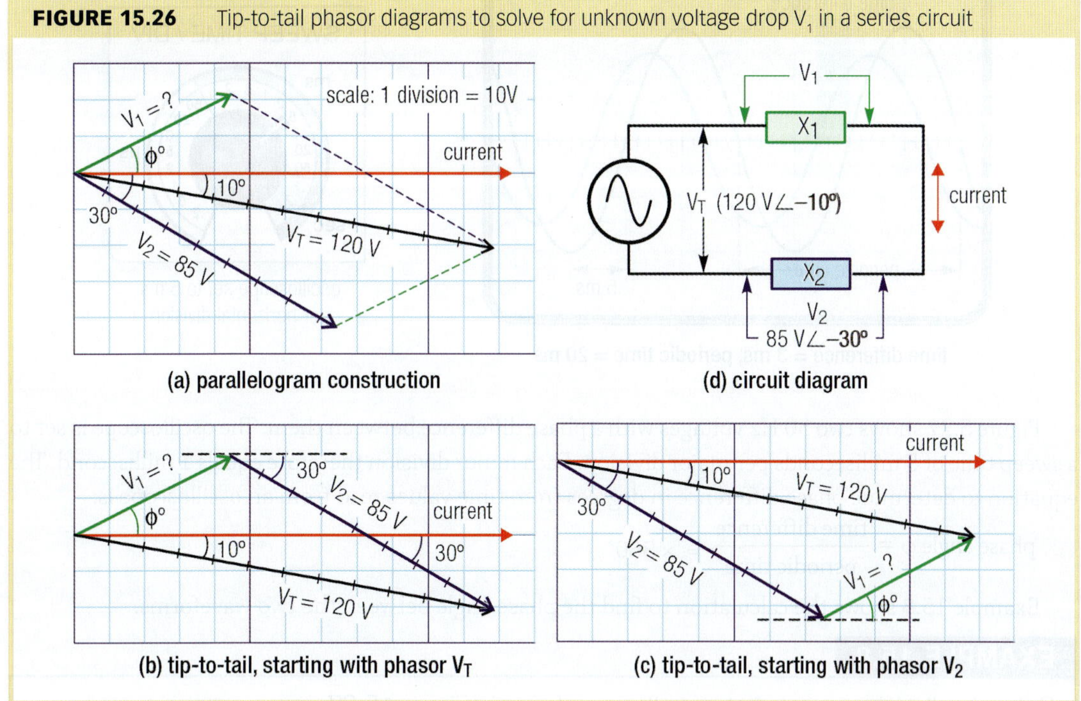

The tip-to-tail phasor diagrams in Figure 15.26 are for a series circuit. The aim is to find the value and phase angle of voltage V_1. As it's a series circuit, current is the reference. Again, it doesn't matter which known phasor value is drawn first.

The phasor diagram in Figure 15.26(b) starts with lagging phasor V_T, so it is rotated clockwise by 10°. The V_2 phasor lags by 30° (also rotated clockwise) and is drawn so its tip is at the tip of V_T. This phasor replaces the blue construction line in (a). The phasor for the unknown value (V_1) completes the construction.

In Figure 15.26(c), the phasor for V_2 is drawn first, rotated 30° clockwise. After drawing the phasor for V_T, the phasor for the unknown voltage V_1 completes the construction.

By measurement from any of the three phasor diagrams, the unknown voltage drop V_1 in Figure 15.26(d) is 49.6 V with a leading phase angle of 25.9° (49.6 V ∠25.9°). Phasor diagrams are generally accurate to one decimal place, depending on the scale used to draw the phasors. The larger the scale, the higher the accuracy.

Page xxxix shows how to draw the phasor diagram in Figure 15.26(c)

15.6 Phase Angle from Oscilloscope

Phase angle between alternating voltages and currents can be measured with a phase angle meter. This type of meter is often expensive, although lower cost hand-held units are sometimes used in the field. Phase angle between waveforms can also be measured using a dual-trace oscilloscope connected to display the waveforms of two voltages. The setup to do this depends on the circuit, especially if the circuit under test is at mains potential.

In principle, the two voltage waveforms with a phase difference are displayed on the screen, in which the oscilloscope is set to display about two cycles of each waveform. A current is converted to a voltage by passing the current through a resistance. The oscilloscope is adjusted so both waveforms are positioned vertically with both half cycles equally around the graticule centre line. The horizontal shift control is adjusted so both waveforms move to a position where one of the waveforms passes through zero at a graduation on the horizontal axis.

FIGURE 15.27 Phase difference is displayed by an oscilloscope as a time value, in this case 3 ms

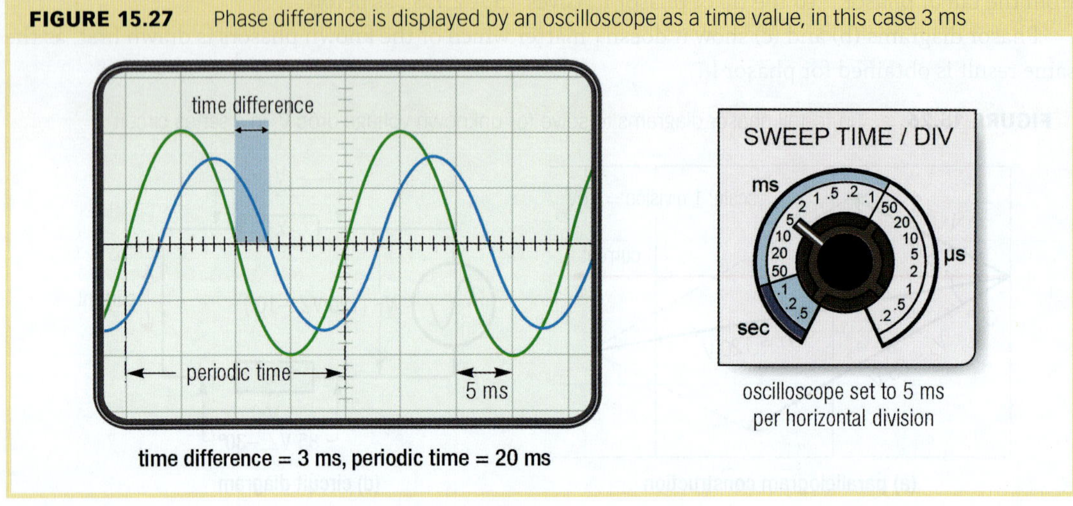

Figure 5.27 shows two 50 Hz voltages with a phase difference between them. The oscilloscope is set to a sweep time of 5 milliseconds per major division. Each minor division therefore equals 1 millisecond. The equation to determine phase difference in degrees from time values read from an oscilloscope is:

$$\text{phase angle } \phi = \frac{\text{time difference}}{\text{periodic time}} = \times 360°$$

Example 15.8 shows the calculation to find the phase angle between the two waveforms.

EXAMPLE 15.8

Determine the phase angle between the waveforms in Figure 15.27.

Solution

Values periodic time = 20 ms
time difference between waveforms = 3 ms
phase angle = ?

Equation $\text{phase angle} = \frac{\text{time difference}}{\text{periodic time}} \times \frac{0.5 \times 10^{-3}}{3.5 \times 10^{-3}} \times 360$

Answer $\phi = 54°$

CHAPTER SUMMARY

- Electrical power in Australia and New Zealand is a sinusoidal alternating voltage with a frequency of 50 hertz. The standard supply voltage is 230 V RMS.
- An RMS value has the same heating power as that value of DC.
- AC voltages and currents are generally given as RMS values. Most digital multimeters display an RMS value.
- The sinewave is a fundamental waveshape, and all other waveforms (square, triangular, complex) are combinations of sinewaves.
- The periodic time (or period) of a waveform is the time taken for a full cycle.
- The frequency of an alternating voltage or current is the reciprocal of its periodic time, measured in hertz.

- The RMS value of a sinewave is 0.707 V_{max}, the average value of half a cycle is 0.637 V_{max}.
- The value at any angle (θ) in a sinewave is:

  $v = \sin\theta\, V_{max}$. Transposed, $V_{max} = \dfrac{V}{\sin\theta}$
- If AC voltages or currents are in phase or 180° out of phase, the total value is their algebraic sum. When out of phase, the total value is found with a phasor diagram.
- Phase difference between two waveforms is denoted with the Greek letter phi (ϕ). Angle of rotation is denoted with the Greek letter theta (θ).
- A phasor diagram always has a reference, which is current in a series circuit and voltage in a parallel circuit. The reference is a line drawn horizontally across the page.
- A phasor is a line with a length representing a voltage or current. Its angle to the reference is the value's phase angle.
- A leading phasor is rotated clockwise, a lagging phasor is rotated anticlockwise.
- Phase angle is shown with an angle symbol (∠) followed by the angle. A minus sign indicates a lagging angle, e.g. 10 A ∠–40°.

REVIEW EXERCISES

Check your answers at the back of the book.

1 For voltage 1 in Figure 15.28, determine its:
   a maximum voltage
   b peak-to-peak voltage
   c RMS voltage
   d instantaneous voltage at 70°.
2 For voltage 1 in Figure 15.28, determine its:
   a periodic time
   b frequency.
3 For voltage 2 in Figure 15.28, determine its:
   a maximum voltage
   b peak-to-peak voltage
   c RMS voltage
   d average voltage (half cycle).
4 For the two waveforms in Figure 15.28, determine the:
   a time difference between the waveforms
   b phase difference between the waveforms.
5 Which waveform is the leading voltage in Figure 15.28?

FIGURE 15.28

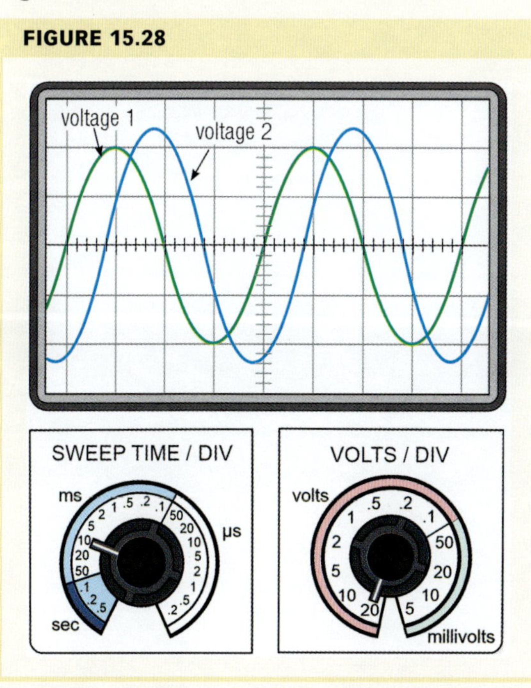

FIGURE 15.29

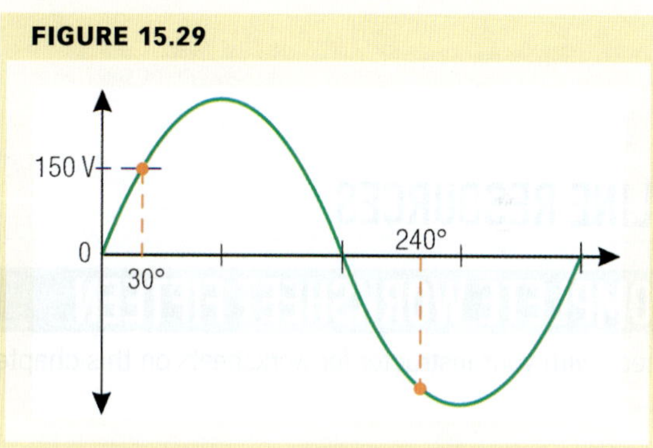

6 For the waveform in Figure 15.29, determine its:
   a maximum value
   b RMS value
   c instantaneous voltage at 240°.

FIGURE 15.30

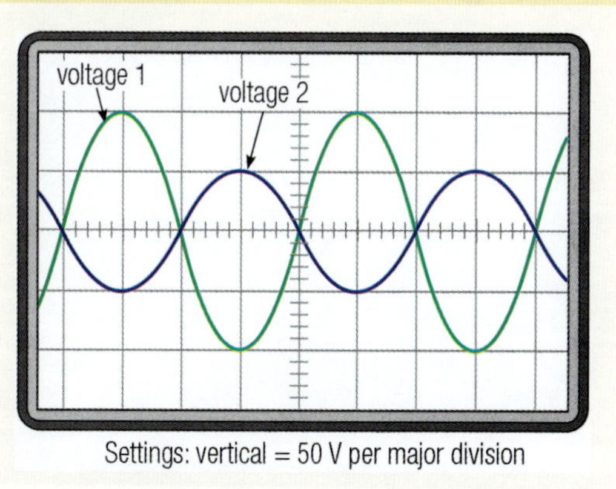

Settings: vertical = 50 V per major division

7 For the waveforms in Figure 15.30, determine their:
   a maximum values
   b RMS values
   c total RMS value.

FIGURE 15.31

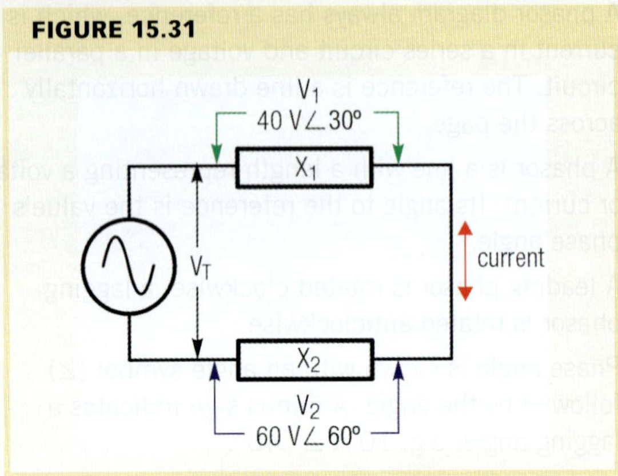

8 For the circuit in Figure 15.31, use a phasor diagram to find the value of the applied voltage V_T and its phase angle to the current.

FIGURE 15.32

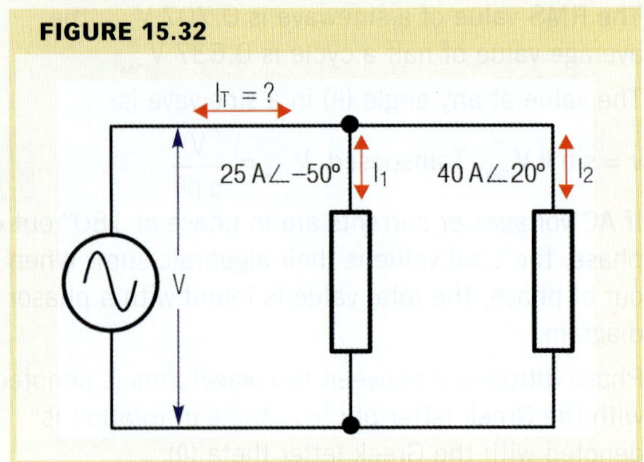

9 For the circuit in Figure 15.32, use a phasor diagram to find the value of the total current I_T and its phase angle to the applied voltage.

10 The total current in a parallel circuit with two branches is 37.75 A ∠13°. The current in branch 1 (I_1) is 20 A ∠40°. Use a phasor diagram to find the current in branch 2 (I_2) and its phase angle.

ONLINE RESOURCES

COMPLETE WORKSHEET FIFTEEN

Check with your instructor for worksheets on this chapter.

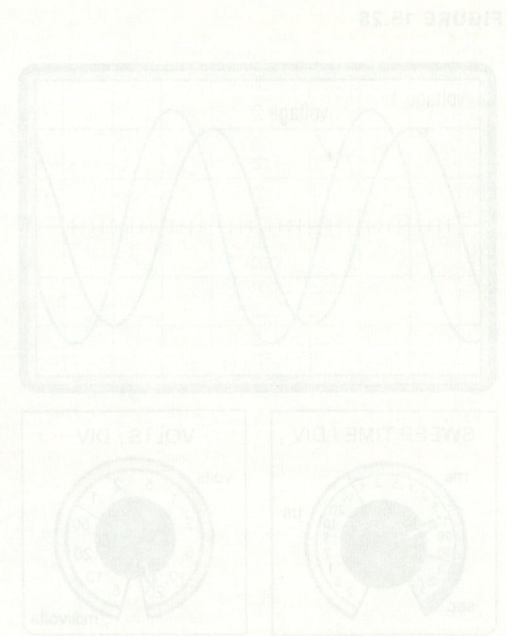

CHAPTER 16

PURE R, L OR C IN AN AC CIRCUIT

This chapter explains the behaviour of AC circuits that are purely resistive, purely capacitive and purely inductive. An AC circuit with resistance is no different to the DC resistive circuit, but there are considerable differences if the circuit has capacitance or inductance. In an AC circuit, a capacitor or an inductor causes a phase shift between voltage and current. Because of a property called reactance, a term introduced in this chapter, these components also determine the amount of current flowing in a given AC circuit.

CHAPTER OUTLINE

16.1 Power in AC Resistive Circuits

Examples of an AC circuit in which the load is a pure resistance include lighting circuits with incandescent lamps and power circuits for stoves, ovens or heaters. As outlined in Chapter 15, a resistance does not cause a phase shift between the voltage and the current in the circuit, which simplifies the circuit calculations. The waveforms of current and voltage in a purely resistive circuit are shown in Figure 16.1, in which both waveforms are in phase with each other.

FIGURE 16.1 Current and voltage are in phase in an AC resistive circuit

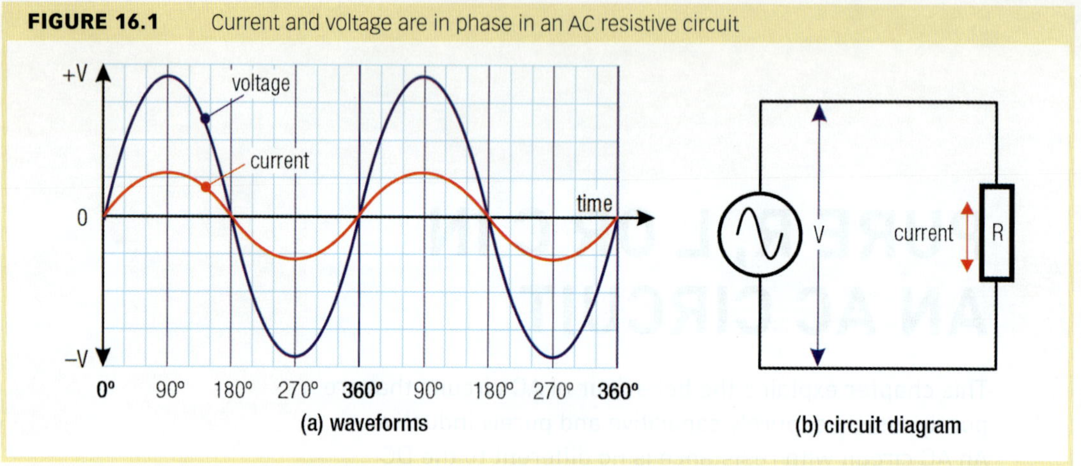

Chapter 15 explained that Ohm's law is used in an AC resistive circuit in the same way as for a DC circuit. That is, current equals the applied voltage divided by the resistance of the circuit. Power is the product of the applied RMS voltage and the RMS current, as in a DC circuit.

Because the circuit current and voltage are periodically changing polarity, the power dissipated by the resistance also changes periodically. When the voltage and current are zero, the power is zero. When the voltage and current are at their maximum value, the power is a maximum. Therefore, the power dissipated by the resistor has a waveform.

The shape of the power waveform can be found by multiplying instantaneous values of the voltage and the current, and plotting the result as a waveform. As explained in Chapter 15, the equation to find an instantaneous value of voltage of a sinewave is:

$v = \sin\theta\, V_{max}$

where:

v = the instantaneous voltage

$\sin\theta$ = the sine of the angle where the instantaneous value is during the cycle

V_{max} = the maximum voltage of the sinewave.

The equation to find an instantaneous value of current of a sinewave is:

$i = \sin\theta\, I_{max}$

where:

i = the instantaneous current

$\sin\theta$ = the sine of the angle where the instantaneous value is during the cycle

I_{max} = the maximum current value of the sinewave.

Table 16.1 lists the instantaneous values of a voltage with a maximum value of 10 V and a current with a maximum value of 2 A, at 15° intervals over a complete 360° cycle. It also shows the instantaneous power at these intervals.

TABLE 16.1

POWER IN A RESISTIVE AC CIRCUIT							
Instantaneous current and voltage values where V_{max} = 10 V and I_{max} = 2 A							
positive half cycle				negative half cycle			
θ°	V volts	I amperes	P = V × I watts	θ°	V volts	I amperes	P = V × I watts
0	0.00	0.00	**0.00**				
15	2.59	0.52	**1.34**	195	−2.59	−0.52	**1.34**
30	5.00	1.0	**5.00**	210	−5.00	−1.0	**5.00**
45	7.07	1.41	**10.00**	225	−7.07	−1.41	**10.00**
60	8.66	1.73	**15.00**	240	−8.66	−1.73	**15.00**
75	9.66	1.93	**18.66**	255	−9.66	−1.93	**18.66**
90	10.00	2.00	**20.00**	270	−10.00	−2.00	**20.00**
105	9.66	1.93	**18.66**	285	−9.66	−1.93	**18.66**
120	8.66	1.73	**15.00**	300	−8.66	−1.73	**15.00**
135	7.07	1.41	**10.00**	315	−7.07	−1.41	**10.00**
150	5.00	1.0	**5.00**	330	−5.00	−1.0	**5.00**
165	2.59	0.52	**1.34**	345	−2.59	−0.52	**1.34**
180	0.00	0.00	**0.00**	360	0.00	0.00	**0.00**

The waveforms in Figure 16.2 are derived from the values in Table 16.1. The values for instantaneous power are positive in both half cycles because multiplying a negative number by another negative number gives a positive result.

FIGURE 16.2 Power, voltage and current waveforms in a resistive circuit

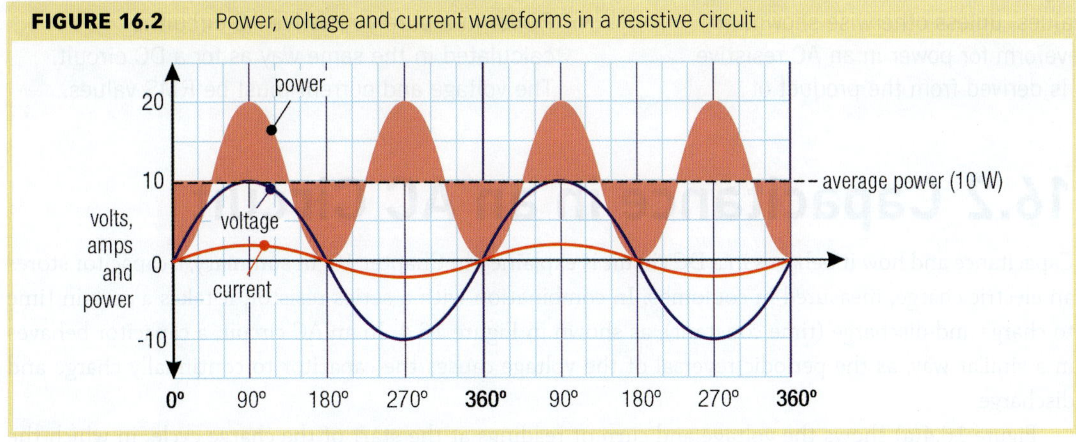

There are several things to notice about the waveform for power.

1 It has the shape of a sinewave.

2 It is always positive. This means power is consumed on both halves of the voltage and current cycles.

3 It is twice the frequency of the voltage and current waveforms. For a 50 Hz system, power frequency is therefore 100 Hz.

4 The average of the waveform gives the average power being dissipated (10 W in Figure 16.2).

Power in an AC resistive circuit is the product of the RMS values of current and voltage. In Figure 16.2, the RMS values are found by multiplying their maximum values by 0.707 ($1/\sqrt{2}$). The maximum voltage is 10 V and the maximum current is 2 A, giving RMS values of 7.07 V and 1.41 A.

 FYI

Because the power frequency is 100 Hz, the hum and flicker in some types of lamps is at 100 Hz

Multiplying these values gives a power dissipation of 10 W. This is called the *true* power taken by the circuit.

AC power in a resistive circuit can be measured with a wattmeter and registers on a kilowatt hour (energy) meter. Although the power consumption is continually changing from a maximum to zero, when a heater is operated from a 50 Hz supply, the thermal inertia of the element causes it to maintain a constant heat output.

EXAMPLE 16.1

Find the circuit current and the power being dissipated by the resistor in Figure 16.3.

Solution

Values $V = 230\ V_{RMS}$

$R = 50\ \Omega$

Equation $I_{RMS} = \frac{V_{RMS}}{R} = \frac{230}{50}$

Answer $\mathbf{I_{RMS} = 4.6\ A}$

Equation $P = VI$ (RMS values) $= 230 \times 4.6$

Answer $\mathbf{P = 1058\ W}$

FIGURE 16.3

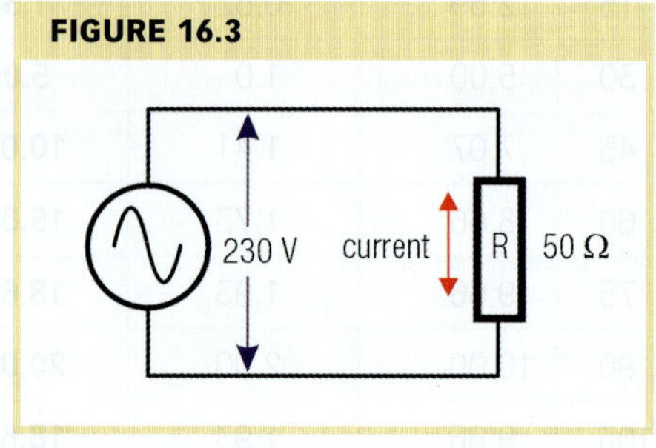

FYI

If power is calculated from peak (maximum) values, the true power is always half the peak (maximum) power value

In a resistive AC circuit, true power can be found using all the equations given in Chapter 4 for a DC circuit. For example, $P = I^2R$. The important thing is to convert current and voltages to RMS values where necessary. In most cases, AC voltages and currents are given as RMS values, and it's not necessary to write RMS. Where a value is not RMS, then write what type of value it is, for example, maximum, peak-to-peak or average.

KEY POINTS...

- Voltage and current are always in phase in a purely resistive AC circuit.
- AC voltages and currents are assumed to be RMS values, unless otherwise shown.
- The waveform for power in an AC resistive circuit is derived from the product of instantaneous values of current and voltage and shows that power has a frequency twice that of the voltage or current.
- Power in a purely resistive AC circuit is calculated in the same way as for a DC circuit. The voltage and current must be RMS values.

16.2 Capacitance in an AC Circuit

Capacitance and how it behaves in a DC circuit is explained in Chapter 10. In summary, a capacitor stores an electric charge, measured in coulombs. In combination with a series resistor, it takes a certain time to charge and discharge (time constant), as shown in Figure 16.4. In an AC circuit, a capacitor behaves in a similar way, as the periodic reversal of the voltage causes the capacitor to continually charge and discharge.

Figure 16.4(a) shows the voltage and current readings at the start of the charge cycle, in which the charge current starts at a maximum value and the capacitor voltage starts at zero. As shown in (b), eventually the current is zero and the voltage is at the maximum value of 10 V. The graphs in (c) show how the current and voltage change. In terms of a phase shift, the current is 90° out of phase to the voltage, as it is a maximum when the voltage is a minimum. As well, the current is leading the voltage because it reaches its maximum before the voltage.

FIGURE 16.4 When a capacitor is first connected to a DC voltage, the charge current immediately rises to its maximum value, while the voltage across the capacitor starts at zero. When charged, the current is zero and the voltage is a maximum value.

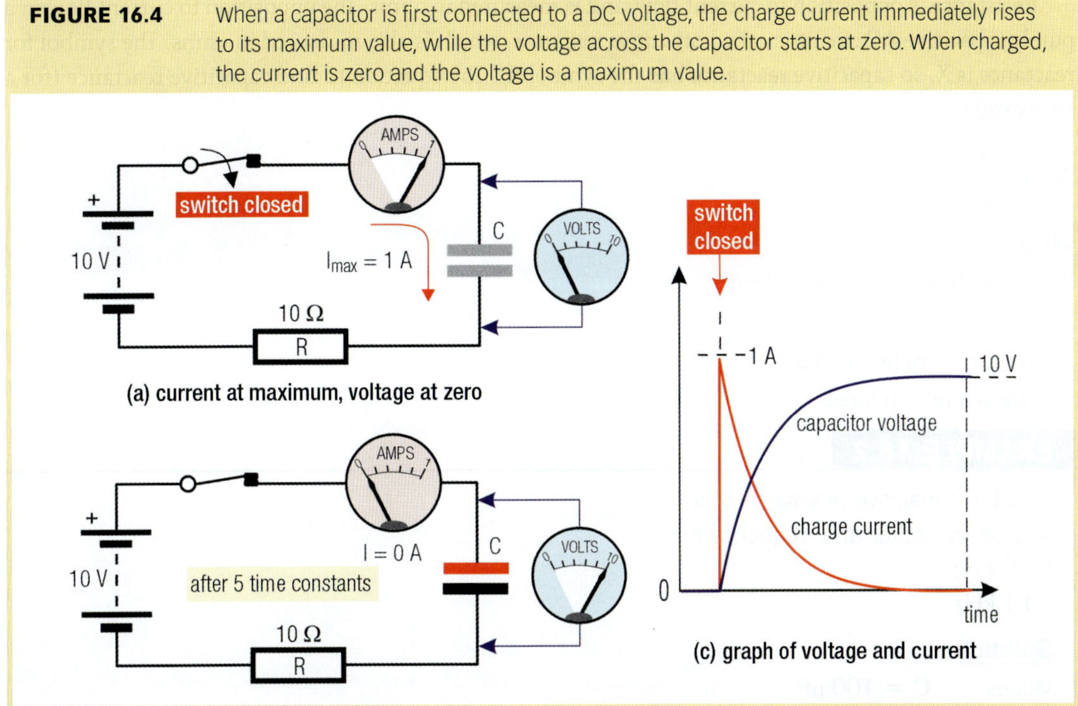

Phase shift between current and voltage

A capacitor in an AC circuit, therefore, causes a phase shift between the current and the voltage across it. For a pure capacitor, this phase shift is 90° in which the current *leads* the voltage, as shown in Figure 16.5. The phasor diagram is also shown, in which the reference is assumed to be the voltage across the capacitor.

KEY CONCEPT

Current through a pure capacitance leads the voltage across the capacitance by 90°

FIGURE 16.5 Current leads the voltage by 90° in a purely capacitive circuit

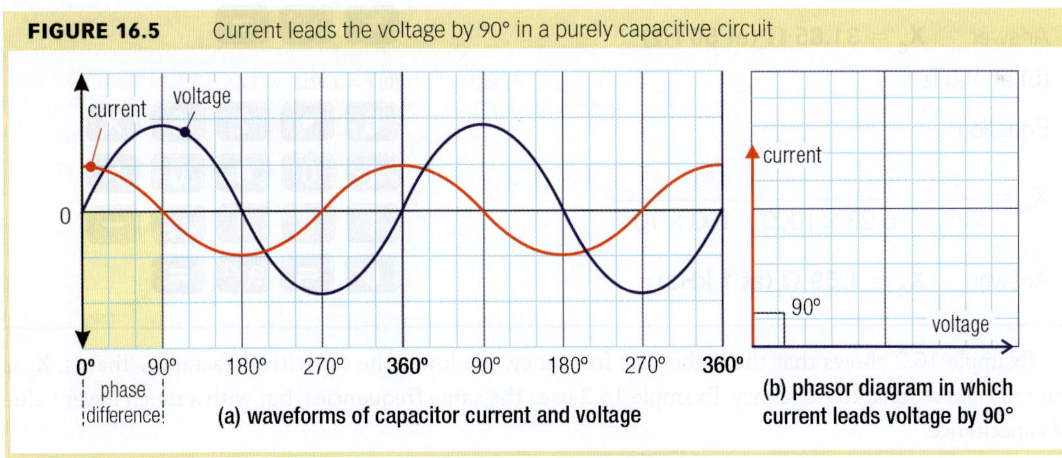

16.3 Capacitive Reactance

A current flows continually in a purely capacitive AC circuit because the capacitance in the circuit is constantly charging and discharging. The larger the capacitance, the higher the value of the current if the capacitor is the only component in the circuit. Therefore, the current flowing in a purely capacitive AC circuit depends on the applied voltage and the value of the capacitor. However, there is another factor to consider: the *frequency* of the AC voltage.

The more cycles there are per second, the more often the capacitor charges and discharges. This means the current flowing in the circuit increases if the frequency increases. Therefore, for a given AC voltage, the current in a purely capacitive circuit depends on the value of the capacitance and the frequency of the AC voltage.

Resistance is opposition to current flow, and is measured in ohms. The opposition to current flow in a purely capacitive AC circuit is called *capacitive reactance*, which is also measured in ohms. The symbol for reactance is X, so capacitive reactance has the symbol X_C. The equation to find capacitive reactance (for a sinewave) is:

$$X_C = \frac{1}{2\pi fC}$$

where:

X_C = capacitive reactance in ohms
$2\pi = 6.28$ (a close approximation)
f = frequency in hertz (Hz)
C = capacitance in farads.

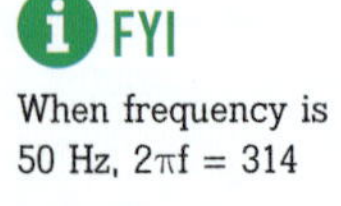

FYI

When frequency is 50 Hz, $2\pi f = 314$

EXAMPLE 16.2

Find the capacitive reactance of a 100 μF capacitor at a frequency of
(a) 50 Hz
(b) 1 kHz

Solution

Values C = 100 μF
f_1 = 50 Hz
f_2 = 1 kHz

(a) at 50 Hz

Equation

$$X_C = \frac{1}{2\pi fC} = \frac{1}{6.28 \times 50 \times 100 \times 10^{-6}}$$

Answer **X_C = 31.85 Ω (at 50 Hz)**

(b) at 1 kHz

Equation

$$X_C = \frac{1}{2\pi fC} = \frac{1}{6.28 \times 1000 \times 100 \times 10^{-6}}$$

Answer **X_C = 1.59 Ω (at 1 kHz)**

Keystrokes for Casio calculator

(a) f = 50 Hz:

3 1 4 × 1 0 0 ×10ˣ − 6 = x^{-1} =

(b) f = 1 kHz:

6 • 2 8 × 1 0 0 0 × 1 0 0 ×10ˣ − 6 = x^{-1} =

Example 16.2 shows that the higher the frequency, the lower the capacitive reactance. That is, X_C is *inversely proportional* to frequency. Example 16.3 uses the same frequencies, but with a much lower value of capacitance.

EXAMPLE 16.3

Find the capacitive reactance of a 0.022 μF capacitor at a frequency of:
(a) 50 Hz
(b) 1 kHz

FIGURE 16.6

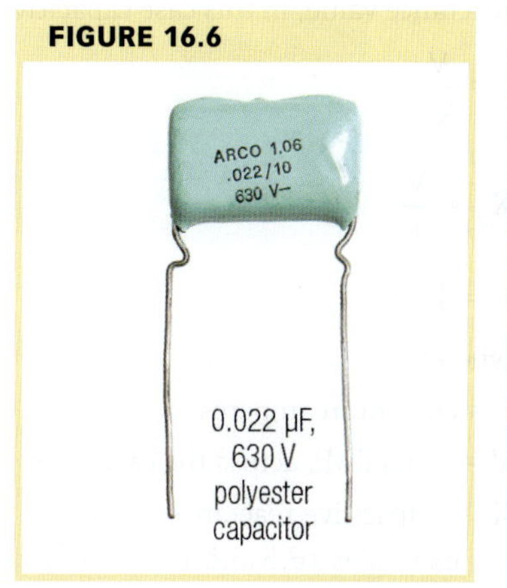

0.022 μF, 630 V polyester capacitor

Solution

Values $C = 0.022\ \mu F$
$f_1 = 50\ Hz$
$f_2 = 1\ kHz$

(a) at 50 Hz

Equation $X_C = \frac{1}{2\pi fC} = \frac{1}{6.28 \times 50 \times 0.022 \times 10^{-6}}$

Answer $\mathbf{X_C = 144\,800\ \Omega}$ **or 144.8 kΩ (at 50 Hz)**

(b) at 1 kHz

Equation $X_C = \frac{1}{2\pi fC} = \frac{1}{6.28 \times 1000 \times 0.022 \times 10^{-6}}$

Answer $\mathbf{X_C = 7238\ \Omega}$ **or 7.24 kΩ (at 1 kHz)**

Example 16.3 shows that the smaller the capacitance, the higher the capacitive reactance. Put another way, X_C is *inversely proportional* to capacitance.

The equation for X_C can be transposed to find the frequency or capacitance:

MATH
For help with transposing this equation, see page xxiv

$$f = \frac{1}{2\pi C X_C}$$

$$C = \frac{1}{2\pi f X_C}$$

Example 16.4 shows how to use these two equations.

EXAMPLE 16.4

Find the:
(a) capacitance of a capacitor whose capacitive reactance is 300 Ω at 250 Hz
(b) frequency that gives a capacitive reactance of 200 Ω for a 10 nF capacitor

FIGURE 16.7

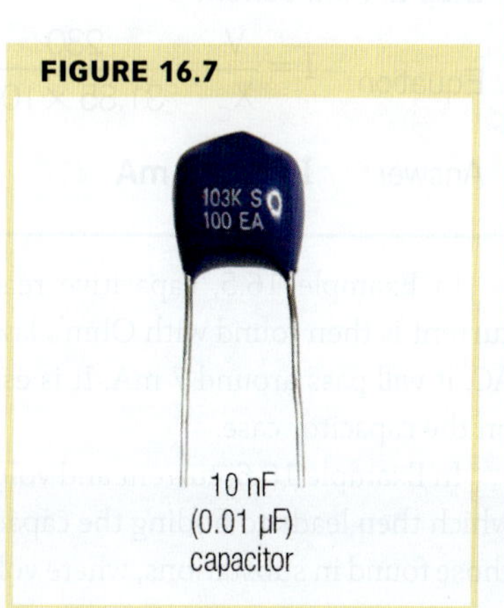

10 nF (0.01 μF) capacitor

Solution

(a) Values $X_C = 300\ \Omega$
$f = 250\ Hz$

Equation $C = \frac{1}{2\pi f X_C} = \frac{1}{6.28 \times 250 \times 300}$

Answer $\mathbf{C = 2.12\ \mu F}$

(b) Values $X_C = 200\ \Omega$
$C = 10\ nF\ (= 10 \times 10^{-9} F)$

Equation $f = \frac{1}{2\pi C X_C} = \frac{1}{6.28 \times 10 \times 10^{-9} \times 200}$

Answer **f = 79 617.8 Hz** or **79.62 kHz**

Capacitive reactance and Ohm's law

Ohm's law is used to find the current in a purely capacitive AC circuit. Resistance (R) is replaced with a reactance value, in this case capacitive reactance (X_C), giving the following equations:

$$I = \frac{V}{X_C}$$

$$X_C = \frac{V}{I}$$

$$V = IX_C$$

where:

I = current in amperes

V = volts RMS across the capacitor

X_C = capacitive reactance in ohms.

Examples 16.5 and 16.6 use these equations.

EXAMPLE 16.5

A 0.1 μF capacitor is connected across a voltage of 230 V AC 50 Hz. How much current flows in the circuit?

Solution

Values $C = 0.1\ \mu F$

$V = 230$ V AC

$f = 50$ Hz

Step 1: Find capacitive reactance

Equation $X_C = \frac{1}{2\pi fC}$

$$X_C = \frac{1}{6.28 \times 50 \times 0.1 \times 10^{-6}}$$

$$X_C = 31.85\ k\Omega$$

Step 2: Find current

Equation $I = \frac{V}{X_C} = \frac{230}{31.85 \times 10^3}$

Answer **I = 7.22 mA**

FIGURE 16.8

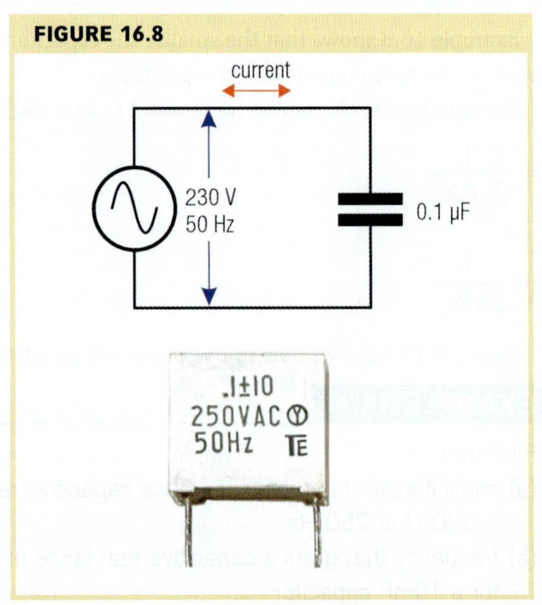

In Example 16.5, capacitive reactance is calculated using capacitance and frequency values. The current is then found with Ohm's law. If the actual capacitor shown in Figure 16.8 is connected to 230 V AC, it will pass around 7 mA. It is essential that the capacitor is rated to withstand 230 V AC, as shown on the capacitor case.

In Example 16.6, current and voltage are known, so Ohm's law is used to find the capacitive reactance, which then leads to finding the capacitance. This example refers to a high voltage capacitor bank such as those found in substations, where voltages and current values are much higher than in other installations.

EXAMPLE 16.6

A capacitor bank connected across an 11 kV, 50 Hz busbar in a substation takes a current of 300 A. What is its capacitance value?

Solution

Values $V = 11\text{ kV} = 11 \times 10^3\text{ V}$
$f = 50\text{ Hz}$
$I = 300\text{ A}$

Step 1: Find capacitive reactance

Equation $X_C = \frac{V}{I} = \frac{11 \times 10^3}{300}$

$X_C = 36.67\ \Omega$

Step 2: Find capacitance

Equation $C = \frac{1}{2\pi f X_C} = \frac{1}{6.28 \times 50 \times 36.67}$

Answer **$C = 86.86\ \mu F$**

FIGURE 16.9

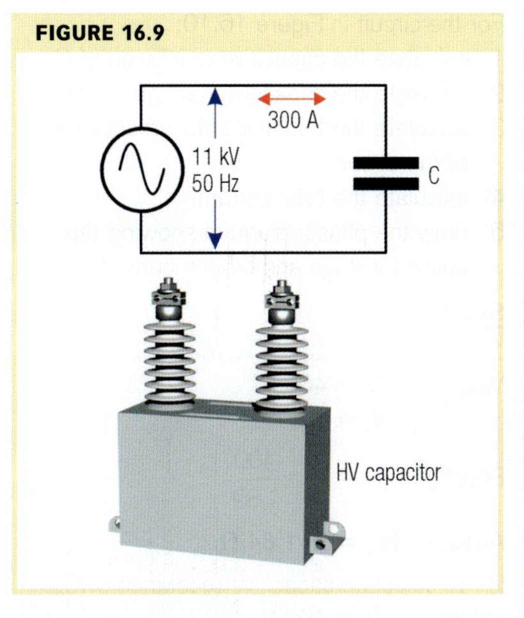

Capacitive reactance in parallel

When capacitors are connected in parallel, the total capacitance is the sum of the individual capacitor values. The total capacitance increases as more capacitors are connected in parallel, which means the capacitive reactance *decreases* with each additional capacitor (see Chapter 10). This is similar to a parallel resistive circuit, in which adding another resistance in parallel reduces the total resistance. Therefore, in a purely capacitive parallel circuit, because the capacitor currents are in phase with each other, total capacitive reactance can be found in the same way as resistance in a resistive parallel circuit. The difference is, instead of writing R (for resistance), write X_C for capacitive reactance. That is:

$$\frac{1}{X_{CT}} = \frac{1}{X_{C1}} + \frac{1}{X_{C2}} + \frac{1}{X_{C3}} + \ldots$$ for three or more values of X_C (ohms)

$$X_{CT} = \frac{X_{C1} \times X_{C2}}{X_{C1} + X_{C2}}$$ for two values of X_C (ohms).

Capacitive reactance in series

When capacitors are connected in series (see Chapter 10), the total capacitance reduces with each additional capacitor. Therefore, the total capacitive reactance will increase. This is similar to a series resistive circuit, in which the total resistance of the circuit increases with each additional resistor. In a purely capacitive series circuit, the voltage drops across each capacitor are in phase with each other, and are all out of phase with the current by 90°. This means they can be added algebraically. It also means the total capacitive reactance of the circuit is the sum of the individual reactance values. That is:

$$X_{CT} = X_{C1} + X_{C2} + X_{C3} \ldots$$

EXAMPLE 16.7

For the circuit in Figure 16.10:

1 calculate the capacitive reactance of C_1
2 calculate the capacitive reactance of C_2
3 calculate the total capacitive reactance of the circuit
4 calculate the total current
5 draw the phasor diagram showing the applied voltage and branch currents

FIGURE 16.10

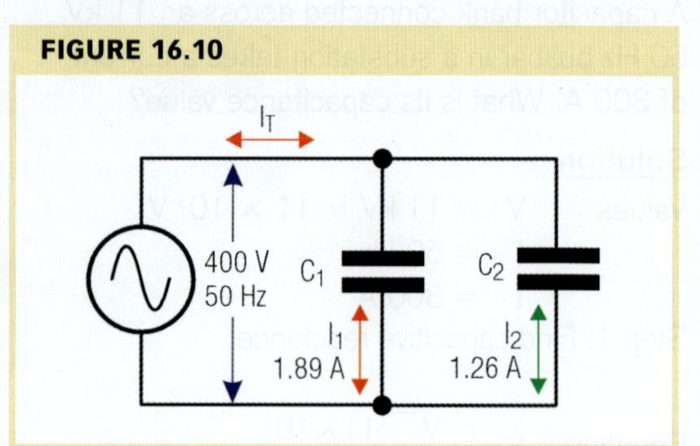

Solution

1 To calculate capacitive reactance of C_1

Values $V = 400$ V

$I_1 = 1.89$ A

Equation $X_C = \frac{V}{I} = \frac{400}{1.89}$

Answer $\mathbf{X_{C1} = 211.64\ \Omega}$

2 To calculate capacitive reactance of C_2

Values $V = 400$ V

$I_2 = 1.26$ A

Equation $X_C = \frac{V}{I} = \frac{400}{1.26}$

Answer $\mathbf{X_{C2} = 317.46\ \Omega}$

3 To calculate total capacitive reactance

Values $X_{C1} = 211.64\ \Omega$

$X_{C2} = 317.46\ \Omega$

Equation $X_{CT} = \frac{X_{C1} \times X_{C2}}{X_{C1} + X_{C2}} = \frac{211.64 \times 317.46}{211.64 + 317.46}$

Answer $\mathbf{X_{CT} = 126.98\ \Omega}$

4 To find total current

Total current can be found by adding the individual currents as they are both out of phase with the applied voltage by 90°. Total current can also be found with Ohm's law using the values already calculated:

Values $I_1 = 1.89$ A

$I_2 = 1.26$ A

Equation $I_T = I_1 + I_2 = 1.89 + 1.26$

Answer $\mathbf{I_T = 3.15\ A}$

or

Values $V = 400$ V

$X_{CT} = 126.98\ \Omega$

Equation $I = \frac{V}{X_C} = \frac{400}{126.98}$

Answer $\mathbf{I_T = 3.15\ A}$

FIGURE 16.11 Phasor diagram for Example 16.7

5 Phasor diagram

The phasor diagram is shown in Figure 16.11. Because it is a parallel circuit, the reference is the applied voltage and both currents are drawn at 90° (leading) to the reference, starting with the phasor for I_1. The tail of phasor I_2 starts at the tip of phasor I_1. Total current is shown by the length of the combined current phasors.

16.4 Power in a Capacitive AC Circuit

The power waveform for a purely capacitive circuit is developed from instantaneous values as for a resistive AC circuit. Table 16.2 lists the instantaneous values of voltage, current and power at 15° intervals for a voltage with a maximum value of 10 V, and a current with a maximum value of 4 A that is leading the voltage by 90°.

TABLE 16.2

POWER IN A PURELY CAPACITIVE AC CIRCUIT							
Instantaneous current and voltage values where V_{max} = 10 V and I_{max} = 4 A							
positive half cycle of voltage				negative half cycle of voltage			
θ°	V volts	I amperes	P = V × I watts	θ°	V volts	I amperes	P = V × I watts
0	0.00	4	**0**				
15	2.59	3.86	**10.00**	195	−2.59	−3.86	**10.00**
30	5.00	3.46	**17.32**	210	−5.00	−3.46	**17.32**
45	7.07	2.83	**20.00**	225	−7.07	−2.83	**20.00**
60	8.66	2.00	**17.32**	240	−8.66	−2.00	**17.32**
75	9.66	1.04	**10.00**	255	−9.66	−1.04	**10.00**
90	**10.00**	**0.00**	**0.00**	**270**	**−10.00**	**0.00**	**0.00**
105	9.66	−1.04	**−10.00**	285	−9.66	1.04	**−10.00**
120	8.66	−2.00	**−17.32**	300	−8.66	2.00	**−17.32**
135	7.07	−2.83	**−20.00**	315	−7.07	2.83	**−20.00**
150	5.00	−3.46	**−17.32**	330	−5.00	3.46	**−17.32**
165	2.59	−3.86	**−10.00**	345	−2.59	3.86	**−10.00**
180	**0.00**	**−4.00**	**0.00**	**360**	**0.00**	**4.00**	**0.00**

Table 16.2 shows that the values for instantaneous power are positive in the first quarter of the cycle (0° to 90°), then negative for the next quarter (to 180°). In the second half of the cycle, the power values are positive from 180° to 270°, then negative for the last quarter of the cycle. The waveforms in Figure 16.12 are from the values in Table 16.2.

FIGURE 16.12 Power, voltage and current waveforms in a purely capacitive circuit

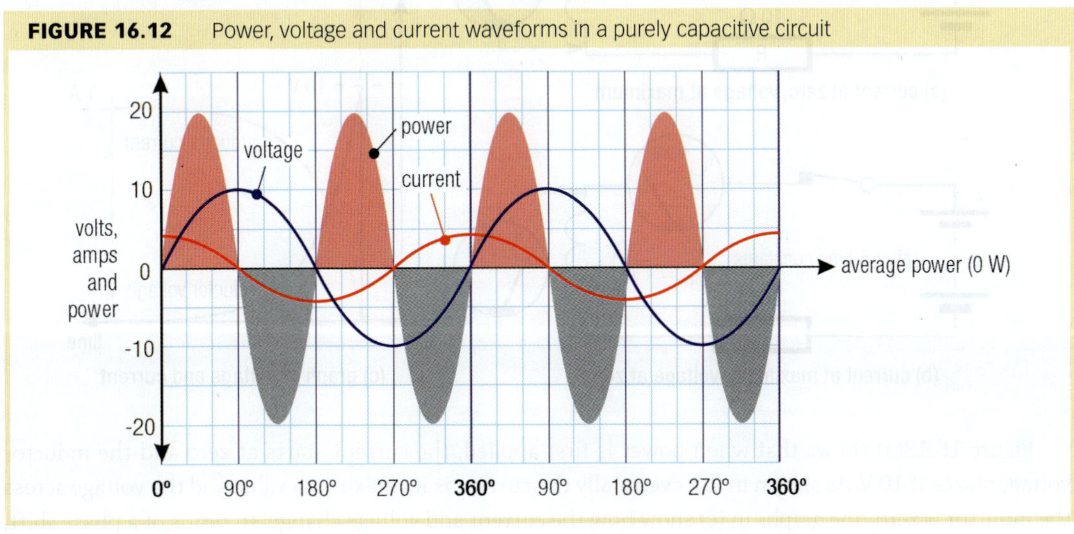

Figure 16.12 shows that power is consumed during the first quarter cycle when the capacitor charges, and that the power is returned during the next quarter cycle when the capacitor discharges. This happens twice over a full cycle of the supply in which the power taken from the supply is returned. Like a resistive AC circuit, the power frequency is twice that of the supply, and the waveform is sinusoidal. This is called *reactive* power, in which no true power is consumed, and the average power is zero.

KEY POINTS...

- Capacitive reactance is inversely proportional to frequency and capacitance. Symbol is X_C, measured in ohms and can be calculated with the equation $X_C = \frac{1}{2\pi fC}$
- Total capacitive reactance in a purely capacitive series or parallel circuit is found the same way as total resistance in a series or parallel circuit.
- If voltage and current are known, capacitive reactance can be found with Ohm's law where $X_C = \frac{V}{I}$
- When capacitive reactance is known, current and voltage in a purely capacitive AC circuit are calculated using Ohm's law where $I = \frac{V}{X_C}$ and $V = IX_C$
- The average power taken by a purely capacitive AC circuit is zero because the same amount of power taken from the supply during one part of the cycle is returned to the supply in the next part of the cycle.

16.5 Inductance

Inductance, as explained in Chapter 12, is the property of a component or conductor that opposes a change in the value of an electric current. The unit of inductance is the henry (H). Figure 16.13 shows a pure inductance in a DC circuit, with graphs of the current flowing in the inductor and the voltage across the inductor when the switch is closed.

FIGURE 16.13 When an inductor is first connected to a DC voltage, the voltage across the inductor immediately rises to its maximum value, while the current in the inductor starts at zero

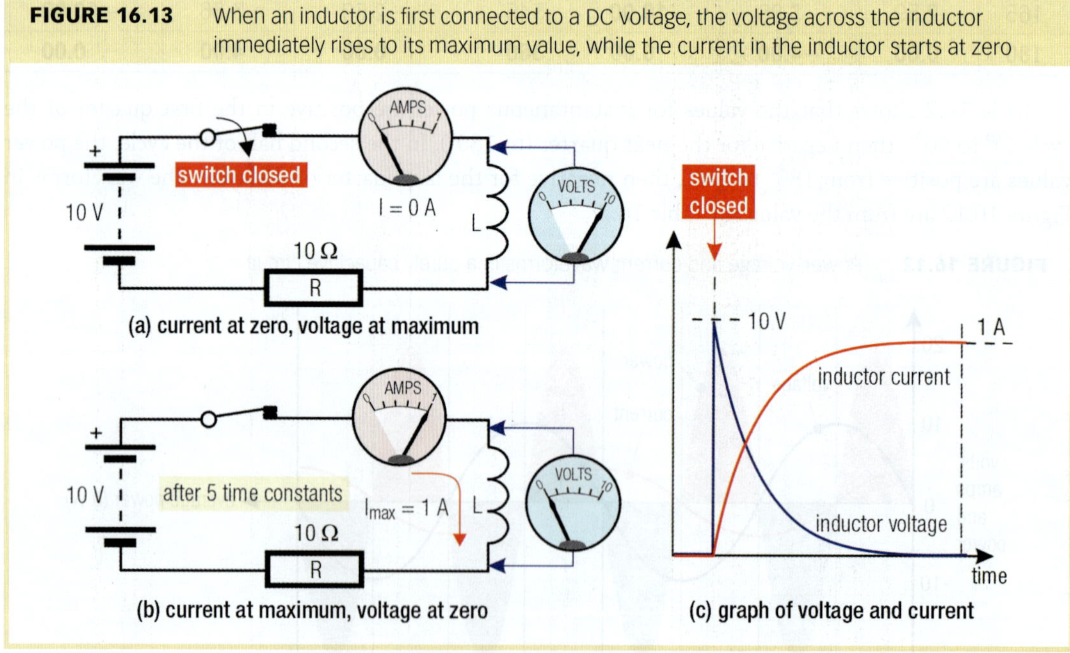

Figure 16.13(a) shows that when power is first applied, the current starts at zero and the inductor voltage starts at 10 V. As shown in (b), eventually the current is its maximum value and the voltage across the inductor is zero. The graphs in (c) show how the current and voltage change. In terms of a phase shift, the voltage is 90° out of phase to the current, as it is at a maximum when the current is zero. The voltage is leading the current because it reaches its maximum before the current. Or, the current is lagging the voltage, if voltage is the reference.

Phase shift between current and voltage

When an inductor is in an AC circuit, the periodic reversal of the current will cause the inductor to continually oppose the change in the value and direction of the current. This is due to the self-induced EMF in the coil caused by the changing magnetic field. This means an inductor in an AC circuit causes a phase shift between the current flowing in the inductor and the voltage across it. For a pure inductance, this phase shift is 90° in which the current lags the voltage, as shown in Figure 16.14(a) and also in the phasor diagram in (b).

KEY CONCEPT

Current through a pure inductance lags the voltage across the inductor by 90°

FIGURE 16.14 Current lags the voltage by 90° in a purely inductive circuit

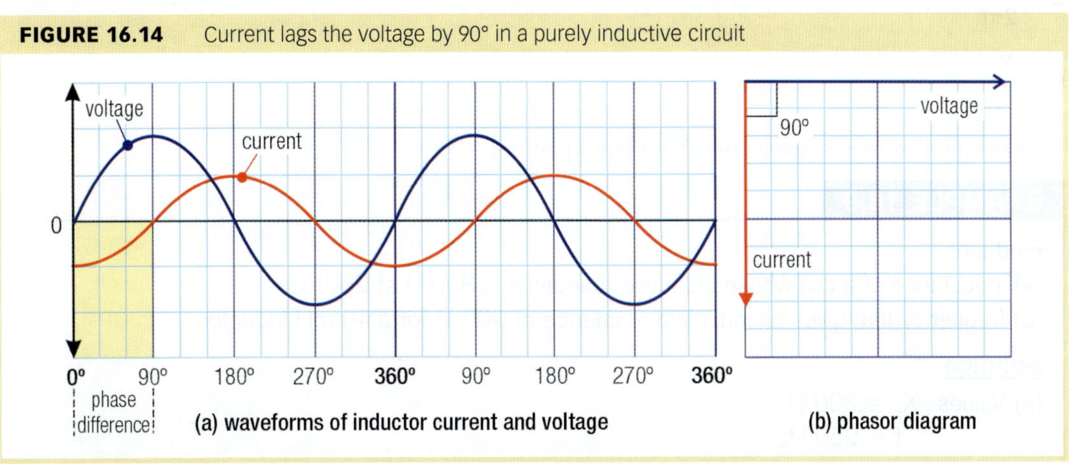

(a) waveforms of inductor current and voltage

(b) phasor diagram

16.6 Inductive Reactance

Lenz's law states that when current in an inductor changes in value or direction, the self-induced EMF in the inductor will have a polarity and value that opposes the change. This opposition is called inductive reactance. It is measured in ohms and has the symbol X_L. The higher the inductance, the higher the inductive reactance. Because inductive reactance is only associated with a changing current, the faster the change, the higher the inductive reactance. This means inductive reactance is directly proportional to the frequency of the AC supply and the inductance of the component. For a sinewave, the equation to find inductive reactance is:

$X_L = 2\pi fL$

where:

X_L = inductive reactance in ohms

$2\pi = 6.28$ (a close approximation)

f = frequency in hertz

L = inductance in henrys.

EXAMPLE 16.8

Find the inductive reactance of the inductors in Figure 16.15 at a frequency of 50 Hz.

Solution

Values $L_1 = 2.8$ H
$L_2 = 10$ mH
$f = 50$ Hz

1 For 2.8 H inductor

Equation $X_L = 2\pi fL = 6.28 \times 50 \times 2.8$

Answer $\mathbf{X_L = 879.2\ \Omega}$

2 For 10 mH inductor

Equation $X_L = 2\pi fL = 6.28 \times 50 \times 10 \times 10^{-3}$

Answer $\mathbf{X_L = 3.14\ \Omega}$

FIGURE 16.15

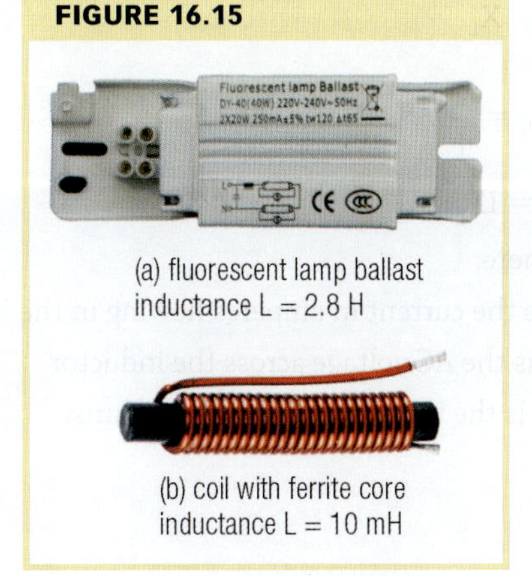

(a) fluorescent lamp ballast inductance L = 2.8 H

(b) coil with ferrite core inductance L = 10 mH

MATH

For help with transposing this equation, see page xxiv

Example 16.8 shows that inductive reactance is directly proportional to the inductance value. If the frequency of the AC supply is increased, the inductive reactance also increases. For example, the inductive reactance of the 10 mH inductor at a frequency of 1 kHz becomes 62.8 Ω, or 20 times its value at 50 Hz. This is to be expected, as 1 kHz is 20 times 50 Hz.

The equation for X_L can be transposed to find frequency or inductance:

$$L = \frac{X_L}{2\pi f}$$

$$f = \frac{X_L}{2\pi L}$$

Example 16.9 shows how to use these two equations.

EXAMPLE 16.9

Find the:

(a) inductance of a coil whose inductive reactance is 300 Ω at 250 Hz

(b) frequency that gives an inductive reactance of 200 Ω for a 40 mH inductor.

Solution

(a) Values $X_L = 300\ \Omega$

$f = 250$ Hz

Equation $L = \frac{X_L}{2\pi f} = \frac{300}{6.28 \times 250}$

Answer $\mathbf{L = 0.19\ H}$

(b) Values $X_L = 200\ \Omega$

$L = 40$ mH

Equation $f = \frac{X_L}{2\ L} = \frac{200}{6.28 \times 40 \times 10^{-3}}$

Answer $\mathbf{f = 796.2\ Hz}$

Inductive reactance and Ohm's law

A pure inductance has no resistance, so the only opposition to current in an AC circuit is its inductive reactance. In practice, it's impossible to have a pure inductance, as all inductors have some resistance because of the resistance of the winding. How this resistance is taken into account is explained in Chapter 17. At this stage, we'll assume an inductor with zero resistance, which means Ohm's law equations can be used in which R is replaced with X_L, giving:

$$I = \frac{V}{X_L}$$

$$X_L = \frac{V}{I}$$

$$V = IX_L$$

where:

I is the current in amperes flowing in the inductor

V is the AC voltage across the inductor

X_L is the inductive reactance in ohms.

Examples 16.10 and 16.11 use these equations.

EXAMPLE 16.10

The inductor in Figure 16.16 has negligible resistance, an inductance of 0.75 H and is connected to a 230 V 50 Hz supply. How much current flows in the inductor?

Solution

Values $L = 0.75$ H

$V = 230$ V AC

$f = 50$ Hz

Equation 1 $X_L = 2\pi fL$

$= 6.28 \times 50 \times 0.75$

Answer $\mathbf{X_L = 235.5\ \Omega}$

Equation 2 $I = \frac{V}{X_L} = \frac{230}{235.5}$

Answer $I =$ **0.98 A**

FIGURE 16.16

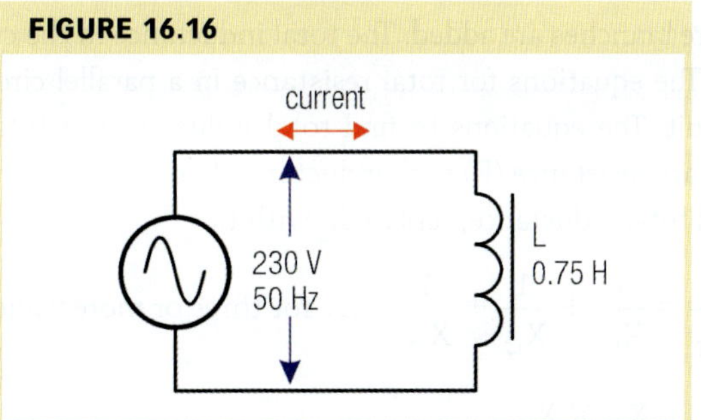

EXAMPLE 16.11

What value of inductance do you need for a coil of negligible resistance that will cause a current of 500 mA when the coil is connected across a 100 V AC 400 Hz supply?

Solution

Values $V = 100$ V

$f = 400$ Hz

$I = 0.5$ A

Equation 1 $X_L = \frac{V}{I} = \frac{100}{0.5}$

Answer $\mathbf{X_L = 200\ \Omega}$

Equation 2 $L = \frac{X_L}{2\pi f} = \frac{200}{6.28 \times 400}$

Answer $L =$ **79.6 mH or 0.08 H** (to 2 decimal places)

FIGURE 16.17

Inductors in series

When pure inductors are connected in series each inductor produces a back-EMF, which collectively opposes the current flow. As for resistors in series, the total inductive reactance in a series inductive circuit is the sum of the individual reactance values. Also, the total inductance in the circuit is the sum of the individual inductance values. The equations are:

$X_{Ltotal} = X_{L1} + X_{L2} + X_{L3} + \ldots$ where X_L = inductive reactance in ohms

$L_{total} = L_1 + L_2 + L_3 + \ldots$ where L = inductance in henrys

Inductors in parallel

When pure inductors are connected in parallel, each inductor takes current from the supply. Because each current has the same phase difference to the supply (90°), the individual currents can be added to obtain the total current. Like the resistive parallel circuit, increasing the number of branches in an inductive parallel circuit causes the total current to increase. Therefore, the total inductive reactance reduces as more branches are added. The total inductance in the circuit will also be reduced.

The equations for total resistance in a parallel circuit therefore apply to a purely inductive parallel circuit. The equations to find total inductance or total inductive reactance in a parallel circuit simply replace resistance (R) with inductance L, or inductive reactance X_L. The equations below are for X_L, so to find total inductance, replace X_L with L.

$$\frac{1}{X_{LT}} = \frac{1}{X_{L1}} + \frac{1}{X_{L2}} + \frac{1}{X_{L3}} + \ldots \text{ for three or more values of } X_L \text{ (ohms)}$$

$$X_{LT} = \frac{X_{L1} \times X_{L2}}{X_{L1} + X_{L2}} \text{ for two values of } X_L \text{ (ohms)}$$

EXAMPLE 16.12

Find the total inductive reactance of the circuit in Figure 16.18.

Solution

Find X_{L1}

Values $L = 64 \text{ mH} = 0.064 \text{ H}$

$f = 50 \text{ Hz}$

Equation $X_L = 2\pi fL$

$= 6.28 \times 50 \times 0.064$

$X_{L1} = 20.1\ \Omega$

Find X_{L2}

Values $V = 400 \text{ V}$

$I_2 = 25 \text{ A}$

Equation $X_L = \frac{V}{I} = \frac{400}{25}$

$X_{L2} = 16\ \Omega$

Find X_{LT}

Equation $X_{LT} = \frac{X_{L1} \times X_{L2}}{X_{L1} + X_{L2}} = \frac{20.1 \times 16}{20.1 + 16}$

Answer $\mathbf{X_{LT} = 8.91\ \Omega}$

FIGURE 16.18

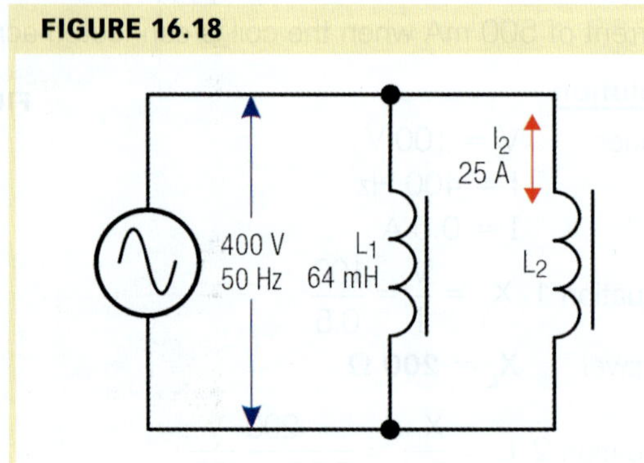

16.7 Power in a Purely Inductive AC Circuit

The power waveform for a purely inductive circuit is developed from instantaneous values as for a resistive AC circuit. Table 16.3 lists the instantaneous values of voltage, current and power at 15° intervals for a voltage with a maximum value of 10 V, and a current with a maximum value of 4 A that is lagging the voltage by 90°.

TABLE 16.3

POWER IN A PURELY INDUCTIVE AC CIRCUIT							
Instantaneous current and voltage values where V_{max} = 10 V and I_{max} = 4 A							
positive half cycle of voltage				negative half cycle of voltage			
θ°	V volts	I amperes	P = V × I watts	θ°	V volts	I amperes	P = V × I watts
0	0.00	−4	**0**				
15	2.59	−3.86	**−10.00**	195	−2.59	3.86	**−10.00**
30	5.00	−3.46	**−17.32**	210	−5.00	3.46	**−17.32**
45	7.07	−2.83	**−20.00**	225	−7.07	2.83	**−20.00**
60	8.66	−2.00	**−17.32**	240	−8.66	2.00	**−17.32**
75	9.66	−1.04	**−10.00**	255	−9.66	1.04	**−10.00**
90	**10.00**	**0.00**	**0.00**	**270**	**−10.00**	**0.00**	**0.00**
105	9.66	1.04	**10.00**	285	−9.66	−1.04	**10.00**
120	8.66	2.00	**17.32**	300	−8.66	−2.00	**17.32**
135	7.07	2.83	**20.00**	315	−7.07	−2.83	**20.00**
150	5.00	3.46	**17.32**	330	−5.00	−3.46	**17.32**
165	2.59	3.86	**10.00**	345	−2.59	−3.86	**10.00**
180	**0.00**	**4.00**	**0.00**	**360**	**0.00**	**−4.00**	**0.00**

Table 16.3 shows that the values for instantaneous power are negative in the first quarter of the cycle (0° to 90°), then positive for the next quarter (to 180°). In the second half of the cycle, the power values are negative from 180° to 270°, then positive for the last quarter of the cycle. The waveforms in Figure 16.19 are from the values in Table 16.3.

FIGURE 16.19 Power, voltage and current waveforms in a purely inductive circuit

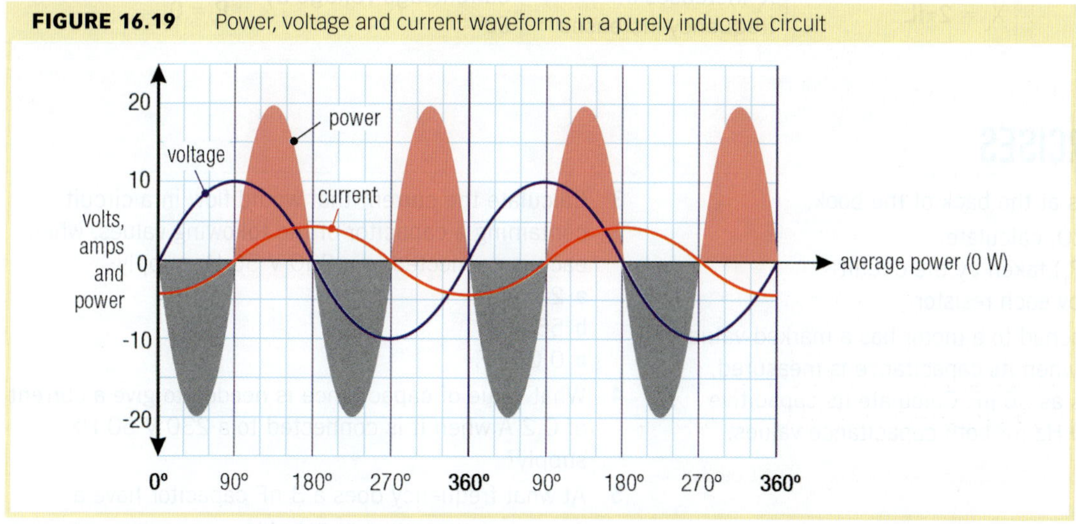

The waveforms in Figure 16.19 show that power is repeatedly consumed during one quarter of the supply cycle, then returned in the next quarter. This is because the magnetic field established in one quarter of the cycle collapses during the next quarter. This causes the inductor to return the power required to establish the magnetic field every quarter cycle. The power frequency is twice that of the supply. Like a capacitive AC circuit, this is reactive power in which, over one cycle, power is repeatedly taken from, then returned to, the supply.

Unlike a resistor, an inductor (and a capacitor) can be used to limit the current flowing in an AC circuit, but without consuming true power. An application is the ballast in a low or high intensity discharge lamp, such as a fluorescent lamp. A ballast is an inductor in series with the lamp that limits the current flowing through the lamp. The only real power it takes is due to the resistance of the ballast.

CHAPTER SUMMARY

- Resistance, capacitance and inductance are fundamental electrical properties.
- Resistance has the same properties in an AC or DC circuit. Current and voltage are always in phase and true power is always consumed by a resistive circuit.
- Capacitance and inductance exhibit reactance in an AC circuit.
- Reactance is opposition to an alternating current caused by an electric or magnetic field that acts to oppose the changing current.
- A circuit of pure capacitance or pure inductance does not consume true power, as power is periodically taken from and returned to the supply. This is called reactive power.
- Inductance (L in henrys) and capacitance (C in farads) have a reactance (X in ohms) that depends on the frequency and the value of the component.
- The reactance of a pure inductor or capacitor can be found with Ohm's law, in which $X = \frac{V}{I}$, where V = voltage across the component and I the current flowing in the component.
- In a purely reactive series or parallel circuit, the total reactance is found the same way as for a series or parallel resistive circuit.
- When voltage and current are known, reactance can be found with Ohm's law where $X = \frac{V}{I}$.
- When reactance is known, current and voltage in a purely reactive AC circuit are calculated using Ohm's law where $I = \frac{V}{X}$ and $V = IX$.

TABLE 16.4 Summary of R, C and L in an AC circuit

COMPONENT	VALUE IN AN AC CIRCUIT (OHMS)	EFFECT ON VALUE WITH CHANGE IN FREQUENCY	PHASE SHIFT BETWEEN CURRENT AND VOLTAGE	TRUE POWER CONSUMED (WATTS)
Resistor (R)	R	nil	0°	$P = I^2R$
Capacitor (C)	$X_c = \frac{1}{2\pi fC}$	X_C decreases if frequency increases	current leads voltage by 90°	P = 0
Inductor (L)	$X_L = 2\pi fL$	X_L increases if frequency increases	current lags voltage by 90°	P = 0

REVIEW EXERCISES

Check your answers at the back of the book.

1 For Figure 16.20, calculate:
   a total power (P_T) taken by the circuit
   b power taken by each resistor.
2 A capacitor attached to a motor has a marked value of 100 μF, but when its capacitance is measured, the value shows as 36 μF. Calculate its capacitive reactance at 50 Hz for both capacitance values.

FIGURE 16.20

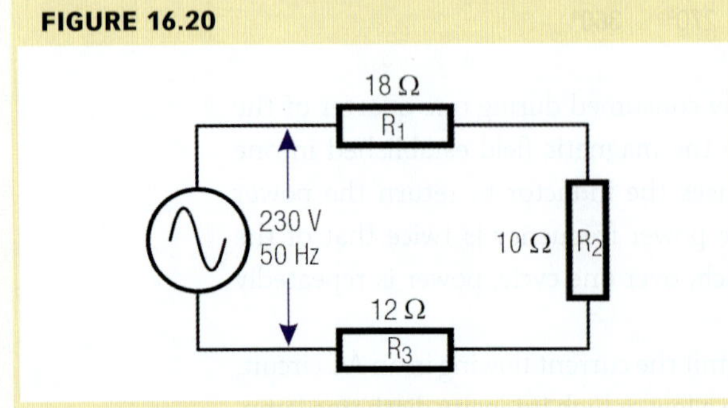

3 Calculate the current that would flow in a circuit containing a capacitor of the following values, when each is connected to a 230 V 50 Hz supply:
   a 2.2 μF
   b 56 μF
   c 0.01 μF.
4 What value of capacitance is needed to give a current of 0.2 A when it is connected to a 230 V 50 Hz supply?
5 At what frequency does a 3 nF capacitor have a capacitive reactance of 500 Ω?
6 If the total capacitance of an AC circuit is increased, how does this affect the capacitive reactance of the circuit and the current flowing in the circuit?
7 Determine the inductive reactance of a coil with an inductance of 620 mH at a frequency of 400 Hz.
8 What value of inductance has an inductive reactance of 60 Ω at a frequency of 1 kHz?

9 At what frequency does a 100 mH coil have an inductive reactance of 100 Ω?

10 Calculate the following from the circuit in Figure 16.21:
- **a** the total inductance
- **b** the total inductive reactance
- **c** the voltage drop across inductor L_1. Include its phase relationship to the current in the inductor.

FIGURE 16.21

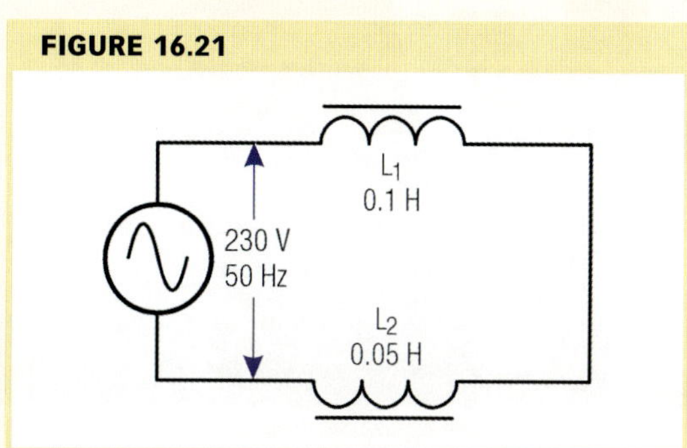

ONLINE RESOURCES

COMPLETE WORKSHEET SIXTEEN

Check with your instructor for worksheets on this chapter.

CHAPTER 17

SERIES COMBINATIONS OF R, L AND C

As Chapter 16 explained, AC circuits that contain either inductance or capacitance have a phase shift between the current and voltage of 90°. When resistance is added in series with an inductor or capacitor, the phase shift is no longer 90°. This chapter explains how to determine phase angle and other unknowns in a series circuit that has inductance, capacitance and resistance. It also introduces the terms 'impedance' and 'resonance', and explains how to determine their values.

CHAPTER OUTLINE

17.1 Series RL AC Circuits

A basic series RL circuit with a single coil and resistor is shown in Figure 17.1, along with its circuit diagram. At this stage, the inductor is assumed to have zero resistance, so the resistance of the circuit is due to resistor R.

FIGURE 17.1 A series RL circuit

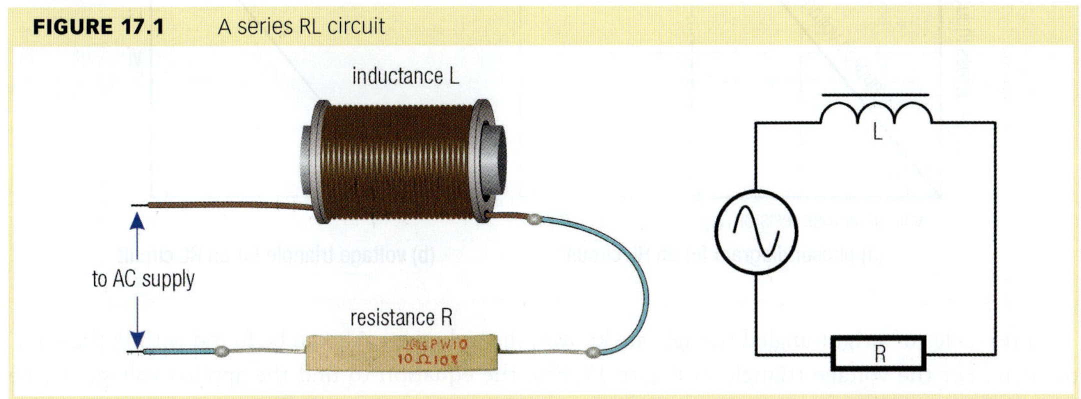

Chapter 16 explained that the current in a pure inductor lags the voltage by 90° or, put another way, the voltage across the inductor leads the current by 90°. For a resistor, the current and voltage are always in phase. In a series circuit, current is the reference, as it's the same in all parts of the circuit. Figure 17.2 shows waveforms of the current flowing in an RL circuit and the voltage drops across the resistor and the inductor. The waveforms of voltage across the resistor and current are in phase, while the waveform for voltage across the inductor leads the other two waveforms by 90°. That is, the voltage across the inductor reaches its maximum positive value before the other two waveforms.

FIGURE 17.2 In a series RL circuit, the voltage across the inductor leads the current by 90°

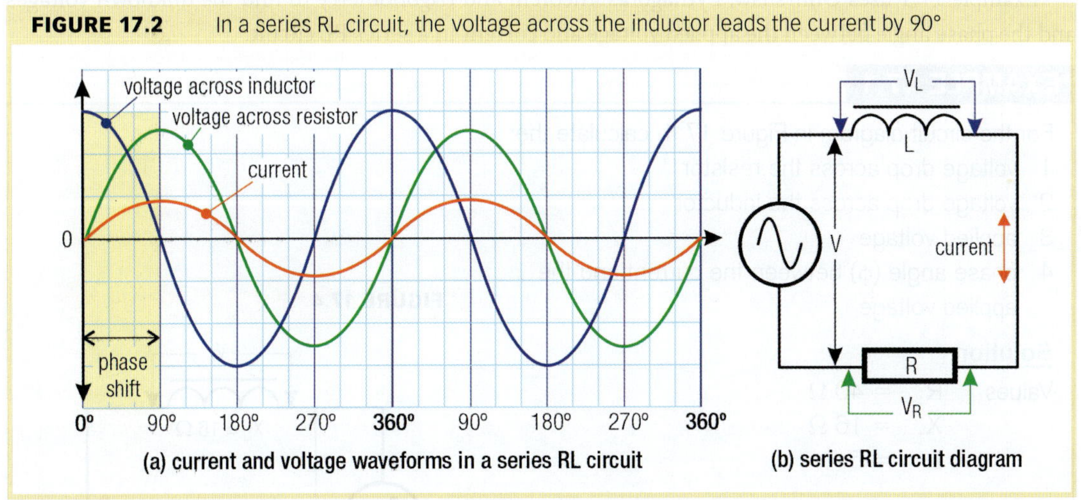

The value and phase angle of the applied voltage (V) in the RL circuit in Figure 17.2(b) is found with the phasor diagram shown in Figure 17.3(a). The reference phasor is current, as it's a series circuit, and the phasor for the voltage across the resistor is drawn on top of the current phasor, as both are in phase. The phasor for the voltage drop across the inductor is drawn at 90° leading the reference. The voltage phasors are drawn to scale to represent the voltage values. The phasor diagram in Figure 17.1(a) looks like a right-angled triangle, and is redrawn as a triangle in Figure 17.3(b). Because each side represents a voltage, it is called a voltage triangle.

FIGURE 17.3 The phasor diagram of an RL circuit can be redrawn as a right-angled triangle

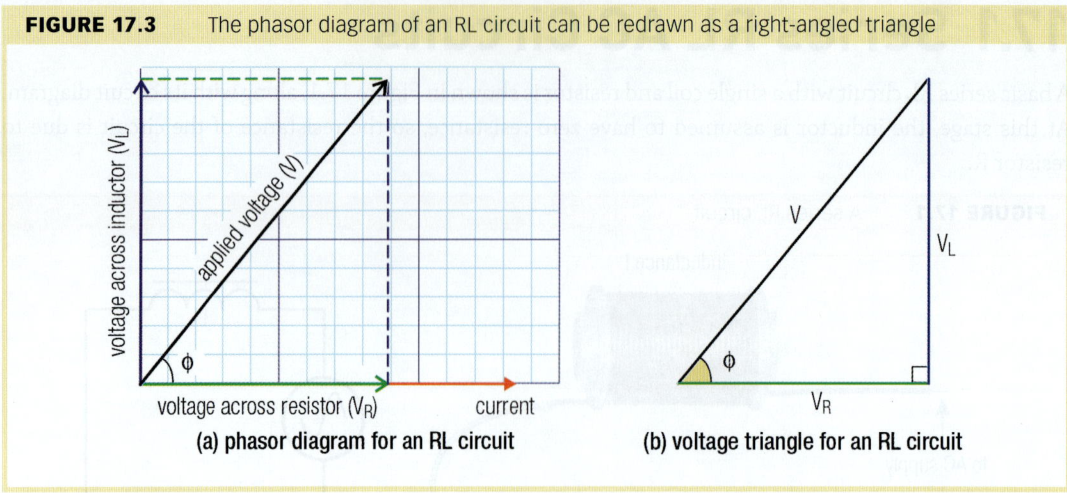

If two sides of a right-angled triangle are known, the unknown side can be found with Pythagoras' theorem. For the voltage triangle in Figure 17.3(b), the equation to find the applied voltage V (the hypotenuse) is:

MATH

Pythagoras' theorem is explained on page xxx

$$V = \sqrt{V_R^2 + V_L^2}$$

where:

V = applied voltage

V_R = voltage across resistor

V_L = voltage across inductor.

Example 17.1 uses Ohm's law, Pythagoras' theorem and trigonometry to find the unknown voltages and the phase angle between the applied voltage and current in a series RL circuit.

EXAMPLE 17.1

For the circuit diagram in Figure 17.4, calculate the:

1 voltage drop across the resistor
2 voltage drop across the inductor
3 applied voltage
4 phase angle (ϕ) between the current and the applied voltage

FIGURE 17.4

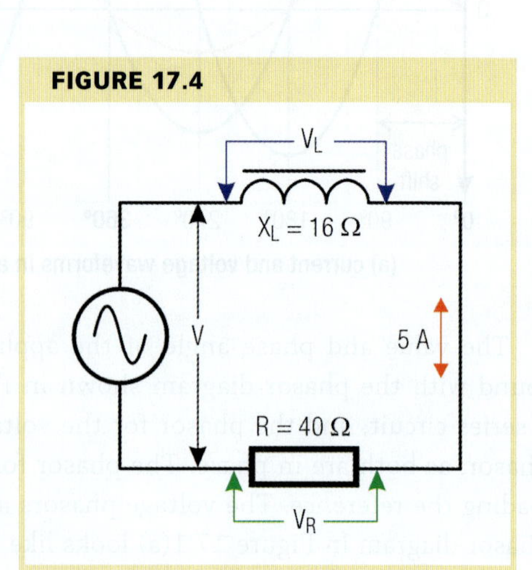

Solution

Values $R = 40\ \Omega$
$X_L = 16\ \Omega$
$I = 5$ A

1 Calculate the voltage across R

Equation $V_R = IR = 5 \times 40$

Answer $\mathbf{V_R = 200\ V}$

2 Calculate the voltage across the inductor

Equation $V_L = IX_L = 5 \times 16$

Answer $\mathbf{V_L = 80\ V}$

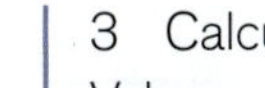

3 Calculate the applied voltage

Values $V_R = 200$ V

$V_L = 80$ V

Equation $V = \sqrt{V_R^2 + V_L^2} = \sqrt{200^2 + 80^2} = \sqrt{46400}$

Answer **V = 215.4 V**

Keystrokes for Casio calculator

4 Calculate the phase angle ϕ

Values $V_R = 200$ V

$V_L = 80$ V

$V = 215.4$ V

Equation Figure 17.5 shows the circuit's voltage triangle. Because all sides of the voltage triangle are known, any of the three trigonometry functions can be used to find the phase angle ϕ, such as the cos function:

$$\cos\phi = \frac{adj}{hyp} \text{ or}$$

$$\phi = \cos^{-1}\frac{adj}{hyp}$$

$$\phi = \cos^{-1}\frac{V_R}{V} = \cos^{-1}\frac{200}{215.4}$$

Keystrokes for Casio calculator

Answer **ϕ = 21.8° (voltage leading current)**

FIGURE 17.5 Voltage triangle with values found in Example 17.1

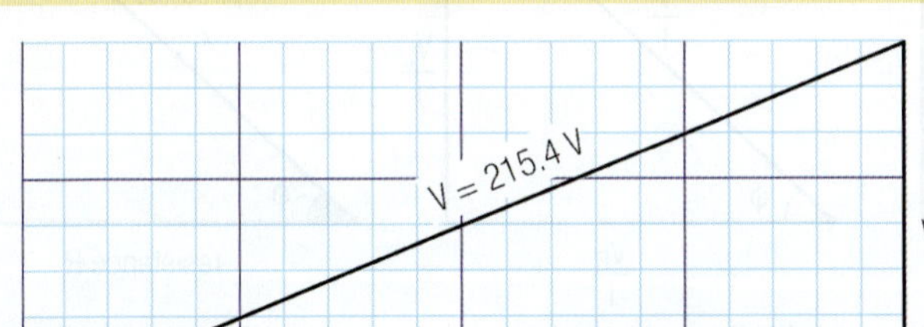

Impedance

In Example 17.1 the circuit current is 5 A, and the calculated applied voltage is 215.4 V. That is, the resistance and the inductive reactance limits the current to 5 A. The combined effect of reactance and resistance is called *impedance*, from the word 'impede. Because impedance is a combination of reactance and resistance, it only applies to AC circuits.

Impedance is measured in ohms, and has the symbol Z. The impedance of any AC circuit always equals the applied voltage divided by the circuit current. That is:

KEY CONCEPT

Impedance is the combination of reactance and resistance in an AC circuit

$$Z = \frac{V}{I}$$

where:

Z = circuit impedance in ohms

V = applied voltage in volts

I = circuit current in amps.

EXAMPLE 17.2

What is the impedance of the circuit in Figure 17.4?

Solution

Values $V = 215.4$ V (from Example 17.1)
$I = 5$ A

Equation $Z = \frac{V}{I} = \frac{215.4}{5}$

Answer $\mathbf{Z = 43.08\ \Omega}$

The impedance of an AC circuit can also be found from the resistance and reactance values in the circuit by using Pythagoras' theorem. To do this requires a right-angled triangle, which can be derived from the voltage triangle. In a series circuit the current is the same in all components. By Ohm's law, resistance equals voltage divided by current, and inductive reactance also equals voltage divided by current.

Therefore, as shown in Figure 17.6, each voltage in a voltage triangle can be divided by the circuit current. Instead of voltages, the triangle is now representing resistance, inductive reactance and impedance, all of which are measured in ohms. The shape and size of the triangle is not altered, nor is the phase angle. This means the phase angle ϕ between the applied voltage and the circuit current depends only on the values of the resistance and the inductive reactance in the circuit.

FIGURE 17.6 An impedance triangle for an RL circuit is derived from its voltage triangle by dividing each voltage by the circuit current

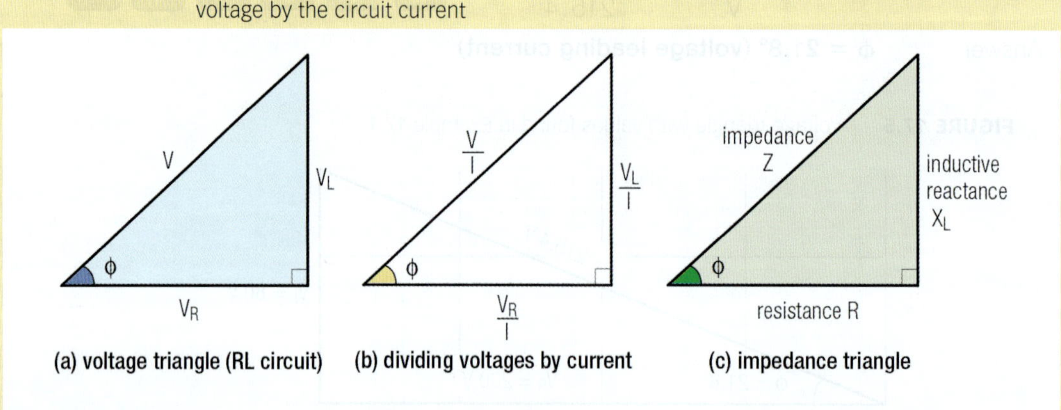

Because an impedance triangle is also a right-angled triangle, Pythagoras' theorem can be used in the same way as with the voltage triangle. This gives a way of finding impedance from the resistance and inductive reactance values. The equation is:

$$Z = \sqrt{R^2 + X_L^2}$$

where:

Z = impedance in ohms

R = total circuit resistance in ohms

X_L = total inductive reactance in ohms.

The total resistance of a series AC circuit is the sum of the individual resistances in the circuit. This includes the resistance of the inductor and any other resistances in the circuit. The total inductance is the sum of the individual inductance values, and the total inductive reactance is the sum of the individual reactance values.

EXAMPLE 17.3

Find the impedance of the circuit in Figure 17.7.

FIGURE 17.7

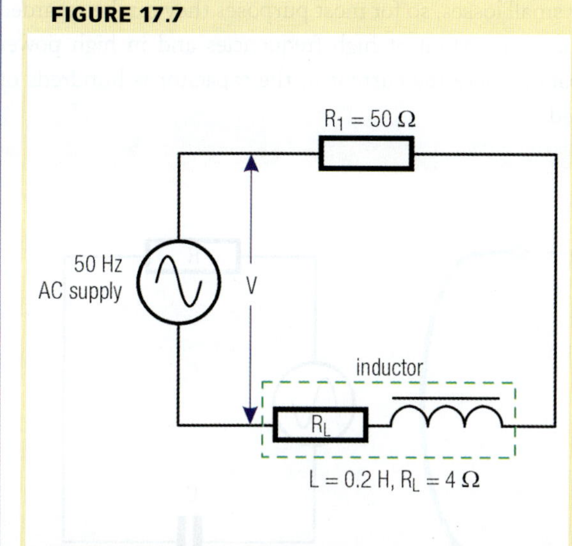

Solution

Values $R_1 = 50\ \Omega$

$R_L = 4\ \Omega$

$L = 0.2\ \text{H}$

$f = 50\ \text{Hz}$

1 find X_L

Equation $X_L = 2\pi fL$

$= 6.28 \times 50 \times 0.2$

$\mathbf{X_L = 62.8\ \Omega}$

2 find Z

Equation $Z = \sqrt{R^2 + X_L^2}$ where

$R = R_1 + R_L = 50 + 4 = 54\ \Omega$

$Z = \sqrt{54^2 + 62.8^2} = \sqrt{6859.8}$

Answer $\mathbf{Z = 82.8\ \Omega}$

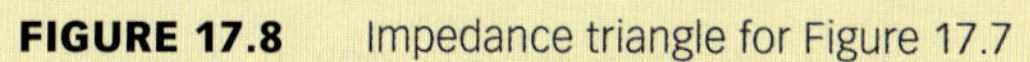

FIGURE 17.8 Impedance triangle for Figure 17.7

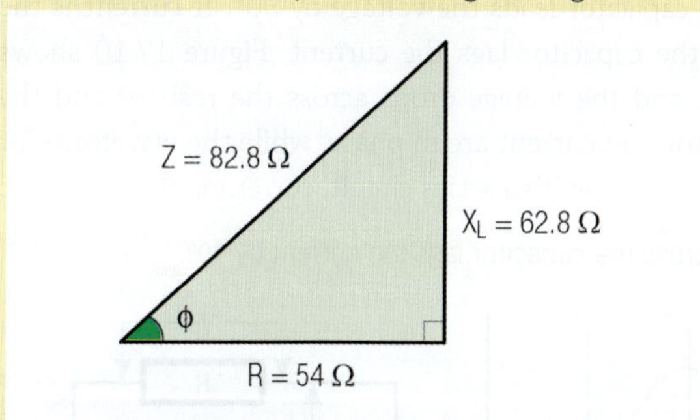

The impedance triangle for Example 17.3 is shown in Figure 17.8. The phase angle ϕ between the applied voltage and current can be found from this triangle using trigonometry, as in Example 17.1. Because the values of all sides of the triangle are known, any of the three trigonometry functions can be used. To find the angle, first determine the number that represents the sine, cosine or tangent of the angle. The angle is then found using the inverse trigonometric function, such as $\sin^{-1}$, $\cos^{-1}$ or $\tan^{-1}$.

EXAMPLE 17.4

Use the impedance triangle in Figure 17.8 to find the phase difference between the applied voltage and current in the circuit of Figure 17.7.

Solution

Values $R = 54\ \Omega$

$X_L = 62.8\ \Omega$

$Z = 82.8\ \Omega$

Equation $\phi = \cos^{-1}\dfrac{R}{Z} = \cos^{-1}\dfrac{54}{82.8}\left(\text{or } \phi = \sin^{-1}\dfrac{X_L}{Z}, \text{or } \phi = \tan^{-1}\dfrac{X_L}{R}\right)$

Answer ϕ **= 49.3° (voltage leading current)**

MATH

Trigonometry is explained on pages xxxi–xxxiv

In Example 17.4, the only loss in the inductor is its resistance, called the copper loss. In a practical inductor there are also losses due to the magnetisation curve of the iron core, called the iron losses. These losses can be regarded as resistive losses, as they generate heat. It is therefore usual to add these losses together and to regard them as a resistance value.

17.2 Series RC AC Circuits

A series RC circuit containing a capacitor and a resistor is shown in Figure 17.9, along with its circuit diagram. Unlike an inductor, capacitors have very small losses, so for most purposes they can be regarded as a pure capacitance. Losses in capacitors become important at high frequencies and in high power installations, such as in a substation capacitor bank, where the current in the capacitor is hundreds of amperes. At this stage, pure capacitance is assumed.

FIGURE 17.9 A series RC circuit

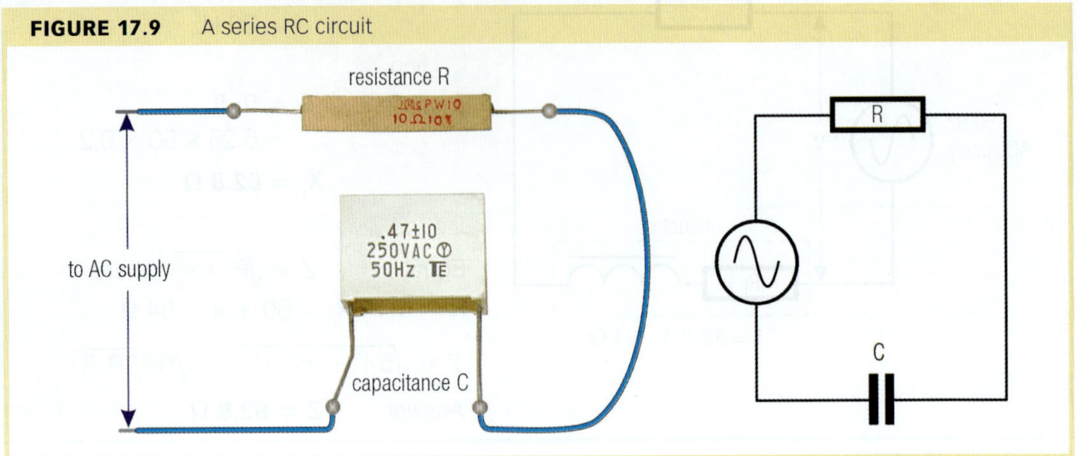

Chapter 16 explained that the current in a pure capacitor leads the voltage by 90°. If current is the reference as in a series circuit, the voltage across the capacitor lags the current. Figure 17.10 shows waveforms of the current flowing in an RC circuit and the voltage drops across the resistor and the capacitor. The waveforms of voltage across the resistor and current are in phase, while the waveform for voltage across the capacitor lags the other two by 90°. Because it's a series circuit, current is the reference.

FIGURE 17.10 In a series RC circuit, the voltage across the capacitor lags the current by 90°

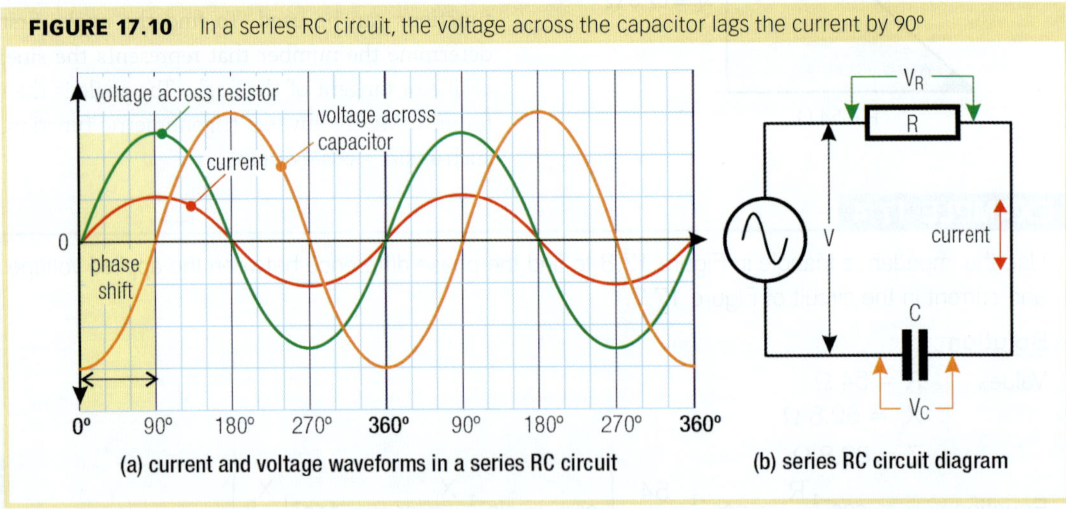

The voltage applied to an RC circuit can be found with a phasor diagram in the same way as for an RL circuit. As shown in Figure 17.11, the reference phasor is current, and the phasor for the voltage across the resistor is drawn on top of the current phasor because they are in phase with each other. The phasor for the voltage drop across the capacitor is drawn at 90° *lagging* the reference. The voltage phasors are drawn to scale to represent the voltage values. From the phasor diagram, a voltage triangle can be drawn for the circuit, shown in Figure 17.11(b).

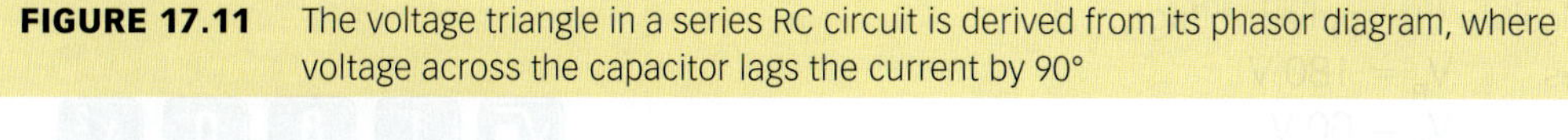

FIGURE 17.11 The voltage triangle in a series RC circuit is derived from its phasor diagram, where voltage across the capacitor lags the current by 90°

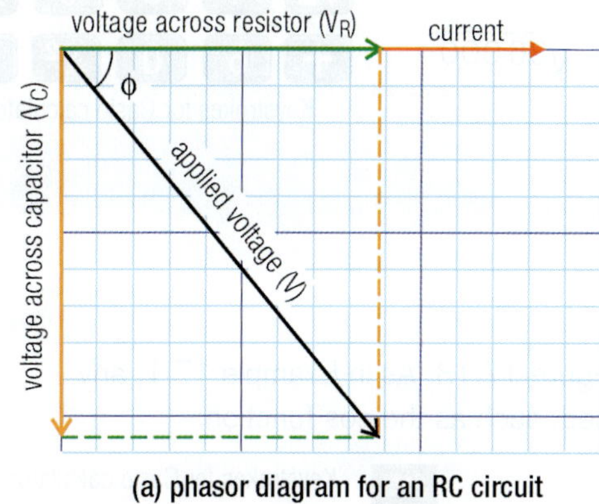

(a) phasor diagram for an RC circuit

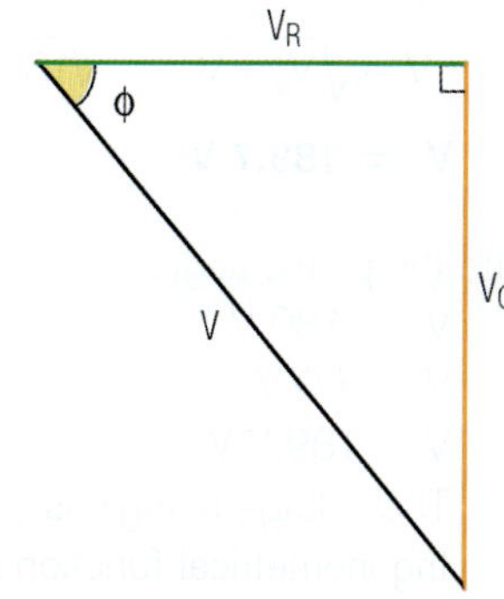

(b) voltage triangle for an RC circuit

From the voltage triangle, the applied voltage in an RC circuit can be found using the same equation used for an RL circuit. The difference is the equation refers to the voltage across a capacitor instead of an inductor. The equation becomes:

$$V = \sqrt{V_R^2 + V_C^2}$$

where:

V = applied voltage

V_R = voltage across resistor

V_C = voltage across capacitor.

To find the voltage drops in an RC circuit, calculate the capacitive reactance of the capacitor and use Ohm's law to find the voltage drop across the resistor and the capacitor. The phase angle (ϕ) between the applied voltage and current can be found with trigonometry, as shown in Example 17.5.

EXAMPLE 17.5

For the circuit diagram in Figure 17.12, calculate the:

FIGURE 17.12

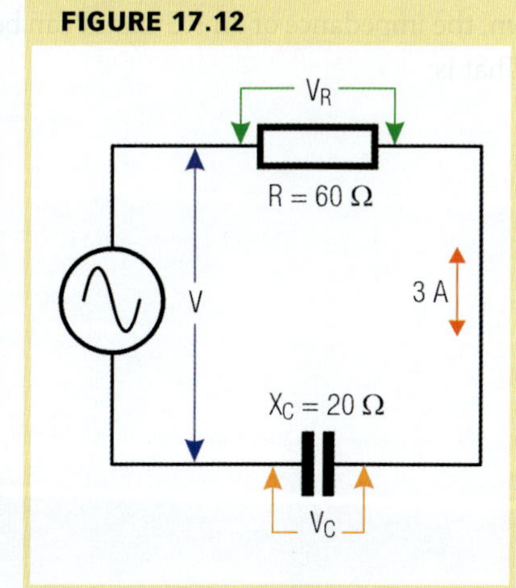

1 voltage drop across the resistor
2 voltage drop across the capacitor
3 applied voltage
4 phase angle (ϕ) between the current and the applied voltage

Solution

Values $R = 60\ \Omega$

$X_C = 20\ \Omega$

$I = 3\ A$

1 Calculate the voltage across R

Equation $V = IR = 3 \times 60$

Answer $\mathbf{V_R = 180\ V}$

2 Calculate the voltage across the capacitor

Equation $V = IX_C = 3 \times 20$

Answer $\mathbf{V_C = 60\ V}$

3 Calculate the applied voltage

Values $V_R = 180\text{ V}$

$V_C = 60\text{ V}$

Equation $V = \sqrt{V_R^2 + V_C^2} = \sqrt{180^2 + 60^2} = \sqrt{36\,000}$

Answer $\mathbf{V = 189.7\text{ V}}$

Keystrokes for Casio calculator

4 Calculate the phase angle ϕ

Values $V_R = 180\text{ V}$

$V_L = 60\text{ V}$

$V = 189.7\text{ V}$

Equation The voltage triangle is shown in Figure 17.13. As in Example 17.1, any trigonometrical function can be used, such as the cos function:

$$\phi = \cos^{-1}\frac{\text{adj}}{\text{hyp}}$$

$$\phi = \cos^{-1}\frac{180}{189.7}$$

Keystrokes for Casio calculator

Answer ϕ **=–18.4° (voltage lagging current)**

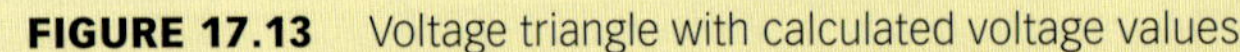

FIGURE 17.13 Voltage triangle with calculated voltage values

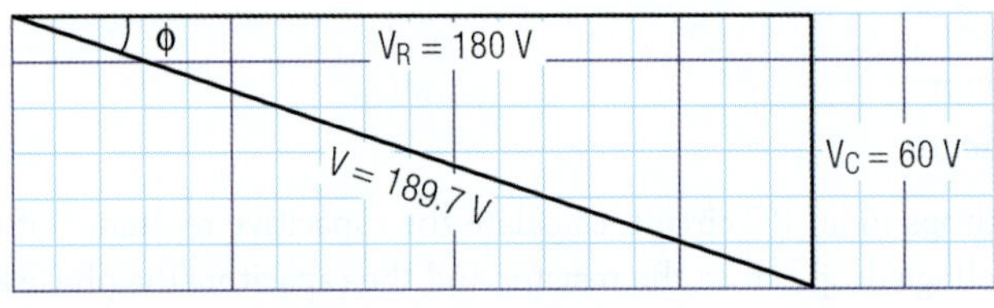

Impedance

Like an RL circuit, if the current and the applied voltage are known, the impedance of an RC circuit can be found with Ohm's law, using the same equation as given before. That is:

$$Z = \frac{V}{I}$$

where:

Z = circuit impedance in ohms

V = applied voltage in volts

I = circuit current in amps.

EXAMPLE 17.6

What is the impedance of the circuit in Figure 17.12?

Solution

Values $V = 189.7$ V (from Example 17.5)
$I = 3$ A

Equation $Z = \frac{V}{I} = \frac{189.7}{3}$

Answer $\mathbf{Z = 63.23\ \Omega}$

As in the RL circuit, Pythagoras' theorem is another way to find the impedance in an RC circuit. The impedance triangle is derived in the same way, in which all voltages in the voltage triangle are divided by the current, as shown in Figure 17.14. Because the voltage across the capacitor lags the current by 90°, the triangle is upside down compared to that for the RL circuit.

FIGURE 17.14 An impedance triangle for an RC circuit is derived from its voltage triangle by dividing each voltage by the current

(a) voltage triangle (RC circuit) **(b) dividing voltages by current** **(c) impedance triangle (RC circuit)**

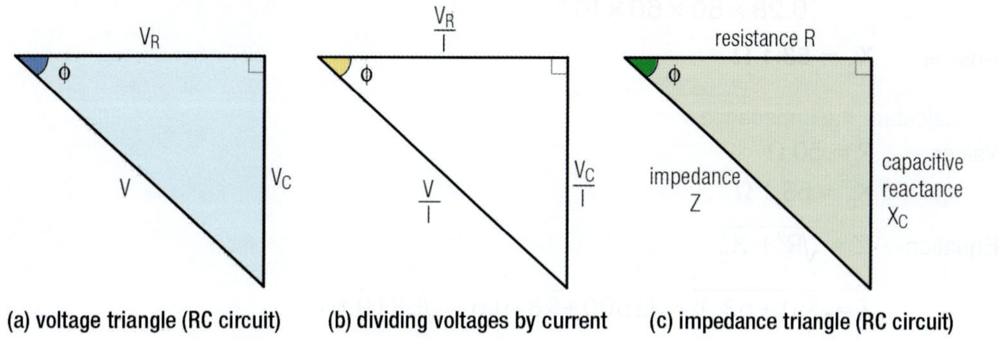

The impedance triangle gives the same equation as for an RL circuit, except it now has the term capacitive reactance (X_C) instead of inductive reactance (X_L):

$$Z = \sqrt{R^2 + X_C^2}$$

where:

Z = impedance in ohms

R = total circuit resistance in ohms

X_C = total capacitive reactance in ohms.

Example 17.7 uses Pythagoras' theorem to find the impedance of an RC circuit, which first requires finding the capacitive reactance. From this, the circuit current can be found.

EXAMPLE 17.7

For the circuit in Figure 17.15, calculate the:

1 capacitive reactance
2 impedance
3 current

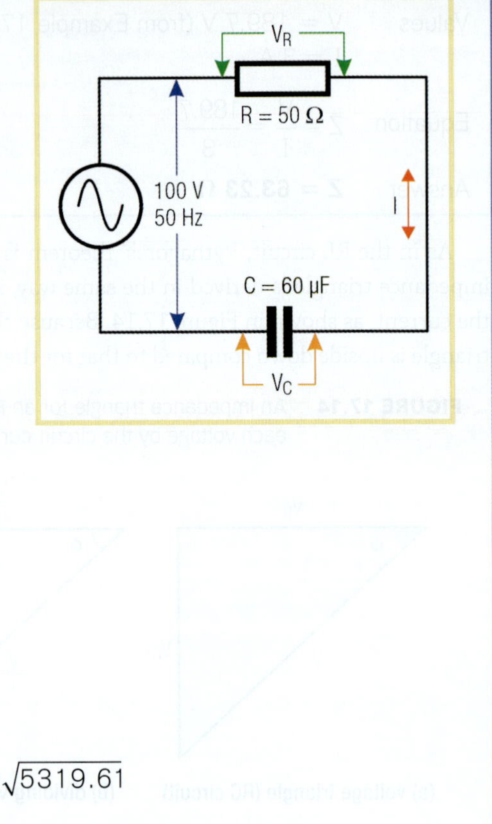

Solution

Values $C = 60\ \mu F$

$R = 50\ \Omega$

$V = 100\ V$

$f = 50\ Hz$

1 Calculate the capacitive reactance

Equation $X_C = \frac{1}{2\pi fC}$

$$= \frac{1}{6.28 \times 50 \times 60 \times 10^{-6}}$$

Answer $\mathbf{X_c = 53.1\ \Omega}$

2 Calculate the impedance

Values $R = 50\ \Omega$

$X_C = 53.1\ \Omega$

Equation $Z = \sqrt{R^2 + X_C^2}$

$$Z = \sqrt{50^2 + 53.1^2} = \sqrt{2500 + 2819.61} = \sqrt{5319.61}$$

Answer $\mathbf{Z = 72.94\ \Omega}$

3 Calculate the current

Values $Z = 72.94\ \Omega$

$V = 100\ V$

Equation $I = \frac{V}{Z} = \frac{100}{72.94}$

Answer $\mathbf{I = 1.37\ A}$

The examples given so far have introduced the letter X as standing for reactance. This term only applies to an AC circuit, and is either due to inductance (X_L) or capacitance (X_C). The combination of reactance and resistance gives an impedance (Z) value. Reactance, resistance and impedance are measured in ohms, so Ohm's law applies to an AC circuit in a similar way to a DC circuit.

KEY POINTS...

- In an RL series AC circuit, the applied voltage leads the current by an angle determined by the values of resistance and inductive reactance in the circuit.
- In an RC series AC circuit, the applied voltage lags the current by an angle determined by the values of resistance and capacitive reactance in the circuit.
- In any series AC circuit containing reactance (either inductive or capacitive), the applied voltage is $V = \sqrt{V_R^2 + V_X^2}$, where V_X is the voltage across either the inductor or the capacitor.
- The impedance (Z) of any AC circuit equals the applied voltage divided by the circuit current: $Z = \frac{V}{I}$ ohms.
- Impedance applies only to an AC circuit and is a phasor combination of the reactance and resistance in the circuit.
- In any series AC circuit containing reactance (either inductive or capacitive) the impedance is $Z = \sqrt{R^2 + X^2}$, where X is the reactance of the inductor or the capacitor.
- The voltage triangle of an AC series reactive circuit is derived from its phasor diagram. The impedance triangle for the circuit is derived from the voltage triangle by dividing each side by the value of the current.
- In a voltage triangle, the phase angle ϕ between current and applied voltage in any series circuit is the angle between the reference (V_R) and the side representing the applied voltage (V). In an impedance triangle, it is the angle between the sides for resistance and impedance.
- The phase angle between current and applied voltage in any series AC circuit containing reactance (either inductive or capacitive) can be calculated using trigonometry from the values shown by a voltage or impedance triangle.

17.3 Series RLC AC Circuits

In an AC circuit, capacitance behaves in a similar but opposite way to inductance. In summary:

1 **Frequency** – when frequency is increased, X_L increases and X_C decreases. Or, X_L is proportional to frequency and X_C is inversely proportional to frequency.

2 **Component value** – when component value is increased, X_L increases and X_C decreases. That is, X_L is proportional to component value (in henrys) and X_C is inversely proportional to component value (in farads).

3 **Phase angle** – in an inductive series circuit, voltage leads the current; in a capacitive series circuit, voltage lags the current.

These differences are important when looking at what happens when capacitance and inductance are both in a series AC circuit, such as the circuit shown in Figure 17.16.

FIGURE 17.16 RLC series circuit

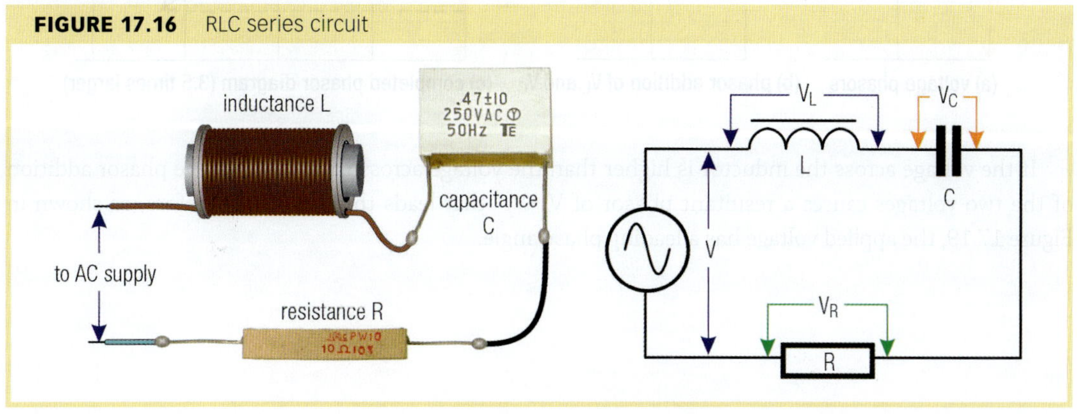

By Ohm's law, the current in the circuit will cause a voltage drop across each component. Each voltage drop will be proportional to the reactance of the inductor, the reactance of the capacitor and the resistance of the resistor. The voltage across the inductor will lead the current by 90°, the voltage across the capacitor will lag the current by 90° and the voltage across the resistor will be in phase with the current. The waveforms of these three voltages are shown in Figure 17.17. The voltages across the capacitor and the inductor are 180° out of phase with each other, which means they are in opposition and therefore tend to cancel each other.

FIGURE 17.17 The voltages across a capacitor and inductor in a series circuit are 180° out of phase with each other

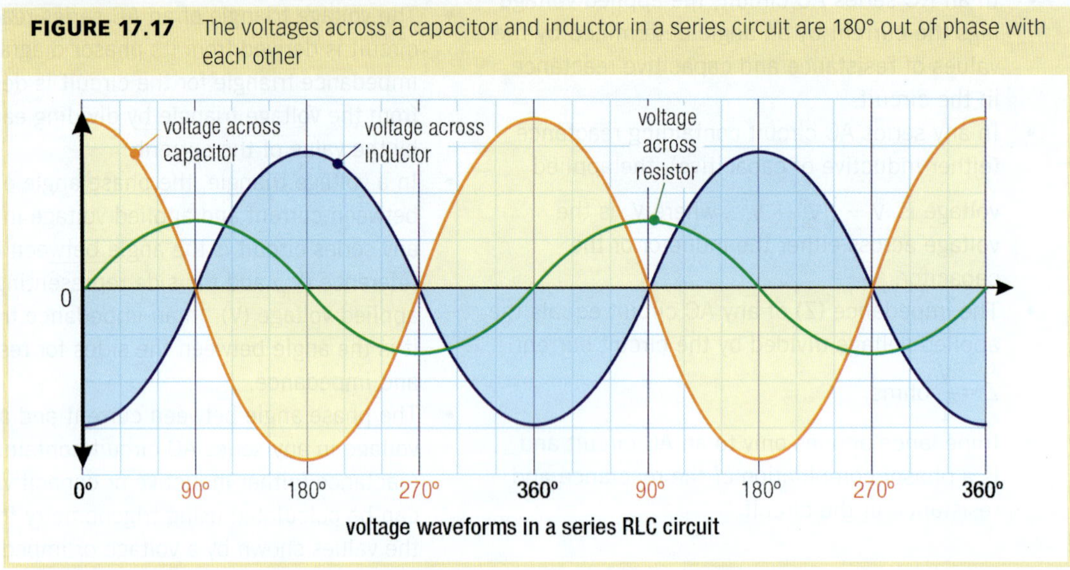

Figure 17.18(a) shows the phasors drawn to scale for the three voltages displayed in Figure 17.17. Phasor V_L leads the current by 90°, phasor V_C lags the current by 90° and phasor V_R is in phase with the current. The algebraic addition of the phasors for V_L and V_C is shown in (b), in which phasor V_C has the largest value. The completed phasor diagram in (c) shows that the applied voltage lags the current because the voltage across the capacitor is larger than the voltage across the inductor.

FIGURE 17.18 The phasors for voltages across a capacitor and inductor tend to cancel each other, giving a resultant phasor called $V_L–V_C$

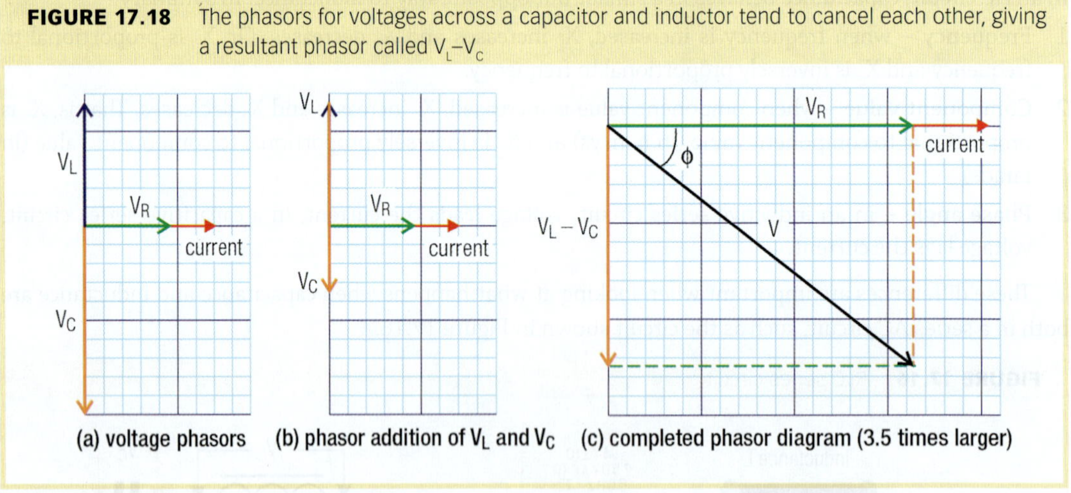

If the voltage across the inductor is higher than the voltage across the capacitor, the phasor addition of the two voltages causes a resultant phasor of $V_L–V_C$ that leads the current. Therefore, as shown in Figure 17.19, the applied voltage has a leading phase angle.

FIGURE 17.19 In a series AC circuit, when the voltage across the inductor is greater than the voltage across the capacitor, the applied voltage leads the current

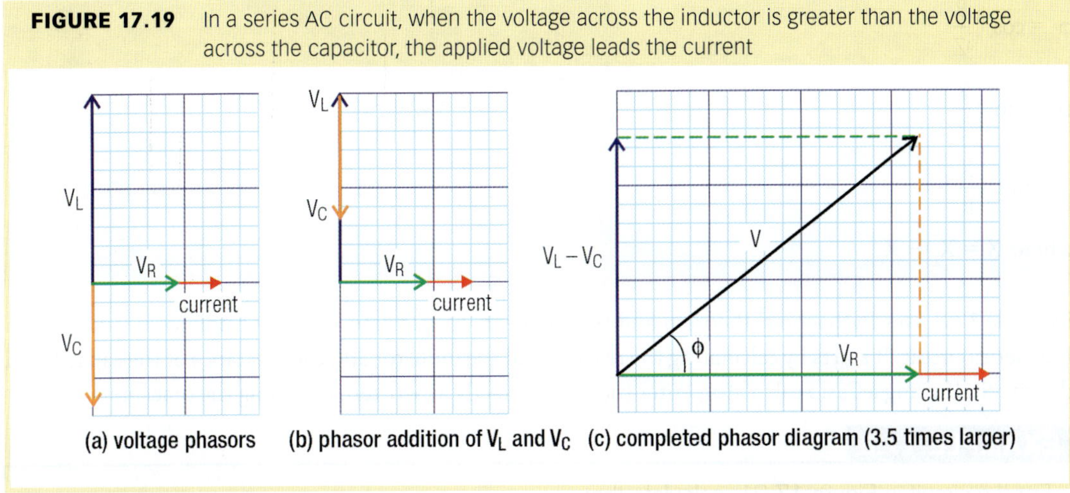

Impedance

The completed phasor diagrams in Figure 17.18(c) and 17.19(c) show that a voltage triangle can be developed for a series RLC circuit in the same way as for a series RL or RC circuit. Also, an impedance triangle can be developed by dividing the voltages in the voltage triangle by the circuit current. If the RLC circuit has a higher voltage drop across the *inductor* compared to the capacitor, the triangles will be like those for an RL circuit, as in Figure 17.6. If the voltage drop across the *capacitor* exceeds the voltage drop across the inductor, the triangles will be the same as for an RC circuit, shown in Figure 17.14.

FIGURE 17.20 Impedance triangles for RLC circuits, where reactance is the difference between inductive reactance and capacitive reactance

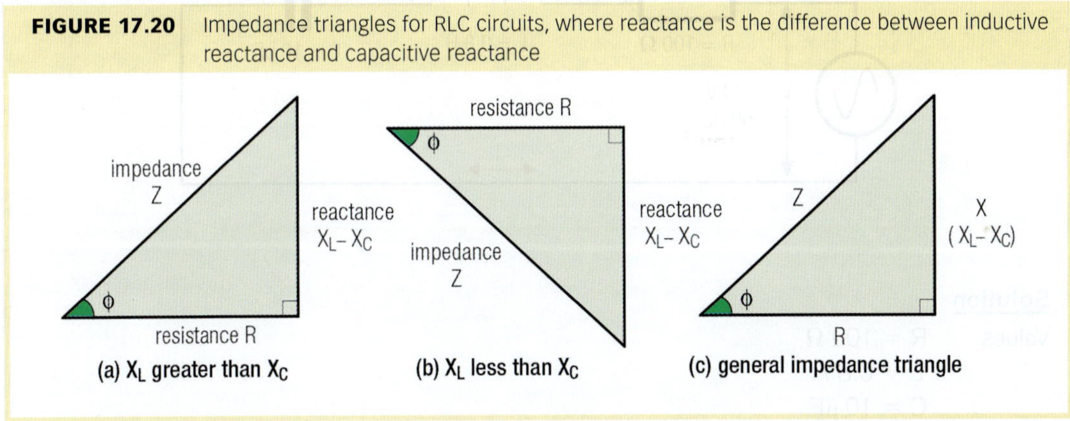

The impedance triangles in Figure 17.20 show that for an RLC circuit, the side representing reactance is the *difference* between the reactance of the inductor (X_L) and the reactance of the capacitor (X_C). From this, a general impedance triangle, shown in Figure 17.20(c) can be constructed, in which reactance (X) is inductive reactance X_L minus capacitive reactance X_C. Because it's a right-angled triangle, Pythagoras' theorem can be used to find the impedance of a series RLC circuit if the reactance and resistance values are known. That is:

FYI

If X_C is greater than X_L, their difference is a negative number. When squared, it becomes a positive value.

$$Z = \sqrt{R^2 + (X_L - X_C)^2}$$

where:

Z = impedance in ohms

R = total circuit resistance in ohms

X_L = total inductive reactance in ohms

X_C = total capacitive reactance in ohms.

The phase angle between current and applied voltage can be found with any of the three basic trigonometrical functions, depending on which values of the impedance triangle are known. These are summarised below, and apply to all the circuits described in this chapter.

$$\phi = \cos^{-1}\frac{R}{Z}$$

$$\phi = \sin^{-1}\frac{X}{Z}$$

$$\phi = \tan^{-1}\frac{X}{R}$$

where: $X = X_L - X_C$.

Example 17.8 uses Pythagoras' theorem and trigonometry to calculate a range of unknown values in an RLC series circuit. It starts by calculating the reactance of the inductor and the reactance of the capacitor in the circuit. If using a Casio calculator, enter the values exactly as written, including the brackets.

EXAMPLE 17.8

For the circuit in Figure 17.21, calculate the:

1 impedance
2 current
3 voltage drops across the resistor, the inductor and capacitor
4 phase angle between the current and the applied voltage

FIGURE 17.21

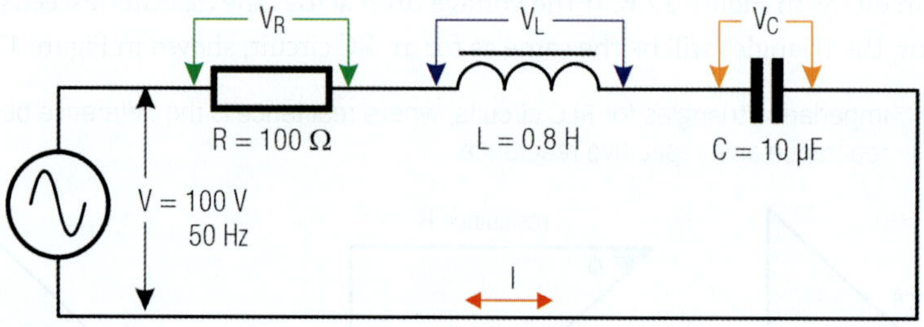

Solution

Values $R = 100\ \Omega$
$L = 0.8\ H$
$C = 10\ \mu F$
$V = 100\ V$
$f = 50\ Hz$

1 To find the impedance, first calculate the inductive and capacitive reactance values:

Equation $X_L = 2\pi fL = 6.28 \times 50 \times 0.8$

$X_L = 251.2\ \Omega$

Equation $X_C = \dfrac{1}{2\pi fC} = \dfrac{1}{6.28 \times 50 \times 10 \times 10^{-6}}$

$X_C = 318.5\ \Omega$

Equation $Z = \sqrt{R^2 + (X_L - X_C)^2} = \sqrt{100^2 + (251.2 - 318.5)^2}$

$Z = \sqrt{10\,000 + (-67.3)^2} = \sqrt{14\,529.3}$

Answer $\mathbf{Z = 120.5\ \Omega}$

2 Calculate the current:

Values $Z = 120.5\ \Omega$

$V = 100\ V$

Equation $I = \frac{V}{Z} = \frac{100}{120.5}$

Answer $\mathbf{I = 0.83\ A}$

3 Calculate the voltage drops:

Values $R = 100\ \Omega$

$X_L = 251.2\ \Omega$

$X_C = 318.5\ \Omega$

$I = 0.83\ A$

Equation $V_R = IR = 0.83 \times 100$

Answer $\mathbf{V_R = 83\ V}$

Equation $V_L = IX_L = 0.83 \times 251.2$

Answer $\mathbf{V_L = 208.5\ V}$

Equation $V_C = IX_C = 0.83 \times 318.5$

Answer $\mathbf{V_C = 264.4\ V}$

4 Calculate the phase angle (ϕ) using cos function:

Values $Z = 120.5\ \Omega$

$R = 100\ \Omega$

Equation $\phi = \cos^{-1}\frac{R}{Z} = \cos^{-1}\frac{100}{120.5} = \cos^{-1}0.83$

Answer $\mathbf{\phi = 33.9°}$ **(voltage lagging current)**

In Example 17.8, the voltage drops across the capacitor and inductor are higher than the applied voltage. This is what happens in practical RLC circuits, in which the voltages across reactive components can be considerably higher than the supply voltage.

The diagrams in Figure 17.22 show in (a) the relative size of the voltage drops for the circuit in Figure 17.21, and in (b) the resulting phasor due to the voltages across the capacitor and the inductor. As a result, the applied voltage lags the current, as shown in (c).

SAFETY

Voltages across reactive components in an AC circuit can far exceed the supply voltage

FIGURE 17.22 Phasor diagrams for Figure 17.21. The voltages across the capacitor and the inductor are both much higher than the supply voltage

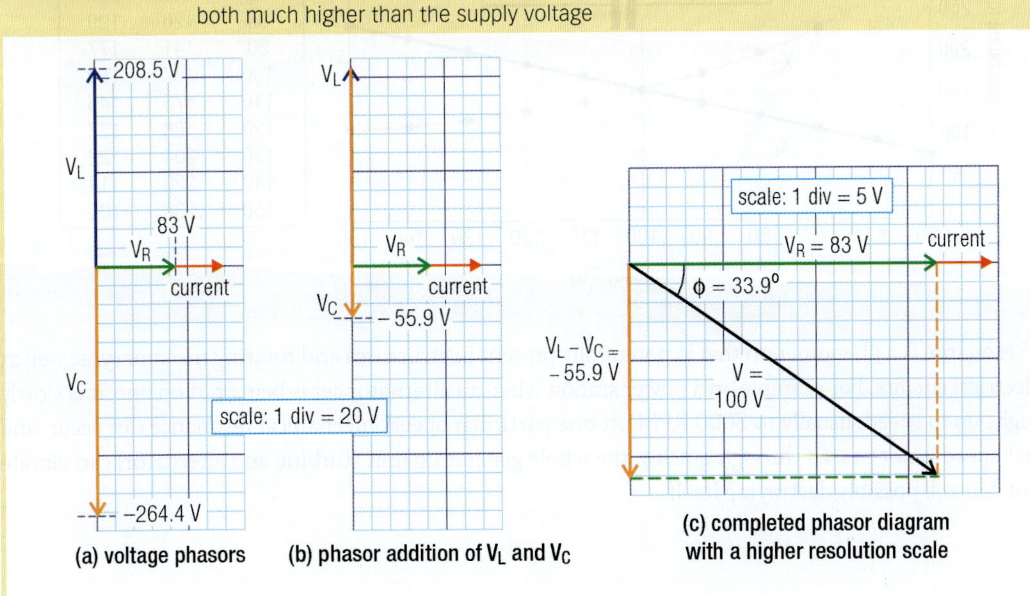

(a) voltage phasors (b) phasor addition of V_L and V_C (c) completed phasor diagram with a higher resolution scale

KEY POINTS...

- In an RLC series AC circuit, the voltages across the capacitor and inductor tend to cancel each other because they are 180° out of phase with each other.
- If the inductive reactance in a series RLC circuit is greater than the capacitive reactance, the circuit is inductive and the supply voltage leads the current.
- If the capacitive reactance is the greater quantity, the circuit is capacitive, and the supply voltage lags the current.
- If the capacitive and inductive reactance are equal, the circuit is resistive and the supply voltage and current are in phase. (This occurs at the circuit's resonant frequency which is described in section 17.4.)
- The impedance of an RLC circuit $Z = \sqrt{R^2 + (X_L - X_C)^2}$ ohms
- The voltage across the reactive components in a series RLC circuit can exceed the supply voltage.
- Phase angle between current and supply voltage can be found with trigonometry, e.g. $\phi = \cos^{-1}\frac{R}{Z}$

17.4 Series Resonance

In a series RLC circuit, inductive reactance and capacitive reactance tend to cancel each other. As well, if the frequency of the supply voltage is changed, inductive reactance changes in one way and capacitive reactance changes in the opposite way. In any RLC circuit, there is a particular frequency that results in the inductive reactance in the circuit being equal to its capacitive reactance. This is the circuit's *resonant* frequency.

Figure 17.23 shows a table listing the reactance at difference frequencies of the inductor and capacitor for the circuit shown. The graphs are plotted from the table, and show that at around 100 Hz the inductive and capacitive reactance values are almost equal.

FIGURE 17.23 An RLC circuit with a resonant frequency close to 100 Hz

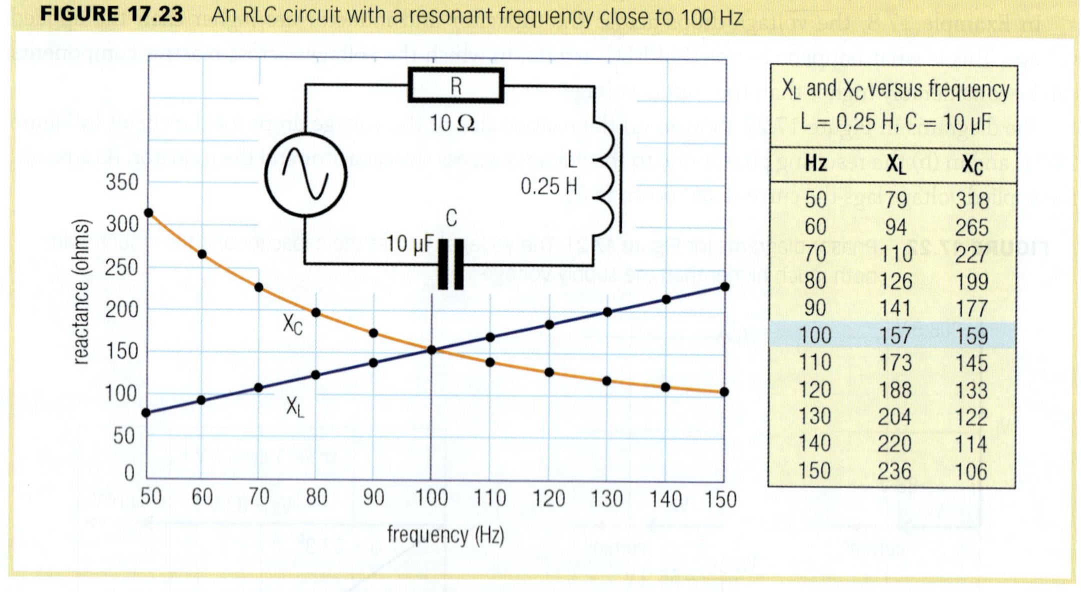

X_L and X_C versus frequency L = 0.25 H, C = 10 μF		
Hz	X_L	X_C
50	79	318
60	94	265
70	110	227
80	126	199
90	141	177
100	157	159
110	173	145
120	188	133
130	204	122
140	220	114
150	236	106

Resonance is a phenomenon that is potentially present in structures and rotating machinery, as well as in electrical circuits. For example, in a power station when an alternator set is being put on line, it is slowly brought up to speed, usually to 3000 RPM. At one particular speed, mechanical resonance can occur, and if that speed is not passed through quickly, the whole generating unit (turbine and alternator) can vibrate out of control, possibly destroying itself.

In an electrical circuit, resonance can cause a number of effects which can be destructive, particularly in a circuit handling high power levels. This is shown after Example 17.9. The exact frequency of an RLC circuit can be calculated with an equation, which is derived as follows.

> **FYI**
> Mechanical resonance can occur in many types of rotating machinery, and can be destructive if not controlled

At resonance, $X_L = X_C$. Therefore, because $X_L = 2\pi fL$ and $X_C = \frac{1}{2\pi fC}$, at resonance $2\pi fL = \frac{1}{2\pi fC}$. Transposing to make f the subject gives the frequency at which $X_L = X_C$ and an equation for the resonant frequency f_r:

$$f_r = \frac{1}{2\pi\sqrt{LC}}$$

where:

f_r = resonant frequency in Hz

2π = 6.28 (a close approximation)

L = inductance in henrys

C = capacitance in farads.

EXAMPLE 17.9

FIGURE 17.24

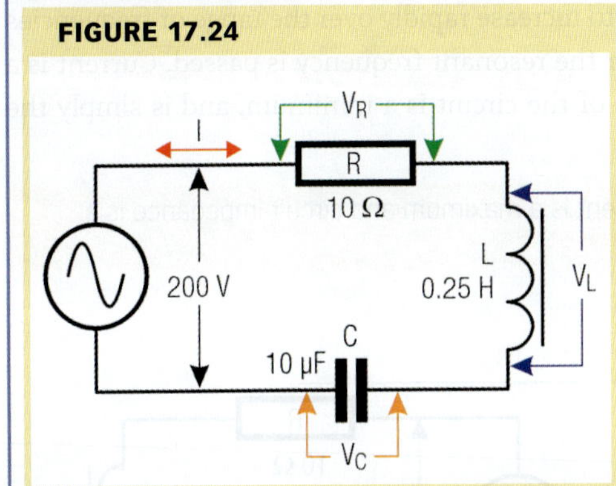

Calculate the resonant frequency of the circuit in Figure 17.24.

Solution

Values L = 0.25 H

C = 10 μF

Equation

$$f_r = \frac{1}{2\pi\sqrt{LC}} = \frac{1}{6.28\sqrt{0.25 \times 10 \times 10^{-6}}}$$

Answer **f_r = 100.71 Hz**

The effects of resonance in the series circuit in Figure 17.24 can be seen with the following calculations. At resonance:

1 The reactance of the capacitor and inductor will be the same:
  $X_C = 1/2\pi fC = X_L = 2\pi fL$ = **158.11 ohms**
2 The impedance of the circuit is the value of the resistor, as the two reactance values cancel each other:
  Z = 10 ohms
3 The current is found with Ohm's law:
  I = V/Z=200/10 = **20 A**
4 The voltage across the inductor and the capacitor will be the same, as they have the same reactance value:
  $V_C = IX_C = V_L = IX_L = 20 \times 158.11 = 3162.2$ = **3.16 kV**

TABLE 17.1

IMPEDANCE, CURRENT AND VOLTAGES FOR FIGURE 17.24						
frequency (Hz)	Z (ohms)	I (amps)	V_R (volts)	V_L (volts)	V_C (volts)	phase angle
50	250	0.8	8	62.9	254.9	−87.7°
80	83.3	2.4	24	301.8	477.8	−83.1°
100	12.1	16.6	166	2601.7	2636	−34.1°
100.71	**10**	**20**	**200**	**3162**	**3162**	**0°**
120	65.9	3	30	572.4	402.7	81.3°
150	139.5	1.43	14.3	337.8	152.1	85.9°

At resonance, the circuit is purely resistive, so current and voltage are in phase. Either side of the resonant frequency causes the phase angle to change from lagging to leading. So, below the resonant frequency, the circuit looks capacitive; above resonance, it looks inductive.

The graph in Figure 17.25 shows the circuit current at frequencies either side of the resonant frequency for the circuit shown. The graph shows that current starts to increase rapidly over the range of frequencies close to the resonant frequency, then reduces rapidly after the resonant frequency is passed. Current is a maximum at the resonant frequency, and the impedance of the circuit is a minimum, and is simply the resistance of the circuit.

FIGURE 17.25 At resonance, in a series RLC circuit, current is a maximum and circuit impedance is a minimum

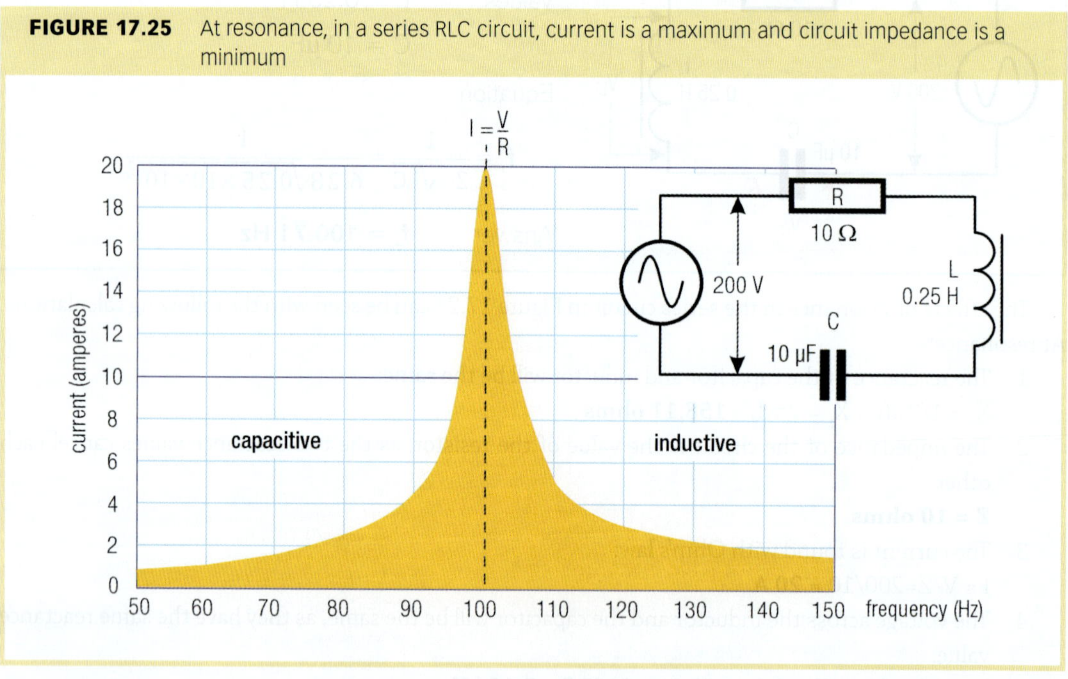

If the resistance of the circuit in Figure 17.24 is increased from 10 Ω to 30 Ω, the current at resonance will be 6.67 A, a third of its previous value of 20 A. Also, the voltages across the capacitor and inductor will be one-third of their previous values. That is, the resistance of the circuit determines how it responds at resonance. The response can be determined by comparing the voltage drops at resonance across the resistance and either of the reactive components. This gives a measure of the amplification effect caused at resonance, or the Q factor (quality factor). That is:

$Q = \frac{V_L}{V}$ or $\frac{V_C}{V}$ where voltages are those occurring at resonance, when $V_L = V_C$

The voltage across the resistor at resonance is the supply voltage V, and V_L is proportional to X_L. Therefore Q is also equal to the ratio of the reactance of the capacitor or inductor to the resistance. That is:

$Q = \frac{X_L}{R}$ or $\frac{X_C}{R}$ where reactance values are those occurring at resonance, when $X_L = X_C$

For the circuit in Figure 17.25, either equation for Q gives a Q factor of 15.8. A circuit with a higher resistance would have a lower Q factor and a wider but flatter resonance response curve. This means it has a wider bandwidth, which is that range of frequencies when the current is greater than 71 per cent of its maximum value. The difference between a high and low Q circuit is shown by the response curves in Figure 17.26.

FIGURE 17.26 Bandwidth is that range of frequencies where the circuit current is greater than 70.7 per cent of the maximum value

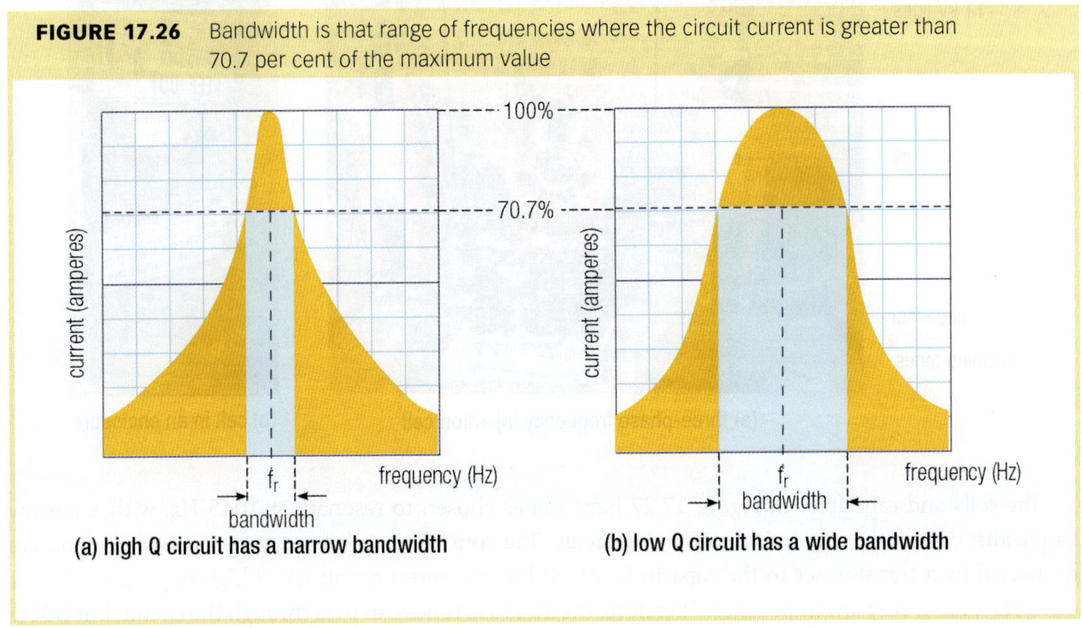

Bandwidth can be found with this equation:

$$BW = \frac{f_R}{Q}$$

where:

BW = bandwidth in Hz

f_R = resonant frequency in Hz

Q = Q factor.

The response curve in Figure 17.26(a) is for the circuit under discussion, which, from the above equation, has a bandwidth of 6.4 Hz, as Q = 15.8 and the resonant frequency is 100.7 Hz. The frequency at the lower point of the bandwidth is the resonant frequency minus half the bandwidth, or 100.7 – 3.2, which is 97.5 Hz. At the higher point of the bandwidth, it is 100.7 plus 3.2, giving 103.9 Hz. At these points, the circuit current will be about 71 per cent of its maximum value.

FYI

A current or voltage value of 70.7 per cent means the power is 50 per cent of its maximum value

The response curve in Figure 17.26(b) is for a circuit with the same values of capacitance and inductance, but with a resistance of 30 Ω, three times the previous value. The circuit has a lower Q factor of one-third the previous value, and the bandwidth is higher by a factor of three. The bandwidth is therefore 19.2 Hz.

Applications of series resonance

Resonance has considerable application in radio communications by use of a 'tuned circuit'. This is an inductor and a capacitor with values that give a resonant circuit at the frequency being received. So if a receiver can pick up a range of frequencies, either the inductor or the capacitor is variable, such as in a typical FM or AM radio.

In the electrical power industry, a common use for a series RLC tuned circuit is to provide a means of coupling control tones to high voltage distribution cables. These tones are used by electrical distributors to control relays fitted to consumer switchboards or to operate street lights. The frequency of the control tone depends on the power company, but is always higher than 50 Hz. An example is 1045 Hz. The photos in Figure 17.27 show a typical frequency injection cell in a substation, and its enclosure.

FIGURE 17.27 A series resonant circuit is used in substations to couple control tones to high voltage distribution cables

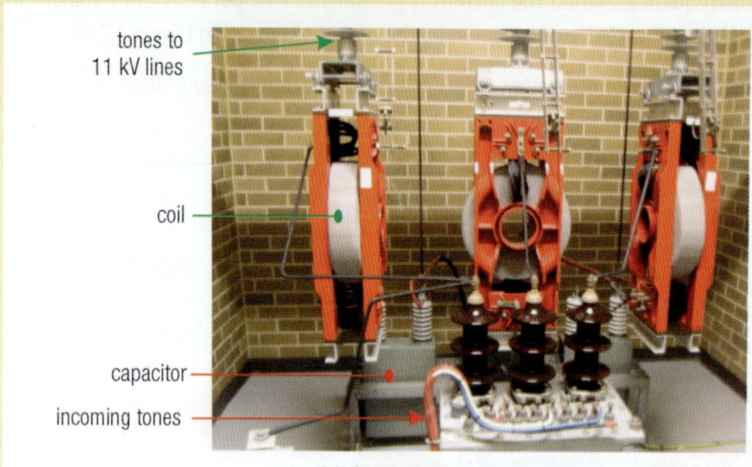

(a) three-phase frequency injection cell

(b) cell in an enclosure

The coils and capacitors in Figure 17.27 have values chosen to resonate at 1045 Hz, with a narrow bandwidth due to the low resistance of the circuit. The control tones are generated externally, and are connected by a transformer to the capacitors. At 50 Hz, the series circuit has a high impedance, but at 1045 Hz it has a very low impedance. Therefore, the 1045 Hz tones can pass through the circuit, but 50 Hz currents are blocked.

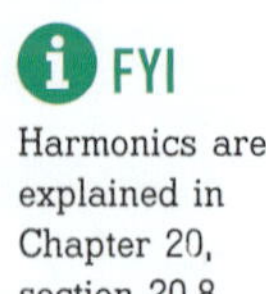
FYI

Harmonics are explained in Chapter 20, section 20.8

Resonance sometimes occurs by accident in the electrical power system. This can happen at 50 Hz when certain circuit conditions arise, such as the capacitance of a long underground cable forming a resonant circuit with the inductance of the transformer it is supplying. Another possibility is the effect of industrial and commercial loads that operate by electronic switching. All computers have a switching type power supply, and many industrial motor control systems operate by rapidly switching the current to the motor.

The effect on the electrical power supply system is to create additional frequencies, called harmonics, which are multiples of the system frequency. For example, the third harmonic of 50 Hz is 150 Hz, the eleventh harmonic is 550 Hz. Under certain conditions, harmonics generated by a switching type load can cause resonance to occur between capacitors and inductors associated with the supply. This will occur at a particular harmonic frequency, and can result in extremely high voltages that cause component breakdown and considerable heat due to the high current.

When a series RLC circuit is at resonance:

- Reactive components cancel each other.
- The only opposition to current flow is the resistance of the circuit.
- Impedance of the circuit equals the resistance of the circuit.
- Current flowing in the circuit equals the applied voltage divided by the resistance.
- Impedance of the circuit is a *minimum* at resonance, so the current is a maximum.
- Current is in phase with the applied voltage.
- Voltages across the inductor and across the capacitor are equal and will usually be considerably higher than the applied voltage.
- Voltage across the capacitor is 180° out of phase with the voltage across the inductor.

CHAPTER SUMMARY

The equations introduced in this chapter are summarised below.

- Applied voltage: $V = \sqrt{V_R^2 + V_L^2}$ (RL circuit),

  $V = \sqrt{V_R^2 + V_C^2}$ (RC circuit),

  $V = \sqrt{V_R^2 + (V_L - V_C)^2}$ (RLC circuit).
- Impedance: $Z = \sqrt{R^2 + X_L^2}$ (RL circuit),

  $Z = \sqrt{R^2 + X_C^2}$ (RC circuit),

  $Z = \sqrt{R^2 + (X_L - X_C)^2}$ (RLC circuit).
- Ohm's law for any AC circuit: $Z = \frac{V}{I}$ ohms,

  $I = \frac{V}{Z}$ amps, $V = IZ$ volts.
- phase angle (any AC circuit): $\phi = \cos^{-1}\frac{R}{Z}$

  $\phi = \sin^{-1}\frac{X}{Z}$, $\phi = \tan^{-1}\frac{X}{R}$.
- phase angle from voltages: $\phi = \cos^{-1}\frac{V_R}{V}$

  $\phi = \sin^{-1}\frac{V_X}{V}$, $\phi = \tan^{-1}\frac{V_X}{V_R}$.
- resonant frequency: $f_R = \frac{1}{2\pi\sqrt{LC}}$.
- Q factor of a series tuned circuit:

  $Q = \frac{V_L}{V}$ or $\frac{V_C}{V}$, also $Q = \frac{X_L}{R}$ or $\frac{X_C}{R}$.
- bandwidth of a tuned circuit: $BW = \frac{f_R}{Q}$.

REVIEW EXERCISES

Check your answers at the back of the book.

1 A large electrical contactor has a coil with an inductance of 0.3 H and a resistance of 20 Ω. It is connected to 230 V AC 50 Hz. Calculate the coil's:
  - a inductive reactance
  - b impedance
  - c current

2 Calculate the phase angle (also identify if lead or lag) between the applied voltage and the current flowing in the coil of Question 1.

3 A high intensity discharge lamp has a ballast (coil) with an inductance of 1.1 H and a resistance of 36 Ω. Calculate its impedance at 50 Hz.

4 The circuit in Figure 17.28 is in series with a resistive AC load. Calculate:
  - a the total inductance of the circuit
  - b the inductive reactance of each coil at 50 Hz
  - c total inductive reactance of the circuit
  - d the impedance of the circuit at 50 Hz
  - e the voltage drop V
  - f the phase angle between the current and voltage V.

FIGURE 17.28

each inductor = 10 mH, resistance = 2 Ω

1.5 A

V

5 A relay coil has an inductance of 160 mH and a resistance of 80 Ω. Calculate the:
  - a coil's impedance at 50 Hz
  - b phase angle between current and voltage when the coil is connected to a 50 Hz supply.

6 A 3.3 μF capacitor is in series with a 1 kΩ resistor. Determine the impedance of the circuit at:
  - a 50 Hz
  - b 1 kHz.

7 Calculate the phase angle between the applied voltage and current in the circuit in Question 6. Do this for both frequencies.

8 The circuit in Figure 17.29 is connected to a 50 Hz AC supply. Calculate the:
  - a voltage drops across the resistor and the capacitor
  - b applied voltage
  - c the phase angle between current and the applied voltage.

FIGURE 17.29

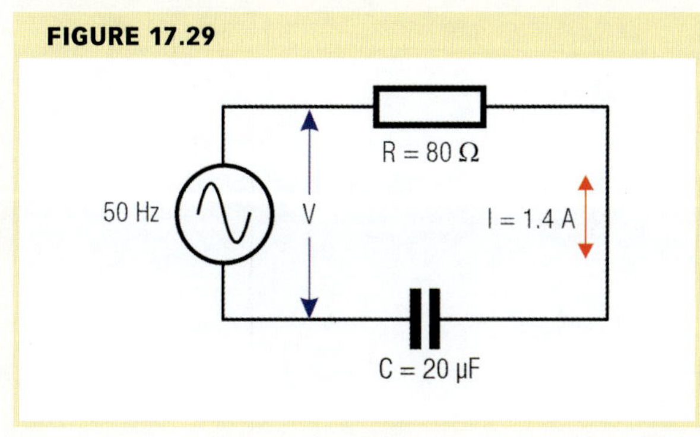

9 For the circuit in Figure 17.30, calculate the:
 a inductive reactance
 b capacitive reactance
 c impedance
 d current flowing in the circuit
 e voltage drops across each component
 f phase angle between the current and voltage V.

FIGURE 17.30

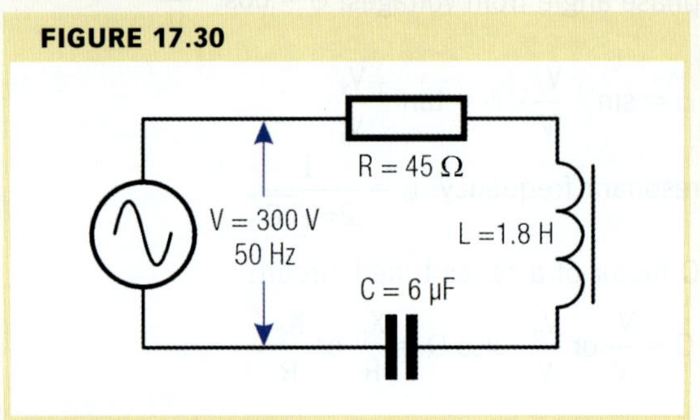

10 For the circuit in Figure 17.31, calculate the circuit's:
 a resonant frequency
 b impedance at resonance
 c Q factor
 d bandwidth.

FIGURE 17.31

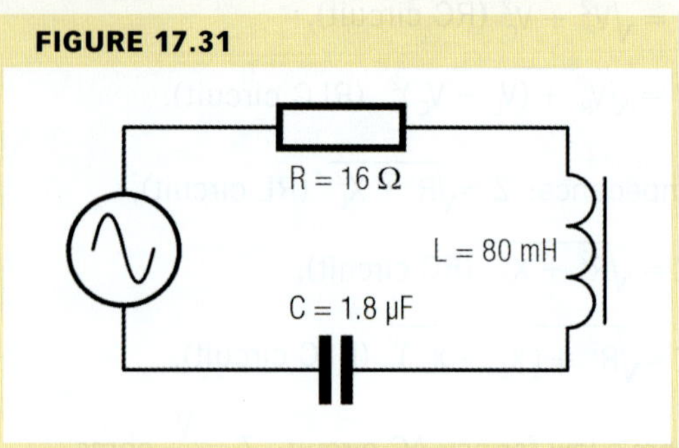

ONLINE RESOURCES

COMPLETE WORKSHEET SEVENTEEN

Check with your instructor for worksheets on this chapter.

CHAPTER 18

PARALLEL AC CIRCUITS

Like a series circuit, a parallel AC circuit often has inductance, resistance and capacitance, sometimes in various combinations. Therefore, the circuit current and applied voltage are unlikely to be in phase. This chapter explains how to use phasor diagrams to find the total current in a parallel AC circuit, its phase difference to the applied voltage and the impedance of the circuit. It also describes parallel resonance and its effects.

CHAPTER OUTLINE

18.1 Parallel RL AC Circuits

In most electrical wiring installations the appliances, lighting, machinery and so on are all connected in parallel. Therefore, the total load on the installation could include capacitance, resistance and inductance. The main difference with a parallel AC circuit compared to a series AC circuit is the applied voltage is the same across each branch. In a series circuit, the current is the same in all parts of the circuit. Figure 18.1 shows a simple RL parallel circuit in which a coil is connected in parallel with a resistor.

FIGURE 18.1 A parallel RL circuit

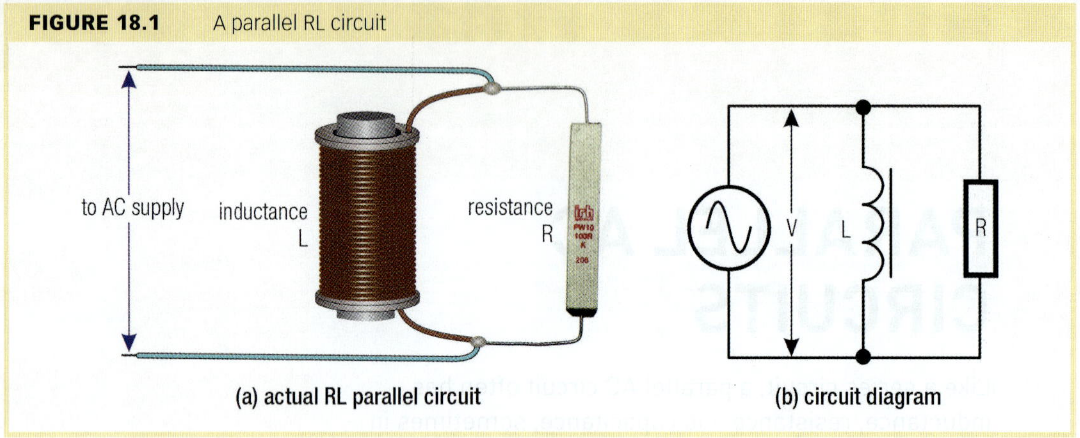

When an inductance is in parallel with a resistance, the current through the inductive branch will lag the applied voltage, while the current in the resistive branch will be in phase with the applied voltage. If the inductive branch has no resistance, the current through it will lag by 90°. Inductors always have some resistance; at this stage, the inductor is assumed to have zero resistance. The waveforms in Figure 18.2 show that current IL lags the applied voltage by 90°.

FIGURE 18.2 In a parallel RL circuit, the current flowing in the inductor lags the applied voltage by 90°

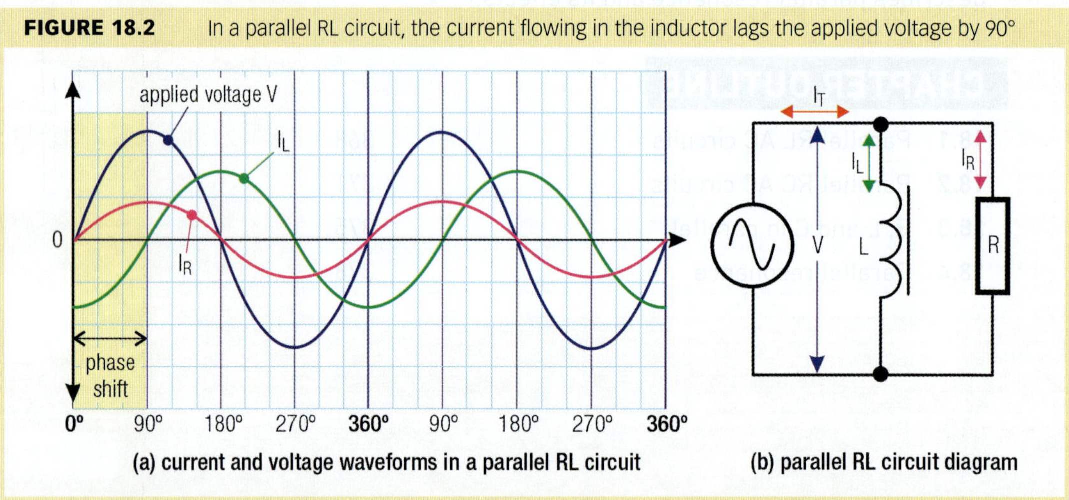

To find the total current in a parallel RL circuit, start by calculating each branch current using Ohm's law. This requires calculating the inductive reactance of the inductor. A phasor diagram can then be constructed to find the total current and its phase difference to the applied voltage. This is shown in Example 18.1.

EXAMPLE 18.1

For the circuit in Figure 18.3:

1 calculate the branch currents
2 draw a phasor diagram to find the total current and its phase difference (ϕ) to the applied voltage
3 calculate the circuit impedance (Z)

Solution

Values $L = 0.25\ \text{H}$
$R = 100\ \Omega$
$V = 400\ \text{V}$
$f = 50\ \text{Hz}$

FIGURE 18.3

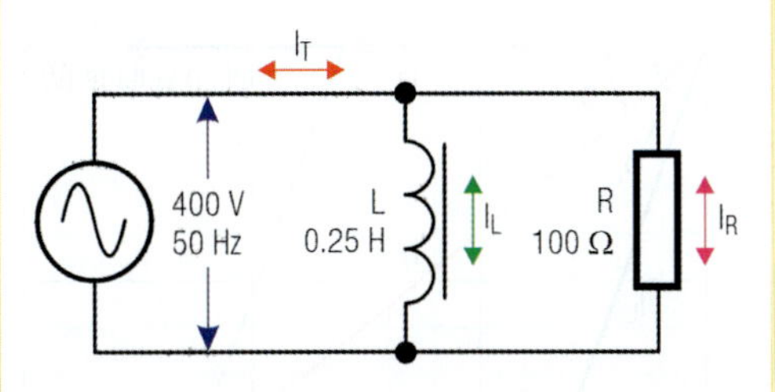

I_T | 400 V 50 Hz | L 0.25 H | I_L | R 100 Ω | I_R

1 Calculate the branch currents

Equation 1 $X_L = 2\pi fL$
$X_L = 6.28 \times 50 \times 0.25$
$X_L = 78.5\ \Omega$

Equation 2 $I_L = \dfrac{V}{X_L} = \dfrac{400}{78.5}$ and $I_R = \dfrac{V}{R} = \dfrac{400}{100}$

Answer **I_L = 5.1 A (lagging by 90°)**
I_R = 4 A (in phase)

2 Calculate the total current and phase difference (ϕ)

The phasors for both currents are drawn to scale with I_L lagging the applied voltage by 90° and I_R in phase. The applied voltage is the reference. See Figure 18.4.

FIGURE 18.4 Phasor diagram for Example 18.1, also a mathematical solution using Pythagoras' theorem and trigonometry

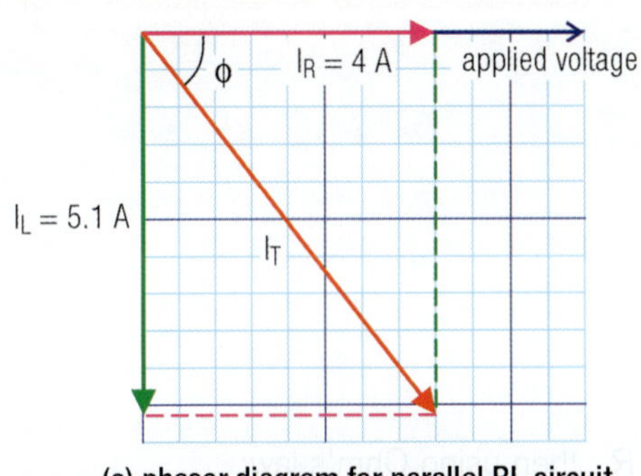

(a) phasor diagram for parallel RL circuit

Pythagoras' theorem to find I_T:
$I_T = \sqrt{I_R^2 + I_L^2} = \sqrt{4^2 + 5.1^2}$
$I_T = 6.48$ A

Trigonometry to find ϕ:
$\phi = \cos^{-1}\dfrac{I_R}{I_T} = \cos^{-1}\dfrac{4}{6.48}$
$\phi = 51.88°$

(b) mathematical solution

By measurement from the phasor diagram:

Answer **I_T = 6.5 A**
ϕ = 51.9° (lagging, as it is an inductive circuit)

3 Use Ohm's law to find impedance Z

Equation $Z = \dfrac{V}{I_T} = \dfrac{400}{6.5}$

Answer **Z = 61.5 Ω**

Including the resistance of the inductor

In a practical circuit, the inductor has resistance due to the resistance of the windings and iron losses. Therefore, the phase angle between the inductor current and the applied voltage is less than 90°. The circuit in Figure 18.5 shows a practical inductor in parallel with a resistor, and a phasor diagram for this type of circuit. The inductor current I_L has a phase shift shown as ϕ_L.

To find the total current in this type of circuit, first calculate the value and phase angle of the current flowing in the inductor. This is done with Pythagoras' theorem and trigonometry. A phasor diagram is then drawn to find the total current. Example 18.2 is based on Example 18.1, except the inductor now has a resistance of 30 Ω.

FIGURE 18.5 Practical circuit of L and R in parallel and its phasor diagram

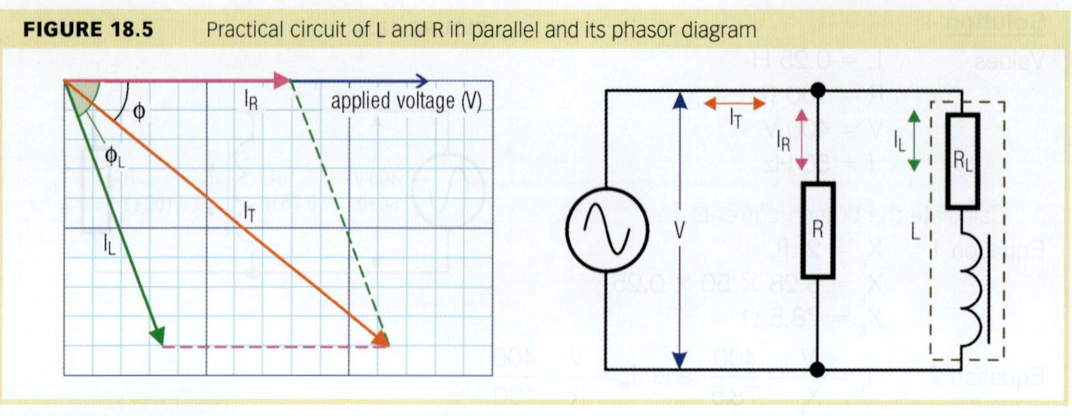

EXAMPLE 18.2

A resistor of 100 Ω is connected in parallel with a 0.25 H coil that has a resistance of 30 Ω. If the applied voltage is 400 V 50 Hz, determine the:

1 current in the inductive branch (I_L)
2 phase angle between the applied voltage and I_L (ϕ_L)
3 total circuit current (I_T)
4 phase angle between the total current and the applied voltage (ϕ)
5 circuit impedance (Z)

Solution

Values $X_L = 78.5\ \Omega$ (from Example 18.1)
$I_R = 4$ A (from Example 18.1)
$R_L = 30\ \Omega$
$R = 100\ \Omega$
$V = 400$ V
$f = 50$ Hz

1 Calculate I_L

This is done by finding the impedance of L and R_L, then using Ohm's law:

Equation 1 $Z_L = \sqrt{R_L^2 + X_L^2} = \sqrt{30^2 + 78.5^2}$

$Z_L = 84.04\ \Omega$

Equation 2 $I_L = \dfrac{V}{Z_L} = \dfrac{400}{84.04}$

Answer $\mathbf{I_L = 4.8\ A}$

2 Calculate the phase angle between applied voltage and I_L (ϕ_L)

The inductor's impedance and resistance are known, so use the inverse cos function:

Equation 3 $\phi_L = \cos^{-1}\dfrac{R_L}{Z_L}$

$\phi_L = \cos^{-1}\dfrac{30}{84.04} = \cos^{-1} 0.357$

Answer $\boldsymbol{\phi_L = 69.1°}$ **(current lagging the voltage as branch is inductive)**

3 Determine the total current (I_T)

The branch currents (I_R and I_L) have already been found, so to find I_T, draw a phasor diagram of the currents, and find I_T by measurement. The branch currents are:

$I_L = 4.8$ A (lagging by 69.1°)
$I_R = 4$ A (in phase with the voltage as branch is resistive)

To draw the phasor, make voltage (V) the reference then draw phasors for the branch currents to scale. Complete the phasor diagram and measure the length of the phasor to find I_T. Measure the angle between this phasor and the reference to find the phase angle (ϕ) between the applied voltage and the total current.

The phasor diagram is shown in Figure 18.6. The scale is one division of the graph paper equals 0.25 A. A tip-to-tail phasor diagram is used, as it is easier to construct. It is otherwise the same as the parallelogram construction in Figure 18.5. By measurement of phasor I_T:

FIGURE 18.6 Tip-to-tail phasor diagram for Example 18.2

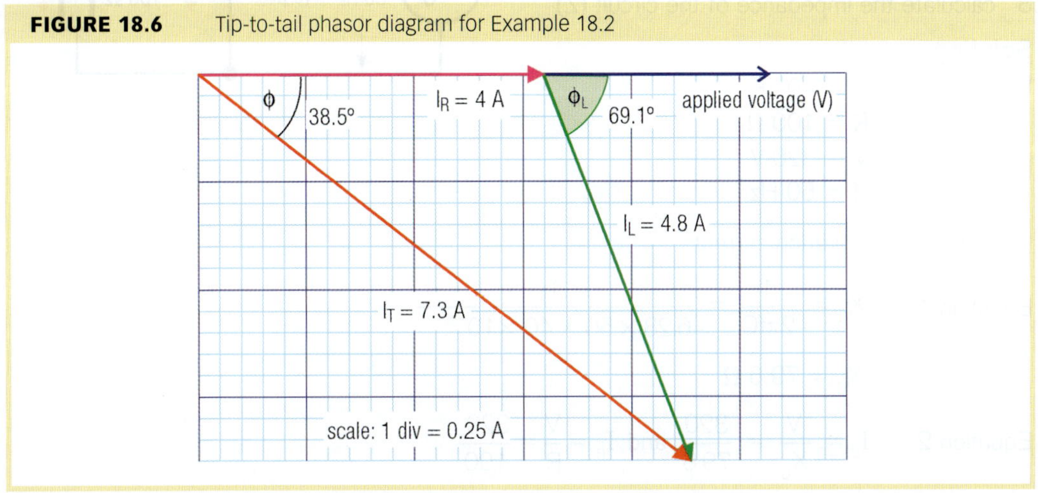

Answer $\quad I_T$ **= 7.3 A**

4 Determine the phase angle ϕ between the total current (I_T) and the applied voltage by measurement from the phasor diagram:

Answer $\quad \phi$ **= 38.5° (lagging)**

5 Use Ohm's law to find impedance (Z)

Equation $\quad Z = \frac{V}{I_T} = \frac{400}{7.3}$

Answer $\quad$ **Z = 54.8 Ω**

MATH Tip-to-tail phasor construction is explained on page xxxix

18.2 Parallel RC AC Circuits

When capacitance is in parallel with resistance, the current through the capacitive branch leads the applied voltage. The current in the resistance will be in phase with the applied voltage. If the capacitive branch has no resistance, the branch current will lead by 90°. This happens in practice, as capacitors usually have very low resistance. If resistance is in series with the capacitor, the current will lead by less than 90°. Figure 18.7(a) shows a motor start capacitor in parallel with resistor, while (b) shows the circuit diagram.

FIGURE 18.7 In an RC circuit, the total circuit current and its phase difference to the applied voltage is found with a phasor diagram. This is shown in Example 18.3.

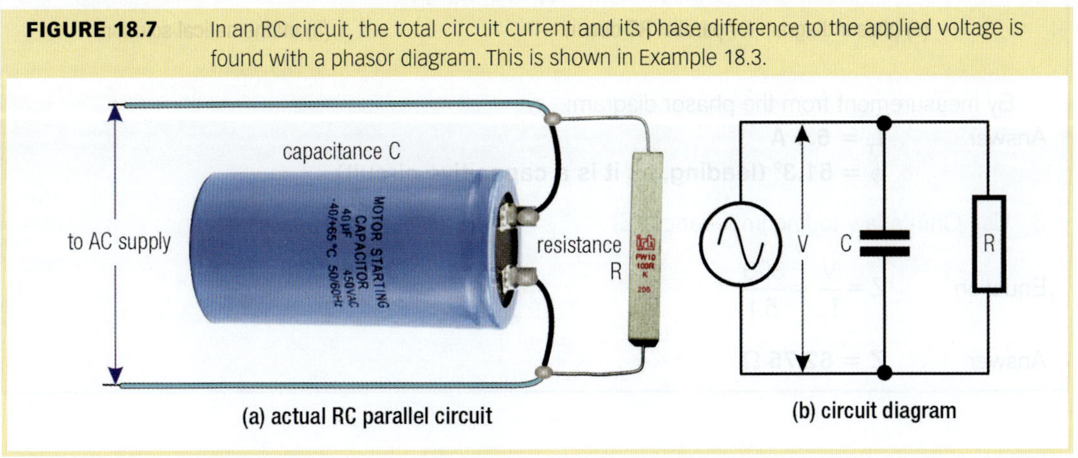

EXAMPLE 18.3

For the circuit in Figure 18.8:

1 calculate the branch currents
2 draw a phasor diagram to find the total current and its phase difference (ϕ) to the applied voltage
3 calculate the impedance of the circuit (Z)

FIGURE 18.8

320 V 50 Hz; C 40 µF; R 100 Ω; I_T, I_C, I_R

Solution

Values $C = 40\ \mu F$

$R = 100\ \Omega$

$V = 320\ V$

$f = 50\ Hz$

1 Calculate the branch currents

Equation 1 $$X_C = \frac{1}{2\pi fC} = \frac{1}{6.28 \times 50 \times 40 \times 10^{-6}}$$

$$X_C = 79.6\ \Omega$$

Equation 2 $$I_C = \frac{V}{X_C} = \frac{320}{79.6} \text{ and } I_R = \frac{V}{R} = \frac{320}{100}$$

Answer **I_C = 4 A (leading by 90°)**

I_R = 3.2 A (in phase)

2 Determine the total current (I_T) and phase difference (ϕ)

The phasors for both currents are drawn to scale with I_C leading the applied voltage by 90° and I_R in phase. The applied voltage is the reference.

FIGURE 18.9 Phasor diagram for Example 18.3, also a mathematical solution using Pythagoras' theorem and trigonometry

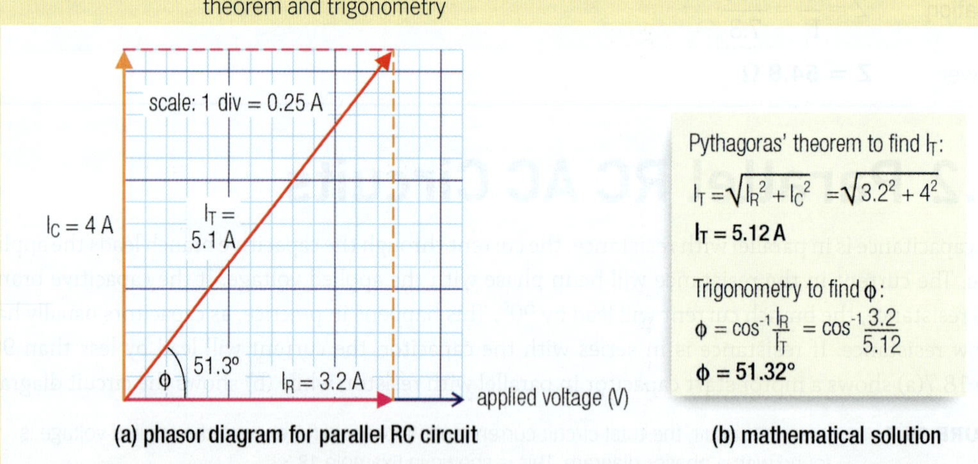

By measurement from the phasor diagram:

Answer **I_T = 5.1 A**

ϕ = 51.3° (leading, as it is a capacitive circuit)

3 Use Ohm's law to find impedance (Z)

Equation $$Z = \frac{V}{I_T} = \frac{320}{5.1}$$

Answer **Z = 62.75 Ω**

Adding resistance in series with the capacitor

A parallel RC circuit could have a branch containing a series resistance and capacitor. This means the phase angle between the current in that branch and the applied voltage is less than 90°. Figure 18.10 shows this type of circuit and the circuit's phasor diagram.

FIGURE 18.10 An RC parallel circuit with resistance in series with the capacitor, and the phasor diagram for the circuit

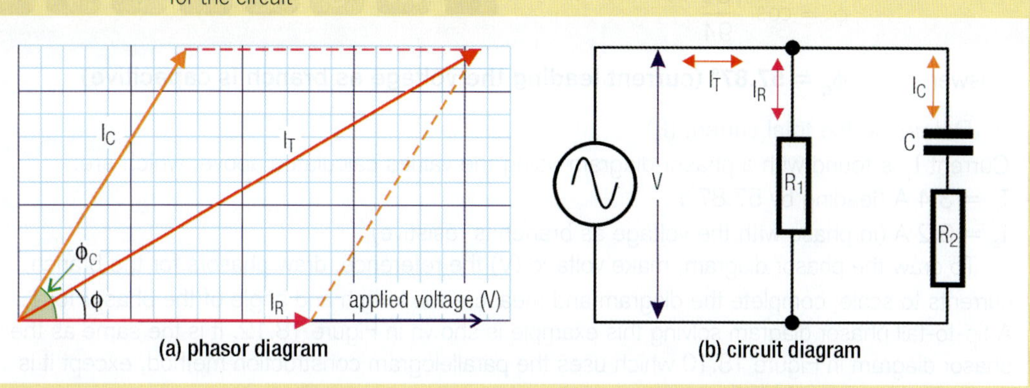

To find current values in this type of circuit, start with the capacitive-resistive branch to find the branch current and its phase angle to the applied voltage. Then, after calculating the current in the resistive branch, a phasor diagram can be constructed to find the total circuit current and its phase angle. Example 18.4, based on Example 18.3, demonstrates this.

EXAMPLE 18.4

For the circuit in Figure 18.11 calculate the:

1 current in the capacitive branch (I_C)
2 phase angle between the applied voltage and I_C (ϕ_C)
3 total current (I_T)
4 phase angle between the total current and the applied voltage (ϕ)
5 impedance (Z) of the circuit

FIGURE 18.11

I_T
I_R
I_C
C
40 μF
320 V
50 Hz
100 Ω
R_1
50 Ω
R_2

Solution

Values $X_C = 79.6\ \Omega$ (from Example 18.3)
$I_R = 3.2$ A (from Example 18.3)
$R_1 = 100\ \Omega$
$R_2 = 50\ \Omega$
$V = 320$ V
$f = 50$ Hz

1 Calculate I_C

This is done by finding the impedance of the series combination of C and R_2, then using Ohm's law:

Equation 1 $Z_C = \sqrt{R_2^2 + X_C^2} = \sqrt{50^2 + 79.6^2}$

$Z_C = 94\ \Omega$

Equation 2 $I_C = \frac{V}{Z_C} = \frac{320}{94}$

Answer $\mathbf{I_C = 3.4\ A}$

2 Calculate the phase angle between applied voltage and I_C (ϕ_C)

Because the impedance and resistance of the branch are known, we can use the inverse cos function:

Equation 3 $\phi_C = \cos^{-1}\dfrac{R_2}{Z_C}$

$\phi_C = \cos^{-1}\dfrac{50}{94}$

Answer ϕ_C **= 57.87° (current leading the voltage as branch is capacitive)**

3 Determine the total current (I_T)

Current I_T is found with a phasor diagram using the values calculated above, which are:

I_C = 3.4 A (leading by 57.87°)

I_R = 3.2 A (in phase with the voltage as branch is resistive)

To draw the phasor diagram, make voltage (V) the reference, draw phasors for the branch currents to scale, complete the diagram and measure the length and angle of the phasor for I_T. A tip-to-tail phasor diagram solving this example is shown in Figure 18.12. It is the same as the phasor diagram in Figure 18.10 which uses the parallelogram construction method, except it is drawn to a larger scale.

FIGURE 18.12 Tip-to-tail phasor diagram for Example 18.4

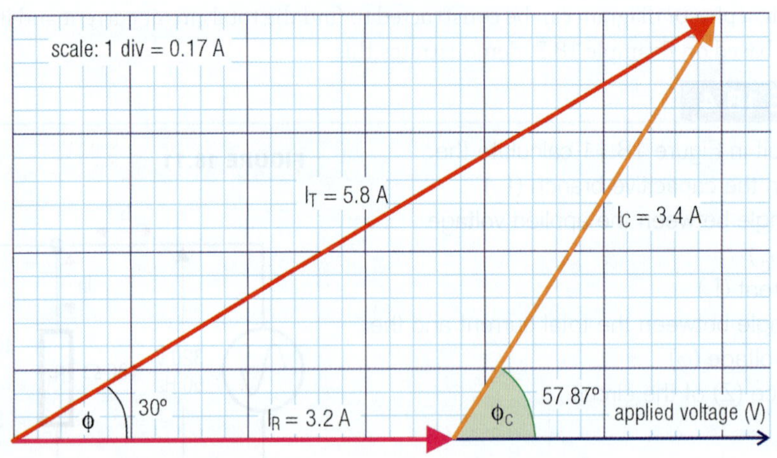

By measurement from the phasor diagram:

Answer I_T **= 5.8 A**

4 Measure the phase angle ϕ between the total current and applied voltage (V)

Answer ϕ **= 30° (with current leading the voltage as circuit is capacitive)**

5 Use Ohm's law to find impedance (Z)

Equation $Z = \dfrac{V}{I_T} = \dfrac{320}{5.8}$

Answer **Z = 55.17 Ω**

In this example, a tip-to tail phasor diagram has been used as the construction method, and is generally easier than the parallelogram method. Whichever method is used, best accuracy is obtained with the largest scale.

KEY POINTS...

- In an AC parallel resistive circuit, all currents and voltages are in phase and the circuit is treated in the same way as for DC.
- When an inductor is in parallel with a resistor, the total circuit current will lag the applied voltage by an angle less than 90°.
- If a branch in a parallel AC circuit is purely inductive, the current in that branch lags the applied voltage by 90° and the total current in the circuit can be found with Pythagoras' theorem or a phasor diagram.
- If an inductive branch in a parallel AC circuit contains resistance, the current in that branch lags the applied voltage by less than 90° and the total current is found with a phasor diagram.
- When a capacitor is in parallel with a resistance, the total circuit current leads the applied voltage by less than 90°.
- If a branch in a parallel AC circuit is purely capacitive, the current in that branch leads the applied voltage by 90° and the total current in the circuit can be found with Pythagoras' theorem or a phasor diagram.
- If a capacitive branch in a parallel AC circuit contains resistance, the current in that branch leads the applied voltage by less than 90° and the total current is found with a phasor diagram.

18.3 R, L and C in Parallel

When a circuit, like that in Figure 18.13 has resistance, pure inductance and pure capacitance in parallel, current in the:

- resistive branch is in phase with the applied voltage
- inductive branch lags the applied voltage by 90°
- capacitive branch leads the applied voltage by 90°.

FIGURE 18.13 A parallel RLC circuit

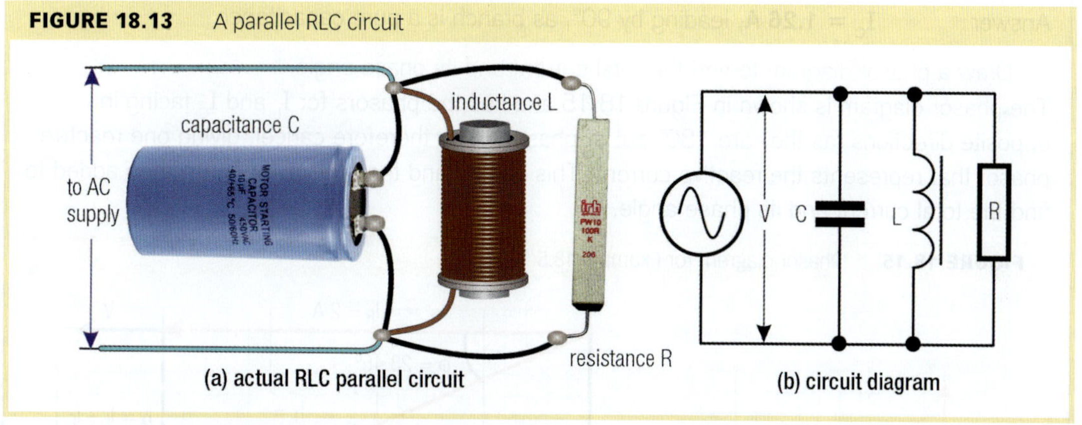

The leading and lagging currents in a pure RLC circuit will tend to cancel each other, as they are in perfect opposition. If one is greater than the other, there will be a reactive current in the circuit, which when combined with the resistive current will give a total circuit current that either leads or lags the applied voltage by a certain angle. Example 18.5 shows how to determine the branch currents, the total current and its phase angle, and the impedance of a pure RLC parallel circuit.

EXAMPLE 18.5

For the circuit in Figure 18.14, find the:

1 value and phase angle of each branch current
2 total circuit current and its phase angle to the applied voltage
3 impedance of the circuit

Solution

Values	$R = 200\ \Omega$
	$L = 0.6$ H
	$C = 10\ \mu$F
	$V = 400$ V
	$f = 50$ Hz

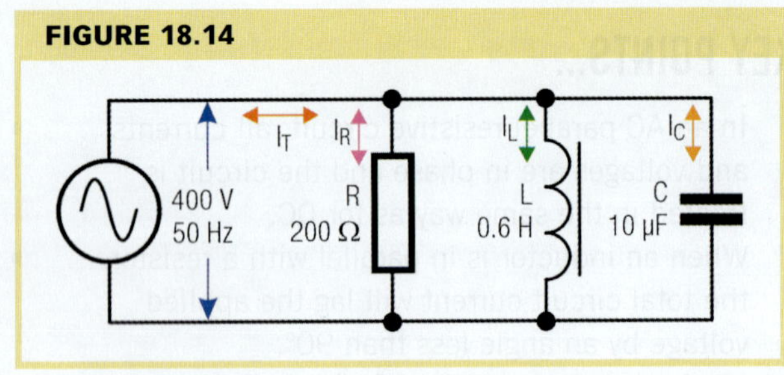

FIGURE 18.14

1 Calculate the value and phase angle of each branch current

The currents are found with Ohm's law. The phase angles will be either 0° or 90° as the components are either purely resistive, or purely reactive.

Equation 1 $$I_R = \frac{V}{R} = \frac{400}{200}$$

Answer $\mathbf{I_R = 2\ A}$, in phase as circuit is resistive

Equation 2 $X_L = 2\pi fL = 6.28 \times 50 \times 0.6$

Answer $\mathbf{X_L = 188.4\ \Omega}$

Equation 3 $$I_L = \frac{V}{X_L} = \frac{400}{188.4}$$

Answer $\mathbf{I_L = 2.12\ A}$, lagging by 90°, as branch is a pure inductance

Equation 4 $$X_C = \frac{1}{2\pi fC} = \frac{1}{6.28 \times 50 \times 10 \times 10^{-6}}$$

Answer $\mathbf{X_C = 318.5\ \Omega}$

Equation 5 $$I_C = \frac{V}{X_C} = \frac{400}{318.5}$$

Answer $\mathbf{I_C = 1.26\ A}$, leading by 90°, as branch is a pure capacitance

2 Draw a phasor diagram to find the total current and its phase angle.

The phasor diagram is shown in Figure 18.15 and has the phasors for I_L and I_C facing in opposite directions (as they are 180° out of phase). They therefore cancel, giving one reactive phasor that represents the reactive current. This phasor and the phasor for I_R are then added to find the total current and its phase angle.

FIGURE 18.15 Phasor diagram for Example 18.5

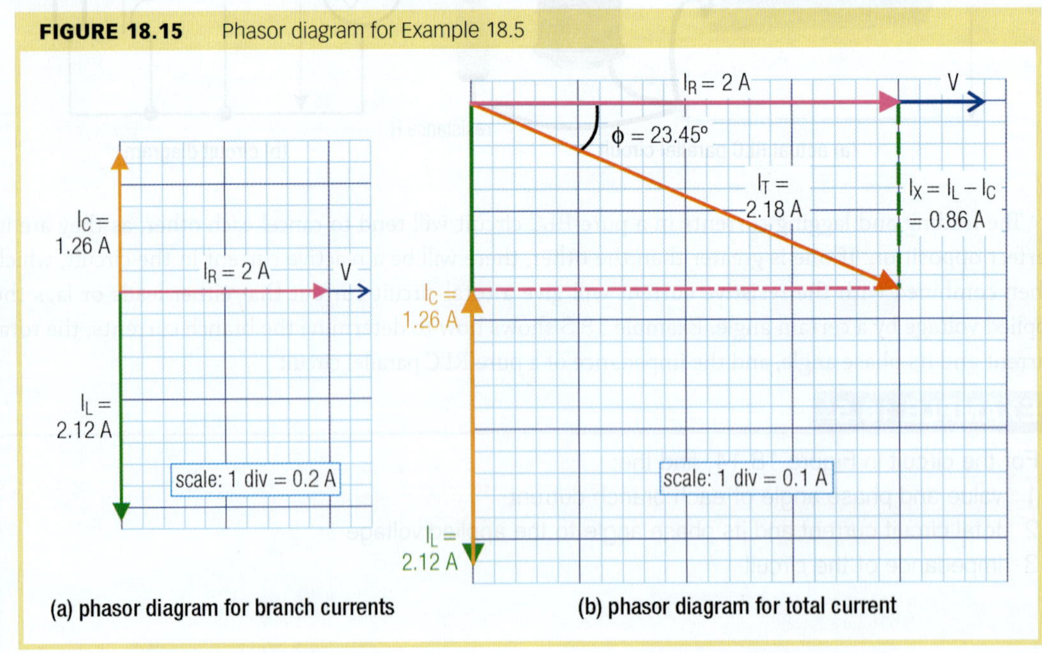

By measurement from the phasor diagram:

Answer I_T = **2.18 A**

ϕ = **23.45°** (lagging)

3 Use Ohm's law to find impedance (Z)

Equation $Z = \frac{V}{I_T} = \frac{400}{2.18}$

Answer **Z = 183.49 Ω**

FIGURE 18.16

Mathematical solution

Pythagoras' theorem to find I_T:

$I_T = \sqrt{I_R^2 + I_X^2} = \sqrt{2^2 + 0.86^2}$

$I_T = 2.18\,A$

Trigonometry to find ϕ:

$\phi = \cos^{-1}\frac{I_R}{I_T} = \cos^{-1}\frac{2}{2.18}$

$\phi = 23.45°$

Including the resistance of the inductor

In most cases the resistance of an inductor means the inductor current is less than 90° out of phase. Therefore, it does not exactly oppose the current in the capacitive branch. To find the total current and its phase angle requires a phasor diagram. Example 18.6 explains the process, and is based on Example 18.5, in which the inductance now has a resistance of 50 Ω.

EXAMPLE 18.6

For the circuit in Figure 18.17, use a phasor diagram to find the total current.

FIGURE 18.17

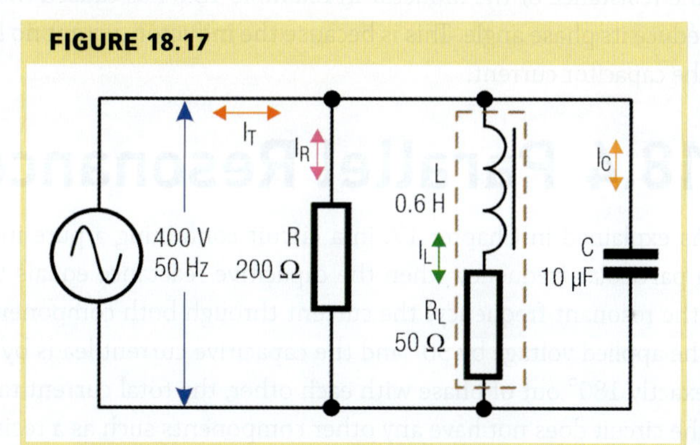

Solution

Values from Example 18.5

$R = 200\ \Omega$

$I_R = 2\ A$

$X_L = 188.4\ \Omega$

$I_C = 1.26\ A\ \angle 90°$

$V = 400\ V$

New values

$R_L = 50\ \Omega$

Step 1 Calculate the impedance of the inductance

Equation 1 $Z_L = \sqrt{R_L^2 + X_L^2} = \sqrt{50^2 + 188.4^2}$

Answer $\mathbf{Z_L = 194.92\ \Omega}$

Step 2 Calculate the current in the inductance and its phase angle

Equation 2 $I_L = \frac{V}{Z_L} = \frac{400}{194.92}$

Answer $\mathbf{I_L = 2.05\ A}$

Equation 3 $\phi = \cos^{-1}\frac{R_L}{Z_L} = \frac{50}{194.92}$

Answer **ϕ = 75.14° (lagging)**

Step 3 Draw the phasor diagram, which is shown in Figure 18.18

FIGURE 18.18 Tip-to-tail phasor diagram to find total current in Figure 18.17

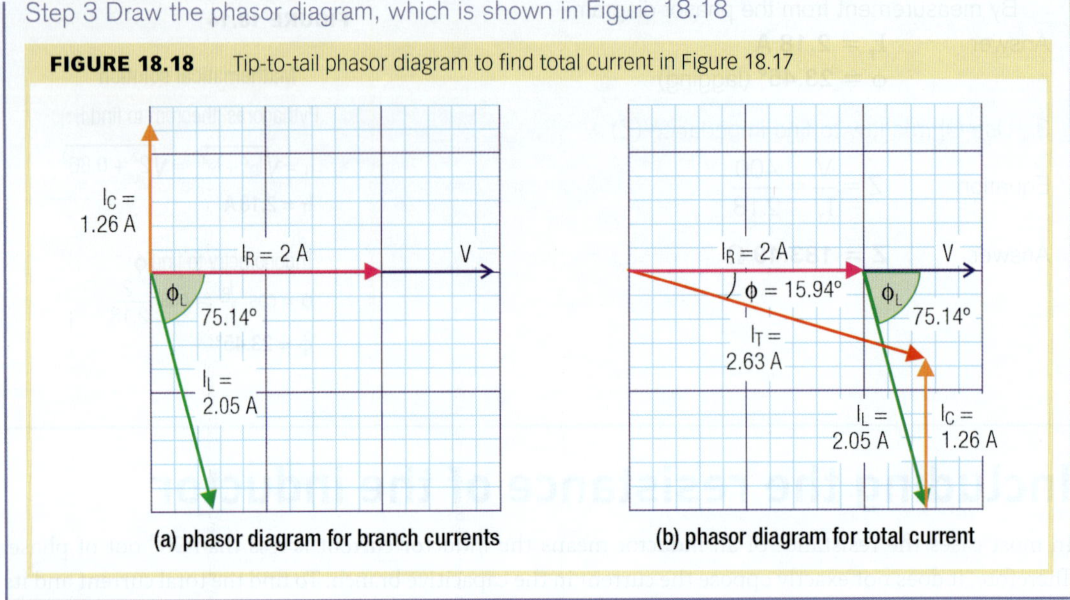

(a) phasor diagram for branch currents (b) phasor diagram for total current

The phasor diagram Figure 18.18(a) shows the three current phasors, and (b) shows their phasor addition using a tip-to-tail phasor diagram. By measurement, the total current is 2.63 A at a lagging phase angle of 15.9°. In Example 18.5, the total current was found to be 2.18 A at an angle of 23.45° (lagging). The resistance of the inductor in Example 18.6 has caused the total current to *increase* by 0.45 A and to reduce its phase angle. This is because the inductor current no longer exactly opposes and thereby cancels the capacitor current.

18.4 Parallel Resonance

As explained in Chapter 17, in a circuit containing a pure inductance and a pure capacitance, there is a particular frequency when the capacitive reactance equals the inductive reactance. At this frequency (the resonant frequency) the current through both components is the same. The inductive current lags the applied voltage by 90° and the capacitive current leads by 90°. Because both currents are equal, but exactly 180° out of phase with each other, the total current taken from the supply is zero. This assumes the circuit does not have any other components such as a resistance. The circuit and the phasor diagram for a parallel resonant circuit without resistance are shown in Figure 18.19.

FIGURE 18.19 When a pure inductor and capacitor in parallel have the same reactance (at resonance), the total current in the circuit falls to zero

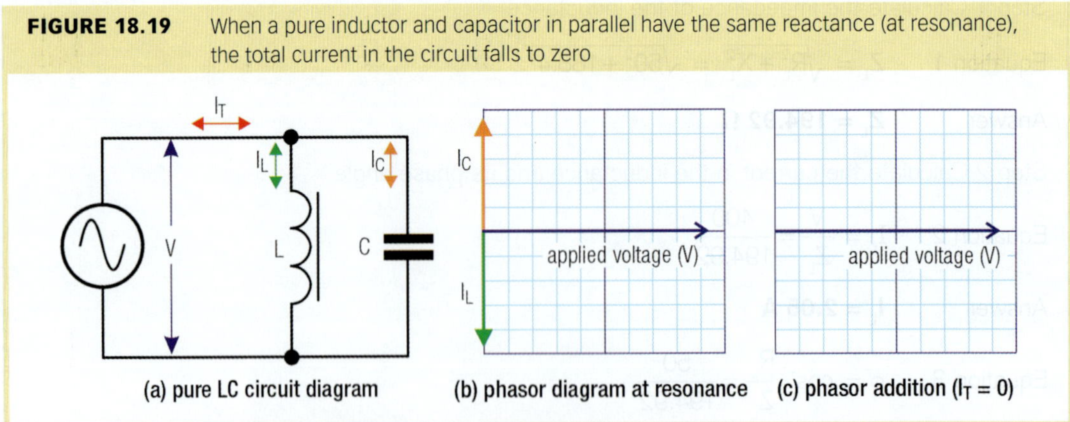

(a) pure LC circuit diagram (b) phasor diagram at resonance (c) phasor addition ($I_T = 0$)

If a resistance is in parallel with the capacitor and the inductor, current will flow through the resistance, regardless of what is happening between the capacitor and the inductor. However, at resonance, this is the only current taken from the supply. The circuit and its phasor diagram are shown in Figure 18.20.

FIGURE 18.20 In a parallel circuit containing resistance, inductance and capacitance, at resonance the only current taken from the supply is the current in the resistance

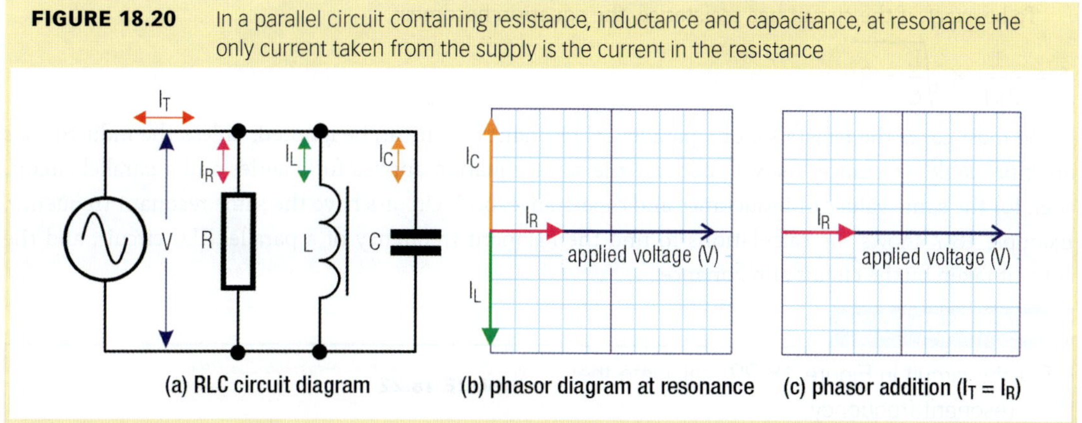

Resonance in any circuit containing capacitance and inductance can occur as a result of multiple frequencies, called harmonics, that can happen in the 50 Hz power supply network. Harmonics are caused by non-linear loads, such as computers or motor control systems that operate by rapidly switching power to the motor.

In a parallel RLC circuit, if resonance occurs, a circulating current flows between the capacitor and the inductor. This is because an inductor and a capacitor both store energy. In an inductor, energy is stored in the magnetic field built up by the current. In a charged capacitor, energy is stored as an electrostatic field in its dielectric.

Figure 18.21(a) shows an inductance with a magnetic field about to collapse. As it collapses, it causes an induced current to flow which charges the capacitor. When the magnetic field has collapsed, as shown in (b), the fully charged capacitor discharges through the coil. This builds up a magnetic field around the inductor, and so the cycle continues in which energy is transferred back and forth between the capacitor and the inductor.

FIGURE 18.21 In a parallel resonant circuit, energy constantly swaps back and forth between the capacitor and the inductor

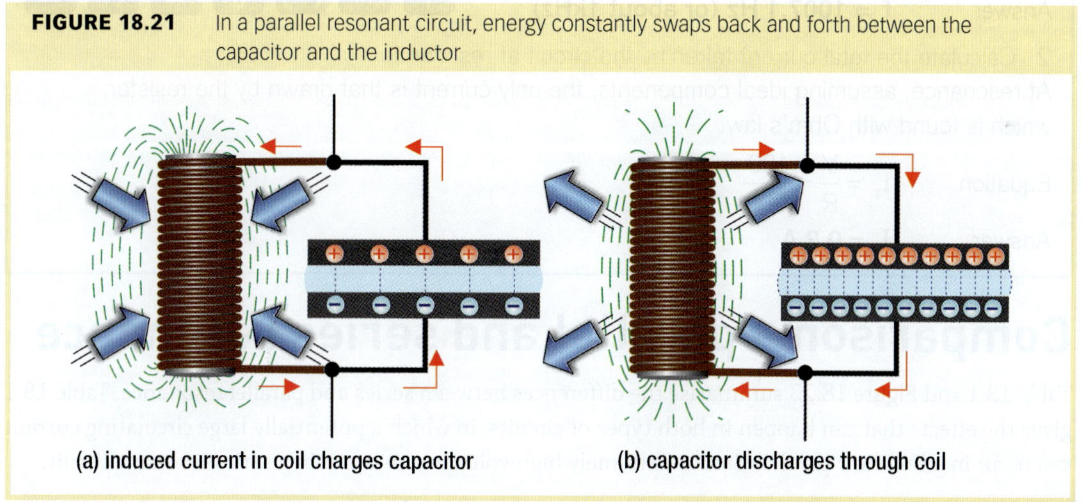

This energy 'swapping' goes on as long as there's an external voltage to sustain it and thereby overcome the losses in the circuit. This action is like water sloshing back and forth in a tank. For this reason, the parallel LC circuit is often called a 'tank circuit'.

In a practical resonant circuit, there's always some resistance. Therefore, a practical parallel resonant circuit takes some energy from the supply to overcome the resistive losses in the circuit. However, there will still be a circulating current in the LC circuit, which can have a very high value limited only by the reactance of the capacitor and the inductor. The equation to find the resonant frequency of a parallel resonant circuit is the same as for a series resonant circuit:

$$\text{resonant frequency } f_r = \frac{1}{2\pi\sqrt{LC}}$$

FYI

If resonance occurs in a parallel power circuit, very high circulating currents can occur

To be exact, if the circuit has resistance, the equation becomes:

$$f_r = \frac{1}{2\pi L} \times \sqrt{\frac{L}{C} - R^2}$$

In most cases the resistance of a parallel LC resonant circuit can be ignored, unless the inductor has an appreciable resistance. As well, because the same equation applies for a series and a parallel circuit, then for the same values of inductance and capacitance, both circuits have the same resonant frequency. Example 18.7 shows the calculations to find the resonant frequency of a parallel RLC circuit, and the current taken by the circuit at resonance.

EXAMPLE 18.7

For the circuit in Figure 18.22, calculate the:

1 resonant frequency

2 total current taken by the circuit at resonance.

FIGURE 18.22

100 V, I_T, R 500 Ω, L 25 mH, C 1 μF

Solution

Values R = 500 Ω

L = 25 mH

C = 1 μF

V = 100 V

1 Calculate the resonant frequency

Equation $f_r = \frac{1}{2\pi\sqrt{LC}}$

$$f_r = \frac{1}{6.28\sqrt{25 \times 10^{-3} \times 1 \times 10^{-6}}}$$

Answer **f_r = 1007.1 Hz (or about 1kHz)**

2 Calculate the total current taken by the circuit at resonance

At resonance, assuming ideal components, the only current is that drawn by the resistor, which is found with Ohm's law.

Equation $I_T = \frac{V}{R} = \frac{100}{500}$

Answer **I_T = 0.2 A**

Comparison – parallel and series resonance

Table 18.1 and Figure 18.23 summarise the differences between series and parallel resonance. Table 18.1 gives the effects that can happen in both types of circuits, in which a potentially large circulating current can occur in a resonant parallel circuit. Extremely high voltages can occur in a series resonant circuit.

TABLE 18.1

COMPARING SERIES AND PARALLEL RESONANCE

	Parallel resonance	Series resonance
Impedance (Z)	maximum	minimum
Supply current (I)	minimum	maximum
Phase angle of I	0°	0°
Effect	circulating current between L and C	voltage across L and C can exceed the supply voltage
Equation for resonance	$f_r = \frac{1}{2\pi\sqrt{LC}}$ (ideal circuit)	$f_r = \frac{1}{2\pi\sqrt{LC}}$

Figure 18.23 shows that a series circuit has its minimum impedance at resonance (equals the DC resistance of the circuit), while a parallel circuit has a maximum impedance. Therefore, at resonance, the current taken from the supply is a maximum for the series circuit and a minimum for the parallel circuit.

FIGURE 18.23 Curves showing how the current and circuit impedance change in a series and a parallel resonant circuit

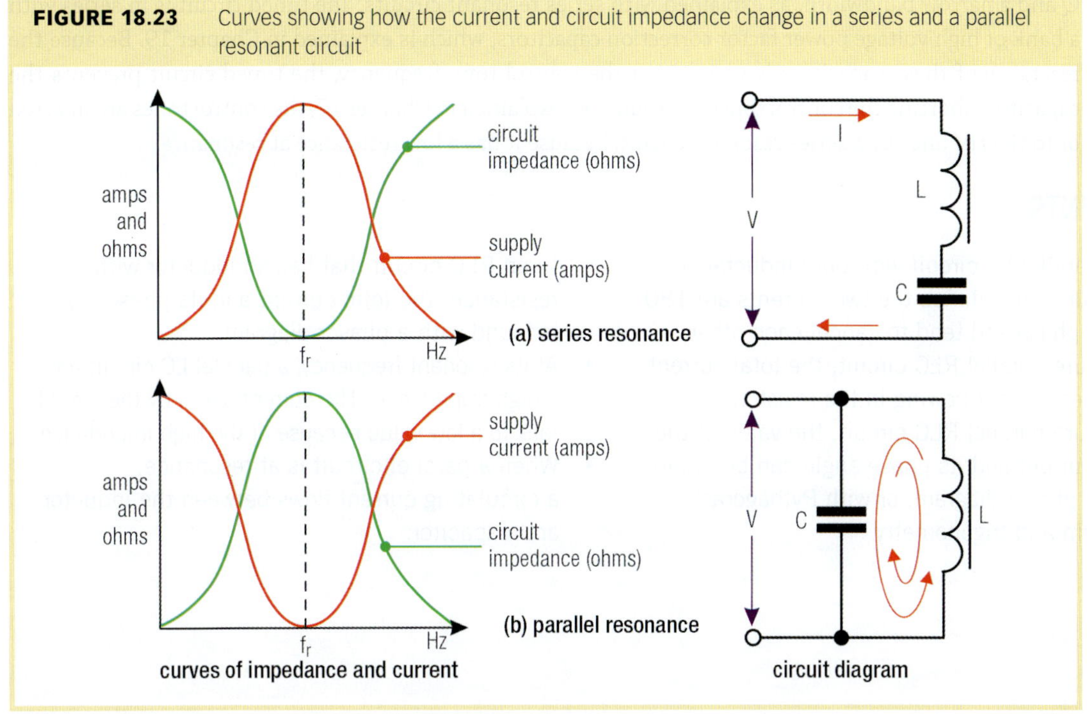

Resonance is therefore generally a factor to be avoided in a power circuit. As already mentioned, some types of loads can generate harmonics that cause resonance in a circuit that has inductance and capacitance. These are often present in some form in many power circuits, and protection against resonance is often needed in substations due to the high power levels.

Applications of parallel resonance

The parallel resonant circuit has many applications, particularly in communications where a tank circuit forms the output of radio transmitters. In the electrical power industry, a common use for a parallel LC tuned circuit is to prevent control tones injected onto high voltage distribution cables from being shorted out by capacitors associated with the substation. The frequency of the control tone depends on the power company, which is 1045 Hz for the equipment in Figure 18.24.

FIGURE 18.24 High power, high voltage inductor (reactor) and capacitor forming a parallel circuit with a resonant frequency of 1045 Hz, creating a high impedance to 1045 Hz control tones

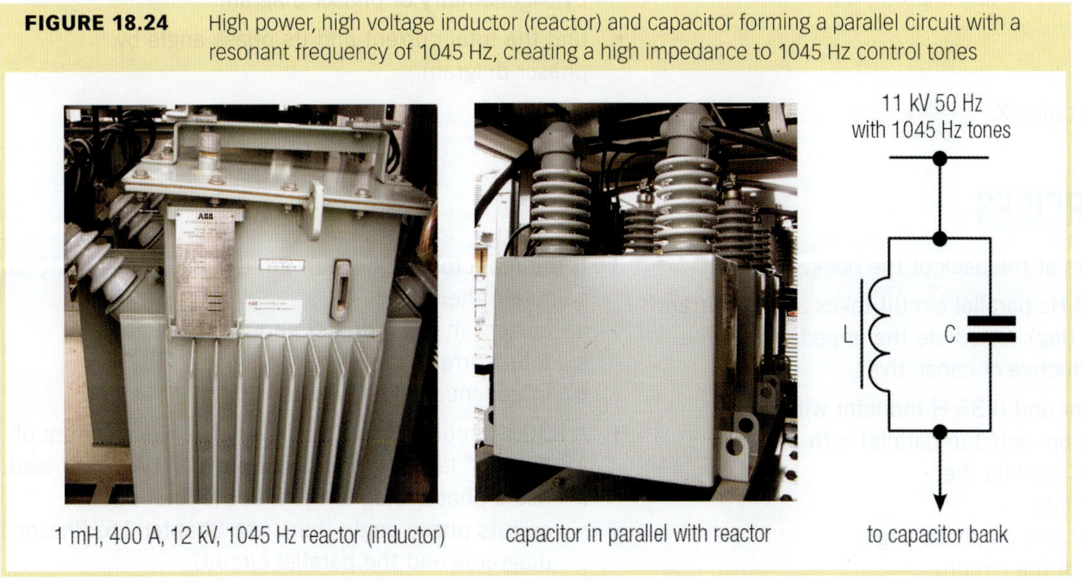

The LC parallel circuit in Figure 18.24 is tuned to a resonant frequency that equals the control tone frequency used at the substation. At this frequency the circuit has a high impedance, as the resistance of the inductor is very small, so it can pass up to 400 A at 50 Hz. Therefore, the tuned circuit has a high Q and a narrow bandwidth, as explained with series resonant circuits. The tuned circuit is in series with a bank of high voltage power factor correction capacitors, which is explained in Chapter 19. Because the reactance of these capacitors is very low at the control tone frequency, the tuned circuit prevents the capacitors shorting the control tones to ground. As explained in Chapter 17, the control tones are injected onto the HV lines by a series resonant circuit, because it has a low resistance at resonance.

KEY POINTS...

- In a parallel LC circuit with pure inductance and pure capacitance the two currents are 180° out of phase and tend to cancel each other.
- In a pure parallel RLC circuit, the total current includes current flowing in the resistor.
- In a pure parallel RLC circuit, the value of the total current and its phase angle can be found with a phasor diagram, or with Pythagoras' theorem and trigonometry.
- In an RLC circuit that has an inductor with resistance, the total current and its phase angle is found with a phasor diagram.
- At its resonant frequency, a parallel LC circuit has a high impedance. The current taken by the circuit falls to a low value because of the high impedance.
- When a parallel circuit is at resonance, a circulating current flows between the inductor and capacitor.

CHAPTER SUMMARY

Equations used in this chapter:

- Inductive reactance $X_L = 2\pi fL$
- Inductor current $I_L = \frac{V}{X_L}$
- Impedance of a parallel circuit $Z = \frac{V}{I_T}$
- Impedance of an inductor with resistance $Z_L = \sqrt{R_L^2 + X_L^2}$
- Phase angle of current in an inductor with resistance $\phi_L = \cos^{-1}\frac{R_L}{Z_L}$
- Capacitive reactance $X_C = \frac{1}{2\pi fC}$
- Capacitor current $I_C = \frac{V}{X_C}$
- Resonant frequency of a parallel circuit $f_r = \frac{1}{2\pi\sqrt{LC}}$

General procedure to find currents, phase angles and impedance in any parallel RLC circuit:

- calculate impedance of each branch
- calculate each branch current
- find the phase angle of each branch current, by trigonometry or phasor diagram
- find the total current and its phase angle by phasor diagram.

REVIEW EXERCISES

Check your answers at the back of the book.

1 A 230 V AC 50 Hz parallel circuit takes a total current of 36 A at 22° (lag). Calculate the impedance of the circuit. Is it inductive or capacitive?

2 A 150 Ω resistor and 0.35 H inductor with zero resistance are connected in parallel with a 100 V 50 Hz supply. Calculate the:
   a branch currents
   b total current and its phase angle
   c impedance of the circuit.

3 If the inductor in Question 2 has a resistance of 30 Ω, calculate the:
   a current in the inductor and its phase angle
   b total current and its phase angle
   c impedance of the circuit.

4 A 230 V circuit has two branches. One has a current of 12 A at 25° lag, the other a current of 10 A at 50° lead.
   a use a phasor diagram to find the total current and its phase angle (hint, see Chapter 15, Phasor diagrams and the parallel circuit)
   b calculate the impedance of the circuit.

5 A parallel circuit takes a total circuit current of 0.72 A at 27° lag. It has two branches, one resistive that takes 0.45 A, the other resistive-inductive. Use a phasor diagram to find the current and its phase angle in the resistive-inductive branch.

6 A 10 μF capacitor and a 500 Ω resistor are in parallel with a 100 V 50 Hz supply. Calculate or use a phasor diagram to find the:
   a branch currents
   b total current and its phase angle
   c impedance of the circuit.

7 For the circuit in Figure 18.25, find the:
   a value and phase angle of each branch current
   b total current and its phase angle
   c impedance of the circuit.

FIGURE 18.25

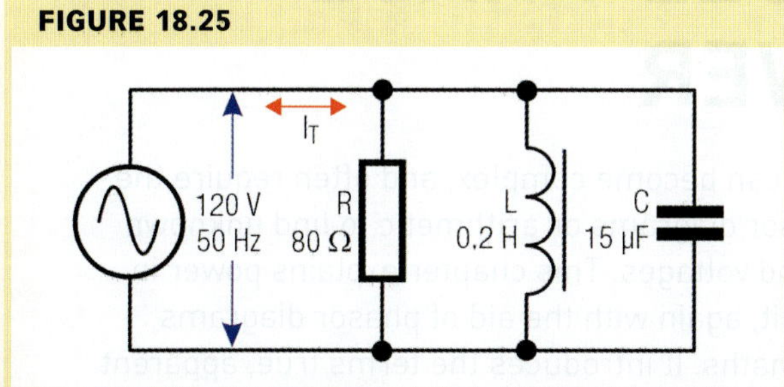

8 Calculate the resonant frequency of the circuit in Figure 18.25.

9 If the circuit in Figure 18.25 is operating at its resonant frequency, what is the total current and its phase angle?

10 For the circuit in Figure 18.26, find the:
   a branch currents and their phase angle
   b total current and its phase angle
   c impedance of the circuit.

FIGURE 18.26

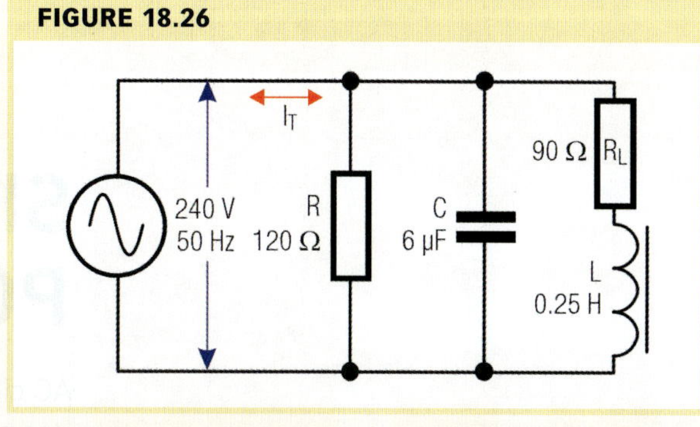

ONLINE RESOURCES

COMPLETE WORKSHEET EIGHTEEN

Check with your instructor for worksheets on this chapter.

CHAPTER 19

SINGLE-PHASE POWER

AC circuits can become complex, and often require the use of phasor diagrams or arithmetic to find unknown currents and voltages. This chapter explains power in an AC circuit, again with the aid of phasor diagrams and basic maths. It introduces the terms true, apparent and reactive power, and explains the problems caused by reactive power in terms of the losses it can cause. This chapter also introduces the term power factor, and the importance of maintaining the power factor of an electrical supply system above a certain value.

CHAPTER OUTLINE

19.1 Introduction

Chapter 16 explained that a purely resistive circuit consumes electrical power, called true power. True power performs work. A purely reactive circuit does not consume true power, even though current and voltage are present in the circuit. That is, no useful work is done. This is because the reactive component takes power on one quarter of a cycle, then returns it to the supply on the next. The waveforms in Figure 19.1 summarise power, voltage and current in the three types of AC circuits: resistive, pure capacitive and pure inductive.

FIGURE 19.1 Power is consumed in a resistive AC circuit, but not in a purely reactive AC circuit

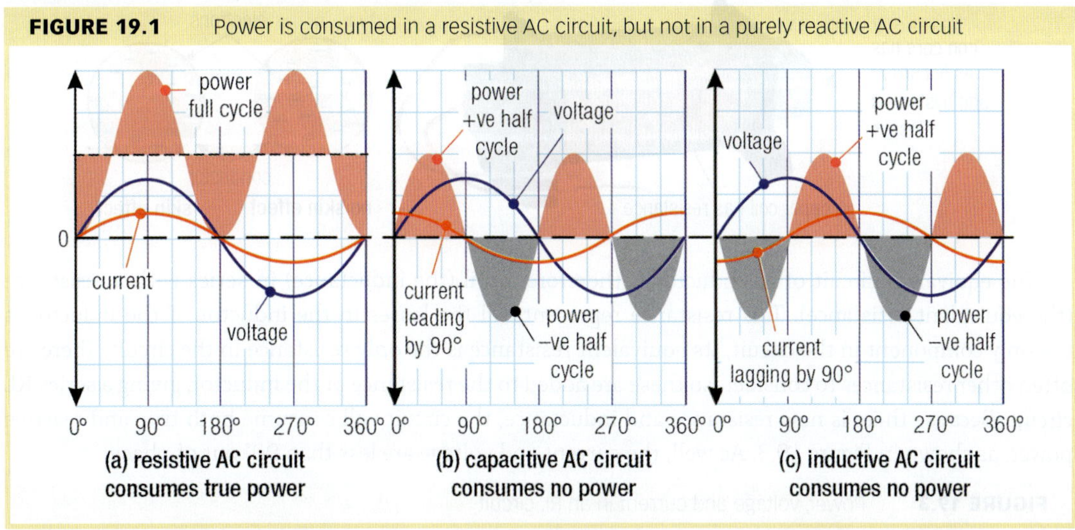

Most AC circuits have a combination of reactance and resistance. Therefore, this type of circuit will take true power (the resistive component), and reactive power (the reactive component). As the waveforms in Figure 19.1(b) and (c) show, reactive power is the result of current and voltage being 90° out of phase, while true power has both current and voltage in phase. Note that the power waveforms in Figure 19.1(b) and (c) are out of phase with each other by 180°.

Reactive power does no useful work, yet requires current flow. If this current is fed to the load over a long power transmission line, the resistance of the line will cause an I^2R loss, which is a true power loss in the form of heat. This means power is consumed (by the transmission line) to deliver a current to a reactive circuit that does no useful work. Therefore, keeping the reactive current in a circuit relatively low is necessary, as less losses occur.

To minimise reactive current, it's usual to counteract the inductive reactance of a load or circuit with a certain amount of capacitive reactance. However, if the capacitive reactance equals the inductive reactance, resonance can occur, so there is usually some reactance in a circuit. Reducing reactive current is called *power factor* correction, which is of importance to electrical power distributors and users.

19.2 Power in Reactive–Resistive Circuits

Most AC circuits contain a mix of reactance and resistance, for example, a motor which has inductance and resistance. A circuit could have a combination of resistance and capacitance, or a combination of resistance, inductance and capacitance. Power and the RL and RC circuit are described in this section.

Power in an RL circuit

In a purely inductive circuit, current lags the voltage by 90°. However, all practical inductors have resistance, which can be measured with an ohmmeter. This is the copper loss. There are also other losses which make the effective resistance of an inductor higher than its measured resistance. These include:

- magnetic hysteresis loss in the core
- eddy current losses in the core
- skin effect.

Figure 19.2 shows the losses in an inductor. Skin effect is the result of electrons flowing near the surface, or 'skin' of a conductor, instead of using the entire conductor. Because less conducting material is being used to carry the current, the conductor resistance increases. This effect increases with frequency. It can be ignored at power frequencies like 50 Hz, but becomes significant at higher frequencies. All these losses consume true power.

FIGURE 19.2 Inductors have various losses that consume true power

The equivalent circuit of an inductor is therefore a coil (the inductance) in series with a resistance (the equivalent resistance). This resistance represents all the losses in the inductor. If the inductor is the only component in the circuit, its equivalent resistance is the only resistance in the circuit. There are often other resistances to consider, so these are added to the resistance of the inductor, giving a series RL circuit. Because there is now resistance and inductance, the circuit will consume both true and reactive power, as shown in Figure 19.3. As well, the current and voltage are less than 90° out of phase.

FIGURE 19.3 Power, voltage and current in an RL circuit

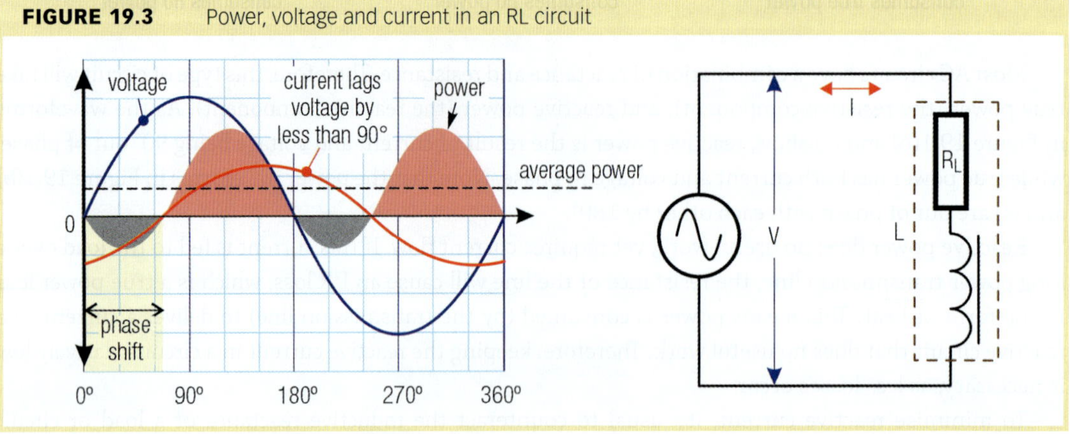

The waveforms in Figure 19.3 show that the power waveform is twice the frequency of the voltage, and that it has a larger positive cycle than a negative cycle. The power is zero when either the current or voltage is zero. The circuit is therefore consuming more power than it's returning, because of the resistance in the circuit.

Power in an RC circuit

Like an inductor, a capacitor has an equivalent resistance to represent losses. The main loss in a capacitor is that due to the dielectric, called the dielectric loss. In most cases, these losses are small and can be ignored. However, they become important in power capacitor installations because of the large values of current flowing in the capacitors. Even so, the losses are smaller compared to those of an inductor.

If resistance is added in series with a capacitor, the resistance takes true power, and the capacitor takes reactive power. So, like an RL circuit, in an RC circuit the power waveform will not be symmetrical around the zero axis, as both true and reactive power are taken by the circuit. This is shown in Figure 19.4.

FIGURE 19.4 Power, voltage and current in an RC circuit

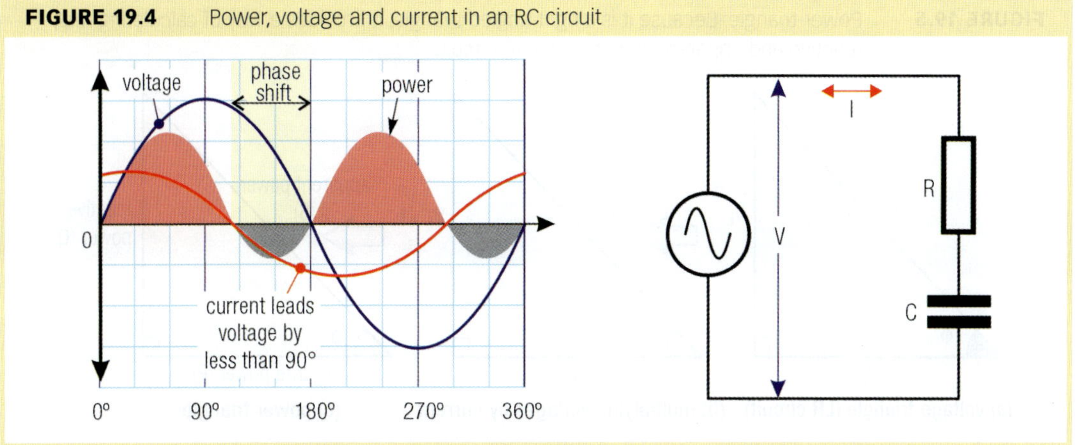

As with the RL circuit, the waveforms in Figure 19.4 show that:

- current and voltage are out of phase by less than 90°
- the power waveform is twice the frequency of the voltage
- power is zero whenever current or voltage are zero
- the power waveform is not central over the zero axis and therefore has an average value.

This again means that the circuit is consuming more power than it's returning, caused by the resistance in the circuit.

KEY POINTS...

- The resistive component of a circuit takes true power from the AC source. This power does work in the circuit, such as producing heat or mechanical motion. It is the power registered by a wattmeter or a kilowatt hour meter in a fuse box.
- Reactance in a circuit takes reactive power from the AC source. Reactive components are inductance and capacitance. Reactive power does no work in a circuit.
- Reactance in a circuit causes a phase shift between the current and the applied voltage, and increases the current flowing in the conductors between the source and the circuit.
- Power loss in a conductor equals I^2R, where I is the total current taken from the supply and R the conductor resistance. Therefore, reactance in an AC circuit causes a power loss between the source and the circuit. This means that the reactive component of a circuit should be kept as low as possible.

19.3 Power in Any AC Circuit

In any AC circuit, the power taken by resistance in the circuit is called true power. Reactive components (inductance and capacitance) in the circuit take reactive power. The phasor addition of these two power values gives a third value called the *apparent* power, which is the power that seems to be taken from the supply. These three types of power can be represented by a triangle, similar to the voltage triangles developed in Chapter 17.

Dividing all sides of the triangle by the circuit current gives an impedance triangle. Because the voltage drop across each component is proportional to its impedance (or resistance), the dimensions of the triangle remain unaltered. In a similar way, by multiplying each side of a voltage triangle by the circuit current, a power triangle is developed.

This is shown in Figure 19.5 in which the voltage triangle for a series LR circuit is developed into a power triangle. This triangle gives a way of determining the values of the three types of power associated with any RL or RC (resistive–reactive) AC circuit.

FIGURE 19.5 Power triangle. Because it is a right-angled triangle, it provides a way of calculating real, reactive and apparent power in any AC circuit.

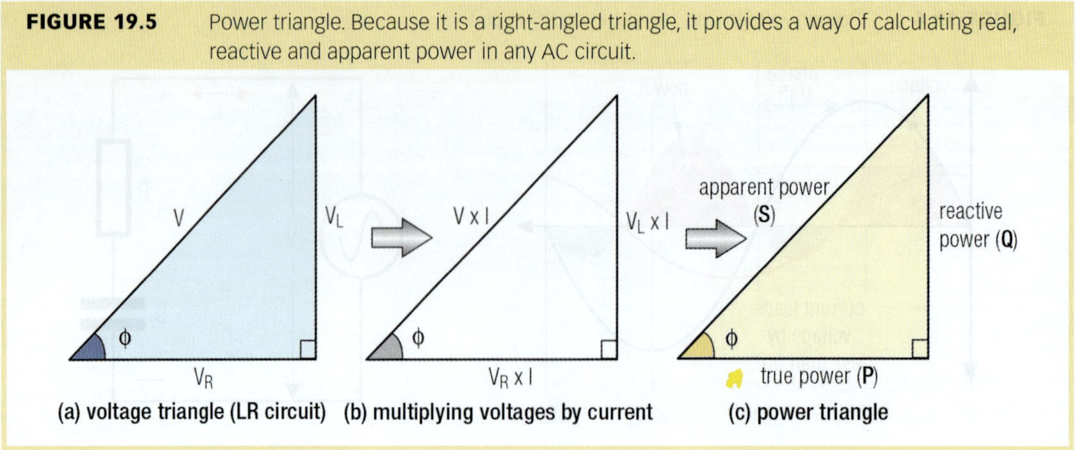

True power (P)

True power can be found by multiplying the current in the resistive component of a circuit by the voltage across the resistance. It has the symbol P and is measured in watts. So:

$P = V_R I_R$

where:

P = true power in watts

V_R = voltage across the resistance in volts (for a parallel circuit, V_R equals the supply voltage)

I_R = current through the resistance in amperes (for a series circuit, I_R equals the supply current).

For a circuit with reactance, V_R and I_R can be found with a phasor diagram or, depending on what is known about the circuit, with trigonometry. The triangles in Figure 19.5 show that the phase angle ϕ is between the two sides representing current (V_R) and the applied voltage (V). It is also the angle between the sides representing true power and apparent power.

In the power triangle, true power is the side adjacent to angle ϕ. The hypotenuse is the product of the total current and the supply voltage (V × I). Dividing the adjacent side of a right-angled triangle by its hypotenuse gives the cosine of the angle between these two sides. Therefore, the cosine function is used to find true power (P) when phase angle, supply voltage and circuit current are known. That is:

$P = VI \cos \phi$

where:

P = true power in watts

V = supply voltage in volts (RMS)

I = current taken from the supply in amperes (RMS)

ϕ = phase difference between the supply voltage and the total circuit current.

The term cos ϕ is called *power factor*. In a purely resistive circuit, ϕ is 0°, so the power factor is 1 (as cos 0° = 1). Therefore, in a purely resistive AC circuit, P = VI.

Other equations to find true power (P) in any circuit are:

$$P = I_R^2 R \text{ or } P = \frac{V_R^2}{R}$$

where:

I_R is the current in the resistance

V_R is the voltage across the resistance

R is the value of the resistance.

Also, by Pythagoras' theorem from the power triangle

$$P = \sqrt{S^2 - Q^2}$$

Reactive power (Q)

This is the product of the current flowing through the reactive component of the circuit and the voltage across the reactance. It is given the symbol Q and is measured in volt-amperes reactive (VA_R or var). That is:

$Q = V_X I_X$

where:

Q = reactive power in VA_R

V_X = voltage across the reactance in volts

I_X = current through the reactance in amperes.

If you know the phase angle (ϕ) between supply current I and supply voltage V, reactive power can also be found with this equation:

$Q = VI \sin \phi$

where:

ϕ = phase angle between the supply voltage and the supply current.

In a purely reactive circuit the phase angle ϕ between current and voltage is 90°, giving a power factor of zero, as cos 90° = 0. This means true power (P) must be zero, as $V \times I \times \cos 90° = 0$.

Reactive power does no work, and is sometimes called wattless power. It is present in any circuit that has inductance or capacitance. Other equations to find reactive power Q are:

$$Q = I_X^2 X \text{ or } Q = \frac{V_X^2}{X}$$

where:

X = inductive or capacitive reactance

I_X = current in the reactance

V_X = voltage across the reactance

$Q = \sqrt{S^2 - P^2}$ (Pythagoras' theorem) where S is the apparent power in volt amperes.

Apparent power (S)

Apparent power is the product of the supply voltage and the current taken from the supply. Its symbol is S and the measurement unit is the volt-ampere (VA).

$S = VI$

where:

S = apparent power in volt-amperes

V = supply voltage in volts (RMS)

I = supply current in amperes (RMS).

Other equations to find apparent power are:

$S = I^2 Z$ or $S = \frac{V^2}{Z}$ where Z = the impedance of the circuit

$S = \sqrt{P^2 + Q^2}$ Pythagoras' theorem, when true and reactive powers are known.

AC power sources (transformers and alternators) are usually rated in volt-amperes rather than watts, as this specifies the maximum current they can supply. For example, a 2000 VA 200 V alternator can supply no more that 10 A, regardless of power factor.

KEY POINTS...

- *True power* does work and is the power consumed by the resistive component of a circuit. Symbol is P, unit is the watt.
  $P = VI \cos \phi$
  $\cos \phi$ is the circuit's power factor
- *Reactive power* does no work and is the power associated with the reactive component of a circuit. Symbol is Q, unit is the volt-ampere reactive (VA_R).
  $Q = VI \sin \phi$
- *Apparent power* is the total power that appears to be consumed by a circuit. Symbol is S, unit is the volt ampere (VA).
  $S = VI$

EXAMPLE 19.1

An AC motor is taking 10 A from a 230 V 50 Hz supply. The phase angle between the motor current and its supply voltage is 20° lagging. Calculate the:

1 true power taken by the motor
2 reactive power taken by the motor
3 apparent power provided by the supply source

Solution

Values $I = 10$ A
$V = 230$ V
$\phi = 20°$

1 Calculate true power (P)

Equation $P = VI \cos \phi = 230 \times 10 \times 0.94$

Answer **$P = 2161.3$ W**

Keystrokes for Casio calculator

2 3 0 × 1 0 × cos 2 0 =

2 Calculate reactive power (Q)

Equation $Q = VI \sin \phi = 230 \times 10 \times 0.34$

Answer **$Q = 786.65$ VA_R**

Keystrokes for Casio calculator

2 3 0 × 1 0 × sin 2 0 =

3 Calculate apparent power (S)

Equation $S = VI = 230 \times 10$

Answer **$S = 2300$ VA**

If the true power and the reactive power are known, the apparent power taken by a circuit can be calculated with Pythagoras' theorem, as shown in Example 19.2.

EXAMPLE 19.2

A circuit is taking 2 kW of true power and 600 VA_R of reactive power. How much apparent power is the circuit taking?

Solution

Values $P = 2$ kW $= 2000$ W
$Q = 600$ VA_R

Equation $S = \sqrt{P^2 + Q^2} = \sqrt{2000^2 + 600^2}$

Answer **$S = 2088.1$ VA**

19.4 Power Factor

Power factor is a way of describing the phase difference between the supply voltage and supply current. It is often given the symbol λ (Greek letter *lambda*), and is found with the equation:

power factor (λ) = cos ϕ, where ϕ is the phase difference between the supply voltage and supply current.

As the power triangle in Figure 19.5(c) shows, power factor can be found from the ratio of true and apparent power. That is:

$$\cos\phi = \frac{\text{true power}}{\text{apparent power}} = \frac{P}{S}$$

where:

P = true power in watts

S = apparent power in volt-amperes.

The value of the power factor in a circuit, therefore, depends on the phase difference between the supply voltage and current. In a purely resistive circuit, the phase angle is zero, so the power factor is one. In a purely reactive circuit, the phase angle is 90°, so the power factor is zero. If the phase angle is 45°, the power factor is 0.71. In summary:

- an inductive circuit has a lagging power factor (current lags voltage)
- a capacitive circuit has a leading power factor (current leads voltage)
- a resistive circuit has a power factor of one (unity power factor, as current and voltage are in phase).

A power factor can only range between 0 and 1, but can be either leading or lagging. It's usual to make voltage the reference, so a lagging power factor means the current is lagging the voltage. In some cases power factor is multiplied by 100 to express the value as a percentage. A power factor of unity is therefore a percentage power factor of 100 per cent.

FYI

Power factor is denoted in various ways, including p.f., cos ϕ and λ

Effects of a low power factor

Power factor is of great importance in electrical power transmission and distribution. In general, the lower the power factor, the higher the current to supply the same true power. Example 19.3 shows the calculations to determine the apparent power needed to power a 2.2 kW motor from a 230 V supply when the power factor is 0.8.

EXAMPLE 19.3

An AC motor takes a current of 12 A from a 230 V 50 Hz supply at a lagging power factor of 0.8. Calculate the:

1 true power taken by the motor
2 phase angle ϕ between supply current and supply voltage
3 apparent power provided by the supply source
4 reactive power taken by the motor.

Solution

Values $I = 12$ A
$V = 230$ V
p.f. $= 0.8 = \cos\phi$

Equation 1 $P = VI\cos\phi = 230 \times 12 \times 0.8$

Answer (1) **P = 2208 W = 2.2 kW**

Equation 2 $\cos\phi$ = power factor (λ) = 0.8. $\phi = \cos^{-1}0.8$

Answer (2) **ϕ = 36.87°**

Equation 3 $S = VI = 230 \times 12$

Answer (3) **S = 2760 VA = 2.76 kVA**

Keystrokes for Casio calculator to find sin ϕ

sin 3 6 • 8 7 =

Equation 4 $Q = VI\sin\phi = 230 \times 12 \times 0.6$

Answer (4) **Q = 1656 VA_R = 1.66 kVA_R**

Example 19.3 shows that the supply source has to provide 2.76 kVA of apparent power to supply a load that only requires 2.2 kW of true power. The power values from Example 19.3 are shown as a power triangle in Figure 19.6.

FIGURE 19.6 Power triangle for Example 19.3

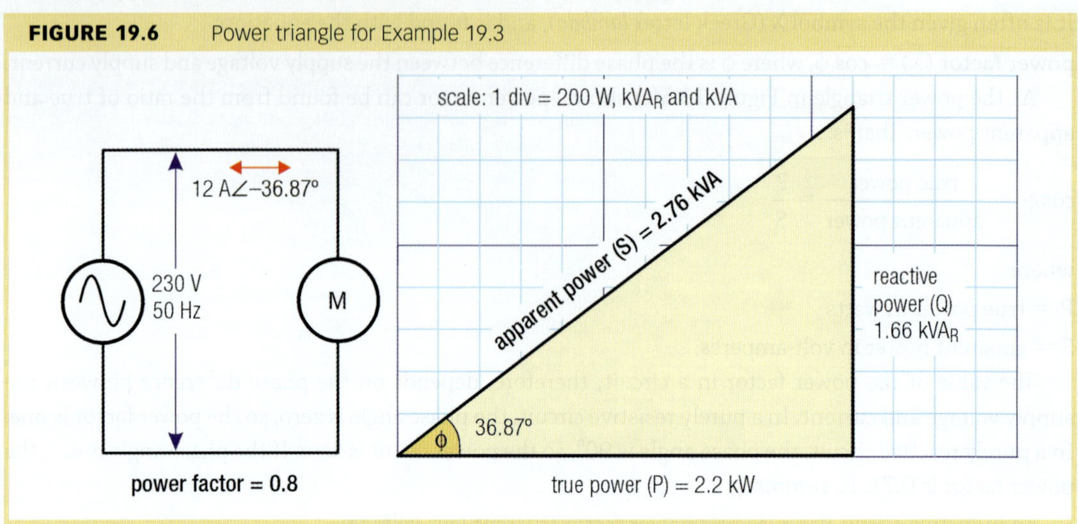

The apparent power taken by the motor in Example 19.3 is 2.76 kVA, requiring 12 A to be supplied by the source. The current doing 2.2 kW of useful work equals the true power (P) divided by the supply voltage (V). It also equals the supply current (I) multiplied by the power factor. In Example 19.3, the power factor is 0.8, so the current doing useful work is 12 × 0.8 = 9.6 A. Example 9.4 shows the effect of increasing the power factor to 0.9 for the same motor as in Example 19.3.

EXAMPLE 19.4

For the 2.2 kW motor in Example 19.3, if the power factor is 0.9, calculate the:

1 supply current (I)
2 apparent power (S)
3 phase angle ϕ
4 reactive power (Q)

Solution

Values P = 2.2 kW or 2208 W
V = 230 V
p.f = 0.9 = cos ϕ

Equation 1 $P = VI\cos\phi$

$$I = \frac{P}{V\cos\phi} = \frac{2208}{230 \times 0.9}$$ (after transposing Equation 1)

Answer (1) **I = 10.67 A**

Equation 2 $S = VI = 230 \times 10.67$

Answer (2) **S = 2453 VA or 2.45 kVA**

Equation 3 $\phi = \cos^{-1}$ p.f. $= \cos^{-1} 0.9$

Answer (3) **ϕ = 25.84°**

Equation 4 $Q = VI\sin\phi = 230 \times 10.67 \times \sin 25.84$

Answer (4) **Q = 1068 VA$_R$**

Effects of a low power factor

The power triangle for this improved power factor is in Figure 19.7, and shows that as the power factor increases, the apparent power becomes closer to the true power. The supply current is now 10.7 A compared to 12 A for a motor with a power factor of 0.8.

FIGURE 19.7 Power triangle for Example 19.4 when the motor power factor is 0.9

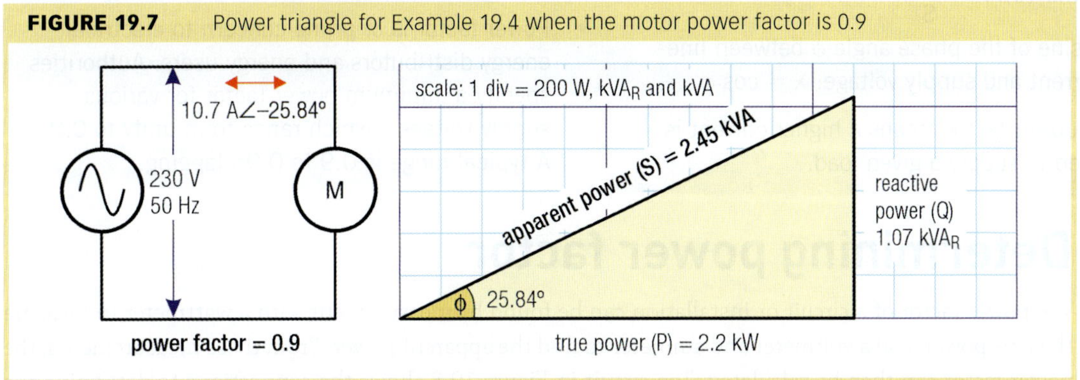

A low power factor means more supply current is needed for the same true power to be delivered to a load. This higher value of current requires:

- larger conductors (greater cross-sectional area)
- larger supply transformers
- higher rated fuses
- higher rated switch gear.

A low power factor causes higher losses in conductors because of their resistance. The power loss is generally known as an I^2R loss, in which doubling the current causes four times the power loss in the conductors. The voltage drop across the conductors is another loss, resulting in reduced voltage at the load.

There are also higher generating costs, which are passed on to the consumer. An industry purchasing power from an energy distributor could be paying for each volt-ampere of energy, not each watt of energy. It therefore makes commercial sense for that industry to keep its power factor close to unity, so every volt-ampere purchased is converted to true power.

Supply authorities, including the national Australian Energy Market Commission (AEMC) and state authorities, specify a range of power factors that must not be exceeded. The allowable range depends on the supply voltage. The Service and Installation Rules of NSW specify a power factor of at least 90 per cent, or 0.9 (lagging). Essential Services Commission (ESC) in Victoria specifies a minimum of 0.75 (lagging) or 0.8 (leading) for a supply voltage of less than 6.6 kV, for loads up to 100 kVA. Ergon in Queensland allows a power factor no less than 0.8 (lagging but not leading) for a supply voltage less than 1 kV.

Causes of a low power factor

A low power factor is typically due to inductive reactance in a circuit, which can come from:

- electric motors that are lightly loaded
- transformers that are lightly loaded
- the inductance (ballast) in series with low and high intensity discharge lamps, such as fluorescent lamps (low intensity), mercury-vapour and metal-halide lamps (high intensity).

Capacitive reactance is usually not a factor that causes a low power factor. Instead, as explained further on, capacitance is used to counteract the effect of inductive reactance. An exception is the capacitance of a long power transmission line, which under certain circumstances can cause a leading power factor. It can also cause the voltage at the load end to be higher than the voltage source (called the Ferranti effect).

KEY POINTS...

- The power factor (λ) for any AC circuit is the:
  - ratio of the true power (P) to apparent power (S): $\lambda = \frac{P}{S}$
  - cosine of the phase angle ϕ between line current and supply voltage: $\lambda = \cos\phi$
- A low power factor means a higher current is required to supply a given load.
- A low power factor causes power losses in power lines, and requires increased ratings of switch gear and protection equipment.
- Power factor is of prime concern to electrical energy distributors and energy users. Authorities specify a minimum power factor for various supply voltages, which range from unity to 0.8. A typical range is 0.9 to 0.95 lagging.

Determining power factor

The power factor of a circuit or installation can be found by measurement with a wattmeter to measure the true power, and a voltmeter and ammeter to find the apparent power. From these measurements, the power factor can then be calculated. The circuit in Figure 19.8 shows the connections to determine the power factor of an AC motor or any reactive load. The operation and use of a wattmeter is explained in Chapter 25.

FIGURE 19.8 Circuit to measure power factor of a motor, where the wattmeter measures true power and the ammeter/voltmeter measure apparent power

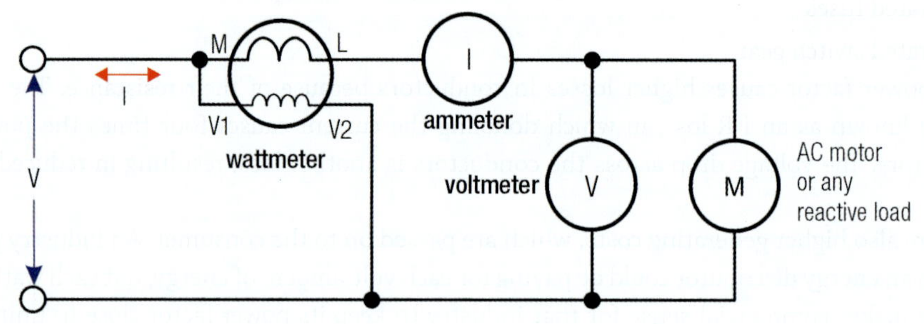

The power factor is calculated from the readings of the meters:

$$\text{power factor } (\lambda) = \frac{\text{true power (watts)}}{\text{volt-amperes (VA)}} = \frac{P}{S}$$

EXAMPLE 19.5

A motor connected to a 230 V 50 Hz supply takes a current of 3 A as measured by an ammeter. A wattmeter measures 500 W. What is the power factor?

Solution

Values $V = 230\ \text{V}$

$I = 3\ \text{A}$

$P = 500\ \text{W}$

Equation 1 $S = VI$

$S = 230 \times 3$

$S = 690\ \text{VA}$

Equation 2 $\text{power factor } (\lambda) = \frac{P}{S} = \frac{500}{690}$

Answer **power factor = 0.73**

Power factor can also be measured with a power factor meter, which has four connections like a wattmeter. A single-phase analog power factor meter has a fixed coil that carries the load current, and crossed coils that are connected to the load voltage. There is no spring to restrain the moving system, which takes a position to indicate the angle between the current and voltage. The scale can be marked in degrees or in power factor. A power factor meter is only accurate at one frequency. An example is shown in Figure 19.9.

FIGURE 19.9 Typical analog power factor meter, in which a factor of 1 is mid scale

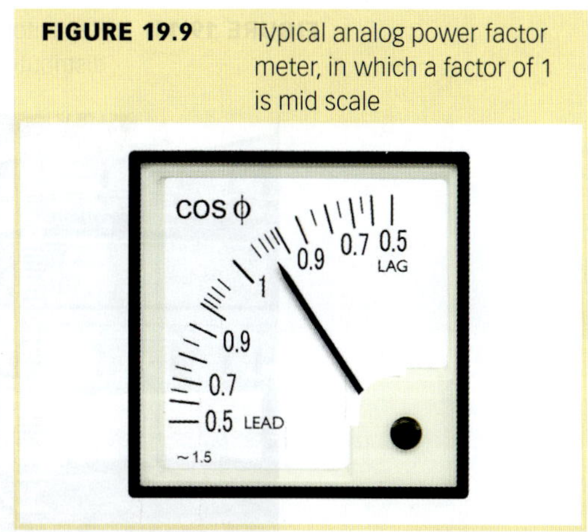

19.5 Power Factor Correction

Maintaining a high power factor can be achieved in various ways. One method is to ensure motors and transformers in an installation are correctly rated so they operate at their rated capacity. An electric motor running at full load appears more resistive, which keeps the power factor high. If the motor is lightly loaded, it appears more inductive, lowering the power factor. The same applies to a transformer.

Another method is to add capacitive reactance to the circuit, assuming the circuit is inductive, which is typically the case. A common situation is the inductance of a bank of fluorescent lamps or any form of gas discharge lamp, due to the ballast in each light fitting. To counteract the inductance, a capacitor is connected as shown in Figure 19.10. A 36 W single tube fitting could have a capacitor of around 4.5 μF. The reactance of the ballast keeps the current in the tube at the correct value. Without it, the tube would take excessive current and be destroyed.

FIGURE 19.10 Single tube fluorescent lamp with a power factor correction capacitor connected across its terminals

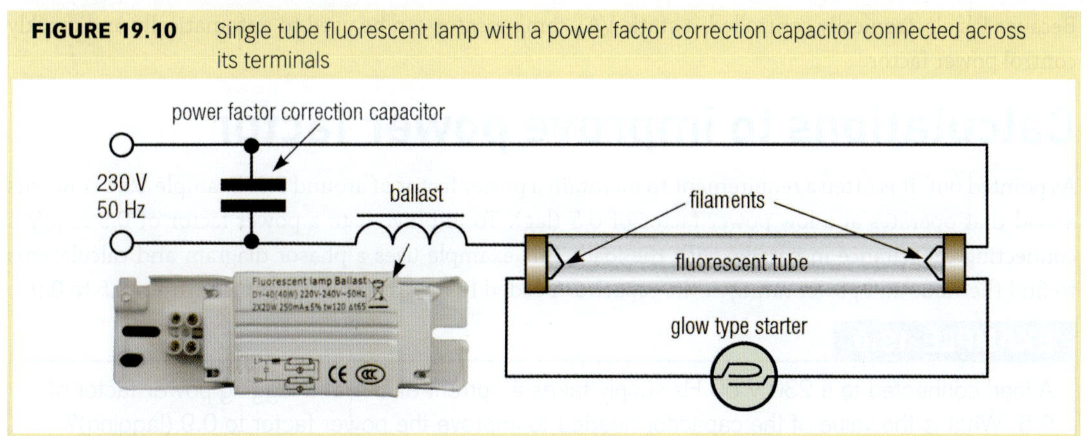

A common method of controlling the power factor of power being transmitted over long high voltage transmission lines is by connecting capacitor banks to the lines. Capacitor banks are usually located in a substation and are switched in or out by a central control authority. For example, the large capacitor bank shown in Figure 19.11 connects to the main electrical power grid. The smaller bank connects to a localised 36 kV power distribution system.

Capacitor banks are rated by their reactive power, written as a kVA_R or MVA_R rating. The capacitance of the bank is not usually given on the nameplate, but is of a value that, at the rated voltage, causes a certain value of reactive current to flow. This current value, when multiplied by the rated voltage, gives the reactive power rating of the bank.

Large installations, such as a factory, or home units with more than 10 metered customers are required to control the power factor of their installation so it is never less than a specified value, such as 0.9 lagging. While capacitors or capacitor banks are often used, other methods that provide better control are also used.

SUSTAINABILITY

A high power factor means lowest electrical losses, greater network and generator efficiency and less impact on the environment

FIGURE 19.11 Capacitor banks are connected at various points in the electrical power transmission and distribution system to help control power factor

36 kV 18 MVA_R capacitor banks

132 kV 96 MVA_R capacitor bank

One method is with a synchronous motor, which is explained in Chapter 24. This type of motor has the interesting characteristic that it takes a lagging current under certain conditions, and a leading current under a different set of conditions. When used for power factor correction, this type of motor is sometimes called a synchronous capacitor. It has the advantage that the amount of leading reactive current created by the motor is easily and quickly adjusted as required.

Another method is with a static VA_R compensator. It is called static as it has no moving parts and comprises a combination of capacitance and inductance and solid-state switches that are electronically controlled. Coarse control of power factor is achieved by switching various values of capacitors in and out of circuit. Fine control is achieved by switching various amounts of inductance in and out of circuit. Because it's electronically controlled, a static VA_R compensator can be used to automatically and quickly control power factor.

Calculations to improve power factor

As pointed out, it is often a requirement to maintain a power factor of around 0.9. Example 19.6 concerns a load that operates at a low power factor of 0.5 (lag). To improve it to a power factor of 0.9 requires connecting capacitance in parallel with the load. This example uses a phasor diagram and calculations to find the value and power rating of the capacitor needed to increase the power factor from 0.5 to 0.9.

EXAMPLE 19.6

A load connected to a 230 V 50 Hz supply takes a current of 5 A at a lagging power factor of 0.5. What is the value of the capacitor needed to improve the power factor to 0.9 (lagging)? What must its VA_R rating be?

Solution

Find the phase angle for both power factors where:

$\lambda_1 = 0.5$ and $\lambda_2 = 0.9$

$\phi_1 = \cos^{-1} 0.5 = 60°$ (lagging)

$\phi_2 = \cos^{-1} 0.9 = 26°$ (lagging)

Draw a phasor diagram as shown in Figure 19.12. The steps are:

1 Draw the reference (V) and draw to scale a phasor for $I = 5$ A at 60° lagging the reference.
2 Draw a vertical line from and at a right angle to the reference to the tip of the phasor for I.
3 Draw a phasor to represent the current at λ_2. This is a line drawn from the origin to the vertical line at an angle of 26° ($I\angle{-26°}$ in Figure 19.12).
4 Measure the distance between the tips of the two phasors just drawn to determine the capacitor current (I_C) needed to correct the power factor. From the diagram I_C = **3.1 A**.

FIGURE 19.12 Phasor diagram for Example 19.6. Draw the phasor for the line current at –60°. Draw the phasor for the new line current at -26° to the vertical line from the tip of the first phasor. The phasor between their tips gives the required capacitor current of 3.1 A.

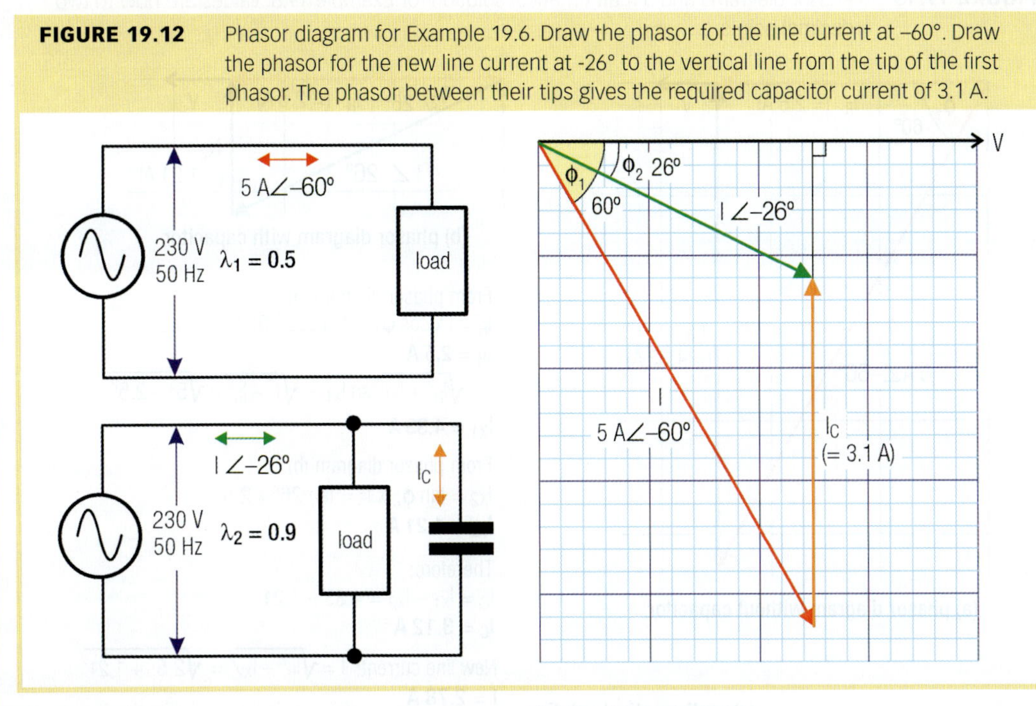

Calculations:

1 Calculate the value of X_C that passes a current of 3.1 A at 230 V 50 Hz.

Equation $X_C = \frac{V}{I_c} = \frac{230}{3.1}$

$X_C = 74.2\ \Omega$

2 Find the value of C that has an X_C of 74.2 Ω at 50 Hz.

Equation $X_C = \frac{1}{2\pi fC}$

$\therefore C = \frac{1}{2\pi fX_C} = \frac{1}{6.28 \times 50 \times 74.2}$

Answer **C = 42.9 μF**

3 Find the VA_R rating of the capacitor

Equation $VA_R = V \times I_C$

$VA_R = 230 \times 3.1$

Answer **VA_R rating = 713 VA**

The new line current taken by the circuit can be determined from the phasor diagram, which by measurement of green phasor I∠–26° to one decimal point is 2.8 A. Prior to adding the capacitor, the in-phase current (the current doing the work) was 2.5 A, which is found by multiplying the line current (5 A) by the power factor (0.5). Adding the capacitor gives a line current of 2.8 A, so the in-phase current is 2.8 A multiplied by the new power factor of 0.9, which also gives 2.5 A.

Adding the capacitor has reduced the line current from 5 A to 2.8 A. Adding more capacitance to further increase the power factor will reduce the supply current even more. If the power factor is increased to 1, the inductive and capacitive currents cancel, creating a resonant circuit at 50 Hz. Depending on circuit values, this should be avoided because a high circulating current between the load and the added capacitance could occur.

Figure 19.13 uses Pythagoras' theorem and trigonometry to solve Example 19.6.

FIGURE 19.13 Phasor diagrams and a mathematical solution for Example 19.6. Values are now to two decimal points.

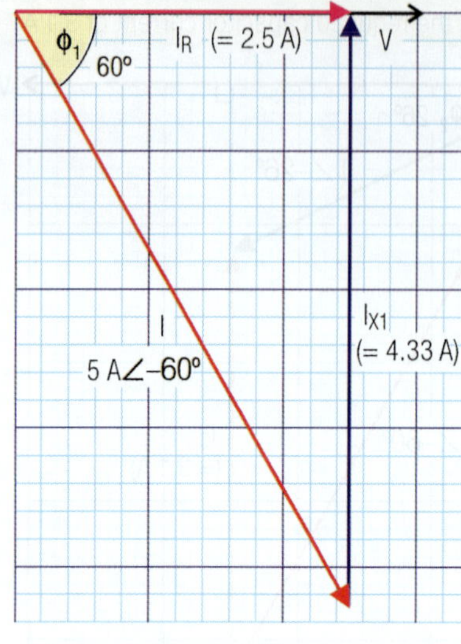

(a) phasor diagram without capacitor

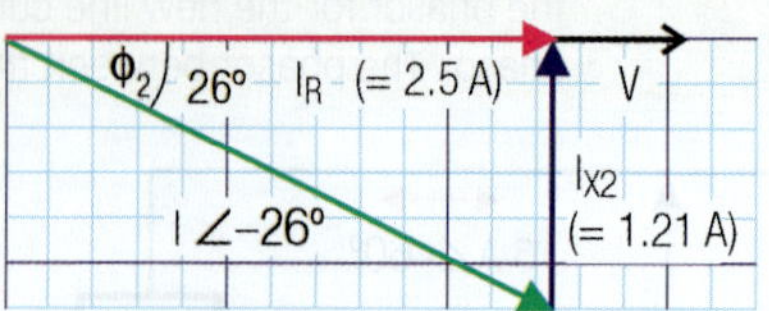

(b) phasor diagram with capacitor

From phasor diagram (a):

$I_R = I \times \cos \phi_1 = 5 \times \cos^{-1} 60° = 5 \times 0.5$

$\mathbf{I_R = 2.5\ A}$

$I = \sqrt{I_R^2 + I_{X1}^2}$ so $I_{X1} = \sqrt{I^2 - I_R^2} = \sqrt{5^2 - 2.5^2}$

$\mathbf{I_{X1} = 4.33\ A}$

From phasor diagram (b):

$I_{X2} = \tan \phi_2 \times I_R = \tan 26° \times 2.5$

$\mathbf{I_{X2} = 1.21\ A}$

Therefore:

$I_C = I_{X1} - I_{X2} = 4.33 - 1.21$

$\mathbf{I_C = 3.12\ A}$

New line current: $I = \sqrt{I_R^2 + I_{X2}^2} = \sqrt{2.5^2 + 1.21^2}$

$\mathbf{I = 2.78\ A}$

(c) mathematical solution

CHAPTER SUMMARY

- True power can be measured with a wattmeter. Apparent power can be measured with an ammeter and voltmeter, where apparent power is the product of the ammeter and voltmeter readings.
- Capacitors, synchronous motors running at a leading power factor and static VAR compensators are methods used to control power factor to 0.9 to 0.95 in power supply systems.
- Power factor correction capacitors are rated by the reactive power they can handle. Reactive power is the product of the applied voltage and the current flowing in the capacitor.
- True power P does work and is measured in watts. It is due to current flow in the resistive part of a load.
- Current in the resistive part of a load is in phase with the supply voltage.
- Equations to find true power:
  - $P = VI \cos \phi$ ($\cos \phi$ = power factor)
  - $P = V_R I_R$ where $I_R = I \times \cos \phi$
  - $P = \sqrt{S^2 - Q^2}$
- Power factor: p.f. or $\lambda = \cos \phi$ where ϕ is the phase angle between supply voltage and line current.
  - $\lambda = \cos \phi = \dfrac{\text{true power}}{\text{apparent power}} = \dfrac{P}{S}$
  - $\cos \phi = \dfrac{R}{Z}$
  - $\phi = \cos^{-1}$ of power factor
- Reactive power Q does no work, and is measured in volt-ampere reactive VA_R.
  - $Q = VI \sin \phi$
  - $Q = \sqrt{S^2 - P^2}$
- Apparent power S is measured in volt-ampere (VA).
  - $S = VI$
  - $S = \sqrt{P^2 + Q^2}$

REVIEW EXERCISES

Check your answers in the back of the book.

1 An inductive load connected to a 230 V 50 Hz AC circuit takes 15 A from the supply. The phase angle between the load current and supply voltage is 27° lagging. Calculate:
   a the true power
   b the reactive power
   c the apparent power taken by the load
   d the circuit power factor.

2 A 230 V 50 Hz contactor coil has a resistance of 15 Ω and takes a current of 3.56 A. Find the power factor and phase angle of the current supplying the coil.

3 An AC power source is delivering 32 kW of true power and 12 kVA_R of reactive power to a load.
   a How much apparent power is the source providing?
   b What is the power factor?

4 An AC motor on full load takes a current of 26 A from a 400 V 50 Hz supply at a lagging power factor of 0.85. Calculate the:
   a true power taken by the motor
   b phase angle between supply current and voltage
   c apparent power provided by the supply source
   d reactive power taken by the motor.

5 A wattmeter measures a value of 12.5 kW taken by a load connected to a 400 V supply. An ammeter measuring the current to the load registers 36.25 A.
   a Determine the power factor of the circuit
   b Calculate the phase angle between the current and supply voltage.

6 A 400 V 50 Hz supply is providing power to two industrial loads. Load 1 is purely resistive and takes 15 kW of true power. Load 2 takes 10 kW of true power at a power factor of 0.85. Determine the supply current and power factor when:
   a only load 1 is on
   b only load 2 is on
   c both loads are on (need a phasor diagram).

7 An electrical circuit is taking 15 kW of true power and 6 kVA_R of reactive power.
   a How much apparent power is the supply source providing?
   b What is the power factor?

8 A 12 kV power distribution system in a factory is supplying a current shown by an ammeter of 450 A, at a lagging power factor of 0.75.
   a How much apparent power is the factory consuming?
   b How much true power is the factory using?

9 An ideal 0.8 H inductor is connected to a 400 V 50 Hz supply.
   a How much true power does it consume?
   b How much apparent power does the supply source need to provide?
   c What is the circuit's power factor?

10 A load connected to a 600 V 50 Hz supply takes a current of 25 A at a lagging power factor of 0.6.
   a Determine the value and reactive power rating of a capacitor needed to improve the power factor to 0.95 (lagging).
   b The current taken by the load after the capacitor is connected.

ONLINE RESOURCES

COMPLETE WORKSHEET NINETEEN

Check with your instructor for worksheets on this chapter.

CHAPTER 20

THREE-PHASE POWER

So far we have looked at single-phase AC circuits, where the electrical supply is provided by two lines. However, most electrical power is multiphase, typically having three separate phases. The first three-phase power transmission took place in the early 1890s and is attributed to Nicola Tesla who sold his patents to Westinghouse in the USA. The simplicity of a three-phase motor is one significant advantage of three-phase power; others include improved efficiency in power transmission. This chapter presents the principles of three-phase power, and explains how to determine voltages, currents, phase angles and power dissipation in three-phase loads.

CHAPTER OUTLINE

20.1 Introduction

Chapter 13 described the DC generator, which is actually an alternator with modifications to make it produce a direct current. In principle, an alternator has a coil rotating in a magnetic field with slip rings to connect the coil to its electrical load. Alternators are covered in more detail in Chapter 24, and are the prime method of generating electrical power. As previously explained, an alternator produces a sinusoidal voltage, which is an important characteristic. The sinewave is a fundamental waveform that can be transformed from one voltage to another (with a transformer) without its shape being changed.

Figure 20.1 shows a basic alternator as a 3D drawing and also as a 2D illustration. A coil is rotating in a fixed magnetic field, and the output is a sinewave. This is a single-phase alternator and supplies its load over two lines.

FIGURE 20.1 A basic single-phase alternator has two output terminals and produces a sinusoidal voltage

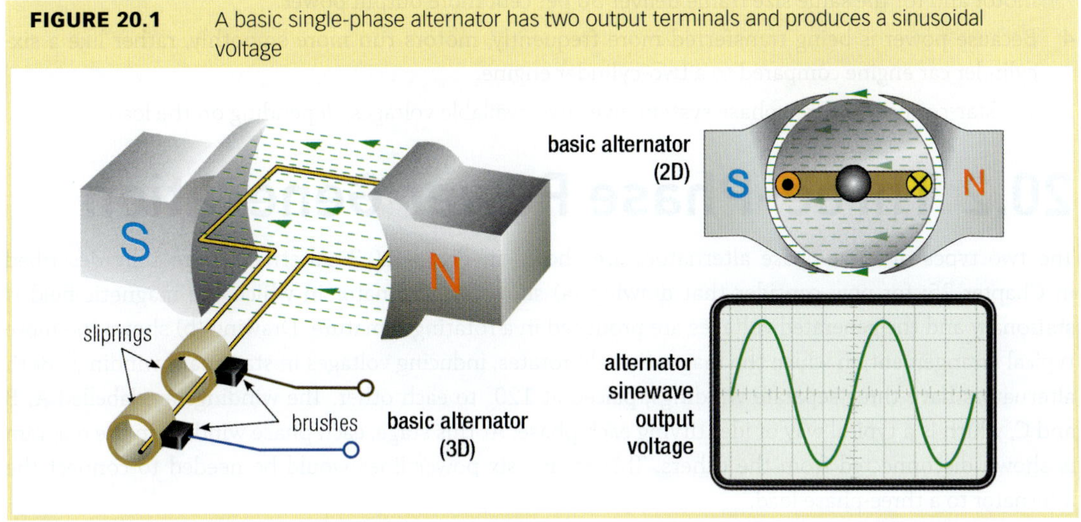

The power produced in an AC circuit has a frequency twice the frequency of the voltage, so at 50 Hz the power cycle frequency is 100 Hz. That is, a single-phase alternator provides power 'pulses' to its load 100 times a second. This can be compared to a single cylinder petrol engine, where it's obvious that the more cylinders, the smoother the power transfer. The first attempts at achieving smoother electrical power transfer were with a two-phase system.

While rarely used today, it formed the basis of early power generation. An advantage of two-phase compared to single-phase power is that a two-phase motor is self-starting and is therefore simpler to construct. As well, a two-phase motor (and alternator) gives around 25 per cent more power for the same size single-phase machine. Figure 20.2 shows a simple two-phase alternator, in which two coils are placed at 90° to each other and are rotating in the same magnetic field. Each coil supplies its output through its own set of slip rings.

FIGURE 20.2 A two-phase alternator has two windings physically displaced by 90° which means the two generated phases are 90° out of phase.

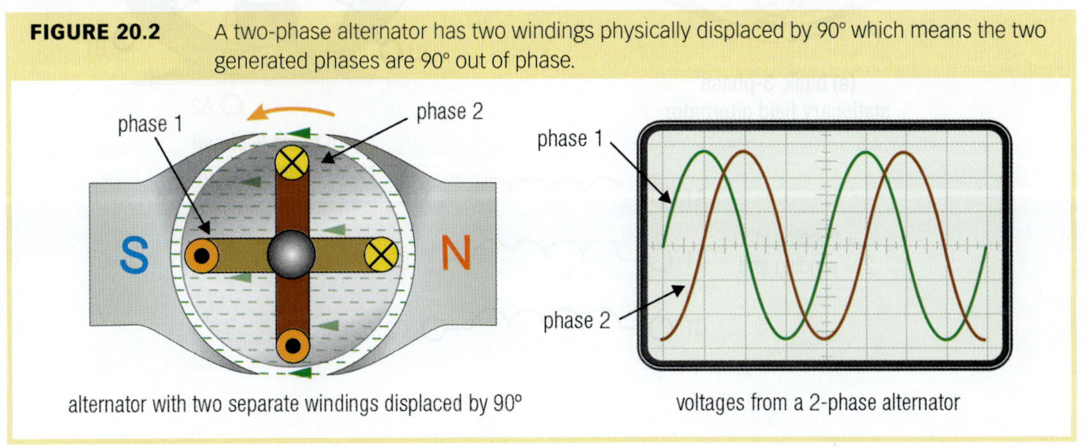

Most electrical power generation and distribution systems are three-phase. A multiphase system with more than three phases gives small improvements in efficiency, but the additional costs make the three-phase system the most economical. Its advantages are:

1 The available output per kilogram of iron in a three-phase alternator is over 50 per cent more compared to the same frame size single-phase alternator. Economically, when compared to a single- or two-phase system, a three-phase system gives the greatest output, and yet only requires one more conductor.
2 It allows smaller power transmission lines, transformers and associated equipment compared to the same amount of single-phase power being transmitted.
3 A three-phase system provides a rotating magnetic field, which means three-phase motors are self-starting, and therefore simpler to construct. They deliver a more constant torque than a single-phase motor and for the same size frame deliver 30 per cent more output power.
4 Because power is being transferred more frequently, motors run more smoothly, rather like a six-cylinder car engine compared to a two-cylinder engine.
5 The star-connected three-phase system gives two available voltages, depending on the load.

20.2 Three-Phase Power Generation

The two types of three-phase alternators are shown in Figure 20.3. Alternators are fully described in Chapter 25; for now consider that drawing (a) shows an alternator in which the magnetic field is stationary and the generated voltages are produced in a rotating armature. Drawing (b) shows the more typical arrangement, in which the magnetic field rotates, inducing voltages in stationary windings. Both alternators have three separate windings, placed at 120° to each other. The windings are labelled A, B and C, which is a typical way of identifying each phase. At this stage, each phase winding in the diagram is shown disconnected from the others. This means six power lines would be needed to connect the alternator to a three-phase load.

FYI

In Australia and New Zealand, the standard conductor colours are red for A phase, white for B phase and blue for C phase

FIGURE 20.3 A three-phase alternator has three windings labelled A, B and C physically displaced by 120°

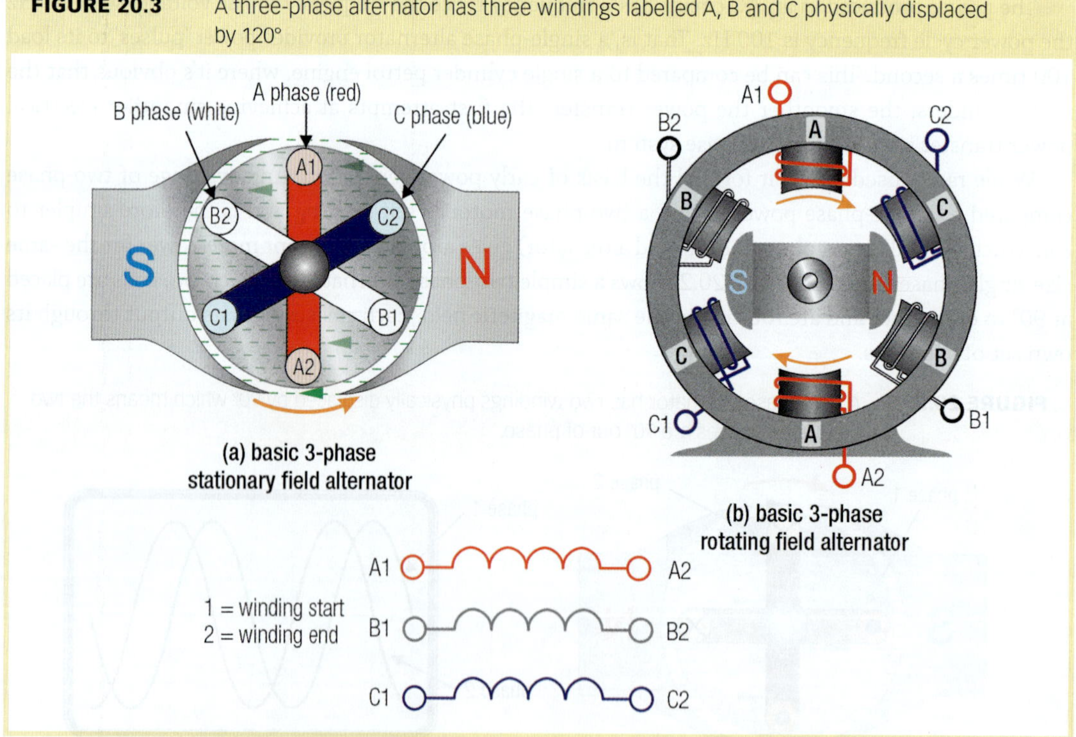

When either alternator in Figure 20.3 is running, each winding produces a sinusoidal voltage. Because the windings are displaced by 120°, each voltage is also displaced by 120°, giving the waveforms shown in Figure 20.4. That is, there's a 120° phase difference between each phase voltage. The filled in (yellow) section of the waveforms illustrates the power transfer characteristic of a three-phase system. The voltage of each phase is the same, as each phase winding in the alternator is identical.

FIGURE 20.4 Voltage waveforms produced by a three-phase alternator, with phase A leading phase B by 120°, which leads phase C by 120°

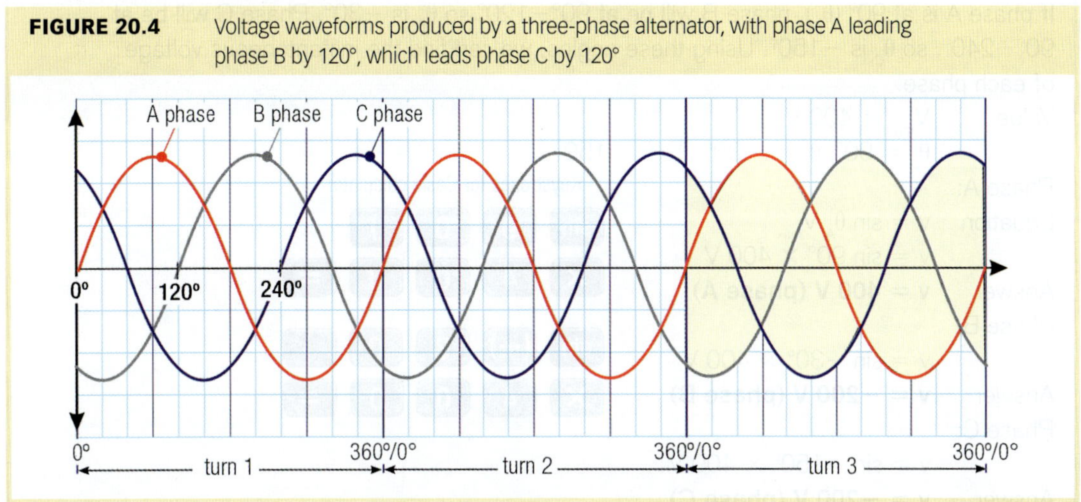

Figure 20.5 shows the phasor diagram of the three voltages generated by a three-phase alternator, in which each phasor has a phase difference of 120°. In this case, phase A is the reference phasor. In the diagram, the phase sequence is phase A, followed by phase B, followed by phase C. That is, phase A reaches a maximum voltage, and 120° later in the cycle phase B reaches a maximum value, with phase C reaching its maximum after a further 120°. This is the called the phase sequence, which in this case is A-B-C.

FIGURE 20.5 Phasor diagram of a three-phase alternator that is rotating anticlockwise

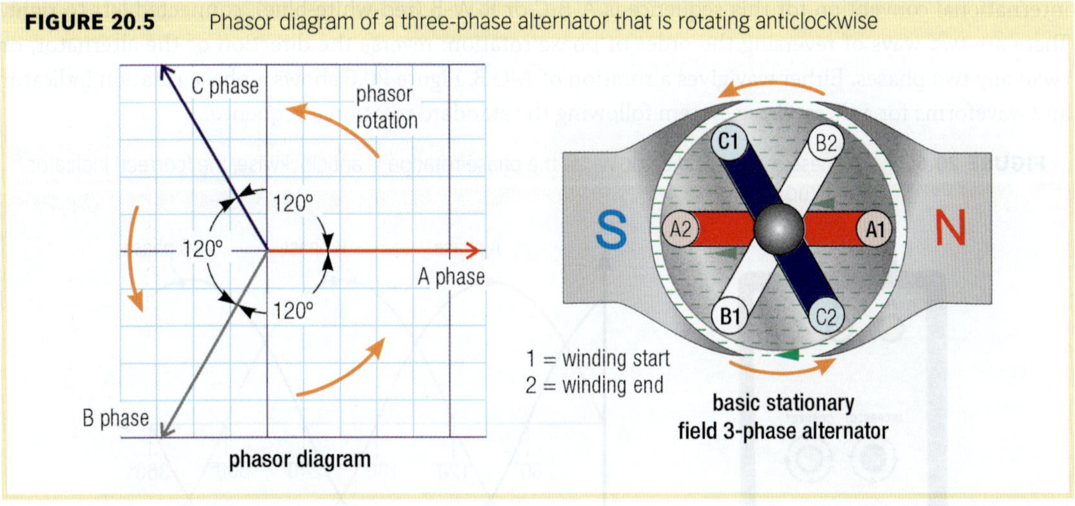

The phasor diagram in Figure 20.5 shows the three voltage phasors and their phase relationships. However, unlike an oscilloscope, a phasor diagram cannot show how the voltages change as the alternator rotates. For example, if phase A has a maximum voltage of 400 V, what are the voltages of the other two phases when phase A is at its maximum positive value? The waveforms in Figure 20.4 show that when phase A is maximum positive, the other two phase voltages are negative, and equal in value. Example 20.1 shows this by calculating the instantaneous voltages for the three phasors when phase A is at 90°.

EXAMPLE 20.1

A three-phase alternator produces a maximum voltage of 400 V. Determine the voltage of each phase when phase A is at a rotational angle of 90°.

Solution

If phase A is at 90° (θ_A), phase B will be at 90°−120° so θ_B is −30°. Phase C will be at 90°−240°, so θ_C is −150°. Using these angles, we can find the instantaneous voltage of each phase.

Value $V_{max} = 400$ V

$\theta_A = 90°$, $\theta_B = -30°$, $\theta_C = -150°$

Phase A:

Equation $v = \sin \theta_A \, V_{max}$

$v = \sin 90° \times 400$ V

Answer **v = 400 V (phase A)**

Keystrokes for Casio calculator

sin 9 0) × 4 0 0 =

Phase B:

$v = \sin -30° \times 400$ V

Answer **v = −200 V (phase B)**

sin (−) 3 0) × 4 0 0 =

Phase C:

$v = \sin -150° \times 400$ V

Answer **v = −200 V (phase C)**

Phase sequence

The phasor diagram in Figure 20.5 shows the phase sequence of each phase voltage depicted as rotating in an anticlockwise direction, which is the standard in Australia and New Zealand. The international convention for this sequence is A-B-C or R-W-B (red-white-blue), connected left to right. There are two ways of reversing the order of phase rotation: reverse the direction of the alternator, or swap any two phases. Either way gives a rotation of A-C-B. Figure 20.6 shows a phase rotation indicator and waveforms for a three-phase system following the standard rotational sequence.

FIGURE 20.6 A phase rotation indicator. When the phase rotation is anticlockwise, the 'correct' indicator lights up.

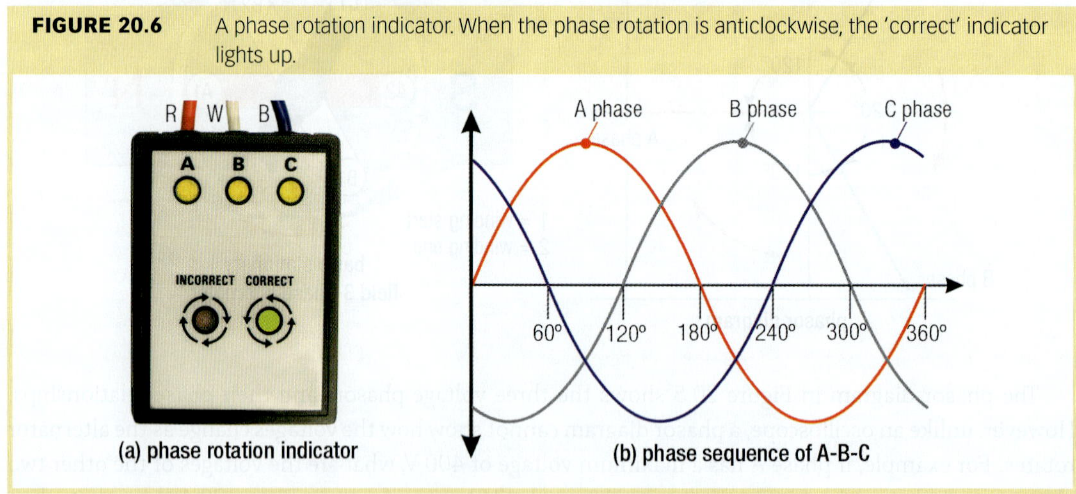
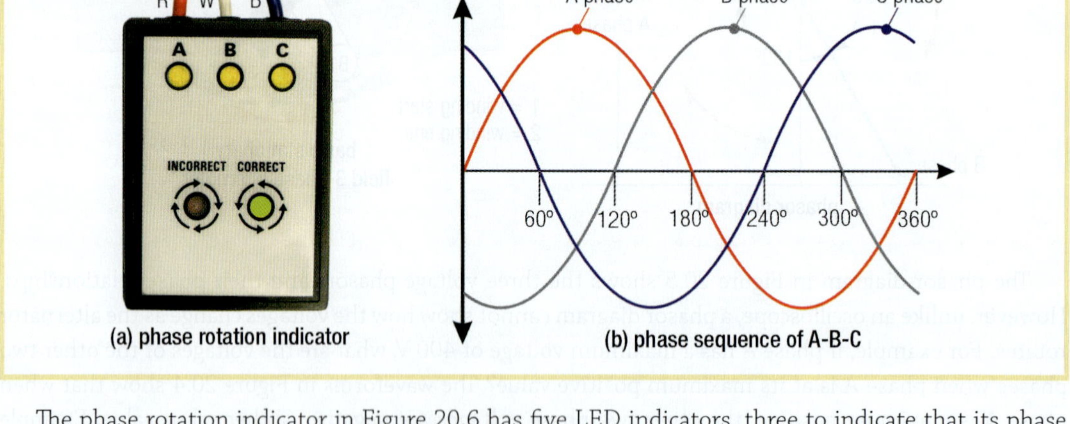

(a) phase rotation indicator

(b) phase sequence of A-B-C

FYI

When disconnecting a three-phase motor, take note of the connections to ensure the motor runs in the correct direction when reconnected

The phase rotation indicator in Figure 20.6 has five LED indicators, three to indicate that its phase is 'live', and one each to indicate correct (anticlockwise) or incorrect (clockwise) phase rotation. If an installation's three-phase supply is to be disconnected and later reconnected, the phase sequence should be recorded before disconnection. It doesn't matter what phase rotation is indicated, it's important to reconnect it with the same phase rotation. Otherwise, all three-phase motors in the installation will run backwards. Other phase-sequence reliant equipment will also be affected.

There is often confusion with phase rotation indicators, as some types light the 'correct' indicator for a clockwise phase sequence, or suggest in some way that clockwise rotation is the correct sequence. Therefore, it's important to use the same type of phase rotation indicator for both disconnection and reconnection of the supply.

Although most three-phase installations today use red, white and blue cable colours, in Australia and New Zealand, active conductors can be any colour except green/yellow, green, yellow, black or light blue. Yellow is no longer permitted in the 2018 revision of AS/NZS 3000:2018, but may be found in old installations.

KEY POINTS...

- Three-phase power is more economical and provides a smoother power transfer than a single phase system.
- A three-phase alternator has three separate windings placed at an angle of 120° to each other. This arrangement produces three sinewave voltages that are out of phase with each other by 120°.
- The instantaneous value of a phase voltage in a three-phase system can be found with the equation $v = \sin\theta\, V_{max}$.
- In a three-phase system, phase voltages reach their maximum in a particular order, called phase sequence. In Australia and New Zealand, a system is considered 'correct' if its phase voltages 'rotate' anticlockwise.

20.3 Star Connection

A three-phase alternator has three separate coil windings, one per phase, as shown in Figure 20.7. Each phase produces the same value of voltage, except each voltage is 120° out of phase with the others. If a load was connected to the alternator in Figure 20.7, six connecting lines would be needed, which is uneconomical and impractical. There are two ways of connecting a three-phase alternator to its load: the *star* connection and the *delta* connection.

FIGURE 20.7 A three-phase alternator has six connections, made up of three windings each with a start and end connection

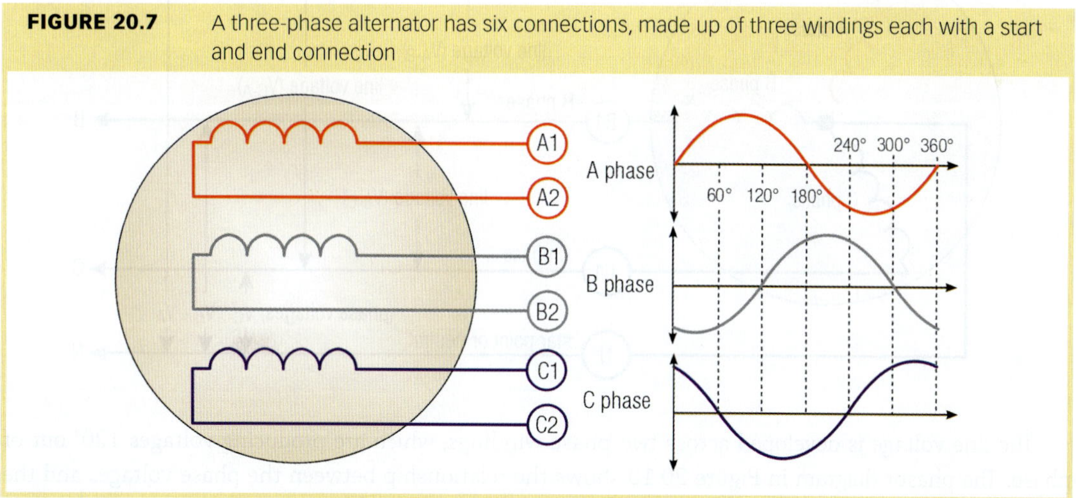

In the star connection, the three starts or ends of each coil winding are connected together. Two ways of drawing the circuit are shown in Figure 20.8, in which (a) creates a circuit layout in the shape of a star and (b) the shape of an inverted Y. The common point of the connection is called the star point, or the neutral point (N). In Figures 20.8(a) and (b), the ends of each winding are connected to the star point, and the start of each winding is the point of connection for that phase.

The star connection has four terminals, and is referred to as a 4-wire system. The common connection provides the fourth output of a three-phase alternator, called the neutral. If the alternator is connected to a load in which the current in each phase of the alternator is different (unbalanced), the neutral carries the out-of-balance current. Otherwise, the neutral is at zero potential, hence its name. It is generally connected to earth, as later explained.

FIGURE 20.8 The star or Y connection, in which the ends (or starts) of each winding are joined at a common point called the star or neutral point

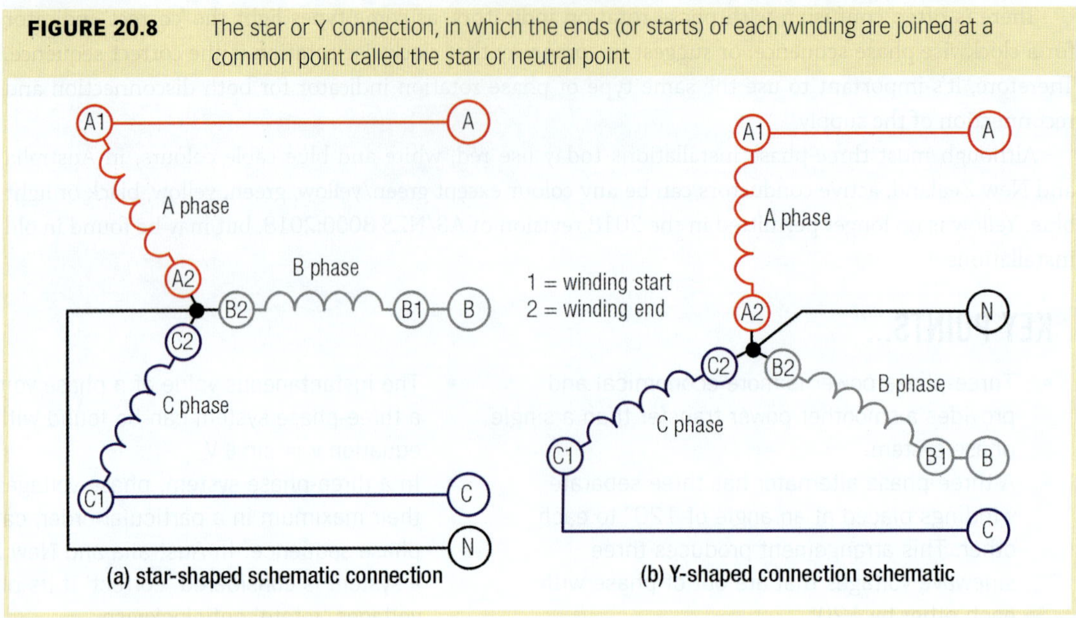

Voltages (star)

Figure 20.9 shows the voltages in a star-connected system. The voltage between any two lines is called the *line* voltage. The voltage between the neutral (star point) and any line is the *phase* voltage. This is the voltage developed by each phase winding in the alternator.

FIGURE 20.9 Line and phase voltages in a star-connected three-phase power system

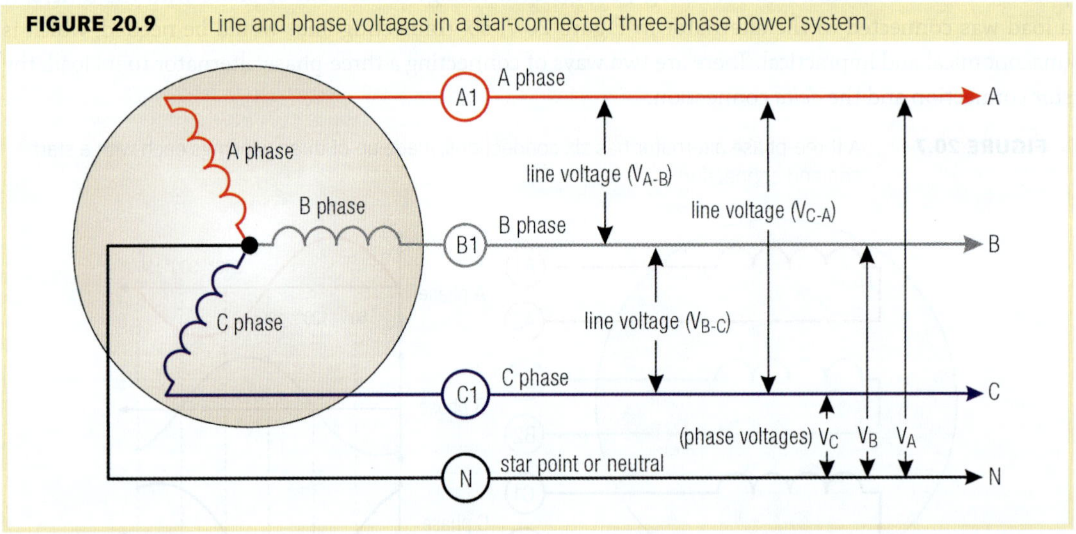

The line voltage is developed across two phase windings, which are producing voltages 120° out of phase. The phasor diagram in Figure 20.10 shows the relationship between the phase voltages and the line voltages in a star-connected system. The line voltage is shown by the phasor joining each phase voltage.

The phasor diagram in Figure 20.11 also shows the relationships between phase and line voltages. In this diagram, each line voltage is the phasor addition of one phase voltage, and the inverted phasor of another phase voltage. This is known as phasor subtraction.

From the phasor diagrams in Figures 20.10 and 20.11, in a star-connected system:

1 There is a phase difference of 30° between line and phase voltages. In Figure 20.11, the line voltage phasors are identified by letters. The first letter is the reference phase voltage, the second is the phase voltage being subtracted from that reference. So phasor V_{A-B} is the phasor difference between phase voltages A and B. Because phase A is the reference, line voltage V_{A-B} leads V_A by 30°.

2 The line voltage is greater than the phase voltage.

FIGURE 20.10 Phasor diagram of line and phase voltages in a three-phase star-connected system

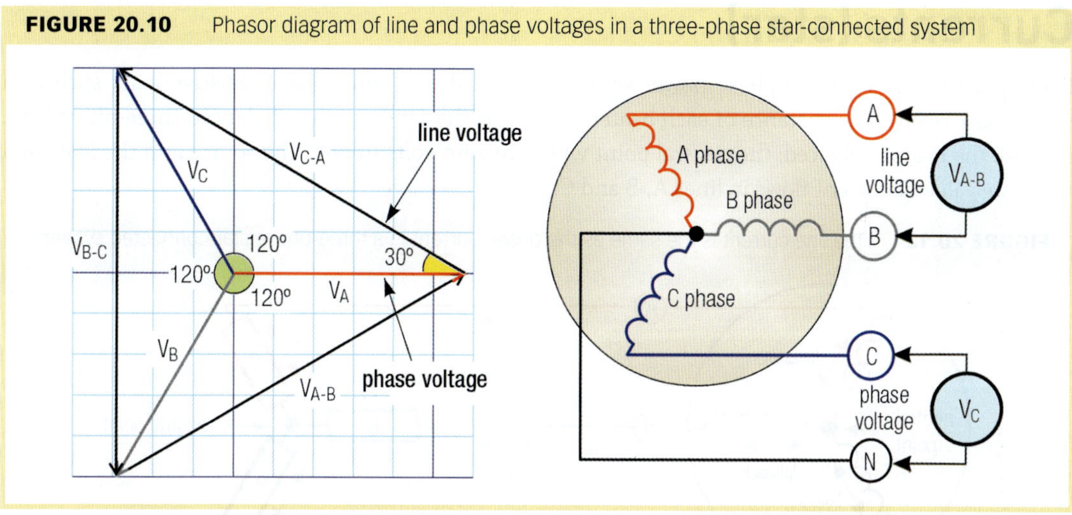

FIGURE 20.11 Phasor diagram showing relationship of line and phase voltages in a three-phase star-connected system in which phase voltages lag line voltages by 30°

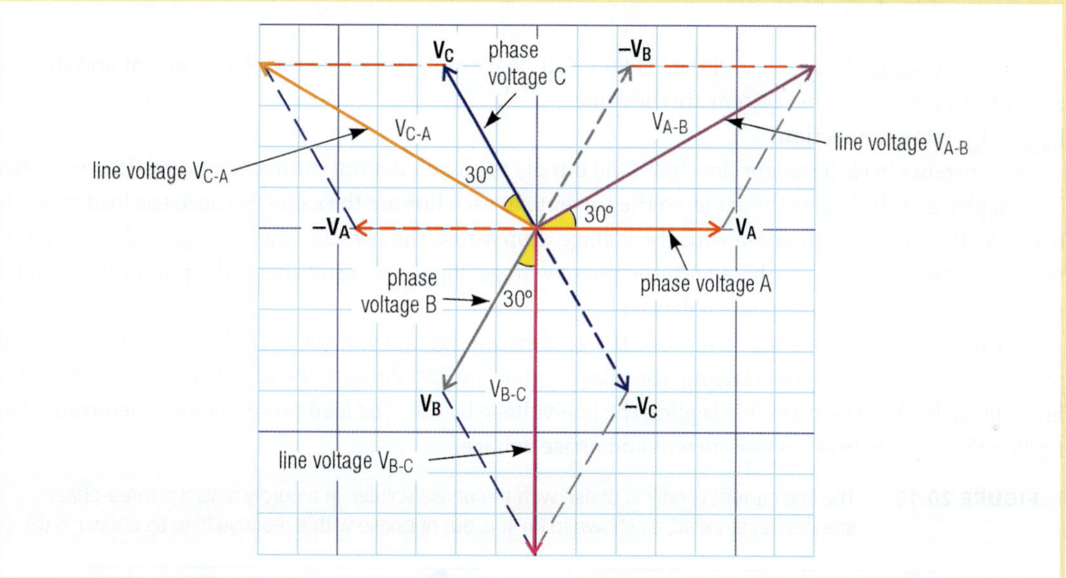

Equations for voltages in a star-connected system are:

$V_{line} = \sqrt{3} \times V_{phase}$ (star connection).

Transposing the above equation to make phase voltage the subject gives:

$V_{phase} = \frac{V_{line}}{\sqrt{3}} = 0.577 \times V_{line}$ (star connection).

A star-connected power system therefore provides two different voltages: a line voltage between any two phases and a lower phase voltage between any phase and neutral.

FYI

$\sqrt{3}$ is used a lot in three-phase calculations. A value of 1.732 is usually accurate enough. The reciprocal is 0.577, also accurate enough.

EXAMPLE 20.2

The line voltage of a star-connected three-phase supply is 398 V. Find the phase voltage.

Solution

Values $V_{line} = 398$ V

Equation $V_{phase} = \frac{V_{line}}{\sqrt{3}} = \frac{398}{\sqrt{3}} = \frac{398}{1.732} = 0.577 \times 398$

Answer $\mathbf{V_{phase} = 230\ V}$ (rounded value)

Currents (star)

Figure 20.12 shows a three-phase power source connected to a three-phase resistive load, such as a three-phase oven. The resistance of each heater element is the same, giving a balanced three-phase load. Because the load is balanced, the neutral point will be at zero volts in both the source and the load, and the same value of current flows in lines A, B and C.

FIGURE 20.12 The line current is the same as the phase current in a three-phase star-connected system

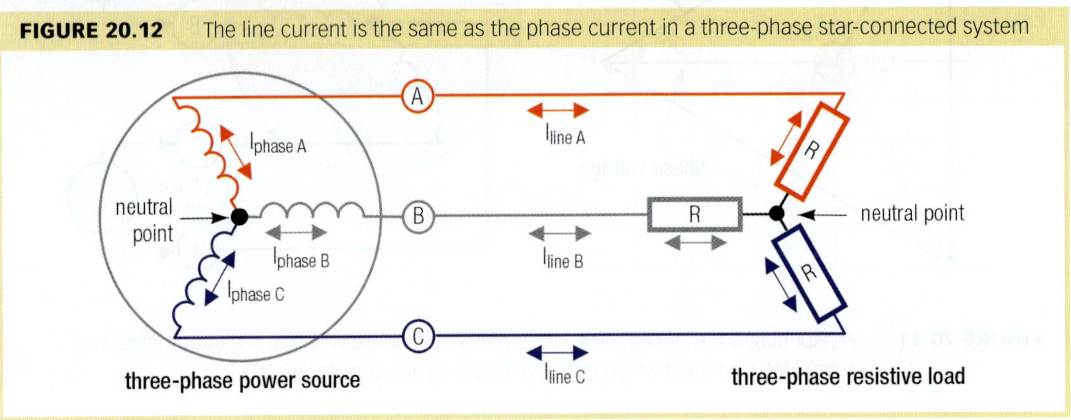

The windings in the source and the load for that phase are in series, so the phase current and the line current in a star-connected system are the same.

$I_{phase} = I_{line}$ (star connection)

The currents in each resistor flow into (and out of) the load's neutral connection point. The resistance of each phase of the load is the same, so the currents in each line are the same. Because the load is purely resistive, the current is in phase with the voltage drop across the resistor. This voltage is also the phase voltage, so the current is in phase with the phase voltage. Figure 20.13(a) shows the phasor diagram for currents in a resistive three-phase balanced load.

FYI

In a star system: Line and phase currents are the same ($I_{line} = I_{phase}$)

For voltages

$V_{line} = V_{phase} \times 1.732$

If the load is inductive (e.g. a motor), the current will lag the voltage as in Figure 20.13(b). In this phasor diagram, the current is lagging the phase voltage by 25°. Because the phase voltage is lagging the line voltage by 30°, the current is lagging the line voltage by 55°. The load power factor is determined by finding the angle between phase current and phase voltage.

FIGURE 20.13 The line current is only in phase with the phase voltage in a purely resistive three-phase star-connected load, as shown in (a). It is out of phase with a reactive load as shown in (b).

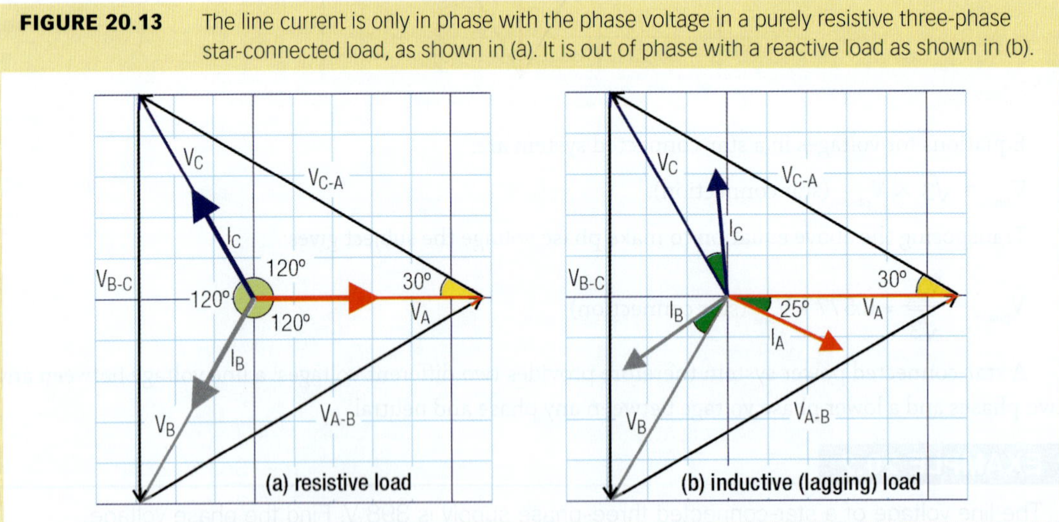

Figure 20.14 shows phasor diagrams of the currents in a balanced star-connected load. In Figure 20.14(a), the load is resistive, so the currents are in phase with the phase voltages. The phasor addition of the currents in the A and C phases gives a phasor called $I_C + I_A$. When this phasor is added to phasor I_B the result is zero.

FIGURE 20.14 The phasor addition of the currents in any balanced star-connected circuit is zero

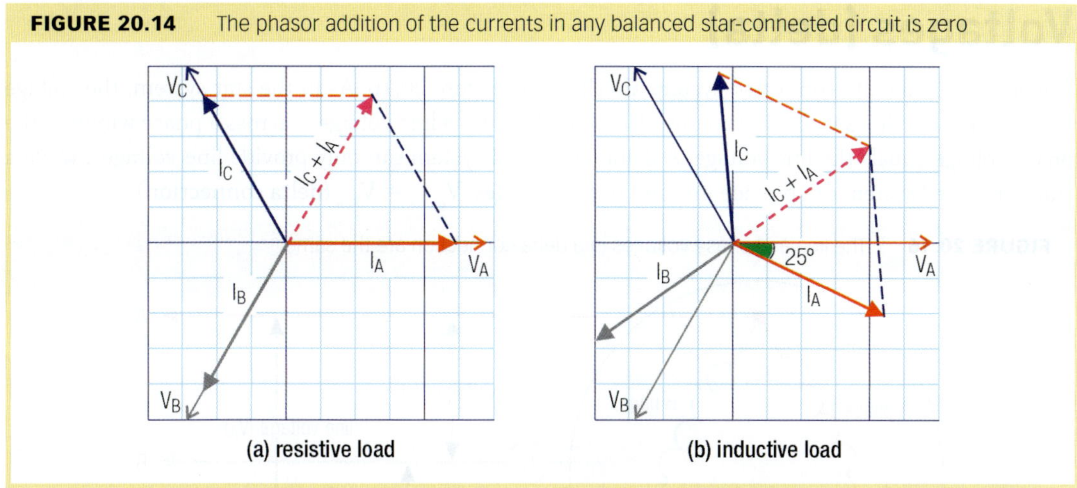

The same effect occurs with a balanced reactive load, as shown in Figure 20.14(b), in which the currents lag the phase voltage by 25°. Although current is passing through the neutral point, the sum of the currents at any one time is zero. Therefore, the neutral point remains at zero volts. This only applies to a balanced three-phase load.

Summary of star connection voltages and currents

- $V_{line} = 1.732 \times V_{phase}$
- $V_{phase} = \frac{V_{line}}{\sqrt{3}} = 0.577 \times V_{line}$ (star connection)
- $I_{line} = I_{phase}$
- V_{line} leads V_{phase} by 30°
- in a balanced star system, the star point (neutral) = 0 V

20.4 Delta Connection

The other way of connecting the windings of a three-phase power source is the mesh or delta configuration, so called because the circuit looks like the Greek letter delta (Δ). Figure 20.15 shows two circuit layouts, in which both diagrams have the start of a phase winding connected to the end of the next. This gives three output connections, labelled as A, B and C. (This book uses A, B and C; other methods are L_1, L_2 and L_3, or R, W and B.)

FIGURE 20.15 The delta connection has a maximum of three lines to transmit three-phase power

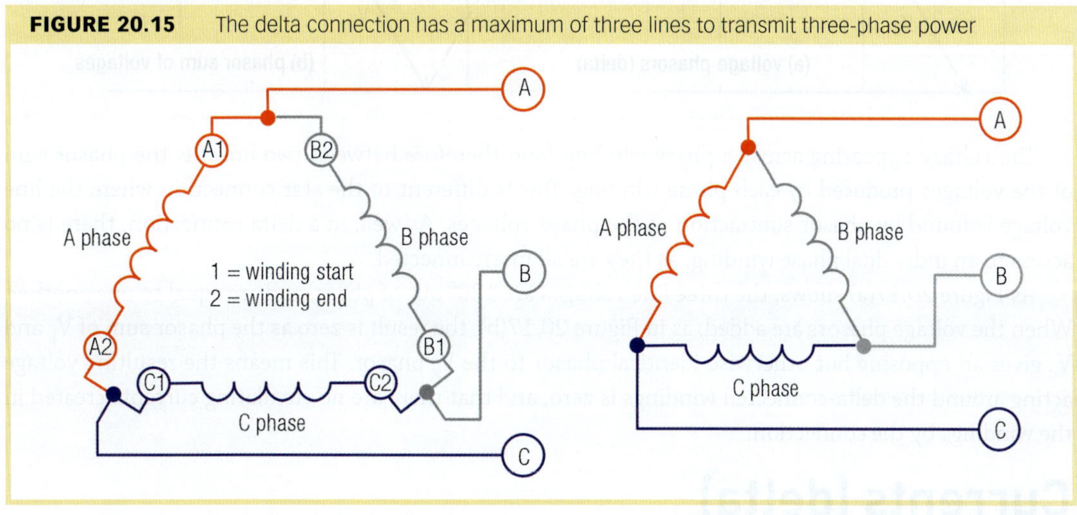

Voltages (delta)

The voltages in a delta-connected system are shown in Figure 20.16. As in the star system, the voltage between any two lines is the line voltage. However, because a line voltage is across a phase winding, the phase voltage equals the line voltage. A delta-connected system can only provide line voltages, while a star-connected system can provide line and phase voltages. $V_{phase} = V_{line}$ (delta connection)

FIGURE 20.16 The line and phase voltages in a delta connection are the same

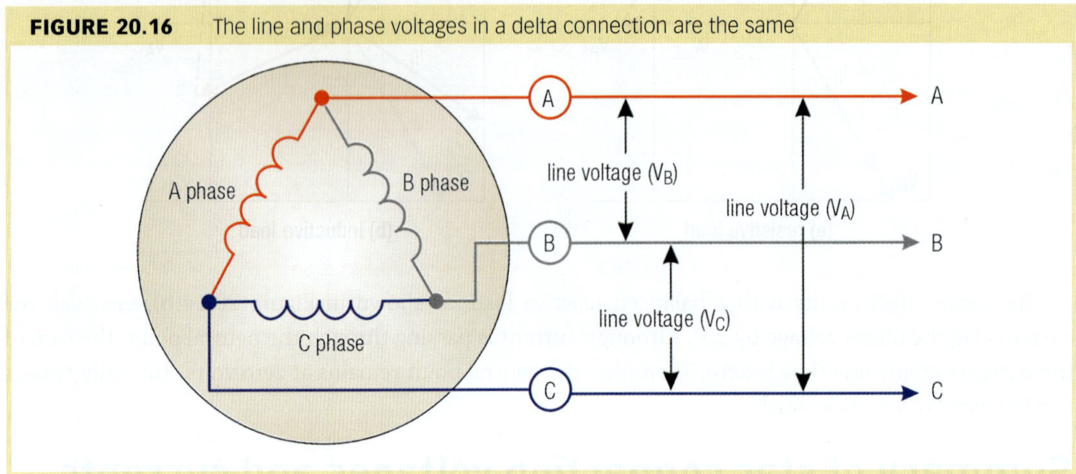

The voltages in Figure 20.16 are labelled by the phase winding they are across, so, for example, the voltage between lines A and C is the phase voltage produced by phase winding A. Therefore, unlike the star connection, in a delta circuit the voltages are simply V_A, V_B and V_C. The phasor diagram of the voltages in a delta-connected system is shown in Figure 20.17(a).

FIGURE 20.17 Phasor diagram of voltages in a delta connection. When the phasors are added, their sum is zero because phasor $V_B + V_C$ is the same length but in opposite direction to phasor V_A.

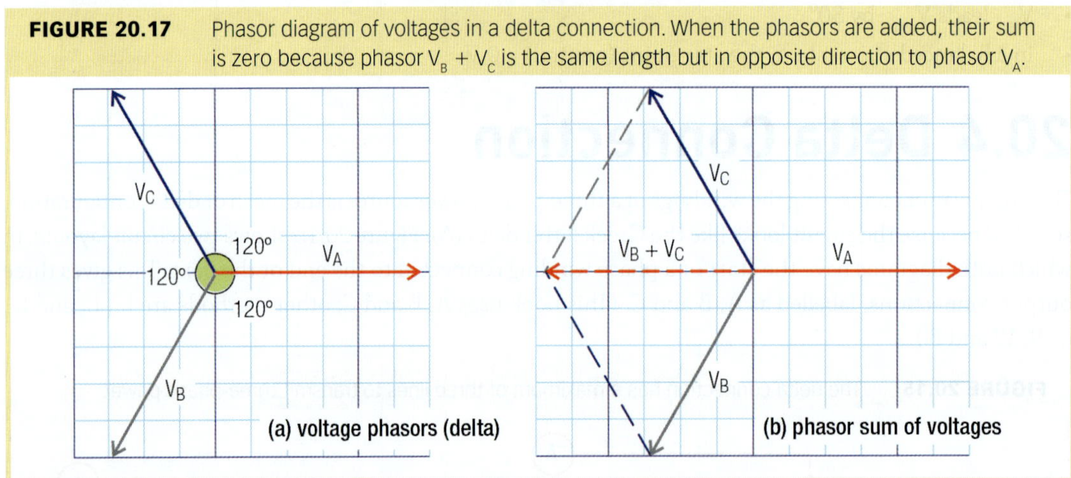

The voltage appearing across a phase winding (and therefore between two lines) is the phasor sum of the voltages produced by each phase winding. This is different to the star connection where the line voltage is found by phasor subtraction of the phase voltages. As well, in a delta connection, there is no access to an individual phase winding, as they are all interconnected.

As Figure 20.17(a) shows, the three line voltages are 120° out of phase, and are equal to one another. When the voltage phasors are added, as in Figure 20.17(b), the result is zero as the phasor sum of V_B and V_C gives an opposing but otherwise identical phasor to the V_A phasor. This means the resulting voltage acting around the delta-connected windings is zero, and that there are no circulating currents created in the windings by the connection.

Currents (delta)

Figure 20.18 shows a delta-connected power source supplying power to a delta-connected resistive load. As the diagram shows, at each phase winding connection there are two currents contributing to the line current. At the load end, the line current supplies two paths in the load.

FIGURE 20.18 The line currents in a delta-connected system are the phasor sum of the individual phase currents contributing to that line current

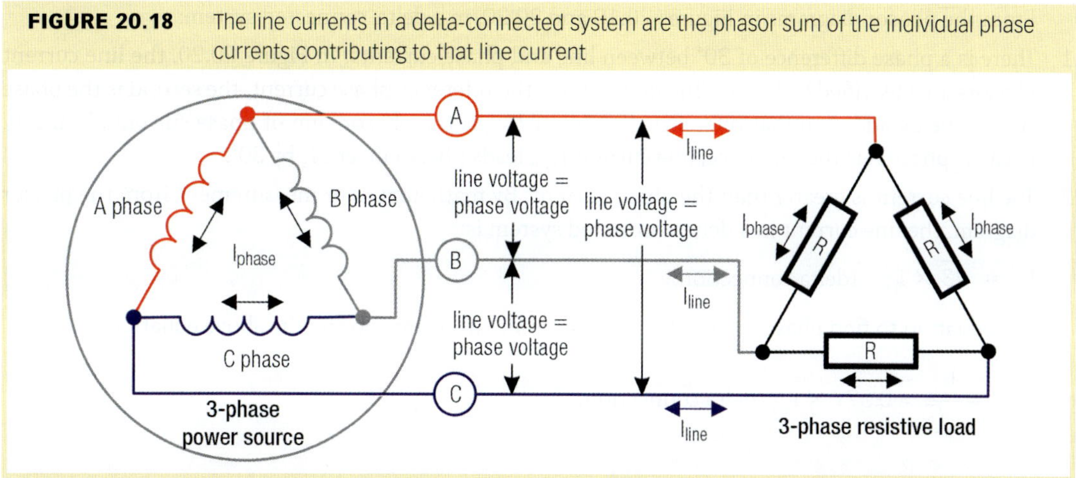

The load in Figure 20.18 is purely resistive, so the current flowing in each resistor is in phase with the voltage across the resistor, which is the line voltage. However, each current is out of phase with the other by 120°, so the current entering the connection of any two resistors is the phasor sum of those leaving the connection. The same thing applies at the alternator, where each line current is the phasor sum of two phase currents. Figure 20.19 shows the phase and line currents in a delta-connected system and the phasor diagram for the currents.

FIGURE 20.19 Line currents are 30° out of phase with phase currents in a delta system

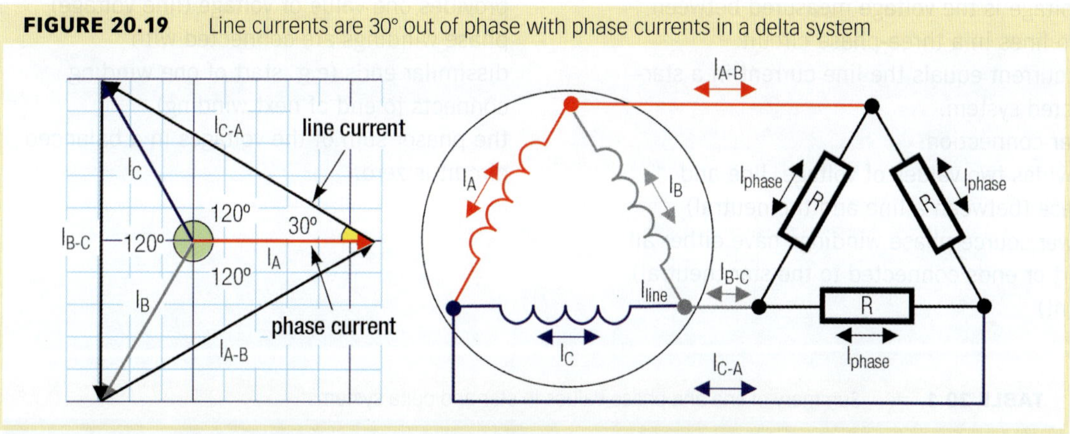

The phasor diagram in Figure 20.20 shows similar information to that in Figure 20.19, except it also shows the relationships between the line voltage, line current and phase current.

FIGURE 20.20 Phasor diagram showing relationship of line and phase currents in a three-phase delta-connected system with a unity power factor (resistive load)

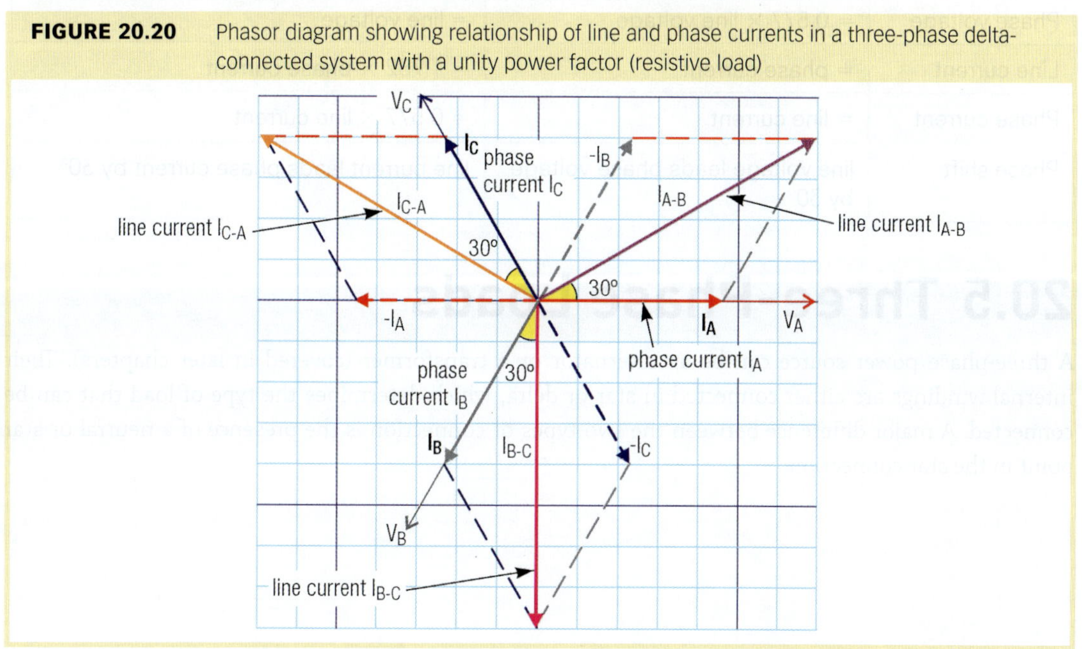

FYI

In a delta system: Line and phase voltages are the same ($V_{line} = V_{phase}$)

For currents $I_{line} = I_{phase} \times 1.732$

From the phasor diagrams in Figures 20.19 and 20.20, in a delta-connected system:

1 There is a phase difference of 30° between line and phase currents. In Figure 20.20, the line current phasors are identified by letters. The first letter is the reference phase current, the second is the phase current being added to that reference. So phasor I_{A-B} is the phasor sum of phase currents I_A and I_B. Because phase A is the reference, line current I_{A-B} leads phase current I_A by 30°.

2 The line current is greater than the phase current. By mathematics or measurement from the phasor diagram, the line current in a delta-connected system is:

$I_{line} = \sqrt{3} \times I_{phase}$ (delta connection).

The equation to find phase current from line current, after transposing the above equation is:

$I_{phase} = \frac{I_{line}}{\sqrt{3}} = 0.577 \times I_{line}$ (delta connection).

KEY POINTS...

- The phase windings of a three-phase power source (e.g. alternator or transformer) can be connected in star (Y) or delta (mesh).
- Phase voltage equals the line voltage in a delta-connected system.
- Line voltage is the voltage measured between any two lines in a three-phase circuit.
- Phase current equals the line current in a star-connected system.
- The star connection:
  - provides two values of voltage, line and phase (between a line and the neutral)
  - power source phase windings have either all start or ends connected to the star (neutral) point)
  - the neutral point (star point) is usually connected to earth
  - the phasor sum of the currents in a balanced circuit is zero.
- The delta connection:
  - provides one value of voltage (line voltage)
  - phase windings are connected with dissimilar ends (e.g. start of one winding connects to end of next winding)
  - the phasor sum of the voltages in a balanced circuit is zero.

TABLE 20.1 Summary of line and phase values in star and delta systems

Values	Star connection	Delta connection
Line voltage	= 1.732 × phase voltage	= phase voltage
Phase voltage	= 0.577 × line voltage	= line voltage
Line current	= phase current	= 1.732 × phase current
Phase current	= line current	= 0.577 × line current
Phase shift	line voltage leads phase voltage by 30°	line current leads phase current by 30°

20.5 Three-Phase Loads

A three-phase power source can be an alternator or a transformer (covered in later chapters). Their internal windings are either connected in star or delta, which determines the type of load that can be connected. A major difference between the two types of connection is the presence of a neutral or star point in the star connection.

In general, a three-phase load can be balanced or unbalanced, reactive or resistive. In a balanced load, each phase has the same impedance and power factor. An unbalanced load has different impedances and power factors in each phase. Loads can be connected in star or delta, and in the case of a star-connected supply, loads can be single phase as well as three phase.

Loads connected to a star supply

A star-connected power source provides three supply lines (A, B and C) and a neutral point (N). The presence of the neutral allows a wider range of loads to be connected, compared to a delta-connected supply. For example, it can supply a single-phase load between one line and the neutral point. A home or any enterprise with a single-phase connection is an example of a load connected to a star supply. A home or a commercial enterprise connected to a three-phase supply is another example of a load powered by a three-phase star-connected supply, as there is a neutral connection for 230 V appliances.

Balanced load

Figure 20.21 shows a star-connected power source connected to a balanced, resistive star-connected load, such as any three-phase load with heater elements. Each element is the same, with one per phase joined in a star configuration.

FIGURE 20.21 In a balanced three-phase load, the sum of the currents at the star point is zero

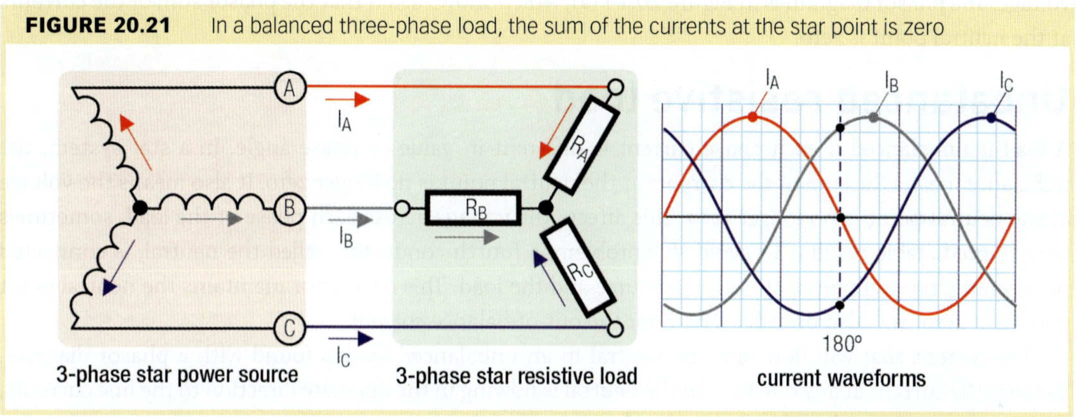

When each phase (line) current is equal in a star-connection, their sum at the star point is zero, as shown by the phasor diagram in Figure 20.14, and also from the waveforms in Figure 20.21. For example, at 180° the A phase current is zero, and the other two phase currents have equal and opposite values. These cancel each other, giving a total of zero for all three currents. This applies at all points along the waveforms.

Each phase of a star-connected load has a current equal to the line current. Because it's a balanced circuit, the line currents all have the same value. The voltage across each phase of the load is the phase voltage produced by the generating source. The current flowing in each phase of the load is found with Ohm's law, as for all electrical circuits. That is:

$$I_{phase} = \frac{V_{phase}}{Z_{phase}} \text{ or } I_p = \frac{V_p}{Z_p} \text{ amps}$$

EXAMPLE 20.3

The industrial three-phase heater in Figure 20.21 has three elements each with an impedance (actually a resistance, as there is no reactive component) of 11.5 Ω. The line voltage of the supply is 400 V. Calculate the:

1 voltage across each heater element (phase voltage)
2 current in each heater element (phase current)
3 line current

Solution

Values $V_{line} = 400\ V$

$Z_A = Z_B = Z_C = 11.5\ \Omega$

Equation (1) $V_{phase} = \frac{V_{line}}{\sqrt{3}} = \frac{400}{\sqrt{3}} = \frac{400}{1.732} = 0.577 \times 400$

Answer (1) $\mathbf{V_{phase} = 230\ V}$ (rounded value)

Equation (2) $I_p = \frac{V_p}{Z_p} = \frac{230}{11.5}$

Answer (2) $\mathbf{I_{phase} = 20\ A}$

In a star-connection, line and phase currents are the same, so:

Answer (3) $\mathbf{I_{line} = 20\ A}$

The phase angle between the current and the phase voltage is 0° because the load is purely resistive. As mentioned previously, the phase voltage lags the line voltage by 30°. If a load is reactive, but with each phase balanced, the only difference to a resistive load is that the current is out of phase with the phase voltage. The phasor diagram is in Figure 20.14 (a), which shows again that the phasor sum of the currents at the neutral point is zero.

Unbalanced resistive load

FYI

A star supply can supply an unbalanced star load if a neutral is connected; it cannot supply an unbalanced delta load

A load is unbalanced if each phase current is different in value or phase angle. In a star system, the unbalance means the sum of the currents at the neutral point is no longer zero. It also means the voltage at the neutral point is no longer zero. This affects the voltage across each phase of the load, sometimes with unpredictable results. To avoid this problem, a fourth conductor, called the neutral, is connected between the neutral points of both the source and the load. This conductor maintains the neutral point voltage at zero, and also provides a path for the out-of-balance current.

The current that will flow into the neutral in an unbalanced load is found with a phasor diagram. Because the current at any one time in the neutral is flowing in the opposite direction to the line currents, the neutral current is therefore *minus* the phasor sum of the three currents.

Figure 20.22 shows an unbalanced resistive load connected to a three-phase supply, with a neutral conductor connected between the neutral points to carry the out-of-balance current. Example 20.4 shows how to find the neutral current in this circuit.

FIGURE 20.22 In an unbalanced three-phase load, the sum of the currents at the star point is no longer zero and causes a neutral current flow that is equal to minus the phasor sum of the phase currents

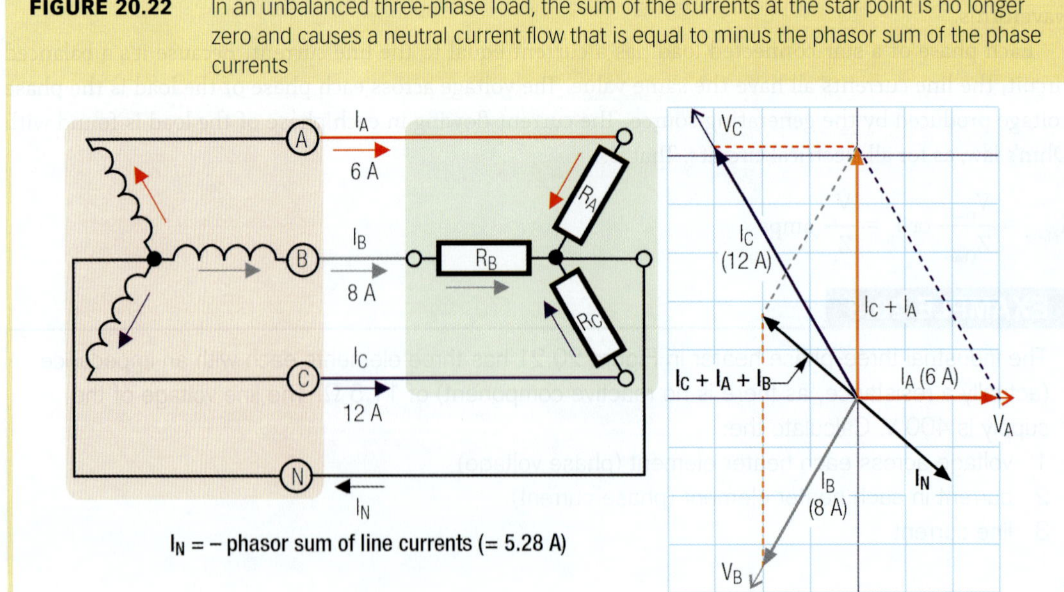

EXAMPLE 20.4

The three-phase resistive load in Figure 20.22 is taking the currents shown. Determine the value of the neutral current.

Solution

Values $I_A = 6$ A
$I_B = 8$ A
$I_C = 12$ A

Method (Refer to the phasor diagram in Figure 20.22)

1 Construct a phasor diagram on a suitable scale (e.g. 1 A = 10 mm) by drawing each current phasor so it is in phase with its phase voltage (because the load is resistive).
2 Construct a parallelogram to find the phasor sum of any two of the line currents. We found the sum of I_A and I_C first, giving a phasor called $I_C + I_A$.
3 Construct a parallelogram to find the phasor sum of the remaining current phasor and the phasor just found. In this case, we are adding phasor I_B with the $I_C + I_A$ phasor. (Gives $I_C + I_A + I_B$.)
4 Because the neutral current is minus the phasor sum of the currents, draw another phasor equal in length and at 180° to phasor $I_C + I_A + I_B$, giving phasor I_N.
5 Measuring the length of phasor I_N gives a neutral current of 5.3 A.

Answer I_N = **5.3 A**

The phase angle between the neutral current and the line currents can also be found by measurement from the phasor diagram. In this example, I_N is lagging I_A by a measured angle of 40.7°. (Note: Phasor $I_C + I_A$ is not always at 90°, as happened in Figure 20.22.)

Unbalanced reactive load

The currents in Example 20.4 are all in phase with their phase voltages. If an unbalanced load has reactance, it could be that each phase current not only has a different value, but a different phase angle to its phase voltage (different power factor). The process to find the neutral current is the same as for Example 20.4, except each current phasor is drawn at an angle to its phase voltage. This angle is derived from the power factor for that phase, where the angle is the inverse cosine of the power factor (λ).

FIGURE 20.23 Unbalance in a star-connected three-phase load can be due to differing power factors in each phase as well as differing phase currents

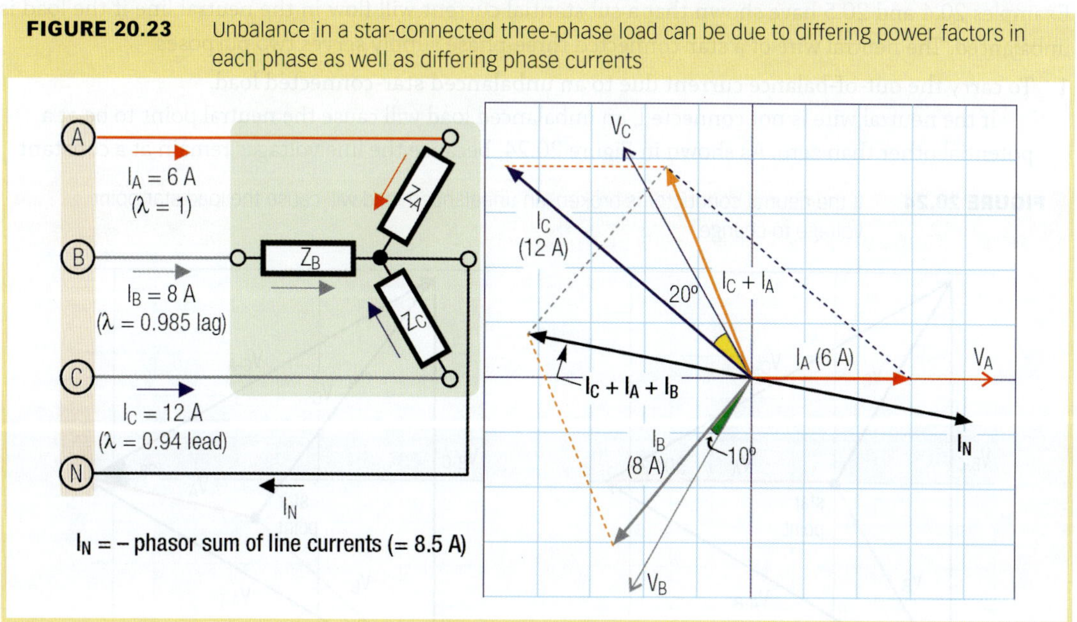

EXAMPLE 20.5

The three-phase reactive load in Figure 20.23 is taking the currents shown. Determine the value of the neutral current.

Solution

Values $I_A = 6$ A ($\lambda_A = 1$)
$I_B = 8$ A ($\lambda_B = 0.985$ lag)
$I_C = 12$ A ($\lambda_C = 0.94$ lead)

Method (Refer to the phasor diagram in Figure 20.23)

1 Determine the phase angles from the power factor values. In this case, I_A is in phase with V, so $\phi_A = 0°$. I_B has a lagging power factor (λ_B) of 0.985 so $\phi_B = 10°$ lag behind V_B. I_C has a leading power factor (λ_C) of 0.94 so $\phi_C = 20°$ lead before V_C.
2 Construct a phasor diagram on a suitable scale (e.g. 1 A = 10 mm) by drawing each current phasor so it is at the calculated phase angle to its phase voltage.
3 Construct a parallelogram to find the phasor sum of any two of the line currents. We found the sum of I_A and I_C first, giving a phasor called $I_C + I_A$.
4 Construct a parallelogram to find the phasor sum of the remaining current phasor and the phasor just found. In this case, we are adding phasor I_B with the $I_C + I_A$ phasor. (Gives $I_C + I_A + I_B$.)
5 Because the neutral current is minus the phasor sum of the currents, draw another phasor equal in length and at 180° to phasor $I_C + I_A + I_B$, giving phasor I_N.
6 Measuring the length of phasor I_N gives a neutral current of 8.5 A.
Answer I_N = **8.5 A**

The phase angle between the neutral current and the line currents can also be found by measurement from the phasor diagram. In this example, I_N is lagging I_A by a measured angle of 10.6°.

Examples 20.4 and 20.5 have the same unequal phase currents, but in Example 20.5 the power factors in each phase are also different. This shows up in the value of the neutral current, which in Example 20.5 is larger than in Example 20.4.

Importance of the neutral connection

Examples 20.4 and 20.5 have shown that a substantial current will flow in the neutral line if the load is unbalanced. The neutral wire of a star-connected three-phase supply serves two purposes.

1 To carry the out-of-balance current due to an unbalanced star-connected load.

If the neutral wire is not connected, an unbalanced load will cause the neutral point to be at a potential other than zero. As shown in Figure 20.24, because the line voltages remain at a constant

FIGURE 20.24 If the neutral conductor is broken, an unbalanced load will cause the load star point voltage to change

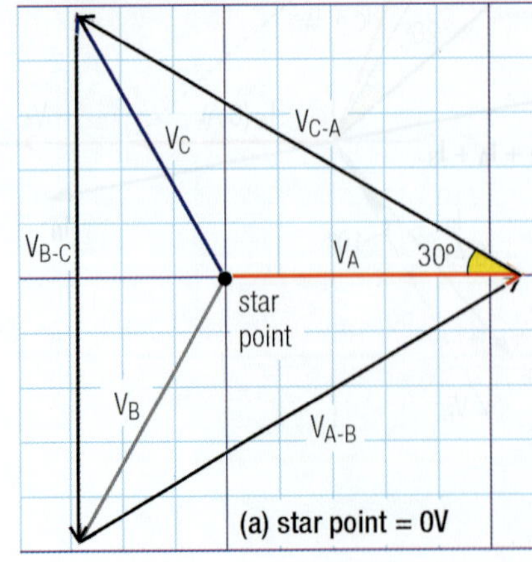

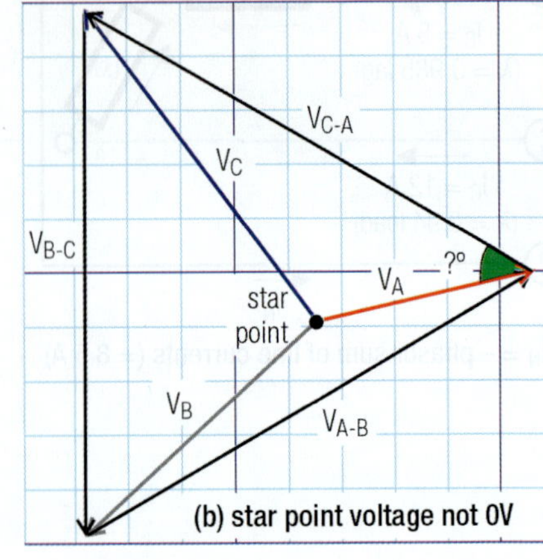

value, a shift in the neutral point voltage causes the voltages across each phase of the load to change. This means the current in each phase changes, further upsetting the imbalance. The extent of the potential problem depends on how much imbalance there is, but in any case, without a neutral conductor any imbalance can cause serious instability and potentially high voltages across one or more phases of the load.

2 **To provide a single-phase power supply service from a three-phase service.**

A three-phase supply with a nominal line voltage of 400 V has a nominal 230 V phase voltage. Figure 20.25 shows a typical arrangement for a three-phase power distribution system. Single-phase loads (e.g. homes) are connected between a line and the neutral, giving a 230 V single-phase supply. To distribute the load between phases, each power line is connected to one-third of the dwellings in a street.

FIGURE 20.25 Typical types of loads that are connected to a star-connected three-phase supply. A single-phase load could be a home, a three-phase unbalanced load could be a small factory.

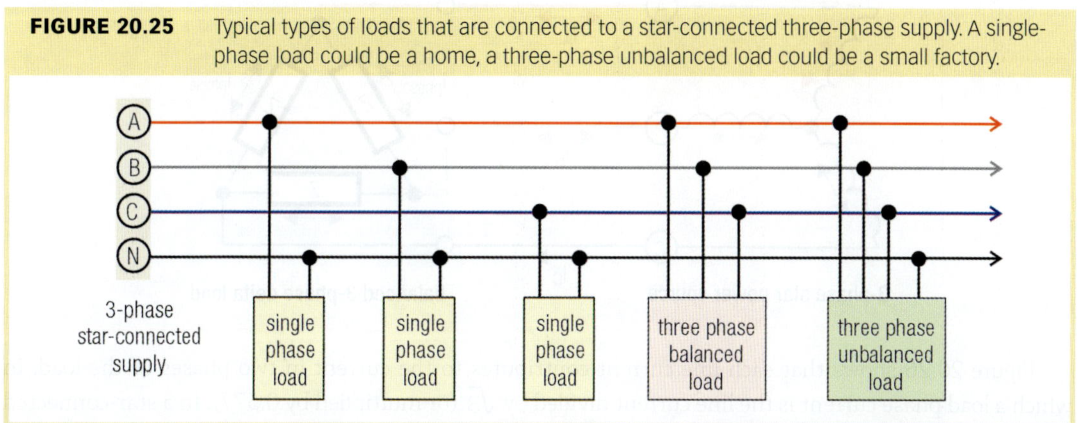

Three-phase balanced loads (e.g. a delta-connected three-phase motor) do not need a neutral, while three-phase unbalanced loads must be connected to the neutral. To accommodate the out-of-balance currents likely to occur in a power distribution system, Clause 3.5.2 of AS/NZS 3000:2018 specifies that the maximum out-of-balance current must be included in the calculation to determine the size of the neutral conductor. Another consideration is the value of harmonic currents generated by the installation, which can be caused by, among other things, variable speed drive systems.

As mentioned, if the neutral line becomes open-circuit or high-impedance, instability and potentially dangerous situations can arise. For these reasons, AS/NZS 3000:2018 requires that a switch or circuit breaker should not be inserted in the neutral conductor of consumer mains. Exceptions are given in clause 2.3.2.1.2.

SAFETY

Fitting switches or circuit-breakers in the neutral is not permitted (the few exceptions are in AS/NZS 3000:2018 2.3.2.1.2)

Effects of an open-circuit neutral

An open-circuit neutral can occur at a connection point, or the neutral conductor could be accidentally severed at a point either within or external to the installation. In Figure 20.25, if the neutral connection became open-circuit at the N terminal of the supply, all loads would be affected other than the balanced load not connected to the neutral.

In the case of a home or any single-phase power installation, the effects depend on the state of the installation's earth connection, a topic that is not covered here. In any event, the effects range from the installation appearing to work normally, through a 'brown out' to where nothing works. It becomes more complex if the neutral conductor is broken between two installations, with the effects again depending on factors such as earth connections.

In the case of a single three-phase unbalanced load, because the neutral point voltage has changed, the voltage across each phase of the load changes. This can cause a significantly higher voltage to appear across one or more phases of the load, and can introduce instability where voltages are no longer predictable.

In many cases, the break in the neutral conductor will have a high voltage across it. In a power distribution system, there is never an exact balance between phases, so there is always some value of current flowing in the neutral conductor. As in any circuit, the highest voltage is across an open-circuit.

FYI

In most electrical installations, the neutral conductor is connected to earth at the installation's fuse box

SAFETY

Never assume a neutral conductor is at zero potential. Test first.

Delta-connected load

Delta-connected loads include three-phase motors and industrial three-phase heaters as these are balanced loads. In the delta connection, each phase of the load has the line voltage across it (e.g. 400 V). The same load connected in star has the system phase voltage across it (e.g. 230 V). Figure 20.26 shows a star-connected power source connected to a balanced delta-connected resistive load. It's important that the load is balanced, as there is no star point and therefore no neutral conductor to carry any out-of-balance current.

FIGURE 20.26 A delta-connected load has the line voltage across each phase of the load and must be balanced to operate successfully on a star-connected supply

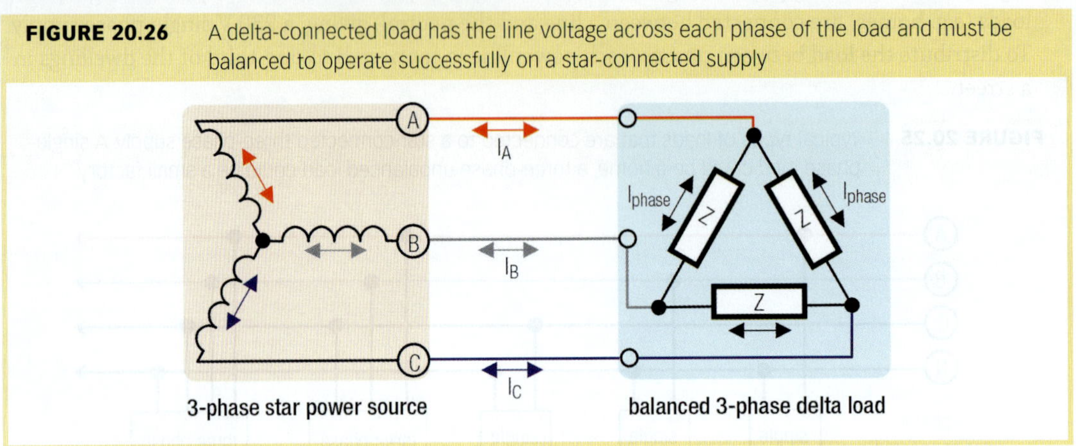

Figure 20.26 shows that each line current contributes to the current in two phases of the load, in which a load phase current is the line current divided by $\sqrt{3}$ (or multiplied by 0.577). In a star-connected supply, the line and phase currents are the same. Example 20.6 shows the calculations to determine line and phase current for the connection in Figure 20.26.

EXAMPLE 20.6

The three-phase load in Figure 20.26 has an impedance in each phase of 20 Ω and is connected to a 400 V, three-phase star-connected supply. Calculate the:

1 phase current in each phase of the load
2 line current in power lines A, B and C
3 voltage across each phase of the load

Solution

Values $V_L = 400$ V

$Z_A = Z_B = Z_C = 20\ \Omega$

1 Calculate the phase currents

Equation $I_P = \dfrac{V_L}{Z_P} = \dfrac{400}{20}$

Answer $\mathbf{I_P = 20\ A}$ (in each phase as load is balanced)

2 Calculate the line currents

Equation $I_L = \sqrt{3} \times I_P = 1.732 \times 20$

Answer $\mathbf{I_L = 34.64\ A}$

3 The phase voltage in a delta-connected load is the same as the line voltage so:

Answer $\mathbf{V_P = 400\ V}$

If the same load is connected in star, each phase of the load will have 230 V across it, not the 400 V calculated above. The phase current is 230 V divided by 20 Ω, which equals 11.5 A. In delta, the phase current was 20 A.

When a load is connected in delta, the phase current is 1.732 ($\sqrt{3}$) times greater than when the same load is connected in star.

Loads connected to a delta supply

The delta-connected supply provides only three supply lines (A, B and C). While this limits its versatility, it has greater stability in the event of an unbalanced load. Figure 20.19 shows a delta source connected to a delta load. If one phase of the load has a different impedance to the other phases, its current will be different to the other two phase currents. However, the only voltage present in a delta supply is the line voltage, which remains constant, regardless of the load conditions. This means the line currents associated with the affected phase current will change, but without affecting anything else in the system.

However, because there is no neutral conductor, only balanced star-connected loads can be connected to a delta supply source. This is similar to the limitation that only balanced delta loads can be connected to a star-connected supply source.

When a star load is connected to a delta supply, the phase current in the load is the line current. However, the phase current in the supply is the line current divided by 1.732 ($\sqrt{3}$).

FYI

A delta supply can only supply an unbalanced delta load, not an unbalanced star load

KEY POINTS...

- A three-phase load can be connected in star (Y) or delta (mesh).
- Loads are either balanced or unbalanced. A balanced load has the same current and power factor in each phase of the load. An unbalanced load does not.
- A star-connected source can supply an unbalanced star load, providing there is a neutral connection. It cannot supply an unbalanced delta load.
- A delta-connected source can supply an unbalanced delta load, but not an unbalanced star load.
- For a star-connected load:
  - load phase current $I_P = \frac{V_P}{Z_P}$ amps (also equals the supply line current I_L)
  - voltage across each phase of the load $V_P = \frac{V_L}{\sqrt{3}}$ where $\sqrt{3} = 1.732$
  - an unbalanced star-connected load requires a neutral connection between the load and the supply to carry the out-of-balance current
  - a break in the neutral connection can cause instability and high voltages to appear across the load
  - the importance of the neutral connection and its continuity cannot be overstated.
- For a delta-connected load:
  - load phase current $I_P = \frac{V_P}{Z_P}$ amps, it also equals $\frac{I_L}{\sqrt{3}}$ amps
  - line current equals the phase current multiplied by 1.732 ($I_{line} = \sqrt{3} \times I_{phase}$)
  - voltage across each phase of the load equals the line voltage.

20.6 Three-Phase Power

In a single-phase circuit, true power is found with the equation P = VI × power factor, where power factor is the cosine of the angle between V and I. That is:

$P = VI \cos \phi$ (single phase)

where:

P = true power in watts

V = supply voltage in volts (RMS)

I = current taken from the supply in amperes (RMS)

ϕ = phase difference between the supply voltage and current.

In a three-phase circuit, a load can be balanced or unbalanced. We start by looking at a balanced load, and derive a general equation to find the power consumed.

Balanced load

Figure 20.27(a) shows a balanced star-connected load, in which the phase current equals the line current, identified as I_A, I_B and I_C. The voltage across each phase of the star load, shown across Z_A as V_{PA}, is the line voltage divided by $\sqrt{3}$ (or multiplied by 0.577). Figure 20.27(b) shows a balanced delta-connected load, in which the line voltage equals the phase voltage, and the phase current is line current divided by $\sqrt{3}$.

FIGURE 20.27 A delta-connected load has the line voltage across each phase of the load, a star-connected load has the line current flowing in each phase

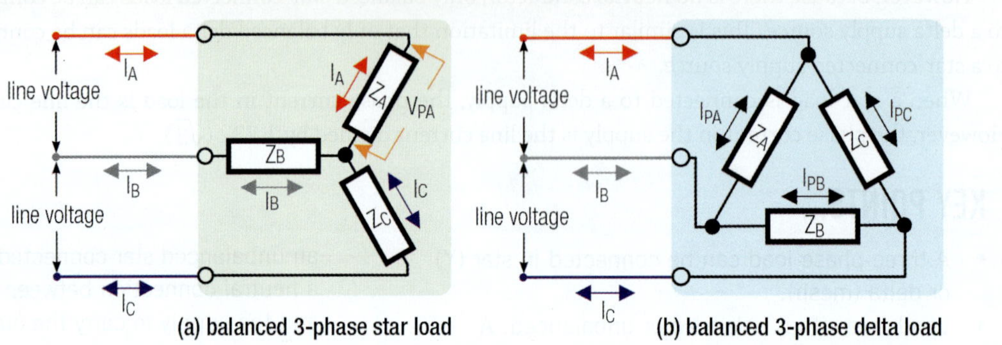

(a) balanced 3-phase star load

(b) balanced 3-phase delta load

The true power taken by a single phase in either of the circuits in Figure 20.27 is found using the same equation as for single-phase power. That is, for phase A:

(1) $P_A = V_{PA} I_{PA} \cos \phi$

where:

P_A is the power taken by phase A of the load

V_{PA} is the voltage across that phase

I_{PA} is the current in that phase.

The total true power taken by either of the balanced circuits in Figure 20.27 is the sum of the power taken by each phase of the load. Because each phase of the load takes the same amount of true power, the equation to find the total power is equation (1) multiplied by 3:

(2) $P = 3 \times V_{PA} I_{PA} \cos \phi$

where:

P is the total true power taken by a balanced three-phase load.

Total power in a balanced three-phase load can also be found if the line current and line voltage are known. As explained above, there is a common factor of 0.577 or $1/\sqrt{3}$ in both types of load. Multiplying 0.577 by 3 gives 1.732 or $\sqrt{3}$. Therefore, equation (2) can be rewritten in terms of line current and line voltage to give a general equation to find power taken by a balanced star- or delta-connected load. That is:

(3) $P = \sqrt{3}\, V_L I_L \cos \phi$

where:

P = total true power taken by a balanced star- or delta-connected load (watts)

V_L = line voltage

I_L = line current

$\cos \phi$ = power factor of one phase, which for a balanced load, is the same in each phase of the load.

KEY CONCEPT

True power taken by a three-phase balanced load uses the same equation as for single-phase, except multiplied by $\sqrt{3}$ (1.732)

EXAMPLE 20.7

A three-phase load draws a line current of 20 A at a lagging power factor of 0.8 from a 400 V three-phase supply. Calculate the power consumed by the load.

Solution

Values $V_L = 400$ V

$I_L = 20$ A

$\cos \phi = 0.8$

Equation $P = \sqrt{3} V_L I_L \cos \phi = 1.732 \times 400 \times 20 \times 0.8$

Answer **P = 11 084 W = 11.1 kW**

Example 20.7 shows how to find the true power in a balanced load (star or delta connected) where all load phase currents and power factors are the same. If the true power is known, rearranging the equation for power so either V_L, I_L or $\cos\phi$ is the subject lets you find the unknown term. That is:

$$V_L = \frac{P}{\sqrt{3}\, I_L \cos\phi} \text{ and } I_L = \frac{P}{\sqrt{3}\, V_L \cos\phi} \text{ and } \cos\phi = \frac{P}{\sqrt{3} V_L I_L}.$$

Unbalanced load

A load is unbalanced if the phase currents or phase power factors are different. In this type of load, the power taken by each phase is different. To find the total power taken by the load requires calculating the power taken by each phase, then adding them together. That is:

$P = P_A + P_B + P_C$

where:

P_A = true power taken by A phase in watts

P_B = true power taken by B phase in watts

P_C = true power taken by C phase in watts.

EXAMPLE 20.8

Calculate the total true power taken by the circuit in Figure 20.28.

Solution

Values $V_L = 400$ V

$I_A = 6$ A, $\lambda_A = 1$

$I_B = 8$ A, $\lambda_B = 0.9$ lag

$I_C = 12$ A, $\lambda_C = 0.8$ lag

Equation 1 Power taken by one phase

$P = V_P I_P \lambda_P$

Equation 2 phase voltage

$$V_P = \frac{V_L}{\sqrt{3}} = \frac{400}{\sqrt{3}} = 230 \text{ V}$$

Equation 3 total power $P = P_A + P_B + P_C$

$P_A = 230 \times 6 \times 1 = 1380$ W

$P_B = 230 \times 8 \times 0.9 = 1656$ W

$P_C = 230 \times 12 \times 0.8 = 2208$ W

$P = 1380 + 1656 + 2208$

Answer **P = 5244 W = 5.2 kW**

FIGURE 20.28

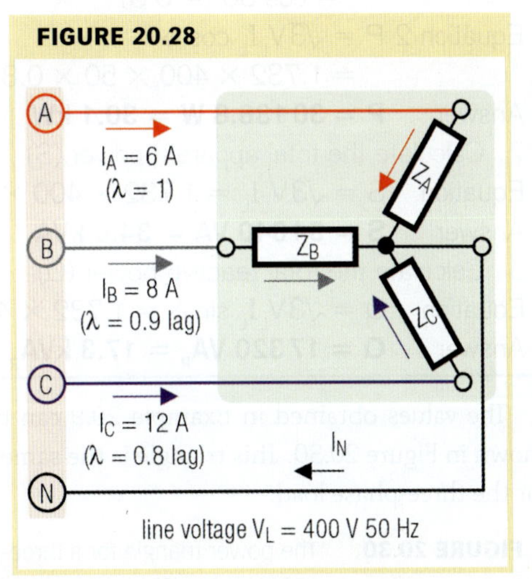

Apparent and reactive power

In a single-phase circuit, as explained in Chapter 19, the reactive power (Q) consumed by a load is the product of the line voltage, line current and the sine of the phase angle. That is:

1 $Q = VI \sin\phi$ (single-phase circuit).

For a single-phase circuit, the apparent power (S) supplied to a load is the product of the line voltage and the line current. That is:

2 $S = VI$ (single-phase circuit).

We can adapt Equations 1 and 2 to find reactive and apparent power in a balanced three-phase load in the same way as for true power by introducing the $\sqrt{3}$ multiplier into the equations. Therefore, for any balanced three-phase load:

3 $Q = \sqrt{3} V_L I_L \sin\phi$ (three-phase balanced load)

FYI

The equations to find reactive and apparent power for a three-phase balanced load are the same equations for single phase, except multiplied by $\sqrt{3}$ (1.732)

4 $S = \sqrt{3}V_L I_L$ (three-phase balanced load), where:
Q = total reactive power (VA_R)
S = apparent power (VA)
V_L = line voltage of the three-phase supply (volts)
I_L = line current
ϕ = phase angle.

EXAMPLE 20.9

The three-phase balanced load in Figure 20.29 takes a line current of 50 A at a lagging phase angle of 30° when connected to a 400 V three-phase supply. Calculate the:

1 total true power taken by the load
2 total apparent power
3 total reactive power

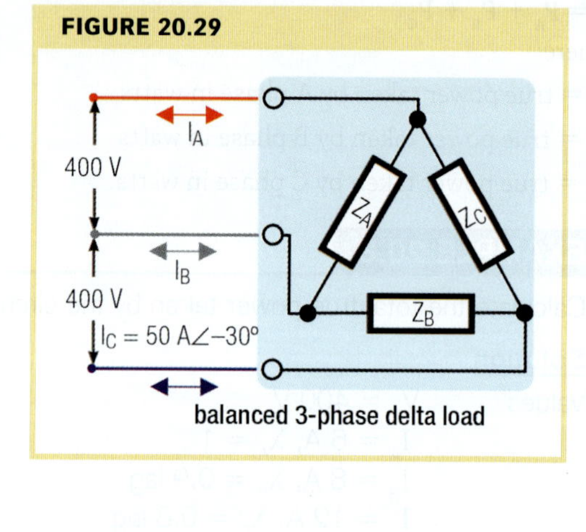

FIGURE 20.29

Solution

Values $V_L = 400$ V
$I_L = 50$ A
$\phi = 30°$

1 To find total true power (P)

Equation 1 power factor $\lambda = \cos\phi$
$= \cos 30° = 0.87$

Equation 2 $P = \sqrt{3}V_L I_L \cos\phi$
$= 1.732 \times 400 \times 50 \times 0.87$

Answer **P = 30 136.8 W = 30.1 kW**

2 Calculate the total apparent power (S)

Equation $S = \sqrt{3}V_L I_L = 1.732 \times 400 \times 50$

Answer **S = 34 640 VA = 34.6 kVA**

3 Calculate the total reactive power (Q)

Equation $Q = \sqrt{3}V_L I_L \sin\phi = 1.732 \times 400 \times 50 \times \sin 30°$

Answer **Q = 17 320 VA_R = 17.3 kVA_R**

The values obtained in Example 20.9 can be used to construct a power triangle for the circuit, as shown in Figure 20.30. This triangle is the same as for single-phase circuits, except each value is the total for the three-phase load.

FIGURE 20.30 The power triangle for a three-phase balanced load shows the total true, reactive and apparent power taken by the load

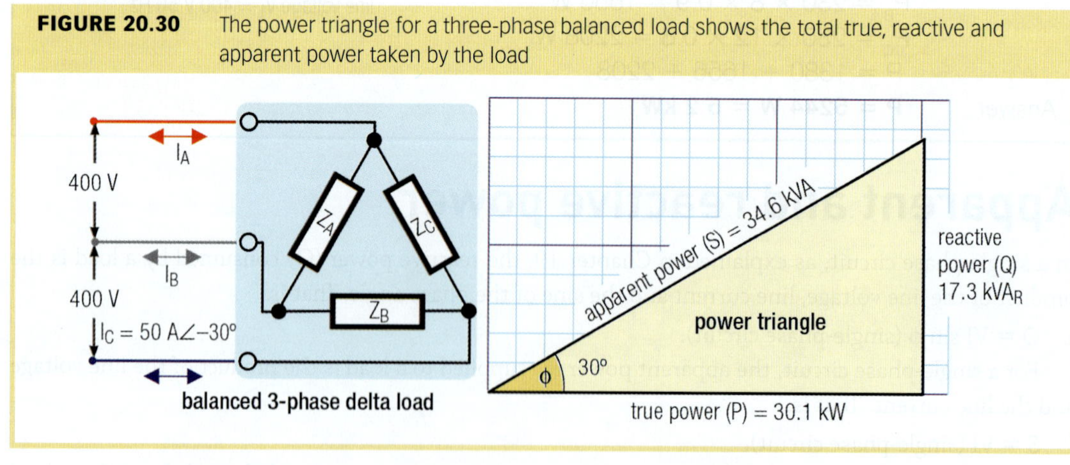

Chapter 19 explained that Pythagoras' theorem can be used with the power triangle for a single-phase load. It also applies to the power triangle for a balanced three-phase load. That is:

$S = \sqrt{P^2 + Q^2}$ when true and reactive powers are known

$Q = \sqrt{S^2 - P^2}$ when true and apparent powers are known

$P = \sqrt{S^2 - Q^2}$ when apparent and reactive powers are known.

Power factor

As for a single-phase load, the power triangle for a three-phase balanced load, as in Figure 20.30, shows the phase angle (ϕ) of the circuit. Therefore, the power factor of a balanced three-phase circuit can be found from its power triangle, in which the power factor is the ratio of the apparent power to the true power. This only applies to a balanced load, in which the current and power factor of each phase are the same. Power factor in an unbalanced load is likely to be different in each phase, so power factor for an unbalanced three-phase circuit becomes meaningless.

$$\text{power factor } (\lambda) = \frac{\text{true power (watts)}}{\text{volt-amperes (VA)}} = \frac{P}{S}$$

EXAMPLE 20.10

A delta-connected motor takes 100 kW of true power when operating from a 3.3 kV three-phase supply. If the line current is 22 A, calculate the:

1 power factor of the motor
2 phase angle of the motor
3 reactive power taken by the motor

Solution

Values $V_L = 3.3\text{ kV} = 3300\text{ V}$
$I_L = 22\text{ A}$
$P = 100\text{ kW}$

1 To calculate power factor, first find the apparent power (S)

Equation 1 $S = \sqrt{3}V_L I_L = 1.732 \times 3300 \times 22$
$S = 125\,743\text{ VA} = 125.7\text{ kVA}$

Equation 2 $\text{power factor } (\lambda) = \frac{P}{S} = \frac{100\text{ kW}}{125.7\text{ kVA}} = \frac{100}{125.7}$

Answer **power factor (λ) = 0.795 (0.8 rounded up)**

2 Calculate the phase angle

Equation $\phi = \cos^{-1}\lambda = \cos^{-1} 0.795$

Answer $\boldsymbol{\phi = 37.3°}$

3 Calculate the reactive power (Q)

Equation $Q = \sqrt{3}V_L I_L \sin\phi = 1.732 \times 3300 \times 22 \times \sin 37.3°$

Answer $\mathbf{Q = 76\,198.9\text{ kVA}_R = 76.2\text{ kVA}_R}$

Example 20.10 shows that power factor is an important consideration in a three-phase supply. In Example 20.10, the poor power factor means over 125 kVA of apparent power is needed to provide 100 kW of true power. To improve the power factor, capacitor banks are connected in parallel with the supply. These are usually connected in a star configuration, as in Figure 20.31, and often as a double-star connection, either grounded or ungrounded. Detecting the unbalanced current with a current transformer provides protection for the bank. Protection systems include detecting an unbalance between phases or, in a double-star configuration, between two star-connected banks.

FIGURE 20.31 Capacitor banks are connected in parallel (shunt) with three-phase power sources to improve power factor

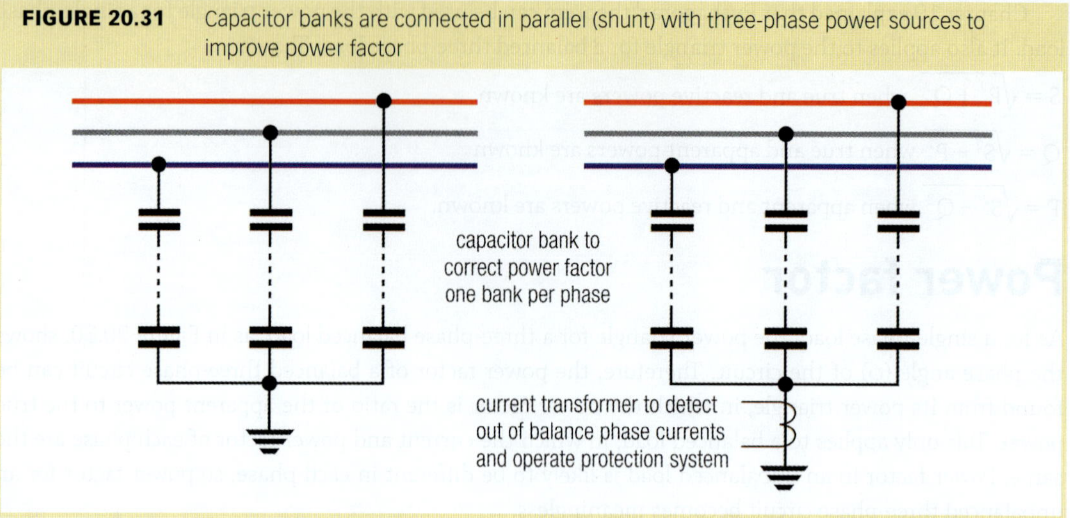

KEY POINTS...

- The true power taken by one phase of a three-phase load can be found with the same equation used in single-phase power: $P_{phase} = V_{phase} I_{phase} \cos\phi$
- The equation to find true power taken by a balanced star- or delta-connected three-phase load is: $P = \sqrt{3}\, V_L I_L \cos\phi$.
- The total true power taken by an unbalanced three-phase load is the sum of the powers taken by each phase: $P = P_A + P_B + P_C$.
- The total apparent power taken by a three-phase balanced load: $S = \sqrt{3} V_L I_L$ (VA).
- The total reactive power taken by a three-phase balanced load: $Q = \sqrt{3} V_L I_L \sin\phi$ (VA_R).
- True, apparent and reactive power in a three-phase circuit can be shown as a power triangle.
- Power factor in a balanced three-phase circuit: $\lambda = \frac{\text{true power (watts)}}{\text{volt-amperes (VA)}} = \frac{P}{S}$.

20.7 Measuring Three-Phase Power

True power, regardless of the power factor, can be measured with a wattmeter as it simultaneously measures current and voltage. Three-phase supplies are either three-wire or four-wire systems, and loads can be balanced or unbalanced, so different techniques are used to measure three-phase power, all involving one or more wattmeters.

One wattmeter, four-wire supply

For a balanced three-phase load, a single wattmeter can be connected to measure the power taken by one phase of the load, and its reading multiplied by three to obtain the total true power. Figure 20.32(a) shows a balanced star load with the current coil of a wattmeter measuring the current in line A. The voltage coil of the wattmeter is connected between line A and the neutral. This means the wattmeter is reading a phase voltage and a phase current.

For a balanced load, total power is:

1 $P = 3P_A = 3P_B = 3P_C$ (balanced load).

If the load is unbalanced, the wattmeter is connected into each phase in turn and the total power is the sum of the individual readings. That is:

2 $P = P_A + P_B + P_C$ watts (balanced or unbalanced load).

FIGURE 20.32 One wattmeter can be used to measure three-phase power in a balanced circuit. If a neutral is not available, an artificial neutral point can be established as shown in (b).

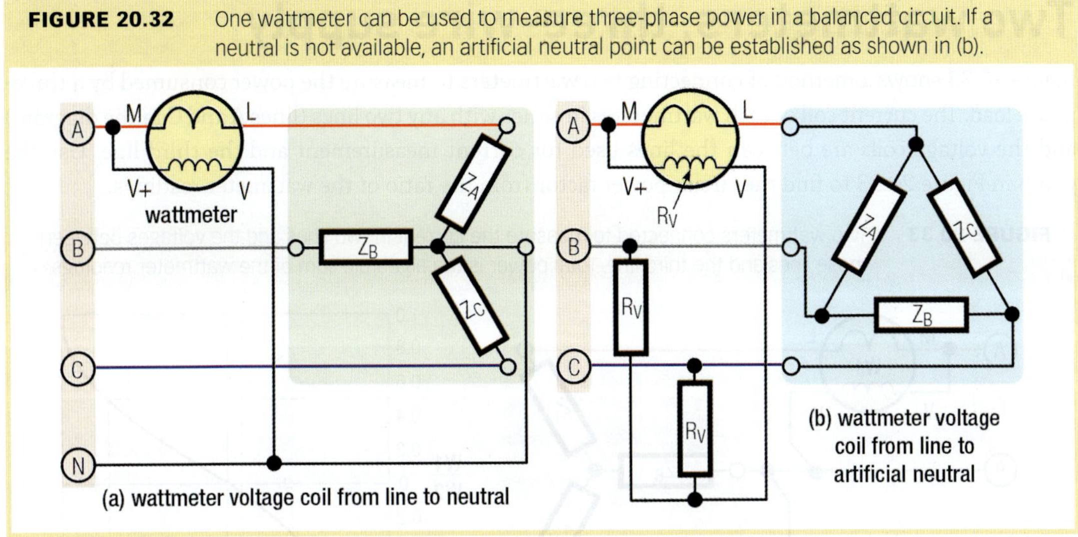

This method of power measurement has the advantages of needing one wattmeter only, and can be used with balanced or unbalanced loads. However, it requires a neutral connection, which is not always available.

One wattmeter, three-wire supply

When there is no neutral available, an artificial neutral point can be made using two resistors with the same resistance as the voltage coil circuit in the wattmeter, as shown in Figure 20.32(b). This forms a star circuit with all impedances equal, so the voltage at the star point will be zero. This is needed, as there is a 30° phase shift between line and phase voltages (see Figure 20.11), so it's important to measure the phase voltage, which is made possible by the artificial neutral point.

The total power in a balanced circuit is found with equation 1 above, and for an unbalanced circuit, with equation 2. This method of measurement has the same advantages as for the four-wire system, but requires a 'Y box' to provide an artificial neutral point. A disadvantage with the one-wattmeter method of power measurement is inaccuracy with fluctuating loads.

EXAMPLE 20.11

A single wattmeter shows 1600 W when connected as shown in Figure 20.32(a) to measure power dissipated by a 400 V balanced, star-connected heating load. Calculate the:

1 total power dissipated by the heater
2 impedance of each phase of the heater

Solution

Values $P_A = 1.6$ kW
$V_L = 400$ V

1 Calculate total power dissipated

Equation $P = 3P_A$

Answer **P = 4.8 kW**

2 To find impedance, first find the current in each phase of the load. Because the load is resistive, this can be found from a power equation. Assume the phase voltage of a 400 V supply is 230 V.

Equation 1 $I_{ph} = \frac{P_{ph}}{V_{ph}} = \frac{1600}{230} = 6.96 \text{ A}$

Equation 2 $Z = \frac{V_{ph}}{I_{ph}} = \frac{230}{6.96}$

Answer **Z = 33.1 Ω**

Two wattmeters, three-wire supply

Figure 20.33 shows a method of connecting two wattmeters to measure the power consumed by a three-phase load. The current coil of each wattmeter is in series with any two lines (lines A and C in the diagram) and the voltage coils are between the lines used for current measurement and the third line. Use the graph in Figure 20.33 to find the circuit power factor from the ratio of the wattmeter readings.

FIGURE 20.33 Two wattmeters connected to measure the current in two lines and the voltages between those lines and the third line. Total power is the algebraic sum of the wattmeter readings.

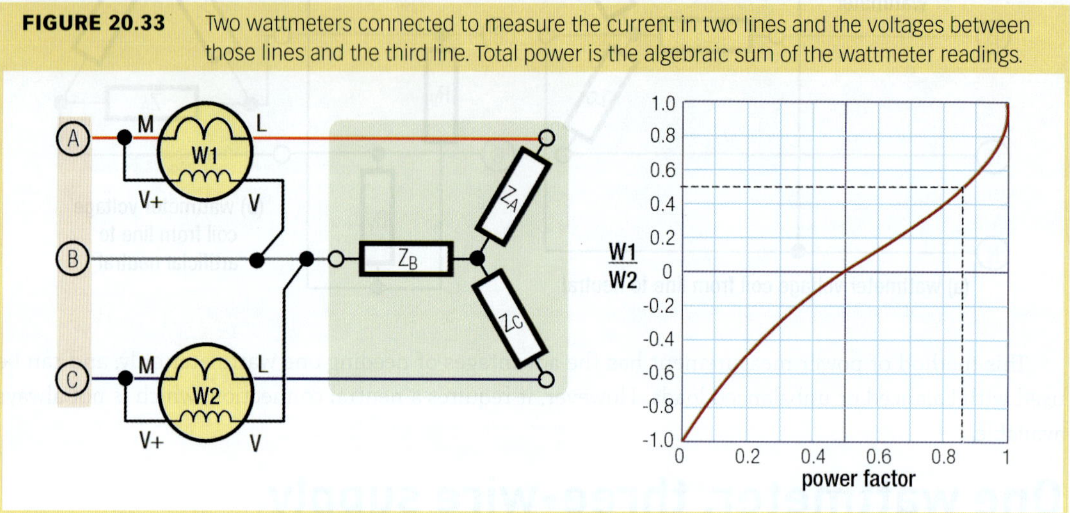

The total power is the algebraic sum of the two wattmeter readings, which for a balanced load at unity power factor will be equal. For all other conditions, the meter readings will be different, and can even be negative.

- $P = W_1 + W_2$ (balanced or unbalanced, star- or delta-connected load, any power factor, no neutral), where W_1 and W_2 are the wattmeter readings in watts.

When a negative reading occurs, the current or voltage connections to that wattmeter can be reversed to obtain the reading. However, the reading must be considered to have a negative value, so the total power is the difference between the meter readings.

The only restriction with the two-wattmeter method is there must *not* be a neutral conductor. This is because there could be current flowing in the neutral from the line that is not in series with a wattmeter's current coil.

If the load is balanced and has pure reactance (inductance or capacitance), one wattmeter will show a positive value and the other a negative reading of the same value. Adding the two readings will give zero, indicating the circuit is not taking any true power.

For a balanced load, the power factor of the load can be found with the two-wattmeter method by using this equation:

$$\tan\phi = \sqrt{3} \times \left(\frac{W_2 - W_1}{W_2 + W_1}\right).$$

Angle ϕ is found from the inverse tan function ($\tan^{-1}$), and power factor is found from the cosine of the angle. Another method is to use the graph shown in Figure 20.33.

EXAMPLE 20.12

Two wattmeters connected to a three-phase motor read 2 kW (W_1) and 4 kW (W_2). Calculate the:

1. power taken by the motor
2. power factor

Solution

Values $W_1 = 2$ kW
$W_2 = 4$ kW

1 Calculate the power

Equation $P = W_1 + W_2 = 2\text{ kW} + 4\text{ kW}$

Answer **$P = 6\text{ kW}$**

2 Calculate the power factor – from equation

Equation 1 $\tan\phi = \sqrt{3} \times \left(\frac{W_2 - W_1}{W_2 + W_1}\right) = \sqrt{3} \times \left(\frac{4-2}{4+2}\right) = \sqrt{3} \times \frac{2}{6}$

$\tan\phi = 0.58,\ \phi = \tan^{-1} 0.58$

$\phi = 30°$

Equation 2 $\lambda = \cos\phi = \cos 30°$

Answer **$\lambda = 0.866$**

From graph in Figure 20.33:

- The ratio of the wattmeter readings $W_1/W_2 = 2/4 = 0.5$
- From the graph, the power factor = **0.87**

Three wattmeters

Four-wire supply

Figure 20.34 shows how three wattmeters can be connected to measure the power being delivered to a three-phase load. The load can be balanced or unbalanced, and connected in star or delta. This method has a wattmeter measuring all three line currents and phase voltages, which gives better accuracy when measuring power to a fluctuating unbalanced load.

FIGURE 20.34 Connections for three wattmeters and a three-phase four-wire system

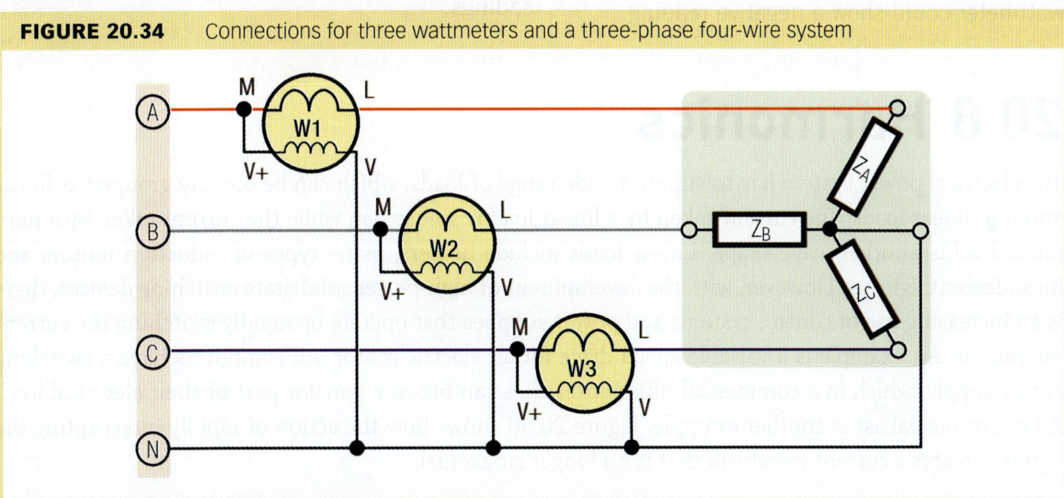

Three-wire supply

If there is no neutral connection available, the wattmeters can be connected as shown in Figure 20.35, assuming they are identical. This is similar to the circuit in Figure 20.32, where a neutral point was established with two resistors and one wattmeter voltage coil.

For both connection methods, the total power is the algebraic sum of the individual wattmeter readings. That is:

- $P = W_1 + W_2 + W_3$ (balanced or unbalanced, star- or delta-connected load, any power factor), where W_1, W_2 and W_3 are the wattmeter readings in watts.

FIGURE 20.35 Connections for three wattmeters and a three-wire system

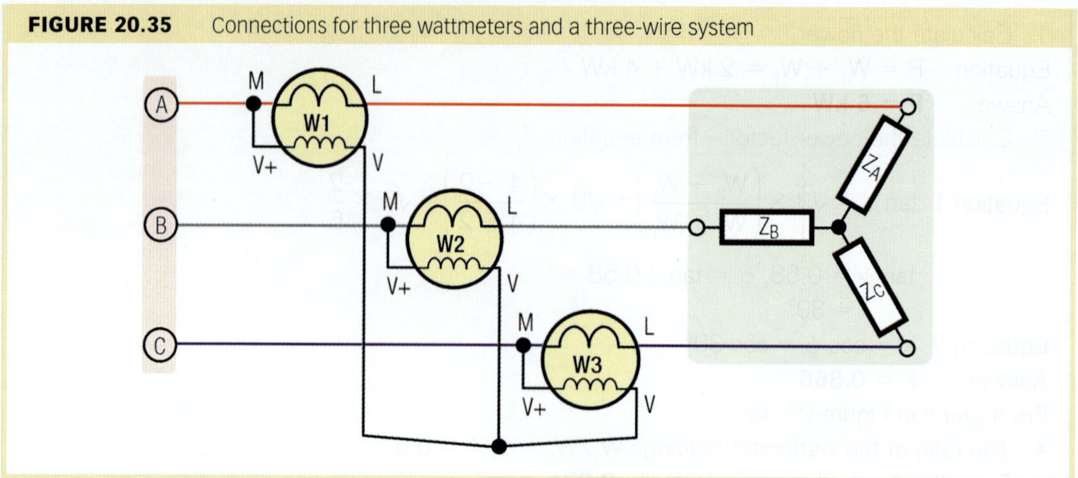

KEY POINTS...

- Power in a three-phase balanced load can be measured with a single wattmeter, where the total power is three times the wattmeter reading. That is $P = 3P_A = 3P_B = 3P_C$.
- Total power using the two-wattmeter method is the algebraic sum of the wattmeter readings: $P = W_1 + W_2$. Depending on the power factor, either wattmeter could show a negative reading.
- Power factor for a balanced load can be derived from the two-wattmeter method: $\tan\phi = \sqrt{3} \times \left(\frac{W_2 - W_1}{W_2 + W_1}\right)$.
- Total power using the three-wattmeter method is the algebraic sum of the three wattmeter readings.

20.8 Harmonics

The electrical power system has to supply a wide range of loads, which can be broadly grouped as linear and non-linear loads. The current taken by a linear load is sinusoidal, while the current taken by a non-linear load is another wave shape. Linear loads include heaters, many types of induction motors and incandescent lighting. However, with the development of high-power solid-state switching devices, there is an increasing use of control systems and power supplies that operate by rapidly switching the current on and off. An example is a variable speed drive for an electric motor. All computers have a switching power supply, which in a commercial office enterprise can become a major part of their electrical load. An electronic ballast is another example. Figure 20.36 shows how the action of rapidly interrupting the current creates a current waveform that is no longer sinusoidal.

FIGURE 20.36 The switching action of power control systems can generate additional current frequencies, called harmonics

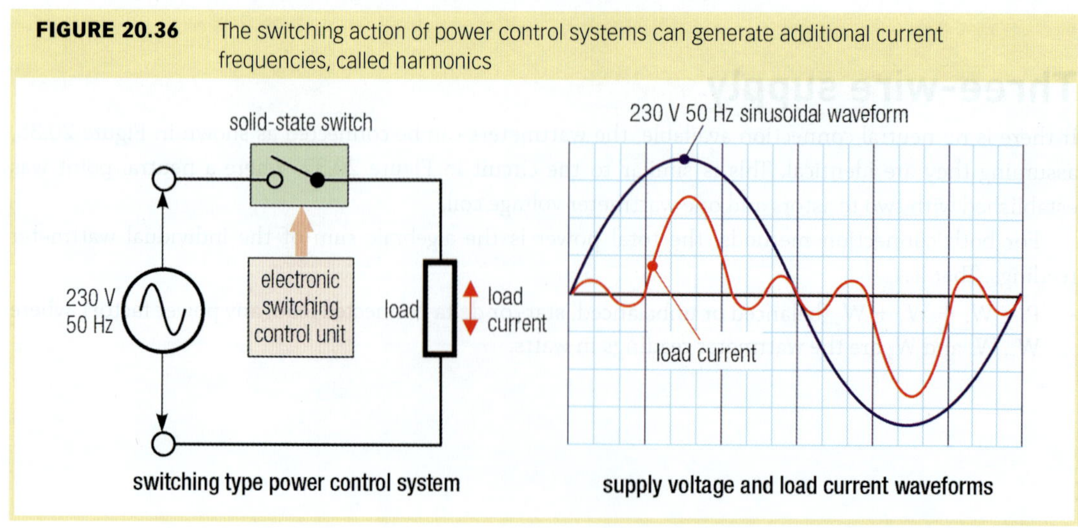

The sinewave is a fundamental waveform and all waveforms, of whatever shape, are made up of numerous sinewaves of various frequencies and various amplitudes. Therefore, if a waveform is not sinusoidal, there are other components making up that waveform. In many cases the other components are multiples of the fundamental frequency. These are called harmonic frequencies, or harmonics. For example, the third harmonic of any frequency is three times that frequency, so the third harmonic of 50 Hz is 150 Hz. Figure 20.37 shows the effect on the waveshape when a 50 Hz sinewave and its third harmonic are combined.

FIGURE 20.37 3rd harmonic distortion caused by the combination of a fundamental waveform and its 3rd harmonic

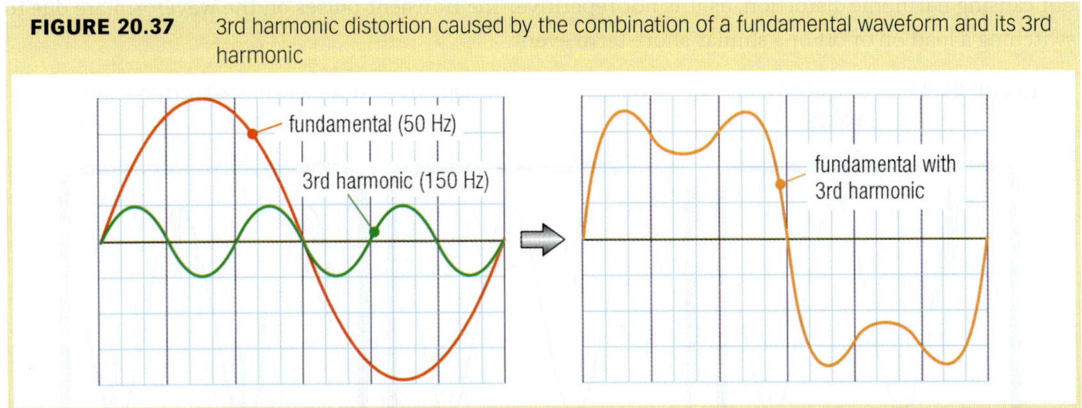

The effect of a 5th harmonic of a 50 Hz sinewave is shown in Figure 20.38. Like the waveform in Figure 20.37, it is no longer sinusoidal due to *harmonic distortion*.

FIGURE 20.38 5th harmonic distortion caused by the combination of a fundamental waveform and its 5th harmonic

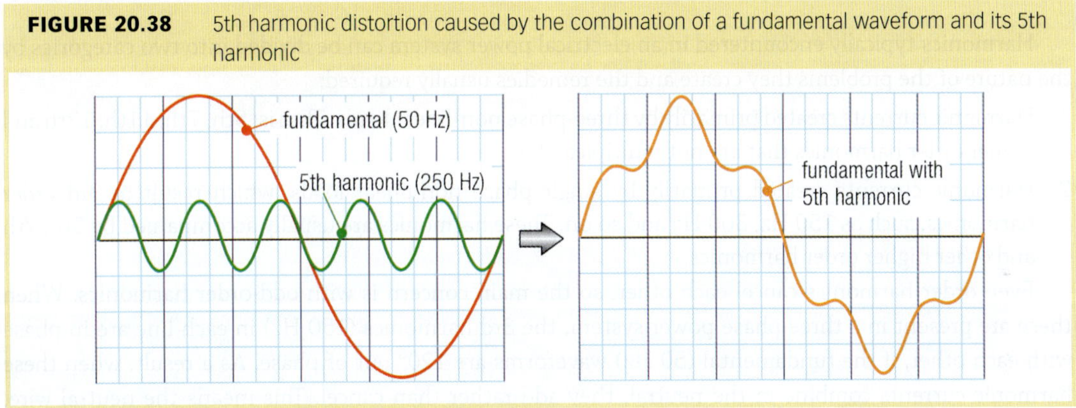

Harmonic classification

Harmonics are grouped into positive, negative and zero sequence components. In a motor, positive sequence harmonics produce magnetic fields and currents rotating in the same direction as the fundamental frequency. Negative sequence harmonics develop magnetic fields and currents that rotate in the opposite direction. Zero sequence harmonics do not develop useable torque, but produce additional losses in the machine.

The interaction between the positive and negative sequence magnetic fields and currents produces torsional oscillations of the motor shaft, causing it to vibrate. If the oscillating frequency coincides with the natural mechanical frequency of the shaft, the vibrations are amplified and severe damage to the motor can occur. Table 20.2 shows the first nine harmonics and their rotational sequence.

TABLE 20.2

HARMONIC SEQUENCE									
Harmonic	F	2nd	3rd	4th	5th	6th	7th	8th	9th
Frequency	50	100	150	200	250	300	350	400	450
Sequence	+ve	−ve	0	+ve	−ve	0	+ve	−ve	0

Problems caused by harmonics

Under some conditions the harmonic currents caused by a non-linear load can cause distortion of the supply voltage waveform, which is the responsibility of the supply utility. Customers are responsible for maintaining current distortion within acceptable limits, but the problem is now reaching the point where the Institute of Electrical and Electronics Engineers (IEEE) is developing a harmonic-distortion-based penalty structure. Examples of supply voltage distortion are shown in Figure 20.39. Waveform (a) is not causing harmonic distortion, as it is not repetitive. The transient pulses on the waveform are due to switching a load on or off or a similar short-term event.

FIGURE 20.39 Distortion of the supply voltage waveform can cause equipment breakdown and other effects

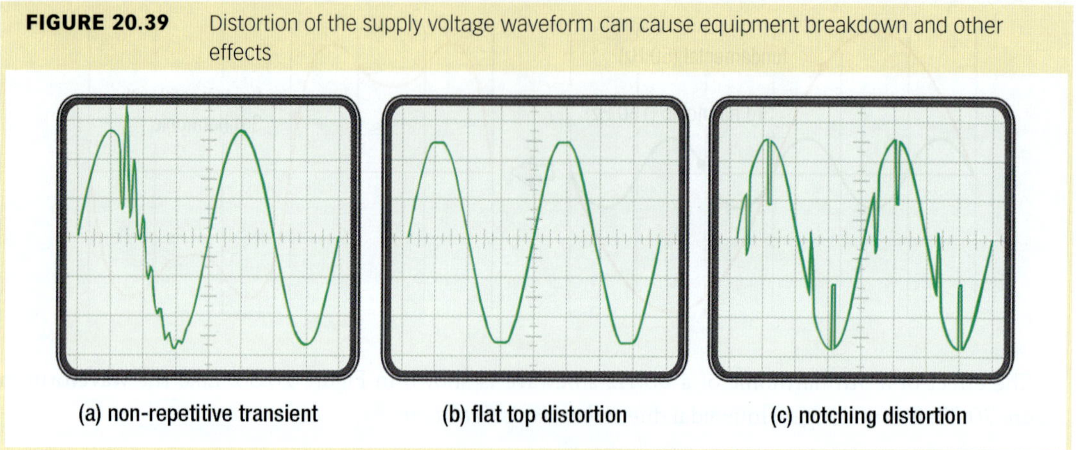

Harmonics typically encountered in an electrical power system can be divided into two categories by the nature of the problems they create and the remedies usually required:

1 Harmonic currents created primarily by three-phase non-linear loads. That is: 5th, 7th, 11th, 13th and higher order harmonics that are not multiples of 3.

2 Harmonic currents created primarily by single-phase non-linear loads, which result in 3rd order harmonics, such as 150 Hz, 300 Hz and so on. These harmonics are usually accompanied by 5th, 7th and other higher order harmonics.

Even-order harmonics cancel each other, so the main concern is with odd-order harmonics. When these are present in a three-phase power system, the 3rd harmonics (150 Hz) in each line are in phase with each other, if the fundamental (50 Hz) waveforms are 120° out of phase. As a result, when these harmonic currents combine in the neutral, they add rather than cancel. This means the neutral wire, which is typically the same size as the phase conductors, is now carrying more current than it is specified to handle.

The effects depend on the magnitude of the harmonic currents, but include heating the neutral conductor and overheating the supply transformer. Other problems are circuit-breaker nuisance tripping, malfunctioning equipment connected to the line, metering problems and over-voltages. In a substation with capacitor banks, the presence of harmonics can cause resonance between the bank and a nearby inductance, such as a transformer.

FYI

Requirements regarding harmonic currents and size of the neutral conductor are in AS/NZS 3000:2018 clause 3.5.2

Harmonic currents are generally not detected by a standard clamp-on ammeter, as these only work at 50 Hz or so. Newer types can handle frequencies up to several thousand hertz, and provide a true RMS reading. These types should be used where harmonic currents are suspected. The difference in the readings from a standard clamp-on ammeter and the newer type gives an indication of the magnitude of the harmonic currents. Old-style mechanical watt-hour meters only read sine wave power, not the harmonics. Smart watt-hour meters read everything.

Reducing the effects of harmonics

Where harmonics are present, or likely to be present, in a three-phase power system, over-sized neutrals can be installed to handle the additional current load. On three-phase branch circuits, running separate neutral conductors for each phase conductor is another way. Transformers and generators supplying a non-linear load are often oversized to prevent overheating due to harmonic currents.

Special types of transformers that can handle harmonics are also used. These have specially designed cores and windings. A delta-star transformer (explained in the next chapter) traps the triplen harmonics (e.g. 3rd, 9th, 15th, 21st and so on). Additional special winding connections can further reduce other harmonics.

The effects of harmonics on a power factor correction capacitor bank can be reduced by connecting each phase in series with an inductance, called a reactor. By selecting the correct value of inductance, a harmonic filter is created. In a three-phase system, each phase of a capacitor bank has a separate reactor, as shown in Figure 20.40. Series reactors are used with capacitor banks for two main reasons.

1 To dampen the effect of transients during capacitor switching.
2 To control the natural frequency of the capacitor bank and the system impedance to avoid resonance, or to sink harmonic currents.

In the second case listed above, a reactor of a chosen inductance value is connected in series with a capacitor bank to give a tuned circuit that acts as a filter at a certain frequency. There could be several LC circuits, involving additional reactors and capacitor banks, allowing specific frequencies to selectively filter out certain harmonic currents. In other cases, the LC circuit could be de-tuned away from a particular frequency to protect the capacitor bank from being overloaded or damaged by a harmonic current.

Harmonic filters are also used in power distribution systems. These are LC circuits tuned to a particular frequency and connected to the three-phase bus. There could be several filters to remove 5th, 7th, 11th, etc. order harmonics.

FIGURE 20.40 Adding reactance to a capacitor bank to reduce the effects of harmonics

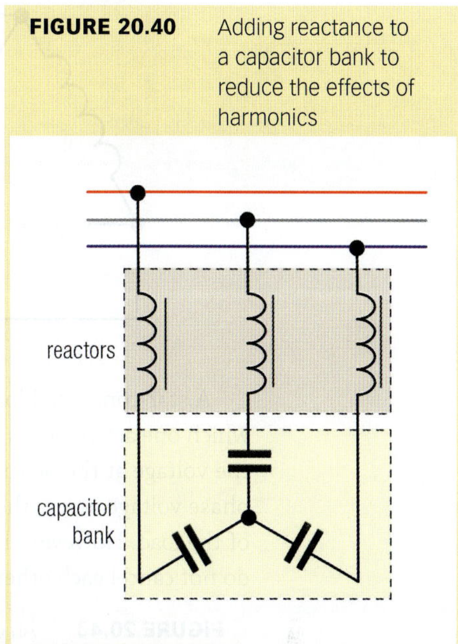

20.9 Faults in Three-Phase Power Systems

A common type of fault is the loss of one phase in a three-phase supply. This could be due to a protection device operating or an open-circuit in a supply line. Another possibility is an open-circuit phase winding in the generating source. Figure 20.41 shows a delta supply connected to a delta load. One phase winding in the generating source is open-circuit.

FIGURE 20.41 Loss of one phase winding in a delta-connected generating source does not affect the voltages across a three-phase load

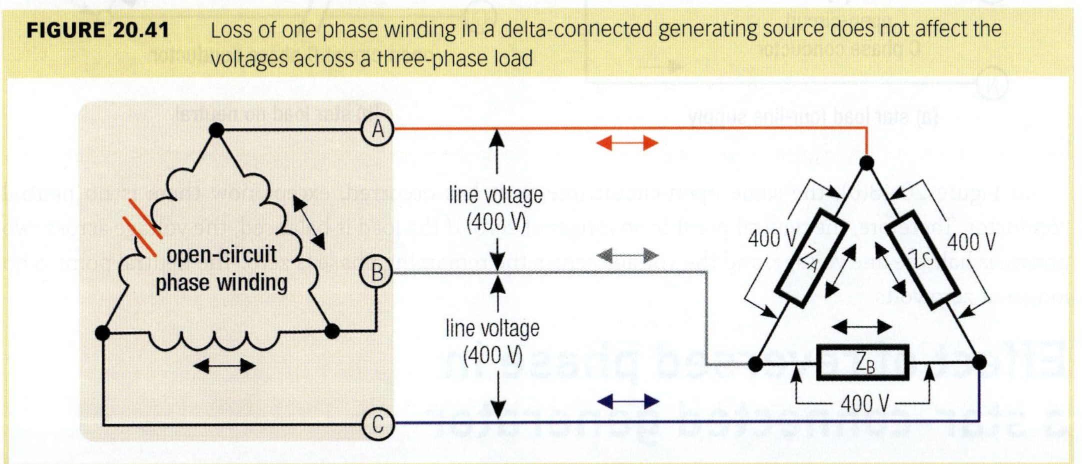

In Figure 20.41, each phase of the load has the full line voltage across it, so the current in each phase will be the same as it was before the fault occurred. The only difference is that the remaining phase windings in the generating source will have increased current flowing in them. Under certain conditions, if the increased current does not exceed the generator current rating, the load will continue to operate normally. This is a distinct advantage of a delta-connected supply source.

Figure 20.42 shows a condition where one supply line is open-circuit. In this case, one phase of the load is not affected, but the voltage across the other two phases of the load is now half the line value, assuming a balanced load. Because of the reduced voltage, the affected phases of the load have reduced current flowing in them.

FIGURE 20.42 An open-circuit supply line affects the voltage across and the current flowing in two phases of a delta-connected load

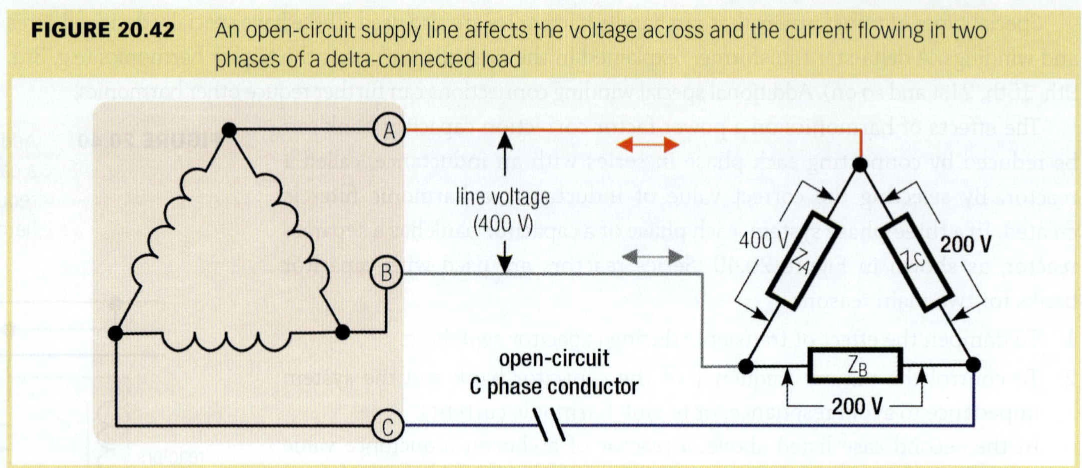

A star-connected load operating from a four-line three-phase supply is shown in Figure 20.43(a) in which one active line is open-circuit. Because there's a neutral connection to the star point of the source, the voltage at the neutral point is zero, as it should be. Therefore, two phases of the load have the usual phase voltage across them of 230 V (assuming a 400 V supply) and there is zero voltage across phase Z_C of the load. However, the current in the neutral will no longer be zero, as the currents in A and B phases do not cancel each other.

FIGURE 20.43 An open-circuit supply line to a star-connected load causes zero volts across one load phase and current to flow in the neutral line. If there is no neutral line, the load star point will be above zero volts.

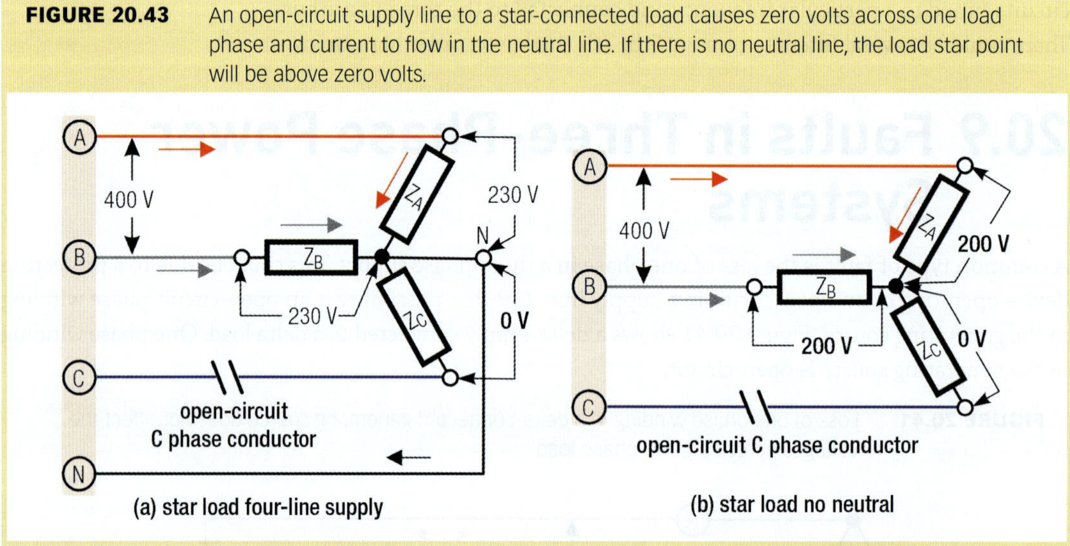

In Figure 20.43(b), the same open-circuit line fault has occurred, except now there is no neutral conductor. Therefore, the neutral point is no longer at zero. If the load is balanced, the voltage across two phases is half the line voltage, and the voltage across the remaining phase is zero. The neutral point is no longer at zero volts.

Effect of reversed phase in a star-connected generator

Three-phase supply sources are usually either an alternator or a transformer. Both sources have phase windings, typically with accessible terminals so the windings can be connected in star or delta. When connected correctly, each phase has a phase difference of 120°. But what happens if a phase winding is connected with reverse polarity?

The effect of a reversed phase in a star-connected source is shown in the phasor diagrams in Figure 20.44. Phasor diagram (a) shows the phasors for line and phase voltages when the windings are correctly connected. Phasor diagram (b) shows the effect of reversing the connections of phase winding C. Line voltage $V_{A\text{-}B}$ is not affected, but line voltages $V_{C\text{-}A}$ and $V_{B\text{-}C}$ are now equal to the phase voltage. The result is the load is no longer balanced.

FIGURE 20.44 Reversing a phase winding in a three-phase star-connected generator causes the line voltages of two phases to become equal to their phase voltage as shown in (b)

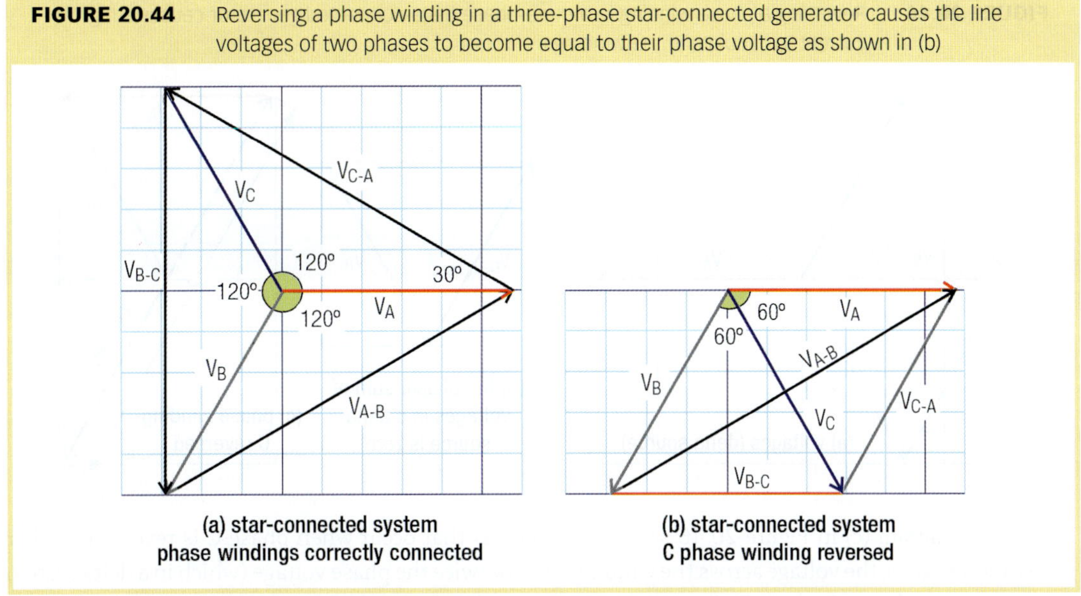

The waveforms in Figure 20.45(a) show that with C phase reversed, there is a 120° phase difference between phase voltages A and B, and 60° phase difference between phase voltages A and C, and between C and B. Therefore the phase sequence is no longer A-B-C, instead it is A-C-B, which will cause three-phase motors to run backwards if they are still able to run.

FIGURE 20.45 Reversing a phase winding in a three-phase star-connected generator changes the rotational sequence and causes two line voltages to become equal to the phase voltage of the generator

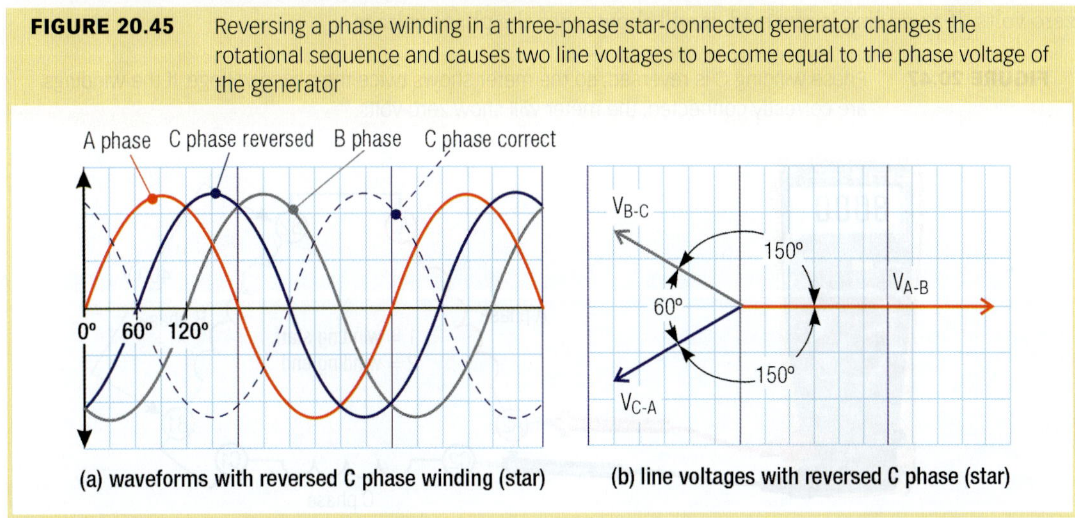

To test for a reversed phase winding in a star-connected source, check that the phase voltages are all the same. Also check that the line voltages have the same value and are 1.732 times higher than the phase voltages.

Effect of reversed phase in a delta-connected generator

A reversed phase winding in a delta-connected three-phase generator has a significant effect. When connected correctly, the three phase voltages cancel each other, so there is no circulating current in the phase windings, as shown in Figure 20.46(a). The three phase voltages are also shown in phasor diagram (b), where V_B is added to the tip of V_A and V_C is added to the tip of V_B. The resulting voltage is the distance between the tip of V_C and the tail of V_A, which is zero.

FIGURE 20.46 Reversing a phase winding in a three-phase delta-connected generator causes a high current to circulate in the windings

Phasor diagram (c) in Figure 20.46 shows the voltages that occur when phase C is reversed. Rather than cancel to zero, the voltage across the windings is now twice the phase voltage (which in a delta system is also the line voltage). In a 400 V three-phase generator, this means there is now 800 V generated within the closed circuit of the windings, which causes a high circulating current to flow in the windings. This will cause them to quickly overheat and burn out.

A method of ensuring correct phase winding connections is shown in Figure 20.47. If the windings are correctly connected, a voltmeter connected between any two open junctions of the circuit will show zero volts. If a winding is reversed, it will show twice the phase voltage.

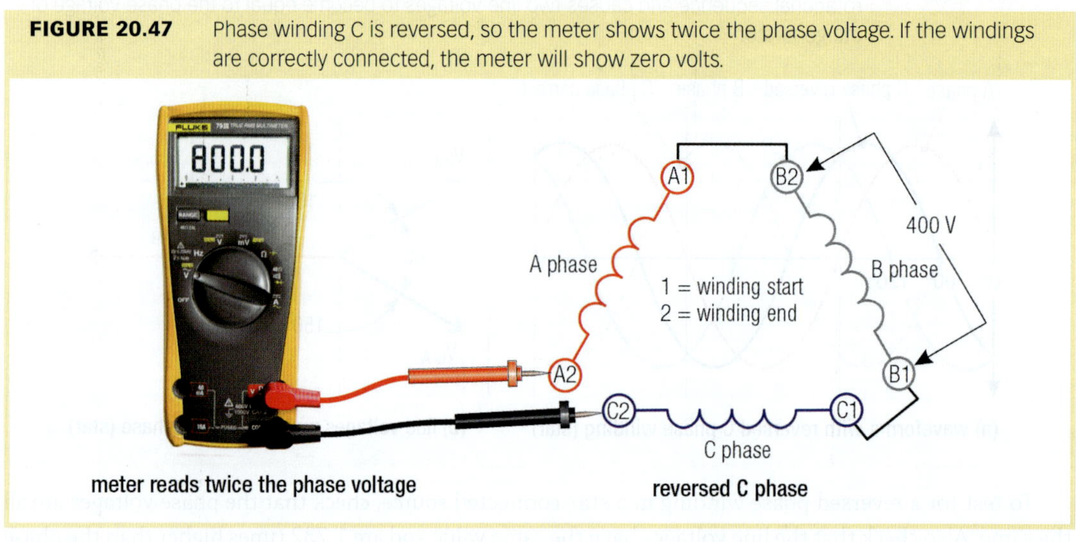

FIGURE 20.47 Phase winding C is reversed, so the meter shows twice the phase voltage. If the windings are correctly connected, the meter will show zero volts.

CHAPTER SUMMARY

- The equations introduced in this chapter are:
  - $V_{line} = \sqrt{3} \times V_{phase}$ (star connection)
  - $V_{phase} = \frac{V_{line}}{\sqrt{3}}$ (star connection)
  - $I_{phase} = I_{line}$ (star connection)
  - $V_{phase} = V_{line}$ (delta connection)
  - $I_{line} = \sqrt{3} \times I_{phase}$ (delta connection)
  - $I_{phase} = \frac{I_{line}}{\sqrt{3}}$ (delta connection)
  - load phase current $I_p = \frac{V_p}{Z_p}$
  - $P_A = V_{PA} I_{PA} \cos \phi$ where P_A is the power taken by phase A of a three-phase load
  - $P = 3 \times V_{PA} I_{PA} \cos \phi$ where P is the total true power taken by a balanced three-phase load

- $P = \sqrt{3}\,V_L I_L \cos\phi$ where P is the total true power taken by a balanced star- or delta-connected load
- $P = P_A + P_B + P_C$ where P_A, P_B and P_C are the true power in watts taken by phases A, B and C
- $Q = \sqrt{3}\,V_L I_L \sin\phi$ (three-phase balanced load) where Q is the total reactive power (VA_R)
- $S = \sqrt{3}\,V_L I_L$ (three-phase balanced load) where S is the total apparent power (VA)
- $\tan\phi = \sqrt{3} \times \left(\frac{W_2 - W_1}{W_2 + W_1}\right)$ where W_1 and W_2 are wattmeter readings from the two-wattmeter method of three-phase power measurement.

- Non-linear loads, such as solid-state variable speed drives, can cause harmonic currents to flow in the electrical supply lines.
- A harmonic is a multiple of the fundamental frequency (e.g. 50 Hz, so 3rd harmonic is 150 Hz). In electrical power systems, even-order harmonics cancel.
- Harmonics can cause supply cables to overheat, in particular the neutral conductor, as the harmonics in each phase do not cancel. Other effects include transformer heating, incorrect operation of motors, distortion of the mains voltage, over-voltages and resonance within the supply system.
- Faults in a three-phase system include an open-circuit in a generator phase winding, an open-circuit conductor and a reversed generator phase winding.

REVIEW EXERCISES

Check your answers at the back of the book.

1 A three-phase alternator produces a maximum voltage of 550 V. Determine the voltage of each phase when phase A is at a rotational angle of 30°.

2 A star-connected power source has a phase voltage of 110 V and a line current of 25 A.
 a What is the line voltage?
 b What is the phase current?

3 A delta-connected supply has a phase voltage of 16 kV and a line current of 15 A.
 a What is the line voltage?
 b What is the phase current?

4 A commercial espresso coffee machine has three heating elements, each with a resistance of 46 Ω and connected in star to a 400 V three-phase supply. Determine the:
 a current in each element
 b line current in each phase of the supply.

5 A motor connected to a three-phase 400 V supply takes a line current of 25 A at a lagging power factor of 0.85. How much power is being consumed by the motor?

6 An unbalanced star-connected resistive load has 30 A in A phase, 20 A in B phase and 40 A in C phase. Determine the neutral current.

7 A three-phase delta load has an impedance in each phase of 150 Ω and is connected to a 600 V, three-phase star-connected supply. Calculate the:
 a current in each phase of the load
 b line current
 c voltage across each phase of the load.

8 A three-phase motor consumes 20 kW at a power factor of 0.75 when connected to a 400 V three-phase supply. Calculate:
 a the line current
 b apparent power taken from the supply
 c the reactive power taken by the motor.

9 Two wattmeters connected to a three-phase motor read 11.2 kW (W_1) and 7.6 kW (W_2). Find the:
 a load true power
 b load power factor.

10 A three-phase star-connected unbalanced load is supplied by a 400 V three-phase supply. The current in phase A is 10 A (p.f. = 0.75 lag); in phase B current is 6.5 A (p.f. = 0.9 lag), and phase C has a current of 12 A (p.f. = 1). Determine the total power consumed by the load.

ONLINE RESOURCES

COMPLETE WORKSHEET TWENTY

Check with your instructor for worksheets on this chapter.

CHAPTER 21

TRANSFORMERS

The transformer is an essential part of today's power transmission and distribution. Efficient, practical transformer designs did not appear until the 1880s, but within a decade the transformer and the AC power system had triumphed over the DC power system. Transformer design still relies on the principles established over 100 years ago, including many of the construction techniques. This chapter examines the operating principles of single- and three-phase transformers, and describes some of the tests and measurements associated with transformers.

CHAPTER OUTLINE

21.1 Introduction

A transformer changes an alternating voltage from one value to another. It does this so efficiently that it is a main reason electrical power systems are alternating current. A transformer allows AC power generated at one voltage to be transformed up to a much higher voltage for transmission, then transformed down to voltages such as 400 V/230 V for domestic and industrial use. Figure 21.1 shows two typical power transformers found in a substation. The power transformer in (a) transforms an incoming 66 kV supply to 11 kV for distribution by way of overhead or underground cables. At various points in the distribution network, pole-or pad-mounted auxiliary transformers, shown in (b), reduce the 11 kV to 400 V/230 V.

FIGURE 21.1 The power transformer transforms the incoming 66 kV supply to 11 kV for distribution by way of overhead or underground cables

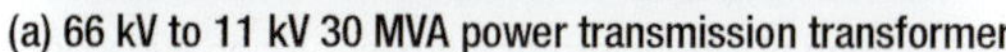

(a) 66 kV to 11 kV 30 MVA power transmission transformer

(b) 11 kV to 400 V auxiliary transformer

As well as forming a major and essential part of the electrical power supply network, transformers are found in nearly every electronic appliance, in battery chargers, mains-powered test equipment and so on. In electronics, transformers handle high frequency signals in communications equipment. They are used in industrial sound systems to power loudspeakers that could be many hundreds of metres from the amplifier. In this book we are looking mainly at power transformers, but whatever the size or purpose of a transformer, its operating principle and basic construction are the same.

21.2 The Ideal Transformer

The two main parts of any transformer are its ferromagnetic core and the windings wrapped around the core. A basic construction is shown in Figure 21.2(a) in which coils are wound on opposite limbs of a hollowed rectangular-shaped laminated core.

The two coil windings are called the primary and secondary windings. The primary is connected to the supply and the secondary to the load. The choice of which is the primary and which is the secondary depends on the application, as a transformer can operate either way. It's common to refer to transformer windings as the high voltage side and low voltage side. In a power transmission system, the primary is often the high voltage winding, although at a power station the output voltage of an alternator is stepped up for transmission, so the primary is the low voltage side.

Operating principles

Transformers operate on the principle of mutual inductance, which occurs when one coil induces a voltage in another coil, as explained in Chapter 12. When the primary winding of a transformer is connected to an AC supply, the current flowing through the winding produces an alternating magnetic flux in the core. If the supply voltage is sinusoidal, the current and the flux produced by the current are also sinusoidal. This in turn will produce a sinusoidal induced voltage in all windings on the transformer, including the winding connected to the supply, as shown in Figure 21.2(b) as voltage e_1.

FIGURE 21.2 Transformers typically have a laminated iron core and two or more windings

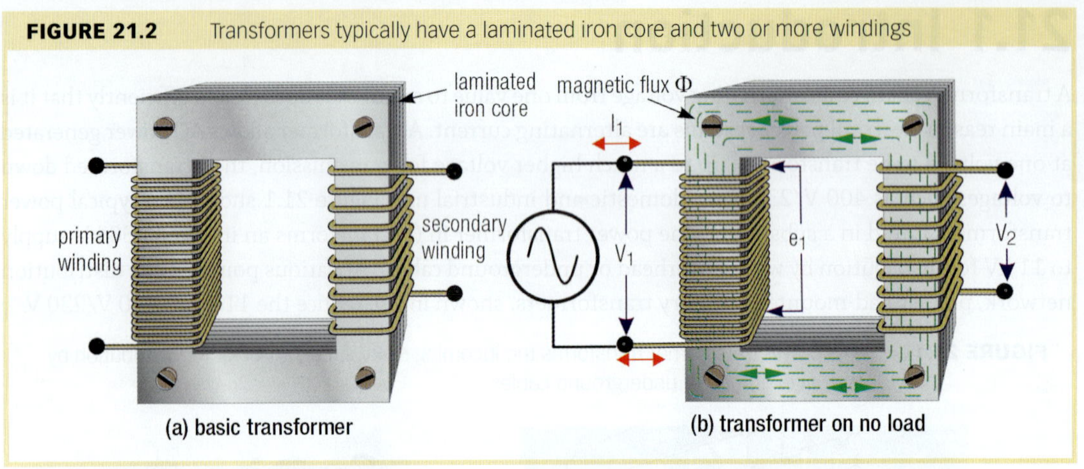

This happens because the magnetic field is expanding and contracting, causing relative motion between the magnetic field and the windings of the transformer. The resulting back-EMF in the primary winding (e_1) opposes the applied voltage, and limits the primary current to a value enough to create the magnetic flux. This current is the *magnetising* (excitation) current of the transformer. It is generally quite low and flows even when a load is not connected to the transformer. In a large power transformer, it can be as low as 1 per cent of the rated primary current.

If the secondary is connected to a load, as shown in Figure 21.3, the voltage induced in the secondary winding (V_2) will cause a current to flow in the load. This current will generate a magnetic flux in the core that opposes the flux produced by the primary winding. As a result, the core flux reduces, which in turn reduces the voltage induced in the primary (e_1). This causes the primary current to increase, which increases the magnetic flux, restoring the flux to its previous value.

FIGURE 21.3 The current flowing in the primary winding of a transformer increases when a load is connected to the secondary and energy is transferred from the supply to the load.

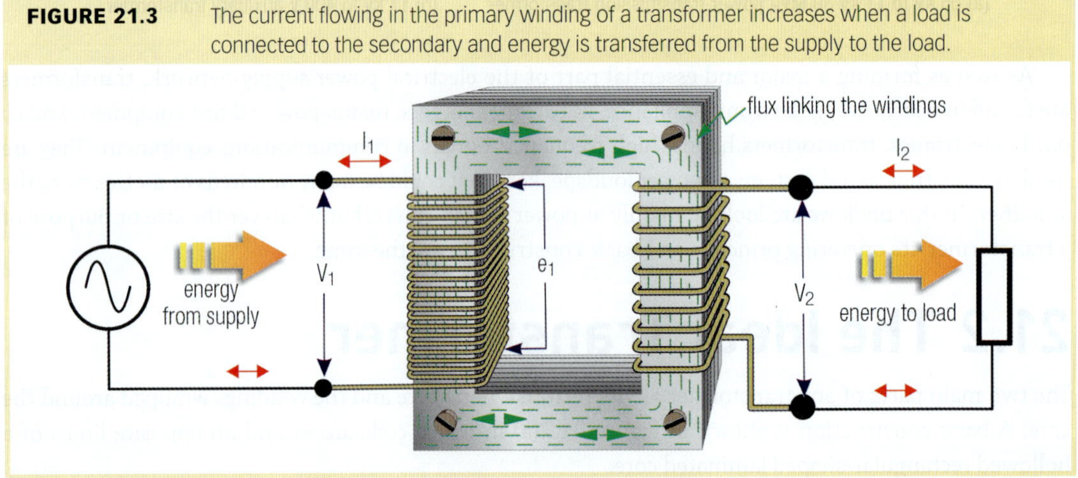

The magnetic flux in the core of a transformer, therefore, tends to remain constant from no load to full load. The primary current depends mainly on the load connected to the secondary, which means electrical energy is transferred from the supply to the load, even though there is no electrical connection between the two windings. This is how electrical isolation is achieved.

Circuit symbols for a transformer are shown in Figure 21.4. The standard symbol has two coils as in Figure 21.4(a), with the type of core material identified by a line (iron core), dashed line (ferrite core) or no line for an air-cored transformer. Simplified symbols for a single- or three-phase transformer are shown in Figure 21.4(b) and (c).

FIGURE 21.4 Transformer circuit symbols

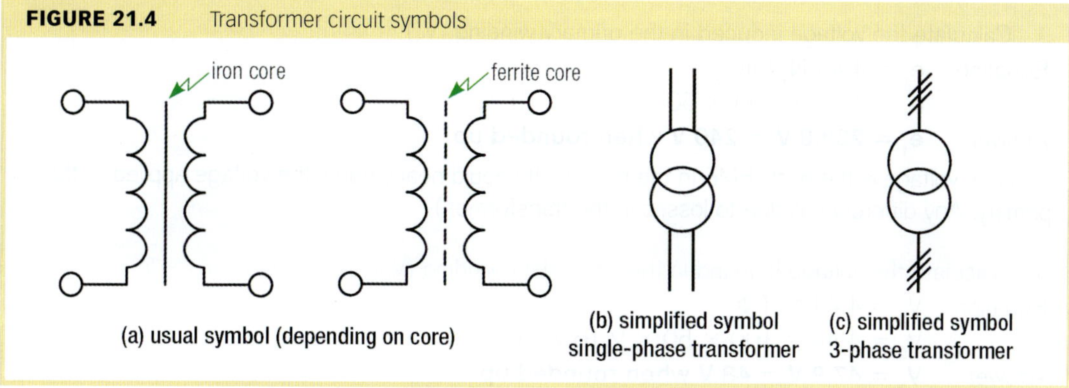

Induced voltage value

In an ideal transformer, the windings have no resistance, the core is never saturated, the magnetic flux produced by the primary completely links with the secondary and the input power to the transformer equals the power dissipated in the load. We are assuming an ideal transformer at this stage as a way of understanding the operation of a transformer.

Chapter 12 explained the factors that determine the magnitude of the induced voltage in a coil. These are the:

- length of the conductor
- strength of the magnetic field
- speed of the relative motion between the conductor and the field.

These factors also determine the value of the voltage induced in the windings of a transformer, where the:

- length of the conductor is related to the number of turns in a winding
- strength of the magnetic field is related to the flux (Φ) in the core
- relative motion between the conductor and the field depends on the frequency of the supply.

For sinusoidal waveforms, these three factors can be combined in an equation to find the RMS voltage induced in the primary or secondary windings of a transformer:

$V = 4.44\ Nf\ \Phi_{max}$

where:

V = RMS value of the induced voltage

N = number of turns on the coil

f = frequency of the voltage in hertz

Φ_{max} = maximum value of the magnetic flux in webers.

This equation is derived from Faraday's law, which says:

- the voltage induced in a coil in a changing magnetic field is proportional to the number of turns on the coil and the rate of change of the flux.

Example 21.1 uses this equation.

EXAMPLE 21.1

A transformer has 500 turns on its primary winding (N_1) and 100 turns on the secondary (N_2). The maximum value of the flux in the core is 2.16 milliweber and the frequency is 50 Hz. Calculate the voltage induced in the:

1 primary winding (e_1)
2 secondary winding (V_2)

Solution

Values N_1 = 500 turns f = 50 Hz
N_2 = 100 turns Φ_{max} = 2.16 milliweber

1 Calculate the voltage induced in the primary winding (V_P)

Equation $e_1 = 4.44\ N_1\ f\ \Phi_{max}$

$e_1 = 4.44 \times 500 \times 50 \times 2.16 \times 10^{-3}$

Answer **$e_1 = 239.8$ V = 240 V when rounded up**

(This voltage is the back-EMF in the primary. It should nearly equal the voltage applied to the primary. Any difference is due to losses in the transformer.)

2 Calculate the voltage induced in the secondary winding (V_2)

Equation $V_2 = 4.44\ N_2\ f\ \Phi_{max}$

$V_2 = 4.44 \times 100 \times 50 \times 2.16 \times 10^{-3}$

Answer **$V_2 = 47.9$ V = 48 V when rounded up**

KEY CONCEPT

Voltage ratio of a transformer equals the turns ratio of primary and secondary windings

Transformation ratio

In Example 21.1, the only difference between calculations (1) and (2) is the number of turns in the windings, in which the primary has five times the number of turns as the secondary. In (1), the primary induced voltage is 240 V and in (2) the secondary induced voltage is five times smaller at 48 V. This shows a relationship between the turns ratio and the voltage ratio, which can be found by expressing the two equations used in Example 21.1 as a ratio.

$$\frac{V_1}{V_2} = \frac{4.44 N_1 f \Phi_{max}}{4.44 N_2 f \Phi_{max}}$$

Because the frequency (f) and the maximum flux (Φ_{max}) are the same in both equations, these cancel, leaving an equation that relates the turns ratio to the voltage ratio:

$\frac{V_1}{V_2} = \frac{N_1}{N_2}$ where $\frac{V_1}{V_2}$ is the transformer voltage ratio and $\frac{N_1}{N_2}$ is the turns ratio.

This relationship is the transformation ratio, in which the turns ratio equals the voltage ratio. Transposing the equation gives two more equations to find either the primary or secondary voltage if one voltage and the turns ratio of the transformer are known.

$$V_1 = V_2 \times \frac{N_1}{N_2}$$

$$V_2 = V_1 \times \frac{N_2}{N_1}$$

where:

V_1 = primary voltage in volts RMS

V_2 = secondary voltage in volts RMS

N_1 = number of turns on the primary winding

N_2 = number of turns on the secondary winding.

EXAMPLE 21.2

A transformer has 960 turns on its primary winding (N_1) and 48 turns on the secondary (N_2). Calculate the:

1 turns ratio of the transformer

2 secondary voltage when 230 V is applied to the primary

Solution

Values N_1 = 960 turns

N_2 = 48 turns

V_1 = 230 V RMS

1 Calculate the turns ratio

Equation turns ratio $= \frac{N_1}{N_2} = \frac{960}{48} = \frac{20}{1}$

Answer **turns ratio = 20:1**

2 Calculate the secondary voltage

Equation $V_2 = V_1 \times \frac{N_2}{N_1}$

$V_2 = 230 \times \frac{48}{960} = 230 \times \frac{1}{20}$

Answer **secondary voltage V_2 = 11.5 V**

Current ratios

An ideal transformer is 100 per cent efficient, so the apparent power into the transformer equals the apparent power out. Apparent power (S) in any single-phase AC circuit is:

S = VI, measured in volt amps (VA)

Because the apparent power in equals the apparent power out:

$V_1I_1 = V_2I_2$

Transposing this equation gives equations relating turns, voltage and current ratios:

$$\frac{V_1}{V_2} = \frac{I_2}{I_1} = \frac{N_1}{N_2}$$

This means the:

- ratio of the primary and secondary voltages in a transformer is *directly* proportional to the turns ratio (N)
- ratio of the currents in the primary and secondary windings of a transformer is *inversely* proportional to the turns ratio (N).

EXAMPLE 21.3

The transformer in Figure 21.5 has 720 turns on its primary winding (N_1) and 100 turns on the secondary (N_2). The transformer primary is connected to 230 V AC and the secondary is connected to a 30 Ω load. Calculate the:

1 secondary voltage
2 secondary current
3 primary current

FIGURE 21.5

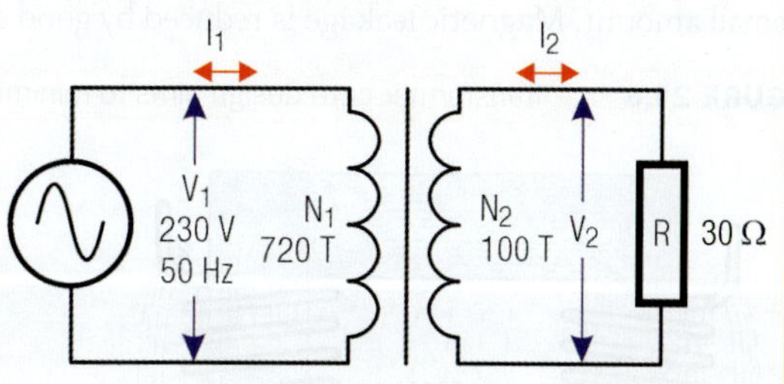

Solution

Values N_1 = 720 turns
N_2 = 100 turns
V_1 = 230 V RMS
R = 30 Ω

1 Calculate the secondary voltage

Equation

$V_2 = V_1 \times \frac{N_2}{N_1} = 230 \times \frac{100}{720}$

Answer **V_2 = 31.9 V = 32 V when rounded up**

2 Calculate the secondary current

Equation $I_2 = \frac{V_2}{R} = \frac{32}{30}$

Answer **I_2 = 1.067 A = 1.07 A**

3 Calculate the primary current

Equation $I_1 = I_2 \times \frac{N_2}{N_1} = 1.1 \times \frac{100}{720}$

Answer **I_1 = 0.149 A**

In Example 21.3 the turns ratio is 720:100 or 7.2:1, giving a *step-down* transformer in which the secondary voltage is 7.2 times less than the primary voltage. However, the secondary current is 7.2 times greater than the primary current. That is, the current is stepped up.

A step-down transformer therefore steps up the current because it has stepped down the voltage. A *step-up* transformer steps up the voltage and steps down the current. This has to be, so the apparent power in equals the apparent power out.

KEY POINTS...

- In an ideal transformer:
  - the voltage ratio equals the turns ratio
  - the current ratio is the inverse of the voltage (or turns) ratio
  - there are no losses and the transformer consumes no power
  - the apparent power (S = VI) in to an ideal transformer equals the apparent power out.

21.3 The Practical Transformer

No transformer is 100 per cent efficient, which means there are losses that need to be kept to a minimum by good design. All transformers take a primary current, even when there is no load on the secondary: a case of power in, but no power out. Losses in a transformer cause it to become hot, which in large power transformers must be managed by cooling methods. Losses are made up of winding (copper) and core (iron) losses.

Core construction

In an ideal transformer, all of the magnetic flux links with the windings. However, in a practical transformer, a certain amount of flux will bypass the magnetic circuit and be ineffective. This is called magnetic leakage, which is shown in Figure 21.6(a) and described in Chapter 11. Magnetic leakage increases as the transformer load current increases, which reduces the secondary voltage and therefore lowers the transformation ratio by a small amount. Magnetic leakage is reduced by good design, but is never entirely eliminated.

FIGURE 21.6 Transformer core design aims to minimise magnetic leakage

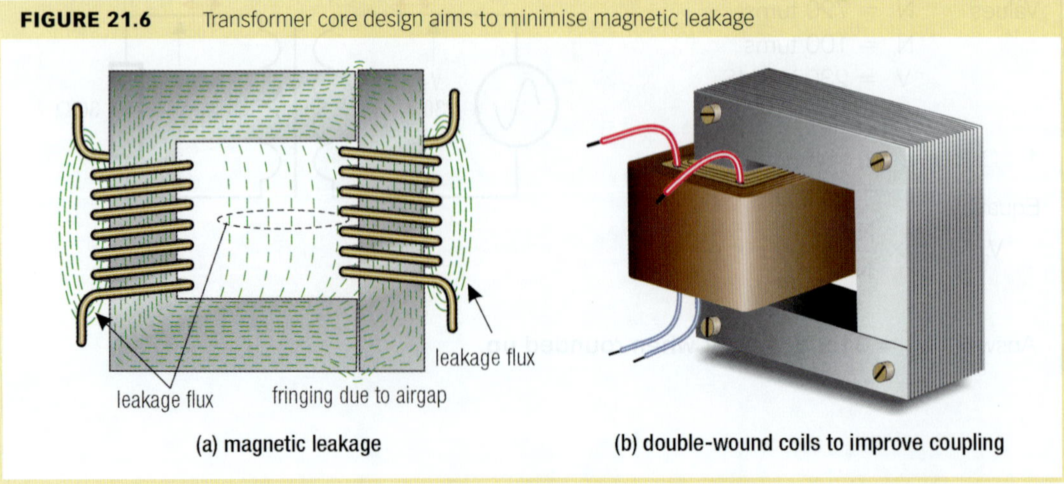

To improve the coupling between windings, transformers often have the primary and secondary windings on the same limb of the core. Most transformers are double-wound, which means one winding is wound concentrically over the other, as shown in Figure 21.6(b). A single-wound transformer has windings placed next to each other on one limb of the core, or the primary winding might be on one limb and the secondary on another limb.

Electrical power transformers typically have a laminated iron core, while smaller types might have a ferrite core. Transformers operating at frequencies greater than 1 kHz usually have a ferrite core or a powdered iron core that has fine insulated iron particles moulded into the required shape. Air-cored transformers are used mainly in communications equipment.

A laminated core is made from layers of thin, high grade, grain-oriented silicon steel (97 per cent iron and 3 per cent silicon), usually 0.25 mm to 1 mm thick. This material is chosen to minimise the *hysteresis loss* in the core (explained in Chapter 12). This loss is caused by the changing magnetic field produced by the alternating current in the primary winding. Because the lines of force in the magnetic field are repeatedly changing direction, the millions of magnetic domains in the core are continually being magnetised in one direction, then the other. This causes molecular friction, which causes a power loss to overcome the friction, producing heat. This is a resistive loss, and the power to overcome it comes from the source supplying the transformer.

Another loss is caused by *eddy currents* that flow in the core. The magnetic field that induces a voltage in the secondary of a transformer also induces a voltage in the core. This causes circulating (or eddy currents) in the core. Making the laminations as thin as possible reduces this loss. The laminations are coated with a thin insulating material to prevent current flowing between them.

Single-phase transformer cores

Transformers are often classified by the type of core. The two basic core designs are the shell-type and the core-type. A *core-type* transformer has the limbs of the core surrounded concentrically by the main windings, as shown in Figure 21.7. The core is made up of U- and I-shaped laminations interleaved to form a stack that is bolted, riveted or otherwise bonded together. The cross-sectional area of the core is uniform throughout to ensure the same flux density in all parts of the core.

FIGURE 21.7 The core-type transformer is commonly used in power transmission and distribution

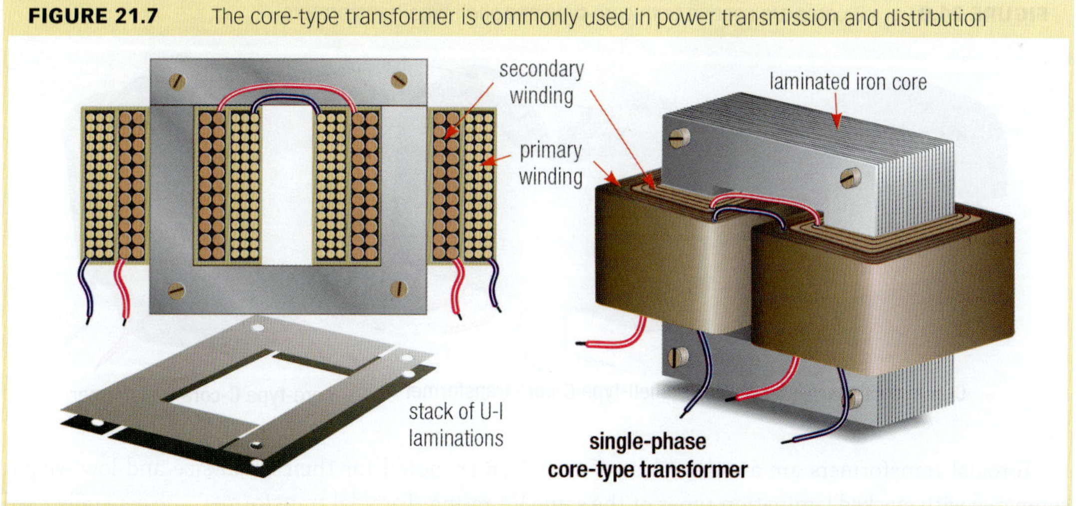

Double-wound windings are usually arranged with the low voltage winding placed nearer the core, to reduce the amount of insulation required between the winding and the earthed core. The insulation provided by the low voltage winding serves to further insulate the high voltage winding from the core. In a single-wound transformer, the high voltage winding could be on one limb of the core and the low voltage winding wound on the other limb.

FYI

When windings surround the core, the transformer is core type; when windings are surrounded by the core, the transformer is shell type

In a *shell-type* transformer the flux-return paths of the core are external and wrap around the windings, as shown in Figure 21.8. This design provides better magnetic shielding than the core-type, and is used in transformers that provide a low-voltage, high current supply, such as in an arc furnace.

The central limb of the core is twice the cross-sectional area of the outer limbs because all the magnetic flux flows in the centre limb, which then divides to flow in the two outer limbs. The windings in Figure 21.8 are double-wound and shown with the primary nearest the core, although in many cases the secondary is nearest the core, for the same reasons as in the core-type transformer. Low power shell-type transformers are often single-wound, with the windings placed side by side.

FIGURE 21.8 The shell-type transformer has its windings wound around the centre limb of the core

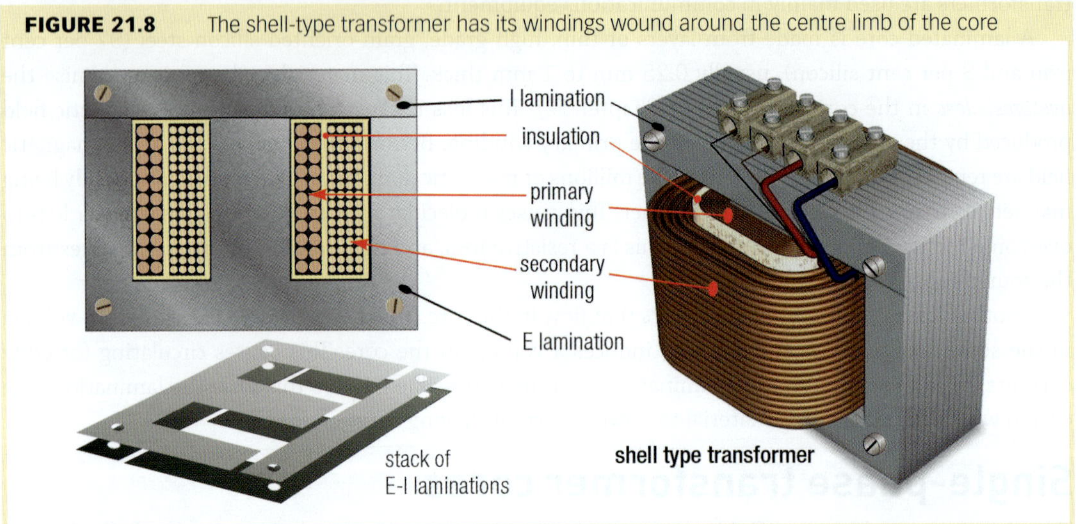

C-cores are made by tightly rolling a continuous ribbon of thin metal alloy strip around a former to give it the desired shape. After bonding under pressure, the consolidated mass is cut in half. To ensure the best possible magnetic coupling (no air gap), the cut ends are machined and polished as a pair. The core halves are held together with banding, as shown in Figure 21.9.

FIGURE 21.9 C-cores are used in both core-type and shell-type transformers

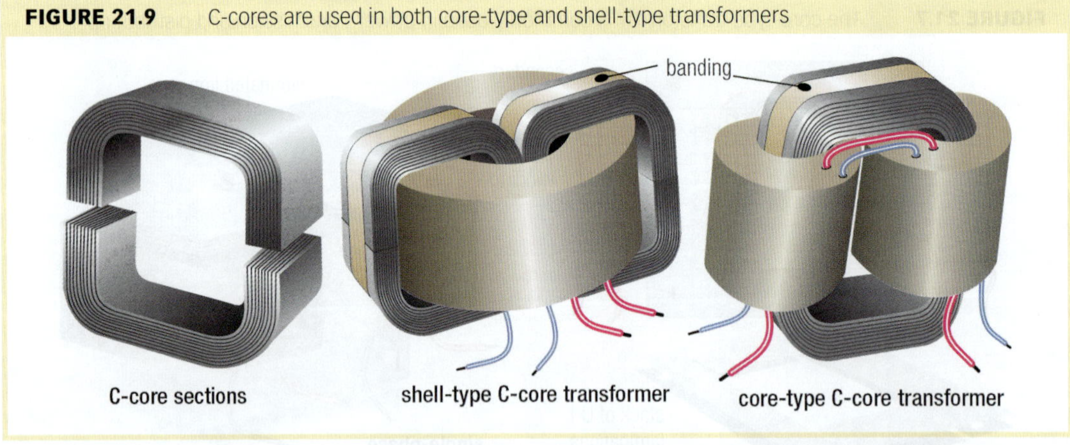

Toroidal transformers are another class of transformer, noted for their small size and low weight compared with stacked lamination types of the same VA rating. Toroidal transformers are generally used for power applications of no more than a few kVA. The core is ring-shaped, and is made from ferrite or rolled silicon steel strip wound into a ring. This type of construction eliminates any air gaps, which can occur in laminated cores. The cross-section of the ring is usually square or rectangular, sometimes circular. The primary and secondary windings are wound concentrically over the entire surface of the core. Figure 21.10 shows a typical toroidal transformer and cores.

FIGURE 21.10 Toroidal transformers have a low level of radiated magnetic field and a high efficiency

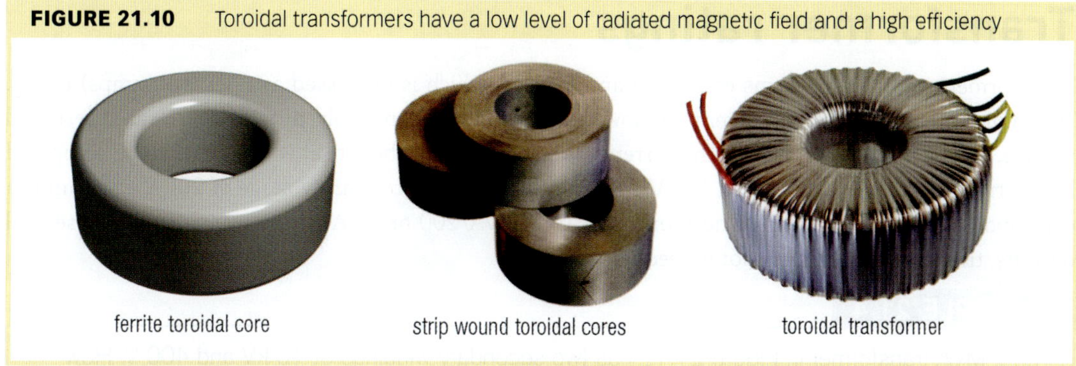

Transformer windings

Small transformers are wound with wire coated with an insulating polyurethane, polyamide or polyester resin. This type of wire is sometimes referred to as enamel-coated winding wire. The primary and secondary coils are wound on a bobbin, and after winding, lacquering and baking the complete coils, the laminations forming the core are assembled.

Transformers used in power transmission and distribution are wound with copper rectangular strip conductors insulated by oil-impregnated paper and blocks of pressboard (compressed layers of paper). Large power transformers often have multiple-stranded conductors, in which interleaved copper strips insulated from each other make up the conductor. This construction reduces eddy currents in the copper windings caused by leakage flux.

High-frequency transformers operating above 10 kHz often have windings made of braided Litz wire to minimise the skin-effect, in which current tends to flow on the outside of the conductor. Litz wire has numerous strands of insulated wire formed into a braid.

Transformers have to conform to various standards. For example, AS/NZS 3000:2018 clause 2.7.2 requires 'the provision of adequate insulation, screening or separation of windings'. To achieve the required level of insulation and safety requirements, transformer windings are arranged in various ways, depending on the operating voltage and power rating of the transformer. Three methods are shown in Figure 21.11.

FIGURE 21.11 Winding arrangements depend on the voltage and power ratings of a transformer

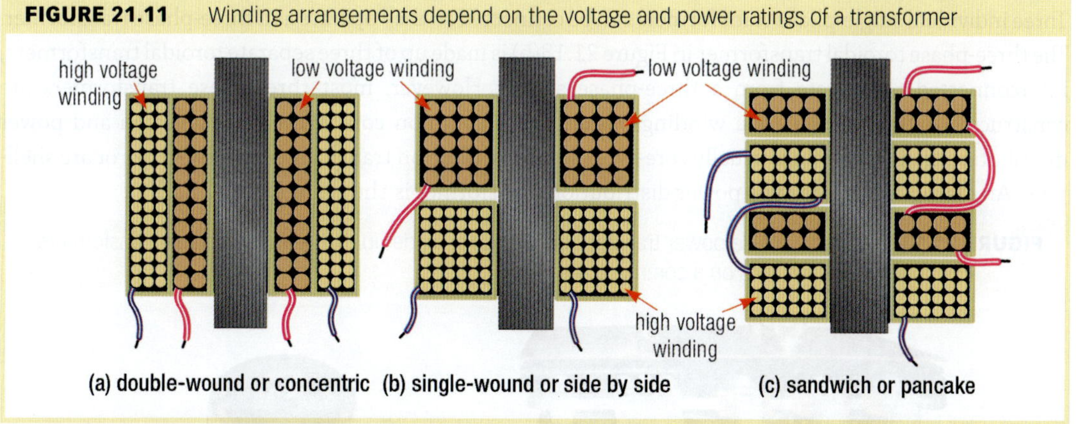

Screening refers to placing a metal shield around a transformer winding. In some transformers, the screen acts as a shield against an electrostatic field, preventing capacitive coupling between windings. In regard to the standards, some types of transformers are required to have a metal shield between the primary and secondary windings, with the shield connected to earth. An example is a 12 V (or 24 V) pool lighting transformer. If the transformer develops a short-circuit between the primary and secondary windings, the earthed metal shield between them will become live. This causes current to flow to earth and thereby operate the residual current device protecting the circuit. Without the shield, the secondary winding voltage could reach 230 V, and make the pool water 'live' and potentially electrocute those in the pool.

Requirements regarding safety and transformers are in AS/NZS 61558.1.2008

Transformer ratings

Transformers are rated in terms of voltage and current, which is expressed as a VA (volt-amps) rating. This indicates the maximum current the low voltage winding can safely provide at its rated terminal voltage, regardless of power factor. The current rating depends on the rate at which heat can be dissipated by the transformer. For example, a 10 kVA transformer with a secondary voltage of 200 V can handle a continuous secondary current of no more than 10 000 VA/200 or 50 A. If the transformer has several outputs, the total output must not exceed its VA rating.

EXAMPLE 21.4

The 2 MVA transformer in Figure 21.12 has two secondary windings of 10 kV and 400 V. How much current can the 10 kV winding provide if the 400 V winding is supplying 100 A to a load?

FIGURE 21.12

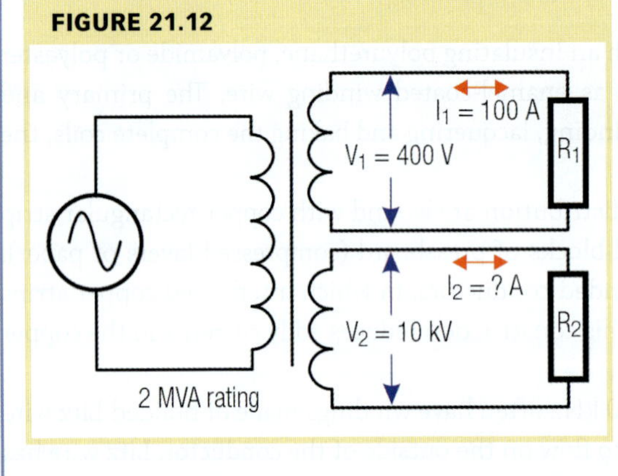

Solution

Values
$V_1 = 400$ V
$I_1 = 100$ A
$V_2 = 10$ kV
VA = 2 MVA

apparent power supplied by 400 V winding = 400 × 100 = 40 kVA

apparent power available from 10 kV winding = transformer rating − 40 kVA = 1.96 MVA

Equation $I_2 = \dfrac{1.96\text{ MVA}}{10\text{ kV}} = \dfrac{1.96 \times 10^6}{10 \times 10^3}$

Answer $\mathbf{I_2 = 196\text{ A}}$

Power transmission transformers are given an MVA rating, such as 60 MVA, which is 60×10^6 VA. Distribution transformers have a lower rating such as 200 kVA, or 200×10^3 VA.

Three-phase transformers

Three individual single-phase transformers can be connected so they operate as a three-phase transformer. The three-phase toroidal transformer in Figure 21.13(b) is made up of three separate toroidal transformers, interconnected to operate from a three-phase supply. However, most three-phase transformers are constructed with three separate windings wound on a common core. Power transmission and power distribution transformers are usually core-type; some distribution transformers have C-cores or are shell-type. An example of a core-type power distribution transformer is shown in Figure 21.13(a).

FIGURE 21.13 A three-phase power transformer is typically made up of three single-phase transformers with the coils on a common core

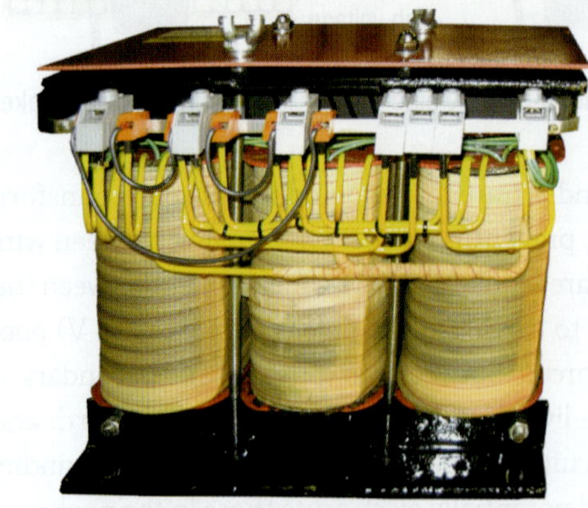

(a) three-phase 200 kVA core-type transformer

(b) three-phase toroidal transformer

The same factors affecting the windings and cores of a single-phase transformer apply to the three-phase transformer. Design features of a three-phase transformer depend on its application. For example, the magnetic coupling between windings is sometimes purposely reduced to limit the secondary current in the event of a short-circuited load. Winding connections are described later in this chapter.

Transformer cooling

All transformers dissipate heat when operating, due to their copper (I^2R) and iron losses. Small transformers have sufficient surface area to remove the heat at a rate that keeps the transformer from overheating under full-load conditions. Transformers such as those used in power transmission and distribution require additional means of cooling. A commonly used method relies on circulating mineral oil through the transformer, and cooling the oil by way of heat exchanging radiators.

The model transformer in Figure 21.14(a) shows how windings are enclosed by the transformer tank, which is filled with mineral oil. The oil impregnates the paper insulation, improving its insulating qualities. It also helps cool the transformer by flowing due to natural convection, removing heat from the windings and core, and transferring the heat to the casing. The fins on the casing provide increased surface area, and cooling is also by natural convection of the air surrounding the casing. This type of cooling is called oil natural, air natural, abbreviated as ONAN.

FIGURE 21.14 Power transformers are usually cooled by circulating mineral oil through the windings, and cooling the oil by convection or forced air flow

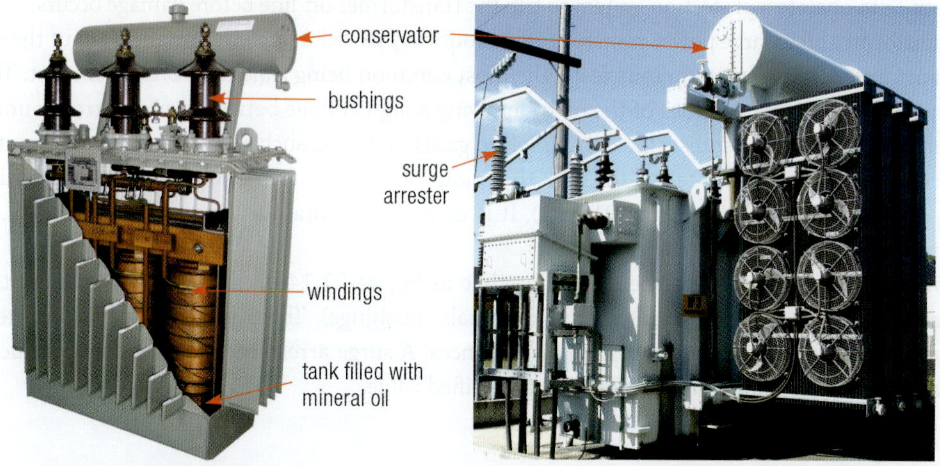

(a) model of a transformer with oil natural, air natural (ONAN) cooling

(b) 35 MVA transformer with oil forced and air forced cooling (OFAF)

Large power transformers often have an associated oil pump to force oil flow around the windings, with fans to cool the heated oil as it flows through a heat exchanger, as shown in Figure 21.14(b). This type of cooling is referred to as oil forced, air forced, or OFAF. Other variations are oil forced, air natural (OFAN) or oil natural, air forced (ONAF).

Transformers that are cooled only by air circulation are either air natural (AN) or air forced (AF). If oil is forced through oil ducts built into the windings, the cooling is called OFD (oil forced directed). Power transformers are often given MVA ratings at two or three different forms of cooling, with the lowest at ONAN, the highest at OFDAF (oil forced directed air forced).

Oil- or liquid-filled transformers often have a conservator tank that acts as a reservoir for the coolant during expansion or contraction of the main transformer tank due to temperature changes. Some conservators are fitted with rubber bladders as a barrier against moisture. While there is always some moisture present in transformer oil, this has to be kept below a certain level to avoid the mineral oil breaking down due to the high voltages present in the windings. To reduce moisture content in the oil, transformers usually have a silica gel breather, such as those in Figure 21.15. The silica gel beads are blue when dry, and become pink when moist, indicating the beads need to be dried by heating.

FIGURE 21.15 The condition of transformer oil is indicated by its colour

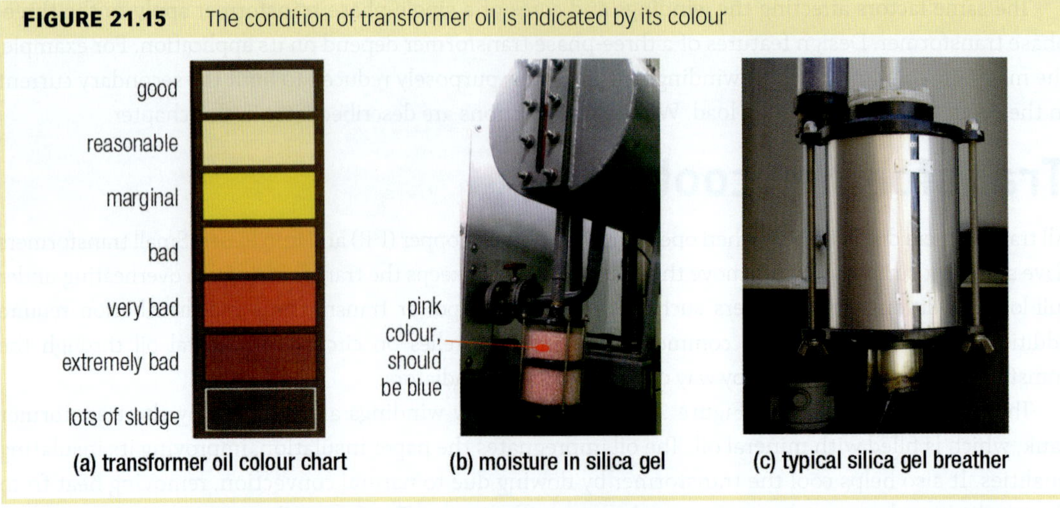

Oil-filled transformers with conservators are also fitted with a device called a Buchholz relay that detects the presence of gases such as acetylene being produced within the transformer tank. Gases are produced by arcing within the transformer, and indicate potential failure of the insulation and the oil. If an arc forms, gas rapidly builds up, causing oil to quickly flow into the conservator. The relay is actuated either by the pressure waves from oil flow, by gas accumulation or by loss of oil below the relay level. Contacts are fitted to operate an alarm or to operate a protection system to trip the transformer off-line before damage occurs.

Transformer oil in large power transformers is routinely tested to gain information about the state of the transformer. There are numerous tests, the most common being a *dielectric breakdown* test. This test measures the insulating qualities of the oil by applying a high voltage between two electrodes immersed in a sample of the oil. Other tests look for dissolved gases in the oil, such as hydrogen, methane, ethylene and acetylene. A simple field test is to check the colour of the transformer oil against a standard colour chart, such as that shown in Figure 21.15(a). It is essential to obtain a clean sample of the oil, as any contamination can affect the test results.

Power transmission transformers, such as that in Figure 21.14(b), are generally fitted with *surge arresters*, located at the high voltage input terminals (bushings). These protect the transformer from voltage surges due to lightning or system disturbances. A surge arrester is a device that becomes a low resistance when the voltage across it exceeds a specified value.

KEY POINTS...

- In a practical transformer:
  - losses are made up of iron and copper (I^2R) losses
  - losses cause heat, which is removed in a large transformer by circulating oil through the windings and the core
  - eddy currents in the core are reduced by laminating the core
  - cores can be shell or core-type; power transformers are usually core-type, and in some cases C-cores are used
  - windings can be double-wound, side by side or pancake, depending on the ratings of the transformer
  - the volt-amp rating of the transformer determines the value of continuous secondary current the transformer can provide without overheating
  - a three-phase transformer has a single multi-limb core with three single-phase windings.

21.4 Transformer Operation

A practical single-phase transformer can be represented by the equivalent circuit shown in Figure 21.16. The circuit assumes the transformer itself is ideal, with the transformer losses represented by the reactive and resistive components. This diagram allows us to examine the operation of a transformer, starting with the no-load condition.

FIGURE 21.16 Equivalent circuit of a single-phase transformer on no-load

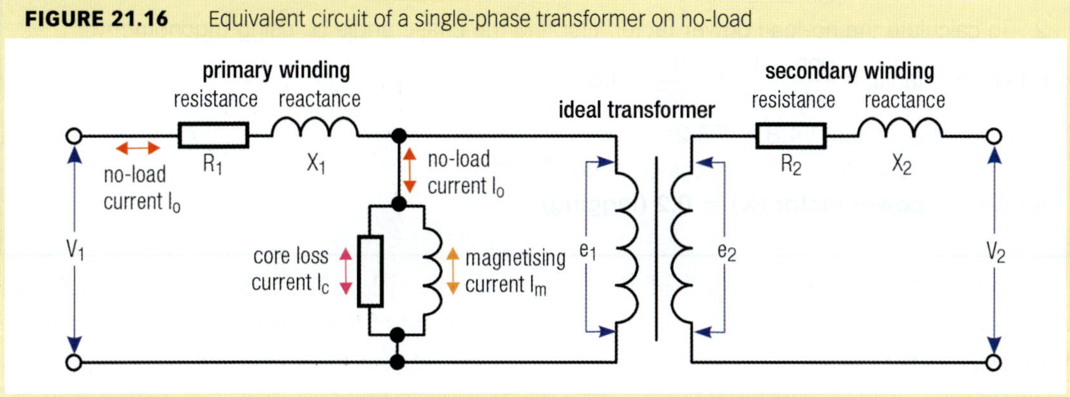

Transformer on no-load

When a transformer is operating without a load, the no-load current taken by the transformer is made up of two components, shown as currents I_c and I_m in Figure 21.16. The core loss current I_c is in phase with the primary voltage (V_1), so this loss is shown as a resistance, because heat is generated. The magnetising current I_m lags the primary voltage by 90°, so this loss is shown as a reactance. The no-load current of a transformer is a few per cent of the full-load current, which means the primary winding resistance R_1 and reactance X_1 can be ignored under no-load conditions.

The phasor diagram of a transformer with equal primary and secondary turns operating on no-load is shown in Figure 21.17. Because the magnetic flux (Φ) is common to both windings, it becomes the reference phasor, drawn to any convenient length. The induced voltages V_2, e_2 and e_1 have the same value because the turns ratio is 1:1, and these voltages all lag the flux by 90°. At no-load there is very little voltage drop across R_1 and X_1, so the phasor for V_1 can be drawn equal in length to and opposite the phasor for V_2, e_2 and e_1.

The magnetising current I_m causes the flux, so it is in phase with the reference phasor Φ. The core loss current I_c is in phase with the primary voltage V_1 so its phasor is drawn over the V_1 phasor. The phasor sum of these two currents is the no-load current I_0. The angle ϕ_0 between I_0 and V_1 is the phase difference between the primary voltage and the transformer's no-load current.

FIGURE 21.17 Phasor diagram for a transformer on no-load

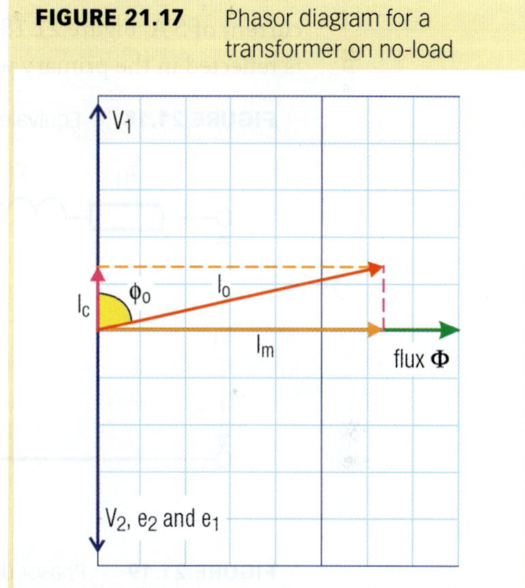

EXAMPLE 21.5

A transformer takes a magnetising current of 6 A and a core loss current of 1.25 A. Determine the transformer's:

1 no-load current
2 no-load power factor

Solution

This problem can be solved with a phasor diagram, as shown in Figure 21.17. It can also be solved mathematically, as there is a right angle between I_c and I_m:

Values $I_c = 1.25$ A
$I_m = 6$ A

1 To calculate I_0, use Pythagoras' theorem

Equation $I_0 = \sqrt{I_m^2 + I_c^2} = \sqrt{1.25^2 + 6^2}$
Answer $\mathbf{I_0 = 6.13\ A}$

2 To calculate the no-load power factor, first find the phase angle ϕ_0 using trigonometry

Equation $\tan\phi_0 = \frac{opp}{adj} = \frac{I_m}{I_c} = \frac{6}{1.25} = 4.8$

$\phi_0 = \tan^{-1} 4.8 = 78.2°$

power factor $(\lambda_0) = \cos \phi_0 = \cos 78.2°$

Answer **power factor (λ) = 0.2 (lagging)**

Example 21.5 shows that on no-load, the power factor is very low. This is because the transformer appears as a highly inductive load, with a very small component of the no-load current being in phase with the primary voltage.

Transformer on load

When a transformer is operating with a load, the primary current taken by the transformer has two components, the no-load current and the current needed to compensate for the flux cancellation caused by the current flowing in the secondary. This current is sometimes called a *reflected* current. For example, if a transformer has a 1:1 turns ratio, a load taking 10 A from the secondary winding causes a primary winding current of 10 A, in addition to the no-load current. The total primary current I_1 is the phasor sum of these two currents. If the turns ratio is 2:1, a load current of 10 A requires a reflected primary current of 5 A. Figure 21.18 shows the equivalent circuit of a transformer under load. The load current I_2 is reflected in the primary winding, shown as I_2'.

FIGURE 21.18 Equivalent circuit of a single-phase transformer on load

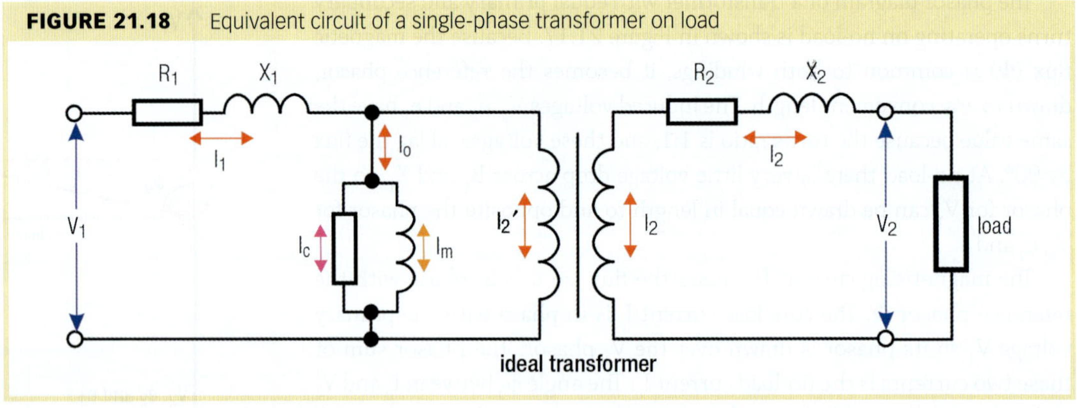

FIGURE 21.19 Phasor diagram for a transformer supplying a resistive load

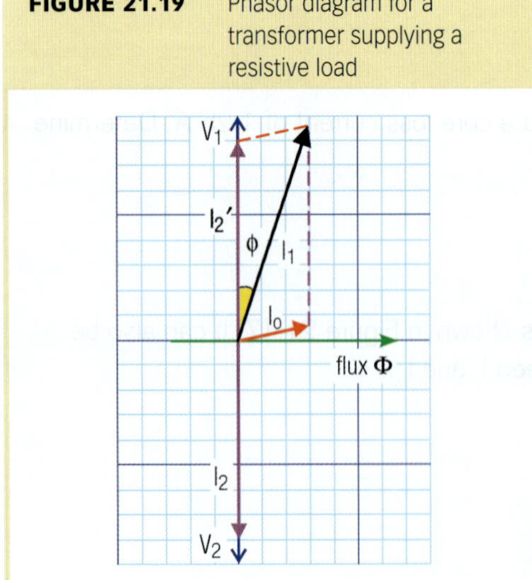

The phasor diagram for a 1:1 turns ratio transformer supplying a resistive load is shown in Figure 21.19. The secondary current I_2 is in phase with the secondary terminal voltage, and the reflected current I_2' in the primary winding is drawn equal in length and opposite to current I_2. (If the transformer turns ratio was 2:1, current I_2' would be half the length of current I_2.) The phasor addition of I_0 and I_2' gives the total primary current I_1. The angle ϕ between I_1 and V_1 is the power factor of the loaded transformer.

EXAMPLE 21.6

The transformer in Example 21.5 is supplying a current of 16 A to a resistive load. Determine the transformer's:

1 primary current
2 power factor

Solution

This problem is solved with a phasor diagram, which is shown in Figure 21.19. The no-load current I_0 was found in Example 21.5 to be 6.13 A with a phase angle ϕ_0 of 78.2°. Secondary current I_2 of 16 A is in phase with the secondary voltage V_2, as the load is purely resistive. Current I_2' is therefore in phase with the primary voltage V_1, and because the transformer has a 1:1 turns ratio, its phasor is also representing 16 A. By measurement from the phasor diagram we find:

1 primary current $\mathbf{I_1 = 18.3\ A}$
2 power factor = cos ϕ = cos 19° = **0.94 (lagging)**

Example 21.6 shows that the power factor for a loaded transformer is significantly higher than for an unloaded transformer. A more general case is a transformer supplying an inductive load, as most industrial loads are inductive. The phasor diagram for a transformer supplying an inductive load is shown in Figure 21.20. In this diagram the same transformer is used as in Example 21.6, except the load current of 16 A is lagging the secondary terminal voltage by 20°.

The secondary current I_2 is drawn so it lags the secondary voltage phasor by angle ϕ_2, in this case 20°. The phasor for current I_2' (reflected current) is drawn opposite the I_2 phasor and has the same length as I_2 because the turns ratio is 1:1. The current I_1 taken by the transformer is the phasor sum of I_2' and the no-load current I_0. The phase angle ϕ_1 can be measured from the phasor diagram, so the power factor of the loaded transformer can be found by finding the cosine of ϕ_1.

As the phasor diagram shows, the power factor of the load ϕ_2 and the power factor of the transformer ϕ_1 are not the same. As well, the inductive load will mean the power factor of the loaded transformer will be lower than for a resistive load of the same current.

The phasor diagrams in Figures 21.19 and 21.20 do not take into account the resistance or the reactance of the primary and secondary windings. These values are usually low enough to ignore for practical purposes. The value of the no-load current I_0 in Figure 21.20 is exaggerated for clarity in the diagram, and would normally be only a few per cent of the load current I_2.

FIGURE 21.20 Phasor diagram for a transformer supplying an inductive load

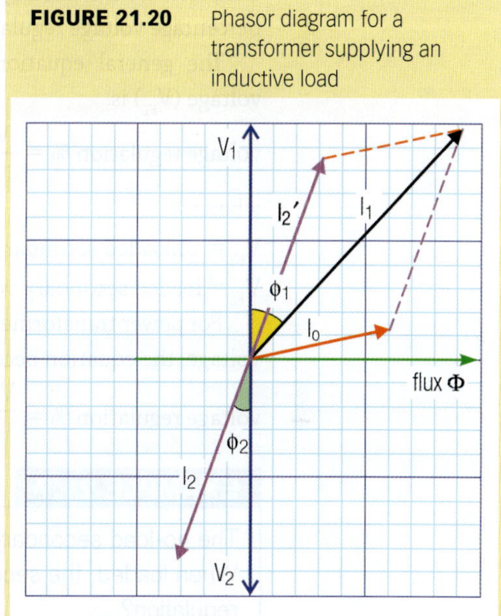

KEY POINTS...

- In a practical transformer:
  - the no-load current is the phasor sum of the magnetising current and the core loss current
  - the primary current of a loaded transformer is the phasor sum of the no-load current and the current reflected in the primary by the secondary current
  - the power factor of an unloaded transformer is usually very low at between 0.1 and 0.2
  - loading a transformer improves the power factor.

21.5 Voltage Regulation

The secondary voltage of a practical transformer changes if the primary voltage or the load current changes. When a load is connected to a transformer, its secondary voltage will fall because of the resistance and reactance of the primary and the secondary windings. The resistance will cause a voltage drop in phase with the current, and the reactance causes a voltage drop 90° out of phase with the current. Calculations to determine the drop in secondary voltage under load are therefore complex, so measurements are usually required. The change in the secondary voltage due to a change in the load current gives a measure of the transformer's *voltage regulation*.

The method to determine voltage regulation depends on the standard being applied. In Australia and New Zealand, standards for power transformers are based on International Electrotechnical Commission (IEC) Standard 60076. However, in the USA and other countries, different standards apply, called ANSI/IEEE standards. A difference between the IEC and ANSI standards is how the voltage regulation of a power transformer is determined. IEC standards state that the reference is the *unloaded* secondary voltage. ANSI standards state that the *loaded* secondary voltage is used when determining voltage regulation of a power transformer.

The difference between the standards can sometimes cause confusion. The IEC standard (as in Australia and New Zealand) specifies the unloaded secondary voltage to ensure the correct voltage rating of insulation requirements for cabling and ancillaries connected to the secondary. However, many small transformers are rated by their loaded output voltage, giving two variations of the equation used to find percentage voltage regulation of a transformer.

The general equation to find voltage regulation when the reference is the *loaded* secondary voltage (V_{FL}) is:

$$\text{voltage regulation \%} = \frac{V_{NL} - V_{FL}}{V_{FL}} \times 100$$

where:

V_{NL} = unloaded secondary voltage

V_{FL} = loaded secondary voltage

For power transformers in Australia and New Zealand, because the reference is the *unloaded* secondary voltage, the equation becomes:

$$\text{voltage regulation \%} = \frac{V_{NL} - V_{FL}}{V_{NL}} \times 100$$

EXAMPLE 21.7

The no-load secondary voltage of a power transformer is specified by IEC standards as 400 V. When loaded, the secondary voltage is 390 V. What is the transformer's percentage voltage regulation?

Solution

Values $V_{NL} = 400$ V

$V_{FL} = 390$ V

Equation $\text{voltage regulation \%} = \frac{V_{NL} - V_{FL}}{V_{NL}} \times 100 = \frac{400 - 390}{400} \times 100$

Answer **voltage regulation = 2.5 per cent**

The load power factor has a considerable effect on the voltage regulation of a transformer. Therefore, a transformer's percentage voltage regulation is given at a specified power factor, usually 0.8 lagging, as this is a typical value. If the load has a leading power factor (capacitive), the loaded secondary voltage can rise above its unloaded value. In general, the voltage regulation of a power transformer is relatively small and is always calculated with secondary winding voltages.

Tap changers

Electrical supply authorities need to maintain the supply voltage to homes and industry within specified limits, despite large variations in the load. This is achieved in various ways, in particular with a device called a *tap changer*. In principle, a tap changer is a type of selector switch connected to various points of a primary or secondary transformer winding, called tappings. This arrangement allows the turns ratio of a transformer to be changed, which changes the secondary voltage.

Figure 21.21 shows the principles of changing tappings to change the secondary voltage. In Figure 21.21(a), the secondary winding has five tappings, with a selector switch to connect the load to one of the five tappings. The lower the number of secondary turns, the lower the secondary voltage. The tappings are marked as a percentage change in the secondary voltage for each switch position, with the centre position the reference or principal tapping.

FIGURE 21.21 A tap changer is a form of selector switch connected to tappings of either the primary or secondary windings of a transformer

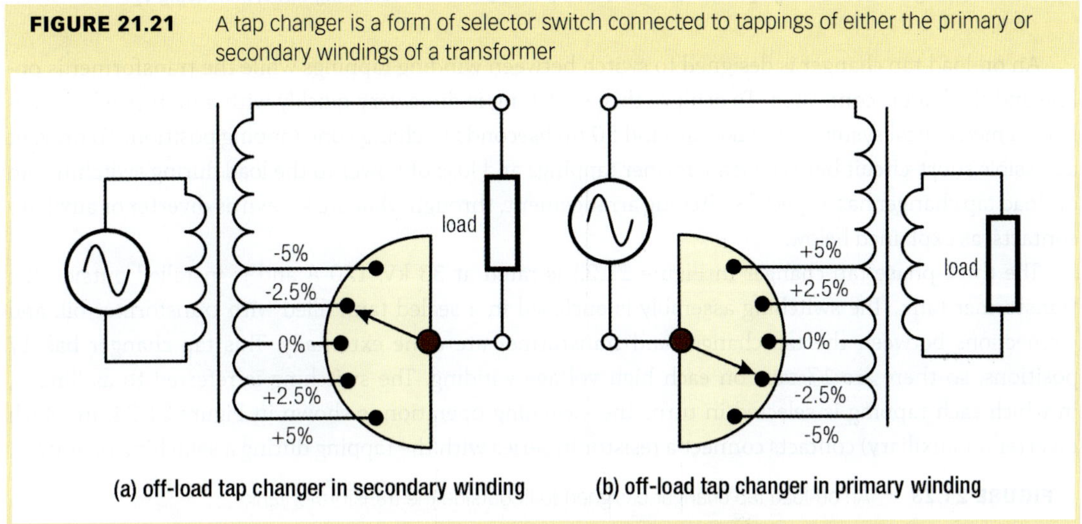

A tap changer is often fitted to the high voltage winding of a transformer, as the current being switched is lower than on the low voltage side. As well, in many transformers, the high voltage winding is the outside winding, making it easier to run tappings to a tap changer. The principle is shown in Figure 21.21(b), in which reducing the number of effective primary turns increases the secondary voltage, again as a percentage of the nominal (or rated) output voltage of the transformer.

Off-load tap changer

Both of the arrangements in Figure 21.21 have the same limitations. When a tapping position is being changed, there is an instant where the switch is open-circuit, causing a momentary power failure to the load. The other problem is arcing during switching, particularly if the tap changer is in the low voltage winding. Therefore, the tap changer in Figure 21.21(a) can only be operated when the load is disconnected, and in (b) only when the transformer is de-energised.

Tap changers that work under unloaded or de-energised conditions are called off-line or off-load tap changers. They are used if tap changing is done manually and at irregular intervals. Power distribution transformers often have an off-line tap changer, which is set during commissioning, and changed perhaps months later due to changes in the load. The switching arrangements used in an off-line tap changer vary, and an example is shown in Figure 21.22.

On-load tap changer

Virtually all power transmission transformers have an on-load tap changer, fitted externally to the transformer or inside the transformer tank. These have a motor-driven mechanism to move the contacts and can be operated automatically by a voltage control system, or operated manually either locally or remotely. In the event of power failure, most tap changers can be operated with a crank handle.

FIGURE 21.22 An off-load tap changer that bridges tappings between two halves of the primary winding

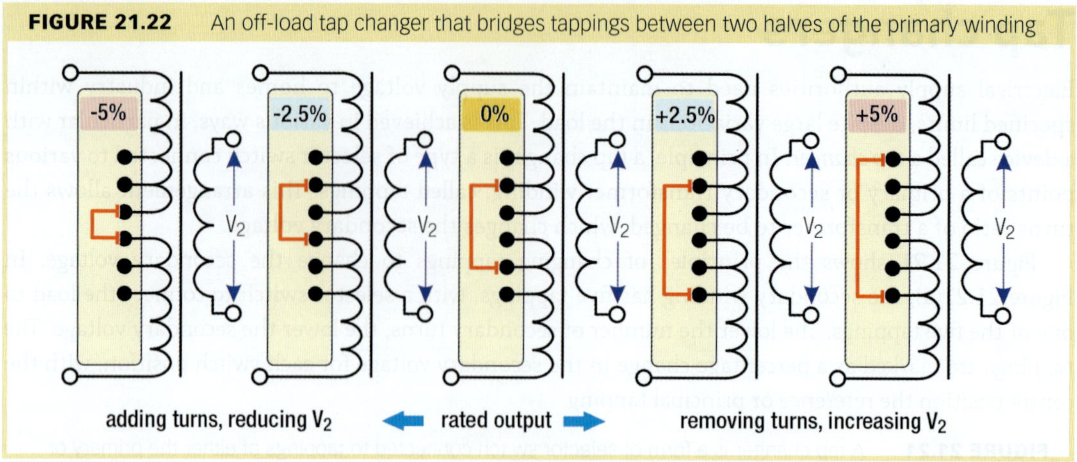

An on-load tap changer is designed to switch between winding tappings while the transformer is on-line and the load is connected. To achieve this, switching is done very quickly with a motorised, spring-loaded mechanical system that takes around 50 milliseconds to change one tapping position. To prevent a possible short-circuit between transformer tappings and loss of power to the load during switching, an on-load tap changer has a special switching arrangement, through what are known as diverter or auxiliary contacts, as explained below.

The three-phase tap changer in Figure 21.23 is rated at 33 kV, 150 A and is installed outside the transformer tank. The switching assembly is enclosed in a sealed tank filled with transformer oil, and connections between the tap changer and transformer are done externally. This tap changer has 17 positions, so there are 17 taps on each high voltage winding. The switching is referred to as 'linear', in which each tapping is selected in turn. The switching operation is shown in Figure 21.24, in which diverter (or auxiliary) contacts connect a resistor in series with the tapping during a switching operation.

FIGURE 21.23 An on-load tap changer designed to fit outside the transformer tank

To achieve rapid switching, the motor-driven mechanism uses a Geneva gear to change rotary motion into a stepping motion. Mechanical energy is stored in tension springs, and the spring energy is released when the mechanism is ready to change positions. Although the 'wind-up' process takes several seconds, the switching action lasts less than three cycles (50 milliseconds or so) and makes a lot of noise. The switching sequence shown in Figure 21.24 occurs very quickly.

As shown, during a switching operation, diverter contacts act as an intermediary to maintain current to the primary of the transformer. This current passes through a resistor which helps reduce arcing at the contacts. The value of the resistor is chosen to limit the arcing current without causing a significant drop in the voltage to the transformer. During switching, a diverter contact maintains current flow through either of the two series resistors. At position 3 in Figure 21.24, two transformer tappings are connected together and the circulating current that will occur is limited by both resistors.

FIGURE 21.24 Diverter contacts maintain the connection during switching by passing the load current through a wire wound transition resistor

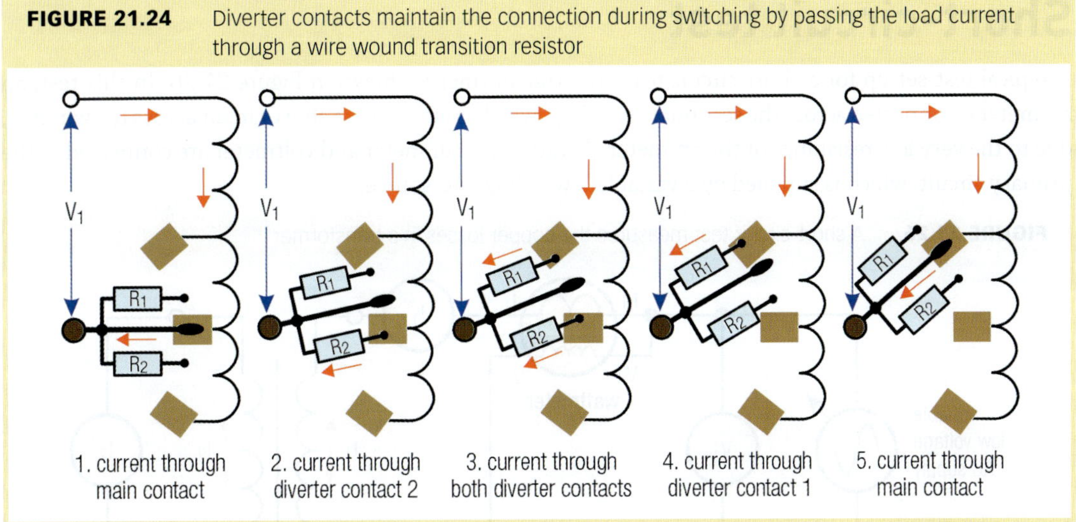

Figure 21.25(a) shows a typical connection of a three-phase tap changer. The tappings are located near the star point (if the transformer is star-connected), as this reduces the voltage levels the tap changer will experience. Some transformers have their tappings at the line end or in the middle of the winding. Figure 21.25(b) shows the connection terminals for the tap changer shown in Figure 21.23. Terminal 9 is the principal terminal, and is located in the middle of the tapping range. The variation range depends on the transformer, and could be from +10 per cent to −10 per cent, −15 per cent to +15 per cent and so on.

FIGURE 21.25 Tappings in a star-connected primary winding are located near the star or neutral point

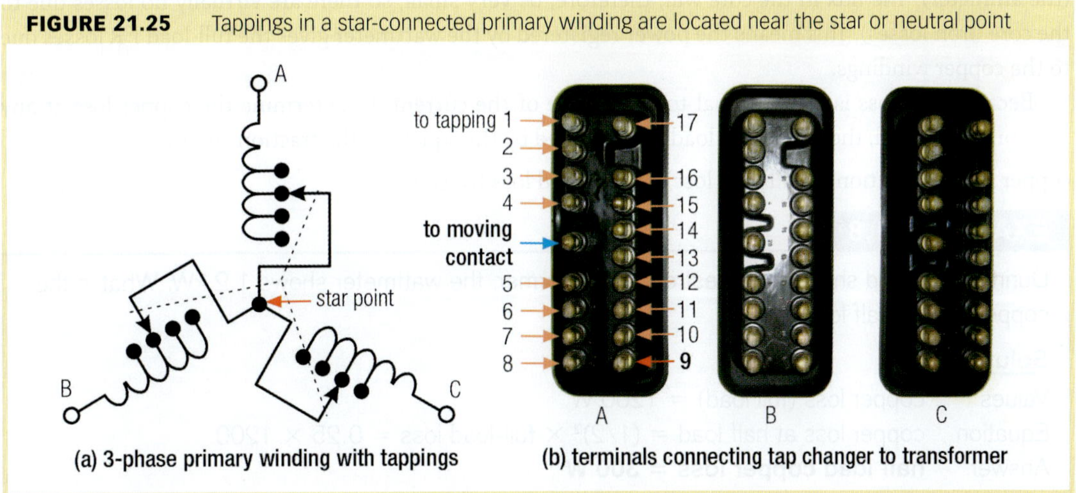

While resistive type on-load tap changers are the most common in Australia and New Zealand, another type uses a reactance in place of the resistor. This type is often used when the tap changer is connected to the secondary of a transformer. There are also variations in the type of switching. The tap changer in Figure 21.23 is for linear switching (one tapping after another), while another type is plus/minus switching in which an additional selector switch changes the polarity of a tapped section of the winding. This gives twice the range for a given number of tapping positions. There are many other variations associated with tap changers that are beyond the scope of this book, but their basic operating principles are similar to those we have described.

21.6 Transformer Performance

There are two tests that can be done on a transformer that allow its performance to be evaluated. These are the *open-circuit test* and the *short-circuit test*. From the results of these tests, a number of important performance characteristics can be determined, such as the transformer's efficiency, its impedance and prospective short-circuit current.

FYI

Geneva drive is a mechanism that has a rotating gear which intermittently engages with another specially shaped gear, rapidly moving it by part of a revolution

Short-circuit test

A typical test set-up for a short-circuit test on a transformer is shown in Figure 21.26. In this test, an ammeter is connected across the secondary winding, which is effectively a short-circuit across the winding, due to the very low resistance of the ammeter. A wattmeter, ammeter and voltmeter are connected in the primary circuit, which is supplied by a variable low voltage AC source.

FIGURE 21.26 A short-circuit test measures the copper losses in a transformer

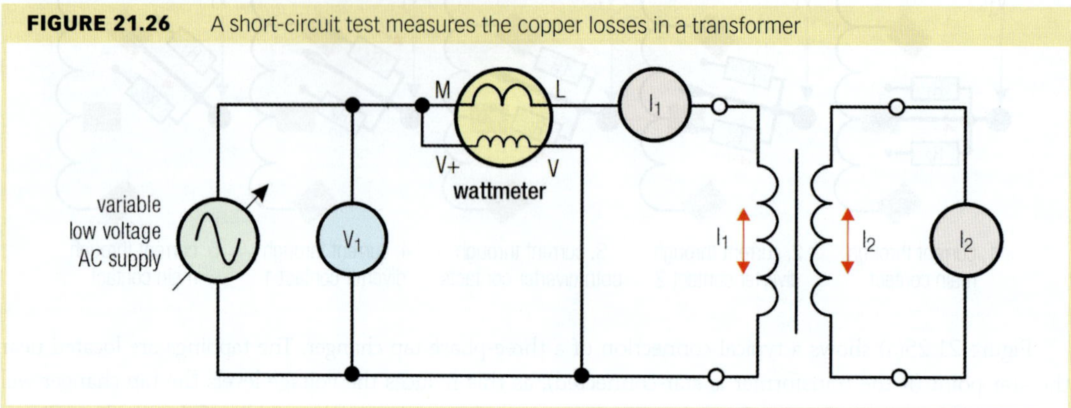

To carry out the test, the supply voltage is slowly increased until the ammeters show the transformer's rated full-load currents. Because the secondary winding is short-circuited by the ammeter, the secondary voltage is virtually zero, so negligible power is consumed by the load (the ammeter). The flux in the core will, therefore, be very small, so there are virtually no losses due to the core (iron losses). This means the power registered by the wattmeter gives the full-load I^2R losses due to the copper windings.

Because this loss is proportional to the square of the current, to determine the copper loss at any fraction of full load, the loss at full load is multiplied by the square of the fraction. That is:

copper loss = (fraction of full-load loss)^2 × full-load loss (watts).

EXAMPLE 21.8

During a full-load short-circuit test on a transformer, the wattmeter shows 1.2 kW. What is the copper loss at half load?

Solution

Values	copper loss (full load) = 1200 W
Equation	copper loss at half load = $(1/2)^2$ × full-load loss = 0.25 × 1200
Answer	**half load copper loss = 300 W**

The short-circuit test on a large transformer requires considerable care, due to the large test currents that are likely to be involved. It also requires good measuring equipment due to the low power factor, plus good connections to minimise errors.

Open-circuit test

The aim of the open-circuit test is to determine the iron losses in a transformer. The test set-up is shown in Figure 21.27, in which the secondary winding is left open-circuit and full rated voltage is applied to the primary winding.

As previously explained, when a transformer's secondary winding is open-circuit, the only current in the primary winding is the magnetising current, called the no-load or excitation current. Because this current is very low compared to the rated current, the transformer's copper loss can be ignored. Therefore, the power indicated by the wattmeter is the power due to the iron losses in the core.

FIGURE 21.27 An open-circuit test measures the iron losses in a transformer

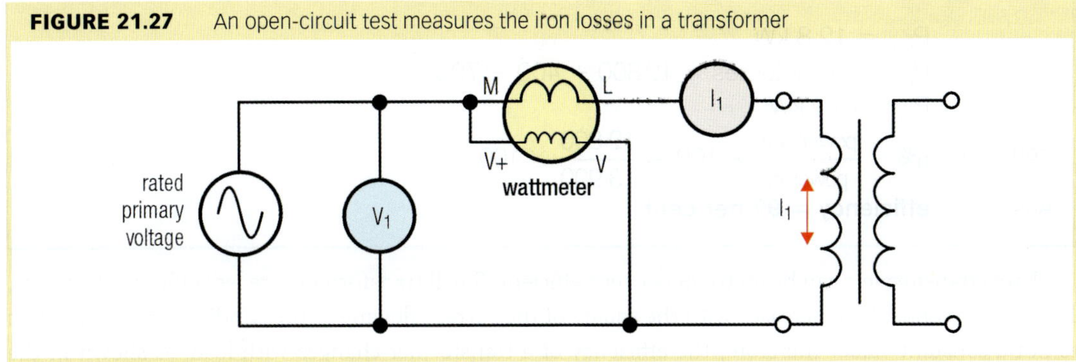

Unlike the copper loss in a transformer, iron losses remain relatively constant over the full-load range, assuming the primary voltage and supply frequency remain constant. By determining the iron and copper losses, the efficiency of a transformer can be determined.

Efficiency

Figure 21.28 shows the power flow through a transformer, in which some of the input power is consumed by the transformer and dissipated as heat, the rest being supplied to the load. Because efficiency is the ratio of the output power to the input power, if we know the input power and the losses, the efficiency of the transformer can be calculated with the following equation. The symbol η (Greek letter *eta*) is for efficiency, which is normally expressed as a percentage:

$$\eta\% = \frac{\text{power out}}{\text{power in}} \times 100$$

FIGURE 21.28 The power loss in a transformer is the sum of the iron and copper losses, and is dissipated as heat

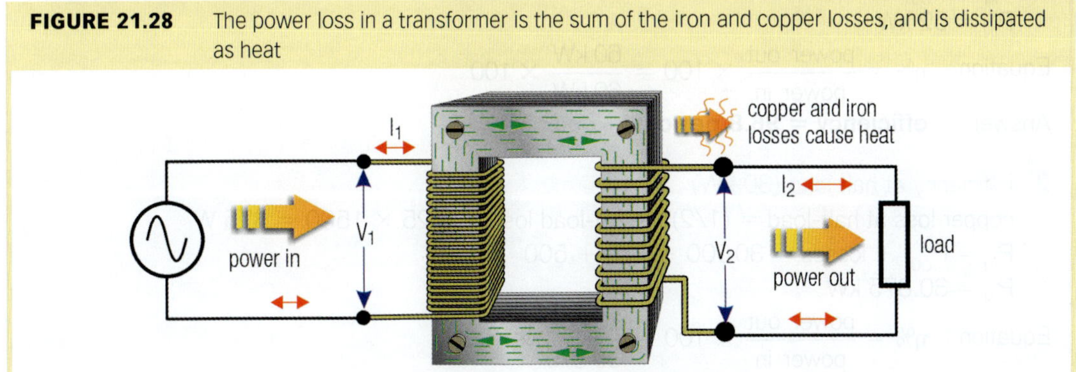

The power taken by any single-phase load is the product of the current, voltage and power factor. Multiply this value by $\sqrt{3}$ for a three-phase load. That is, for:

single-phase: $P_{IN} = V_1 I_1 \lambda_1$, and $P_{OUT} = V_2 I_2 \lambda_2$

three-phase: $P_{IN} = \sqrt{3} V_1 I_1 \lambda_1$, and $P_{OUT} = \sqrt{3} V_2 I_2 \lambda_2$.

EXAMPLE 21.9

An 11 kV/400 V single-phase transformer is supplying a load current of 40 A at a power factor of 0.8 lag. The transformer has iron losses of 400 W and at this load, the copper loss is 700 W. Determine the efficiency of the transformer.

Solution

Values copper loss = 700 W, iron loss = 400 W

$I_2 = 40$ A, $V_2 = 400$ V, $\lambda_2 = 0.8$

$P_{OUT} = V_2 I_2 \lambda_2 = 400 \times 40 \times 0.8$

$P_{OUT} = 12.8$ kW
$P_{IN} = P_{OUT} + \text{losses} = 12800 + 400 + 700$
$P_{IN} = 13.9$ kW

Equation $\eta\% = \frac{\text{power out}}{\text{power in}} \times 100 = \frac{12\,800}{13\,900} \times 100$

Answer **efficiency = 92 per cent**

Large transformers can be up to 99 per cent efficient. Small transformers are less efficient. However, because the copper loss increases with the square of the current flowing in the windings, the copper loss is not a constant value. Therefore, the efficiency of a transformer changes with load, as shown in the following example.

EXAMPLE 21.10

A 60 kVA transformer has a full-load copper loss of 1.5 kW and an iron loss of 500 W.
If the transformer is supplying a purely resistive load, calculate the efficiency of the transformer at:

1 full load
2 half load
3 one quarter load

Solution

Values full load copper loss = 1500 W, iron loss = 500 W
full load S = 60 kVA and because $\lambda = 1$, $P_{OUT} = 60$ kW

1 Efficiency at full load

$P_{IN} = P_{OUT} + \text{losses} = 60000 + 1500 + 500$
$P_{IN} = 62$ kW

Equation $\eta\% = \frac{\text{power out}}{\text{power in}} \times 100 = \frac{60 \text{ kW}}{62 \text{ kW}} \times 100$

Answer **efficiency = 96.8 per cent**

2 Efficiency at half load (30 kW)

copper loss at half-load = $(1/2)^2 \times$ full-load loss = $0.25 \times 1500 = 375$ W
$P_{IN} = P_{OUT} + \text{losses} = 30000 + 375 + 500$
$P_{IN} = 30.875$ kW

Equation $\eta\% = \frac{\text{power out}}{\text{power in}} \times 100 = \frac{30\,000}{30\,875} \times 100$

Answer **efficiency = 97.17 per cent**

3 Efficiency at quarter load (15 kW)

copper loss at quarter load = $(1/4)^2 \times$ full-load loss = $0.0625 \times 1500 = 93.75$ W
$P_{IN} = P_{OUT} + \text{losses} = 15000 + 93.75 + 500$
$P_{IN} = 15.593$ kW

Equation $\eta\% = \frac{\text{power out}}{\text{power in}} \times 100 = \frac{15\,000}{15\,523} \times 100$

Answer **efficiency = 96.63 per cent**

In Example 21.10, notice that the highest efficiency occurs at half load. In fact, the highest efficiency is reached when the copper losses equal the iron losses. This is nearly occurring in the example at half-load. Otherwise, the efficiency drops, and becomes increasingly lower as the load current reduces. At zero load current, efficiency is also zero. Because of these variations, a more useful measurement for a power transformer is an average efficiency taken over a period of time, such as 24 hours.

All-day efficiency

To determine the all-day efficiency of a transformer, its input energy and output energy are measured over a 24-hour period. A transformer's all-day efficiency is found with this equation:

$$\text{all-day efficiency } \eta\% = \frac{\text{kWh out}}{\text{kWh in}} \times 100, \text{ where kWh} = \text{kW} \times \text{time.}$$

EXAMPLE 21.11

Measurements taken on a power transformer show that over a 24-hour period it delivered 3500 kWh of energy. The energy losses over this period were measured at 300 kWh. What is the transformer's all-day efficiency?

Solution

Values energy out = 3500 kWh, power losses = 300 kWh (over 24 hour period)
energy in = energy out + energy losses = 3500 kWh + 300 kWh
energy in = 3800 kWh

Equation $\text{all-day efficiency } \eta\% = \frac{\text{kWh out}}{\text{kWh in}} \times 100 = \frac{3500}{3800} \times 100$

Answer **all-day efficiency = 92.11 per cent**

FYI

The highest all-day efficiency is achieved when a transformer is operating at or near full load at all times

Transformer impedance

The impedance of a transformer determines the amount of fault current that can flow in the windings should a short-circuit occur across its terminals. If a transformer is to work in parallel with another transformer in a load sharing arrangement, it's important that both transformers have the same impedance. The impedance of a transformer is calculated from the measurements taken during a short-circuit test.

As Figure 21.26 shows, a voltmeter measures the voltage applied to the transformer that causes full rated currents to flow in both windings. This voltmeter reading is referred to as the short-circuit test voltage, or V_{SC}. It is also called the impedance voltage, and is defined as the voltage required to cause rated current to flow through the impedance of the windings. To find the effect impedance has on voltage regulation requires also knowing the power factor of the primary circuit and the secondary circuit, which is beyond the scope of this book. We can say, however, that the higher the impedance, the greater the effect.

While impedance is measured in ohms, it's usual to refer to transformer impedance as a ratio of the voltage V_{SC} to the transformer's rated voltage V. Multiplying by 100 gives the impedance as a percentage. That is:

$$\%Z = \frac{V_{SC}}{V} \times 100$$

where:

V_{SC} is the voltage applied to the winding to achieve full rated current
V is the winding's rated voltage.

EXAMPLE 21.12

During a short-circuit test on a three-phase 10 MVA 66 kV/11 kV transformer, it was found that 2.8 kV was required to cause full rated current to flow in the windings. Calculate the transformer's:

1 rated primary current
2 percentage impedance

Solution

Values S = 10 MVA
V_1 = 66 kV, V_2 = 11 kV, V_{SC} = 2.8 kV

1 Calculate rated primary current

Equation $I_1 = \frac{S}{\sqrt{3} \times V_1} = \frac{10 \times 10^6}{1.732 \times 66 \times 10^3} = \frac{10 \times 10^3}{114.32}$

Answer **I_1 = 87.47 A**

2 Calculate percentage impedance

Equation $\%Z = \frac{V_{SC}}{V_1} \times 100 = \frac{2.8\,\text{kV}}{66\,\text{kV}} \times 100 = \frac{2.8}{66} \times 100$

Answer **%Z = 4.24 per cent**

Transformer manufacturers are required to state a transformer's short-circuit impedance on its nameplate. This value is given along with the temperature it applies to (typically 75 °C), the apparent power rating it refers to and, where appropriate, the tapping position. Apparent power rating of a transformer is often related to how it is cooled, and in many cases several power ratings are given for a transformer for each method of cooling.

The section of a transformer nameplate in Figure 21.29 shows a percentage impedance value of 14.8 per cent. This value was determined at the maximum rating of the transformer, which occurs when the cooling is oil forced directed, air forced (OFDAF).

FIGURE 21.29 Section of a 35 MVA 66 kV/11 kV power transformer nameplate showing its impedance as a percentage voltage value

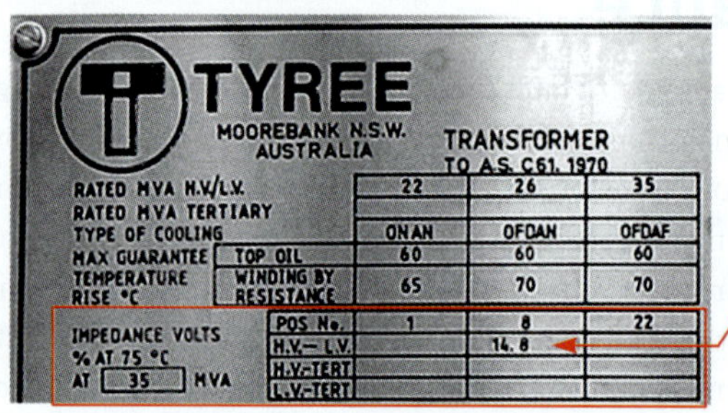

Prospective short-circuit current

When the percentage impedance of a transformer is known, the maximum value of a possible short-circuit current in the low voltage winding can be determined. This can occur if a fault condition arises that places a short-circuit across the low voltage winding of a power transformer. This value of current would occur during a short-circuit test if the test voltage was raised to the transformer's rated voltage. The equation to find the prospective short-circuit current is:

$$I_{SC} = \frac{100}{\%Z} \times I_{rated}$$

where:

I_{sc} is the prospective short-circuit current

%Z is the percentage impedance of the transformer

I_{rated} is the maximum rated current of the low voltage winding.

EXAMPLE 21.13

A three-phase 66 kV/11 kV, 35 MVA power transformer has a percentage impedance of 14.8 per cent. Calculate the transformer's:

1 rated secondary line current
2 prospective short-circuit current

Solution

Values $S = 35$ MVA
$V_1 = 66$ kV, $V_2 = 11$ kV
$\%Z = 14.8$

1 Calculate rated secondary line current (I_2)

Equation $I_2 = \frac{S}{\sqrt{3} \times V_2} = \frac{35 \times 10^6}{1.732 \times 11 \times 10^3} = \frac{35 \times 10^3}{19.1}$

Answer $\mathbf{I_2 = 1837\ A}$

2 Calculate prospective short-circuit current

Equation $I_{SC} = \frac{100}{\%Z} \times I_{rated} = \frac{100}{14.8} \times 1837$

Answer $\mathbf{I_{SC} = 12\,412.5\ A}$ **or 12.4 kA**

The short-circuit current found in Example 21.13 is a worst-case value, and assumes no other resistance or impedance in the circuit except that of the transformer. In a typical fault scenario, the impedance of conductors on both sides of the transformer and other circuit components would reduce the value of the short-circuit current.

KEY POINTS...

- Voltage regulation of a power transformer is a measure of how the secondary voltage changes under load. When multiplied by 100, the value becomes % voltage regulation.
- A tap changer is a type of switch that changes the turns ratio of a transformer and therefore the secondary voltage by selecting tappings on either the primary or secondary windings.
- Distribution transformers usually have an off-load tap changer, while transmission transformers have an on-load tap changer in the primary winding that is controlled by a voltage regulation system.
- An on-load tap changer has a complex switching arrangement that ensures current is maintained at all times to the load during a switching operation.
- A short-circuit test measures the copper losses in a transformer. It also gives the voltage required to establish full-load current in the windings, thereby allowing the transformer's impedance to be calculated.
- An open-circuit test measures the iron losses of a transformer, which remain virtually constant regardless of load. Copper loss varies with the square of the winding current.
- Efficiency of a transformer $\eta\% = \frac{\text{power out}}{\text{power in}} \times 100$.
- Transformer impedance is a ratio of the voltage required to establish full-load current (V_{SC}) to the transformer's rated voltage (V). Multiplying by 100 gives the impedance as a percentage, expressed as Z%.
- The prospective short-circuit current (I_{SC}) of a transformer $= \frac{100}{\%Z} \times$ secondary rated current.

21.7 Transformer Connections

Australian Standard AS 60076.1 *Power Transformers* sets out the standards for marking the terminals of each winding in a transformer. The basic rules are:

- Letters are assigned to windings in which the same letter is used for all windings on one limb of the core.
- The high voltage winding is identified with capital letters (e.g. A, B or C), and the low voltage winding on the same core is identified with a corresponding small letter (e.g. a, b or c).

- An auxiliary (or third) winding is identified by a capital letter for that phase, preceded by the numeral 3 (e.g. 3A, 3B or 3C).
- The start of a winding is identified by the number 1 written as a subscript (e.g. A_1 or a_1), the end by number 2 as a subscript (A_2 or a_2).

Figure 21.30 shows the markings for a single-phase and a three-phase transformer. The lower set of symbols are typical of those used in the standards and shown on transformer nameplates. Although the high voltage winding is usually the primary winding, in a step-up transformer the low voltage winding becomes the primary. It is identified as the low voltage winding, with small letters.

FIGURE 21.30 Transformer winding markings as specified by standards

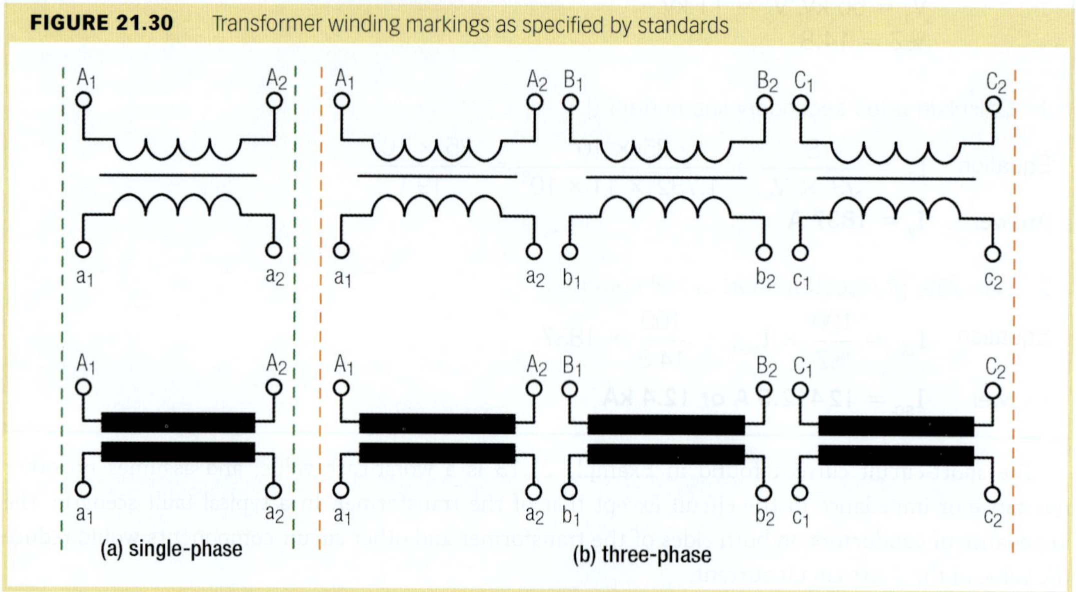

Three-phase transformers

Examples of three-phase transformers are shown in Figures 21.13 and 21.14. As these photos show, a three-phase transformer is essentially three single-phase transformers sharing a common core. The core typically has three limbs, like those in the photos. Large transformers might have two outer limbs as well, to provide a parallel path for the flux.

A three-phase transformer is smaller, cheaper, lighter and more efficient than three, separate single-phase transformers connected as a three-phase transformer. The disadvantage is the need for an expensive spare three-phase transformer, compared to a spare single-phase transformer. For transport reasons, very large power transformers might have to be made of three separate single-phase units.

There are three ways to connect each winding of a three-phase transformer: star, delta and a third called inter-star or zigzag. This connection is not commonly used, and we are not covering it here. We described the star and delta connections in Chapter 20, showing various ways of drawing each connection. There are four ways of combining the star and delta connections in a transformer:

1. star-connected primary and star-connected secondary (star–star)
2. star-connected primary and delta-connected secondary (star–delta)
3. delta-connected primary and delta-connected secondary (delta–delta)
4. delta-connected primary and star-connected secondary (delta–star).

A star–star connected transformer is shown in Figure 21.31(a), alongside the typically used symbol depicting the connection on a transformer nameplate. A delta–star transformer is in Figure 21.31(b), also with the symbols used on a transformer nameplate.

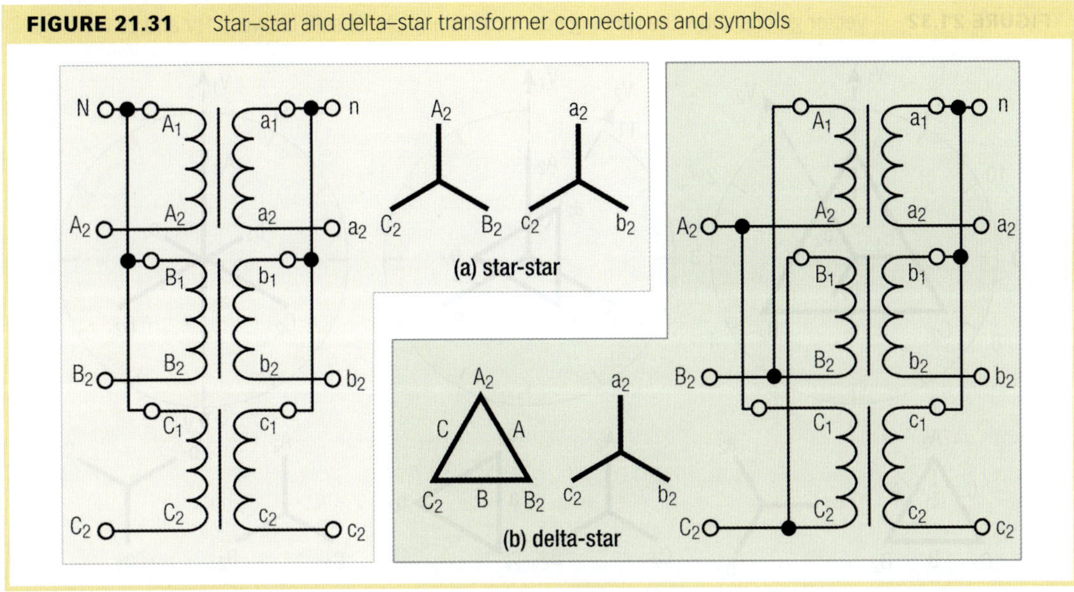

FIGURE 21.31 Star–star and delta–star transformer connections and symbols

Each of the four connections listed above has its particular advantages and disadvantages. For example, a distribution transformer would have a star-connected secondary to provide a neutral connection. A delta-connected secondary can continue supplying three-phase power if one phase of the transformer is open-circuit. Another consideration is the effect of a connection on harmonics, which we discuss later.

An important aspect to consider if two transformers are to be connected in parallel is the phase difference between the primary and secondary voltages. Some of these are:

1 star–star – 0° phase shift
2 star–delta – 30° phase shift, V_2 leading V_1
3 delta–delta – 0° phase shift
4 delta–star – 30° phase shift, V_2 lagging V_1.

It's common practice to connect power transformers in parallel, so a standard way of identifying the high voltage and secondary voltage winding connections and the phase shift between them is specified by AS 60076.1. This is referred to as a transformer's vector group.

Vector group

The method for describing a vector group is a code consisting of two or three letters, followed by one or two digits. Capital letters are for the high voltage (HV) windings and small letters are for the low voltage (LV) windings. The letters are:

- Y or y – indicates a star connection
- D or d – indicates a delta connection
- N or n – indicates that the neutral point is brought out.

The phase difference is indicated by using a clock face notation. The phasor representing the high voltage winding is taken as the reference and is set at 12 o'clock. All references are taken from phase-to-neutral and assume an anticlockwise phase rotation. The neutral point may be real (as in a star connection) or imaginary (as in a delta connection).

The numeral 0 refers to 12 o'clock, so because the high voltage winding is 'set' to 12 o'clock, the phase difference is 0°. Numeral 1 (1 o'clock) means the phase difference is 30° lagging, 2 means 60° lagging and so on around to numeral 6, in which the windings are 180° out of phase. Numeral 11 (11 o'clock) indicates a 30° leading phase shift. Three examples are shown in Figure 21.32.

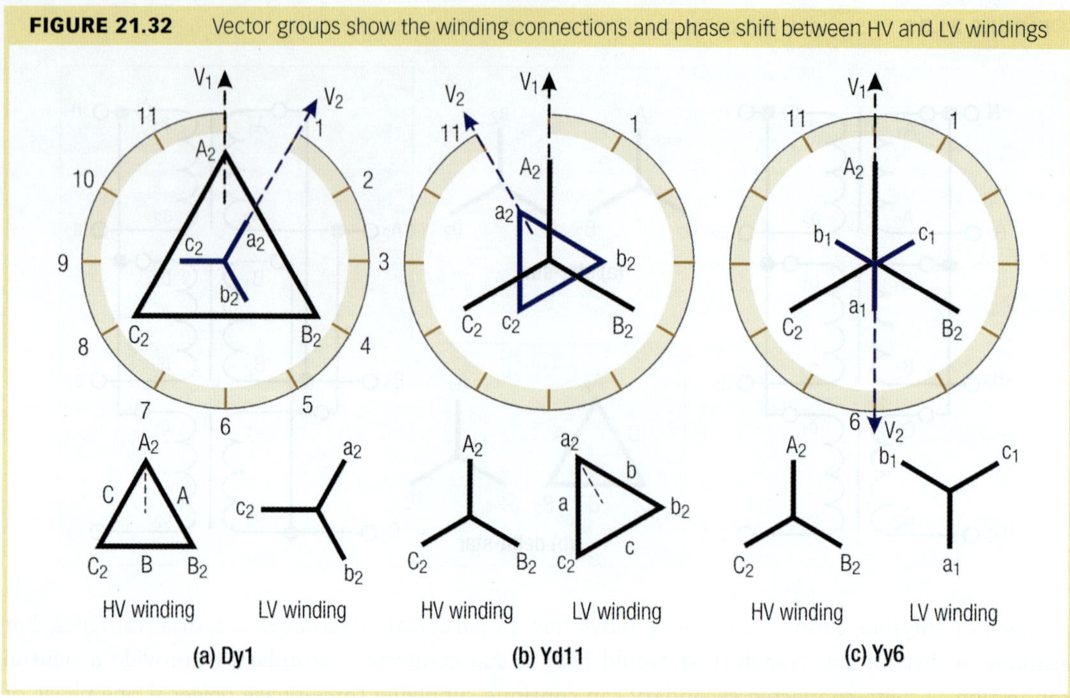

FIGURE 21.32 Vector groups show the winding connections and phase shift between HV and LV windings

The vector group in Figure 21.32(a) shows the delta–star connection, which has a phase shift of 30° lagging, so it is notated as Dy1 because vector V_2 is at 'one oclock'. The star–delta connection has a phase shift of 30° leading, and is notated as Yd11. The star–star connection in Figure 21.32(c) is for the condition where the LV winding is 180° out of phase with the high voltage winding. If the windings were in phase, the vector group would be Yy0. Because they are out of phase, the vector group is Yy6. A delta–delta connected transformer can be connected to give a 180° phase difference between windings, so its vector group is Dd6, otherwise Dd0. The standards also give a method of showing a vector group with symbols, in which the star or delta symbol of the LV winding is shown rotated if there's a phase shift. Examples are shown in Figure 21.32.

Sections of a power transformer nameplate showing the vector group and winding information are shown in Figure 21.33. The complete nameplate has more details such as ratings of the transformer. Notice that this transformer has a delta-connected tertiary winding, as explained next. The transformer vector group is YNyn0d1, where N or n indicates neutral is brought out, Y or y indicates star connection, 0 gives the phase difference (0°) and d1 indicates the delta-connected tertiary winding with a phase displacement of +30°.

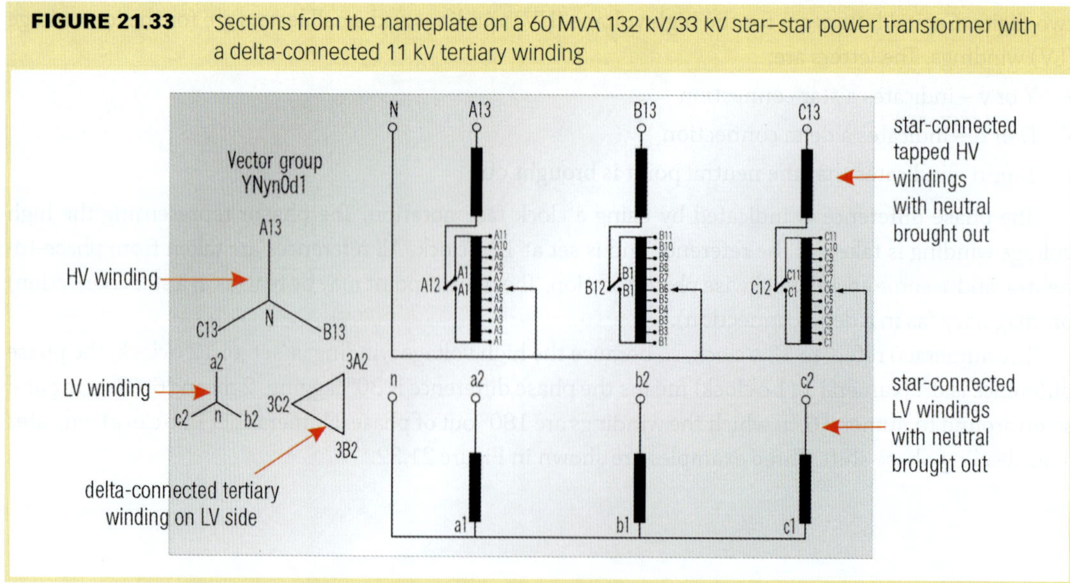

FIGURE 21.33 Sections from the nameplate on a 60 MVA 132 kV/33 kV star–star power transformer with a delta-connected 11 kV tertiary winding

There are four basic vector groups. Group 1 has a phase displacement between primary and secondary of 0°, group 2 a phase displacement of 180°, group 3 a phase displacement of −30° (secondary lags primary) and group 4, where the phase displacement is +30° (secondary leads primary). Transformers being connected in parallel must belong to the same vector group.

Tertiary winding and harmonics

The B–H curve of a transformer core is not linear, which means the magnetising current of a transformer is not purely sinusoidal (assuming a sinusoidal primary winding voltage). Therefore, the magnetising current contains harmonics. Harmonics are explained in Chapter 15. The third-order harmonic (150 Hz) is the most prominent harmonic in a transformer, although others occur, but in smaller magnitudes.

The third-order harmonic component of the magnetising current flows through the neutral of a star-connected winding, or around any delta-connected winding. If there is no path for the harmonic current, the output voltage will contain distortion, and no longer be a pure sinewave. A circulating third-harmonic current in the neutral can cause interference with telecommunications and electronic equipment, and other effects.

To provide a path for third-order harmonics, some star–star transformers have a third winding connected in delta, called a tertiary winding or a stabilising winding. This winding could have its terminals brought out so it can provide power to an auxiliary load, or the winding terminals may not be accessible. In any case, the winding must be rated to handle the maximum circulating current which can flow as a result of worst case system unbalance. This situation occurs if there's a line-to-earth short-circuit of the secondary winding with the secondary neutral point connected to earth.

21.8 Transformers in Parallel

Power transformers are connected in parallel to provide increased capacity to supply a load that a single transformer cannot. There are a number of requirements that have to be met before two transformers can be connected in parallel.

1. The transformers should have the same percentage impedance (or regulation).
2. The voltage ratio must be the same.
3. The instantaneous polarity of the secondary voltages must be the same.
4. The phase sequence must be the same (three-phase transformer).
5. The transformers must have the same inherent phase angle between primary and secondary terminals (three-phase transformer).

Requirements 1, 2 and 3 apply to both single-phase and three-phase transformers.

Percentage impedance (or regulation)

Percentage impedance of a transformer gives a guide to its voltage regulation. That is, it gives a guide as to how much the terminal voltage drops as load current is increased. The relationship between percentage impedance and voltage drop can be complex due to the power factor of a load, so we are assuming a resistive load only.

Figure 21.34 shows two single-phase transformers connected in parallel supplying a 500 kVA resistive load. Both transformers are rated at 250 kVA, but transformer A (Tx_A) has an impedance of 4 per cent and Tx_B has an impedance of 2 per cent. Therefore, they will not share the total load equally.

FIGURE 21.34 If two transformers in parallel have different impedances, one will supply more power to the load than the other

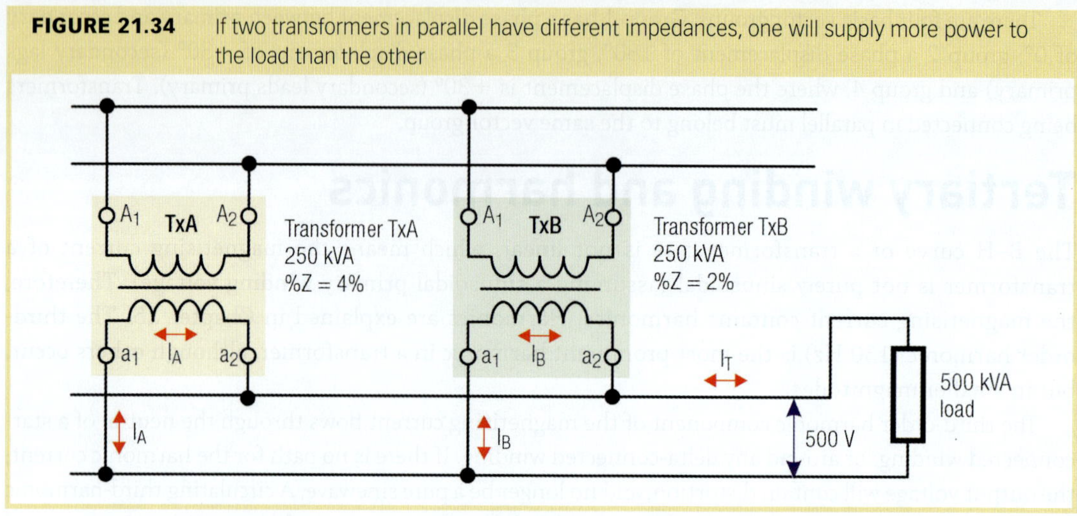

The method to determine how each transformer in Figure 21.34 shares the load uses these equations:

$$\text{Load taken by Tx}_A = \frac{Z_B}{Z_A + Z_B} \times \text{total load}$$

$$\text{Load taken by Tx}_B = \frac{Z_A}{Z_A + Z_B} \times \text{total load}$$

where:

Z_A = % impedance of transformer A

Z_B = % impedance of transformer B.

EXAMPLE 21.14

For the circuit in Figure 21.34, calculate the:

1 load taken by each transformer

2 current supplied by each transformer

Solution

Values VA rating of both transformers = 250 kVA

V_2 = 500 V, load = 500 kVA or 500 kW as load is resistive

$\%Z_A = 4$, $\%Z_B = 2$

1 Calculate load taken by each Tx

Equation $\text{Load taken by Tx}_A = \frac{Z_B}{Z_A + Z_B} \times \text{total load} = \frac{2}{4+2} \times 500\text{ kVA}$

Answer **Tx_A load = 166.67 kVA (or kW)**

Equation $\text{Load taken by Tx}_B = \frac{Z_A}{Z_A + Z_B} \times \text{total load} = \frac{4}{4+2} \times 500\text{ kVA}$

Answer **Tx_B load = 333.33 kVA (or kW)**

2 Calculate current supplied by each transformer

Equation $I_A = \frac{\text{kVA supplied by Tx}_A}{\text{load voltage}} = \frac{166.67}{500}$

Answer **I_A = 333.3 A**

Equation $I_B = \frac{\text{kVA supplied by Tx}_B}{\text{load voltage}} = \frac{333.33}{500}$

Answer **I_B = 666.7 A**

Example 21.14 shows that:

- transformer A is supplying one-third of the load kVA, and is not working at its rated capacity
- transformer B is supplying two-thirds of the load kVA and is operating at 33 per cent overload.

If a load is reactive, it becomes more complex to determine the load sharing between two transformers. The important thing is that when connecting transformers in parallel, either single-phase or three-phase, their impedance values should be equal.

Voltage ratio

When transformers are connected in parallel, the high voltage windings are connected to the same high voltage bus. Therefore, to achieve the same output voltage, the turns ratio of the transformers must be the same. If one is higher than the other, the transformer with the higher output voltage will cause a current to flow in the other transformer, called a circulating current. It is sometimes impossible to have two transformers provide exactly equal output voltages, which means there is often a small value of circulating current flowing between them. This can be acceptable if the circulating current is relatively small and the transformers are not fully loaded.

Polarity

The polarity of the windings *must* be the same. Figure 21.35(a) shows two single-phase transformers with the same voltage polarity at both secondary windings, indicated by the voltmeter showing zero volts. These transformers can be connected in parallel. If the windings have opposite polarity, the voltmeter will show twice the secondary voltage as in Figure 21.35(b). Connecting these transformers in parallel would cause a dangerously high circulating current between the secondary windings.

FIGURE 21.35 If two voltages have the same polarity and the same value, there is no potential difference between them as in (a); otherwise the potential difference is twice the individual voltages as in (b)

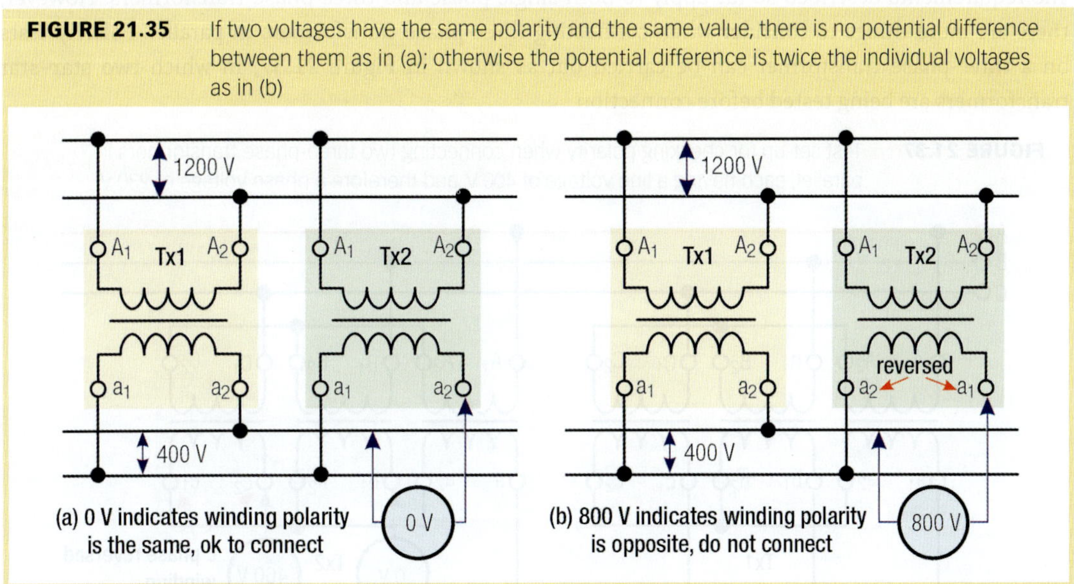

Transformer windings, as explained above, are marked in a standard way in which, for example, A_1 is the start of an HV winding and A_2 the end of the winding. The LV winding is marked as a_1 for the start of the winding, and a_2 for the end. However, one transformer manufacturer might wind the HV winding in a clockwise direction and wind the LV winding in an anticlockwise direction, and another manufacturer might have both windings wound in a clockwise direction.

This means the markings on the transformer, although correctly identifying the start and finish of each winding, do not identify its polarity. That is, the output voltage of both transformers, while identical in value, are 180° out of phase. The transformers, therefore, have different polarities.

Figure 21.36 shows how to determine the polarity of a transformer. The transformers in Figure 21.36(a) and (b) are connected to an AC supply of a suitable voltage, and terminals A_1 and a_1 are joined. That is, the start of the primary winding and the start of the secondary winding are connected together. A voltmeter is connected to measure the voltage between the ends of the windings, marked as A_2 and a_2.

FIGURE 21.36 The polarity of a transformer can be determined with this test set-up

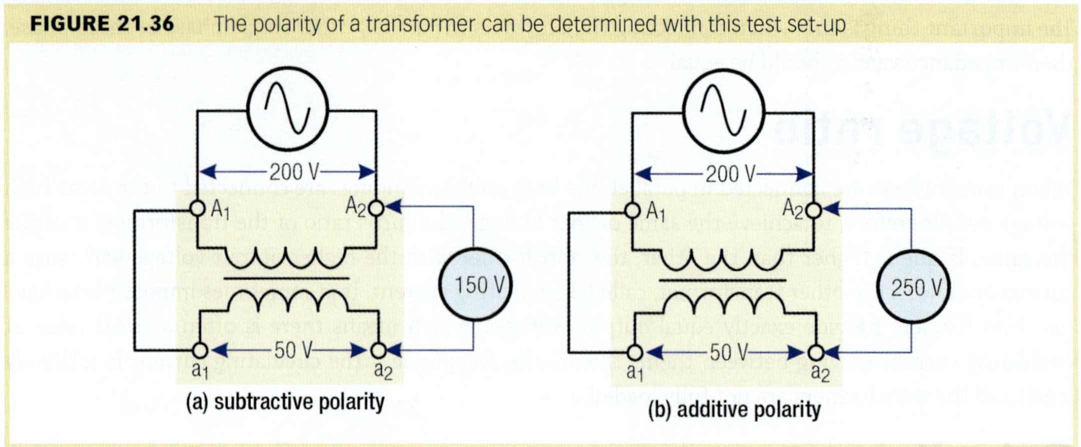

If the voltage between A_2 and a_2 is less than the primary voltage, the transformer has a *subtractive* polarity. This is the case with most power transformers. If the transformer has an *additive* polarity, the voltmeter will read more than the primary voltage. When connecting transformers in parallel, it is, therefore, important to confirm their polarity is the same, either with the test in Figure 21.36, or better still, the test in Figure 21.35.

Three-phase transformers in parallel

The requirements described so far apply to both single-phase and three-phase transformers. However, there are more things to consider when connecting three-phase transformers in parallel. Polarity tests on a three-phase transformer can be carried out as shown in Figure 21.37, in which two star–star transformers are being tested before connection.

FIGURE 21.37 Test set-up for checking polarity when connecting two three-phase transformers in parallel, each having a line voltage of 400 V, and therefore a phase voltage of 230 V

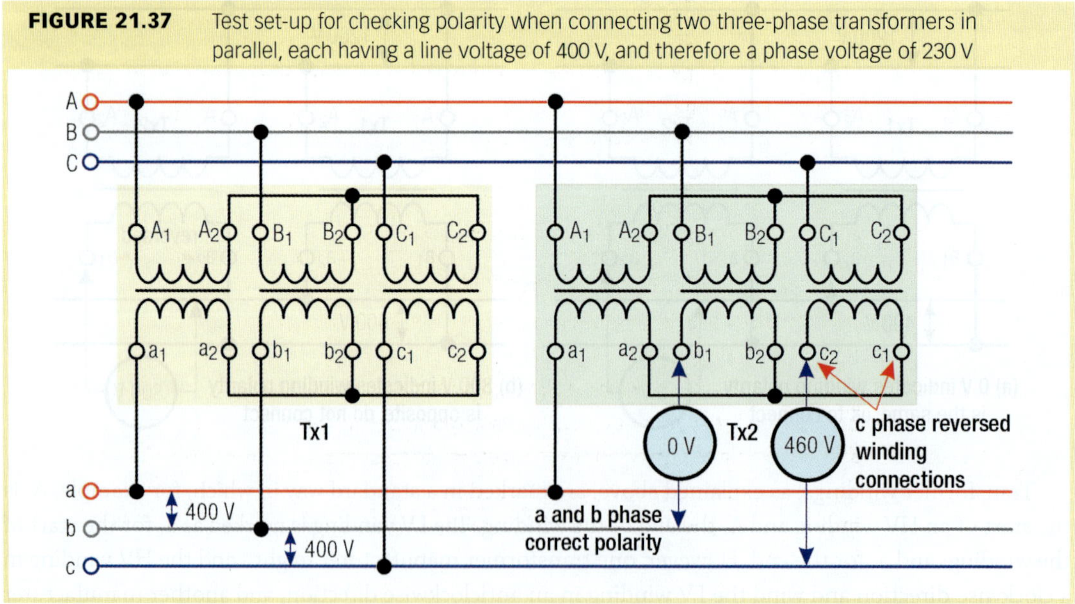

In Figure 21.37, the polarity of both transformers is the same, but the c phase secondary winding of the second transformer (Tx2) is incorrectly wired, and is reversed. This reverses the polarity of the voltage compared to the c phase voltage of the other transformer, indicated by the voltmeter reading twice the phase voltage. The voltmeter readings would all be high if one of the transformers has subtractive polarity and the other has additive polarity, assuming all windings to be otherwise correct.

Phase sequence

In single-phase transformers, phase sequence does not apply, as it's a characteristic of polyphase transformers only. Phase sequence refers to the order in which each phase reaches its maximum voltage, which can be A-B-C or A-C-B. The order of phase sequence is not important, simply that both transformers must have the same phase sequence.

Inherent phase angle

The phase difference between the primary and secondary terminals of a three-phase transformer depends on how the windings are connected, which is identified by the transformer's vector group. For example, a star–star connection has 0° phase shift between windings, while a star–delta connection has a 30° phase shift. This is called the *inherent phase angle*, which must be the same in both transformers if they are to be connected in parallel.

Transformers that belong to the same vector group can be connected in parallel, all other requirements having been met. For example, transformers in group 1 and group 2 have a 180° phase difference between the two secondary windings. A group 3 and a group 4 transformer have a 60° phase difference between their secondary voltages. Transformers belonging to different vector groups cannot be operated in parallel without altering the internal connections of one of them.

Table 21.1 shows the combinations of transformer winding connections that allow parallel operation. Knowing the vector group is important, as each connection has two possible inherent phase angles. For example, Yd11 and Yd1 are both star–delta connected, but one has a −30° phase shift (group 3), the other a +30° phase shift (group 4). Also shown are the winding combinations that cannot be paralleled without changing the internal connections of the transformer.

Table 21.1

PHASE RELATIONSHIPS AND CONNECTING TRANSFORMERS IN PARALLEL						
TRANSFORMER 1			TRANSFORMER 2			PARALLEL CONNECTION
HV	LV	group	HV	LV	group	
delta	delta	1 or 2	delta	delta	1 or 2	✓
star	star	1 or 2	star	star	1 or 2	✓
delta	delta	1 or 2	star	star	1 or 2	✓
star	star	1 or 2	delta	delta	1 or 2	✓
star	delta	3 or 4	star	delta	3 or 4	✓
star	delta	3 or 4	delta	star	3 or 4	✓
delta	delta	1 or 2	delta	star	3 or 4	✗
star	star	1 or 2	delta	star	3 or 4	✗
star	delta	3 or 4	delta	delta	1 or 2	✗
star	delta	3 or 4	star	star	1 or 2	✗
delta	star	3 or 4	star	star	1 or 2	✗
delta	star	3 or 4	delta	delta	1 or 2	✗

21.9 High Voltage Safety

All electricity supply organisations have certain requirements in regard to working on high voltage equipment. A high voltage is defined as any voltage greater than 1 kV AC or 1.5 kV DC. The requirements may vary from one organisation to another, but all require certain procedures to be followed and certain protective clothing to be worn. Supply organisations usually provide training to ensure all electricians and those working in a high voltage environment are fully aware of the dangers, and are able to work safely. This includes rescue techniques, such as a bucket rescue where a person working aloft on high voltage lines is carried safely to ground by other workers in the event of a problem (e.g. electrocution).

In Australia, each state and territory produces safety guides for electrical workers based on these regulations. New Zealand has similar legislation and regulations. AS/NZS 3000:2018 covers high voltage electrical installations in section 7.6.

In all cases, before working on or near high voltage equipment or exposed conductors, you will need an access permit. This will require that the equipment or lines being worked on are isolated, and proven to be de-energised by an approved method. It should be impossible for the site to become accidentally energised, which means circuit breakers might be removed from their cabinets, and lines and equipment are earthed, as shown in Figure 21.38. Danger tags are fitted at the isolation points and the work area is identified by taping around the area. Apprentices and permit holders are required to be supervised by an authorised person.

FIGURE 21.38 When working on high voltage equipment, an access permit is required, the equipment is isolated, danger tagged and various points of the equipment are earthed

(a) danger tag fitted to HV equipment that has been isolated for maintenance

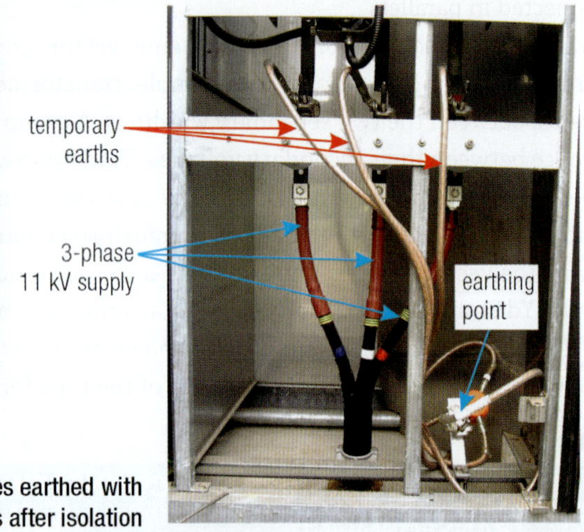

temporary earths

3-phase 11 kV supply

earthing point

(b) 11 kV supply lines earthed with temporary earths after isolation

KEY POINTS...

- Transformer winding terminals are marked with letters and subscript numbers in accordance with AS 60076.1 to identify the start and end of each winding.
- Three-phase transformer windings are generally connected in either delta or star configurations. Others include zigzag and open-delta, but these are not commonly used.
- The vector group of a transformer identifies the phase difference between the primary and secondary voltages, which is typically 0° (group 1), 180° (group 2), −30° (group 3) or +30° (group 4).
- To provide a path for third order harmonics, some transformers have a delta-connected tertiary winding.
- Single-phase transformers being connected in parallel must have the same percentage impedance (or regulation), voltage ratio and polarity.
- Three-phase transformers being connected in parallel must have all of the above plus the same phase sequence and inherent phase angle between primary and secondary terminals.

21.10 Auto-Transformers

A conventional transformer has separate primary and secondary windings which are electrically isolated from each other. Their only connection is through the magnetic flux in the core. An auto-transformer has one continuous winding, shared by the primary and secondary. The winding is wound on a laminated silicon steel core, with the primary and secondary sections in the same magnetic circuit.

The circuit diagrams in Figure 21.39 show in (a) a step-down transformer, in which a tapping from the winding provides one secondary output terminal, with its other output terminal connected to one of the primary input terminals. The secondary voltage V_2 is lower than the primary voltage. A step-up autotransformer is shown in (b).

FIGURE 21.39 An auto-transformer has an electrical connection between its primary and secondary windings, and can be either step down or step up

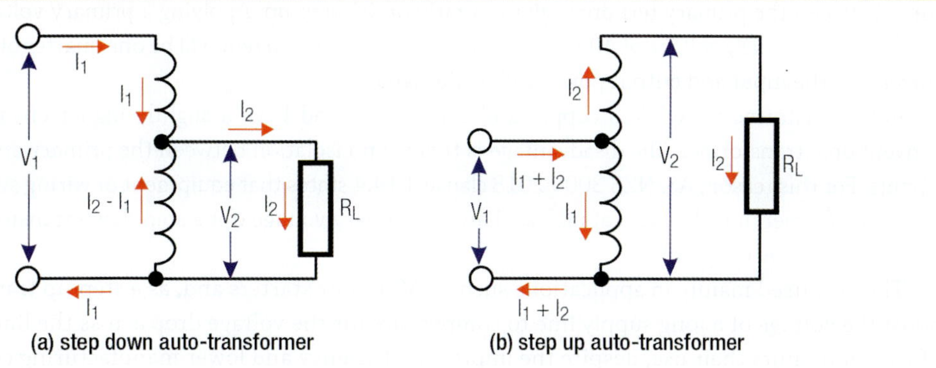

As for a conventional transformer, the ratio of the primary and secondary voltage is the same as the turns ratio and current is the inverse ratio. The turns ratio is the total number of turns between the input terminals compared to the total number of turns between the output terminals. Example 21.15 will help show the operation of the step down auto-transformer.

EXAMPLE 21.15

The transformer in Figure 21.39(a) has a total of 400 turns. A tapping is taken so the secondary winding has 100 turns. The primary voltage is 200 V and the load on the secondary is 25 Ω (resistive). Calculate the:

1 secondary voltage and current
2 primary current
3 transformer input and output power

Solution

Values Primary turns $N_1 = 400$, secondary turns $N_2 = 100$
$V_1 = 200$ V
$R_L = 25\ \Omega$

1 Calculate V_2 and I_2

Equation $V_2 = V_1 \times \frac{N_2}{N_1} = 200 \times \frac{100}{400}$

Answer $\mathbf{V_2 = 50\ V}$

$I_2 = \frac{V_2}{R_L} = \frac{50}{25}$

Answer $\mathbf{I_2 = 2\ A}$

2 Calculate primary current (I_1)

Equation (ignoring losses) $I_1 = I_2 \times \frac{N_2}{N_1} = 2 \times \frac{100}{400}$

Answer $\mathbf{I_1 = 0.5\ A}$

3 Calculate input and output power

Equation $P_{in} = V_1 \times I_1 = 200 \times 0.5$

Answer $\mathbf{P_{in} = 100\ W}$

$P_{out} = V_2 \times I_2 = 50 \times 2$

Answer $\mathbf{P_{out} = 100\ W}$

Because the direction of the current in the secondary is 180° out of phase with the current in the primary, the current in the common part of the winding is the difference between the two currents. Therefore, this part of the winding can be wound with smaller gauge wire, saving copper and costs.

A step-up auto-transformer is shown in Figure 21.39(b). In this type, there are more secondary turns than primary turns. For example, there could be 200 turns between the secondary terminals, and 50 turns between the primary terminals giving a ratio of 4:1 step up. Applying a primary voltage of 100 V produces a secondary voltage of 400 V. However, the secondary current will be one quarter of the primary current, so the input and output power will be the same.

Auto-transformers have less copper and core losses, and have a slightly higher efficiency than a conventional transformer. The disadvantage is there is no isolation between the primary and secondary circuits. For this reason, AS/NZS 3000:2018 clause 4.14.4 states that equipment or wiring supplied by an auto-transformer must be rated at no less than the primary voltage (for a step-down transformer) or the secondary voltage (step up).

They are used mainly in applications such as AC motor starters and, as a step-up transformer, to boost the voltage of a long supply line to compensate for the voltage drop across the line. Their lack of isolation limits their use, despite the improved efficiency and lower manufacturing cost.

FYI

The term 'Variac' was a US trademark, but is now applied to all such transformers

Variable auto-transformers

Figure 21.40 shows a variable auto-transformer, often called a variac. These are found in electrical workshops and test laboratories. As the circuit diagram in Figure 21.40(b) shows, depending on the position of the wiper, this transformer can be either step up or step down. The unit in Figure 21.40 has a maximum output voltage of around 260 V for a 230 V primary voltage.

FIGURE 21.40 A variable auto-transformer (variac) provides a variable output voltage determined by the position of the wiper

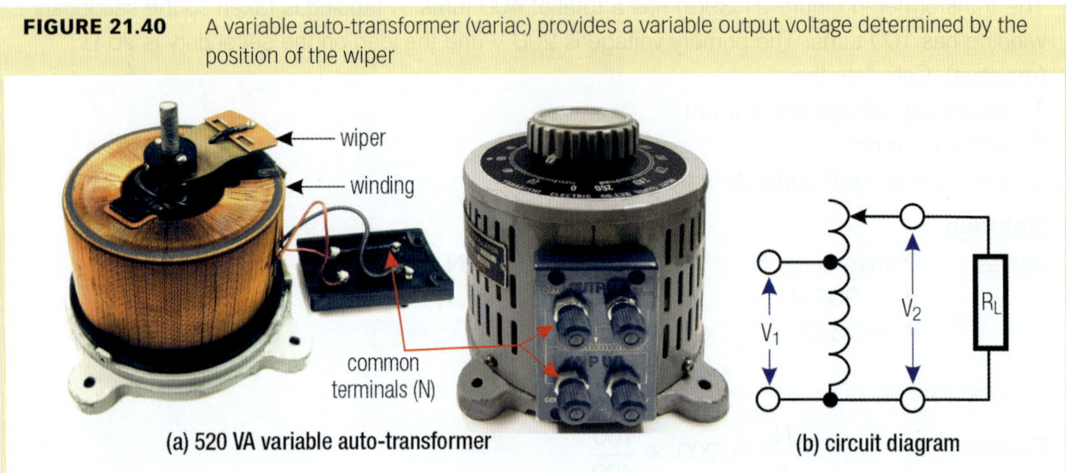

SAFETY

The neutral of a supply should be connected to the common terminal of an autotransformer

There are two important issues to consider with a variable auto-transformer. The first is the polarity of the input voltage. In a typical application, a single-phase unit is connected to the active and neutral of a 230 V supply. The neutral conductor is earthed at the switchboard, and is therefore at zero potential. The neutral should be connected to the common terminal, as shown in Figure 21.40(b). Otherwise, this terminal is at 230 V above earth, presenting a shock hazard.

The other issue is the maximum value of the secondary current. The unit shown in Figure 21.40 has a rating of 520 VA, which means the primary current cannot exceed 2 A. Because the winding is common to both primary and secondary circuits, the current in any section of the winding is limited to 2A.

This means the apparent power able to be supplied to a load will decrease below a certain output voltage, which for the unit in Figure 21.40 is 130 V, half its maximum output of 260 V. For example, if the output voltage is set to 50 V while connected to a 520 VA load, the load current will be 10.4 A. The input apparent power is also 520 VA, which at 230 V gives a primary current of 2 A.The current in the shared part of the winding is 10.4 A – 2 A, giving a current exceeding 8 A. This current will burn out that part of the winding because it is rated at only 2 A.

21.11 Instrument Transformers

Instrument transformers fall into two categories: voltage transformers (VT) and current transformers (CT). A voltage transformer is sometimes referred to as a potential transformer (PT). Instrument transformers are used extensively in electrical power generation and distribution to measure and monitor the high voltages and currents that are otherwise impossible to measure. They are also used in industrial power installations for the same reason. They usually form part of a network's metering, protection and control systems.

Voltage transformers

Voltage transformers (VT) can be either of two types, inductive (as in a conventional transformer) or capacitive, in which capacitors provide a potential divider network. Inductive voltage transformers are most economical up to a system voltage of 145 kV and capacitor voltage transformers above 145 kV. Inductive VTs are similar to small power transformers, but are designed for a controlled ratio accuracy over the specified range of output. Because the load on a VT is typically a voltmeter, a protection relay or the voltage coil of a wattmeter, they have a low VA rating.

Figure 21.41 shows a three-phase VT rated at 33 kV (primary). The windings are connected in star–star, and the secondary voltage (phase to neutral) is $110/\sqrt{3}$ (63.5 V). The neutral connection on the secondary side is connected to earth, along with an earth screen between the windings.

FIGURE 21.41 Three-phase VT in a 33 kV substation, monitoring the bus voltage to a power transformer

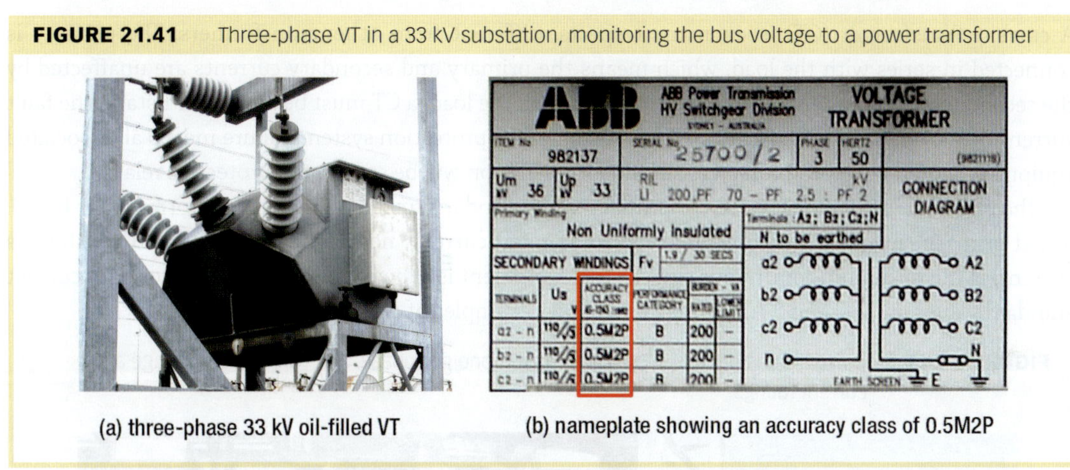

(a) three-phase 33 kV oil-filled VT

(b) nameplate showing an accuracy class of 0.5M2P

A VT is designed to operate at a low core flux and with a specified 'burden'. IEC and other standards state that accuracy should be maintained from 25 per cent to 100 per cent of the rated burden. In some types, two or more secondary windings are provided, in which one is connected to metering equipment and the other to protection relays. Standard values of rated output VA, at a power factor of 0.8 lagging, range from 10 VA to 500 VA, where the rated output is the rated output per phase.

Metering accuracy classes range from around 0.15, which means the error at a metered load with a power factor of 0.6 to 1.0 will be 0.15 per cent for a burden ranging from 0 to 100 per cent with a primary voltage ranging from 90 per cent to 110 per cent of rated voltage. When used as part of a protection system, a VT is given an accuracy class of, for example, 3P, which means accuracy is fulfilled between 5 per cent to 80 per cent of the primary rated voltage, and between 120 per cent to over 150 per cent of rated voltage. A VT might have a combined class, such as 0.5/3P. Standard burdens range from 35 VA (at 0.2 power factor) to 400 VA at 0.85 power factor.

A voltage transformer might be protected by fuses on the primary, up to 66 kV, but must always be protected on the secondary side to protect against secondary short-circuit currents. Voltage transformers should never be shorted at the secondary terminals. Figure 21.42 shows the connections of a three-phase VT. It is common practice for the primary and secondary windings to be earthed at the star point. The voltmeters would be calibrated to give a direct readout of the HV line voltage.

FIGURE 21.42 A three-phase voltage transformer (VT) is usually connected in star–star with the neutral points earthed

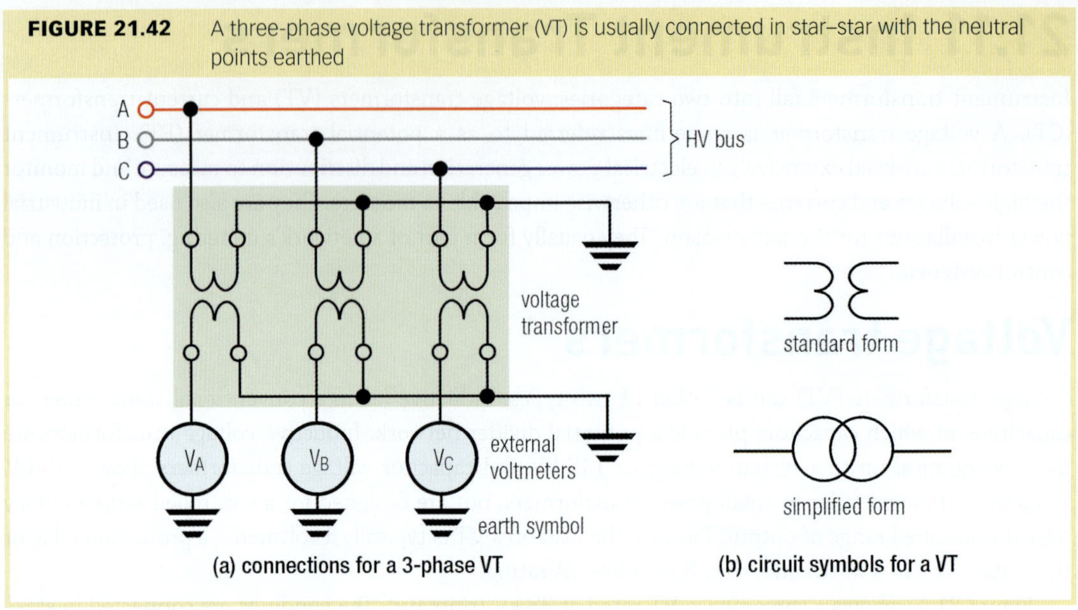

Current transformers

A current transformer (CT) is, in many respects, different from other transformers. The primary is connected in series with the load, which means the primary and secondary currents are unaffected by the secondary load (burden). Because it's in series with the load, a CT must be able to withstand the fault current that may occur at its location. If a CT is part of a protection system, failure means all associated equipment would be left unprotected, since no information will be sent to the protective relays.

The rated currents are the values of primary and secondary currents on which performance is based. Rated primary current (sometimes referred to as rated current, nominal current or rated continuous current) is the maximum continuous current the equipment is allowed to carry. Rated secondary current standard values are 1, 2 and 5 A. Figure 21.43 shows examples of current transformers.

FIGURE 21.43 Current transformers (CTs) are arranged one per phase, and vary in size, voltage and current ratings

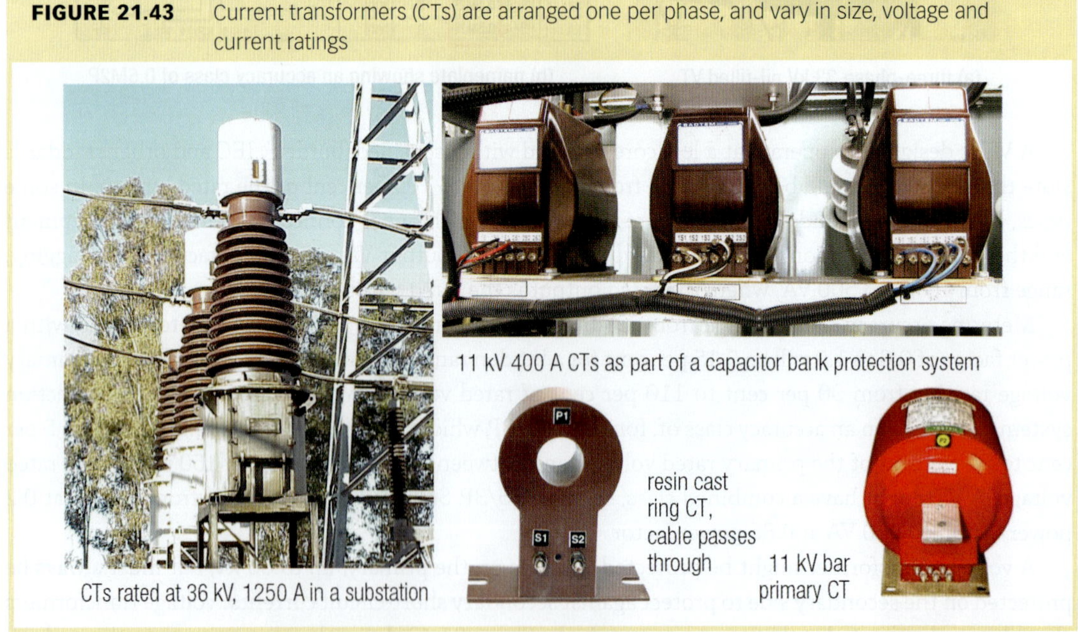

CTs are also given a burden rating in VA. This is the external impedance in the secondary circuit in ohms at the specified power factor. It is usually expressed as the apparent power which is taken up at rated secondary current. The accuracy class for measuring windings is given as 0.2, 0.5 or 1.0, depending on the application.

SAFETY

An in-service current transformer should always have a load across the secondary, to prevent very high voltages occurring in the secondary winding

A CT is designed to work at a low flux density, and as shown in Figure 21.43, the primary is sometimes the actual conductor, with the secondary coil arranged around the conductor. Because the primary and secondary current are determined by the line current, they cannot become excessive when a fault occurs, so fuses are not needed to protect a CT.

However, it is extremely important to connect a short-circuit across the secondary terminals of a CT if its load is removed. Otherwise, there will be no secondary current to produce a demagnetising effect in the core, and the flux will increase well above its normal value. As a result, a very high voltage will appear across the secondary terminals, creating a hazard and possibly causing the CT to break down.

Figure 21.44 shows how CTs are connected to protect a three-phase power factor correction capacitor bank. Three CTs monitor the line current to the bank, operating protection relays if the current exceeds a certain value. A single CT monitors the unbalance current between each bank, again operating a protection relay if the unbalance current is excessive. Ideally, there should be no unbalance current if both banks are working correctly.

FIGURE 21.44 CTs providing information to a protection system in a three-phase double star capacitor bank

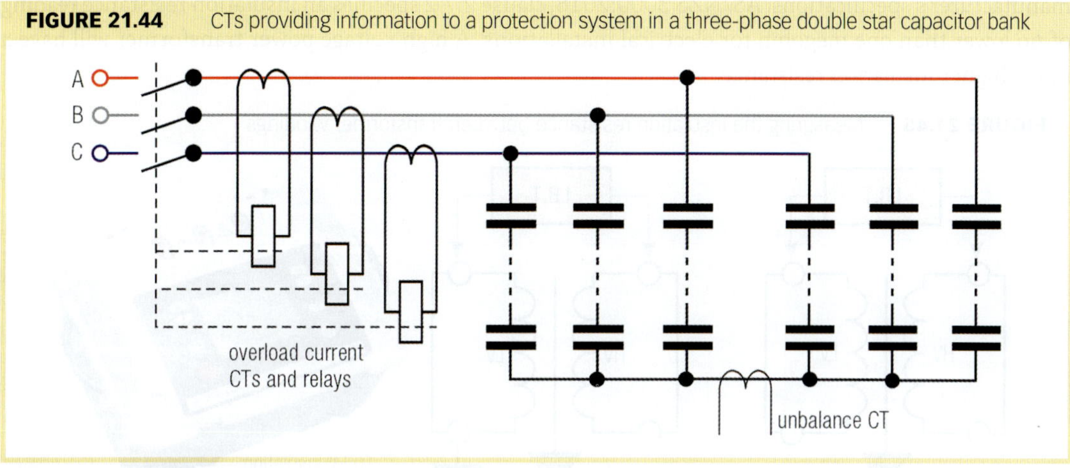

EXAMPLE 21.16

A current transformer has a ratio of primary to secondary current of 1500:5 A and is connected to monitor the current taken by an induction heater. What is the current flowing in the secondary circuit of the transformer if the primary current is 950 A?

Solution

Values Current ratio = 1500:5

$I_1 = 950$ A

Equation $I_2 = I_1 \times \frac{N_2}{N_1} = 950 \times \frac{5}{1500}$

Answer $\mathbf{I_2 = 3.17\ A}$

21.12 Insulation Resistance Test

When a transformer of any type is being put into service, either for the first time or after repairs, it is given a number of tests before being connected to the supply. Many of these tests are carried out by trained personnel, and some are also carried out by maintenance electricians. One of these tests is an insulation resistance (IR) test.

The instrument used to carry out this test is called an *insulation resistance tester*, commonly called a megger, as the Megger company invented this type of instrument. It operates by applying a high DC voltage to the equipment under test. If there is any leakage current flowing as a result of the high voltage, a reading in ohms is displayed, indicating the value of the leakage resistance. The tests applied to a transformer are between:

- HV and LV windings
- HV winding and earth (frame of transformer, which will be earthed)
- LV winding and earth
- HV winding and both the LV winding and earth
- LV winding and both the HV winding and earth.

Figure 21.45 shows the connections for two of these tests. The primary and secondary windings are usually short-circuited during the test. Ideally, the resistance reading should be no lower than the manufacturer's specifications. AS/NZS 3000:2018, clause 2.7.2 specifies an insulation resistance reading of no lower than one megohm for electrical installations. A high voltage power transformer will have a much higher insulation resistance.

FIGURE 21.45 Measuring the insulation resistance between transformer windings

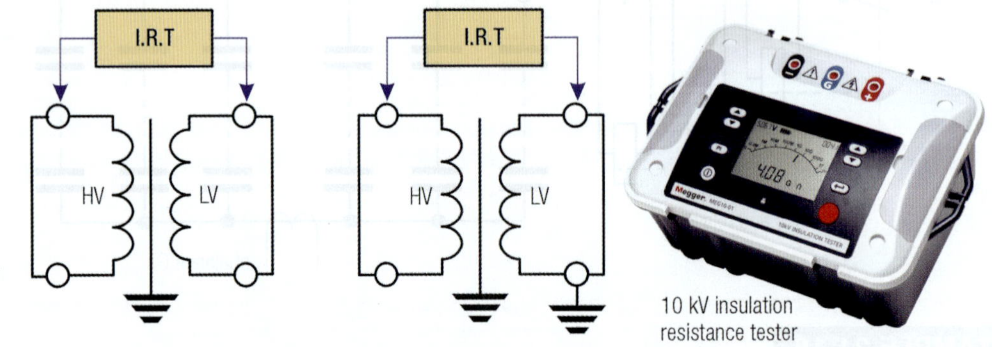

(a) between HV and LV windings (b) between HV and earthed LV windings

SAFETY
An insulation resistance tester can build up a lethal charge in an item under test

When measuring insulation resistance on an inductive or capacitive load, such as long cables or transformers, an IR tester can produce a lethal charge within half a second of pressing the test button. Also, even when the test points are shorted together, a charge can build up after removing the short-circuit.

In review of this section:

- An auto-transformer can be step up, step down or variable and has a single winding shared by the primary and secondary terminals. There is no isolation between each side.
- Voltage transformers (and current transformers) are used to measure and monitor high values of voltage and current. They are used in metering, protection and control systems.
- An insulation resistance test is carried out on a transformer to test the leakage resistance between the windings, and the windings and the earthed frame of the transformer.

CHAPTER SUMMARY

Equations in this chapter:

- $V = 4.44\,Nf\,\Phi_{max}$ where V = RMS value of induced voltage, N = the number of turns, f = supply voltage frequency, Φ_{max} = maximum value of magnetic flux (Wb)
- Transformer voltage, current and turns ratios: $\frac{V_1}{V_2} = \frac{I_2}{I_1} = \frac{N_1}{N_2}$
- IEC standard for transformer voltage regulation (Australia and New Zealand): voltage regulation $\% = \frac{V_{NL} - V_{FL}}{V_{NL}} \times 100$
- Efficiency $\eta\% = \frac{\text{power out}}{\text{power in}} \times 100$
- all-day efficiency $\eta\% = \frac{\text{kWh out}}{\text{kWh in}} \times 100$ where kWh = kW × hours
- $\%Z = \frac{V_{SC}}{V} \times 100$ where V_{sc} = voltage applied to a winding to obtain full rated current in both windings. is the winding's rated voltage.
- $I_{SC} = \frac{100}{\%Z} \times I_{rated}$ where I_{sc} is the prospective short-circuit current, %Z is the percentage impedance of the transformer, and I_{rated} the maximum rated current of the low voltage winding
- For two transformers in parallel, the load taken by $Tx_A = \frac{Z_B}{Z_A + Z_B} \times$ total load and the load taken by $Tx_B = \frac{Z_A}{Z_A + Z_B} \times$ total load, where Z_A = % impedance of transformer A, Z_B = % impedance of transformer B.

REVIEW EXERCISES

Check your answers at the back of the book.

1 A transformer has 200 turns on its primary winding and 48 turns on the secondary winding. The maximum value of the flux in the core is 10 mWb and the frequency is 50 Hz. Calculate the voltage induced in the:
   a primary winding
   b secondary winding.

2 A transformer connected to 230 V AC has 460 primary turns and 96 secondary turns. The secondary supplies a 20 Ω load. Calculate the:
   a secondary voltage
   b secondary current
   c primary current.

3 A transformer rated at 200 kVA has two secondary windings, one producing 3.3 kV the other producing 400 V. If 35 A of current is being supplied by the 3.3 kV secondary, calculate how much current the 400 V secondary can supply.

4 A three-phase 33 kV/11 kV power transformer has a power rating of 10 MVA when the type of cooling is ONAN, and 16 MVA when the cooling is OFDAF. Calculate the full-load line current, for both types of cooling, in the:
   a high voltage winding
   b low voltage winding.

5 A transformer takes a magnetising current of 4.5 A and a core loss current of 1 A. Calculate the transformer's:
   a no-load current
   b no-load power factor.

6 The no-load secondary voltage of a power transformer is specified as 33.9 kV (IEC rated). When fully loaded at a power factor of 0.8, the secondary voltage is 33 kV. What is the transformer's percentage voltage regulation?

7 A 3.3 kV/200 V single phase transformer is supplying a full load current of 50 A at a lagging power factor of 0.85. The transformer has an iron loss of 200 W and a full load copper loss of 600 W. Find the full-load efficiency of the transformer.

8 A short-circuit test on a three-phase 5 MVA 66 kV/3.3 kV transformer showed that 4.1 kV was required to cause full rated current to flow in the windings. Calculate the transformer's:
   a rated primary current
   b percentage impedance.

9 Two transformers rated at 2 MVA are connected in parallel to supply a 3.5 MVA load. Transformer A has an impedance of 8.5 per cent and transformer B an impedance of 7 per cent. Determine the load taken by each transformer.

10 An auto-transformer winding has 400 turns with a tapping giving a secondary winding of 80 turns. The primary voltage is 300 V and a resistive load of 15 Ω is connected to the secondary. Determine the:
   a secondary voltage
   b secondary current
   c primary current
   d current in the common part of the windings.

ONLINE RESOURCES

COMPLETE WORKSHEET TWENTY-ONE

Check with your instructor for worksheets on this chapter.

CHAPTER 22

THREE-PHASE MOTORS

The three-phase induction AC motor is attributed to a number of people, including Nikola Tesla. Today this type of motor is the workhorse of industry, noted for its simplicity, reliability and smooth running. This chapter explains the operation of three-phase induction and wound rotor motors. It also explains the principles of motor protection and starting systems, and typical faults associated with these types of motors.

CHAPTER OUTLINE

22.1 Three-Phase Induction Motor

There are two main types of AC motors, induction or synchronous. The induction motor is the most common and is the first type of motor we will discuss. AC motors can be either single-phase or three-phase, the simplest being the three-phase motor. Figure 22.1 shows examples of typical three-phase motors. These differ in size, power rating and factors such as mounting method, and whether they are totally enclosed, drip proof and so on. Single-phase motors often look similar to the three-phase types, and both share the same basic construction.

FIGURE 22.1 Examples of typical three-phase motors

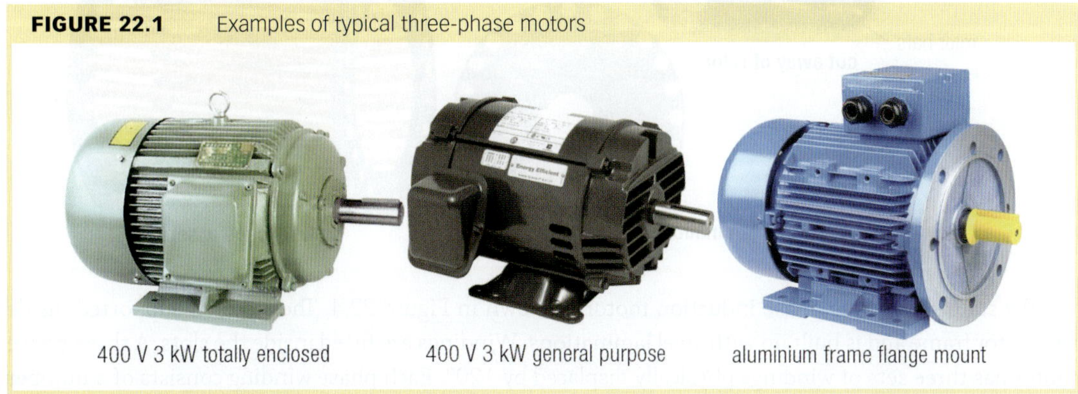

Construction

The construction of a three-phase induction motor is shown in Figure 22.2. When compared to a DC motor, it is much simpler, as there is no need for a commutator or brushes. The frame might be made of aluminium or steel. End plates provide support for the rotor bearings and some motors have cooling fins as part of the frame.

FIGURE 22.2 Construction of a three-phase induction motor

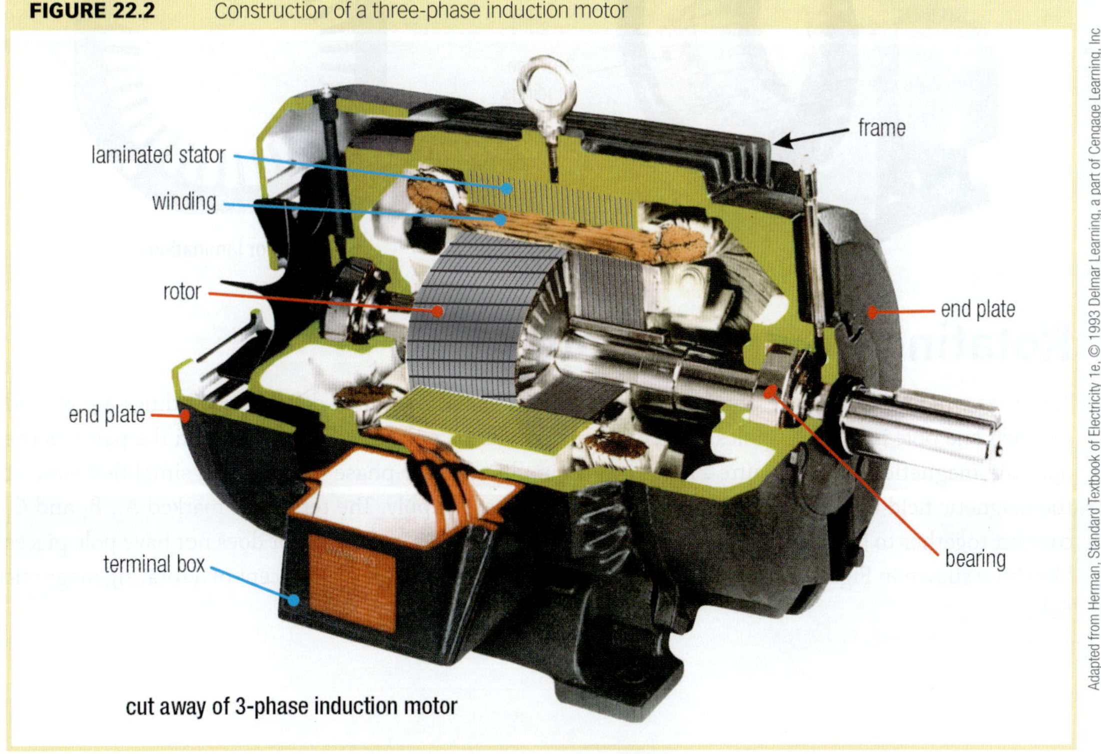

The rotor in a typical induction motor is shown in Figure 22.3. The main component is the so-called squirrel cage assembly. This consists of bars of copper or aluminium supported and connected electrically by end rings. This assembly is embedded in a laminated iron core. The rotor has no electrical connections.

FIGURE 22.3 Construction of a squirrel cage rotor

The stator of a three-phase induction motor is shown in Figure 22.4. The stator is supported inside the motor frame and is built up with steel laminations. Windings are fitted inside the slots. A three-phase motor has three sets of windings physically displaced by 120°. Each phase winding consists of a number of series-connected coils that produce the required number of poles (typically 2, 4 or 6).

FIGURE 22.4 Stator construction of a three-phase induction motor

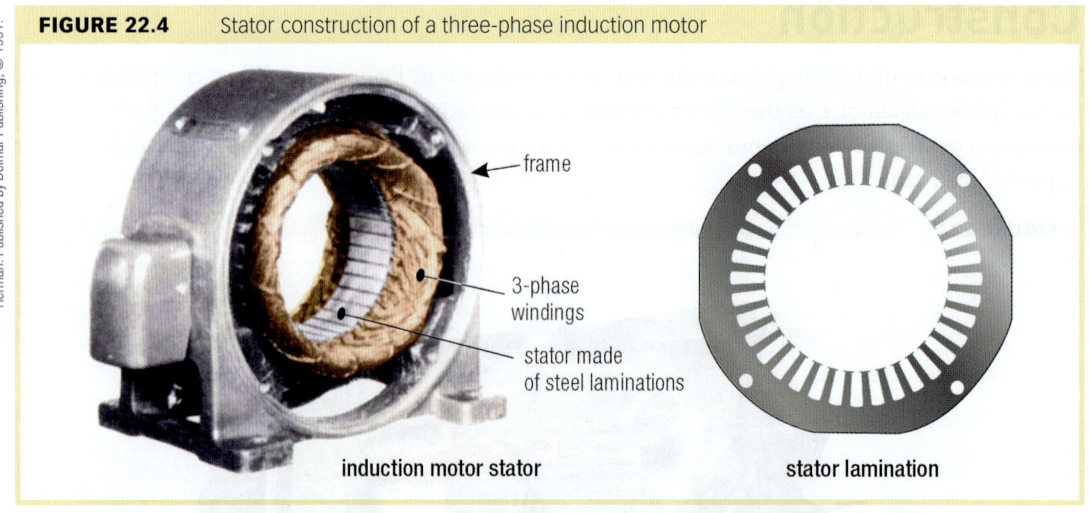

Rotating magnetic field

The three windings of a three-phase motor can be connected in star or delta. The windings are wound so when one pole in a pair of poles has a particular magnetic polarity, the other pole in the pair has the opposite magnetic polarity. Figure 22.5 shows a simplified three-phase motor and a simplified view of the magnetic field that results from applying a three-phase supply. The terminals marked A_1, B_1 and C_1 connect together to give a star connection. As Figure 22.4 shows, an AC motor does not have pole pieces like those shown in Figure 22.5, which are drawn this way to illustrate the concept of a rotating magnetic field.

FIGURE 22.5 The windings of a three-phase motor are arranged to provide a rotating magnetic field when connected to a three-phase supply

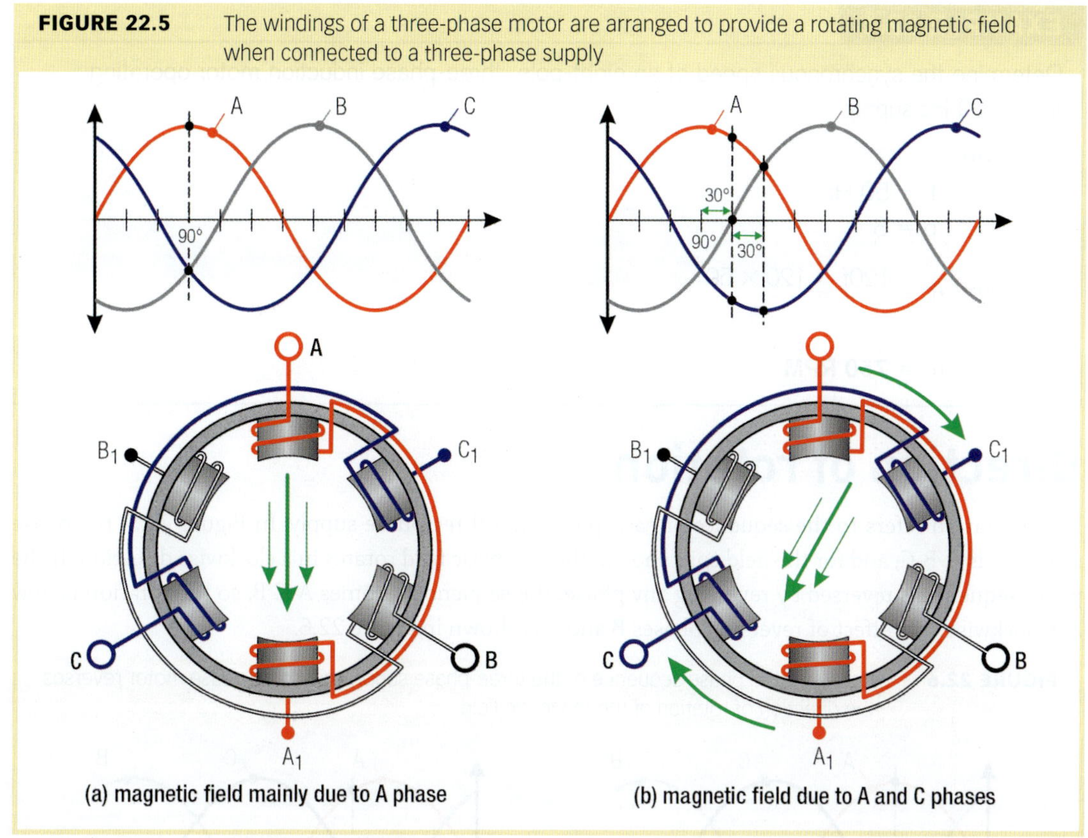

In Figure 22.5(a), phase A is at maximum positive and phases B and C are both equally negative and at a lower voltage than phase A. The magnetic field is therefore concentrated between the A-phase magnetic poles. The windings are assumed to be wound so the top winding produces a north pole and the bottom winding a south pole.

Figure 22.5(b) shows that 30° later, B phase is passing through zero, and phases A and C are equal, although of opposite polarity. The magnetic field strengths produced by the A and C phases are therefore equal, so the magnetic field is concentrated between the A- and C-phase poles. After a further 30°, phase C is maximum negative, while phases A and B are both equally positive. Therefore, the magnetic field now aligns with the C phase windings.

Speed of rotation

For the two-pole machine shown in Figure 22.5, it takes one full cycle of phase A for the magnetic field to rotate through 360°. For a 50 Hz supply, the field will rotate 50 times each second, as one hertz is the same as one cycle per second. Motor speed is usually given in revolutions per minute (RPM), so multiplying 50 revolutions per second by 60 gives 3000 RPM.

However, motors can have more than two poles. For example, with a four-pole machine, one cycle of phase A will cause the magnetic field to rotate by 180°, so two cycles are needed for one complete revolution. Therefore, the speed is now 1500 RPM. Because we are only referring to the rotation of the magnetic field, and not necessarily that of the motor, this speed is called the *synchronous speed*. It can be found for a motor with any number of poles with this equation:

$$n_s = \frac{120f}{p}$$

where:

n_s = synchronous speed in RPM

f = supply frequency in hertz

p = number of poles.

FYI

Three-phase induction motors are widely used in electric cars, in which speed of rotation is determined by the frequency of the supply to the motor

EXAMPLE 22.1

Determine the synchronous speed of an eight-pole, three-phase induction motor operating from a 50 Hz supply.

Solution

Values $f = 50$ Hz

$p = 8$

Equation $n_s = \frac{120f}{p} = \frac{120 \times 50}{8}$

Answer $n_s =$ **750 RPM**

Direction of rotation

Phase rotation refers to the sequence of each phase in a three-phase supply. In Figure 22.5, the phase sequence is A-B-C, and for the field coils shown, the magnetic field rotates in a clockwise direction. If the phase sequence is reversed by reversing any phase, the sequence becomes A-C-B, so the rotation is now anticlockwise. The effect of reversing phases B and C is shown in Figure 22.6.

FIGURE 22.6 Reversing the phase sequence of the three-phase supply to a three-phase motor reverses the direction of rotation of the magnetic field

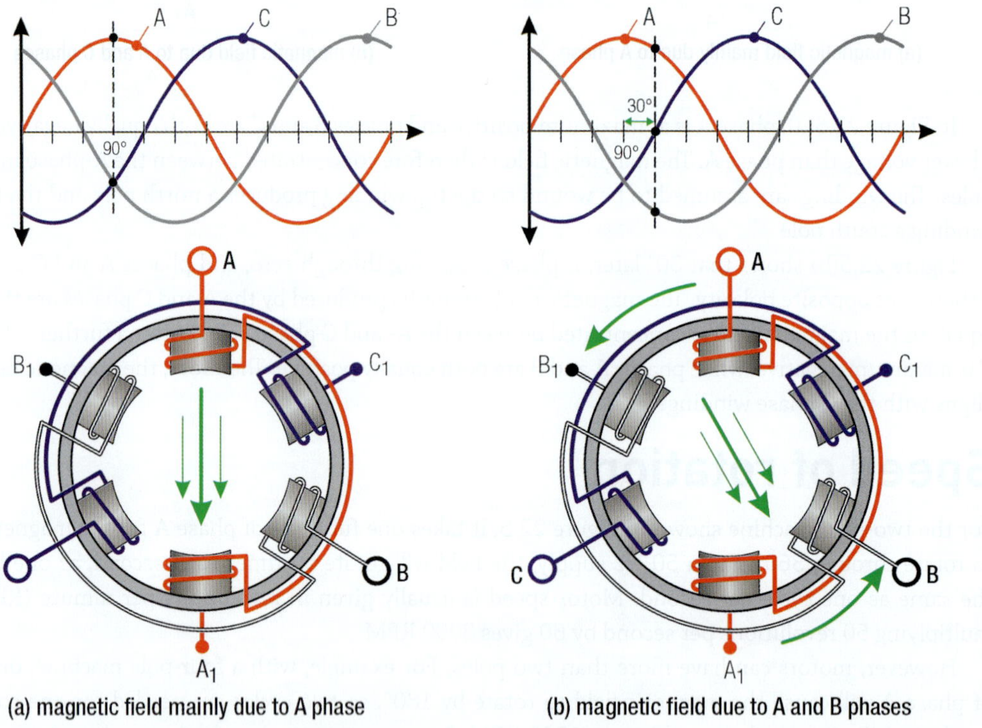

(a) magnetic field mainly due to A phase

(b) magnetic field due to A and B phases

Reversing the supply phase sequence from A-B-C to A-C-B causes the magnetic field to change direction, as phase C is now at zero volts after 30°, compared to being a negative value, as in Figure 22.5(b). Phase B is now a negative value, causing the magnetic field to rotate in an anticlockwise direction. The same thing happens if phases A and B, or A and C are swapped.

Induction motor operating principles

To take advantage of a rotating magnetic field to create rotational torque requires a magnetic field in the rotor. This is achieved by induction. As shown in Figure 22.7(a), when a rotating magnetic field is established in the stator, the field crosses the air gap and cuts the rotor conductors. Because there is relative motion and a magnetic field, we can apply Fleming's right-hand rule and determine the direction of the induced current. Figure 22.7(b) shows that for the given stator field, the induced current flows towards you, creating its own magnetic field, which flows anticlockwise.

FIGURE 22.7 Currents induced in the rotor conductors produce magnetic fields that have a direction to cause the rotor to spin in the direction of the stator's rotating magnetic field

(a) rotor in a rotating magnetic field (b) stator and rotor magnetic fields (c) fields repel each other

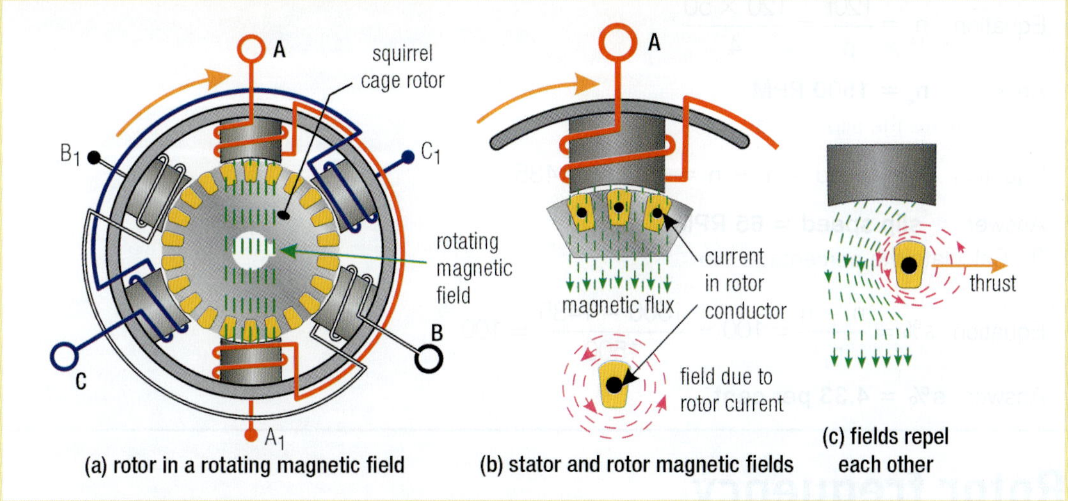

The effect of the interaction between the two fields is shown in Figure 22.7(c), where the stator field and the field induced in the rotor repel each other. This creates a rotational thrust which causes the rotor to turn in the direction of the rotating stator field. If it is free to rotate, the rotor will accelerate towards but never quite reach synchronous speed.

The reason is that to produce torque, there has to be rotor flux, which is produced by the relative motion between the stator magnetic field and the rotor conductors. If the rotor is spinning at synchronous speed, there is no longer any relative motion between its conductors and the rotating magnetic field. Therefore, there will be no current induced in the rotor conductors and no magnetic field needed to interact with the stator field.

This means an induction motor can never run at synchronous speed. Instead, it has to run slightly slower to maintain the necessary rotational torque between the two magnetic fields. The difference between this speed and synchronous speed is the *slip speed* or *slip*. That is:

slip speed = synchronous speed – rotor speed.

Slip is expressed as a percentage of the synchronous speed, and can be found with this equation:

$$s\% = \frac{n_s - n}{n_s} \times 100$$

where:

n_s is synchronous speed

n is rotor speed.

Practical values of slip are between 2 per cent and 5 per cent for small induction motors (less than 10 kW), and as low as 0.5 per cent for very large motors. The greater the load on the motor, the higher the percentage slip needed to create the necessary torque.

EXAMPLE 22.2

A three-phase four-pole induction motor on load runs at 1435 RPM when operating from a 50 Hz supply. Calculate the motor's:

1 synchronous speed
2 slip speed
3 percentage slip

Solution

Values $n = 1435$ RPM

$f = 50$ Hz

$p = 4$

1 Calculate the synchronous speed

Equation $n_s = \frac{120f}{p} = \frac{120 \times 50}{4}$

Answer $\mathbf{n_s = 1500}$ **RPM**

2 Calculate the slip speed

Equation slip speed $= n_s - n = 1500 - 1435$

Answer **slip speed = 65 RPM**

3 Calculate the percentage slip

Equation $s\% = \frac{n_s - n}{n_s} = 100 = \frac{1500 - 1435}{1500} = 100$

Answer **s% = 4.33 per cent**

Rotor frequency

When a rotor is stationary and the stator is connected to a 50 Hz supply, the stator field cuts each rotor conductor at a rate of 50 times per second. Therefore, when the rotor is stationary the frequency of the voltage induced in the rotor conductors is 50 Hz. This is referred to as the *rotor frequency*. If the rotor was turning at synchronous speed, the rotor frequency will be zero. At points in between standstill and synchronous speed, the rotor frequency will be somewhere between the supply frequency (f) and zero. This frequency depends on the slip speed, and can be found with this equation:

$f_r = \frac{s\% \times f}{100}$ where f_r = rotor frequency, and f = supply frequency

The relationship between rotor frequency and slip speed is shown in Figure 22.8. As the rotor accelerates, the rotor frequency decreases and, as a result, the inductive reactance of the rotor conductors decreases. That is, as the rotor frequency varies, so too does the rotor inductive reactance, which affects the starting and running characteristics of the motor, as explained later in this chapter.

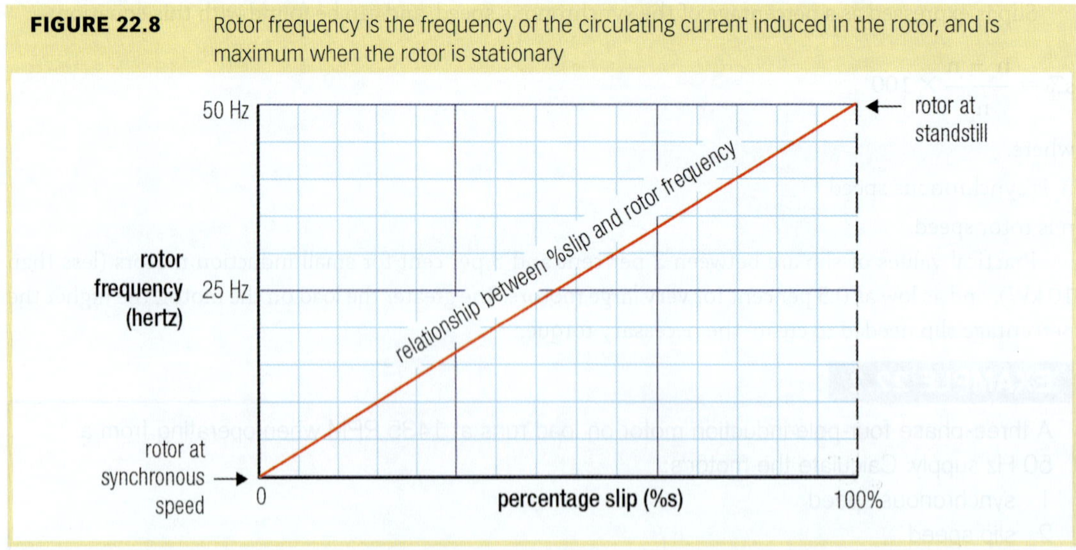

FIGURE 22.8 Rotor frequency is the frequency of the circulating current induced in the rotor, and is maximum when the rotor is stationary

Motor terminal block markings

A three-phase motor has three sets of identical windings. These can be connected in star or delta, and like transformer windings, must be connected correctly to avoid a reversed winding. The international standard for identifying motor windings is to label them as U, V and W in which the:

- start of each winding is labelled as U_1, V_1 and W_1
- end of each winding is labelled as U_2, V_2 and W_2.

If a motor is internally connected in either star or delta, the motor's terminal block will have three studs that connect to the three-phase supply. If the motor is not internally connected, its terminal block has six terminals, marked as described above. A typical motor terminal block and terminal box are shown in Figure 22.9. Notice the links between the terminal studs. The terminal block is housed in a terminal box, which is typically sealed against dust and moisture.

FIGURE 22.9 Three-phase motor terminal blocks usually have six terminals with links to change connections

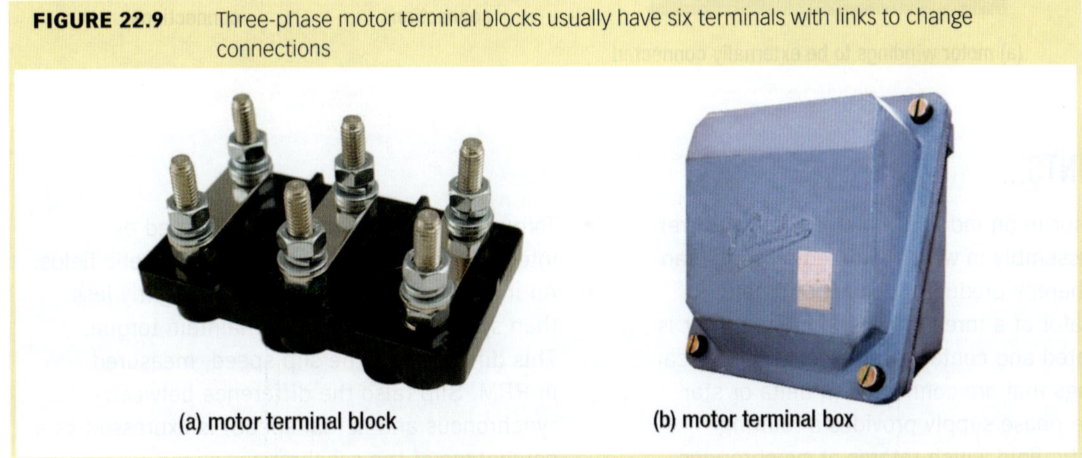

If a motor is internally connected, three leads are brought out to the terminal block as shown in Figure 22.10. Links between the motor winding connections and the supply are through links connecting the studs.

FIGURE 22.10 Terminal block connections for internally connected three-phase motors

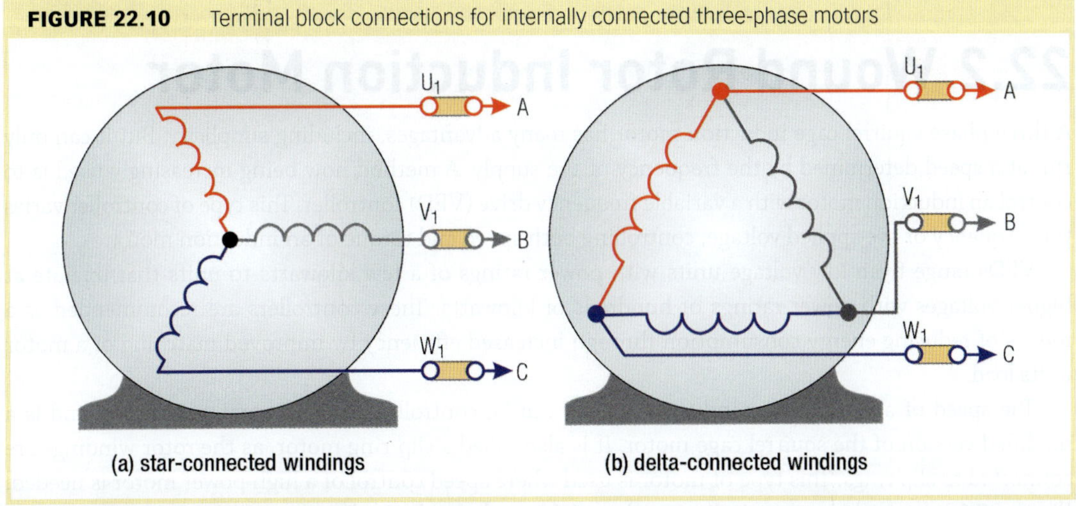

Some motors have both ends of the windings brought out to its terminal block. These can be connected with links, as shown in Figure 22.11. If the motor is to be connected to a star–delta starter (explained later), all six connections will go directly to the starter.

FIGURE 22.11 Terminal block connections for externally connected three-phase motors

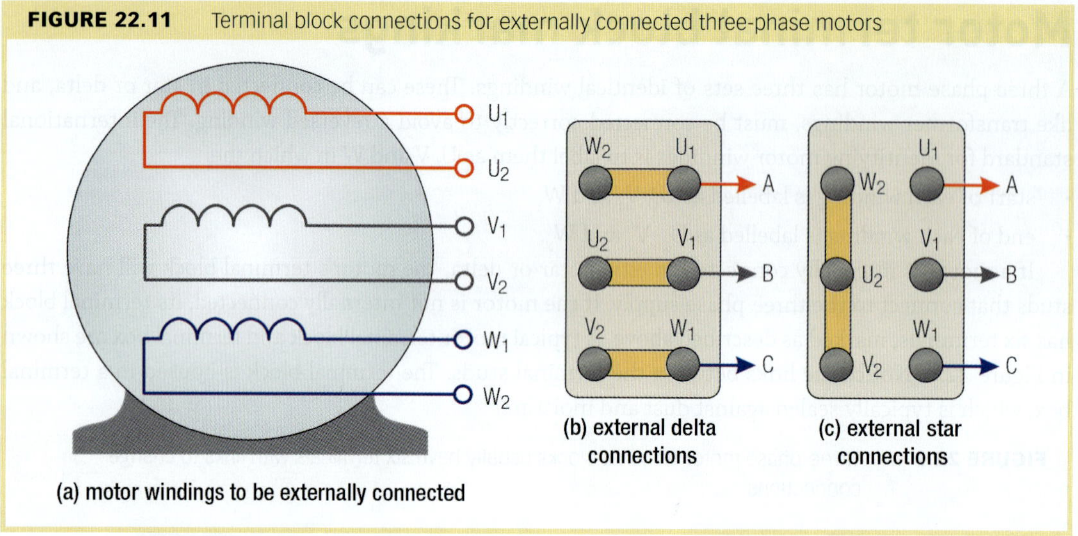

KEY POINTS...

- The rotor in an induction motor has a squirrel cage assembly in which induced currents can flow, thereby producing a magnetic field.
- The stator of a three-phase induction motor is laminated and contains three sets of identical windings that are connected in delta or star.
- A three-phase supply provides a rotating magnetic field which rotates at synchronous speed (n_s). This speed equals 120 times the supply frequency, divided by the number of poles in the motor.
- Torque in an induction motor is caused by interaction of the stator and rotor magnetic fields.
- An induction motor has to run at slightly less than synchronous speed to maintain torque. This difference is the slip speed, measured in RPM. Slip (also the difference between synchronous and actual speed) is expressed as a percentage of the synchronous speed.
- Motor windings are identified with the letters U, V and W; the winding starts with a subscript 1, ends with a subscript 2.

22.2 Wound Rotor Induction Motor

VFD motor control is a significant way to achieve energy savings

A three-phase squirrel cage induction motor has many advantages, including simplicity. But it can only run at a speed determined by the frequency of the supply. A method now being increasingly used is to control an induction motor with a variable frequency drive (VFD) controller. This type of controller varies the frequency of the applied voltage, controlling both speed and torque of an induction motor.

VFDs range from low voltage units with power ratings of a few kilowatts to units that operate at higher voltages with power ratings of hundreds of kilowatts. These controllers are recommended as a means of reducing energy consumption through increased efficiency by improved matching of a motor to its load.

The speed of a wound rotor induction motor can be controlled with external resistances, and is a modified version of the squirrel cage motor. It is also called a slip-ring motor, as the rotor windings are connected to slip rings. This type of motor is used where speed control of a high power motor is needed. Figure 22.12 shows the stator windings and rotor from a large three-phase wound rotor motor.

FIGURE 22.12 Wound rotor motors have a rotor with a set of star-connected three-phase windings that connect to three slip rings

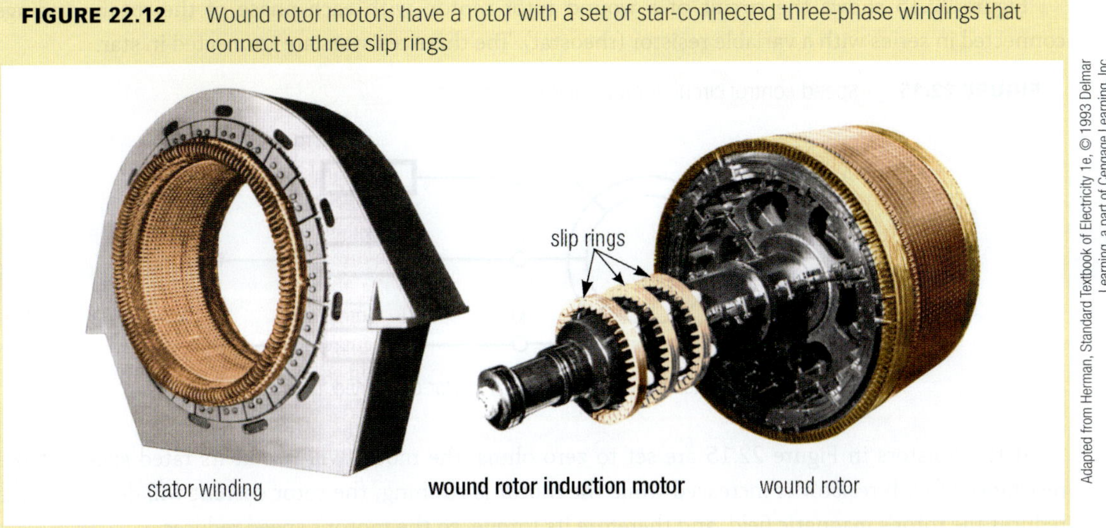

The stator of a slip-ring motor is the same as the stator of a squirrel cage induction motor. However, the rotor has a set of three single-phase windings embedded in its laminated core, instead of a squirrel cage assembly. The windings are 120 electrical degrees apart, and are grouped to give the same number of poles as in the stator. The windings are star-connected and terminate at three slip rings fitted to the shaft. Carbon brushes rub against the slip rings and provide a connection to an external speed control circuit.

FIGURE 22.13 Connections of a wound rotor motor. Stator windings are marked U, V and W, rotor connections as K, L and M.

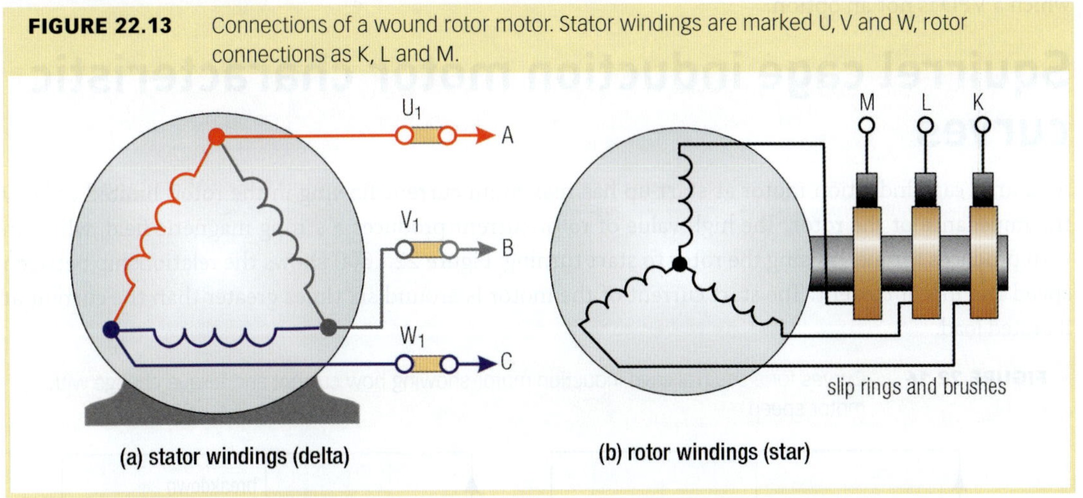

Figure 22.13 shows the connections of a wound rotor motor. The stator windings are connected in delta and the rotor windings are connected in star, and terminated at three slip rings. The stator windings are identified as for a squirrel cage motor. The rotor windings are marked M, L and K, with M usually nearest the rotor. The circuit symbol for a wound rotor motor is in Figure 22.14.

Operation of wound rotor motor

Because the rotor windings are now accessible, the value of the induced current flowing in the windings can be controlled with an external bank of variable resistors (rheostats). If terminals K, L and M are open-circuit, no rotor current can flow and there is no magnetic field produced by the rotor. Therefore, there is no rotational torque produced, and the rotor remains stationary. If terminals K, L and M are short-circuited, the motor behaves like a squirrel cage motor.

FIGURE 22.14 Circuit symbol for a wound rotor motor

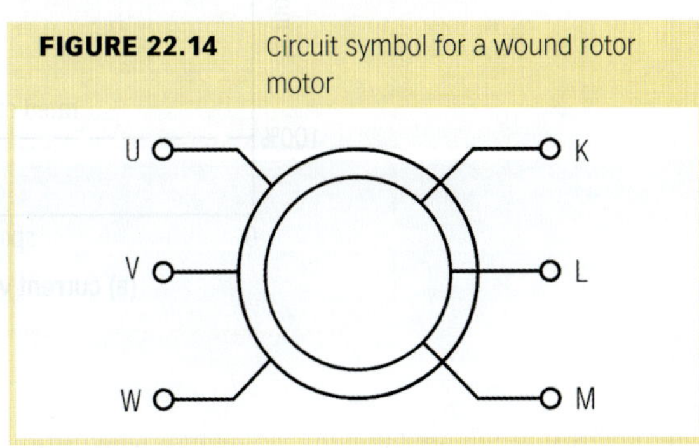

Figure 22.15 shows the circuit of a wound rotor motor, with each phase of the rotor windings connected in series with a variable resistor (rheostat). The three resistors are connected in star.

FIGURE 22.15 Speed control circuit for a wound rotor motor

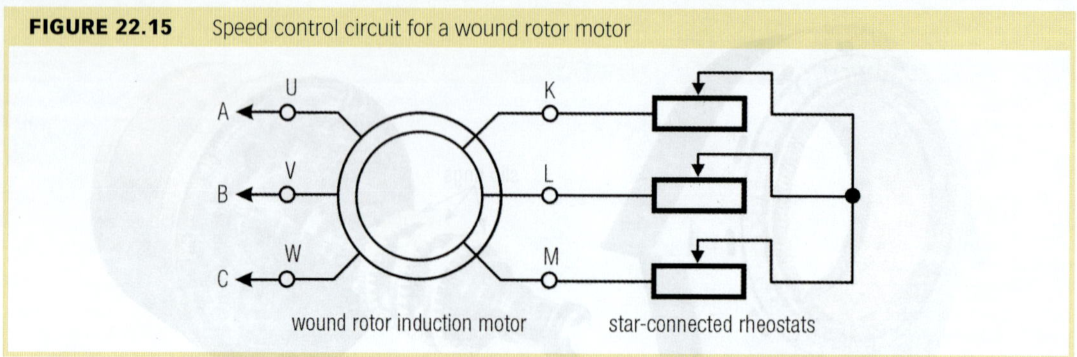

If the resistors in Figure 22.15 are set to zero ohms, the motor will run at its rated speed. If the resistance of each resistor is increased while the motor is running, the rotor current will decrease. This reduces the rotor's magnetic field, and therefore its torque, so the motor's speed reduces.

Because the speed has reduced, the stator current increases, which increases the induced voltage in the rotor windings. This increases the rotor current, which increases the torque. This sequence occurs almost instantly, and the motor now runs at a slower speed, but with enough torque to drive the load. The higher the resistance, the lower the speed of the motor. Although this type of speed control is relatively simple and inexpensive, a wound rotor motor is expensive. They are therefore usually high power machines in which a VFD is not an option.

Squirrel cage induction motor characteristic curves

A squirrel cage induction motor at start-up has maximum current flowing in the rotor, limited only by the impedance of the rotor. The high value of rotor current produces a strong magnetic field, which in turn produces torque, causing the rotor to start turning. Figure 22.16(a) shows the relationship between speed and motor current. The start current of the motor is around six times greater than the current at its rated load.

FIGURE 22.16 Curves for a squirrel cage induction motor showing how current and torque change with motor speed

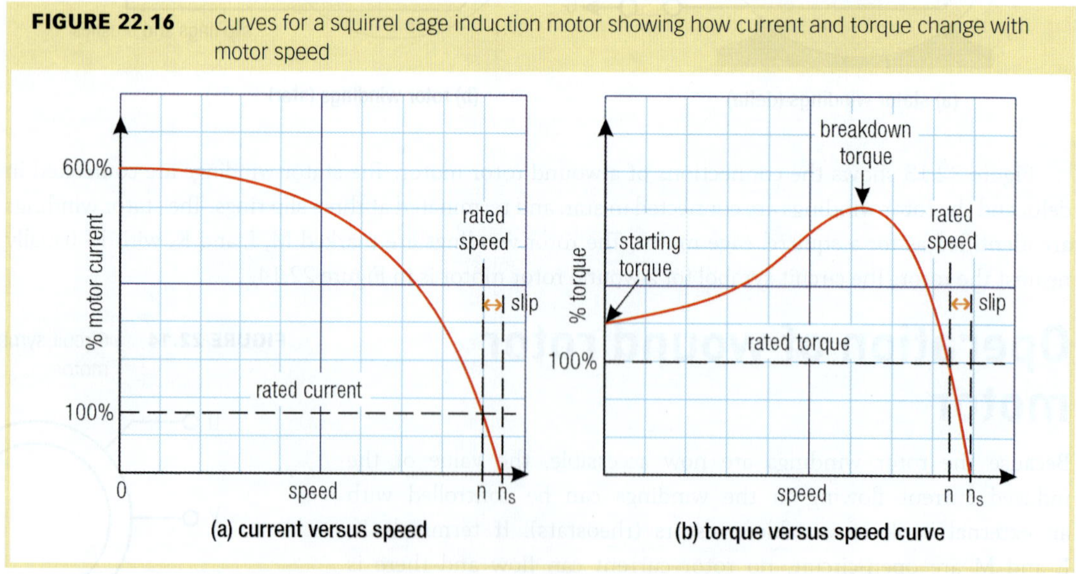

As the rotor accelerates, the voltage generated in the rotor decreases, which reduces the rotor current and the stator current. When the motor reaches its rated speed, at its rated load, the current falls to its rated value.

Figure 22.16(b) shows the relationship between torque and speed for a squirrel cage motor. The starting torque, also called the locked-rotor torque, is higher than the rated torque, and the motor current at this point is referred to as the 'locked rotor current'. At this stage, the rotor reactance is higher than its resistance, as the rotor frequency is at the supply frequency.

As the motor accelerates, it reaches its 'breakdown torque' when the resistance of the rotor equals its reactance, the condition required for maximum torque. The breakdown torque is also the stalling torque, as increasing the load too far beyond this point will stall the motor.

The rated torque is reached when the motor is running at its rated speed, voltage and load. At this point, the rotor reactance is less than its resistance, as the rotor frequency is low. The motor's slip speed allows it to develop the required torque to rotate the load.

KEY CONCEPT

Maximum torque occurs when the resistance of the rotor equals its reactance

Wound rotor motor characteristic curve

As explained, maximum torque occurs when the resistance and reactance of the rotor are equal. The resistance of a wound rotor remains constant. Its reactance depends on the rotor frequency, where rotor frequency is equal to the supply frequency multiplied by the percentage slip (see Figure 22.8).

When the rotor is at standstill, the rotor frequency equals the supply frequency (50 Hz). At rated speed, the rotor frequency is around 2 Hz for a slip speed of 4 per cent. Therefore, the reactance of the rotor is a maximum at standstill, falling to a minimum at rated speed, and reaching zero at synchronous speed. Therefore, if the resistance of the rotor circuit is changed to keep it around the same value of the rotor reactance, the available torque will be increased.

Because of the external resistors in the rotor circuit, the resistance of the rotor circuit can be changed as required to improve the torque characteristics. To achieve maximum torque at start-up, the rotor resistance should equal the rotor reactance at start-up. If the external resistance in the rotor circuit is set to achieve this value, the motor will have a torque/speed curve as shown in Figure 22.17(a). This curve shows that the torque is maximum at start-up, but at rated speed, the torque is considerably less than the torque available from a standard squirrel cage motor.

FIGURE 22.17 Curves comparing a squirrel cage induction motor and a wound rotor motor

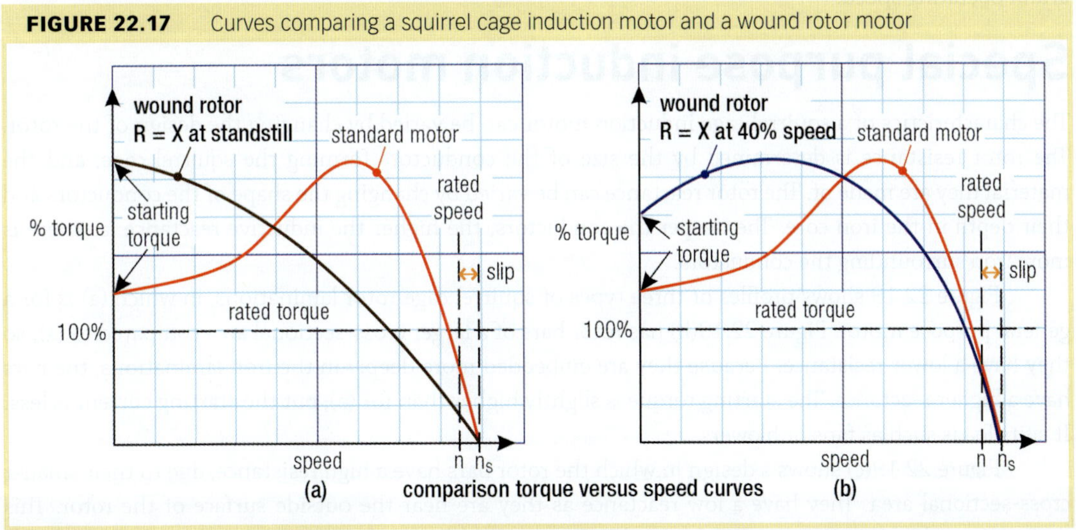

Curve (b) in Figure 22.17 shows the torque versus speed curve for a wound rotor motor in which its rotor resistance equals its rotor reactance when the motor is running at around 40 per cent of its rated speed. The starting torque is reduced compared to that in curve (a), but is higher than a standard motor. The torque at rated speed is slightly less than a standard motor.

By varying the rotor resistance of a wound rotor motor during start-up, a relatively constant and high torque can be achieved. This also gives a lower starting current. Figure 22.18 shows the torque versus speed curves for a squirrel cage motor and a wound rotor motor in which the rotor resistance is changed as the motor speeds up. During stage 1, the external resistors are set so the rotor resistance equals the rotor reactance at start-up. During stage 2, the external resistance is reduced, and in stage 3 the resistance is set to zero (short-circuit).

FIGURE 22.18 Torque/speed curves comparing a standard induction motor with a wound rotor motor where its rotor resistance is decreased in stages during start-up

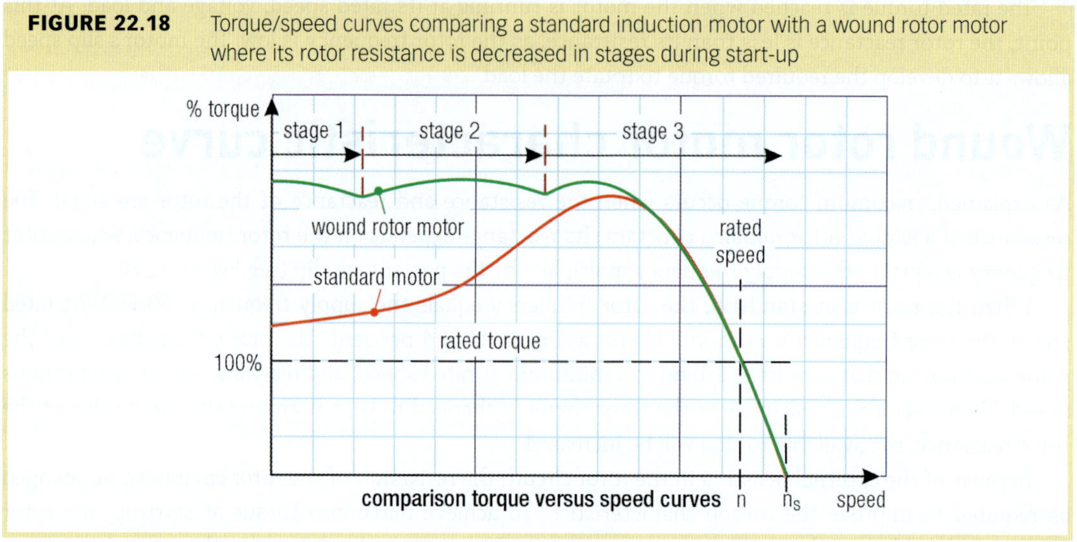

The curve in Figure 22.18 shows that a wound rotor motor can be controlled so it produces a high torque during its start-up phase, which makes it suitable for driving high inertia loads. Another advantage is reduced starting current, which means the motor can drive a load that could take several minutes to attain operating speed.

Special purpose induction motors

The characteristics of a squirrel cage induction motor can be varied by changing the design of the rotor. The rotor resistance is determined by the size of the conductors forming the squirrel cage, and the material they are made of. The rotor reactance can be varied by changing the shape of the conductors and their depth in the iron core. The deeper the conductors, the higher the inductive reactance, as there is more iron surrounding the conductors.

Figure 22.19 shows profiles of three types of squirrel cage rotor laminations, in which (a) is for a general purpose motor. Figure 22.19(b) has rotor bars of a larger cross-sectional area compared to (a), so they have a lower resistance. Because they are embedded more deeply in the iron laminations, the bars have a higher reactance. The starting torque is slightly higher than for (a), but the starting current is less. It suits loads such as fans or blowers.

Figure 22.19(c) shows a design in which the rotor bars have a high resistance, due to their smaller cross-sectional area. They have a low reactance as they are near the outside surface of the rotor. This design gives a high starting torque of around 280 per cent of the rated torque, and a relatively low starting current. However, its rated speed is lower, due to the increased slip speed needed to produce the rated torque.

FIGURE 22.19 Three-phase induction motor torque/speed curves for three types of squirrel cage rotors

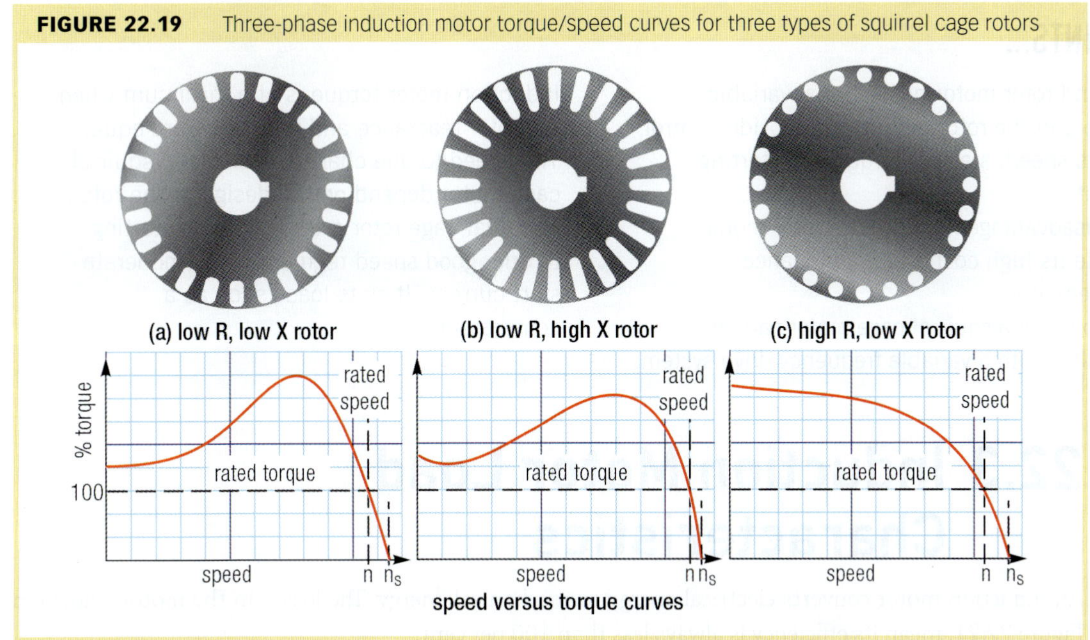

A variation is the dual cage rotor, shown in Figure 22.20. The outer cage is near the surface of the rotor, so it has a low reactance. It is also made from a resistive material, so it has a relatively high resistance. The inner cage is made of copper, so it has a low resistance, and is embedded more deeply in the rotor core, giving the inner cage a relatively high reactance.

At start-up, because the rotor frequency is high, the inner cage has a high reactance, so most of the rotor current flows in the outer cage. Because this cage has a high resistance, the starting torque is high and the starting current is acceptably low. As the motor accelerates, the reactance of both cages reduces because the rotor frequency reduces. When the motor reaches its rated speed, the reactance of both cages is low, so most of the rotor current now flows in the low resistance inner cage.

The torque/speed curve for the motor shows that the torque remains high as the motor accelerates, and the rated speed is only a few per cent lower than the synchronous speed. The starting current is about five times the rated current. Its high starting torque makes this type of motor suitable for compressors and loads needing a high starting torque.

FIGURE 22.20 Double cage rotor, which provides a high starting torque, low starting current and good speed regulation

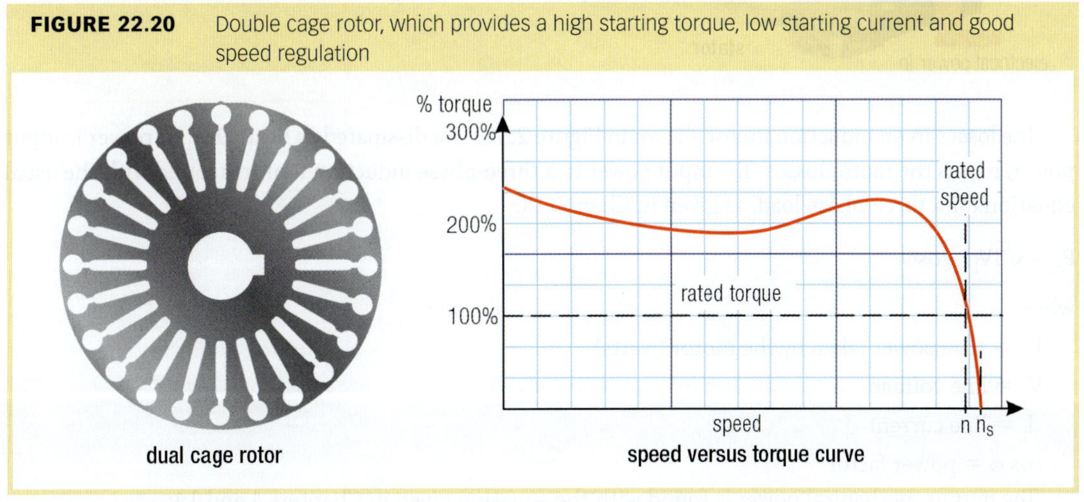

KEY POINTS...

- A wound rotor motor has external variable resistors in the rotor circuit that provide control over its speed, starting torque and starting current.
- The disadvantages of a wound rotor motor include its high cost and maintenance requirements.
- The speed of a squirrel cage motor can be controlled with a variable frequency drive system.
- Induction motor torque is at a maximum when the rotor reactance and resistance are equal.
- The speed/torque characteristics of a squirrel cage motor depend on the design of the rotor.
- The dual cage rotor provides a high starting torque, good speed regulation and moderate start current. It suits loads such as a compressor.

22.3 Induction Motor Load Characteristics

An induction motor converts electrical energy to mechanical energy. The losses in the motor, shown in Figure 22.21, mean its efficiency is always less than 100 per cent.

FIGURE 22.21 Losses in an induction motor include iron and copper losses in the stator and the core, and mechanical losses

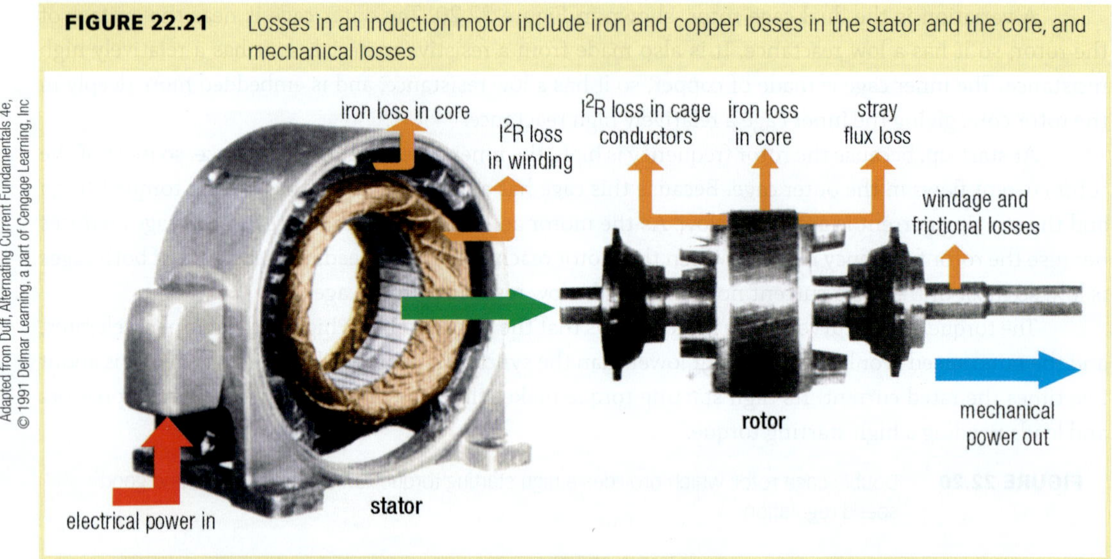

The losses in an induction motor shown in Figure 22.21 are dissipated as heat. Output power is input power, minus the motor losses. The input power to a three-phase induction motor is found with the usual equation for a three-phase load, as given in Chapter 20:

$$P_{in} = \sqrt{3} V_L I_L \cos\phi$$

where:

P_{in} = true power taken by the motor (watts)
V_L = line voltage
I_L = line current
$\cos\phi$ = power factor.

The output mechanical power is found with the equation given in Chapters 4 and 14:

$$P_{out} = \frac{2\pi nT}{60} = \frac{nT}{9.55}$$

where:

P_{out} = output power in watts
n = speed in RPM
T = torque in newton metres.

The efficiency of a three-phase induction motor depends on its losses. A motor running at full load is generally more efficient than a motor on a light load, as the losses remain relatively constant. The equation to find percentage efficiency is the same as that for a transformer:

$$\%\eta = \frac{P_{out}}{P_{in}} \times 100$$

where:

η = efficiency

P_{out} = output power in watts

P_{in} = input power in watts.

EXAMPLE 22.3

The nameplate of a three-phase induction motor is shown in Figure 22.22. Calculate the full-load efficiency of the motor.

Solution

Values $V_L = 400$ V

$I_{FL} = 19.2$ A

$P_{out} = 11$ kW

$\cos\phi = 0.9$

Equation $P_{in} = \sqrt{3} V_L I_L \cos\phi$

$P_{in} = 1.732 \times 400 \times 19.2 \times 0.9$

$P_{in} = 11.97$ kW

Equation $\%\eta = \frac{P_{out}}{P_{in}} \times 100$

$= \frac{11\,\text{kW}}{11.97\,\text{kW}} \times 100$

Answer **η = 91.9 per cent**

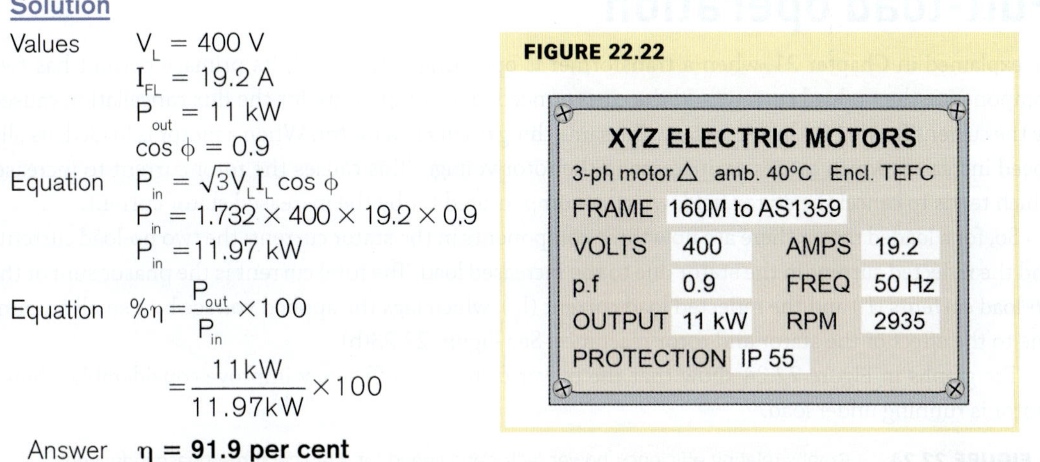

FIGURE 22.22

The efficiency of three-phase induction motors with an output rated from 0.73 kW to 185 kW must comply with Minimum Energy Performance (MEPS) requirements which are set out in AS/NZS 1359.5-2004. These requirements were introduced in 2001, and have become more stringent. This has led to the development of the high efficiency motor (HEM). For example, a standard 10 kW motor has a typical efficiency of 88 per cent, where its HEM equivalent has an efficiency of up to 93 per cent.

In the interests of sustainability, the efficiency of many types of three-phase induction motors must meet MEPS requirements (star rating)

No-load operation

When a three-phase induction motor is operating without a load, the power into the motor has to establish the rotating magnetic field and overcome the motor losses. The current taken by the motor on no-load is made up of two components that are outlined below.

- Magnetising current, which is higher than that taken by a transformer because of the air gap between the rotor and the stator. This current is typically around 30 per cent of the full load current and lags the applied voltage by 90°. Because the stator flux remains almost constant from no-load to full load, the magnetising current is also a relatively constant value.
- Motor loss current (iron and mechanical losses), which is typically around 10 per cent to 15 per cent of the full-load current. This current is in phase with the applied voltage.

The current taken by a motor on no-load is the phasor sum of these two currents. See phasor diagram (a) in Figure 22.23. Because the magnetising current is the main component, the no-load current lags the applied voltage by a large amount, giving a poor power factor, typically around 0.2.

FIGURE 22.23 Phasor diagrams of a three-phase induction motor on (a) no-load and (b) full load

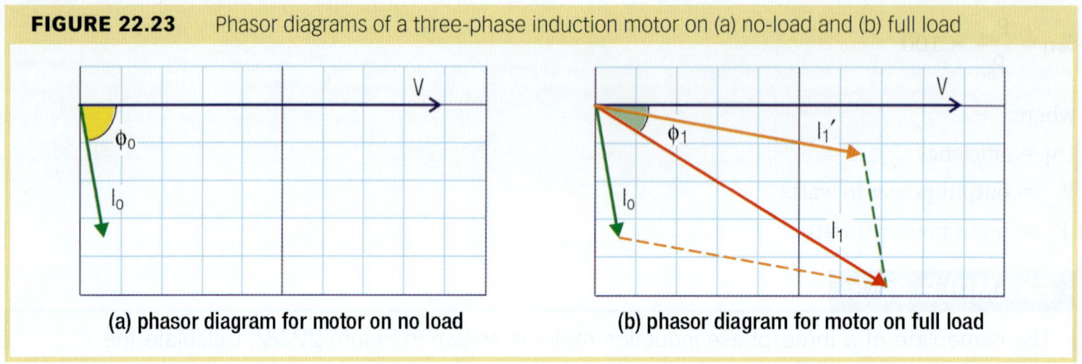

Full-load operation

As explained in Chapter 21, when a transformer is operating with a load, its primary current has two components, the no-load current and the current needed to compensate for the flux cancellation caused by the current flowing in the secondary. The same thing occurs in a motor. When a motor is loaded, its slip speed increases, which causes an increase in the rotor voltage. This causes the rotor current to increase, which tends to cancel the stator field, which is compensated for by the increased stator current.

So, for a loaded motor there are now three components in the stator current: the two no-load currents and the reflected current in the stator due to the increased load. The total current is the phasor sum of the no-load currents (I_0) and the reflected load current (I_1'), which lags the applied voltage by a small amount due to the effect of the stator and rotor reactance. See Figure 22.23(b).

The graphs in Figure 22.24 show that the power factor and efficiency improve considerably when a motor is running under load.

FIGURE 22.24 Graphs relating efficiency, power factor and speed for a typical three-phase induction motor from no-load to full load

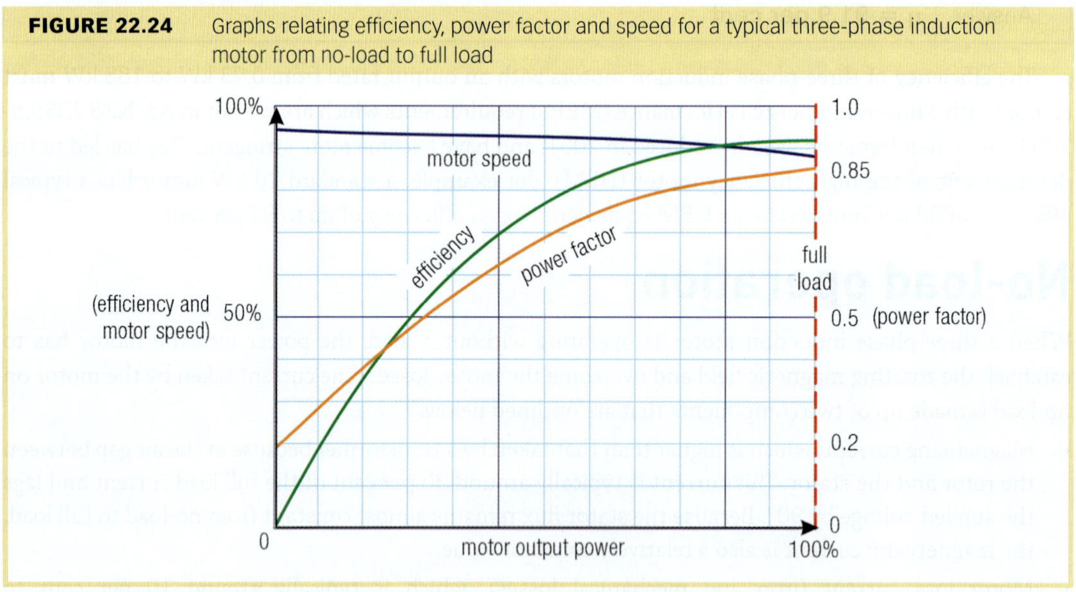

The efficiency of a motor is zero when it is operating with no-load, rising to around 90 per cent for motors rated at 10 kW or more. Low-power motors can have efficiencies of less than 50 per cent. The power factor is also low at no-load, shown as 0.2 in Figure 22.24. When loaded, the power factor increases to around 0.9 for a large motor, less for small motors. The power factor curve in Figure 22.24 indicates a full-load value of 0.85. The speed regulation of a motor depends also on its size and type, which is typically a few per cent.

Fault-finding a three-phase induction motor

Three-phase induction motors, while generally reliable, are also subject to failure. Motor repair and rewinding is a specialist area that is beyond the scope of this book. We instead look at the types of faults and situations that can generally be dealt with or analysed on-site.

When dismantling a motor, it's important to make sure the motor is reassembled in its original form. Before removing the end plates, mark them in a way that ensures you know which way up the plates go, and how they are aligned to the frame. A method is to use a centre punch to make adjacent punch marks on the top of the frame and the end plates. Another way is to scribe a line to show the alignment. Similar methods can be used when removing other parts, such as bearing covers.

Removing the end plates requires care as the rotor will no longer be supported by the bearings, and will drop onto the stator as each end plate is removed. Removing the rotor should also be done carefully to avoid damage to the stator, the windings and the rotor. Large rotors are heavy and could need lifting equipment or someone to assist with their removal.

Aspects to consider include mechanical failure, such as worn bearings, loose couplings, rotor loose on the shaft, rust, water ingress and other forms of damage. Electrical problems include three-phase supply problems and electrical problems with the motor windings or stator cage.

Mechanical problems

Worn bearings can cause the rotor to ride on the stator, resulting in very noisy operation. If the bearings are excessively worn, the rotor will rest on the stator, making it impossible for it to rotate. Figure 22.25 shows how to test for a worn bearing. In a large motor, it may be possible to use feeler gauges to check if the air gap between the stator and rotor is the same at all points. In a small motor, try moving the shaft up and down. If the shaft can move, even by a small amount, the bearings should be replaced, or the shaft itself may be worn.

FIGURE 22.25 Worn bearings will cause the air gap between the stator and rotor to no longer be consistent

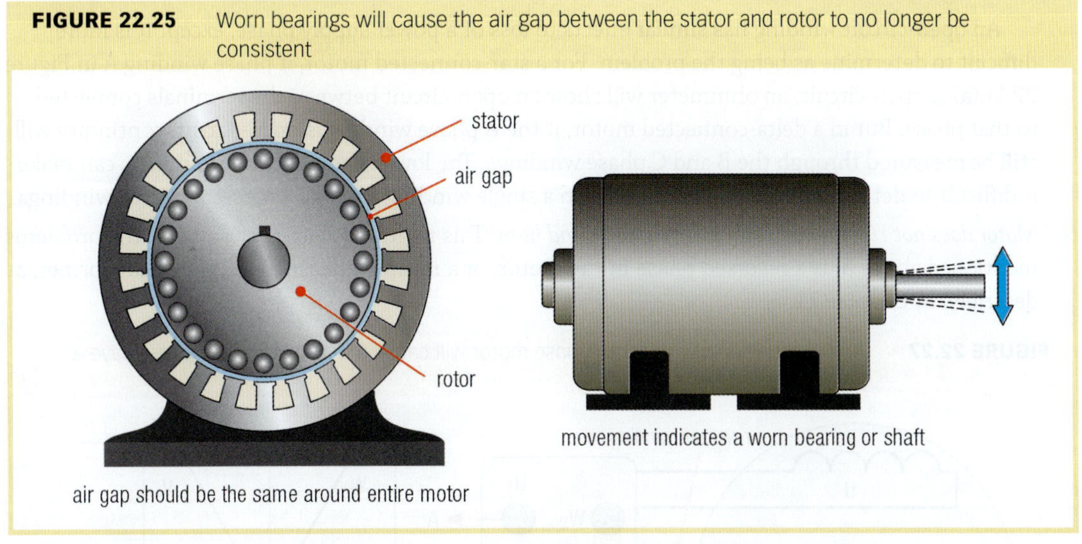

If a motor is excessively hot but is otherwise running normally, there may be a mechanical problem with the load, such as tight bearings or other forms of failure. Or the motor bearings may be tight. This is best checked by disconnecting the load, if possible, and confirming the motor runs smoothly and quietly and that it can be turned by hand. If a motor fails to start it can be due to a seized bearing, assuming the electrical supply is correct. A rattling sound might be due to a loose bar in the rotor.

Electrical problems

Symptoms indicating a motor problem do not always indicate an electrical problem, as a mechanical or an electrical failure can give similar symptoms. The following symptoms assume mechanical failure has been ruled out.

1 *Motor fails to start.* This can be due to the operation of a fuse or overload in one or more phases supplying the motor. It can also be caused by a defective controller (e.g. variable frequency drive). An open-circuit phase winding in the motor or open-circuit bar in the rotor, incorrect internal connections or a short-circuited winding are other reasons. Figure 22.26 shows some of the effects due to failure of one phase. Note that failure of one phase while the motor is running could mean it still runs, but eventually it will overheat.

FIGURE 22.26 Failure of one phase of the supply while a motor is running will cause it to remain running but to overheat

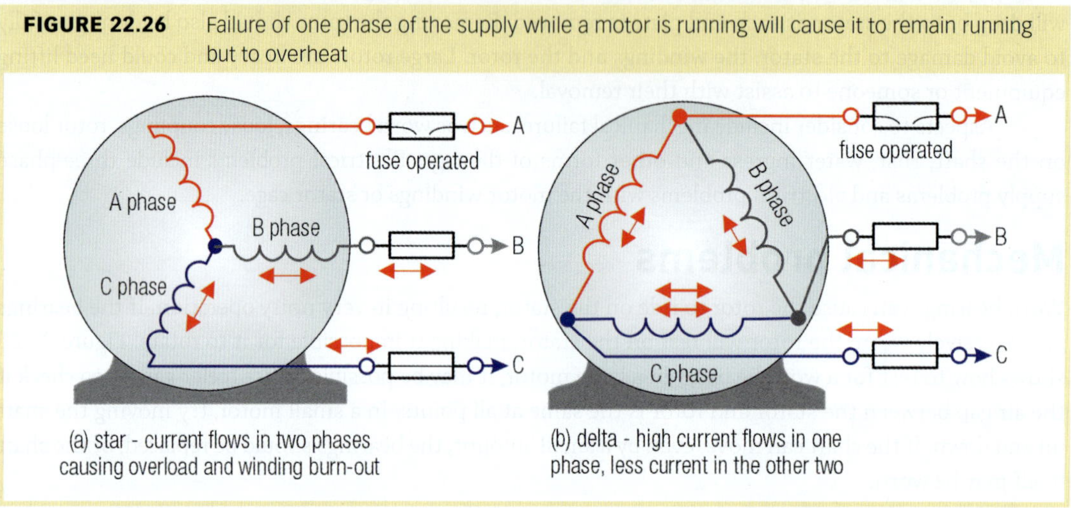

An open-circuit winding has similar effects to loss of a power supply phase, except it is more difficult to determine as being the problem. For a star-connected motor, if phase winding A in Figure 22.26(a) is open-circuit, an ohmmeter will show an open-circuit between the terminals connected to that phase. But in a delta-connected motor, if the A phase winding is open-circuit, continuity will still be measured through the B and C phase windings. The low resistance of the windings can make it difficult to determine the difference between a single winding and two series-connected windings.

2 *Motor does not run properly and exhibits noise and hum.* This might be caused by some of the problems mentioned above plus a reversed phase in the motor, or a reversed phase in a supply transformer, as described in Chapter 21.

FIGURE 22.27 A reversed winding in a three-phase motor will cause the currents to no longer have a phase difference of 120°

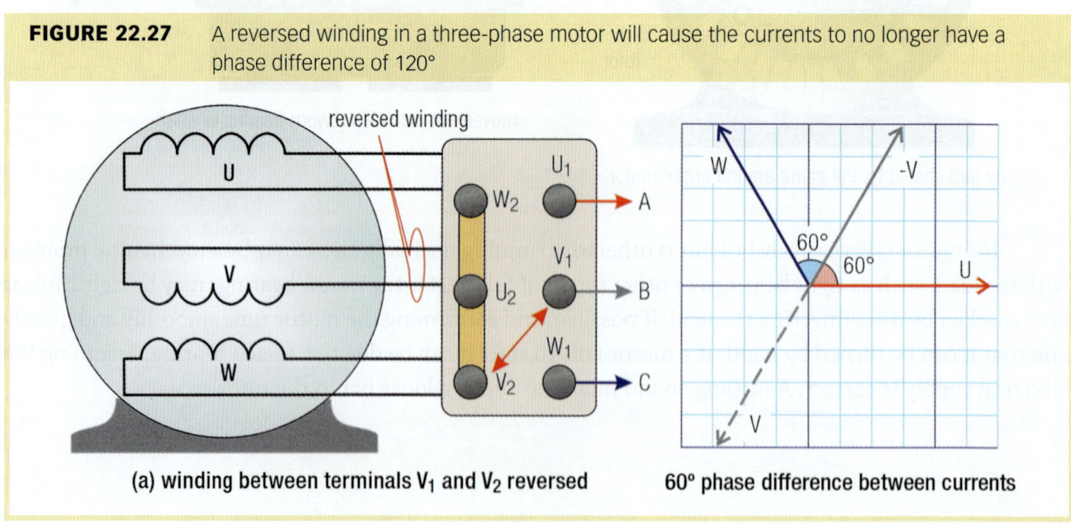

If one winding in a three-phase motor is reversed, the phase relationship between the currents is no longer 120°. Figure 22.27(a) shows a star-connected motor with winding V reversed. The effect of a reversed phase winding will cause the motor to rotate slowly, if at all, and to emit a growling sound. The current in each phase will remain high, and each phase will have a different value of current.

If the line voltages to the motor are not equal, for whatever reason, the phase currents will be unequal. This can cause heating due to internal circulating currents, and loss of torque.

3 *Motor runs slowly.* An open-circuit bar in a squirrel cage rotor will cause the motor to lose power. It will run below the nameplate speed even on a light load, and the increased slip will cause the motor to overheat due to the excessive stator current. Cages made of cast aluminium are often impossible to repair. Cages in large motors are often made of brass or copper bars which can work loose or become open-circuit at the end rings. There must be a solid connection between rotor bars and the end rings. The connections can sometimes be repaired, and broken bars can sometimes be replaced.

4 *Motor becomes hot.* Possibilities include loose rotor bars, failure of one phase of the supply and a short-circuited phase winding. To check this, measure the line current of each phase, usually with a clamp ammeter. Currents should be the same and be approximately equal to the nameplate value. If not, suspect a short-circuited winding. This can be hard to test with an ohmmeter, due to parallel paths in a delta-connected motor and also the low resistance of a typical winding. Visual inspection might show a burnt section of the winding.

Insulation resistance

A motor can also have failing winding insulation, particularly in older motors. If left unattended, the leakage current between phases and between each phase and the metal core will cause the insulation to become hot and eventually break down. An insulation resistance test should be carried out on motors being serviced. The test voltage will depend on the voltage rating of the motor. Figure 22.28 illustrates testing the insulation resistance between windings and motor frame of a three-phase 400 V motor with a test voltage of 500 V. A test voltage of 1000 V would be used to measure the insulation resistance between each phase winding.

FIGURE 22.28 The resistance of the insulation between windings and motor core should exceed one megohm, and is measured with an insulation resistance tester

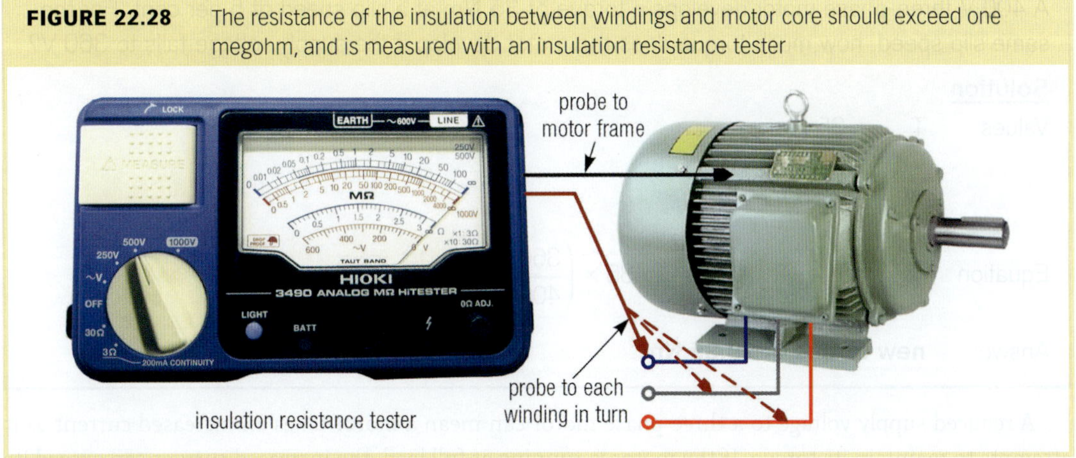

As shown in Figure 22.28, one probe of the tester is connected to the motor's metal frame, making sure it contacts metal, and not paint or other insulation. The other probe is connected in turn to each phase winding terminal (in the terminal box) and the resistance reading shown by the meter is taken. If a value close to or below one megohm is shown, the motor insulation is suspect. The test should then be repeated at the motor's normal operating temperature.

FIGURE 22.29 Torque changes with the square of the voltage applied to the motor

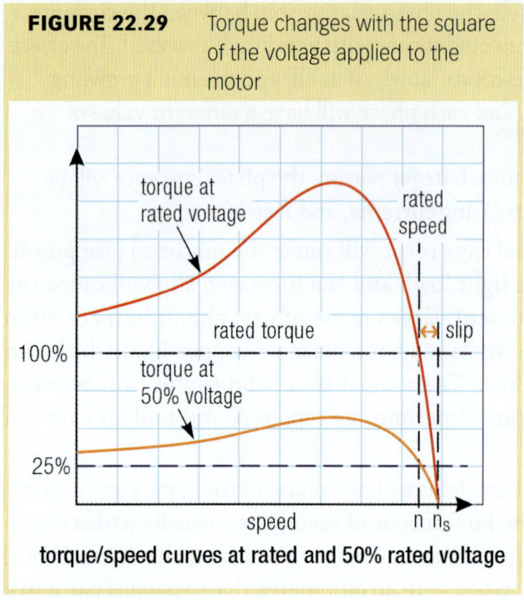

Effects of reduced voltage

The supply voltage to a three-phase motor can change for various reasons, such as an overload on the supply lines or problems with the electrical supply. When this happens, the motor no longer delivers its rated torque. As an example, assume the supply voltage falls to half the rated voltage of the motor. The effects are to reduce the stator current to half its normal value, which will reduce the rotor current to half its normal value. Because power is proportional to the product of current and voltage, and torque is proportional to power, the power available to produce torque has dropped to one quarter of the correct value.

Therefore, torque varies with the square of the applied voltage (for a given value of %slip). For any value of slip, the torque produced by a motor when the applied voltage has fallen to a new voltage can be found with this equation:

$$\text{new value of torque} = \text{rated torque} \times \left(\frac{\text{new voltage}}{\text{rated voltage}}\right)^2$$

Figure 22.29 shows the torque/speed curves for a standard squirrel cage motor at rated voltage and half rated voltage. The torque at the same slip speed is 25 per cent or one quarter of its rated value.

EXAMPLE 22.4

A 400 V three-phase motor develops a torque of 35 Nm at a slip speed of 5 per cent. For the same slip speed, how much torque does the motor develop if the supply voltage falls to 360 V?

Solution

Values $T_{rated} = 35$ Nm
$V_{rated} = 400$ V
$V_{new} = 360$ V

Equation $T_{new} = T_{rated} \times \left(\frac{V_{new}}{V_{rated}}\right)^2 = 35 \times \left(\frac{360}{400}\right)^2 = 35 \times 0.81$

Answer **new torque = 28.35 Nm**

A reduced supply voltage to a three-phase motor can mean a motor takes an increased current as it attempts to maintain its torque. If the motor is running at full load, the increased current can exceed its rated current, causing it to possibly overheat. As well, the motor could stall, allowing a high current to flow, enough to make it burn out. An increased supply voltage can cause the magnetic field to saturate the iron core, again causing an increased current.

KEY POINTS...

- The losses in an induction motor are made up of stator I^2R and iron losses, rotor I^2R and iron losses, and mechanical losses due to friction and windage.
- The input power to a three-phase induction motor $P_{in} = \sqrt{3}V_L I_L \cos\phi$. The output power developed by the motor $P_{out} = \frac{2\pi nT}{60} = \frac{nT}{9.55}$.
- The percentage efficiency of a three-phase induction motor $\%\eta = \frac{P_{out}}{P_{in}} \times 100$.
- Efficiency of three-phase induction motors is subject to Minimum Energy Performance (MEPS) requirements (AS/NZS 1359.5-2004).
- An unloaded motor has a low power factor and zero efficiency; a loaded motor has a higher power factor and an efficiency determined by the design of the motor and its size.
- An open-circuit phase winding or loss of one phase of the supply causes a three-phase motor to operate as a single-phase motor, which will overheat the motor. As well, the motor will not start from rest.
- Faults in three-phase motors include short-circuited coils, shorts between windings and frame, low leakage resistance, loose or open-circuit bars in the rotor cage and worn bearings or shafts.
- Torque varies with the square of the applied voltage, so a voltage reduction of 50 per cent will reduce the torque to 25 per cent of its rated value. A low supply voltage to a motor can also cause it to stall and subsequently burn out.

22.4 Motor Protection

Motor protection and isolation requirements are specified by AS/NZS 3000:2018 in section 4.13. In general, motor protection covers overload, overheating, over-temperature and, where required, protection against restarting or reversal of rotation. Isolation includes starting and stopping a motor with correctly rated devices. AS/NZS 3000:2018 (sub clause 4.13.2) requires that all motors with a rating greater than 370 W be provided with control equipment that incorporates protection against overload, unless the motor is part of a machine that complies with an appropriate standard.

Motor starters

When an induction motor is started, it takes a certain value of starting current. The actual value depends on the design of the motor and its power rating. Starting currents can be many times the rated current, although the value quickly drops as the motor accelerates. AS/NZS 3000:2018 gives a guide to determining the value of a motor's starting current, in the absence of manufacturer data, as being eight times the rated current of the motor (4.13.1.2 (i)). Supply authorities impose limits on motor starting currents, although the limits depend on the authority.

The reason for applying limits is to prevent significant fluctuations of the supply voltage. For example, the *Service and Installations Rules of NSW* sets down a maximum starting current of 53 A for a three-phase 400 V motor in a domestic installation. The limit is 45 A for a single-phase motor in a domestic installation. In other types of installations, the general guideline is that a starting current will not cause the supply voltage to the installation to fall by more than 5 per cent over a period of 20 milliseconds. To keep motor starting currents within the specified limits requires a unit called a motor starter. Where the motor starting current is already within limits, the starter simply connects it directly to the supply lines. This is called a direct-on-line or DOL starter.

Motors with a starting current that exceeds the limits are started with a more sophisticated type of starter. These reduce the supply voltage to the motor when it is first energised, and increase the voltage as the motor accelerates.

In addition to a means of starting and stopping a motor, it must also be protected against overload conditions. While fuses or thermal/magnetic circuit breakers in the supply will achieve this to some extent, a more usual method is with a device called a *thermal overload*. Another requirement is to prevent a motor restarting after stopping due to an overload or an under-voltage condition. This applies wherever the unexpected restarting of a motor could cause a danger.

DOL motor starter

The simplest form of DOL starter is a device called a *contactor*. This is an electromechanical device in which contacts are closed or opened by a mechanism operated by the magnetic effect of a coil. Two types of contactors are shown in Figure 22.30. Contactor (b) shows the coil that, when energised, attracts an armature that closes the contacts. Contactor (a) has a similar arrangement, except the coil and armature mechanism are enclosed. The circuit symbol for a contactor is in Figure 22.31.

FIGURE 22.30 Two types of contactors used to start and stop a three-phase motor

main contacts

coil and armature

auxiliary contacts

(a) three-phase 400 V contactor

(b) three-phase contactor for large loads

To provide current overload protection, a separate thermal overload unit is usually fitted to a contactor, if one is not already integrated. In some cases, the thermal overload device may be separate from the contactor. A thermal overload unit (relay) is designed to cause the contactor to open if the motor current exceeds a certain limit over a period of time. Typically, the unit is adjusted during installation to a setting equal to the full-load current of the motor. If the current increases and remains high for a certain period, the unit will operate and trip the contactor. Thermal overload units and the circuit symbol are shown in Figure 22.32.

FIGURE 22.31 Circuit symbol for a three-phase contactor

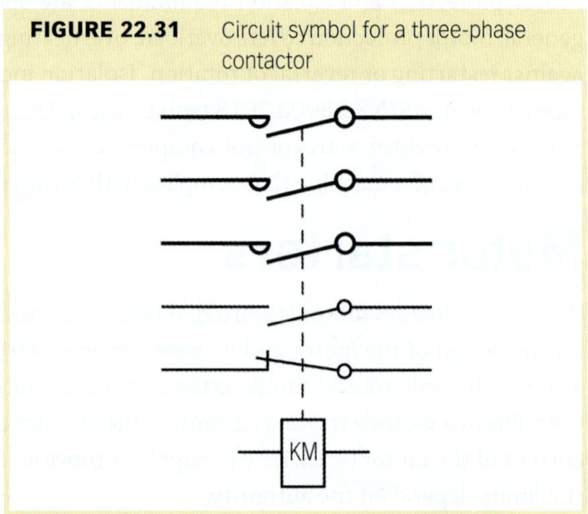

FIGURE 22.32 Thermal overload units with adjustments to set the load current which, when exceeded, operates the unit

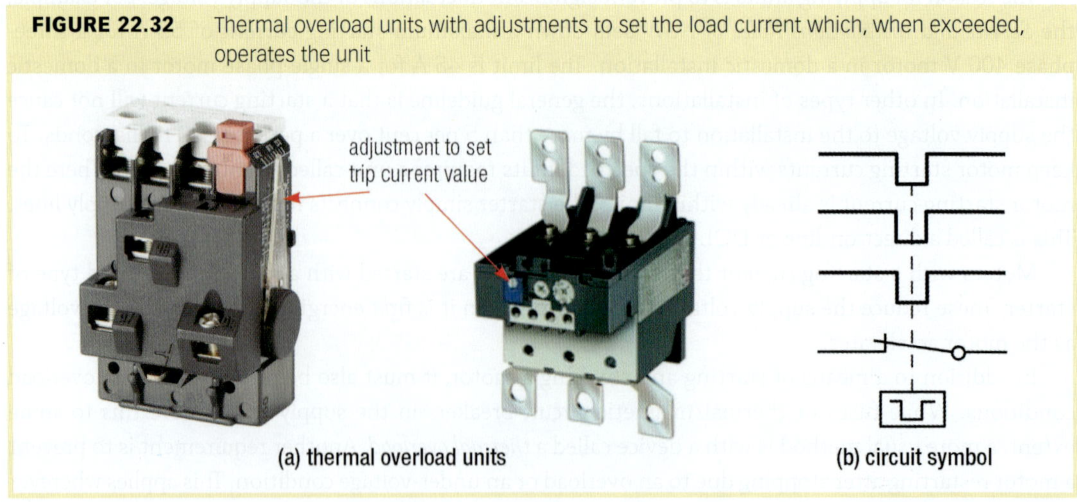

(a) thermal overload units

(b) circuit symbol

Figure 22.33 shows a circuit with a DOL motor starter. The motor supply is shown in black and the control circuit is shown in red. When main contacts KM1 are closed, current flows from the supply lines, through the main contacts and the thermal overload relay (TOL) to the motor. In some installations, the thermal overload is connected in only two of the three phases.

FIGURE 22.33 Circuit diagram of a DOL three-phase motor starter with thermal overload protection

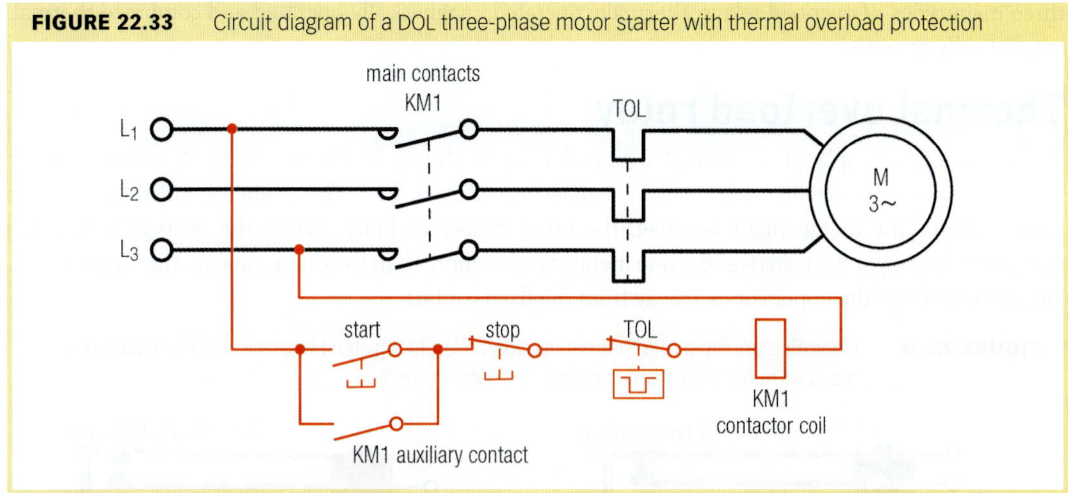

The control circuit of the starter has two push-buttons, labelled start and stop. These are mounted at a convenient point in the installation, in which the start button is usually green and the stop button is red. When the start button is pressed, contactor coil KM1 is energised, as there is now a circuit from L_1, through the start button, the stop button and the TOL contacts to the coil, which has its other terminal connected to L_3. As a result, the main contacts close, allowing the motor to start. When the contactor operates, it also closes an auxiliary contact, which is connected across the start button. This allows the start button to be released as the contactor coil is now supplied through the auxiliary contact.

When the stop button is pressed, the circuit to the contactor coil KM1 is broken, and the contactor releases, opening the main contacts and stopping the motor. The auxiliary contact KM1 also opens, so the motor can only be restarted by pressing the start button.

If the current to the motor exceeds its rated value, the thermal overload unit will operate by opening the contact marked TOL, which opens the circuit to the contactor coil, causing the contactor to open. The motor then needs to be restarted by pressing the start button, assuming the thermal overload has reset after cooling down. If the supply to the motor fails, the contactor coil will be de-energised and the contactor will open, requiring the start button to be pressed to restart the motor when the supply is restored.

AS/NZS 3000:2018 specifies that the switching apparatus (in this case a contactor) must be capable of interrupting a motor's locked rotor current, not just the normal rated current. A motor with a full-load current of 5 A could have a starting current of 40 A, so the contactor must be rated at 40 A, or higher if a higher starting current is specified by the manufacturer.

The position of the start and stop buttons is also required to be located to allow easy operation. This often means the start and stop station is mounted remotely from the contactor. Any number of emergency stop buttons can be added to the control circuit by connecting them in series with the stop button and the TOL unit. These, like the stop button, are push-buttons with contacts that are opened when the push-button is pressed (called normally closed contacts).

The control circuit shown in Figure 22.33 is connected between two lines, giving line voltage across the start push-button when the contactor is in its off position. In some cases the control circuit is powered from a separate low voltage supply, such as 24 V AC. This means high line voltages are not present in the control circuit, for safety reasons.

Overload protection

The circuit in Figure 22.33 has one form of overload protection, a thermal overload relay. As explained, this type of relay is designed to operate if the motor current exceeds its full-load current. There are three main types of overload relays: thermally operated, magnetically operated and combined thermal-magnetically operated.

Thermal overload relay

The operating principle of a thermal overload relay is shown in Figure 22.34. A resistive coil is wound around a bimetallic strip, which is made up of two dissimilar metals bonded together to form a strip. One metal might be steel, the other copper or brass. When the strip is heated due to current flowing in the resistive coil, one metal expands more than the other, causing the strip to bend. This action pushes the upper contact away from the fixed contact.

FIGURE 22.34 Operating principle of a thermal overload relay, in which a bimetallic strip is heated by a resistive heating element carrying the motor current

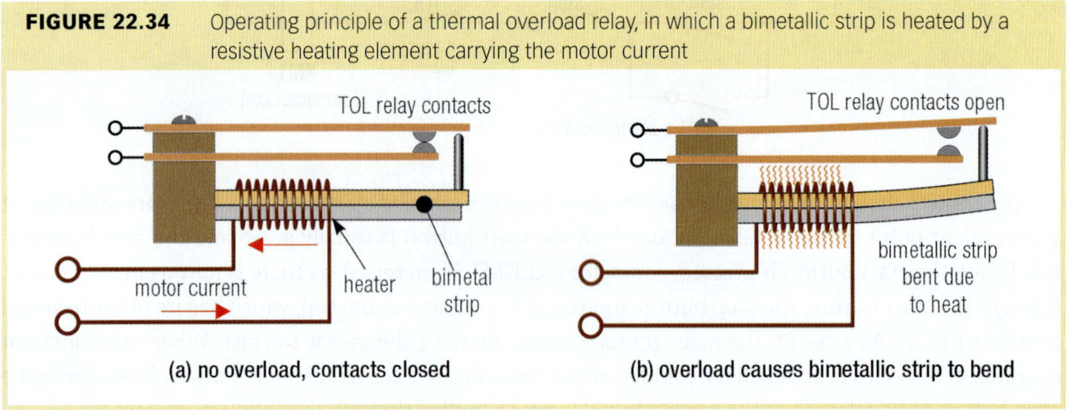

This type of thermal overload has a time delay, caused by the time it takes to heat the bimetallic strip. The higher the overload current, the faster the unit will operate. This is called an inverse-time characteristic. Because of the inherent time delay, the starting current of a motor does not remain high for long enough to operate the relay. Being temperature-sensitive, this type of overload unit will operate at a lower current if the ambient temperature increases. This can be a good thing if the motor being protected is exposed to the same ambient temperature. Compensated thermal overload relays are designed to be unaffected by ambient temperature changes.

Thermal overload relays are available as three-pole units for three-phase loads, which can also be used for single phase loads. Most types have an adjustable current setting over a range, such as from 4 A to 8 A. The trip current should be set to the motor's full-load line current.

A differential thermal overload relay causes the relay to operate when the current in each phase is different (unbalanced). The greater the unbalance condition, the faster the relay operates. Phase current imbalance occurs if a fuse in one phase operates.

Magnetic overload relay

A conventional magnetic relay operates too quickly for use as an overload sensing device, as it would trip every time the protected motor is started. To slow down the action, a piston inside an oil-filled dashpot is attached to the plunger of a solenoid. As shown in Figure 22.35, the piston has a bleed hole to allow oil to flow slowly through as the solenoid attempts to raise the piston. The motor current passes through the solenoid coil, which generates a magnetic field with a strength proportional to the motor current. When set correctly, the time delay caused by the oil-immersed piston will prevent the relay operating during motor start-up, but will operate if the motor current exceeds the setting by even a small amount.

FIGURE 22.35 Operating principle of a time delay magnetic overload relay, in which a piston attached to the solenoid plunger is slowed by travelling in an oil-filled dashpot

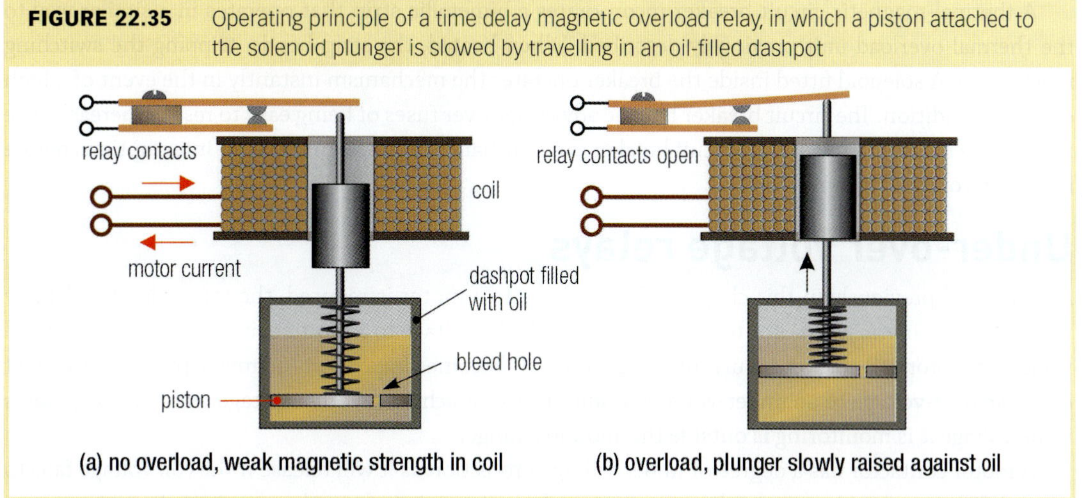

A combined thermal-magnetic overload relay incorporates a fast acting solenoid and the time delay due to the bimetallic strip action. This type of overload relay will trip instantaneously under high overload conditions, but will not trip when the motor is started.

An advantage of all these types of overload relays is they automatically reset. This is particularly important in situations where access is difficult. Typically someone will need to press the start button to restart the motor, unless the control circuit causes the motor to restart when the overload resets. In that case, it is important that an unexpected motor start does not pose a hazard to others.

Fuses

The fuse is the simplest form of overload protection, in which a fusible element inside a casing melts when the current rating is exceeded. The main purpose of a fuse is to protect a circuit, not necessarily a load connected to that circuit. If a short-circuit occurs in the circuit, the short reaction time of the fuse provides optimal protection. When used in conjunction with a motor, the fuse must have a current rating high enough to withstand the motor starting current, so it cannot offer the same level of protection as an overload relay.

The most common type of fuse in this application is the high rupturing capacity (HRC) fuse. Examples are shown in Figure 22.36. This type of fuse has the element enclosed in a ceramic tube filled with powdered silica. This ensures rapid heating of the fusible element, arc quenching and containment of the energy. A standard HRC fuse is not suitable for use in a motor start circuit. Instead, motor start HRC fuses are available that incorporate a special fusible element that provides a delay. These are sometimes referred to as 'gM' fuses, depending on the manufacturer. They often have a larger barrel than the standard HRC fuse.

FIGURE 22.36 HRC fuses operate quickly and contain the heat produced during operation

Circuit breakers

A motor circuit is often protected by a miniature circuit breaker, such as those shown in Figure 22.37. Circuit breakers incorporate an on-off switch as well as in-built overload protection. They operate in a similar way to a thermal or magnetic overload unit, in which an overload condition operates the switch mechanism. All three phases are simultaneously interrupted, unlike a fuse, where only one phase might be interrupted.

FIGURE 22.37 Thermal magnetic circuit breakers operate quickly under high overload; less quickly under low to moderate overload

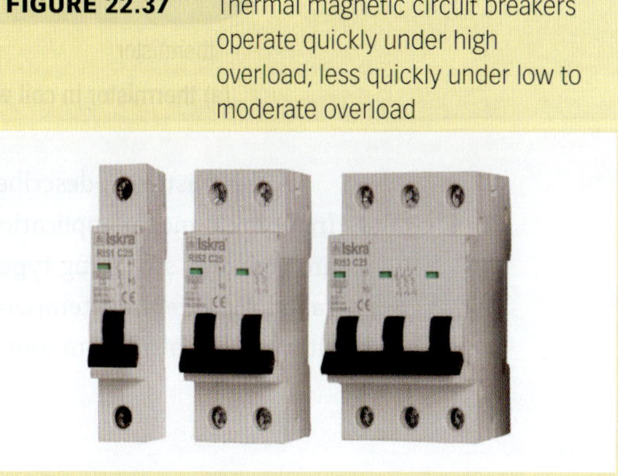

A thermal-magnetic circuit breaker incorporates a bimetallic strip that operates in a similar way to the thermal overload unit previously described. When heated, the strip bends, tripping the switching mechanism. A solenoid fitted inside the breaker operates the mechanism instantly in the event of a high overload condition. The circuit breaker has the advantage over fuses of being easy to reset, whereas a fuse needs to be replaced. However, a circuit breaker can only handle a certain number of interruptions before its switch contacts become worn.

Under-over voltage relays

As explained previously in this chapter, if the voltage to a motor is reduced, the torque is considerably reduced. This can cause a motor to overheat or stall. If the voltage to a motor is increased above its rated value, the motor will take more current and overheat. It becomes important to provide protection against either an over-voltage or an under-voltage condition. This is achieved with a voltage relay which operates if the voltage it is monitoring is outside the allowable range.

A motor contactor has a degree of under-voltage protection as it will release if the coil voltage falls to around 75 per cent of its rated voltage, assuming the coil supply is from the same supply as the motor. However, a voltage relay provides a more reliable form of protection and is connected into the control circuit of the motor. This type of device is not always included in a motor control circuit.

Over-temperature protection

AS/NZS 3000:2018 section 4.13.3 requires that all motors rated greater than 2.25 kW and unattended motors above 240 VA, such as commercial and communal refrigeration units, have some form of over-temperature protection. Exceptions are motors that are part of a fire-protection system, or where opening the motor circuit could cause a hazard. Protection devices are specified in AS/NZS 60947.

A thermal overload, as previously explained, protects a motor against excessive current causing it to overheat. Another method is built-in protection, in which the temperature of the motor, regardless of the cause, operates the protective device when the motor temperature reaches a certain value. A common method is with a thermistor embedded in the motor windings, as shown in Figure 22.38(a).

FIGURE 22.38 Thermistors embedded in motor windings have a positive temperature coefficient (PTC) in which the thermistor resistance increases quickly after the reference temperature is reached

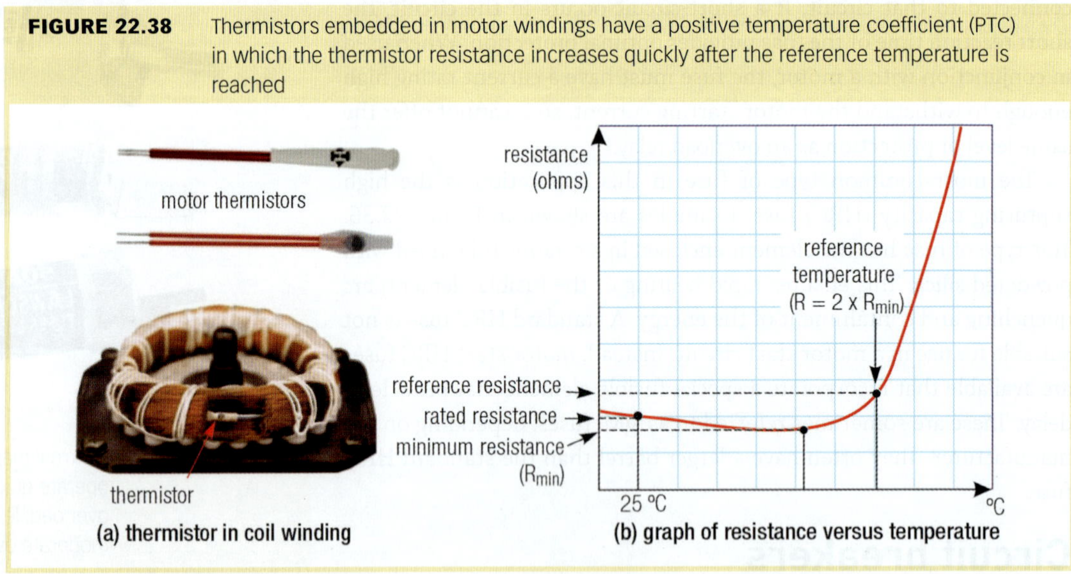

A thermistor, as described in Chapter 5, is a component whose resistance changes with temperature. In a typical motor application, thermistors with a positive temperature coefficient (PTC) are used. These are generally switching types, in which their resistance rises suddenly at a certain critical temperature, called the reference temperature. They are made of a doped polycrystalline ceramic containing barium titanate and other compounds.

Figure 22.38(b) shows a graph of resistance versus temperature for a motor thermistor. The rated resistance is its resistance at 25 °C and the reference resistance is the point when switching starts. The switching temperature might be 60 °C, depending on the application.

In a three-phase motor, thermistors are usually embedded in each phase winding, although AS/NZS 3000:2018 requires that two phases are protected as a minimum. When the winding temperature reaches the switching temperature, the thermistor resistance will increase significantly after a further small increase in temperature. The thermistors are usually connected in series and are typically connected to a motor control circuit with separate wiring, often with screened cables. If one or more of the thermistors exceeds its switching temperature, its high resistance will cause a relay in the motor control circuit to operate, isolating supply to the motor.

Adverse environments

Motors are often required to operate in hot, humid, wet or dusty environments. Motors manufactured to meet these needs are identified by an IP Code (ingress protection rating). This code is stamped on the motor nameplate, and consists of the letters IP followed by two digits. The first digit rates the level of protection the motor enclosure provides against access to live or moving parts, and the ingress of dust or similar foreign bodies. The numbers range from 0 (no protection) to 6, which is dust tight.

The second digit refers to the level of protection a motor enclosure provides against the ingress of water. The numbers range from 0 (no protection) to 8, in which the motor can be fully and continuously immersed in water at a depth greater than one metre.

A typical industrial motor has a protection rating of IP55, which means its enclosure provides dust protection and protection against the ingress of water projected onto the motor. A motor rated at IP56 has an even higher level of protection against the ingress of water.

Motor enclosures are designed to match the motor to its working environment. The most common type of motor in industry is the totally enclosed, fan-cooled motor or TEFC motor. This type of motor has a fan fitted externally on the non-drive end of the shaft that draws air over the motor fins. Figure 22.39(a) shows an example of a TEFC motor.

FIGURE 22.39 A motor's enclosure determines its suitability to different operating environments

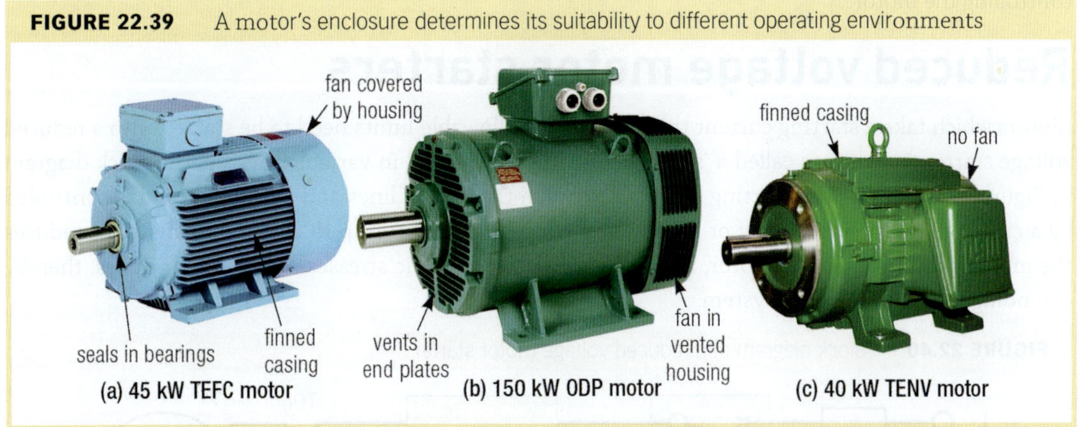

The open drip proof (ODP) motor typically has a protection rating of IP22 or IP23. This type of enclosure has no dust protection and allows access to the motor parts with objects such as a finger. An IP22 rated motor has limited water ingress protection and an IP23 motor can withstand low pressure spraying water. It has vent openings in the end plates and an internal fan to blow air over the motor parts. Figure 22.39(b) shows a 150 kW ODP motor with a flat steel frame (no fins) and vents in the end plates.

When the operating environment could cause a motor fan to become clogged, the totally enclosed non-ventilated (TENV) motor is used. This type of motor does not have a fan, and relies on the increased surface area of the motor frame to cool the motor by convection. The motor shown in Figure 22.39(c) is rated at IP55, and has V-ring seals between casing parts to maintain a good seal. A TENV motor is often built with special high temperature insulation, since the motor is designed to run hot.

The totally enclosed air over (TEAO) motor is designed to be used in the air stream of the fan or blower it is driving. The motor has a dust-tight cover and an aerodynamic body. The totally enclosed blower cooled (TEBC) motor is commonly used with a variable speed drive system. These motors are often rated as 'inverter duty' or 'vector duty' and are more expensive than similarly rated TEFC motors. The motor is constructed with a dust-tight, moderately sealed enclosure which rejects a degree of water. A constant speed blower pulls air over the motor fins to keep the motor cool at all operating speeds.

Motors designed for use in hazardous locations include the explosion proof motor (XPRF). This type of motor is designed to prevent it igniting surrounding materials, liquids or gases in the event the motor has an internal explosion, sparking or breakdown. It should contain the internal pressure caused by the problem, and flames should not be able to pass from inside the motor through any part of its casing and connections. These are given a maximum surface temperature rating from T1 (450 °C) down to T6 (85 °C). A hazardous location is defined as a place where concentrations of flammable gases, vapours or dusts occur.

Starting and reversing

The life of a motor is affected by the number of times it is started and stopped, or reversed. Each start-stop or forward-reverse cycle imposes stress on the insulation, the motor shaft, its bearings and general condition. Each time a motor is reversed during operation, considerable current needs to flow in the windings to make it reverse. This stresses the windings, which can flex at their connections, and cause mechanical stress in the stator insulation. The number of start-stop cycles per hour depends on the motor design. For example, a typical three-phase 1 kW motor might be able to do this 10 to 15 times an hour, whereas a large motor (25 kW or more) would be limited to less than five starts an hour. Each start-stop or forward-reverse cycle also causes wear on switching contacts, in particular the main contacts controlling the motor.

Reduced voltage motor starters

Motors which take a starting current that exceeds the allowable limits need to be started with a reduced voltage starter, sometimes called a 'soft' starter. This is achieved in various ways. As the block diagram in Figure 22.40 shows, the starting circuit is between the supply lines and the motor, and is controlled by a circuit that either manually or automatically brings the motor up to speed. A soft starter reduces the mechanical stress on the motor, as well as the electrodynamic stresses on the power cables, thereby extending the lifespan of the system.

FIGURE 22.40 Block diagram of a reduced voltage motor starter

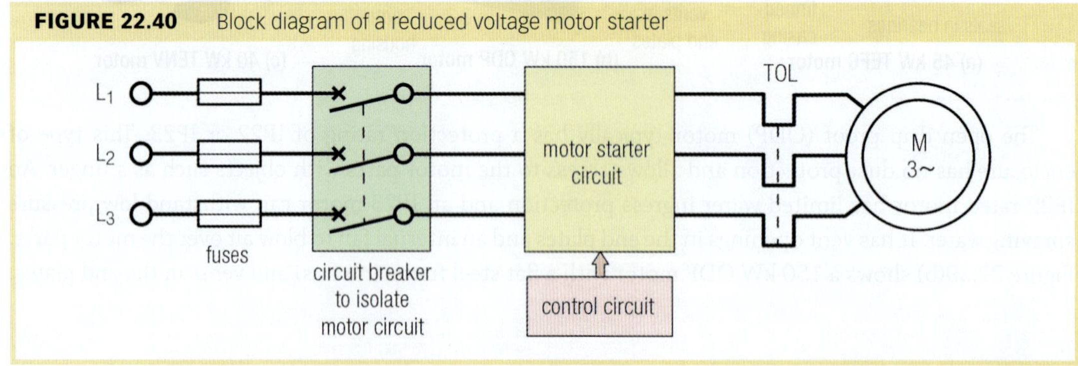

Electronic starter

In principle, an electronic starting system can be compared to a light dimmer circuit, except the starting system deals with substantially larger currents. Electronic components called silicon controlled rectifiers (SCRs) are connected in series with each motor winding. These are controlled by an electronic circuit that turns each SCR on for a part of the AC cycle. As in a light dimmer, a wide range of control is available. The control circuit provides a ramp signal to progressively apply more power to the motor as it accelerates. When the motor is up to speed, a contactor closes to bypass the starter.

Star–delta starter

Figure 22.41 shows the motor winding switching circuit of an open transition star–delta starter. This type of starter has two main stages, in which during the first stage the motor windings are connected in star caused by contactor KM2 closing. During this stage, the voltage applied to each winding is 1/√3 or 58 per cent of the line voltage, which for a 400 V supply means each winding has 230 V across it. The starting current and torque is therefore considerably reduced (to 33 per cent of full-load torque), limiting this type of starter to where the motor load can be started with a low torque.

FIGURE 22.41 Open transition star–delta three-phase motor starter circuit

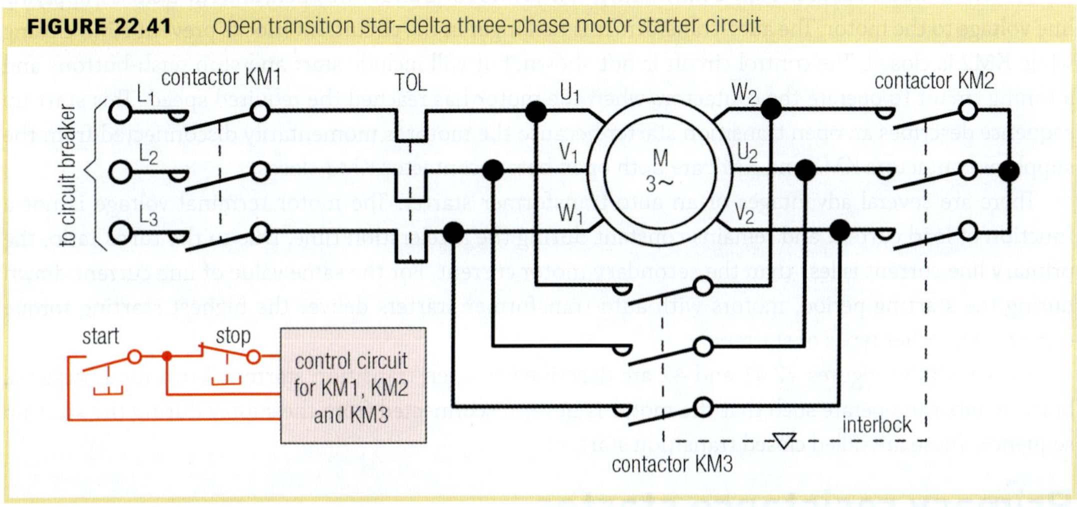

When the motor has reached around 80 per cent of its full-load speed, contactor KM2 opens and contactor KM3 closes, connecting the motor windings in delta. The timing of this action could be with a timing relay or a current-sensing circuit. It is critical that both contactors are never closed at the same time, as this would short-circuit the supply lines. Therefore, an electrical and mechanical interlock is usually incorporated. When the changeover from star to delta occurs, a high current peak can occur, as well as high mechanical stress due to the surge in torque.

In effect, this type of starter has four states. State 1 has all contactors open, state 2 (star state) has KM1 and KM2 closed, with KM3 open. State 3 is an open state, as KM1 is closed but momentarily KM2 and KM3 are open. The final state, called the delta state, has KM1 and KM3 closed. This type of starter is therefore called an open transition starter, due to state 3.

Auto-transformer starter

This type of starter applies a reduced voltage to the motor using an auto-transformer. An example circuit is in Figure 22.42. During the start-up phase, contactors KM1 and KM2 are closed, and KM3 is open. KM2 connects the auto-transformer in a star configuration, and KM1 applies line voltage to the auto-transformer. The motor is now connected to the auto-transformer tappings, which provide reduced voltage to the motor. Tapping positions are generally either 80 per cent, 65 per cent or 50 per cent of the line voltage.

FIGURE 22.42 Auto-transformer three-phase motor starter circuit

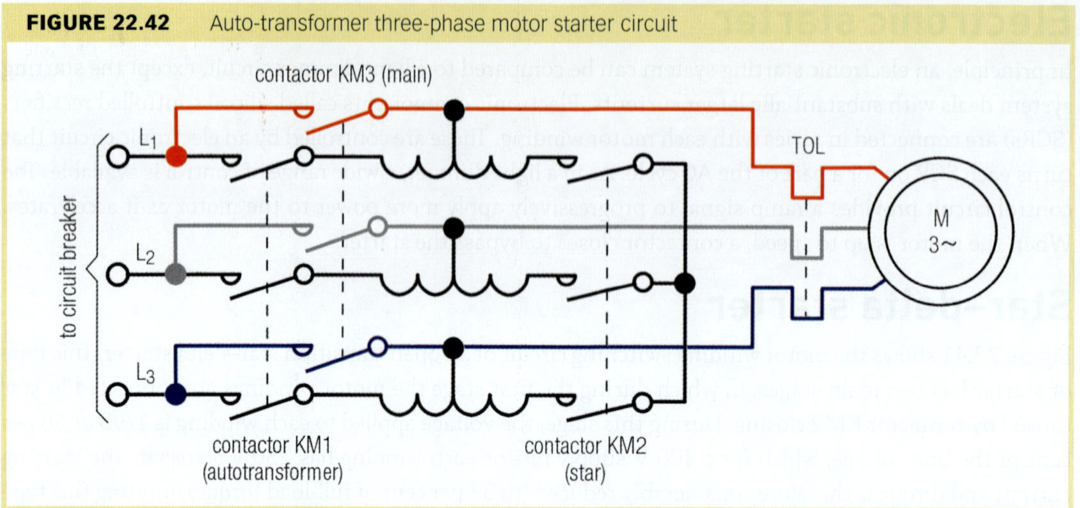

When the motor has accelerated sufficiently, contactors KM1 and KM2 open, and KM3 closes, applying line voltage to the motor. The auto-transformer is no longer in the circuit. Interlocks prevent KM3 closing while KM2 is closed. The control circuit is not shown, but will include start and stop push-buttons and a timing circuit to operate the contactors when the motor has reached the required speed. This start-up sequence describes an open transition starter because the motor is momentarily disconnected from the supply as contactors KM1 and KM2 are both open before contactor KM3 closes.

There are several advantages of an auto-transformer starter. The motor terminal voltage is not a function of load current and remains constant during the acceleration time. Due to the turns ratio, the primary line current is less than the secondary motor current. For the same value of line current drawn during the starting period, motors with auto-transformer starters deliver the highest starting torque compared to other types of starters.

The circuits in Figures 22.41 and 42 are described as open transition starters. With modifications, both circuits can operate such that the motor is never disconnected from the supply during the start-up sequence. These are called closed transition starters.

Primary resistance starter

Figure 22.43 shows the circuit of a primary resistance starter. The resistors are typically made from a resistive metal, such as cast iron grid resistors or wire wound types. For large motors, liquid resistors are sometimes used, in which the resistance is between electrodes in a tank filled with an electrolyte, such as salt water.

FIGURE 22.43 Primary resistance three-phase motor starter circuit

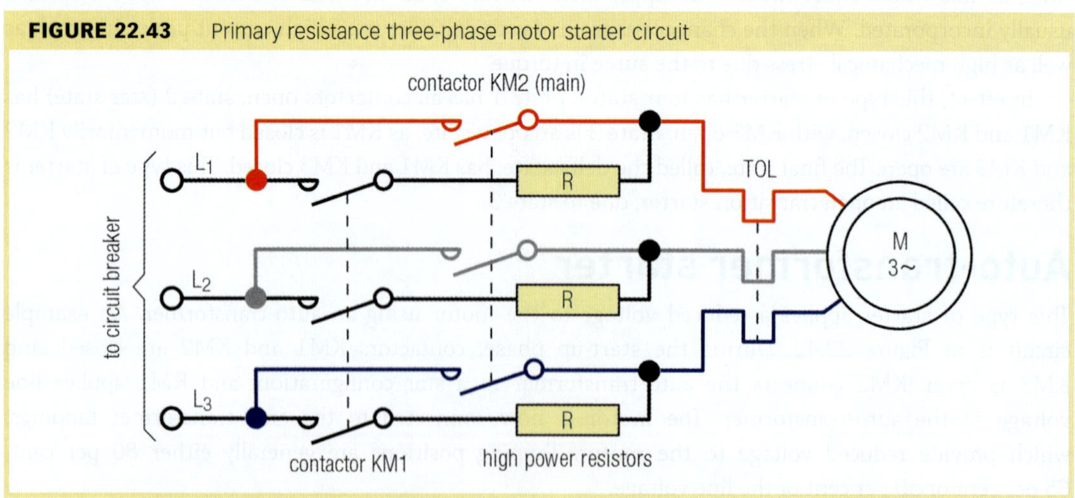

During start-up, contactor KM1 is closed with KM2 open. The motor is now in series with the resistors and draws current from the supply. The voltage drop across the resistors is initially high, but as the motor gains speed the starting current reduces, which in turn reduces the voltage drop across the resistors. This raises the terminal voltage of the motor, allowing it to gain even more speed. After a preset time interval, contactor KM2 closes, followed by KM1 opening. Because there is voltage applied at all times to the motor, this gives a closed transition starter. This type of starter is only suited to low inertia loads, as the torque at start-up is very low.

KEY POINTS...

- A DOL starter applies full voltage to a motor during its start-up phase. This type of starter is permitted if the motor starting current does not exceed specified limits.
- A DOL starter can be a manually operated switch, a contactor controlled by start and stop push-buttons or a contactor controlled automatically by another circuit.
- When the starting current of a motor is excessive, a reduced voltage starter is used. These include the star–delta starter, electronic starter, auto-transformer starter and primary resistance starter.
- A contactor or switch controlling a motor should be rated at the motor's starting current.
- AS/NZS 3000:2018 requires that all motors with a rating greater than 370 W be protected against overload. Unattended motors above a certain rating are required to have over-temperature devices fitted to the motor (provisos apply).
- A thermal overload relay senses the current taken by a motor and opens when the current exceeds the set value. Over-temperature protection is often provided by thermistors embedded in the motor windings.
- Fuses in a motor circuit must be capable of handling the motor starting current. Motor start HRC fuses have a delay incorporated in the fusible element.
- A miniature circuit breaker with thermal or magnetic (or both) overload protection is preferred over fuses, as all three phases of the supply are interrupted together under an overload condition.
- Motor protection circuits often have a voltage relay to guard against an over- or under-voltage condition, which can cause a motor to overheat.
- Motors are rated to suit their operating environment and are given an ingress protection rating (IP rating). Motor enclosures range from the open drip proof type (ODP) to explosion proof (XPRF).

CHAPTER SUMMARY

The following equations were introduced in this chapter:

- $n_s = \dfrac{120f}{p}$

  where n_s = synchronous speed in RPM, f = supply frequency in hertz, p = number of poles

- $s\% = \dfrac{n_s - n}{n_s} \times 100$ and $n = n_s - \dfrac{n_s \times \%s}{100}$

  where s% = percentage slip, ns = synchronous speed, n = rotor speed

- $f_r = \dfrac{s\% \times f}{100}$

  where f_r = rotor frequency, f = supply frequency in hertz

- new value of torque

  $$= \text{rated torque} \times \left(\frac{\text{new voltage}}{\text{rated voltage}}\right)^2$$

Other equations that apply to three-phase motors are:

- $P_{in} = \sqrt{3}\, V_L I_L \cos\phi$

  where P_{in} = true power taken by the motor (watts), V_L = line voltage, I_L = line current, $\cos\phi$ = power factor

- $P_{out} = \dfrac{2\pi nT}{60} = \dfrac{nT}{9.55}$

  where P_{out} = output power in watts, n = speed in RPM, T = torque in newton metres

- $\%\eta = \dfrac{P_{out}}{P_{in}} \times 100$

  where η = efficiency, P_{out} = output power in watts, P_{in} = input power in watts.

REVIEW EXERCISES

Check your answers at the back of the book.

1 A three-phase eight pole induction motor on load runs at 720 RPM when operating from a 50 Hz supply. Determine the motor's:
   - a synchronous speed
   - b slip speed
   - c percentage slip.

2 A four-pole three-phase motor is powered by a variable frequency drive with a frequency range of 5 Hz to 130 Hz. The percentage slip is 4 per cent. Calculate the minimum and maximum speeds of the motor.

3 A two-pole 50 Hz three-phase induction motor is running at 2850 RPM. Calculate its:
   - a percentage slip
   - b rotor frequency.

4 Calculate the rotor frequency of a motor operating from a 50 Hz supply that has a percentage slip of 5.6 per cent.

5 A three-phase motor connected to a supply with a line voltage of 400 V takes 6 A per phase at a power factor of 0.85. It delivers 21 Nm of torque at a speed of 1430 RPM. Calculate the efficiency of the motor.

6 A two pole, three-phase induction motor connected to a 50 Hz supply is running at a percentage slip of 4.5 per cent and delivers 40 Nm of torque. It has an efficiency of 89 per cent and a power factor of 0.9. Calculate the motor's:
   - a synchronous speed
   - b slip speed
   - c rotor speed
   - d output power
   - e input power
   - f line current for a line voltage of 400 V.

7 An 11 kV three-phase motor develops a torque of 80 Nm at a slip speed of 4.6 per cent. For the same slip speed, how much torque does the motor develop if the supply voltage falls to 9 kV?

8 A three-phase induction motor takes 45 kW of power, and delivers 42.5 kW of output power. What is its percentage efficiency?

9 A 400 V three-phase induction motor rated at 150 Nm is started with a star delta starter. What is its approximate starting torque?

10 When comparing the star–delta, autotransformer and primary resistance motor starters, which type provides the best starting torque characteristics?

ONLINE RESOURCES

COMPLETE WORKSHEET TWENTY-TWO

Check with your instructor for worksheets on this chapter.

CHAPTER 23

SINGLE-PHASE MOTORS

Single-phase motors are used extensively in homes, industry and commerce. They operate in a similar way to the three-phase induction motor, but have an additional winding to make them start. There are various types of single-phase motors, including capacitor motors and shaded-pole motors. This chapter describes the various types of single-phase motors, their protection requirements and typical faults.

CHAPTER OUTLINE

23.1 Single-Phase Induction Motors

A three-phase induction motor, as explained in Chapter 22, has a rotating magnetic field due to each phase being spaced by 120° (electrical degrees). This means a motor starts rotating as soon as it is connected to a three-phase supply. A single-phase supply does not provide a rotating magnetic field; instead, as shown in Figure 23.1, it produces a magnetic field that reverses polarity with each half cycle. This creates a pulsing effect in an induction motor, in which the rotor remains stationary.

FIGURE 23.1 A single-phase motor produces a pulsating magnetic field that holds the rotor stationary

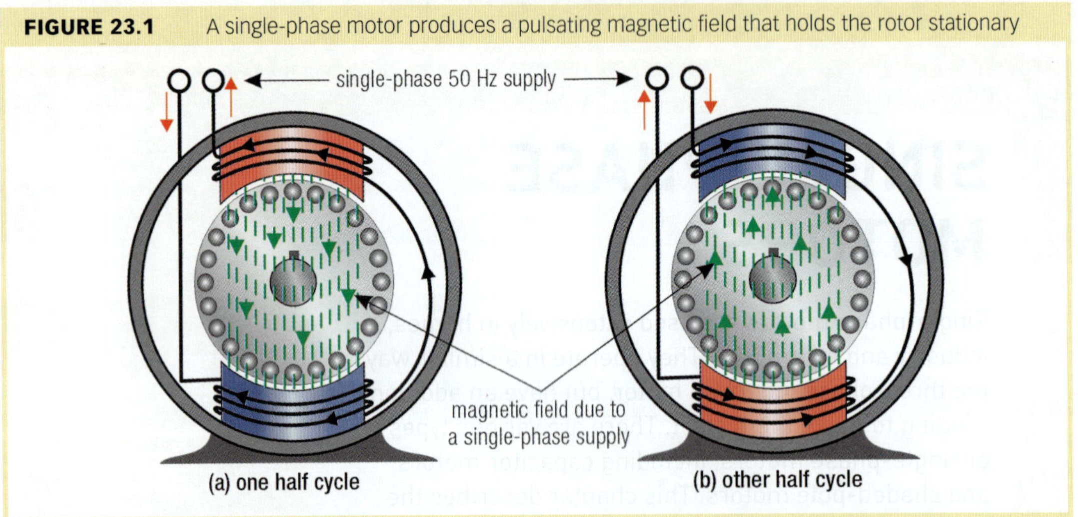

If the rotor can be brought up to a sufficient speed by an external force, the action between the rotor and stator magnetic fields will keep the motor running. A rotating magnetic field sufficient to get an induction motor running can be achieved in various ways.

Split-phase induction motor

Figure 23.2 shows a split-phase motor that has two sets of stator windings to create the effect of a rotating magnetic field from a single-phase supply. The main winding, also called the run winding, is wound with a relatively heavy gauge insulated wire, and is placed at the bottom of the stator slots. The other winding, called the auxiliary or start winding, is wound with a smaller gauge wire. It has around 25 per cent less turns than the run winding, and is placed at the top of the slots. This results in the start winding having a higher resistance and a lower reactance than the run winding.

FIGURE 23.2 A split-phase induction motor has two sets of windings, in which the run winding has a higher inductance and a lower resistance than the start (auxiliary) winding

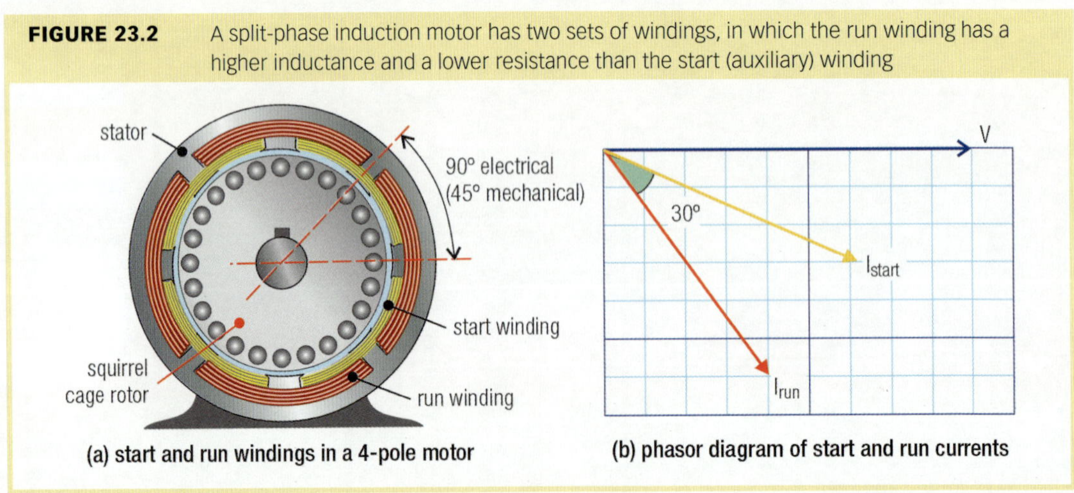

The phasor diagram in Figure 23.2 shows the phase relationship between the currents in both windings. Because the start winding has a higher resistance and lower inductance than the run winding, the start winding current lags the applied voltage by a smaller amount than the current in the run winding. A typical phase difference between these two currents is 30° to 40°, which means the magnetic fields produced by the two windings have this same phase difference between them.

The phasor addition of the two fields produces a rotating magnetic field, although the field is not uniform and is more elliptical than circular. However, its action on the squirrel cage rotor produces enough torque to start the rotor turning and to bring it up to speed. The vibration and hum associated with a split-phase motor is due to the non-uniform magnetic field.

The start (or auxiliary) winding can only be left in circuit for around 10 seconds, as it is made of relatively thin wire and has a low inductance. Therefore, the split-phase induction motor needs a way of disconnecting the start winding from the supply when the motor has reached sufficient speed. The method shown in Figure 23.3 uses a centrifugal switch mechanism that opens a contact and disconnects the start winding from the supply when the motor is around 75 per cent of its rated speed.

FIGURE 23.3 Centrifugal switch in which contacts are held closed by a collar that moves inward as the motor increases, causing the contacts to spring open

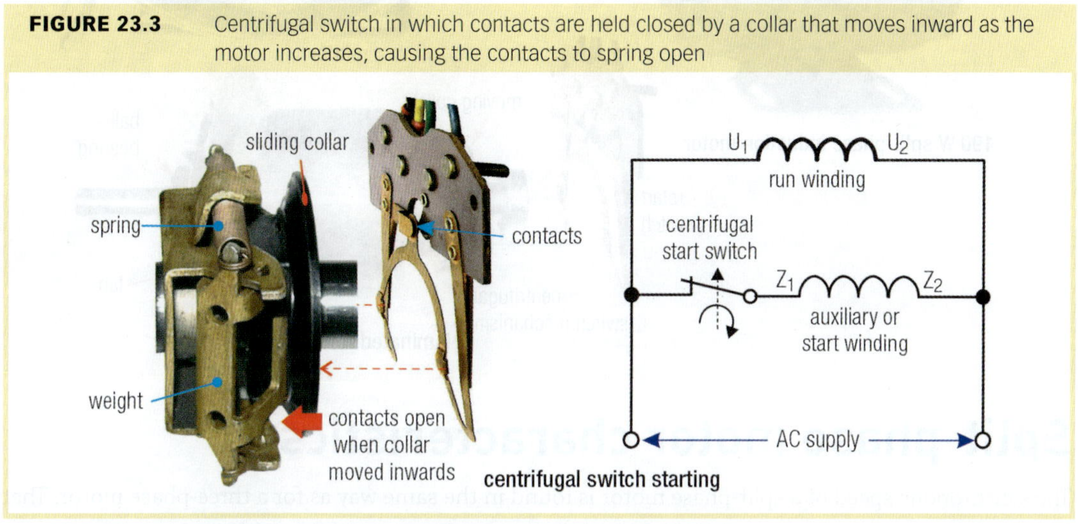

A centrifugal switch mechanism has two weights that fly outwards as the rotor gains speed. This action causes a sliding collar to move inwards and allow the spring-loaded contacts to open and disconnect the start winding. Another method of achieving automatic disconnection of the auxiliary winding is with a motor starting relay. A motor starting relay and the circuit diagram are shown in Figure 23.4. The relay coil is in series with the motor as shown in Figure 23.4(b).

FIGURE 23.4 Motor start relay, in which the run winding is in series with the relay coil

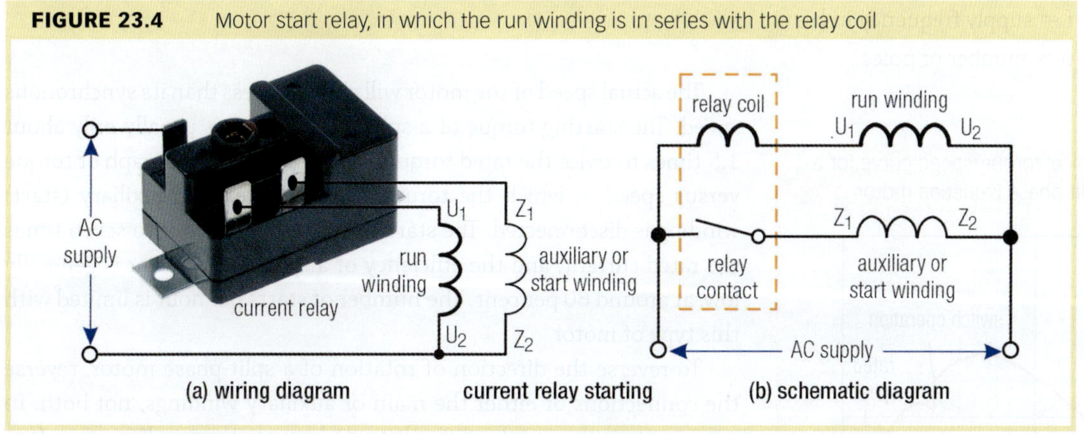

A motor start relay can be mounted on the motor. When the motor is started, the large starting current causes the relay to operate and close its contacts, connecting the auxiliary winding to the supply. This allows the motor to start. When the current through the relay coil reduces to its normal value the relay armature opens, opening the contacts and disconnecting the start winding from the supply.

Figure 23.5 shows the main parts of a typical split-phase induction motor. The rotor usually has a cast aluminium cage and a core made up of steel laminations. A fan mounted on the shaft provides cooling. Bearings are often ball bearings, sometimes sleeve bearings. If a centrifugal switch is fitted, it will also be on the rotor shaft. The switch contact operated by the centrifugal switch is attached to an end plate and connections are made at terminals on the switch contact assembly.

FIGURE 23.5 A typical split-phase induction motor with a centrifugal switch in the start circuit

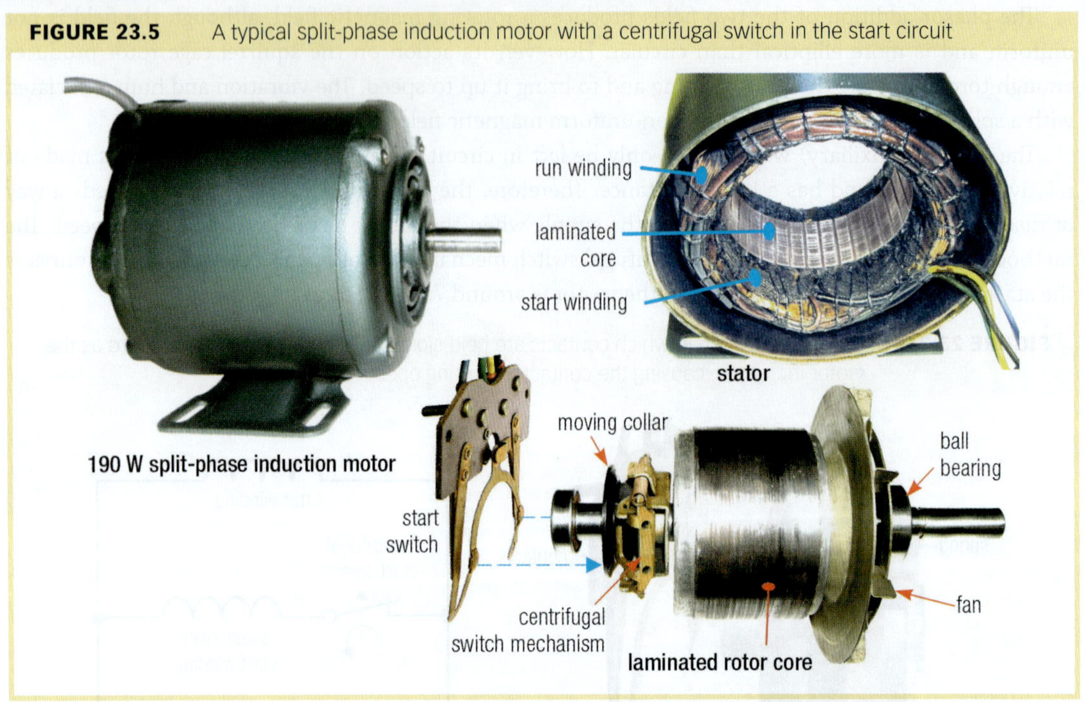

Split-phase motor characteristics

The synchronous speed of a split-phase motor is found in the same way as for a three-phase motor. That is:

$$n_s = \frac{120f}{p}$$

where:

n_s = synchronous speed in RPM

f = supply frequency in hertz

p = number of poles.

FIGURE 23.6 Typical torque/speed curve for a split-phase induction motor

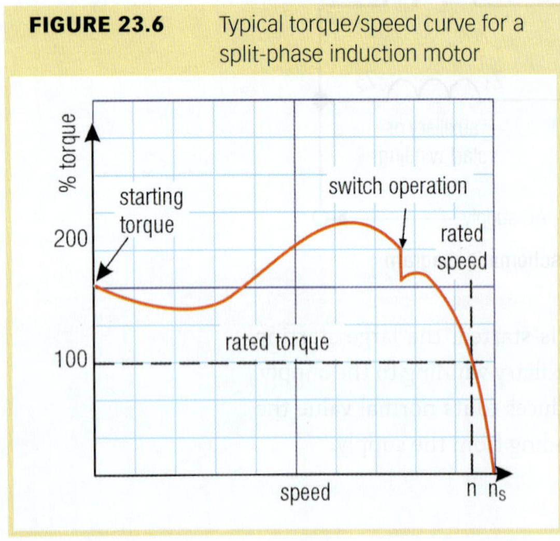

The actual speed of the motor will always be less than its synchronous speed. The starting torque of a split-phase motor is usually only about 1.5 times to twice the rated torque. Figure 23.6 shows a graph of torque versus speed in which the torque changes when the auxiliary (start) winding is disconnected. The start current is around five to seven times the rated current, and the efficiency of a split-phase motor is generally low, at around 60 per cent. The number of starts per hour is limited with this type of motor.

To reverse the direction of rotation of a split-phase motor, reverse the connections of either the main or auxiliary windings, not both. In some motors this can be difficult to do because the windings are often internally connected at their common point, with only three leads accessible at the motor terminals.

Split-phase motors are normally rated at less than 1 kW, more typically less than 500 W. Applications include powering grinders, sanders, small lathes and drilling machines. They are also used in washing machines, fans and some types of pumps.

23.2 Capacitor Motors

There are three types of capacitor motors: capacitor start, in which a single capacitor is in the circuit only during starting; permanently split capacitor motor, which has one capacitor in the circuit during starting and running; and capacitor start, capacitor run, in which there are two different values of capacitance for starting and running. Example capacitor motors are shown in Figure 23.7. Capacitor motors range in size from less than 30 W to over 10 kW. The capacitors are typically mounted on the motor frame, but are sometimes mounted separately to the motor.

FIGURE 23.7 Typical capacitor motors

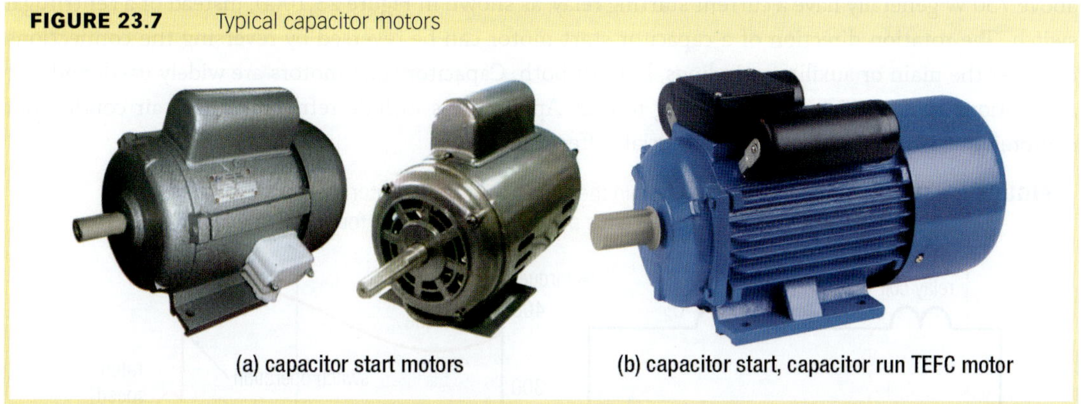

(a) capacitor start motors (b) capacitor start, capacitor run TEFC motor

Capacitors for use with a capacitor motor range in size from a few microfarads to over 200 μF. They are given an AC voltage rating, such as 230 V AC, 440 V AC, 660 V AC and so on. Capacitors in the starting circuit of a motor are typically non-polarised electrolytic types. Capacitors connected in circuit while the motor is running are usually oil-filled paper or plastic film dielectric types. A typical plastic film is metallised polypropylene. Some types have two capacitors in the same can. Example capacitors are shown in Figure 23.8.

FIGURE 23.8 Typical capacitors used with capacitor motors

Capacitor start motor

The capacitor start motor is similar to a split-phase motor, except it has a capacitor in series with the auxiliary (start) winding. This winding may have heavier gauge wire and more turns than for a split-phase motor. The circuit is shown in Figure 23.9(a). As shown in the phasor diagram in Figure 23.9(b), the addition of the capacitor increases the phase angle between the start winding and run winding currents, which, therefore, gives increased torque during start-up.

FIGURE 23.9 Circuit of a capacitor start, induction run motor in which the phase difference between the start and run currents approaches 90°

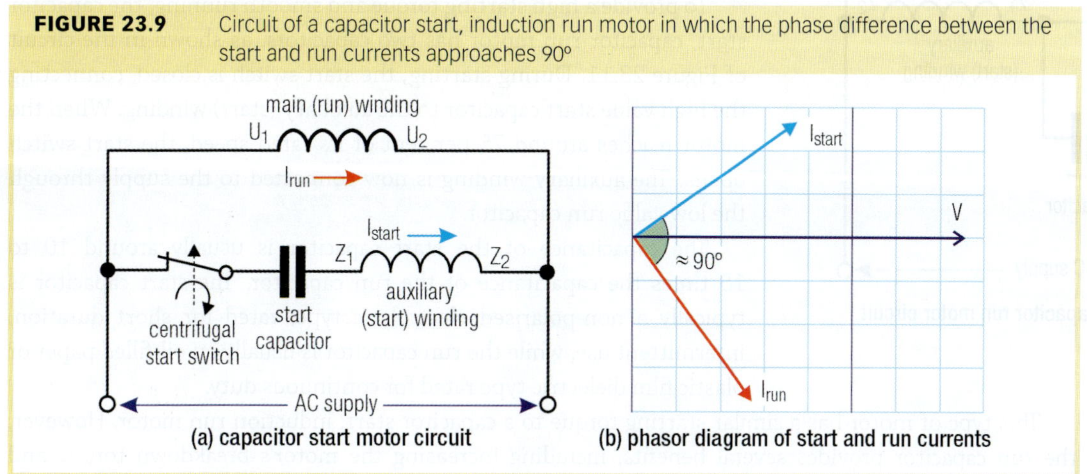

(a) capacitor start motor circuit (b) phasor diagram of start and run currents

A typical value of capacitance for a 250 W motor is around 180 μF. The capacitor is usually a non-polarised electrolytic type (AC voltage rated), with a duty time of less than 10 seconds. Some types are rated at less than 5 seconds, which means the motor must reach 75 per cent of its rated speed in that time, to prevent overloading the capacitor.

The starting torque of a capacitor start motor can be up to four times the rated torque, as shown in Figure 23.10(b). However, when the start winding is disconnected by the start switch, the torque is no greater than for a split-phase motor.

The start switch of any induction motor has to interrupt a high current. Capacitor start motors rated above 750 W generally have a current starting relay, as shown in Figure 23.10(a), instead of a centrifugal switch. The rotation direction of a capacitor start motor can be reversed by reversing the connections of either the main or auxiliary windings, but not both. Capacitor start motors are widely used, and suit applications that require a high starting torque. Applications include refrigerator and air conditioner compressors, machine tools and commercial coffee grinders.

FIGURE 23.10 (a) A current relay is used in the start circuit of capacitor motors rated above 750 W (b) capacitor start motors have a starting torque up to four times the rated torque

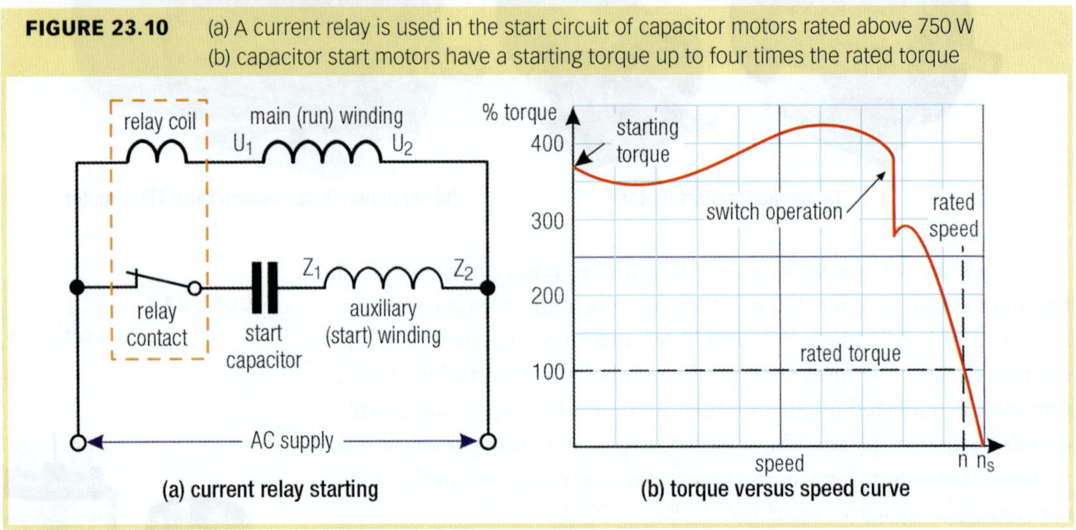

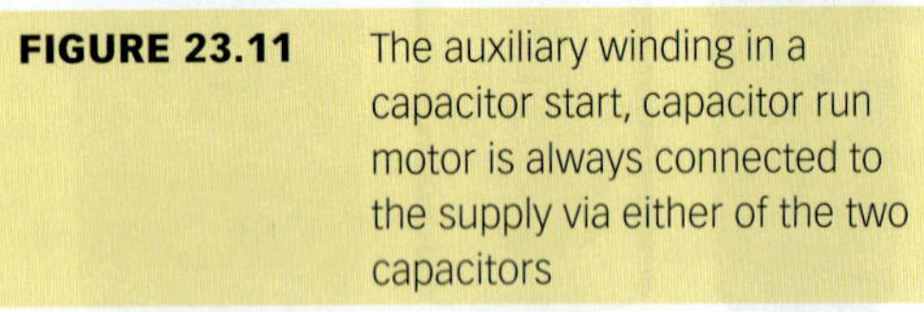
FIGURE 23.11 The auxiliary winding in a capacitor start, capacitor run motor is always connected to the supply via either of the two capacitors

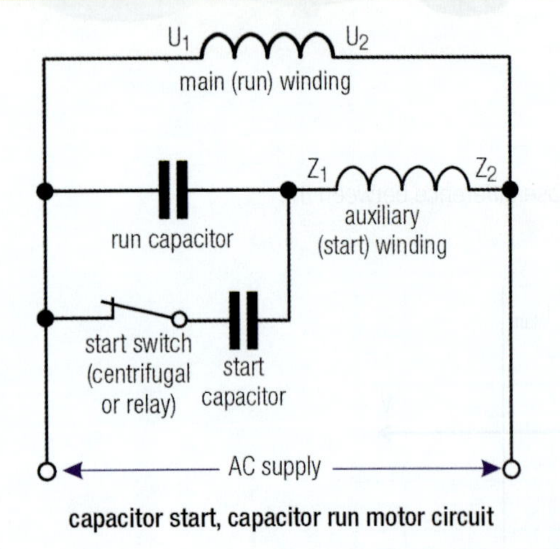

Capacitor start, capacitor run motor

Adding a capacitor in series with the auxiliary (start) winding not only increases the starting torque of the motor, it also provides a more uniform magnetic field during start-up. To achieve a high starting torque, the start capacitor should have a high value, while to achieve a uniform magnetic field during running requires a small value capacitor.

To provide a high starting torque and smooth running, the capacitor start, capacitor run motor has two capacitors, as shown in the circuit of Figure 23.11. During starting, the start switch is closed, connecting the high value start capacitor to the auxiliary (start) winding. When the motor reaches around 75 per cent of its rated speed, the start switch opens. The auxiliary winding is now connected to the supply through the low value run capacitor.

The capacitance of the start capacitor is usually around 10 to 15 times the capacitance of the run capacitor. The start capacitor is typically a non-polarised electrolytic type, rated for short duration, intermittent use, while the run capacitor is usually an oil-filled paper or plastic film dielectric type rated for continuous duty.

This type of motor has a similar starting torque to a capacitor start, induction run motor. However, the run capacitor provides several benefits, including increasing the motor's breakdown torque and improving its efficiency and power factor. It also provides a more uniform, circular rotating magnetic field, which reduces motor noise and vibration. This makes the motor suitable for use where a high starting torque is needed, along with relatively silent running. They are used extensively in air-conditioning systems, compressors and many types of industrial equipment. The motor can be reversed by reversing the connections of either winding, but not both.

Permanently split capacitor motor

This type of motor has two identical windings with a capacitor permanently connected in series with one of them. The circuit is shown in Figure 23.12(a), in which winding 1, for the switch position shown, is connected to the supply via a capacitor. Winding 2 is connected directly across the supply.

FIGURE 23.12 A permanently split capacitor motor does not have a starting switch. It is easily reversed by adding an external switch or relay that selects which of the identical windings has a capacitor in series with it and the supply.

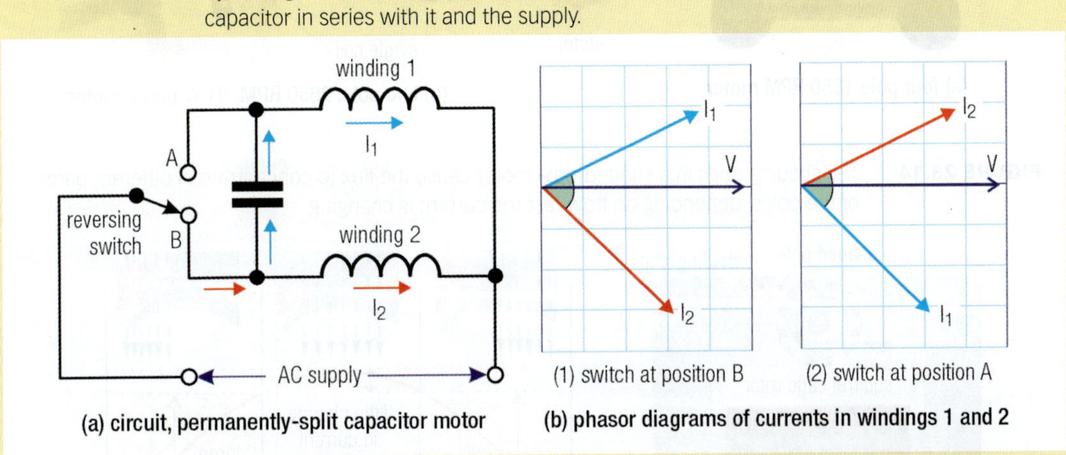

The phasor diagrams in Figure 23.12(b) show the phase relationship between the winding currents. In (1), when the switch is at position B, as shown in the circuit diagram, current I_1 leads current I_2. When the switch position is changed to A, the two currents are as shown in (2). As a result, the motor runs in a reverse direction.

The permanently split capacitor motor has a low starting torque, but it features quiet, smooth running. It also has a simple construction, is easy to reverse and suits electronic speed control. It is used extensively in ceiling fans, some types of blowers and in applications where the motor has to be reversed, such as opening and closing air conditioning vents.

23.3 Shaded-Pole Induction Motors

The shaded-pole motor, shown in Figure 23.13, gets its name from how it achieves rotation. A shading ring is made from one or two copper loops embedded into a slot at the tip of each stator pole. A two-pole motor has two shading rings and one coil, while a four-pole motor has four shading rings and two coils. Like all induction motors, it has a squirrel cage rotor.

When energised, the alternating magnetic field induces a current in the shading rings. By Lenz's law, the current opposes the change in the magnetic field, to an extent that depends on how rapidly the current is changing. The effect is shown in Figure 23.14, in which the opposing action causes the field to compress on one side of the pole, then appear across the entire pole, finally to compress on the other side. This gives a moving field that causes the rotor to turn.

FIGURE 23.13 Shaded-pole motors are used in many types of fans, small pumps and loads that have low starting torque

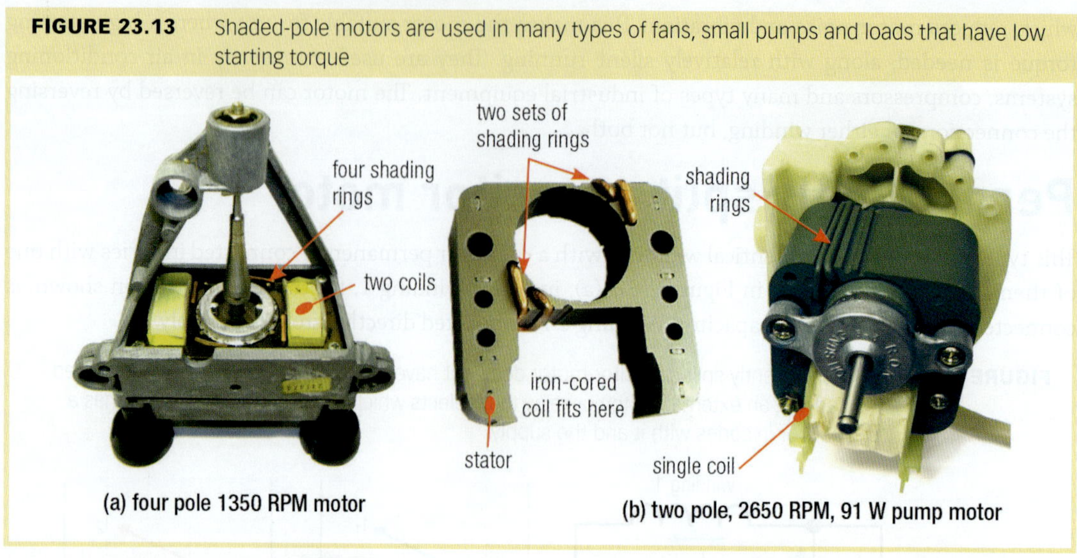

FIGURE 23.14 The shading rings in a shaded-pole motor cause the flux to concentrate in different parts of the poles, depending on how fast the current is changing

The starting torque of a shaded-pole motor is very low, and the motor runs with a higher amount of slip compared to a conventional squirrel cage induction motor. However, it has a very simple construction, which makes the motor useful in many applications, providing a high starting torque is not required. It is impossible to reverse a standard shaded-pole motor other than to physically reverse the stator because the direction of rotation is always towards the shading rings. Reversible shaded-pole motors have two sets of shade 'windings', rather than rings. By opening one set of windings and closing the other, the motor can be electrically reversed.

The shaded-pole motor is inefficient. For example, the motor in Figure 23.13(b) is rated at 91 W at 240 V, 0.8 A. The power input is therefore 192 W, giving an efficiency of 47 per cent. Despite this, shaded-pole motors are widely used in many types of appliances.

Although generally reliable, a common fault in a shaded-pole motor is a 100 Hz buzz, caused by a loose shading coil or lamination. Because of its low starting torque, dry or sticky bearings can cause the motor to remain stalled, although this does not usually cause the coil to burn out. Other than cleaning the motor and oiling its bearings, this type of motor is generally not repairable and replacement cost is usually cheap.

23.4 Universal Motors

The universal motor is not an induction motor, as it has a wound armature like those in a DC motor. They are called universal because this type of motor operates from AC as well as DC. A typical universal motor is shown in Figure 23.15. The universal motor is used in mains-powered hand tools (drills, saws, planers, etc.), vacuum cleaners, lawn mowers, kitchen appliances and applications where a high speed, relatively small motor is required.

FIGURE 23.15 Universal motors have their field coils in series with the armature, and operate at a high speed when unloaded

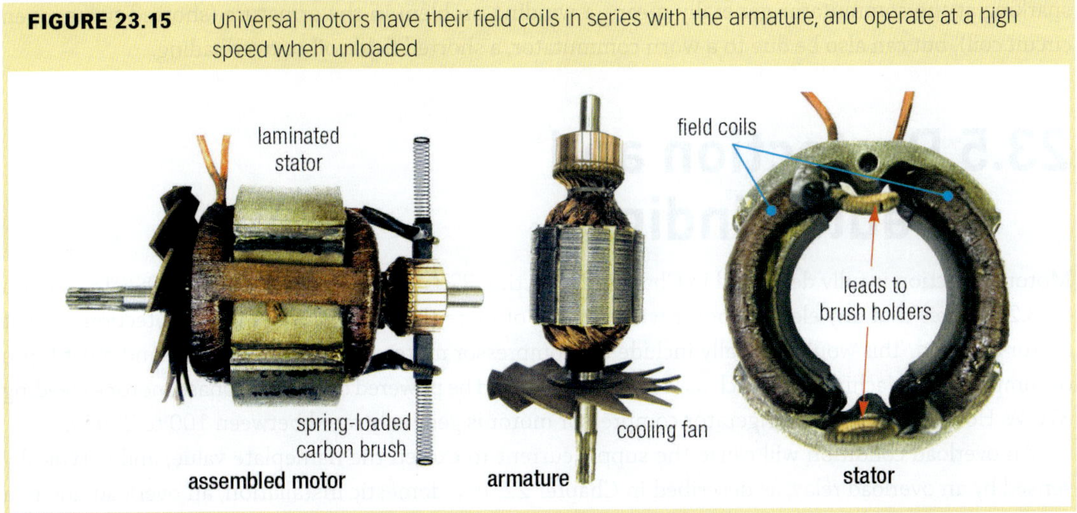

A universal motor generally has two field coils fitted to a laminated stator, as Figure 23.15 shows. Some types have a field winding distributed around the stator, as in an AC induction motor. The field coils are in series with the armature, so the motor behaves like a series DC motor (see Chapter 14). As a result, these motors have a high speed when unloaded, upwards of 20 000 RPM, requiring a suitably designed armature to withstand this type of stress.

The series connection causes the magnetic fields in the stator and the armature created by an alternating current to simultaneously reverse direction. This maintains the torque in the same direction, regardless of the polarity of the current. This can only be achieved with a wound armature, in which the commutator reverses the armature current to maintain the same direction of magnetic field interaction.

The circuit diagram and torque/speed curve of a universal motor are shown in Figure 23.16. When the motor is first started, it takes a high current, limited by the impedance of the armature and field coils. As it gains speed, the voltage induced in the armature opposes the applied voltage, so the motor current reduces. Because there is less current, the magnetic fields produce less torque. Therefore, the maximum torque is produced at start-up, which reduces significantly with speed.

FIGURE 23.16 Circuit diagram and torque speed curve for a universal motor

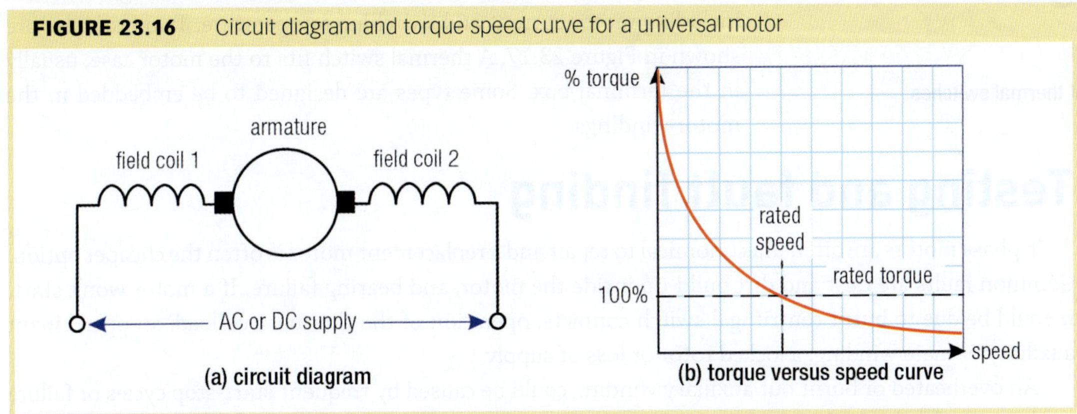

Universal motors have a relatively poor efficiency, and are generally limited to a power rating of 1 kW or less. There is usually sparking at the brushes, which creates radio interference and ozone. Their main advantage is a high power to weight ratio. A 1 kW universal motor is considerably smaller than most other motors of the same rating, making them suitable for portable tools and the like.

The direction of rotation is changed by reversing the current flow through either the field coils or the armature. In many motors, this is difficult to do, due to the construction of the motor. Some types of power tools have a motor fitted with a double-pole reversing switch to reverse the armature connections. These motors are designed to operate in either direction, where other types function best in one direction only.

The universal motor is far less robust than an induction motor. The armature has to withstand a high speed, and is often a cause of failure. The commutator can wear, and brushes will always wear. A common fault is brushes worn down to their copper connection which can then damage the commutator. Excessive sparking at the commutator generally means a winding problem in the armature (shorted turns, open circuit coil), but can also be due to a worn commutator, a shorted field coil or overloading.

23.5 Protection and Fault-Finding

Motor protection is fully described in Chapter 22, section 22.4. In summary, AS/NZ 3000:2018, section 4.13.2 requires that any electric motor with a rating of more than 370 W must have protection against motor overload. This would typically include the compressor motor in an air conditioner and most types of compressors. Machine tools such as lathes or drills could be powered by a single-phase motor exceeding 370 W. However, a typical refrigerator compressor motor is generally rated between 100 to 250 W.

An overload condition will cause the supply current to exceed the nameplate value, and is typically sensed by an overload relay, as described in Chapter 22. In a domestic installation, an overload due to a motor will trip a circuit breaker, although the rating of the circuit breaker is for the whole circuit, not that of the motor. A protection device for a single-phase motor with one supply conductor earthed (the neutral) must interrupt the active conductor.

FIGURE 23.17 Bi-metal thermal switches are specified by trip temperature and current rating. An auto reset switch will reset when the temperature has dropped by a specified amount.

Section 4.13.3 refers to protection against over-temperature, in which unattended shaded-pole motors with a rating of 480 VA or more must be fitted with an over-temperature device; this is similar for any unattended motor of all other types with a power rating greater than 240 VA.

Single-phase induction motors are generally limited in power to around five kilowatts, although higher rated motors are available. A method of over-temperature protection in single-phase motors is a bi-metal thermal switch. This type of sensor uses a snap action disc-shaped bi-metallic switch that reverses its curvature in the event of excessive heat. A replacement must match the specifications of the original switch, including motor current and switching temperature. Two examples are shown in Figure 23.17. A thermal switch fits to the motor case, usually in the terminal box. Some types are designed to be embedded in the motor windings.

Testing and fault finding

Split-phase motors are often uneconomical to repair and a replacement motor is often the cheaper option. Common faults are dust and dirt build-up inside the motor, and bearing failure. If a motor won't start, it could be due to burnt centrifugal switch contacts, operation of the thermal overload, an open circuit auxiliary or main winding, a locked rotor or loss of supply.

An overheated or burnt out auxiliary winding could be caused by frequent start-stop cycles or failure in the centrifugal switch or starting relay, which prevents the windings being disconnected from the supply. It could also be due to a low supply voltage or an overload that prevents the motor reaching sufficient speed to operate the start switch, therefore leaving the start winding connected.

Typical tests include measuring the resistance of the motor between its supply terminals and an insulation resistance test, as shown in Figure 23.18. The insulation resistance between the windings and the frame (earth) of the motor should be no less than one megohm. A more usual reading is infinity, or at least 20 MΩ. Older motors will usually have a lower insulation resistance. Winding resistance will be low, but never zero ohms.

As well as the faults that can occur in any induction motor, a capacitor motor can fail due to a faulty capacitor. Capacitor handling and testing are explained in Chapter 10, section 10.8. In summary, to check a capacitor, measure its capacitance and leakage resistance. It's important to replace a motor capacitor with the same type, same capacitance value and same voltage rating.

FIGURE 23.18 Tests include measuring the DC resistance of both windings and measuring the insulation resistance between windings and motor frame, which should be earthed

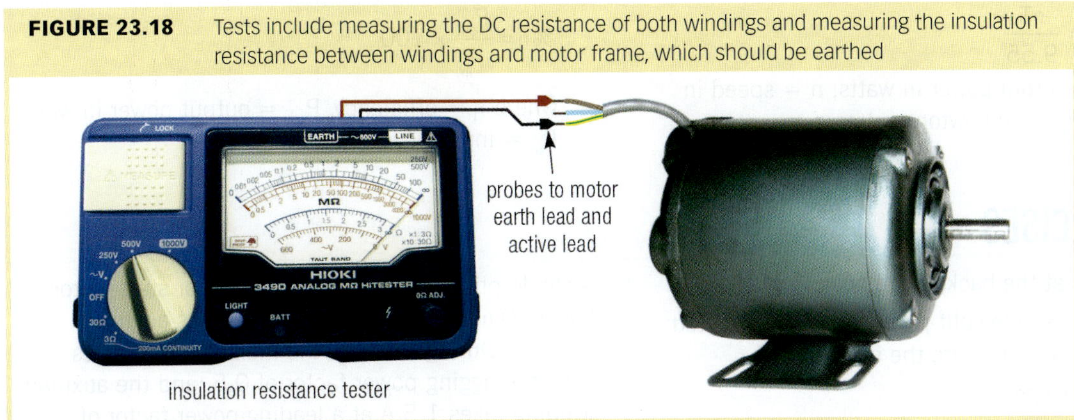

CHAPTER SUMMARY

- All single-phase induction motors require some means of creating a rotating magnetic field to start the rotor turning. Once rotating above a certain speed, the rotor and stator magnetic fields produce the rotational torque.
- All induction motors have a squirrel cage laminated steel rotor. The cage is usually made of aluminium.
- A split-phase induction motor has two windings, one with a higher resistance and lower inductance (auxiliary winding) than the other (main winding). This creates a phase shift between the winding currents, giving a rotating field that can start the motor turning. The auxiliary winding is disconnected from the supply when the motor reaches around 75 per cent of its rated speed.
- The auxiliary winding is disconnected from the supply by a centrifugal switch mechanism, or a current relay in series with the main winding.
- A capacitor start motor has a large value capacitor in series with the auxiliary winding that provides additional phase shift between the winding currents during starting, giving the motor an increased starting torque.
- A capacitor start, capacitor run motor (dual capacitor motor) has a large value capacitor in series with the auxiliary winding during starting, and a low value capacitor in series during running. This gives improved efficiency, torque, power factor and smoother running.
- A permanently split capacitor motor has two identical windings with a capacitor permanently in series with one of them. It does not have a start switch, it is easily reversed and runs smoothly and relatively quietly.
- A shaded-pole motor has a copper shading coil fitted to the trailing tips of the stator poles. It has limited starting torque, low efficiency and cannot be easily reversed.
- A universal motor runs on AC and DC, and has a wound armature in series with the field coils. The highest torque occurs at start-up (zero speed), it can reach dangerously high speeds when unloaded and features a high power-to-weight ratio.
- Tests on a motor to prove serviceability include winding resistance tests, confirming correct operation of the start switch mechanism or relay, and measuring the insulation resistance between windings and motor frame. This should not be lower than 1 MΩ.
- The following equations are used with single-phase motors:
  - $n_s = \frac{120f}{p}$
    where n_s = synchronous speed in RPM,
    f = supply frequency in hertz,
    p = number of poles
  - $s\% = \frac{n_s - n}{n_s} \times 100$
    where s% = percentage slip,
    n_s = synchronous speed, n = rotor speed
  - $f_r = \frac{s\% \times f}{100}$ where f_r = rotor frequency
  - $P_{in} = VI \cos \phi$ where P_{in} = true power taken by the motor (watts), V = supply voltage, I= motor current, $\cos \phi$ = power factor

- $P_{out} = \frac{2\pi nT}{60} = \frac{nT}{9.55}$

  where P_{out} = output power in watts, n = speed in RPM, T = torque in newton metres

- $\%\eta = \frac{P_{out}}{P_{in}} \times 100$

  where η = efficiency, P_{out} = output power in watts, P_{in} = input power in watts.

REVIEW EXERCISES

Check your answers at the back of the book.

1 A 230 V, 50 Hz six-pole split-phase motor has a rated speed of 960 RPM. Calculate the motor's:
   (a) synchronous speed
   (b) slip speed
   (c) percentage slip.

2 A split-phase motor operating from a 50 Hz supply has a percentage slip of 4.8%. What is the motor's rotor frequency?

3 A single-phase motor takes a current of 5 A at a lagging power factor of 0.82 from a 230 V 50 Hz supply. How much true power is the motor consuming?

4 A single-phase motor running at 1425 RPM takes 1050 W of true power when delivering a torque of 5.8 Nm. Calculate the motor's:
   (a) output power
   (b) efficiency.

5 A single-phase 50 Hz, 230 V capacitor start motor has a 20 μF capacitor in series with its auxiliary winding. During starting, the main winding takes 2 A at a lagging power factor of 0.5, and the auxiliary winding takes 1.5 A at a leading power factor of 0.866. Determine the:
   (a) phase difference between the currents in both windings
   (b) total current taken by the motor during starting
   (c) the voltage across the start capacitor during start-up.

ONLINE RESOURCES

COMPLETE WORKSHEET TWENTY-THREE

Check with your instructor for worksheets on this chapter.

CHAPTER 24

SYNCHRONOUS MACHINES

Synchronous machines run at a speed that locks to a frequency, typically 50 Hz. Synchronous motors come in various sizes, from a few watts to many megawatts. Alternators are also synchronous machines and when driven by a prime mover provide most of our electricity supply. These machines and their operating principles are described in this chapter.

CHAPTER OUTLINE

24.1 Alternators

An alternator is an AC machine that when coupled to a prime mover generates an alternating voltage. Example machines are shown in Figure 24.1. Some types are single-phase such as those used on building sites to provide AC power. These are usually driven by a petrol engine. Larger machines are driven by a diesel engine, and typically produce three-phase power.

FIGURE 24.1 Alternators come in a wide range of power ratings and output voltages

The largest alternators are those in thermal power stations where generating units are often rated at more than 500 MVA at an output voltage of around 20 kV. Alternators in hydro-electric power stations are smaller, and are rated at 100 MVA or less. Emergency supply alternators, powered by diesel engines, are always on standby in hospitals and wherever it's critical to maintain electrical power. These are power-rated to suit the situation.

On the lower end of the power scale is the automotive alternator, which is used in virtually every type of road vehicle. These are typically rated at less than 1 kVA, are usually three-phase and have diodes to convert the generated alternating current to direct current in order to charge the vehicle battery. A car alternator and typical small alternator stator windings are shown in Figure 24.2.

FIGURE 24.2 Automotive alternators are rated at less than 1 kVA, are usually 3-phase and are fitted with diodes to convert AC to DC

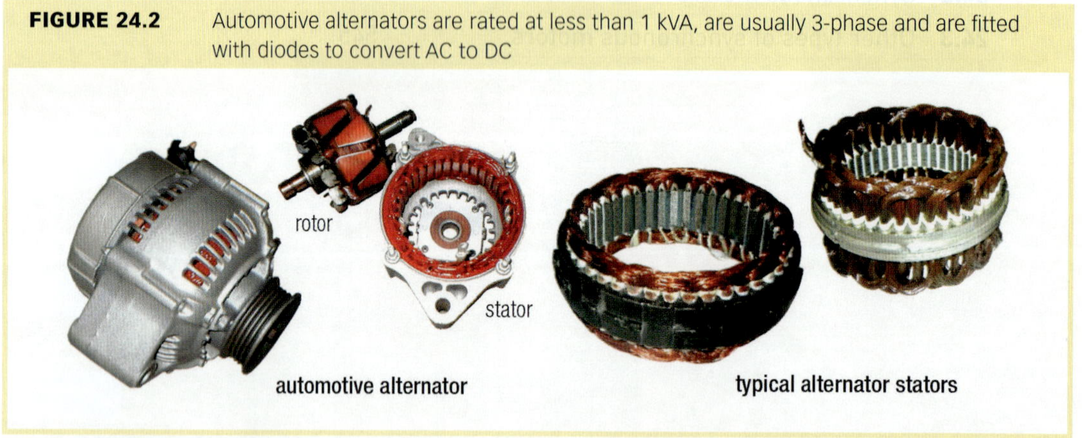

FYI

The rotor of an automotive alternator has a 'claw-pole' field construction which produces a multi-pole field from a single coil winding

The operating principles of an alternator have been mentioned in previous chapters. As the simplified diagram of a basic alternator in Figure 24.3 illustrates, an alternator has three main components: a magnetic field and a coil, in which one of them rotates relative to the other, and a set of slip rings. In Figure 24.3, the coil is rotating inside a field produced by permanent magnets.

FIGURE 24.3 A basic alternator produces a non-sinusoidal alternating voltage

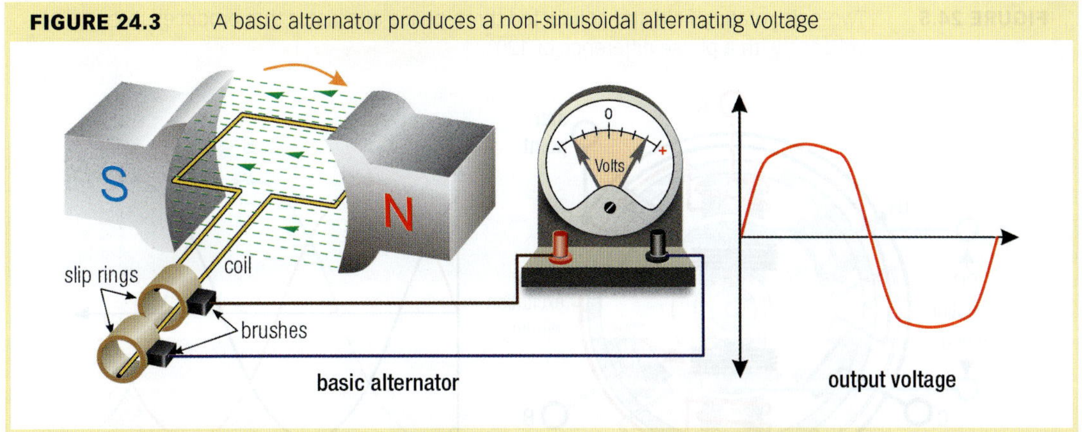

The voltage waveform in Figure 24.3, although alternating, is not a pure sinewave. A sinusoidal waveform is obtained by appropriate design of the alternator, such as specially shaped pole pieces. The simple alternator shown in Figure 24.3 has a coil rotating in a magnetic field and the voltage induced in the coil is connected to an external circuit with brushes and slip rings. A practical construction is shown in Figure 24.4, in which an armature rotates inside a magnetic field established by a DC current flowing in four field coils. The armature is wound to produce a three-phase, star-connected output via four slip rings.

FIGURE 24.4 A three-phase alternator with an armature rotating in a magnetic field created by a DC excitation current

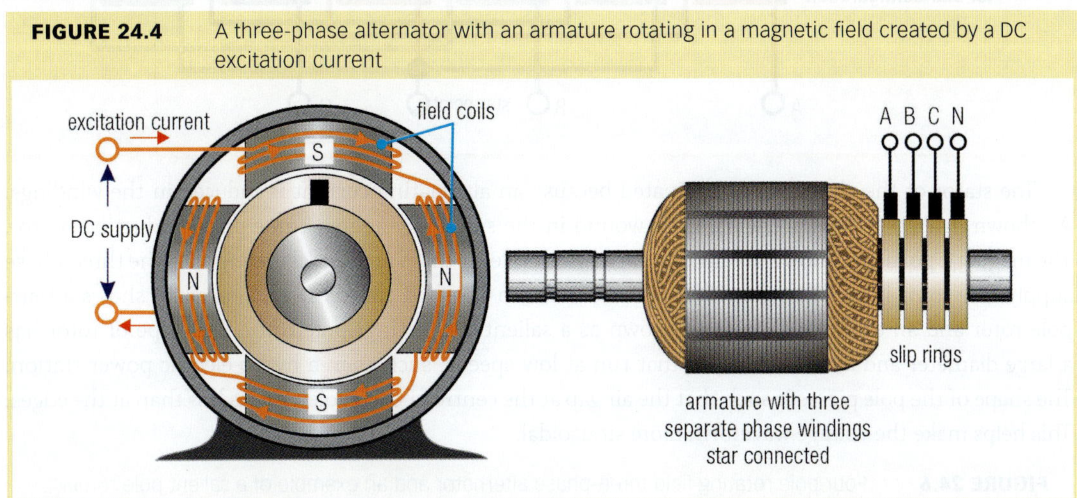

The magnetic field is developed by passing an excitation current through the field windings, which are wound to give the magnetic polarities shown. The armature coils are laid in slots in a laminated core, and the induced current is supplied to a three-phase load by brushes contacting the four slip rings.

However, this arrangement of a stationary magnetic field and rotating armature is only suited to small alternators. Instead, most alternators have a rotating magnetic field and the armature in Figure 24.4 is wound in the stator of the alternator. This design has only two slip rings, and the advantage is the slip rings are passing the excitation current instead of the much larger generated current.

Rotating field alternator

A simplified drawing of a rotating field three-phase alternator is shown in Figure 24.5. The rotor winding is connected via slip rings and brushes to a DC supply that supplies a field excitation current. In operation, this arrangement provides a rotating magnetic field. The armature winding of Figure 24.4 is wound on the stator in Figure 24.5. This winding is referred to as the armature, despite being wound in the stator.

FIGURE 24.5 Two-pole rotating field star-connected three-phase alternator that produces three voltages with a phase difference of 120°

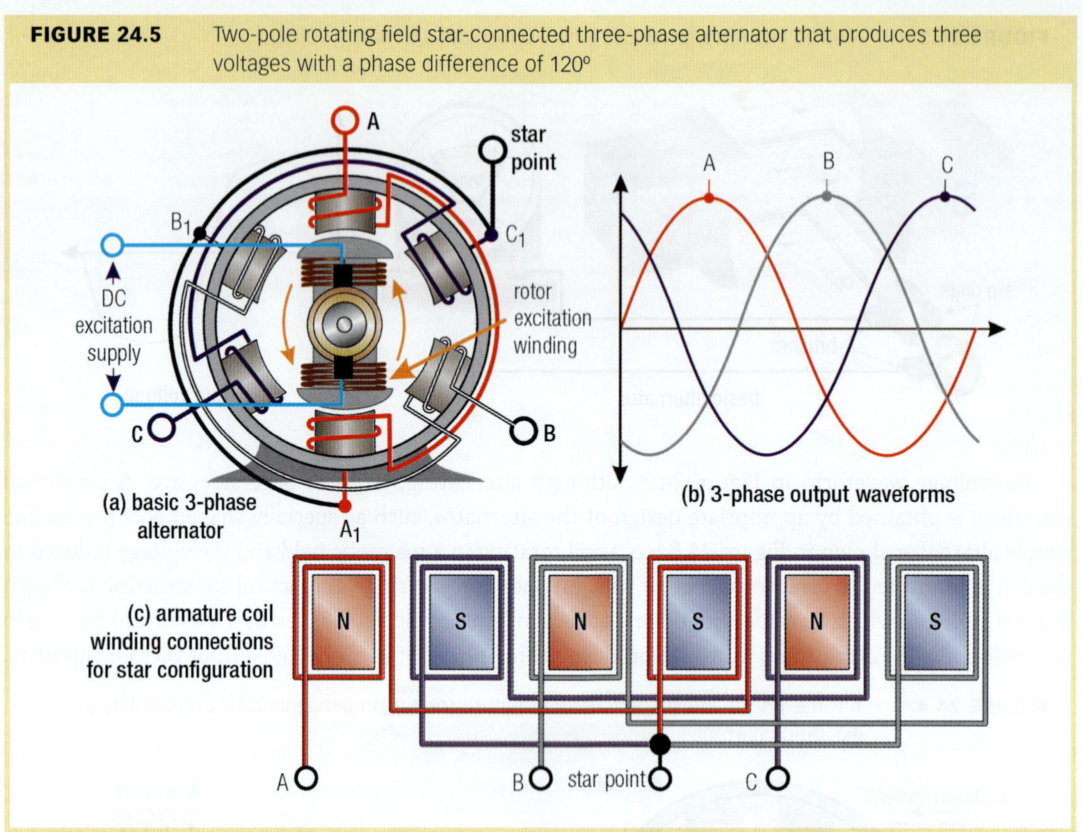

The stator of the alternator is laminated because an alternating current is induced in the windings. As shown in Figure 24.5(c), the stator is wound in the same way as in a three-phase induction motor. The magnetic polarities shown are for one part of the cycle, as these are ever changing with the three-phase supply. The rotor is usually solid steel, as the current in the field coils is DC. Figure 24.6 shows a four-pole rotor and an example of what is known as a salient (projecting) pole rotor. This type of rotor has a large diameter and suits alternators that run at low speeds, such as in a hydro-electric power station. The shape of the pole pieces is such that the air gap at the centre of each pole piece is less than at the edges. This helps make the voltage waveform more sinusoidal.

FIGURE 24.6 Four-pole rotating field three-phase alternator and an example of a salient pole rotor

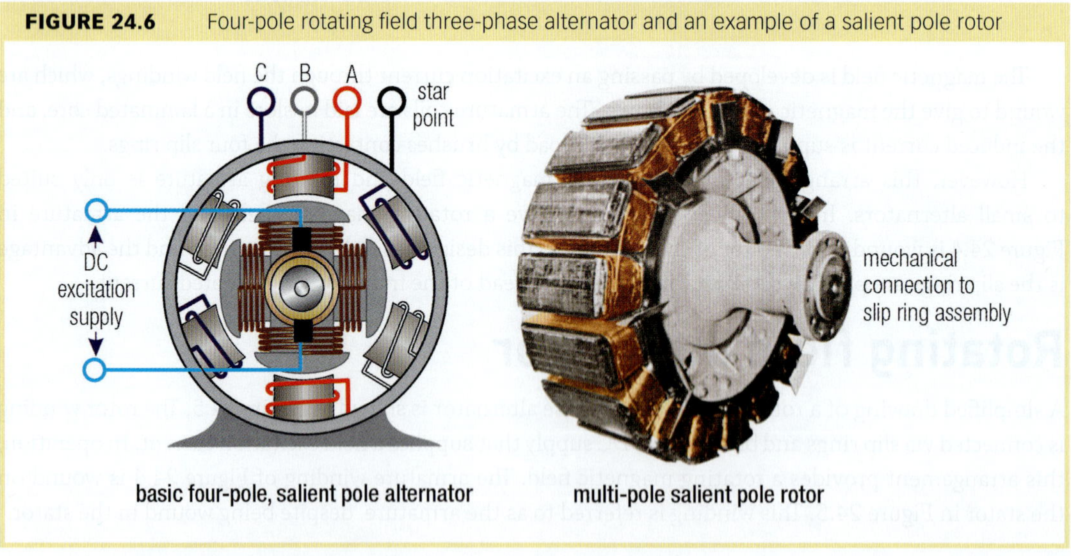

Machines that rotate at high speeds have a long cylindrical rotor. For example, a turbine driven alternator in a thermal power station rotates at 3000 RPM. The large diameter of a salient pole rotor would mean a very high peripheral speed, so diameter is traded for length. This type of rotor is also easier to balance, and makes less noise due to its smoother surface.

Frequency and number of poles

Alternators are designed to generate a specified voltage at a specified frequency. In Australia and New Zealand, this is typically 50 Hz, although units designed for higher frequencies include 'gen sets', such as that shown in Figure 24.7. This unit produces control tones for distribution over the power lines to control various functions within a power distribution network. It is an example of frequency conversion, where a 50 Hz motor drives a multi-pole alternator to produce a different frequency, in this case 1050 Hz.

FIGURE 24.7 Three-phase 50 Hz motor coupled to an alternator that produces a single-phase output voltage at a frequency of 1050 Hz

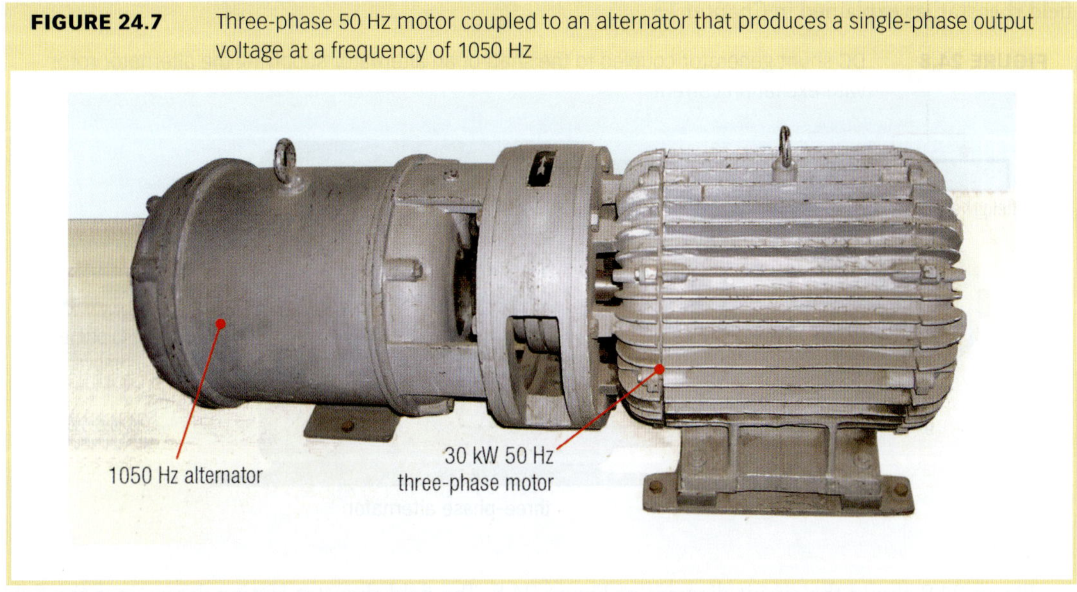

As in any AC machine, the relationship between frequency, rotational speed and number of poles for an alternator is found with the equation:

$$n_s = \frac{120f}{p}$$

where:

n_s = speed in RPM

f = frequency in hertz

p = number of poles.

The rotational speed of an alternator is determined by the type of prime mover. A steam turbine prime mover operates at high speed, typically 3000 RPM, while a diesel engine prime mover runs at a much lower speed. Rearranging the above equation in terms of number of poles (p) gives a means of determining how many poles an alternator should have to produce an output at a specified frequency for a given rotational speed. That is:

$$p = \frac{120f}{n_s}$$

From this equation, an alternator coupled to a steam turbine running at 3000 RPM can only have two poles to produce a 50 Hz output. It would have a cylindrical rotor, due to its high rotational speed. An alternator coupled to a diesel engine running at 750 RPM would need eight poles to produce a 50 Hz output and would have a salient pole rotor.

A critical aspect is maintaining a constant rotational speed, as any variations in speed mean a variation in the frequency of the output voltage. The correct speed produces the required frequency and is called the synchronous speed. For example, the synchronous speed of a 24-pole alternator producing a 50 Hz supply is 250 RPM. Any speed above or below is no longer the synchronous speed.

Excitation

An alternator needs a DC supply to provide excitation current to establish the alternator's magnetic field. The DC source can be a separate DC generator coupled to the shaft of the alternator or derived by rectifying an AC supply with diodes. The DC excitation current supplied to an alternator is controlled in some way. A method is to connect a DC shunt generator to the shaft of the alternator, as shown in Figure 24.8. The DC generator output voltage and therefore the excitation current is controlled with a field rheostat, as explained in Chapter 13.

FIGURE 24.8 DC shunt generator coupled to the shaft of an alternator supplying the alternator rotor with excitation current

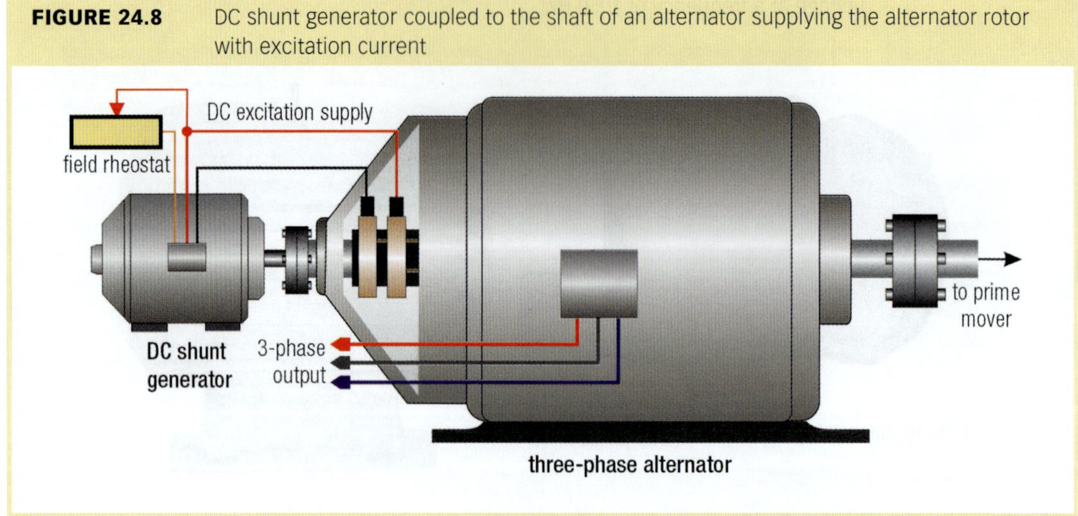

Figure 24.9 shows the circuit diagram of Figure 24.8. The field rheostat setting determines the field current in the DC shunt generator. Increasing this current by reducing the rheostat resistance raises the DC generator's output voltage. This increases the excitation current flowing in the alternator's rotor windings, which increases the output voltage of the alternator. The rheostat is usually replaced with an automatic voltage control system, which controls the exciter field current to maintain the alternator output voltage at a constant level.

FIGURE 24.9 Circuit diagram of a shunt wound exciter supplying excitation current to a three-phase alternator

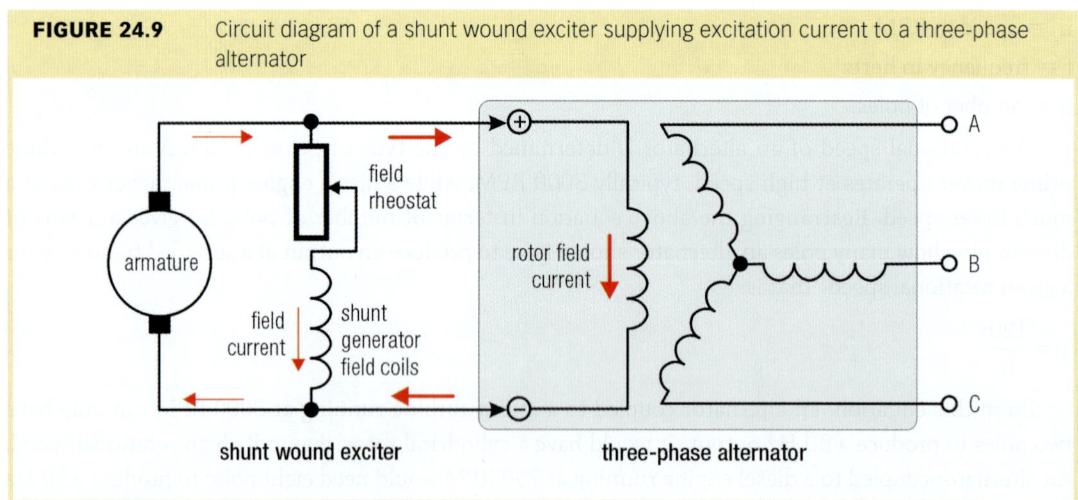

Brushless alternator

The brushless alternator derives its DC excitation current from another, smaller three-phase alternator attached to the same shaft as the main alternator. The DC excitation current is produced by rectifying the AC output of the exciting alternator with a set of six diodes (see Figure 24.10). The rotor of the exciter alternator and the diode set are mounted on the shaft of the main alternator, and rotate as an assembly. The output of the diodes (DC voltage) connects directly to the excitation windings on the rotor of the main alternator.

FIGURE 24.10 Circuit diagram of a brushless excitation system, in which solid state diodes convert the output of a three-phase rotating armature alternator to provide excitation current for the main alternator

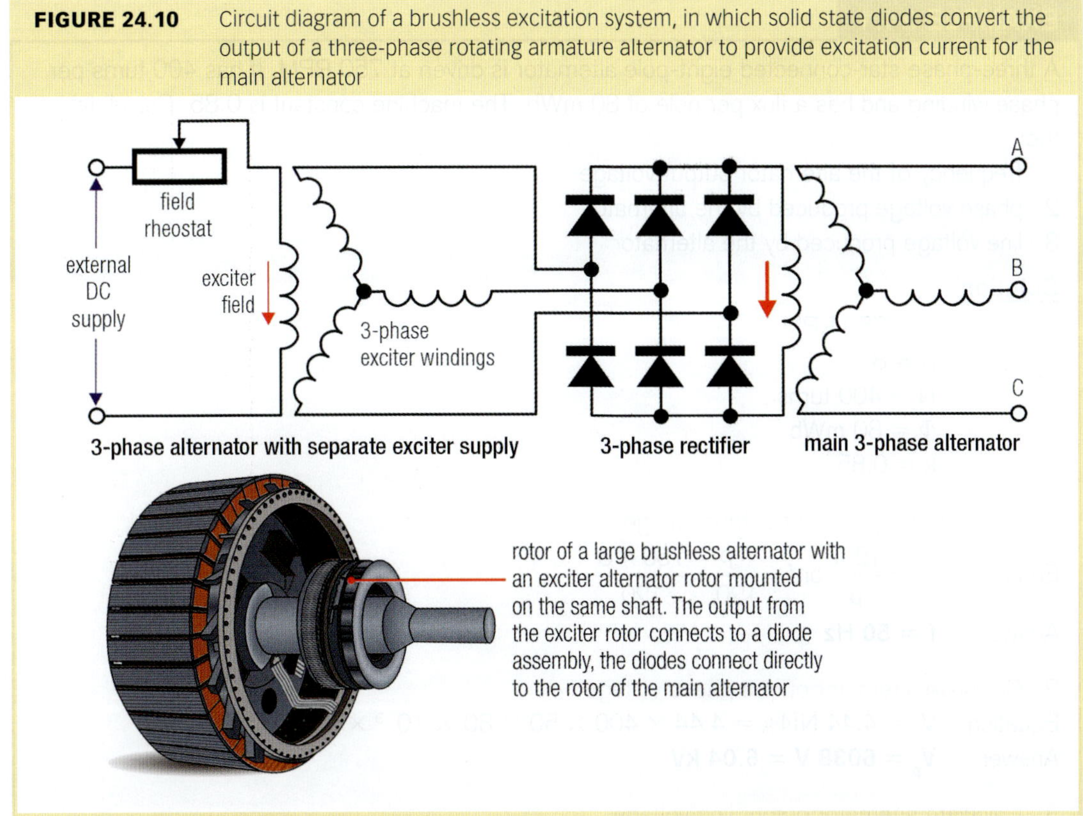

The three-phase output voltage of the exciter is controlled with a field rheostat that determines the current in its field windings. The higher the exciter field current, the higher the exciter's output voltage and therefore the greater the excitation current in the rotor of the main alternator, thereby increasing its output voltage. The advantage of this arrangement is there are no slip rings or brushes, giving less maintenance and higher reliability.

Permanent magnet alternators

The availability of high field strength rare earth magnets has meant the development of alternators in which the rotor is magnetised by permanent magnets, instead of field coils. They feature simpler construction, lower cost and less maintenance as there are no slip rings or brushes. A problem is regulating the alternator's output voltage, as the magnetic field strength is fixed. They are used in small wind turbines and are limited to a few hundred kilowatts.

Alternator output voltage

The phase voltage developed by an alternator can be found with an equation similar to that used to find the voltage induced in transformer windings. It requires knowing a machine constant which takes into account how the alternator is wound and other factors:

$$V_p = 4.44\, Nf\,\Phi k$$

where:

V_p = phase voltage in volts produced by the alternator

N = number of armature turns per phase

f = frequency of the generated voltage in hertz

Φ = magnetic flux per pole in webers

k = machine constant.

EXAMPLE 24.1

A three-phase star-connected eight-pole alternator is driven at 750 RPM. It has 400 turns per phase winding and has a flux per pole of 80 mWb. The machine constant is 0.85. Calculate the:

1 frequency of the alternator output voltage
2 phase voltage produced by the alternator
3 line voltage produced by the alternator

Solution

Values $n = 750$ RPM
$p = 8$
$N = 400$ turns
$\Phi = 80$ mWb
$k = 0.85$

1 Calculate frequency of output voltage

Equation $n_s = \frac{120f}{p}$ or $f = \frac{n_s p}{120} = \frac{750 \times 8}{120}$

Answer **$f = 50$ Hz**

2 Calculate alternator output phase voltage

Equation $V_p = 4.44\ Nf\Phi k = 4.44 \times 400 \times 50 \times 80 \times 10^{-3} \times 0.85$

Answer **$V_p = 6038$ V = 6.04 kV**

3 Calculate alternator output line voltage

Equation $V_L = \sqrt{3} \times V_p = \sqrt{3} \times 6038$

Answer **$V_L = 10\,458$ V = 10.46 kV**

FYI

The term 'armature' refers to that part of an alternator that generates the output voltage, and is typically the stationary part (stator)

If the flux per pole in Example 24.1 is halved to 40 mWb, the alternator output voltage will drop by half because flux per pole is the only variable in the equation. If the alternator is slowed, the voltage will remain the same but the frequency will be lower. Therefore, this shows that the output voltage for a given alternator can be controlled by the excitation current in the field winding.

Open-circuit characteristics

The equivalent circuit of an alternator can be simplified to an AC generator in series with a resistor and an inductor. The resistor represents the winding resistance, iron and copper losses, and the inductance represents the alternator's winding inductance and magnetic leakage. Figure 24.11 shows the equivalent circuit for a single-phase alternator, or one phase of a three-phase alternator.

FIGURE 24.11 Equivalent circuit of a single-phase alternator

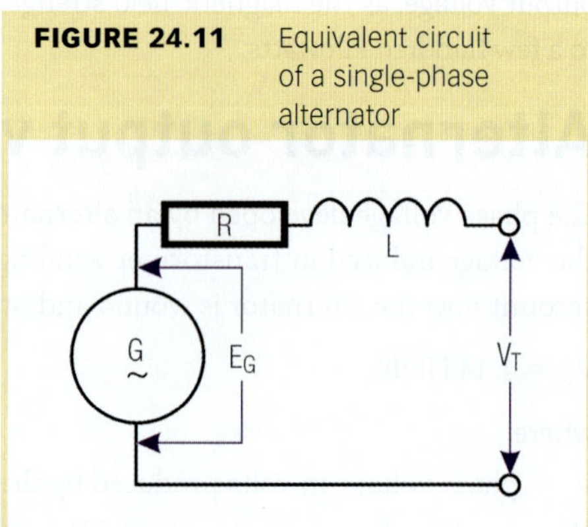

Figure 24.11 shows that if there is no load connected to the alternator, its terminal voltage will equal the EMF generated in the armature (E_G). The value of this voltage depends on the magnetic field created by the rotor field coils, as explained above.

The open-circuit characteristic curve of an alternator shows the relationship between its terminal voltage and field current (see Figure 24.12). To obtain this curve, the alternator is driven at its synchronous speed while the field current is increased from zero to its maximum value. The terminal voltage of the alternator is measured for each value of field current. Because the alternator is not supplying a load, the terminal voltage and generated EMF are the same.

FIGURE 24.12 Open-circuit characteristic curve of an alternator

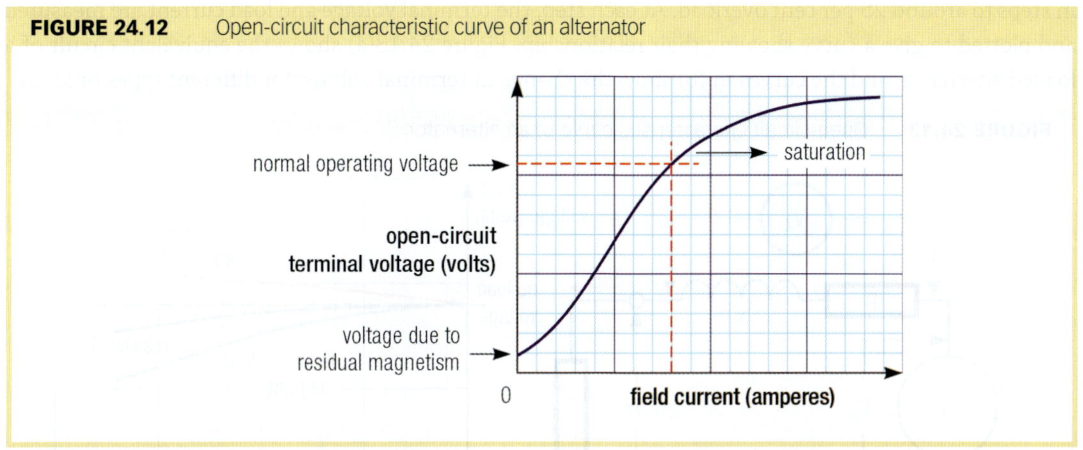

The curve in Figure 24.12 is similar to that for a DC generator, in that it is also the machine's magnetisation curve. As in a DC generator, the straight or linear part of the curve shows that the terminal voltage of the alternator is directly proportional to the field current. Above a certain value of field current, the terminal voltage no longer rises in a linear way, indicating the field poles are becoming magnetically saturated. As in a DC generator, when the field current is zero, there is a small generated voltage due to the residual magnetism of the field poles.

Alternator rating

An alternator is rated by its:

- terminal voltage at full load current
- apparent power capability (VA rating)
- frequency, which determines its rotational speed
- number of phases.

Because the manufacturer has no control over the type of load the alternator will be connected to, it is given a VA rating, which is used to determine the maximum current the alternator can deliver, regardless of power factor. From previous chapters, the equations to find the apparent power rating of an alternator are:

$S = VI$ (single-phase alternator)

$S = \sqrt{3}\, V_L I_L$ (three-phase alternator).

EXAMPLE 24.2

What is the full-load current of a three-phase 400 V alternator rated at 80 kVA?

Solution

Values $V_L = 400$ V

$S = 80$ kVA

Equation $S = \sqrt{3}\ V_L I_L$ so $I_L = \dfrac{S}{\sqrt{3} \times V_L} = \dfrac{80 \times 10^3}{1.732 \times 400}$

Answer $\mathbf{I_L = 115.47\ A}$

Loaded characteristics

The effect of a load on the terminal voltage of an alternator is found by measuring its terminal voltage for different values of load current. The field current and speed are kept constant, and the load is increased in steps to around 25 per cent overload. At each step, the terminal voltage and load current are measured and plotted to give a curve showing their relationship. Figure 24.13(a) shows the equivalent circuit of a loaded alternator, and the curves in (b) show the change in terminal voltage for different types of loads.

FIGURE 24.13 Open-circuit characteristic curve of an alternator

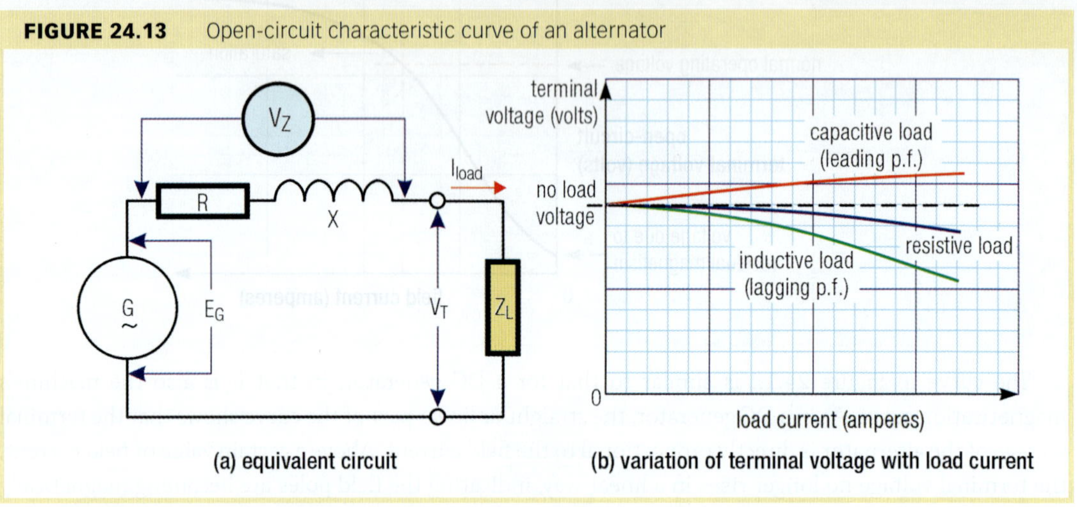

FYI

Armature reaction is explained in Chapter 14. Armature leakage reactance is additional reactance caused by flux leakage.

The curves in Figure 24.13(b) show that:

- for a resistive load (unity power factor), the terminal voltage falls as the load current is increased, following a slightly drooping curve
- an inductive load (lagging power factor) causes the terminal voltage to fall more dramatically
- a capacitive load (leading power factor) causes the terminal voltage to rise.

There are two factors that contribute to the change in terminal voltage: armature leakage reactance (X_L) and the effect of armature reaction. These two factors can be represented by a reactance X, along with the relatively small resistance of the armature windings and other in-phase losses, giving the equivalent circuit shown in Figure 24.13(a). The effect of the impedance of an alternator with different types of loads and therefore power factors is shown by the phasor diagrams in Figure 24.14.

FIGURE 24.14 Phasor diagrams of an alternator for loads taking the same current, at the same terminal voltage, but at different power factors

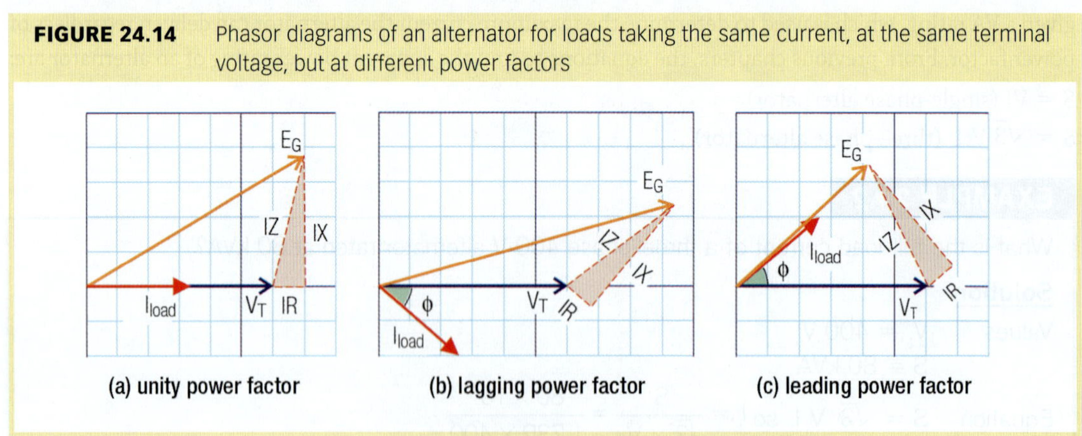

Resistive load (unity power factor)

Phasor diagram (a) is for a resistive load, in which the load current is in phase with the terminal voltage. The voltage drop across the alternator impedance R and X in Figure 24.13(a) has two components, the resistive voltage drop IR which is in phase with the load current, and the voltage drop IX across the reactance, which is 90° out of phase. The triangle shows these two voltage drops and their phasor addition (IZ). The alternator's generated voltage (E_G) is the phasor sum of the terminal voltage (V_T) and voltage drop IZ. As the diagram shows, the generated voltage is higher than the terminal voltage because of the voltage drop across R and X.

As explained in Chapter 13, armature reaction opposes the rotational force being applied by the prime mover and has the effect of changing how the magnetic flux lines are distributed through a machine's magnetic circuit. In an alternator, for a resistive load current, armature reaction distorts the magnetic field, requiring more power from the prime mover to overcome the additional load. However, the distorted field remains constant, and its effect on the terminal voltage is minimal. This is also shown by the curve for a resistive load in Figure 24.13(b).

Inductive load (lagging power factor)

A lagging load current gives the phasor diagram in Figure 24.14(b). The load current is the same as for the resistive load, so the only change is the phase difference between the terminal voltage and the load current. As before, the triangle representing the voltage drops across R and X has the IR component in phase with the load current, which tilts the triangle clockwise. Again, the generated voltage (E_G) is the phasor sum of the terminal voltage (V_T) and voltage drop IZ.

This shows that to produce the same full load terminal voltage for an inductive load requires a higher generated voltage than for a resistive load. As shown in Figure 14.13(b), an inductive load will therefore cause the terminal voltage of an alternator to drop by more than a resistive load taking the same load current.

Capacitive load (leading power factor)

The phasor diagram in Figure 24.14(c) shows the effect of a leading power factor. Using the same values of load current and full load terminal voltage, the triangle is now rotated anticlockwise. The phasor addition of V_T and IZ now gives a generated voltage (E_G) that is *less* than the terminal voltage V_T. As shown in Figure 24.13(b), the terminal voltage therefore *rises* when the load current leads the terminal voltage. This is because a leading load current tends to strengthen the magnetic field, which causes the generated voltage and therefore the terminal voltage to rise.

Voltage regulation

The terminal voltage of an alternator is rated at its full-load value. The regulation of an alternator can be defined as the percentage rise in terminal voltage when its full rated load is removed. This assumes the excitation current was adjusted to give rated output voltage at full load. The equation to find the percentage voltage regulation of an alternator is:

$$\text{voltage regulation \%} \ \frac{V_{NL} - V_{FL}}{V_{FL}} \times 100$$

where:

V_{NL} = no-load voltage

V_{FL} = full-load voltage.

The voltage regulation of an alternator depends on the design of the machine and the power factor of the load. The curves in Figure 24.15 show how the terminal voltage changes for loads of different power factors. These curves assume the excitation current was adjusted at the full-load current for each type of load. Notice that for a load with a leading power factor (capacitive load), the terminal voltage drops when the load is removed.

FIGURE 24.15 Change in alternator terminal voltage when loads of different power factors are removed

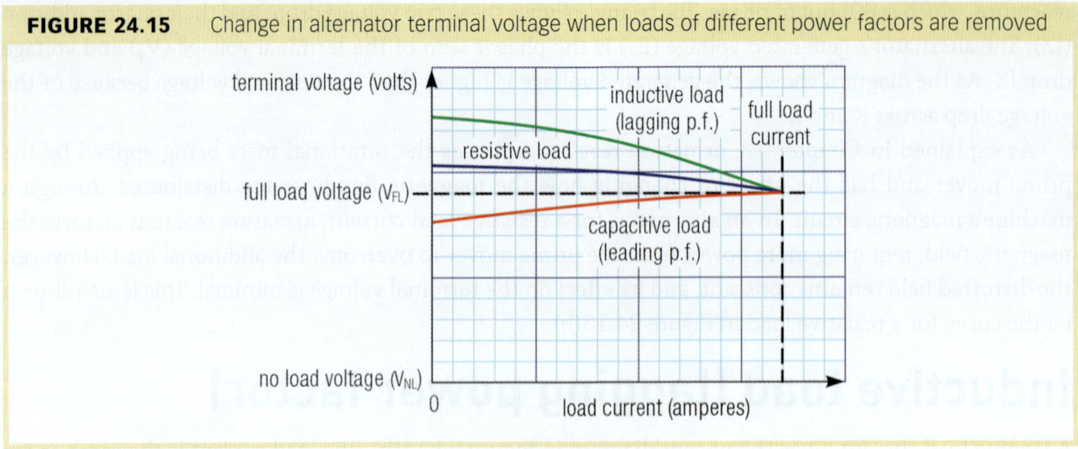

As the curves in Figure 24.15 show, the type of load has a significant effect on the unloaded terminal voltage when that load is removed. For this reason, manufacturers often include the power factor when stating the voltage regulation of an alternator.

EXAMPLE 24.3

What is the percentage voltage regulation of an alternator rated at 600 V, if its terminal voltage is 650 V when the load is removed?

Solution

Values $V_{FL} = 600$ V
$V_{NL} = 650$ V

Equation voltage regulation $\% = \frac{V_{NL} - V_{FL}}{V_{FL}} \times 100 = \frac{650 - 600}{600} \times 100$

Answer **voltage regulation = 8.33%**

If the load in Example 24.3 has a different power factor, the voltage regulation will be different from the value found. Practical values are around 10 per cent for a unity power factor load, while for a load with a power factor of 0.8 lagging, the voltage regulation could be as high as 35 per cent.

Alternators are often designed to have a relatively high impedance because this limits the short-circuit current in the event of a fault. The terminal voltage of the alternator is maintained under normal load conditions with an automatic voltage regulator that changes the excitation current with load.

Efficiency

Like all machines, the efficiency of an alternator is the ratio of the true power output and the true power input. That is:

$$\eta\% = \frac{P_{out}}{P_{in}} \times 100$$

where:

$\eta\%$ = efficiency expressed as a percentage

P_{out} = output power in watts

P_{in} = input power to the machine when the machine is producing its rated output power.

There are a number of losses associated with an alternator. These include rotational losses such as bearing friction, windage and brush friction at the rotor slip rings, plus the usual iron and copper losses associated with AC machines. There are also losses due to the exciter, which is powered by the prime mover driving the alternator. In large alternators, there are also stray load losses due to the non-uniform distribution of current in the conductors. For alternators rated above 200 kVA, this loss is generally around 1 per cent of the output of the alternator.

EXAMPLE 24.4

A 5 MVA, 3.3 kV three-phase alternator when operating at full load is supplying current with a lagging power factor of 0.9. It has friction and windage losses of 40 kW, an iron loss of 75 kW, a copper loss of 60 kW, an excitation loss of 8 kW and stray losses of 50 kW. Calculate the alternator's:

1 full-load current
2 total losses
3 input power
4 percentage efficiency

Solution

Values $S = 5$ MVA
$V_L = 3.3$ kV
$\lambda = 0.9$ lag

1 Calculate the full-load current

Equation $S = \sqrt{3}\ V_L I_L$ so $I_L = \dfrac{S}{\sqrt{3} \times V_L} = \dfrac{5 \times 10^6}{1.732 \times 3.3 \times 10^3}$

Answer $\mathbf{I_L = 874.8\ A}$

2 Total losses = 40 kW + 75 kW + 60 kW + 8 kW + 50 kW

Answer **total losses = 233 kW**

3 Calculate input power

Equation $P_{in} = P_{out} + \text{losses}$
$P_{out} = S \times \lambda = 5 \times 10^6 \times 0.9 = 4.5$ MW
$P_{in} = 4.5$ MW + 233 kW = 4.5 MW + 0.233 MW

Answer $\mathbf{P_{in} = 4.733\ MW}$

4 Calculate efficiency

Equation $\eta\% = \dfrac{P_{out}}{P_{in}} \times 100 = \dfrac{4.5\ \text{MW}}{4.733\ \text{MW}} \times 100$

Answer **efficiency = 95.1%**

Alternators in parallel

Alternators are connected in parallel under two scenarios: connecting to the power grid as in a power station, or connecting to a power supply system that is isolated from the power grid. Bringing an alternator on line in a power station is an automated process. Paralleling an alternator with an isolated system might be done manually. In this case, the reason could be to increase power to a load or to shift the load to the incoming machine without interrupting supply. When connecting an alternator in parallel with any system, the following conditions need to be met:

- Both line voltages must be equal and sinusoidal in shape. The voltage of the alternator being connected is controlled by adjusting its excitation current.
- The frequencies must be identical. This is controlled by adjusting the governed speed of the prime mover driving the alternator being connected.
- Both voltages must have the same phase sequence and be in phase with each other. Phase sequence is the order in which each phase reaches its maximum value, which can be A-B-C or A-C-B.

There are several ways of synchronising an alternator to another supply system. Providing the phase sequence is correct, an accurate method that responds to phase and frequency difference is with an instrument called a synchroscope. This device has a pointer that rotates over a dial face and is connected as shown in Figure 24.16.

FIGURE 24.16 A synchroscope indicates the difference in frequency and phase relationship between two three-phase supply systems

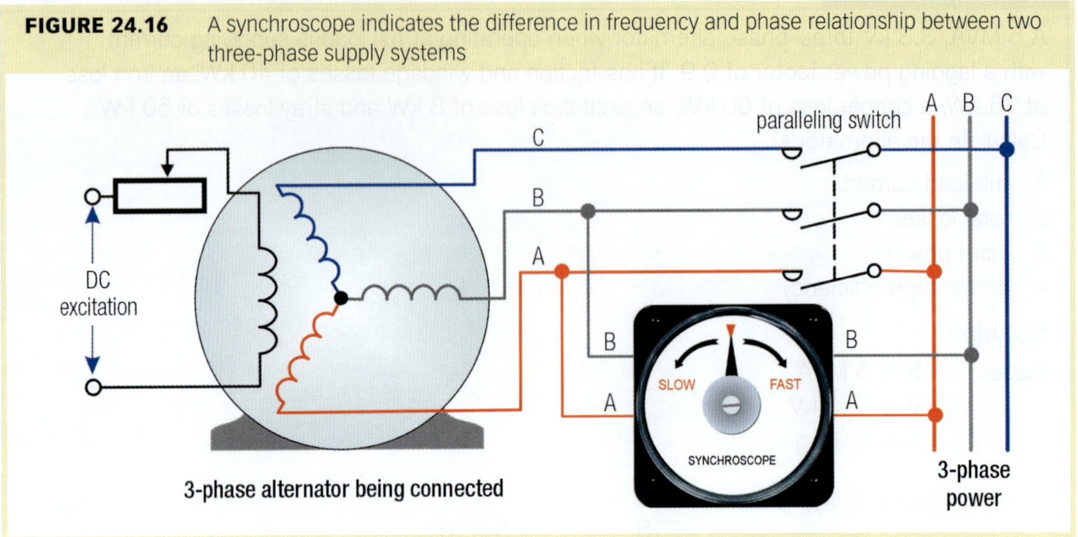

The moving pointer of a synchroscope rotates in either direction to indicate a difference between the alternator and the system frequency and phase difference. The faster the pointer rotates, the greater the difference. A slowly moving pointer rotating clockwise indicates the alternator frequency is slightly higher than the system frequency. By adjusting the governed speed of the alternator's prime mover, it can be brought into synchronism and connected to the system when the pointer is stationary and pointing vertically.

A less accurate method is with the 'three lamps dark' connection shown in Figure 24.17. This method is not suited to paralleling large alternators but, unlike a synchroscope, has the advantage of checking phase sequence. Lamps must be rated at twice the system voltage. If the phase sequence is correct, the three lamps will glow and darken at the same time. If the phase sequence is incorrect, the lamps will glow and darken in a sequence.

FIGURE 24.17 If all lamps glow and darken together, the phase sequence is the same between the alternator and the system

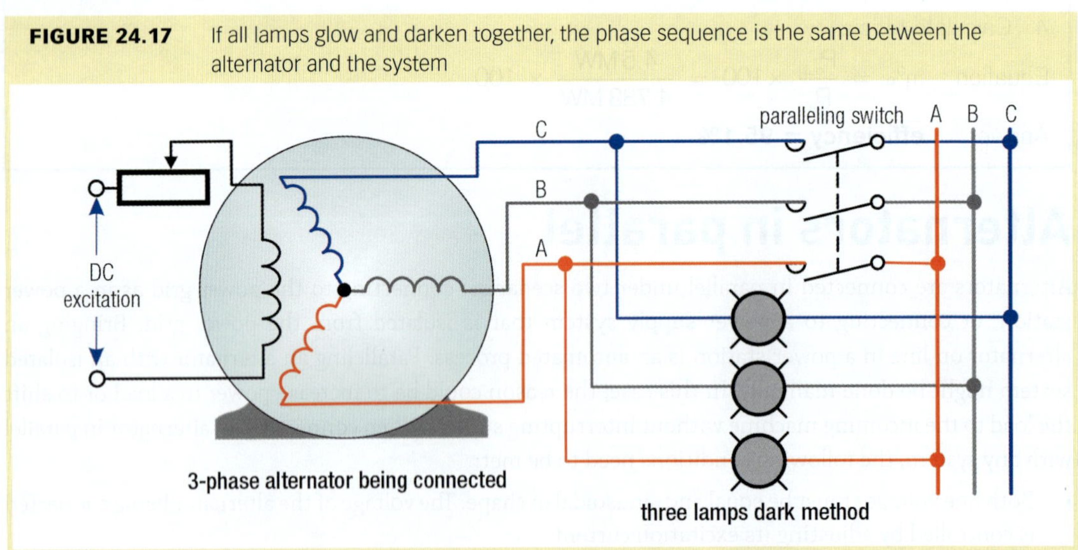

As the alternator is brought up to synchronous speed, the lamps will flicker at a rate that indicates the difference between its frequency and the system frequency. When all lamps are dark, both frequencies are the same and in phase with each other, allowing the alternator to be paralleled with the system. A limitation is that although lamps may look dark, there could still be a voltage difference across them, which is why this method only suits small alternators.

A more accurate method is the 'two bright, one lamp dark' connection, shown in Figure 24.18. The difference is that two of the lamps are connected between different phases, and one is connected between the same phase. A limitation is that this connection does not indicate an incorrect phase sequence, which can be checked with the 'three lamps dark' connection.

FIGURE 24.18 If L1 is dark and L2 and L3 are glowing, the alternator and the system are synchronised

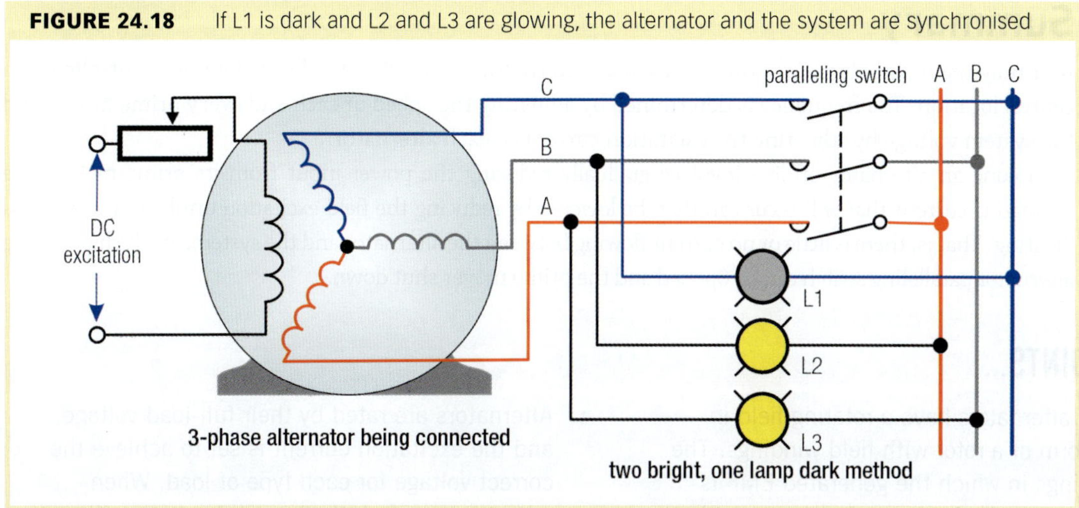

If the frequency of the incoming alternator is below that of the system, the lamps will glow in a sequence of L3-L2-L1, indicating the alternator speed is too slow. If the alternator is running too fast, the lamps will glow in the reverse sequence of L1-L2-L3. Arranging the lamps in a triangle then gives an indication rather like that of a synchroscope. The best synchronism is shown when L1 is dark and L2 and L3 are glowing brightly.

Load sharing

When an alternator is first synchronised with another alternator or power system, it will not be taking any of the load and may even be drawing power from the system (motoring). The voltage and frequency of both will be the same, as they are connected in parallel. Assuming two paralleled alternators of the same size, adjustments to one alternator will affect the operation of the other.

Torque-speed adjustment

To increase the load taken by the just-connected alternator requires adjusting the speed of its prime mover, which in effect increases the torque applied to the alternator. However, changing the speed of one alternator only will cause it to change the system frequency. The aim is to make the adjustment without changing the system frequency, which requires adjusting the applied torque of both alternators together. That is, while adjusting the governor controlling the speed of the just-connected alternator to increase its share of the load, the governor controlling the speed of the other alternator must be adjusted so its share of the load is decreased. Changing the system frequency requires adjusting both governors in the same direction.

Excitation adjustment

When an alternator is paralleled with a three-phase power system, it cannot cause the system voltage to increase unless it is providing most of the power. The system voltage is changed by adjusting the excitation current on both alternators. Increasing the excitation current of one alternator causes no appreciable change in the true power it is delivering; instead it produces a lagging reactive current that creates a circulating current between the two alternators. The correct amount of field excitation for alternators operating in parallel is the excitation each alternator would need if it were carrying its load alone at the same voltage and frequency.

Summary

In a fully regulated electrical power system all alternators connected to the system are controlled at a central location. The frequency is determined by adjusting the speed of each and every prime mover and the system voltage by adjusting the excitation current of each alternator.

Taking an alternator off-line involves gradually reducing the power input from its prime mover. The circulating current that will occur can then be lowered by reducing the field excitation until the alternator is 'floating'. That is, there is little or no current flowing between the alternator and the system, at which point the alternator paralleling switch can be opened and the prime mover shut down.

KEY POINTS...

- Most alternators have a rotating field in the form of a rotor with field windings. The windings in which the generated EMF is induced are in the form of coils wound in the stator. This winding is called the armature winding.
- Alternators are designed to produce a sinusoidal voltage waveform at a rated voltage and frequency. They can be single-phase or multi-phase.
- The magnetic field in a rotating field alternator is produced by passing a DC excitation current through the rotor windings. The terminal voltage of an alternator is determined by the value of the excitation current.
- The DC excitation source can be supplied to the rotor slip rings from a DC generator coupled to the alternator shaft. The brushless alternator has a second alternator and a three-phase rectifier built onto the shaft. This removes the need for slip rings.
- Alternators are rated by their full-load voltage, and the excitation current is set to achieve the correct voltage for each type of load. When a load is removed, the terminal voltage will either rise or fall, depending on the type of load.
- If the load is resistive (unity power factor) or inductive (lagging power factor), the terminal voltage will rise when the load is removed. It will rise more for an inductive load than a resistive load. If the load is capacitive (leading power factor), it will fall.
- An alternator can be connected in parallel with another alternator or power bus bars only when both are in synchronism in terms of frequency and phase difference.
- When an alternator is operating in parallel with another power source, it will run at the synchronous speed determined by the power source, either as a motor or as a generator. It will run as an alternator when its prime mover is supplying sufficient input power.

24.2 Synchronous Motors

As previously explained, under the right conditions, an alternator will run as a motor. Furthermore, it will run at a synchronous speed set by the frequency of its supply. Therefore, an alternator and a synchronous motor have a similar design and construction.

The two main categories of synchronous motors are those that need an excitation source and those that don't need external excitation, called non-excited motors. These motors have a permanent magnet field and are described at the end of this chapter.

FIGURE 24.19 The ship propulsion motor has a similar construction to an alternator. The smaller motors have a permanent magnet rotor.

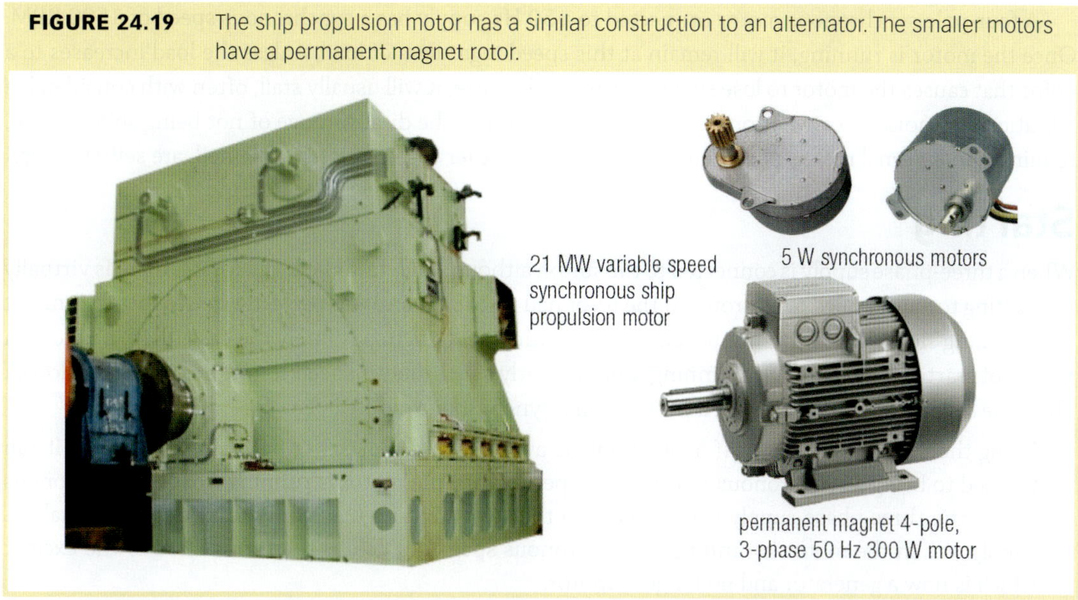

Large synchronous motors require external excitation in the same way as an alternator and, as shown in Figure 24.19, are used in high power applications. They have a higher efficiency than induction motors and, as later explained, can provide power factor correction.

Their construction is similar to an alternator, with modifications to improve their characteristics as a motor. These include the length of the air gap and number of field winding turns. They usually have a salient pole rotor, which limits their speed to 1500 RPM; higher speed motors have a cylindrical rotor. The stator is wound in the same way as in a three-phase induction motor. The general arrangement of a three-phase synchronous motor is shown in Figure 24.20, in which the stator winding is supplied from a three-phase power source and the rotor windings from an external DC source.

FIGURE 24.20 Synchronous motor with a separate DC exciter

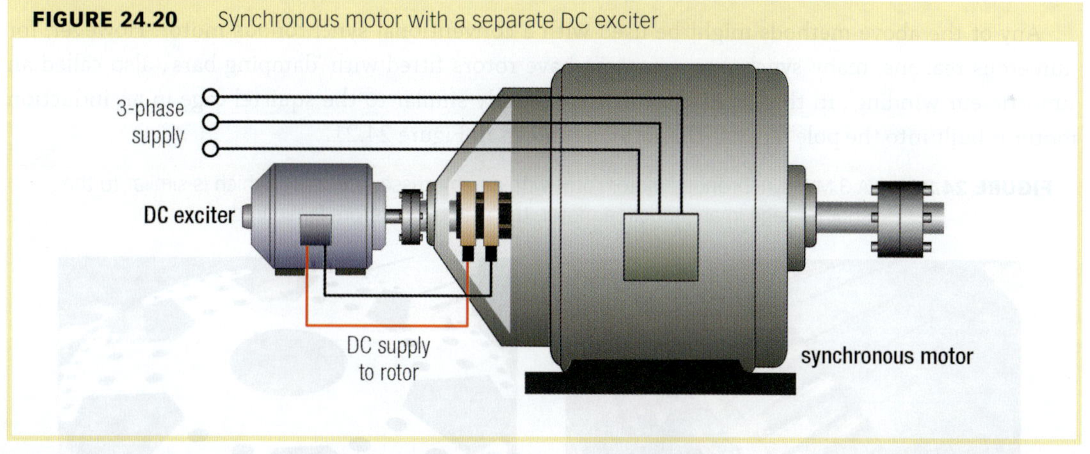

Operating principle

When a three-phase supply is connected to the stator of a synchronous motor, a rotating field is set up in the stator. The field travels at synchronous speed, which is found with the usual equation:

$$n_s = \frac{120f}{p}$$

where:

n_s = synchronous speed in RPM

f = supply frequency in hertz

p = number of poles.

A four-pole synchronous motor connected to a 50 Hz supply has a synchronous speed of 1500 RPM. Once the motor is running, it will remain at this speed regardless of load, unless the load increases to a point that causes the motor to lose synchronism. In this case, it will usually stall, often with considerable vibration and noise. A conventional synchronous motor has the disadvantage of not being self-starting, requiring an external means of bringing it up to speed. Other types, as later explained, are self-starting.

Starting

When a three-phase supply is connected to the stator, although the stator field will rotate, there is virtually no starting torque applied to the rotor. If the rotor had no inertia and was perfectly free to rotate, it might start turning slowly in an attempt to follow the rotating field. As this is rarely the case, there has to be a means of starting the motor and running it up to nearly synchronous speed so torque can be developed. There are several ways used to start a conventional synchronous motor.

1 Using the exciter as a motor. If a DC supply is available and the exciter has sufficient power, it can be used to bring a synchronous motor up to speed. Once the motor is running close to synchronous speed, the three-phase supply is connected to the stator and its field connected to the external DC supply. When the motor is running at synchronous speed, its field is then connected to the exciter, which is now a generator and no longer a motor.

2 Using a pony motor. This arrangement has a small three-phase induction motor, called a pony motor, coupled to the synchronous motor shaft. It will have two less poles than the synchronous motor, so it can reach the required speed. Once up to speed, AC power and DC excitation can be applied to the synchronous motor.

3 Using a variable frequency drive (VFD). If a synchronous motor is supplied by a VFD, which is often the case, it can be started by setting the frequency of the drive system to a very low value and gradually increasing the frequency to the desired value. The low frequency allows the rotor to start turning slowly as the rotating field is slow enough for the rotor to follow and overcome its inertia. During the starting phase, the voltage applied to the motor is reduced to compensate for the reduced reactance of the motor at a low frequency.

Any of the above methods might be used with a conventional synchronous motor. However, for numerous reasons, many synchronous motors have rotors fitted with 'damping bars', also called an 'amortisseur winding'. In this type of rotor, an assembly similar to the squirrel cage in an induction motor is built into the pole faces of the rotor, as shown in Figure 24.21.

FIGURE 24.21 A 3 MW synchronous motor rotor with an amortisseur winding, which is similar to the squirrel cage in an induction motor. The rotor is therefore laminated.

Synchronous motors with damping bars in the rotor can be started as an induction motor. The rotating field in the stator cuts the damping winding, which behaves like the cage in the rotor of an induction motor. As a result, torque is developed and the machine runs up to nearly synchronous speed as an induction motor. As the speed increases, the motor's excitation current is increased until the rotor locks into synchronism with the rotating field.

With this method of starting, a soft starter is usually required, due to the large starting current. The DC excitation current is isolated during the start-up phase, as a high voltage will be induced in the excitation windings by transformer action. For this reason, the field windings are insulated to withstand a much higher voltage than the DC excitation voltage.

Loading the motor

Induction and DC motors slow down when their mechanical load is increased. The reduction in speed reduces the motor's induced or counter-EMF, which allows more current to flow through the motor windings to provide the required increase in torque to compensate. However, a synchronous motor cannot slow down, as it would lose its torque and stall. Instead, under load, a synchronous motor rotor changes its relative *position* to the rotating field, while still following the field.

Figure 24.22(a) shows the rotor position relative to the rotating magnetic field when the synchronous motor is unloaded. The centre of the field aligns with the centre of a pole piece, and remains that way as the rotor follows the rotating field. Figure 24.22(b) shows that when load is applied, the rotor falls behind the field, stretching the magnetic lines of force. The angle (α) between the rotor's unloaded and loaded positions relative to the centre of the magnetic field is the load or torque angle.

FIGURE 24.22 At no load, a synchronous motor rotor pole is central with the rotating stator field; when loaded, it lags the field by torque angle α

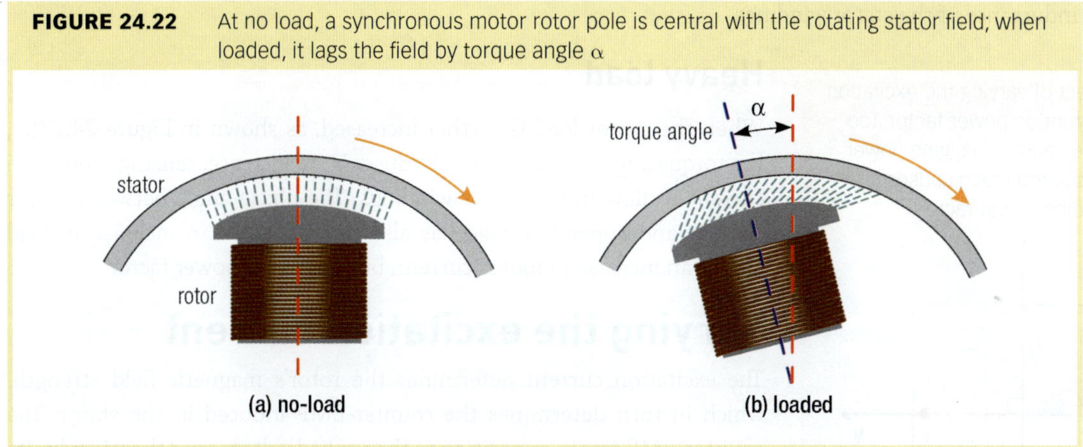

The magnetic lines of force in Figure 24.22 behave like elastic, and try to pull the rotor back into its central position. This is referred to as elastic magnetic coupling and is broken only when the motor is overloaded. The larger the load, the higher the value of the torque angle and the greater the magnetic stretch. Once the maximum torque angle is exceeded (called the 'pull-out' torque, and occurs at around 90°), torque reduces and the motor stalls, often dramatically. If it has amortisseur windings, it will attempt to run as an induction motor, but unless the mechanical overload is reduced, it will be unable to reach synchronous speed.

No load

The counter-EMF (E_G) in a synchronous motor is induced in its stator winding by the motion of the rotating field poles of the rotor. Under no-load conditions, the counter-EMF is almost equal and opposite to the applied voltage, so the stator current is just enough to make up for the losses in the motor. In an ideal motor (has no losses), the counter-EMF would exactly equal the applied voltage and no current would flow, as shown in Figure 24.23(a).

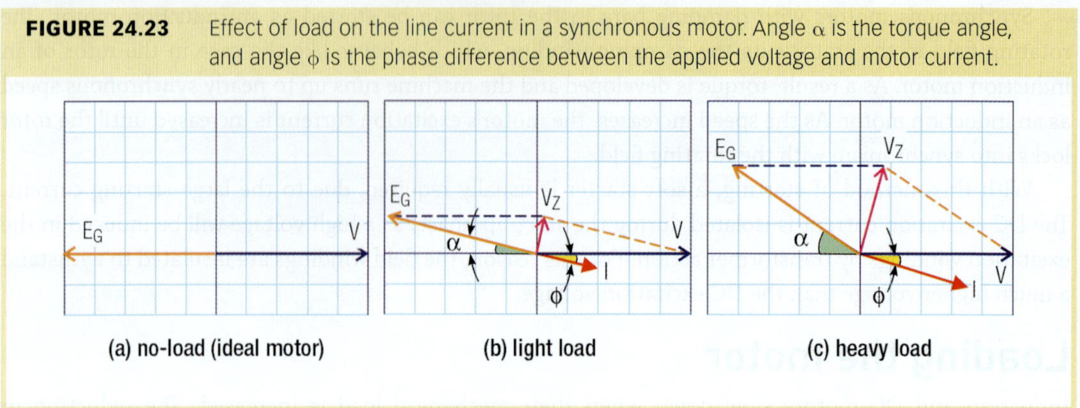

FIGURE 24.23 Effect of load on the line current in a synchronous motor. Angle α is the torque angle, and angle ϕ is the phase difference between the applied voltage and motor current.

Light load

When a light load is applied to the motor, the rotor falls behind the rotating magnetic field by the torque angle (α). The counter-EMF (E_G) also moves in relation to the applied voltage by the same angle. That is, instead of being 180° out of phase, it is now (180 − α)° out of phase. It no longer counteracts the applied voltage by as much as under no load. The phasor addition of the applied voltage V and the counter-EMF (E_G) results in voltage V_Z, which overcomes the motor impedance and causes current to flow in the stator windings. That is, the increase in load has caused the motor current to increase, which in turn increases the strength of the rotating magnetic field. The motor current is almost 90° lagging V_Z because of the high inductance of the stator windings.

FIGURE 24.24 Effect of varying the excitation current on power factor. Too little causes a lagging power factor, too much causes a leading power factor.

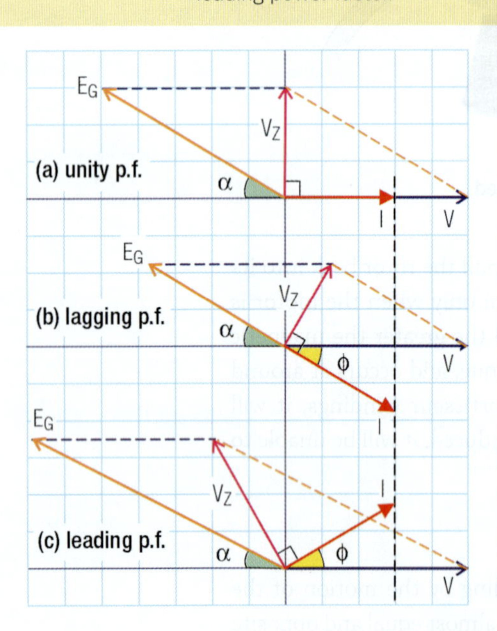

Heavy load

When the motor load is further increased, as shown in Figure 24.23(c), the torque angle increases, which causes V_Z to increase, causing even more current to flow in the stator windings. The phase angle between motor current and applied voltage has also increased, so an increase in load causes an increase in motor current, but at a lower power factor.

Varying the excitation current

The excitation current determines the rotor's magnetic field strength, which in turn determines the counter-EMF induced in the stator. The counter-EMF acts in opposition to the applied voltage, and therefore limits the current taken by the motor. However, we have just shown that motor current is determined by the torque angle, not the excitation current. It turns out that there is an optimum value of excitation current for a given load. It is the value that causes the stator current to be a minimum for that load. We'll use the phasor diagrams in Figure 24.24 to explain.

Normal excitation

Phasor diagram 24.24(a) shows the condition when the rotor excitation current is adjusted so the stator current I is in phase with the applied voltage V. This is the optimum value, known as normal excitation, and for a given load, the torque angle α will be as needed for that load. The voltage V_Z is the phasor addition of V and the counter-EMF (E_G). As explained previously, V_Z is the voltage that causes current to flow in the stator, and leads the current by 90° because the motor impedance is mainly inductive.

Below normal excitation

Phasor diagram 24.24(b) shows the effect of reducing the excitation current below its normal value. The torque angle remains the same because the load has not varied. However, the counter-EMF (E_G) reduces because the field strength is reduced. The phasor addition of E_G and V gives a smaller value of V_Z, which, compared to (a), is now phase-shifted in a lagging direction. Current I remains at 90° to V_Z, so its phase angle also shifts, giving a lagging power factor.

The motor current I is larger than in (a) because it now has two components, the in-phase current (which is doing the work of rotating the load) and a component that is at 90° (magnetising current, which does no work). The phasor addition of these two currents gives the resultant current I shown in Figure 24.24(b), which is now lagging the applied voltage V.

Above normal excitation

Phasor diagram (c) shows the effect of increasing the excitation current. The torque angle is not changed as the load is the same, but the counter-EMF (E_G) increases due to the increased field strength. The phasor addition of E_G and V gives an increased value of V_Z, which is now rotated in a leading direction compared to Figure 24.24(a). Current I is at 90° to V_Z, so it too rotates and leads the applied voltage V. The motor now has a leading power factor.

As for phasor diagram (b), the motor current I in (c) has two components. It has an in-phase component of the same value as before, and a magnetising current that tends to offset the increased field strength caused by the increased excitation current.

Controlling power factor

If the excitation current of a synchronous motor is increased from a small value to a higher value, two things happen:

1 The stator current will decrease until it reaches a minimum value (normal setting), and then it will increase as the excitation current exceeds the normal setting.

2 The power factor will at first be lagging, then reach unity at the normal value of the excitation current. When the excitation current exceeds this value, a leading power factor will result.

It follows, therefore, that the power factor of a synchronous motor can be controlled by varying the field current. The range of variation depends on the load. Figure 24.25 shows how power factor and motor current change with field current for a constant load. These curves are sometimes referred to as V-curves, and a more complete set would show the variations for different load values. To maintain unity power factor with an increase in motor load requires increasing the excitation current.

FIGURE 24.25 V-curves of a synchronous motor showing how power factor and motor current change with a change in field excitation current

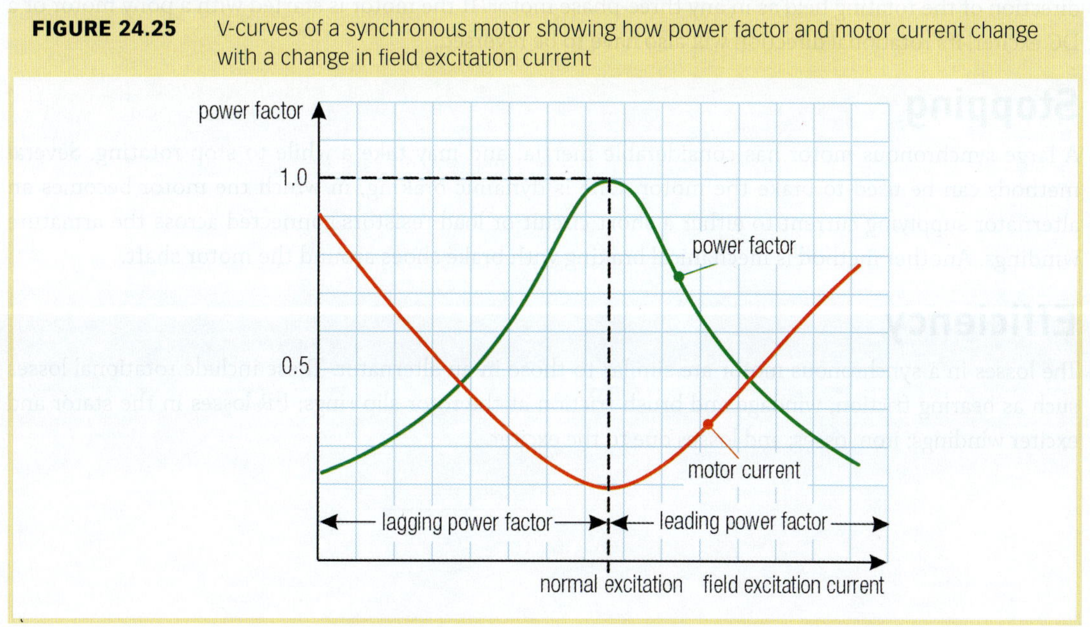

There are practical limits concerning over- and under-excitation. If the field current is reduced too far, the magnetic bond between the rotating field and the rotor field can become too weak to hold the rotor in synchronism with the rotating field. This can cause the motor to stall or, if it has amortisseur windings, to operate as an induction motor. Over-excitation can result in the motor current exceeding its rated value. It can also cause the magnetic bond between the stator and rotor fields to be so strong and stiff that changes in the load cause severe mechanical stress on the shaft of the motor.

Hunting

As Figure 24.22 shows, the torque angle depends on the load. Therefore, when the load changes from, say, a heavy load to a much lighter load, the torque angle will reduce, causing the rotor to take up a new position relative to the rotating field. In so doing, its inertia will cause it to overshoot the correct position. It will then be pulled back by the rotating field, this time overshooting that position. That is, it will oscillate around the correct torque angle position. This effect is called *hunting* and can possibly cause mechanical resonance to be set up in the machine, which in turn increases the amount of hunting. The resulting vibrations will generate lots of noise, and eventually the machine will loose synchronism, causing the effects already described.

In any case, it will be shown by ammeters measuring the motor current, in which analog meter pointers will oscillate or a digital meter will show an ever-changing value. Hunting is caused typically by a sudden change in the load, but can also be due to a change in the supply frequency or problems associated with the excitation circuit: for example, a loose connection, chattering brushes or similar problems. A reciprocating load such as a piston-type pump or compressor can also cause the motor to hunt.

Hunting can be reduced or prevented by increasing the excitation current to make the magnetic bond stronger. Another technique is the use of damper bars, or amortisseur windings, fitted to the rotor as shown in Figure 24.21. As explained previously, these can also provide a means of starting the motor if the bars are interconnected with end rings in the form of a squirrel cage. The hunting effect induces a current in the bars, which by Lenz's law will act to prevent the motion causing the induced current. This dampens the swaying action of the rotor, and establishes smoother running. If there is no swaying action occurring and the rotor is running uniformly, there are no induced currents in the bars.

Hunting can also occur in an alternator, and amortisseur windings are often fitted to the rotor for the same reasons as for a motor. These might only be in each pole face, with each set of bars independent from the others.

Reversing

A synchronous motor can be reversed by swapping any two phases supplying the stator. This reverses the direction of the rotating field as in any three-phase motor. If the motor is started with a pony motor or a DC exciter, its rotational direction will also have to be reversed.

Stopping

A large synchronous motor has considerable inertia, and may take a while to stop rotating. Several methods can be used to brake the motor. One is dynamic braking, in which the motor becomes an alternator supplying current to either a short circuit or load resistors connected across the armature windings. Another method is mechanical braking with brake shoes around the motor shaft.

Efficiency

The losses in a synchronous motor are similar to those in an alternator. These include rotational losses such as bearing friction, windage and brush friction at the rotor slip rings; I^2R losses in the stator and exciter windings; iron losses; and losses due to the exciter.

EXAMPLE 24.5

A three-phase 50 Hz synchronous motor nameplate rates the motor output power at 3 MW and states a full load current of 162 A at 11 kV for a unity power factor. Calculate the efficiency of the motor.

Solution

Values $P_{out} = 3\ \text{MW}$
$V_L = 11\ \text{kV}$
$I_L = 162\ \text{A}$
$\lambda = 1$

Equation $P_{in} = \sqrt{3}\, V_L I_L \cos\phi = 1.732 \times 11 \times 10^3 \times 162 \times 1$
$P_{in} = 3\,086\,424\ \text{W or } 3.086\ \text{MW}$

Equation $\eta\% = \frac{P_{out}}{P_{in}} \times 100 = \frac{3\ \text{MW}}{3.086\ \text{MW}} \times 100$

Answer **efficiency = 97.2%**

Applications

Synchronous motors are used extensively in heavy industrial applications, such as mining drag lines, large ball mills, cement plants and compressors. They have a higher efficiency than an equivalent size induction motor, particularly at low speeds. Their higher initial cost is therefore offset by reduced power consumption.

The high efficiency of a reluctance motor suits energy minimisation standards

They are also used in power factor correction applications. When run either unloaded or loaded, by over-exciting the rotor windings, the motor behaves as a capacitor and causes the current to lead the applied voltage. In a typical industrial environment, most of the plant is inductive. Therefore, a synchronous motor can serve two simultaneous purposes: to operate a load and to improve the system power factor. In some cases, the only task a synchronous motor has is power factor correction. In this application, the motor is called a synchronous capacitor.

24.3 Other Types of Synchronous Motors

A synchronous motor is any motor that runs at a speed synchronised with the supply frequency. Speed control can only be achieved with a variable frequency drive system. There are numerous types of synchronous motors, as described below.

Single-phase synchronous motors

These motors are generally low power, usually less than 1 kW and often only a few watts. There are three main types: reluctance, hysteresis and permanent magnet motors.

Synchronous induction motor

The synchronous induction motor is similar in construction to the wound rotor induction motor described in Chapter 22. The stator is the same as any induction motor and the rotor slip rings are connected to load resistors during start-up. When synchronous speed is approached, the resistors are disconnected and a DC excitation supply is connected, creating a magnetised rotor, which pulls it into synchronism. This type of motor is self-starting, and can start and pull into synchronism against more than full load torque. It can operate with a leading power factor by changing the excitation current.

FYI

Switched reluctance motors are used in various makes of electric cars

Reluctance motors

The term 'reluctance' refers to the principle of a magnetic material aligning itself to a magnetic field so the magnetic flux lines flow through a path of least reluctance. Reluctance motors fall into two categories; switched or synchronous.

A switched reluctance motor has a solid iron rotor with salient poles and a stator with two more poles than the rotor. Figure 24.26(a) shows a six-pole stator with field coils and a four-pole rotor. Coil pair 1 is shown energised, which will cause the rotor to move so the magnetic field has the path of least reluctance. Making the rotor continually rotate requires switching each pair of coils so the rotor is never locked between a pair of stator poles. This is achieved using an electronic drive circuit with a position sensor attached to the motor shaft. Switched reluctance motors have a low starting current and high torque, with a fast-dynamic response. They also have a high efficiency and a simple construction. They are used in electric vehicles and many industrial applications.

FIGURE 24.26 Reluctance motors have specially designed rotors that turn to provide a low reluctance path for the rotating magnetic field

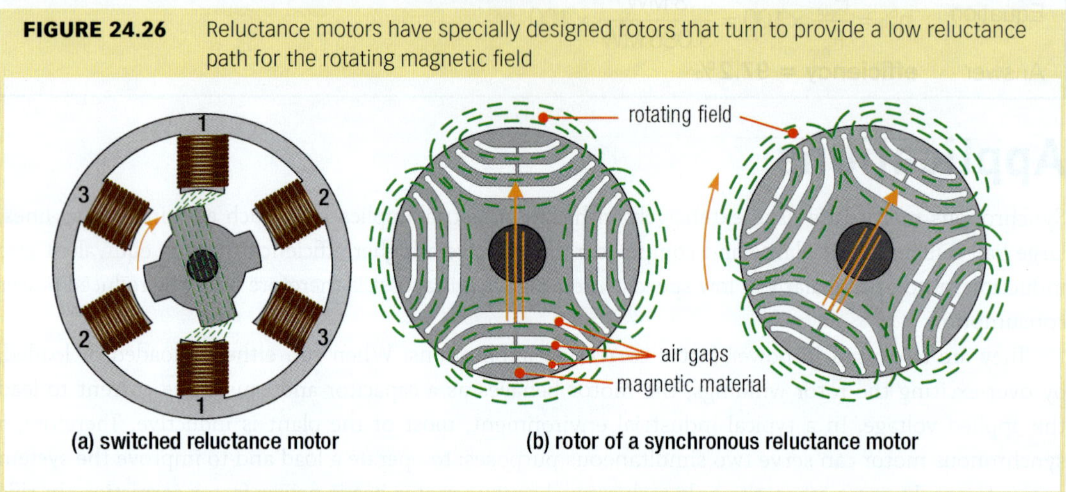

The synchronous reluctance motor has a rotor as shown in Figure 14.26(b). It is made of iron laminations, and has air gaps that offer a high reluctance magnetic path. When it is in a rotating magnetic field, as provided by a three-phase supply, the rotor will follow the field so the magnetic field only flows in the low reluctance part of the rotor. To make it self-starting, the rotor could also have a squirrel cage as part of its construction. This type of motor is efficient, and can run from a three-phase supply or, for speed control, a variable frequency drive. They are used to operate pumps and compressors, and range in size up to hundreds of kilowatts.

Hysteresis motor

The rotor of a hysteresis motor is a smooth cylinder made of a hardened magnetic alloy. Hysteresis refers to how the magnetic flux in the metal rotor lags behind the rotating magnetising force. These motors have a stator like any induction motor. On start-up, when slip is sufficiently reduced, the rotor becomes magnetised by the stator field and the rotor's magnetised poles stay in place. The motor then runs at synchronous speed as if the rotor were a permanent magnet. They are silent and smooth in operation, and suit low power applications.

Permanent magnet motor

Permanent magnet synchronous motors (PMSM) are similar to a brushless DC motor in that both motors have a permanent magnet rotor and a rotating magnetic field. In a DC motor, electronics sequentially switch power to the stator coils to create a rotating field. In a PMSM, a rotating field is either from an AC supply or a variable frequency drive. Without the electronics, they are not self-starting and have various arrangements to make them start, such as a shaded-pole. Once up to speed, the magnetic rotor locks into the rotating field in the same way as for a conventional synchronous motor. They have a relatively low torque and efficiency, and are used in low power applications, such as instrumentation and turntables.

KEY POINTS...

- A synchronous motor runs at a fixed speed determined by the frequency of the AC supply and the number of poles in the motor. The speed of all types of synchronous motors can be controlled with a variable frequency drive.
- Conventional synchronous motors have a wound rotor supplied by an external DC supply to magnetise the rotor.
- A conventional synchronous motor is not self-starting, and various methods are used to bring the motor up to near synchronous speed, after which it will lock into the rotating stator field and maintain synchronous speed.
- Amortisseur windings, also called damping bars, are fitted to the rotor of a conventional synchronous motor to prevent it hunting, and to also provide a means of starting the motor. In this case, the bars allow the motor to start as an induction motor.
- When the load on a synchronous motor is increased, the rotor speed remains constant, but the rotor shifts backwards by an amount called the torque angle.
- The power factor of a conventional synchronous motor depends on the level of DC excitation. An under-excited motor has a lagging power factor, while an over-excited motor has a leading power factor. The optimum exciting current gives the minimum stator current for that load at unity power factor.
- Because power factor can be determined by excitation, conventional synchronous motors are sometimes used to improve power factor. In this application, the motor is a synchronous capacitor.
- A switched reluctance motor is like a stepper motor in which field coils are switched in a rotating sequence, causing a specially shaped rotor to synchronise with the rotating stator field.
- A synchronous reluctance motor has a rotor with pathways of low reluctance that cause it to synchronise with a rotating stator field.
- Reluctance motors have a high efficiency and suit energy minimisation standards, making them a preferred choice in many applications.

CHAPTER SUMMARY

The following equations apply to this chapter:

- $n_s = \dfrac{120f}{p}$ where n_s = synchronous speed in RPM, f = supply frequency in hertz, p = number of poles
- $p = \dfrac{120f}{n_s}$ which is the above equation in terms of number of poles
- $V = 4.44\ Nf\Phi k$ where V = phase voltage in volts produced by an alternator, N = number of armature turns per phase, f = frequency of the generated voltage in hertz, Φ = magnetic flux per pole in webers, k = machine constant
- apparent power S = VI (single-phase alternator), $S = \sqrt{3}\ V_L I_L$ (three-phase alternator)
- for an alternator, voltage regulation $\% = \dfrac{V_{NL} - V_{FL}}{V_{FL}} \times 100$ where V_{NL} = no-load voltage, V_{FL} = full-load voltage
- $\eta\% = \dfrac{P_{out}}{P_{in}} \times 100$ where $\eta\%$ = percentage efficiency, P_{out} = output power in watts, P_{in} = input power to the machine when the machine is producing its rated output power
- $P_{in} = P_{out}$ + losses
- $T = \dfrac{P_{out} \times 9.55}{n}$ where T = torque in newton metres (Nm), n = motor's rotational speed in RPM.

REVIEW EXERCISES

1 What is the synchronous speed of a 12-pole, 50 Hz alternator?

2 A three-phase star-connected 50 Hz alternator has 300 turns per phase, a flux per pole of 80 mWb and a machine constant of 0.88. Calculate the alternator's:
  - a phase voltage
  - b line voltage.

3 A three-phase alternator is rated at 20 kVA at a full-load line voltage of 400 V. How much line current can the alternator supply?

4 The full-load terminal voltage of an alternator is 400 V, rising to 430 V when the load is disconnected. What is its percentage voltage regulation?

5 A 200 kVA, 600 V three-phase alternator is supplying a full load current at a lagging power factor of 0.86. It has friction and windage losses of 2.5 kW, an iron loss of 4 kW, a copper loss of 3 kW, an excitation loss of 600 W and stray losses of 2.8 kW. Calculate the alternator's:
  - a full load current
  - b total losses
  - c input power
  - d percentage efficiency.

6 A 100 kVA alternator has an efficiency of 89 per cent when operating at full load with a power factor of 0.85 lag. How much power does the prime mover need to provide under these conditions?

7 An 8-pole synchronous motor is operating from a 50 Hz supply. What is its rotational speed?

8 Give three methods used to start a synchronous motor.

9 A three-phase 50 Hz synchronous motor is rated at 250 kW. Its nameplate states a voltage rating of 3.3 kV and a current of 46.5 A at unity power factor. Calculate the efficiency of the motor.

10 A three-phase six-pole 400 V 50 Hz synchronous motor when delivering 40 kW to a load takes a current of 62 A with normal excitation applied to the motor. Calculate:
  - a the input power to the motor
  - b the efficiency of the motor
  - c rotational speed of the motor
  - d torque produced by the motor.

ONLINE RESOURCES

COMPLETE WORKSHEET TWENTY-FOUR

Check with your instructor for worksheets on this chapter.

CHAPTER 25

TEST EQUIPMENT

This chapter describes the oscilloscope and typical electrical test equipment used in the field.

CHAPTER OUTLINE

25.1 The Oscilloscope

An oscilloscope is an instrument that can show a lot of information about electrical waveforms, such as shape and phase differences. Waveform measurements such as voltage and period can be read either directly or indirectly from the display, making the oscilloscope a very important tool when working on electrical circuits. There are various types of oscilloscopes as explained later, but all work on the same principle.

Safety

Oscilloscopes that are mains-powered have probe sockets in which one terminal is connected to the earth of the three-pin power plug, which connects to the mains earth through the power outlet. Therefore, it is not possible to use this type of oscilloscope to monitor a mains voltage because connecting the earth lead of the probe to the neutral connection will cause the RCD protection device to trip. Accidentally contacting the active line with the earth lead of the probe will also cause an RCD to trip, assuming there is one. If not, excessive current will flow until a fuse or circuit breaker operates, possibly damaging the instrument and causing an injury.

Some types of oscilloscopes are designed to work with mains power, in which the probe connections are isolated from earth. These should have a Cat rating of at least 600 V Cat III, and the probes should have the same Cat rating. An instrument is rated by whichever has the lowest Cat rating: the oscilloscope or the probe.

Operating principle

You could try this demonstration yourself. Hold a pencil on a piece of paper, and practise moving your hand up and down to draw a vertical line. Then practise moving your hand across the page at a constant speed to draw a horizontal line. Now combine the two movements. That is, move your hand at a constant speed up and down the page while you also move it across the page. You should get a drawing like that shown in Figure 25.1.

FIGURE 25.1 Simple demonstration to show how an oscilloscope traces a changing voltage

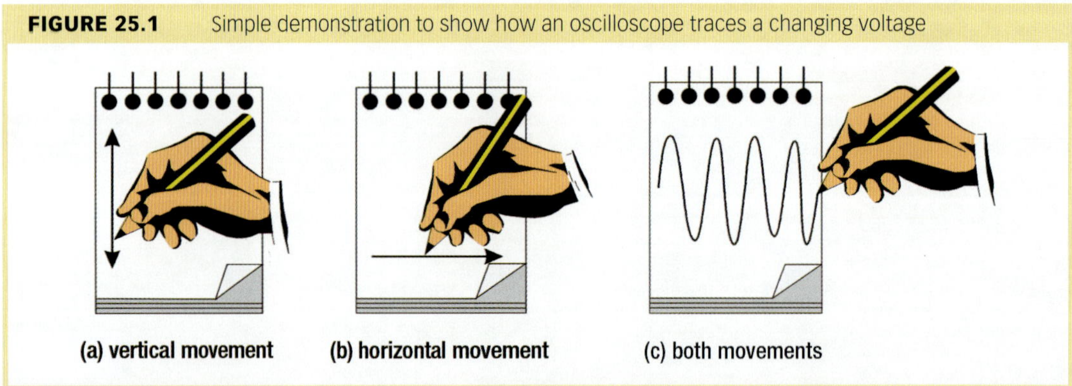

An oscilloscope moves a trace across a screen in a similar way. The movement across the screen is controlled by circuitry called a sweep generator, also called a time base. Front panel controls let you set the speed at which the trace moves. The vertical movement is caused by applying the voltage waveform under examination to circuitry within the oscilloscope that causes the trace to move up and down in accordance with the incoming voltage. Front panel controls are used to set how far the trace moves vertically per graticule division, often called the volts/division switch. We explain the horizontal and vertical controls further on.

Types of oscilloscopes

There are three basic types of oscilloscopes, which are explained below. While each type operates with different technologies, they all have similar controls and are used in much the same way. Each type of oscilloscope is available in different models with different features, depending on the cost of the instrument. We look at typical models only, as many of the features on more expensive instruments are not generally needed in electrical work.

Cathode ray oscilloscope (CRO)

The earliest type is the cathode ray oscilloscope, which has a 'trace' produced by an electron beam that strikes a phosphor surface inside an evacuated glass tube, called a *cathode ray tube*. The beam, which is attracted to the phosphor surface by a high voltage, is moved both horizontally and vertically by electrical signals that connect to electrodes within the tube. A limitation of the CRO is that it can only display repetitive waveforms, as the trace only lasts a short time after the electron beam has passed. This type of 'scope has been superseded by newer technologies, but many such instruments remain in use.

FIGURE 25.2 A typical general purpose cathode ray oscilloscope displaying a sinewave

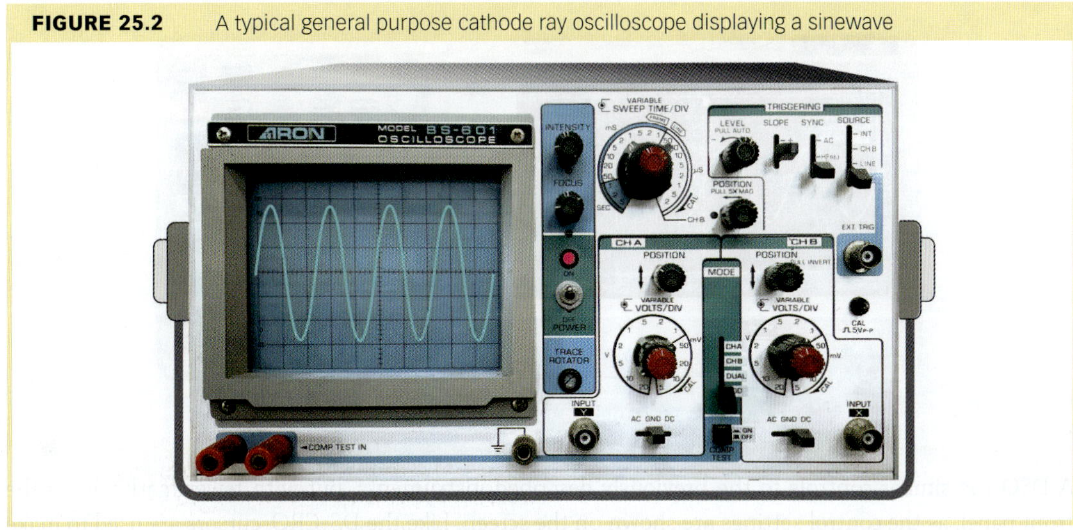

The oscilloscope in Figure 25.2, like most oscilloscopes, has two traces (or beams), allowing two waveforms to be viewed at the same time. This is needed when measuring or observing phase difference between waveforms, or when making comparisons between them, such as shape or frequency differences.

Digital storage cathode ray oscilloscope (DS-CRO)

This type of oscilloscope started appearing in the 1990s, when solid-state memory and integrated circuits had reached sufficient development. While still using a cathode ray tube, this instrument incorporates circuitry that can store a waveform in memory, allowing it to be continually displayed on the cathode ray tube screen. Storage allows a single capture of a waveform that can then be displayed as long as necessary.

FIGURE 25.3 A digital storage oscilloscope has similar controls to the cathode ray type, but with extra controls to capture a waveform in memory

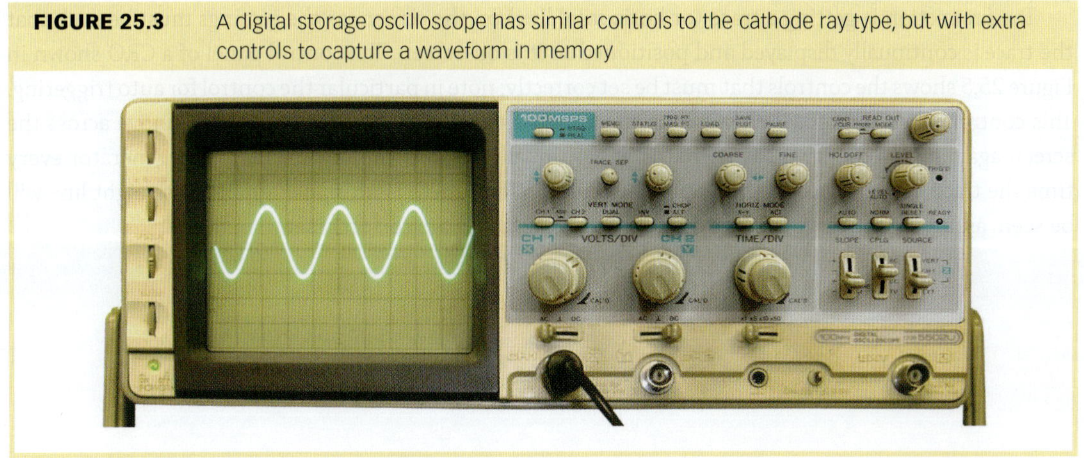

Because of the digital circuitry in the instrument, readings such as period and voltage can be read directly from the screen by moving cursors to relevant parts of the waveform, as explained later. Apart from the storage capability, the DS-CRO is operated in a very similar way to the cathode ray oscilloscope.

Digital storage oscilloscope (DSO)

Most oscilloscopes now in use are based on computer technology. These include bench models and hand-held versions. The screen is a display like those in mobile phones, which means it can show a range of colours and display written information. Signals being examined are either fed directly to the screen or they can be stored in memory in a similar way to the DS-CRO.

FIGURE 25.4 A low cost digital oscilloscope that is typical of those found in colleges and electrical workshops

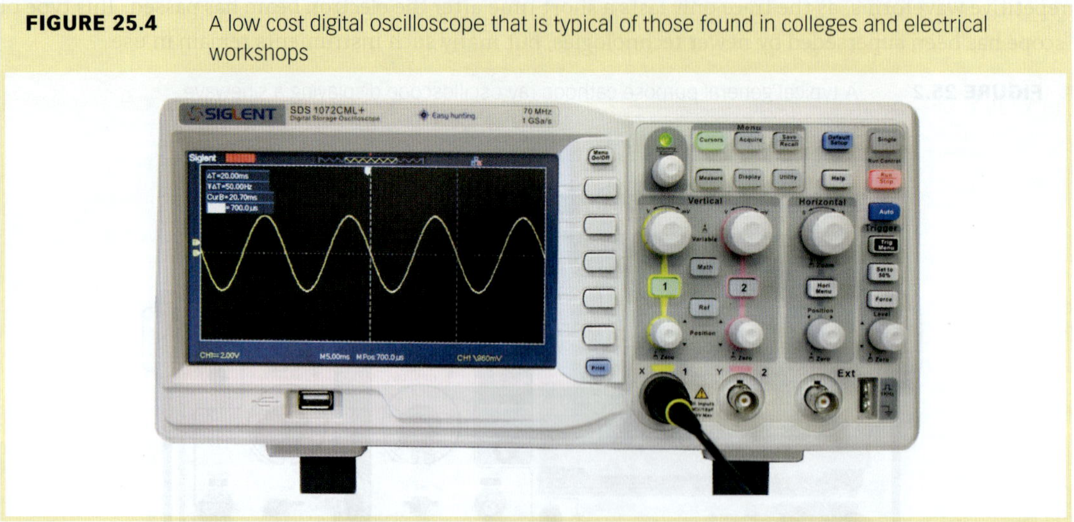

A DSO has similar controls to the previously described instruments, but with fewer markings on the front panel, as the control settings are shown on the screen. Like the DS-CRO, cursors are used to make measurements of a waveform. The instrument in Figure 25.4 has two traces, while more expensive instruments have four or more traces, making them useful in three-phase work. The features and specifications of a DSO depend on price, but for general electrical work on equipment isolated from the mains, an instrument costing no more than $500 is usually satisfactory.

Using a cathode ray oscilloscope

We start by explaining how to use a dual-beam cathode ray oscilloscope, in which you need to derive measurements indirectly from the screen, as there are no cursors or values shown on the screen. Understanding how to use a CRO will allow you to use any type of oscilloscope.

The first action is getting a trace (or horizontal line) on the screen, as the settings must be such that the trace is continually displayed and positioned so it is on-screen. The control panel of a CRO shown in Figure 25.5 shows the controls that must be set correctly; note in particular the control for auto triggering. This control, when on, causes the trace to move across the screen, fly back quickly and move across the screen again, on a repetitive basis. That is, it continually triggers the horizontal sweep generator every time the trace has returned to the left of the screen. When there is no signal applied, a straight line will be seen, assuming the brightness and trace position controls are correctly set.

FIGURE 25.5 Controls to adjust to get a trace on the screen

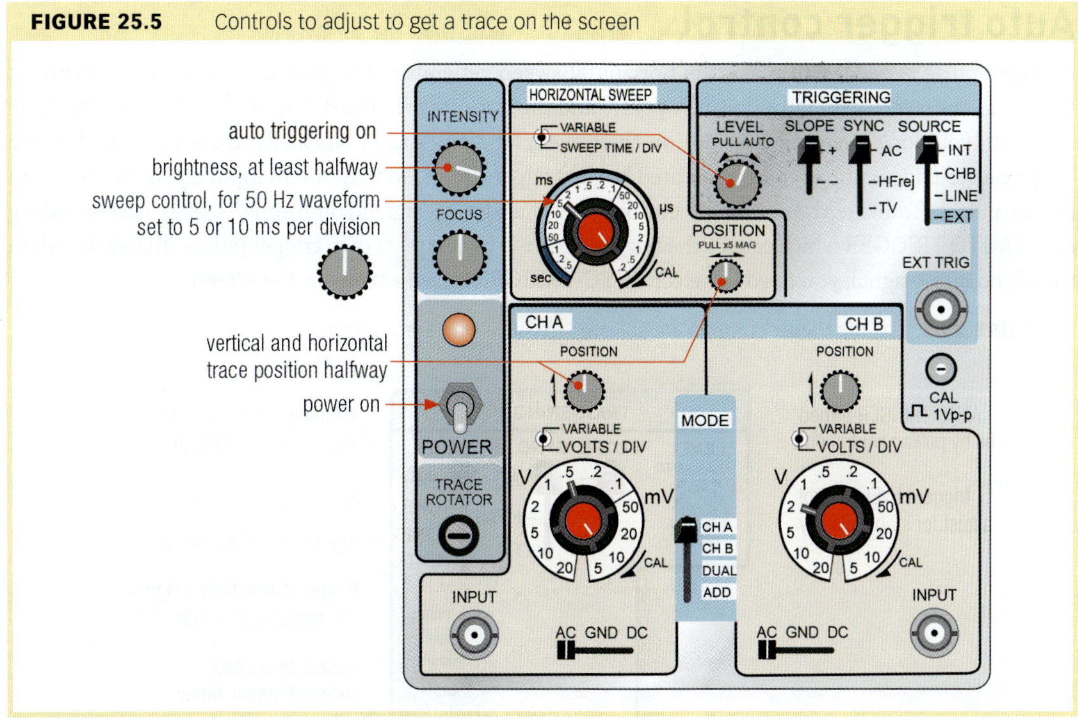

Triggering

To explain what *triggering* means, we return to the simple demonstration of tracing a waveform on a piece of paper with a pencil. Imagine that the tracing made with the pencil fades away very quickly. To keep the trace visible all the time, you have to keep redrawing it. But to do this, you have to start at the same point each time, and also move the pencil in the same direction. If not, you get pencil traces like those in Figure 25.6.

FIGURE 25.6 To repeat the trace exactly, you have to start at the same point and move the pencil in the same direction each time

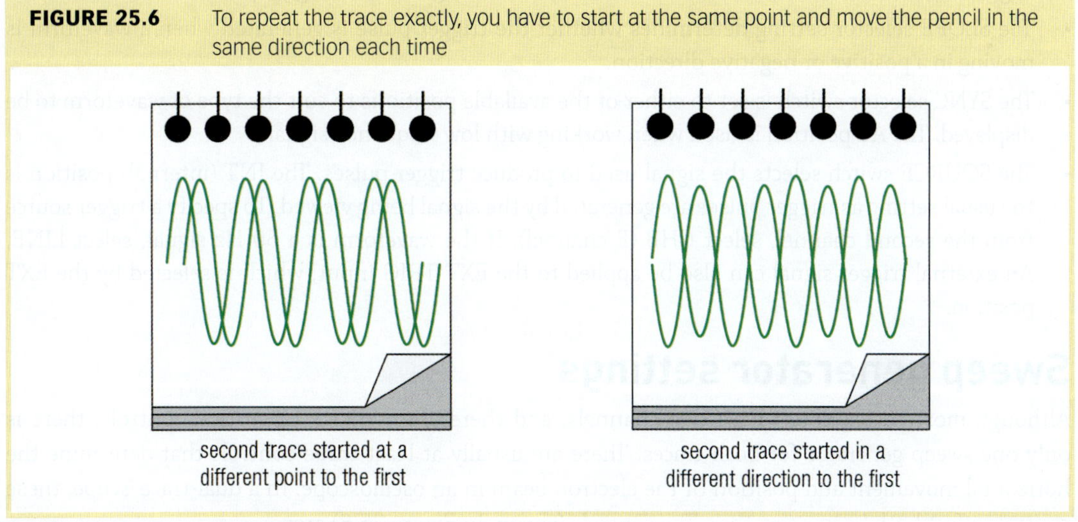

Because the phosphor on a cathode ray tube screen stays lit for a very short time, to show a waveform the beam has to keep tracing it over and over again. Therefore, the electron beam has to start at *exactly* the same point on the screen, and move in the same direction each time. This is achieved by correctly adjusting the 'trigger' controls. This assumes the signal being examined is repetitive, such as the sinewave signal from the secondary of a transformer operating at 50 Hz.

Auto trigger control

A trigger pulse is needed to make the horizontal sweep generator start its trace-retrace cycle. Without this pulse, the sweep generator doesn't operate and the trace stays in the one spot. Also, because the trace is cut off, you don't even see the spot on the screen, let alone a line. Normally, the input signal to the 'scope causes trigger pulses to be generated by the trigger circuit, so without an input signal, there are no trigger pulses. To make the sweep generator operate under no-signal conditions, 'scopes have a setting called AUTO TRIGGER which causes the trigger circuit to generate its own trigger pulses. Therefore, when there's no input signal, you need to select AUTO TRIGGER to see a trace on the screen.

FIGURE 25.7 The trigger controls are adjusted to give a stationary display

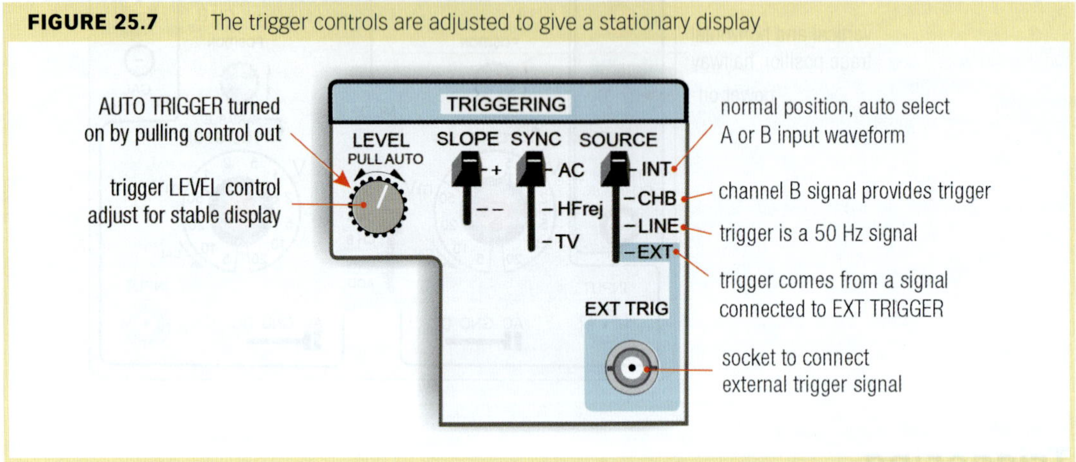

The trigger controls are often the least understood part of an oscilloscope. The idea is to set the controls so the trigger circuit can produce reliable trigger pulses to get a stable display. This is easy to do with some waveforms, but very hard with others, particularly if the waveform is continually changing its shape. Most oscilloscopes have several trigger controls to give a range of settings, such as those in Figure 25.7.

- The LEVEL (Pull Auto) controls are a switch combined with a potentiometer to set or disable auto triggering, and to set the point on the waveform that will generate a trigger pulse.
- The SLOPE selector setting determines whether the trigger pulse is generated when a waveform is moving in a positive or negative direction.
- The SYNC selector switch is set to either of the available positions to suit the type of waveform to be displayed. The AC position is used when working with low frequency signals.
- The SOURCE switch selects the signal used to produce trigger pulses. The INT (internal) position is the usual setting as trigger pulses are generated by the signal being viewed. To specify a trigger source from the second channel, select CHB (B channel). If the waveform is a 50 Hz signal, select LINE. An external trigger signal can also be applied to the EXT TRIG input, which is selected by the EXT position.

Sweep generator settings

Although most oscilloscopes have two channels, and therefore two sets of vertical controls, there is only one sweep generator for both traces. There are usually at least three controls that determine the horizontal movement and position of the electron beam in an oscilloscope. In a dual-trace 'scope, these controls affect both beams together. The controls (shown in Figure 25.8) are:

- horizontal POSITION control, to move the trace left or right on the screen
- SWEEP TIME/DIV selector switch, which sets the time taken for the trace to move by one graticule position
- VARIABLE sweep time adjustment.

FIGURE 25.8 Controls that determine the horizontal position and movement of the trace

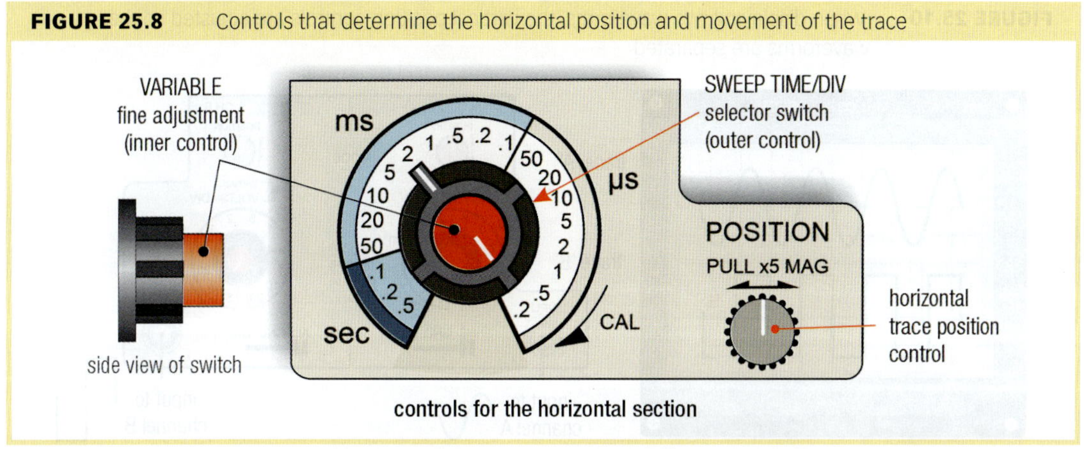

Like the VOLTS/DIV switch in the vertical section, the SWEEP TIME/DIV switch is calibrated, and the VARIABLE control is part of the switch assembly. Again, the calibrations are only true if the VARIABLE control is set to the CAL (calibrate) position. The SWEEP TIME/DIV is the time it takes for the trace to travel from one graticule division to the next. In Figure 25.8, the slowest time is 0.5 seconds, and the fastest is 0.2 microseconds.

For this 'scope, the POSITION control is also a switch for a ×5 magnifier. When this control is pulled to the 'on' position, the trace travels five times faster, which magnifies the width of the waveform by five times. As shown in Figure 25.9, a part of a waveform can be looked at in detail by switching on the magnification and moving the waveform with the horizontal position control.

FIGURE 25.9 Details of a waveform can be examined more closely with the magnify function

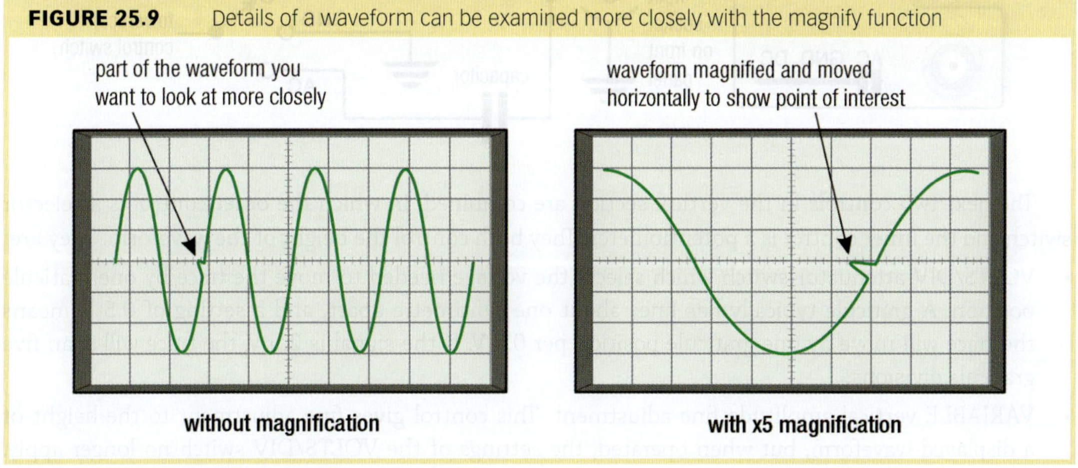

Vertical controls

There are at least four controls in the vertical section of any oscilloscope. In a dual-trace 'scope, there are two identical sets of vertical controls.

- Vertical POSITION control, which moves the trace up and down the screen. This control is useful when displaying two waveforms, as they can be positioned on the screen so they are not overlapping each other. Figure 25.10 shows two waveforms of the same frequency but different shapes and amplitude displayed on a dual-trace oscilloscope. The MODE switch is set to DUAL when displaying two waveforms.

FIGURE 25.10 When displaying two waveforms, the vertical position controls are adjusted so the waveforms are separated

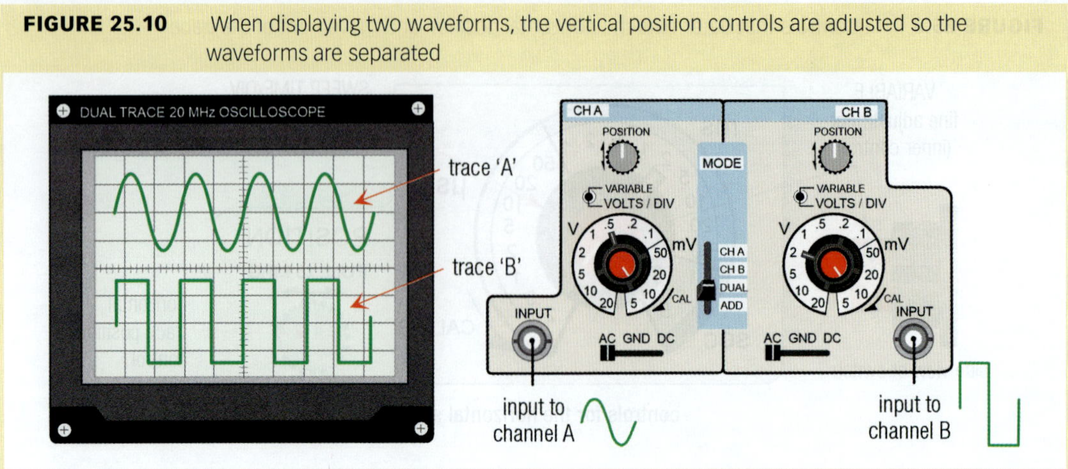

- AC-GND-DC switch. The circuit of this switch is shown in Figure 25.11. In the AC position, the input signal is connected to the oscilloscope through a capacitor. In the DC position, the signal connects directly to the oscilloscope. In the GND position, the input signal is isolated and the oscilloscope input is connected to ground (0 V). If an AC waveform has a DC component, select the AC position so the capacitor blocks the DC voltage. Otherwise, the DC component can cause the trace to disappear off the screen. In this case, selecting GND will usually cause the trace to return on-screen.

FIGURE 25.11 Circuit of the AC-GND-DC selector switch

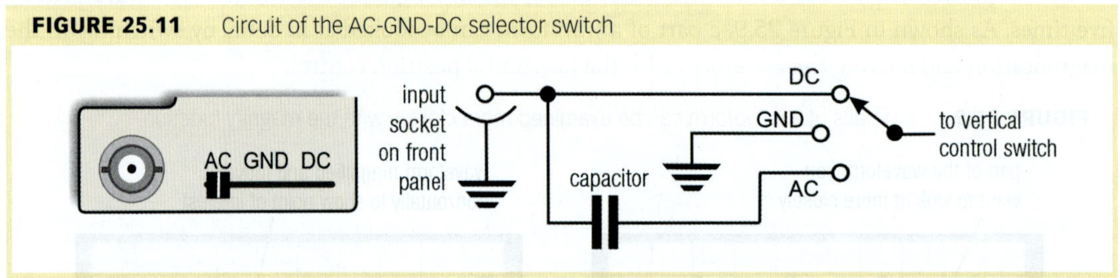

The next two controls in the vertical section are combined, in which the outer control is a selector switch and the inner control is a potentiometer. They both control the height of the waveform. They are:

- VOLTS/DIV attenuator switch which selects the voltage needed to move the trace by one graticule position. A graticule typically has lines about one centimetre apart, and a setting of 0.5 V means the trace will move by one graticule position per 0.5 V. If the signal is 2.5 V, the trace will span five graticule divisions.
- VARIABLE vertical amplitude fine adjustment. This control gives fine adjustment to the height of a displayed waveform, but when operated, the settings of the VOLTS/DIV switch no longer apply. When taking measurements of a waveform, the fine adjustment control must be in the CAL (calibrate) position, when it will have no effect.

Figure 25.12 shows the controls in one channel of the vertical section of a CRO, with details of the VOLTS/DIV switch and the VARIABLE fine adjustment control.

FIGURE 25.12 Controls that affect the vertical position and height of a displayed waveform

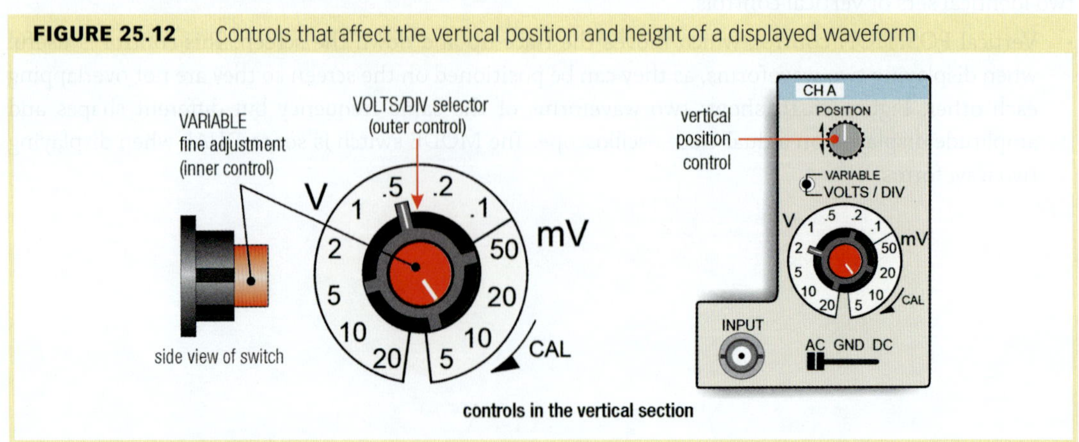

Measurements with an oscilloscope

An oscilloscope can show the shape of an electrical signal, but an important use is making measurements. There are many cases where an oscilloscope is the only way to take a particular measurement, such as a waveform's maximum value. Three common types of measurements that can be made using an oscilloscope are:

- voltage (peak-to-peak and maximum)
- frequency and period
- phase difference between waveforms.

Measuring AC voltages

In Figure 25.13, the VOLTS/DIV switch is set to the 10 V per graticule division and the fine control is set to the CAL position. This allows voltage readings such as peak-to-peak and maximum values to be read from the display. The peak-to-peak voltage is its overall height in graticule divisions multiplied by the setting of the volts/div switch. For a sinewave, the RMS value is calculated with the usual equations:

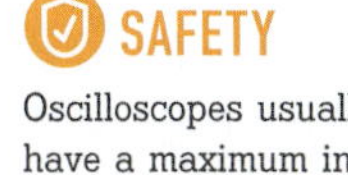

SAFETY

Oscilloscopes usually have a maximum input voltage rating of 400 V

- $V_{RMS} = 0.707 \times V_{max}$, or $V_{RMS} = 0.707 \times \frac{V_{p-p}}{2}$

FIGURE 25.13 The graticule allows an oscilloscope to be used to measure voltages of a waveform. In this case, the waveform has a peak-to-peak voltage of 40 V, or a maximum value of 20 V.

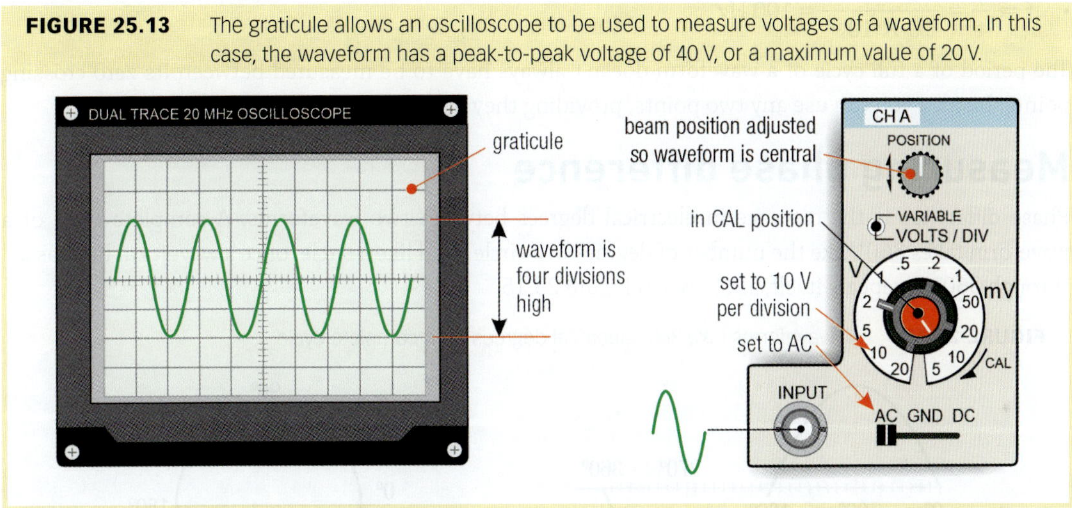

Measuring frequency

The frequency of a waveform is found with a CRO by measuring the periodic time (t) of the waveform, which is the time taken for a waveform to go through a complete cycle. The equation to find frequency is 1/t. To measure the period of a waveform, first display the waveform as already described for voltage measurement. Then:

1. Set the horizontal sweep VARIABLE control to its CAL position, so the calibrations of the SWEEP TIME/DIV control apply.
2. Set the SWEEP TIME/DIV switch to display one or two cycles of the waveform.
3. Position the waveform with the horizontal POSITION control so a zero crossing point is aligned with a vertical graticule line. Make sure the magnification control (if any) is switched off.
4. Multiply the width of *one cycle* of the waveform in graticule divisions by the setting of the SWEEP TIME/DIV switch.

FIGURE 25.14 The period of a waveform is the number of divisions taken by one cycle multiplied by the SWEEP TIME/DIV setting

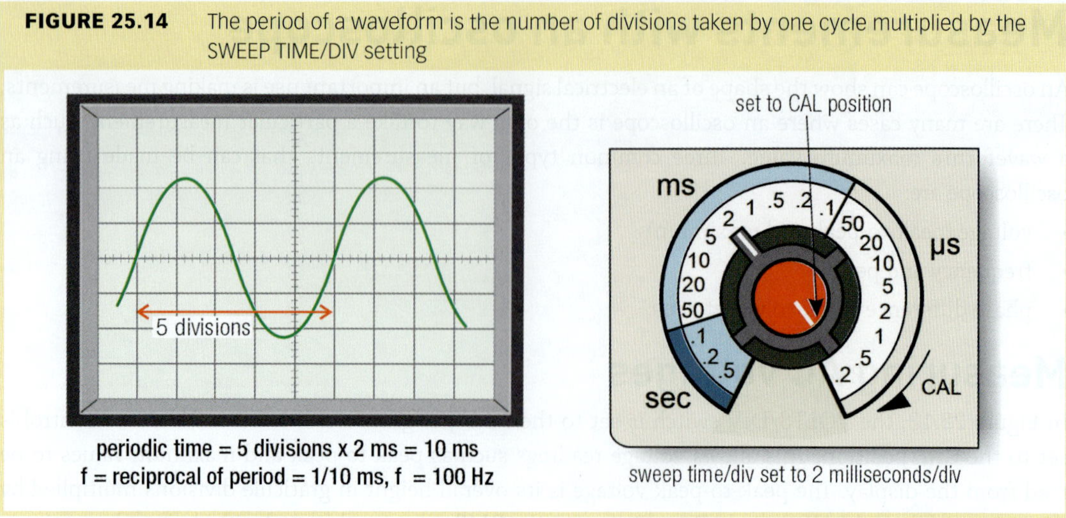

For the waveform in Figure 25.14, the periodic time is 10 milliseconds. Frequency (f) is the reciprocal of the period (t), so:

- $f = \frac{1}{t} = \frac{1}{10 \times 10^{-3}} = 100\text{ Hz}$

The period of a full cycle of a waveform doesn't always have to be measured between its zero crossing points. Instead you can use any two points, providing they are the same in each cycle.

Measuring phase difference

Phase difference is the distance in electrical degrees between two waveforms. A complete cycle of a waveform takes 360°, like the number of degrees in a circle. And like a circle, once a waveform has passed through 360°, it repeats itself, as shown in Figure 25.15.

FIGURE 25.15 All waveforms take 360° electrical degrees for a complete cycle

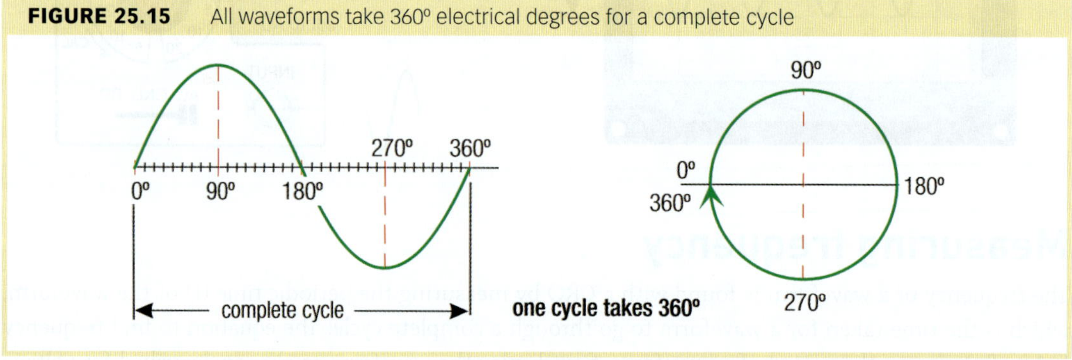

Because there are always 360° in a complete cycle, the distance over one cycle can be divided into 36 equal parts, giving divisions every 10°. As explained in Chapter 15, for a sinewave, the maximum positive voltage is at 90°, it passes through zero at 180° and the maximum negative voltage is at 270°.

Figure 25.16 shows two waveforms with a phase difference of 60°. The difference is measured between two similar points on each waveform, in this case where they pass through zero as they head towards their positive peak. In this diagram, waveform 1 leads waveform 2 by 60°.

FIGURE 25.16 Phase difference is measured in electrical degrees

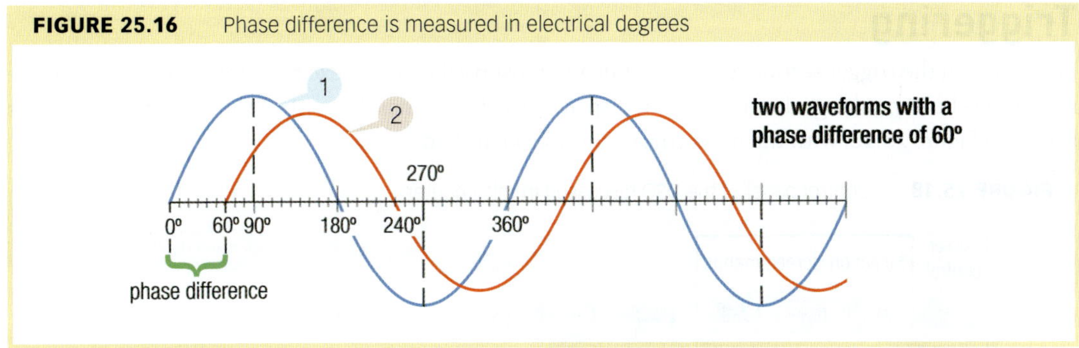

To measure phase difference on an oscilloscope, use both channels because two waveforms have to be displayed. Unlike the procedure given in Chapter 15, the method below does not require converting values into time:

1 Connect one signal to channel A and the other to channel B. Select DUAL with the MODE switch. Adjust the 'scope so about two cycles of each waveform are displayed, as shown in Figure 25.17.

FIGURE 25.17 Phase difference between the waveform is 60°

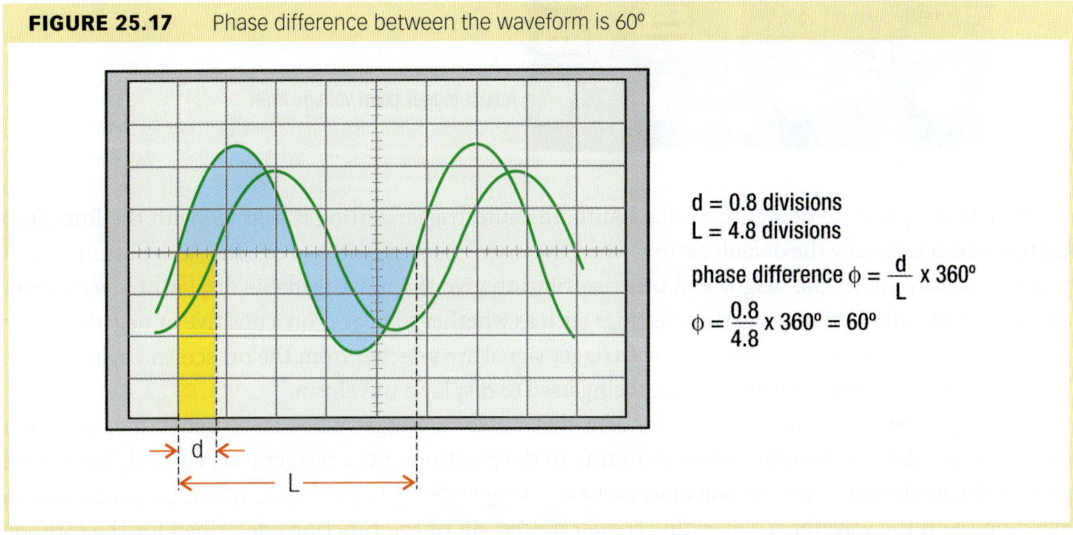

2 Find the length (L) of one cycle in graticule divisions. In the diagram, L = 4.8 divisions.

3 Find the distance (d) between two equal points of the waveforms. In the diagram the distance (d) between the zero crossing points of the waveforms is 0.8 divisions.

4 Find the phase difference with the equation $\phi = \frac{d}{L} \times 360°$. In the diagram $\phi = \frac{0.8}{4.8} \times 360° = 60°$

A typical example of phase measurement is between an applied voltage and the circuit current. In this case, a low value resistance is inserted into the current-carrying conductor so a voltage that is in phase with the current is produced across the resistor. In this situation, the earth lead of both probes must be connected to the same point in the circuit. As explained later, some types of oscilloscopes have inputs that are isolated from each other and isolated from earth, allowing greater flexibility as to where probes are connected in a circuit. This type of instrument is suitable for making phase measurements in mains-powered circuits.

Using a digital storage oscilloscope (DSO)

Because DSOs are based on computer technology, they typically incorporate capabilities that are only possible with this type of technology. A DSO is often easier to use than a CRO, as measurements are displayed on-screen through the use of cursors. (This also applies to a DS-CRO.) In this chapter we look only at the basic capabilities of a DSO, but where possible, we recommend you examine the manual for a DSO, as it can be very instructive.

Triggering

We start with the trigger section, which is similar to most oscilloscopes, except many of the functions are selected via buttons alongside an on-screen display of the functions. The control panel of a typical DSO is shown in Figure 25.18, with the main trigger controls identified.

FIGURE 25.18 Control panel of the DSO discussed in this chapter

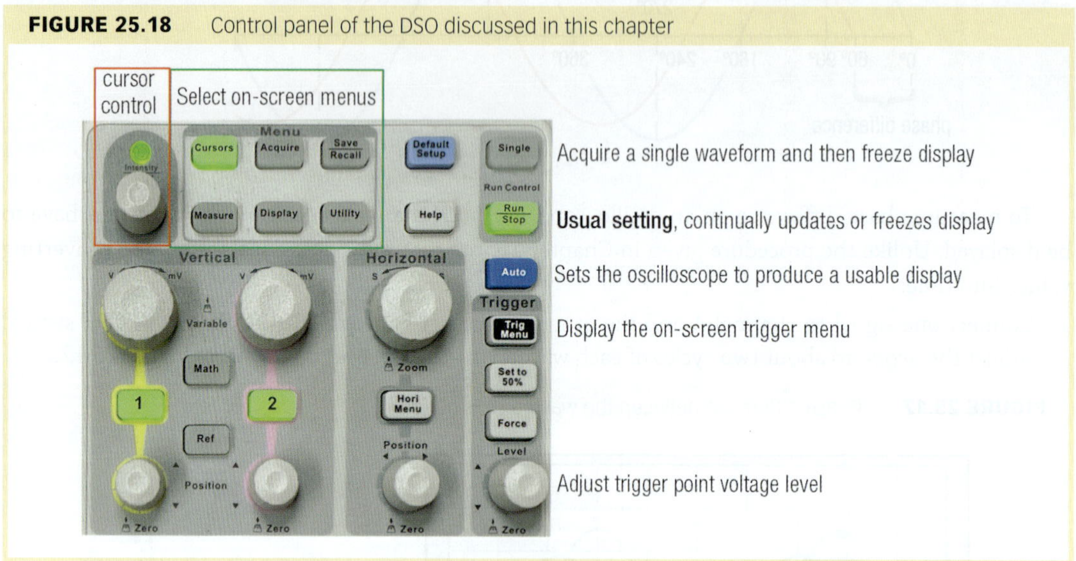

Despite the presence of a button called Auto, the auto trigger setting is selected with the Run/Stop button, which is usually the default setting when the instrument is switched on. The Auto button is used to automatically adjust the height and other settings to give the most readable display of a waveform, and is not just dedicated to triggering. Settings such as whether to trigger on a positive or negative going part of the waveform and the source of the trigger signal are selected from the on-screen trigger menu. Typically, the trigger source is the channel being used to display a waveform.

The trigger level (voltage of the waveform that causes a trigger pulse to be generated) is shown on-screen as a dotted line that moves according to the position of the trigger level control. This useful feature shows the point on the waveform where a trigger pulse is produced. It can be positioned to avoid noise on the waveform generating trigger pulses. All of the functions described for the cathode ray oscilloscope are available in a DSO, with other functions also provided. Figure 25.19 shows the two pages of the on-screen trigger menu for this DSO.

FIGURE 25.19 The trigger menu has two pages of options, and buttons alongside select submenus for each option

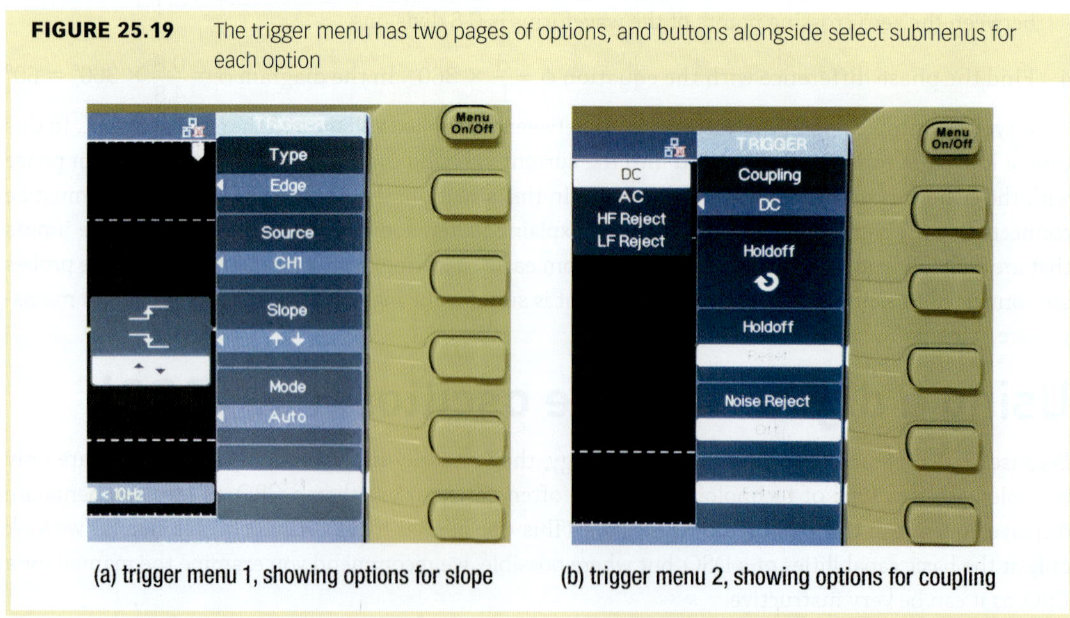

(a) trigger menu 1, showing options for slope

(b) trigger menu 2, showing options for coupling

To display a stationary sinewave connected to channel 1, on menu page 1, select Edge trigger (other settings may also work), CH1 as the source, either positive or negative slope (usually positive) and Auto. On page 2, select AC coupling (DC coupling will work if there is no DC component in the waveform) and leave other settings at the default values.

Measuring AC voltages

The two vertical controls for each channel set the voltage per graticule division (as in a CRO), and the lower control moves the trace vertically as required for that channel. The voltage per graticule division is displayed on-screen and has settings from 2.0 mV to 10.0 V, giving a maximum peak-to-peak input voltage of 80 V, as there are eight vertical divisions. (Higher values can be displayed with a 10:1 probe, as explained further on.) Pressing a position control will place the trace for that channel so the waveform is centrally positioned on the screen. Pressing AUTO (see Figure 25.18) will usually adjust the vertical amplification automatically to give the best display. Figure 25.20 shows a 50 Hz sinewave being displayed with cursors set to measure the peak-to-peak voltage of the waveform.

FIGURE 25.20 Cursors A and B are switched on via the cursor menu, and adjusted with the top left-hand control shown in Figure 25.18

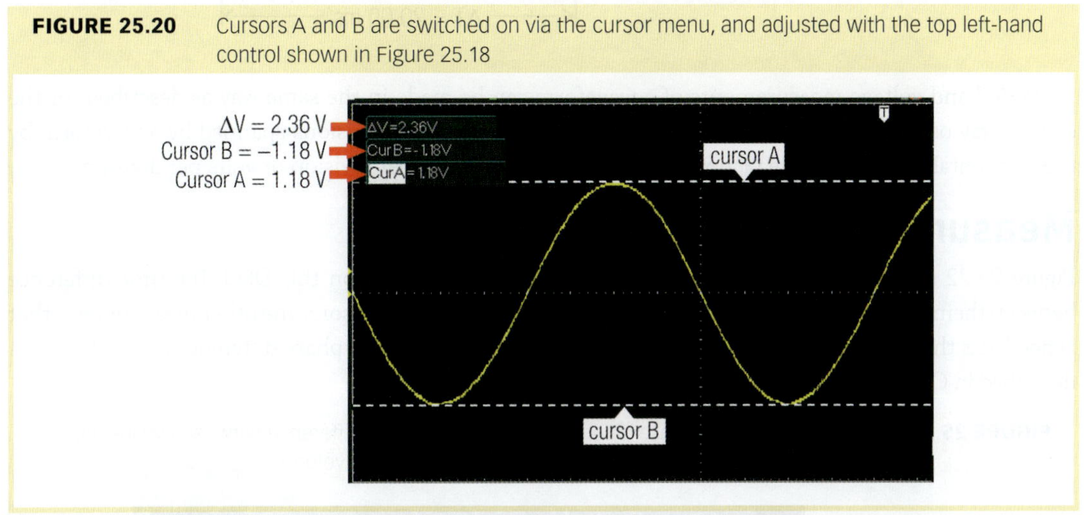

The cursor menu has an option called Type, which is either Voltage or Time. In Figure 25.20, this option is set to Voltage. Because the graticule is no longer needed, it is switched off to more clearly see the cursors. It can be turned on by pressing the Display button (Figure 25.18) and selecting Grid from the menu. The values shown on-screen are the change in voltage (ΔV) between each cursor position, in this case 2.36 V, which is the peak-to-peak voltage of the waveform. The values for cursors A and B are the maximum voltages, as these measure from the centre of the waveform (zero crossing) to the cursor position. If the waveform has a DC component, these voltages would be different.

Measuring period and frequency

There are two controls associated with the horizontal section of this DSO: time per graticule division and horizontal position of the trace. The time per division ranges from 5.0 nanoseconds to a very slow 50 seconds per division. As there are 16 horizontal divisions, a scan at 50 seconds per division will take nearly a quarter of an hour. Only a digital oscilloscope has this capability, and it is used with very slow changing waveforms. The 5.0 nanoseconds per division is extremely fast and might be used to view part of a waveform, as shown in Figure 25.13.

Figure 25.21 shows a 50 Hz sinewave being displayed with cursors set to two identical points of the waveform. The on-screen values show the time between the cursors, which is the period of the waveform, and also the frequency as calculated by the DSO. To cause the cursors to show time values, select Time from the cursor menu. Adjusting the time/division setting will show more or fewer cycles of a waveform. For highest accuracy, a single cycle of a waveform should be displayed. The time/division setting is displayed on screen (not shown in Figure 25.21).

FIGURE 25.21 Cursors A and B are positioned at two identical points of the waveform. The DSO displays the period and frequency of the waveform.

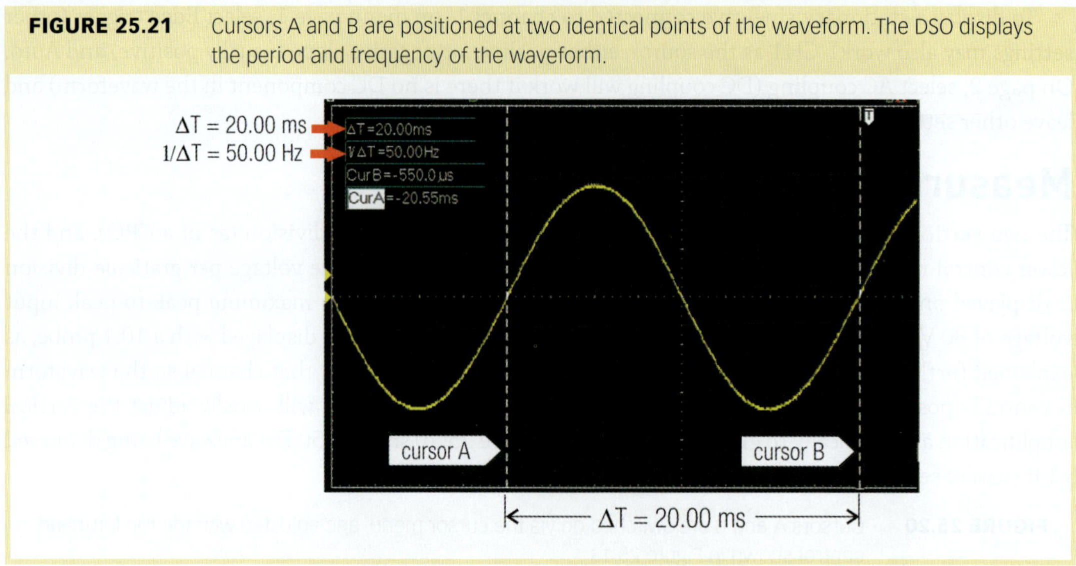

Period and voltage measurements of a waveform can be made in the same way as described for the cathode ray oscilloscope. That is, multiply the number of graticule divisions spanned by a waveform by the horizontal or vertical scale factor, which is either sweep time per division or volts per division.

Measuring phase difference

Figure 25.22 shows two out-of-phase 50 Hz sinewaves as displayed on this DSO. The time difference between them is shown to be 4.5 ms as determined by adjusting the cursors. The other measurement that is needed is the periodic time of the waveforms (both will be same). The phase difference is then found as described in Chapter 15, Figure 15.29, and summarised below.

FIGURE 25.22 When determining phase difference, measure the time difference between two identical points of the waveforms, and the periodic time of the waveforms

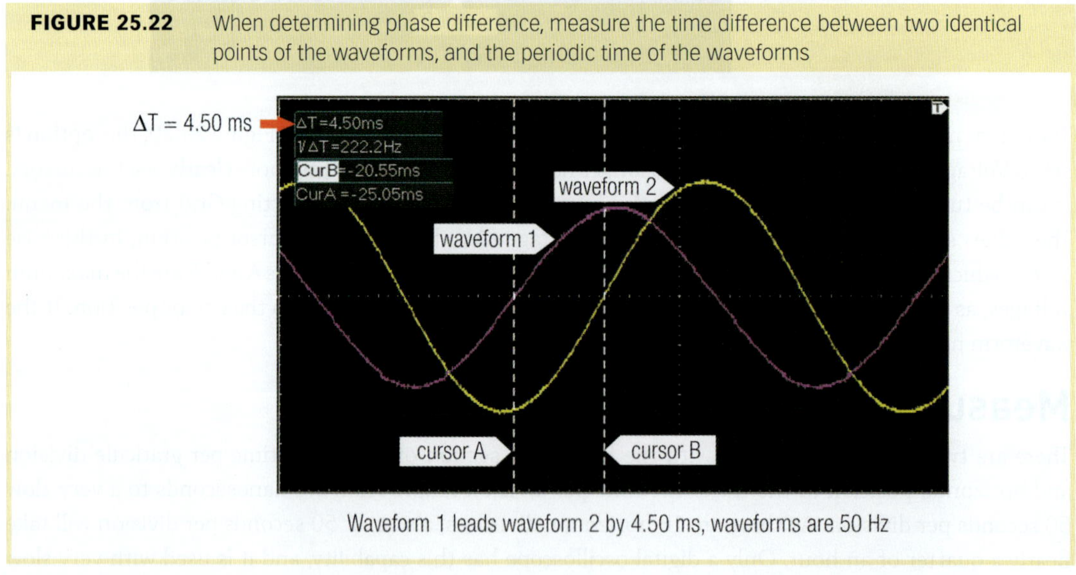

The phase difference between waveforms 1 and 2 is found with the equation $\phi = \frac{\text{time difference}}{\text{periodic time}} \times 360°$.

In Figure 25.22, the periodic time is 20 ms (because the frequency of the waveforms is 50 Hz), the time difference is 4.5 ms, so the phase difference is 81°.

Storing waveforms

A key feature of a DSO is the ability to store waveforms in memory or to freeze a displayed waveform. On this DSO, pressing the Run/Stop button will cause the display to stop updating and show only the waveform being displayed at the time the Run/Stop button is pressed. The horizontal and vertical controls remain effective, which means a captured waveform can be expanded or contracted by adjusting the horizontal and vertical settings. As well, the memory will hold a greater sample of the waveform than the display will show. For example, if four cycles of a 50 Hz waveform are captured at a setting of 5.0 ms per time division, selecting 10.0 ms per division will display eight cycles. A time base setting of 100 μs will show a small portion of one cycle, giving a huge magnification. The vertical settings are also effective, giving vertical magnification of any part of a waveform.

Waveforms can be stored in memory for later recall, or they can be saved to an external USB drive for viewing on a computer. Real time recordings can be made of an incoming signal, limited in duration by the available memory. The recording can be played back at any time.

In summary, a DSO is a versatile instrument with many features that include mathematical functions, such as adding or subtracting waveforms, as well as Fast Fourier Transform analysis of a waveform, which is useful in identifying harmonics in a waveform. Because of the many functions and capabilities, just playing with the instrument can be a good way to learn about it.

Handheld oscilloscopes

A handheld oscilloscope is especially useful when working in the field. There are many brands and models on the market, including some that do not have a Cat rating and are therefore a potential safety hazard. These should not be used when working with single- or three-phase power. In some industries (such as mines), the minimum allowable Cat rating for portable test equipment is Cat IV 1000 V. Figure 25.23 shows two handheld oscilloscopes made by Fluke, one of the few companies that produce Cat IV rated handheld oscilloscopes.

FIGURE 25.23 Handheld oscilloscopes suitable for industrial troubleshooting. Both are battery powered and offer numerous functions with inbuilt intelligence.

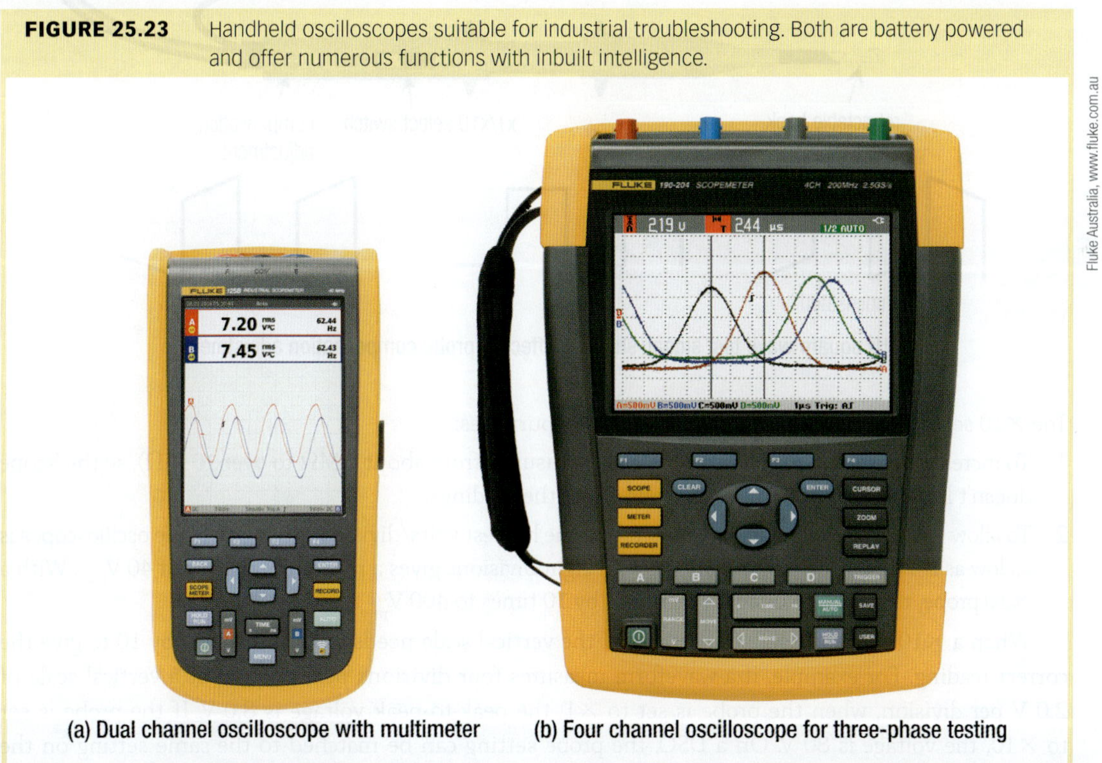

(a) Dual channel oscilloscope with multimeter (b) Four channel oscilloscope for three-phase testing

Fluke Australia, www.fluke.com.au

The instrument shown in (a) is battery-powered, has two independent, isolated oscilloscope inputs and inputs for digital multimeter probes. It can therefore be used safely to monitor mains power, as the oscilloscope probe input sockets are electrically separated and isolated from earth. It has many of the features of the DSO described previously, plus the advantage of portability and safety.

The example shown in (b) has four independent, isolated oscilloscope inputs, and is designed for use with three-phase systems. It is also battery-powered and Cat IV rated. Instruments such as these are comparatively expensive and find their greatest application when troubleshooting automated systems, motor drive systems and complex control systems. Because the inputs are isolated, these instruments can be used in single- and three-phase mains powered circuits, in particular when measuring phase differences.

Oscilloscope probe

An oscilloscope probe connects the instrument to the circuit under test. A typical 'scope probe has a shielded lead about one metre long, with a BNC plug on one end and the probe on the other. The lead has two conductors: an earth braid that connects to the metal case of the 'scope and also the mains earth, and an inner conductor that conducts the signal to the vertical amplifier input of the oscilloscope. The earth braid around the inner conductor reduces the amount of electrical noise that could otherwise be picked up by the inner conductor. The same type of probe is used with oscilloscopes that have isolated inputs, such as the handheld instruments described above.

The probe itself usually has a retractable hook to connect to a test point, and an earth lead to connect to the circuit's common point. Most oscilloscope probes also have a feature called a ×10 attenuator, as shown in Figure 25.24.

FIGURE 25.24 Most oscilloscope probes have a ×1/×10 select switch

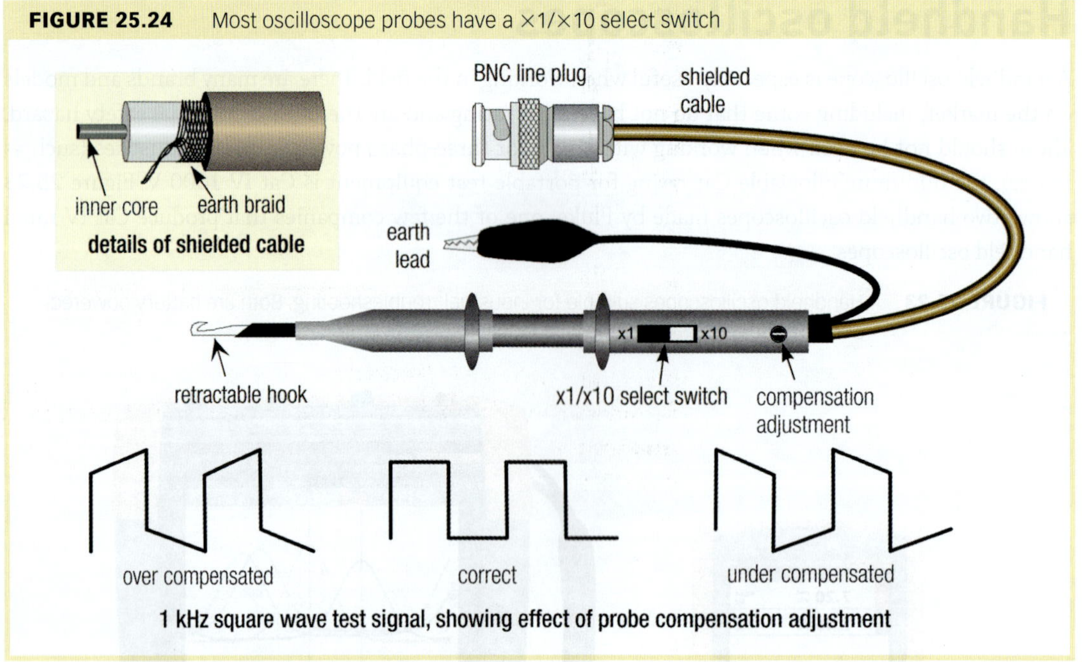

The ×10 setting on an oscilloscope probe has two purposes:

1. To increase the input impedance of the 'scope (usually from about 1 MΩ to over 10 MΩ), so the 'scope doesn't load the circuit being tested and affect the readings.
2. To allow the 'scope to display higher voltages. The highest volts/division setting on some oscilloscopes is as low as 5 V/div, which for a graticule with eight divisions gives a maximum display of 40 V_{p-p}. With a ×10 probe, the display voltage is increased by 10 times to 400 V_{p-p}.

When a ×10 probe is used, the setting of the vertical scale needs to be multiplied by 10 to give the correct reading. For example, if a waveform measures four divisions peak to peak at a vertical scale of 2.0 V per division, when the probe is set to ×1, the peak-to-peak voltage is 8.0 V. If the probe is set to ×10, the voltage is 80 V. On a DSO, the probe setting can be matched to the same setting on the instrument, thereby ensuring the displayed voltage values are correct.

As shown in Figure 25.24, probes with a ×10 setting have an adjustment to compensate for the capacitance of the lead. This is done by connecting the probe to a 1 kHz (or so) square wave, which is provided on the front panel of most oscilloscopes. The adjustment is only done at the ×10 setting, and when correct should cause the display to be a square wave without top and bottom tilt or distortion.

25.2 Electrical Test Equipment

This section describes the following electrical test instruments:

- voltage indicators
- voltage testers
- insulation resistance testers
- clip-on ammeters
- wattmeters.

Voltage indicators and testers

Voltage indicators include a wide range of devices, the simplest being an electrician's test lamps. These have been superseded by numerous testing devices, and some organisations no longer approve their use. Voltage testers fall into two categories: those that indicate the presence of a voltage and those that indicate the presence of and measure the voltage. There are two basic types: those that need direct contact with a live terminal and a path to earth, either through the user or an external connection, and those that respond to an electric field (proximity tester).

Safety

A problem with lamps and many voltage testers is the potential for the tester to fail just as you are about to use it, thereby causing it to indicate that a line or circuit is de-energised when it is still active. It is important for your safety to always prove that a voltage testing device is working before and after use. That is:

1. test the tester
2. use the tester
3. test the tester.

Step 3 is especially important if the tester indicates a terminal is dead. Some organisations provide a 'proving unit' for voltage testers, which is a device that outputs a voltage suitable for confirming a tester is functional.

A simple indicator that has been around for a long time is the neon test pencil, which has a neon lamp inside a clear body, generally with a screwdriver blade attached to the body. These operate by passing enough current to earth through the person using the indicator to light the internal neon lamp. They are an unsafe indicator for these reasons:

- In normal light, the neon lamp can be difficult to see when it is lit.
- The screwdriver tip is not insulated and is broad enough to span two close-by connections.
- If the user does not provide a sufficient path to earth, the neon lamp will not light.
- Most test pencils do not have an IP rating and will not prevent water entering the case. If water does enter, using the test pencil could cause a fatal shock.
- These devices do not usually have a Cat rating, despite being rated at 500 V. They would not be permitted in some workplaces.

FIGURE 25.25 Test lamps and neon test pencils are no longer permitted in many workplaces as they do not have a suitable Cat rating, and are generally unsafe to use

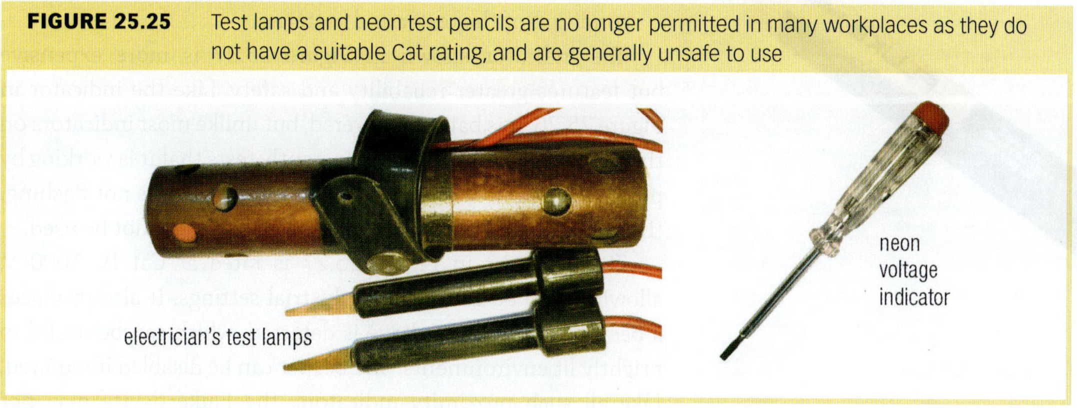

Test lamps are also an unsafe voltage indicator for these reasons:

- If test lamps are connected between an active and an earth terminal (for example, at a GPO), the RCD protection device will trip due to the earth current.
- Lamp filaments can become open-circuit due to vibrations such as when climbing a ladder.
- A lower-than-usual voltage may not cause the lamps to light, even though a lethal voltage is present.
- It is unlikely test lamps have an IP rating, allowing water to enter the lamp sockets.
- Test lamps are unlikely to be Cat rated.

Voltage indicators

Voltage indicators are a family of devices that indicate the presence of a voltage. The most common type is the proximity indicator, which senses the electric field associated with a voltage. A typical device is shown in Figure 25.26.

FIGURE 25.26 A typical low-cost battery-powered proximity voltage indicator

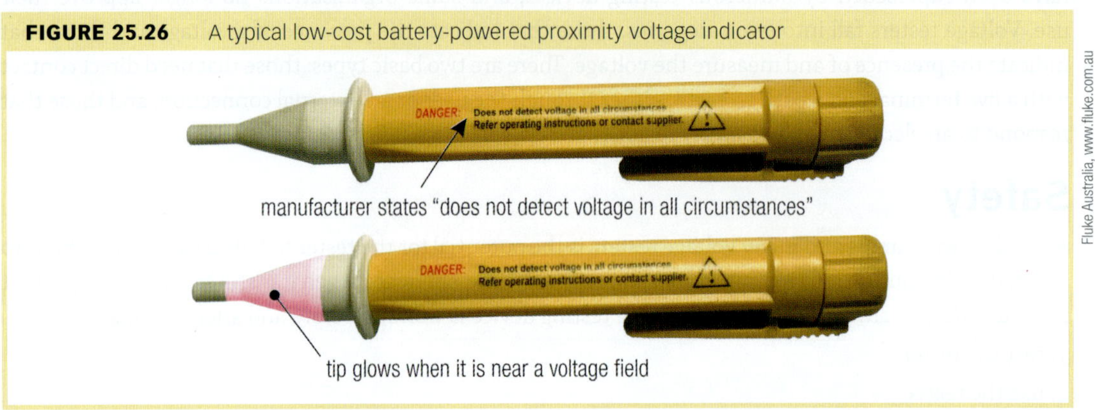

Proximity indicators can be especially useful in troubleshooting a problem in an electrical installation. For example, holding the indicator near a light switch can indicate if power is available at the switch. Most proximity indicators can respond when the tip is placed near the switch, so there is no need to remove it from a wall cavity.

But these devices are not reliable enough to depend on. The manufacturer of the device in Figure 25.26 warns the user that it may not respond in all cases. If the tip is held near a cable with an active and neutral pair, the indicator will not respond. Sometimes nearby cables can cause the indicator to light, or they can prevent the indicator from lighting. Reasons a proximity indicator might not light include:

- shielded cables in which the voltage field is contained by the shielding
- thickness and type of insulation around a cable
- distance from the voltage source, requiring the probe to be held closer to the cable
- fully isolated users that prevent an effective ground, such as being on an insulated ladder while using the indicator
- receptacles in recessed sockets, preventing the indicator from being close enough to the electric field
- condition of the tester and batteries.

FIGURE 25.27 A Cat IV rated proximity tester that has a self-test function to confirm it is working

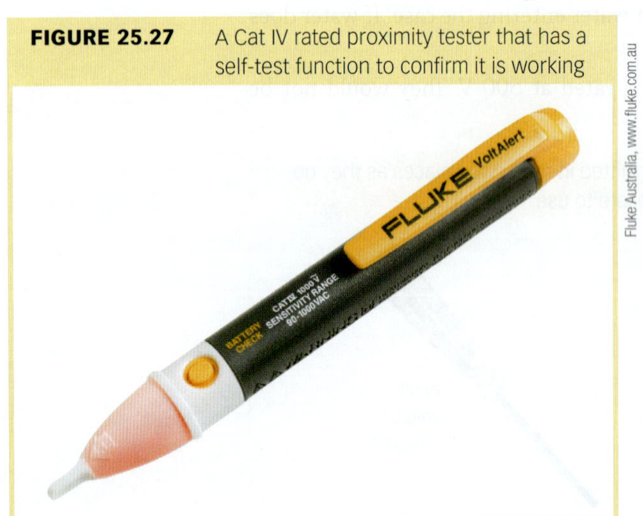

The indicator shown in Figure 25.27 is more expensive but features greater reliability and safety. Like the indicator in Figure 25.26, it is battery-powered, but unlike most indicators on the market, the Fluke device constantly tests that it is working by producing a double-flash every two seconds. If it is not flashing, the manufacturer states that the indicator should not be used.

The indicator in Figure 25.27 is rated at Cat IV 1000 V, allowing it to be used in all industrial settings. It also produces a beeping sound if a voltage is detected, which can be useful in brightly lit environments. The beeper can be disabled if required. Like all such proximity indicators, the Fluke device may not operate because of any of the conditions listed above. As a general rule, if the indicator does not light, do not assume the circuit under test is isolated. Use other tests to confirm this.

Voltage testers

A voltage tester displays a measurement of a voltage. These devices are sometimes called voltage probes, as most models have a probe that contacts the terminal under test. There are numerous examples of voltage testers on the market, all with various features. Figure 25.28 shows examples of two types of voltage testers.

The voltage tester in Figure 25.28(a) operates in a similar way to using a multimeter, except the tester is more compact. The device shown has been superseded by newer models in the Fluke range. Features of these testers include LED indicators, a digital display to show measured values and an audible or vibration indication of a voltage.

FIGURE 25.28 Voltage testers display the value of a voltage, giving a more reliable indication of the presence of a voltage

(a) Voltage tester with probes (b) Voltage tester with field sensing technology

Fluke Australia, www.fluke.com.au

The tester shown in Figure 25.28(b) is more sophisticated because it detects and measures AC voltage, frequency and current using field-sensing technology. It also has two probes for measuring DC voltages and resistance. To measure an AC voltage, the tester is placed so a single cable sits at the bottom of the fork opening. It will not operate on twin-pair cables carrying an active and a neutral conductor.

Both testers in Figure 25.28 are Cat rated, which is the main reason Fluke devices are included in this book. Other brands of testers are also suitable, providing they are Cat rated. In general, a Cat IV 1000 V rating means the device is acceptable in all electrical work environments. Cat III 600 V are also permitted in many workplaces.

Insulation resistance tester

When electrical power wiring is first installed, it has to be checked before it's connected to a 400 V/230 V supply. There are quite a few tests, and one of these is a test of insulation resistance between conductors, and between conductors and earth. Insulation resistance testing of a 400 V/230 V electrical installation is covered in clause 8.3.6 of AS/NZS 3000:2018. Testing insulation resistance of a high voltage installation, such as in a substation, is outlined in clause 7.6.3 of AS/NZS 3000:2018. These two situations differ in that the test voltage applied by an insulation resistance tester needs to be considerably higher when testing high voltage equipment. The test instrument explained here is typical of those used by electricians testing a 400 V/230 V wiring installation.

An insulation resistance tester is an item of test equipment that displays a resistance reading of a circuit by applying a suitably high DC voltage across the circuit and registering the leakage current that occurs. Test voltages are typically 250 V (for circuits with electronic equipment attached), 500 V for a 230 V circuit and 1000 V for 400 V installations. Insulation resistance between conductors, and between live conductors and earth, should be no less than 1 MΩ.

Because the minimum allowable value is 1 MΩ, it makes sense for this value to be about mid-scale. And to make the instrument more versatile, it's usual for it to include an ohmmeter function for measuring low resistance values. A typical insulation resistance tester is shown in Figure 25.29.

FIGURE 25.29 Typical solid-state insulation resistance tester with selectable test voltages of 250 V, 500 V or 1000 V DC

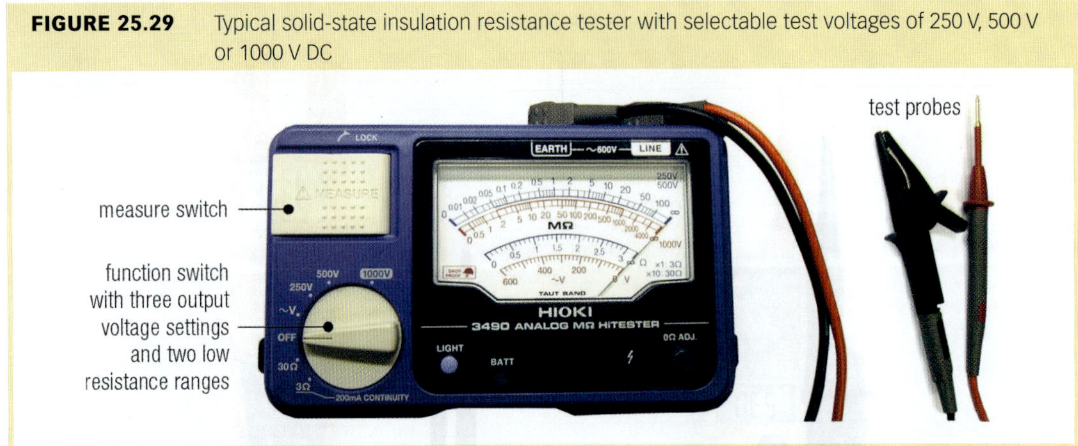

Two ways of producing the high voltage needed by an insulation tester are:

- with a hand-cranked generator, where you turn a crank handle that turns a generator built inside the tester
- with a battery-powered solid-state electronic circuit called an inverter.

Hand-cranked insulation tester

The hand-cranked type has been mostly replaced by the solid-state version as it's easier to use, smaller and lighter. However, the hand-cranked tester has the advantages of being rugged and not needing batteries. The meter movement in this type of tester has two coils. One coil takes the place of the hair-springs normally used in a moving-coil meter movement. The basic circuit is shown in Figure 25.30.

FIGURE 25.30 The meter movement in a hand-cranked insulation tester has two coils: a control coil instead of control springs and a deflection coil to move the pointer

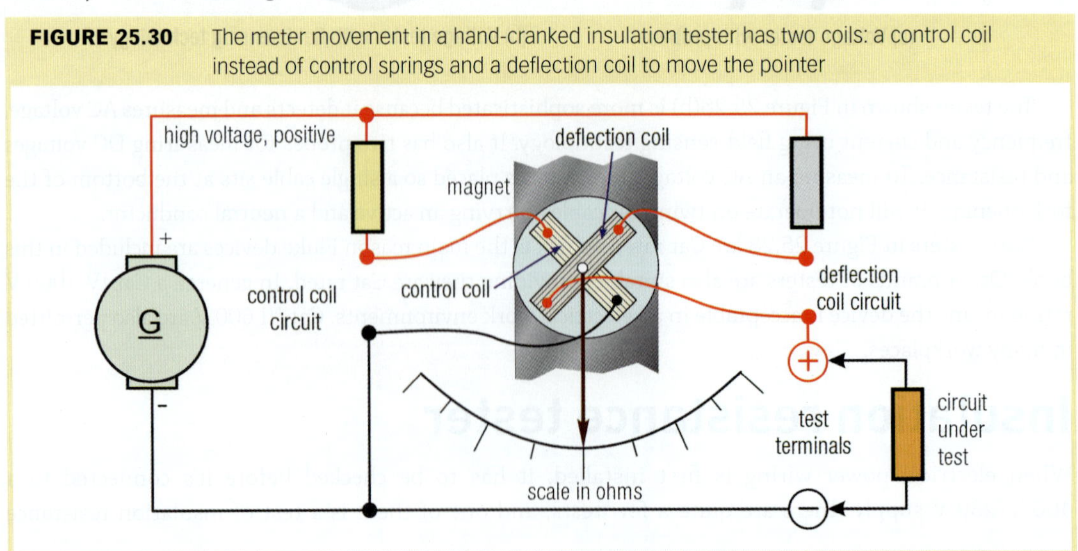

The circuit in Figure 25.30 is like that of an ohmmeter, except that instead of a battery there's a DC generator. As in an ohmmeter, the voltage source (generator) is in series with the meter movement and the circuit being tested. However, instead of control springs to return the movement to its zero position, the assembly has a second coil, called the control coil, fixed at right angles to the deflection coil. This coil is also connected to the voltage source through a resistor.

The deflecting torques of both coils oppose each other, and the assembly will take up a position depending on the current flowing in both coils. The current in the control coil depends on the test voltage. The current in the deflection coil depends on the test voltage and the resistance of the circuit being tested. The reason for this arrangement is that the resistance reading is relatively independent of the test voltage. This is important in a hand-cranked tester because the generator speed is likely to vary, giving a varying output voltage.

Solid-state insulation tester

The insulation tester in Figure 25.29 is a solid-state type. An electronic circuit, called an inverter, replaces the hand-powered generator. Most solid-state resistance testers are powered by up to six size AA 1.5 V cells. It's also usual for the meter movement to have a taut band suspension, as these are more rugged than jewel and pivot assemblies. The unit shown is claimed to be able to survive a metre-high drop onto concrete, at least once.

An insulation resistance test can cause a circuit or some types of appliances to become charged to a lethal voltage. Discharging the test circuit or appliance is essential to prevent the possibility of an electric shock. As well, don't rely on the discharge circuit of the tester. You should also short the two connection points together before touching either of them. To use the tester in Figure 25.29:

- select the test voltage (usually 500 V)
- make sure the circuit or device being tested is not 'live'
- connect the probes to the circuit, press the measure switch to activate the high voltage circuit and read the insulation resistance
- while the probes are still connected to the test circuit or device, release the measure switch so the in-built discharge circuit can remove any charge built up in the circuit or device. When fully discharged, the meter pointer will read infinity (∞).

AS/NZS 3000:2018 sub clause 8.3.6.2 specifies that an insulation resistance tester must be able to maintain its terminal voltage within +20 per cent or −10 per cent of its nominal open-circuit output voltage when measuring a resistance of 1 MΩ on the 500 V range or 10 MΩ on the 1000 V range. This can be tested by connecting a 1 MΩ or 10 MΩ resistor across the tester's output terminals, and measuring the DC voltage with a voltmeter. Hand-cranked units are likely to produce a far less stable output voltage than the solid-state types.

Clip-on ammeter

A clip-on ammeter (also called a clamp ammeter or a tong tester) can be either analog or digital. Its main advantage is that it can measure current without having to cut the current-carrying conductor. Instead, it's clamped around the conductor, as shown in Figure 25.31.

FIGURE 25.31 Analog and digital clip-on ammeters

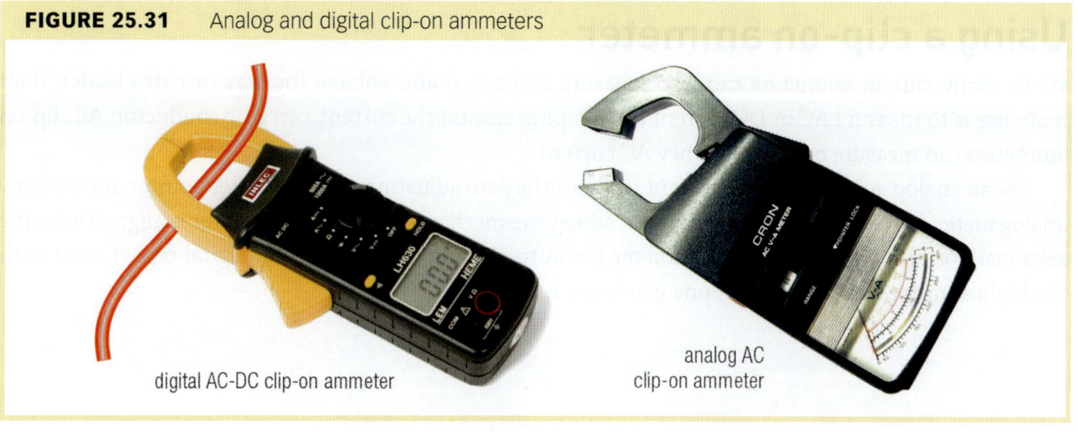

There are two types of clip-on ammeters: those that can measure an alternating current only, and those that can measure both direct (DC) and alternating currents. An AC only clip-on ammeter has a current transformer.

Current transformer type clip-on ammeter

The current transformer (CT) in a clip-on ammeter has a core formed by two laminated iron or ferrite jaws. One jaw is hinged, the other fixed. The moveable jaw is opened by pressing a lever, so the current-carrying conductor can be passed through the centre of the core. Figure 25.32 shows the basic construction of this type of clip-on ammeter. Notice the circuit that converts the AC current from the coil to a DC current. This circuit is called a rectifier, and has four semiconductor diodes.

FIGURE 25.32 Basic construction of a CT type clip-on ammeter

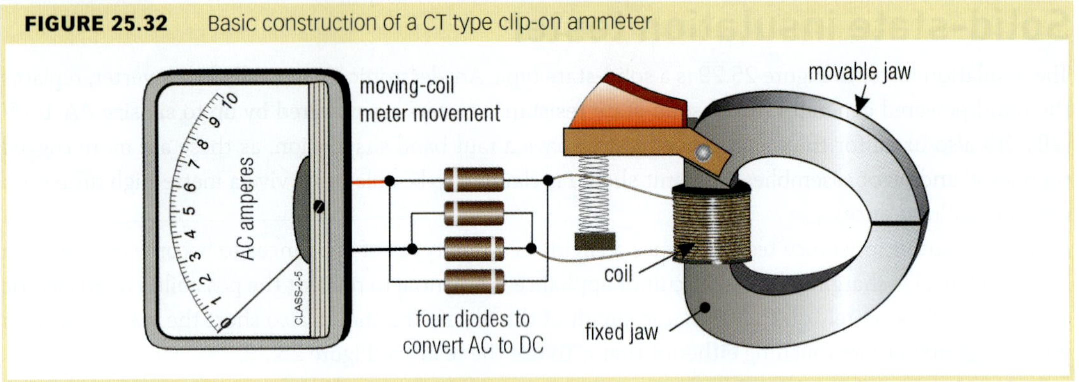

To get a number of current ranges, tappings are usually brought out from the coil and selected by a range switch. The greater the number of turns, the smaller the current the meter can read.

Hall effect clip-on ammeter

A clip-on ammeter that can read DC as well as AC has a semiconductor called a Hall effect device, rather than a transformer. The Hall effect was discovered in 1879 at the Johns Hopkins University (USA) by Edward Hall, who used a piece of pure gold to demonstrate the effect. These days, a semiconductor material is used instead of gold. The Hall effect is this:

- If a constant value current flows through a wafer of semiconductor, and if a magnetic field is brought near the semiconductor, then a voltage will be produced across the edges of the semiconductor that are parallel with the current flow. This voltage is directly proportional to the strength of the magnetic field, and its polarity depends on the direction of the magnetic field.

A Hall effect device is built into the jaws of a clip-on ammeter, and circuitry in the meter supplies the constant value current for the device. In most cases, a Hall effect ammeter has a digital display, as the voltage produced by the device is too small to drive a moving coil meter.

This type of clip-on ammeter can read AC as well as DC because the Hall effect works for moving and stationary magnetic fields. As well, a Hall effect meter can often measure alternating currents with a frequency of 10 kHz or more, making the instrument useful in electronics.

Using a clip-on ammeter

While many clip-on ammeters can also measure resistance and voltage (by way of extra leads), their main use is to measure AC or DC current by clamping around the current-carrying conductor. All clip-on ammeters can measure mains frequency AC current.

For an analog ammeter, it's important to check the zero adjustment of the meter pointer. As with any analog meter, before taking a measurement, adjust the mechanical zero so the pointer is aligned with the zero mark on the scale. Unless the ammeter has auto-ranging (usually only on digital types), start with the highest range setting, or select one you know is suitable.

AC current measurement

The most typical use of a clip-on ammeter is to measure the current taken by a load. To do this, open the moveable jaw and position the meter so the current-carrying conductor is encircled by the jaws. For the reading to be accurate, the jaws must shut properly.

When measuring current taken by a mains-operated appliance, pass one power lead through the ammeter jaws. A convenient way to do this is with an adaptor plug-socket, in which both the active and the neutral leads are looped outside so they're accessible. To measure the current taken by the appliance, plug the adaptor into a power outlet and plug the appliance into the adaptor. Then clip the meter around one of the extended leads, as shown in Figure 25.33.

FIGURE 25.33 An adaptor plug makes it easier to measure the current taken by a mains appliance

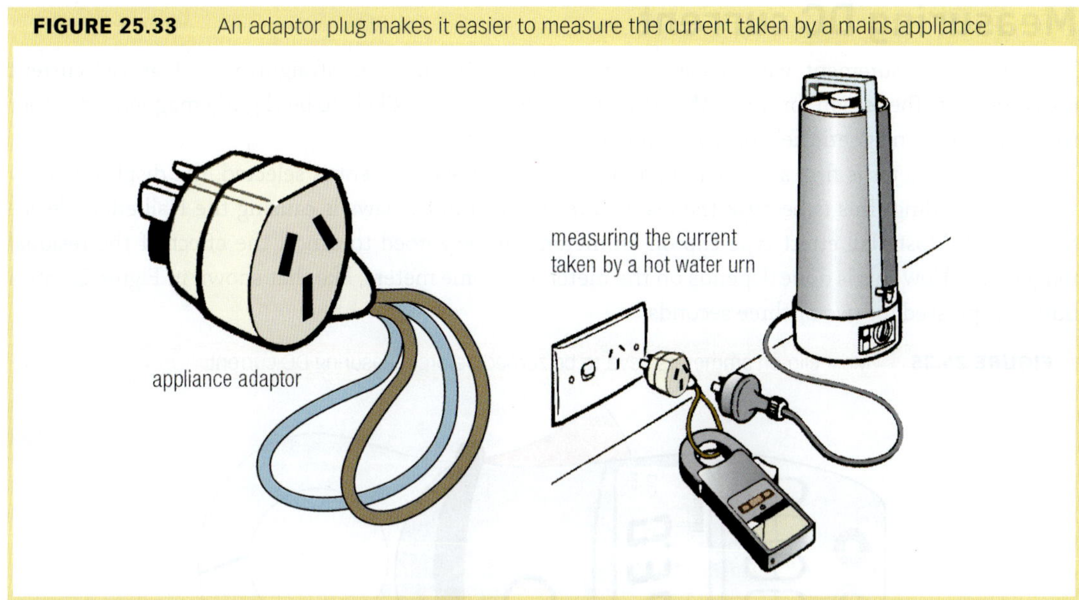

When measuring current with a clip-on ammeter, it's important to remember the voltage of the current-carrying conductor. As a general rule, don't use a clip-on ammeter to measure current in a conductor that's operating at 600 V or more. Also, like all test instruments used in electrical measurement, a clip-on ammeter should have a category rating, typically Cat III, 600 V.

Measuring leakage current

Leakage current of an electrical appliance is the current flowing in the earth conductor. Ideally, this current should be zero. If it's more than 30 mA it will trip a residual current device (RCD). If there's any leakage current, it usually means the appliance is faulty, and dangerous.

A clip-on ammeter can be used to measure leakage current of an appliance in two ways. The first is to measure the current flowing in the earth conductor. Another way is to clamp the meter around both the active and neutral conductors of the appliance. An easy way to do this is with the appliance adaptor described before. Rather than clamping the meter around one lead, both leads pass through the jaws of the meter, as in Figure 25.34.

FIGURE 25.34 To measure leakage current, clamp the meter around both leads of the adaptor

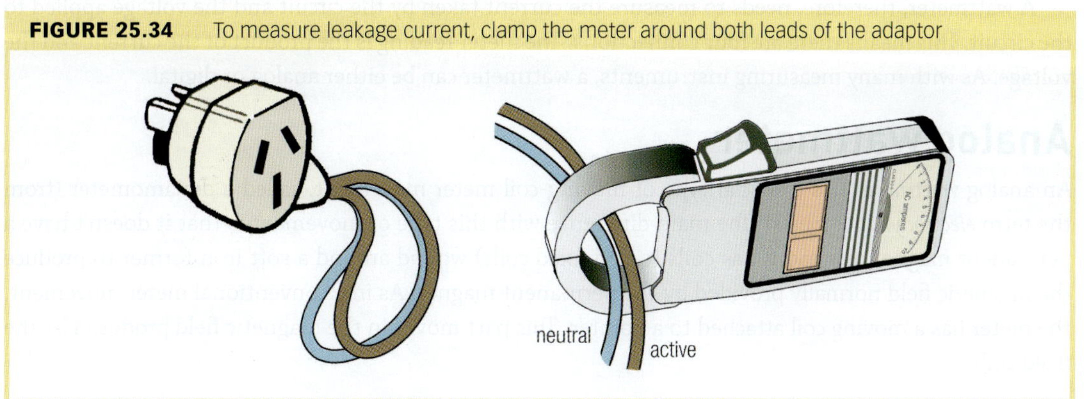

Because the currents in the active and neutral are flowing in opposite directions, their magnetic fields cancel. If the two currents are exactly equal, there will be no magnetic field, and the clip-on ammeter will read zero. But when there's leakage current flowing in the earth conductor of the appliance, the neutral current is less than the current in the active conductor, by an amount equal to the current in the earth conductor. This means the magnetic fields around the active and neutral don't cancel completely, making the ammeter show a reading. This reading is the difference in the active and neutral currents, which is the leakage current of the appliance.

The leakage current of a complete power circuit can be found in the same way. Here the ammeter is clipped around the main active and neutral conductors of the circuit.

Measuring DC current

DC current measurement with a clip-on ammeter is not quite as straightforward as AC current measurement. The reason for this is that the jaws of the meter are likely to be slightly magnetised, either from the Earth's magnetic field or from a previous measurement.

When the meter is first switched on, and DC current measurement is selected, the display is likely to show a reading. This is because the residual magnetism in the jaws is causing the Hall effect device to respond. Most Hall effect type clip-on ammeters can be zeroed to cancel the effect of the residual magnetism. How this is done depends on the meter. For some meters, like that shown in Figure 25.35, a button is pressed for two or three seconds.

FIGURE 25.35 Most clip-on ammeters have to be zeroed before measuring DC current

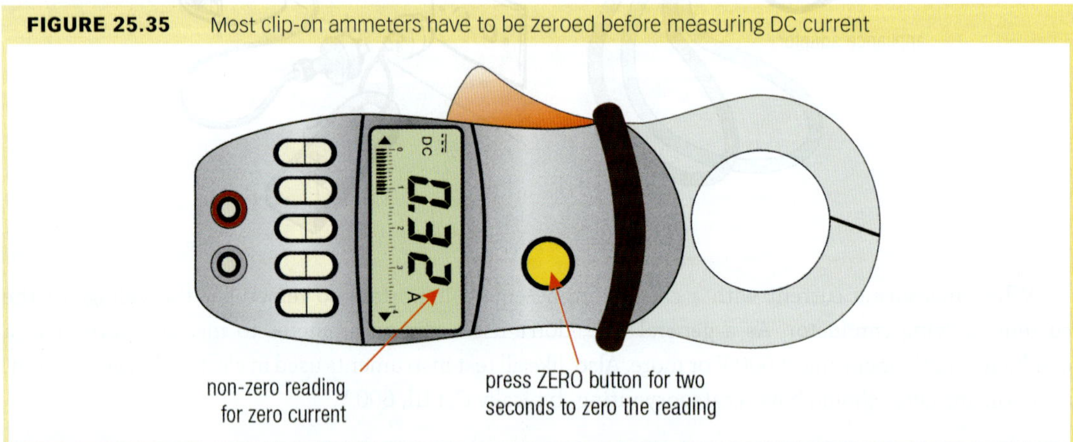

Once the meter is zeroed, the procedure is the same as for AC current measurement, except the polarity of the current will be shown. For best accuracy the meter should be positioned so the conductor is in the centre of the jaws. Some instruments have alignment marks to help you judge the centre position.

The wattmeter

A wattmeter measures electrical power. As you know, power in a DC circuit or in an AC resistive circuit can be calculated by multiplying the voltage of the circuit by the current flowing in the circuit. That is, $P = VI$ watts.

A wattmeter, therefore, needs to measure the current taken by the circuit and the voltage applied to the circuit. This means there are four connections. The meter reading is the product of the current and the voltage. As with many measuring instruments, a wattmeter can be either analog or digital.

Analog wattmeter

An analog wattmeter has a special type of moving-coil meter movement, called a dynamometer (from the term *electrodynamic meter*). The main difference with this type of movement is that it doesn't have a permanent magnet. Instead it has coils (called fixed coils) wound around a soft iron former to produce the magnetic field normally provided by the permanent magnet. As in a conventional meter movement, the meter has a moving coil attached to a pointer. This part moves in the magnetic field produced by the fixed coils.

When the wattmeter is connected to measure power, the circuit current flows through the fixed coils, and the voltage is applied to the moving coil (through a multiplier resistor, as in a voltmeter). The deflection of the movement is therefore determined by the strength of the magnetic fields produced by the fixed coils and the moving coil. The meter scale is then calibrated in watts, as shown in Figure 25.36.

FIGURE 25.36 A wattmeter has fixed coils for the current and a moving coil for the voltage

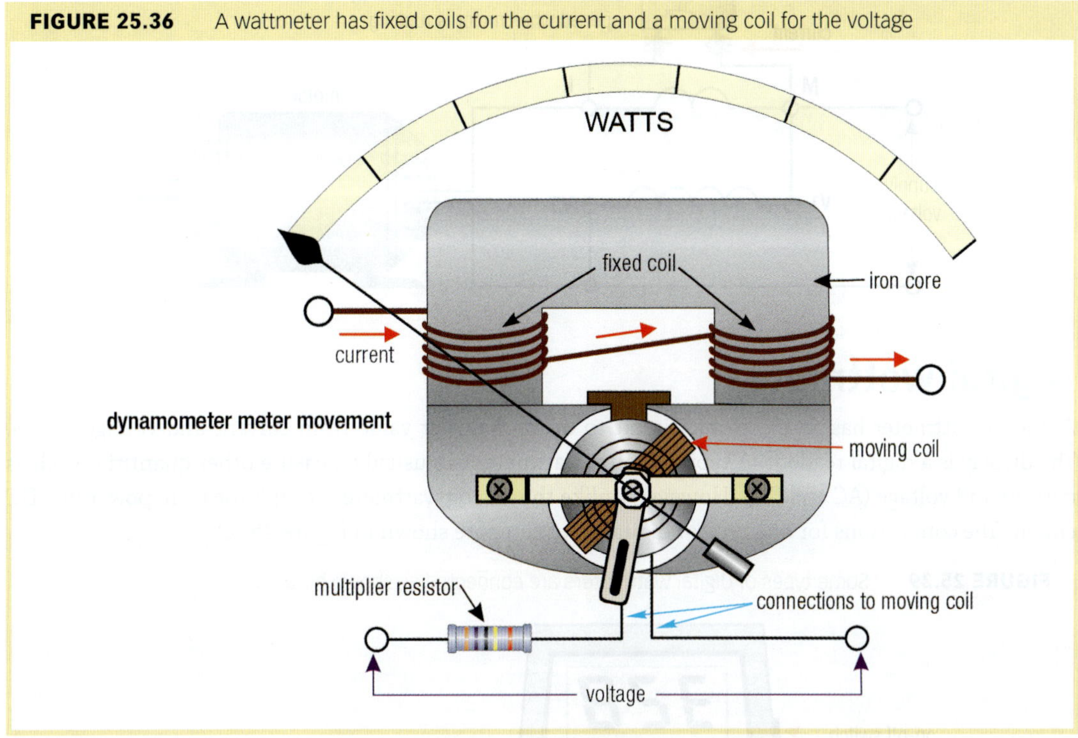

Because the direction of the current in either coil determines the direction of deflection, it's important to get the polarity of both connections correct. The terminals of most analog wattmeters are marked as shown in Figure 25.37, in which the:

- current coil is marked M for connection to the supply side and L for the side that connects to the load
- voltage coil is marked V1 or V+ on one side and V2 or V on the other.

FIGURE 25.37 A wattmeter is labelled in a standard way and is connected as shown

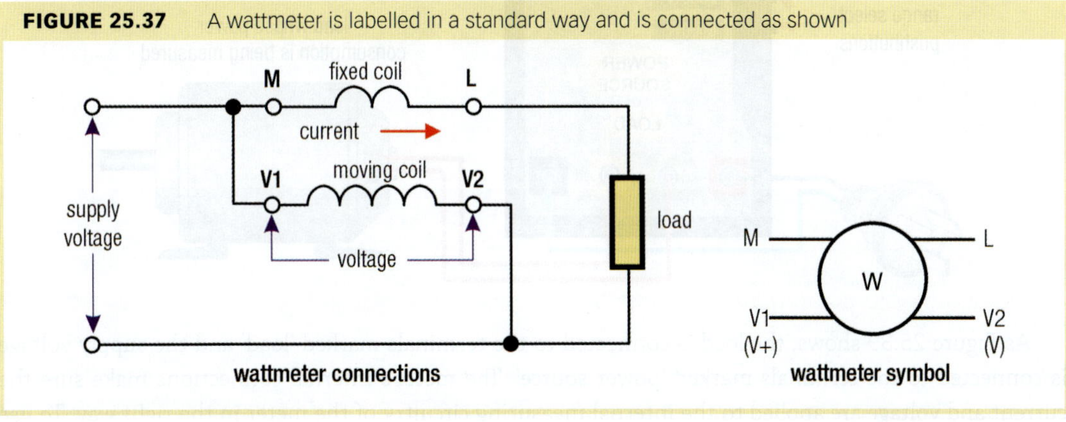

Some wattmeters are fitted with a switch that connects between the M and L terminals. When the switch is on, the current flows through the switch rather than the current coil, as in Figure 25.38. This is needed if a load, like an electric motor, takes a large current when it's first switched on. Without the switch, the high value of current might burn out the current coil. Once the current has fallen to its normal value, the switch is opened so the current flows through the current coil, and the wattmeter can show the power taken by the motor.

FIGURE 25.38 The switch bypasses the current coil so it is not overloaded when the motor is started

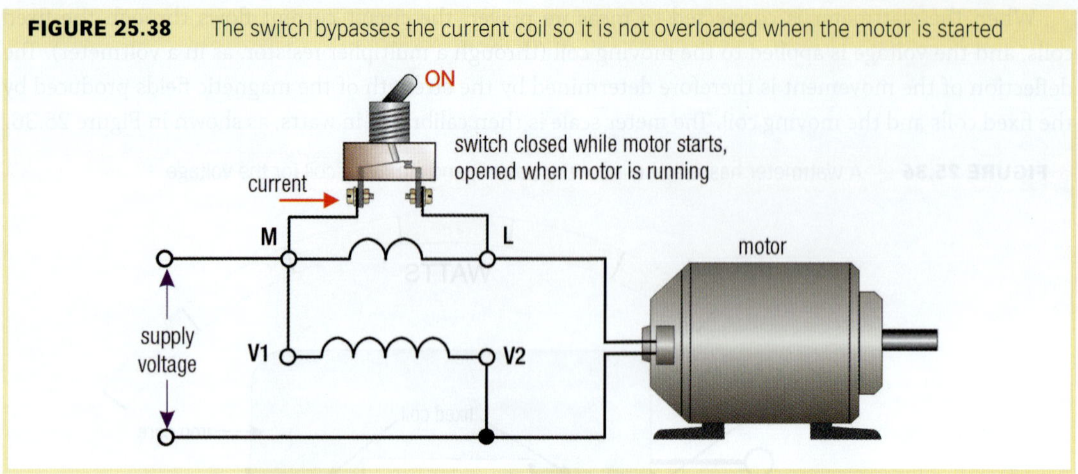

Digital wattmeter

A digital wattmeter has an electronic circuit to derive a power value from current and voltage values. The display is a digital readout. A typical digital wattmeter can usually measure other quantities such as current and voltage (AC and DC). However, unlike the analog wattmeter, it can't measure power in a DC circuit. The connections for one type of digital wattmeter are shown in Figure 25.39.

FIGURE 25.39 Some types of digital wattmeters are connected as shown here

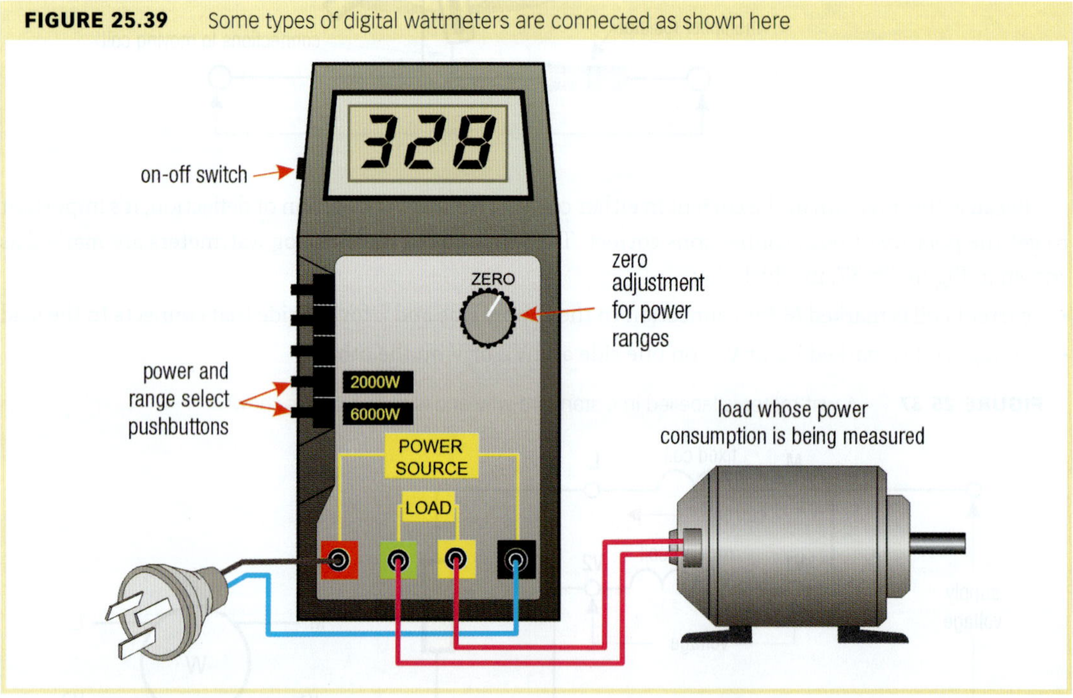

As Figure 25.39 shows, the load is connected to the terminals marked 'load' and the supply voltage is connected to the terminals marked 'power source'. The meter's internal connections make sure the current and voltage are applied to the internal measuring circuitry of the meter in the right way. To use this meter:

1 switch on the meter and adjust the zero control so the display shows zero
2 select the 6000 W range
3 connect the load to the meter as shown
4 connect the meter to the supply source as shown
5 switch on the supply
6 select the 2000 W range if the power reading is less than 2000 W.

To prevent damage to the meter from a high start-up current, the manufacturer recommends connecting a 10 A fuse in series with the load. This meter also lets you measure the load current and the supply voltage during power measurement.

Using a wattmeter

A wattmeter, whether it's digital or analog, is connected so the:

- circuit current flows through the meter
- supply voltage (or load voltage) is read by the meter.

The digital wattmeter shown in Figure 25.39 has these specifications:

- maximum current that can flow through the meter is 10 A (AC or DC)
- maximum voltage that can be applied to the meter is 600 V (AC).

Because P = VI, the maximum AC power the meter can measure is 10 × 600, which is why the highest power range for the meter is 6000 W. But what if the circuit voltage is less than 600 V? Let's say it is 300 V. For a power consumption of 6000 W, the load current is 20 A. But because the wattmeter can only handle a current up to 10 A, the meter will be damaged.

Or, what if the load current is less than its rated 10 A, say 5 A? For a power consumption of 6000 W, the voltage is now 1200 V. This voltage is twice the wattmeter's voltage rating, so obviously it will damage the meter and create a serious hazard.

In both these cases the wattmeter will be damaged, even though the power consumption of the circuit is within the range of the meter. So unless you're certain the current and voltage values are within range of the meter's specifications, measure them *before* a wattmeter is connected into a circuit, and *while* it's connected. Otherwise, the meter can be damaged because of a voltage or current overload. Figure 25.40 shows a typical power measurement setup.

FIGURE 25.40 So the wattmeter is not damaged by overload, the circuit current and voltage should be separately measured

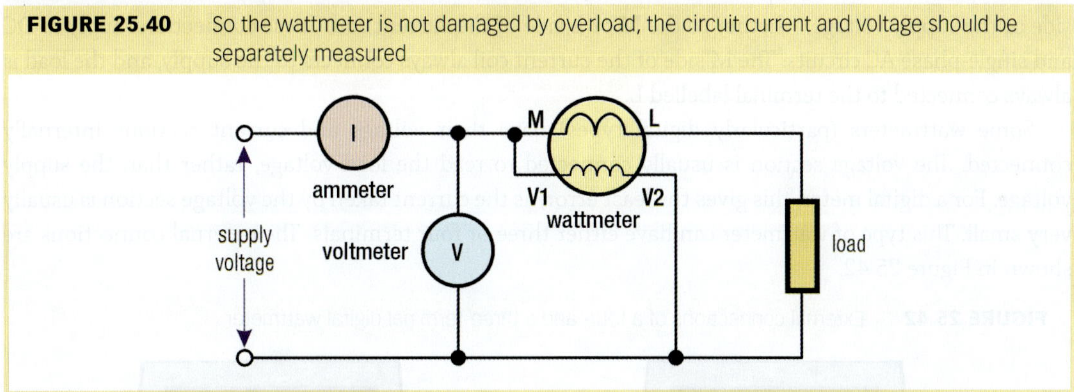

The setup in Figure 25.40 can also be used to find the power factor of a circuit, as the wattmeter measures true power, and apparent power is measured with the voltmeter (V) and the ammeter (I). Apparent power (S) = VI. Power factor, as explained in Chapter 19, is found with the equation:

$$\cos\phi = \frac{\text{true power}}{\text{apparent power}} = \frac{P}{S}$$ where P = true power in watts, S = apparent power in volt-amperes.

Wattmeter connections

If the voltage and current sections of a wattmeter are independent (as in many analog wattmeters), they must be externally connected. There are two ways of connecting the voltage section: across the supply (called the *long-shunt* connection) or across the load (*short-shunt* connection). These are shown in Figure 25.41.

FIGURE 25.41 Two ways of connecting the voltage and current sections of a wattmeter

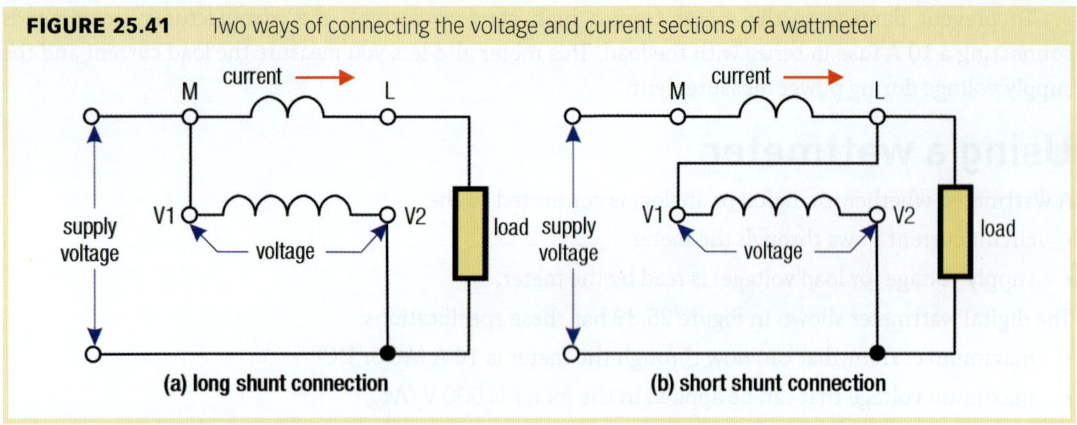

The long-shunt connection is in (a), where the supply voltage rather than the load voltage is measured by the wattmeter. The difference between the two is the voltage drop across the current coil. This gives an error in the power reading that depends on the resistance of the ammeter coil of the wattmeter and the value of the circuit current.

The short-shunt connection is in (b), where the wattmeter reads the load voltage (not the supply voltage). However, the current read by the wattmeter includes that taken by the voltage coil. The error in the power reading depends on the current the voltage coil is taking compared to the load current.

Either way gives an error, however small. The best connection to use depends on the wattmeter characteristics, the load resistance (and therefore load current) and the supply voltage.

An important thing to remember is that the two sections of a wattmeter must be connected with the right polarity. Notice in Figure 25.41 that the voltage coil is shown with two markings, as wattmeter manufacturers use both. The point marked V1 (or V+) always connects to the positive (or active for AC) side of the supply voltage, whether to the M or L side of the current coil. These connections apply to DC and single-phase AC circuits. The M side of the current coil always connects to the supply, and the load is always connected to the terminal labelled L.

Some wattmeters (particularly digital types) have their voltage and current sections internally connected. The voltage section is usually connected to read the load voltage, rather than the supply voltage. For a digital meter, this gives the least error, as the current taken by the voltage section is usually very small. This type of wattmeter can have either three or four terminals. The external connections are shown in Figure 25.42.

FIGURE 25.42 External connections of a four- and a three-terminal digital wattmeter

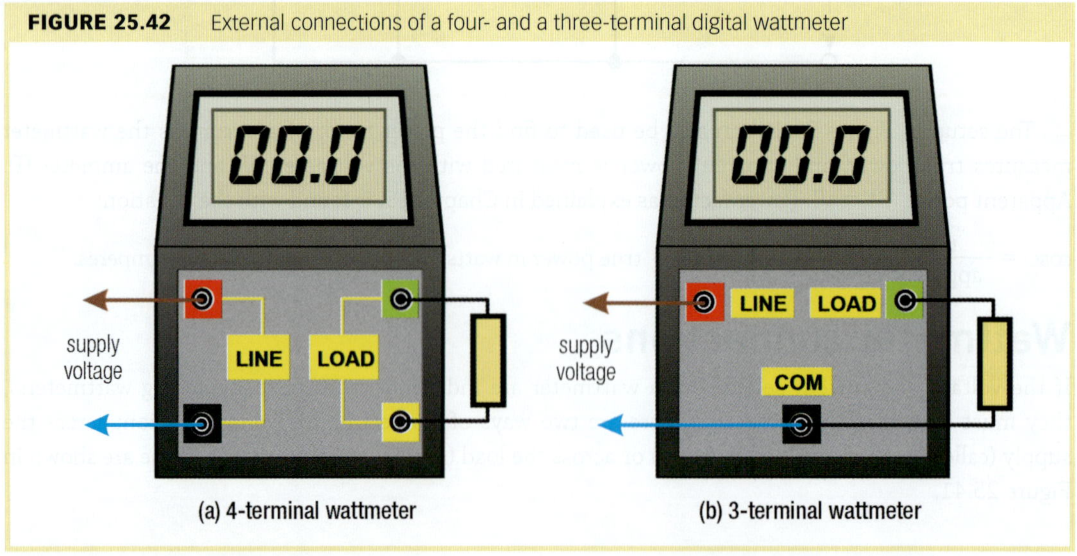

Because some digital wattmeters can't measure DC power, their terminals are not marked with a polarity sign. An AC supply can be connected either way because it doesn't have a polarity.

Finally, remember that the current section of a wattmeter can be damaged by the start-up current of some types of loads, particularly electric motors. For these loads, use a shorting switch between the line and load terminals, or connect a suitably rated fuse in series with the supply and the wattmeter.

Digital wattmeters come in a range of packages. The unit shown here is just one example. Some wattmeters are able to measure three-phase power, including power factor. A commonly available wattmeter plugs into a standard GPO, and the appliance plugs into the wattmeter. The current taken by the appliance is measured, as is the voltage at the GPO, allowing true power to be measured and displayed. An issue is accuracy with low power appliances, such as a computer in sleep mode. The current waveform may be non-sinusoidal, causing a much higher reading than it should be. This is also a problem with some types of loads, such as a lightly loaded motor, in which reactive power might be displayed, rather than true power.

REVIEW EXERCISES

Check your answers at the back of the book.

1. An oscilloscope is displaying a waveform that measures 5.6 divisions high. The vertical attenuator is set to 2 V/division and the signal is coupled to the 'scope through a ×10 probe. What's the peak-to-peak voltage of the waveform?
2. Two 50 Hz out-of-phase sinewaves are displayed on an oscilloscope, in which one cycle takes 9.2 graticule positions. The distance between the positive going zero crossing points of both waveforms measures 1.3 divisions. What's the phase difference between the waveforms?
3. A DSO is showing two 100 Hz out-of-phase sinewaves. The cursors are positioned at the positive going zero crossing points of both waveforms, and the DSO displays a time difference of 1.6 ms. Determine the phase difference between the waveforms.
4. How can you be sure a mains circuit is isolated when using a proximity voltage tester?
5. When testing the insulation resistance of a 230 V stove element, what test voltage should be used?
6. How can you use a clamp ammeter to read earth leakage current due to an appliance?
7. How can you be sure a voltage tester is giving the correct result?
8. Give three reasons electrician's test lamps should not be used.
9. A wattmeter has a maximum range of 5 kW and a maximum current rating of 10 A. How much power can the wattmeter safely read if the applied voltage is 230 V?
10. How can the true power and the apparent power taken by an appliance be measured?

ONLINE RESOURCES

COMPLETE WORKSHEET TWENTY-FIVE

Check with your instructor for worksheets on this chapter.

SOLUTIONS

CHAPTER 1

REVIEW EXERCISES

1 Two parts are renewable energy and energy efficiency.
2 An electric current is a flow of electrons.
3 A voltage and a path for the current are required.
4 Voltage symbol is V, measured in volts. Current symbol is I, measured in amperes. Resistance symbol is R, measured in ohms.
5 Copper atoms have one electron in the outer shell, called a valency of 1. This allows the outer electron to easily detach from an atom and move to another atom.
6 20 coulombs per second equals 20 amperes (20 A).
7 The circuit symbols are (a) fuse (b) lamp (c) open switch.
8 A closed-circuit means current can flow, an open-circuit means current cannot flow.
9 The current will decrease.
10 An ammeter is connected in series so all circuit current flows through the meter.

CHAPTER 2

REVIEW EXERCISES

1 Sources of energy are mechanical, chemical, heat and light.
2 Magnetic effect.
3 To produce pure copper that has a low resistance to an electric current.
4 The anode has a positive potential.
5 Zinc is more reactive than steel (see Figure 2.17), so the zinc coating is sacrificed to protect the steel.
6 The three components of a basic electric cell are: the electrolyte (acid or alkali), the anode (positive) and the cathode (negative electrode).
7 The two main chemical effects are electrolysis and the voltaic effect (the electric cell).
8 Advantages of LED lighting systems include high efficiency, long life, consistent light output over wide temperature range, low voltage operation.
9 The value of current flowing through a human body is determined by the resistance of your body, and the voltage causing the electric shock.
10 Protection against direct contact with a live part of a circuit under normal service conditions, and protection against indirect contact with a live part due to a fault condition.

CHAPTER 3

TASK 3.1

1 24 ohms
2 40 V
3 0.5 ohms
4 30 V
5 46 ohms

TASK 3.2

1 2.5 megohms, or 2.5 M ohms, or 2M5 ohms
2 4 milliamps or 4 mA
3 2 200 megawatts
4 6.53 V
5 3 micro ohms (3 $\mu\Omega$)

TASK 3.3

1 1800 MW
2 1.25 V
3 0.47 volts
4 3 333 330 ohms
5 4.25 GW

REVIEW EXERCISES

1 The current will drop in value by half.
2 R = 1/G so resistance is 1/12 S = 0.08 ohms
3 R = V/I, resistance = 230/12 = 19.17 ohms
4 I = V/R, so I = 12/120 = 0.1 A
5 **(a)** 1.2×10^6 **(b)** 8.8 **(c)** 20×10^3
6 I = V/R, so current = 25/10 = 2.5 A
7 **(a)** 330×10^3 V **(b)** 22×10^{-3} A **(c)** 4.7×10^6 ohms
8 **(a)** 1.2 millivolts **(b)** 9.6 μA **(c)** 0.0356 amperes
9 8 V
10 250×10^{-6} ohms, or 250 μohms

CHAPTER 4

TASK 4.1

1 245 N
2 734 joules
3 300 N
4 357.7 W (force = 715.4 N, work done = 7154 joules)
5 Efficiency = 90 per cent, losses = 50 W

TASK 4.2

1 4600 W or 4 kW
2 15.65 A
3 2 420 000 W or 2.42 MW
4 10 kW, 240 kWh
5 $72

TASK 4.3

1 1125 W or 1.125 kW
2 25 ohms
3 400 V
4 93.75 W
5 47.43 A

REVIEW EXERCISES

1 Work is done when energy is transformed from one form to another. Work and energy are both measured in joules.
2 141.38 kW
3 92 per cent
4 6.52 A
5 (a) 46.8 Mj (b) 13 kW (c) $3.90
6 10 kW
7 **(a)** R = 46 Ω, P = 1.15 kW
(b) R = 3 kΩ, P = 192 mW
(c) R = 2 kΩ, P = 80 W
8 **(a)** R = 266.67 Ω **(b)** I = 7.75 A **(c)** V = 230.04 V
9 **(a)** 3.91 A **(b)** 7.83 A **(c)** 14.35 A
10 **(a)** 58.78 Ω **(b)** 29.39 Ω **(c)** 16.03 Ω

CHAPTER 5

TASK 5.1

1 2.58 ohms
2 1.22 ohms
3 1.87 ohms
4 10.75×10^{-6} m², or 10.75 mm²
5 60 ohms

TASK 5.2

1 Values from left to right: 220 kΩ ±10%, 4.7 kΩ ±5%, 62 Ω ±2%, 4.3 kΩ ±1%, 1 kΩ ±5%

REVIEW EXERCISES

1 Use the 2.5 mm² cable, as its resistance is 1.376 ohms. A larger cable size is more expensive and unnecessary.
2 The csa of the 2.5 mm diameter cable is 4.91 mm², so it has the lowest resistance.
3 50 A
4 negative temperature coefficient
5 2 m
6 minimum csa of copper cable is 5.19×10^{-6} m² or 5.19 mm²
7 1.43×10^{-3} ohms, or 1.43 milliohms
8 A surge arrestor acts as a short-circuit for the duration of the lightning strike and passes the lightning strike current to ground, thereby protecting the transmission line and equipment connected to the line.
9 No, the highest allowable within tolerance value is 220 ohms.
10 **(a)** 2k7 Ω ±10% **(b)** 680 k Ω ±5%
(c) 39 k Ω ±2% **(d)** 240 k Ω ±1%
(e) 12 Ω ±5%

CHAPTER 6

TASK 6.1

1 Current = 2 A
2 R_2 = 36 Ω
3 Voltage across R_2 = 192 V
4 Supply voltage V_T = 50 V
5 R_2 = 5.2 Ω, R_3 = 8 Ω, R_4 = 7.8 Ω, R_T = 24 Ω

REVIEW EXERCISES

1 **(a)** R_T = 50 ohms **(b)** I = 4.6 A **(c)** voltage across 15 ohm elements = 69 V **(d)** voltage across 10 ohm elements = 46 V **(e)** total power = 1058 W
2 **(a)** 6.96 A **(b)** 8.26 ohms
3 **(a)** voltage across faulty element is 230 V, and 0 V across the other elements
(b) voltage across faulty element is 0 V, and 76.67 V across the other elements
4 **(a)** 20 ohms **(b)** 80 W
5 **(a)** 11.76 ohms **(b)** 19.57 A **(c)** 2.94 ohms **(d)** 1125 W or 1.125 kW **(e)** 57.5 V
6 **(a)** 76.67 V **(b)** 26.08 A **(c)** 5997.73 or 6 kW
7 **(a)** current = 50 mA **(b)** R_6 = 330 Ω **(c)** circuit resistance (R_T) = 2 kΩ **(d)** R_2 = 600 Ω

CHAPTER 7

TASK 7.1

1 **(a)** V = 120 V **(b)** I_2 = 1.2 A **(c)** I_T = 7.2 A **(d)** R_T = 16.67 Ω **(e)** R_1 = 120 Ω
2 **(a)** R_1 = 80 Ω **(b)** I_1 = 2.5 A **(c)** I_2 = 7.5 A **(d)** R_T = 12.5 Ω **(e)** R_2 = 26.67 Ω **(f)** R_3 = 33.33 Ω
3 **(a)** 23.23 Ω **(b)** 169.3 Ω **(c)** 1.86 kΩ **(d)** 8 Ω

REVIEW EXERCISES

1 **(a)** 7.5 A **(b)** P_1 = 345 W, P_2 = 690 W, P_3 = 460 W and P_4 = 230 W **(c)** 1725 W or 1.75 kW
2 **(a)** 2200 W or 2.2 kW **(b)** 700 W appliances, I = 3.04 A, 400 W appliances I = 1.74 A **(c)** 9.56 A **(d)** 24.06 ohms
3 **(a)** lathe current = 4.35 A, drill press current = 2.17 A, compressor current = 6.52 A, grinder current = 1.3 A **(b)** 14.35 A **(c)** lathe = 52.87 ohms, drill press = 106 ohms, compressor = 35.28 ohms, grinder = 176.92 ohms **(d)** 16.04 ohms
4 **(a)** 225 V **(b)** I_2 = 3 A **(c)** I_3 = 5 A **(d)** R_3 = 45 Ω **(e)** total power = 2.14 kW
5 **(a)** R_T = 24.73 Ω **(b)** I_1 = 2.3 A, I_2 = 5 A, I_3 = 2 A **(c)** P_1 = 529 W, P_2 = 1150 W, P_3 = 460 W **(d)** P_T = 2138.9 W or 2.14 kW
6 **(a)** 15.52 ohms **(b)** 2.56 A **(c)** 4.6 A **(d)** 14.82 A (rounding might give 14.84 A)

CHAPTER 8

TASK 8.1

1 **(a)** 50 Ω **(b)** 100 Ω

TASK 8.2

1 **(a)** R_T = 160 ohms **(b)** I_{R1} = 0.25 A **(c)** V_{R4} = 100 V
2 **(a)** R_T = 100 ohms **(b)** V_{R2} = 60 V **(c)** I_{R6} = 0.5 A

REVIEW EXERCISES

1 **(a)** I_T = 10 A **(b)** I_1 = 6 A **(c)** I_2 = 2.5 A **(d)** R_T = 1.2 ohms **(e)** $I_{lamp\ 1}$ = 4 A, $I_{lamp\ 2}$ = 3.5 A, $I_{lamp\ 3}$ = 2.5 A **(f)** total power = 120 W **(g)** power dissipated in the cable = 9.85 W

2 **(a)** $R_{eq\ LED}$ = 24 Ω, R_T = 8.12 Ω **(b)** I_T = 1.48 A **(c)** voltages: lamp 1 = 11.93 V, lamp 2 = 11.83 V, lamp 3 = 11.73 V **(d)** total power = 17.73 W **(e)** power dissipated in the cable = 0.25 W

3 **(a)** R_T = 83.72 Ω **(b)** V_{R2} = V_{R4} = 13.33 V **(c)** ?V = 0 V

CHAPTER 9

TASK 9.1

1 1 milliohm

2 1 megohm

3 20 megohm

REVIEW EXERCISES

1 100 kilohms

2 R_{shunt} = 0.0083 Ω or 8.33 milliohms

3 200 V

4 Analog meter causes least loading because it has a resistance of 20 MΩ, DVM has a resistance of 10 MΩ.

5 Use the short shunt connection (voltmeter across load) as the voltmeter current will be insignificant compared to the load current, and the voltage drop across the ammeter is accounted for.

6 V_{R2} = 33.33 V before meter is connected, = 25 V when meter is connected.

7 V_{R2} = 33.22 V (assumes DVM resistance is 10 MΩ)

8 Test current is not high enough, also lead and contact resistance adds to the measured value.

9 Cat III meter can handle a fault current six times that of a Cat II meter.

10 One meter is loading the circuit more than the other; the highest voltage reading is the most correct reading.

CHAPTER 10

TASK 10.1

1 Q = 2 millicoulombs (2 mC)

2 V = 600 V

3 Energy = 18.15 joules

4 C = 0.55 μF

5 Energy = 1.8 joules

TASK 10.2

1 **(a)** Total capacitance = 573.8 nF **(b)** Charge on C_1 = 0.6 μC **(c)** Total charge = 28.69 μC

2 **(a)** Total capacitance = 0.8 μF **(b)** Total charge = 32 μC **(c)** Charge on C_1 = 32 μC **(d)** Voltages across each capacitor: V_1 = 20 V, V_2 = 16 V, V_3 = 4 V.

REVIEW EXERCISES

1 **(a)** C_T = 113.3 nF or 0.113 μF **(b)** Q_T = 22.66 μC **(c)** Q_1 = 20 μC, Q_2 = 2 μC, Q_3 = 660 nC or 0.66 μC

2 **(a)** C_T = 122.79 nF or 0.123 μF **(b)** Q_T = 49.12 μC **(c)** V_{C1} = 223.26 V, V_{C2} = 104.51 V, V_{C3} = 72.23 V

3 C_T = 0.27 μF, voltage across each capacitor = 166.67 V

4 W = 199.65 joules

5 C = 0.295 μF

6 V_C = 13.33 kV

7 200 secs or 3 mins, 20 secs.

8 The lowest intolerance value is 29.7 μF, so the capacitor should be replaced because it is out of tolerance and likely to deteriorate even further.

9 100 capacitors are needed.

10 4255.32 Ω or 4.26 kΩ.

CHAPTER 11

TASK 11.1

1 50 000 lines of force

2 B = 0.1 T or 100 mT

3 force over a metre length = 250 N

4 F_M = 12 500 ampere turns

5 H = 50 000 At/m

TASK 11.2

1 $\mu = 3.28 \times 10^{-3}$ H/m

2 B = 7.2 T (H = 1600 At/m)

3 R_M = 500 000 or 500 x10³ At/Wb

4 B = 0.33 T

5 relative permeability (μ_r) = 1763.67

REVIEW EXERCISES

1 **(a)** south pole **(b)** south pole

2 A ferromagnetic material is strongly attracted to a magnetic field, a paramagnetic material is weakly attracted to a magnetic field and a paramagnetic material is weakly repelled by a magnetic field.

3 A magnetic shield causes magnetic flux to flow in the shield and bypass the shielded object. Shields can be of any magnetic material, although special materials are made for the purpose.

4 B = 0.025 T or 25 mT

5 Factors that reduce the magnetic strength of a permanent magnet include exposure to a high temperature, hammering or jarring the magnet, exposing it to another magnetic field, and poor storage such as not using a keeper across the magnet's poles.

6 The conductors repel each other with a force of 0.0067 N.

7 Conductors are attracted to each other with a force of 88 N.

8 F = 1920 N, conductors repel each other.

9 A south pole.

10 F_m = 840 ampere turns

11 H = 4000 At/m

12 B = 1.63 tesla, $\mu_r = 323$

13 **(a)** A = 0.000314 m^2 or 314 x 10^{-6} m^2 **(b)** Φ = 0.51 mWb

14 $\mu = 44.1 \times 10^{-3}$ H/m

15 B = 0.314 T or 314 mT

16 Φ = 110 μWb

17 $R_m = 2 \times 10^{-6}$ At/Wb

18 Residual flux is a measure of the remaining magnetism in a material when the magnetising force is zero. Coercive force refers to the amount of magnetising force needed to reduce the residual flux to zero.

19 Magnetic losses include hysteresis loss in the material, air gap loss, fringing and leakage flux, and eddy current losses in magnetic circuits operating from alternating current.

20 1.27 x 10^{-6} At/Wb

CHAPTER 12

REVIEW EXERCISES

1 Current flow is anticlockwise.

2 Point A is positive with respect to point B.

3 e = 1.8 mV

4 e = 44 V (rate of flux change = 0.1 Wb/5 = 0.02 Wb/sec, e = 2200 × 0.02 = 44 V)

5 induced voltage = –60 V.

6 L = 154.6 μH

7 L = 0.74 H

8 **(a)** steady state current = 0.1 A **(b)** time constant = 60 ms

9 When the switch is first set to position B the current flows in a clockwise direction, or the same direction it flowed before.

10 L = 2.4 H

CHAPTER 13

REVIEW EXERCISES

1 To provide a magnetic field

2 The output voltage reduces

3 1875 V or 1.88 kV

4 The output voltage increases

5 The output voltage will be zero or very small and will not build up

6 378.4 V

7 **(a)** 300 V **(b)** 1.08 kW **(c)** 318 V **(d)** 636 W **(e)** 961 W **(f)** 18 kW **(g)** 20.68 kW

8 91.68 per cent

9 0.25 ohms

10 V_{reg} =7.6 per cent

CHAPTER 14

REVIEW EXERCISES

1 **(a)** 393.75 N **(b)** 59.1 Nm

2 393.6 V

3 13.61 kW

4 A series motor has the greatest torque because the torque is proportional to the square of the armature current due to this current also flowing in the field coils.

5 **(a)** 16 kW **(b)** 14.28 kW **(c)** 1.72 kW

6 P_T = 2.11 kW (P_{SH} = 781.25 W, P_{SE} = 540 W, P_A = 792 W)

7 718.75 W (= $P_{in} - P_{SH}$)

8 93 per cent

9 1002.5 A or 1 kA

10 $P_{rheostat}$ = 22.5 kW

CHAPTER 15

REVIEW EXERCISES

1 **(a)** V_{max} = 40 V **(b)** $V_{p\text{-}p}$ = 80 V **(c)** V_{RMS} = 28.28 V **(d)** v = 37.59 V

2 **(a)** 40 ms **(b)** 25 Hz

3 **(a)** V_{max} = 48 V **(b)** $V_{p\text{-}p}$ = 96 V **(c)** V_{RMS} = 33.94 V **(d)** V_{av} = 30.58 V

4 **(a)** 8 ms **(b)** 72°

5 Voltage 1 leads voltage 2

6 **(a)** V_{max} = 300 V **(b)** V_{RMS} = 212.13 V **(c)** v at 240° = –259.71 V

7 **(a)** V_{max1} = 100 V, V_{max2} = 50 V **(b)** V_{RMS1} = 70.71 V, V_{RMS2} = 35.36 V **(c)** V_{total} = 35.36 V_{RMS}

8 V_T = 96.7 V, phase angle (ϕ) = 48.1° (leading)

9 I_T = 54 A, phase angle (ϕ) = –5.8° (lagging)

10 I_2 = 21.9 A, phase angle (ϕ) = –11.5° (lagging)

CHAPTER 16

REVIEW EXERCISES

1 **(a)** P_T = 1322.5 W or 1.32 kW **(b)** P_{R1} = 595.125 W, P_{R2} = 330.625 W, P_{R3} = 396.75 W

2 When C is 100 μF, X_C = 31.85 Ω, when C is 36 μF, X_C = 88.46 Ω

3 **(a)** I = 0.159 mA [X_C = 1447.6 Ω] **(b)** I = 4.04 A [X_C = 56.9 Ω] **(c)** I = 722.1 μA [X_C = 318.5 kΩ]

4 C = 2.77 μF

5 f = 106.16 kHz

6 The capacitive reactance is reduced and the current increases.

7 X_L = 1557.44 Ω

8 L = 9.55 mH

9 f = 159.2 Hz

10 **(a)** L_T = 0.15 H **(b)** X_{LT} = 47.1 Ω **(c)** V_{L1} = 153.33 V/90° leading.

CHAPTER 17

REVIEW EXERCISES

1 **(a)** $X_L = 94.2\ \Omega$ **(b)** $Z = 96.3\ \Omega$ **(c)** $I = 2.39$ A

2 $\phi = 78°$ (voltage leading current)

3 $Z = 347.27\ \Omega$

4 **(a)** total inductance = 30 mH **(b)** $X_L = 9.42\ \Omega$ **(c)** $X_{LT} = 28.26\ \Omega$ **(d)** $Z = 28.89\ \Omega$ **(e)** $V = 43.33$ V **(f)** $\phi = 77.74°$ (voltage leading current)

5 **(a)** $Z = 94.47\ \Omega$ **(b)** $\phi = 32.79°$ (voltage leading current)

6 **(a)** Z at 50 Hz = 1389.73 Ω **(b)** Z at 1 kHz = 1001.16 Ω

7 **(a)** at 50 Hz $\phi = 43.98°$ **(b)** at 1 kHz $\phi = 2.76°$ (voltage lagging current in both cases, as circuit is capacitive)

8 **(a)** $V_R = 112$ V, $V_C = 222.93$ V **(b)** $V = 249.49$ V **(c)** phase angle = 63.33° (voltage lagging current)

9 **(a)** $X_L = 565.2\ \Omega$ **(b)** $X_C = 530.79\ \Omega$ **(c)** $Z = 56.65\ \Omega$ **(d)** $I = 5.3$ A **(e)** $V_R = 238.31$ V, $V_L = 2995.56$ V or 3 kV, $V_C = 2813.19$ V or 2.81 kV **(f)** phase angle = 37.41° (voltage lagging current)

10 **(a)** $f_r = 419.62$ Hz **(b)** $Z = R = 16\ \Omega$ **(c)** $Q = 13.18$ **(d)** bandwidth = 31.85 Hz

CHAPTER 18

REVIEW EXERCISES

1 6.39 Ω, circuit is inductive as current lags the supply voltage

2 **(a)** $I_L = 0.91\ A\angle{-90°}$ (lag), $I_R = 0.67\ A\angle 0°$ **(b)** $I_T = 1.13\ A\angle{-53.64°}$ (lag) **(c)** $Z = 88.5\ \Omega$

3 **(a)** $I_L = 0.88\ A\angle{-74.73°}$ (lag) **(b)** $I_T = 1.24\ A\angle{-43.2°}$ (lag) **(c)** $Z = 80.6\ \Omega$

4 **(a)** $17.5\angle 8.48°$ (lead) **(b)** $Z = 13.14\ \Omega$

5 $I_L = 0.38\ A\angle{-60°}$ (lag)

6 **(a)** $I_C = 0.314\ A\angle 90°$ (lead), $I_R = 0.2$ A **(b)** $I_T = 0.37\ A\angle 57.28°$ (lead) **(c)** $Z = 270\ \Omega$

7 **(a)** $I_R = 1.5\ A\angle 0°$, $I_L = 1.91\ A\angle{-90°}$ (lag), $I_C = 0.57\ A\angle 90°$ (lead) **(b)** $I_T = 2.01\ A\angle{-41.73°}$ (lag) **(c)** 59.7 Ω

8 $f_r = 91.9$ Hz

9 current at resonance = $1.5\ A\angle 0°$

10 **(a)** branch currents: $I_R = 2\ A\angle 0°$, $I_{L\text{-}R} = 2\ A\angle{-41°}$ (lag), $I_C = 0.45\ A\angle 90°$ (lead) **(b)** $I_T = 3.6\ A\angle{-13.8°}$ (lag) **(c)** impedance = 66.7 Ω

CHAPTER 19

REVIEW EXERCISES

1 **(a)** $P = 3074$ W or 3.07 kW **(b)** $Q = 1566.3\ VA_R$ or 1.57 kVA_R **(c)** $S = 3450$ VA or 3.45 kVA **(d)** p.f. = 0.89.

2 p.f. = 0.23, $\phi = 76.7°$

3 **(a)** $S = 34.18$ kVA **(b)** p.f. = 0.94

4 **(a)** $P = 8.84$ kW **(b)** $\phi = 31.79°$ **(c)** $S = 10.4$ kVA **(d)** $Q = 5.48\ kVA_R$

5 **(a)** p.f. = 0.86 **(b)** $\phi = 30.68°$

6 **(a)** $I_1 = 37.5$ A p.f. = 1 **(b)** $I_2 = 29.4$ A p.f. = 0.85 **(c)** $I_T = 64.3$ A p.f. = 0.97

7 **(a)** $S = 16.16$ kVA **(b)** p.f. = 0.93

8 **(a)** 5.4 MVA **(b)** 4.05 MW

9 **(a)** $P = 0$ W as circuit is purely inductive **(b)** $S = 636.94$ VA **(c)** p.f. = 0 as circuit is purely reactive.

10 **(a)** $C = 80\ \mu$F, rating is 9042 VA_R or 9 kVA_R **(b)** $I = 15.79$ A.

CHAPTER 20

REVIEW EXERCISES

1 $V_A = 275$ V, $V_B = -550$ V, $V_C = 275$ V

2 **(a)** $V_{line} = 190.5$ V **(b)** $I_{phase} = 25$ A

3 **(a)** $V_{line} = 16$ kV **(b)** $I_{phase} = 8.66$ A

4 **(a)** $I_{phase} = 5$ A **(b)** $I_{line} = 5$ A

5 $P = 14.72$ kW

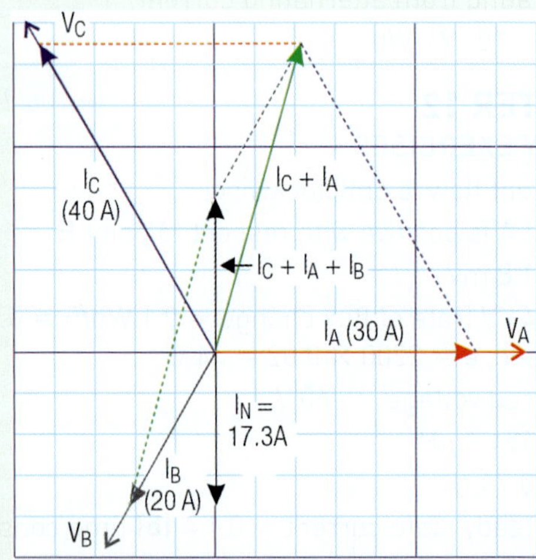

Phasor diagram for Question 6

6 $I_N = 17.3$ A approx

7 **(a)** $I_P = 4$ A **(b)** $I_L = 6.93$ A **(c)** $V_L = 600$ V

8 **(a)** $I_{line} = 38.5$ A **(b)** $S = 26.7$ kVA **(c)** $Q = 17.7\ kVA_R$

9 **(a)** $P = 18.8$ kW **(b)** p.f. = 0.95

10 $P_T = 5831$ W or 5.83 kW

CHAPTER 21

REVIEW EXERCISES

1 **(a)** $e_1 = 444$ V **(b)** $V_2 = 106.56$ V

2 **(a)** 48 V **(b)** 2.4 A **(c)** 0.5 A

3 211.25 A

4 **(a)** at 10 MVA, $I_{HV} = 303$ A, $I_{LV} = 909.1$ A **(b)** at 16 MVA, $I_{HV} = 484.85$ A, $I_{LV} = 1454.55$ A or 1.45 kA

5 **(a)** I_o = 4.6 A **(b)** p.f. = 0.22 lagging (as transformer on no-load is inductive)
6 voltage regulation = 2.73 per cent
7 efficiency = 91.4 per cent
8 **(a)** I_1 = 43.74 A **(b)** %Z = 6.2
9 Tx_A takes 1.58 MVA, Tx_B takes 1.92 MVA
10 **(a)** V_2 = 60 V **(b)** I_2 = 4 A **(c)** I_1 = 0.8 A **(d)** I in common winding = 3.2 A

CHAPTER 22

REVIEW EXERCISES

1 **(a)** synchronous speed = 750 RPM
(b) slip speed = 30 RPM
(c) percentage slip = 4 per cent.
2 minimum speed = 144 RPM, maximum speed = 3744 RPM
3 **(a)** s% = 5% **(b)** fr = 2.5 Hz
4 f_r = 2.8 Hz
5 efficiency = 89 per cent
6 **(a)** n_s = 3000 RPM **(b)** slip speed = 135 RPM **(c)** rotor speed = 2865 RPM
(d) P_o = 12 kW **(e)** P_{in} = 13.48 kW
(f) I_L = 21.62 A
7 53.6 Nm
8 efficiency = 94.44%
9 Starting torque = 50 Nm
10 The auto-transformer starter provides the highest starting torque for comparable values of line current.

CHAPTER 23

REVIEW EXERCISES

1 **(a)** 1000 RPM **(b)** 40 RPM **(c)** 4 per cent.
2 2.4 Hz
3 943 W
4 P_{out} = 865.5 W (b) efficiency = 82.4%
5 **(a)** phase difference = 90° [60° lag in main winding, 30° lead in auxiliary winding] **(b)** total current = 2.5 A **(c)** V_c = 238.9 V

CHAPTER 24

REVIEW EXERCISES

1 500 RPM
2 **(a)** V_{ph} = 4688.6 V or 4.69 kV
(b) V_{line} = 8120.9 V or 8.12 kV
3 I_{line} = 28.87 A
4 voltage regulation = 75 per cent
5 **(a)** 192.5 A **(b)** 12.9 kW **(c)** 184 900 W or 185 kW
(d) 93 per cent.
6 P_{in} = 95.51 kW
7 750 RPM
8 (1) Using the exciter as a motor (2) using a pony motor (3) power the synchronous motor from a variable frequency drive, starting at a low frequency.
9 efficiency = 94.1 per cent
10 **(a)** P_{in} = 42.95 kW **(b)** efficiency = 93.1 per cent **(c)** n = 1000 RPM
(d) torque = 382 Nm

CHAPTER 25

REVIEW EXERCISES

1 112 Vp-p
2 50.87°
3 57.6°
4 Proximity testers may not always indicate a live circuit. Use a suitably Cat rated voltage probe or meter to get direct contact. If there is no indication, test the probe or meter, then test the circuit again.
5 500 V setting
6 Measure the current flowing in the earth lead, or put the clamp ammeter around the active and neutral wires to the appliance.
7 Test the tester, then use the tester, then test the tester, especially if it does not register a voltage.
8 Reasons include:
- If test lamps are connected between an active and an earth terminal (for example, at a GPO), the RCD protection device will trip due to the earth current.
- Lamp filaments can become open-circuit due to vibrations such as when climbing a ladder.
- A lower-than-usual voltage may not cause the lamps to light, even though a lethal voltage is present.
- It is unlikely test lamps have an IP rating, allowing water to enter the lamp sockets.
- Test lamps are unlikely to be Cat rated.

9 2300 W
10 True power is measured by the wattmeter, apparent power is measured with an ammeter measuring the current and a voltmeter measuring the voltage. The apparent power is the product of the voltmeter and ammeter readings.

ANSWERS TO MATHS EXERCISES

EXERCISE 1

1. Calculation simplifies to $5 \times 20 = 100$
2. Calculation simplifies to $4 \times 5 = 20$
3. Calculation simplifies to $12 + 18 - 5 = 25$

EXERCISE 2

1. 0.7
2. 0.58333 (or 0.58)
3. R = 12.76595 (or 12.77)

EXERCISE 3

1. 6.45
2. 234.34
3. 12.99
4. 10.62

EXERCISE 4

1. $V = \frac{P}{I}$
2. $I = \frac{P}{V}$
3. $C = \frac{Q}{V}$
4. $H = \frac{B}{\mu}$

EXERCISE 5

1. $C = \left(\frac{A}{D}\right) - B$ or $C = \frac{A\text{-}BD}{D}$
2. $T = \frac{60P}{2\pi n}$
3. $L = \frac{X_L}{2\pi f}$
4. $V = \frac{Q}{RST}$

EXERCISE 6

1. $P = \frac{V^2}{R}$
2. $I = \sqrt{\frac{P}{R}}$
3. $R = \frac{V^2}{P}$
4. $P = I^2R$

EXERCISE 7

1. $a = \sqrt{c^2 - b^2}$
2. 27.98
3. $X = \sqrt{Z^2 - R^2}$
4. 45 mm

EXERCISE 8

1. **(a)** 0.87 **(b)** 0.5 **(c)** 1.73
2. **(a)** 30° **(b)** 60° **(c)** 26.57°
3. **(a)** 36.87° **(b)** 28 mm **(c)** 53.13°
4. **(a)** 29.06° **(b)** 41.18 mm **(c)** 60.94°
5. **(a)** 30.72 mm **(b)** 38.98 mm **(c)** 38°

INDEX